Space Almanac
Second Edition

Gulf Publishing Company
Houston, London, Paris, Zurich, Tokyo

Space Almanac
Second Edition

Anthony R. Curtis

Space Almanac
Second Edition

First edition © 1990 by Arcsoft Publishers.

Printed in the United States of America.

Library of Congress Cataloging-in-Publication Data

Curtis, Anthony R., 1940–
  Space almanac/Anthony R. Curtis.—2nd ed.
    p. cm.
  Includes bibliographical references and index.
  ISBN 0-88415-039-9—ISBN 0-88415-030-5 (pbk.)
    1. Outer space—Exploration. 2. Space stations. 3. Artificial satellites.
  4. Rockets (Aeronautics) 5. Solar system. I. Title.
  QB500.C87   1992
  629.4—dc20   91-40370
                                                                   CIP

Contents
****

Maps
****

Glossaries
****

Preface

****

If you're like me, you scan the local newspaper everyday, hoping for a tidbit about the latest deep space discovery by astronomers, radio signals from unmanned probes beyond the Solar System, American and Soviet space shuttle flights, scientific satellite launches, space stations in orbit, or the unexpected supernova. You check out TV news for scraps of information on men in space from Russia or the latest Chinese, Japanese, Indian, Israeli and European launches. But you find remarkably little to satisfy that deep hunger to know what's going on Out There.

An X-ray glow surrounds the heart of our galaxy. Bricks can be made from Moon dust. The edge of the Universe we can see has been extended by the continuing discovery of ever-more-distant quasars. Some say Russians are a decade ahead in manned spaceflight. They've been living for years in orbiting space stations. Japanese scientists are flying Moon probes and designing interplanetary spacecraft. European and Japanese engineers are moving ahead on their own space shuttles. Foreigners pay China to launch satellites. Brazil, India, Israel, Iraq and other would-be spacefaring countries are trying to fire satellites to orbit. Radio hams around the globe have sent up a small flotilla of exciting new high-tech communications and science satellites, with interplanetary plans for the future.

Do TV networks report all of this? Do newspapers keep the interested consumer up-to-date? Frustratingly, no! Unless you're one of a precious few space agency insiders or military top brass, most space news just whizzes by overhead. You never hear about all the exciting happenings in and about space.

And that's why this book exists. There continue to be so many developments affecting mankind's future, they simply are bursting to be let out of the news box. This book rounds up all the news about space stations, astronauts, cosmonauts, shuttles, satellites, rockets, planetary explorers, the Moon, our Sun, Venus, Mars, Jupiter, Saturn, Uranus, Neptune, Pluto, comets, asteroids, our Solar System, the Milky Way galaxy and neighboring galaxies, pulsars, quasars, black holes.

This reference database is a collection of facts, figures, names, dates, places, lists, tables, maps and charts covering all of Space from Earth to the edge of the Universe. Data has been gathered from many sources around the world. The chapters are original, written and edited for this volume. Much of the First Edition material has been rewritten for this Second Edition and a great deal of new data has been added. I thank for their valuable assistance the National Aeronautics and Space Administration, Goddard Space Flight Center, Johnson Space Flight Center and other NASA centers, Jet Propulsion Laboratory, NORAD, Smithsonian Astrophysical Observatory, European Space Agency, Arianespace, governments of the USSR, Japan, China, Great Britain, France, West Germany, Brazil, India and other governments around the world, as well as satellite owners, commercial launch companies and a wide variety of other sources.

Space Almanac is designed to be a user-friendly database which you'll pick up and use easily, over and over at the slightest urge; not a cold, formal coffee-table book to look at once and put away. We've included hundreds of updates and background checks on thousands of items too small for the popular press to report. Plus plenty of reference tables, charts, maps, histograms and quick look-up lists.

Space Almanac will bring you up-to-date with what's happening just now Out There— covering space from Earth to the edge of the Universe. Here's the most detailed history anywhere and all the latest news of importance from and about space. As can be seen in the Table of Contents, we divide articles into seven categories: Astronauts & Cosmonauts, Space Stations, Space Shuttles, Rockets, Satellites, Solar System and Deep Space. News from Deep Space can be sporadic as it is slowed by the time necessary for astronomers and scientists to make and confirm discoveries. Solar System news is as exciting as Deep Space, but often slowed by the years it takes for probes to get to other planets. For example, the Galileo probe is sailing six years to Jupiter. It may make news along the way, but at one-year to two-year intervals. In the interim, men will continue to plod toward other important achievements. The Russians will add to their space station and America may build one. Probes will fly to the Moon, Venus and Mars from Japan, the USSR and the U.S.

America might return to the Moon. Both the U.S. and the USSR are likely to send men and women to Mars before 2010—possibly including Europeans and Japanese in the flight crew. It's as if we have a slow motion view of human progress outward, punctuated by an occasional rapid discharge of important news from other parts of our Solar System, the Milky Way and Deep Space beyond.

Each small step is one square in a patchwork quilt. We see the whole quilt in our mind's eye and are impatient to finish the job. A particular Soyuz flight to Mir or an American shuttle's week in orbit may seem relatively insignificant in the big picture...but it's taking us Out There, slowly but surely.

Man has studied the heavens since ancient times but the view has always been made fuzzy by the ocean of air between Earth's hard surface and the near-vacuum of space. Only just now are we able to put our observatories in space where the view is not blurred by an atmosphere. American shuttles are ferrying NASA's Great Observatories to Earth orbit—four high-tech instruments with strange-sounding acronyms—Hubble Space Telescope (HST), Gamma Ray Observatory (GRO), Advanced X-Ray Astrophysics Facility (AXAF) and Space Infrared Telescope Facility (SIRTF).

Much news will flow from discoveries made by astronomers using these magnificent instruments, but it won't end there. In time, a town on the backside of the Moon will be home to mankind's best-ever observatory for deep-space astronomy. Then there will be a flow of news beyond anything we know now. It will be a heyday for folks like you and me who see a significance in pulsars, quasars, galaxies, black holes, supernovas and unknown planets. And, when those men and women build the first human town on Mars, every single one of Earth's five billion people will know about it and be electrified by it.

—*Anthony R. Curtis*

Astronauts & Cosmonauts
★★★★★★★★★★★★★★★★★★★★★★★★★★★★★★★★★★★★★★★

Man's Journey To The Stars

This chronology is a timetable of selected highlights of 20th Century space history. Major events scheduled for the last decade of the 20th Century are included. Additional details on the events recorded here may be found elsewhere in this Space Almanac.

Year	Date	Launch	Event
1903-1926		rockets	Russian rocket pioneer Konstantin Tsiolkovskii wrote of liquid-fuel rockets.
1926	Mar 16	rocket	Robert Goddard, talking of firing vehicles to the Moon someday, blasted off world's first liquid fuel rocket on a farm at Auburn, Massachusetts. It flew to altitude of 184 feet, landed in a cabbage patch.
1927		space travel	Verein fur Raumschiffahrt, the Society For Space Travel, was founded at Breslau, Germany. Wernher von Braun was an early member.
1930	Feb 18	Pluto planet	U.S., the planet Pluto was discovered.
1930	Dec 30	rocket	At Roswell, New Mexico, Robert Goddard fired a liquid-fuel rocket to 2,000 feet altitude.
1931	Feb 21	rocket	Johannes Winkler, a member of Germany's Verein fur Raumschiffahrt, the Society For Space Travel, launched Europe's first liquid-fuel rocket.
1932	Apr	military rocket	German army assigned Walter Dornberger to start work on liquid-fuel military rocket. Small rockets were built with help of Verein fur Raumschiffahrt.
1936	Oct	A-4 rocket	German engineers designed the bigger liquid-fuel A-4 rocket. It evolved into V-2 military rocket.
1936	Oct 31	rocket	California Institute of Technology students and friends blasted off a small liquid-fuel rocket in the dry Arroyo Seco riverbed, near where today's Jet Propulsion Laboratory stands outside Pasadena.
1937		asteroid	The asteroid Hermes passed less than 500,000 miles from Earth.

1940's

Year	Date	Launch	Event
1942	Jun	A-4 rocket	Germany, the first flight test of the A-4 rocket at Peenemunde by Wernher von Braun, Walter Dornberger. It would evolve into the Vergeltungswaffe Zwei or V-2 WWII missile.
1942	Oct 3	V-2 rocket	First successful V-2 launch. The Vergeltungswaffe Zwei rocket prototypes were painted dark green, leading to the nickname "cucumber." Production models were green and brown camouflage colors. V-2 was the first mass-produced long range rocket. Some 6,000 were made during WWII with 3,225 fired at England, France, Belgium, killing tens of thousands and destroying even more buildings.
1943	Nov 20	rocket	Jet Propulsion Laboratory founded to expand rocket research beyond work already done in the dry Arroyo Seco riverbed outside Pasadena, Calif.
1945		V-2 rocket	65 captured Peenemunde rockets, data files and rocket experts moved by the U.S. to New Mexico, in 300 railroad freight cars.
1945		space satellites	Arthur C. Clarke wrote of communication sats.

1945	Oct	rocket	U.S. Navy, Bureau of Aeronautics, started planning to build a liquid-fuel space rocket to launch a man-made artificial Earth satellite. Naval Research Laboratory started planning sounding rockets to send TV pictures from high altitudes.
1945	Oct 11	WAC-Corporal	Jet Propulsion Laboratory's first sounding rocket launched from new U.S. Army site, White Sands, New Mexico. JPL was developing 3 military missiles: Private, Corporal, Sergeant. The version of Corporal for science research was called WAC, Women's Army Corps, because it was thought of as the military Corporal's little sister. Reaching 33 miles altitude, WAC-Corporal was the highest flier to that time.
1946		rocket	James A. Van Allen, a cosmic ray physicist at Applied Physics Lab, Johns Hopkins Univ, fired "rockoons," small sounding rockets carried to high altitudes by balloons, then blasted above the atmosphere to measure ozone and cosmic rays.
1946	Jan 10	Moon	U.S. Army Signal Corps bounced radar off Moon.
1946	Feb 27	space science	The V-2 Panel, including Harvard astronomer Fred Whipple, Johns Hopkins' James A. Van Allen, Princeton's M.H. Nichols, Michigan's W.G. Dow, formed to design equipment for cosmic ray, solar and atmosphere research, riding in empty nosecones of V-2 test flights.
1946	Apr 16	V-2 rocket	Wernher von Braun launched a captured German V-2 rocket at White Sands, New Mexico, for U.S. Army. It flew up about 3 miles, then a fin fell off.
1946	May 10	V-2 rocket	Wernher von Braun successfully launched a captured German V-2 at White Sands for U.S. Army, to an altitude of 71 miles. Some 64 V-2 launches were conducted by Sep 1952. The first primates in space, monkeys Albert 1 and Albert 2, died in nose cones during early V-2 test flights.
1946		Aerobee rocket	U.S. Navy, Bureau of Ordinance, designed the successful Aerobee high-altitude sounding rocket.
1946		Viking rocket	Naval Research Laboratory designed the Viking high-altitude sounding rocket.
1946		space satellites	U.S. Navy started a space satellite program.
1946-1950		V-2 Aerobee	V-2 plus Aerobee upper stage, U.S. Air Force high altitude research rockets fired from Wallops Island.
1947	May 29	V-2 rocket	Gyroscope broke during test flight from White Sands, New Mexico, of a captured V-2. The missile streaked over El Paso, Texas, crossed the Rio Grande River, into a hill beyond Juarez, Mexico.
1947	Sep	V-2 rocket	U.S. Navy V-2 launched from the carrier Midway.
1948	Feb 16	Uranus planet	U.S., Uranus' moon Miranda found.
1948	May 13	Bumper-WAC	First flight of WAC-Corporal rocket mounted atop V-2 rocket to gain altitude. Launch from White Sands, New Mexico, failed. Next 3 flights failed. Fifth flight a success, Feb 24, 1949, reached 244 mi. altitude, set record. First rocket in outer space.

1948	Jun 3	Hale telescope	U.S., 200-in. telescope placed in service.
1949	Feb 24	Bumper-WAC	Fifth flight, first successful flight of Bumper-WAC, a WAC-Corporal atop a V-2. Launched from White Sands, New Mexico, reached 244 mi. altitude, set new record. First rocket in outer space.
1949	May 1	Neptune planet	USA, Neptune's moon Nereid found.
1949		rockets	Joint Long Range Proving Ground established at deserted, remote Cape Canaveral, Florida.

1950-1956

1950	Apr	rockets	von Braun's U.S. Army rocket team moved to Redstone Arsenal, Huntsville, Alabama.
1950	July 24	Bumper-WAC	A two-stage Bumper-WAC was the first rocket launched from The Cape.
1951		Atlas rocket	U.S. Air Force started Atlas ICBM design.
1951	Sep 20	Aerobee rocket	Earlier in 1951, monkey and mice shot to edge of space on a U.S. Aerobee rocket died when their parachute failed to open. In the first successful spaceflight for living creatures, a monkey and 11 mice shot to the edge of space on an Aerobee rocket Sep 20 were recovered alive.
1953	Aug 20	Redstone rocket	U.S., first test launch.
1955		Titan rocket	U.S. Air Force started Titan ICBM design.
1956	Sep 20	Jupiter-C rocket	U.S. first launch.

1957

1957	Aug 3	R-7 rocket	USSR, first ICBM flight.
1957	Sep 20	Thor rocket	U.S., first launch.
1957	Oct 4	Sputnik 1 satellite	Start of the Space Age, USSR orbited the first man made, artificial Earth satellite.
1957	Nov 3	Sputnik 2 satellite	USSR, first higher life form in space: Laika was a live dog on a life-support system in Sputnik 2. Capsule remained attached to the converted intercontinental ballistic missile rocket. The dog captured hearts around the world as life slipped away from Laika a few days into her journey. Sputnik burned in atmosphere 4/14/1958. Six other dogs were launched to space by the end of the Sputnik satellite series in 1961 as the Russians prepared to send men to orbit: Laika (Barker in Russian), Belka (Squirrel in Russian), Strelka (Little Arrow), Pchelka (Little Bee), Mushka (Little Fly), Chernushka (Blackie) and Zvezdochka (Little Star). Laika, Pchelka, and Mushka died in flight. In 1966, the dogs Veterok and Ugolek were launched in Voskhod 3 to be observed in orbit for 23 days via TV and biomedical telemetry.
1957	Dec 6	Vanguard rocket	Navy failed first attempt to beat the Army in launching the first U.S. satellite. In case it were to fail, the Navy had referred to the rocket before launch as TV-3 or Test Vehicle No. 3. With world wide media tuned in, Vanguard lost thrust two seconds after launch, just four feet off the pad. It fell back and exploded. The six-in.-diameter

			satellite popped out of the flames and rolled away, transmitting its radio signal on the ground. Newspapers called it Kaputnik, Stayputnik.
1957-1958		sounding rockets	International Geophysical Year (IGY), July 1, 1957, to Dec 31, 1958. Dozens of sounding rockets fired to measure atmosphere, X-ray, ultraviolet and other energy from Sun, solar flares, including many U.S. Navy Aerobee and U.S. Army Nike-Cajun rockets. Nike-Cajuns were Nike-Ajax air defense missiles with Ajax stage replaced by Cajun stage to lift science instruments to 145 mi. altitude. V-2 Panel, renamed Rocket and Satellite Research Panel, conducted IGY experiments.

1958

1958	Jan 25	Vanguard rocket	U.S. Navy tried again to beat the Army to space with America's first satellite, but Vanguard fizzled within 14 seconds of ignition.
1958	Jan 31	Explorer 1 satellite	Army successfully launched first U.S. satellite to orbit from Cape Canaveral on a Jupiter-C rocket, making the U.S. the 2nd nation with space rocket. The Explorer 1 satellite had been designed by Jet Propulsion Laboratory.
1958	May 1	Moon probe	USSR launch failed in an unannounced secret attempt to send an unmanned probe to the Moon.
1958	Mar 17	Vanguard 1 sat	Navy launched U.S. satellite from Cape Canaveral.
1958	Mar 26	Explorer 3 sat	Army launched U.S. satellite from Cape Canaveral.
1958	May 15	Sputnik 3 sat	USSR science satellite, solar powered.
1958	July 26	Explorer 4 sat	Army launched U.S. satellite from Cape Canaveral. James Van Allen, analyzing data from Explorer sats 1,3, & 4, discovered belts of radiation trapped in the magnetic field surrounding planet Earth.
1958	Jul 29	Space Act	U.S., the National Aeronautics and Space Act of 1958 legislation enabled birth of National Aeronautics and Space Administration (NASA) which then was born Oct 1. Dept of Defense fought the Space Act legislation, wanting to keep space research entirely military.
1958	Aug 2	Atlas rocket	U.S., first successful launch.
1958	Aug 17	Thor-Able 1	U.S. Thor-Able 1 rocket exploded during an attempt to send an unmanned probe to the Moon.
1958	Sep 24	Moon probe	USSR launch failed in an unannounced secret attempt to send an unmanned probe to the Moon.
1958	Oct 1	NASA	National Aeronautics and Space Administration (NASA), the modern U.S. space agency, was founded. It included the old National Advisory Committee for Aeronautics (NACA).
1958	Oct 11	Pioneer 1 probe	U.S. failed in an attempt to send an unmanned probe to the Moon when rocket did not achieve necessary velocity, unable to leave Earth's gravity, but sent down 43 hours of data.
1958	Nov 8	Pioneer 2 probe	U.S. failed attempt to send unmanned probe to Moon. Rocket did not achieve necessary velocity and was unable to leave Earth's gravity.

1958	Nov 26	Moon probe	USSR launch failed in an unannounced secret attempt to send an unmanned probe to the Moon.
1958	Dec 6	Pioneer 3 probe	U.S. failed in an attempt to send an unmanned probe to the Moon when rocket did not achieve necessary velocity, unable to leave Earth's gravity, but measured radiation bands around Earth as it passed through.
1958	Dec 15	Moonflight	NASA received recommendations from Wernher von Braun and others to start planning for launching men to the Moon by 1967. Later, von Braun designed the Saturn V Moon rocket.
1958	Dec 18	Score satellite	U.S., Pres. Eisenhower taped Christmas message, broadcast from space.

1959

1959	Jan 2	Luna 1 probe	USSR finally succeeded in sending an unmanned probe to the Moon. It was the first craft to leave Earth's gravity. Luna 1 flew by within 3,500 mi. of the Moon, then on to orbit the Sun.
1959	Mar 3	Pioneer 4 probe	U.S. finally succeeded in sending an unmanned probe to the Moon. It was the second craft to leave Earth's gravity. Pioneer 4 flew by within 37,500 mi. of the Moon, then on to orbit the Sun. Sent back radiation data.
1959	Apr 27	astronauts	NASA selected its first class of astronauts, known as the Mercury Seven: Malcolm Scott Carpenter, Leroy Gordon "Gordo" Cooper Jr., John Herschel Glenn Jr., Virgil Ivan "Gus" Grissom, Walter Marty "Wally" Schirra Jr., Alan Bartlett Shepard Jr., Donald Kent "Deke" Slayton. The astronauts and their wives became media stars after they signed a deal giving Life rights to their stories and the magazine published a series of cover stories.
1959	Sep 12	Luna 2 probe	USSR, unmanned lunar probe, crashed into the Moon Sep 13 to gain science data. It was the first craft to impact on Moon, first man-made object to land on another celestial body, first Moon-junk. Wreckage still there. Measured solar wind enroute.
1959	Sep 18	Vanguard 3 sat	U.S. Navy sat to measure Earth's magnetic field.
1959	Oct 4	Luna 3 probe	On second anniversary of Sputnik 1, USSR launched the unmanned, solar-powered Luna 3 to circle Moon, radio back first pictures of the far side of Moon on Oct 7. Measured solar wind.
1959	Oct 13	Explorer 7 sat	U.S. satellite measured effects of magnetic storm on Sun on Earth radiation belts.
1959	Nov 26	Atlas-Able	U.S. failed to send unmanned probe to orbit Moon. Payload shroud broke off 45 seconds into flight.
1959	Dec 4	Mercury capsule	Monkey Sam made a suborbital flight in U.S. Mercury Little Joe 3.

1960

| 1960 | Jan 21 | Mercury capsule | Monkey, Miss Sam, made a suborbital flight in U.S. Mercury Little Joe 4. |
| 1960 | Mar 11 | Pioneer 5 probe | U.S. observatory probe to the orbit the Sun, |

			between Earth and Venus, measuring plasma streaming toward Earth.
1960	Mar 14	cosmonauts	The USSR's first cosmonauts, known as the Star City 12, were selected: Pavel I. Belyayev, 28; Valeri F. Bykovsky, 26; Yuri A. Gagarin, 26; Viktor V. Gorbatko, 26; Yevgeny V. Khrunov, 27; Vladimir M. Komarov, 33; Alexei A. Leonov, 26; Andrian G. Nikolayev, 31; Pavel R. Popovich, 30; Georgi S. Shonin, 25; Gherman S. Titov, 25; Boris V. Volynov, 26.
1960	Apr 4	Tiros 1 satellite	U.S., Television Infra-Red Orbital Satellite (TIROS), first weather satellite, cloud photos.
1960	May 15	Sputnik 4 sat	Unmanned Vostok capsule test failed. Fragments fell near Manitowoc, Wisc., Sep 5, 1962.
1960	Aug 12	Echo 1 satellite	U.S., 100-ft.-diameter balloon. A passive communications satellite which bounced signals transmitted from Earth back down to receivers.
1960	Aug 19	Sputnik 5 sat	Second unmanned test of Vostok capsule carried dogs Belka and Strelka, 40 mice. Tape of Russian choral group played over radio during descent Aug 20 with TV pictures of a dummy in a spacesuit. Dogs ejected after re-entry, parachuting to Earth. First successful recovery of a satellite from space.
1960	Sep 25	Atlas-Able	U.S. Atlas-Able failed again to send an unmanned probe to orbit the Moon, when the rocket's second stage malfunctioned.
1960	Oct 4	Courier 1B sat	U.S., first active communications satellite, first store-and-forward message board in space.
1960	Oct 10	Mars probe	USSR rocket failed to achieve Earth orbit in an unannounced secret attempt to send an unmanned probe on a Mars fly-by. First of 2 1960 attempts.
1960	Oct 14	Mars probe	USSR rocket failed to achieve Earth orbit in an unannounced secret attempt to send an unmanned probe on a Mars fly-by. 2nd of 2 1960 attempts.
1960	Dec 1	Sputnik 6 sat	Third unmanned Vostok capsule test, dogs Pchelka and Mushka. Dogs died in the burning capsule when it re-entered at too steep angle Dec 2.
1960	Dec 19	Mercury 1 capsule	U.S., unmanned test, flew 235 mi., alt. 130.7 mi.

1961

1961	Jan 31	Mercury capsule	Ham, a chimp, made suborbital flight in U.S. Mercury MR-2.
1961	Feb 4	Sputnik 7 sat	First launch of unmanned Venera to explore planet Venus, but A1 rocket reached only Earth orbit. Also known as Tyazheily Sputnik 4.
1961	Feb 12	Venera 1 probe	USSR, second launch of unmanned Venera to explore planet Venus, first interplanetary probe to escape Earth orbit, flew within 62,000 miles of Venus May 19. Radio failed 14 million miles from Venus, but probe was tracked by radar. Sometimes referred to as Sputnik 8 or Tyazheily Sputnik 5.
1961	Mar 9	Sputnik 9 sat	Fourth unmanned Vostok test carried dog Chernushka and dummy in spacesuit. Descended after 1 orbit, dog ejected and parachuted to Earth.

1961	Mar 25	Sputnik 10 sat	Fifth and last unmanned Vostok test carried dog Zvezdochka and dummy in spacesuit. Descended after 1 orbit, dog ejected and parachuted to Earth.
1961	Apr 12	Vostok 1 capsule	USSR cosmonaut Yuri Alekseyevich Gagarin was the first human in space. One orbit of Earth. Also, the first person to eat and drink in Earth orbit. His first-ever manned flight made Gagarin's trip the world's best-known space voyage.
1961	May 5	Mercury 3 capsule	U.S. astronaut Alan Shepard was the first American in space, on a 15 minute 22 second suborbital flight in Mercury MR-3 capsule Freedom 7. It was the shortest stay in space for a U.S. astronaut. Mercury astronauts called their tiny 6 ft. by 9-ft. capsules "garbage cans." Shepard splashed down into the Atlantic Ocean.
1961	May 6	Moonflight	NASA and Defense Dept recommend to President John F. Kennedy U.S. astronauts fly to the Moon.
1961	May 25	Moonflight	President John F. Kennedy, in a speech on "Urgent National Needs" to joint session of Congress, said he would send astronauts to the Moon by 1970.
1961	Jul 21	Mercury 4 capsule	U.S. astronaut Gus Grissom was the second American in space, on a 15 minute 37 second suborbital flight in Mercury MR-4 capsule Liberty Bell 7. It was the second shortest stay in space for a U.S. astronaut, 15 seconds longer than Alan Shepard's flight. When Grissom splashed down into the Atlantic Ocean, explosive bolts blew open a hatch, flooding the capsule with seawater. He floated out in a buoyant spacesuit, but water seeped through an open vent into his suit. He was about to go under when he finally grabbed a sling dangling from a rescue helicopter and was yanked to safety. The water-filled capsule, too heavy for the helicopter, was abandoned. The chopper cut the capsule loose and reeled in Grissom. Today, Liberty Bell 7 is on the ocean floor near the Bahamas beneath 3 miles of water.
1961	Aug 6	Vostok 2 capsule	USSR cosmonaut Gherman Titov spent 25 hours 18 minutes in 16 trips around the globe. He was the second person in orbit, the first to sleep in space, the first beyond 24 hours in space, and, at age 26, the youngest person to travel in space.
1961	Aug 23	Ranger 1 probe	Ranger was a series of Moon probes in the early 1960's, the first U.S. space program to investigate another Solar System body. Ranger 1 failed Aug 23, 1961, in its attempt to crash into the Moon when its Agena rocket failed to restart, sending Ranger into the wrong orbit. Ranger 1 was to have bounced onto the lunar surface with a seismometer.
1961	Nov 18	Ranger 2 probe	U.S. failed to send an unmanned probe to crash into the Moon. It entered the wrong orbit when the Agena rocket attitude control system failed.
1961	Nov 29	Mercury 5 capsule	First primate in orbit: chimp Enos flew two orbits

of Earth in U.S. Mercury MA-5.

1961 Dec 12 OSCAR 1 satellite U.S., first amateur radio satellite launched to orbit.

1962

1962 Jan 26 Ranger 3 probe U.S. finally successfully sent an unmanned Ranger to crash into the Moon, but missed target by 20,000 miles in a hard landing.

1962 Feb 20 Mercury 6 capsule U.S., John H. Glenn Jr. was the first American to orbit Earth, flying Mercury capsule Friendship 7. First manned flight in a Florida winter. Glenn ate applesauce during his three orbits. Perth, Australia, residents turned on all lights in town to signal Glenn as he passed overhead. "As if I were walking through a field of fireflies," Glenn radioed after luminous yellow green particles drifted past his window. The fireflies remained a mystery 3 months until Scott Carpenter found they were frost particles flaking off the outside of his capsule. Glenn also was first astronaut in public office, elected to U.S. Senate from Ohio in 1974.

1962 Mar 7 OSO 1 sat U.S., Orbiting Solar Observatory, measured solar flares from Earth orbit.

1962 Apr 23 Ranger 4 probe U.S. unmanned probe crashed on the far side of the Moon, returned no data.

1962 Apr 26 Cosmos 4 sat First USSR weather satellite.

1962 Apr 26 Ariel 1 sat British satellite launched by NASA from Cape Canaveral, first international satellite, observed X-rays, cosmic rays.

1962 May 24 Mercury 7 capsule U.S. astronaut M. Scott Carpenter was 2nd American in orbit. Three orbits in five hours in Mercury capsule Aurora 7. Pillsbury supplied Carpenter with new high-protein cereal snacks and Nestle sent "bone-bones" of cereals with raisins and almonds. Today we call these granola bars. Carpenter was to splash down in the Atlantic near Puerto Rico, but firing of landing rockets was delayed 3 seconds so he overshot the target by 250 miles. He was out of radio contact 41 minutes; NASA was worried. Finally, he flashed a passing boat with a hand mirror. He was rescued after 3 hours bobbing among the waves.

1962 Jul 10 Telstar 1 satellite U.S., with a transponder making it the first active real-time communications satellite. First direct TV to Europe.

1962 Jul 22 Mariner 1 probe U.S., failed, Venus fly-by probe. The rocket flew off course and was destroyed by Cape Canaveral range safety officer.

1962 Aug 11 Vostok 3 & 4 USSR cosmonaut Andrian Nikolayev was launched to Earth orbit in the Vostok 3 capsule Aug 11. Pavel Popovich was launched in Vostok 4 Aug 12. The first team flight. In the first space rendezvous, Vostok 3 and Vostok 4 approached within three miles of each other in space in August 1962. Their closeness was due to launch trajectory. They were

			unable to maneuver in space.
1962	Aug 14	Mariner 2 probe	U.S. Venus probe, flyby within 22,000 miles of Venus on Dec 14, 1962, first successful interplanetary probe, first flyby of any planet, first spacecraft to visit another planet. Measured atmosphere, pressure, temperature.
1962	Aug 25	Venus	USSR rocket failed to leave Earth orbit in a secret attempt to send unmanned probe on Venus fly-by.
1962	Aug 31	comm sats	U.S., Comsat created.
1962	Sep 1	Venus	USSR rocket failed to leave Earth orbit in a secret attempt to send unmanned probe on Venus fly-by.
1962	Sep 1	astronauts	The first civilian astronauts were aeronautical engineer Neil Armstrong and U.S. Maritime Academy engineer Elliott See, selected in NASA's second class. Armstrong flew in Gemini 8 in 1966 and in Apollo 11 in 1969 to become the first man on the Moon. See was killed in an airplane crash in 1966 without having made a spaceflight.
1962	Sep 12	Venus	USSR rocket failed to leave Earth orbit in a secret attempt to send unmanned probe on Venus fly-by.
1962	Sep 29	Alouette 1 sat	Canada satellite launched by NASA from Cape Canaveral, second international satellite, measured ionosphere.
1962	Oct 3	Mercury 8 capsule	U.S. astronaut Walter M. "Wally" Schirra Jr. orbited in his capsule Sigma 7. Mercury astronauts squeezed food from toothpaste-style tubes. Schirra lunched on beef, vegetables and peaches, spent 9 hours in space, then became first to splashdown in the Pacific, landing within 4.5 miles of target.
1962	Oct 18	Ranger 5 probe	U.S. unmanned probe crashed into the Moon for seismometer readings, but power failed and little data was sent.
1962	Oct 24	Mars probe	USSR rocket failed to leave Earth orbit in a secret attempt to send unmanned probe on Mars fly-by. First of three 1962 Mars attempts.
1962	Nov 1	Mars 1 probe	USSR finally gets a Mars probe away from Earth. First successful USSR Mars probe, but radio failed 69 million miles from Mars. This second 1962 attempt was known officially as Mars 1.
1962	Nov 4	Mars probe	USSR rocket failed to leave Earth orbit in a secret attempt to send unmanned probe on Mars fly-by. Third of three 1962 Mars attempts.
1962	Dec 13	Relay 1 satellite	U.S., communications satellite; 12 simultaneous telephone circuits, one TV channel.

1963

1963	Jan	quasars	Quasi-stellar objects calculated most distant objects from Earth, by Maarten Schmidt, Mt. Wilson Observatory. Quasars, powerful faraway energy sources, had been discovered by Cambridge Univ. radio astronomer Martin Ryle in 1950's.
1963	Jan 4	Moon	USSR rocket failed to leave Earth orbit in a secret attempt to send an unmanned probe to soft-land on the Moon.

1963	Feb 3	Moon	USSR rocket launch failed in a secret attempt to send an unmanned probe to soft-land on the Moon.
1963	Feb 14	Syncom 1 satellite	U.S., first attempt to send a communications satellite to a high stationary orbit; failed.
1963	Feb 21	Telstar 1 satellite	Electronics in the orbiting communications satellite knocked out by atom bomb test on Earth.
1963	Apr 2	Explorer 17 sat	U.S. satellite, measured atmosphere, density, pressure, zone of neutral helium around Earth.
1963	Apr 2	Luna 4 probe	USSR, Moon flyby. The USSR finally was successful in sending an unmanned probe toward a soft-landing on the Moon, but missed the Moon by 5,300 miles.
1963	May 7	Telstar 2 satellite	U.S. communications satellite, first trans-Atlantic color TV.
1963	May 9	West Ford	U.S. needles scattered in space as dipoles for passive communications antennas.
1963	May 15	Mercury 9 capsule	U.S., orbit, Gordon Cooper.
1963	Jun 14	Vostok 5 capsule	USSR cosmonaut Valeri Bykovsky made the longest solo space flight ever. He stayed in space four days (119 hours 6 minutes) making 81 revolutions around Earth in Vostok 5. In the USSR's second team flight, he passed within 3 miles of Valentina Tereshkova in Vostok 6.
1963	Jun 16	Vostok 6 capsule	USSR, Valentina V. Tereshkova, the first woman in space, spent 71 hours in "Sea Gull" orbiting Earth 48 times. At 26, she was the youngest woman to fly in space. She also was the first ordinary person in space: a textile mill worker who enjoyed the hobby of parachute jumping when she was selected to train with a group of Soviet women. Premier Nikita Khrushchev wanted a spectacular, so, by age 25, Tereshkova was a cosmonaut. Although she had made 126 parachute jumps, her spaceflight training of less than one year didn't fully prepare her for the stress of a high-G re-entry. Tereshkova also was the first politician from space. In 1967, she was elected a member of the Supreme Soviet. She became a member of the Presidium of the Supreme Soviet in 1974. Tereshkova and cosmonaut Andrian G. Nikolayev became the first couple married after spaceflight, in November 1963, five months after her June 16, 1963, flight in Vostok 6. He had flown in Vostok 3 August 11, 1962, and later flew in Soyuz 9 in 1970. She and Nikolayev had a romance before her flight. Khrushchev was toastmaster at their wedding. Their daughter, Yelena, born in June 1964, was the first child of cosmonauts.
1963	Jul 26	Syncom 2 satellite	U.S., communications sat, first sat in stationary orbit, TV from 1964 Olympics games in Japan.
1963	Nov 11	Venus/Cosmos 21	USSR rocket failed to leave Earth orbit in a secret attempt to send unmanned probe to Venus.

			Announced as Earth satellite Cosmos 21.
1963	Nov 26	Explorer 18 sat	U.S. satellite, Interplanetary Monitoring Platform A, measured geomagnetic field from 200 to 200,000 miles above Earth. Reported interaction of solar plasma stream with Earth geomagnetic field. Found bow shock where solar wind breaks upon magnetic field like surf on a beach.

1964

1964	Jan 25	Echo 2 satellite	U.S., first joint space project with USSR.
1964	Jan 30	Ranger 6 probe	U.S. unmanned probe crashed into Moon. Transmitted some TV pictures before Ranger 6 bounced onto the surface for seismometer reading, but camera failed and no seismic data was sent.
1964	Jan 30	Elektron sats	First USSR double launch, from Baikonur Cosmodrome, blasted Elektron 1 and Elektron 2 satellites to Earth orbit to study radiation belts.
1964	Apr 2	Zond 1 probe	USSR, Venus flyby.
1964	Apr 8	Gemini 1 capsule	U.S., unmanned test.
1964	Jul 11	Elektron sats	Second USSR double launch blasted Elektron 1 and Elektron 2 satellites to Earth orbit.
1964	July 28	Ranger 7 probe	U.S. Ranger series, redesigned for last-minute close-up photography as a probe neared lunar surface. Ranger 7 crashed into Moon July 31 in area later named Mare Cognitum, "the sea which has become known," a level mare without craters chosen as a potential Apollo landing site. Six sequential TV cameras radioed 4,316 photos, first close-ups of an extraterrestrial body. Widest angle covered square mile, showed 30-ft. craters. Most detailed revealed 3-ft. craters in 100x160 ft. area.
1964	Aug 28	Nimbus 1 sat	U.S., first of Nimbus polar-orbit weather sats. High-resolution TV of cloud cover in visible light, infrared. Automatically transmitted pictures.
1964	Sep 4	OGO 1 sat	U.S., Orbiting Geophysical Observatory.
1964	Oct 12	Voskhod 1 capsule	Cosmonauts Komarov, Yegorov, Feoktistov in the first three-man capsule. The USSR put extra seats in their one-man Vostok capsule and called it Voskhod, sunrise in Russian, then crammed the 3 cosmonauts in for a 24-hour flight, sixteen orbits of Earth—so cramped they couldn't even take along spacesuits for safety, the first crew to fly without spacesuits. First live TV from space.
1964	Nov 5	Mariner 3 probe	U.S., failed, Mars fly-by probe. The payload shroud covering the probe failed to blow away, trapping the spacecraft. No data was received.
1964	Nov 28	Mariner 4 probe	U.S. Mars probe, first successful Mars flyby 6,000 miles above the Red Planet July 14, 1965. Radioed 22 photos, first close-ups of another planet. Revealed cratered surface. Winter ice on crater rims in southern hemisphere. Found no canals.
1964	Nov 30	Zond 2 probe	USSR unmanned fly-by, passed within 1,000 mi. of Mars in 1965, radio failed, sent no data. The flight was known officially as Zond 2. During the

1964 launch window, a second USSR rocket may have failed to reach Earth orbit in a second attempt to send an unmanned probe on a Mars fly-by.

1964   Dec 15   San Marco A sat   Italy became third nation to send satellite to space. Fired by Italian launch team on U.S. Scout rocket from platform in Indian Ocean off Kenya. First NASA rocket launch by foreign technicians.

1965

1965   Jan   cosmic noise   Bell Telephone Labs physicists Arno Penzias and Robert Wilson discovered cosmic background radio noise coming to Earth from all directions. Some theorize it is a remnant of the Big Bang explosion they say formed the Universe.

1965   Jan 19   Gemini 2 capsule   U.S., unmanned test.

1965   Feb 17   Ranger 8 probe   U.S. Moon hard lander, crashed within 15 mi. of planned bulls-eye in southwest Mare Tranquillitatis. Transmitted 7,137 last-minute photos before impact.

1965   Feb 11   Les 1 satellite   U.S., communications sat.

1965   Mar 9   OSCAR 3 satellite   U.S., first active real-time amateur communications satellite.

1965   Mar 12   Moon/Cosmos 60   USSR rocket failed to leave Earth orbit in a secret attempt to send unmanned soft lander to Moon. Announced as Earth satellite Cosmos 60.

1965   Mar 18   Voskhod 2 capsule   USSR cosmonauts Leonov and Belyayev spent 26 hours in space. Leonov took the first-ever spacewalk (extravehicular activity or EVA), 10 minutes on a 10-ft. tether outside the capsule. They returned to Earth March 19, but waited more than a day to be picked up after they parachuted into a forest in the snowy Ural Mountains. A helicopter found them in three hours, but couldn't land on the rugged terrain. Belyayev and Leonov spent the night on the hillside, eating and keeping warm with supplies dropped by the helicopter. The nearest suitable landing site was 12 miles away and a party descended there the next day. But they didn't reach the cosmonauts until after midday and had to call for another supply drop to spend that night. It took the entire third day to hike back to the helicopter landing area. When Belyayev and Leonov reached civilization March 21, they had spent three times as long in the Urals as in space. Voskhod 2 was the first two-man flight. It was followed five days later by the U.S. two-man Gemini 3. However, the USSR already had launched three persons in Voskhod 1 in 1964.

1965   Mar 21   Ranger 9 probe   U.S. Moon hard lander, crashed as planned between the central peaks and rilles on the floor of the Alphonsus Crater near the crater's east wall. Transmitted 5,814 last-minute photos before impact. It was last in the Ranger series, which showed that mare were free of small craters, cracks and rock fields, and would support the weight of a manned spacecraft. They accurately measured the Moon's radius and mass.

1965	Mar 23	Gemini 3 capsule	U.S. astronauts Virgil I. "Gus" Grissom, John W. Young in orbit. Grissom ate first corned beef sandwich in orbit during 5-hr flight. Astronaut Wally Schirra made the sandwich before launch and gave it to Young who passed it to Grissom. The sandwich caused a congressional fuss and new rules later prevented such casual dining. Grissom and Young were the first to maneuver a ship in space, flying Gemini 3 around space during three orbits. While all the Mercury capsules had nicknames, only this first capsule in the Gemini series of two man spacecraft had a nickname, Molly Brown. After that, it became too hard for government bureaucrats to approve astronaut nicknames.
1965	Apr 6	Early Bird satellite	U.S., 1st commercial comsat, known as Intelsat 1.
1965	Apr 23	Molniya 1A sat	first USSR active real-time communications sat, elliptical orbit. Molniya 1B launched 10/14/65.
1965	May 9	Luna 5 probe	USSR successfully got an unmanned soft lander to the Moon, but it crashed on the surface.
1965	May 29	Explorer 28 sat	U.S., Interplanetary Monitoring Platform 3, radiation 120 to 163,000 miles above Earth.
1965	Jun 3	Gemini 4 capsule	U.S. astronauts White, McDivitt fly in orbit. White takes the first U.S. spacewalk outside the Gemini 4 capsule on a tether for 20 minutes.
1965	Jun 8	Luna 6 probe	USSR successfully got an unmanned soft lander off toward the Moon, but missed the Moon by 100,000 miles.
1965	Jul 16	Proton rocket	USSR, heavy-lifter launches 50,000 lbs. to orbit.
1965	Jul 18	Zond 3 probe	USSR Mars fly-by attempt outside of the usual Mars launch window, failed Mars, passed Moon at distance of 5,717 miles, sent back 25 photos of far side of Moon as communications test.
1965	Aug 21	Gemini 5 capsule	U.S., orbit, Cooper, Conrad.
1965	Oct 4	Luna 7 probe	USSR successfully got an unmanned soft lander to the Moon, but it crashed on the surface.
1965	Nov 12	Venera 2 probe	USSR, third launch of unmanned Venera to fly by planet Venus. Radio failed, probe tracked by radar as it passed within 15,000 miles on Feb 27, 1966.
1965	Nov 16	Venera 3 probe	USSR, fourth launch of unmanned Venera to planet Venus. On Mar 1, 1966: the first craft to impact Venus, the first human artifact to reach the surface of another planet. Carried USSR flag to a spot within 280 miles of planet center, but had a dead radio which failed just before entering atmosphere.
1965	Nov 23	Venus/Cosmos 96	USSR rocket failed to leave Earth orbit in a secret attempt to send unmanned atmosphere probe to Venus. Announced as Earth satellite Cosmos 96.
1965	Nov 26	Asterix 1 satellite	France's first satellite launch to orbit, Diamant rocket, 3rd nation with space rocket.
1965	Dec 3	Luna 8 probe	USSR successfully got an unmanned soft lander to

			the Moon, but it crashed on the surface.
1965	Dec 4	Gemini 7 & 6	In the first U.S. space rendezvous, astronauts Frank Borman and James Lovell flew to Earth orbit in Gemini 7 on Dec 4, followed by Wally Schirra and Thomas Stafford in Gemini 6 on Dec 15. The two craft met in orbit. It was the first time four men were in orbit at the same time.
1965	Dec 16	Pioneer 6 probe	U.S. probe of the Sun, launched into solar orbit. Measured Sun and effects on interplanetary space.
1965	Dec 21	OSCAR 4 satellite	U.S., high-altitude hamsat.

1966

1966	Jan 31	Luna 9 probe	USSR launched unmanned probe which successfully soft-landed on Moon Feb 3 at Oceanus Procellarium. The first soft-landing on Moon, first TV from lunar surface, showed rubble-strewn lava field. Transmitted 3 days, including radiation data.
1966	Feb 3	ESSA 1 sat	U.S., advanced weather satellite, photographed entire sunlit portion of Earth daily.
1966	Mar 1	Moon/Cosmos111	USSR rocket failed to leave Earth orbit in a secret attempt to send unmanned probe to the Moon. Announced as Earth satellite Cosmos 111.
1966	Mar 16	Gemini 8 capsule	U.S. astronauts Neil Armstrong and David Scott completed the first space docking, flying Gemini 8 to dock with an orbiting Agena rocket.
1966	Mar 31	Luna 10 probe	USSR launched an unmanned probe which successfully went into orbit around the Moon Apr 3, 1966. The first satellite to orbit the Moon. It had a gamma ray spectrometer to study surface chemistry. Luna 10 sent no photos.
1966	May 30	Surveyor 1 probe	U.S. Moon probe, soft landing on Moon June 2 at Oceanus Procellarum. Tested soil. Sent 11,237 surface photos of undulating maria surface, texture of freshly-plowed field, landing gear and shadow.
1966	Jun 3	Gemini 9 capsule	U.S., orbit, Stafford, Cernan.
1966	Jul 18	Gemini 10 capsule	U.S., orbit, Young, Collins.
1966	Aug 10	Lunar Orbiter 1	U.S., first of 5 satellites launched between Aug 1966 and Aug 1967 to orbit and map the Moon. L.O. 1 was the first U.S. craft to circle the Moon. After radioing 211 high resolution photos, L.O. 1 was crashed into surface to avoid conflict with L.O. 2. Lunar Orbiters 1, 2 and 3 sought Apollo landing sites, while L.O. 4 and 5 did global mapping. Lunar Orbiter series discovered excess concentrations of mass under the maria, known as mascons, photographed 99% of lunar surface.
1966	Aug 17	Pioneer 7 probe	U.S. probe of the Sun, launched into solar orbit. Found Earth's magnetosphere tail while measuring the Sun and its effects on interplanetary space.
1966	Aug 24	Luna 11 probe	USSR unmanned probe launched into orbit around the Moon. Sent no photos, but did send other data.
1966	Sep 12	Gemini 11 capsule	U.S. astronauts Pete Conrad, Richard Gordon flew Gemini 11 to a record altitude of 739.2 miles. Photographed stars, galaxies, took 2 spacewalks.

1966	Sep 20	Surveyor 2 probe	U.S. Moon probe was to soft land on the Moon, but crashed into the surface instead.
1966	Oct 22	Luna 12 probe	USSR unmanned probe launched into orbit around the Moon. Transmitted photos and other data.
1966	Nov 6	Lunar Orbiter 2	U.S., second of 5 satellites sent to orbit and map Moon. After radioing photos of 13 possible Apollo landing sites, Lunar Orbiter 2 was crashed into surface to avoid Lunar Orbiter 3.
1966	Nov 11	Gemini 12 capsule	U.S., orbit, Aldrin, Lovell.
1966	Dec 21	Luna 13 probe	USSR unmanned probe soft-landed on the Moon, transmitted pictures, used a mechanical arm digger, measured chemistry of surface.

1967

1967	Jan 27	Apollo 1 capsule	The first manned test flight of America's powerful new Saturn 1B rocket was to be Feb. 21. On Jan. 27, 25 days before scheduled lift-off, astronauts Gus Grissom, Edward White, Roger Chaffee were killed when fire broke out inside their Apollo 204 capsule while they were practicing during a simulated countdown on the Cape Canaveral, Fla., launch pad. The Apollo 204 capsule was renamed Apollo 1. A 20-month delay in manned flight followed the tragedy.
1967	Feb 4	Lunar Orbiter 3	U.S., third of 5 satellites sent to orbit and map Moon. After sending Apollo landing sites photos, L.O. 3 was crashed into surface to avoid L.O. 4.
1967	Mar 27	Mars probe	USSR rocket launch failed in a secret attempt to send an unmanned probe to land on Mars. There may have been other unannounced attempts which failed in the 1967-68 Mars launch window.
1967	Apr 17	Surveyor 3 probe	U.S. probe, soft landing on Moon. Tested soil, transmitted 6,315 surface photos.
1967	Apr 24	Soyuz 1 capsule	USSR cosmonaut Vladimir Komarov was the first to die in space flight. He was killed Apr 24 as his Soyuz 1 capsule crashed on the central-Asian plain. He had been launched on Apr 23 for 17 orbits of Earth in a flight of 26 hours 40 minutes. The test flight seemed successful, until it crashed after re-entry. A parachute opened to slow the descending capsule, but lines became snarled at 23,000 feet altitude. The parachute twisted and deflated. Soyuz 1 hit the ground at 200 miles per hour and burst into flames.
1967	May 4	Lunar Orbiter 4	U.S., fourth of 5 satellites sent to orbit and map Moon. Made first photos of Moon's south pole. Measured gravity, radiation. After radioing Apollo landing sites photos, L.O. 4 was crashed into surface to avoid L.O. 5.
1967	Jun 12	Venera 4 probe	USSR, Venus atmosphere probe, arrived Oct 18, capsule parachuted toward night side surface, radioing down to an altitude of 15 miles, finding the atmosphere to be 90-95% carbon dioxide. The first return of data from within the atmosphere.

1967	Jun 14	Mariner 5 probe	U.S., Venus flyby probe, arrived Oct 19, day after Venera 4, flew by at distance of 2,440 miles. Confirmed carbon dioxide atmosphere. Found ionosphere, but no magnetic field. Atomic hydrogen cloud at 11,800 miles altitude.
1967	Jun 17	Venus/Cosmos167	USSR rocket failed to leave Earth orbit in a secret attempt to send an unmanned atmosphere probe to Venus. Announced as Earth satellite Cosmos 167.
1967	Jun 25	various satellites	Our World TV program, five satellites bring together Beatles, London; Marc Chagall, Joan Miro, Paris; Van Cliburn, Leonard Bernstein, NY.
1967	Jul 14	Surveyor 4 probe	U.S. Moon probe was to soft land on the Moon, but crashed into the surface instead.
1967	Aug 1	Lunar Orbiter 5	U.S., fifth of 5 satellites sent to orbit and map Moon. First photos of Moon's south pole. Photos and measurements of both sides of the Moon. Radioed 212 photos, then crashed into surface.
1967	Sep 8	Surveyor 5 probe	U.S. Moon probe, made a soft landing Sep 10 at Mare Tranquillitatis (Sea of Tranquility), carrying cameras and a soil chemical analyzer in a small gold box. It found basalt, like volcanic rock on Earth, showing the Moon is a partially-evolved planet, not an undifferentiated meteorite as had been thought. Transmitted 18,006 surface photos.
1967	Nov 7	Surveyor 6 probe	U.S. Moon probe, made a soft landing, made the first lunar lift off when it hopped 10 feet.
1967	Nov 9	Apollo 4 capsule	U.S., unmanned test launch.
1967	Dec 13	Pioneer 8 probe	U.S. probe of the Sun, launched into solar orbit to measure solar radiation and its effects.

1968

1968	Jan 6	Surveyor 7 probe	U.S. Moon probe, made a soft landing near crater Tycho. Photos, soil sample analysis.
1968	Jan 22	Apollo 5 capsule	U.S., unmanned lunar module to low Earth orbit.
1968	Feb 24	pulsar	Cambridge Univ. astronomers announced finding in 1967 the first pulsating star (pulsar), known as CP 1919. It emitted pulses of naturally-generated radio energy every 1.337 seconds.
1968	Mar 2	Zond 4 probe	USSR Moon spacecraft failed in an apparent unmanned test of manned lunar landing equipment. Destroyed around Mar. 9.
1968	Mar 4	OGO-5 sat	NASA's most sophisticated Orbiting Geophysical Observatory carried science gear from Britain, Netherlands, France. Found electrical fields in bow shock where solar wind hits geomagnetic field.
1968	Apr 4	Apollo 6 capsule	U.S., last test of Saturn 5 and capsule.
1968	Apr 7	Luna 14 probe	USSR unmanned lunar orbiter. Photos and data.
1968	Sep 14	Zond 5 probe	USSR Moon spacecraft testing manned lunar landing equipment. Circled the Moon and returned to Earth in the first-ever flight to the Moon and back. Carried turtles, worms, flies, plants, seeds.
1968	Oct 11	Apollo 7 capsule	U.S. astronauts Schirra, Eisele, Cunningham flew in Earth orbit. Schirra came down with a cold. Then, during 10 days in space, Eisele and

Cunningham caught it from him, leading to a decision not to wear helmets during re-entry to keep air pressure on ear drums equalized as cabin pressure changed during descent. Seventeen years, later Schirra appeared in TV ads for a cold remedy.

1968	Oct 26	Soyuz 3 capsule	USSR, first manned rendezvous, Beregovoi.
1968	Nov 8	Pioneer 9 probe	U.S. probe of the Sun, launched into solar orbit to explore space around the Sun, measure solar radiation and its effects on interplanetary space.
1968	Nov 10	Zond 6 probe	Second USSR Moon spacecraft testing manned lunar landing equipment. Circled the Moon and returned to Earth, including a double-dip re-entry.
1968	Dec 7	OAO 2 sat	U.S., Orbiting Astronomical Observatory, Earth orbiting telescopes observed stars in ultraviolet and infrared light, gamma rays, X-rays. At 4,453 lbs., it was the heaviest American science satellite orbited up to that time.
1968	Dec 21	Apollo 8 capsule	U.S. astronauts Borman, Lovell and Anders in the first manned flight of the Saturn 5 rocket, first manned craft to leave Earth's gravity, first manned craft to orbit Moon, the first persons to see all of planet Earth at one time, the first men to travel beyond the radiation protection of Earth's magnetic field, the first men to fly to the Moon. Astronauts experienced light flashes when they closed their eyes, maybe as cosmic rays hit the optic nerve. The 147 hour trip included 10 orbits of the Moon. They did not land. During their six day flight, longest on record to that time, they photographed landing sites. Back at Earth they plunged into the atmosphere at 24,696 miles per hour, a speed no human had reached. First to spend Christmas in space. On Christmas Eve, telecast a view of Earth from 250,000 miles out, then delivered a worldwide broadcast reading from the Holy Bible, book of Genesis. "Up to that point it was the largest audience that had ever listened to a human voice," Lovell said.

1969

1969	Jan 5	Venera 5 probe	USSR, Venus atmosphere probe, arrived May 16, dropped capsule which parachuted to surface, transmitted 53 mins. down to altitude 16 miles above surface. Signals went up to Venera 5 which relayed to Earth as it continued on into solar orbit. 100 atmospheres pressure, 932 F. temperature.
1969	Jan 10	Venera 6 probe	USSR, Venus atmosphere probe, arrived May 17, dropped capsule which parachuted to surface, transmitted 51 mins. down to altitude 7 miles above surface. Signals went up to Venera 6 which relayed to Earth as it continued on into solar orbit. 100 atmospheres pressure, 932 F. temperature.
1969	Jan 14	Soyuz 4 & 5	USSR's first space docking, first USSR manned launch during a Soviet winter, first change of

clothes in space: cosmonaut Vladimir Shatalov flew to Earth orbit in Soyuz 4 on Jan 14. Boris Volynov, Alexei Yeliseyev and Yevgeny Khrunov flew Jan 15 in Soyuz 5. They rendezvoused in orbit and docked. In the first change of clothes, Yeliseyev and Khrunov donned spacesuits for a one-hour spacewalk, pulling themselves along handrails away from Soyuz 5 to Soyuz 4 airlock. It was the first spacecraft-to-spacecraft personnel exchange in orbit. Yeliseyev, Khrunov and Shatalov flew home in Soyuz 4. Volynov flew down to Earth in Soyuz 5.

1969	Feb 24	Mariner 6 probe	U.S.Mars probe, flyby July 31, over Martian equator at altitude of 2,106 miles, sending 75 photos showing fewer craters than Mariner 4 found, but a chain of huge craters, including one 310 miles diameter, had frosty rims.
1969	Mar 3	Apollo 9 capsule	U.S. astronauts McDivitt, Schweikart and Scott tested the lunar excursion module (LEM) in Earth orbit, following the first health delay in American spaceflight history. Apollo 9 blast off had been delayed 4 days from Feb 28 to Mar 3 when the crew came down with sniffles.
1969	Mar 27	Mariner 7 probe	U.S.Mars probe, flyby Aug 5 over southern hemisphere, sending 126 photos including south pole ice cap, evidence of volcanos, atmosphere density about 0.7 of Earth at sea level, mostly carbon dioxide.
1969	May 18	Apollo 10 capsule	U.S. astronauts Stafford, Cernan, and Young orbited the Moon in a full dress rehearsal for a lunar landing. In the first American change of clothes in space, Stafford and Cernan donned spacesuits in the command service module for a ride down to within ten miles of the Moon's surface in the lunar excursion module.
1969	Jun	Biosatellite 3	U.S. Biosat 3, three weeks before 1st men landed on Moon. Biosat monkey passenger supposed to orbit for a month, but was brought down ill from loss of body fluids after 9 days, to die shortly after landing. Later USSR and U.S. biological sats carried rats, monkeys, newts, flies, fish.
1969	Jun 14	Moon probe	USSR launch failed in an unannounced secret attempt to send an unmanned lunar rover and sample return mission to the Moon.
1969	Jun 21	IMP-7/Explorer 41	U.S., Interplanetary Monitoring Platform 7 (IMP 7), Explorer 41. In 1977, from ratio of beryllium 9 and 10 isotopes in cosmic ray particles, found cosmic rays 20 million years old.
1969	Jul 13	Luna 15 probe	Apollo 11 was to be launched Jul 15 to make the first manned landing on the Moon Jul 20. In an attempt to dull Apollo 11 publicity, USSR sent Luna 15 on Jul 13 on a mission to land unmanned lunar rover on Moon, pick up soil sample, return

			to Earth. Luna 15 arrived in lunar orbit 48 hours before Apollo 11, but attempt failed. Probe orbited Moon 52 times, then crashed onto surface Jul 21.
1969	Jul 15	Apollo 11	To protect the astronauts from germs, a NASA doctor called off a dinner President Richard M. Nixon wanted to give for astronauts Armstrong, Aldrin, Collins the night before they were to leave for the first Moon landing. Later Collins said, "I'm sure presidential germs are benign."
1969	Jul 16	Apollo 11 capsule	U.S. Apollo 11 astronauts Neil Armstrong, Buzz Aldrin, and Michael Collins fly to the Moon.
1969	Jul 20	LEM Eagle	Apollo 11 astronauts Armstrong and Aldrin landed the lunar excursion module on the Moon's Sea of Tranquility. "Houston, Tranquility Base here. The Eagle has landed," Armstrong said. "Roger, Tranquility. You've got a bunch of guys about to turn blue," replied Houston. Armstrong stepped outside the LEM, onto the lunar surface, and said, "That's one small step for man...one giant leap for mankind." On the Moon, Armstrong and Aldrin drank hot coffee with their hot dogs, bacon squares, canned peaches, sugar cookies. They left behind a gold olive branch, cameras, tools, boots, bags, containers, armrests, brackets, a U.S. flag and flagstaff, a solar wind experiment mast, a TV camera and power cable, a Moonquake detector, a laser light reflector, the Eagle lander, life-support backpacks, an Apollo 1 shoulder patch remembering three dead astronauts, medals commemorating two dead cosmonauts, a 1.5-in. silicon disk with goodwill speeches from heads of 23 nations, and a plaque, "Here Men from the Planet Earth first set foot upon the Moon, July 1969 A.D. We came in peace for all mankind."
1969	July 21	CSM Columbia	Armstrong and Aldrin blasted off from the lunar surface on a return trip to the command service module Columbia in lunar orbit, toppling the American flag they had planted in the lunar soil into the dirt as they blasted.
1969	Jul 22	Apollo 11	Astronauts Armstrong, Aldrin, Collins became the first men to be launched from the Moon to Earth. Armstrong's Moon landing left him the world's best-known spaceman.
1969	Aug 7	Zond 7 probe	Third USSR Moon spacecraft testing manned lunar landing equipment. USSR's first deep space color photos. Circled the Moon and returned to Earth.
1969	Aug 27	Pioneer E probe	U.S., Sun probe to be in solar orbit, but rocket malfunctioned, destroyed by range safety officer.
1969	Sep 23	Moon/Cosmos300	USSR rocket failed to leave Earth orbit in attempt to send to the Moon an unmanned lunar rover to bring back a soil sample. Announced as Earth satellite Cosmos 300.
1969	Oct 11	Soyuz 6, 7, 8	For the first time three spacecraft and seven

cosmonauts were in Earth orbit at the same time; the USSR launched three capsules to orbit. Georgi Shonin and Valeri Kubasov flew in Soyuz 6 on Oct 11. Anatoli Filipchenko, Vladislav Volkov and Viktor Gorbatko flew Oct 12 in Soyuz 7. Vladimir Shatalov and Alexei Yeliseyev flew Oct 13 in Soyuz 8. Each craft flew down to Earth after 118 hours in space, landing a day apart just as they had taken off. While in orbit, Shonin and Kubasov did the first welding of metal in space.

1969   Oct 22   Moon/Cosmos305   USSR rocket failed to leave Earth orbit in attempt to send to the Moon an unmanned lunar rover to bring back a soil sample. Announced as Earth satellite Cosmos 305.

1969   Nov 14   Apollo 12 capsule   U.S. astronauts Conrad, Bean, Gordon to Moon. Conrad, Bean landed at Oceanus Procellarum Nov 19 in second manned lunar landing. Hiked to Surveyor 3 which landed in 1967, finding probe merely dusty. Retrieved 24 lbs. of glass and metal from Surveyor and 75 lbs. of soil for analysis.

1970

1970   Feb 11   Ohsumi satellite   Japan launched its first satellite to orbit, making it the fourth nation with a space rocket.

1970   Feb 19   Moon probe   USSR rocket failed to leave Earth orbit Feb 19 in a secret attempt to send an unmanned lunar rover and sample return mission to the Moon. Altogether, the USSR had failed in four secret attempts to launch a huge Moon rocket. It wasn't until 1991 the USSR revealed its N1 booster designed to send cosmonauts to the Moon. The rocket had four stages with 43 engines, while America's Saturn 5 had three stages with 11 engines. One N1 managed to climb to an altitude of 70,000 feet before failing. The Moon program then was scrapped.

1970   Apr 11   Apollo 13 capsule   U.S. astronauts Lovell, Swigart and Haise attempted a flight to the Moon. Swigart had replaced Mattingly at the last moment, following the first space measles scare. Apollo 13 flight was interrupted on the way to the Moon when an oxygen tank in the command service module ruptured. There was an explosion and the flight was aborted. Apollo 13 swung around the Moon, without landing, and headed back to Earth. Swigart, Lovell and Haise used the lunar excursion module as a lifeboat for oxygen and power. The pre-flight scare: Apollo 13 launch was not delayed because a back-up crew was ready. Pilot Mattingly had to be replaced April 6 by Swigart when doctors realized Mattingly had been in contact with an astronaut who had come down with German measles. Mattingly was judged not immune. During the aborted flight, Mattingly was safe at home where he did not develop measles. Later,

Mattingly piloted Apollo 16 to Moon in 1972. In Apollo 13, Lovell became the first man to fly twice to the Moon. He had flown in Apollo 8 which didn't land, and Apollo 13 didn't land.

1970	Apr 24	Mao 1 satellite	China, first satellite launch to orbit, Long March rocket, 5th nation with space rocket.
1970	Jun 1	Soyuz 9 capsule	USSR cosmonaut Vitali Sevastyanov flew a record 18 days in space. During the five years after his flight, he became a popular TV commentator on the USSR program "Man, Earth and Universe." In May 1975, he returned to space in Soyuz 18B on a flight to the Salyut 4 space station for a record 63 days in space.
1970	Aug 17	Venera 7 probe	USSR probe, arrived Venus Dec 15, 1970, dropped instrument capsule for first soft landing on surface. Radio signals weakened after 35 minutes, indicating the capsule reached planet's night-side surface. Computer enhancement extracted another 26 minutes of data. Temperatures were recorded as 887 deg. Fahrenheit, pressure at 90 atmospheres.
1970	Aug 22	Venus/Cosmos359	USSR rocket carrying Venus lander failed to leave Earth orbit. Announced as Earth sat Cosmos 359.
1970	Sep 12	Luna 16 probe	USSR Moon probe landed at Mare Foecunditatis Sep 20, picked up 101 grams of soil, sealed it in a box, launched itself back to Earth. First automated sample retrieval from another celestial body.
1970	Oct 14	Intercosmos 4	USSR science satellite launched from Kapustin Yar, carrying Czech, East German and Russian gear to monitor solar ultraviolet light and X-rays.
1970	Oct 20	Zond 8 probe	Fourth and last USSR Moon spacecraft testing manned lunar landing equipment. USSR's first deep space color photos of Earth. Circled the Moon and returned to splashdown in the Indian Ocean.
1970	Nov 10	Luna 17 probe	USSR Moon probe, landed at Mare Imbrium Nov 17, sent out instrumented Lunokhod 1 lunar rover vehicle to get soil samples and analyze density and composition. Lunokhod, first wheeled vehicle on the Moon, broadcast TV over its 2.5 mile range and observed stars with an X-ray telescope.
1970	Dec 11	NOAA 1 sat	U.S. weather satellite in polar orbit. Camera for continuous day and night pictures.
1970	Dec 12	Uhuru/Explorer 42	Marjorie Townsend, the first woman to launch a spacecraft, sent the U.S. Explorer 42 satellite (SAS-1) to Earth orbit from Italy's San Marco Platform in the Indian Ocean off Kenya. Townsend named the satellite Uhuru, a Swahili word for "freedom," in honor of Kenya's independence day Dec 12. Uhuru detected a black hole in the constellation Cygnus on Jan 30, 1971, when the small astronomy satellite received X-rays and natural radio waves from the direction of a faint blue star which had a tiny, unseen, dark companion, more massive than a neutron star.

1971

1971   Jan 31   Apollo 14 capsule   U.S. astronauts Shepard, Mitchell, Roosa fly to Moon. Shepard, Mitchell land Fra Mauro Formation hills 110 miles east of Apollo 12 site, picking up 96 lbs. of rocks. Third manned landing on Moon. Shepard hit golf ball hundreds of yards with jury-rigged six-iron. Set off 2 small "Moonquakes" to be read by seismic monitors planted by earlier Apollo visitors, finding 28-ft. thick powdery layer. Organic compounds were in soil, but no micro-organisms were found so Apollo 14 astronauts were the last quarantined upon return to Earth.

1971   Apr 19   Salyut 1 station   The USSR launched the first-ever space station, into 200-mi.-high Earth orbit from Baikonur Cosmodrome. Cosmonauts Nikolai Rukavishnikov, Vladimir Shatalov, Alexei Yeliseyev were the first men to go to a space station, leaving the USSR April 22 in Soyuz 10. They docked at Salyut 1 Apr 24, stayed in dock less six hours, then hurried home. Soyuz 11 crew went in June, satyed 23 days. Salyut 1 then fell from orbit Oct 11.

1971   May 8   Mariner 8 probe   U.S. Mars probe, Centaur rocket failed.

1971   May 10   Mars/Cosmos419   USSR rocket carrying Mars orbiter and lander failed to leave Earth orbit. Announced as Earth sat Cosmos 419.

1971   May 19   Mars 2 probe   USSR unmanned Mars orbiter and lander, launched just 11 days before U.S. Mariner 9. However, Mariner 9 arrived in orbit over Mars Nov 13 while Mars 2 went into low orbit over the planet Nov 27, so Mars 2 was not the first spacecraft to orbit another planet. Mars 2 did study surface and atmosphere from orbit, but its lander crashed.

1971   May 28   Mars 3 probe   USSR unmanned orbiter with lander. First landing on Mars. Went into low orbit over planet Dec 2, 1971. Orbiter studied surface and atmosphere. Lander was first to soft land on Mars, sent first panoramic image of Martian surface, stopped transmitting after two minutes.

1971   May 30   Mariner 9 probe   U.S. Mars orbiter Mariner 9 was launched May 30, just 11 days after the USSR's unmanned Mars 2 orbiter/lander was launched May 19. However, Mariner 9 arrived in orbit over Mars Nov 13 while Mars 2 went into orbit there Nov 27, making Mariner 9 the first spacecraft to orbit another planet. As Mariner 9 arrived, planet-wide dust storm was raging below. After the storm, Mariner 9 sent 7,000 photos of plains, volcano mountains larger than on Earth, a 2,500-mi. Grand Canyon along equator. River and stream beds showed past erosion. Some said life may have evolved earlier. Mapped 85% of Mars. First photos of Martian

1971	Jun 6	Salyut 1 station	moons Deimos and Phobos. Cosmonauts Georgi Dobrovolsky, Vladislav Volkov, Viktor Patsayev flew in Soyuz 11 to become the first crew from Earth to work in a space station. They ended as the USSR's second space tragedy. They stayed in the station 23 days. Then, not wearing spacesuits, all three died on the flight home as air escaped Soyuz 11 during re-entry.
1971	Jun 30	Soyuz 11 capsule	The second, third and fourth men to die during space flight were USSR cosmonauts Georgi Dobrovolsky, Vladislav Volkov and Viktor Patsayev killed June 30. They had been launched to orbit June 6 in Soyuz 11 for a visit to the new Salyut 1 space station where they stayed a record 23 days. But, the three were not wearing spacesuits on the flight home as air escaped the Soyuz 11 capsule during re-entry. They were found dead when the capsule was opened on the ground after an automatic landing. It was the last USSR three-man flight for nearly a decade.
1971	Jul 26	Apollo 15 capsule	U.S. astronauts Scott, Irwin and Worden flew to Moon; Scott, Irwin landed Jul 30 at Hadley Rille. Apollo 15 made Scott the second man to fly twice to Moon: previously Apollo 9. Scott, Irwin drove the first Moon rover, a four-wheel battery-powered vehicle, 17 miles along front of Apennine Mountain. They sunk probes which revealed a hot interior of Moon, maybe radioactive decay. Picked up 169 lbs. soil, rocks. Fourth manned Moon landing. Worden took first deep-space spacewalk on their way home from the Moon. At 197,000 miles from Earth, he stepped outside Apollo 15 for 16 minutes to retrieve two film cassettes with pictures he had made of the Moon from lunar orbit.
1971	Sep 2	Luna 18 probe	USSR Moon lander and soil sample return mission. Lander crashed on the Moon.
1971	Sep 28	Luna 19 probe	USSR probe launched to low orbit over Moon for photography and gravity, gamma ray and magnetic field measurements.
1971	Oct 3	Luna 19 probe	USSR probe, Moon orbiter.
1971	Oct 28	Black Knight 1 sat	Great Britain, first satellite launch to orbit, 6th nation with space rocket.

1972

1972	Feb 14	Luna 20 probe	Second USSR sample return mission. Landed on Mare Crisium, collected 5 oz. sub-surface soil with hollow drill, put soil in box, returned to USSR.
1972	Mar 3	Pioneer 10 probe	U.S. probe launched from Cape Canaveral on a Jupiter flyby, arriving at Jupiter Dec 3, 1973. First spacecraft to explore planets in the outer Solar System, and interplanetary space beyond.
1972	Mar 22	Venera 8 probe	USSR, Venus atmosphere probe and lander, transmitted data 50 minutes, on way down and from day side surface, analyzed soil.

1972 Mar 31 Venus/Cosmos482 USSR rocket carrying Venus atmosphere probe and lander failed to leave Earth orbit. Announced as Earth satellite Cosmos 482.

1972 Apr 16 Apollo 16 capsule U.S. astronauts Young, Duke, and Mattingly flew to the Moon; Young and Duke landed in lunar highlands at Cayley Formation Apr 20. Fifth manned lunar landing. Second lunar rover. Collected 207 lbs. of rock, including one rock 4.25 billion years old. Apollo 16 made Young the third man to fly twice to the Moon: he previously flew in Apollo 10.

1972 Jul 23 Landsat/ERTS-1 U.S., Earth Resources Technology Satellite 1, later called Landsat 1, in polar orbit, shoots photos and scans Earth 14 times a day, recording forests, oceans, deserts, croplands, urban areas.

1972 Aug 21 Copernicus/OAO-3 U.S., Orbiting Astronomical Observatory 3, also known as Copernicus. Earth orbiting telescope observed stars in ultraviolet light. Heavier than OAO-2, at 4,860 lbs. OAO-3 was the heaviest American astronomy satellite orbited up to that time. It contained more than 328,000 parts. In 1973, Copernicus found a low ratio of heavy hydrogen to hydrogen in the interstellar medium, pointing to a continuously expanding Universe. Found heavier elements more in stars than in gas clouds, suggesting elements condensed into dust grains. In 1974, found supernova explosions leave low-density cavities in space filled with higher temperature gas. In 1976, found most neutral interstellar gas to be in small dense clouds, not uniformly distributed.

1972 Sep 23 IMP-9/Explorer 47 U.S., Interplanetary Monitoring Platform 9 (IMP 9), Explorer 47. In 1977, from ratio of beryllium 9 and 10 isotopes in cosmic ray particles, found cosmic rays 20 million years old.

1972 Dec 7 Apollo 17 capsule U.S. astronauts Cernan, Schmitt, Evans flew to the Moon. Cernan, Schmitt made man's sixth and last landing Dec 11 in the Littrow Valley at the foot of the Taurus Mountains. They took longest ride in a Moon car, driving lunar rover 21 miles at up to nine mph. Cernan snagged a hammer and ripped the fender on the rover. He patched it with plastic Moon maps. He dented the rover's tires driving over rocks. Schmitt was first geologist in space. Most colorful stuff found on the Moon was orange glass near Shorty Crater, suggesting possible ice within the Moon. They collected 243 lbs. rocks. In Apollo 17, Cernan became the fourth man to fly twice to the Moon: previously in Apollo 10. On Dec 14, Cernan was the last man on the Moon. He left a plaque, "Here Man completed his first exploration of the Moon December 1972 A.D. May the spirit of peace in which we came be reflected in the lives of all mankind."

1973

1973	Jan 8	Luna 21 probe	USSR, unmanned soft landing on the Moon. Prospected the surface with the rover Lunokhod-2 and returned data until June 3.
1973	Apr 3	Salyut 2 station	USSR launched 2nd first-generation space station to 200-mi.-high Earth orbit from Baikonur Cosmodrome. Station broke into 25 pieces Apr 14; was ever manned; descended May 22.
1973	Apr 5	Pioneer 11 probe	U.S. probe, twin of Pioneer 10, launched from Cape Canaveral to fly by Jupiter and Saturn, arrived Jupiter Dec 2, 1974, then Saturn Sep 1, 1979. Discovered Saturn's 11th moon, two new rings, magnetic field.
1973	May 11	Cosmos 557 stn	USSR, first-generation space station launch failed.
1973	May 14	Skylab 1 station	Skylab was America's first and only space station, launched unmanned to Earth orbit on the last Saturn 5 Moon rocket. The station was 84.2 feet long and weighed 84 tons, four times the size of the USSR's Salyut space station. Including orbital workshop, airlock, multiple docking adaptor and Apollo telescope mount, it weighed 220,460 lbs. After three astronaut visits in 1973-74, Skylab re-entered atmosphere and burned July 11, 1979.
1973	May 25	Skylab 2 capsule	U.S. astronauts Charles Conrad, Joseph Kerwin and Paul Weitz were the first persons to go to a U.S. space station when they opened Skylab May 25 and stayed 28 days. They spacewalked for 5 hrs 41 mins to repair outside of station. They set manned flight duration record of 28 days 49 mins.
1973	Jul 21	Mars 4 probe	USSR unmanned Mars orbiter. Braking rocket failed, preventing going into orbit, making it a fly by as it overshot Mars in Feb 1974.
1973	Jul 25	Mars 5 probe	USSR unmanned Mars orbiter to relay radio signals from landers Mars 6 and Mars 7. Mars 5 went into orbit around the planet Feb 12, 1974. Photos of Mars. Stopped transmitting days later.
1973	Jul 28	Skylab 3 capsule	Bean, Garriott, Lousma were the second team to go to the U.S. space station. They photographed a solar flare and spacewalked for 13 hrs 44 mins to repair outside of station. They set manned spaceflight duration record of 59 days 11 mins.
1973	Aug 5	Mars 6 probe	USSR unmanned Mars fly by and lander. Flew past Mars Mar 1974, dropped lander. Signals from lander stopped moments before landing.
1973	Aug 9	Mars 7 probe	USSR unmanned Mars fly by and lander. Flew by Mars on Mar 1974, dropped lander, but rocket failure caused it to miss the planet.
1973	Nov 3	Mariner 10 probe	U.S. probe launched to fly by Venus and Mercury, passed Venus on Feb 5 1974, then on to make the first of 3 encounters with Mercury Mar 29 1974. The first double planet reconnaissance mission.
1973	Nov 16	Skylab 4 capsule	U.S., astronauts Carr, Gibson, Pogue to the space

station. They observed Kohoutek, the first comet photographed from orbit, snapped 200,000 photos of the Sun, and spacewalked 4 times for total of 22 hrs 21 mins to make repairs. Their zero-gravity game involved a dry-roasted nut escaping its tin, floating away in the 84-ft. station. When it wandered by, an astronaut would open his mouth and push off a wall toward it. If he didn't aim just right, the peanut would bounce off his face, necessitating a chase. They had Velcro darts and board, but darts wobbled toward the target. They had 3 balls which seemed never to stop bouncing, often becoming lost. They set a manned spaceflight duration record of 84 days, still the record for American astronauts. After they left, Feb. 8, 1974, the station was not used again. Skylab burned in the atmosphere July 11, 1979.

1973   Dec 3   Pioneer 10 probe   U.S. Jupiter probe, launched Mar 2, 1972, made the first Jupiter fly-by Dec 3, passing within 80,530 mi. of the giant planet's cloud tops. Enroute, discovered low density of debris in Asteroid Belt which had been feared as hazard to space travel. Snapped first close-up photos of Jupiter. Found Great Red Spot to be a hurricane. Found no solid surface, planet mostly liquid hydrogen. Tinted bands of clouds from air masses rising and falling in atmosphere. Found huge magnetic field extending out to 420 times the planet's radius. The field trapped radiation so intense it interfered with Pioneer 10 instruments. Found Jupiter gives off more energy than it receives from Sun. Found powerful electrical field linked Jupiter's moon Io to planet Jupiter. Pioneer 10 was first to cross the Asteroid Belt, first to fly by Jupiter, first to radio pictures of that giant gas planet to Earth, first to chart Jupiter's intense radiation belts, locate Jupiter's magnetic field, show that the planet is mostly liquid, and measure the mass of Jupiter's four planet-size moons Europa, Ganymede and Callisto. Gravity flung Pioneer 10 out of Solar System. In June 1983 Pioneer 10 became first interstellar probe. Pioneer 11 was twin to Pioneer 10.

1973   Dec 18   Soyuz 13 capsule   During the first time five persons were in orbit at the same time, the USSR manned flight Soyuz 13 was in space while the U.S. space station Skylab was manned. Cosmonauts Pyotr Klimuk and Valentin Lebedev were not on a trip to a space station, but on an astronomy flight using an ultraviolet camera in Soyuz to photograph solar X-rays and stars obscured by Earth atmosphere. Astronauts in Skylab were Carr, Gibson, Pogue. Klimuk and Lebedev were the first cosmonauts to spend Christmas in space. They made 10,000 spectrograms of 3,000 stars. Found no unusual stars on Christmas Eve.

1974

1974	Feb 5	Mariner 10	U.S. probe launched Nov 3, 1973, to fly by Venus and Mercury, passed within 2,980 mi. of Venus on Feb 5, 1974. Photographed Venus clouds in visible and ultraviolet light. Top layer of clouds rapidly circling planet. Found abundance of atomic oxygen above atmosphere, suggesting water vapor in the past, indicating Venus once had an ocean. Confirmed 95% carbon dioxide atmosphere, surface air pressure 95 to 100 atmospheres, surface temperature 842 deg. Fahrenheit. Mariner 10 then used a gravity assist from Venus to fly on toward the first of its three encounters with Mercury on Mar 29, 1974. The first double planet reconnaissance mission.
1974	Mar 29	Mariner 10	U.S. probe launched Nov 3, 1973, to fly by Venus and Mercury, passed Venus on Feb 5, 1974, used gravity assist from Venus to fly on toward the first of its three encounters with Mercury, flying by at 437 mi. above Mercury on Mar 29, 1974. Photos of heavily cratered, lunar-like surface, one large basin like Mare Imbrium on Moon, sharp cliffs. Dayside temperature of 950 deg. Fahrenheit. Nightside temp of −346 deg. F. Mariner 10 returned to Mercury Sep 21 and Mar 16, 1975.
1974	Apr	Westar 1 satellite	U.S., first sat for inside-U.S. communications.
1974	May 29	Luna 22 probe	USSR unmanned lunar orbiter launched to photograph Moon. Sent data for more than a year.
1974	May 30	ATS-6 sat	U.S., Applications Technology Satellite 6, communications satellite in stationary orbit. Transmitted educational and informational TV, plus telephone, telegraph, fax, to small portable ground stations in developing countries. Relay for Apollo-Soyuz U.S.-USSR joint flight in 1975.
1974	Jun 24	Salyut 3 station	USSR launched 3rd first-generation space station from Baikonur Cosmodrome, 200-mi.-high orbit. Stayed in orbit until Aug 24, 1975.
1974	Sep 21	Mariner 10	Second fly by of Mercury. U.S. probe launched Nov 3, 1973, passed Venus Feb 5, 1974, encountered Mercury, Mar 29, 1974, Sep 21, 1974, and Mar 16, 1975.
1974	Oct 28	Luna 23 probe	USSR unmanned soft landing on the Moon. Drill was damaged, but soil sample returned to Earth.
1974	Dec 2	Pioneer 11 probe	Jupiter fly by of U.S. probe, twin of Pioneer 10, launched Apr 5, 1973. Flew by at 67,800 mi. above planet. Colorful photos of the biggest planet and large moons Europa, Ganymede, Callisto. Found 4 new small satellites, bringing Jupiter's total to 16. Found Jupiter radiates nearly twice the heat it receives from the Sun, also is the

strongest emitter of natural radio signals in the Solar System after the Sun. Pioneer 11 then flew on toward Saturn where it arrived Sep 1, 1979.

1974	Dec 10	Helios 1 probe	West German Sun observatory, U.S. launched from Cape Canaveral into solar orbit.
1974	Dec 26	Salyut 4 station	USSR launched its fourth first-generation space station, weighing 41,666 lbs., into 200-mi.-high Earth orbit from Baikonur Cosmodrome. Stayed in orbit until Feb 2, 1977.

1975

1975	Mar 16	Mariner 10	Third fly by of Mercury. U.S. probe launched Nov 3, 1973, at Venus Feb 5, 1974, at Mercury Mar 29, 1974, Sep 21, 1974, Mar 16, 1975.
1975	Jun 8	Venera 9 probe	USSR Venus orbiter with lander. First successful Venus orbiter. Lander dropped Oct 22, first photos ever from Venus surface. Earth-like amount of uranium, potassium, thorium.
1975	Jun 14	Venera 10 probe	USSR Venus orbiter with lander. Arrived, dropped lander Oct 25, high-resolution photos from surface, transmitted 65 mins. Earth-like amount of uranium, potassium, thorium.
1975	Jul 15	Apollo-Soyuz	Apollo 18 astronauts Stafford, Brand and Slayton, and Soyuz 19 cosmonauts Leonov and Kubasov in the first U.S.-USSR space rendezvous. They docked at noon EDT July 17. It was a political maneuver during a time of Cold War detente. The astronauts and cosmonauts held the first press conference from space. They lunched on borscht, chicken and turkey. Slayton was the first person over 50 in space. Apollo-Soyuz carried ultraviolet light and X-ray telescopes for astronomy, and electrophoresis and crystal growth equipment. An extreme-ultraviolet telescope saw white dwarf stars and a low density of gas around the Sun. A total of 7 persons were in orbit: three in Apollo 18; two in Soyuz 19; plus cosmonauts Klimuk and Sevastyanov at the USSR's Salyut 4 space station. Seven had been in space at one time before, in Soyuz 6, Soyuz 7, and Soyuz 8, in 1969. Probably the least-known space trip: while millions watched the Apollo-Soyuz joint flight, no one paid attention to Klimuk and Sevastyanov at Salyut 4. Launched May 24, 1975, in Soyuz 10-B, they set a record of 63 days stay in space.
1975	Aug 9	Cos-B sat	European Space Agency X-ray satellite launched by NASA. In 1978, it recorded X-rays from quasar OX-169 with intensity changing every 6 hours. At that rate, if the X-rays were matter falling into a black hole, the black hole would have a mass one million times the mass of the Sun.
1975	Aug 20	Viking 1 probe	U.S. Mars probe, lander parachuted to the Chryse Planitia desert on the Martian surface Jul 20, 1976, with weather station, inorganic and organic

chemistry labs to analyze soil, seismology station and TV transmitter. Result of search for live microbes in the soil was ambiguous. Orbiter photographed, scanned from above. First U.S. landing on another planet.

1975   Sep 9   Viking 2 probe   U.S. Mars probe, lander parachuted to the Utopia Planitia plain Sep 3, 1976. Equipped like Viking 1. Result of microbes search also ambiguous. Orbiter photographed, scanned planet. Second U.S. landing on another planet.

1975   Oct 16   GOES-1 sat   U.S., Geostationary Operational Environmental Satellite, 1st weather sat in geosynchronous orbit.

1975   Dec   Raduga satellite   USSR stationary comm sat.

1976

1976   Jun 22   Salyut 5 station   USSR launched 5th first-generation space station from Baikonur Cosmodrome, 200-mi.-high orbit. Stayed in orbit until Aug 8, 1977.

1976   Aug 9   Luna 24 probe   USSR's third and last unmanned lunar soil sampler, soft landed at Mare Crisium Aug 18, automatic drill penetrated 6.6 ft., retrieved 3.9 oz. subsoil, sealed it in a box and blasted off for Earth.

1976   Oct 26   Ekran satellite   USSR stationary communications satellite.

1977

1977   Mar 10   Uranus planet   Cornell Univ astronomers Elliott, Dunham, Mink, while flying in NASA's Kuiper Airborne Observatory (KAO) and looking at a faint star intermittently blocked by the planet, spotted rings around the planet Uranus.

1977   Aug 3   Bio-Cosmos 936   USSR biosatellite launched from Plesetsk, carried experiments from USSR, U.S., France, Poland, Romania, Czechoslovakia, Bulgaria, Hungary, East Germany. Landed Aug 22 in Siberia.

1977   Aug 12   HEAO-1 sat   U.S., the heavy 6,945-lb. High Energy Astronomical Observatory 1 (HEAO-1) was launched on a powerful Atlas-Centaur rocket, carrying telescopes to map X-ray and gamma ray sources in deep space. The X-ray telescope found 1,500 sources and a universal background gas enveloping galaxies. The gamma ray telescope showed the strange galaxy Centaurus A emits gamma rays at one million electron volts energy.

1977   Aug 12   Enterprise shuttle   The space shuttle prototype was called OV-101 as construction started in 1974. After 100,000 Star Trek TV fans wrote in, President Gerald Ford named it Enterprise in 1976. NASA didn't like the name, wanting to call the shuttle Constitution. The test bed space shuttle Enterprise had no engines and never went to space. The 130-ton prototype only flew bolted to the back of a Boeing 747 jet. The plane dropped Enterprise to test gliding. Astronauts Fred Haise and Charles Fullerton were at controls for the first flight Aug 12, 1977. Five test flights were made over two years. Haise and Fullerton piloted flights one, three, five. Joe Engle and Richard Truly piloted flights two, four. Later, Enterprise was used to get launch pads ready for real shuttles. Then the vehicle was flown city to city for publicity. It was a hit with the public. Enterprise was to have gone back to the workshop for engines, but it became cheaper to make a new ready-to-fly structure-test orbiter. Enterprise was stripped and displayed at Kennedy Space Center, then was flown to Washington, D.C., for the Smithsonian Institution to build a museum around it. The museum has not been built and, today, Enterprise sits, paint peeling, among weeds in the outback of Washington's Dulles Int'l Airport.

1977	Aug 20	Voyager 2 probe	U.S., outer Solar System probe, was to fly by Jupiter, Saturn; extended after Saturn to Uranus, Neptune for Grand Tour. Flew by Jupiter July 9, 1979, Saturn Aug 25, 1981, Uranus Jan 24, 1986, Neptune Aug 24, 1989. Now on its way out of Solar System to interstellar space.
1977	Sep 5	Voyager 1 probe	U.S., outer Solar System probe, flew by Jupiter Mar 5, 1979, Saturn Nov 12, 1980. Now on its way out of Solar System to interstellar space. In 1990, first portrait of Solar System.
1977	Sep 29	Salyut 6 station	USSR, 1st second-generation space station from Baikonur Cosmodrome, 200-mi.-high orbit. Redesign had dock at each end so station could be refueled by visiting unmanned Progress freighter with Soyuz capsule in other dock. Radiotelescope with 32.8-ft. dish was deployed. Salyut 6 stayed in orbit until Jul 29, 1982.
1977	Oct 22	ISEE-1/2 sats	International Sun-Earth Explorers 1 & 2 (ISEE). Pair of U.S./European solar observatories launched into long looping Earth orbits on 1 Delta rocket, swinging out to 85,000 mi., back to 170 mi., each revolution. The larger ISEE-1 managed by NASA; the smaller, maneuverable ISSE-2 by European Space Agency. Both observed Sun during 11 year sunspot cycle which had started June 1976. Measured effects of solar activity on Earth's upper atmosphere, ionosphere and magnetosphere.
1977	Nov 1	Chiron asteroid	An asteroid named Chiron, minor planet 2060, was found. Orbiting mostly beyond Saturn, the 200-mi.-wide body takes 50 years to circle Sun.

1978

| 1978 | Jan 26 | IUE sat | International Ultraviolet Explorer (IUE) satellite in stationary Earth orbit. Built by U.S., U.K., ESA, controlled from Goddard Space Flight Center, Maryland, and ESA's Cillafranca Station, Spain. Maneuvers to point at deep space astronomy targets. Discovered a hot, 180,000 deg. Fahrenheit, halo of oxygen, sulphur, iron, silicon |

and carbon atoms around the Milky Way galaxy. Found star Capella has chromosphere like the Sun.

1978 Jan   Salyut 6 station   First refueling in orbit from an unmanned Progress tanker to Salyut 6 space station. Soyuz 26 cosmonauts Grechko and Romanenko at station.

1978 Mar 2   Salyut 6 station   When Czech cosmonaut Vladimir Remek, the son of the Czechoslovakian defense minister, flew in Soyuz 28 to visit Salyut 6 space station, he was the first non-USA/non-USSR person in space. Remek was the first Intercosmos cosmonaut.

1978 Mar 16   Salyut 6 station   Old U.S. man-in-space record of 84 days from Skylab in 1974 was broken by USSR cosmonauts Yuri Romanenko, Georgi Grechko who did a 96 day tour at Salyut 6.

1978 May 20   Pioneer-Venus 1   U.S. Venus atmosphere probe and mapper, also known as Pioneer 12. In orbit around Venus Dec 4. Radar and gravity maps of the planet surface topography included continents Aphrodite and Ishtar. Also studied solar wind. Paired with Pioneer-Venus 2 (Pioneer 13).

1978 Jun 22   Charon moon   U.S. astronomer James W. Christy at the U.S. Naval Observatory, Flagstaff, Arizona, found Pluto's moon Charon, 48 years after Clyde Tombaugh discovered the planet. Charon orbits the planet every 6.4 Earth days.

1978 Jun 27   Salyut 6 station   Miroslaw Hermaszewski, first Polish cosmonaut, flew to space station in the Soyuz 30 capsule.

1978 Jun 27   Seasat-1 sat   U.S. ocean observer satellite in polar orbit with synthetic aperture radar. High resolution photos day and night through clouds.

1978 Aug 8   Pioneer-Venus 2   U.S. Venus multiple-purpose probe, also known as Pioneer 13 and Pioneer-Venus Multi-Probe. In orbit around Venus Dec 9, dropped 4 probes, then main spacecraft also descended into atmosphere. Measured wind, circulation, pressure, temperature, composition of atmosphere. Found extra heavy hydrogen, relative to hydrogen, indicating an ocean of water may have evaporated. Paired with Pioneer-Venus 1 (Pioneer 12).

1978 Aug 12   ISEE-3/ICE   International Sun-Earth Explorer 3 (ISEE-3). Solar observatory launched by NASA into halo orbit at Earth-Sun libration point where the pull of gravity from Earth and Sun are equal, one million miles from Earth toward the Sun. ISEE-3 warned of solar flares before magnetic disturbances reached Earth. Renamed International Cometary Explorer (ICE) when repointed to explore the inrushing Comet Giacobini-Zinner on Sep 11, 1985. ICE was used again, on Mar 25, 1986, to study Comet Halley at a distance of 17.46 million miles.

1978 Aug 26   Salyut 6 station   Sigmund Jahn, first East German cosmonaut, flew to space station in the Soyuz 31 capsule.

1978 Sep 9   Venera 11 probe   USSR Venus flyby craft. Arrived at Venus Dec 21 and dropped a soft-lander through the dense atmosphere. Found lightning in the clouds, tape recorded sound of thunder. Lander reached the surface, transmitted 95 minutes. No photos.

1978 Sep 14 Venera 12 probe — USSR Venus flyby craft. Arrived at Venus Dec 25, dropped a soft-lander through dense atmosphere. Transmitted 110 minutes. Lightning in clouds. Lander reported temperature on surface 860 deg. F.

1978 Nov 13 Einstein/HEAO-2 — U.S., the Einstein Observatory or High Energy Astronomical Observatory 2 (HEAO-2), launched by NASA into 300-mi. circular orbit, carried X-ray telescope, TV camera. Spotted X-rays from quasars billions of lightyears away. X-rays found from sources in deep space 1,000 times farther than previously seen. X-rays from Jupiter found.

1978 Dec 19 Gorizont satellite — USSR stationary communications satellite.

1979

1979 Mar 5 Voyager 1 probe — U.S. outer Solar System probe launched Sep 5, 1977, flew by Jupiter Mar 5, 1979. Passing 216,865 mi. above the planet, it snapped color photos of clouds, rings, and moons Io, Ganymede and Callisto. Io volcanos made it first volcanic body found beyond Earth. Ganymede, Callisto scarred by meteorite craters. Ganymede a mix of rock, ice. Voyager 1 reached Saturn Nov 12, 1980. Now leaving Solar System.

1979 Apr 10 Salyut 6 station — Georgi Ivanov in Soyuz 33 was the first Bulgarian cosmonaut. Engine failed on Soyuz 33 while docking at Salyut 6. Capsule and station were separating at 60 mph. Ground control used backup engine to bring Soyuz home immediately.

1979 Jul 9 Voyager 2 probe — U.S. outer Solar System probe, launched Aug 20, 1977, flew by Jupiter July 9, 1979. At 404,000 mi. above Jupiter's clouds, it photographed cloud weather patterns and moons Europa, Callisto, Ganymede, Io. Planet rings were 4,040 mi. wide with particles down to cloud tops. Massive volcano erupting on moon Io. Large impact basin on moon Callisto. Voyager 2 reached Saturn Aug 25, 1981, Uranus Jan 24, 1986, Neptune Aug 24, 1989. Now leaving Solar System.

1979 Jul 11 Skylab station — Skylab, America's first and only space station, launched to Earth orbit in 1973, used by astronauts only in 1973-74, re-entered the atmosphere July 11, 1979, broke up at 18.6 mi. over western coast of Australia. Perth residents saw brilliant light show as chunks rained on hundreds of miles of the Outback. No injuries or property damage.

1979 Sep 1 Pioneer 11 probe — First-ever Saturn fly by, U.S. probe, twin of Pioneer 10, launched Apr 5, 1973, from Cape Canaveral, arrived Jupiter Dec 2, 1974, then Saturn Sep 1, 1979. Passed within 13,300 mi. of cloud tops. Colorful photos of ringed planet.

Discovered 2 new rings, F and G. Photographed big moon Titan for first time. Found Saturn radiates 2.5 times as much heat as it receives from the Sun. Pioneer 11 then headed on out of the Solar System to interstellar space.

1979	Sep 20	HEAO-3 sat	High Energy Astronomical Observatory 3, launched by NASA, two cosmic ray telescopes.
1979	Dec 24	Ariane 1 rocket	First Ariane launched by European Space Agency; from Kourou, French Guiana; CAT satellite.

1980

1980   Feb 14   Solar Max sat   U.S. Solar Maximum Mission satellite, SMM or Solar Max, had 7 instruments to observe Sun. Solar flares observed in visible and ultraviolet light, gamma rays and X-rays. Photographed 12,500 flares; 250,000 pictures of Sun's corona; 100,000 ultraviolet light images of Sun; discovered 10 Sun-grazing comets. Solar Max was first to detect gamma rays from Supernova 1987a. On Apr 6, 1984, astronauts Nelson and Van Hoften in shuttle Challenger flight STS-41C made first satellite repair in orbit. They grabbed Solar Max, did the first in-space satellite repair in the cargo bay, then released Solar Max back to orbit. Solar Max fell from orbit Dec 2, 1989, burning over the Indian Ocean near Sri Lanka.

1980   May 26   Salyut 6 station   Bertalan Farkas, first Hungarian cosmonaut, flew to space station in the Soyuz 36 capsule.

1980   Jun 2   Pluto planet   MIT astronomers Lupo and Lewis calculated the density of Pluto at only 14% higher than water, and the planet's diameter at 1583 miles. Pluto had been thought to be as dense as mercury.

1980   Jul 18   Rohini 1 satellite   India launches first satellite to orbit using rocket known as Satellite Launch Vehicle (SLV). India is the seventh nation with a space rocket.

1980   Jul 23   Salyut 6 station   When cosmonaut Pham Tuan of Vietnam flew in Soyuz 37 to visit the Salyut 6 space station during Summer Olympic Games in Moscow, he was the first Asian, the first non-Warsaw Pact and the first third-world person in space.

1980   Sep 18   Salyut 6 station   Cuban cosmonaut Arnaldo Tomayo Mendez flew in Soyuz 38. He was the 1st Hispanic person, 2nd non-Warsaw Pact, 2nd third-world person in space.

1980   Oct 11   Salyut 6 station   USSR Soyuz 35 cosmonauts Valery Ryumin and Leonid Popov return to Earth, setting a new endurance record of 185 days.

1980   Nov 12   Voyager 1 probe   U.S. outer Solar System probe launched Sep 5, 1977, at Jupiter Mar 5, 1979, arrived at Saturn Nov 12, 1980. Passing 78,300 mi. above Saturn's cloud tops, it snapped 17,500 color photos revealing six additional moons raising total to 18. Confirmed Pioneer 11 finding that Saturn radiates more heat than it gets from Sun. Clouds revealed 1,100 mph winds. Saturn's major A, B and C rings found to have ringlets. Shepherding moons found in F ring. Mimas, Tethys, Dione and Rhea moons were heavily cratered. Largest moon, Titan, had nitrogen atmosphere 1.5 times denser than Earth's atmosphere. Voyager 1 now leaving Solar System.

1981

1981	Apr 12	Columbia shuttle	STS-1 was the maiden flight of the first U.S. space shuttle. Astronauts John Young, Robert Crippen went to space in Columbia for 54 hours 20 minutes. Columbia became the first spacecraft to land like an airplane on a runway, landing on Rogers Dry Lake bed at Edwards Air Force Base, California. First of 4 test flights.
1981	Mar 22	Salyut 6 station	Jugderdemidiyn Gurragcha, first Mongolian cosmonaut, flew in the Soyuz 39 capsule.
1981	May 14	Salyut 6 station	Dumitru Prunariu, first Romanian cosmonaut, flew in the Soyuz 40 capsule.
1981	Jul 26	Pioneer 10 probe	Passed the distance of 25 astronomical units (AU) from Earth. Between Saturn and Uranus, on its way out of Solar System, radioed solar wind reports. Pioneer 10 launched Mar 3, 1972, eventually left the Solar System June 13, 1983. On Feb 23, 1990, it was 4.5 billion mi. away, the most distant man-made object.
1981	Aug 25	Voyager 2 probe	U.S. outer Solar System probe, launched Aug 20, 1977, at Jupiter July 9, 1979, arrived at Saturn Aug 25, 1981. At 62,760 mi. above Saturn's cloud tops, measured thickness of rings, found spokes in rings which may be clouds of micrometeoroids. Moons Mimas, Tethys, Dione and Rhea were heavily cratered. Evidence of motion of crust on moon Enceladus. Largest moon, Titan, had nitrogen atmosphere 1.6 times denser than Earth's atmosphere, methane in abundance, surface temperature of −292 deg. F., density twice water. Voyager 2 reached Uranus Jan 24, 1986, Neptune Aug 24, 1989, now leaving Solar System.
1981	Oct 30	Venera 13 probe	USSR, Venus orbiter dropped off landers. Landed Mar 3, 1982. Transmitted color surface photo.
1981	Nov 4	Venera 14 probe	USSR, Venus orbiter dropped off landers. Landed May 5, 1982. Transmitted color surface photo.

1982

1982	Apr 19	Salyut 7 station	USSR launched second and last second-generation space station from Baikonur Cosmodrome, 200 mi.-high orbit. It had an X-ray telescope and biomedical center. Salyut 7 fell Feb 7, 1991.
1982	Jun 24	Salyut 7 station	French cosmonaut Jean-Loup Chretien was the first Westerner, from a Western industrial democracy, to fly in a Soviet spacecraft. He took the first French cuisine to orbit in Soyuz T-6.
1982	Aug 19	Salyut 7 station	USSR cosmonaut Svetlana Savitskaya was the second woman in orbit, flying in Soyuz T-7.

It was the first mixed male/female crew in space: Savitskaya, Popov and Serebrov. Savitskaya returned to space in July 1984 in Soyuz T-12.

1982 Nov 11 Columbia shuttle Flight STS-5 was the first operational shuttle mission. Four astronauts, Brand, Overmyer, Lenoir and Allen were the largest crew up to that time. Lenoir and Allen were the first mission specialists. A first-ever shuttle spacewalk was cancelled when spacesuits didn't work.

1983

1983 Apr 4 Challenger shuttle Flight STS-6, the maiden voyage of U.S. shuttle Challenger. Astronauts Peterson and Musgrave donned new spacesuits and went outside into the open cargo hold for 4 hours 17 minutes on the first shuttle spacewalk (EVA). Challenger was NASA's second operational shuttle.

1983 Jun 2 Venera 15 probe USSR Venus radar mapping orbiter. High-resolution photos, including north pole.

1983 Jun 6 Venera 16 probe USSR Venus radar mapping orbiter. High-resolution photos, including planet surface regions not visible from Earth.

1983 Jun 13 Pioneer 10 probe U.S. probe Pioneer 10 passed the orbit of the most distant planet and left the Solar System June 13, 1983. Pioneer 10 was supposed to have a working life of only 21 months when it was launched Mar 3, 1972. By Jun 13, 1990, it was 4.7 billion mi. away, the most distant man-made object. Its twin, Pioneer 11, which became the fourth man-made object to leave the Solar System on Feb 23, 1990, traveling in the opposite direction.

1983 Jun 18 Challenger shuttle U.S. astronaut Sally K. Ride, aboard flight STS-7, was the first American woman in space and, at 32, the youngest U.S. woman ever in orbit. She was part of first American mixed male/female crew in orbit: Ride, Crippen, Hauck, Fabian, Thagard. Ride returned to space in Challenger in 1984.

1983 asteroid Comet IRAS-Araki-Alcock passed within 3 million miles of Earth.

1983 Jun 27 Salyut 7 station After they arrived at the Salyut 7 space station June 29, USSR cosmonauts Alexandrov and Lyakhov were forced to bail out, into their Soyuz T-9 capsule, until it was found that no life threatening damage had been done by a tiny micrometeorite stone which had pinged loudly against a half-inch thick pane in a station porthole. It did not penetrate. Later, while taking on fuel from a Progress tanker, a station oxidizer line ruptured. Alexandrov and Lyakhov again bailed out. After all-clear, they by-passed the line.

1983 Aug 30 Challenger shuttle U.S. shuttle flight STS-8 carried the first black man in space, astronaut Guion S. Bluford Jr.

1983 Sep 26 Soyuz T-10A USSR cosmonauts Vladimir Titov and Gennadi Strekalov were strapped into their Soyuz T-10A capsule September 26, 1983, ready for launch to the Salyut 7 space station, when fire erupted at the base of their A-2 rocket 90 seconds before launch. The Soyuz escape rocket was triggered. The capsule separated from the booster, lifting off seconds before the A-2 exploded. Titov and Strekalov landed, uninjured, several miles away.

1983   Nov 28   Columbia shuttle   The first ham radio operator to chat from space with friends on the ground, from the first ham shack in space, was U.S. astronaut Owen K. Garriott, callsign W5LFL, flight STS-9. West German Ulf Merbold, American Byron K. Lichtenberg were first payload specialists. Merbold was first non-American to fly in U.S. spacecraft. STS-9 commander John Young set a record for most rides to space, having flown six times on four different kinds of spacecraft: he flew in Gemini 3 in 1965, Gemini 10 in 1966, Apollo 10 in 1969, Apollo 16 to land on the Moon in 1972, the maiden voyage of shuttle Columbia in 1981, Columbia in 1983.

1984

1984   Feb 7   Challenger shuttle   U.S. astronauts Stewart and McCandless took the first untethered spacewalk, using the Manned Maneuvering Units (MMU) for the first time. Flying free around Challenger, without ties to the shuttle, they were the first human satellites.

1984   Apr 3   Salyut 7 station   The first person from India in space was Rakesh Sharma who flew in Soyuz T-11 to visit the Salyut 7 space station. The fourth third-world person to fly in a USSR spacecraft, he used yoga to combat effects of weightlessness. With him in Soyuz T-11 were cosmonauts Malyshev and Strekalov. In addition, Atkov, Kizim and Solovyev were in residence at the station. Then U.S. shuttle Challenger was launched April 6. For the first time, 11 persons were in orbit at one time: five astronauts in Challenger and six cosmonauts at Salyut 7, the most people ever in space at one time, up to that time.

1984   Apr 6   Challenger shuttle   Flight STS-41C (STS-11), the first satellite repair in orbit: astronauts Nelson and Van Hoften rescued the Solar Max astronomy satellite which had been launched Feb 14, 1980, made the first-ever in space satellite repair in the shuttle cargo bay, then released Solar Max back to orbit.

1984   Jul 17   Salyut 7 station   The first woman to go to space twice was USSR cosmonaut Svetlana Savitskaya who flew in Soyuz T-7 to the Salyut 7 space station in August 1982, and again to Salyut 7 in Soyuz T-12 July 17, 1984. During the 1984 visit, she made the first female spacewalk and the first male-female spacewalk, 3.5 hrs. with Vladimir Dzhanibekov on July 25.

1984	Aug 30	Discovery shuttle	Flight STS-12 (STS-41D) was the maiden flight of U.S. shuttle Discovery. It carried the second American woman in orbit, Judith A. Resnick. First paying spaceflight passenger: Charles D. Walker was the first American private industry representative in space. McDonnell Douglas purchased his ticket as the first commercial payload specialist for drug research.
1984	Oct 5	Challenger shuttle	In U.S. shuttle flight STS-41G (STS-13), astronaut Kathryn Sullivan was the first American woman to walk in space. Sullivan and David Leestma took the first American male-female spacewalk. The flight also was the first time two women were in space at same time: Sullivan and Sally Ride. Ride became the first U.S. woman to go to space twice, in Challenger in June 1983 (STS-7) and again in Oct 1984. Flight STS-41G carried Paul Scully Power, the first oceanographer to observe the oceans from space, as well as the first Canadian in space, Marc Garneau.
1984	Nov 8	Discovery shuttle	The first mother in space was U.S. astronaut Anna L. Fisher, M.D., flight STS-51A.
1984	Dec 15	Vega 1 probe	USSR probe, flew by planet Venus in June 1985, then on to fly by Comet Halley, within 5,523 miles, on Mar 6, 1986.
1984	Dec 21	Vega 2 probe	USSR probe, flew by planet Venus in June 1985, dropped a lander and atmosphere balloon probe, then on to fly by within 4,989 miles of Comet Halley on Mar 9, 1986.

1985

1985	Jan 8	Sakigake probe	Japan's probe flew by Comet Halley, within 3.73 million miles of the comet, on Mar 11, 1986. Measured effect of solar wind on comet's tail.
1985	Apr 12	Discovery shuttle	First politician in space: U.S. Senator Edwin "Jake" Garn orbited Earth in STS-51D.
1985	Apr 29	Challenger shuttle	U.S. shuttle flight STS-51B (STS-17) carried the first Dutch astronaut, Lodewijk van den Berg of the European Space Agency.
1985	Jun 6	Salyut 7 station	USSR space station Salyut 7, launched in 1982, malfunctioned while unoccupied in 1985, leaving the station out of ground control. Cosmonauts Dzhanibekov and Savinykh flew Soyuz T-13 to the station June 6, finding it frozen from power loss and tumbling in space. They recharged batteries, repaired water pipes, and warmed living quarters. It was 10 days before they could live in the station.
1985	Jun 17	Discovery shuttle	U.S. shuttle flight STS-51G (STS-18) carried France's Patrick Baudry, of the French space agency CNES, who treated fellow crew members to gourmet dining. It also carried the first royalty in orbit and the first Arab in space: Saudi Arabia's Prince Sultan Salman Abdel Aziz Al-Saud.
1985	Jul 2	Giotto probe	European Space Agency probe made the closest

approach of any probe to Comet Halley, within 372 miles of nucleus Mar 13, 1986. Giotto survived a battering by high-speed particles in the comet's tail. Its radio fell silent, but returned after 20 minutes. A month later, European Space Operations Center, Darmstadt, Germany, put Giotto in hibernation, rounding the Sun every 10 months. Giotto was parked in that orbit from Apr 1986 to Feb 1989, using minimum power, radio transmissions shut down. Giotto's camera was broken during the Halley fly-by, but European scientists said it still could relay valuable information so ESA decided to send Giotto to a July 10, 1992, rendezvous with Comet Grigg Skjellrup. Giotto swung by Earth July 2, 1990, for a gravity boost to catch the comet.

1985	Jul 29	Challenger shuttle	U.S. shuttle STS-51F astronauts Fullerton, Bridges, Acton, Musgrave, England, Henize, Bartoe quenched thirsts with special cans of the first Coke and Pepsi in orbit. Henize, at 58, was the oldest person to fly in space up to that time. At 59, Vance Brand broke the record in 1990.
1985	Aug 18	Suisei probe	Japan's probe, Comet Halley flyby. Passed within 93,827 miles of the comet Mar 8, 1986. Transmitted ultraviolet light photos of the comet's 12 million-mi.-long coma.
1985	Sep 11	ICE/ISEE-3	International Sun-Earth Explorer 3 (ISEE-3) was renamed International Cometary Explorer (ICE) and turned to explore the inrushing Comet Giacobini-Zinner. ICE was used again, on Mar 25, 1986, to study Comet Halley at a distance of 17.46 million miles. ISEE-3 had been launched by NASA on Aug 12, 1978, as a solar observatory in a halo orbit at the Earth-Sun libration point where pull of gravity from Earth and Sun are equal, one million mi. from Earth toward the Sun. ISEE-3 used to warn of solar flares, before being reassigned as ICE.
1985	Oct 3	Atlantis shuttle	Secret military flight STS-21 (STS-51J) was the maiden flight of U.S. shuttle Atlantis.
1985	Nov 26	Atlantis shuttle	U.S. shuttle flight STS-61B (STS-23) carried the first Mexican in space, Rudolfo Neri Vela.
1985	Nov	Salyut 7 station	In the first space station illness, cosmonaut Vladimir Vasyutin became ill with a 104 degree fever while orbiting in Salyut 7. He had flown there Sep 17 in Soyuz T-14 with USSR cosmonauts Grechko and Volkov to continue repairs to the station. His mission ended suddenly when he became ill in November. He flew home and was hospitalized into December with a full recovery.

1986

1986	Jan 12	Columbia shuttle	U.S. Representative William "Bill" Nelson rode to Earth orbit in STS-61C.
1986	Jan 24	Voyager 2 probe	U.S. probe made its first solo planet flyby, within

50,600 mi. of Uranus' cloud tops, discovering a magnetic field. Planet's average temperature was 350 degrees Fahrenheit. Uranus radiated much ultraviolet light, known as dayglow. Voyager 2 found 10 unknown moons, the largest only 90 mi. diameter, bringing the total to 15. Astronomers previously had seen five large ice-and-rock moons: innermost is Miranda, found by Voyager to be one of the strangest bodies in the Solar System, with fault canyons up to 12 mi. deep, terraced layers and mixed young and old surfaces. Titania was marked by huge faults and deep canyons. Ariel had the brightest and youngest surface of Uranian moons, with many deep valleys and broad ice flows. Dark-surfaced Umbriel and Oberon looked older. Voyager found 9 rings around Uranus quite different from Jupiter and Saturn rings. Uranus' rings may be young remnants of a shattered moon.

1986	Jan 28	Challenger shuttle	U.S. astronauts aboard Challenger shuttle flight STS-51L—Francis R. "Dick" Scobee, Michael J. Smith, Judith A. Resnik, Ellison S. Onizuka, Ronald E. McNair, Gregory B. Jarvis and Concord, New Hampshire, high school social studies teacher Sharon Christa McAuliffe—were the first Americans to die during a spaceflight, the first persons to die enroute to space. The 7 were killed during lift off from a Cape Canaveral, Fla., launch pad Jan. 28, 1986. "Obviously a major malfunction. We have no downlink. The vehicle has exploded," intoned NASA mission control at one minute thirteen seconds into lift off.
1986	Feb 20	Mir station	USSR's first third-generation space station launched from Baikonur Cosmodrome to a 200-mi.-high orbit. Cosmonauts Kizim, Solovyev flew Soyuz T-15 to Mir Mar 15. In what the USSR called the first "space taxicab," they left Mir in Soyuz T-15 May 5, cruised to the old station Salyut 7, docked, entered, mothballed Salyut 7, leaving it parked in orbit near Mir. They flew back to Mir June 26, then down to Earth July 16.
1986	Mar 25	ICE/ISEE-3	International Cometary Explorer (ICE) studied Comet Halley at a distance of 17.46 million miles. ICE was launched by NASA Aug 12, 1978, as International Sun-Earth Explorer 3 (ISEE-3), a solar observatory in a halo orbit at the Earth-Sun libration point where pull of gravity from Earth and Sun are equal. ISEE-3 warned of solar flares, before being renamed International Cometary Explorer (ICE) Sep 11, 1985, and turned to explore Comet Giacobini-Zinner.

1987

| 1987 | Feb 6 | Mir station | USSR cosmonauts Yuri Romanenko and Alexander |

			Laveikin launched to Mir in Soyuz TM-2 Feb 6 to start permanent occupancy Feb 8, the first permanent-manning crew of any space station. Laveikin had to fly home after six months when he showed an unusual heartbeat. Doctors on the ground in July ordered him home. Romanenko stayed to set a multi-trip record 430 days in space.
1987	Feb 23	Supernova 1987a	The previously-noted, but unremarkable, star Sanduleak 69.202, in the nearby Large Magellanic Cloud galaxy, exploded in a fiery death, the spectacular Supernova 1987a.
1987	May 15	Energia rocket	USSR, first test flight of the world's largest, most powerful space booster.
1987	Jun 6	various satellites	Pope John Paul II prays in 35 languages in a broadcast over 23 communications satellites.
1987	Jul 22	Mir station	USSR, Soyuz TM-3. Syrian cosmonaut Mohammed Faris, 2nd Arab in space, 1st Syrian at Mir, carried Syria's tricolor flag, a four-color Pan-Arab banner, a portrait of Syrian President Hafez Assad, and samples of Damascus soil

1988

1988	Jun 7	Mir station	A blooming orchid grown on Earth from seeds previously planted in the space station were carried back to Mir by Soyuz TM-5 capsule cosmonauts Solovov, Savinykh, Alexandrov. Alexandrov was second Bulgarian in space.
1988	Jul 7	Phobos 1 probe	USSR probe to orbit Mars and drop lander on moon Phobos, travelled 12 million miles of 111 million-mile route, but was accidentally turned off by ground controller Aug 29, 1988, now aimless in solar orbit.
1988	Jul 12	Phobos 2 probe	USSR probe to orbit Mars and drop lander on moon Phobos, arrived in Mars orbit Jan 29, 1989, mapped the planet, found water vapor in atmosphere, transmitted photos of moon Phobos, but radio contact was lost Mar 27, 1989, unable to drop hopping lander on Phobos in April 1989.
1988	Aug	astronaut	Ukrainian cosmonaut Anatoly Levchenko, 47, who had been a cosmonaut since 1981, training to fly the USSR's space shuttle Buran, still was a rookie pilot when he flew Soyuz TM-4 to Mir Dec 21, 1987. He returned to Earth Dec 29. In 1988, Moscow doctors found he had a brain tumor. A summer operation was unsuccessful and Levchenko died of a brain tumor in August 1988, eight months after his eight-day stay at Mir station.
1988	Aug 29	Mir station	Abdul Ahad Mohmand first cosmonaut from Afghanistan, flew in Soyuz TM-6.
1988	Sep 19	Horizon 1 sat	Israel orbited first satellite, Horizon 1 or Ofek 1, on 3-stage Shavit rocket from military launch pad in Negev Desert—joining China, France, Great Britain, India, Japan, USSR, USA as 8th nation with space rocket. Shavit is Hebrew for comet.

			Horizon 1 fell from orbit after 4 months. A 2nd satellite was launched Apr 3, 1990.
1988	Sep 29	Discovery shuttle	U.S. shuttle Discovery flight STS-26 returned the American flag to space, 975 days after the Jan 1986 Challenger disaster.
1988	Oct 20	Mir station	The first amateur radio contacts from a space station were made by the first all-ham space station crew. USSR cosmonauts Vladimir Titov and Musa Manarov went outside Mir Oct 20 to mount a ham radio antenna. Titov, Manarov and Dr. Valery Polyakov called CQ in November, chatting with hams on the ground. Callsign: Titov U1MIR; Manarov U2MIR; Polyakov U3MIR.
1988	Nov 15	Buran shuttle	The first USSR shuttle, Buran No. 1, lifted off from Baikonur Cosmodrome on the world's most powerful space rocket, Energia. The unmanned Buran flew two orbits at 155 miles above Earth to a picture-perfect automated landing on a runway just eight miles from its original launch pad.
1988	Nov 31	Mir station	French cosmonaut Soyuz TM-7 passenger Jean Loup Chretien played a portable organ at Mir as USSR cosmonauts Vladimir Titov, Musa Manarov, Alexander A. Volkov, Sergei M. Krikalev and Dr. Valery Polyakov dined on 23 gourmet foods from a French chef, including compote of pigeon with dates and dried raisins, duck with artichokes, oxtail fondue with tomatoes and pickles, beef bourguignon, saute de veau Marengo, ham and fruit pates, bread, rolls, cheeses, nuts, coffee and chocolate bars. The delicacies were canned. The chef, saying French cuisine is inconceivable without wine, regretted a bottle could not go to space when he learned wine and sauces would float away without gravity. Dishes had to be small without bones, or sauces to become flying droplets. Meat was made to absorb sauce to guard against dryness. Chretien relaxed with music taped by the British rock group Pink Floyd. Chretien became the first non-USSR cosmonaut/non-USA astronaut to make a spacewalk when he went outside Mir to build an experimental plastic web frame for science gear.
1988	Dec 21	Mir station	USSR cosmonauts Titov and Manarov set a 366 day one-trip spaceflight record (Dec 21, 1987-Dec 21, 1988) at Mir. French cosmonaut Chretien set a record for the longest stay at a USSR station by a non-USSR cosmonaut, 25 days at Mir. As Titov, Manarov and Chretien left Mir in Soyuz TM-6 Dec 21, trouble developed. Telemetry showed a computer overload, delaying Soyuz descent 3 hours; they switched to manual back-up.

1989

| 1989 | Jan 29 | Phobos 2 probe | USSR unmanned Phobos 2 orbits Mars, 6-months |

111 million miles from Earth. Launched July 12, 1988, Phobos 2 transmitted photos of the Martian moon Phobos February 21. Contact was lost unexpectedly March 27.

1989	Mar 13	Discovery shuttle	First artwork in space: America's first official space artwork was a seven-pound cube-shaped sculpture Boundless Aperture, created by Boston artist Lowry Burgess, a professor at Massachusetts College of Art. He conceived Boundless Aperture in the 1960's as part of what he called The Quiet Axis. By the time Boundless Aperture went to space, Burgess already had buried a sculpture in the floor of the Pacific Ocean off Easter Island and lugged another into central Afghanistan's Hindu Kush mountains. Shuttle flight STS-29.
1989	Mar 23	asteroid	Earth had a close call March 23 when the passing half-mile-diameter asteroid 1989fc rushed by at a distance of 465,000 mi., closest approach in 300 years. Other 1989 close calls: June 3 and Aug 24.
1989	Mar 27	Phobos 2 probe	Contact was lost with the USSR's unmanned Phobos 2, orbiting Mars. Launched July 12, 1988, Phobos 2 arrived at Mars Jan 29 and transmitted photos of the moon Phobos February 21.
1989	Apr 27	Mir station	USSR Mir space station unmanned Apr 27-Sep 8. Cosmonauts reopened Mir Sep 6.
1989	May 4	Magellan probe	Magellan was launched to Earth orbit in U.S. shuttle Atlantis flight STS-30 from which astronauts fired the radar-mapping probe to Venus. It went into orbit around Venus August 10, 1990. Started full-time mapping September 15, 1990, with radar peering down through Venus' thick clouds. It was the first shuttle launch of an interplanetary spacecraft. The unmanned Magellan had radar-mapped the entire planet twice during some 3,000 orbits of Venus by the end of 1991.
1989	Jun 3	asteroid	The two-mile-wide asteroid 1989ja passed within 8 million miles of Earth. Other 1989 close calls were Mar 23 and Aug 24.
1989	Jun 14	Titan 4 rocket	Currently the most powerful unmanned rocket in the U.S. inventory, Titan 4 thundered to space from Cape Canaveral in its maiden voyage.
1989	Jun 17	astronaut	U.S. astronaut Navy Reserve Rear Adm. S. David Griggs, 49, assigned to pilot shuttle Discovery in November, was killed June 17 while stunt flying in a vintage World War II trainer plane in Arkansas. John E. Blaha, just back from his first trip as pilot of Discovery Mar 13, replaced Griggs.
1989	Aug 8	Hipparcos sat	The High Precision Parallax Collecting Satellite (Hipparcos) was launched into an elliptical transfer orbit ranging from 125 miles to 22,304 miles altitude by European Space Agency (ESA) Ariane 4 rocket flight V-33. The star-mapper's kick motor, which should have ignited to

circularize the satellite's orbit into a stationary position 22,300 miles over Africa, failed Aug 10, stranding Hipparcos in the elliptical orbit. Engineers later salvaged much of the project by maneuvering the satellite in its lower orbit and reprogramming it. The solar-powered Hipparcos has a small, but very sophisticated reflecting telescope designed to measure accurately the positions of 120,000 stars. The satellite spins slowly in orbit so the telescope can scan the entire sky. In its imperfect orbit, Hipparcos has been able to chart the positions of about 80 percent of the 120,000 stars. Astronomers will use this astrometry data to draw up precision star maps, showing distances to stars and their directions of movement, by the late 1990's. Hipparcos is controlled from ESA's European Space Operations Centre at Darmstadt, near Frankfurt, Germany, and monitored by ground stations in Perth, Australia, Odenwald, Germany and Kourou, French Guiana.

1989 Aug 24 asteroid

A dumbbell-shaped mile-wide mountain of rock, asteroid 1989pb, came within 2.5 million miles of Earth. Other 1989 close calls: Mar 23 and June 3.

1989 Aug 24 Voyager 2 probe

U.S. probe flew within 3,000 mi. of Neptune while that large gas planet was the most distant in the Solar System. Voyager found six moons at Neptune, bringing total to eight. The found moons were named Naiad, Thalassa, Despina, Galatea, Larissa, and Proteus. It also found large dark spots on Neptune like Jupiter's Great Red Spot hurricane. The largest was named the Great Dark Spot. A small irregular cloud flitting across cloud tops was named Scooter. The strongest winds of any planet were measured on Neptune, with winds near the Great Dark Spot blowing at 1,200 mph. The probe found an odd magnetic field, as it had at Uranus. Voyager found four complete rings around Neptune—1989N1R, 1984N4R, 1989N2R and 1989N3R—and no partial ring arcs. The fine material in the rings is so diffuse they can not be seen from Earth. The largest moon, Triton, turned out to be one of the most intriguing satellites in the Solar System, with erupting geysers spewing invisible nitrogen gas and dark dust particles high in the extremely thin atmosphere which extends 500 miles above Triton's surface. Triton may not always have been a moon of Neptune. If so, tidal heating might have melted Triton, leaving the moon liquid for a billion years after its capture by Neptune. Today it may have thin clouds of nitrogen snowflakes a few miles above its firm surface. With a surface temperature of −391 deg. Fahrenheit, Triton is the coldest body known in the Solar System. Triton's atmospheric pressure was 1/70,000th the surface pressure on Earth. Voyager 2 now is headed out of the Solar System. Pluto once again will become the most distant planet in 1999.

1989		Endeavour shuttle	NASA chose the name Endeavour, suggested by students in a nationwide competition, for the Challenger replacement shuttle.
1989	Oct 18	Galileo probe	Unmanned interplanetary spacecraft launched to Earth orbit in U.S. shuttle Atlantis STS-34. Galileo's inertial upper stage rocket then fired it on six-year, 2.4 billion mile trip to Jupiter. It was the second shuttle launch of an interplanetary spacecraft. Using planet gravity for speed boosts, Galileo swooped around Venus Feb 10, 1990, and Earth Dec 8, 1990. It grazed the Asteroid Belt, passing close to the asteroid 951 Gaspra Oct 29, 1991, then by Earth again Dec 8, 1992. On July 10, 1995, Galileo will separate into orbiter and atmosphere probe. On Dec 7, 1995, orbiter will arrive over Jupiter and probe will parachute 370 miles down into the atmosphere, measuring clouds and atmosphere for 75 minutes, until pressure and temperature crush and vaporize it. The orbiter above will transmit data about Jupiter and its moons for 22 months.
1989	Nov 18	COBE telescope	U.S., Cosmic Background Explorer (COBE) launched on Delta rocket to find out why the Universe has a lumpy consistency. COBE performed splendidly, looking out across the Universe from Earth orbit, searching for explanations of why the far flung star-clouds we call galaxies are distributed unevenly across the cosmos. Many scientists think the Universe exploded into existence 15 billion years ago in a hot, dense fireball which instantly started expanding in all directions—the so-called Big Bang. It would seem the Universe should be expanding smoothly, with an even distribution of matter everywhere. The problem is, the Universe looks uneven or lumpy, with great clusters of galaxies dotted across giant dark voids. Orbiting 570 miles above Earth, COBE's 3 receivers were supposed to record weak microwave radiation left over from the Big Bang, to find out why the Universe is lumpy, but failed to find a trace of any significant energy release after the first huge explosion. COBE did send down the best pictures ever seen of the center of our Milky Way galaxy. NASA combined COBE's infrared images to make individual pictures, clear of the dust that usually blocks the view of the heart of the galaxy.
1989	Nov 22	Discovery shuttle	First woman aboard a U.S. military flight was

1989   Dec 2   Solar Max sat

1989   Dec 5   Iraq satellite

astronaut Kathryn C. Thornton, flight STS-33.
High energy from the Sun heated Earth's
atmosphere in 1989, dragging down satellites,
including Solar Max, the Solar Maximum Mission
observatory which burned in atmosphere Dec 2.
Iraq launched a 48-ton three-stage rocket from the
Al-Anbar Space Research Center 50 miles west of
Baghdad on a six-orbit flight. It was not a separate
satellite, but the 3rd stage of a rocket which swept
around the Earth for 6 revolutions before falling
out of orbit. A U.S. nuclear-attack-warning
satellite spotted fiery rocket exhaust as it blasted
off. North American Aerospace Defense Command
tracked the 3rd stage around the globe. Iraq became
the 9th nation with a space rocket. The 75-ft., 3-
stage rocket was similar to the U.S. Scout rocket
used to send small satellites to low orbits. The
booster may have been a modified version of
Argentina's Condor ballistic missile. Such
a ballistic missile could carry a nuclear warhead
1,240 miles. The launch was the first time Iraq
exposed its space research. It already had a 600-
mi. missile built around the USSR Scud. Scuds were
launched against Iranian cities in the eight-year
Persian Gulf war in the 1980's and Israel and
Saudi Arabia in the 1991 Persian Gulf War.

1990

1990   Jan 9   Columbia shuttle

1990   Jan 22   amateur radio sats

1990   Jan 24   Hiten/Muses-A

Flight STS-32 extended by weather and computer
problems in orbit to record 10 days 21 hrs, half
day longer than previous. Retrieved Long Duration
Exposure Facility (LDEF) with 57 science
experiments including 12.5 million tomato seeds
which NASA mailed to schools so 3.5 million
students could study the effect of space on growth.
Eight new amateur radio satellites were fired to
Earth orbit in 1990, the biggest proliferation
since 1981 when Russian hams sent 6 in 1 flight.
The 8 were OSCARs 14, 15, 16, 17, 18 and 19
launched Jan 22, Fuji-OSCAR 20 Feb 7, Badr-A
July 16. European Space Agency launched
OSCARs 14, 15, 16, 17, 18 and 19 on 1 rocket.
OSCAR 15 failed in orbit. China launched Badr-A,
Pakistan's first amateur satellite; it fell Dec 9.
Japan launched unmanned Muses-A Moon probe to
Earth orbit; renamed it Hiten, or Spaceflyer, in
orbit. Hiten looped out from Earth and around the
Moon, passing within 9,100 miles of the lunar
surface Mar 19, 1990. Hiten dropped off a small
lunar satellite, Hagoromo, initiating Japan into
the small club of nations having spacecraft
circling Earth's natural satellite. Only the U.S. and
the USSR had done it before. Hiten made a second
lunar swing-by July 10, 1990, and a third Aug 4,

1990. Lunar gravity boosts its speed, enlarging its long elliptical orbit around Earth. Altogether, Hiten will loop Earth and Moon 8 times, 4 to speed up and 4 to slow down. Eventually, Hiten will be travelling 600,000 miles from Earth. That apogee distance will hold for the rest of Hiten's life. Hiten carries a German micrometeorite counter. Hagoromo will remain in orbit around the Moon, despite a broken radio. Hagoromo's rocket fired March 19 to pull away from Hiten at 12,500 miles from the Moon, but Hagoromo's transmitter failed, leaving a record of rocket firing but no signal. Astronomers used optical telescopes to see Hagoromo orbiting the Moon.

1990	Feb 3	Pioneer 11	The U.S. interplanetary probe Pioneer 11 crossed the orbit of Neptune 2.8 billion miles from Earth February 3, leaving behind forever the known parts of the Solar System. It joined three other U.S. spacecraft, Pioneer 10, Voyager 1 and Voyager 2, in the unexplored wide open spaces beyond our Solar System.
1990	Feb 4	Long March	After no Long March rocket launches in 1989, the People's Republic of China was busy in 1990 with launches including the nation's 5th communications satellite Feb 4; ASIAsat 1 Apr 7; Pakistan's Badr-A July 16; Feng Yun 1B weather satellite Sep 3; a Chinese Academy of Sciences satellite carrying animals and plants Oct 5; a suborbital space probe to study the depletion of the ozone layer up to 19 miles above the Earth Oct 5; and a remote-sensing satellite which dropped a film cannister to Earth. The animals, launched to assess response to weightlessness, returned to Earth after 8 days. ASIAsat 1 was the former Westar 6 ferried to space in a U.S. shuttle in 1984. It failed and was recaptured by shuttle in 1984.
1990	Feb 10	Galileo probe	Unmanned interplanetary spacecraft Galileo, launched Oct 18, 1989, swoops around Venus, passing within 9,300 miles of the planet for a gravity assist on its way to Jupiter in 1995.
1990	Feb 14	Voyager 1 probe	America's Voyager 1, floating through space 3.7 billion miles from Earth, looked back toward home and snapped its two aging cameras to make a sweeping portrait of the planets of the Solar System and the Sun. It was the first opportunity astronomers have had to take a picture of the planets from outside the Solar System.
1990	Feb 23	Pioneer 11 probe	U.S. probe became the fourth man-made object to leave the Solar System on Feb 23, 1990, at a distance of 2.8 billion miles from the Sun. It was flying the opposite direction from its twin Pioneer 10, which passed the orbit of the most distant planet and left the Solar System June 13,

			1983. Pioneer 11 was supposed to have a working life of only 21 months at launch Apr 5, 1973.
1990	Feb 28	Atlantis shuttle	A sore throat and head cold of commander John Creighton was the first time a U.S. shuttle had to be postponed due to crew illness. Atlantis finally was launched on the sixth try after five delays. Later Creighton said, "I probably had the world's most famous cold."
1990	Apr 3	Horizon 2 sat	Israel launched its 2nd satellite, Horizon 2 or Ofek 2, on a three-stage Shavit rocket from a military base launch pad in the Negev Desert south of Tel Aviv and Jerusalem to an elliptical orbit from 125 to 923 miles altitude. Shavit is Hebrew for comet. Horizon 2 fell into the atmosphere July 9, 1990. Israel launched 1st satellite Sep 19, 1988.
1990	Apr 5	Pegasus rocket	First flight the 15-ton Pegasus space rocket was launched from a 36-year-old Air Force B-52 bomber. To orbit lightweight satellites up to 600 lbs., it was first all-new unmanned U.S. launch vehicle since 1960's. Winged, three-stage, solid fuel Pegasus, 50 ft. long, 50 in. diameter, strapped under the wing of a B-52 bomber, ferried 7 miles above Earth, dropped free and blasted to orbit, carrying to polar orbit Pegsat containing a small Navy communications satellite, a package of instruments and two barium canisters.
1990	Apr 7	astronaut	Apollo 17 command module pilot Ronald E. Evans, 56, died. His Dec 1972 trip was NASA's last manned flight to the Moon. Evans didn't land, said he felt fortunate to have had the opportunity to travel within 80,000 feet of the Moon.
1990	Apr 25	Hubble telescope	Hubble Space Telescope, one of NASA's Great Observatories, a powerful astronomy satellite 380 mi. above Earth, orbited by shuttle Discovery. Discovery flew to a shuttle record 380 mi. altitude. Two months later, NASA was embarrassed by blurry Hubble pictures. The 12-ton Cassegrain telescope's 94-in. mirror was curved incorrectly. Hubble sent 2 "first light" pictures May 20. Astronauts will fly in Nov 1993 to replace the wide-field camera to correct spherical aberration. Astronauts also will service Hubble in 1996, 1999, 2002, 2005, 2008.
1990	Jun 1	Rosat telescope	Germany's Roentgen Satellite (Rosat) launched on a Delta 2 rocket from Cape Canaveral to circular orbit 360 mi. above Earth to map sources of extreme-ultraviolet light and X-ray across the entire sky. Rosat was the largest X-ray telescope ever built. Rosat is a cooperative program between U.S., West Germany and Great Britain.
1990	Jul 10	asteroid	After 3 close calls in 1989, Planet Earth again had asteroids visits in 1990. Asteroid 1990mf streaked by July 10 at the relatively-short distance of 3

million miles, one of the closest in 50 years. The 300-1,000 ft. diameter rock flashed past at 22,000 mph. Another 1990 close call was Aug 17.

1990	Jul 10	astronauts

NASA yanked shuttle commanders Robert L. "Hoot" Gibson and David M. Walker from 1991 flights for violating Johnson Space Center rules against risky flying. Gibson was involved in a crash fatal to another pilot. Walker flew too close to an airliner. It was the first time NASA had removed crew for discipline. Gibson also was barred from T-38 jet trainer flights one year. Walker grounded from T-38 flights 60 days. NASA policy prohibits high-risk recreational activities by those named to crews.

1990	Jul 15	COBE satellite

Cosmic Background Explorer (COBE) finished all sky survey of background microwave radiation left over from Big Bang, leaving cosmologists pondering how Universe became complex.

1990	Jul 25	CRRES satellite

U.S. Combined Release/Radiation Effects Satellite (CRRES) launched by Atlas-Centaur rocket to study Earth's trapped radiation belts.

1990	Aug 10	Magellan probe

Radar probe Magellan arrived in orbit around Venus. First high-resolution mapping of planet Aug 28. Started full-time mapping Sep 15, with radar peering down through thick clouds. Mapped entire planet twice during 3,000 orbits of Venus by the end of 1991.

1990	Aug 17	asteroid

Australian astronomers found quarter-mile asteroid 1990mu after it crossed Earth's orbital path around the Sun in June. It came within 1.2 million miles, and may cross Earth path closer in 1992. Astronomers said chances are slim of 1990mu hitting Earth within a million years. Another 1990 close call was July 10.

1990	Sep 22	Pioneer 10

Pioneer 10, the first interplanetary probe to leave the Solar System, floated past an important milestone September 22 when it was 50 times farther from the Sun than our planet Earth is from the Sun. Earth stands off from the Sun at an average distance of about 93 million miles. Astronomers refer to that distance as an astronomical unit, or AU. On September 22, Pioneer 10 was 50 AU from the Sun—4.65 billion miles. Since it was launched from Cape Canaveral on March 2, 1972, Pioneer 10 has travelled farther than any other human-made object. It left the Solar System planets behind on June 13, 1983.

1990	Oct 6	Ulysses probe

European-American solar probe Ulysses, once known as International Solar Polar Mission (ISPM), was ferried 160 mi. above Earth in U.S. shuttle Discovery. Its Inertial Upper Stage rocket fired it away from Earth orbit toward Jupiter on a 4 year voyage across the Sun's north and south poles. Ulysses will swing around Jupiter February 8, 1992. Planet gravity will fling it above the plane in which planets of the Solar System orbit the Sun, boosting Ulysses over the Sun's south pole in 1994, then over the Sun's north pole in 1995. Ulysses will measure magnetic field, plasma, electrons, protons, ions, interstellar neutral gas, gravity waves, solar particles, cosmic rays, plasma waves, radio signals, X-rays, gamma rays, cosmic dust. It will be the first probe to explore the Solar System's "third dimension." The probe's radio and plasma-wave receiver has 2 boom antennas, one 24.3 ft. long, the other 238 ft. During deployment of the 24.3-ft. boom Nov 4, 1990, a wobble appeared in the spacecraft's spin, but the shake seemed gone by 1991. One radio stopped working for a time in June 1991. As Ulysses passes within 280,000 miles of Jupiter in Feb 1992, it will sweep down the huge planet's northern hemisphere across its southern hemisphere, measuring magnetic field and interaction with solar wind for 2 weeks.

1990	Dec 2	Mir station	USSR cosmonaut Musa Manarov launched in Soyuz TM-11. He stayed 175 days at Mir, to May 26, 1991. With 366 days in space in 1987-88, Manarov's world record total time was 541 days.
1990	Dec 2	Mir station	First Japanese, first journalist in space, second paying spaceflight passenger. Tokyo Broadcasting System paid more than $10 million to the USSR to have TV journalist Toyohiro Akiyama spend a week at Mir as a "cosmoreporter," launched in Soyuz TM-11. Akiyama took along Japanese green tree frogs. The USSR rushed reporters from the newspapers Krasnaya Zvezda, Delovoi Mir and Sovietskaya Molodyozh of Riga, Latvia, and from the Ukrainian TV Film Studio and the magazine Literaturnaya Gazeta, to space training camp in 1989 after journalists griped about their country sending a Japanese journalist to space first. Advertisements painted on the 150-ft. Soyuz launch rocket which carried the manned capsule Soyuz TM-11 to space, brought the Soviets hard currency. The booster was plastered with Japanese ads for Unicharm women's hygiene products, Sony electronic products and Ohtsuka Chemicals. After he kicked a four-pack-a-day habit for the flight, Akiyama admitted he was looking forward desperately to his first cigarette after landing.
1990		deep space	The largest galaxy ever seen was found a billion lightyears away in the Abell 2029 cluster of galaxies. The galaxy at the center of the cluster has a diameter of 6 million lightyears—5 times

wider than the galaxy previously thought to be largest and 60 times larger than our own Milky Way galaxy—one of the most luminous, shining with 2 trillion times the light of the Sun. More than a quarter of all the light radiated by the cluster is contributed by the one galaxy.

| 1990 | Dec 2 | Columbia shuttle | U.S. shuttle STS-35 commander Vance Brand, at 59, was the oldest person ever to travel in space. The first dress shirt and tie worn in space was donned by U.S. astronaut Jeffrey Hoffman on Dec 7 as he dressed up to teach elementary pupils on the ground by TV from the shuttle. With shuttle in orbit, most people ever in space at the same time included 12 astronauts and cosmonauts: 5 at the USSR's Mir space station and 7 in Columbia. Columbia carried Astro-1 Spacelab for astronomy. It was first flight in 5 years of non-NASA astronauts, since 1986 Challenger disaster. |
| 1990 | Dec 8 | Galileo probe | Unmanned interplanetary spacecraft Galileo, launched Oct 18, 1989, swoops around Earth at 620 mph on its way to Jupiter in 1995. |

1991

1991	Apr 5	GRO telescope	U.S. shuttle Atlantis STS-37 astronauts launched Gamma Ray Observatory into Earth orbit, one of NASA's Great Observatories. Also, it was first all ham space shuttle crew: amateur radio operators Kenneth Cameron, KB5AWP; Jay Apt, N5QWL; Linda Godwin, N5RAX; Steve Nagel, N5RAW; Jerry Ross, N5SCW.
1991	May 18	Mir station	Helen Sharman, 1st British cosmonaut, in Soyuz TM-12, for 8 days studying how orchids, dwarf magnolia vines improved atmosphere aboard Mir.
1991	Aug 30	Solar-A telescope	Japan's Institute of Space and Astronautical Science (ISAS) launched a solar flare observation satellite from the Kagoshima Space Center in southern Japan on a M-3S2 three-stage solid-fuel rocket. The small, boxy, 860-lb. Solar-A, Japan's 14th scientific satellite, will capture high-energy X-rays and gamma rays released from flares on the Sun for three years. Four instruments aboard are NASA's Soft X-Ray Telescope, Great Britain's Bragg Crystal Spectrometer, Japan's Hard X-Ray Telescope and Japan's Wide-Band Spectrometer.
1991	Oct 2	Mir station	First Austrian cosmonaut Franz Viehboeck. Soyuz TM-13 had unusual crew with 2 researcher cosmonauts—Viehboeck and Kazakhstan's Takhtar Aubakirov. Past USSR guest cosmonaut flights had a Soviet commander and engineer with 1 foreign researcher. Soyuz TM-13 had commander, no flight engineer, 2 researchers.
1991	Oct 29	Galileo probe	Unmanned interplanetary spacecraft Galileo, launched Oct 18, 1989, grazed the Asteroid Belt, passing close to 20-mile-wide asteroid 951 Gaspra

			on Oct 29, 1991, on its way to Jupiter in 1995.
1991	Dec	EUVE telescope	U.S., Extreme Ultraviolet Explorer (EUVE) was launched on a Delta 2 rocket to Earth orbit to map ultraviolet light sources in the sky. Shuttle astronauts will retrieve the payload in Feb 1995.

1992

1992	Feb	Discovery shuttle	U.S. flight STS-42 is to carry the first Canadian woman in space, Roberta L. Bondar, M.D.
1992	Feb 8	Ulysses probe	European-American solar probe passes Jupiter. Ulysses was launched Oct 6, 1990, to Jupiter on a 4-year voyage to the Sun's north and south poles. Passing within 280,000 miles of Jupiter, Ulysses sweeps down the northern hemisphere and across the southern hemisphere, measuring magnetic field, charged particles, and interaction with the solar wind for 2 weeks. Then Jupiter's gravity flings it above the plane of the Solar System and over the Sun's south pole in 1994, then over the Sun's north pole in 1995.
1992	Apr	Mir station	The USSR planned launch in 2nd quarter 1992 large Mir add-on module, Spektr, with telescopes to photograph Earth. Spektr, Russian for Optical, is a large capsule like Kvant-2 and Kristall expansion modules added to Mir in 1989-90.
1992	May	Atlantis shuttle	U.S. flight STS-45 is to carry the first female payload commander, Kathryn D. Sullivan.
1992	May	Mir station	USSR cosmonaut Sergei Krikalev is to become the 3rd person to spend a year in space on one trip.
1992	May	Endeavour shuttle	U.S. flight STS-49 is to be the maiden flight of U.S. shuttle Endeavour, replacing Challenger.
1992	Jul 10	Giotto probe	European Space Agency probe,which made closest approach of any to Comet Halley Mar 13, 1986, is to rendezvous with Comet Grigg-Skjellrup.
1992	Sep	Endeavour shuttle	Endeavour's second flight, STS-47, is likely to carry the first married couple in space, astronauts Mark Lee and Jan Davis. It also is to carry the first black woman in space, Mae C. Jemison, M.D.
1992	Oct	Mir station	The USSR planned to launch the international ecological module Priroda to expand Mir in 4th quarter 1992. Priroda is a large capsule like Kvant-2 and Kristall modules added in 1989-90.
1992	Dec 8	Galileo probe	Unmanned interplanetary spacecraft Galileo, launched Oct 18, 1989, was to swoop around Earth again on its way to Jupiter in 1995.
1992	Sep	Mars Observer	U.S. orbiter launched on Titan 3 rocket to study surface, climate, gravity, magnetic field. First U.S. probe to Mars since Viking in 1976. Mars Observer will orbit 272-mi. above the surface, sending photos and data for 2 years. The lightweight Mars Observer, designed after a communications satellite with weather satellite parts added, will have a moving-mirror infrared camera to peer into atmospheric dust. For the best

pictures ever of Mars from orbit, its camera will have a compact telescope to show two-mi. squares of landscape. It will have radar altimeter to map hills, valleys; electronics to measure magnetic field; gamma-ray scanner for uranium, iron, potassium, calcium, magnesium.

1992	Oct	ISY 1992/Mars

For the International Space Year 1992, a six-nation fleet of spacecraft powered by large, whispy solar sails may race to Mars. Columbus 500 Space Sail Cup race, sponsored by the Christopher Columbus 500 Quincentenary Jubilee Commission, celebrating International Space Year 1992, is to begin Columbus Day as the spacecraft, with sails folded in canisters, are launched by rocket to Earth orbit. The gossamer sails range from a small disk like sunflower petals to a 3,000 ft. sheet. Unfurled in space, they will catch the flow of photons from the Sun, blowing the spacecraft toward Mars. The voyage past the Moon to Mars will take one to five years. Sailcraft can travel as fast between planets as rocket ships, and they can carry heavy cargo, however the 1992 racers weigh under 1,100 lbs each. Representing the Americas to which Columbus sailed, Europe from which he sailed, and Asia where he was headed, racers are to include the U.S., Canada, Great Britain, Italy, the USSR and China. NASA will provide advice and tracking. There may be winners in four categories: getting to Earth orbit and unfurling; leaving Earth's gravity; passing closest to the Moon or Mars within 6,000 miles; and reaching Mars.

1992		ISY 1992/Moon

For the International Space Year 1992, Lunar Exploration Inc. (LEI), a U.S. non-profit group of engineers, scientists, marketers and lawyers, had planned to launch a lunar orbiter called Lunar Prospector. LEI had hoped to pay for the probe by selling advertising on the side of the USSR rocket used to launch the 600-lb. Lunar Prospector from the USSR. The USSR would donate a rocket, but LEI would pay launch costs. The satellite would orbit the Moon, mapping the surface for a year, searching for ice deposits, measuring the magnetic field, counting frequency and location of gas releases and producing the first complete map of the lunar gravity field. Lunar Prospector's five science experiments were to be donated by aerospace contractors and NASA. The project was delayed in 1991 when fast-food restaurants, soft drink bottlers and other businesses were not interested in advertising on the side of a Soviet rocket. If LEI finds funds at NASA or elsewhere, it would take 18 to 24 months to launch Lunar

Prospector. If it were to fly soon, LPP might be the first American spacecraft sent to the Moon since the Apollo 17 manned flight in Dec 1972. LPP then also would be the first lunar orbiter from any nation since the Japanese spacecraft Hiten dropped the small Hagoromo satellite into lunar orbit Mar 19, 1990. LPP then also would be the first private lunar mission. Japan launched Hiten Jan 24, 1990. The USSR sent its last 4 probes to the Moon from 1973-1976.

| 1992 | ISY 1992/Moon | For the International Space Year 1992, non-profit Space Studies Institute wants to send a Lunar Polar Probe (LPP) to orbit the Moon, searching for water ice, oxygen, iron, uranium, silicon, and carbon resources which might be used to build lunar towns. The 300-lb. remote-sensing LPP would be built from left-over spare parts left over from past space missions. For instance, one such item of government surplus would be an Apollo gamma ray spectrometer warehoused by NASA's at the Jet Propulsion Laboratory, Pasadena, California. Space Studies Institute was trying to raise money for LPP from businesses and private donations; then NASA or European Space Agency could carry LPP to Earth orbit free. LPP would have its own fuel and rocket to boost it away from Earth's gravitational pull to the Moon. If it were to fly soon, LPP might be the first American spacecraft sent to the Moon since the Apollo 17 manned flight in Dec 1972. LPP then also would be the first lunar orbiter from any nation since the Japanese spacecraft Hiten dropped the small Hagoromo satellite into lunar orbit Mar 19, 1990. LPP then also would be the first private lunar mission. Japan launched Hiten Jan 24, 1990. The USSR sent its last 4 Moon probes in 1973-1976. |

1993

| 1993 | U.S. shuttles | The first Hispanic woman in space, U.S. astronaut Ellen Ochoa, is likely to fly in 1993. The first U.S. military females in space are likely to be Army Capt. Nancy J. Sherlock and Air Force Capt. Susan J. Helms flying U.S. shuttles in 1993. |

1994

| 1994 Feb | Hubble telescope | U.S. shuttle astronauts will fly to Hubble Space Telescope 380 mi. above Earth to replace the satellite's wide-field camera to correct blurry pictures caused by spherical aberration. |
| 1994 | Ulysses probe | European-American solar probe Ulysses, launched Oct 6, 1990, passes over the Sun's south pole in 1994, then on to the Sun's north pole in 1995. |

1995

| 1995 Jul 10 | Galileo probe | Unmanned interplanetary spacecraft Galileo, |

			launched Oct 18, 1989, will separate into orbiter and atmosphere probe, 150 days ahead of Dec 7 arrival at the planet Jupiter.
1995		Ulysses probe	European-American solar probe, launched Oct 6, 1990, passes over the Sun's north pole in 1995.
1995	Nov	Freedom station	NASA starts building U.S.-international space station Freedom in Earth orbit. Some 18 shuttle flights are planned, through Dec 1999. Use of Freedom station will start with a May 1997 flight.
1995		Zenit rocket	USSR's Zenit space booster may ferry American satellites from northeastern Australia to Earth orbit starting in 1995—if Aborigines permit United Technologies Corp., Hartford, Connecticut, to build the spaceport for the Cape York Space Agency on their lands in northern Queensland. The U.S. government gave informal approval in 1990 for the launch site on Australia's northernmost tip.
1995	Dec 7	Galileo probe	Unmanned interplanetary spacecraft Galileo, launched Oct 18, 1989, will complete its six-year, 2.4 billion mile trip to the giant planet Jupiter. Having separated July 10, 1995, Galileo's atmosphere probe will parachute 370 miles down, measuring clouds and atmosphere for 75 mins, until pressure and temperature crush and vaporize the probe. The Galileo orbiter above will transmit data about Jupiter and its moons for 22 months.
1995	Dec	Cassini probe	NASA launches a Titan 4/Centaur rocket carrying a robot explorer to Saturn. NASA's spacecraft Cassini will fly by the asteroid 66 Maja in 1997 and the planet Jupiter in 1999 on its way to rendezvous with Saturn in Oct 2002. At Saturn, Cassini will drop a European Space Agency probe, Huygens, through the atmosphere to a soft landing on Saturn's large moon Titan. Cassini then will study Saturn, its rings and moons, over four years, completing 30 orbits around the giant planet.

1996

1996	Feb	CRAF probe	Comet Rendezvous Asteroid Flyby (CRAF) may visit asteroid and study Comet Kopff for 2.5 years as comet approaches Sun between orbits of Jupiter and Mars. CRAF will find composition of comet dust, ice. CRAF and Saturn probe Cassini will be identical Mariner Mark II interplanetary craft, except for unique science experiments.
1996	Mar	Planet B probe	Japan plans to launch a probe called Planet B to Venus atop an improved MU-3S-2 solid rocket.
1996		Moon lander	Japan plans to launch an unmanned Moon lander on an M-5 rocket. Probes aboard the Institute of Space and Astronautical Science (ISAS) spacecraft would penetrate lunar soil.
1996	Nov	Hubble telescope	U.S. shuttle astronauts will fly to service the Hubble Space Telescope 380 mi. above Earth.

1997

| 1997 | May | Freedom station | NASA starts using the U.S.-international space station Freedom in Earth orbit. Utilization shuttle flights will continue through 1999, even as construction flights continue to Dec 1999. |
| 1997 | | Cassini probe | NASA's robot explorer will fly through the Asteroid Belt on its way to a rendezvous in 2002 with Saturn. In 1997, Cassini will fly near and examine the asteroid 66 Maja. |

1998

| 1998 | Mar | AXAF telescope | Advanced X-ray Astrophysics Facility (AXAF), one of NASA's Great Observatories, to be launched to Earth orbit from where it will receive X-rays from quasars, galaxy clusters, starburst galaxies, supernova remnants, neutron stars. Its imaging spectrometer, 1,000 times more sensitive, will produce X-ray images 10 times as sharp. Shuttle astronauts would visit AXAF during its 15 year life, servicing the telescope in 2002, 2007, 2012. |

1999

| 1999 | | Cassini probe | NASA's unmanned spacecraft will fly by Jupiter, picking up a gravity boost on its way to Saturn in 2002. Cassini will send back photos and data from Jupiters and its moons. |
| 1999 | Nov | Hubble telescope | U.S. shuttle astronauts will fly to service the Hubble Space Telescope 380 mi. above Earth. |

2000

| 2000 | Jun | SIRTF telescope | Space Infrared Telescope facility (SIRTF), one of NASA's Great Observatories, may be launched to Earth orbit where it will receive infrared light from distant objects with 1,000 times more sensitivity. |
| 2000 | Oct | Lunar Observer | U.S. lunar orbiter to study the Moon. |

2002

2002	Apr	AXAF telescope	U.S. shuttle astronauts will fly to Earth orbit to service the Advanced X-ray Astrophysics Facility.
2002	May	Solar Probe	U.S. spacecraft to study unexplored regions near the Sun, measure electromagnetic fields, particles.
2002	Oct	Cassini probe	NASA's robot explorer Cassini will arrive at Saturn. It will drop the European Space Agency probe Huygens through Titan's clouds to a soft landing for soil analysis. Cassini will relay Huygens' photos, data to Earth, then study Saturn, its rings and moons for 4 years, completing 30 orbits around the giant planet. Cassini, the first spacecraft from Earth to visit Saturn since Voyager 2 in 1981, will use radar to map Titan's cloud shrouded surface as Magellan mapped Venus in 1990-91. Cassini and Comet Rendezvous Asteroid Flyby (CRAF), versions of NASA Mariner Mark II.
2002	Nov	Hubble telescope	U.S. shuttle astronauts will fly to service the Hubble Space Telescope 380 mi. above Earth.

Cosmonauts

All men and women who have flown, or are preparing to fly, in USSR spacecraft are cosmonauts in this list. Astronauts are those who have flown in U.S. spacecraft. A few persons from various nations are both cosmonauts and astronauts. No other nations have originated manned flights; Europe, Japan and China are preparing to do so.

The very first cosmonauts, known as the Star City 12, were selected March 14, 1960: Pavel I. Belyayev, 28; Valeri F. Bykovsky, 26; Yuri A. Gagarin, 26; Viktor V. Gorbatko, 26; Yevgeny V. Khrunov, 27; Vladimir M. Komarov, 33; Alexei A. Leonov, 26; Andrian G. Nikolayev, 31; Pavel R. Popovich, 30; Georgi S. Shonin, 25; Gherman S. "Herman" Titov, 25; and Boris V. Volynov, 26. All 12 flew in space. Gagarin became the first human being in space in his Vostok 1 capsule April 12, 1961.

Name	Flight
Afanasyev, Viktor	Soyuz TM-11
Akiyama, Toyohiro	Soyuz TM-11
Aksyonov, Vladimir V.	Soyuz 22, Soyuz T-2
Alexandrov, Alexander (USSR)	Soyuz T-9, Soyuz TM-3
Alexandrov, Alexander (Bulg.)	Soyuz TM-5
Artsebarsky , Anatoly	Soyuz TM-12
Artyukhin, Yuri	Soyuz 14
Atkov, Oleg	Soyuz T-10B
Aubakirov, Takhtar	Soyuz TM-13
Balandin, Alexander N.	Soyuz TM-9
Belyayev, Pavel I.	Voskhod 2
Beregovoi, Georgi T.	Soyuz 3
Berezovoi, Anatoli	Soyuz T-5
Bykovsky, Valeri F.	Vostok 5, Soyuz 22, Soyuz 31
Chretien, Jean-Loup	Soyuz T-6, Soyuz TM-7
Clemens, Lothaller	Soyuz TM-13-backup
Demin, Lev S.	Soyuz 15
Dobrovolsky, Georgi T.	Soyuz 11
Dzhanibekov, Vladimir A.	Soyuz 27, Soyuz 39, Soyuz T-6,Soyuz T-12, Soyuz T-13
Faris, Mohammed	Soyuz TM-3
Farkas, Bertalan	Soyuz 36
Feoktistov, Konstantin P.	Voskhod 1
Filipchenko, Anatoli V.	Soyuz 7, Soyuz 16
Gagarin, Yuri A.	Vostok 1
Glazkov, Yuri N.	Soyuz 24
Gorbatko, Viktor V.	Soyuz 7, Soyuz 24, Soyuz 37
Grechko, Georgi M.	Soyuz 17, Soyuz 26, Soyuz T-14
Gubarev, Alexei	Soyuz 17, Soyuz 28
Gurragcha, Jugderdemidiyn	Soyuz 39
Hermaszewski, Miroslaw	Soyuz 30
Ivanchenkov, Alexander S.	Soyuz 29, Soyuz T-6
Ivanov, Georgi	Soyuz 33
Jahn, Sigmund	Soyuz 31
Khrunov, Yevgeny V.	Soyuz 5
Kizim, Leonid	Soyuz T-3, Soyuz T-10B, Soyuz T-15
Klimuk, Pyotr I.	Soyuz 13, Soyuz 18B, Soyuz 30
Komarov, Vladimir M.	Voskhod 1, Soyuz 1
Kovalyonok, Vladimir	Soyuz 25, Soyuz 29, Soyuz T-4
Krikalev, Sergei M.	Soyuz TM-7, Soyuz TM-12
Khrunov, Yevgeny V.	Soyuz 5
Kubasov, Valeri N.	Soyuz 6, Soyuz 19, Soyuz 36
Laveikin, Alexander I.	Soyuz TM-2
Lazarev, Vasili G.	Soyuz 12, Soyuz 18A
Lebedev, Valentin	Soyuz 13, Soyuz T-5

Leonov, Alexei A.	Voskhod 2, Soyuz 19
Levchenko, Anatoly	Soyuz TM-4
Lyakhov, Vladimir	Soyuz 32, Soyuz T-9, Soyuz TM-6
Makarov, Oleg G.	Soyuz 12, Soyuz 18A, Soyuz 27, Soyuz T-3
Malyshev, Yuri V.	Soyuz T-2, Soyuz T-11
Manakov, Gennadi	Soyuz TM-10
Manarov, Musa K.	Soyuz TM-4, Soyuz TM-11
Mendez, Arnaldo Tomayo	Soyuz 38
Mohmand, Abdul Ahad	Soyuz TM-6
Musabaev, Talgat	Soyuz TM-13-backup
Nikolayev, Andrian G.	Vostok 3, Soyuz 9
Patsayev, Viktor I.	Soyuz 11
Polyakov, Valery	Soyuz TM-6
Popov, Leonid I.	Soyuz 35, Soyuz 40, Soyuz T-7
Popovich, Pavel R.	Vostok 4, Soyuz 14
Prunariu, Dumitru	Soyuz 40
Remek, Vladimir	Soyuz 28
Romanenko, Yuri V.	Soyuz 26, Soyuz 38, Soyuz TM-2
Rozhdestvensky, Valeri I.	Soyuz 23
Rukavishnikov, Nikolai	Soyuz 10, Soyuz 16, Soyuz 33
Ryumin, Valeri V.	Soyuz 25, Soyuz 32, Soyuz 35
Sarafanov, Gennadi	Soyuz 15
Savinykh, Viktor P.	Soyuz T-4, Soyuz T-13, Soyuz TM-5
Savitskaya, Svetlana	Soyuz T-7, Soyuz T-12
Serebrov, Alexander A.	Soyuz T-7, Soyuz T-8 Soyuz TM-8
Sevastyanov, Vitali	Soyuz 9, Soyuz 18B
Sharma, Rakesh	Soyuz T-11
Sharman, Helen	Soyuz TM-12
Shatalov, Vladimir A.	Soyuz 4, Soyuz 8, Soyuz 10
Shonin, Georgi S.	Soyuz 6
Solovov, Anatoly	Soyuz TM-5, Soyuz TM-9
Solovyev, Vladimir	Soyuz T-10B, Soyuz T-15
Strekalov, Gennadi	Soyuz T-3, Soyuz T-8, Soyuz T-10A,Soyuz T-11,Soyuz TM-10
Tereshkova, Valentina V.	Vostok 6
Titov, Gherman S. "Herman"	Vostok 2
Titov, Vladimir G.	Soyuz T-8, Soyuz T-10A, Soyuz TM-4
Tuan, Pham	Soyuz 37
Vasyutin, Vladimir	Soyuz T-14
Viehboeck, Franz	Soyuz TM-13
Viktorenko, Alexander S.	Soyuz TM-3, Soyuz TM-8, Soyuz TM-13-backup
Volk, Igor P.	Soyuz T-12
Volkov, Alexander A.	Soyuz T-14, Soyuz TM-7, Soyuz TM-13
Volkov, Vladislav N.	Soyuz 7, Soyuz 11
Volynov, Boris V.	Soyuz 5, Soyuz 21
Yegorov, Boris B.	Voskhod 1
Yeliseyev, Alexei S.	Soyuz 5, Soyuz 8, Soyuz 10
Zholobov, Vitali	Soyuz 21
Zudov, Vyacheslav	Soyuz 23

Vostok

Vostok, meaning East in Russian, was the USSR's first man-in-space program, lasting from 1961-63. Yuri Gagarin in 1961 rode Vostok 1 to become the first man in space. Valentina Tereshkova, in Vostok 6 in 1963, was the first woman to fly in space and the first woman to orbit Earth.

The cosmonauts gave their capsules nicknames: Swallow, Eagle, Falcon, Golden Eagle, Hawk and Sea Gull. All ejected from their capsules and parachuted to Earth.

VOSTOK 1

Cosmonaut: Yuri A. Gagarin
Rocket: A-1
Earth orbits: 1

Capsule callsign/nickname: Swallow
Launch: April 12, 1961
Flight duration: 1 hrs 48 mins

Events: First human being in space and first person in Earth orbit. Made one complete orbit. First to eat and drink in orbit.

VOSTOK 2

Cosmonaut: Gherman S. Titov
Rocket: A-1
Earth orbits: 16

Capsule callsign/nickname: Eagle
Launch: August 6, 1961
Flight duration: 25 hrs 18 mins

Events: Second person in Earth orbit. First spaceflight of more than 24 hours. Youngest person to fly in space, age 26. First to sleep in space. Capsule heater malfunctioned, allowing temperature inside to drop to 43 deg. Fahrenheit. Titov suffered nausea throughout the flight.

VOSTOK 3

Cosmonaut: Andrian G. Nikolayev
Rocket: A-1
Earth orbits: 64

Capsule callsign/nickname: Falcon
Launch: August 11, 1962
Flight duration: 94 hrs 22 mins

Event: First space rendezvous, with Vostok 4. First team flight. Nikolayev and Popovich in Vostok 4 approached within three miles of each other in space. Their closeness was due to launch trajectory. They were unable to maneuver in space.

VOSTOK 4

Cosmonaut: Pavel R. Popovich
Rocket: A-1
Earth orbits: 48

Capsule callsign/nickname: Golden Eagle
Launch: August 12, 1962
Flight duration: 70 hrs 57 mins

Event: First space rendezvous, with Vostok 3. First team flight. Popovich, on his first orbit, and Nikolayev in Vostok 3 approached within three miles of each other in space. Their closeness was due to launch trajectory. They were unable to maneuver in space.

VOSTOK 5

Cosmonaut: Valeri F. Bykovsky
Rocket: A-1
Earth orbits: 81

Capsule callsign/nickname: Hawk
Launch: June 14, 1963
Flight duration: 119 hrs 06 mins

Events: 2nd team flight, passing within 3 miles of Vostok 6. Longest solo flight ever, 4 days.

VOSTOK 6

Cosmonaut: Valentina V. Tereshkova
Rocket: A-1
Earth orbits: 48

Capsule callsign/nickname: Sea Gull
Launch: June 16, 1963
Flight duration: 70 hrs 50 mins

Events: First woman in space. First woman in Earth orbit. Youngest woman to fly in space, age 26. Second team flight, passing within about 3 miles of Vostok 5. Tereshkova was the first ordinary person in space. She had been a textile mill worker with a hobby of parachute jumping. She and cosmonaut Andrian G. Nikolayev became the first couple married after spaceflight, in November 1963, five months after her June 16 Vostok 6 flight. Nikolayev had flown Vostok 3 August 11, 1962. USSR Premier Nikita Khrushchev was toastmaster at the wedding. The 1st cosmonaut progeny was Yelena, daughter of Tereshkova, Nikolayev, born June 1964. She was 1st politician from space. In 1967, elected to Supreme Soviet, then to Presidium of the Supreme Soviet in 1974.

Voskhod

Voskhod, meaning Sunrise in Russian, was the USSR's second man-in-space program. It featured three-man capsules.

USSR Premier Nikita Khrushchev, wanting to upstage America's two-man Gemini flights scheduled for 1965, had Soviet engineers remove life-support and safety equipment and bolt three seats into what otherwise would have been a two-man Vostok capsule. The converted capsule was named Voskhod and flew in 1964. There wasn't room for cosmonauts to wear bulky spacesuits so they wore shirt-sleeve-style work coveralls. Cosmonauts landing from previous Vostok flights had ejected and parachuted to Earth. But Voskhod didn't have room for ejection equipment and individual cosmonaut parachutes, so one large parachute was designed to float the entire Voskhod capsule to a hard landing. The cosmonauts even dieted to reduce launch weight. When Voskhod 1 carried three men to space October 12, 1964, it weighed only 1,300 lbs. more than the older two-man Vostok. In Moscow that day, the newspaper Pravda referred to future American three-man capsules in the headline, "Sorry, Apollo!" A day later, the three Voskhod 1 cosmonauts suddenly and mysteriously were brought back to Earth. Khrushchev had been kicked out of office October 13.

America's first two-man Gemini flight didn't blast off until March 1965. The first U.S. three-man flight was Apollo 7 in October 1968.

VOSKHOD 1
Cosmonauts: Konstantin P. Feoktistov, Vladimir M. Komarov, Boris B. Yegorov
Flight callsign/nickname: Ruby

Rocket: A-2	Launch: October 12, 1964
Earth orbits: 16	Flight duration: 24 hrs 17 mins

Events: First multi-person crew in space. First three-man crew. First live TV pictures from space. First to fly without spacesuits. Maybe the most uncomfortable space trip ever with two extra seats stuffed in a one-man Vostok to make it into Voskhod. Three cosmonauts were crammed in and sent to space for 24 hours. So cramped they couldn't even take along spacesuits for safety.

VOSKHOD 2
Cosmonauts: Pavel I. Belyayev, Alexei A. Leonov
Flight callsign/nickname: Diamond

Rocket: A-2	Launch: March 18, 1965
Earth orbits: 17	Flight duration: 26 hrs 2 mins

Event: After discomfort of three cosmonauts in Voskhod 1, Voskhod 2 carried only two. It was the first ever two-man crew on a spaceflight, just five days before two U.S. astronauts were launched in Gemini 3, but five months after the USSR launched three in Voskhod 1. Leonov made the first-ever spacewalk (extravehicular activity EVA), 10 minutes on a 10-ft. tether outside. Belyayev and Leonov spent 26 hours in space and returned to Earth March 19. Unfortunately, they had to wait more than a day to be picked up after they parachuted into a forest deep in the snow-covered Ural Mountains. A helicopter found them in less than three hours, but couldn't land on the rugged terrain. Belyayev and Leonov spent the night, eating and keeping warm with supplies dropped on the hillside by the chopper. The nearest suitable landing site was 12 miles away and a party descended there the next day. But they didn't reach the stranded cosmonauts until after midday and had to call for another helicopter supply drop to spend that night. It took the entire third day to hike back to the helicopter landing area. When Belyayev and Leonov reached civilization March 21, they had spent almost three times as long in the Urals as in space.

VOSKHOD 3 or COSMOS 110
Passengers: the dogs Veterok and Ugolek

Rocket: A-2	Launch: February 22, 1966
Flight duration: 23 days	

Events: The last Voskhod was a long-duration test preparing for manned Soyuz capsule flights. The dogs Veterok and Ugolek were observed in orbit 23 days via TV and biomedical telemetry. Previously, seven Russian dogs had been flown to orbit in Sputniks between November 1957 and March 1961: Laika (Barker in Russian), Belka (Squirrel in Russian), Strelka (Little Arrow), Pchelka (Little Bee), Mushka (Little Fly), Chernushka (Blackie) and Zvezdochka (Little Star).

Soyuz

Soyuz spacecraft have been a series of capsules used by the USSR to transport men and women to Earth orbit. Soyuz is the USSR's third man-in-space program, after the early Vostok and Voskhod. Soyuz has featured two-man and three-man capsules. From 1967 to the present, independent trips to space have been made, as well as trips to the USSR's Salyut and Mir space stations.

Design. Blueprints for a new manned spacecraft probably were on USSR drawing boards by 1962. The new capsule was to be named Soyuz, Russian for Union.

Ready for its first flight April 23, 1967, the new Soyuz design took some elements from the USSR's earlier Vostok and Voskhod manned capsules, and may have borrowed heavily from an American man-in-space capsule design by General Electric Co. (GE) for the Apollo program. Press reports in the 1980's held that Soyuz designer Sergei P. Korolev, the father of the USSR space program, could have had access to technology reports and design proposals published in the U.S.

Soyuz had three main parts: service module, nearly-spherical orbital module, and bell-shaped reentry capsule. GE similarities included instrument positions, orbital module, side hatch, torus propellant tank, bell-shaped reentry capsule stocked for three cosmonauts, and heatshield to separate before landing. Soyuz was similar to the GE total weight of 16,000 lbs., and capsule weight of 5,000 lbs. Soyuz weighed 14,550 lbs. and was 30 feet long from rear antenna to docking probe and 33 feet wide with solar arrays deployed with 318 cubic feet livable volume.

Soyuz capsules have been blasted to low Earth orbit atop the same A-2 rocket used for Voskhod launches.

Moon. The original Soyuz version, flown from 1966-70, was designed to carry a man on a flight around the Moon. Known as Soyuz B-V, the circumlunar mission was abandoned as the Soviets fell behind in the Moon Race and they switched to a more complex lunar landing plan to keep up with America's Apollo program.

Generations. There have been four generations of basically-the-same capsules from 1967 to the present: Soyuz, Soyuz Ferry, Soyuz T and Soyuz TM. The first generation Soyuz capsules flew from 1967 into the 1970's.

The second generation was Soyuz Ferry, a modified original Soyuz used to carry cosmonauts to and from Salyut space stations through the 1970's. Soyuz 12 in 1973 was the shakedown flight of Soyuz Ferry. It had solar panels removed to save weight, limiting flights to two days on internal batteries. There was a new launch escape rocket, rebuilt air valves and new spacesuits.

Only two-man missions were flown for six years, until introduction of the Soyuz-T capsule version in 1979. The T in Soyuz-T stood for troika, for third-generation Soyuz. The first unmanned flight of a Soyuz-T capsule was in 1979. The first Soyuz-T manned flight was in 1980.

The fourth generation was Soyuz TM. The M in Soyuz TM was for modified, as in modified Soyuz T. The first unmanned flight of Soyuz TM was in 1986. The first Soyuz TM manned flight was in 1987.

The Progress series of unmanned cargo freighters used to supply space stations are modified Soyuz capsules. The first Progress flight was in 1978.

Components. Dimensions and capacities have changed some over the years, but a Soyuz capsule today is very much like the first Soyuz in the 1960's.

The forward third of a Soyuz is the orbital module—a 2,600-lb. ball seven feet in diameter and ten feet long used as sleeping quarters, lab, air lock and cargo hold.

The middle third is the 6,200-lb., 7-ft.-long, 7-ft.-diameter reentry module, shaped like a bell for aerodynamic lift. A capsule can land on water, but a ground landing in the USSR makes recovery easier.

The rear third of a Soyuz is a 6,000-lb., 8-ft.-long, 8-ft.-diameter service module with a 9-ft.-diameter flared base. It has an unpressurized engine compartment and a pressurized forward instrumentation and battery area. The engine compartment has four large fuel tanks of nitric acid and UDMH (unsymmetrical dimethal hydrazine) and four small tanks of hydrogen peroxide for the main and backup engines. Small fuel tanks of hydrogen peroxide for the attitude control system are forward in the service module.

Crew. Soyuz crews can include commander, flight engineer and research cosmonaut. The commander flies the spacecraft during docking or emergencies. The flight engineer monitors life support and communications. Soyuz flies itself most of the time so its instruments—a globe showing spacecraft position, indicator lights, TV to watch docking and display a checklist, and periscope for manual rendezvous and docking— mostly keep cosmonauts up-to-date on performance and malfunctions. Carbon dioxide and water are filtered from the 68 degree Fahrenheit air.

Landing. Cosmonauts today wear pressure suits during launch and landing. A Soyuz capsule heatshield can withstand high forces and heat of re-entering Earth's atmosphere, even if returning from lunar orbit at 25,200 mph. During reentry, temperature in the capsule reaches 77-86 degrees F.

After reentry, a parachute opens at an altitude of 23,000 feet, the heatshield separates, and landing radar fires solid-fuel braking rockets at seven feet off the ground for a soft landing at less than 7 mph.

On the ground, the capsule flashes three beacon lights for its recovery team. Capsule emergency supplies include insulating suits, sea survival suits, raft, radio, flares, knife, medicine and food.

SOYUZ 1
Cosmonaut: Vladimir M. Komarov
Rocket: A-2   Launch: April 23, 1967
Flight duration: 26 hrs 40 mins for 17 orbits
Events: First Soyuz. First person to die in space flight. Komarov died April 24 as his Soyuz 1 capsule crashed on the central-Asian plain. He had been launched 26 hours earlier. The test flight seemed successful, until it crashed after re-entry. A parachute to slow the capsule opened, but lines became snarled at 23,000 feet altitude. The parachute twisted and deflated. Soyuz 1 hit the ground at 200 miles per hour and burst into flames.

SOYUZ 2
Rocket: A-2   Launch: October 25, 1968
Flight duration: 3 days
Events: First unmanned Soyuz, rendezvous target for Soyuz 3.

SOYUZ 3
Cosmonaut: Georgi T. Beregovoi
Rocket: A-2   Launch: October 26, 1968
Flight duration: 94:51 for 64 orbits
Events: Rendezvous by ground control to within 600 ft. of Soyuz 2. Beregovoi flew to within 3 feet but was unable to complete docking.

SOYUZ 4
Cosmonaut launched: Vladimir A. Shatalov
Cosmonauts landed: Shatalov, Yeliseyev & Khrunov
Rocket: A-2   Launch: January 14, 1969
Flight duration: 71:14 for 45 orbits
Events: First USSR manned launch in winter. First USSR space docking, with Soyuz 5. First change of clothes in space as Yeliseyev, Khrunov donned spacesuits for a one-hour spacewalk. First crew transfer between spacecraft. Yeliseyev, Khrunov left Soyuz 5, pulled themselves along handrails into the Soyuz 4 airlock. The capsules remained docked 4.5 hours. Shatalov, Yeliseyev & Khrunov flew home in Soyuz 4.

SOYUZ 5
Cosmonauts launched: Boris V. Volynov, Alexei S. Yeliseyev, Yevgeny V. Khrunov
Cosmonaut landed: Volynov
Rocket: A-2   Launch: January 15, 1969
Flight duration: 72:46 for 46 orbits
Events: First USSR space docking, with Soyuz 4. First crew transfer between spacecraft. Yeliseyev & Khrunov went outside in spacesuits, pulled themselves along handrails and into Soyuz 4. The spacewalk took about an hour. The two capsules remained docked over 4 hours. Volynov flew home in Soyuz 5, a day after Soyuz 4.

SOYUZ 6
Cosmonauts: Georgi S. Shonin, Valeri N. Kubasov
Rocket: A-2   Launch: October 11, 1969
Flight duration: 118:42 for 79 orbits
Events: Soyuz 6 was the first of three launches in three days. The Soviets made a precision launch of three spacecraft to orbit with Shonin, Kubasov in Soyuz 6 on October 11. Filipchenko, Volkov, Gorbatko flew October 12 in Soyuz 7. Shatalov, Yeliseyev flew October 13 in Soyuz 8. It was the first time three manned spacecraft and seven cosmonauts were in Earth orbit at the same time. Each

craft flew down to Earth after 118 hours in space, landing one day apart just as they had taken off. First welding of metal in space using the Vulcan module.

SOYUZ 7
Cosmonauts: Anatoli V. Filipchenko, Vladislav N. Volkov, Viktor V. Gorbatko
Rocket: A-2   Launch: October 12, 1969
Flight duration: 118:41 for 79 orbits
Events: Soyuz 7 was the second of three launches in three days. Soyuz 7 was the docking target for Soyuz 8. With Soyuz 6 and Soyuz 8, Soyuz 7 was the first time three spacecraft and seven cosmonauts were in orbit at the same time. First space lab construction.

SOYUZ 8
Cosmonauts: Vladimir A. Shatalov, Alexei S. Yeliseyev
Rocket: A-2   Launch: October 13, 1969
Flight duration: 118:41 for 79 orbits
Events: Soyuz 8 was the third of three launches in three days. Soyuz 8 failed to dock with Soyuz 7. With Soyuz 6 and Soyuz 7, Soyuz 8 was the first time three manned spacecraft and seven cosmonauts were in orbit at the same time.

SOYUZ 9
Cosmonauts: Andrian G. Nikolayev, Vitali Sevastyanov
Rocket: A-2   Launch: June 1, 1970
Flight duration: a record 18 days
Events: Broke 5-year-old U.S. Gemini 7 record of 14 days. Modified Soyuz carried science gear in place of docking apparatus. Inside: portable TV camera, with controls and lights, film cameras, science equipment, food, water, waste facilities, 4 portholes around middle. Soyuz TV transmitted on 625 MHz, 625 lines/frame, 25 frames/second. Outside: voice and telemetry radio antenna belted the middle of the capsule. After the flight, Sevastyanov became a popular TV commentator on the USSR program "Man, Earth and Universe." He flew again in Soyuz 18B. Nikolayev had been husband of cosmonaut Valentina V. Tereshkova since November 1963.

SOYUZ 10
Cosmonauts: Nikolai Rukavishnikov, Vladimir A. Shatalov, Alexei S. Yeliseyev
Rocket: A-2   Launch: April 22, 1971
Flight duration: 2 days
Events: First men to go to a space station. The Russians had sent the world's first space station, Salyut 1, to Earth orbit April 19. Rukavishnikov, Shatalov and Yeliseyev left the USSR April 22 in Soyuz 10 to dock at Salyut 1, but stayed in dock only 5.5 hours, then hurried home.

SOYUZ 11
Cosmonauts: Georgi T. Dobrovolsky, Vladislav N. Volkov, Viktor I. Patsayev
Rocket: A-2   Launch: June 6, 1971
Flight duration: a record 23 days (569 hrs 40 mins) for 360 orbits
Events: The first crew from Earth to work in a space station, but also the second USSR space tragedy. The second men to die in spaceflight. Soyuz 11 docked with the new Salyut 1 space station. The cosmonauts stayed in the station a record 23 days. Then, the three cosmonauts were not wearing spacesuits when all three died on the flight home June 30 as air escaped the Soyuz 11 capsule during re-entry. Dobrovolsky, Volkov, Patsayev were found dead when the capsule was opened on the ground after an automatic landing. It was the USSR's last three-man flight for 8 years. The Soyuz design was changed after Soyuz 11, with cosmonauts wearing pressure suits for launch and landing, with only enough room for 2 cosmonauts, rather than 3, and spacesuit environmental controls.

SOYUZ 12
Cosmonauts: Vasili G. Lazarev, Oleg G. Makarov
Rocket: A-2   Launch: September 27, 1973
Flight duration: 2 days
Events: Shakedown flight of a new version of the Soyuz capsule. The Soyuz Ferry version was designed to fly cosmonauts to Salyut space stations. Solar panels removed to save weight, limiting flights to 2 days on internal batteries. New launch escape rocket, rebuilt air valves, new spacesuits. Starting with Soyuz 12, only two-man missions were flown for 6 years, until introduction of the Soyuz-T capsule version in December 1979.

SOYUZ 13
Cosmonauts: Pyotr I. Klimuk, Valentin Lebedev
Rocket: A-2   Launch: December 18, 1973
Flight duration: 8 days

Events: Not a trip to the space station, but an independent astronomy flight in solar-panel version of Soyuz, using Orion-2 ultraviolet camera to photograph Sun X-rays and stars obscured by Earth atmosphere, making 10,000 spectrograms of 3,000 stars. The first Soviets to spend Dec 25 in space, they did not report any unusual stars on Christmas Eve. Soyuz 13 was in orbit while the U.S. space station Skylab was manned. The two cosmonauts and three Skylab astronauts were the first time five persons had been in space at the same time. Klimuk and Lebedev landed Dec 26.

SOYUZ 14
Cosmonauts: Yuri Artyukhin, Pavel R. Popovich
Rocket: A-2   Launch: July 3, 1974
Flight duration: 16 days
Events: First operational flight of the lighter Soyuz ferrying cosmonauts to Salyut 3. Soyuz equipped for water recovery.

SOYUZ 15
Cosmonauts: Lev S. Demin, Gennadi Sarafanov
Rocket: A-2   Launch: August 26, 1974
Flight duration: 2 days
Events: Failed to dock with Salyut 3, a night landing in bad weather.

SOYUZ 16
Cosmonauts: Anatoli V. Filipchenko, Nikolai Rukavishnikov
Rocket: A-2   Launch: December 2, 1974
Flight duration: 6 days
Events: Backup Apollo-Soyuz crew testing equipment before Soyuz 19. U.S. tracking stations took part. Photos of Sun, stars. Fish, plants, fungus carried along for study.

SOYUZ 17
Cosmonauts: Georgi M. Grechko, Alexei Gubarev
Rocket: A-2   Launch: January 11, 1975
Flight duration: 30 days
Events: First cosmonaut ferry to Salyut 4. Returned in snow storm.

SOYUZ 18A
Cosmonauts: Vasili G. Lazarev, Oleg G. Makarov
Rocket: A-2   Launch: April 5, 1975
Flight duration: did not make it to orbit
Events: Flight aborted to suborbital when the A-2 rocket stages didn't separate properly. Landing in Siberian cold, the crew climbed out and built a fire to stay warm.

SOYUZ 18B
Cosmonauts: Pyotr I. Klimuk, Vitali Sevastyanov
Rocket: A-2   Launch: May 24, 1975
Flight duration: a record 63 days
Events: Klimuk and Sevastyanov set a USSR record, for that time, of 63 days in space. Their effort may have been the least known space trip. While millions watched the Apollo 18-Soyuz 19 linkup in July, no one paid attention to Klimuk and Sevastyanov at Salyut 4 station. Sevastyanov, who had flown before in Soyuz 9, had been a popular TV commentator on the USSR program "Man, Earth and Universe."

SOYUZ 19 or SOYUZ-APOLLO
Cosmonauts: Alexei A. Leonov, Valeri N. Kubasov
USSR rocket: A-2   USSR launch: July 15, 1975
USSR flight duration: 143 hrs 31 mins, including 96 orbits of Earth.
Astronauts: Thomas P. Stafford, Vance D. Brand, Donald K. "Deke" Slayton.
U.S. rocket: Saturn   U.S. launch: July 15, 1975
U.S. flight duration: 217 hrs 28 mins 23 secs, including 136 orbits of Earth.
Events: The first USSR/U.S. joint space mission, a show of international cooperation in Earth orbit as Soyuz 19 docked with Apollo 18. Stafford and Leonov met in the dock airlock, shaking hands over Metz, France, July 17. Crews spent 7 hours in each others' craft. The five men conducted experiments, shared meals, held a news conference. They lunched on borscht, chicken and turkey. Also known as Apollo Soyuz Test Program (ASTP), the craft separated after 48 hours joint flight, docked again, separated finally July 19. Slayton was the first person over 50 in space. First time seven people were in orbit at the same time: Stafford, Brand, Slayton, Leonov, Kubasov plus Klimuk and Sevastyanov at Salyut 4 space station. Soyuz 19 landed July 21. Apollo, July 24.

SOYUZ 20

Rocket: A-2   Launch: November 17, 1975
Flight duration: 91 days
Events: Second unmanned Soyuz, last Soyuz to dock with Salyut 4. Testing 3 month shutdown and restart, prior to launch of Salyut 5. Held special climate with turtles, cactus, corn, legumes, vegetable seeds, drosophila, gladioli bulbs.

SOYUZ 21
Cosmonauts: Boris V. Volynov, Vitali Zholobov
Rocket: A-2   Launch: July 6, 1976
Flight duration: 49 days
Events: First flight to Salyut 5, ended by Salyut environment problem.

SOYUZ 22
Cosmonauts: Vladimir V. Aksyonov, Valeri F. Bykovsky
Rocket: A-2   Launch: September 15, 1976
Flight duration: 8 days
Events: The last Soyuz to fly free, not to a space station, had East German MKF-6 spy camera for 10-meter-resolution photos in visible and infrared light, the first foreign equipment to fly in a USSR spacecraft. Had fish and plants for Biokat studies.

SOYUZ 23
Cosmonauts: Valeri I. Rozhdestvensky, Vyacheslav Zudov
Rocket: A-2   Launch: October 14, 1976
Flight duration: 2 days
Events: Docking failure, like Soyuz 15. Upon return home, landed a mile from shore in a lake in a blizzard with four-below-zero temperatures. Rescue rafts were blown away, helicopters had to tow the capsule to shore.

SOYUZ 24
Cosmonauts: Yuri N. Glazkov, Viktor V. Gorbatko
Rocket: A-2   Launch: February 7, 1977
Flight duration: 18 days
Events: The last flight to Salyut 5. Manual maneuvering to overcome problems with automatic docking on earlier flights.

SOYUZ 25
Cosmonauts: Vladimir Kovalyonok, Valeri V. Ryumin
Rocket: A-2   Launch: October 9, 1977
Flight duration: 2 days
Events: Failed first docking with Salyut 6 second-generation station.

SOYUZ 26
Cosmonauts: Georgi M. Grechko, Yuri V. Romanenko
Rocket: A-2   Launch: December 10, 1977
Flight duration: 37 days for Soyuz 26 capsule, 96 days for cosmonauts
Events: To Salyut 6, first second-generation space station. Crew broke U.S. Skylab 4 84-days-in-space record. First USSR spacewalk in 9 years. First refueling in orbit from a Progress cargo freighter. Grechko and Romanenko received two crews of visitor-cosmonauts: Soyuz 27 and Soyuz 28. The Soyuz 27 visitor-crew used Soyuz 26 for trip home, leaving behind Soyuz 27. The Soyuz 28 crew flew home in Soyuz 28. That pattern of station crew rotation continues today: a long-time crew occupies the station, receiving short-time visitors. When a short-time crew leaves, it takes the old Soyuz home, leaving its newer Soyuz behind. Until recently, Soyuz was considered reliable for restart after shutdown up to about 90 days. Grechko and Romanenko flew home in Soyuz 27 after a record 96 days in orbit.

SOYUZ 27
Cosmonauts: Vladimir A. Dzhanibekov, Oleg G. Makarov
Rocket: A-2   Launch: January 10, 1978
Flight duration: 6 days for cosmonauts, 65 days for Soyuz 27 capsule
Events: The first short-time visitor-crew to Salyut 6 switched their individually-contoured seats from Soyuz 27 to Soyuz 26 for the flight home, leaving Soyuz 27 behind for Grechko and Romanenko.

SOYUZ 28
Cosmonauts: Alexei Gubarev, Vladimir Remek
Rocket: A-2   Launch: March 2, 1978
Flight duration: 8 days

Events: Remek, the son of the Czechoslovakian defense minister, was the first non-USSR/non-U.S. person to go to space. As the first Intercosmos or non-USSR cosmonaut, he went to space in Soyuz 28 as a short-time visitor to Salyut 6. Gubarev and Remek flew home in Soyuz 28, leaving Soyuz 27 still at the station for use by long-timers Grechko and Romanenko.

SOYUZ 29
Cosmonauts: Alexander S. Ivanchenkov, Vladimir Kovalyonok
Rocket: A-2   Launch: June 15, 1978
Flight duration: 80 days for Soyuz 29 capsule, 140 days for cosmonauts
Events: The USSR moved ahead of U.S. man-hours in space for the first time since 1965 when Ivanchenkov and Kovalyonok spent 140 days at Salyut 6 station. During the time, Ivanchenkov and Kovalyonok received two sets of short-time visitor-cosmonauts in Soyuz 30 and Soyuz 31 capsules, unloaded three Progress resupply freighters, and made a spacewalk. Ivanchenkov and Kovalyonok flew home November 2, 1978, in Soyuz 31.

SOYUZ 30
Cosmonauts: Miroslaw Hermaszewski, Pyotr I. Klimuk
Rocket: A-2   Launch: June 27, 1978
Flight duration: 8 days
Events: Hermaszewski, the first Polish cosmonaut, made a short visit to Salyut 6. The Sirena materials processor produced cadmium-mercury telluride. Photos were made of southern Poland. Klimuk and Hermaszewski flew home in Soyuz 30.

SOYUZ 31
Cosmonauts: Valeri F. Bykovsky, Sigmund Jahn
Rocket: A-2   Launch: August 26, 1978
Flight duration: 8 days for cosmonauts, 68 days for Soyuz 31
Events: Jahn, the first East German cosmonaut, made a short visit to Salyut 6. The Berolina materials processor was used to make ampules of bismuth-antimony and lead-telluride. Jahn used East German MFK-6M and KATE-140 cameras. Bykovsky and Jahn flew home in Soyuz 29, leaving Soyuz 31 behind for Ivanchenkov and Kovalyonok.

SOYUZ 32
Cosmonauts: Vladimir Lyakhov, Valeri V. Ryumin
Rocket: A-2   Launch: February 25, 1979
Flight duration: 108 days for Soyuz 32 capsule, 175 days for cosmonauts.
Events: Lyakhov and Ryumin set a new space endurance record aboard Salyut 6 station. They refurbished the 1.5-year-old station, including replacing a ruptured fuel tank. They received three Progress cargo freighters with food, fuel, water, supplies and mail from home. The last of the three carried the KRT-10 radiotelescope. Its antenna was dropped out the rear docking port, later to become tangled when the crew tried to jettison it as they departed. Lyakhov and Ryumin had to make a spacewalk to clear the fouled wire. They received no visitor-cosmonauts during their stay in Salyut 6 as an engine failed on Soyuz 33. Soyuz 32 was sent home unmanned with materials processing results June 13, 1979. Soyuz 34 was sent unmanned to the station June 6, 1979, and Lyakhov and Ryumin flew home in it August 9, 1979.

SOYUZ 33
Cosmonauts: Georgi Ivanov, Nikolai Rukavishnikov
Rocket: A-2   Launch: April 10, 1979
Flight duration: 2 days
Events: Ivanov was the first Bulgarian cosmonaut. A major engine failed on Soyuz 33 while docking at Salyut 6. The capsule and station were separating at 60 miles per hour. Ground control used a backup engine to bring Soyuz home immediately. Ivanov and Rukavishnikov made it safely. Soyuz 34 had to be sent unmanned to Salyut to bring home Lyakhov and Ryumin.

SOYUZ 34
Rocket: A-2   Launch: June 6, 1979
Flight duration: 74 days for Soyuz 34
Events: The third unmanned Soyuz. Soyuz 32 was docked at the station, but ground control knew it had the same kind of engine as the ill-fated Soyuz 33. Also, Soyuz 32 had been shut down more than 90 days. Soyuz 32 was sent home unmanned with materials processing results June 13, 1979. Soyuz 34 was sent unmanned June 6, 1979, for two days of tests in space before docking. It brought Lyakhov and Ryumin home August 9, 1979.

SOYUZ T-1
Rocket: A-2   Launch: December 16, 1979

Flight duration: 100 days
Events: The fourth unmanned Soyuz was the final unmanned test of the third generation of Soyuz. Most systems were improved from the second-generation. Solar panels replaced internal chemical batteries. Capsule interior was redesigned for three cosmonauts in spacesuits. Digital electronics were installed. Re-entry and rescue equipment were modified for safety. Fuel system was changed so capsule main engine and maneuvering engines used hydrazine, the same fuel as Salyut and Progress. Over the following two years, second-generation capsules were phased out, replaced by Soyuz-T spacecraft. Soyuz T-1 returned from Salyut 6 to Earth on March 25, 1980.

SOYUZ 35
Cosmonauts: Leonid I. Popov, Valeri V. Ryumin
Rocket: A-2   Launch: April 9, 1980
Flight duration: 185 days for cosmonauts, 55 days for Soyuz 35 capsule
Events: During a record 185 days orbiting in Salyut 6, Popov and Ryumin received four groups of visitor-cosmonauts—Soyuz 36, Soyuz 37, Soyuz 38 and Soyuz T-2—and four Progress supply freighters. Salyut 6 passed 1,000 days in orbit. After growing plants in greenhouse and making 40,000 Earth-atmosphere measurements, Popov and Ryumin flew Soyuz 37 home October 11, 1980, setting a new endurance record of 185 days in space.

SOYUZ 36
Cosmonauts: Bertalan Farkas, Valeri N. Kubasov
Rocket: A-2   Launch: May 26, 1980
Flight duration: 8 days for cosmonauts, 66 days for Soyuz 36 capsule
Events: Farkas was the first Hungarian cosmonaut. Experiments in Salyut 6 produced gallium-arsenide-chromium and aluminum-chromium alloys. Farkas and Kubasov flew home June 3, 1980, in Soyuz 35.

SOYUZ T-2
Cosmonauts: Vladimir V. Aksyonov, Yuri V. Malyshev
Rocket: A-2   Launch: June 5, 1980
Flight duration: 4 days
Events: First manned test-flight of the third-generation Soyuz-T spacecraft. T stands for troika or three, for third generation. Manually docked with Salyut 6. New re-entry program tested on the way back to Earth June 9, 1980.

SOYUZ 37
Cosmonauts: Viktor V. Gorbatko, Pham Tuan
Rocket: A-2   Launch: July 23, 1980
Flight duration: 8 days for cosmonauts, 80 days for Soyuz 37 capsule
Events: Pham Tuan, from Vietnam, was the first non-Warsaw Pact and first third-world Intercosmos cosmonaut. The flight to the Salyut 6 station took place during Summer Olympic Games in Moscow. Vietnamese-USSR experiment called Halong made bismuth-tellurium-selenium and gallium phosphide ampules. Photos of Vietnam for agriculture and resource searches. Pham Tuan and Gorbatko flew home July 31, 1980, in Soyuz 36.

SOYUZ 38
Cosmonauts: Arnaldo Tomayo Mendez, Yuri V. Romanenko
Rocket: A-2   Launch: September 18, 1980
Flight duration: 8 days
Events: To the Salyut 6 station. Mendez, from Cuba, was the second non-Warsaw Pact and second third-world Intercosmos cosmonaut. Grew organic monocrystals in space from Cuban sugar. Mendez and Romanenko flew home September 26, 1980, in Soyuz 38.

SOYUZ T-3
Cosmonauts: Leonid Kizim, Oleg G. Makarov, Gennadi Strekalov
Rocket: A-2   Launch: November 27, 1980
Flight duration: 13 days
Events: The first USSR three-man flight in a decade, to test Soyuz-T systems and repair the Salyut 6 station controls, fuel, telemetry and heating systems. They shot holograms, grew plants, made cadmium-mercury-telluride monocrystals and flew home December 10, 1980.

SOYUZ T-4
Cosmonauts: Vladimir Kovalyonok, Viktor P. Savinykh
Rocket: A-2   Launch: March 12, 1981
Flight duration: 75 days for cosmonauts, 75 days for Soyuz T-4 capsule
Events: During their 75 days at Salyut 6, Kovalyonok and Savinykh received cosmonaut-visitors in

Soyuz 39 and Soyuz 40, and one Progress resupply vessel. Kovalyonok and Savinykh flew home in Soyuz T-4 May 26, 1981.

SOYUZ 39
Cosmonauts: Vladimir A. Dzhanibekov, Jugderdemidiyn Gurragcha
Rocket: A-2   Launch: March 22, 1981
Flight duration: 8 days
Events: Gurragcha, from Mongolia, a short-time visitor at Salyut 6, was the third third-world Intercosmos cosmonaut. Dzhanibekov and Gurragcha flew home in Soyuz 39 March 30, 1981.

SOYUZ 40
Cosmonauts: Leonid I. Popov, Dumitru Prunariu
Rocket: A-2   Launch: May 14, 1981
Flight duration: 8 days
Events: To the Salyut 6 station. Last of second-generation Soyuz capsules. Last of the first phase of Intercosmos cosmonaut flights. Last non-USSR cosmonaut in a second-generation Soyuz, Prunariu was from Romania. Short-time visitors, Popov and Prunariu flew home in Soyuz 40 May 22, 1981.

SOYUZ T-5
Cosmonauts: Anatoli Berezovoi, Valentin Lebedev
Rocket: A-2   Launch: May 13, 1982
Flight duration: 211 days for cosmonauts, 106 days for Soyuz T-5
Events: First manned craft to dock at new Salyut 7 station. Berezovoi and Lebedev set a 211-day record, received 2 three-cosmonaut visitor crews in Soyuz T-6 and T-7, received 4 Progress cargo tugs, took a spacewalk and flew home in Soyuz T-7 December 10, 1982.

SOYUZ T-6
Cosmonauts: Jean-Loup Chretien, Vladimir A. Dzhanibekov, Alexander S. Ivanchenkov
Rocket: A-2   Launch: June 24, 1982
Flight duration: 8 days
Events: The first three-man flight in two years went to Salyut 7 for a short visit. Second phase of Intercosmos cosmonaut program started with Chretien, from France, first from a Western industrial democracy at a USSR station. France was the first nation to have both astronauts and cosmonauts. Patrick Baudry flew in U.S. shuttle STS-51G. The trio flew Soyuz T-6 home July 2, 1982.

SOYUZ T-7
Cosmonauts: Leonid I. Popov, Svetlana Savitskaya, Alexander A. Serebrov
Rocket: A-2   Launch: August 19, 1982
Flight duration: 8 days for cosmonauts, 113 days for Soyuz T-7
Events: Second woman in orbit. Savitskaya, the first woman to go to orbit since 1963, visited Salyut 7 station. First mixed male/female space crew. Popov, Savitskaya and Serebrov flew home in Soyuz T-5 August 27, 1982. The first woman to go to space twice, Savitskaya flew again to Salyut 7 in Soyuz T-12 in 1984.

SOYUZ T-8
Cosmonauts: Alexander A. Serebrov, Gennadi Strekalov, Vladimir G. Titov
Rocket: A-2   Launch: April 20, 1983
Flight duration: 2 days
Events: The fourth docking failure for Soyuz craft. Soyuz radar antenna damaged, crew couldn't tell how fast they were closing with the Salyut 7 space station, flew home April 22, 1983.

SOYUZ T-9
Cosmonauts: Alexander Alexandrov, Vladimir Lyakhov
Rocket: A-2   Launch: June 27, 1983
Flight duration: 149 days, both for cosmonauts and Soyuz T-9 capsule
Events: Three mishaps at Salyut 7 space station: A micrometeorite pinged a half-inch-thick pane in a station porthole, but did not penetrate. Alexandrov and Lyakov bailed out temporarily, into Soyuz, until it was found that no life-threatening damage had been done by the meteorite. While taking on fuel from a Progress tanker, a station oxidizer line ruptured. Alexandrov and Lyakhov again bailed out temporarily. After an all-clear, the cosmonauts by-passed the line. Soyuz T-10A blew up at launch leaving the Salyut crew with Soyuz T-9 which had been shut down longer than past experience. Three craft were docked together for six weeks when Cosmos 1443 joined Salyut 7 and Soyuz T-9, a major step in docking craft with large masses and volumes. Alexandrov and Lyakhov went on two spacewalks and received three Progress supply ships, before flying Soyuz T-9 home safely November 23, 1983.

SOYUZ T-10A

Cosmonauts: Vladimir G. Titov, Gennadi Strekalov
Rocket: A-2   Launch: September 26, 1983
Flight duration: rocket exploded on the launch pad
Events: Titov and Strekalov had failed to reach Salyut 7 in Soyuz T-8. In Soyuz T-10A, they were set to go again to Salyut 7 when fire erupted at the base of the A-2 rocket 90 seconds before launch. The Soyuz escape rocket was triggered. The capsule separated from the booster, lifting off seconds before the A-2 exploded. Titov and Strekalov landed, uninjured, several miles away.

SOYUZ T-10B
Cosmonauts: Oleg Atkov, Leonid Kizim, Vladimir Solovyev
Rocket: A-2   Launch: February 8, 1984
Flight duration: 237 days for cosmonauts, 63 days for Soyuz T-10B capsule
Events: At Salyut 7, the first three-cosmonaut long-duration station crew set a 237-days-in-space record. During the time they received visitor-cosmonauts in Soyuz T-11 and Soyuz T-12, plus five Progress cargo ships. Dr. Oleg Atkov was a cardiovascular medical specialist. During nearly 8 months at the station, Kizim and Solovyev took 6 spacewalks and the visiting Soyuz T-12 crew made a spacewalk. Altogether, a record 35 hours of spacewalk EVA's or extravehicular activities. Five spacewalks were to repair the fuel line ruptured during Soyuz T-9. Unified Manual Instrument, with the Russian acronym URI, was used on two walks for cutting, welding, spraying and soldering. Solar panels were added during the sixth spacewalk. The T-12 crew also used URI during their spacewalk. Inside, Insparitel-M was used to melt metals and plastics to coat objects. Atkov, Kizim and Solovyev flew home October 2, 1984, in Soyuz T-11, showing a third-generation Soyuz could work after six months' shutdown in space.

SOYUZ T-11
Cosmonauts: Yuri V. Malyshev, Rakesh Sharma, Gennadi Strekalov
Rocket: A-2   Launch: April 3, 1984
Flight duration: 8 days for cosmonauts, 182 days for Soyuz T-11 capsule
Events: Sharma, from India, as a short-time visitor at Salyut 7 was the fourth third-world Intercosmos cosmonaut. Yoga was used to combat effects of weightlessness. Strekalov, who couldn't make it to Salyut in Soyuz T-8 and T-10A, made it in T-11. The Insparitel-M apparatus was used to make a molten silver-germanium alloy. With 5 astronauts in U.S. shuttle Challenger and 6 cosmonauts at Salyut 7, it was the first time 11 people had been in orbit at the same time. Malyshev, Sharma and Strekalov flew home in Soyuz T-10B April 11, 1984.

SOYUZ T-12
Cosmonauts: Vladimir A. Dzhanibekov, Svetlana Savitskaya, Igor P. Volk
Rocket: A-2   Launch: July 17, 1984
Flight duration: 8 days
Events: Savitskaya returned to space to become the first woman to make a spacewalk. The first male-female spacewalk. She and Dzhanibekov went outside Salyut 7 for about 3.5 hours on July 25. They used URI to cut and weld stainless steel and titanium, to solder lead and tin, and to coat anodized aluminum with molten silver. Dzhanibekov, Savitskaya and Volk flew Soyuz T-12 home July 29, 1984.

SOYUZ T-13
Cosmonauts: Vladimir A. Dzhanibekov, Viktor P. Savinykh
Rocket: A-2   Launch: June 6, 1985
Flight duration: 168 days Savinykh, 111 days Dzhanibekov and Soyuz T-13
Events: Salyut 7, launched in 1982, malfunctioned while unoccupied early in 1985, leaving the station out of contact with ground control. Dzhanibekov and Savinykh flew Soyuz T-13 to the station June 6, finding it frozen from power loss and tumbling in space. They recharged batteries, repaired water pipes, warmed living quarters. It was 10 days before they could live in the station. Two Progress freighters flew to the station with fuel and equipment. Dzhanibekov and Savinykh spacewalked to install a third solar panel for electricity. Soyuz 14 arrived at the station in September. Grechko and Dzhanibekov flew Soyuz T-13 home September 25, 1985. Savinykh stayed on with Vasyutin and Volkov in Salyut 7. On May 6, 1986, cosmonauts left Mir in Soyuz T-15, flew to Salyut 7, entered, mothballed the Salyut station and returned to Mir June 26, 1986. The USSR considered flying its space shuttle Buran to Salyut 7 to retrieve the station, but the Salyut plunged into the atmosphere February 7, 1991, and burned over Argentina. Chunks were found 12 miles from Buenos Aires.

SOYUZ T-14
Cosmonauts: Georgi M. Grechko, Vladimir Vasyutin, Alexander A. Volkov

Rocket: A-2   Launch: September 17, 1985
Flight duration: 8 days Grechko, 65 days Vasyutin, Volkov & Soyuz T-14
Events: To continue repair and use of Salyut 7 station. A Progress cargo freighter was received. Cosmos 1686, a large habitable module, was sent to Salyut 7 September 27, 1985 and attached to the station's forward dock. It doubled the station length and increased living space by 50 percent. The Soyuz T-14 mission ended suddenly when Vasyutin became ill with a 104º fever. Savinykh, Vasyutin and Volkov flew Soyuz T-14 home November 21. Vasyutin was hospitalized until December 20 with a full recovery. Salyut 7 plunged into the atmosphere February 7, 1991, and burned.

SOYUZ T-15
Cosmonauts: Leonid Kizim, Vladimir Solovyev
Rocket: A-2   Launch: March 15, 1986
Flight duration: 51 days Mir, 52 days Salyut 7, 20 days Mir, total 123 days
Events: Kizim and Solovyev were a temporary crew to check-out the new Mir third-generation space station. In what the USSR called the first "space taxicab," Kizim and Solovyev left Mir in Soyuz T-15 on May 5, 1986, cruised through space to Salyut 7, docked and entered the 4-year-old second-generation space station. They mothballed Salyut 7, leaving it parked in orbit near Mir. Kizim and Solovyev flew back to Mir June 26, 1986, in Soyuz T-15. After 20 days, they flew home July 16, 1986, in Soyuz T-15. Salyut 7 re-entered the atmosphere and burned Feb 7, 1991.

SOYUZ TM-1
Rocket: A-2   Launch: May 21, 1986
Flight duration: 9 days
Events: Test of restyled Soyuz capsule. Descended May 30, 1986.

SOYUZ TM-2
Cosmonauts: Yuri V. Romanenko, Alexander I. Laveikin
Rocket: A-2   Launch: February 6, 1987
Flight duration: 326 days Romanenko, 174 days Laveikin, 174 days Soyuz TM-2
Events: Arrived at Mir February 8, 1987, as first permanent-manning crew. Romanenko and Laveikin made three 1987 spacewalks: April 11 to clear a dock for the small astronomy module Kvant-1 which was attached to the station in April. Kvant is Russian for Quantum. In a spacewalk June 12, they started installing a third solar panel to the station, and June 17 they finished installing the solar panel. Kvant-1 was launched March 31 and was attached to Mir April 11, after Romanenko and Laveikin spacewalked to remove a small white cloth bag obstructing docking of the station and the module. After six months in space, Laveikin showed an unusual heartbeat. Doctors on the ground ordered him to fly home in Soyuz TM-2 July 30, 1987, with short-time Soyuz TM-3 visitors Viktorenko and Faris. Alexandrov, from TM-3, stayed on with Romanenko in Mir. Ground control called an impromptu fire drill August 31, 1987. Romanenko and Alexandrov rushed into Soyuz TM-3, wriggled into spacesuits and ran through a checklist of preparations as if they were going to make an emergency separation from the Mir station. They did not take Soyuz out of the dock. Alexandrov already had Soyuz T-9 bail out experience. Romanenko and Alexandrov, in Kvant-1 astrophysics module in Aug 1987, studied Supernova 1987a in the Large Magellanic Cloud galaxy. Romanenko set 326-days endurance record as he and Alexandrov flew home Dec 29, 1987, in Soyuz TM-3. Altogether, in various flights, Romanenko had spent a total of 430 days in space.

SOYUZ TM-3
Cosmonauts: Alexander S. Viktorenko, Alexander Alexandrov, Mohammed Faris
Rocket: A-2   Launch: July 22, 1987
Flight duration: 8 days for Faris, 160 days for Viktorenko, Alexandrov, and Soyuz TM-3
Events: To the Mir station. Faris, from Syria, was the second Arab in space. The first had been a Saudi Arabian prince who flew in 1985 in a U.S. shuttle. Viktorenko, Faris and Laveikin flew home in Soyuz TM-2 July 30, 1987. Alexandrov stayed in Mir with Romanenko, flying home December 29, 1987, in Soyuz TM-3.

SOYUZ TM-4
Cosmonauts: Vladimir G. Titov, Musa K. Manarov, Anatoly Levchenko
Rocket: A-2   Launch: December 21, 1987
Flight duration: 366 days for Titov, Manarov, 10 days Levchenko, 179 days Soyuz TM-4
Events: Titov and Manarov made two 1988 spacewalks outside Mir: June 30 to start repair of Kvant telescope and October 20 to complete the telescope repair. Vladimir Titov is not related to Gherman Titov who, in 1961, was the second man to orbit Earth and the first to stay in space more than a day. Levchenko flew Romanenko and Alexandrov home in Soyuz TM-3 December 29, 1987. Levchenko,

47, from the Ukraine, died of a brain tumor 8 months after returning to Earth. A cosmonaut since 1981, Levchenko had not flown to space before Soyuz TM-4. Titov and Manarov received three groups of cosmonaut visitors in 1988: Soyuz TM-5, TM-6 and TM-7. Titov and Manarov went on to set a 366-days-in-space record. They left for home with Chretien in TM-6 December 21, 1988. Trouble developed briefly for Titov, Manarov and Chretien on the flight to Earth when telemetry showed an on-board computer overload. Descent was delayed three hours as they switched manually to a back-up program. Unlike the Soyuz TM-5 return flight, TM-6 carried three cosmonauts. TM-6 landed safely December 21, 1988. The mission of the TM-4 crew was the longest in history. The longest American space flight was 84 days by three astronauts aboard Skylab in 1973.

SOYUZ TM-5
Cosmonauts: Anatoly Solovov, Viktor P. Savinykh, Alexander Alexandrov
Rocket: A-2   Launch: June 7, 1988
Flight duration: 10 days for cosmonauts, 92 days for Soyuz TM-5
Events: The USSR's 63rd manned spaceflight. Alexandrov, from Bulgaria, was not the same cosmonaut as the Soviet with the same name who stayed 160 days in Mir in 1987. Soyuz TM-5 was the second Bulgarian flight in nine years. The first had been the ill-fated Soyuz 33 attempt to go to the Salyut 6 space station. Solovov, Savinykh and Alexandrov took to Mir a blooming orchid grown on Earth from seeds planted in the space station. Solovov, Savinykh and Alexandrov flew home in TM-4 June 17, 1988.

SOYUZ TM-6
Cosmonauts: Abdul Ahad Mohmand, Valery Polyakov, Vladimir Lyakhov
Rocket: A-2   Launch: August 29, 1988
Flight duration: 9 days for Mohmand, Lyakhov, 241 days for Polyakov, 114 days Soyuz TM-6
Events: Flight to Mir of Mohmand, from Afghanistan, the eighth manned-spaceflight country among developing nations. Previous USSR flights: Cuba, India, Mongolia, Syria, Vietnam. U.S. flights: Mexico, Saudi Arabia. Cardiovascular specialist Dr. Polyakov was sent to monitor Titov and Manarov health. He stayed on in Mir with Titov and Manarov as Lyakhov and Mohmand left for home September 6, 1988, in Soyuz TM-5. Trouble developed for Lyakhov and Mohmand as equipment and human error delayed their landing by a day. The Soyuz life-support system was good for at least two days, maybe a week. They ran low on food and air before landing safely September 7, 1988.

SOYUZ TM-7
Cosmonauts: Alexander A. Volkov, Sergei M. Krikalev, Jean-Loup Chretien
Rocket: A-2   Launch: November 26, 1988
Flight duration: 25 days Chretien, 152 days Volkov, Krikalev and Soyuz TM-7
Events: Back in 1982, Chretien, from France, had ridden Soyuz T-6 to spend 8 days at the Salyut 7 station as the first cosmonaut from a Western industrial democracy at a USSR station. At Mir station in 1988, Chretien's 25 days at Mir set a record for non-USSR cosmonaut stay at a USSR space station. Chretien also became the first non-USSR cosmonaut/non-USA astronaut to spacewalk when he and Volkov went outside December 9, 1988, to build an experimental plastic web frame for science gear. The Soyuz TM-7 crew took along a music tape by British rock group Pink Floyd. Chretien carried a portable organ to play in Mir as the cosmonauts dined on 23 gourmet foods from a French chef, including compote of pigeon with dates and dried raisins, duck with artichokes, oxtail fondue with tomatos and pickles, beef bourguignon, saute de veau Marengo, ham and fruit pates, bread, rolls, cheeses, nuts, coffee and chocolate bars. The delicacies were canned. The chef, saying French cuisine is inconceivable without wine, regretted a bottle could not go to space when he learned wine and sauces would float away without gravity. Dishes had to be small without bones, or sauces to become flying droplets. Meat was made to absorb sauce to guard against dryness. It wasn't the first French cuisine to go to orbit. Chretien took it in 1982 to Salyut 7 and French astronaut Patrick Baudry treated in U.S. shuttle Discovery in 1985. Chretien, Titov and Manarov left for home December 21, 1988, in Soyuz TM-6. Trouble developed briefly for the trio on the flight to Earth when telemetry showed an on-board computer overload. Descent was delayed three hours as they switched manually to a back-up program. Soyuz TM-6 landed safely December 21. Dr. Polyakov, sent to Mir in TM-6, remained at Mir to monitor health of Volkov and Krikalev. The trio flew home in TM-7 April 27, 1989.

SOYUZ TM-8
Cosmonauts: Alexander S. Viktorenko, Alexander A. Serebrov
Rocket: A-2   Launch: September 5, 1989
Flight duration: 167 days Viktorenko, Serebrov and Soyuz TM-8

Events: Commander Viktorenko with Alexander N. Balandin as flight engineer had been scheduled to blast off April 19, 1989, in Soyuz TM-8 to dock at Mir April 21 to relieve Polyakov, Volkov, Krikalev, but the flight was cancelled two weeks before launch because two station expansion modules, Kvant-2 and Kristall, were not ready to be launched to Mir on schedule in 1989. Also, the Mir solar power system developed problems. Polyakov, Volkov and Krikalev flew Soyuz TM-7 home April 27, 1989, leaving Mir temporarily (134 days) with no cosmonauts in residence. Mission commander Viktorenko and engineer Serebrov finally were launched in Soyuz TM-8 September 5, 1989. Balandin, the original TM-8 flight engineer became backup engineer for the Sept 6 flight. The other backup was Anatoly Solovov who then prepared to fly Soyuz TM-9 in Feb 1990. Viktorenko and Serebrov received first of two 21-ton add-on modules, Kvant-2 Dec 6, 1989, doubling size of Mir station and adding a space motorcycle (manned maneuvering unit or MMU) for spacewalks, new space shower, more comfortable hatch for spacewalks, remote manipulator arm for repositioning the module after docking. Kvant-2, the same size as the original Mir, was launched on a Proton rocket Nov 26, 1989, and docked with Mir December 6. Viktorenko and Serebrov took 5 spacewalks outside the station, on Jan 9, 11 and 26 and Feb 1 and 5. The other large module, Kristall, was launched later, on May 31, 1990. Balandin and Solovov in Soyuz TM-9 arrived Feb 11. Viktorenko and Serebrov flew home in Soyuz TM-8 Feb 19, 1990.

SOYUZ TM-9

Cosmonauts: Anatoly Solovov and Alexander N. Balandin
Rocket: A-2   Launch: February 11, 1990
Flight duration: 179 days for Solovov, Balandin and Soyuz TM-9
Events: Second crew to occupy Mir since reopened Sep 1989, boarded Mir Feb 13 to relieve Viktorenko and Serebrov. They were USSR's sixth main expedition to the station and 22nd and 23rd cosmonauts to go to Mir. Balandin and Solovov received the second of the two 21-ton add-on station modules, Kristall, on June 6, 1990. Kristall was the third major building-block addition to the space station. The same size as Kvant-2 and the original Mir, Kristall was launched on a Proton rocket May 31, 1990, and docked June 9. With the addition of Kristall, the entire Mir complex in Earth orbit weighed 90 tons. Kristall, called a microfactory in orbit, had a lab to grow crystals in weightlessness for use in electronics, optical equipment and batteries. Solovov and Balandin took several spacewalks after Kristall arrived. Solovov and Balandin, using a furnace to melt materials, produced 23 pure crystals worth $1 million each for use in semiconductors. The USSR hoped to make millions of dollars in profits from sales of the manufactured crystals. The cosmonauts also produced cell cultures in zero gravity for pharmaceutical companies. The Russian space program was under pressure to show a profit while the USSR faced consumer shortages. Officials said Soyuz TM-9 mission cost $134 million, but brought in revenues of $155 million in sales of the pure crystals for computer chips and of high-definition photos of Earth's surface. Sales were said to show the Soviet space program was profitable. Balandin and Solovov spent a grueling 7 hours in space in July trying to repair torn thermal insulation on the Soyuz TM-9 capsule, only to discover they could not close completely the airlock hatch leading back into the space station. Soviet media reported they were dangerously close to running out of oxygen. They finally were able to close the hatch on a subsequent space walk, but it continued to cause problems. Solovov and Balandin flew Soyuz TM-9 home Aug 9, 1990.

SOYUZ TM-10

Cosmonauts: Gennadi Manakov and Gennadi Strekalov
Rocket: A-2   Launch: August 1, 1990
Flight duration: 131 days for Manakov, Strekalov and Soyuz TM-10
Events: Flight commander Manakov and flight engineer Strekalov docked Soyuz TM-10 at Mir Aug 3. Manakov was a rookie; Strekalov had made 5 launches. They carried five live Japanese quail to Baikonur. Manakov and Strekalov were 24th and 25th cosmonauts to visit Mir and 7th primary crew to occupy Mir. On Oct 1, they entered the Soyuz TM-10 capsule docked at Mir, fired up its engine and boosted the entire station complex to a slightly higher altitude. After the maneuver, Mir was at an apogee or high point in its orbit of 273 miles. The perigee or low point was 236 miles. It was circling Earth every 92.4 minutes. In November, Manakov and Strekalov unloaded TV gear from the Progress M-5 cargo freighter, preparing for a December visit and daily broadcasts by Japanese TV journalist Toyohiro Akiyama in Soyuz TM-11. Manakov and Strekalov replaced Mir's heater with a new heating loop sent up in Progress M-5. They did lab work to continue earning money for the USSR space program, using their Gallar electric furnace for 240-hour experiments to grow gallium arsenide and zinc oxide crystals, semiconductor materials. They used the KAP-350 topographic camera, the Priroda-5 camera and other instruments on a remote-controlled hydrostabilized platform

in Kvant-2 to photograph and spectrograph Kharkov, Volgograd, the Aral Sea, the Caucasus, the Caspian Depression and other USSR areas. Using the Buket telescope and Granat spectrometer in the Kristall module, they measured X-rays, gamma rays and neutron radiation arriving at Earth from deep space. They used Kvant-2 to observe infrared and visible light from deep space. Manakov and Strekalov took spacewalks. On Oct 19, they repaired a broken joint on the outer hatch of the airlock in the Kvant-2 module. The previous crew, Solovov and Balandin, had broken the joint at the start of their July 17 spacewalk. Strekalov, Manakov and Soyuz TM-11's Akiyama flew home Dec 10, 1990, in Soyuz TM-10.

SOYUZ TM-11

Cosmonauts: Viktor Afanasyev, Musa K. Manarov and Toyohiro Akiyama
Rocket: A-2   Launch: Dec 2, 1990
Flight duration: 8 days for Akiyama, 175 days for Afanasyev, Manarov and Soyuz TM-11
Events: Ads for Unicharm women's hygiene products, Sony electronic products and Ohtsuka Chemicals painted on launch rocket. At launch, flight engineer Manarov shared the record for longest single trip in space. Flight commander Afanasyev was a rookie spaceflier who had flown 40 different kinds of aircraft. Afanasyev and Manarov were the 8th main crew to occupy Mir. Guest cosmonaut Akiyama, veteran Japanese TV correspondent and former Washington bureau chief for the Tokyo Broadcasting System, was first Japanese and first journalist in space. TBS paid $10 million for the flight. Akiyama took along six green tree frogs from Japan as well as amulets given to him by the Japanese prime minister, the president of TBS and his own father. He observed the frogs, which usually move with vacuum suckers, in weightlessness. Akiyama broadcast a daily TV show from space. He photographed Tokyo, Moscow, New York and Paris from 200 miles overhead. Akiyama, who had kicked a four-pack-a-day habit for the flight, said he was looking forward to a cigarette after landing. Japanese backup for Akiyama was TBS camerawoman Ryoko Kikuchi. After coming down with appendicitis days before launch, Kikuchi said she hoped Krikalev and Artsebarsky would take her along "as a personal item" in Soyuz TM-12. Noting Manarov and Afanasyev were taking along the same doll Yuri Gagarin had carried on his maiden spaceflight in 1961, Kikuchi told reporters she wished they would take her along "like that little doll." Gagarin's doll, used as a gravity indicator, had begun floating free when his Vostok 1 capsule reached zero gravity. Some 100,000 Japanese live in the South American country of Peru. On Dec 6, Akiyama broadcast from Mir to Peru. Japanese-Peruvians beamed up a TV greeting and conversed by radio with Akiyama. After visiting Mir 8 days, cosmoreporter Akiyama joined Soyuz TM-10 cosmonauts Strekalov and Manakov in Soyuz TM-10 on Dec 10, landing in a snowy field in the USSR Khazakhstan Republic. Afanasyev and Manarov stayed at Mir, making four spacewalks. They spent Russian Christmas, Jan 7, 1991, on a spacewalk repairing Kvant-2 airlock hatch. They did astronomy research with the Buket telescope and the Granat spectrometer, and used the Gallar furnace to grow zinc oxide monocrystals. The most people ever in space at one time included 12 astronauts and cosmonauts during the week of December 2, 1990. Five were at Mir space and seven were in U.S. shuttle Columbia. When Afanasyev, Manarov and Soyuz TM-12 cosmonaut Helen Sharman flew home in Soyuz TM-11 May 26, 1991, Manarov gained the record for most time accumulated in space at 541 days, while continuing to share the 366-day Soyuz TM-4 record for longest flight.

SOYUZ TM-12

Cosmonauts: Anatoly Artsebarsky, Sergei M. Krikalev and Helen Sharman
Rocket: A-2   Launch: May 18, 1991
Flight duration: 8 days Sharman, 145 days Artsebarsky & Soyuz TM-12, ~365 days Krikalev
Events: Mission commander Artsebarsky and flight engineer Krikalev were the 9th main crew at Mir. Guest cosmonaut Sharman, 27, a former Mars candy company worker and a chemist from Sheffield, England, with no flying experience, was the first woman at Mir. She had been chosen from 13,000 applicants who replied to a newspaper ad by a private British company which was to have raised $10 million from the public for Sharman's training and flight. When Britons wouldn't donate the money, the USSR gave Sharman a free flight, but dropped British science experiments. Sharman took along one of the blue space passports issued to all cosmonauts for identification and requesting assistance if they were to be forced off course in landing. USSR President Mikhail Gorbachev welcomed Sharman by radio after Soyuz TM-12 docked at Mir. Sharman, having studied Russian while preparing for space travel, replied to Gorby in fluent Russian. Later, she did science experiments and spoke by radio with English school children. Sharman operated the Mir amateur radio station, using callsign GB1MIR. Sharman used an "elektrotopograf" to study degradation of dielectric materials in space. In an experiment called Vazon, she studied how orchids and dwarf

magnolia vines improved the psychological atmosphere aboard Mir. Tass News Agency said the four male crew members gave Sharman "the best living quarters aboard the station, a room with a view of Earth." After 8 days, Sharman on May 26, 1991, joined Soyuz TM-11 cosmonauts Manarov and Afanasyev in Soyuz TM-11, landing on the barren Kazakhstan steppe near Dzhezkazgan. Artsebarsky stayed on at Mir for 145 days, landing in Soyuz TM-12 Oct 10, 1991, accompanied by Soyuz TM-13's Viehboeck and Aubakirov. Krikalev was to stay 365 days, expecting to land in Soyuz TM-13 around May 17, 1992.

SOYUZ TM-13

Cosmonauts: Alexander A. Volkov, Takhtar Aubakirov, Franz Viehboeck
Rocket: A-2   Launch: October 2, 1991
Flight duration: 8 days for Viehboeck, Aubakirov, ~228 days for Volkov and Soyuz TM-13
Events: Unusual crew with 2 researcher cosmonauts—Viehboeck from Austria and Aubakirov from Kazakhstan. Past guest cosmonaut flights had a Soviet commander and engineer with one foreign researcher. Soyuz TM-13 had a commander, no flight engineer, and 2 researchers. Aubakirov had been honored in 1988 with the USSR's top distinction, Hero of the Soviet Union, for aircraft test pilot work. TM-13 backup crew: Austria's Lothaller Clemens, Kazakhstan's Talgat Musabaev and Alexander S. Viktorenko. Preparing for TM-13, Aubakirov and Musabaev received Soyuz/Mir docking training at Baikonur Cosmodrome. After the 1991 coup, USSR officials extended Mir spaceflight research agreements to the newly sovereign republics, such as Kazakhstan. Aubakirov, the first guest cosmonaut from within USSR, in space searched his native Kazakhstan's Kulunda Steppe and Aral Sea areas for natural resources and potential oil fields. Viehboeck operated a package of 16 Austrian space experiments, AustroMir 91. It included Austrian Amateur Radio Experiment Aboard Mir (AreMir) with equipment by Radio Club for Communication and Wave Propagation, Gratz, Austria. AreMir receivers were in Austrian and Russian schools. Amateur radio callsigns: Krikalev U5MIR, Volkov U4MIR. Artsebarsky, Viehboeck and Aubakirov flew down to Earth on Oct 10, 1991, in Soyuz TM-12. Krikalev and Volkov, with their Soyuz TM-13 transport, stayed in space until about May 17, 1992—about 365 days in orbit for Krikalev. Volkov and Krikalev will be the 10th main crew at Mir.

SOYUZ TM-14

Cosmonauts: to be announced, may include Alexander S. Viktorenko, Talgat Musabaev
Rocket: A-2   Launch: ~May 1992
Flight duration: ~8 days for guest cosmonaut, ~180 days for main crew and Soyuz TM-14
Events: The USSR space program had severe budget problems in 1991-92. On July 18, 1991, officials postponed launch of Soyuz TM-14 and laid off its crew. They said it would be necessary for cosmonaut Krikalev, at Mir station since May 1991, to extend his stay to a year. Some members of the Soyuz TM-13 backup crew—Alexander S. Viktorenko, Talgat Musabaev from Kazakhstan, Lothaller Clemens from Austria—could be in the Soyuz TM-14 crew launched around May 1992. In May 1991, deputy chief flight controller Viktor Blagov said the USSR expected to launch in the second quarter of 1992 the large Mir add-on module, Spektr, with telescopes to photograph Earth. Spektr is Russian for Optical. The USSR also planned to launch the international ecological module Priroda to expand Mir in the fourth quarter of 1992. Spetkr and Priroda are large capsules like Kvant-2 and Kristall expansion modules added to Mir in 1989-90.

USSR Space Stations

The USSR has launched a total of eight space stations to Earth orbit, including five first-generation space stations in the 1970's, two second-generation stations in the 1970's and 1980's, and a third-generation space station in 1986.

The first- and second-generation space stations were called Salyut—Russian for Salute—a name given to the first station in 1971 to honor the tenth anniversary of the first man in space. Cosmonauts lived for record periods of time aboard Salyuts.

The third-generation station was called Mir, Russian for Peace. Cosmonauts have lived for record periods of time aboard Mir.

SALYUT 1

First-generation station launched: April 19, 1971
Duration: fell from orbit and burned October 11, 1971
Events: First USSR space station.

Soyuz visits:
Soyuz 10 cosmonauts Nikolai Rukavishnikov, Vladimir A. Shatalov, Alexei S. Yeliseyev
Soyuz 11 cosmonauts Georgi T. Dobrovolsky, Vladislav N. Volkov, Viktor I. Patsayev

SALYUT 2
First-generation station launched: April 3, 1973
Duration: fell from orbit and burned May 22, 1973
Events: Failed, military.
Soyuz visits: none

SALYUT 3
First-generation station launched: June 24, 1974
Duration: fell from orbit and burned August 24, 1975
Events: First military version actually used.
Soyuz visits:
Soyuz 14 cosmonauts Yuri Artyukhin, Pavel R. Popovich
Soyuz 15 cosmonauts Lev Demin, Gennadi Sarafanov

SALYUT 4
First-generation station launched: December 26, 1974
Duration: fell from orbit and burned February 2, 1977
Events: Civilian.
Soyuz visits:
Soyuz 17 cosmonauts Georgi Grechko, Alexei Gubarev
Soyuz 18B cosmonauts Pyotr I. Klimuk, Vitali Sevastyanov

SALYUT 5
First-generation station launched: June 22, 1976
Duration: fell from orbit and burned August 8, 1977
Events: Last solely-military.
Soyuz visits:
Soyuz 21 cosmonauts Boris V. Volynov, Vitali Zholobov
Soyuz 23 cosmonauts Valeri Rozhdestvensky, Vyacheslav Zudov
Soyuz 24 cosmonauts Yuri Glazkov, Viktor V. Gorbatko

SALYUT 6
Second-generation station launched: September 29, 1977
Duration: fell from orbit and burned July 29, 1982
Events: First of second generation.
Soyuz visits:
Soyuz 25 cosmonauts Vladimir Kovalyonok, Valeri Ryumin
Soyuz 26 cosmonauts Georgi Grechko, Yuri V. Romanenko
Soyuz 27 cosmonauts Vladimir Dzhanibekov, Oleg Makarov
Soyuz 28 cosmonauts Alexei Gubarev, Vladimir Remek
Soyuz 29 cosmonauts Alexander Ivanchenkov, Vladimir Kovalyonok
Soyuz 30 cosmonauts Miroslaw Hermaszewski, Pyotr Klimuk
Soyuz 31 cosmonauts Valeri F. Bykovsky, Sigmund Jahn
Soyuz 32 cosmonauts Vladimir Lyakhov, Valeri Ryumin
Soyuz 33 cosmonauts Georgi Ivanov, Nikolai Rukavishnikov
Soyuz 34 Unmanned.
Soyuz T-1 Unmanned.
Soyuz 35 cosmonauts Leonid Popov, Valeri Ryumin
Soyuz 36 cosmonauts Bertalan Farkas, Valeri N. Kubasov
Soyuz T-2 cosmonauts Vladimir Aksyonov, Yuri Malyshev
Soyuz 37 cosmonauts Viktor V. Gorbatko, Pham Tuan
Soyuz 38 cosmonauts Arnaldo Tomayo Mendez, Yuri V. Romanenko
Soyuz T-3 cosmonauts Leonid Kizim, Oleg Makarov, Gennadi Strekalov
Soyuz T-4 cosmonauts Vladimir Kovalyonok, Viktor P. Savinykh
Soyuz 39 cosmonauts Vladimir Dzhanibekov, Jugderdemidiyn Gurragcha
Soyuz 40 cosmonauts Leonid Popov, Dumitru Prunariu

SALYUT 7
Second-generation station launched: April 19, 1982
Duration: fell from orbit and burned February 7, 1991
Events: The last Salyut.

Soyuz visits:
Soyuz T-5 cosmonauts Anatoli Berezovoi, Valentin Lebedev
Soyuz T-6 cosmonauts Jean-Loup Chretien, Vladimir Dzhanibekov, Alexander Ivanchenkov
Soyuz T-7 cosmonauts Leonid Popov, Svetlana Savitskaya, Alexander Serebrov
Soyuz T-8 cosmonauts Alexander Serebrov, Gennadi Strekalov, Vladimir Titov
Soyuz T-9 cosmonauts Alexander Alexandrov, Vladimir Lyakhov
Soyuz T-10B cosmonauts Oleg Atkov, Leonid Kizim, Vladimir Solovyev
Soyuz T-11 cosmonauts Yuri Malyshev, Rakesh Sharma, Gennadi Strekalov
Soyuz T-12 cosmonauts Vladimir Dzhanibekov, Svetlana Savitskaya, Igor Volk
Soyuz T-13 cosmonauts Vladimir Dzhanibekov, Viktor P. Savinykh
Soyuz T-14 cosmonauts Georgi Grechko, Vladimir Vasyutin, Alexander A. Volkov
Soyuz T-15 cosmonauts Leonid Kizim, Vladimir Solovyev
MIR
Third-generation station launched: February 20, 1986
Duration: still in orbit, in permanent full-time use.
Events: Eighth space station, first of third generation.
Soyuz visits:
Soyuz T-15 cosmonauts Leonid Kizim, Vladimir Solovyev
Soyuz TM-2 cosmonauts Yuri V. Romanenko, Alexander I. Laveikin
Soyuz TM-3 cosmonauts Alexander Viktorenko, Alexander Alexandrov, Mohammed Faris
Soyuz TM-4 cosmonauts Vladimir Titov, Musa Manarov, Anatoly Levchenko
Soyuz TM-5 cosmonauts Anatoly Solovov, Viktor P. Savinkyh, Alexander Alexandrov
Soyuz TM-6 cosmonauts Abdul Ahad Mohmand, Valery Polyakov, Vladimir Lyakhov
Soyuz TM-7 cosmonauts Alexander A. Volkov, Sergei M. Krikalev, Jean-Loup Chretien
Soyuz TM-8 cosmonauts Alexander S. Viktorenko, Alexander A. Serebrov
Soyuz TM-9 cosmonauts Alexander N. Balandin and Anatoly Solovov
Soyuz TM-10 cosmonauts Gennadi Manakov and Gennadi Strekalov
Soyuz TM-11 cosmonauts Viktor Afanasyev, Toyohiro Akiyama and Musa Manarov
Soyuz TM-12 cosmonauts Anatoly Artsebarsky , Sergei M. Krikalev and Helen Sharman
Soyuz TM-13 cosmonauts Takhtar Aubakirov, Franz Viehboeck and Alexander A. Volkov

Non-USSR Cosmonauts

Many trips have been made by persons from various nations to USSR space stations. Men and women who have flown, or are preparing to fly, in USSR spacecraft are considered cosmonauts in this list. A few persons from various nations are both cosmonauts and astronauts.

Since 1978, the USSR has earned foreign-exchange hard currency by selling to other governments trips for non-USSR cosmonauts to the space stations Salyut 6, Salyut 7 and Mir. Future guest-cosmonaut flights might include persons from Germany, Malaysia, Israel, China and even a U.S. astronaut.

All candidates are prepared for spaceflight by the Yuri Gagarin Memorial Cosmonaut Training Center at Star City outside Moscow. All manned flights to space have been launched from Baikonur Cosmodrome in the USSR's Republic of Kazakhstan.

From	Cosmonaut	Flight		Craft	Station
Czech	Vladimir Remek	Mar	1978	Soyuz 28	Salyut 6
Poland	Miroslaw Hermaszewski	Jun	1978	Soyuz 30	Salyut 6
E German	Sigmund Jahn	Aug	1978	Soyuz 31	Salyut 6
Bulgaria	Georgi Ivanov	Apr	1979	Soyuz 33	Salyut 6
Hungary	Bertalan Farkas	May	1980	Soyuz 36	Salyut 6
Vietnam	Pham Tuan	Jul	1980	Soyuz 37	Salyut 6
Cuba	Arnaldo Tomayo Mendez	Sep	1980	Soyuz 38	Salyut 6
Mongolia	Jugderdemidiyn Gurragcha	Mar	1981	Soyuz 39	Salyut 6
Romania	Dumitru Prunariu	May	1981	Soyuz 40	Salyut 6
France	Jean-Loup Chretien	Jun	1982	Soyuz T-6	Salyut 7
India	Rakesh Sharma	Apr	1984	Soyuz T-11	Salyut 7

Syria	Mohammed Faris	Jul	1987	Soyuz TM-3	Mir
Bulgaria	Alexander Alexandrov	Jun	1988	Soyuz TM-5	Mir
Afghan	Abdul Ahad Mohmand	Aug	1988	Soyuz TM-6	Mir
France	Jean-Loup Chretien	Nov	1988	Soyuz TM-7	Mir
Japan	Toyohiro Akiyama	Dec	1990	Soyuz TM-11	Mir
G Britain	Helen Sharman	May	1991	Soyuz TM-12	Mir
Austria	Franz Viehboeck	Oct	1991	Soyuz TM-13	Mir
Future	**Cosmonaut**	**Flight**			
USSR	journalist to be named	1992, six training for one slot			
German	R. Ewald or K.D. Flade	1992, second half of year			
Malaysia	to be named	1992-94 date to be announced			
France	M. Tognini or J.L.Chretien	1992-93, France wants flight every two years			
Israeli	to be named	1993-94 date to be announced			
American	to be named	1993-94 date to be announced			
France	J.L.Chretien or M. Tognini	1993-94 date to be announced			
China	to be named	pending decision to train for flight			

USSR Man-In-Space Summary

Vostok: One-person capsules, 6 flights, 6 cosmonauts. First: April 12, 1961. Last: June 16-17, 1963. Shortest: the first by Yuri A. Gagarin, 1 hr 48 min, 1 orbit of Earth, first person in space. Longest: next to last by Valeri F. Bykovsky, 119 hrs 6 min, 81 orbits of Earth.

Voskhod: Two-man/three-man capsules, 3 flights, 5 cosmonauts, 2 dogs. Between Oct 12, 1964, and Feb 22, 1966. Longest cosmonaut Voskhod flight: 26 hrs 2 min., 17 orbits of Earth.

Soyuz: Three-man capsules, 70 flights, 90 cosmonauts. Flights started April 23, 1967, and continue today. Longest cosmonaut trip to space in Soyuz was 366 days for cosmonauts Vladimir Titov and Musa Manarov. They flew to space Dec 21, 1987, in Soyuz TM-4 and landed Dec 21, 1988, in Soyuz TM-6.

Salyut/Mir: Salyut 1, the first of eight USSR space stations, was launched in 1971. Mir continues in use today. Cosmonaut flights to space stations: 51. Longest cosmonaut stay: 366 days in one trip to Mir by Musa Manarov and Vladimir Titov in 1987-88. Second longest trip: 326 days in one stay at Mir by Yuri Romanenko in 1987. Most time accumulated in space: 541 days by Musa Manarov in two trips to Mir, including 366 days in 1987-88 and and 175 days in 1990-91. Second most accumulated time in orbit: 430 days in three trips to space by Yuri Romanenko, including 96 days at Salyut 6 in 1977, 8 days at Salyut 6 in 1980 and 326 days at Mir in 1987.

Soyuz-Apollo: U.S.-Soviet link-up in space. One flight, 2 cosmonauts, 3 astronauts, July 15-24, 1975.

Space Shuttles: Buran No. 1 reusable space vehicle flown unmanned Nov 15, 1988. Buran No. 2 under construction.

Totals: By the end of 1991, 79 flights, 99 crew members.

Chinese Cosmonauts

People's Republic of China reportedly is designing a capsule to ferry four persons to Earth orbit late in the 1990's. China's Long March 2E rocket has sufficient power to lift a manned payload to space. China is said to have been preparing since the late 1970's to launch astronauts. In 1986, the official newspaper People's Daily said the first flight was "not far off." Life-support systems and a crew cabin reportedly were ready. Spacesuits and food had been in the works since 1980. China has been bringing satellites back to Earth successfully since the 1970's. Astronauts training in a mock spaceship appeared in the Chinese magazine Science Life in 1980. A first-flight crew was being selected, according to the People's Daily newspaper. A four-place spacecraft would be twice the size of the U.S. Gemini two-man capsules of the 1960's. A shuttle also may be on the drawing board. In 1989, the government announced plans to build its own space station. They may be planning a 24-ft., 22-ton manned station to be sent to a low Earth orbit by 1998.

Astronauts

Following common usage of the title, all men and women who have flown, or are preparing to fly, in U.S. spacecraft are considered astronauts in this list. They include mission commanders, pilots, payload commanders, mission specialists, payload specialists and space flight participants. NASA, however, refers to payload specialists and space flight participants as members of a shuttle flight crew, not as astronauts.

Cosmonauts, listed on previous pages, are men and women who have flown in USSR spacecraft. A few persons from various nations are both astronauts and cosmonauts. No other nations have originated manned flights, but at least Europe, Japan and China are preparing to do so in a few years.

First astronauts. Most early astronauts were selected from the military. The very first astronauts, known as the Mercury 7, were selected April 27, 1959: Malcolm Scott Carpenter, Leroy Gordon "Gordo" Cooper Jr., John Herschel Glenn Jr., Virgil Ivan "Gus" Grissom, Walter Marty "Wally" Schirra Jr., Alan Bartlett Shepard Jr., and Donald Kent "Deke" Slayton.

On May 5, 1961, Shepard became the first American in space. The first civilian astronauts were aeronautical engineer Neil Armstrong and U.S. Maritime Academy engineer Elliot See, selected in 1962 in NASA's second group of astronauts.

Newest astronauts. The most-recently-trained astronauts were in NASA's class of 1990, including seven pilots and sixteen mission specialists. Eleven of them were American civilians and twelve were U.S. military officers. Among the five women in the group, three were military officers, including the first woman pilot. The class completed training in July 1991 and became eligible for assignment to space shuttle flights.

Name	Flight
Acton, Loren W.	STS-51F Challenger
Adamson, James C.	STS-28 Columbia, STS-43 Atlantis
Akers, Thomas D.	STS-41 Discovery, STS-49 Endeavour
Al-Saud, Salman Abdel Aziz	STS-51G Discovery
Aldrin, Edwin E. "Buzz", Jr.	Gemini 12, Apollo 11, Moon
Allen, Andrew M.	STS-46 Atlantis
Allen, Joseph P.	STS-5 Columbia, STS-51A Discovery
Anders, William A.	Apollo 8
Apt, Jerome "Jay"	STS-37 Atlantis, STS-47 Endeavour
Armstrong, Neil A.	Gemini 8, Apollo 11, Moon
Bagian, James P.	STS-29 Discovery, STS-40 Columbia
Baker, Ellen S.	STS-34 Atlantis
Baker, Michael A.	STS-43 Atlantis
Bartoe, John-David F.	STS-51F Challenger
Bassett, Charles	died in 1966 crash of T-38 jet trainer
Baudry, Patrick	STS-51G Discovery
Bean, Alan L.	Apollo 12, Moon, Skylab 3
Blaha, John E.	STS-29 Discovery, STS-33 Discovery, STS-43 Atlantis
Bluford, Guion S. "Guy", Jr.	STS-8 Challenger, STS-61A Challenger, STS-39 Discovery
Bobko, Karol J. "Bo"	STS-6 Challenger, STS-51D Discovery, STS-51J Atlantis
Bolden, Charles F., Jr.	STS-61C Columbia, STS-31 Discovery, STS-45 Atlantis
Bondar, Roberta L.	STS-42 Discovery
Borman, Frank	Gemini 7, Apollo 8
Bowersox, Kenneth D.	STS-50 Columbia
Brand, Vance D.	Apollo-Soyuz (Apollo 18), STS-5 Columbia, STS-41B Challenger, STS-35 Columbia
Brandenstein, Daniel C.	STS-8 Challenger, STS-51G Discovery, STS-32 Columbia, STS-49 Endeavour
Bridges, Roy D., Jr.	STS-51F Challenger
Brown, Mark N.	STS-28 Columbia, STS-48 Discovery
Brummer, Renate Luise	trained for STS-55 Columbia

Buchli, James F.	STS-51C Discovery, STS-61A Challenger, STS-29 Discovery, STS-48 Discovery
Bursch, Daniel W.	eligible to be assigned
Cabana, Robert D.	STS-41 Discovery
Cameron, Kenneth D.	STS-37 Atlantis
Carpenter, Malcolm Scott	Mercury MA-7 Aurora 7
Carr, Gerald P.	Skylab 4
Carter, Manley Lanier "Sonny", Jr.	STS-33 Discovery
Casper, John H.	STS-36 Atlantis, STS-50 Columbia
Cenker, Robert J.	STS-61C Columbia
Cernan, Eugene A.	Gemini 9, Apollo 10, Apollo 17, Moon
Chaffee, Roger B.	Apollo 1
Chang-Diaz, Franklin R.	STS-61C Columbia, STS-34 Atlantis, STS-46 Atlantis
Chiao, Leroy	eligible to be assigned
Chilton, Kevin P.	STS-49 Endeavour
Cleave, Mary L.	STS-61B Atlantis, STS-30 Atlantis
Clifford, Michael R. U.	eligible to be assigned
Coats, Michael L.	STS-41D Discovery, STS-29 Discovery, STS-39 Discovery
Cockrell, Kenneth D.	eligible to be assigned
Collins, Eileen	eligible to be assigned
Collins, Michael	Gemini 10, Apollo 11
Conrad, Charles P. "Pete", Jr.	Gemini 5, Gemini 11, Apollo 12, Moon, Skylab 2
Cooper, Leroy Gordon "Gordo", Jr.	Mercury MA-9 Faith 7, Gemini 5
Covey, Richard O.	STS-51I Discovery, STS-26 Discovery, STS-38 Atlantis
Creighton, John O. "J.O."	STS-51G Discovery, STS-36 Atlantis, STS-48 Discovery
Crippen, Robert L.	STS-1 Columbia, STS-7 Challenger, STS-41C Challenger, STS-41G Challenger
Culbertson, Frank L., Jr.	STS-38 Atlantis
Cunningham, R. Walter	Apollo 7
Davis, N. Jan	STS-47 Endeavour
DeLucas, Lawrence J.	STS-50 Columbia
Duffy, Brian	STS-45 Atlantis
Duke, Charles M., Jr	Apollo 16, Moon
Dunbar, Bonnie J.	STS-61A Challenger, STS-32 Columbia, STS-50 Columbia
Durrance, Samuel T.	STS-35 Columbia
Eisele, Donn F.	Apollo 7
England, Anthony W. "Tony"	STS-51F Challenger
Engle, Joe H.	STS-2 Columbia, STS-51I Discovery
Evans, Ronald E.	Apollo 17
Fabian, John W.	STS-7 Challenger, STS-51G Discovery
Fisher, Anna L.	STS-51A Discovery
Fisher, William F.	STS-51I Discovery
Foale, C. Michael	STS-45 Atlantis
Franco Malerba	trained for STS-46 Atlantis
Freeman, Theodore	died in 1964 crash of T-38 jet trainer
Frimout, Dirk D.	STS-45 Atlantis
Fullerton, Charles Gordon	STS-3 Columbia, STS-51F Challenger
Furrer, Reinhard	STS-61A Challenger
Gaffney, Francis Andrew "Drew"	STS-40 Columbia
Gardner, Dale A.	STS-8 Challenger, STS-51A Discovery
Gardner, Guy S.	STS-27 Atlantis, STS-35 Columbia
Garn, Edwin "Jake"	STS-51D Discovery
Garneau, Marc	STS-41G Challenger
Garriott, Owen K.	Skylab 3, STS-9 Columbia
Gemar, Charles D. "Sam"	STS-38 Atlantis, STS-48 Discovery
Gibson, Edward G.	Skylab 4
Gibson, Robert L. "Hoot"	STS-41B Challenger, STS-61C Columbia, STS-27 Atlantis
Glenn, John Herschel, Jr.	Mercury MA-6 Friendship 7
Godwin, Linda M.	STS-37 Atlantis, STS-60 Endeavour

Gordon, Richard F., Jr.	Gemini 11, Apollo 12
Grabe, Ronald J.	STS-51J Atlantis, STS-30 Atlantis, STS-42 Discovery
Gregory, Frederick D.	STS-51B Challenger, STS-33 Discovery, STS-44 Atlantis
Gregory, William G.	eligible to be assigned
Griggs, S. David	STS-51D Discovery
Grissom, Virgil Ivan "Gus"	Mercury MR-4 Liberty Bell 7, Gemini 3, Apollo 1
Guidoni, Umberto	trained for STS-46 Atlantis
Gutierrez, Sidney M.	STS-40 Columbia
Haise, Fred W., Jr.	Apollo 13
Halsell, James D., Jr.	eligible to be assigned
Hammond, L. Blaine, Jr.	STS-39 Discovery
Harbaugh, Gregory J.	STS-39 Discovery
Harris, Bernard A., Jr.	eligible to be assigned
Hart, Terry J.	STS-41C Challenger
Hartsfield, Henry W., Jr.	STS-4 Columbia, STS-41D Discovery, STS-61A Challenger
Hauck, Frederick H. "Rick"	STS-7 Challenger, STS-51A Discovery, STS-26 Discovery
Hawley, Steven A.	STS-41D Discovery, STS-61C Columbia, STS-31 Discovery
Helms, Susan J.	eligible to be assigned
Henize, Karl G.	STS-51F Challenger
Hennen, Thomas J.	STS-44 Atlantis
Henricks, Terence T. "Tom"	STS-44 Atlantis
Hieb, Richard J. "Rick"	STS-39 Discovery, STS-49 Endeavour
Hilmers, David C.	STS-51J Atlantis, STS-26 Discovery, STS-36 Atlantis, STS-42 Discovery
Hoffman, Jeffrey A.	STS-51D Discovery, STS-35 Columbia, STS-46 Atlantis
Hughes-Fulford, Millie	STS-40 Columbia
Irwin, James B.	Apollo 15, Moon
Ivins, Marsha S.	STS-32 Columbia
Jarvis, Gregory B.	STS-51L Challenger
Jemison, Mae C.	STS-47 Endeavour
Jernigan, Tamara E. "Tami"	STS-40 Columbia
Jones, Thomas D.	eligible to be assigned
Kerwin, Joseph P.	Skylab 2
Lampton, Michael L.	trained for STS-45 Atlantis, didn't fly for medical reason
Lee, Mark C.	STS-30 Atlantis, STS-47 Endeavour
Leetsma, David C.	STS-41G Challenger, STS-28 Columbia, STS-45 Atlantis
Lenoir, William B.	STS-5 Columbia
Lichtenberg, Byron K.	STS-9 Columbia, STS-45 Atlantis
Lind, Don L.	STS-51B Challenger
Lounge, John M. "Mike"	STS-51I Discovery, STS-26 Discovery, STS-35 Columbia
Lousma, Jack R.	Skylab 3, STS-3 Columbia
Lovell, James A., Jr.	Gemini 7, Gemini 12, Apollo 8, Apollo 13
Low, G. David	STS-32 Columbia, STS-43 Atlantis
Lucid, Shannon W.	STS-51G Discovery, STS-34 Atlantis, STS-43 Atlantis
Malerba, Franco	trained for STS-46 Atlantis
Mattingly, Thomas K., 2nd	Apollo 16, STS-4 Columbia, STS-51C Discovery
McArthur, William S., Jr.	eligible to be assigned
McAuliffe, Sharon Christa	STS-51L Challenger
McBride, Jon A.	STS-41G Challenger
McCandless, Bruce, II	STS-41B Challenger, STS-31 Discovery
McCulley, Michael J.	STS-34 Atlantis
McDivitt, James A.	Gemini 4, Apollo 9
McMonagle, Donald R.	STS-39 Discovery
McNair, Ronald E.	STS-41B Challenger, STS-51L Challenger
Meade, Carl J.	STS-38 Atlantis, STS-50 Columbia
Melnick, Bruce E.	STS-41 Discovery, STS-49 Endeavour
Merbold, Ulf D.	STS-9 Columbia, STS-42 Discovery
Messerschmid, Ernst	STS-61A Challenger
Mitchell, Edgar D.	Apollo 14, Moon

Mohri, Mamoru	STS-47 Endeavour
Mullane, Richard M. "Mike"	STS-41D Discovery, STS-27 Atlantis, STS-36 Atlantis
Musgrave, F. S. "Story"	STS-6 Challenger, STS-51F Challenger, STS-33 Discovery, STS-44 Atlantis
Nagel, Steven R.	STS-51G Discovery, STS-61A Challenger, STS-37 Atlantis
Nelson, George D. "Pinky"	STS-41C Challenger, STS-61C Columbia, STS-26 Discovery
Nelson, William "Bill"	STS-61C Columbia
Neri Vela, Rodolfo	STS-61B Atlantis
Newman, James H.	eligible to be assigned
Nicollier, Claude	STS-46 Atlantis
O'Connor, Bryan D.	STS-61B Atlantis, STS-40 Columbia
Ochoa, Ellen	eligible to be assigned
Ockels, Wubbo J.	STS-61A Challenger
Onizuka, Ellison S.	STS-51C Discovery, STS-51L Challenger
Oswald, Stephen S.	STS-42 Discovery
Overmyer, Robert F.	STS-5 Columbia, STS-51B Challenger
Pailes, William A.	STS-51J Atlantis
Parise, Ronald A.	STS-35 Columbia, STS-69 Columbia
Parker, A. Robert	STS-9 Columbia, STS-35 Columbia
Payton, Gary E.	STS-51C Discovery
Peterson, Donald H.	STS-6 Challenger
Pogue, William R.	Skylab 4
Precourt, Charles J.	eligible to be assigned
Readdy, William F.	STS-42 Discovery
Reightler, Kenneth S., Jr.	STS-48 Discovery
Resnik, Judith A.	STS-41D Discovery, STS-51L Challenger
Richards, Richard N. "Dick"	STS-28 Columbia, STS-41 Discovery, STS-50 Columbia
Ride, Sally K.	STS-7 Challenger, STS-41G Challenger
Roosa, Stuart A.	Apollo 14
Ross, Jerry L.	STS-61B Atlantis, STS-27 Atlantis, STS-37 Atlantis, STS-55 Columbia
Runco, Mario, Jr.	STS-44 Atlantis
Schirra, Walter Marty "Wally", Jr.	Mercury MA-8 Sigma 7, Gemini 6, Apollo 7
Schlegel, Hans-Wilhelm	trained for STS-55 Columbia
Schmitt, Harrison H.	Apollo 17, Moon
Schweickart, Russell L.	Apollo 9
Scobee, Francis R. "Dick"	STS-41C Challenger, STS-51L Challenger
Scott, David R.	Gemini 8, Apollo 9, Apollo 15, Moon
Scully-Power, Paul D.	STS-41G Challenger
Searfoss, Richard A.	eligible to be assigned
Seddon, Margaret Rhea	STS-51D Discovery, STS-40 Columbia
See, Elliott	died in 1966 crash of T-38 jet trainer
Sega, Ronald M.	eligible to be assigned
Shaw, Brewster H., Jr.	STS-9 Columbia, STS-61B Atlantis, STS-28 Columbia
Shepard, Alan Bartlett, Jr.	Mercury MR-3 Freedom 7, Apollo 14, Moon
Shepherd, William M.	STS-27 Atlantis, STS-41 Discovery
Sherlock, Nancy J.	eligible to be assigned
Shriver, Loren J.	STS-51C Discovery, STS-31 Discovery, STS-46 Atlantis
Slayton, Donald Kent "Deke"	Apollo-Soyuz (Apollo 18)
Smith, Michael J.	STS-51L Challenger
Spring, Sherwood C.	STS-61B Atlantis
Springer, Robert C.	STS-29 Discovery, STS-38 Atlantis
Stafford, Thomas P.	Gemini 6, Gemini 9, Apollo 10, Apollo-Soyuz (Apollo 18)
Stewart, Robert L.	STS-41B Challenger, STS-51J Atlantis
Sullivan, Kathryn D.	STS-41G Challenger, STS-31 Discovery, STS-45 Atlantis
Swigart, John L., Jr.	Apollo 13
Thagard, Norman E.	STS-7 Challenger, STS-51B Challenger, STS-30 Atlantis, STS-42 Discovery
Thiele, Gerhard	trained for STS-55 Columbia

Thomas, Donald A.	eligible to be assigned
Thornton, Kathryn C.	STS-33 Discovery, STS-49 Endeavour
Thorton, William E.	STS-8 Challenger, STS-51B Challenger
Thuot, Pierre J.	STS-36 Atlantis, STS-49 Endeavour
Trinh, Eugene H.	STS-50 Columbia
Truly, Richard H.	STS-2 Columbia, STS-8 Challenger
Umberto Guidoni	trained for STS-46 Atlantis
Van Hoften, James D. "Ox"	STS-41C Challenger, STS-51I Discovery
van den Berg, Lodewijk	STS-51B Challenger
Veach, Charles Lacy	STS-39 Discovery
Voss, James S.	STS-44 Atlantis
Voss, Janice E.	eligible to be assigned
Walker, Charles D.	STS-41D Discovery, STS-51D Discovery, STS-61B Atlantis
Walker, David M.	STS-51A Discovery, STS-30 Atlantis
Walter, Ulrich	trained for STS-55 Columbia
Walz, Carl E.	eligible to be assigned
Wang, Taylor G.	STS-51B Challenger
Weitz, Paul J.	Skylab 2, STS-6 Challenger
Wetherbee, James D.	STS-32 Columbia, STS-46 Atlantis
White, Edward H., 2nd	Gemini 4, Apollo 1
Wilcutt, Terrence W.	eligible to be assigned
Williams, Clifton "C.C."	died in 1967 airplane crash
Williams, Donald E.	STS-51D Discovery, STS-34 Atlantis
Wisoff, Peter J. K.	eligible to be assigned
Wolf, David A.	eligible to be assigned
Worden, Alfred M.	Apollo 15
Young, John W.	Gemini 3, Gemini 10, Apollo 10, Apollo 16, Moon, STS-1 Columbia, STS-9 Columbia

Mercury

America's first men in space went there in Mercury capsules. There were 23 unmanned and manned launches in the first U.S. manned spaceflight program—Project Mercury. Mercury astronauts made two suborbital and four orbital flights in the tiny 6-ft. by 9-ft. capsules they called "garbage cans."

Americans in space. After World War II, manned spaceflight was studied in the U.S., but not planned until the USSR launched Sputniks 1 and 2 in October and November 1957. When the Russians indicated they were rushing toward manned flight, the U.S. Congress started planning to send Americans to orbit.

In March 1958, the U.S. National Advisory Committee for Aeronautics (NACA), suggested a wingless satellite carrying a person on a ballistic path to re-enter the atmosphere without subjecting the astronaut to damaging acceleration or temperatures.

The National Aeronautics and Space Act of 1958 enacted July 29, 1958, enabled the birth of National Aeronautics and Space Administration (NASA) which then was born October 1. NASA included the highlights of the NACA proposal into Project Mercury.

More than two million men and women worked on Project Mercury from 1959-63, developing capsule, rocket, and a tracking network. Project Mercury was followed in the 1960's by Gemini and Apollo programs on the way to the Moon.

Mercury capsule. Project Mercury featured one-person, bell-shaped capsules fired to 100-176 mile-high orbits. Weighing less than 3,000 lbs., the capsule was 9 feet 7 inches from the blunt-end heat shield to the top of the cylindrical recovery section. The spacecraft had three sections:

A cylindrical recovery section held a small drogue parachute, large ringsail parachutes, radio antennas, and automatic control sensors.

The cabin, modeled after a fighter plane cockpit, was pressurized with oxygen. Its reclining contour couch protected the astronaut from acceleration up to 20 gravities. Mercury astronauts squeezed their food in orbit from toothpaste-style tubes.

The astronaut usually rode through space backward with heat shield pointing in the direction of flight. The capsule could be turned around and flown forward, its position (attitude) controlled by 18 thrusters squirting hydrogen-peroxide gas. The thrusters rolled the capsule, pitched it down or up, and yawed it right or left. An autopilot could control attitude or the astronaut could fire thrusters manually with a hand controller.

Behind the cabin was a bell-shaped heat shield and six solid-fuel rockets. Three 400-lb.-thrust rockets fired to separate capsule from rocket or change the orbit of the capsule. Three 1,000-lb.-thrust rockets fired to brake the capsule, slowing it to fall from orbit into the atmosphere.

Once in orbit, the most critical part of the flight was firing the braking rockets. The astronaut had to position the capsule precisely for retrofire to land near the ocean recovery ships. A Mercury capsule re-entered the atmosphere ballistically and parachuted to an ocean splashdown to be recovered by helicopters from U.S. Navy ships.

Mercury rockets. A total of 23 rockets were used for the various Mercury launches, including seven Little Joe boosters, one Blue Scout, five Redstone, three Atlas, six Atlas D, and one Atlas Big Joe.

The less-powerful Redstone rockets were used to launch the first two manned capsules which only reached suborbital speed and altitude.

Mercury capsule size and weight were dictated by the lifting capability of the more-powerful Atlas-D rockets used to launch the four manned capsules all the way to Earth orbit.

Atlas-D was a converted Atlas intercontinental ballistic missile (ICBM), the only U.S. rocket in the 1960 arsenal with sufficient reliability and power to loft a heavy manned satellite to a 100-mile-high orbit. Time between launches ranged from 3 to 224 days, averaging 61 days.

Escape tower. Before manned flight, capsules were launched on a series of suborbital flights testing structural integrity and the steeple-shaped launch escape tower.

For astronaut safety, the launch escape tower was built around a small rocket to pull the capsule up and away from the top of a large space rocket failing during liftoff from launch pad. After an escape, the capsule would float to the ground on a parachute.

Mercury astronauts: The original Mercury 7 astronauts were selected by NASA in 1959. The astronauts and their wives, became media superstars after they signed a deal giving Life exclusive rights to their stories. The magazine published a series of cover stories about their lives. The astronauts:

Alan B. Shepard Jr., Virgil I. "Gus" Grissom, John H. Glenn Jr., M. Scott Carpenter, Walter M. "Wally" Schirra Jr., Leroy Gordon "Gordo" Cooper Jr. and Donald K. "Deke" Slayton. Slayton did not fly in a Mercury capsule, but went to space in 1975 as docking-module pilot in the joint U.S.-USSR Apollo-Soyuz Test Project (ASTP).

Callsign/nicknames. As did the USSR cosmonauts, the first American astronauts gave their capsules nicknames which were used as radio callsigns:

Nickname/callsign	Astronaut	Flight
Freedom 7	Alan B. Shepard Jr.	Mercury MR-3
Liberty Bell 7	Virgil I. "Gus" Grissom	Mercury MR-4
Friendship 7	John H. Glenn Jr.	Mercury MA-6
Aurora 7	M. Scott Carpenter	Mercury MA-7
Sigma 7	Walter M. "Wally" Schirra Jr.	Mercury MA-8
Faith 7	Leroy Gordon "Gordo" Cooper Jr.	Mercury MA-9

Of the 23 launches in the Mercury series, only about a quarter carried men. Thirteen carried no crew, four carried animals, six carried astronauts.

Sixteen flights were successful. Seven launches failed. There were no flight failures during manned launches.

MERCURY Big Joe
Crew: none
Rocket: Atlas Big Joe
Recovery: Sept. 9, 1959

Previous Mercury launch: none
Launch: Sept. 9, 1959
Event: suborbital capsule test

MERCURY Little Joe 1
Crew: none
Rocket: Little Joe
Recovery: Oct. 4, 1959

Previous Mercury launch: 25 days
Launch: Oct. 4, 1959
Event: suborbital capsule test

MERCURY Little Joe 2
Crew: none
Rocket: Little Joe
Recovery: failed

Previous Mercury launch: 31 days
Launch vehicle failure: Nov. 4, 1959
Events: suborbital capsule test

MERCURY Little Joe 3
Passenger: Sam, a monkey
Rocket: Little Joe
Recovery: Dec. 4, 1959

Previous Mercury launch: 30 days
Launch: Dec. 4, 1959

Event: suborbital capsule, biomedical and escape system test

MERCURY Little Joe 4
Passenger: Miss Sam, a monkey
Rocket: Little Joe
Recovery: Jan. 21, 1960

Previous Mercury launch: 48 days
Launch: Jan. 21, 1960

Event: a suborbital capsule flight downrange. Biomedical tests were made. It also tested an escape system.

MERCURY MA-1
Crew: none
Rocket: Atlas
Recovery: failed July 29, 1960

Previous Mercury launch: 190 days
Launch: July 29, 1960
Event: launch vehicle failure

Capsule location: recovered in 1981, now stored at Kissimme, Florida

MERCURY Little Joe 5
Crew: none
Rocket: Little Joe
Recovery: failed

Previous Mercury launch: 102 days
Launch: Nov. 8, 1960
Event: escape rocket fired prematurely

MERCURY MR-1A
Crew: none
Rocket: Redstone
Recovery: Dec. 19, 1960

Previous Mercury launch: 41 days
Launch: Dec. 19, 1960

Event: In a suborbital test, Mercury-Redstone 1A launched unmanned on 235-mi. flight down Atlantic Missile Range. Capsule reached altitude of 130.7 miles at 4,909 mph.
Capsule location: now at McDonnell-Douglas, St. Louis, Missouri.

MERCURY MR-2
Passenger: Ham, a chimp
Rocket: Redstone
Recovery: Jan. 31, 1961

Previous Mercury launch: 43 days
Launch: Jan. 31, 1961

Event: In a 16-min. suborbital flight, Mercury-Redstone 2 carried a 37-lb. chimpanzee named Ham in a Mercury capsule on a suborbital flight. Ham was recovered unharmed.
Capsule location: now at Kennedy Space Center, Cape Canaveral, Florida

MERCURY MA-2
Crew: none
Rocket: Atlas
Recovery: Feb. 21, 1961

Previous Mercury launch: 21 days
Launch: Feb. 21, 1961
Event: suborbital capsule test

Capsule location: now at World Trade Center, Houston, Texas

MERCURY Little Joe 5A
Crew: none
Rocket: Little Joe
Recovery: failed

Previous Mercury launch: 25 days
Launch: March 18, 1961
Event: escape system failed

MERCURY MR-BD
Crew: none
Rocket: Redstone
Recovery: March 24, 1961

Previous Mercury launch: 6 days
Launch: March 24, 1961
Event: vehicle test without capsule

MERCURY MA-3

Crew: none
Rocket: Atlas
Recovery: failed

Previous Mercury launch: 32 days
Launch: April 25, 1961
Event: launch vehicle failure

MERCURY Little Joe 5B

Crew: none
Rocket: Little Joe
Recovery: failed

Previous Mercury launch: 3 days
Launch: April 28, 1961
Event: launch vehicle failure

MERCURY MR-3 "Freedom 7"

Crew: Alan B. Shepard Jr.
Rocket: Redstone
Recovery: May 5, 1961

Previous Mercury launch: 7 days
Launch: May 5, 1961
Flight duration: 15 mins 22 secs, less than 1 orbit

Event: First U.S. manned spaceflight. First American in space. Shepard is said to have wet his spacesuit as he sat strapped in the Freedom 7 capsule atop a Redstone rocket, waiting to become the first American in space. Then, in a 15-min. suborbital flight, Mercury-Redstone 3 carried Shepard to space and 303 mi. downrange from the Cape Canaveral launch site. The capsule reached 116 miles altitude. Shepard tested controls and reported "AOK" (all okay). Shepard flew again in Apollo 14 in 1971. Capsule location: National Air & Space Museum Smithsonian Institution, Washington, D.C.

MERCURY MR-4 "Liberty Bell 7"

Crew: Virgil I. "Gus" Grissom
Rocket: Redstone
Recovery: July 21, 1961

Previous Mercury launch: 77 days
Launch: July 21, 1961
Flight duration: 15 mins 37 secs, less than 1 orbit

Events: Second U.S. manned spaceflight. Second American in space. Second and final manned suborbital flight. In a 15-min. flight, Mercury-Redstone 4 carried Grissom 302 miles downrange from Cape Canaveral. The capsule, Liberty Bell 7, reached 118.3 miles altitude, but sank after splashdown in the Atlantic Ocean near the Bahamas. Explosive bolts blew open a hatch, flooding the capsule with seawater. He floated out in his buoyant spacesuit, but water seeped through an open vent into his suit. He was about to go under when he finally grabbed a sling dangling from a rescue helicopter and was yanked to safety. The water-filled capsule, too heavy for the helicopter, was abandoned. The chopper cut the capsule loose and reeled in Grissom. Liberty Bell 7 sank three miles to the ocean floor. Grissom later was pilot of Gemini 3 in 1965. He died in the Apollo 1 launch pad fire in 1967. Capsule location: Atlantic Ocean floor near the Bahamas in 3 miles of water.

MERCURY MA-4

Crew: none
Rocket: Atlas D
Recovery: Sept. 13, 1961

Previous Mercury launch: 54 days
Launch: Sept. 13, 1961
Event: first orbital flight test

MERCURY MS-1

Crew: none
Rocket: Blue Scout
Recovery: failed

Previous Mercury launch: 49 days
Launch: Nov. 1, 1961
Event: launch vehicle failure

MERCURY MA-5

Passenger: Enos, a chimp
Rocket: Atlas D
Earth orbits: 2

Previous Mercury launch: 28 days
Launch: Nov. 29, 1961
Recovery: Nov. 29, 1961

Event: Mercury-Atlas 5 carried the chimpanzee Enos in a Mercury capsule for two orbits of Earth. Enos was recovered unharmed, showing a man could do the same.

MERCURY MA-6 "Friendship 7"

Crew: John H. Glenn Jr.
Rocket: Atlas D
Earth orbits: 3

Previous Mercury launch: 83 days
Launch: Feb. 20, 1962
Recovery: Feb. 20, 1962

Flight duration: 4 hrs 55 mins 23 secs

Events: First U.S. manned orbital flight. First American in orbit. Third person to go to orbit. Third U.S. manned spaceflight. Third American in space. First Florida winter launch. Following the pair of Mercury-Redstone manned suborbital test flights and the Mercury-Atlas flight of Enos the chimp, an Atlas-D rocket carried Glenn in the Mercury-Atlas 6 capsule Friendship 7 from Cape Canaveral to space for 3 revolutions around the planet. Perth, Australia, residents turned on all lights in town to signal Glenn as he passed overhead. Glenn said a swarm of yellowish-green luminous particles drifting past the capsule window looked like fireflies. They remained a mystery until Mercury 7's Scott Carpenter discovered they were frost particles flaking off the outside of the capsule. Glenn ate applesauce in orbit. He splashed down in the Atlantic Ocean. Americans gave Glenn the same hero's welcome as after Charles A. Lindbergh's 1927 New York-Paris flight. Glenn was elected to the U.S. Senate from Ohio in 1974, the first astronaut in public office.

Capsule location: National Air & Space Museum Smithsonian Institution, Washington, D.C.

MERCURY MA-7 "Aurora 7"

Crew: M. Scott Carpenter

Rocket: Atlas D

Earth orbits: 3

Flight duration: 4 hrs 56 mins 5 secs

Previous Mercury launch: 93 days

Launch: May 24, 1962

Recovery: May 24, 1962

Events: Second U.S. manned orbital flight. Second American in orbit. Fourth U.S. manned spaceflight. Fourth American in space. An Atlas-D rocket carried Carpenter's Mercury-Atlas 7 capsule Aurora 7 from Cape Canaveral to space for 3 orbits. First granola bars in orbit: Pillsbury had supplied Carpenter with new high-protein cereal snacks and Nestle sent along "bone-bones" of cereals with raisins and almonds. He was supposed to splashdown in the Atlantic Ocean near Puerto Rico, but firing of landing rockets was delayed three seconds and the capsule was in the wrong position. Carpenter overshot the target by 250 miles. He was out of voice radio contact 41 minutes; NASA was worried. Navy search aircraft homed in on his emergency radio beacon until he flashed a passing boat with a hand mirror. After three hours bobbing among the waves in a rubber raft outside Aurora 7, he was rescued. Capsule location: now at Hong Kong Space Museum, Hong Kong.

MERCURY MA-8 "Sigma 7"

Crew: Walter M. "Wally" Schirra Jr.

Rocket: Atlas D

Earth orbits: 6

Flight duration: 9 hrs 13 mins 11 secs

Previous Mercury launch: 132 days

Launch: Oct. 3, 1962

Recovery: Oct. 3, 1962

Events: Third U.S. manned orbital flight. Third American in orbit. Fifth U.S. manned spaceflight. Fifth American in space. Atlas-D rocket lofted Mercury-Atlas 8 capsule Sigma 7 for 6 orbits, doubling previous U.S. flight time in orbit. Like all Mercury astronauts, Schirra squeezed his lunch of beef, vegetables and peaches from a toothpaste-style tube. Schirra made the first Pacific Ocean splashdown, just 4.5 miles from the waiting U.S. aircraft carrier Kearsarge. Schirra flew again, Gemini 6 in 1965 and Apollo 7 in 1968.

Capsule location: NASA Space & Rocket Center, Huntsville, Alabama.

MERCURY MA-9 "Faith 7"

Crew: L. Gordon "Gordo" Cooper Jr.

Rocket: Atlas D

Earth orbits: 22

Flight duration: 34 hrs 19 mins 49 secs

Previous Mercury launch: 224 days

Launch: May 15, 1963

Recovery: May 16, 1963

Events: Fourth U.S. manned orbital flight. Fourth American in orbit. Sixth U.S. manned spaceflight. Sixth American in space. First American to stay a day in space. First live TV from U.S. spacecraft. Final Mercury flight. Atlas-D rocket lifted Mercury-Atlas 9 capsule Faith 7 to space for 22 orbits. When Cooper made the first pilot-controlled re-entry and splashdown, the first American manned space program was completed less than five years after Congress established NASA in 1958. Cooper flew again in Gemini 5 in 1965.

Capsule location: now at NASA Johnson Space Center, Houston, Texas.

Gemini

Project Gemini was America's second man-in-space program, following Project Mercury. Gemini launches from 1964-66 featured two-person capsules boosted to low Earth orbit on Titan 2 rockets. There were 12 flights, including two unmanned tests.

Gemini's purpose. A bridge between Mercury and Apollo was authorized by Congress in 1961 as a step toward landing on the Moon. Gemini, with a two-person crew, was named for the third constellation of the zodiac which is said to have twin stars, Castor and Pollux. Gemini is Latin for twins.

Project Gemini was designed to test men and hardware in flights of up to two weeks in Earth orbit, to develop and practice rendezvous and docking with spacecraft in orbit, to move around space using the docking-target rocket's propulsion, to polish atmospheric re-entry techniques, and to touch down on dry land. Ocean splashdowns were continued and the idea of landing on solid ground was dropped in 1964.

Mercury had shown astronauts could live in orbit up to 34 hours. Now NASA needed to know if a sophisticated life-support system would work and astronauts could endure the length of weightless freefall time needed for a round trip to the Moon. Gemini used NASA's global tracking and communications network built for Mercury.

Gemini capsule. The spacecraft, sometimes referred to as Gemini-Titan because of its launching rocket, was like the bell-shaped, blunt-end Mercury capsule enlarged to 19 feet in length and 10 feet in diameter at the base, weighing 8,400 lbs. Gemini had 50 percent more cabin space, but weighed twice as much. Maintenance was simplified.

The cabin had side-by-side couches facing a cylindrical nose housing a docking mechanism and parachutes. Gemini 10 pilot Michael Collins compared the cabin space and couches to the front seats of a Volkswagen beetle. Gemini 7 astronauts Frank Borman and James A. Lovell Jr. lived two weeks in that tiny space.

Plastic bags of freeze-dried foods, liquified by water pistol, were used, culminating in the first shrimp cocktail in space. The astronauts urinated into cups which were dumped overboard. For solid waste, they used stick-on plastic bags carried back to Earth.

The retrograde section behind the cabin held four retro-rockets fired in quick succession to slow the spacecraft for reentry. Gemini was more maneuverable than Mercury. A rear adapter section held 16 orbital attitude and maneuvering system (OAMS) engines, fuel-cell batteries for electricity, and extra oxygen.

To land, the crew separated forward re-entry cabin from rear adapter section and ignited four 2,500 lb.-thrust retro-rockets. Retro-rockets and adapter section were jettisoned during re-entry.

Eight re-entry control system (RCS) thrusters stabilized the capsule which gained lift from the atmosphere. Within a small range, the astronauts could control Gemini's landing point by rolling the capsule. At the end, the capsule parachuted to splashdown in the ocean.

Titan rocket. Two-stage Titan 2 intercontinental ballistic missiles (ICBM), more powerful than Redstone rockets and Atlas ICBMs used with Mercury capsules, were converted to boost Gemini capsules to low Earth orbit.

Gemini astronauts: Edwin E. "Buzz" Aldrin Jr., Neil A. Armstrong, Frank Borman, Eugene A. Cernan, Michael Collins, Charles "Pete" Conrad Jr., Leroy Gordon "Gordo" Cooper Jr., Richard F. Gordon Jr., Virgil I. "Gus" Grissom, James A. Lovell Jr., James A. McDivitt, Walter M. "Wally" Schirra Jr., David R. Scott, Thomas P. Stafford, Edward H. White 2nd, and John W. Young.

The manned flights were Gemini 3 through Gemini 12. Cooper, Grissom and Schirra were original Mercury 7 astronauts. Conrad, Lovell, Stafford and Young flew twice in Gemini capsules. Grissom and White were killed later in Apollo 1.

Aldrin, Armstrong and Collins later flew Apollo 11 to the Moon. Borman and Lovell flew Apollo 8. Lovell flew Apollo 13. Cernan flew Apollo 10 and Apollo 17. Gordon and Conrad flew Apollo 12.

Conrad flew to the first U.S. space station in the Skylab 2 flight. McDivitt and Scott flew Apollo 9. Later Scott flew Apollo 15. Schirra flew Apollo 7.

Stafford flew Apollo 10 and the U.S.-USSR Apollo-Soyuz Test Project (ASTP). Young flew Apollo 10 and Apollo 16 and shuttle Columbia flights STS-1 and STS-9.

Altogether, Young flew to space six times: in Gemini 3 in 1965, Gemini 10 in 1966, Apollo 10 in 1969, Apollo 16 to land on the Moon in 1972, the maiden voyage of shuttle Columbia in 1981, and Columbia again in 1983.

Callsign/nickname. Only the first manned flight, Gemini 3, had a radio callsign like the Mercury capsule nicknames. Gemini 3 command pilot Virgil I. "Gus" Grissom gave his capsule the nickname Molly Brown in honor of his Mercury capsule which sank after splashdown in the Atlantic Ocean. After that, it was said it became too hard for government bureaucrats to approve astronaut nickname choices.

GEMINI 1
Rocket: Titan 2   Launch: April 8, 1964
Event: Unmanned test launch of Gemini capsule on Titan 2.
Capsule location: Gemini 1A, Hall of Sciences, Queens, New York; Gemini 1B El Kabong, Michigan Space Center, Jackson Michigan

GEMINI 2
Rocket: Titan 2   Launch: January 19, 1965
Event: Unmanned test launch of Gemini capsule heat shield on Titan 2.
Capsule location: Gemini 2, USAF Space Museum, Cape Canaveral, Florida; Gemini 2A, Kansas Cosmosphere & Discovery Center, Hutchinson, Kansas

GEMINI 3
Astronauts: Virgil I. "Gus" Grissom, John W. Young
Capsule callsign/nickname: Molly Brown
Rocket: Titan 2   Launch: March 23, 1965
Earth orbits: 3   Splashdown: March 23, 1965
Flight duration: 4 hours 53 mins
Event: First Gemini to orbit. First manned Gemini flight. First U.S. two-man flight. First manned craft to change orbit. First lifting re-entry of a manned spacecraft. Grissom ate the first corned beef sandwich in orbit. Astronaut Wally Schirra made the sandwich before launch and gave it to Young who passed it to Grissom. The sandwich caused a congressional fuss and new rules later prevented such casual dining. Capsule location: Grissom Memorial Museum, Mitchell, Indiana

GEMINI 4
Astronauts: James A. McDivitt, Edward H. White II
Rocket: Titan 2   Launch: June 3, 1965
Spacewalk: one for 22 minutes   Earth orbits: 62
Splashdown: June 7, 1965   Flight duration: 97 hrs 56 min 11 sec (4 days 2 hr)
Event: Second Gemini to orbit. First U.S. spacewalk (extravehicular activity or EVA). White made a 22-minute EVA, wearing a 31-lb. spacesuit and a 25-ft. tether attached to the Gemini capsule. He maneuvered by firing a twin-barrel gas gun. McDivitt remained inside. USSR cosmonaut already had made first spacewalk, 10 minutes on a 10-ft. tether March 18, 1965.
Capsule location: National Air & Space Museum, Smithsonian Institution, Wash. D.C.

GEMINI 5
Astronauts: Leroy Gordon "Gordo" Cooper Jr., Charles "Pete" Conrad Jr.
Rocket: Titan 2   Launch: August 21, 1965
Earth orbits: 120   Splashdown: August 29, 1965
Flight duration: 190 hrs 55 mins 14 secs (8 days 21 hrs)
Event: Third Gemini to orbit. First to track a typhoon from space. First fuel cells for electricity. First person to fly in space with radar, checking out navigation and guidance for future rendezvous. American space-endurance record was extended to eight days. Set a new space endurance record of 3,312,993 miles in less than 191 hours. Moved the U.S. ahead of the USSR in manhours in space with 225 hours 15 mins. Performed 17 science experiments. Observed weather, ocean currents. Cooper was 1st astronaut to make a 2nd orbital flight, having flown in 1963 in Mercury capsule.
Capsule location: NASA, Johnson Space Center, Houston, Texas

GEMINI 6
Astronauts: Walter M. "Wally" Schirra Jr., Thomas P. Stafford
Rocket: Titan 2   Launch: December 15, 1965
Earth orbits: 16   Splashdown: December 16, 1965
Flight duration: 25 hrs 51 mins 24 secs
Event: First space rendezvous. To continue progress in the Moon Race, the Gemini program in 1965 needed to achieve rendezvous and docking with a satellite in orbit. NASA planned for a Gemini spacecraft to dock with an Agena rocket upper stage launched to orbit on an Atlas rocket. The first planned Agena blew up after launch on an Atlas, so the first rendezvous attempt was cancelled Oct 25, 1965. A second rocket fizzled on the launch pad. Gemini 7 had been launched Dec 4 so NASA sent Gemini 6 to orbit Dec 15 to meet Gemini 7. Schirra flew Gemini 6 within 1 foot of Gemini 7 Dec 15—the 1st successful space rendezvous. The 2 capsules practiced station-keeping in orbit for 5 hours at distances from 1 to 295 feet. Gemini 6 splashed down Dec 16.
Capsule location: Gemini 6A, McDonnell Planetarium, St. Louis, Missouri

GEMINI 7
Astronauts: Frank Borman, James A. Lovell Jr.

Rocket: Titan 2
Earth orbits: 206
Launch: December 4, 1965
Splashdown: December 18, 1965
Flight duration: 333 hrs 35 mins 31 secs (13 days 18 hrs 35 mins)
Event: Gemini 7 was intended to test whether man could live two weeks in space. It was given an additional assignment at the last minute as a substitute target after Gemini 6 was scrubbed when an Agena rendezvous target failed to reach orbit. After performing the first space rendezvous with Gemini 6 on December 15, Borman and Lovell set an endurance record of 13 days 18 hrs 35 mins in orbit, longest of all Gemini flights. Flying 20 times the distance to the Moon while staying in Earth orbit, Gemini 7 showed astronauts could make a round trip to the Moon. Gemini 7 splashed down in a controlled landing in the Atlantic Ocean December 18.
Capsule location: National Air & Space Museum, Smithsonian Institution, Wash. D.C.

GEMINI 8
Astronauts: Neil A. Armstrong, David R. Scott
Rocket: Titan 2
Earth orbits: 6.5
Launch: March 16, 1966
Splashdown: March 16, 1966
Flight duration: 10 hrs 41 min 26 sec
Event: Rendezvous and first docking with another spacecraft, an unmanned orbiting Agena target rocket. A broken control in the OAMS thrusters threw Gemini 8 into an uncontrollable spin, ending the flight suddenly. The crew undocked from the Agena and made the first emergency landing for a U.S. spacecraft, splashing down in the western Pacific Ocean. Armstrong had been a civilian pilot when chosen as an astronaut. In 1969, he became the first man to walk on the Moon.
Capsule location: Neil Armstrong Museum, Wapakoneta, Ohio

GEMINI 9
Astronauts: Thomas P. Stafford, Eugene A. Cernan
Rocket: Titan 2
Spacewalk: 2 hours
Splashdown: June 6, 1966
Launch: June 3, 1966
Earth orbits: 44
Flight duration: 72 hrs 21 mins (3 days 21 hours)
Event: An Agena docking target failed to reach orbit May 17, 1966, so Gemini 9 postponed while an augmented target docking adapter (ATDA) was rigged from Gemini pieces. Stafford and Cernan then were launched, attempted docking with ATDA June 3, but were prevented when faulty ATDA shroud did not completely separate. Three rendezvous were completed and 2 hours spacewalking.
Capsule location: Gemini 9A, NASA Kennedy Space Center, Cape Canaveral, Florida

GEMINI 10
Astronauts: John W. Young, Michael Collins
Rocket: Titan 2
Spacewalks: 2, total 88 mins
Splashdown: July 21, 1966
Launch: July 18, 1966
Earth orbits: 43
Flight duration: 70 hrs 46 mins 39 secs (3 days)
Event: Rendezvous and docking. First use of target rocket for propulsion. Young and Collins docked Gemini 10 with an unmanned orbiting Agena target on July 18 and fired the Agena's engine to boost the joined spacecraft to 475 miles altitude. Collins made two EVAs, 49 minutes standing in the Gemini 10 hatch and 39 minutes retrieving a detector from the Agena.
Capsule location: Swiss Museum of Transport, Lucerne, Switzerland

GEMINI 11
Astronauts: Charles "Pete" Conrad Jr., Richard F. Gordon Jr.
Rocket: Titan 2
Spacewalks: 2, total 2 hrs 33 min
Splashdown: September 15, 1966
Launch: September 12, 1966
Earth orbits: 44
Flight duration: 71 hrs 17 min 8 sec (5 days 8 hrs)
Event: Conrad and Gordon rendezvoused and docked with the Agena target on their first orbit, then set the Gemini altitude record of 739.2 miles by using the Agena target rocket to boost Gemini higher. Highest manned flight ever for any type of U.S. or USSR capsule or shuttle in Earth orbit (not including trips away from Earth to the Moon). Gordon made two EVAs, a 33-minute spacewalk to attach a tether to Agena and a two-hour standup in the hatch. They photographed stars and galaxies. Capsule location: NASA, Ames Center, Mountain View, California

GEMINI 12
Astronauts: James A. Lovell Jr., Edwin E. "Buzz" Aldrin Jr.
Rocket: Titan 2
Spacewalks: 3, total 5 hrs 30 min
Splashdown: November 15, 1966
Launch: November 11, 1966
Earth orbits: 59
Flight: 94 hr 34 min 31 sec (3 days 22 hrs 34 min)
Event: Rendezvoused and docked by visual means with its target Agena and kept station with Agena

while Aldrin made three EVAs, setting an EVA record of 5 hours 30 minutes including one 2 hr 9 min spacewalk and two stand-up-in-the-hatch exercises. The first of Aldrin's EVAs were disrupted when he suffered overheating and his spacesuit face-plate fogged. He was able to overcome the problem. The Agena rocket was fired, while docked with Gemini 12, to push Gemini 12 to a higher altitude. It was the final Gemini flight. Capsule location: NASA, Goddard Spaceflight Center, Greenbelt, Md.

Apollo

Between July 1969 and December 1972, American astronauts used Apollo, the first manned interplanetary transportation system, to fly to the Moon, land there, explore the lunar landscape and pick up samples at six sites on the near side of the Moon.

Project Apollo. U.S. President John F. Kennedy on May 25, 1961, set the national goal of landing a man on the Moon within the decade and returning him safely to Earth. Project Apollo's goals were to establish technology to meet American interests in space, to achieve preeminence in space for the U.S., to carry out scientific exploration of the Moon, and to develop a capability to work in the lunar environment.

Saturn rocket. The plan to use a lunar orbit rendezvous technique for reaching the Moon and landing astronauts on the lunar surface was set in 1962. It required an extraordinarily-powerful space rocket, a high-tech capsule to carry astronauts, and a vehicle to fly down from lunar orbit to the surface of the Moon.

The Saturn family of space rockets developed for Project Apollo included Saturn 1, Saturn 1B and Saturn V (Saturn 5). Saturn 1B was used for flights in Earth orbit. Saturn V was used for lunar flights.

Manned capsule. The 1964-75 Apollo program followed the 1959-63 Mercury project and the 1965-66 Gemini program in which orbiting, docking and spacewalking techniques were developed. While Gemini featured two-man capsules, Apollo brought three-man capsules, two-man lunar landers and Saturn rockets.

An Apollo Moonship had three parts: command module (CM) with quarters for three crewmen and flight controls, service module (SM) with propulsion and support systems, and lunar excursion module—sometimes referred to as LM, LEM or lunar lander—to land two astronauts on the Moon, house and support them on the lunar surface, and fly them back up to lunar orbit to the CSM.

CM and SM modules were connected to each other and generally referred to as CSM. The CSM was 10 feet in diameter at the blunt end and 34 feet long.

The crew rode on three couches in the CM which was pressurized with oxygen at 5 lb/sq. in. A major Apollo advance over earlier spacecraft was its inertial guidance system developed by Massachusetts Institute of Technology (MIT).

The SM, attached behind the CM, housed the oxygen and hydrogen tanks, fuel-cell batteries, environmental control system, reaction control system, and the 21,500-lb.-thrust main engine used for course corrections, major changes in orbit, injection of the combined CSM/LEM into lunar orbit and CSM escape from lunar orbit for return to Earth.

When re-entering Earth's atmosphere, the CM separated from the SM. The CM was protected by a heat shield and stabilized by a reaction control system, until low enough to deploy its drogue and three main parachutes.

Lunar lander. The LEM included a descent stage and an ascent stage, attached during flight down to the Moon's surface. The descent module provided a platform from which the ascent module blasted off on a return flight up to a CSM in lunar orbit.

The electric 4-wheel-drive Lunar Roving Vehicle (LRV), first used on the Moon by Apollo 15, was carried to the surface, with its wheels folded, in the LEM descent stage.

Apollo 204 tragedy. There was a rush to beat the Russians to the Moon by the end of the 1960's. Pre-flight testing of Apollo capsules started May 28, 1964. The U.S. sent the powerful new Saturn 1B rocket on its first unmanned suborbital test flight February 26, 1966, carrying a dummy Apollo capsule 5,500 miles downrange.

In a second unmanned test flight July 5, the rocket second stage went into orbit, but carried no other satellite. A third Saturn 1B test that year was a suborbital flight August 25. The first manned Apollo test was to follow on February 21, 1967, but it didn't happen.

On January 27, 1967, 25 days before their scheduled lift-off, Grissom, White and Chaffee had taken an early lunch and were practicing in Apollo-Saturn 204 during a simulated countdown rehearsal on the pad at Cape Canaveral's launch complex 34. It was routine—what NASA called a "plugs out" test. A cumbersome hatch locked in the astronauts, who were breathing pure oxygen, and the sealed capsule was disconnected from outside electrical power to simulate launch.

Problems had dropped the simulation behind schedule near sunset when a communications problem caused a 10-minute delay. Somehow a broken, bruised or frayed wire contacted metal, short-circuited and sparked beneath Grissom's couch, triggering a fire in the cabin's 100-percent-oxygen atmosphere. The electrical arcing ignited usually-fire-resistant plastic which turned flammable in the pure oxygen atmosphere of the CM. Dense acrid smoke from burning plastic suffocated the crew.

Blockhouse controllers and workers in the launch pad clean-area white room heard a radio cry from the capsule, "There is a fire in here." As fire engulfed the capsule interior, the lead worker on the pad shouted to a co-worker, "Get them out of there."

Seconds later flame burst from Apollo-Saturn 204. Gasping white room workers tugged at the hatch to open the capsule while controllers in the blockhouse were paralyzed by panic. Apollo-Saturn 204 did not have explosive bolts on its main hatch.

A short 5.5 minutes after the fire broke out, the hatch was opened. Fourteen minutes after the first outcry, physicians Allan Harter and G. Fred Kelly reached the astronauts, already dead from asphyxiation of toxic gases. The charred Apollo-Saturn 204 shell was stored in a locked container at NASA's Langley Research Center in Hampton, Virginia.

20-month delay. In 1967, NASA renamed Apollo-Saturn 204 to Apollo 1. Before the change of designation, the previous Apollo-Saturn 201 and Apollo-Saturn 202 flights, carrying Apollo CMs, had been known unofficially as Apollo 1 and Apollo 2. Apollo-Saturn 203 carried only an aerodynamic nose cone. Officially, there were no Apollo 2 or Apollo 3 flights.

A 20-month delay in manned flight followed the Apollo-Saturn 204 tragedy. There were three unmanned Saturn tests in 1967 and 1968, known as Apollo 4, 5 and 6, before a Saturn 1B hefted the manned Apollo 7 to space October 11, 1968.

The first unmanned Saturn V launch, November 9, 1967, was called Apollo 4. The Saturn 1B rocket originally designated to carry an Apollo CM as Apollo-Saturn 204 was renamed Apollo 5 and launched January 22, 1968, carrying a LEM as payload.

The second Saturn V launch, known as Apollo-Saturn 502 or Apollo 6, carrying a CM, was April 4, 1968. It was a success despite two first-stage engines shutting down prematurely and the third stage engine failure to re-ignite in orbit. For Apollo 7, a Saturn 1B carried a CSM with three astronauts October 11, 1968.

Manned flights. NASA launched 15 Apollo Saturn rockets carrying manned Apollo spacecraft between October 1968 and July 1975.

Eleven were in the lunar landing project, including two test flights in low Earth orbit, two test flights in lunar orbit, six landings, and one aborted flight which resulted in a circumlunar flight.

Four Apollo flights were made after lunar landings ended. Three in 1973-74 ferried astronauts to and from America's Skylab space station. One was the 1975 Apollo-Soyuz Test Project joint flight with USSR cosmonauts.

Apollo astronauts. There were 36 Apollo astronauts and 12 manned Apollo missions. Three astronauts died in a launch pad fire. Six flew only in Earth orbit. Eighteen flew to the Moon. Twelve landed on the Moon. In the list, mission commander is listed first, command module pilot is second followed by lunar excursion module.

Apollo 1	Virgil I. "Gus" Grissom, Edward H. White 2nd, Roger B. Chaffee
Apollo 7	Walter M. "Wally" Schirra Jr., Donn F. Eisele, R. Walter Cunningham
Apollo 8	Frank Borman, James A. Lovell Jr., William A. Anders
Apollo 9	James A. McDivitt, David R. Scott, Russell L. Schweickart
Apollo 10	Eugene A. Cernan, John W. Young, Thomas P. Stafford
Apollo 11	Neil A. Armstrong, Michael Collins, Edwin E. "Buzz" Aldrin Jr.
Apollo 12	Charles P. "Pete" Conrad Jr., Richard F. Gordon Jr., Alan L. Bean
Apollo 13	James A. Lovell Jr., John L. Swigart Jr., Fred W. Haise Jr.
Apollo 14	Alan B. Shepard Jr., Stuart A. Roosa, Edgar D. Mitchell
Apollo 15	David R. Scott, Alfred M. Worden, James B. Irwin
Apollo 16	John W. Young, Thomas K. Mattingly 2nd, Charles M. Duke Jr.
Apollo 17	Eugene A. Cernan, Ronald E. Evans, Harrison H. Schmitt
Apollo 18	Thomas P. Stafford, Vance D. Brand, docking pilot Donald K. "Deke" Slayton

The 27 Apollo astronauts who left and flew to the Moon have been farther away than anyone else from Earth. First to fly to the Moon were Apollo 8 astronauts Borman, Lovell and Anders in December 1968. They made ten orbits of the Moon, but did not land on the lunar surface.

The all-time best known spaceman is Apollo's Neil Armstrong, the first man to walk on the Moon. His memorable remark on July 20 1969, was, "That's one small step for (a) man...one giant leap for mankind." NASA reinserted the radio-garbled "a" later.

A NASA doctor called off a dinner President Richard M. Nixon wanted to give for Armstrong and his fellow Apollo 11 astronauts Buzz Aldrin and Michael Collins on July 15, 1969, the night before they left for the first-ever Moon landing. Later Collins said, "I'm sure presidential germs are benign."

Along with Armstrong, those who walked on the Moon were Aldrin, Conrad, Bean, Shepard, Mitchell, Scott, Irwin, Young, Duke, Cernan and Schmitt. Fourteen astronauts flew to the Moon, but didn't get to land: Borman, Lovell, Anders, Stafford, Young, Collins, Gordon, Lovell, Swigart, Haise, Roosa, Worden, Mattingly and Evans.

First to fly twice to the Moon was Lovell in Apollo 8 and 13. Later, Scott flew in Apollo 9 and 15. Young flew in Apollo 10 and 16. Cernan flew in Apollo 10 and 17.

Apollo Flight	Crews	Astronauts	Result
1	1	3	Died in launch pad fire
7, 9	2	6	Flew in Earth orbit only
8, 10, 13	3	9	Flew to Moon, didn't land
11, 12, 14, 15, 16, 17	6	18	Two landed on Moon, with one only in lunar orbit

Young rode the most spaceships, flying to space six times—Gemini 3 in 1965, Gemini 10 in 1966, Apollo 10 in 1969, Apollo 16 to land on the Moon in 1972, the maiden voyage of shuttle Columbia in 1981, and shuttle Columbia again in 1983. Grissom, Schirra, Shepard and Slayton were original Mercury 7 astronauts

Callsign/nicknames. The practice of using popular names for radio callsigns to distinguish separated LEM from CSM started when the Apollo 9 CSM was referred to as Gumdrop and the LEM as Spider.

Apollo 9	CSM: Gumdrop	LEM: Spider
Apollo 10	CSM: Charlie Brown	LEM: Snoopy
Apollo 11	CSM: Columbia	LEM: Eagle
Apollo 12	CSM: Yankee Clipper	LEM: Intrepid
Apollo 13	CSM: Odyssey	LEM: Aquarius
Apollo 14	CSM: Kitty Hawk	LEM: Antares
Apollo 15	CSM: Endeavor	LEM: Falcon
Apollo 16	CSM: Casper	LEM: Orion
Apollo 17	CSM: America	LEM: Challenger

Moon rocks. The 12 men who walked the Moon after six Apollo landings picked up 2,000 samples of rock and soil weighing 842 pounds. As a result, scientists now have a much clearer picture of the origin and history of the Moon.

After forming, the Moon's outer layers probably melted because of heat from decay of radioactive elements or the accretion of material that formed the planet. As molten layers cooled, the Moon was bombarded by huge meteorites.

As the bombardment slowed four billion years ago, heat from decay of radioactive elements warmed the interior bringing great floods of lava to the surface, filling the large impact craters, creating the dark maria visible today. The Moon hasn't changed much in the last 3 billion years. Today, the surface changes so slowly astronaut footprints are likely to remain clearly defined for millions of years.

The Apollo astronauts found the Moon covered by a layer, from 3 to 66 feet deep, of fine soil and rock fragments called regolith. Water, wind and life, which change Earth soil, are not found on the Moon. Lunar soil built up on the airless surface over billions of years of bombardment by meteorites, most of which probably are so small they would have burned in Earth's atmosphere. The meteorites shattered solid rock and scattered debris widely, mixing the soil in a process scientists call gardnering.

The astronauts found no sedimentary rocks. All were igneous—solidified volcanic lava. Minerals in Moon rocks were mostly the same as in Earth lava, but three new minerals were discovered. One was named Tranquillityite for the Apollo 11 landing site. Another was labeled Armalcolite for Apollo 11 astronauts Armstrong, Aldrin and Collins. The third new mineral was named Pyroxferroite.

Ancient moon. Maria on the Moon are broad flatlands, erroneously thought by early astronomers to be seas. Rocks and soil collected there were the youngest found on the Moon. Even so, they were some 3.8 billion years. That's as old as the oldest rocks on Earth.

Rocks collected in the lunar highlands were more than 4 billion years old. Some very small green rock chips collected by Apollo 17 astronauts turned out to be 4.6 billion years. They might have been first material solidified on a once-molten Moon.

The 3.5 billion-year-old rocks were dry, unlike Earth rocks. They were well preserved having no water to cause rust or form clay. Crystals in the old rock seemed fresher than those in water-bearing lava erupted recently from an Earth volcano.

Apollo astronauts left behind seismometers which detected about 3,000 Moonquake tremors a year. Seismometers on Earth sense hundreds of thousands a year. The Moonquakes were very weak compared with Earthquakes, popping off only as much energy as a big firecracker.

Scientists say the Moon may be structured like Earth, in layers with a 37-mile thick crust over a 500-mi.-deep mantle of denser rock. The Moon's core remains a mystery. It could be hot or molten. Today, NASA keeps the Moon rocks at the Planetary Materials Laboratory, Johnson Space Center, Houston, Texas.

Moon race. There really was a Moon Race. The Soviet Union failed in at least four secret attempts to fly a Moon rocket around 1970. It wasn't until 1991 that the USSR gave the West a peek at its massive N1 booster, designed to send cosmonauts to the Moon. The Soviet Moon rocket had four stages with 43 engines, while America's successful Saturn V Moon rocket had three stages with 11 engines. One N1 managed to climb to an altitude of 70,000 feet before failing. The Soviet man-on-the-Moon program then was scrapped.

End of Apollo. There were supposed to be 20 Apollo flights in the lunar program, but the last Moon flight was Apollo 17 in 1972. In 1969, NASA had hoped to build on public enthusiasm for America's success.

A 50-person space station was proposed for completion in Earth orbit by 1980, leading to a flight by astronauts to Mars in 1983. U.S. Vice President Spiro Agnew was the most prominent person to endorse the project, while U.S. President Richard M.

Nixon, as well as most of his administration, Congress and the press, were not interested. NASA's 1971 budget was cut to the lowest level in nine years. Lunar flights Apollo 18 and Apollo 19 were postponed to 1974, and then not flown to the Moon. Apollo 20 was cancelled. The space station was delayed past 1972. Saturn V, the world's most powerful rocket, was terminated.

No more Apollo Moon flights were launched. The last Saturn V rocket carried the Skylab station to low Earth orbit in 1973.

Three Apollo capsules were launched on smaller Saturn 1B rockets in 1973, but only to carry astronauts to the new Skylab space station.

The last Apollo capsule was launched on a Saturn 1B in 1975 in the joint U.S.-USSR Apollo-Soyuz Test Project.

APOLLO 1
Astronauts: Virgil I. "Gus" Grissom, Edward H. White 2nd, Roger B. Chaffee
Date: January 27, 1967
Events: Not a launch, the crew was killed in launch-pad fire during ground test.
CSM capsule location: NASA, Langley Research Center, Hampton, Virginia

APOLLO 4
Rocket: Saturn V   Launch: November 9, 1967
Events: The first unmanned flight of the mighty Saturn V rocket.
CSM capsule location: North Carolina Museum of Life & Science, Durham N.C.

APOLLO 5
Rocket: Saturn 1B   Launch: January 22, 1968
Events: First unmanned test of the lunar excursion module (LEM) in Earth orbit.

APOLLO 6
Rocket: Saturn V   Launch: April 4, 1968
Events: Second unmanned Saturn V launch; final capsule check-out before manned flight.
CSM capsule location: Fernbank Science Center, Atlanta, Georgia

APOLLO 7
Astronauts: cmdr Walter M. "Wally" Schirra Jr., CM pilot Donn F. Eisele,
            LM pilot R. Walter Cunningham
Rocket: Saturn 1B   Launch: October 11, 1968
Orbits of Earth: 163   Splashdown: October 21, 1968.
Duration: 260 hrs 09 mins 03 secs (10 days 20 hours)
Events: Apollo-Saturn transportation system first full space test. First manned Apollo flight, in Earth orbit only, of the Apollo moonflight Command Service Module (CSM). First head colds suffered in orbit. First live TV from U.S. manned spacecraft. Work load friction with Mission Control, Houston. Cunningham had been a civilian test pilot. As CSM had only minor problems, the flight showed it spaceworthy for duration of a lunar mission. Schirra came down with a cold in orbit. Eisele and Cunningham then caught it from him, leading to a decision not to wear helmets during re-entry to keep air pressure on ear drums equalized as cabin pressure changed during descent. Seventeen years, later Schirra appeared on TV advertising a cold remedy.
CSM capsule location: National Museum Science & Technology, Ottawa Canada

APOLLO 8
Astronauts: cmdr Frank Borman, CM pilot James A. Lovell Jr., LM pilot William A. Anders
Rocket: Saturn V   Launch: December 21, 1968
Arrived at Moon: December 24, 1968   Moon orbits: 10 by CSM over 20 hours
Flight duration: 147:00:42 (6 days 3 hrs) Splashdown on Earth: December 27, 1968
Events: Second manned Apollo-Saturn flight. First manned flight around Moon. Six-day space flight was the longest up to the that time. Orbited the Moon ten times, but did not land there. Reconnaissance from lunar orbit for future landing sites. Photos and live TV pictures of Earth and Moon. Anders, Borman and Lovell made a worldwide telecast Christmas Eve 1968 reading passages from the Holy Bible, Book of Genesis, with a stunning view of Earth from 250,000 miles. "Up to that point it was the largest audience that had ever listened to a human voice," Lovell said 20 years later as he and Borman recalled helping the U.S. out of trying times. "The time...was the end of 1968...not a very stellar year...Bobby Kennedy was assassinated, Martin Luther King...the Vietnam war, the Democratic convention in Chicago...riots...it was sort of a down year," Lovell said. Apollo 8 was an "achievement everybody could look up to...you could go outside and see the

Moon and know the United States had finally put a spacecraft around the Moon...it couldn't have happened at a better time," Lovell said. Apollo generated "a nationalistic fervor" that faded as the public became blase about spaceflight, Borman recalled. Back at Earth, Apollo 8 carrying Borman, Lovell and Anders plunged into the atmosphere at 24,696 miles per hour, a speed no human had reached before.

CSM capsule location: Chicago Museum of Science & Technology, Chicago, Illinois

APOLLO 9

Astronauts: cmdr James A. McDivitt, CM pilot David R. Scott, LM pilot Russell L. Schweickart

CSM: Gumdrop   LEM: Spider
Rocket: Saturn V   Launch: March 3, 1969
Orbits of Earth: 151   Splashdown: March 13, 1969
Total flight duration: 241:00:54 (10 days 1 hour)

Events: First manned flight of all lunar hardware, including Lunar Excursion Module (LEM), only in Earth orbit. Tested LEM in Earth orbit, practiced docking with it. Schweickart, who had been a civilian pilot, took a 37 minute spacewalk. The first health delay in American spaceflight history had postponed the blastoff of Apollo 9 four days from February 28 to March 3 when McDivitt, Scott and Schweickart came down with sniffles.

CSM location: Michigan Space Center, Jackson, Michigan.
LEM location: NASA Kennedy Space Center, Cape Canaveral, Florida.

APOLLO 10

Astronauts: cmdr Eugene A. Cernan, CM pilot John W. Young, LM pilot Thomas P. Stafford

CSM: Charlie Brown   LEM: Snoopy
Rocket: Saturn V   Launch: May 18, 1969
LEM flight over Moon: May 22, 1969   CM Moon orbits: 31 during 61 hours 36 mins
Duration: 192:3:23 (8 days 3 mins)   Splashdown on Earth: May 26, 1969

Events: Apollo 10 was a full dress rehearsal for a Moon landing, with the first live color TV from space. Orbiting above the Moon on May 22, Young remained in the CSM as Stafford and Cernan donned spacesuits, climbed into the LEM and separated from the CSM. They flew down to within 8.4 nautical miles (50,000 feet) of the lunar surface, skimmed across the large lunar basin known as the Sea of Tranquility and among the mountain peaks on the Moon, exploring landing places for Apollo 11. Stafford and Cernan jettisoned the LEM descent stage in Moon orbit and flew the LEM ascent stage back up to dock with the CSM. Safely back inside the CSM, the astronauts fired the LEM out of lunar orbit, away from the Moon and Earth, into a path of nearly-endless circles around the Sun. Stafford had said the LEM was named Snoopy because they were going to the Moon to "snoop around." Snoopy is a beagle dog belonging to a boy named Charlie Brown in the Peanuts comic strip. Cartoonist Charles M. Schulz permitted the astronauts to use his characters. NASA employees doing good work have been rewarded with Silver Snoopy lapel pins depicting the dog in a spacesuit.

CSM capsule location: Science Museum, London, England

APOLLO 11

Astronauts: cmdr Neil A. Armstrong, CM pilot Michael Collins,LM pilot Edwin E. "Buzz" Aldrin Jr.

CSM: Columbia   LEM: Eagle
Rocket: Saturn V   Launch: July 16, 1969
Coasting to Moon: July 18   Arrival in lunar orbit: July 19
LEM lands: July 20, 16:17:42 EDT   Man steps on Moon: July 20
LEM on surface: 21 hrs 36 mins 21 secs   Moonwalk: one for 2 hours 31 mins
CSM leaves lunar orbit: July 22   CSM Moon orbits: 30 by CSM (59 hours 30 mins)
Duration: 195:18:35 (8 days 3 hrs 18 m)   Splashdown back on Earth: July 24

Events: First landing of men on Moon July 20. LEM landed at Mare Tranquillitatis (Sea of Tranquility), "Houston, Tranquility Base here. The Eagle has landed." Armstrong first outside, Aldrin second. Armstrong stepped off ladder to say, "That's one small step for (a) man...one giant leap for mankind." NASA later found the "a" lost in transmission. Collins stayed in Moon orbit in CSM. Armstrong, Aldrin set out U.S. flag, science instruments, including laser beam reflector, seismometer which transmitted evidence of a moonquake, and sheet of aluminum foil to trap solar wind particles, and a plaque on the LEM descent stage, "Here Men From Planet Earth First Set Foot Upon the Moon. July 1969 A.D. We Came In Peace For All Mankind." They ate hot dogs, bacon squares, canned peaches, and sugar cookies, and drank hot coffee, on the Moon. They photographed and collected 48.5 lbs. of rocks, soil. Armstrong, Aldrin left on the Moon a gold olive branch, cameras, tools, boots, bags, containers, armrests, brackets, a U.S. flag and flagstaff, a solar wind experiment mast, a TV camera and power cable, a Moonquake detector, a laser light reflector, a plaque, their Eagle lander, life-support backpacks, an Apollo 1 shoulder patch remembering three dead astronauts, medals commemorating two dead cosmonauts, and a 1.5-in.silicon disk with goodwill speeches from heads of 23 Earth nations. They flew back to the CSM, leaving the LEM ascent stage in lunar orbit. Made man's first return from a celestial body. Splashdown in Pacific Ocean. CSM capsule location: Nat'l Air & Space Museum, Smithsonian Inst'n, Washington, D.C.

APOLLO 12
Astronauts: cmdr Chas. P. "Pete" Conrad Jr., CM pilot Richard F. Gordon Jr., LM pilot Alan L. Bean
CSM: Yankee Clipper   LEM: Intrepid
Rocket: Saturn V   Launch: November 14, 1969
LEM lands Moon: November 19, 1969   Moonwalks: 2 totaling 7 hours 50 minutes
LEM on Moon surface: 31 hrs 31 mins CSM Moon orbits: 45, during 89 hours
Duration: 244:36:25 (10 days 4 hr 36 m) Splashdown on Earth: November 24, 1969
Events: Lightning struck Saturn twice at liftoff. Conrad and Bean in LEM made man's second Moon landing, at Oceanus Procellarum (Ocean of Storms), a very large young area of the Moon previously visited by unmanned Luna 9, Luna 13, Surveyor 1 and Surveyor 3. Hiked half mile to retrieve 25 lbs. of parts from Surveyor 3 which had landed in April 1967. Set out Apollo Lunar Surface Experiments Package (ALSEP). Astronauts collected 74.7 lbs. of rocks & soil. Gordon stayed in CSM. LEM Intrepid intentionally crashed into Moon to create first artificial moonquake.
CSM capsule location: NASA, Langley Research Center, Hampton, Virginia

APOLLO 13
Astronauts: cmdr James A. Lovell Jr., CM pilot John L. Swigart Jr., LM pilot Fred W. Haise Jr.
CSM: Odyssey   LEM: Aquarius
Rocket: Saturn V   Launch: April 11, 1970
Explosion: April 13, 1970   Splashdown on Earth: April 17, 1970
Duration: 142:54:41 (5 days 22 hours 54 minutes)
Events: Third lunar landing flight was disrupted two days after launch when SM oxygen tank ruptured, onboard explosion, crippled power and life-support systems, caused flight abort. Radio: "Houston, we have a problem!" A thermostatically-controlled switch had failed and allowed the oxygen tank to overheat. Crew returned to Earth safely using LEM Aquarius as a lifeboat for oxygen and power. They used LEM descent engine to accelerate CSM around the Moon and back to Earth. Back near Earth, crew entered CM to land. NASA said it was a "successful failure" due to crew rescue experience gained. Apollo 13 LEM was to have landed at Moon's Fra Mauro formation. The spent rocket upper stage crashed into the Moon. Before the Apollo 13 flight, CSM pilot Thomas Mattingly had been replaced April 6 by Swigart after Mattingly contacted astronaut Charles Duke who had German measles. Mattingly was judged not immune. Apollo 13 was launched with Swigart, but the flight was aborted. Safe at home, Mattingly did not develop measles. He and Duke flew later in Apollo 16. CSM capsule location: Musee de l'Air, Paris, France

APOLLO 14
Astronauts: cmdr Alan B. Shepard Jr., CM pilot Stuart A. Roosa, LM pilot Edgar D. Mitchell
CSM: Kitty Hawk   LEM: Antares
Rocket: Saturn V   Launch: January 31, 1971
LEM lands on Moon: February 5, 1971 LM on Moon surface: 33 hours 31 minutes
Moonwalks: 2 totaling 9 hours 25 mins CSM Moon orbits: 34 (67 hours)
Duration: 216:01:57 (9 days) Splashdown on Earth: February 9, 1971
Events: Shepard and Mitchell made man's third Moon landing in the uplands at Fra Mauro formation cone crater. They set out ALSEP and instruments, including second laser reflector, and collected 96 lbs. of rock, soil samples. They towed the first two-wheeled lunar "rickshaw" hand cart, containing tools and instruments, to the edge of the crater to transport rocks. In the first golf shot on Moon, Shepard hit balls hundreds of yards with a jury-rigged six-iron. Roosa stayed in lunar orbit in the CSM. First two man-made Moonquakes were set off to be read by seismic monitors planted by earlier Apollo Moonwalkers: the Saturn V rocket third stage was fired into the Moon for seismic readings, releasing the explosive power of 11 tons of TNT, the Moon rang like a bell, vibrating up 3 hours at depths to 25 miles; after Shepard and Mitchell returned to the CSM, the LEM Antares was crashed into Moon, tremors were recorded by seismic stations left by Apollos 12 and 14. NASA's 14-day quarantine at Houston of returned lunar astronauts ended after Apollo 14.
CSM capsule location: Los Angeles County Museum, Los Angeles, California.

APOLLO 15
Astronauts: cmdr David R. Scott, CM pilot Alfred M. Worden, LM pilot James B. Irwin

CSM: Endeavor
Rocket: Saturn V
LEM lands on Moon: July 30, 1971
Moonwalks: 3 totaling 18 hrs 36 mins
Duration: 295:11:53 (12 days17 hrs12 m)

LEM: Falcon
Launch: July 26, 1971
LM on Moon surface: 66 hours 54 minutes
CSM Moon orbits: 74 (145 hours)
Splashdown on Earth: August 7, 1971

Events: Scott and Irwin made man's fourth landing, at Hadley Rille on the Apennine Mountains front. First to use the electric 4-wheel-drive Lunar Roving Vehicle (LRV), driving it 18.6 miles to collect 169 lbs. of samples, take measurements and set out ALSEP. LRV had been carried to the Moon with its wheels folded in the LEM descent stage. Improved spacesuits gave Scott and Irwin increased mobility and let them stay longer. The amount of science gear landed on Moon doubled over previous flights. Worden stayed in CSM. Apollo 15 was first to carry orbital sensors in SM. For first time, a small satellite, to measure gamma rays and X-rays from the lunar surface, was dropped off in lunar orbit, by Worden. The LEM Falcon was sent crashing into Moon for seismic readings. In the first deep-space spacewalk (extra-vehicular activity, EVA), on his way home from the Moon, at a distance of 197,000 miles from Earth, Worden stepped outside Apollo 15 for 38 minutes to retrieve two film cassettes with pictures he had made of the Moon from lunar orbit.
CSM location: USAF Museum, Wright-Patterson Air Force Base, Dayton, Ohio

APOLLO 16

Astronauts: cmdr John W. Young,CM pilot Thomas K. Mattingly 2nd,LM pilot Charles M. Duke Jr.
CSM: Casper
Rocket: Saturn V
LEM lands on Moon: April 21, 1972
Moonwalks: totaling 20 hours 14 mins
Duration: 265:51:05 (11 days 1 hr 51 m)

LEM: Orion
Launch: April 16, 1972
LM on Moon surface: 71 hours 2 minutes
Moon orbits: 64 by the CSM (126 hours)
Splashdown on Earth: April 27, 1972

Events: The last two lunar landings, Apollo 16 and 17, penetrated the lunar highlands. Young and Duke in Apollo 16 made man's fifth landing, at Cayley-Descartes formation in the highlands of the Moon's southern hemisphere for the first study of lunar highlands. They set out science experiments on the lunar surface and collected 213 lbs. of samples. Drove the second electric 4-wheel-drive Lunar Roving Vehicle (LRV) 16.8 miles. For the first time, they used an ultraviolet camera and spectrograph on the Moon. Mattingly in CSM. 90-lb. satellite released into lunar orbit, crashed after 5 weeks. LEM was crashed into Moon. Mattingly took a one-hour spacewalk.
CSM capsule location: NASA, Alabama Space & Rocket Center, Huntsville, Alabama

APOLLO 17

Astronauts: cmdr Eugene A. Cernan, CM pilot Ronald E. Evans, LM pilot Harrison H. Schmitt
CSM: America
Rocket: Saturn V
LEM lands on Moon: December 11, 1972
Moonwalks: 3 totaling 22 hours 4 mins
Duration: 301:51:59 (12 days 13hrs 52m)

LEM: Challenger
Launch: December 7, 1972
LM on Moon surface: 75 hours
Moon orbits: 75 by CSM
Splashdown on Earth: December 19, 1972

Events: For the last manned flight to the Moon, the Saturn V rocket blasted off in the first U.S. manned launch at night. Cernan and Schmitt made man's sixth Moon landing, in the Taurus Mountains, near the Littrow crater, a highlands and valley area on the border of Mare Serenitatis. Evans stayed in in lunar orbit in the CSM. Schmitt, a geologist, was the first scientist to land on the Moon. They collected 243 lbs. of samples and set up the sixth automated research station. They found orange soil near Shorty Crater. Cernan and Schmitt drove the third LRV lunar rover Moon car 22 miles at speeds up to nine miles per hour. Cernan snagged a hammer and ripped the fender on the rover, patched it with plastic Moon maps, then dented the rover's tires while driving over Moon rocks. Cernan was the last man on the Moon as he left December 14. Cernan left behind a plaque on the LEM descent stage, "Here Man completed his first exploration of the Moon December 1972 A.D. May the spirit of peace in which we came be reflected in the lives of all mankind." After Cernan and Schmitt flew it back up to the CSM, the LEM ascent stage was crashed into Moon. Evans took a 1 hour 6 minute spacewalk on the way back to Earth.
CSM capsule location: NASA, Johnson Space Center, Houston, Texas

APOLLO-SOYUZ or APOLLO 18

Apollo astronauts: cmdr Thomas P. Stafford, Vance D. Brand,
     docking pilot Donald K. "Deke" Slayton
Soyuz cosmonauts: cmdr/pilot Alexei Leonov, flight engineer Valeri Kubasov
Capsule: Apollo 18
U.S. rocket: Saturn 1B

Capsule: Soyuz 19
USSR rocket: A-2

U.S. launch: July 15, 1975
Soyuz docks with Apollo: July 17
Apollo splashdown: July 24
U.S. orbits of Earth: 136 orbits
U.S. duration: 217:28:23 (9 days)

USSR launch: July 15, 1975
Soyuz and Apollo separate: July 19
Soyuz landing: July 21
USSR orbits of Earth: 96 orbits
USSR duration: 143:31

Events: A demonstration of U.S.-USSR detente in space. Joint orbital flight of U.S. Apollo and USSR Soyuz manned spacecraft called Apollo-Soyuz Test Project (ASTP). Flight proposed by U.S. in 1969, approved by USSR during the 1972 Nixon-Kosygin summit conference in Moscow. As the first US/USSR joint space mission, it required construction of a new universal docking module. First American visit to the USSR's secret Baikonur Cosmodrome space launch site. Western press went inside the USSR flight-control center at Kaliningrad. Apollo-Soyuz was not a Moon flight, but a show of international cooperation in Earth orbit. They were the first simultaneous American and Russian space flights. Brand was a civilian. Slayton, 51, the first person over 50 in space, was a civilian and one of the original Mercury astronauts. The two-man Soyuz 19 was smaller than the three-man Apollo 18. Leonov was Soyuz commander and pilot and Kubasov was flight engineer. Soyuz 19 was launched from Baikonur at 8:20 a.m. EDT July 15 with Apollo launched later at 3:50 p.m. EDT from Cape Canaveral. With Soyuz as target, flying in tandem at more than 18,000 mph, the Apollo astronauts completed a rendezvous two days later, docking with Soyuz at 12:09 p.m. EDT on July 17 over the Atlantic Ocean 640 miles west of Portugal. Stafford shook hands with Leonov at 3:19 p.m. EDT July 17 in the connecting airlock. Millions around the globe watched via TV as U.S. President Gerald Ford and USSR Communist Party Secretary Leonid Brezhnev congratulated the spacemen. The project demonstrated crew transfer between spacecraft and space rescue in low Earth orbit. The five men transferred between spacecraft four times during two days, completed five experiments, shared meals, held a news conference. After separating July 19, the Soyuz carrying Leonov and Kubasov landed in the USSR's Kazakhstan Republic July 21. Apollo splashed down in the Pacific Ocean 270 miles west of Hawaii on July 24. The Apollo crew suffered discomfort from inhaling nitrogen tetroxide fumes entering the cabin after a landing-procedure error. It was the last Apollo mission. When the five Apollo-Soyuz crew members reunited 15 years later in July 1990 at Johnson Space Center, Houston, Texas and at Kennedy Space Center, Florida, Stafford said, "We'd like to go back and do it again." Leonov, who had been the first person to walk in space and then the USSR's chief of cosmonaut training, said he hoped for a joint Mars mission. Slayton called a joint Mars flight "inevitable...we're probably 15 or 20 years...it will be a major international venture." The two nations agreed in 1991 to joint flights in the 1990's with a cosmonaut flying in a U.S. shuttle and an astronaut flying in a Soyuz capsule to the Mir space station. Brand said the Apollo-Soyuz joint mission was a "crack in the door through the Iron Curtain which had been closed for so long."
Apollo-Soyuz CSM location: Museum of Science, Seattle, Washington

Skylab Space Station

The U.S. space agency started talking in 1965 about a space station in earth orbit to follow the Apollo man-on-the-Moon project. A working space station would test man's ability to survive extended periods in space in preparation for very long multi-year trips beyond the Moon to Mars.

NASA received large Moon Race budgets through the 1960's and couldn't foresee the financial freefall ahead in the 1970's. Planners imagined a space station in Earth orbit by the mid-1970's, serviced by an inexpensive, reusable space shuttle to ferry men and materials to orbit. The space station would be a convenient waystation enroute to colonizing the Moon by the 1980's and exploring Mars by the 1990's.

In 1969, NASA sought to build on public enthusiasm for America's Moon landing success. A 50-person space station was proposed for completion in Earth orbit by 1980, leading to a flight by astronauts to Mars in 1983.

But the well dried up. There were to be 20 Apollo flights in the lunar program, but suddenly the last Moon flight was Apollo 17 in 1972. While U.S. Vice President Spiro Agnew was the most prominent person to endorse NASA's plan, U.S. President Richard M. Nixon, most of his administration, Congress and the press, were not interested.

NASA's 1971 budget was the lowest in nine years. Apollo 18, 19 and 20 did not fly to the Moon. The space station was delayed past 1972. Saturn V was terminated and the last Moon rocket only carried the Skylab station to low Earth orbit in 1973.

Three Apollo capsules were launched on smaller Saturn 1B rockets in 1973, but only to carry astronauts to the new Skylab space station. The last Apollo capsule was launched on a Saturn 1B in 1975 in the joint U.S.-USSR Apollo-Soyuz Test Project.

Skylab station. The space station project, called Apollo Applications Program (AAP) for a time, was renamed Skylab. It was built from modified Apollo equipment and the empty third stage of a Saturn V Moon rocket. The rocket stage was dry of fuel and converted into two stories, living quarters in one, lab in the other.

The station was launched to orbit on the last Saturn V Moon rocket. Over nine months after Skylab was sent to orbit, three teams of astronauts flew to the station in Apollo command modules (CM) launched on smaller Saturn 1B rockets. America could afford only to have three teams of astronauts visit Skylab.

The entire station, including docked CM, was nearly 120 feet long and weighed 84 tons, triple the USSR's Salyut space station. Skylab's primary living and working area, known as the orbital workshop, was 48 feet long and 22 feet in diameter with a volume of about 10,000 cubic feet. That was 150 times the volume of a Gemini capsule, 60 times an Apollo capsule. The other main station components were the Apollo command module used to fly crews to and from Skylab, a multiple docking adaptor, an airlock module and shroud for spacewalking, and an Apollo telescope mount for observing the Sun.

Food and games. The space station had a refrigerator and cook stove so the astronauts could have prime rib, German potato salad made with onions and vinegar, hot chili, scrambled eggs, liquid pepper for spice and ice cream.

Skylab offered the first shower in orbit. It leaked, causing astronauts to waste time cleaning up. The station also had the first private toilet. A funnel collected urine and forced air blew it into a bag. A toilet seat on a wall with a seat belt was the target for solid human waste, which was bagged, dried and carried down to Earth.

Skylab astronauts thought up a fun game to play in zero gravity. They called it Catch the Peanut. A dry-roasted peanut would escape its tin from time to time, floating away in the 84-ft. station. When a hapless peanut would wander by, an astronaut would open his mouth and push off the nearest wall toward it. If he didn't aim just right, the peanut would bounce off his face, necessitating a chase. The astronauts also had a Velcro dartboard, which wasn't much fun as darts wobbled across space to the target. Three balls included in recreation gear seemed never to stop bouncing, often becoming lost in the station.

Space station launch. America's first and only space station left Cape Canaveral May 14, 1973, atop a column of fire pouring from the last of the mammoth Saturn V Moon rockets, bound for orbit 275 miles above Earth.

Despite extraordinary launch vibrations which shook loose a micrometeorite shield, tearing off one solar wing and jamming the other, the house-trailer-sized station reached orbit and was visited successfully by three sets of American astronauts later that year.

The micrometeorite shield was supposed to shade the cabin-in-the-sky from the Sun's heat. Breaking the shield threatened to scuttle the project as inside temperatures reached 190 degrees. But, NASA ground controllers moved the giant satellite so the temperature inside settled down at 110 degrees. And they did it without pointing the operating solar panel away from the Sun which would have cut electrical power to the space station.

Skylab astronauts: Charles P. Conrad Jr., Joseph P. Kerwin, Paul J. Weitz. Alan L. Bean, Owen K. Garriott, Jack R. Lousma, Gerald P. Carr, Edward G. Gibson, and William R. Pogue.

Astronauts at Skylab. There were three more Skylab launches in 1973, to ferry astronaut crews to work in the station. The astronauts flew in Apollo capsules on smaller Saturn 1B rockets.

★Skylab 2 May 25, 1973, 11 days after the station launch, the first team of astronauts sent to man the space station, opened the station, made repairs, tour by Kerwin, Conrad, Weitz lasted 28 days.

★Skylab 3 July 28, 1973, the second crew fixed broken equipment, tour by Bean, Garriott, Lousma lasted 59 days.

★Skylab 4 November 16, 1973, the third team of astronauts, tour by Carr, Gibson, Pogue lasted 84 days. They closed the station permanently February 8, 1974.

The total time spent by nine astronauts at Skylab: 171 days 13 hours. The astronauts brought back from Skylab important information in astrophysics, biomedicine, engineering, materials processing, solar physics, and other technology.

Spacewalks. Altogether, six Skylab astronauts completed 10 spacewalks (extra-vehicular activities or EVAs) totaling 41 hours 56 minutes.

Skylab 2's Conrad and Kerwin did three. First was a stand-up-in-the-hatch exposure to the space environment for 37 minutes May 25. The second was a 3 hour 30 minute spacewalk June 7. The third was a 1 hour 44 minute spacewalk June 19. Total EVA time was 5 hours 51 minutes.

Skylab 3's Garriott and Lousma did three. First was a 6 hour 29 minute spacewalk August 6. Second was a 4 hour 30 minute spacewalk August 24. Third was a 2 hour 45 minute spacewalk September 22. Total EVA time was 13 hours 44 minutes.

Skylab 4's Pogue and Gibson did four. First was 6 hours 33 minutes on November 22. The second was a record 7 hours 1 minute December 25. Third was 3 hours 28 minutes December 29. Fourth was 5 hours 19 minutes February 8. Total EVA time was 22 hours 21 minutes.

A gag in the sky. During a mission, a Skylab crew would talk continuously with, and work closely with, ground controllers. Problems were discussed in detail. Solutions resulted from agreements between engineers, scientists and technicians on the ground, and the Skylab crew. But ground crews were not prepared when Skylab 3 astronaut Owen Garriott played a joke on NASA directors in Houston mission control.

A soft female voice called over the radio from space, "Hello Houston, this is Skylab, are you reading me down there?" There was a long silence in mission control.

"Hello, Houston? Are you reading Skylab?" the voice called again.

After a pause, the chief communicator at Houston replied, "Skylab, this is Houston. I heard you all right, but I had a little difficulty recognizing your voice. Who do we have on the line here?"

"Houston, Roger. I haven't talked with you for awhile. Is that you down there, Bob? This is Helen here in Skylab. The boys hadn't had a home-cooked meal in so long, I thought I'd just bring one up. Over."

A crowd gathered in mission control. Not quite sure what was going on, the controller said, "Roger, Skylab. I think somebody has got to be pulling my leg. Helen, is that really you? Where are you?"

"Just a few orbits ago, we were looking down on the forest fires in California. You know, the smoke sure does cover a lot of territory. And, oh, Bob, the sunrises are just beautiful."

Suddenly, the female voice changed moods, "Oh, oh, I have to cut off now," she said. "I see the boys are floating up toward the command module, and I'm not supposed to be talking to you. See you later, Bob."

Still shaken by the event, the controller muttered, "Bye bye," as the Skylab crew roared with laughter. Garriott had taped his wife's voice before flying to space.

Out of orbit. Skylab was used by three teams of astronauts for a total of only nine months, to February 1974, then the station went unused until 1979 when it was allowed to drop into Earth's atmosphere to burn.

Despite 171 days of crew experience aboard Skylab, NASA, for political reasons, redirected its energy away from the space station. Instead, money was spent on building and flying the large reusable space shuttle. After five years of no visits, technical problems caused the orbit of Skylab to deteriorate. NASA made a weak attempt to keep it in space by remote control from the ground, but no astronauts were sent up for repairs. NASA's hopes failed July 11, 1979, when Skylab re-entered the atmosphere and burned.

On to Mars. The nine Skylab astronauts were reunited in 1988, on the 15th anniversary of the launch, at the Wernher von Braun Civic Center, Hunstville, Alabama, with engineers who worked on Skylab. Skylab 3 commander Bean said, "Space is our frontier, and beginning its exploration may be our generation's greatest contribution to human history."

Skylab 2 pilot Kerwin reported when "someone asked if we would volunteer to go to Mars...hands shot up as rapidly as possible." Skylab 4 commander Carr added, "We would do it in a minute."

SKYLAB 1
Rocket: Saturn V   Launch: May 14, 1973
Events: Unmanned launch of the U.S. Skylab to low Earth orbit. Space station was 84.2 feet long, weighed 84 tons, four times the size of the USSR's Salyut space station. Skylab re-entered the atmosphere July 11, 1979, and burned.

SKYLAB 2
Astronauts: cmdr Charles P. Conrad Jr., scientist-pilot Joseph P. Kerwin, pilot Paul J. Weitz
Rocket: Saturn 1-B   Launch: May 25, 1973
Spacewalks: 3, totaling 5 hrs 51 mins   Crew landed: June 22, 1973
Duration: 28 days 49 min (672 hrs 49 mins 49 secs)
Event: First manned flight in Apollo capsule to space station. Astronauts had to overcome problems which jeopardized the Skylab project. During launch to orbit, a meteorite shield and heat shield were ripped off the station, striking solar panels so they would not unfold correctly. Controllers changed the orbit and built a shield the first crew members took to space 11 days later. Conrad, Weitz and Kerwin, first to visit Skylab, opened the station and made repairs. Conrad, Kerwin did three EVAs. First a stand-up-in-the-hatch exposure for 37 mins May 25. Second a 3 hr 30 min spacewalk June 7. Third a 1 hr 44 min spacewalk June 19. Total EVA time was 5 hrs 51 mins. They salvaged the mission with a spacewalk to adjust the solar panels and erect a makeshift parasol Sun shade, dropping the inside temperature to a comfortable shirt-sleeve level. Set a spacewalk (extra-vehicular activity or EVA) duration record. First pictures of solar flares shot above the atmosphere. Flew down to Earth June 22. CSM location: Naval Aviation Museum, Pensacola, Fl.

SKYLAB 3
Astronauts: cmdr Alan L. Bean, scientist-pilot Owen K. Garriott, pilot Jack R. Lousma
Rocket: Saturn 1-B   Launch: July 28, 1973
Spacewalk total: 13 hrs 44 mins   Crew landed: September 25, 1973
Duration: 59 days 11 hrs (1427 hrs 9 mins 4 secs)
Events: Second manned flight in Apollo capsule to station. After docking at station, 2 of 4 capsule thrusters leaked, had to be shut down. NASA considered aborting, began planning rescue, but decided breakdowns had known causes and Apollo could fly on 2 thrusters. Astronauts were allowed to go ahead. Garriott, Lousma did 3 EVAs. First was 6 hr 29 min spacewalk Aug 6. Second was 4 hr 30 min spacewalk Aug 24. Third was 2 hr 45 min spacewalk Sept 22. Total EVA time was 13 hrs 44 mins. They placed a new Sun shield over the Skylab 2 makeshift parasol, replaced film in cameras, installed panels to measure micrometeorites. Skylab was moved so sensors pointed to most of the U.S. and parts of 33 countries. Ecological, agricultural, forestry, mapping, geological, water resource, fishing, oceanography, mineral prospecting, and meteorological data collected. Sun was very active so more solar observations were made. Astronauts had taken along a common Cross spider known as Arabella to see if a web could be spun in no-gravity "zero-G" environment. The spider was able to spin her web. CSM location: NASA, Ames Center, Mountain View, California

SKYLAB 4
Astronauts: cmdr Gerald P. Carr, scientist-pilot Edward G. Gibson, pilot William R. Pogue
Rocket: Saturn 1-B   Launch: November 16, 1973
Spacewalk total: 22 hrs 21 mins   Single spacewalk record: 7 hrs 1 min
Crew landed: February 8, 1974   Duration: 84 days 1 hr (2017 hrs 16 mins 30 secs)
Events: Third manned flight in Apollo capsule to the space station. Record U.S. man-in-space duration of 84 days. The astronauts had problems, at first, getting experiments started. Morale

suffered as they fell behind on work. The schedule was shuffled for more time off. Things improved. A malfunctioning radio antenna was repaired during a Thanksgiving Day spacewalk. Other equipment was repaired. But, then, the star tracker failed completely and had to be shut down. The crew spotted and photographed Comet Kohoutek. A solar flare was photographed for the first time from space. By coincidence, in space at the same time using an ultraviolet camera to photograph solar X-rays and stars were USSR Soyuz 13 cosmonauts Pyotr Klimuk, Valentin Lebedev, for a total of five persons in orbit at one time. Gibson, Pogue did four EVAs. First 6 hours 33 minutes Nov 22. Second 7 hrs 1 min Dec 25. Third 3 hours 28 minutes Dec 29. Fourth 5 hours 19 minutes Feb 8. Total EVA time 22 hours 21 minutes. When Carr, Gibson, Pogue flew home Feb. 8, no one visited the space station again. It fell from orbit July 11, 1979.
CSM location: National Air & Space Museum, Smithsonian Institution, Washington, D.C.

U.S. Space Shuttles

A U.S. space shuttle is a winged spaceplane with wheeled landing gear and large cargo-carrying capacity. It is launched to Earth orbit by rocket, then leaves orbit to glide to a runway landing. The first shuttle spaceflight was in 1981.

First shuttle. Prototype shuttle Enterprise was called OV-101 when its construction started in 1974. After 100,000 Star Trek TV fans wrote in, U.S. President Gerald Ford named it Enterprise in 1976. NASA didn't like the name, wanting to call the shuttle Constitution.

Glide tests. The test-bed space shuttle Enterprise never went to space. It had no engines. The 130-ton Enterprise only flew bolted to the back of a Boeing 747 jumbo jet. The airplane dropped Enterprise to test its gliding ability.

For the first flight August 12, 1977, astronauts Fred Haise Jr and Charles Fullerton were at the controls. Five test flights were made over two years. Haise and Fullerton piloted flights one, three and five. Astronauts Joe Engle and Richard Truly piloted flights two and four. Engle and Truly later flew shuttle Columbia on the second actual spaceflight.

Publicity tour. Enterprise also was used to get launch pads ready for real shuttles. Then the vehicle was flown city to city, piggyback on its 747 jumbo jet, as a NASA publicity stunt—a big hit with the public. Enterprise was to have gone back to the workshop for rocket engines, but it became cheaper to make a new ready-to-fly structure-test orbiter.

Enterprise was stripped and displayed at Kennedy Space Center, then was flown to Washington, D.C., for the Smithsonian Institution to build a museum around it. The museum has not been built and, today, Enterprise sits among weeds in the outback of Washington's Dulles International Airport, paint peeling, out of sight, out of mind.

Shuttle names. The structure-test orbiter, OV-099, became Challenger, the shuttle which exploded in 1986 after nine successful spaceflights. Orbiter OV-101, to be named Columbia, became the first shuttle to fly in space. Orbiter OV-102 became Discovery and OV-103 became Atlantis. Orbiter OV-105, built to replace Challenger, was named Endeavour.

Launches and landings. All shuttle spaceflights have been launched from Kennedy Space Center, Florida. Most landings have been at Edwards Air Force Base, California, except for the few shown in the list below as landing elsewhere.

Shuttle astronauts. A shuttle crew may include a mission commander, pilot, payload commander, mission specialists, payload specialists and space flight participants. A shuttle payload commander is the lead mission specialist who plans and coordinates shuttle cargo while a mission commander retains overall responsibility for mission success and flight safety. Payload commanders are forerunners of space station mission commanders.

To most people, an astronaut is any man or woman who has flown, or is preparing to fly, in a space shuttle or other spacecraft. NASA, however, refers to payload specialists and space flight participants as members of a shuttle flight crew, not as astronauts.

Food and facilities. Shuttle astronauts eat well, such as smoked turkey, cream of mushroom soup, mixed Italian veggies, vanilla pudding, and freeze-dried strawberries moistened in the mouth.

They chew gum and snack on almond crunch bars, graham crackers, pecan cookies, nuts, and Life Savers. Mostly they drink orange, lemon, orange-grapefruit, orange-pineapple, strawberry and apple drinks and tropical punch.

Space shuttle astronauts share one toilet bowl which can't be flushed. As Skylab space station astronauts did in the 1970's, shuttle astronauts still have to use seat belts to hold themselves on the toilet seat, but their waste matter is dumped overboard. They also have hand and toe holds to keep from floating off their unisex potty. A porthole for a view of Earth is at hand. Solid waste is shredded by something called the slinger, then dried, disinfected, and dumped.

Shuttle astronauts use toothbrush, toothpaste, comb, hairbrush, razor, shaving cream and nail clippers. The clippers don't see much use as fingernails and toenails grow sparingly in weightlessness.

Pre-flight quarantine. To prevent illness in space, all but family and those absolutely necessary are required by NASA to stay six feet or more from crew members for seven days before a shuttle liftoff. Others also wear masks around crew members while NASA has isolation warnings posted.

From the earliest Mercury manned space flights, NASA had trained back up flight crews for each mission to take over in case of illness. NASA stopped training back-up crews in 1982. As the shuttle program speeded up, trained astronauts were on hand for last-minute crew changes.

Flight numbers. In the chronological list of American manned shuttle flights below, STS stands for Space Transportation System.

After STS-9, NASA changed to a numbering scheme indicating flight, year and launch site. For example, STS-41G, the seventh flight planned for the 1984 fiscal year, was to be launched from Cape Canaveral. NASA's fiscal year starts in October so 1985 started in October 1984. G is the seventh letter. The numeral 4 is for 1984. The number 1 indicates Cape Canaveral. Thus, 41G. A number 2 would have indicated Vandenberg Air Force Base, California, but no shuttles were launched from Vandenberg. Some missions, such as STS-41E, were cancelled.

After the Challenger disaster, the 25th flight, numbers were simplified to numerical order. For instance, STS-26, STS-27, STS-28, etc. NASA went back and added the optional numbers STS-10 through STS-25 for flights STS-41B through STS-51L.

Even with a simplified numbering system, the numerical order is disrupted when flights are listed in chronological order. For instance, flights STS-29 and STS-30 were made before STS-28. STS-34 flew before STS-33 which was before STS-32. This is due to assignment of flight numbers 19 months before launch, followed by technical delays which force NASA to juggle the order of flights. Below, tba indicates to be announced.

STS-1
Shuttle: Columbia
Crew: commander John W. Young, pilot Robert L. Crippen (2 persons in the crew)
Launch: April 12, 1981   Duration: 2 days (54:20:52)
Altitude: 172 nautical miles   Inclination: 40.3 degrees
Events: The first manned test flight of a space shuttle. After 37 revolutions around the globe, Columbia became the first manned spacecraft to touchdown on a runway like an airplane. Landed at Edwards Air Force Base, California. In the first decade following STS-1, U.S. shuttle flights in Earth orbit travelled farther than the equivalent 94 million distance from Earth to Sun, flying 123 different astronauts and delivering more than a million pounds of payload to orbit.
STS-2
Shuttle: Columbia
Crew: commander Joe H. Engle, pilot Richard H. Truly (2)

Launch: November 12, 1981   Duration: 2 days (54:13:12)
Altitude: 140 nautical miles   Inclination: 38.0 degrees
Events: The robot arm was tested. Flight cut to 2 days when a fuel cell failed. First time a spacecraft had been reflown with a second crew.

STS-3
Shuttle: Columbia
Crew: commander Jack R. Lousma, pilot Charles Gordon Fullerton (2)
Launch: March 22, 1982   Duration: 8 days (192:05:01)
Altitude: 130 nautical miles   Inclination: 38.0 degrees
Landing site: California weather forced White Sands, New Mexico, landing.
Events: Third flight. Tested remote-manipulator robot arm. Tested thermal tiles by turning orbiter tail, top and bottom toward the Sun. Toilet failed.

STS-4
Shuttle: Columbia
Crew: cmdr Thomas K. Mattingly 2nd, pilot Henry W. Hartsfield Jr. (2)
Launch: June 27, 1982   Duration: 7 days (169:09:40)
Altitude: 162 nautical miles   Inclination: 28.5 degrees
Events: Last test flight. First time a shuttle was launched precisely on scheduled time. Two boosters lost. First classified military payload.

STS-5
Shuttle: Columbia
Crew: commander Vance D. Brand, pilot Robert F. Overmyer,
     mission specialists Joseph P. Allen, William B. Lenoir (4)
Launch: November 11, 1982   Duration: 5 days (122:14:34)
Altitude: 160 nautical miles   Inclination: 28.5 degrees
Events: First operational flight. Largest crew to that time including first mission specialists, Allen and Lenoir. Ferried Anik and SBS commercial communications satellites to orbit. A spacewalk planned for Allen and Lenoir would have been the first from a shuttle, but was cancelled when neither spacesuit worked.

STS-6
Shuttle: Challenger
Crew: commander Paul J. Weitz, pilot Karol J. "Bo" Bobko,
     mission specialists Donald H. Peterson, F. S. "Story" Musgrave (4)
Launch: April 4, 1983   Duration: 5 days (120:23:42)
Altitude: 150 nautical miles   Inclination: 28.5 degrees
Events: Maiden flight of Challenger carried NASA's TDRS-A communications satellite to orbit. Peterson and Musgrave donned new spacesuits to take the first shuttle spacewalks. They went outside into the open cargo hold for 4 hours 17 minutes.

STS-7
Shuttle: Challenger
Crew: commander Robert L. Crippen, pilot Frederick H. "Rick" Hauck,
     mission specialists John W. Fabian, Sally K. Ride, Norman E. Thagard (5)
Launch: June 18, 1983   Duration: 6 days (146:24:20)
Altitude: 160 nautical miles   Inclination: 28.5 degrees
Events: Sally Ride was first and youngest U.S. woman in space, age 32. Carried Anik, Palapa sats to orbit. Released and recovered pallet satellite with robot arm. First American mixed male/female crew in space. STS-7 was first to carry the German-designed payload carrier SPAS (Shuttle Pallet Satellite) a small satellite, outfitted with science instruments, carried into space by shuttle, and deployed for several days in its own independent orbit to gather scientific data and perform technology demonstrations. Before landing, the shuttle returned to SPAS, retrieved it, and returned it to Earth to be available for another mission.

STS-8
Shuttle: Challenger
Crew: commander Richard H. Truly, pilot Daniel C. Brandenstein,
     mission specialists Dale A. Gardner, Guion S. "Guy" Bluford Jr., William E. Thorton (5)
Launch: August 30, 1983   Duration: 6 days (145:08:41)
Altitude: 160 nautical miles   Inclination: 28.5 degrees
Events: NASA's first night shuttle launch and night landing. Challenger ferried India's Indian National Satellite (Insat-1B) to orbit. Bluford was the first black man in space.

STS-9 or STS-41A
Shuttle: Columbia
Crew: commander John W. Young, pilot Brewster H. Shaw Jr., mission specialists Owen K.
   Garriott, A. Robert Parker, payload specialists Ulf D. Merbold, Byron K. Lichtenberg (6)
Launch: November 28, 1983   Duration: 10 days (247:47:41)
Altitude: 135 nautical miles   Inclination: 57.0 degrees
Events: Carried Spacelab-1, a removeable pressurized science laboratory in the cargo hold.
Astronauts floated through a short connecting tube from Columbia's crew quarters into Spacelab.
Merbold from West Germany, first non-American in a U.S. spacecraft. He and Lichtenberg, first
payload specialists. First ham shack in space: Garriott made amateur radio voice contacts, first ham
operator to chat from space with friends on the ground. His callsign: W5LFL. Young had the widest
experience with six flights on four different kinds of spacecraft: Gemini 3 in 1965, Gemini 10 in
1966, Apollo 10 in 1969, Apollo 16 to land on the Moon in 1972, maiden voyage of shuttle
Columbia in 1981, and this 1983 shuttle flight.

STS-10 or STS-41B
Shuttle: Challenger
Crew: commander Vance D. Brand, pilot Robert L. "Hoot" Gibson,
   mission specialists Ronald E. McNair, Robert L. Stewart, Bruce McCandless II (5)
Launch: February 3, 1984   Duration: 8 days (191:15:55)
Altitude: 165 nautical miles   Inclination: 28.5 degrees
Landing Site: Kennedy Space Center, Florida
Events: Carried Indonesia's Palapa-B2 and Western Union's Westar satellites. Both failed after
release, were stranded in a low orbit, and were recovered in flight STS-51A Nov 1984. First use of
Manned Maneuvering Unit (MMU) backpack for untethered spacewalk let McCandless and Stewart
fly free as the first human satellites. McCandless flew 320 feet from Challenger. STS-41B saw the
second use of the German-designed payload carrier SPAS-1A (Shuttle Pallet Satellite), first used on
flight STS-7, a small satellite, outfitted with science instruments, carried into space by shuttle, and
deployed for several days in its own independent orbit to gather data and perform technology
demonstrations. Before landing, the shuttle returned to SPAS, retrieved it, and returned it to Earth
for another mission. Challenger made the first shuttle landing in Florida.

STS-11 or STS-41C
Shuttle: Challenger
Crew: commander Robert L. Crippen, pilot Francis R. "Dick" Scobee,
   mission specialists Terry J. Hart, George D. "Pinky" Nelson, James D. "Ox" Van Hoften (5)
Launch: April 6, 1984   Duration: 7 days (167:40:05)
Altitude: 250 nautical miles   Inclination: 28.5 degrees
Events: Dropped off Long Duration Exposure Facility (LDEF) satellite with 57 experiments in orbit.
Rescued Solar Max astronomy satellite. Nelson and Van Hoften made the first satellite repair in
shuttle cargo bay and released Solar Max back to orbit. By coincidence, 11 persons, including 5 in
Challenger and 6 at the USSR's Salyut 7 space station, were most people ever in space at one time.

STS-12 or STS-41D
Shuttle: Discovery
Crew: cmdr Henry W. Hartsfield Jr., pilot Michael L. Coats, mission specialists Richard M. "Mike"
   Mullane, Steven A. Hawley, Judith A. Resnick, payload specialist Charles D. Walker (6)
Launch: August 30, 1984   Duration: 6 days (144:56:16)
Altitude: 160 nautical miles   Inclination: 28.5 degrees
Events: Maiden flight of Discovery carried Syncom, Telstar and SBS sats. Walker was the first
paying spaceflight passenger, the first commercial payload specialist and the first American private
industry representative in space. The manufacturer McDonnell Douglas purchased Walker's ticket so
he could do drug research. Resnick was second American woman in orbit.

STS-13 or STS-41G
Shuttle: Challenger
Crew: cmdr Robert L. Crippen, pilot Jon A. McBride, mission specialists Sally K. Ride, Kathryn D.
   Sullivan, David C. Leetsma, payload specialists Marc Garneau, Paul D. Scully-Power (7)
Launch: October 5, 1984   Duration: 8 days (197:23:47)
Altitude: 190 nautical miles   Inclination: 57.0 degrees
Landing Site: Kennedy Space Center, Florida
Events: Ferried Earth Radiation Budget Satellite (ERBS) to orbit. It was the first time two women
were in space at the same time. Sullivan made the first U.S. female spacewalk (EVA) as she and

Leetsma spent 3 hours outside. Garneau was first Canadian in space. Scully-Power was the first oceanographer to observe the oceans from space. Second shuttle landing in Florida.

STS-14 or STS-51A
Shuttle: Discovery
Crew: commander Frederick H. "Rick" Hauck, pilot David M. Walker,
      mission specialists Joseph P. Allen, Anna L. Fisher, Dale A. Gardner (5)
Launch: November 8, 1984   Duration: 8 days (191:45:08)
Altitude: 160 nautical miles   Inclination: 28.5 degrees
Landing Site: Kennedy Space Center, Florida
Events: Deployed Leasat and Anik communications satellites, then Allen and Gardner made the first-ever satellite rescue, of Westar and Palapa which had failed after flight STS-41B in Feb 1984. The satellites were returned to Earth in the shuttle. Anna L. Fisher, M.D., was the first mother to go to space. It was the third shuttle landing in Florida.

STS-15 or STS-51C
Shuttle: Discovery
Crew: commander Thomas K. Mattingly 2nd, pilot Loren J. Shriver,
      mission specialists Ellison S. Onizuka, James F. Buchli, payload specialist Gary E. Payton (5)
Launch: January 24, 1985   Duration: 3 days (73:23:26)
Altitude: secret   Inclination: secret
Landing Site: Kennedy Space Center, Florida
Events: First secret Department of Defense mission, carried a military spy satellite. Challenger had been scheduled for this flight; Discovery was substituted after thermal tile problems were found. Fourth landing in Florida. First of 10 shuttle launches in 1985.

STS-16 or STS-51D
Shuttle: Discovery
Crew: commander Karol J. "Bo" Bobko, pilot Donald E. Williams,
      mission specialists Jeffrey A. Hoffman, S. David Griggs, Margaret Rhea Seddon,
      payload specialists Charles D. Walker, U.S. Sen. Edwin "Jake" Garn (7)
Launch: April 12, 1985   Duration: 7 days (167:55:31)
Altitude: 250 nautical miles   Inclination: 28.5 degrees
Landing Site: Kennedy Space Center, Florida
Events: Garn was the first politician, the first elected official and the first U.S. senator to fly to space. Hoffman & Griggs made unplanned spacewalk, using jury-rigged flyswatter devices in attempting to fix their Syncom satellite which had failed. It didn't work, the satellite was left in orbit to be repaired by flight STS-51I Aug 1985. Fifth Florida landing. Brakes locked up during landing in a stiff crosswind as Discovery rolled to a stop, blowing a right main landing gear tire.

STS-17 or STS-51B
Shuttle: Challenger
Crew: commander Robert F. Overmyer, pilot Frederick D. Gregory,
      mission specialists Don L. Lind, Norman E. Thagard, William E. Thorton,
      payload specialists Lodewijk van den Berg, Taylor G. Wang (7)
Launch: April 29, 1985   Duration: 7 days (168:07:55)
Altitude: 190 nautical miles   Inclination: 57.0 degrees
Events: Spacelab-3 in cargo bay for materials processing and life-sciences experiments. Lodewijk van den Berg, of European Space Agency, was first Dutch astronaut.

STS-18 or STS-51G
Shuttle: Discovery
Crew: commander Daniel C. Brandenstein, pilot John O. "J.O." Creighton,
      mission specialists Steven R. Nagel, John W. Fabian, Shannon W. Lucid,
      payload specialists Patrick Baudry, Prince Sultan Salman Abdel Aziz Al-Saud (7)
Launch: June 17, 1985   Duration: 7 days (169:39:05)
Altitude: 190 nautical miles   Inclination: 28.5 degrees
Events: Carried 3 communications sats. Prince Salman Al Saud, first Arab in space, first royalty in orbit. Patrick Baudry of French space agency CNES made France the first nation with both cosmonauts and astronauts; French cosmonaut Jean-Loup Chretien had flown in 1982 to USSR Salyut 7 station. French astronaut Patrick Baudry treated his fellow Discovery crew members to gourmet French dining. Spartan, the Shuttle Pointed Autonomous Research Tool for Astronomy (Sptn-1), was carried aboard STS-51G. Sptn is a small, retrievable, free-flying satellite for X-ray astronomy.

STS-19 or STS-51F
Shuttle: Challenger
Crew: commander Charles Gordon Fullerton, pilot Roy D. Bridges Jr.,
    mission specialists F. S. "Story" Musgrave, Anthony W. "Tony" England, Karl G. Henize,
    payload specialists Loren W. Acton, John-David F. Bartoe (7)
Launch: July 29, 1985   Duration: 8 days (190:45:35)
Altitude: 174 nautical miles   Inclination: 50.0 degrees
Events: First engine failure during flight to space. Spacelab-2 in cargo bay for astronomy. Tony England made amateur radio contacts with hams; callsign WØORE. Henize was the oldest person up to that time to have flown in space, age 58. Astronauts quenched their thirsts by drinking the first Coke and Pepsi in orbit from special cans.

STS-20 or STS-51I
Shuttle: Discovery
Crew: commander Joe H. Engle, pilot Richard O. Covey,
    mission specialists James D. "Ox" Van Hoften, John M. "Mike" Lounge, William F. Fisher (5)
Launch: August 27, 1985   Duration: 7 days (170:17:51)
Altitude: 190 nautical miles   Inclination: 28.5 degrees
Events: Carried 3 communications sats. Fisher & Van Hoften spacewalked to repair Syncom broken in orbit after flight STS-51D in April.

STS-21 or STS-51J
Shuttle: Atlantis
Crew: commander Karol J. "Bo" Bobko, pilot Ronald J. Grabe,
    mission specialists Robert L. Stewart, David C. Hilmers, William A. Pailes (5)
Launch: October 3, 1985   Duration: 4 days (97:45:16)
Altitude: secret   Inclination: secret
Events: Atlantis maiden flight carried 2 military satellites.

STS-22 or STS-61A
Shuttle: Challenger
Crew: commander Henry W. Hartsfield Jr., pilot Steven R. Nagel,
    mission specialists James F. Buchli, Guion S. "Guy" Bluford Jr., Bonnie J. Dunbar,
    payload specialists Reinhard Furrer, Ernst Messerschmid, Wubbo J. Ockels (8)
Launch: October 30, 1985   Duration: 7 days (168:45:00)
Altitude: 175 nautical miles   Inclination: 57.0 degrees
Events: Record 8-person crew. Pressurized Spacelab-D1 for German materials processing experiments. Furrer, Messerschmid from West Germany. Ockels from Holland. Houston shared control with West German Space Operations Center at Oberpfaffenhofen near Munich.

STS-23 or STS-61B
Shuttle: Atlantis
Crew: commander Brewster H. Shaw Jr., pilot Bryan D. O'Connor,
    mission specialists Mary L. Cleave, Sherwood C. Spring, Jerry L. Ross,
    payload specialists Charles D. Walker, Rodolfo Neri Vela (7)
Launch: November 26, 1985   Duration: 7 days (165:05:00)
Altitude: 190 nautical miles   Inclination: 28.5 degrees
Events: Second night launch, after Challenger in August 1983, carried 3 communications satellites. Neri, first Mexican in space. Ross and Spring spacewalked to test in-orbit construction methods.

STS-24 or STS-61C
Shuttle: Columbia
Crew: commander Robert L. "Hoot" Gibson, pilot Charles F. Bolden Jr.,
    mission specialists George D. "Pinky" Nelson, Steven A. Hawley, Franklin R. Chang-Diaz,
    payload specialists Robert J. Cenker, U.S. Rep. William "Bill" Nelson (7)
Launch: January 12, 1986   Duration: 6 days (146:04:09)
Altitude: 175 nautical miles   Inclination: 28.5 degrees
Events: Carried a communications sat, 12 Get Away Special cannisters (GAScans) of experiments, Material Science Lab. Nelson was the first member of the U.S. House of Representatives, as well as the second politician and the second elected official to fly to space. Robert J. Cenker was a commercial passenger.

STS-25 or STS-51L
Shuttle: Challenger
Crew: commander Francis R. "Dick" Scobee, pilot Michael J. Smith,

mission specialists Judith A. Resnik, Ellison S. Onizuka, Ronald E. McNair,
payload specialist Gregory B. Jarvis, space flight participant Sharon Christa McAuliffe (7)
Launch: January 28, 1986   Duration: 13 minutes
Events: The Challenger disaster: seven astronauts died in the fall and crash into the Atlantic ocean after a liftoff explosion. The shuttle era's most heart-stopping quote, "Obviously a major malfunction. We have no downlink. The vehicle has exploded," was intoned by NASA mission control at one minute thirteen seconds into the liftoff of shuttle Challenger from Cape Canaveral. The shuttle orbiter fell into the Atlantic off Cape Canaveral; some parts recovered. McAuliffe was the first school teacher astronaut. Lost in the accident were NASA's Tracking and Data Relay Satellite (TDRS-B) and the Spartan (Sptn-Halley) ultraviolet telescope to study Comet Halley.

STS-26
Shuttle: Discovery
Crew: commander Frederick H. Hauck, pilot Richard O. Covey,
       mission specialists George D. "Pinky" Nelson, John M. "Mike" Lounge, David C. Hilmers (5)
Launch: September 29, 1988   Duration: 4 days (91 hours)
Altitude: 160 nautical miles   Inclination: 28.5 degrees
Events: America's recovery flight 975 days after the Challenger disaster carried NASA's TDRS-C communications satellite. NASA chose five veteran astronauts to return its shuttle fleet to space—the first no-rookie crew.

STS-27
Shuttle: Atlantis
Crew: commander Robert L. "Hoot" Gibson, pilot Guy S. Gardner,
       mission specialist Jerry L. Ross, William M. Shepherd, Richard M. "Mike" Mullane (5)
Launch: December 2, 1988   Duration: 4 days (99:05)
Altitude: secret   Inclination: secret
Events: An all-military crew on a secret Defense Dept. mission carried the huge Lacrosse radar spy satellite to Earth orbit in the cargo bay of Atlantis. The crew dropped Lacrosse off in orbit from where it bounces a powerful radar beam off objects on the ground to create images.

STS-29
Shuttle: Discovery
Crew: commander Michael L. Coats, pilot John E. Blaha,
       mission specialists James P. Bagian, James F. Buchli, Robert C. Springer (5)
Launch: March 13, 1989   Duration: 4 days 23 hrs 39 mins
Altitude: 163 nautical miles   Inclination: 28.5 degrees
Events: Carried NASA's TDRS-D communications sat plus 4 rats, 32 fertilized chicken eggs, plants, crystals. Miles of film were exposed showing Earth's environmental problems. America's first official space artwork, carried to orbit in Discovery, was a seven-pound cube-shaped sculpture called Boundless Aperture, created by Boston artist Lowry Burgess, professor at Massachusetts College of Art. Burgess had conceived Boundless Aperture in the 1960's as part of what he called The Quiet Axis. By the time Boundless Aperture went to space, Burgess already had buried a sculpture in the floor of the Pacific Ocean off Easter Island and lugged another into central Afghanistan's Hindu Kush mountains.

STS-30
Shuttle: Atlantis
Crew: commander David M. Walker, pilot Ronald J. Grabe,
       mission specialists Mary L. Cleave, Mark C. Lee, Norman E. Thagard, M.D. (5)
Launch: May 4, 1989   Duration: 4 days 56 mins
Altitude: 161 nautical miles   Inclination: 28.9 degrees
Events: In Earth orbit, the Atlantis crew lifted the 7,600-lb. interplanetary spacecraft Magellan from the shuttle cargo bay 6 hours 17 minutes after blastoff from Cape Canaveral. Then the inertial upper stage (IUS) rocket attached to the probe fired Magellan on a 466-day trajectory to Venus, the first planet probe launched from a shuttle and the first American planet explorer since two Pioneer craft left in 1978 to study Venus. Magellan, resembling a giant mechanical insect, fell into orbit around Venus August 10, 1990, and mapped the entire surface terrain with cloud-penetrating radar. Each 3.1-day trip around the planet offered a 40-minute period when Magellan was near the surface. Each orbit, the synthetic-aperture radar mapped a swath 15 miles wide and 10,000 miles long. It took 1,853 such fragments over 243 days to piece together a chart of the entire planet by April 1991. The STS-30 crew also used a small video camera for Earth photos. Cleave and Lee grew crystals in weightlessness.

STS-28

Shuttle: Columbia
Crew: commander Brewster H. Shaw Jr., pilot Richard N. "Dick" Richards,
   mission specialists David C. Leestma, James C. Adamson, Mark N. Brown (5)
Launch: August 8, 1989   Duration: 5 days
Altitude: secret   Inclination: secret
Events: Columbia was extensively refurbished for STS-28, its 8th flight since 1981. Columbia had
been NASA's first shuttle to go to space April 12, 1981. It made 2 trips in 1981, 3 in 1982, 1 in
1983. It last had flown January 12, 1986, just 16 days before the Challenger disaster. Since then, it
had been cannibalized for parts to keep Discovery and Atlantis running. Following Challenger, the
three remaining shuttles were rebuilt, each with 250 modifications improving safety and
performance. Discovery and Atlantis returned to space in 1988. Columbia had been scheduled for
July 1989 launch, but was delayed to August to complete safety mods, bumping STS-33 to after the
high-priority October launch of STS-34 with Galileo to Jupiter. With STS-28, NASA had all 3
shuttles back to normal operations. STS-28 may have trucked to space a 10-ton digital-TV photo-
reconnaissance satellite, known as Key Hole-12 (KH-12), to spy on USSR and Middle Eastern
countries. On Columbia's 5th circuit around Earth, astronauts dropped the secret payload overboard
into a 200-mi.-high orbit. It then circled Earth every 91 minutes, probably taking pictures
electronically as commanded and converting them into digital data transmitted to the ground. A KH-
12 would be an enhanced KH-11. The U.S. may have had several KH-11s already in orbit, from
unmanned launches in 1984, 1987 and 1988. With no information on the payload from the Defense
Dept., the spysat ferried in STS-28 also might have been more like AFP-731 launched by STS-36 in
1990. If like AFP-731, it would have been first in that series of new-generation photo spysats,
relieving older KH-11s. The new-style spysat would see infrared images at night. It might even be a
dual-function eye-in-the-sky and ear-in-the-sky, able to snap detailed photos with digital-imaging
reconnaissance cameras, while eavesdropping on electronic communications such as radio, TV and
telephones with signal-intelligence receivers. Its photos would be radioed to Earth as sharp digital
images to be enhanced by computer. KH-12 and AFP satellites probably can make changes in their
orbit, moving over changing events on the ground. Whatever it was, the STS-28 spysat was left in
orbit 100 miles lower than the Lacrosse radar spysat launched by STS-27 in 1988. Shaw had worked
for the presidential commission which investigated the Challenger disaster and was astronaut office
liaison with the Pentagon for classified payloads. It was Richards' first flight, although he had been
an astronaut nine years. Adamson, on his first flight, had worked at Johnson Space Center before
becoming an astronaut in 1984. Brown, on his first spaceflight, had been on the solid-fuel booster
redesign team after Challenger. At blast off August 8, Brown was sitting alone on the lower deck of
Columbia's split-level crew cabin while the other four were seated above on the flight deck; for
landing August 13, Brown and Adamson traded seats. In orbit, a steering thruster turned leaky; the
crew shut it down and used others. STS-28 was the fourth flight on which the Defense Dept. had
imposed a news blackout. However, astronauts always board 2.5 hours before launch and were quite
visible as they were bused to the pad in darkness lit by spotlight and escorted by helicopter.
Colleagues waved goodbye, camera crews filmed and a Russian trawler spiked with radio aerials
eavesdropped from the Atlantic Ocean 50 miles off Cape Canaveral. While Houston mission control
told the American public only that Columbia flew "right down the middle of the pike," hobbyists
along the Eastern Seaboard were able to hear shuttle-to-ground radio transmissions during flight to
orbit on standard NASA ultra-high frequencies (UHF) of 296.8 MHz and 259.7 MHz. NASA refused
to tell reporters where the solid-fuel boosters fell, but USSR spy ships tracked the shuttle flight and
spysat orbit by radar. USSR technicians, using cameras attached to ground telescopes capable of
seeing parts as small as one foot, were able to photograph astronauts high above Earth as well as
the huge satellite in space. Nugzar Ruhadze, a journalist from the USSR was given official
credentials for a front row seat at the Kennedy Space Center press site to observe the STS-28 launch.
Ruhadze, a native of Soviet Georgia on a foreign exchange program with Atlanta, Georgia,
television station WXIA, said the shuttle "roared like a lion."

STS-34

Shuttle: Atlantis
Crew: commander Donald E. Williams, pilot Michael J. McCulley,
   mission specialists Ellen S. Baker, Shannon W. Lucid, Franklin R. Chang-Diaz (5)
Launch: October 18, 1989   Duration: 7 days
Altitude: 168 nautical miles   Inclination: 34.3 degrees
Events: Unmanned interplanetary spacecraft Galileo launched to Earth orbit in Atlantis. Then

Galileo's inertial upper stage (IUS) rocket was fired to send the spacecraft on a 6-year 2.4-billion-mile trip to Jupiter. It was the second shuttle launch of an interplanetary spacecraft, after Magellan in STS-30 in 1989. Using planet gravity for speed boosts, Galileo swooped around Venus Feb 10, 1990, and Earth Dec 8, 1990. NASA radioed a command to Galileo on April 11, 1991, to unfurl its main antenna, but the high-gain antenna stuck partially open. Galileo continued to be able to transmit optical navigation photos and data from its dust detector, magnetometer, and extreme ultraviolet instrument. Engineers tried by radio command, but failed to free the jammed communications antenna, turning it away from the Sun for days to cool and shrink its central mast, which may have snagged ribs on the 16-ft. umbrella-shaped antenna. Galileo grazed the Asteroid Belt, passing close to the asteroid 951 Gaspra Oct 29, 1991, man's first close encounter with a minor planet. The stuck antenna wasn't needed for the asteroid flyby, but would be important when orbiting Jupiter in 1995. After Gaspra, Galileo flew by Earth again Dec 8, 1992. On July 10, 1995, Galileo will separate into orbiter and atmosphere probe. On Dec 7, 1995, orbiter will arrive over Jupiter and probe will parachute 370 miles down into the atmosphere, measuring clouds and atmosphere for 75 minutes, until pressure and temperature crush and vaporize it. The orbiter above will transmit data about Jupiter, the Solar System's largest planet, and its moons and rings for 22 months. The number of photos returned from Jupiter would be limited if the high-gain antenna could not be used. STS-34 astronauts also used the Shuttle Solar Backscatter Ultraviolet Instrument (SSBUV-1) to measure ozone in Earth's atmosphere.

STS-33
Shuttle: Discovery
Crew: commander Frederick D. Gregory, pilot John E. Blaha, mission specialists Manley L.
    "Sonny" Carter Jr., Kathryn C. Thornton, F.S. "Story" Musgrave (5)
Launch: November 22, 1989   Duration: 5 days
Altitude: secret   Inclination: secret
Events: The secret military flight carrying a spysat was the third night shuttle launch after Challenger Aug 1983 and Atlantis Nov 1985. Gregory was the first black shuttle commander. Civilian physicist Thornton was the first female astronaut on a U.S. military mission. Musgrave was a civilian physician which may have been lucky for Gregory who suffered a foot infection in orbit. The astronauts ate Thanksgiving dinner in orbit: turkey, gravy, cranberry sauce.

STS-32
Shuttle: Columbia
Crew: commander Daniel C. Brandenstein, pilot James D. Wetherbee,
    mission specialists Bonnie J. Dunbar, Marsha S. Ivins, G. David Low (5)
Launch: January 9, 1990   Duration: 10 days 21 hrs
Altitude: 178 nautical miles   Inclination: 28.5 degrees
Events: Columbia carried Syncom IV-5 communications sat to orbit, then retrieved Long Duration Exposure Facility (LDEF). Loaded with 57 science experiments, LDEF was a school bus-sized satellite orbited in STS-41C in 1984. On its exterior, NASA and industry experiments gauged effects of atomic oxygen (free atoms of oxygen) on paint and coatings. The exterior and experiments inside also were exposed to six years of ultraviolet radiation from the Sun and to meteoroids and space debris. Intended to stay in orbit one year, the retrieval of LDEF was delayed by NASA schedule changes, then by the Challenger disaster. The 10.5-ton satellite was in danger of falling into the atmosphere when it was plucked from orbit by Dunbar using Columbia's mechanical arm. Among the 57 experiments were 12.5 million tomato seeds which NASA later mailed to schools so 3.5 million students could study the effect of space on growth (Space Exposed Experiment Developed for Students or SEEDS). Chemists dissected materials "weathered" by years of continuous battering by atomic oxygen molecules, ultraviolet rays, and other elements. One coating, a white thermal-control paint, turned brown from exposure to ultraviolet light. Silver-colored Teflon blankets were stretched and stressed as ultraviolet light and atomic oxygen eroded or cracked the surface. Scientists expect LDEF data to help engineers plan the U.S.-international space station Freedom. Flight STS-32 was extended by weather and computer problems in orbit to a record 10 days 21 hours, half a day longer than any previous.

STS-36
Shuttle: Atlantis
Crew: commander John O. "J.O." Creighton, pilot John H. Casper,
    mission specialists David C. Hilmers, Richard M. "Mike" Mullane, Pierre J. Thuot, 34 (5)
Launch: February 28, 1990   Duration: 5 days
Altitude: secret   Inclination: secret

Events: Secret military flight carried AFP-731 spysat to be used by the National Security Agency and the Central Intelligence Agency. Five launch delays caused by computer glitch, weather, Creighton illness. Second American spaceflight health delay after Apollo 9's 4 days in 1968. Creighton's sore throat, head cold was the first time a U.S. shuttle had to be postponed due to crew illness. Creighton later told reporters, "I probably had the world's most famous cold." The dazzling night launch came on the 6th try after 5 delays. Only one shuttle mission had seen more: Columbia STS-61C had 6 postponements for mechanical and weather problems from December 18, 1985, to January 12, 1986. The ill-fated Challenger flight STS-51L was delayed 5 times from January 22 to January 28, 1986, then exploded during lift off. STS-36 was the 4th night launch after Challenger Aug 1983, Atlantis Nov 1985, Discovery Nov 1989. At an inclination of 62 degrees, the STS-36 launch flew the highest angle to the equator ever travelled by a U.S. shuttle so its payload could be dropped off in an orbit covering most of the USSR, including areas of the far north with concentrations of ships, submarines and missile bases. The satellite could fly over every point on the globe between 62 degrees north and south latitude, including most population and industrial centers in the USSR, Europe and Asia. Spectators after the 2:50 a.m. launch saw the shuttle pass north along the U.S. Atlantic Ocean coast 86 miles east of Cape Hatteras, North Carolina, and Cape Cod, Massachusetts, closer to shore than ever before. NASA retrieved the spent solid-fuel boosters where they parachuted into the Atlantic Ocean 190 miles east of Jacksonville, Florida. Cameras had revealed chunks shooting away from the shuttle 1 minute 15 seconds into flight from Kennedy Space Center. Booster-maker Morton Thiokol Inc. said it was hot slag, ash from burned propellant, and chunks routinely blast out of shuttle rockets during ascent. Amateur astronomers on the ground saw the shuttle and spysat 127 miles overhead after the astronauts dropped the reconnaissance payload overboard March 1. AFP-731 might have been second in a series of new-generation photo spysats, relieving older KH-11 models. The first of the new style, capable of infrared night photography, may have been launched by STS-28 in 1989. If AFP-731 has a dual-function capability, it might be able to snoop on most of the world, snapping detailed photos with digital-imaging cameras, while eavesdropping on electronic communications such as radio, TV and telephones with signal-intelligence receivers. Its photos would be radioed to Earth as digital images to be sharpened by computer. Before the secret STS-36 launch, 23 USSR observers, including 10 from the Defense Committee of the Soviet National Legislature and advisers on disarmament, foreign affairs, science and space, were on a 12-day visit to the Pentagon, aerospace factories and military installations. They were taken within 2 miles of Atlantis on launch pad 39A before the countdown. Mission specialist Thuot was a rookie, while Mullane made his 3rd and final trip, retiring from NASA and Air Force July 1. He first had flown on the maiden flight of Discovery, an assignment he said was gratifying because poor eyesight had kept him from becoming a U.S. Air Force fighter pilot. The March 4 landing shuttle flew an unusual route, crossing the Pacific coast north of Los Angeles, flying in just ahead of winds that could have delayed the trip home. As Atlantis approached Edwards Air Force Base, sensors sounded an alarm as pressure dropped in the main hydraulic system which provides power for the landing gear brakes and moves wing flaps and rudder to steer the spacecraft under computer control. Bright red hydraulic fluid may have leaked from the the black-and-white spaceplane as it whisked along the runway at 200 mph, its tires kicking up small clouds of dust, but the touchdown was nominal. Astronaut Stephen Oswald at Houston mission control radioed, "Congratulations on a great flight, guys, and welcome back." Military censors refused to reveal the response from Atlantis. The AFP-731 satellite left behind in space must have malfunctioned almost immediately. By March 7, the USSR was reporting the reconnaissance satellite was disintegrating with four large remnants about to plunge into the atmosphere. The first burned in the sky March 19 over the Pacific Ocean 900 miles north of Midway Islands. The second disintegrated March 20. The two remaining disintegrated in the atmosphere without reaching ground by mid-May.

STS-31

Shuttle: Discovery
Crew: commander Loren J. Shriver, pilot Charles F. Bolden Jr.,
       mission specialists Steven A. Hawley, Bruce McCandless II, Kathryn D. Sullivan (5)
Launch: April 24, 1990   Duration: 5 days
Altitude: 330 nautical miles   Inclination: 28.5 degrees
Events: Discovery flew 380 statute miles above Earth, 70 miles higher than any previous shuttle, to drop off Hubble Space Telescope (HST). For more than a decade, NASA had been touting HST as a powerful astronomy satellite to hunt for the edge of the Universe. After many delays, it finally was dropped off in space in April 1990 by Discovery STS-31. Sullivan, McCandless almost had to take a

spacewalk when a telescope solar panel didn't unfold until third try. Two months later, NASA was embarrassed when Hubble pictures were blurry. The 12-ton telescope's 94-in. mirror was curved incorrectly. In 1991, gyroscopes which stabilize the satellite failed and a faulty power supply cut off a key spectrograph. Astronauts will have to fly Discovery to orbit in Feb. 1994 to replace Hubble's wide-field planetary camera with a new one designed to correct the spherical aberration, fix the spectrograph power supply, replace gyroscopes and tighten loose solar panels. During STS-31, Shriver found a watch lost in Discovery by Sonny Carter during flight STS-33 in Nov 1989.

STS-41
Shuttle: Discovery
Crew: commander Richard N. "Dick" Richards, pilot Robert Cabana,
    mission specialists Thomas D. Akers, Bruce E. Melnick, William M. Shepherd (5)
Launch: October 6, 1990   Duration: 4 days
Altitude: 160 nautical miles   Inclination: 28.5 degrees
Events: The European-American solar probe Ulysses, once known as International Solar Polar Mission (ISPM), was ferried 160 mi. above Earth in Discovery. There, its built-in inertial upper stage (IUS) rocket fired it away from Earth orbit toward Jupiter on a 4-year voyage across the Sun's north and south poles. Ulysses swung around Jupiter February 8, 1992, where planet gravity flung it above the plane in which planets of the Solar System orbit the Sun, boosting Ulysses over the Sun's south pole in 1994, then over the Sun's north pole in 1995. Ulysses will measure magnetic field, plasma, electrons, protons, ions, interstellar neutral gas, gravity waves, solar particles, cosmic rays, plasma waves, radio signals, X-rays, gamma rays, cosmic dust. It will be the first probe to explore the Solar System's "third dimension." The probe's radio and plasma-wave receiver has two boom antennas, one 24.3 ft. long, the other 238 ft. During deployment of the 24.3-ft. boom Nov 4, 1990, a wobble appeared in the spacecraft's spin, but the shake seemed gone by 1991. One radio stopped working for a time in June 1991. As Ulysses passed within 280,000 miles of Jupiter in Feb 1992, it swept down the huge planet's northern hemisphere across its southern hemisphere, measuring magnetic field and interaction with solar wind for 2 weeks. STS-41 astronauts also used the Shuttle Solar Backscatter Ultraviolet Instrument (SSBUV-2) to measure ozone in Earth's atmosphere. SSBUV-1 was carried in STS-34.

STS-38
Shuttle: Atlantis
Crew: commander Richard O. Covey, pilot Frank L. Culbertson Jr.,
    mission specialists Charles D. "Sam" Gemar, Carl J. Meade, Robert C. Springer (5)
Launch: November 15, 1990   Duration: 5 days
Altitude: secret   Inclination: secret
Landing site: High winds in California forced landing at Kennedy Space Center, Florida
Events: Night launch. Secret military flight during Persian Gulf War, Operation Desert Shield. Crew released a spysat into low Earth orbit. Meade celebrated his 40th birthday in orbit, but the mission was so secret the text of his birthday greeting couldn't be revealed. It was the sixth Florida shuttle landing, first since April 1985. Culbertson, Gemar and Meade were spaceflight rookies.

STS-35
Shuttle: Columbia
Crew: commander Vance D. Brand, pilot Guy S. Gardner,
    mission specialists Jeffrey A. Hoffman, John M. "Mike" Lounge, A. Robert Parker,
    payload specialists civilian astronomers Samuel T. Durrance, Ronald A. Parise (7)
Launch: December 2, 1990   Duration: 8 days 23 hrs 5 mins 8 secs
Altitude: 190 nautical miles   Inclination: 28.5 degrees
Events: The first shuttle flight in five years dedicated entirely to science. Four of the seven astronauts were astronomers. Used Astro-1 ultraviolet light telescopes and the Broad Band X-Ray Telescope (BBXRT-1) to study quasars, binary stars, pulsars, black holes, galaxies and high-energy stars. Specific targets included Supernova 1987a, the nearby supergiant star Betelgeuse, radio-quiet quasar Q1821, spiral-poor galaxy cluster Abell 2256, NGC 1633, NGC 1399 in the constellation Fornax, Q1821+64. Despite early telescope control problems in orbit, the Spacelab astronomy instruments in the shuttle cargo bay ended up examining 135 deep space targets during 394 observations. The observations included Hopkins Ultraviolet Telescope, 75 targets, 101 observations; Ultraviolet Imaging Telescope, 64 targets, 89 observations; Wisconsin Ultraviolet Photo Polarimeter Experiment, 70 sources, 88 observations; and Broad Band X-ray Telescope, 76 targets, 116 observations. It was the 15th longest U.S. space mission and the third longest shuttle flight. Previous long-duration shuttle flights were Columbia STS-9, STS-32. Other longer flights:

Gemini 7, Apollo 7, 9, 12, 14, 15, 16 and 17, all 3 Skylab missions, the joint U.S.-USSR Apollo-Soyuz Test Project. During STS-35, Parise used Shuttle Amateur Radio Experiment (SAREX) equipment to make ham radio contact with amateur operators on Earth; his callsign WA4SIR. Hoffman wore the first dress shirt and tie in space when he dressed up to teach elementary pupils on the ground by TV from the shuttle Dec 7. NASA's idea of a Space Classroom became a reality as Durrance, Hoffman, Parise and Parker conducted space-to-ground TV lessons on the invisible and visible Universe, electromagnetic spectrum and how telescopes work in space. The astronauts answered questions from pupils on the ground. Brand, age 59, was the oldest person to have gone to orbit to that time. It was the first time a dozen people were in orbit at the same time. The 12 astronauts and cosmonauts the week of December 2 included 5 at the USSR's Mir space station and 7 in Columbia. Gardner retired from NASA after STS-35 to become commandant of the USAF Test Pilot School at Edwards Air Force Base, California. Parise and Durrance were the first non-NASA astronauts to fly in a U.S. spacecraft in five years, since the Challenger disaster.

STS-37

Shuttle: Atlantis
Crew: commander Steven R. Nagel, pilot Kenneth D. Cameron,
     mission specialists Jerome "Jay" Apt, Linda M. Godwin, Jerry L. Ross (5)
Launch: April 5, 1991   Duration: 6 days
Altitude: 243 nautical miles   Inclination: 28.5 degrees
Events: The first shuttle flight of 1991 ferried to orbit Gamma Ray Observatory (GRO), the second of NASA's four Great Observatories. The 17-ton spacecraft was the largest science satellite ever carried by a shuttle. The 31-ft. GRO filled half of Atlantis' cargo bay. On April 7, Linda Godwin reached into the cargo bay with the 50-foot robot arm, grasped GRO, lifted the observatory out of the hold and dropped it overboard into its own orbit 280 statute miles above Earth. Ross and Apt stood by in spacesuits, prepared to spacewalk to unfold GRO's high-gain antenna or solar panels if needed—and an antenna boom did jam. A thermal blanket on the boom had become hung up. Apt and Ross spacewalked in the cargo bay April 8 to shake loose a jammed 16.5-ft. antenna boom. Later, they spacewalked again to test space station construction gear in the cargo hold. It was the second-ever emergency repair spacewalk, after STS-51D in 1985, and the first U.S. spacewalk in five years, since Ross and Sherwood Spring went into the Atlantis STS-61B cargo hold Dec 1, 1985. While outside STS-37, Apt punctured an outer layer of his spacesuit glove and scraped the skin of his hand until it bled lightly, but was able to return inside safely. While in space, the crew received word of the plane crash death of fellow astronaut Manley L. "Sonny" Carter Jr. STS-37 had the first crew composed entirely of licensed amateur radio operators. Their call letters: Cameron KB5AWP, Apt N5QWL, Godwin N5RAX, Nagel N5RAW, Ross N5SCW. They used Shuttle Amateur Radio Experiment (SAREX) gear to chat from space with school children at nine sites in seven states with pupils listening in from 20 other schools. As Atlantis passed overhead, students had 20 minutes to ask questions, like what is blast off like, how hard is it to eat up there, can you see Oklahoma, and what do you think when you look down on Earth? To the latter, pilot Ken Cameron replied, "When you float up here above the Earth in a little spacecraft and look down at the big spacecraft that we all fly on, it makes us all think about what our priorities are." Mission commander Nagel told the students to "Take school seriously. Worry about your grades right now no matter how young you are. Study hard and get good grades, go to college, study science or engineering, then get into the space program after that." The STS-37 crew heard cosmonaut-hams at the USSR's Mir station calling across space to Atlantis via amateur radio, but couldn't establish contact. High winds in California forced Atlantis to land one day late, on April 11. Nagel said later he had turned too wide in the final approach at Edwards Air Force Base, California. Then windshear pushed the shuttle low in altitude. It landed 600 feet ahead of its target—the first shuttle to touch down short of a runway. A month later, GRO worked well as science operations began with the 17-ton observatory pointed at a Crab Nebula pulsar May 16, 1991. GRO looks for invisible gamma rays, the powerful radiation created when atomic nuclei collide and matter is annihilated in the presence of antimatter. The observatory records high-energy gamma rays flowing from quasars, supernovas, pulsars, neutron stars, black holes, most violent natural processes in the Universe. For GRO's first target of opportunity, controllers at Goddard Space Flight Center pointed the satellite at the Sun on June 7, 1991, to peer at two X-class solar flares. X-class are the largest, most powerful solar flares. Geysers of hot gas, looping 430,000 miles from the Sun into space, generate outbursts of intense radiation which disrupt Earth's magnetic field and interfere with communications and electrical power distribution. By the end of 1991, GRO's Burst and Transient Source Experiment (BATSE) was detecting gamma ray bursts with more sensitivity than previous receivers in more detail than previously possible—

BATSE was observing a gamma ray burst almost every day, a rate of 250 per year. GRO is doing a 15-month full-sky gamma ray survey from its low orbit 280 statute miles above Earth.

STS-39

Shuttle: Discovery
Crew: commander Michael L. Coats, pilot L. Blaine Hammond Jr., mission specialists Guion S.
   "Guy" Bluford Jr., Gregory J. Harbaugh, Richard J. "Rick" Hieb, Donald R. McMonagle,
   Charles Lacy Veach (7)

Launch: April 28, 1991   Duration: 8 days
Altitude: 140 nautical miles   Inclination: 57.0 degrees

Landing site: High winds in California forced a landing at Kennedy Space Center, Florida, May 6.
Events: STS-39's mission was to practice spotting missiles flying through space against a variety of natural backdrops including Earth, the planet's atmosphere and scintillating aurora. Aurora around Discovery was so strong, Coats said it was "like flying through a curtain of light." The Star Wars military flight carried Strategic Defense Initiative (SDI) payload AFP-675 (CIRRIS). Also, IBSS in the German-designed payload carrier SPAS-2 (Shuttle Pallet Satellite). IBSS was the U.S. Strategic Defense Initiative Organization's Infrared Background Signature Survey of Earth's infrared background. SPAS is a small satellite outfitted with science instruments, deployed for several days in its own orbit to gather data. Before landing, the shuttle returned to SPAS-2, retrieved it, and returned it to Earth for another mission. SPAS-2 flew 6.5 miles from Discovery to observe 15 shuttle rocket firings which cartwheeled the orbiter through space. The crew completed 32 of 33 primary observations, 18 of 21 secondary, 41 of 44 third-level. SPAS-2 measured aurora and watched how seven rocket chemicals looked in space. The astronauts jury-rigged 3 experiments with faulty tape recorders and deployed small satellites to simulate a missile defense net across space. Discovery was tracked over South Africa by an unmanned ultraviolet telescope in the SDI satellite LACE launched in 1990. The crew photographed Earth, particularly Bangladesh after it was ravaged by cyclone May 1. On May 6, Discovery dropped off in orbit the last of four small military satellites and retrieved SPAS-2. Then, for the seventh shuttle landing in Florida, Hammond piloted Discovery along an unusual path across Alaska, western Canada and 14 states in a steep, high-speed plunge across America to Kennedy Space Center. Discovery landed at 242 mph, one of the fastest touchdowns. Discovery's right tires contacted the strip 168 feet past the end of the 3-mi. runway. The left tires touched down 215 feet on down the strip. Nose gear touched down 4,723 feet farther along. Discovery rolled 6,316 feet. With Coats applying 37 million ft. lbs. of energy to right-hand brakes and 30 million ft. lbs. to the left, Discovery stopped 9,152 feet down the runway—one of the longer rollouts. Tires were damaged, but none blew out. Landing gear tires have 16 plies; a flat would require eight or more lost. STS-39's right outboard tire was worn most, with three plies gone. Tougher synthetic-rubber tires debuted later in 1991. Discovery STS-51D had blown a tire in April 1985 during a Florida landing in a stiff crosswind. The Florida runway has 1,000 feet of load-bearing overrun at either end. Had Discovery landed short, as Atlantis did in April 1991, the shuttle still would have been safe. Atlantis landed 600 feet short of the runway at Edwards Air Force Base, California, April 11. An hour after the STS-39 landing, the astronauts climbed out to inspect the tire first hand. Three days after the STS-39 landing in Florida, U.S. Vice President J. Danforth Quayle, chairman of the U.S. National Space Council, mistakenly described the shuttle landing as in California. "To know America's history is to understand our love of adventure and exploration, boldness and ingenuity. From the vast travels of Lewis and Clark, from the laboratory of Thomas Edison, to the garage of Henry Ford, to the sands of Kitty Hawk, to the desert of Edwards Air Force Base where the shuttle Discovery landed just three days ago. All of these moments in history illustrate the great American journey of discovery and high achievement," Quayle said.

STS-40

Shuttle: Columbia
Crew: commander Bryan D. O'Connor, pilot Sidney M. Gutierrez,
   mission specialists James P. Bagian, Tamara E. "Tami" Jernigan, Margaret Rhea Seddon,
   payload specialists Francis Andrew "Drew" Gaffney, Millie Hughes-Fulford (7)

Launch: June 5, 1991   Duration: 9 days
Altitude: 160 nautical miles   Inclination: 39.0 degrees

Events: Spacelab Life Sciences (SLS-1) explored how heart, blood vessels, lungs, kidneys and hormone-secreting glands respond to microgravity, the causes of space sickness and changes in muscles, bones and cells during space flight, and readjustment to gravity upon returning to Earth, the most detailed physiological measurements in space since Skylab in 1973-74. Spacelab is a removeable pressurized science lab carried to space and back in Columbia's cargo bay. Astronauts float through a tube into Spacelab. The crew shepherded 30 rats plus 2,478 tiny jellyfish in plastic bags and bottles—the first jellyfish in orbit—for Eastern Virginia Medical School biologist Dorothy Spangenberg who wanted to see how Aurelia aurita (moon jellies) grow in weightlessness. Since moon jellies on Earth mature to free swimming in six days if iodine is added to their water, they were ideal research organisms for Columbia's nine-day spaceflight. In orbit, Jernigan injected iodine into the water of a third of them and videotaped their behavior. The experiments were duplicated the same day in a Norfolk laboratory. After flight, Spangenberg examined them to see how much calcium they lost. GAScans (Get Away Special Cannisters) are small containers carrying experiments to orbit in the cargo hold. A GAS Bridge carried 12 GAScans in STS-40. Crew sleep was interrupted to fix Spacelab cooling equipment. First, a backup cooling pump had to be switched on for the Research Animal Holding Facility. Then, the Spacelab refrigerator-freezer broke down. A shortage nearly required Spacelab power to be shut down for a day to conserve liquid hydrogen and oxygen for fuel cells. Crew observations of Earth included a dust storm from Algeria stretching across the Atlantic Ocean to the Caribbean, and a yellow pall from the 12-mi.-high plume of ash blowing out of the erupting Mt. Pinatubo volcano in the Philippines. It was the first shuttle with three woman in one crew. Bagian, Gaffney and Seddon were MDs. Jernigan, an astronomer. Hughes-Fulford, a biological chemist. A loose seal threatened to require a spacewalk to close the payload bay doors at the end of the flight, but the doors closed tightly without help. After STS-40 landed June 14, Columbia underwent 6 months of major modifications prior to flight STS-50.

STS-43
Shuttle: Atlantis
Crew: commander John E. Blaha, pilot Michael A. Baker,
    mission specialists James C. Adamson, G. David Low, Shannon W. Lucid (5)
Launch: August 2, 1991   Duration: 9 days
Altitude: 160 nautical miles   Inclination: 28.5 degrees
Events: Ferried NASA's fifth Tracking and Data Relay Satellite (TDRS-E) to orbit. Carried two 22-ft.-long heat pipes for space tests. In orbit, the astronauts used the Shuttle Solar Backscatter Ultraviolet Instrument (SSBUV-3) to measure ozone in Earth's atmosphere. SSBUVs had been carried in U.S. shuttles before: SSBUV-1 in STS-34 and SSBUV-2 in STS-41. They developed polymer membranes in microgravity Experimented with fire inside the orbiting shuttle: Adamson flipped a switch to ignite a 4.3-inch piece of ashless paper. It glowed brightly, turned into a slow-moving round flame. It took a minute for the flame to consume the paper. Baker was a spaceflight rookie. Two electricity-generating fuel cells were left on by mistake after landing. They flooded and had to have their power sections overhauled. It was the first Florida landing in 6 years.

STS-48
Shuttle: Discovery
Crew: commander John O. "J.O." Creighton, pilot Kenneth S. Reightler Jr.,
    mission specialists Mark N. Brown, James F. Buchli, Charles D. "Sam" Gemar (5)
Launch: September 12, 1991   Duration: 5 days
Altitude: 292 nautical miles   Inclination: 57.0 degrees
Landing site: Florida clouds forced a landing change to Edwards Air Force Base, California.
Events: Ferried the Upper Atmosphere Research Satellite (UARS) to Earth orbit to study the upper atmosphere on a global scale. Nine sensors in the satellite explore winds, chemical composition and energy of the stratosphere, providing details on how natural processes and man-made pollutants destroy Earth's ozone layer. STS-48's high 57-degree inclination carried Discovery as far south as the tip of South America and as far north as Denmark. Reightler flew Discovery to an altitude of 336 statute miles, then on up to an orbit 354 miles above Earth, one of the highest ever for a shuttle. Spaceflight rookie Reightler looked out the window and reported seeing the entire Nile River, black on either side, but with the Nile and Suez Canal a river of lights. He could see to the coast of Israel in one direction and the burning Kuwait oil fields in another. Astronauts must breathe pure oxygen before a spacewalk to avoid the bends, a painful, sometimes fatal release of nitrogen bubbles into the bloodstream. Buchli and Gemar used special helmets to breathe pure oxygen for an hour. Brown reached into the cargo bay with the 50-foot robot arm Sept 15, grasped UARS, lifted it out of the hold and dropped it overboard into its own orbit 354 miles above Earth. Buchli and Gemar wore spacesuits in preparation for an emergency spacewalk. The oxygen tank for Gemar's suit leaked slightly, but was within specifications. On Sept 17, small thrusters lifted UARS another 18 miles to an altitude of 372 miles. Eight female rats were carried in STS-48 in a middeck cage as a muscle-change experiment. STS-48 was to have been the first nighttime Florida shuttle landing, but low clouds forced a change. Discovery roared across the Oregon coast and down California's San Joaquin

Valley to Edwards Air Force Base for a 200 mph landing on concrete runway 22 shortly after midnight at 12:39 a.m. PDT Sept 18.

STS-44

Shuttle: Atlantis
Crew: commander Frederick D. Gregory, pilot Terence T. "Tom" Henricks,
    mission specialists F.S. "Story" Musgrave, Mario Runco Jr., James S. Voss,
    payload specialist Thomas J. Hennen (6)
Launch: scheduled December 1991   Duration: 10 days
Altitude: 195 nautical miles   Inclination: 28.5 degrees
Events: Military flight carrying Defense Support Program (DSP) satellite to low Earth orbit on its way to higher stationary orbit.

STS-42

Shuttle: Discovery
Crew: commander Ronald J. Grabe, pilot Stephen S. Oswald, payload commander Norman E.
    Thagard, mission specialists David C. Hilmers, William F. Readdy, payload specialists
    Roberta L. Bondar, Ulf D. Merbold (7)
Launch: scheduled February 1992   Duration: 7 days
Altitude: 163 nautical miles   Inclination: 57.0 degrees
Events: International Microgravity Laboratory (IML-1), first microgravity research in a long shuttle Spacelab module, studying effects on material processes and living organisms of very low gravity in space near Earth. With NASA in the project are 5 international organizations: European Space Agency (ESA), Canadian Space Agency (CSA), French National Center for Space Studies (CNES), West German Research and Development Institute for Air and Spacecraft (DLR), and National Space Development Agency of Japan (NASDA). IML Spacelabs are to fly at 17-mo. to 25-mo. intervals; separation allowing scientists to study results and apply them to new research. Spacelab is a removeable pressurized science lab carried to space and back in the cargo bay. Astronauts float through a tube from Discovery crew quarters into Spacelab. They will carry plants and animals, and protein crystal, vapor crystal and mercury iodide crystal growth experiments. Discovery to orbit with its tail pointing down toward Earth in the "gravity gradient" position offering the least gravity during flight. Before the flight, Hilmers replaced Manley L. "Sonny" Carter Jr., who later died in 1991. Carter previously had replaced Mary L. Cleave. Merbold from Germany. Roberta L. Bondar MD is to be the first Canadian woman in space; first non-U.S. astronaut in a U.S. spaceflight since STS-61B in 1985.

STS-45

Shuttle: Atlantis
Crew: commander Charles F. Bolden, Jr., pilot Brian Duffy, payload commander Kathryn D.
    Sullivan, mission specialists C. Michael Foale, David C. Leestma, payload specialists Dirk D.
    Frimout, Byron K. Lichtenberg (7)
Launch: scheduled May 1992   Duration: 8 days
Altitude: 160 nautical miles   Inclination: 57.0 degrees
Events: Sullivan to be the first female payload commander, in charge of Atmospheric Laboratory for Applications and Science (Atlas-1) to study atmospheric phenomena, energy from the Sun and changes in the solar spectrum from a Spacelab aboard Atlantis. The atmospheric lab will be carried in an igloo in the payload bay. Atlas-1 is to be the first of several Atlas missions scheduled for orbit over a decade to study interaction of Earth's atmosphere with the Sun. Atlas is part of NASA's Mission to Planet Earth program. STS-45 astronauts also will use the Shuttle Solar Backscatter Ultraviolet Instrument (SSBUV-A1) to measure ozone in Earth's atmosphere. SSBUVs were carried in U.S. shuttles before: SSBUV-1 in STS-34, SSBUV-2 STS-41, SSBUV-3 STS-43. Duffy and Leestma are to use Shuttle Amateur Radio Experiment (SAREX) to contact ham radio operators on Earth. During pre-flight training in September 1991, Frimout replaced Michael L. Lampton who was disqualified for medical reasons. Lampton remained as alternate payload specialist.

STS-49

Shuttle: Endeavour
Crew: commander Daniel C. Brandenstein, pilot Kevin P. Chilton, mission specialists Thomas D.
    Akers, Richard J. "Rick" Hieb, Bruce E. Melnick, Kathryn C. Thornton, Pierre J. Thuot (7)
Launch: scheduled May 1992   Duration: 7 days
Altitude: 183 nautical miles   Inclination: 28.5 degrees
Events: The maiden flight of Endeavour, which replaced shuttle Challenger destroyed in 1986. In orbit, crew members will attach a new booster to the stranded Intelsat-VI and release it for a new

attempt to boost the communications satellite to stationary orbit. Three spacewalks will test building techniques, strut handling for the proposed U.S.-international space station Freedom.

STS-50

Shuttle: Columbia

Crew: commander Richard N. "Dick" Richards, pilot John H. Casper, payload commander Bonnie J. Dunbar, mission specialists Kenneth D. Bowersox, Carl J. Meade, payload specialists Lawrence J. DeLucas, Eugene H. Trinh (7)

Launch: scheduled June 1992   Duration: 13 days

Altitude: 160 nautical miles   Inclination: 28.5 degrees

Events: U.S. Microgravity Laboratory (USML-1), materials-processing experiments in a Spacelab aboard the first extended-duration mission of the refurbished Columbia, similar to earlier International Microgravity Laboratory IML-1 aboard STS-42. Columbia's 13-day flight to be the longest shuttle trip to date. Trinh and DeLucas to conduct 30 technology and science experiments in materials, fluids, biological processes. Trinh to be first Vietnamese in a U.S. spacecraft, second Vietnam native in orbit; the first in space was North Vietnamese war hero Pham Tuan in USSR's Soyuz-37 to station Salyut 6 in 1980.

STS-46

Shuttle: Atlantis

Crew: commander Loren J. Shriver, pilot James D. Wetherbee, payload commander Jeffrey A. Hoffman, mission specialists Andrew M. Allen, Franklin R. Chang-Diaz, Claude Nicollier, payload specialist from Italy to be Umberto Guidoni or Franco Malerba(7)

Launch: scheduled September 1992   Duration: 7 days

Altitude: 230 nautical miles   Inclination: 28.5 degrees

Events: First flights to Earth orbit of European Retrievable Carrier (Eureca-1), Tethered Satellite System (TSS-1), Evaluation of Oxygen Interaction with Materials (EOIM-3) and Two-Phased Experiment Mounting Plate (TEMP2A-3). Eureca-1, developed by European Space Agency (ESA), is a free-flying reusable platform carrying material-science and life-science experiments. TSS-1 satellite to be released from shuttle payload bay on 12-mile tether collecting electrodynamic data in upper reaches of atmosphere, joint project of NASA and Italian space agency Agenzia Spaziale Italiana. Nicollier, from France, repesenting ESA. Umberto Guidoni or Franco Malerba from Italy, representing Agenzia Spaziale Italiana.

STS-47

Shuttle: Endeavour

Crew: commander tba, pilot tba, payload commander Mark C. Lee, mission specialists N. Jan Davis, Mae C. Jemison, Jerome "Jay" Apt, payload specialist Mamoru Mohri (7)

Launch: scheduled September 1992   Duration: 7 days

Altitude: 160 nautical miles   Inclination: 57.0 degrees

Events: Spacelab-J joint science research project of NASA and the Japanese National Space Development Agency (NASDA). Spacelab is a removeable pressurized science lab carried to space and back in Endeavour's cargo hold. Astronauts float through a tube to and from Endeavour and Spacelab. Mohri will be the second native of Japan in orbit and the first Japanese to fly in a U.S. spacecraft; Japanese cosmoreporter Toyohiro Akiyama flew USSR Soyuz TM-11 to Mir station in 1990. GAScans (Get Away Special Cannisters) are small containers carrying experiments to orbit in the shuttle cargo hold. GAS Bridge, a carrier holding 12 GAScans, will be carried in STS-47. Lee and Davis the first married couple to travel in space. Jemison the first black woman in orbit. Apt would use Shuttle Amateur Radio Experiment (SAREX) equipment; his callsign N5QWL.

STS-52

Shuttle: Columbia

Crew: commander, pilot, payload commander, mission specialists, payload specialist tba (6)

Launch: scheduled November 1992   Duration: 9 days

Altitude: 160 nautical miles   Inclination: 28.5 degrees

Events: To ferry the U.S. Microgravity Payload (USMP-1), Laser Geodynamics Satellite (Lageos-2), and Canadian Experiments (Canex-2) equipment to orbit. USMP is a group of materials-processing experiments using extremely low gravity—microgravity—of Earth orbit. Lageos is a passive sphere encrusted with mirrors to reflect laser beams so scientists can measure tiny movements in Earth's crust. Canex-2 is a package of shuttle-related experiments from Canada.

STS-53

Shuttle: Discovery

Crew: commander, pilot, payload commander, mission specialists, payload specialist tba (5)

Launch: scheduled December 1992   Duration: 4 days
Altitude: 200 nautical miles   Inclination: 57.0 degrees
Events: A Dept. of Defense military flight.

STS-54

Shuttle: Endeavour
Crew: commander, pilot, payload commander, mission specialists, payload specialist tba (5)
Launch: schedule January 1993   Duration: 6 days
Altitude: 160 nautical miles   Inclination: 28.5 degrees
Events: To ferry NASA's Tracking and Data Relay Satellite (TDRS-F now called TDRS-6) to orbit.
The astronauts will use Diffuse X-Ray Spectrometer (DXS) to observe the soft X-ray background of
the Milky Way galaxy to determine abundance and temperature of the hot interstellar medium.

STS-55

Shuttle: Columbia
Crew: commander tba, pilot tba, payload commander Jerry L. Ross,
     two mission specialists tba, two payload specialists tba (7)
Launch: scheduled March 1993   Duration: 9 days
Altitude: 160 nautical miles   Inclination: 28.5 degrees
Events: Spacelab-D2, removeable pressurized science lab carried to space and back in Columbia's
cargo bay. Astronauts will float through a tube between Columbia and Spacelab. Two payload
specialists to be selected from four Germans: female meteorologist Renate Luise Brummer; and
physicists Hans-Wilhelm Schlegel; Gerhard Thiele; and Ulrich Walter. Ross would use Shuttle
Amateur Radio Experiment (SAREX) equipment to make ham radio contact with amateur operators
on Earth; his callsign N5SCW.

STS-51

Shuttle: Discovery
Crew: commander, pilot, payload commander, mission specialists, payload specialist tba (5)
Launch: scheduled April 1993   Duration: 8 days
Altitude: 160 nautical miles   Inclination: 28.5 degrees
Events: Shuttle to ferry to orbit the experimental Advanced Communications Technology Satellite
(ACTS) and the Orbiting and Retrievable Far and Extreme Ultraviolet Spectrometer (ORFEUS-
SPAS), a German telescope to measure distribution across the Solar System of material which might
absorb ultraviolet light and to study ultraviolet light from interstellar deep space. SPAS (Shuttle
Pallet Satellite) is a German-designed payload carrier outfitted with science instruments, carried into
space by shuttle, and deployed for several days in its own independent orbit to gather data. Before
landing, the shuttle returns to SPAS, retrieves it, and returns it to Earth for another mission.

STS-56

Shuttle: Endeavour
Crew: commander, pilot, payload commander, mission specialists, payload specialist tba (5)
Launch: scheduled May 1993   Duration: 9 days
Altitude: 160 nautical miles   Inclination: 57.0 degrees
Events: To carry Atmospheric Laboratory for Applications and Science (Atlas-2) to study
atmospheric phenomena, energy from the Sun and changes in the solar spectrum from a Spacelab
carried in shuttle Endeavour's cargo bay. Atlas-2 is to be the second of several Atlas missions
scheduled for orbit over a decade to study interaction of Earth's atmosphere with the Sun. Atlas-1
was carried to the same altitude and inclination in flight STS-45. Atlas is part of NASA's Mission to
Planet Earth program. STS-56 astronauts also will use the Shuttle Solar Backscatter Ultraviolet
Instrument (SSBUV-A2) to measure ozone in Earth's atmosphere. SSBUVs were carried in U.S.
shuttles before: SSBUV-1 in STS-34, SSBUV-2 STS-41, SSBUV-3 STS-43, SSBUV-A1 STS-45.
Also to be aboard STS-56 is Spartan, the Shuttle Pointed Autonomous Research Tool for Astronomy
(Sptn-201-1). Spartan is a small, retrievable, free-flying satellite for X-ray astronomy.

STS-57

Shuttle: Atlantis
Crew: commander, pilot, payload commander, mission specialists, payload specialist tba (6)
Launch: scheduled July 1993   Duration: 8 days
Altitude: 160 nautical miles   Inclination: 28.5 degrees
Events: Carrying Spacehab-1, a private commercial pressurized man-tended module. STS-57 also is
to retrieve the European Retrievable Carrier (Eureca-1) satellite "platform" of experiments which
was to be left in orbit by Atlantis STS-46 in Sept 1992. Another STS-57 payload known as Shoot is
to demonstrate the use of superfluid helium in orbit. GAScans (Get Away Special Cannisters) are

small containers carrying experiments to orbit in the shuttle cargo hold. GAS Bridge, a carrier holding 12 GAScans, will be carried in STS-57.

STS-58
Shuttle: Columbia
Crew: commander, pilot, payload commander, mission specialists, payload specialist, possible USSR cosmonaut, all tba (7)
Launch: scheduled July 1993   Duration: 13 days
Altitude: 160 nautical miles   Inclination: 28.5 degrees
Events: The second extended-duration flight, carrying Spacelab Life Sciences (SLS-2) exploring how the heart, blood vessels, lungs, kidneys and hormone-secreting glands respond to microgravity, the causes of space sickness and changes in muscles, bones and cells during space flight and in the readjustment to gravity upon returning to Earth. The first Spacelab Life Sciences test (SLS-1) was aboard Columbia STS-40 in 1991. Spacelab is a removeable pressurized science lab carried to space and back in Columbia's cargo bay. Astronauts float through a short connecting tube between Columbia crew quarters and Spacelab. Columbia's earlier 13-day flight was STS-50 in 1992. At a 1991 summit meeting, U.S. President George H.W. Bush and USSR President Michael Gorbachev agreed on a first-time exchange: a Soviet cosmonaut would fly aboard a U.S. shuttle and a NASA astronaut would travel to the USSR's Mir space station. The agreement called for medical research on how humans adapt to weightlessness preparing for lengthy flights to Mars after the year 2000. The cosmonaut might fly 13 days in STS-58. Date for astronaut to fly to Mir has not been set.

STS-59
Shuttle: Discovery
Crew: commander, pilot, payload commander, mission specialists, payload specialist tba (tba)
Launch: scheduled August 1993   Duration: tba
Altitude: tba   Inclination: tba
Events: an available flight with payload to be determined.

STS-60
Shuttle: Endeavour
Crew: commander, pilot, payload commander Linda M. Godwin, mission specialists, payload specialist tba (6)
Launch: scheduled October 1993   Duration: 9 days
Altitude: 130 nautical miles   Inclination: 57.0 degrees
Events: To carry Space Radar Laboratory (SRL-1) making radar images of Earth's surface, to be used to make maps, interpret geological features and locate resources. Godwin would use Shuttle Amateur Radio Experiment (SAREX) to make contact with amateur operators on Earth; her callsign N5RAX.

STS-61
Shuttle: Atlantis
Crew: commander, pilot, payload commander, mission specialists, payload specialist tba (6)
Launch: scheduled December 1993   Duration: 7 days
Altitude: 160 nautical miles   Inclination: 28.5 degrees
Events: To carry to space the second private commercial pressurized man-tended module, Spacehab-2. STS-61 also is to carry Wake Shield Facility (WSF-1) to grow semiconductors; Capillary Pump Loop Experiment (Capl-1) to test heat transfer; Office of Aeronautics, Exploration and Technology (Oaet-1) technology experiments on a platform in the cargo bay; and GAS Bridge, a cargo hold carrier holding 12 GAScans. GAScans (Get Away Special Cannisters) are small containers carrying experiments to orbit in the shuttle.

STS-62
Shuttle: Columbia
Crew: commander, pilot, payload commander, mission specialists, payload specialist tba (5)
Launch: scheduled January 1994   Duration: 9 days
Altitude: 160 nautical miles   Inclination: 28.5 degrees
Events: Carrying the second U.S. Microgravity Payload (USMP-2) for materials processing experiments in the cargo bay. Also, Spartan, the Shuttle Pointed Autonomous Research Tool for Astronomy (Sptn-204), a small, retrievable, free-flying telescope satellite for X-ray astronomy. Also aboard STS-62, a telerobotic servicer test (FTS-DTF-1) and composite materials to be tested for degradation when exposed to the raw environment of space (CMSE-1).

STS-63
Shuttle: Discovery
Crew: commander, pilot, payload commander, mission specialists, payload specialist tba (7)

Launch: scheduled February 1994   Duration: 8 days
Altitude: 330 nautical miles   Inclination: 28.5 degrees
Events: The first Hubble Space Telescope (HST) retrieval and repair in orbit. HST was dropped off in space in 1990 by Discovery STS-31. A solar panel didn't unfold until third try. Two months later, NASA was embarrassed when pictures were blurry; mirror was curved incorrectly. In 1991, gyroscopes which stabilize the satellite failed and a faulty power supply cut off a key spectrograph. STS-63 astronauts will replace the wide-field planetary camera with one designed to correct spherical aberration, fix the power supply, replace gyroscopes and tighten loose solar panels.

STS-64
Shuttle: Endeavour
Crew: commander, pilot, payload commander, mission specialists, payload specialist tba (7)
Launch: scheduled April 1994   Duration: 9 days
Altitude: 160 nautical miles   Inclination: 57.0 degrees
Events: To carry Atmospheric Laboratory for Applications and Science (Atlas-3) to study atmospheric phenomena, energy from the Sun and changes in the solar spectrum from a Spacelab carried in shuttle Endeavour's cargo bay. Atlas-3 is to be the third of several Atlas missions scheduled for orbit over a decade to study interaction of Earth's atmosphere with the Sun. Atlas-1 and Atlas-2 were carried in flights STS-45 and STS-56. Atlas is part of NASA's Mission to Planet Earth program. STS-64 astronauts also will use the Shuttle Solar Backscatter Ultraviolet Instrument (SSBUV-A3) to measure ozone in Earth's atmosphere. SSBUVs were carried in U.S. shuttles before: SSBUV-1 in STS-34, SSBUV-2 STS-41, SSBUV-3 STS-43, SSBUV-A1 STS-45 and SSBUV-A2 STS-56. Also aboard STS-64 will be the Cryogenic Infrared Spectrometer Telescope for Atmosphere (Crista-Spas), a small U.S.-German aeronomy satellite to measure changes in Earth's atmosphere and complement the UARS satellite launched by STS-48. SPAS (Shuttle Pallet Satellite) is a German-designed payload carrier outfitted with science instruments, carried into space by shuttle, and deployed for several days in its own independent orbit to gather data. Before landing, the shuttle returns to SPAS, retrieves it, and returns it to Earth for another mission.

STS-65
Shuttle: Atlantis
Crew: commander, pilot, payload commander, mission specialists, payload specialist tba (5)
Launch: scheduled May 1994   Duration: 7 days
Altitude: 160 nautical miles   Inclination: 28.5 degrees
Events: SPAS-3 (Shuttle Pallet Satellite) is a German-designed payload carrier outfitted with science instruments, carried into space by shuttle, and deployed for several days in its own independent orbit to gather data. Before landing, the shuttle returns to SPAS, retrieves it, and returns it to Earth for another mission. SPAS-3 will carry Strategic Defense Initiative Organization's (SDIO) Star Wars experiments. SPAS variations previously had been aboard STS-7, STS-10, STS-39, STS-51, STS-64. STS-65 astronauts also are to test laser sensors (Lite-1) and test a remote-arm precision controller (DEE).

STS-66
Shuttle: Columbia
Crew: commander, pilot, payload commander, mission specialists, payload specialist tba (7)
Launch: scheduled May 1994   Duration: 13 days
Altitude: 160 nautical miles   Inclination: 28.5 degrees
Events: International Microgravity Laboratory (IML-2), second microgravity research in a long shuttle Spacelab module, studying effects on material processes and living organisms of very low gravity in space near Earth. With NASA in the project are European Space Agency (ESA), Canadian Space Agency (CSA), French National Center for Space Studies (CNES), West German Research and Development Institute for Air and Spacecraft (DLR), and National Space Development Agency of Japan (NASDA). IML Spacelabs are to fly at 17-mo. to 25-mo. intervals; separation allowing scientists to study results and apply them to new research. Spacelab is a removeable pressurized science lab carried to space and back in the cargo bay. Astronauts float through a tube from crew quarters to Spacelab. They will work with plants and animals, and protein, vapor and mercury iodide crystal growth experiments. Columbia to orbit with tail toward Earth in "gravity gradient" position offering least gravity during flight. Columbia's previous 13-day flights were STS-50 and STS-58.

STS-67
Shuttle: Discovery
Crew: commander, pilot, payload commander, mission specialists, payload specialist tba (6)
Launch: scheduled June 1994   Duration: 7 days

Altitude: 160 nautical miles   Inclination: 28.5 degrees
Events: To carry to space the third private commercial pressurized man-tended module, Spacehab-3. Also to carry Spartan, the Shuttle Pointed Autonomous Research Tool for Astronomy (Sptn-201-2). Sptn is a small, retrievable, free-flying telescope satellite for X-ray astronomy. Also to carry the International Extreme-Ultraviolet Far-Ultraviolet Hitchhiker (IEH-1) telescope to study ultraviolet light from deep space. Hitchhiker is a cargo bay holder for small payloads in orbit. Other experiments aboard STS-67 are to be a cadmium telluride growth test (DSCT); and robot handling of thin-film semiconductor samples (Romps-1).

STS-68
Shuttle: Atlantis
Crew: commander, pilot, payload commander, mission specialists, payload specialist tba (tba)
Launch: scheduled August 1994   Duration: tba
Altitude: tba   Inclination: tba
Events: an available flight with payload to be determined.

STS-69
Shuttle: Columbia
Crew: commander tba, pilot tba, payload commander civilian astronomer Ronald A. Parise, mission specialists tba, payload specialist tba (7)
Launch: scheduled September 1994   Duration: 9 days
Altitude: 190 nautical miles   Inclination: 28.5 degrees
Events: To carry Astro-2 telescopes to study quasars, pulsars, black holes, galaxies, high-energy stars. Parise used Astro-1 at same altitude and inclination in STS-35 in 1990. A small satellite referred to as a free-flying platform and known as Office of Aeronautics, Exploration and Technology-Flyer (OAET-Flyer), to be deployed from Columbia STS-69, will contain technology experiments. Parise would use Shuttle Amateur Radio Experiment (SAREX) equipment to make ham radio contact with amateur operators on Earth; his callsign WA4SIR.

Not Just Along For The Ride

The first U.S. astronauts, seven white male military test pilots, were selected in 1959. For 19 years after that first class of astronauts, American spaceflights were exclusively for men.

The USSR, on the other hand, included a woman in its earliest group of cosmonauts. That first woman in space was 26-year-old Soviet cosmonaut Valentina V. Tereshkova. She orbited Earth alone in a small one-person capsule, Vostok 6, on June 16, 1963. It wasn't until 20 years later on June 18, 1983, that a U.S. woman would go to space.

From 1959-1965, U.S. astronauts were almost exclusively military test pilots. Two civilian male test pilots were selected in NASA's second class of astronauts, in 1962. Not until 1965 and 1967 were any male scientists admitted to the exclusive club. Today, the majority of astronauts still come from military service.

Between 1959 and 1992, 175 men have been selected to become career astronauts. Women first were included in 1978. Six women were selected that year. Between 1978 and 1992 a total of 20 women have trained for astronaut careers.

NASA makes a distinction between a career astronaut, who is trained either as a mission specialist or pilot, and a payload specialist, who is defined as "a member of the shuttle crew who is not a NASA astronaut." More than two dozen men have trained as payload specialists while three women have been so trained.

Breaking the barrier. NASA now lists "an affirmative action program goal of having qualified minorities and women" among its astronaut corps. In 1978 NASA first advertised for women and minorities to join its astronaut ranks. "At that time in our country, people were feeling a little bit bad about the way they had treated women," a NASA spokesperson explained.

Sally K. Ride, Ph.D. physics, replied to a newspaper advertisement and was one of the first six women selected in 1978. Five years later, on June 18, 1983, the 32-year-old Ride, on the seventh shuttle flight, became the first U.S. woman in space.

The five other women selected in 1978 included:

★Anna L. Fisher, M.D. In 1984, Fisher became the first mother to go to space;

★Margaret Rhea Seddon, M.D.;

★Shannon W. Lucid, Ph.D. biochemistry, flew a record third flight by a woman in August 1991;

★Kathryn D. Sullivan, Ph.D. geology, the first woman to take a spacewalk. She also is scheduled for a third flight in 1992 on which she will be the first female payload commander;

★Judith A. Resnick, Ph.D. electrical engineering, the second U.S. woman in space, who later died in the 1986 shuttle Challenger explosion.

The class of 1978 included four minority male astronauts. Two more women were admitted in 1980, three in 1984, two in 1985, and two in 1987, including the first black woman, Mae C. Jemison, M.D. Five women were admitted in 1990, including the first Hispanic woman, Ellen Ochoa, Ph.D., and the first woman to train as a shuttle pilot, U.S. Air Force Major Eileen M. Collins. Two other women in the class of 1990 also are military officers. (See the table below, Female Astronauts 1978-1992.)

The primary barrier of having to be a white male military test pilot in order to become an astronaut crumbled under affirmative action and with the advent of the space shuttle. Still, all female astronauts, except for Major Eileen Collins, have been trained only as mission specialists, who carry out scientific experiments, instead of as shuttle pilots or commanders.

To become a pilot "at least 1000 hours pilot-in-command time in high performance jet aircraft" is required. Major Eileen Collins, a jet fighter pilot, was the first woman at the U.S. Air Force Test Pilot School.

The other female astronauts have scientific backgrounds in physics, engineering, physical sciences, or medicine. The pool of women scientists to draw upon is not very large. Reportedly, only 16 percent of U.S. scientists and engineers are women. Because of role expectations, woman often don't pursue math and science careers.

Role models make a difference. At first, Sally Ride said she was uncomfortable with the media coverage about her flight. "I did not come to NASA to make history," she said. But when she retired from her career in 1987 she acknowledged the importance of having been a role model.

"It was important that NASA added that role model to all the other role models for little girls," she said. NASA administrator James Fletcher said her flight as the first American woman in space established an equal role for women in space exploration.

Gloria Steinem, watching the 1983 blastoff said, "It's important because millions and millions of little girls are going to sit in front of the television and know they can become astronauts after this."

Tamara E. Jernigan, Ph.D. physics, was just such a young woman. She said Ride's "acceptance as a mission-specialist astronaut candidate made me realize I had a shot at becoming an astronaut." Jernigan, at 26 years old, became the youngest astronaut candidate when she was selected in 1985. Her first flight was in June 1991.

Questions of gender. When Sally Ride was interviewed before her first flight, the press questioned her in stereotypical fashion, asking whether she planned to wear a bra in outer space. She replied, "There is no sag in zero G." Another reporter wanted to know if she cried when she had a problem.

Newsweek reported NASA modified the space suits "to allow for shoulders that are less broad" and reported "doctors doubt that the menstrual period would cause problems."

When physician Margaret Rhea Seddon helped stitch together a device to retrieve the Solar Maximum Mission satellite for repair during her flight in 1985, astronaut David Hilmers, acting as mission control capsule communicator, complimented Seddon on her "seamstress work." Ride, also in mission control, corrected him saying, "That was the work of a surgeon."

Working in space. By the end of 1991, fourteen U.S. women had orbited in space on 16 successful missions. One flight was unsuccessful—the 1986 Challenger disaster which killed Judith Resnick and Christa McAuliffe.

"Astronauts don't have to be either very feminine or very masculine women or very superhuman males, or any color or anything," Judith Resnick had said when she was training in 1978. "It's about people in space." She said she thought NASA had "come to accept its role more as a routine than an adventure." Those were prophetic words.

In 1984, for the first time, two women were in space together. Ride, who helped NASA develop the shuttle's robot arm, was aboard Challenger when Kathryn D. Sullivan took the first U.S. female spacewalk to test satellite refueling. On her second flight in 1990, Sullivan helped launch NASA's large Hubble Space Telescope in Earth orbit.

In 1985, Mary Cleave, Ph.D. civil and environmental engineering, operated the 50-foot robot arm to help two spacewalking astronauts test platform construction for future space stations. In weightless space, the 5-foot 2-inch Cleave said she didn't have to worry about being too short or something being too heavy. She could do anything the men could do. On her most recent mission in 1989 she helped launch the Magellan probe on its way to Venus.

In 1989, Kathryn C. Thornton, Ph.D. physics, was the first woman on a secret Defense Department flight. In June 1991, for the first time three women were in space at one time, in one shuttle crew: Tamara Jernigan, Rhea Seddon and Millie Hughes-Fulford.

In 1992, Jan Davis and Mark Lee plan to blast off together in shuttle Endeavour. They will be the first married couple ever to fly together in space.

A risky business. Judith Resnick was one of seven astronauts who died January 28, 1986 when their shuttle Challenger exploded 73 seconds into flight. Sharon Christa McAuliffe was aboard as a space flight participant, America's first teacher in space.

When a faulty O-ring seal caused the solid-fuel booster to explode engulfing Challenger in flames, a small cross-section of America died—two women, three men, one black man, and one Japanese-American man.

The media saw Christa McAuliffe, especially, as "a nonprofessional, an innocent" and "her death therefore seemed doubly poignant and unfair." She was only the ninth American woman, eleventh in the world, to attempt to go to space. Space flights resumed September 29, 1988. The next three flights after Challenger had all-male crews.

The future. The first black woman astronaut, Mae C. Jemison, M.D., was accepted as an astronaut in 1987, after the Challenger disaster. She said the "tragedy saddened but did not deter" her and she thinks the space program offers blacks a promising career. "This is one time when we can get in on the ground floor," she said.

Women advanced toward command positions in 1990 with the acceptance of Major Collins for pilot training. It will be several more years, however, before she actually pilots an orbiter or could be considered for the highest position of crew commander.

The appointment of Sullivan as payload commander on a 1992 flight has been another first. Bonnie Dunbar and Linda Godwin also are slated to be payload commanders. However, many more men than women work in space, which will remain true as long as military and engineering fields from which astronauts are recruited remain mostly male.   —*This report on female astronauts was prepared especially for Space Almanac by Judith G. Curtis, managing editor of Space Today newsmonthly.*

Female Astronauts 1978-1992

Mission Specialists	Ed.	Class	Flights	Remarks
Anna L. Fisher	MD	1978	1984	first mother in space
Shannon W. Lucid	PhD	1978	1985, 1989, 1991	
Judith A. Resnick	PhD	1978	1984, 1986	second U.S. woman, died in Challenger

Sally K. Ride	PhD	1978	1983, 1984	at age 32, first U.S. woman, now retired from NASA
Margaret Rhea Seddon	MD	1978	1985, 1991	
Kathryn D. Sullivan	PhD	1978	1984, 1990, 1992s	first female spacewalk, first female payload commander
Bonnie J. Dunbar	PhD	1980	1985, 1990, 1992s	
Mary L. Cleave	PhD	1980	1985, 1989	
Ellen S. Baker	MD	1984	1989	
Marsha S. Ivins	BS	1984	1990	
Kathryn C. Thornton	PhD	1984	1989, 1992s	first woman on military flt
Linda M. Godwin	PhD	1985	1991, 1993s	
Tamara E. Jernigan	PhD	1985	1991	
Nancy Jan Davis	PhD	1987	1992s	
Mae C. Jemison	MD	1987	1992s	first black woman
Susan J. Helms	MS	1990	--	US Air Force captain
Ellen Ochoa	PhD	1990	--	first Hispanic woman
Nancy J. Sherlock	MS	1990	--	US Army captain
Janice E. Voss	PhD	1990	--	
Pilot				
Eileen M. Collins	MS, MA	1990	--	to be first woman pilot, US Air Force major
Payload Specialists				
Roberta L. Bondar	MD	1983	1992s	first Canadian woman
Millie Hughes-Fulford	PhD	1984	1991	
Space Flight Participant				
Sharon Christa McAuliffe	BEd	1985	1986	school teacher, died in Challenger

NASA refers to Payload Specialists and Space Flight Participants as members of a shuttle crew, but not astronauts.
Ed. = education   s = scheduled future flight

Non-U.S. Astronauts

Spaceflights have been made by persons from various nations in U.S. space shuttles. Men and women who have flown, or are preparing to fly, in U.S. spacecraft are considered astronauts in this list. A few persons from various nations are both cosmonauts and astronauts.

Since 1983, the U.S. has flown foreign astronauts on shuttle missions. All candidates for U.S. spaceflights are prepared at NASA's Johnson Space Center, Houston, Texas. All manned shuttle flights are launched from Kennedy Space Center, Florida.

From	Astronaut	Date		Flight	Orbiter
Germany	Ulf D. Merbold	Nov	1983	STS-9/STS-41A	Columbia
Canada	Marc Garneau	Oct	1984	STS-13/STS-41G	Challenger
Netherlands	Lodewijk van den Berg	Apr	1985	STS-17/STS-51B	Challenger
Saudia Arabia	Prince Sultan Salman Abdel Aziz Al-Saud	Jun	1985	STS-18/STS-51G	Discovery
France	Patrick Baudry	Jun	1985	STS-18/STS-51G	Discovery
Netherlands	Wubbo J. Ockels	Oct	1985	STS-22/STS-61A	Challenger
Germany	Reinhard Furrer	Oct	1985	STS-22/STS-61A	Challenger
Germany	Ernst Messerschmid	Oct	1985	STS-22/STS-61A	Challenger
Mexico	Rodolfo Neri Vela	Nov	1985	STS-23/STS-61B	Atlantis
Future	**Astronaut**	**Date**		**Flight**	**Orbiter**
Canada	Roberta L. Bondar	Feb	1992	STS-42	Discovery
Germany	Ulf D. Merbold	Feb	1992	STS-42	Discovery
Italy	Umberto Guidoni or Franco Malerba	Sep	1992	STS-46	Atlantis

France	Claude Nicollier	Sep	1992	STS-46	Atlantis
Japan	Mamoru Mohri	Sep	1992	STS-47	Endeavour
Germany	Renate Luise Brummer or Hans-Wilhelm Schlegel or Gerhard Thiele or Ulrich Walter	Mar	1993	STS-55	Columbia
USSR	tba	tba	1993	tba	tba

U.S. Man-In-Space Summary

Mercury: One-man capsules, 6 flights, 6 astronauts. First was May 5, 1961; last was May 15-16, 1963. Shortest was the first by Alan B. Shepard, 15 minutes, 22 seconds; longest was the last with Gordon Cooper making 22 orbits in 34 hours, 19 minutes.

Gemini: Two-man capsules, 10 flights, 20 astronauts, between March 23, 1965, and November 11-15, 1966. Longest was 14 days by Gemini 7.

Apollo: Three-man capsules, 11 flights, 33 astronauts, 3 flights circled the Moon, 6 landed on the Moon. Flights started October 11-22, 1968, ended December 7-19, 1972. Longest was 12 days, 13 hours, 48 minutes.

Skylab: America's one and only space station. Apollo capsules ferried three men each to the station; 3 flights, 9 astronauts; the first liftoff May 25, 1973; the final landing February 8, 1974. Longest stay was 84 days, 1 hour, 15 minutes.

Apollo-Soyuz: U.S.-Soviet link-up in space. One Apollo flight with 3 astronauts, one Soyuz flight with 2 cosmonauts, July 15-24, 1975.

Space Shuttles: Prototype space shuttle Enterprise in the 1970's did not fly to space. Five reusable space vehicles did blast off to space and return to land on a runway like an airplane: Columbia, Discovery, Challenger, Atlantis, and Endeavour which replaced Challenger, destroyed in 1986. First flight was April 12-14, 1981. At the start of the 25th flight January 28, 1986, shuttle Challenger exploded 73 seconds into liftoff, killing all seven crew members including four who had flown before. Shuttle fleet grounded 975 days. Flights resumed September 29, 1988. By Fall 1991, 42 successful flights, more than 100 astronauts.

Totals: By the end of 1991, 58 flights, 213 crew members.

Astronaut Hall Of Fame

Log books, training equipment, clothing, photos, mementos and souvenirs of America's original Mercury Seven astronauts are on display in the Astronaut Hall of Fame at Space Camp near the Cape Canaveral spaceport.

The astronauts selected in 1959 were Alan B. Shepard Jr., Virgil I. "Gus" Grissom, John H. Glenn Jr., M. Scott Carpenter, Walter M. "Wally" Schirra Jr., Leroy Gordon "Gordo" Cooper Jr. and Donald K. "Deke" Slayton. Grissom died in a launch pad fire in 1967. Slayton did not fly in Mercury, but went to space in 1975 in the joint U.S.-USSR Apollo-Soyuz Test Project. Shepard, the first American to fly in space, is president of the Mercury Seven Foundation founded by America's first astronauts. The foundation operates Space Camp and grants college science, technology and engineering scholarships. The Astronaut Hall of Fame, Space Camp and a 200-room hotel are on 18 acres outside the gates of Kennedy Space Center, Titusville, Florida. Each year, 3,600 children in grades four through seven attend Space Camp's three-day to five-day classes. Shepard said the Hall of Fame explains space travel from the astronauts' point of view.

U.S. Space Capsule Locations

Poking around in a musty old space capsule can be fun. NASA has spread retired U.S. manned spacecraft across the globe. Here's a handy vacation planner for just about anywhere you may be travelling: Mercury, Gemini, Apollo, Skylab, Enterprise, Lunar

Modules, Challenger—all were important manned craft in United States space history. Not all spacecraft manufactured were flown, however. For instance, McDonnell-Douglas Corp. built 22 Mercury capsules. Six took men to space. The others were used for pad-abort tests, suborbital tests, orbital tests, spacecraft pressurization tests, power tests and tracking and training tests. Some never were used. Many used and unused now are retired and on public display.

Where are the capsules? Not all retired space capsules are readily available for public viewing. When in doubt, call ahead to inquire. List includes locations where most U.S. manned craft known to exist can be found. Abbreviations: LM lunar module; NASA National Aeronautics and Space Administration; USAF United States Air Force.

Spacecraft	Location
Mercury 1	Goddard Museum, Roswell, New Mexico
Mercury 9	North Carolina Museum of Life & Science, Durham, North Carolina
Mercury 10	Kansas Cosmosphere & Discovery Center, Hutchinson, Kansas
Mercury 12B	National Luchtvaart Museum, Schipol, The Netherlands
Mercury 14	NASA, Langley Research Center, Hampton, Virginia
Mercury 15B	NASA, Ames Center, Mountain View, California
Mercury 17	Hall of Science, Queens, New York
Mercury 19	Swiss Museum of Transport, Lucerne, Switzerland
Mercury MA-1	Recovered 1981, in storage, Kissimmee, Florida
Mercury MA-2	World Trade Center, Houston, Texas
Mercury MA-6	National Air & Space Museum, Smithsonian Institution, Wash., D.C.
Mercury MA-7	Hong Kong Space Museum, Hong Kong
Mercury MA-8	NASA, Alabama Space & Rocket Center, Huntsville, Alabama
Mercury MA-9	NASA, Johnson Space Center, Houston, Texas
Mercury MR-1A	McDonnell-Douglas Corp., St. Louis, Missouri
Mercury MR-2	NASA Kennedy Space Center, Cape Canaveral, Florida
Mercury MR-3	National Air & Space Museum, Smithsonian Institution, Wash., D.C.
Mercury MR-4	Atlantic Ocean floor near the Bahamas
Mercury ADIOS-MF	Oklahoma Aviation & Space Hall of Fame, Oklahoma City, Oklahoma
Mercury (unknown #)	Junkyard near NASA Kennedy Space Center, Cape Canaveral, Florida
Gemini 1A	Hall of Sciences, Queens, New York
Gemini 1B El Kabong	Michigan Space Center, Jackson, Michigan
Gemini 2	USAF Space Museum, Cape Canaveral, Florida
Gemini 2A	Kansas Cosmosphere & Discovery Center, Hutchinson, Kansas
Gemini 3	Grissom Memorial Museum, Mitchell, Indiana
Gemini 4	National Air & Space Museum, Smithsonian Institution, Wash., D.C.
Gemini 5	NASA, Johnson Space Center, Houston, Texas
Gemini 6A	McDonnell Planetarium, St. Louis, Missouri
Gemini 7	National Air & Space Museum, Smithsonian Institution, Wash., D.C.
Gemini 8	Neil Armstrong Museum, Wapakoneta, Ohio
Gemini 9A	NASA Kennedy Space Center, Cape Canaveral, Florida
Gemini 10	Swiss Museum of Transport, Lucerne, Switzerland
Gemini 11	NASA, Ames Center, Mountain View, California
Gemini 12	NASA, Goddard Spaceflight Center, Greenbelt, Maryland
Gemini (unknown #)	USAF Museum, Wright-Patterson Air Force Base, Dayton, Ohio
Gemini (unknown #)	Florence Air & Missile Museum, Florence, South Carolina
Gemini MSHO #1890	NASA, Alabama Space & Rocket Center, Huntsville, Alabama
Gemini PARAGLIDER	Manchester Air & Space Museum, England
Apollo 1	NASA, Langley Research Center, Hampton, Virginia
Apollo 2	Kansas Cosmosphere & Discovery Center, Hutchinson, Kansas
Apollo 4	North Carolina Museum of Life & Science, Durham, North Carolina
Apollo 6	Fernbank Science Center, Atlanta, Georgia
Apollo 7	National Museum Science & Technology, Ottawa, Ontario, Canada
Apollo 8	Chicago Museum of Science & Technology, Chicago, Illinois
Apollo 9	Michigan Space Center, Jackson, Michigan
Apollo 10	Science Museum, London, England

Apollo 11	National Air & Space Museum, Smithsonian Institution, Wash., D.C.
Apollo 12	NASA, Langley Research Center, Hampton, Virginia
Apollo 13	Musee de l'Air, Paris, France
Apollo 14	Los Angeles County Museum, Los Angeles, California
Apollo 15	USAF Museum, Wright-Patterson Air Force Base, Dayton, Ohio
Apollo 16	NASA, Alabama Space & Rocket Center, Huntsville, Alabama
Apollo 17	NASA, Johnson Space Center, Houston, Texas
LM-2	National Air & Space Museum, Smithsonian Institution, Wash., D.C.
LM-9	NASA Kennedy Space Center, Cape Canaveral, Florida
LM-13 (Apollo 18)	Cradle of Aviation Museum, Garden City, New York
LM-14 (Apollo 19)	Franklin Institute, Philadelphia, Pennsylvania
LTA-1	Cradle of Aviation Museum, Garden City, New York
LTA-3	NASA, Alabama Space & Rocket Center, Huntsville, Alabama
LTA-8	NASA, Johnson Space Center, Houston, Texas
TM-3	Junkyard near NASA Kennedy Space Center, Cape Canaveral, Florida
Skylab 1B	National Air & Space Museum, Smithsonian Institution, Wash., D.C.
Skylab 2	Naval Aviation Museum, Pensacola, Florida
Skylab 3	NASA, Ames Center, Mountain View, California
Skylab 4	National Air & Space Museum, Smithsonian Institution, Wash., D.C.
Skylab (unknown #)	Junkyard near NASA Kennedy Space Center, Cape Canaveral, Florida
Apollo-Soyuz	NASA Kennedy Space Center, Cape Canaveral, Florida
Apollo-Soyuz CSM	Museum of Science, Seattle, Washington
Apollo-Soyuz 1B	National Air & Space Museum, Smithsonian Institution, Wash., D.C.
Shuttle Enterprise	National Air & Space Museum, Smithsonian Institution, Wash., D.C.
Shuttle Challenger	NASA Kennedy Space Center, Cape Canaveral, Florida

Orient Express

American pilots might be zipping across the skies in the year 2000 in Orient Express, a spaceplane the size of a DC-9 jetliner or U.S. space shuttle, flying at 25 times the speed of sound. Also known as X-30 and National Aero-Space Plane, it would travel 17,000 mph, ferrying passengers through space in a three-hour New York-to-Tokyo run.

NASA and the Dept. of Defense will decide in 1993 whether to build the plane. If so, a prototype might be flying by 2000. The spaceplane would take off and land horizontally on an airport runway, yet fly at Mach 25, the speed needed to reach orbit. X-30 would make long hypersonic flights through the atmosphere. Nowhere on Earth would be more than three hours away.

X-30 would not be the same as a shuttle, which blasts off vertically from a launch pad, requires booster rockets and dumps a large expendable fuel tank after launch. Before commercial flights, engineers must invent new motors, construction materials and computers.

The plane would have a skin with sensors, a frame that would be both body and engine, instruments projecting images into pilots' eyes and artificial intelligence. Computer images would be projected through pilots' pupils, using retinas as miniature projection screens, eliminating windows from the plane.

Pilots would fly by voice command and by flexing fingers. Optical sensors outside would replace radar, radio antennas and infrared sensors.

Artificial intelligence would navigate, control the plane and regulate its explosive hydrogen-fuel scram-jets.

France. The French company Aerospatiale in 1990 said it is looking into a successor for the aging supersonic Concorde operated by British Aerospace and Aerospatiale. The German manufacturer AirBus and U.S. aircraft contractors McDonnell Douglas and Boeing also are studying a Concorde replacement.

If feasible, Aerospatiale said a faster, larger, longer-range plane could be flying by the year 2005.

First Man In Space

USSR cosmonaut Yuri A. Gagarin, the first man ever to travel in space, was born in 1934, rode the Vostok 1 capsule to orbit and back April 12, 1961, and was killed in a plane crash March 27, 1968, when other military aircraft approached too closely in bad weather.

Vostok, which means East, was launched atop a Russian A-1 rocket from Tyuratam in the USSR's central-Asian Kazakh Republic, home of the modern Baikonur Cosmodrome. The capsule orbited at an altitude of 112 to 203 miles.

First spaceship. Vostok included two sections, an instrument module and a 98-in.-diameter reentry vehicle. Inside the reentry capsule was an ejection seat for the cosmonaut, three viewing portholes, film and television cameras, space-to-ground radio, control panel, life-support equipment, food and water. Two radio antennas protruded from the top of the capsule. Flight was controlled from the ground.

Man's world changed forever when Gagarin lifted off from Tyuratam at 9:07 a.m. Moscow time in Vostok 1, becoming the first man to orbit Earth.

Two minutes into Gagarin's flight from Tyuratam, four boosters strapped to the A-1 separated and fell away. Half a minute later a protective shroud covering Vostok was jettisoned. At five minutes, the core booster burned out and the final stage rocket ignited. That stage shut down as Vostok reached orbit 11 minutes 16 seconds into the flight.

A small doll Gagarin was carrying as a gravity indicator began floating free as Vostok 1 reached zero gravity. Thirty years later on April 12, 1991, the man with the most space experience at 541 days, Musa Manarov, was carrying Gagarin's little doll again in Earth orbit, this time at the USSR's orbiting space station Mir on the 30th anniversary of Gagarin's 1961 flight.

Gagarin made one orbit of the Earth April 12, 1961. At 44 minutes into the flight, Vostok was turned into position so a braking rocket could fire to slow its speed, causing it to fall back into the atmosphere. At 78 minutes, the retrorocket was fired. The instrument module separated from the capsule. Reentry began ten minutes later. As planned, at 108 minutes, Gagarin ejected himself from the capsule at an altitude of 23,000 ft. He separated from his ejection seat at 13,000 ft., descending via parachute to land in the Saratov area southeast of Moscow near the Volga River, 1,000 miles west of Tyuratam. The empty capsule came down on parachute. The historic mission lasted 118 minutes from launch to landing.

Dinghy from the sky. Peasant women fled as Gagarin parachuted to Earth near the settlement of Uzmoriye on the Volga River. It was only a year after Gary Powers' U-2 spy plane was shot from the sky by Soviet anti-aircraft gunners.

"Mother, where are you running? I am not a foreigner," Gagarin shouted to forester's wife Anna Tahktorovna and her six-year-old grandaughter, Rita, according to the newspaper Komsomolskaya Pravda in 1991. Gagarin's use of the Russian language calmed Tahktorovna.

Rolling up on motorcycles, Uzmoriye villagers buried the cosmonaut's radio and inflatable rubber dinghy. "The dinghy was a genuine gift for the village fishermen...it literally fell down from the sky," Komsomolskaya Pravda explained.

Then the KGB drove up, threatening to arrest the entire village if the equipment was not given back. The villagers said the dinghy was torn, but the KGB captain put it in his car anyway and drove off.

Predecessor rumors. Before launching the first man to orbit in 1961, the USSR sent dummy human figures, wearing tags printed with the name Ivan Ivanovich, to space in Vostok capsule test flights from Baikonur Cosmodrome in Central Asia.

Georgi Grechko, a veteran of three Soviet space flights, said in 1991 he witnessed the dummy flights and all early manned flights as assistant to the Soviet Union's chief space rocket designer Sergei Korolev.

The dummies were dressed in real space suits and the capsules carried tape-recorded messages to simulate two-way radio. The messages were combinations of letters and numbers. The tape transmissions, overheard around the globe, led to rumors that a cosmonaut had called for help from an out-of-control spacecraft.

Grechko said some rockets blew up before the Gagarin flight. Controllers lost one pre-Vostok test capsule in space, he said. It may still be spinning off somewhere in the cosmos. But most of the dummy test capsules landed as commanded, bouncing down at various sites in Central Asia. They were located first by local residents. Seeing lifeless dummies in space suits, those residents spread rumors that cosmonauts had died.

Dogs travelled in place of dummies on some test flights. Grechko, 60 years old at the time of his report in the February 1991 issue of the Russian newspaper Completely Secret, confirmed Gagarin was the first man in space. Korolev died in 1966.

Hero's end. Seven years after his historic 1961 flight, Yuri Gagarin and co-pilot Vladimir Seryogin were training for the flight of Soyuz 3. They had just taken off in a MiG-15 training airplane March 27, 1968, when they were killed in a crash.

Gagarin, 34 years old when he died, was training for a second space flight to include the first docking of two orbiting capsules. Seryogin was a senior test pilot and decorated military hero who had flown more than 200 combat missions during World War II.

Vladimir Shatalov, now head of cosmonaut training, replaced Gagarin on the future docking flight, piloting Soyuz 4 in 1969.

They may have received a bad weather forecast. The wind was extra gusty as they flew the MiG-15 at high speed between two cloud layers at relatively low altitude. The MiG-15 was fitted with two extra fuel tanks which may have made it less stable. There may have been nearly a head-on collision. Marks on a pressure gauge showed the cockpit had depressurized before the MiG-15 hit the ground.

Apparently, a minute after Gagarin and Seryogin took off in their MiG-15, a pair of faster MiG-21 jets took to the air, overtaking the smaller plane. They were followed a minute later by a second MiG-15 whose pilot didn't see the MiG-15 carrying Gagarin and Seryogin. The two planes were brought "in dangerous proximity to each other" less than 1,640 feet apart, according to a team of investigators in 1987.

The radio callsign of Gagarin's plane was 625. "One can conclude that 625 got on the tail of 614 and was following it," the Soviet newspaper Pravda said. "Finding itself in the trailing vortex of the aircraft in front, the plane piloted by Gagarin and Seryogin got into a spin." The crew tried to stabilize the plane but didn't have enough time to avoid the crash.

The clouds probably prevented the pilot from seeing the horizon. His maneuver to prevent dive and spin while leveling off—the downward deflection of the aileron on a dropping wing to prevent the aircraft from going into a spin—led to wing stall and spin. The entire accident happened in a brief moment. Gagarin and Seryogin did what they could to level off in bad weather but low altitude and lack of time didn't allow them to prevent the crash, Moscow's Novosti Press Agency commented in 1989.

A computer simulation of the crash by an official review commission re-enacted the last stage of flight. The commission concluded, "The plane went into a spin characterized by maximal energy loss. Subsequently it recovered from the spin after which the aircraft ploughed into the ground." Either Gagarin's MiG-15 fell into the vortex wake of another aircraft or else it banked sharply to avoid hitting another plane or an instrument probe, the computer simulation showed.

Intoxication rumors. Official silence clouded the circumstances of Gagarin's death for two decades. Rumors had included one in which the pilots were said to be intoxicated at the time of the crash.

A 1987 investigation uncovered air safety violations in the 1968 exercise, but, in 1989, Novosti said, "The commission's findings unequivocally showed that careless flying or lack of discipline on the part of the crew, as well as a ground-control negligent attitude or someone's malicious intent, were ruled out as possible causes for the disaster."

"The crew's actions aimed at regaining level flight were correct in the highest degree," the investigation report noted. "The pilots retained their capacity for work to the end of the flight, skillfully and efficiently piloting the aircraft."

On the far side of the Moon, a crater was named in Yuri Gagarin's honor.

First American In Space

Former Navy test pilot Alan Bartlett Shepard Jr., 37, was the first American in space. On May 5, 1961, the astronaut was launched on a Redstone rocket in a 15 minute suborbital flight in the Mercury-3 capsule he had named Freedom 7.

Mercury Seven. America was heavily into the Space Race with the USSR when Shepard was selected in NASA's first class of astronauts April 27, 1959. Known as the Mercury Seven, the first class also included: Malcolm Scott Carpenter, Leroy Gordon "Gordo" Cooper Jr., John Herschel Glenn Jr., Virgil Ivan "Gus" Grissom, Walter Marty "Wally" Schirra Jr., Donald Kent "Deke" Slayton. The astronauts and their wives became media stars after they signed a deal giving Life rights to their stories and the magazine published a series of cover stories.

Interest in the first manned flight was strong. NASA kept secret the final selection of its first astronaut to fly in space, saying only Glenn, Grissom or Shepard would be pilot. Three days before liftoff, the press found out Shepard was the one. The suborbital flight was to be in the Mercury-Redstone 3 (MR-3) capsule Shepard called Freedom 7.

Mercury-Redstone. Mercury astronauts called their 6 ft. by 9-ft. capsules "garbage cans." The Redstone rocket was a slender black-and-white booster, 83 feet tall including the capsule on top. Redstone provided 78,000 lbs. of thrust. That compares with today's space shuttle which is 184 feet tall and lifts with 7.7 million lbs. of thrust.

Launch day. America held its breath May 5 as Shepard clambered into Freedom 7 for his second launch attempt. The first had been postponed by storms three days earlier. Shepard wet his spacesuit as he waited more than four hours, strapped in the cramped capsule atop the Redstone rocket on launch pad 5, while NASA fixed a ground computer, the electrical system and rocket fuel pressure.

The Redstone finally ignited at 9:34 a.m., tossing Shepard 303 miles downrange from the Cape Canaveral launch site on a 15 minute 22 second suborbital flight through the edge of space. The capsule reached an altitude of 116 miles and a speed of 5,180 mph.

Another Mercury Seven astronaut, Deke Slayton, was capsule communicator on the ground for Shepard's flight. "Roger, liftoff and the clock has started," Shepard radioed to Slayton as the booster slowly climbed off the launch pad toward space. "Reading you loud and clear, this is Freedom 7. The fuel is go. 1.2 G (gravity). Cabin at 14 psi. Oxygen is go. Freedom 7 is still go."

Shepard, reporting from space everything was "AOK," was weightless just five minutes before splashing down in the Atlantic Ocean. It was to be the shortest stay in space for any U.S. astronaut.

On to the Moon. U.S. President John F. Kennedy awarded Shepard NASA's Distinguished Service Medal at the White House three days later. Three weeks later, on May 25, Kennedy announced America would commit to "landing a man on the Moon and returning him safely to the Earth" by the end of the 1960's. Today, Cape Canaveral tourists read a weathered sign on the old launch pad 5 blockhouse, "From this beginning man reached the Moon."

Shepard, backup pilot for the sixth and final Mercury flight, lobbied hard for a seventh. He even went to the President, but the flight was refused in light of the coming two-man Gemini launch schedule.

Shepard had been the first American, but he was the second person to fly in space. USSR cosmonaut Yuri A. Gagarin had flown to orbit 23 days before Shepard's blast off. The space pioneers did not meet and Gagarin was killed in a 1968 plane crash.

Shepard was scheduled for a Gemini flight, but an inner ear problem produced dizziness and nausea and he was tied to a NASA desk job for six years. As second in command of the astronaut corps, he helped chief astronaut Deke Slayton select crews. Shepard later became chief of the astronaut office. After he underwent a successful secret operation in 1968 in Los Angeles to correct the ear disorder, Shepard was reapproved for spaceflight. His second and last flight was in Apollo 14 to land on the Moon.

Apollo 14. While the Moon flight was a triumph for Shepard, Apollo 14 also gave a boost to NASA after the Apollo 13 failure in 1970.

Shepard, along with Stuart A. Roosa and Edgar D. Mitchell, were launched on a Saturn V rocket January 31, 1971, in the Apollo 14 capsule bound for the Moon. Shepard was mission commander; Roosa pilot of the command module Kitty Hawk; Mitchell pilot of the lunar excursion module Antares.

On February 5, Shepard and Mitchell left Kitty Hawk and flew Antares down to man's third Moon landing. Touching down in the uplands at Fra Mauro formation cone crater, they set out science instruments, towed the first two-wheeled lunar "rickshaw" hand cart containing tools and instruments to the edge of the crater to transport rocks, and collected 96 lbs. of rock and soil samples. In 33 hours 31 minutes on the surface, they took two moonwalks totaling 9 hours 25 minutes. As they prepared to leave the lunar surface in Antares, Shepard made the first golf shot on the Moon. He jury-rigged a 6-iron club head to the end of a digging tool and hit a ball hundreds of yards. He described golf balls as travelling "miles and miles and miles" in the weak lunar gravity.

While Shepard and Mitchell were on the surface, Roosa stayed in the command module which circled the Moon 34 times in 67 hours.

The astronauts set off the first two man-made Moonquakes to be read by seismic monitors planted by earlier Apollo moonwalkers. The Saturn V rocket third stage was fired into the Moon for seismic readings, releasing the explosive power of 11 tons of TNT. It made the Moon ring like a bell, vibrating up to three hours at depths to 25 miles. After Shepard and Mitchell returned to the CSM, the ascent stage of their lunar excursion module (LEM) was crashed into the Moon, creating seismic tremors.

The entire Apollo 14 Earth-Moon-Earth roundtrip lasted nine days (216 hours 1 minute 57 seconds). The trio splashed down on Earth February 9, 1971. Although organic compounds were in the lunar soil, no micro-organisms were found so Apollo 14 astronauts were the last to be quarantined for 14 days at Houston upon return to Earth.

Today. The Mercury capsule Freedom 7 is on display at the National Air and Space Museum, Smithsonian Institution, Washington, D.C. The Apollo 14 command module Kitty Hawk is displayed at the Los Angeles County Museum, Los Angeles, California.

Shepard was promoted to admiral in 1971. He resigned from NASA in 1974. Today he's a millionaire businessman in Houston—a commercial property developer, a venture capital group partner, and director of mutual fund companies. Shepard also is chairman of the Mercury Seven Foundation, created by the six living Mercury Seven astronauts and Grissom's widow to raise money for science and engineering scholarships. Most of Shepard's spaceflight mementos are at the United States Astronaut Hall of Fame just 14 miles from the old Mercury launch pad, and other museums.

First American In Orbit

John Herschel Glenn Jr. only made one spaceflight, but it was of singular importance: he was the first American in orbit.

Nine months after Alan B. Shepard Jr. had made America's first spaceflight, a quick up-and-down suborbital ride on May 5, 1961, Glenn was launched all the way to orbit on an Atlas-D rocket February 20, 1962, in the Mercury-6 capsule Friendship 7.

Space Race. America was in a Space Race with the Russians when Glenn was selected in NASA's first class of astronauts April 27, 1959. Known as the Mercury Seven, the first class also included: Malcolm Scott Carpenter, Leroy Gordon "Gordo" Cooper Jr., Virgil Ivan "Gus" Grissom, Walter Marty "Wally" Schirra Jr., Alan Bartlett Shepard Jr., Donald Kent "Deke" Slayton.

First but fourth. Glenn's journey was the first U.S. manned orbital flight, the third manned American spaceflight and the first manned lift off in a Florida winter. Glenn was the third American in space, the fourth person in space, the first American in orbit and the third person in orbit.

USSR cosmonaut Yuri Gagarin had been the first person in orbit, riding Vostok 1 for one revolution around Earth April 12, 1961. Then American astronaut Alan Shepard made a suborbital flight of less than one orbit in a Mercury capsule May 5, 1961. Gus Grissom made a second U.S. suborbital flight of less than one orbit in a Mercury capsule July 21, 1961. Cosmonaut Gherman Titov rode Vostok 2 for 16 orbits August 6, 1961.

Glenn and Grissom had been mentioned for the first Mercury suborbital flight in May 1961, but Shepard was chosen to pilot the first manned "garbage can" as the Mercury astronauts called their 6 ft. by 9-ft. capsules. Glenn called his Mercury-Atlas 6 (MA-6) capsule Friendship 7.

In orbit. Following the pair of U.S. Mercury-Redstone suborbital test flights by Shepard and Grissom in May and July 1961, and the two-orbit Mercury-Atlas flight of Enos the chimp in November 1961, NASA decided to go for it.

An Atlas-D rocket, launched from Cape Canaveral, carried Glenn in Friendship 7 to space for three revolutions around the planet February 20, 1962.

While circling Earth three times, Glenn ate applesauce and watched for Perth, Australia, residents who turned on all the lights in town to signal as he passed overhead.

"As if I were walking through a field of fireflies," was how Glenn described a flock of luminous yellow-green particles which drifted past his window in orbit. The fireflies remained a mystery for three months until another Mercury Seven astronaut, Scott Carpenter, saw them and discovered they were frost particles flaking off the outside of his capsule.

Glenn splashed down in the Atlantic Ocean. The flight lasted 4 hours 55 minutes 23 seconds. Americans gave him the same hero's welcome as after Charles A. Lindbergh's 1927 New York-Paris flight.

Today. Glenn's Mercury-6 capsule Friendship 7 is on display at the National Air and Space Museum, Smithsonian Institution, Washington, D.C. Glenn went on to become the first astronaut elected to public office. He was elected to the U.S. Senate from Ohio in 1974. He still holds the office. Glenn is on the board of the Mercury Seven Foundation, created by the six living Mercury Seven astronauts and Grissom's widow to raise money for science and engineering scholarships.

Mercury Seven. America's first astronauts all flew in space. Carpenter rode in the Mercury-Atlas 7 capsule Aurora 7. Cooper flew in Mercury MA-9 Faith 7 and Gemini 5. Glenn flew in the Mercury-Atlas 6 capsule Friendship 7. Grissom travelled in Mercury-Redstone 4 capsule Liberty Bell 7 and in Gemini 3, then was killed in Apollo 1. Schirra flew in the Mercury-Atlas 8 capsule Sigma 7, in Gemini 6 and in Apollo 7. Shepard rode in Mercury-Redstone 3 capsule Freedom 7 and went to walk on the Moon in Apollo 14. Slayton flew in the Apollo 18 part of the joint-U.S./USSR Apollo-Soyuz Test Project.

First Woman In Space

Soviet cosmonaut Valentina V. Tereshkova was not only the first woman in orbit, but also the first ordinary person in space. Born March 6, 1937, at Maslennikovo, she had been a tire factory worker, then a cotton mill worker who enjoyed the hobby of parachute jumping when she was picked for a class of women to train for spaceflight.

As Yuri Gagarin made man's first space voyage in 1961, Valentina V. Tereshkova was head of the local cotton mill workers parachuting club and secretary of the local Young Communist League. It was said she dreamed of spaceflight after Gagarin's Vostok 1 flight. After Vostok 2, she wrote to Moscow asking for assignment as a cosmonaut.

However it happened, she was selected February 16, 1962, and started training March 12 with five other women: Tatyana D. Kuznetsova, Irina B. Solovyova, Valentina L. Ponomaryova, and Zhanna D. Yorkina. Among rigorous preparations, they made hundreds of parachute jumps and learned to pilot jets.

USSR Premier Nikita Khrushchev wanted a space spectacular, so Tereshkova was selected the first female cosmonaut. She spent 70 hours 50 minutes orbiting Earth 48 times on June 16, 1963, in her Vostok 6 capsule Sea Gull. At age 26, Tereshkova was the youngest woman to fly in space. Her flight was the second team flight—she came within three miles of cosmonaut Valeri F. Bykovsky orbiting in Vostok 5.

She had made 126 parachute jumps, but her brief spaceflight training had not prepared her for high-G re-entry. She ejected as planned from Vostok 6 and floated by parachute to Earth. Reportedly, while hanging from the parachute on the way down, she looked up and was hit in the face with loose metal. Extra make-up was required for several days. After landing in the Southern Urals, she was reported in "pitiful condition," but recovered quickly and was rushed to Moscow for the celebration. Premier Khrushchev boasted, "It is our girl who is first in space."

Interestingly, no other "girls" flew in space for 19 years until USSR cosmonaut Svetlana Savitskaya flew in 1982. Meanwhile, Sally Ride, who in 1983 would become the first American woman in orbit, was a 12-year-old sixth grader in Los Angeles when Tereshkova was in orbit in 1963.

Tereshkova became the first politician from space when she was elected a member of the Supreme Soviet in 1967. Later she became a member of the Presidium of the Supreme Soviet in 1974, then was elected to the Congress of People's Deputies in 1989.

Tereshkova and cosmonaut Andrian Nikolayev became the first couple married after spaceflight when they joined in November 3, 1963, five months after her flight in Vostok 6. Nikolayev had flown in Vostok 3 in August 1962 and later flew in Soyuz 9 in 1970. The couple was said to have had a romance before her flight, but then there was speculation the marriage was arranged for scientific reasons. At any rate, Premier Khrushchev was toastmaster at the wedding.

The first cosmonaut progeny was Yelena, daughter of Tereshkova and Nikolayev, born in June 8, 1964. Yelena became a medical student.

None of Valentina Tereshkova's fellow women cosmonauts flew to space. When a planned Voskhod flight with an all-woman crew was cancelled after their training, the group of five female cosmonauts was disbanded in 1969. Tereshkova and Nikolayev were reported in June 1983 to be divorced.

First American Woman In Space

Astronaut Sally K. Ride, aboard shuttle Challenger flight STS-7 June 18, 1983, was the first American woman in space and, at 32, the youngest U.S. woman ever in orbit. Ride and astronauts Robert L. Crippen, Frederick H. "Rick" Hauck, John W. Fabian and Norman E. Thagard composed the first American mixed male/female crew in orbit. Ride compared her first shuttle launch to a Disneyland ride.

Ride became the first U.S. woman to go to space twice when she returned in Challenger flight STS-41G (STS-13) on October 5, 1984. It also was the first time two women were in space at same time. Kathryn D. Sullivan rode along with Ride and Robert L. Crippen, Jon A. McBride, David C. Leestma, Marc Garneau and Paul D. Scully-Power.

Sullivan became the first American woman to take a spacewalk when she went outside for three hours with Leetsma, the first male-female spacewalk.

Tereshkova. The first woman in space had been Valentina V. Tereshkova, a Soviet textile mill worker and hobby parachute jumper. At age 25, cosmonaut Tereshkova spent 71 hours in Vostok 6 orbiting Earth 48 times on June 16, 1963.

Sally Ride, born in Los Angeles and growing up nearby in Encino, was a 12-year-old California sixth grader when Tereshkova was in orbit. Ride's father was a college administrator and her mother a psychologist.

They introduced Sally to tennis to keep her from playing football in the streets. By age 13, she was working toward becoming a nationally ranked tennis player. She received tennis scholarships, joking later she would have been a professional if her backhand had been better.

Shakespeare. Shakespeare and science interested Ride. At Stanford University in the 1970's, she earned bachelor's degrees in English and physics, then a master's and doctorate in physics, graduating as NASA was opening the astronaut corps to women.

Ride was one of six women in a group of 35 persons selected for the astronaut corps in 1978. Kathryn D. Sullivan was in the group. Another was Judith A. Resnick who died in Challenger. The others were Anna L. Fisher, Shannon W. Lucid and Margaret Rhea Seddon.

One of the men in the 1978 astronaut class was astronomer Steven A. Hawley. He and Ride were married in 1982, but they were divorced in 1987. Hawley was deputy chief of the astronaut office at NASA's Johnson Space Center after flying to space in Discovery STS-41D, Columbia STS-61C, and Discovery STS-31.

Challenger. Ride helped develop the space shuttle remote manipulation arm. She was ground-to-space capsule communicator for four shuttle flights.

After Challenger exploded during lift off in January 1986, Ride spent days with the family of Judy Resnik, her close friend and one of seven crew members killed.

U.S. President Ronald Reagan appointed Ride to the Rogers Commission which investigated the accident. She represented the astronaut corps and was a key member of the commission in 1986 and 1987.

The Ride Report. After her Rogers Commission work, NASA administrator James C. Fletcher appointed Ride special associate administrator to study long-range goals for the space agency.

In March 1989, Ride called on Congress to support NASA's proposed Mission to Planet Earth to study the planet's environment from space. The project would use science gear aboard the U.S.-international space station Freedom, as well as orbiting satellites and 1,000 land and sea stations, to gather data about Earth's environment.

Ride promised "significant economic benefits," including an ability to predict droughts and heavy rainfall and find large schools of fish for the world's fishing fleets.

After NASA. Ride quit the astronaut corps and left NASA in 1987 to return to Stanford as a physicist for the university's Center for International Security and Arms Control, Palo Alto, California.

At the time of her departure from NASA, Ride, 36, had been an associate administrator for long range planning at NASA headquarters in Washington, D.C.

In 1989, she became a professor at the University of California, San Diego, and director of the school's California Space Institute.

The California Space Institute, which coordinates University of California space research, was created by the state legislature in 1979 to maintain liaison with the aerospace industry and develop technology.

Fame. Shuttle flights made her a celebrity which the media-shy Ride said was distasteful. She chided the press for highlighting a woman in space, saying people should realize women can do any job they want.

NASA administrator James Fletcher said Ride's flight as the first American woman in space established an equal role for women in space exploration. The assignment of women to shuttle crews today is a routine matter based on ability and need and is no longer a cause for notice, he said.

First Man On The Moon

The world's best-known spaceman must be U.S. astronaut Neil A. Armstrong, who landed on the Moon July 20, 1969, to become the first man to walk the lunar surface.

As his Apollo 11 lunar lander touched down on the Sea of Tranquility, Armstrong radioed, "Houston, Tranquility Base here. The Eagle has landed."

"Roger, Tranquility. You've got a bunch of guys about to turn blue," was the colorful reply from Houston as NASA engineers drew sighs of relief at the successful landing.

Hopping off the lander onto the surface, Armstrong said, "That's one small step for (a) man...one giant leap for mankind." NASA later found the "a" lost in transmission.

First civilian astronaut. Seven years before his historic landing a quarter million miles from home, aeronautical engineer Neil Armstrong had been selected America's first civilian astronaut. The other civilian astronaut in NASA's astronaut class of 1962 was U.S. Maritime Academy engineer Elliot See.

It was NASA's second group of astronauts. Armstrong went on to fly to space in Gemini 8 in 1966. See was killed in an airplane crash in 1966 without having made a spaceflight.

Gemini. Armstrong and David R. Scott flew to space in the Gemini 8 capsule atop a Titan 2 rocket on March 16, 1966. Just 101 minutes before their flight, an Agena rocket had blasted off from Cape Canaveral to be their target in orbit.

Orbiting above Earth, they maneuvered Gemini 8 to dock with the Agena target rocket 6 hours 34 minutes after launch. It was the first space docking, beating the Russians to that milestone in the Space Race by three years.

Armstrong and Scott were supposed to stay three days in space after the docking, but the joined pair of vehicles began tumbling wildly. Armstrong was able to back Gemini away from Agena, but used up 75 percent of his fuel in the maneuver. NASA ended the flight immediately, after Gemini had completed only 6.5 orbits of Earth. The flight lasted 10 hours 41 minutes 26 seconds.

Apollo. To protect the astronauts from carrying germs to the Moon, a NASA doctor called off a dinner U.S. President Richard M. Nixon wanted to give for Armstrong and U.S. Air Force officers Edwin E. "Buzz" Aldrin Jr. and Michael Collins on July 15, 1969, the night before the trio was to leave for the first Moon landing. Later Collins said, "I'm sure presidential germs are benign."

On July 16, Armstrong, Aldrin and Collins blasted off from Cape Canaveral in the Apollo 11 spacecraft atop a mighty Saturn V rocket, bound for the Moon. Armstrong was mission commander, Collins was CSM pilot, and Aldrin was LEM pilot.

Some 75 hours 50 minutes later, Apollo 11 reached lunar orbit on July 19. A day later, on July 20, at 100 hours 12 minutes into their journey, Armstrong and Aldrin entered the lunar excursion module (LEM) Eagle and flew away from the command service module (CSM) Columbia. After 2 hours 17 minutes of descent, the LEM was at an altitude of 26,500 feet and five miles from touchdown in Mare Tranquillitatis when the braking rocket was fired.

Landing. When Eagle was 60 feet above the surface, Armstrong and Aldrin scanned the surface for a level spot to land, finding the right spot on the dusty, desolate Sea of Tranquility (Mare Tranquillitatis). The first landing of men on the Moon came 102 hours 45 minutes after they left Earth. At 109 hours 24 minutes, Armstrong was first to step onto the surface, making that giant leap for mankind.

As Collins stayed above in the CSM in lunar orbit, Armstrong and Aldrin slept seven hours of their 21 hours 36 minutes 21 seconds on the surface. During their one moonwalk lasting 2 hours 31 minutes, Armstrong and Aldrin set out a U.S. flag, science instruments, laser beam reflector, seismometer which transmitted evidence of a moonquake, a sheet of aluminum foil to trap solar wind particles, and a plaque on the LEM descent stage, "Here Men From Planet Earth First Set Foot Upon the Moon. July 1969 A.D. We Came In Peace For All Mankind."

They ate hot dogs, bacon squares, canned peaches, and sugar cookies, and drank hot coffee, then photographed and lugged 48.5 pounds of rocks and soil into the LEM.

They left behind on the Moon a gold olive branch, cameras, tools, boots, bags, containers, armrests, brackets, a U.S. flag and flagstaff, a solar wind experiment mast, a TV camera and power cable, a moonquake detector, a laser light reflector, a plaque, their Eagle lander, life-support backpacks, an Apollo 1 shoulder patch remembering three dead astronauts, medals commemorating two dead cosmonauts, and a 1.5-in.silicon disk with goodwill speeches from heads of 23 Earth nations.

Leaving. At 124 hours 22 minutes since leaving Earth, they lifted off the lunar surface July 21 in the LEM. The downward rocket blast from the LEM ascent stage toppled the American flag the astronauts had planted in the lunar soil.

After an upward bound LEM flight of 3 hours 41 minutes, they arrived at the CSM. The trio left the LEM ascent stage in lunar orbit and blasted out of lunar orbit July 22 to return home in the CSM. The CSM had circled the Moon 30 times in 59 hours 30 minutes. Armstrong, Aldrin and Collins became the first men to be launched from the Moon to Earth.

They splashed down in the Pacific Ocean July 24—man's first return from a celestial body. The complete Earth-Moon-Earth roundtrip had lasted 195 hours 18 minutes 35 seconds—about eight days.

After Apollo. Millions of people back on Earth watched Armstrong's historic first step 238,000 miles away and heard his words via an intricate live TV hookup.

Back home, the 38-year-old Armstrong was hailed as a hero. He was decorated by 17 countries and given dozens of medals and honors.

Armstrong became deputy associate administrator for aeronautics at NASA's Washington, D.C., headquarters office of advanced research and technology. He retired in 1971 to become a professor of aerospace engineering at the University of Cincinnati. Armstrong taught a few classes and researched ways to make everyday use of space technology. He retired from the University of Cincinnati in 1979.

He sat on the boards of corporations, including Gates Learjet, and continued to test pilot commercial jet aircraft. Armstrong served on the Rogers Commission which investigated the 1986 Challenger explosion. There have been 11 Ohio astronauts, including Judith A. Resnik who died in the Challenger disaster.

Museum. While Armstrong was on the Moon, then-Ohio-Gov. James A. Rhodes proposed a museum at the astronaut's hometown of Wapakoneta, Ohio. Three years later, the Neil Armstrong Air and Space Museum opened in the small town. In its first year, the museum attracted 210,000 visitors to its domed theater on Interstate Route 75. Recently, 55,000 persons have visited each year.

Exhibits in the 15,000-sq.-ft. facility include Armstrong's Gemini 8 capsule, his Gemini and Apollo spacesuits and other Apollo gear. A Moon rock is the centerpiece of a Moon geology exhibit. Exhibits from the Smithsonian Institution also are shown from time to time. The Apollo 11 CSM capsule Columbia is on display at the National Air and Space Museum, Smithsonian Institution, Washington, D.C.

Over the years, Armstrong has tried to live a mostly-private life. A few years after his moonwalk, as Armstrong lived with his wife and two children on a farm just 100 miles away near Cincinnati, the director of the Wapakoneta museum said, "The kids think Neil's dead."

20th anniversary. Armstrong, Aldrin and Collins were reunited for the July 1989 anniversary of their Moon flight. Armstrong said, "Twenty years ago this morning...the three of us left on our summer vacation. It was a vacation of going new places, seeing new sights, taking lots of pictures."

NASA Administrator Richard Truly called Apollo 11 "one of the most important events in human history in this entire last thousand years."

Growing up. Armstrong was born August 5, 1930, near the small northwestern Ohio community of Wapakoneta, Ohio, population 7,000. When he was two years old, his father took him to the Cleveland airport. He flew his first airplane ride at six.

Fascinated by flight, he bicycled to the tiny Wapakoneta airport, hanging around to learn airplane mechanics. He paid for flying lessons, earning an airplane pilot's license before he was old enough for an automobile driver's license.

Armstrong flew 78 Navy combat missions in the Korean War, once flying a plane with a shredded wing out of enemy territory to bail out over friendly ground. Later he was a high-speed test pilot, including X-15 rocket planes, for seven years.

He received a B.S. degree in aeronautical engineering from Purdue University, then a master's degree in aerospace engineering from University of Southern California.

Stephen K. Armstrong of Wapakoneta, Armstrong's father, died February 3, 1990, at 82. Just 107 days later, Viola Louise Armstrong of Wapakoneta, Armstrong's mother, died May 21 at 83.

Longest Space Trip

Cosmonauts Vladimir G. Titov, 40, and Musa K. Manarov, 37, captured the single-trip space endurance record of 366 days when they flew home from the USSR's orbiting Mir space station December 21, 1988. The previous record of 326 days in one trip had been set by Yuri Romanenko in 1987.

Manarov, with a total of 541 days in two trips, also holds the personal record for most time accumulated in space. Romanenko is second with 430 days in three trips. The longest American space flight was 84 days by three astronauts aboard Skylab in 1973.

Records. Titov, Manarov and Anatoly Levchenko flew to Mir December 21, 1987, in the Soyuz TM-4 transport. Levchenko flew home in Soyuz TM-3 December 29, as planned. Dr. Valery Polyakov flew to Mir August 29, 1988, to check on the health of Titov and Manarov. He found them physically healthy, although reports later indicated Titov and Manarov had quarreled occasionally in the space station and once went three days without speaking to each other. On November 11, Titov and Manarov surpassed Romanenko's record for a single trip to space.

Sergei M. Krikalev, who was launched May 18, 1991, in Soyuz TM-12 to Mir is expected to come close to meeting or exceeding the Manarov-Titov record. Krikalev, who had planned to fly down to Earth October 10, 1991, had his trip extended to around 365 days, flying home around May 17, 1992.

Krikalev previously flew to space for 152 days in the Soyuz TM-7 flight launched November 26, 1988. He flew home April 27, 1989. If his Soyuz TM-12 flight is extended to 365 days, landing May 17, 1992, he will have accumulated a total of 517 days in space—not as much as Manarov's 541 days but more than Romanenko's 430.

Entertainment. The USSR has sent cosmonauts on six-month and year-long assignments at space stations to see how humans withstand weightlessness and other space conditions in preparation for a three-year manned flight to Mars around 2005.

Vladimir Titov had thought he might want his guitar along in space to while away evening rest periods.

"Do you play your guitar?" mission control asked one day. "Not often. We have little time for it," Titov replied.

Exercise. Cosmonauts have an exercise treadmill in Mir to keep fit. One day, Titov radioed mission control, "I have forgotten to say to you that shoulder straps must be wider. The belt hurts the shoulder. It is difficult to run."

Mission control knew he was talking about rubber straps holding the cosmonauts to the treadmill during training sessions. All such complaints were logged for study by the space station's designers.

Astronomy. Mission control center, one day, radioed a request, "Let us work with Canopus. The star will enter the telescope's field of vision for a long time. It must be kept in a small ring. There is no lens hood on the objective so try to let less light pass." Mission control then recalled, "Romanenko worked with a flashlight. Rule out all illumination inside the station. Recording must be reliable."

At the appointed hour, mission control said, "According to calculations, Canopus must now be entering the telescope's field of vision."

"We don't see Canopus. Just a moment. Here it is," Manarov replied. "Vladimir, record it."

Mission control reminded, "Musa, don't forget to switch on the gas analyzer at 1810." Later, mission control sent up congratulations, "Your work on Canopus has been done well. In the next communication session, let's try to record the star for a longer time. Did the tape on the tape recorder move?"

"It was dark. We could not notice. Incidentally, sometimes the still camera's blind is not closed. Film can be exposed to light," Titov replied. A discussion of how to handle the camera followed.

Spacewalks. Titov and Manarov took two spacewalks outside Mir in 1988: June 30 to start repair of the Kvant-1 telescope and October 20 to complete the repair.

Guest cosmonauts. Titov and Manarov received three groups of cosmonaut visitors in 1988: Soyuz TM-5, TM-6 and TM-7.

Soyuz TM-5 on June 7, 1988, carried Anatoly Solovov, Viktor P. Savinykh and guest cosmonaut Alexander Alexandrov from Bulgaria, the second Bulgarian Soyuz flight in nine years. Solovov, Savinykh and Alexandrov flew home June 17, 1988.

Soyuz TM-6 on August 29, 1988, ferried Dr. Valery Polyakov, Vladimir Lyakhov and guest cosmonaut Abdul Ahad Mohmand from Afghanistan. Cardiovascular specialist Dr. Polyakov was sent to monitor Titov and Manarov health. Lyakhov and Mohmand left September 6, 1988. Dr. Polyakov stayed 241 days to April 27, 1989.

Soyuz TM-7 on November 26, 1988, launched Alexander A. Volkov, Sergei M. Krikalev and guest cosmonaut Jean-Loup Chretien from France. Chretien stayed 25 days. Volkov and Krikalev stayed 152 days. Chretien's 25 days set a record for non-USSR cosmonaut stay at a USSR space station. He also became the first non-USSR cosmonaut/non-USA astronaut to spacewalk when he and Volkov went outside December 9, 1988. The Soyuz TM-7 crew took along a music tape by British rock group Pink Floyd. Chretien treated Manarov and the other cosmonauts to a French chef's gourmet dinner with 23 foods.

Going home. Titov and Manarov on December 21, 1988, handed over the station to a new main crew, Alexander A. Volkov and Sergei M. Krikalev, who had flown to space November 26. Dr. Polyakov remained at Mir with Volkov and Krikalev, while Titov and Manarov left for home with Chretien in Soyuz TM-6.

During the flight down to Earth, trouble developed briefly when telemetry showed an on-board computer overload. Descent was delayed three hours as they switched manually to a back-up program and landed safely December 21.

Landing after 366 days in weightlessness, Titov and Manarov looked pale and drawn as they were lifted from their Soyuz capsule. They had problems maintaining equilibrium when they tried to stand up. Within about three hours, they were able to walk with assistance. Within two days, they were able to walk unassisted.

Titov. Vladimir Titov was born January 1, 1947, at Sretensk, Russia. He worked in oil prospecting, heating a drill, and joined the Communist Party in 1971. Titov became a test pilot and joined the cosmonaut team in 1976. He celebrated his 41st birthday while in orbit. Titov is married and has a daughter and son.

Altogether, Titov has accumulated 368 days in space in three Soyuz missions. His first flight was Soyuz T-8 in April 1983 with Alexander Serebrov and Gennadi Strekalov. They went to space to dock at Salyut 7 space station, but the Soyuz radar was damaged and the crew couldn't tell how fast they were closing with the station. After a very close call in which Soyuz and Salyut whizzed by each other at high speed, the cosmonauts were ordered home immediately. Displeased officials gave Titov only the Order of Lenin for his effort, not the higher honor of Hero of the Soviet Union.

His second flight was to have been with Gennadi Strekalov in Soyuz T-10A to Salyut 7 in September 1983, but fire erupted at the base of their A-2 rocket 90 seconds before launch and the rocket exploded on the pad. They were set to lift off from Baikonur Cosmodrome when, at T-25 seconds, they noticed red-yellow flames licking up past their porthole and black smoke roiling up from the launch pad. They tried a manual abort to fire their escape rocket, but couldn't make it ignite. The rocket was falling over, already tipped 20 degrees off vertical, when frantic ground controllers managed to trigger the escape rocket. The capsule separated from the booster, blasting away six seconds before the A-2 exploded. Titov and Strekalov landed, uninjured, several miles away.

His third flight was Soyuz TM-4 with Manarov in December 1987. Vladimir Titov is not related to Gherman Titov who, in August 1961, became the second man to orbit Earth and the first to stay in space more than a day.

Manarov. Musa Manarov, a rookie cosmonaut who made his first trip to space in Soyuz TM-4, was born March 22, 1951, at Baku in the USSR's Azerbaijan Republic.

He was selected as a cosmonaut in 1978, joined the Communist Party in 1980, and began active training at Yuri Gagarin Cosmonaut Training Center in 1983. Manarov celebrated his 37th birthday in orbit.

Manarov started his second space flight to Mir station December 2, 1990, with Viktor Afanasyev and Toyohiro Akiyama in Soyuz TM-11. Akiyama stayed eight days while Manarov and Afanasyev stayed 175 days. Akiyama, a veteran Japanese TV correspondent and former Washington bureau chief for the Tokyo Broadcasting System, was the first Japanese and the first journalist in space. To celebrate the 30th anniversary of Yuri Gagarin's first-ever manned space flight in 1961, Manarov took along the same small doll Yuri Gagarin had carried as a gravity indicator. The doll floated free when the manned capsule reached zero gravity. Manarov and Afanasyev made four spacewalks outside Mir. They spent Russian Christmas, January 7, 1991, on a spacewalk repairing Kvant-2 airlock hatch. They did astronomy research with the Buket telescope and the Granat spectrometer, and used the Gallar furnace to grow zinc oxide monocrystals. When he flew home in Soyuz TM-11 May 26, 1991, with Afanasyev and British cosmonaut Helen Sharman, Manarov set the record for most time accumulated in space at 541 days, while continuing to share the 366-day Soyuz TM-4 record for longest single spaceflight.

Manarov was part of the largest number of persons ever to orbit Earth at the same time, In December 1990—twelve astronauts and cosmonauts.

Levchenko. Levchenko, 47, from the Ukraine, died of a brain tumor 8 months after returning to Earth in Soyuz TM-3. A cosmonaut since 1981, Levchenko had not flown to space before Soyuz TM-4.

Most Time In Space

USSR cosmonaut Musa K. Manarov, with a total of 541 days on two trips, holds the record for most time accumulated in space.

Manarov, then 37, and Vladimir G. Titov set the single-trip space endurance record of 366 days in space during a visit to the USSR's orbiting Mir space station in 1988. Manarov and Titov were launched in Soyuz TM-4 December 21, 1987. On November 11, 1988, they surpassed Yuri Romanenko's 326-day record for a single trip to space set in 1987. Manarov and Titov returned to Earth December 21, 1988.

Manarov later returned to space for 175 days, launched December 2, 1990, in Soyuz TM-11 to Mir station. He was part of the largest number of persons ever to orbit Earth at the same time—twelve astronauts and cosmonauts. He returned to Earth May 26, 1991.

Romanenko. Yuri Romanenko has the second greatest amount of time accumulated in space with a total of 430 days in three trips to orbit. The longest American space flight was 84 days by three astronauts aboard Skylab in 1973.

Krikalev. Sergei M. Krikalev, who was launched May 18, 1991, in Soyuz TM-12 to Mir is expected to come close to meeting or exceeding the Manarov-Titov record, staying in space to May 1992. He previously flew to space for 152 days in the Soyuz TM-7 flight in 1988. If his Soyuz TM-12 flight extends to 365 days, he will have accumulated 517 days in space—not as much as Manarov's 541 days but more than Romanenko's 430.

Azerbaijan. Musa Manarov, a rookie cosmonaut who made his first trip to space in Soyuz TM-4, was born March 22, 1951, at Baku in the USSR's Azerbaijan Republic. Manarov is of the Lakets nationality, a culture of 100,000 people in the USSR's Daghestan Autonomous Soviet Socialist Republic on the west shore of the Caspian Sea.

He joined the Young Communists League as a teenager, worked as aircraft designer, and applied to become a cosmonaut in 1976. He was selected in 1978, joined the Communist Party in 1980, and began active training at Yuri Gagarin Cosmonaut Training Center in 1983. Manarov celebrated his 37th birthday in orbit. He is married to a first-aid doctor. They have a daughter and son. Manarov visited Johnson Space Center, Houston, Texas, in November 1991.

Most People In Orbit At One Time

There never have been more people from Earth in space at one time than after the launch of Soyuz TM-11 and Columbia STS-35 on December 2, 1990. Altogether, 12 astronauts and cosmonauts were in space at the same time for a week, although not in the same orbits.

At the USSR's Mir space station were Gennadi Strekalov, Gennadi Manakov, Musa K. Manarov, Viktor Afanasyev and Japanese broadcaster Toyohiro Akiyama. In U.S. shuttle Columbia were Vance D. Brand, Guy S. Gardner, John M. "Mike" Lounge, A. Robert Parker, Jeffrey A. Hoffman, Samuel T. Durrance and Ronald A. Parise.

The previous record number of men in space at one time had been 11 in April 1984, when five were in U.S. shuttle Challenger flight STS-41C, and six were at the USSR's Salyut 7 space station.

In Challenger were Robert L. Crippen, Francis R. "Dick" Scobee, Terry J. Hart, George D. "Pinky" Nelson and James D. "Ox" Van Hoften. At Salyut 7 were cosmonauts Oleg Atkov, Leonid Kizim, Vladimir Solovyev, Yuri V. Malyshev, Rakesh Sharma and Gennadi Strekalov.

Most Nationalities In Orbit

People of many nations and ethnic groups have flown in space. The USSR has flown the most nationalities to space, taking cosmonauts from 16 foreign nations to Salyut and Mir stations. The U.S. has flown astronauts from six countries in space shuttles.

The USSR has flown cosmonauts from Czechoslovakia, Poland, Germany, Bulgaria, Hungary, Vietnam, Cuba, Mongolia, Romania, France, India, Syria, Afghanistan, Japan, Great Britain and Austria. The USSR has plans to fly cosmonauts from Malaysia, Israel, China and the U.S.

The U.S. has flown astronauts from Germany, Canada, The Netherlands, Saudia Arabia, France and Mexico, and plans to fly astronauts from Italy, Japan and the USSR.

The first non-U.S./non-USSR person in orbit was cosmonaut Vladimir Remek, the son of the Czechoslovakian defense minister, in 1978 at Salyut 6 space station.

The first Asian in space was Pham Tuan of Vietnam, launched in 1980 for eight days in space. Visiting the Salyut 6 space station during Summer Olympic Games in Moscow, he was the first non-Warsaw Pact and first third-world cosmonaut.

The first Hispanic in space was Cuban cosmonaut Arnaldo Tomayo Mendez launched in 1980 to the Salyut 6 space station. The second non-Warsaw Pact and second third-world cosmonaut, he grew organic monocrystals in space from Cuban sugar. The first Hispanic woman to go to space is to be Ellen Ochoa. She plans to fly in a U.S. shuttle in 1992 or 1993.

The first Westerner in Soviet space was French cosmonaut Jean-Loup Chretien, the first person from a Western industrial democracy at a USSR space station, in 1982 at Salyut 7 station. France was the first nation with both cosmonauts and astronauts. Astronaut Patrick Baudry of CNES, the French space agency, flew in U.S. shuttle Discovery in 1985.

The first black man in space was Guion S. Bluford Jr. He flew in 1983 for six days in U.S. shuttle Challenger. The first black woman, Mae C. Jemison, M.D., is to fly in shuttle Endeavour in 1992.

The first non-American astronaut to fly in an American spacecraft was West German astronaut Ulf Merbold. He flew in shuttle Columbia in 1983.

Rakesh Sharma was the first India in space, in 1984 at the Salyut 7 space station. The fourth third-world cosmonaut, he used yoga to combat effects of weightlessness.

The first Canadian in space was Marc Garneau, in shuttle Challenger in 1984. Roberta L. Bondar, M.D., was to be the first Canadian woman in space, in shuttle Atlantis in 1991.

The first Mexican in space was Rodolfo Neri, in shuttle Atlantis in 1985.

The first Dutch astronaut was Lodewijk van den Berg of the European Space Agency, in shuttle Challenger in 1985.

The first Arab in space was Saudi Arabia's Prince Sultan Salman Abdel Aziz Al-Saud in U.S. shuttle Discovery in 1985. The second Arab in space, Syrian cosmonaut Mohammed Faris, visited Mir space station in 1987.

The first Afghan in space, Abdul Ahad Mohmand, flew in 1988 to Mir space station.

The first Japanese in orbit was Toyohiro Akiyama, the cosmoreporter who flew in 1990 to Mir.

Helen Sharman, first from Great Britain, almost didn't make it to Mir in 1991 when sponsors were unable to raise money to pay for the flight. The USSR let her fly anyway.

The first Austrian person to fly in space was Franz Viehboeck. He visited the orbiting Mir space station in 1991.

First Space Bill Of Rights

The rules for life in future space colonies should include free speech, due process of law, property rights, voting rights, adherence to international law and environmental protection, according to former U.S. Supreme Court Justice William J. Brennan.

Lawyers and legal rules will be needed, he said in 1987, endorsing the "first principles for the governance of space societies." The U.S. is obliged "to insure that the fundamental needs for life, individual freedom, liberty, justice, dignity and the responsibilities inherent in self determination are integral parts of humanity's exploration and settlement of space," according to the first of 11 principles to govern future space colonies. Justice Brennan worked with National Air and Space Museum and Boston University's Center for Democracy, setting down principles, anticipating "enormous new responsibilities" for lawyers when space settlements are established. Justice Brennan asked the U.N. to apply the rules to all countries.

Space Food

The first person to eat in space was, of course, the first person to travel there. USSR cosmonaut Yuri A. Gagarin ate and drank during his one orbit of Earth in the capsule Vostok 1 in 1961.

U.S. astronauts in the first one-man Mercury capsules squeezed most of their food from toothpaste-style tubes. John H. Glenn Jr. swallowed applesauce in the weightlessness of space, during his three-orbit Mercury flight in 1962. Walter M. "Wally" Schirra Jr. lunched in orbit that year on beef, vegetables and peaches.

Bone-bones. For his 1962 flight, Pillsbury supplied Mercury astronaut Scott Carpenter with amazing new high-protein cereal snacks and Nestle sent along "bone-bones" of cereals with raisins and almonds. Today we call these granola bars.

Things improved ever so slightly for U.S. astronauts who orbited in two-man Gemini capsules in the mid-1960's. Plastic bags of freeze-dried foods, liquified by water pistol, were used in Gemini flights, culminating in the first shrimp cocktail in space.

Corned-beef sandwich. Virgil I. "Gus" Grissom ate the first corned beef sandwich in orbit during his five-hour Gemini 3 flight in 1965. Wally Schirra made the sandwich before launch and gave it to Grissom's flight companion, John W. Young, who passed it to Grissom. The sandwich caused a congressional fuss and new NASA rules later prevented such casual dining in space.

Hot food. The first hot food in space came with the three-man Apollo capsules. Astronauts could heat water to 154 degrees Fahrenheit to make hot foods and drinks.

The first chow on the Moon was consumed in 1969 by Apollo 11 astronauts Neil A. Armstrong and Edwin E. "Buzz" Aldrin Jr. They drank hot coffee and ate hot dogs, bacon squares, canned peaches and sugar cookies.

Liquid pepper. In 1973, America's Skylab space station had a refrigerator so the American astronauts could have the first ice cream in orbit. They also ate prime rib, German potato salad made with onions and vinegar, hot chili, scrambled eggs with liquid pepper spice.

Borscht. During the Apollo-Soyuz linkup in 1975, the USSR's Soyuz 19 cosmonauts Alexei Leonov and Valeri Kubasov treated Apollo 18 astronauts Thomas P. Stafford and Donald K. "Deke" Slayton to a lunch of borscht, chicken and turkey.

The first French cuisine in orbit was taken to the USSR's space station Salyut 7 in 1982 by French cosmonaut Jean-Loup Chretien.

Dried strawberries. U.S. space shuttle astronauts in the 1980's and '90's have dined on smoked turkey, cream of mushroom soup, mixed Italian veggies, vanilla pudding, and freeze-dried strawberries moistened in the mouth. They have snacked on almond crunch bars, graham crackers, pecan cookies, nuts, Life Savers and chewing gum. Along with water, they drink orange, lemon, orange-grapefruit, orange-pineapple, strawberry and apple drinks and tropical punch.

French cookin' In 1985, French astronaut Patrick Baudry treated his fellow crew members aboard U.S. space shuttle Discovery to gourmet dining.

That same year, U.S. shuttle Challenger astronauts Charles Gordon Fullerton, Roy D. Bridges Jr., Loren W. Acton, F. S. "Story" Musgrave, Anthony W. "Tony" England, Karl G. Henize and John-David F. Bartoe quenched their thirsts with special cans of the first Coke and Pepsi in orbit.

Probably the best meal ever in space was served at the USSR's Mir space station in 1988. French cosmonaut Jean-Loup Chretien treated USSR cosmonauts Vladimir Titov, Musa Manarov, Alexander A. Volkov, Sergei M. Krikalev and Valery Polyakov. They dined on 23 gourmet foods from a French chef, including compote of pigeon with dates and dried raisins, duck with artichokes, oxtail fondue with tomatoes and pickles, beef bourguignon, saute de veau Marengo, ham and fruit pates, bread, rolls, cheeses, nuts, coffee and chocolate bars.

No wine. The delicacies were canned for the high-pressure ride to Earth orbit. The chef said French cuisine is inconceivable without sauces and wine, but the astronauts had to do without because wine and sauces would float away without gravity. Dishes had to be small, without bones, and without sauces which could become flying droplets. Meat was made to absorb sauce to guard against dryness.

Gourmet cuisine. Do the Russians dream of a gourmet French restaurant floating in Earth orbit at a future time when anybody can visit a space station?

Soviet spacemen spend months in orbit. To spice up the tedium, they feast on delicacies in a diet of 70 dishes, including meat, fish and dairy products. Some of the mouth-watering delicacies found recently on floating space dishes have been so exquisite that lovers of gourmet food on Earth would have been glad to have them.

Tasty morsels have included chicken paste with plums, sturgeon with jellied sauce, processed fruit and all kinds of juices. The delicacies have to be canned, of course, for the trip in a Progress cargo freighter to Mir.

Appetite lost. NASA has found that astronauts eat and drink as much as 70 percent less in space. Space agency scientists, looking over results of a 1991 Spacelab shuttle flight, found the human body starting to adapt to weightlessness even on the launch pad. Seven astronauts orbited nine days with 2,478 tiny jellyfish and 29 lab rats. The jellyfish had gravity receptors something like a human inner ear. On average, the astronauts lost six lbs. during the first few days in orbit, but regained three lbs. before returning to Earth.

Their body fluid levels dropped, decreasing the pumping capacity of their hearts, but their heart muscles seemed unaffected by weightlessness. The lungs were unaffected, but surprisingly, fewer red blood cells were produced in the bone marrow. White blood cells were less responsive in space.

Space Bathrooms

American astronauts report the one question most asked by the public is: how do you use a toilet in space? Their answer: with great difficulty.

U.S. astronaut Alan B. Shepard Jr. reportedly wet his space suit while he waited four hours in 1961, in his Mercury capsule Freedom 7 atop a Redstone rocket, for a 15-minute suborbital ride into history as the first American in space.

Later in the 1960's, Gemini and Apollo astronauts urinated into cups which they then dumped overboard. They used stick-on plastic bags for solid waste which then was carried down to Earth—very messy, smelly, uncomfortable and poor hygiene for the Apollo astronauts travelling to and from the Moon.

Privacy. The American space station Skylab, used in 1973-74, had the first private toilet. A funnel collected urine, and air blew it into a bag. A toilet seat on a wall, with a seat belt, was the target for solid human waste, which then was bagged, dried and flown home with the astronauts for analysis by doctors on the ground.

U.S. space shuttle astronauts share one bowl which can't be flushed. They still have to use a seat belt to hold themselves on the commode, but their waste matter is dumped overboard. They also have hand and toe holds to keep from floating off their unisex potty. There's a porthole for a view of Earth from the bathroom. Solid waste is shredded by something called the slinger, then dried, disinfected, and dumped.

Leaky shower. How to keep clean in space? Skylab offered the first shower in orbit. It leaked, causing astronauts to waste time cleaning up. U.S. space shuttles do not have showers.

Shuttle astronauts do have toothbrushes, toothpaste, combs, hairbrushes, razors, shaving cream and nail clippers. Of course, the clippers don't see much use as fingernails and toenails grow only sparingly in weightlessness.

Stomach gas after meals can be a problem in the weightlessness of outer space.

Passing gas out of the mouth in zero gravity often brings vomit with it. Passing gas from the body in the other direction is not a better solution, so astronauts learn to suffer their gas pains.

Space Sickness

Imagine falling ill hundreds of miles from the nearest medical help. Astronauts orbit more than 100 miles above Earth and the Apollo astronauts on the Moon were a quarter-million miles away from home. During man's first three decades in space, people travelling there suffered varying degrees of illness.

Earth's gravity is very weak to non-existent above and beyond the planet. In that weightless environment, half of all astronauts and cosmonauts suffer something like sea sickness or car sickness—dizziness, nausea, even vomiting—when they first arrive in orbit. Some suffer for a day or two, then adjust. Others suffer through an entire trip.

The ill feeling does not always appear immediately after launch. Symptoms take a day or more to develop in some astronauts and cosmonauts.

USSR cosmonaut Gherman S. Titov, the second person to fly in orbit, suffered nausea throughout his Vostok 2 flight of 25 hours 18 minutes. Titov was the first person to sleep in space as he orbited Earth 16 times in 1961. Sleep did not relieve his nausea.

Space sickness may be caused by the effect of weightlessness on balance and visual reflexes. Doctors have been unable to predict who will suffer with it. Motion sickness drugs used on Earth have not worked in space.

Inner ear confused? The inner ear controls balance. Does the rapid acceleration in a launch to space create unusual pressure on the inner ear of someone lying down?

Three astronauts who flew in U.S. shuttle Challenger in 1985—Ernst Messerschmid and Rheinhard Furrer of Germany and Wubbo Ockels of The Netherlands—suggested in 1990 that launching astronauts flat on their backs might make space sickness worse.

They lay on their backs for 1.5 hours in a centrifuge, which generated gravity three times normal. Moving from the centrifuge to normal gravity, they felt symptoms like space sickness for up to six hours.

Astronauts take off in a supine position because it offers the least resistance to the pressure of gravity and reduces the pooling of blood and fluids in feet and legs. Flight controls in shuttles and manned capsules are positioned for use by reclining astronauts. Of course, astronauts spend only a few minutes in high gravity during launch, much less time than Messerschmid, Furrer and Ockels spent in the centrifuge.

Sniffles. The first health delay in American spaceflight history was the blastoff of Apollo 9, delayed four days in 1968, when crew members James A. McDivitt, David R. Scott and Russell L. Schweickart came down with sniffles.

Later that same year, Walter M. "Wally" Schirra Jr. came down with the first cold caught in space while orbiting Earth in Apollo 7. During their ten days in space, his fellow flyers, Donn F. Eisele and R. Walter Cunningham, caught the cold from Schirra, leading ground controllers to a decision to have the astronauts not wear helmets during re-entry. They wanted to keep air pressure on ear drums equalized as cabin pressure changed during descent. Seventeen years later, Schirra appeared on TV advertising a cold remedy.

Benign germs. In 1969, on the night before they were to leave for the first-ever Moon landing, a White House dinner for Apollo 11 astronauts Neil A. Armstrong, Edwin E. "Buzz" Aldrin Jr. and Michael Collins, given by U.S. President Richard M. Nixon, was called off by a NASA doctor to protect the astronauts from germs. Later Collins said diplomatically, "I'm sure presidential germs are benign."

Launch of Apollo 13 in 1970 was affected by illness, but wasn't delayed because a back-up crew was ready. Command module pilot Thomas K. Mattingly 2nd had to be replaced five days before launch by John L. Swigart Jr. when doctors realized Mattingly had been in contact with astronaut Charles M. Duke Jr. before Duke came down with German measles. Mattingly was judged not to be immune.

Apollo 13 was launched with Swigart rather than Mattingly, but then the flight was interrupted on the way to the Moon when an oxygen tank in the command service module ruptured. There was an explosion and the flight was aborted. Swigart and his fellow astronauts James A. Lovell Jr. and Fred W. Haise Jr. returned to Earth using the lunar module as a lifeboat for oxygen and power. Meanwhile, Mattingly, safe at home, did not develop measles. Later, Mattingly piloted the Apollo 16 command module launched to the Moon in 1972.

Isolation masks. NASA stopped training back-up spaceflight crews in 1982. As the pace of American space shuttle launches speeded up, trained astronauts were on hand for last-minute crew changes. To prevent illness, NASA rules require everybody to stay six feet or more from crew members for seven days before liftoff, except for family members and those workers absolutely necessary. Others also wear masks around crew members while NASA has isolation warnings posted.

In the first space station illness, USSR cosmonaut Vladimir Vasyutin became ill with a 104 degree fever while orbiting in the Salyut 7 space station in 1985. Vasyutin flew home and was hospitalized for a month with a full recovery.

Heart murmur. USSR cosmonaut Alexander I. Laveikin flew to the Mir space station in 1987. After six months in space, Laveikin showed an unusual heartbeat or murmur. Cautious doctors on the ground ordered him to fly home right away.

After flight. Ukrainian cosmonaut Anatoly Levchenko, 47, training to fly the Soviet space shuttle, died of a brain tumor in 1988, eight months after returning to Earth from an eight-day stay at the Mir space station. Levchenko had been a cosmonaut since 1981, but still was a rookie test pilot when he blasted off in Soyuz TM-4 in 1987. After returning home in 1987, the tumor turned up in 1988. An operation by Moscow doctors during the summer of 1988 was unsuccessful.

Pale and drawn. Vladimir Titov and Musa Manarov spent more than a year in space. Landing at the end of 1988 after 366 days in weightlessness, Titov and Manarov looked pale and drawn as they were lifted from their Soyuz capsule. They had problems maintaining equilibrium when they tried to stand up. Within about three hours, they were able to walk with assistance. Within two days, they were able to walk unassisted.

Famous cold. The sore throat and head cold of mission commander John O. "J.O." Creighton led to the first time a U.S. shuttle had to be postponed due to crew illness. The dazzling 1990 night-time launch of Atlantis flight STS-36 came on the sixth try after five delays, including sick leave for mission commander Creighton. After the flight, Creighton told reporters, "I probably had the world's most famous cold."

Space Art

America's first official space artwork, carried to orbit by shuttle Discovery March 13, 1989, was a seven-pound cube-shaped sculpture Boundless Aperture, created by Boston artist Lowry Burgess, a professor at Massachusetts College of Art.

He conceived Boundless Aperture in the 1960's as part of what he called The Quiet Axis. By the time Boundless Aperture went to space, Burgess already had buried a sculpture in the floor of the Pacific Ocean off Easter Island and lugged another into central Afghanistan's Hindu Kush mountains.

Foil Ring. An American artist living in Switzerland crafted a space sculpture—a jumbo ring of foil to inflate in space, become rigid in sunlight and display the message "peace" in various languages. The 20-ft.-diameter ring was to have been ferried to the USSR's orbiting Mir space station where cosmonauts would have thrown it out an airlock to its own orbit. The glint of sunlight off the foil ring was to have been visible to the naked eye on Earth.

Arthur Woods, 39, said in 1988 at Embrach, Switzerland, he had signed an agreement with Dmitri Poletayev of the USSR's Glavkosmos space agency to launch the sculpture to Mir on a Proton rocket in 1990 or 1991. Mir is Russian for peace. Woods said NASA didn't answer when he wrote the U.S. space agency about launching his sculpture.

Woods called his project Orbiting Unification Ring Satellite (OURS). He intended to collect $35 donations from the public, then spend $300,000 on sculpture and launch.

Launch of the OURS project was shelved by the Soviets when scientists around the world blasted Woods' sculpture as space pollution which would interfere with astronomy.

A self-styled painter and sculptor, Woods grew up near Cape Canaveral and worked there two summers while a student. He moved to Switzerland in 1974.

The sculpture's foil technology, known as Inflatable Space Rigidized Structures (ISRS), had been developed by Contraves, a Swiss company. Woods collaborated with Contraves on the project. The 20-ft. ring was to have been a prototype for a gargantuan 3,300-ft. sculpture Woods had wanted to send to Earth orbit by the year 2000.

Space Animals

A variety of large and small animals have been flown to space for science experiments in orbit. The first primates in space were the monkeys Albert 1 and Albert 2. They died in the late 1940's in the nose cones of captured German V-2 rockets during U.S. launch tests.

A monkey and mice died in 1951 when their parachute failed to open after a U.S. Aerobee rocket launch. Then a monkey and 11 mice shot to the edge of space on an Aerobee rocket September 20, 1951, were recovered alive. It was the first successful spaceflight for living creatures.

Laika was a live dog on a life-support system in the second USSR satellite, Sputnik 2, sent to space by the Russians in 1957. Laika's capsule remained attached to the converted intercontinental ballistic missile which rocketed her to orbit. The dog captured the hearts of people around the world as life slipped away from Laika a few days into her journey. Sputnik 2 fell into the atmosphere and burned in 1958.

At least seven other Russian dogs were launched to orbit between November 1957 and March 1961. At the end of the Sputnik series of satellites, the Russians prepared to send men to orbit by sending dogs first: Laika (Barker in Russian), Belka (Squirrel), Strelka (Little Arrow), Pchelka (Little Bee), Mushka (Little Fly), Chernushka (Blackie) and Zvezdochka (Little Star). Laika, Pchelka, and Mushka died in flight.

In February 1966, the dogs Veterok and Ugolek were launched in Voskhod 3 to be observed in orbit for 23 days via TV and biomedical telemetry.

In December 1959, the monkey Sam made a suborbital flight in a Mercury capsule. Another monkey, Miss Sam, flew in January 1960 in a suborbital Mercury capsule. Ham, a chimp, made a suborbital flight in January 1961 in a Mercury capsule.

The first primate in orbit was the chimp Enos, flying two orbits around Earth November 29, 1961, in a Mercury capsule in preparation for manned flight.

The U.S. launched Biosatellite 3 in 1969, three weeks before the first men were to land on the Moon. A monkey passenger was to orbit in Biosatellite 3 for a month, but was brought down, ill from loss of body fluids, after only nine days. It died shortly after landing. The USSR, cooperating with the U.S. and European nations, has flown a number of biosatellites in orbit, testing different kinds of plants and animals in weightlessness. The biological test flights have carried white Czechoslovakian rats, rhesus monkeys, squirrel monkeys, newts, fruit flies, fish and others. In 1990, a Japanese reporter took green tree frogs to the Mir space station.

Space Tragedies

At least 243 astronauts, cosmonauts and spaceport workers have been killed since 1960 in eleven fatal accidents connected with spaceflights and training. Of the 243 persons, eight men and two women were killed in two U.S. spacecraft accidents, five U.S. astronauts were killed in airplane crashes, and 228 men and women died in five USSR disasters.

Some 165 workers were killed in a 1960 rocket explosion on a USSR launch pad. The first three astronauts to die preparing to go to space were killed in 1967. The first cosmonaut to die in flight was killed the same year.

The second cosmonaut in-flight tragedy killed three in 1971. At least nine were killed in a USSR launch-pad rocket explosion in 1973.

At least 50 were killed in a 1980 rocket explosion on a USSR launch pad. The first astronauts to die enroute to space were killed in 1986.

Alan B. Shepard Jr., the first American to fly in space, recalled in 1991, "Had we said thirty years ago that we were going to put man in space for thirty years and we're only going to have two accidents, we would have said, 'Boy, we'll take that right now.' Certainly, pushing out the frontiers as we did and still are doing, and having one accident in flight, the other on the ground, really is remarkable."

ROCKET EXPLOSION
Country: USSR
Accident date: October 24, 1960
Personnel killed: 165 workers
Events: In the worst space-related accident in history, a giant rocket was on the launch pad at Baikonur Cosmodrome on the Central Asian plain in the Kazakh Republic when an explosion killed 165 workers. Field Marshall Mitrofan Nedelin, father of the Soviet Union's Strategic Rocket Forces, was killed. The story was kept secret until 1990 when the USSR army newspaper Red Star reported a new kind of rocket had been erected that October 1960 on the Baikonur launch pad. A first flight test was delayed repeatedly. Launch crew personnel, checking rocket joints and tubing for leaks, found holes burned through their rubber gloves by leaking fuel.

On the morning of October 24, the blast-off signal finally was sent to the rocket, but something went wrong. The first stage failed to ignite. Technicians, who went outside their control bunker to replace a component, apparently forgot to close valves conducting fuel to the second stage. When they attached an umbilical cord, the second stage ignited. Sheets of flame engulfed workers and turned nearby fuel trucks into fireballs. Many victims were burned beyond recognition. No death toll was published in the 1960's, but a monument shows 54 names. The Red Star article reported 165. Western reports claimed up to 300 died. Launch crew chief Stanislav Pavlov was 20 yards from the rocket when it exploded. Burned badly and unconscious three days, he was in a hospital a year. His injury was not recorded officially by the government until 1988. Field Marshall Nedelin, commander in chief of the Rocket Forces, was listed for years by the USSR government as having died elsewhere in a plane crash.

JET CRASH
Country: USA
Accident date: October 31, 1964
Astronaut killed: Theodore Freeman
Events: Freeman was killed in the crash of his T-38 jet trainer at Ellington Air Force Base near Johnson Space Center, Houston.

JET CRASH
Country: USA
Accident date: February 28, 1966
Astronauts killed: Charles Bassett and Elliott See
Events: Bassett and See were killed when their T-38 trainer jet aircraft crashed in heavy fog while approaching St. Louis, Missouri. They had planned to visit McDonnell Douglas Astronautics Company. Elliott See had been a U.S. Maritime Academy engineer when he and aeronautical engineer Neil Armstrong were selected to be America's first civilian astronauts. They were selected in 1962 in NASA's second group of astronauts. Armstrong went on to become the first person to walk on the Moon.

APOLLO 1

Country: USA
Accident date: January 27, 1967
Astronauts killed: Virgil I. "Gus" Grissom, Edward H. White 2nd, Roger B. Chaffee
Events: There was a rush to beat the Russians to the Moon by the end of the 1960's. Pre-flight testing of Apollo capsules started May 28, 1964. The U.S. sent the powerful new Saturn 1B rocket on its first unmanned suborbital test flight February 26, 1966, carrying a dummy Apollo capsule 5,500 miles downrange. In a second unmanned test flight July 5, the rocket second stage went into orbit, but carried no other satellite. A third Saturn 1B test that year was a suborbital flight August 25. The first manned Apollo test was to follow on February 21, 1967, but it didn't happen.

On January 27, 1967, 25 days before their scheduled lift-off, Grissom, White and Chaffee had taken an early lunch and were practicing in Apollo 204 during a simulated countdown rehearsal on the pad at Cape Canaveral's launch complex 34. It was routine—what NASA called a "plugs out" test. A cumbersome hatch locked in the astronauts. They were breathing pure oxygen. The sealed capsule was disconnected from outside electrical power to simulate launch. Problems had pushed the simulation behind schedule near sunset when a communications problem caused a 10-minute delay.

Somehow a broken, bruised or frayed wire contacted metal, short-circuited and sparked beneath Grissom's couch. The spark triggered a fire in the cabin's 100-percent-oxygen atmosphere. The electrical arcing ignited fire-resistant plastic which had been turned flammable by the pure oxygen atmosphere. Dense acrid smoke from burning plastic suffocated the crew.

At 6:31 p.m., blockhouse controllers and workers in the launch pad clean area—the white room—received a dreadful cry by radio from the capsule, "There is a fire in here." As fire engulfed the capsule interior, the lead worker on the pad, Donald Babbit, shouted to his co-worker James Gleaves, "Get them out of there." Seconds later flame burst from Apollo 204. Gasping white room workers tugged at the hatch to open the capsule while controllers in the blockhouse were paralyzed by panic.

Gary Propst, a blockhouse employee of RCA watching on closed-circuit TV, recalled flames rising for three minutes. He saw arms as one of the astronauts inside the capsule wrestled with the hatch. Propst shouted, "Blow the hatch, why don't they blow the hatch?" But Apollo 204 did not have explosive bolts on its main hatch.

A short 5.5 minutes after the fire broke out, the hatch was opened. Fourteen minutes after the first outcry, physicians Allan Harter and G. Fred Kelly reached the astronauts, already dead from asphyxiation of toxic gases.

A 20-month delay in manned flight, and a thorough redesign to improve safety, followed the tragedy. NASA in 1967 renamed the Grissom, White and Chaffee Apollo 204 mission Apollo 1.

There were no Apollo 2 or Apollo 3 flights. There were three unmanned Saturn 1B tests in 1967 and 1968, known as Apollo 4, 5 and 6, before a Saturn 1B hefted the manned Apollo 7 to space October 11, 1968. The first Saturn V launch, November 9, 1967, was called Apollo 4. The eventual Saturn 1B launch designated Apollo 204 became Apollo 5 and carried a LEM as payload, not an Apollo CM. Apollo 201 and Apollo 202, carrying Apollo spacecraft, previously had been unofficially known as Apollo 1 and Apollo 2. Apollo 203 carried only an aerodynamic nose cone. The second Saturn V launch, known as Apollo 502 or Apollo 6, was April 4, 1968. It was a success despite two first-stage engines shutting down prematurely and the third stage engine failure to re-ignite in orbit.

The shell of the charred Apollo 1 capsule has been stored in a locked container at NASA's Langley Research Center in Hampton, Virginia. Until about 1980, the container had been kept in a low-pressure nitrogen atmosphere to minimize corrosion, but that container deteriorated and small leaks developed. The container was repaired, but NASA said its age prevented further maintenance. In 1990, NASA proposed shipping the capsule from Langley to a permanent space tomb inside two abandoned missile silos at Cape Canaveral Air Force Station, Florida, where it would have been dumped on top of debris from the 1986 Challenger explosion.

Besides closing the open-ended maintenance on the containers, NASA wanted the Langley storage area for other purposes. Putting the 17,600 pounds of Apollo 1 hardware in permanent storage in the Florida missile silo would have ended NASA's attention to it.

The capsule was taken out of storage and its heat shield was cut into pieces for shipping. Hardware and investigation materials in 81 cartons were shipped to Kennedy Space Center for burial with the capsule. But, the move was stopped at the last minute when the Smithsonian Institution's National Air and Space Museum indicated it might want the capsule as an historic artifact. The Smithsonian had turned down offers of Apollo 1 materials in 1985 and 1989, but said now it might be interested in the capsule as an historically-significant artifact, not for public display. The Kansas Cosmosphere and Space Center, Hutchinson, Kansas, also wanted it for a memorial. After NASA talked it over with other Apollo astronauts and relatives of the dead astronauts who wanted it kept at Langley indefinitely, the cartons were returned to Langley.

JET CRASH
Country: USA
Accident date: October 15, 1967
Astronaut killed: Clifton "C.C." Williams
Events: Williams died in an airplane crash near Tallahassee, Florida.

SOYUZ 1
Country: USSR
Launched: April 23, 1967, on an A-2 rocket
Accident date: April 24, 1967
Cosmonaut killed: Vladimir M. Komarov
Flight duration: 26 hours 40 minutes, for 17 orbits
Events: USSR cosmonaut Vladimir M. Komarov died April 24, 1967, as his Soyuz 1 capsule crashed on the central-Asian plain. He had been launched in Soyuz 1 on April 23 for 17 orbits of Earth in a flight of 26 hours 40 minutes. It was the first Soyuz multi-man space capsule, but carried only one cosmonaut. The test flight seemed successful, until it crashed after re-entry. A parachute to slow the capsule opened, but lines became snarled at 23,000 feet altitude. The parachute twisted and deflated. Soyuz 1 hit the ground at 200 miles per hour and burst into flames. Komarov was the first casualty of an actual space flight.

SOYUZ 11
Country: USSR
Launched: June 6, 1971, on an A-2 rocket
Accident date: June 30, 1971
Cosmonauts killed: Georgi T. Dobrovolsky, Vladislav N. Volkov, Viktor I. Patsayev
Flight duration: 569 hours 40 minutes, for 360 orbits
Events: Cosmonauts Georgi T. Dobrovolsky, Vladislav N. Volkov and Viktor I. Patsayev flew to space June 6, 1971, in Soyuz 11 to become the first crew from Earth to work in a space station, but also the second USSR space tragedy. They had been launched to orbit June 6 in Soyuz 11 for a visit to the new Salyut 1 space station. They stayed a record 23 days. Then, the three cosmonauts were not wearing spacesuits on the flight home as air escaped the Soyuz 11 capsule during re-entry. They were found dead when the capsule was opened on the ground after an automatic landing. It was the last USSR three-man flight for nearly a decade.

SOYUZ ROCKET
Country: USSR
Accident date: June 26, 1973
Personnel killed: 9 space workers
Events: As a Soyuz space rocket was being fueled with kerosene and liquid oxygen at Plesetsk Cosmodrome—a space launch complex 500 miles north of Moscow—an explosion tore through the launch pad, killing nine. Today, the remains are under red granite slabs displaying pictures of the workers, topped by a space rocket carved of granite. The monument, in the main square of the bedroom community of Mirny in northern Russia, overlooks Lake Plesetskaya surrounded by a birch and fir forest. The USSR made public the story in 1989.

SOYUZ ROCKET
Country: USSR
Accident date: March 18, 1980
Personnel killed: 50 space workers
Events: As a Soyuz rocket was being fueled with kerosene and liquid oxygen at Plesetsk Cosmodrome, an explosion shattered the launch pad, killing 45. Five more died later from burns. Today, the remains are under red slabs displaying pictures of the workers, topped by a granite rocket in the Mirny main square. The USSR made the tragedy public in 1989.

CHALLENGER SHUTTLE
Country: USA
Launch accident date: January 28, 1986
Astronauts killed: Francis R. "Dick" Scobee, Michael J. Smith, Judith A. Resnik,
        Ellison S. Onizuka, Ronald E. McNair, Gregory B. Jarvis, Sharon Christa McAuliffe
Flight duration: 1 minute 13 seconds, did not make it to orbit

Events: Challenger was destroyed 73 seconds into liftoff January 28, 1986, when fuel leaking from the right-side solid-fuel booster rocket ignited. All seven astronauts on board died in the explosion. Killed in the disaster were commander Francis "Dick" Scobee, pilot Michael Smith, mission specialists Ronald McNair, Ellison Onizuka, Judith Resnik, satellite engineer Gregory Jarvis and New Hampshire school teacher Christa McAuliffe. NASA said the seven astronauts probably passed out quickly as the shuttle was breaking up while falling into the Atlantic Ocean. Remains of the astronauts were recovered from the crew cabin, but NASA was unable to say exactly what caused their deaths.

The 215,000 lbs. of salvaged Challenger wreckage, including 20,000 cubic feet of debris, was lowered by crane into waterproofed steel cylinders in two Cape Canaveral silos in 1987. Each of the 5,000 items was catalogued with location noted for future retrieval. Half of one of the 90-ft.-deep 12-ft.-diameter missile silos is filled with Challenger debris while the other silo a quarter-mile away is completely filled, as well as underground equipment rooms at each site. Several major components of the shuttle were left on the Atlantic Ocean floor, including Challenger's left wing.

JET CRASH
Country: USA
Accident date: April 5, 1991
Astronaut killed: Manley Lanier "Sonny" Carter Jr.
Events: Carter died in a commuter plane crash in Georgia.

Astronauts Memorial

Space Mirror, the first American monument to astronauts fallen in the line of duty, was erected in 1991 by 430,000 Florida residents who bought their state's Challenger license plates.

Fifteen glowing names float among the clouds in the Florida sky reflected on a vast curtain of black granite beside a quiet lagoon: Charles Bassett, Manley L. "Sonny" Carter Jr., Roger B. Chaffee, Theodore Freeman, Virgil I. "Gus" Grissom, Gregory B. Jarvis, Sharon Christa McAuliffe, Ronald E. McNair, Ellison S. Onizuka, Judith A. Resnik, Francis R. "Dick" Scobee, Elliott See, Michael J. Smith, Edward H. White 2nd and Clifton "C.C." Williams.

The massive memorial, centerpiece of six acres near the Kennedy Space Center visitors center, is a 42-ft. by 50-ft. rampart of 93 two-in.-thick, 800-lb. black granite panels in a rectangular framework on an immense turntable. Astronaut names are cut through some of the thin granite slabs. The 37-ton wall rotates slowly on a computerized turntable, tracking the Sun which reflects from hidden mirrors through the cut-out astronaut names. At night and on cloudy days, floodlights replace the Sun.

Cracks. Unfortunately, a month before it was opened to the public in May 1991, the thin polished name panels started cracking. The carvings, and clear plastic used to fill the cut-out spaces, may have led to the cracks, which NASA said it would cost $12,000 to repair. Three months later, in August, rotating and pitching of the monument had to be stopped for repairs when the edifice started making odd popping sounds. NASA engineers determined the mirror-pitch controller was off by 1.5 inch.

The 1986 Challenger disaster led Orlando architect Alan Helman to found the Astronauts Memorial Foundation. NASA provided the six acres. There were 756 entries in a national design competition, won by San Francisco architects Paul Holt, Marc Hinshaw, Peter Pfau and Wes Jones. The $6.2 million construction cost was paid by the sale of commemorative license plates. An additional $2 million endowed a perpetual-maintenance trust fund. An inscription on the monument, built by VSL Corp., described the Astronauts Memorial as "a tribute to American men and women who have made the ultimate sacrifice believing the conquest of space is worth the risk of life."

On to Mars. Sales of Challenger plates also was to help pay for trips to Mars. The Florida legislature set aside one-quarter of the money for scholarships and one-quarter for research by the Florida Technical Research and Development Authority into manned flights to Mars. The Challenger license plate depicts a shuttle rising on an orange-and-white cloud into a blue sky. A plate cost an automobile owner an extra $15 in 1987-88. More than 500,000 were sold.

Lunar Crater Memorials

Astronomers on Earth have named 25 craters on the Moon after space-flying pioneers. Such memorials usually commemorate dead persons, but the International Astronomical Union in 1970 paid tribute to a dozen U.S. and USSR space pioneers.

Six living Apollo 8 and Apollo 11 astronauts were honored in 1970 along with three astronauts killed in Apollo 1.

William A. Anders, Frank Borman and James A. Lovell Jr. in December 1968 in Apollo 8 were the first men from Earth to see the far side of the Moon. Most of the craters named after the 16 American space fliers are in and around the 280-mi.-wide Apollo crater on the far side of the Moon.

Neil A. Armstrong and Edwin E. "Buzz" Aldrin Jr. in Apollo 11 were the first men to land on the Moon. Michael Collins was with them in Apollo 11. Craters named for Armstrong, Aldrin and Collins are near the spot in Mare Tranquillitatis on the Moon's near side where the Eagle landed in 1969.

Except for Yuri Gagarin, the craters honoring cosmonauts also are on the far side, in and around the 233-mi.-wide lava basin known as Mare Moscoviense.

Tribute was paid to six living USSR space fliers named along with three dead cosmonauts in 1970.

Seven asteroids also were named for the Challenger astronauts. Since 1919, the IAU's Working Group For Planetary System Nomenclature has named extraterrestrial bodies. Dates in the list below are birth and death, if deceased. The craters:

USSR COSMONAUTS

Belyayev, Pavel I.	1925-1970
Feoktistov, Konstantin P.	1926-
Gagarin, Yuri A.	1934-1968
Komarov, Vladimir M.	1927-1967
Leonov, Aleksei A.	1934-
Nikolayev, Andriyan G.	1929-
Shatalov, Vladimir A.	1927-
Tereshkova, Valentina V.	1937-
Titov, Gherman S.	1935-

USA APOLLO ASTRONAUTS

Aldrin, Edwin E. "Buzz", Jr.	1930-
Anders, William A.	1933-
Armstrong, Neil A.	1930-
Borman, Frank	1928-
Chaffee, Roger B.	1935-1967
Collins, Michael	1930-
Grissom, Virgil I. "Gus"	1926-1967
Lovell, James A., Jr.	1928-
White, Edward H., 2nd	1930-1967

USA CHALLENGER ASTRONAUTS

Jarvis, Gregory B.	1944-1986
McAuliffe, Sharon Christa	1948-1986
McNair, Ronald E.	1950-1986
Onizuka, Ellison S.	1946-1986
Resnik, Judith A.	1949-1986
Scobee, Francis R. "Dick"	1939-1986
Smith, Michael J.	1945-1986

Astronauts/Cosmonauts Glossary

Apollo. The third U.S. man-in-space program, Apollo featured three-man capsules, six trips to the Moon from 1969-72, and a rendezvous in Earth orbit with the Russians.

Apollo-Soyuz. Joint U.S.-USSR space flight and docking in Earth orbit in 1975. Three American astronauts in Apollo 18 and two Soviet cosmonauts in Soyuz 19.

Armstrong, Neil. In 1969, the U.S. Apollo 11 astronaut who was the first man to walk on the Moon. See also: Apollo.

Astronaut. In common usage of the title, all men and women who have flown, or are preparing to fly, in U.S. spacecraft are considered Astronauts. They include mission commanders, pilots, payload commanders, mission specialists, payload specialists and space flight participants. NASA, however, refers to payload specialists and space flight participants as members of a space shuttle flight crew, not as astronauts. A few persons from various nations are both astronauts and cosmonauts. See also: Cosmonauts.

Atlantis. The fourth U.S. space shuttle to fly to orbit. See also: Shuttle.

Atlas. The U.S. rockets which carried the manned Mercury capsules to space. See also: Mercury.

Bluford, Guion. In 1983, U.S. astronaut who was the first black man in space.

Buran. The first orbiter of the USSR's space shuttle fleet. Buran is Russian for snowstorm or blizzard. The first Buran flew to Earth orbit. It will fly to space station Mir. Buran is launched on an Energia rocket. See also: Shuttle.

Capsule. A small pressurized compartment or spacecraft in which astronauts travel. For instance, Mercury capsule, Gemini capsule, Apollo capsule, Soyuz capsule.

Challenger. The second U.S. space shuttle, destroyed in 1986 launch explosion. See also: Endeavour, Shuttle.

Columbia. The first U.S. space shuttle to fly to orbit. See also: Shuttle.

Columbus. Europe's manned module in development for space station Freedom.

Command service module. CSM, the crew compartment of U.S. Apollo Moon spacecraft. See also: Apollo.

Cosmonaut. All men and women who have flown, or are preparing to fly, in USSR spacecraft are Cosmonauts. A few persons from various nations are both cosmonauts and astronauts. See also: Astronauts.

Crew commander. A U.S. space shuttle crew commander is the astronaut who has overall responsibility for mission success and flight safety. See also: Payload Commander.

Discovery. The third U.S. space shuttle to fly to orbit. See also: Shuttle.

Endeavour. The fifth U.S. shuttle to fly to orbit, it replaced shuttle Challenger. See also: Shuttle.

Energia. The newest USSR super-powerful space-launch rocket, used to launch the shuttle Buran to Earth orbit. See also: Buran, Shuttle.

Enterprise. The 1970's prototype U.S. shuttle, it was not flown to space. See also: Shuttle.

EVA. Extra-vehicular activity by an astronaut or cosmonaut, a spacewalk outside a shuttle or space station.

Freedom. The second-generation U.S.-international space station in development.

Gagarin, Yuri. The USSR cosmonaut who, in 1961, was the first man in space and in Earth orbit. See also: Vostok.

Gemini. Project Gemini was America's second man-in-space program. It featured two-person capsules launched on Titan 2 rockets. See also: Capsule.

Glenn, John. In 1962, the third American in space, the first American in Earth orbit. See also: Mercury.

Grissom, Gus. In 1961, the second American in space. See also: Mercury.

Hermes. Europe's manned space shuttle under development. See also: Shuttle.

Hope. Japan's manned space shuttle under development. See also: Shuttle.

JEM. Japan's manned module for the U.S.-international space station Freedom.

Kvant. A manned astronomy and science research cabin module attached to Mir.

LEM. See: Lunar excursion module.

Lunar excursion module. LEM, the U.S. Apollo Moon landing craft used by pairs of astronauts to fly down from their lunar-orbiting command service modules to the surface. See also: Apollo, Armstrong, Command Service Module.

Manarov, Musa. In 1988, USSR cosmonauts Musa Manarov and Vladimir Titov completed the longest one-trip stay in space. See also: Mir.

Manipulator. A mechanical arm used by astronauts to lift cargo from the shuttle bay of a U.S. shuttle; also a mechanical arm used by cosmonauts to move add-on expansion modules between ports on the USSR's orbiting Mir space station.

Mercury. America's first men in space went there in Mercury capsules. There were a total of 23 launches in the first U.S. manned spaceflight program—Project Mercury. Of the 23, only about a quarter carried men. Of the 23, thirteen carried no crew, four carried animals and six carried astronauts. Sixteen of the flights were successful. There were seven launch failures, but no flight failures during manned launches. See also: Capsule.

Mir. The USSR's third-generation space station in Earth orbit. Russian for Peace.

Mission commander. A U.S. space shuttle mission commander is the astronaut who has overall responsibility for mission success and flight safety. See also: Crew Commander.

Mission specialist. An astronaut trained for general work aboard a U.S. space shuttle flight.

NASA. National Aeronautics and Space Administration, the U.S. space agency.

Payload commander. A U.S. space shuttle payload commander is the astronaut who is the lead mission specialist planning and coordinating shuttle cargo. Payload commanders are forerunners of space station mission commanders. See also: Crew Commander.

Payload specialist. An astronaut trained to work with a payload aboard a U.S. space shuttle flight. NASA refers to payload specialists as members of a space shuttle flight crew, not as astronauts.

Pilot. The astronaut especially trained to operate the flight controls of a U.S. space shuttle.

Progress. The series of unmanned cargo-transporting spacecraft which ferry food, fuel, water, science equipment and other supplies to cosmonauts aboard the USSR's orbiting Mir space station.

Progress M. A redesigned Progress. M for modified.

Ride, Sally. In 1983, the first American woman in space.

Romanenko, Yuri. A USSR cosmonaut, the human with longest accumulated time in space.

Salyut. The first-generation and second-generation USSR space stations.

Saturn V. A 1970's series of heavy-lifting U.S. rockets which blasted the manned Apollo capsules to the Moon. See also: Apollo.

Service module. The oxygen, water, and fuel area of a U.S. Apollo moonship.

Shepard, Alan. In 1961, the first American in space. See also: Mercury.

Shuttle. A partly-reusable manned spacecraft which can land on a runway.

Shuttle-C. A proposed unmanned cargo version of a U.S. shuttle.

Skylab. The first-generation U.S. space station, in orbit in the 1970's.

Soyuz. The long series of USSR cosmonaut-transporting spacecraft. USSR Soyuz capsules ferry cosmonauts to and from space, especially to and from the orbiting Mir space station. Soyuz is Russian for Union, symbolizing the spacecraft's rendezvous and docking capability.

Soyuz T. A redesigned Soyuz. The T symbolizes: (1) transport, (2) troika indicating a space transport capable of carrying three cosmonauts to the Mir station, and again (3) troika indicating the third generation of the Soyuz design.

Soyuz TM. A redesign of Soyuz T. The M is for modified and for Mir, its main destination. Soyuz TM is a third generation modified Soyuz.

Space flight participant. An astronaut travelling aboard a U.S. space shuttle flight. Space flight participants include politicians, businessmen and school teachers. NASA refers to space flight participants as members of a space shuttle flight crew, not as astronauts.

Space shuttle. See: Shuttle.

Space station. Structure with quarters for long-duration living & working in orbit. See also: Mir, Salyut, Skylab.

Space station mission commander. See: Payload Commander.

Spacewalk. Extra-vehicular activity, EVA, a trip by an astronaut or cosmonaut outside a shuttle or space station.

Tereshkova, Valentina. In 1963, the USSR cosmonaut who was the first woman in space. See also: Vostok.

Titan. The series of U.S. space launch rockets which carried manned Gemini capsules to Earth orbit in the 1960's. See also: Gemini.

Titov, Gherman. In 1961, the USSR cosmonaut who was the second man to go into Earth orbit. See also: Vostok.

Titov, Vladimir. In 1988, USSR cosmonauts Vladimir Titov and Musa Manarov completed the longest one-trip stay in space. See also: Mir.

Voskhod. The second USSR manned spacecraft series. A redesigned Vostok capsule, with three seats. Voskhod is Russian for Sunrise, following USSR Premier Khrushchev's observation that "the Sun rises in the East." See also: Vostok.

Vostok. The first-ever manned spacecraft. The USSR's first manned spacecraft series, with a single seat. Vostok is Russian for East, symbolizing the East-West competition in international relations. See also: Gagarin, Tereshkova, Titov, Voskhod.

Space Stations
★★★★★★★★★★★★★★★★★★★★★★

History Of Space Stations

Europeans thought up space stations in the 1920's as permanently manned science and technology bases orbiting Earth, Moon or other planets. They were envisioned as large human colonies with research laboratories to study Earth; astronomy observatories to study the Sun, Solar System and deep space; and manufacturing factories.

With experimental rocket launches in the 1920's, Hermann Oberth, referred to as the father of modern astronautics, was first to discuss building a station. A captain in Austria's imperial army, using the pen name Hermann Noordung, in 1929 described a doughnut-shaped station rotating to generate its own gravity. A German rocket scientist in the U.S., Wernher von Braun, promoted the idea of a wheel-shaped station in 1951.

Wet. Von Braun's colleague Krafft Ehricke proposed in the 1950's firing an Atlas rocket to orbit, then cleaning out its interior and outfitting it as a space station. This would have been a "wet" station, meaning the rocket would have been filled with fuel for launch, then cleaned out and converted to a space station in orbit by astronauts.

MOL. The Douglas Aircraft Corp. picked up Ehricke's idea in the 1960's in its proposed Manned Orbiting Laboratory (MOL). Douglas wanted to use the S-IV second stage of the Saturn I rocket. MOL was dropped after astronauts landed on the Moon and NASA wanted to turn Project Apollo into an Apollo Applications Program (AAP).

AAP. Known as a "wet orbital workshop," AAP would have used a Saturn IB rocket's S-IV-B stage filled with fuel for launch, then cleaned out in space and refitted by astronauts. The AAP station in Earth orbit would have been a pit stop for astronauts on their way to the Moon and Mars. Shortage of cash slowed NASA's program, then doubts about doing a wet-to-dry conversion in orbit changed the plan to a dry orbital workshop.

Salyut 1-5. The Soviet Union was participating vigorously in the Space Race in the 1960's, including developing capsules to carry a man or men to the Moon and, from that research, a space station for Earth orbit. While the U.S. won the Moon race in 1969, the USSR won the competition for first space station when Salyut 1 was launched in 1971. Salyut 2 was launched in 1973. Salyut 3, 1974. Salyut 4, 1974. Salyut 5, 1976.

Skylab. Meanwhile, in the U.S., the name AAP was dropped and the dry orbital workshop was called Skylab. It was outfitted on the ground and launched in 1973 on a two-stage Saturn V rocket. Astronauts were shipped to the station separately in groups of three in left-over Apollo Moon capsules on Saturn 1B rockets. In 1973-74 three groups spent a total of 172 days at the station.

Skylab 2. NASA drew up plans for a larger space station to be launched by a two-stage Saturn V, but sharply reduced space budgets forced tough decisions. The first Skylab was not used after just three visits and plans for the second Skylab were dumped. NASA thought a reusable space transportation system was needed to carry men and equipment to orbit, so the limited amount of money available was switched to designing and building space shuttles. Six shuttles were built and dozens of week-long space flights have been made since 1981.

Salyut 6-7. Meanwhile, the USSR's highly successful first-generation space station technology, displayed in Salyuts 1 through 5, was superseded in 1977 by the improved second-generation station Salyut 6. The second-generation station Salyut 7 was sent to Earth orbit in 1982.

Mir. The Salyut series was replaced in 1986 by a third-generation space station called Mir. In the 1990's, Mir will become more than five times its 1986 size. Additions were sent to space and attached to Mir in 1987, 1989 and 1990. Two others are to be orbited in 1992-93.

Space station activities through the 1970's and 1980's gave the USSR the lead in man-hours in orbit and in space biomedicine which it holds today. Several cosmonauts have spent from half a year and to a year and a half in space.

The USSR designed a space shuttle in the 1980's. It made a brief test flight in 1988. Space station visits by USSR shuttles may start in 1992-93.

Mir 2. After expanding the first Mir, the USSR plans a larger station in the 1990's with maneuverable modules which could fly away from the station and return to dock.

Freedom. In 1984, U.S. President Ronald Reagan announced a project to build a permanently-manned space station in Earth orbit within 10 years. The station was named Freedom and Canada, Japan and the European Space Agency joined the project.

The plan was opposed vigorously by some scientists and politicians who said it was unnecessary. Station configurations and construction dates changed several times over subsequent years. NASA still hopes to construct Freedom station from 1995-2000.

Next century. Long-term thinkers continue to look forward to large communities in space, orbiting Earth, Moon or Mars, housing thousands in Earth-style environments aboard space stations built from materials mined on the moons of Earth and Mars and from asteroids.

Launches. So far, nine space stations have been launched to low Earth orbits by the USSR and the U.S. since the first in 1971. This list of stations includes launch date, country of origin, name, date when the station fell from orbit to burn in the atmosphere (down), the station's technology generation and station purpose:

Launch	Country	Station	Down	Generation	Purpose
1971 Apr 19	USSR	Salyut 1	1971 Oct 11	1	science
1973 Apr 3	USSR	Salyut 2	1973 May 22	1	broke up Apr 14
1973 May 14	USA	Skylab	1979 Jul 11	1	science
1974 Jun 24	USSR	Salyut 3	1975 Aug 24	1	military
1974 Dec 26	USSR	Salyut 4	1977 Feb 2	1	science
1976 Jun 22	USSR	Salyut 5	1977 Aug 8	1	military
1977 Sep 29	USSR	Salyut 6	1982 Jul 29	2	science
1982 Apr 19	USSR	Salyut 7	1991 Feb 7	2	science & military
1986 Feb 20	USSR	Mir	still in orbit	3	science

USSR Space Stations

The USSR launched a series of five first-generation space stations to Earth orbit in the 1970's, then launched two second-generation space stations to Earth orbit in the 1970's and 1980's.

All were known as Salyut—Russian for Salute—a name given to the first in 1971 honoring the tenth anniversary of the flight of the first man in space, Yuri Gagarin. Cosmonauts lived for record periods of time in space aboard Salyut stations. All seven have since fallen into the atmosphere and burned.

In 1986, the USSR launched a third-generation space station, Mir, Russian for Peace. Mir is in use today in Earth orbit. Cosmonauts also have lived for record periods of time aboard Mir.

FIRST-GENERATION SPACE STATIONS

SALYUT 1

Rocket: Proton, also known as D-1

Launch: April 19, 1971

Duration: fell from orbit and burned October 11, 1971

Events: First USSR space station, used for meteorology, searches for Earth resources, astronomy with spectrogram telescope Orion 1 and gamma ray telescope Anna III. A hydroponic garden. First-generation stations, which could be used manned or unmanned, each had only one dock to receive another spacecraft.

SALYUT 2

Rocket: Proton, also known as D-1

Launch: April 3, 1973

Duration: fell from orbit and burned May 22, 1973

Events: May have been for military. Catastrophe April 14 tore solar panels, dock and radio transponder from the station, leaving it tumbling uselessly in low orbit.

COSMOS 557
Rocket: Proton, also known as D-1
Launch: May 11, 1973
Duration: fell from orbit and burned May 22, 1973
Events: Failed, unused civilian space station.

SALYUT 3
Rocket: Proton, also known as D-1
Launch: June 24, 1974
Duration: fell from orbit and burned August 24, 1975
Events: First military version of Salyut actually to be used.

SALYUT 4
Rocket: Proton, also known as D-1
Launch: December 26, 1974
Duration: fell from orbit and burned February 2, 1977
Events: Civilian Salyut with biomedical, earth resources and science experiments. Infrared, solar and X-ray telescopes observed Solar System, Milky Way, other galaxies.

SALYUT 5
Rocket: Proton, also known as D-1
Launch: June 22, 1976
Duration: fell from orbit and burned August 8, 1977
Events: Last solely-military space station, like Salyut 3. Splav processor smelted bismuth, tin, lead, cadmium. Kristall processor grew crystals. Fish, plants, fruit flies, algae for biology research.

SECOND-GENERATION SPACE STATIONS

SALYUT 6
Rocket: Proton, also known as D-1
Launch: September 29, 1977
Duration: fell from orbit and burned July 29, 1982
Events: The first USSR second-generation space station, it too could be used manned or unmanned. New propulsion system, water regeneration system, spacesuits. Second-generation stations had two docks—one forward, one aft—to receive other spacecraft. With two docks, Salyut could hold a Soyuz or Progress in each dock, allowing more convenient resupply by freighter and short-term visits by cosmonauts. Refueling and resupply doubled the anticipated 18-month life of the station. Progress unmanned cargo freighters carried food, water, fuel, supplies, mail to the station 12 times from Jan 1978 through Jan 1981. Earth resources searches, astronomy, biomedical tests, materials-processing furnaces produced semiconductors, superconductors, alloys, oxides, glass, pure metals, ionic crystals. Radio telescope deployed outside through rear docking port mapped Milky Way. In four years' service in orbit, Salyut 6 accommodated 33 people, working about 700 days in space.

SALYUT 7
Rocket: Proton, also known as D-1
Launch: April 19, 1982
Duration: fell from orbit and burned February 7, 1991
Events: The last Salyut, a modified version of Salyut 6, had more crew comfort, better food, nicer colors, improved science gear, a 300- lb. materials processing furnace, new X-ray telescope, improved biomedical telemetry equipment. A malfunction while unoccupied in 1985 left the station out of contact with ground control. Cosmonauts flew June 6, 1985, in Soyuz T-13 to the station, finding it frozen from power loss and tumbling in space. They recharged batteries, repaired water pipes, warmed the living quarters. It was 10 days before they could live in the station. Two Progress cargo freighters flew to the station with new equipment. The cosmonauts spacewalked to install a third set of solar panels for electricity. A major step in docking ships with large masses and volumes came when three spacecraft—Cosmos 1443, Salyut 7 and Soyuz T-9—were docked together for six weeks in 1983. Together, the three weighed 94,000 lbs. and were 115 feet long. Three Soyuz T-10B cosmonauts set a 237-days-in-space record in 1984. On May 6, 1986, cosmonauts left Mir in Soyuz T-15, flew to Salyut 7, entered, mothballed the Salyut station and returned to Mir June 26, 1986. The USSR considered flying its space shuttle Buran to Salyut 7 to retrieve the station, but the Salyut plunged into the atmosphere February 7, 1991, and burned over Argentina. Chunks were found 12 miles from Buenos Aires. Over the years, Salyut 7 crews worked more than 800 days in orbit. Altogether, 12 Progress unmanned cargo freighters carried food, water, fuel, supplies and mail to the station from May 1982 to June 1985.

THIRD-GENERATION SPACE STATION
MIR
Rocket: Proton, also known as D-1
Launch: February 20, 1986
Duration: still in orbit, in permanent full-time use.
Events: Mir is the USSR's eighth space station and the first of a new third generation of station hardware. A temporary crew of cosmonauts checked-out Mir from March 15 to July 16, 1986. When an operational crew arrived at Mir February 8, 1987, it became the first permanently-manned space station, with rotating crews. Mir has been manned continuously since then, except for four months in 1989 when it was unattended. The station is a modified and improved version of the Salyut 7 design, with the same mass, outer contours and main dimensions. Mir, 43 feet long and big as a house trailer, has twice the electrical power of Salyut. Crews can adjust the inside temperature for shirt-sleeve work from from 64º to 82º F. The station has two big cruise engines to maneuver in space and 32 small thrusters for attitude control. The usual crew is two or three, but six cosmonauts can work in Mir easily. Six spacecraft can dock at Mir. Cosmonauts have an amateur radio station aboard Mir, using it for recreation to chat with hams around the world.
Mir crews: Several international crews have flown to Mir. Cosmonauts have stayed up to a year at a time. Two cosmonauts, Musa Manarov and Vladimir Titov, hold the human single-trip-in-space endurance record of 366 days. Manarov holds the one-person time in space endurance record with 541 days total in two trips to Mir station.
Mir resupply: Progress unmanned cargo freighters carry food, water, fuel, supplies, science equipment and mail to the station about four times a year.
Mir expansions: The station grows as add-on modules are ferried to space. A small astronomy and science research module, Kvant-1, was attached to the station in April 1987. Kvant is Russian for Quantum. A 19-ton expansion module, Kvant-2, was sent to the station in 1989, relieving overcrowding by doubling the size of the station. Kvant-2, the same size as the original Mir, was launched November 26, 1989, and docked with Mir December 6. Kristall was the third major addition to the station. The same size as Kvant-2 and the original Mir, Kristall was launched May 31, 1990, and docked June 9. Kristall is Russian for Crystal, reflecting the equipment installed in the Kristall module—furnaces to produce semiconductor crystals in the weightless environment of space. The USSR described Kristall as a "microfactory in orbit." Two more large add-on modules to be shipped to Mir in 1992-93, were named Spektr, Russian for Optical, and Priroda.

Salyut 7

Salyut 7 was the second and last of the USSR's second-generation space stations. It was launched to Earth orbit April 19, 1982, on a Proton rocket from Baikonur Cosmodrome in the USSR state of Kazakhstan. The station fell back into the atmosphere to burn February 7, 1991.

The Salyuts. The USSR's first two generations of space stations were known as Salyut—Russian for salute. Salyuts 1 to 5 were of a first technological generation. Salyuts 6 and 7 were a second generation.

Mir is Russian for peace. Launched February 20, 1986, Mir space station was the first of a third generation of stations for the USSR.

Mir was said to have had the first "permanent manning." Previously the USSR had sent cosmonauts only periodically to the seven Salyut stations. The plan for Mir was to keep a crew in the station at all times. However, Mir was unmanned four months in 1989.

Salyut 7. Salyut 7 was a modified version of Salyut 6, the first second-generation station. Salyut 7 had more crew comfort, better food, nicer colors, improved science gear, a 300-lb. materials processing furnace, new X-ray telescope and improved biomedical telemetry equipment.

Mir, in turn, is an improved version of the Salyut 7 design, but with the same mass, outer contours and main dimensions. Salyut 7 or Mir without add-on modules, each weighed 42,000 lbs. and was 43 feet long, big as a house trailer.

Cosmonauts. Salyut 7 was used by eleven groups of cosmonauts, who flew to the space station in Soyuz T-5, T-6, T-7, T-8, T-9, T-10B, T-11, T-12, T-13, T-14, and T-15.

Three cosmonauts, who flew to Salyut 7 in the Soyuz T-10B capsule, set a 237-days-in-space record at the station in 1984. Over the years, Salyut 7 crews worked more than 800 days in orbit.

Two international crews flew to the station. France's Jean-Loup Chretien flew to Salyut 7 in Soyuz T-6 in June 1982 and India's Rakesh Sharma flew to the station in Soyuz T-11 in April 1984.

Those two Salyut 7 flights contrast with nine earlier international-crew flights to the Salyut 6 space station between 1978 and 1981, with cosmonauts from Czechoslovakia, Poland, East Germany, Bulgaria, Hungary, Vietnam, Cuba, Mongolia and Romania.

Trouble in space. A malfunction during a time when it was unoccupied in 1985 left Salyut 7 out of contact with ground control. Cosmonauts flew June 6, 1985, in Soyuz T-13 to the station, finding it frozen from power loss and tumbling in space. They recharged batteries, repaired water pipes, and warmed the living quarters. It was 10 days before they could live in the station. Two Progress cargo freighters flew to the station with new equipment.

The cosmonauts spacewalked to install a third set of solar panels for electricity. Powered by solar energy and chemical batteries, Salyut 7 did not have a nuclear-powered electricity generator.

Expansion. A major step in linking spacecraft with large masses and volumes came when three—Cosmos 1443, Salyut 7 and Soyuz T-9—were docked together for six weeks in 1983. Altogether, the three weighed 94,000 lbs. and were 115 feet long.

Cosmos 1686, an experimental cargo freighter with descent module, was docked with the station September 27, 1985.

Retrieval. Shortly after Mir was launched to orbit near Salyut 7 on February 20, 1986, a check-out crew of cosmonauts flew to space in Soyuz T-15 March 15, 1986, to open the new station.

On May 5, they flew in Soyuz T-15 to the nearby Salyut 7, entered the station, shut it down, mothballed it for a then-planned future retrieval, returned to Mir June 26 and flew to Earth July 16. No one visited Salyut 7 after that, despite the station's location in a parking orbit not far from Mir, 200 miles above Earth.

The Soviets had thought of sending cosmonauts in the shuttle Buran to retrieve Salyut 7. The station could have been returned to Earth, refurbished and sent back to be attached to Mir, increasing room for cosmonauts.

The Sunspot Cycle and a shortage of money defeated that plan. Meanwhile, since 1987, three major add-on modules have been sent to space and attached to Mir, tripling its size.

Sunspots. During peak activity times on the Sun, extra energy blasting outward reaches Earth and heats the atmosphere, causing it to expand. The expansion engulfs high-flying satellites, dragging them downward.

A space station is in a relatively-low orbit, making it especially vulnerable to such drag. Rocket engines attached to a space station must be used from time to time to boost the station back up above 200 miles in altitude.

During a 1988-92 peak in solar activity, the Sun sprayed Earth with an unusual burden of radiation, heating the atmosphere, causing it to expand, increasing drag on satellites of all kinds. The normal slow fall of objects in orbit, such as Salyut 7, was accelerated by the solar activity.

At the same time, the USSR found itself in a cash crunch, unable to fund a special effort to get fuel, cosmonauts or additional engines to Salyut 7 to boost it higher in altitude. As a result, the Sunspot Cycle combined with a shortage of money lead to the demise of the station.

The end. The last of Salyut 7's fuel was burned in 1986 to boost the station to an altitude high enough to keep it in space 10 to 20 years. But, growing sunspot activity in 1988 slowed Salyut 7 in orbit, causing the station to fall faster.

Many satellites fall from orbit each year. Usually, they break up as they are buffeted by the thick lower atmosphere. All the fragments burn. But, some parts of a jumbo spacecraft like Salyut 7, especially with extensive shielding protecting the Cosmos 1686 descent module, would be able to make it all the way down.

The 40-ton station, unused since 1986, was out of control after October 1990, tumbling along its path around the globe, pushed lower by energy from the Sun. By February 5, 1991, Salyut 7 was down to an altitude of 110 miles and falling.

Salyut 7 and the attached Cosmos 1686 cargo module together weighed 40 tons and were 93 feet long, about the size of a railroad car. They entered Earth's atmosphere at 17,000 mph the morning of February 7, 1991.

Argentina. Salyut 7 plunged into Earth's atmosphere over South America's Argentine pampa and "burned out of existence," as the official Soviet news agency Tass put it. However, large chunks of the station did not burn, falling all the way to Earth's surface. Some fell within 12 miles of Argentina's capital of Buenos Aires. One 10.5-ft.-diameter cylindrical fragment was found near the Argentine city of Chaniar.

The official Argentine government news agency Telam reported debris from the spacecraft crashed into the Andes Mountains in a sparsely-populated area near the border between Argentina and Chile. Altogether, as many as 250 small fragments, a total of 1.5 tons, about four percent of the station's weight, made it to the ground. Reportedly, some pieces actually fell in a junkyard.

Eighth to fall. Salyut 7 was the eighth space station to have fallen from Earth orbit over the years. The seven others were the six earlier USSR Salyuts and the one U.S. space station known as Skylab.

The fiery re-entry of Salyut 7 was reminiscent of the 1979 plummet of Skylab into an unpopulated area of Australia. The combined Salyut 7-Cosmos 1686 amounted to only about half the weight of Skylab.

The USSR keeps a search and rescue team on standby for every manned Soyuz launch. The team includes a rigger-engineer, an electrician and a rover driver. The team and its rover, which fly where needed in an IL-76 airplane, were prepared to rush to Argentina. However, no injuries were reported.

Mir Space Station

Mir is a Russian word for peace. It also is the USSR's name for the first of a third technology generation of space stations. Mir was launched February 20, 1986, to a low Earth orbit on a Proton rocket from Baikonur Cosmodrome.

The launch came just 23 days after the fatal explosion of U.S. shuttle Challenger. Today, Mir continues to circle the globe in orbit, in permanent full-time use. Soviet officials say Mir will continue to be manned permanently.

The generations. Russia's first two generations of space stations were known as Salyut. Mir was said to have the first permanent manning because the USSR had sent cosmonauts only periodically to the seven Salyut stations, rather than attempt to keep someone up there all the time.

Mir is an improved version of the last Salyut design—Salyut 7—with the same mass, outer contours and main dimensions as Salyut. Mir alone, without add-on modules, weighed 42,000 lbs. and was 43 feet long—big as a house trailer in orbit. Three add-on modules have expanded Mir's size since 1986.

Mir has twice the electrical power of Salyut 7. The station has two big cruise engines to maneuver in space and 32 small thrusters for attitude control.

Progress. After Mir was placed in orbit February 20, 1986, a check-out crew visited from March 15-July 16, 1986. They turned on equipment, tested hardware and made the station ready for permanent occupancy. They also visited the nearby Salyut 7 space station in May and June to mothball it.

Mir was unmanned when Progress 27, an unmanned cargo freighter loaded with food, fuel and supplies for the space station, was launched in subzero temperatures from Baikonur Cosmodrome January 16, 1987. Progress 27 flew up to 200 miles altitude for rendezvous and automatic docking with Mir. Progress 27 pushed the station to a circular orbit 215 miles above Earth, permitting launch windows from the USSR every two days.

Permanent crew. On February 6, 1987, two cosmonauts, Yuri Romanenko and Alexander Laveikin, were launched in a new-style Soyuz TM-2 capsule to Mir. Two days later, after docking Soyuz at a port on Mir, the cosmonauts went about unloading provisions from Progress 27 into Mir. Thus, permanent manning of the space station began February 8, 1987.

The orbit. Mir is kept 215 miles above Earth. Since its low orbit causes the station to be dragged slowly down into the planet's atmosphere, engines in Mir, and in Progress supply ships attached to Mir, are used from time to time to move it back up to around 215 miles altitude. Progress supply ships arrive with food, water, fuel, supplies and mail for the station every three or four months.

Living quarters. The station has a large galley, several recreation facilities, big bathroom with shower, and private compartments for crew members. Crews adjust the inside temperature for shirt-sleeve work from 64º to 82º F.

Mir station is more like a home than Salyut stations were. The galley, for instance, was made more palatable. Special private compartments for crew members were included, along with nicer shower and bathroom facilities.

The usual Mir permanent crew is two cosmonauts, but six to eight cosmonauts can work in Mir easily.

Expansion. Six spacecraft can dock at Mir. The station has grown as three add-on modules have been sent to space.

A small module—the 19-ft.-long, 13.6-ft.-diameter, astronomy lab Kvant-1—was sent up and attached to Mir in April 1987. Kvant-1 is an astronomy observatory created by British, Dutch, West German and USSR scientists. It has an X-ray telescope with electronic equipment from the European Space Agency. It also has an ultraviolet telescope developed by Switzerland and the USSR. The Soviet Academy of Sciences, the European Space Agency, and West Germany each supplied special spectrometers. Great Britain and the Netherlands supplied a telescope for electronic images.

Kvant-1 is used to gather data about remote reaches of the Universe. Such an observatory above Earth's clouds and dusty atmosphere sees the Moon, Sun, planets, Milky Way galaxy and deep space much more clearly.

The first of two larger add-on modules, Kvant-2, the same size as the original Mir, was launched on November 26, 1989, and docked December 6. The 21-ton module doubled the size of Mir and added a space motorcycle (manned maneuvering unit or MMU) for spacewalks, new shower, more comfortable hatch for spacewalks, and a remote manipulator arm for repositioning the module after docking.

Kristall, second of the two large, 21-ton add-on modules was launched on May 31, 1990, and docked June 9. Kristall was the third major building-block addition to the space station, after Kvant-1 and Kvant-2. Kristall was the same size as Kvant-2 and the original Mir. With the addition of Kristall, the entire Mir complex in Earth orbit weighed 90 tons.

Called a microfactory in orbit, Kristall has a lab to grow crystals in weightlessness for use in electronics, optical equipment and batteries. It has a furnace to melt materials, producing pure crystals worth $1 million each for use in semiconductors. The USSR hopes to make millions of dollars in profits from sales of manufactured crystals. The cosmonauts also produced cell cultures in zero gravity for pharmaceutical companies.

Two more large modules will be added to Mir in 1992-93. The USSR expected to launch in the second quarter of 1992 a large add-on module, Spektr, with telescopes to photograph Earth. Spektr, Russian for Optical, is a large capsule like the Kvant-2 and

Kristall expansion modules added to Mir in 1989-90.

The USSR planned to launch the international ecological module Priroda to expand Mir in the fourth quarter of 1992. Priroda also is large, like Kvant-2 and Kristall.

Guest cosmonauts. Cosmonauts have stayed in Mir up to a year at a time. Two cosmonauts, Vladimir Titov and Musa Manarov, hold the single-trip space endurance record of 366 days. Seven international crews have flown to Mir station between 1987 and the end of 1991.

Recreation. During work hours, crews do science research, commercial development work and military projects. For recreation, cosmonauts have taken various games and musical instruments, as well as audio and video tape recordings of music and movies, to the station for entertainment during rest periods. They have an amateur radio station aboard Mir, using it frequently for recreation to chat with hams around the world.

Docks. Mir was built with a docking port on each end. An additional docking hub with six ports was sent separately to the station and attached in space. The hub is a sphere with six sets of docking holes and collars around its surface.

One of the six ports on the docking hub is attached to the small end of Mir. That leaves the other five ports on the hub available as docks for arriving spacecraft.

The docking ports are needed for Progress unmanned cargo freighters, Soyuz manned spacecraft, and modules of various sizes sent up to be added to the station.

Soyuz ships ferry men to and from the station. Progress ships are unmanned versions of Soyuz-type spacecraft used as radio-controlled cargo carriers. Progress freighters go up with food, fuel, water, equipment and mail from Baikonur every couple of months.

At least one Soyuz spacecraft for carrying cosmonauts to and from Earth is kept at the station at all times, in case of emergency. That Soyuz often is attached to a single port at the large end of Mir.

The Kvant-2 astrophysics module is attached to the single dock on the large end of Mir. In turn, attached to the outer end of Kvant is the current Soyuz manned capsule for return to Earth. Kvant-2 and Kristall are, and Spektr and Priroda will be, attached to the station's large multi-port docking hub.

All of the add-on modules fly unmanned to the space station. An arriving module docks temporarily at the central port facing Mir on the hub. After checkout by the station crew, a manipulator arm grabs the module and swings it around to a port on the side of the docking hub. In that position, it sticks out like a spoke at a right angle from Mir. Ultimately, six 43-ft. Mir clones could be sent up and linked to form one 250,000-lb. six-pointed star, 85-90 feet in diameter. Beyond that, other such "six-packs" could be mated to create an ever growing complex city of Mir stars in space.

Soviet Orbital Station Mir

This report on the USSR's orbiting space station Mir was written in Summer 1987 by physicist and mathematician Dr. Tamara Breus, Ph.D., Head of Sector, Institute of Space Research, USSR Academy of Sciences. The USSR began permanent occupancy in 1987 of its Mir third-generation station in Earth orbit.

A large new-generation scientific laboratory—the Mir orbital station—was launched into space in the Soviet Union on February 20, 1986. It was carried into orbit by the Proton booster rocket which had been earlier used to launch Salyut orbital stations.

The Mir station is designed for fundamentally new research, has much better conditions for the work and rest of the crew, and, most important, its effectiveness has been enhanced perceptibly. This manifests itself specifically in the fact that the crew has been ridded of many auxiliary jobs and can, therefore, pay much more attention to research work.

The fact that the new station and the stations of the previous generation are launched into orbit by means of the same booster rocket predetermined the closeness of some of their major characteristics. Specifically, the same mass, outer geometrical contours and main dimensions. There is similarity also in some basic technical design decisions.

Four compartments. The total length of the Mir station is 13.13 meters (43 ft.). Like the stations of the Salyut family, it consists of four compartments, three of them being air-proof and one not air-tight.

The transition and work compartments and the transition chamber are air-proof. They are linked by hatches. The unit compartment, which is not air-tight, does not stand out in its outer appearance and, together with the work compartment, forms the single external surface of the station.

Two 300-kilogram cruise engines are installed in the unit compartment. They enable the station to maneuver in space and to change its orbit. There are also 32 fourteen-kilogram engines of the attitude control system in the unit compartment.

The Mir station's crew usually consists of two to three members but five to six people can work in it. Transportation of them to the station aboard two transport ships is envisaged.

Multiple docks. To receive transport ships, both manned and automatic, the Mir station, like the Salyut ones, has two docking units—one on the side of the transition compartment and the other on the side of the unit compartment.

Besides that, there are four other docking units in the transition compartment for including scientific modules—a new element of space technology—in the orbital complex.

Modules are launched into orbit by a separate booster and then dock with the station, which enhances the latter's research potential. Modules, like transport ships, tether the main docking unit located in the butt-end of the transition compartment.

After that a mechanical manipulator carries the scientific module to one of the four side docking units. The crew has free access to the equipment installed in the modules. One can get from the transition module into each of them through a hatch with a diameter of just under one meter (three feet).

Mir building block. The very principle of supplementing the scientific, technical and technological equipment of the station with modules brought to it separately and particularly the possibility of replacing them as new research objectives emerge sharply increases the possibilities of carrying out various research in orbit.

The station is called the "basic unit" because it now forms the basis for building a multi-purpose constantly-operating orbital complex and the tasks it accomplishes are largely determined by specialized scientific modules. The Mir station, itself, the basic unit of the orbital complex, mainly ensures necessary conditions for the work and rest of the crew.

The work of the entire complex is controlled from the station, and the latter supplies the complex with electricity. The radiotechnical systems of the station have considerably broadened the possibilities for transmitting scientific and operational information to the Earth. Communication with it can now be maintained, in effect, round the clock due to the use of repeater satellites stationed in a geostationary orbit.

Computer by radio. Reliable round-the-clock communication between the station and the Earth not only creates comforts for the crew but makes it possible to transmit information from the station without forced intervals.

The most important thing is that the station's on-board computer equipment can form an integral complex with powerful ground-based computers. Such direct "machine-to-machine" contact opens up new opportunities for automatic control of the on-board systems during scientific experiments.

The installation of equipment in modules which are separately brought to the station has freed a considerable part of the work compartments from scientific instruments. This has made it possible to provide the crew, for the first time ever, with personal premises which are called cabins, like on ships.

The fact that the crew members can stay alone, if they wish to, is a major element of psychological comfort, especially if one works aboard the station for many months.

Space rendezvous. The turning of the station into a large orbital complex made it expedient to introduce a new system of rendezvous and docking of transport ships.

When a transport ship came up to a Salyut station, the station used its small jet engines to maneuver in space so that its docking unit would be turned toward the approaching transport ship. The calculations have shown that such a docking method is inexpedient as applied to the Mir station.

It is more fuel-efficient for the station's attitude control engines if the transport ship, having approached the station to a distance of several hundred yards, flies around the station and comes up to the docking unit which has been prepared to receive the transport ship. To perform this flying-around and docking maneuver, the station carries necessary automatic systems and radiotechnical equipment which guide and turn the transport ship, leading it to the docking unit.

During the development of the station, its designers paid special attention to the on-board system of controlling the orbital complex, seeking to provide it with certain intellect. For instance, a new strategy of putting the station in the set space position after a long-duration flight without any fixed orientation is used in the system.

The necessary commands to the executing mechanisms are given by the on-board computer. The sensor data on the position of the station during the latest accurate attitude control of it are fed into the computer's memory while the computer learns from the accelerometers in what direction and to what distance the station moved in the past.

Gyrodyne. One more important innovation is the use of installations which do not require reserves of propulsive mass, in the attitude control system, apart from the small-power jet engines. These are gyroscopical stabilizers, or gyrodynes.

The rotating flywheel of a gyrodyne becomes the "fulcrum" around which it is possible to turn the station. Controlling the kinetic moment of gyrodynes, it is possible also to control the turns of the station. That is, to do the same that is done by means of the attitude control jet engines but without consuming propulsive mass (without using fuel).

The power-supply system of the Mir orbital station has considerably changed, compared with Salyuts.

It comprises two solar-battery panels (there were three of them on Salyuts) with a total area of nearly 80 square meters. While deviation, to one or other side, of several volts from the normal supply voltage was allowed on Salyuts, now it is stabilized and can change by not more than 0.5 volts. This makes the operation of the whole electronic, electromechanical, radiotechnical and other equipment more reliable, to say nothing of the fact that the "individual" voltage stabilizers which were indispensable in some instruments become unnecessary.

Improved thermostat. Substantial changes have been made also in the thermo-regulating system which must not only maintain necessary temperatures in the air-tight compartments of the station itself, and in the transport ships and scientific modules, but also ensure the preset thermal conditions of the elements of the construction of the complex, instruments and equipment inside and outside the compartments.

Such a dependable and effective thermo-technical element as heat-exchanger pipes is used in Mir's heat supply system, instead of the traditional coil pipes. This is only one of the innovations, thanks to which the thermo-regulating system has become much more effective.

The crew can preset any temperature within the range of 18 to 28 degrees centigrade. It will be maintained automatically.

First steps in using the Mir station as the basis for a large orbital station have already been made. Its first crew—Leonid Kizim and Vladimir Solovyev—successfully worked aboard it for more than two months, from March 15 to May 5 and from June 26 to July 16, 1986.

Then the second expedition, consisting of Yuri Romanenko and Alexander Laveikin, arrived at the station on February 8, 1987.

Space telescope. The first Kvant orbital module, carrying an astrophysical observatory, docked with the Mir station. This observatory is the largest special-purpose complex ever installed aboard the Soviet orbital space vehicles.

The observatory will have no match in the world in its technical and scientific characteristics, at least up to the year 1990. For instance, the telescope for spectrometric studies of hard X-ray radiation, which forms part of Kvant, has the effective area of detecting devices six times greater than the American HEAO-3 (High Energy Astronomic Observatory) satellite.

New modules, designed to form part of the station, are being made and new scientific equipment for them is being developed.

Progress Space Freighters

Progress cargo freighters have supplied the USSR space stations Salyut and Mir in Earth orbit since 1978, with shipments to Mir every six to eight weeks.

Progress is an unmanned version of the Soyuz capsule used to ferry men to space. The life-support gear, parachutes, re-entry heat shields and solar panels are gone. A docking collar with radar homing transmitter and TV cameras guide the unmanned freighter to a space station dock.

A Progress spacecraft can carry about 5,000 lbs. of goods to the space station from Baikonur Cosmodrome on the Asian steppe. Progress ships supply cosmonauts with food, fuel, water and other necessities as well as scientific experiments to be conducted and mail from home.

A round cargo hold, behind the docking collar, has a special frame with quick-release tie-downs for packages. It usually holds 3,000 lbs. of food, water, clothing, replacement parts, science experiments, personal items, mail and newspapers.

Behind that freight area is a fuel module with four tanks of hydrazine fuel and oxidizer. Compressed air and nitrogen for the station also are in the fuel tank module. Oxygen has to be added from time to time as some is lost when airlocks are opened for spacewalks. Pumps move the fuel, air and nitrogen to space station tanks. About 2,000 lbs. of fuel and gases are carried. Refueling can be done without a crew at the space station.

Beyond the fuel tank area is an instrument module to control Progress. Behind that is the spacecraft's main engine.

Outside. Progress has two television cameras for docking, three docking lights, 14 small docking and orientation thrusters, eight minute thrusters for precision control, plus antennas, sensors and other apparatus.

Both Progress unmanned freighters and Soyuz manned capsules are launched on medium-sized A-2 rockets. Both take about 48 hours to catch the space station in orbit and dock. Progress has sufficient battery power for eight days of flight.

Cosmos 1686, an experimental module, carried cargo to Salyut 7 on September 27, 1985. An improved Progress model M was introduced in August 1989.

Like all Earth satellites, a space station slowly slips down toward the atmosphere. It needs to be lifted back to its original altitude from time to time. A Progress, after being emptied of cargo and after refueling the space station, often is used as a space tug to push the station to a slightly higher orbit.

Cosmonauts at Mir station generate a ton of trash each month. After they unload a freighter, trash is loaded into the empty Progress. The unmanned Progress then is undocked from Mir and commanded to fall into Earth's upper atmosphere where it burns over the South Pacific Ocean.

In September 1990 the USSR launched a recoverable capsule designed to bring results of science experiments down to Earth. About the size of a 30-gal. garbage can, the recoverable capsule can carry 330 lbs. of materials. To send samples home, cosmonauts unload supplies from Progress into Mir, load station trash into the empty freighter, fill the recoverable capsule with materials to go down and install the capsule in Progress's docking port. The unmanned freighter is undocked from Mir 200 miles above Earth and commanded to fall into the atmosphere. Progress drops down to an altitude between 80 and 65 miles where it jettisons the capsule. The freighter burns in the atmosphere over the South Pacific, while the recoverable capsule descends by parachute to a soft landing in the USSR. The list below shows the destination space station beside the year:

Progress	Station	Launch	Docking	Jettison
1978	**Salyut 6**			
Progress 1	Salyut 6	Jan 20	Jan 22	Feb 8
Progress 2	Salyut 6	Jul 7	Jul 9	Aug 4
Progress 3	Salyut 6	Aug 7	Aug 9	Aug 23
Progress 4	Salyut 6	Oct 3	Oct 5	Oct 26
1979	**Salyut 6**			
Progress 5	Salyut 6	Mar 12	Mar 14	Apr 5
Progress 6	Salyut 6	May 13	May 15	Jun 9
Progress 7	Salyut 6	Jun 28	Jun 30	Jul 20
1980	**Salyut 6**			
Progress 8	Salyut 6	Mar 27	Mar 29	Apr 26
Progress 9	Salyut 6	Apr 27	Apr 29	May 22
Progress 10	Salyut 6	Jun 29	Jul 1	Jul 19
Progress 11	Salyut 6	Sep 28	Sep 30	Dec 11
1981	**Salyut 6**			
Progress 12	Salyut 6	Jan 24	Jan 26	Mar 20
1982	**Salyut 7**			
Progress 13	Salyut 7	May 23	May 25	Jun 6
Progress 14	Salyut 7	Jul 10	Jul 12	Aug 13
Progress 15	Salyut 7	Sep 18	Sep 20	Oct 16
Progress 16	Salyut 7	Oct 31	Nov 2	Dec 14
1983	**Salyut 7**			
Progress 17	Salyut 7	Aug 17	Aug 19	Sep 18
Progress 18	Salyut 7	Oct 20	Oct 22	Nov 16
1984	**Salyut 7**			
Progress 19	Salyut 7	Feb 21	Feb 23	Apr 1
Progress 20	Salyut 7	Apr 15	Apr 17	Apr 17
Progress 21	Salyut 7	May 7	May 9	May 26
Progress 22	Salyut 7	May 28	May 30	Jul 15
Progress 23	Salyut 7	Aug 14	Aug 16	Aug 28
1985	**Salyut 7**			
Progress 24	Salyut 7	Jun 21	Jun 23	Jul 15
Cosmos 1686	Salyut 7	Sep 27	Sep 29	Feb 7, 1991
1986	**Mir**			
Progress 25	Mir	Mar 19	Mar 21	Apr 21
Progress 26	Mir	Apr 23	Apr 25	Jun 23
1987	**Mir**			
Progress 27	Mir	Jan 16	Jan 18	Feb 25
Progress 28	Mir	Mar 3	Mar 5	Mar 28

Progress 29	Mir	Apr 21	Apr 23	May 11
Progress 30	Mir	May 19	May 21	Jul 19
Progress 31	Mir	Aug 3	Aug 5	Sep 23
Progress 32	Mir	Sep 23	Sep 25	Nov 19
Progress 33	Mir	Nov 20	Nov 22	Dec 19
1988	**Mir**			
Progress 34	Mir	Jan 20	Jan 22	Mar 4
Progress 35	Mir	Mar 23	Mar 25	May 5
Progress 36	Mir	May 13	May 15	Jun 5
Progress 37	Mir	Jul 18	Jul 20	Aug 12
Progress 38	Mir	Sep 9	Sep 11	Nov 23
Progress 39	Mir	Dec 25	Dec 27	Feb 7, 1989
1989	**Mir**			
Progress 40	Mir	Feb 10	Feb 12	Mar 5
Progress 41	Mir	Mar 16	Mar 18	Apr 25
Progress M-1	Mir	Aug 23	Aug 25	Dec 1
Progress M-2	Mir	Dec 20	Dec 22	Feb 9, 1990
1990	**Mir**			
Progress M-3	Mir	Feb 28	Mar 2	Apr 28
Progress 42	Mir	May 5	May 7	May 27
Progress M-4	Mir	Aug 15	Aug 17	Sep 20
Progress M-5	Mir	Sep 27	Sep 29	Nov 28
1991	**Mir**			
Progress M-6	Mir	Jan 14	Jan 16	Mar 15
Progress M-7	Mir	Mar 19	Mar 21	May 7
Progress M-8	Mir	May 30	Jun 1	Aug
Progress M-9	Mir	Aug 21	Aug 23	Aug 30
Progress M-10	Mir	Oct 17	Oct 21	1991

Guest Cosmonauts

The USSR conducts regular manned spaceflight operations with cosmonauts flying to and from the Mir space station in Earth orbit. Flights to the station happen two to four times a year. Guest cosmonauts frequently go along on trips to the space station.

The USSR says it has invested more than $11 billion in Mir. To recover some of that money, and earn foreign-exchange hard currency, the USSR's commercial space agency, Glavkosmos, sells flights to Mir to any foreign nation for $10 million. Looking to Madison Avenue for guidance, Glavkosmos also has taken to painting foreign company logos on the sides of rockets for a fee.

Many trips have been made by persons from various nations to USSR space stations. At least 18 visitors had flown to USSR space stations by the end of 1991. Nine flew to the Salyut 6 station between 1979-81. Two flew to Salyut 7, in 1982 and 1984. Seven flew to Mir station between 1987 and the end of 1991.

Qualifications. Men and women who fly in USSR spacecraft are cosmonauts. The USSR has liked to train air force pilots as cosmonauts because combat pilots supposedly have better reflexes as well as mental and physical qualifications. However, journalists, scientists and civilians have been trained.

Training. All candidates are prepared for spaceflight by the Yuri Gagarin Memorial Cosmonaut Space Flight and Space Training Center at Star City outside Moscow. Foreign cosmonauts usually receive one to two years training before space flight.

They study Russian, astronomy, advanced mathematics and physics at Gagarin Center. Training includes theory classes and general classroom education about spaceflight. Cosmonauts take copious notes, rather than receive preprinted handouts in classes. They receive 30 to 40 hours of teaching in such subjects as flight mechanics, ballistics, space piloting and navigation, and Soyuz and Mir systems. Preparations include physical training. The trainees practice in simulated weightlessness in 30-second parabolic maneuvers in flights aboard an Ilyushin IL-76 airplane. They also make ten parachute jumps and learn to fly a plane.

Launch. All cosmonauts are launched in Soyuz capsules from Baikonur Cosmodrome on the central-Asian steppe near Tyuratam in the Kazakh Republic, some 1,250 miles southeast of Moscow.

To Mir. A Soyuz capsule carrying two or three cosmonauts takes two days to catch up and dock with the orbiting Mir space station.

At Mir. Space station visits by non-USSR cosmonauts are brief, usually lasting five to ten days. Visitors help the permanent station crew with science and medical experiments, then make a one-day flight back down to a parachute landing in the USSR.

Past Guest Cosmonauts

Czechoslovakia. Czech cosmonaut Vladimir Remek was the first person not from the USA or USSR to go to space. He went to the USSR's Salyut 6 space station, flying in Soyuz 28 in March 1978.

Poland. Miroslaw Hermaszewski from Poland went to Salyut 6 for seven days, blasting off in Soyuz 30 on June 15, 1978.

East Germany. Sigmund Jahn of East Germany traveled to Salyut 6 in Soyuz 31 on August 26, 1978. He stayed seven days and flew back to Earth in Soyuz 29.

Bulgaria. Georgi Ivanov of Bulgaria went to Salyut 6 April 10, 1979, in Soyuz 33. His craft was unable to dock at the station so returned immediately to Earth.

Hungary. Bertalan Farkas of Hungary went to Salyut 6 May 26, 1980, in Soyuz 36. He stayed seven days and returned to Earth in Soyuz 35.

Vietnam. Pham Tuan of Vietnam, the first space flight by a person from outside Europe or North America, went to Salyut 6 July 23, 1980, in Soyuz 37. He stayed seven days and flew down in Soyuz 36.

Cuba. Arnaldo Tamayo Mendez of Cuba, the first person from a Western Hemisphere country outside of the U.S., flew to Salyut 6 September 18, 1980, in Soyuz 38. He stayed seven days.

Mongolia. Jugderdemidiyn Gurragcha of Mongolia flew in Soyuz 39 on March 22, 1981, for seven days at the Salyut 6 station.

Romania. Dumitru Prunariu of Romania flew in Soyuz 40 on May 14, 1981, for seven days at Salyut 6.

France. France's Jean-Loup Chretien flew in Soyuz T-6 to Salyut 7 for eight days starting June 24, 1982. Patrick Baudry, the Soyuz T-6 backup, went on to fly in 1985 aboard U.S. shuttle Discovery STS-51G. Chretien, in turn, served as backup to Baudry on the Discovery flight.

India. Rakesh Sharma, from India, was a cosmonaut aboard Soyuz T-11 when it left April 2, 1984, for eight days at Salyut 7 station.

Syria. Syrian Air Force Lt. Col. Mohammed Faris spent eight days at Mir after blasting off from Baikonur July 22, 1987, in Soyuz TM-3.

Bulgaria. Bulgaria's Alexander Alexandrov flew to Mir June 7, 1988, in Soyuz TM-5. The 10-day flight was the second trip for Bulgaria.

Afghanistan. Afghan Abdul Ahad Mohmand flew August 29, 1988, in Soyuz TM-6 for an eight-day stay at Mir.

France. France's Jean-Loup Chretien returned to space for a record 26 days at Mir starting November 26, 1988, in Soyuz TM-7.

Japan. Cosmoreporter Toyohiro Akiyama, a veteran Japanese TV correspondent and former Washington bureau chief for the Tokyo Broadcasting System, became the first Japanese and the first journalist in space. TBS paid more than $10 million for the flight. He flew in Soyuz TM-11 to Mir station December 2, 1990. He stayed eight days. Akiyama broadcast live to Japan and Peru from the space station. He flew to Earth December 10.

Great Britain. The first woman to visit Mir station was Helen Sharman, 27, a former Mars candy company worker and chemist from Sheffield, England, with no flying experience. She was chosen from 13,000 applicants who replied to a newspaper ad by a private British firm. The company tried to raise $10 million from the public to pay for training and flight. When Britons wouldn't donate the money, the USSR flew Sharman to Mir anyway, but dropped British science experiments. Sharman studied the Russian language, then took off for space May 18, 1991, in Soyuz TM-12. In space, she studied how orchids and dwarf magnolia vines improved the psychological atmosphere. The four male crew members gave Sharman the best living quarters aboard the station, a room with a view of Earth. She stayed eight days and landed May 26.

Austria. After the government of Austria advertised for a scientist to become a cosmonaut and fly to Mir, Franz Viehboeck was selected. He was launched October 2, 1991, in Soyuz TM-13 as the first Austrian in space. His country paid $6.6 million for the flight. Viehboeck spent eight days in space and landed October 10. Austria's Lothaller Clemens was Soyuz TM-13 backup.

Kazakhstan. One of the republics of the USSR is Kazakhstan. Even so, cosmonaut Takhtar Aubakirov was listed as a guest cosmonaut when he blasted off in the Soyuz TM-13 flight to Mir October 2, 1991. It was an unusual crew with two researcher cosmonauts—Viehboeck from Austria and Aubakirov from Kazakhstan. Past guest cosmonaut flights had a Soviet commander and engineer with one foreign researcher. Soyuz TM-13 had a commander, no flight engineer, and 2 researchers. Aubakirov had been honored in 1988 with the USSR's top distinction, Hero of the Soviet Union, for aircraft test pilot work. Kazakhstan's Talgat Musabaev was Soyuz TM-13 backup.

Future Guest Cosmonauts

Plans are under discussion, or in the works, for several guest-cosmonaut flights to Mir station in coming years. Countries include France, Germany, Malaysia and even the United States. Coming trips in Soyuz-TM capsules by guest cosmonauts will include:

Germany. A German cosmonaut was in training for a lift off from Baikonur Cosmodrome in the second half of 1992.

France. The French have asked the USSR for month-long flights aboard Mir every two years through the end of the century. They last flew in 1988. Two previous French cosmonaut flights to Mir were free in exchange for technology, but the next flight will cost France $1 million a day in space. The 1988 flight was made by Jean-Loup Chretien. His back-up, Michel Tognini, could make the next trip. Both French cosmonauts have trained in the USSR.

Chretien, Tognini and Patrick Baudry also are preparing to fly the European Space Agency's Hermes reusable space shuttle by the end of the 1990's. Hermes will be blasted to orbit atop an Ariane 5 rocket and will land on a runway. It might fly to Mir station.

United States. The Reagan Administration in 1988 killed a U.S. Air Force plan for a joint flight with U.S. astronauts and USSR cosmonauts in a Soviet spacecraft to recover a part of an American military satellite. Reversing field three years later, the Bush Administration in 1991 approved an unprecedented exchange: a Soviet cosmonaut would fly aboard a U.S. shuttle and a NASA astronaut would travel to the USSR's Mir space station. The 1991 Bush-Gorbachev summit agreement called for medical research on how humans adapt to weightlessness in preparation for lengthy flights to Mars after the year 2000. A cosmonaut might fly 13 days in shuttle Columbia in 1993. A date for an astronaut to fly in a Soyuz TM capsule to Mir has not been set.

Malaysia. The Malaysians have been in negotiations with the Soviet Union about sending a guest cosmonaut to Mir in the 1990's.

Canada. The USSR invited Canada to take part in experiments at Mir station as well as on missions to Mars. The Canadians have not indicated whether they will accept. U.S. President Bush, vice president at the time of the invitation, said, if he were elected President, America would not have glittering space spectaculars. Not long after Bush's declaration, Canadian and Soviet officials met behind closed doors in Moscow to explore ways the countries could collaborate in space. The first cooperation mentioned was sending a Canadian to Mir. The Canadians decided to think it over. There was no hurry since the USSR was booked up at the time for sightseeing jaunts to Mir. Previously, Canada had chosen a U.S. route to space so it has no launch pads of its own. Canada's space efforts have been linked to the U.S. space shuttle fleet and plans for the U.S.-international space station Freedom. In 1986, Canada contributed an ultraviolet-light camera for a Russian civilian science satellite.

China. Col.-Gen. Chi Haotian, chief of staff of China's People's Liberation Army, visited the Yuri Gagarin Memorial Cosmonaut Training Center at Star City outside Moscow August 7, 1991. He was briefed on spaceflight training at the center.

Israel. An Israeli may fly in a Soyuz capsule to Mir station in the 1990's, according to the USSR government newspaper Izvestia citing the Soviet news agency, Novosti. Israel would pay for the flight, which would symbolize good relations between Israel and the USSR. Diplomatic relations were broken after the 1967 Middle East War. A preliminary letter of intent for a spaceflight reportedly was signed in July 1991 by M. Yakobi of the Israeli "Forum" enterprise.

USSR journalists. There was an outcry in 1989 among Soviet journalists when the 1990 Japanese journalist flight was announced. Russian journalists thought they should have been the first to ride in a USSR spacecraft, but the Soviet Union needed the $10 million it would earn from a Japanese trip to Mir. Six USSR journalists selected as cosmonaut-candidates started training in October 1990 at the Gagarin center for a late-1992 flight in a Soyuz TM capsule. Preparing are A. Andryushkov and V. Baberdin of the newspaper Krasnaya Zvezda; Yu. Krikun of the Ukrainian TV Film Studio; P. Mukhortov of the newspaper Sovietskaya Molodyozh at Riga, Latvia; S. Omelchenko of the newspaper Delovoi Mir; and V. Sharov of the Literaturnaya Gazeta.

From	Cosmonaut	Flight		Craft	Station
Czech	Vladimir Remek	Mar	1978	Soyuz 28	Salyut 6
Poland	Miroslaw Hermaszewski	Jun	1978	Soyuz 30	Salyut 6
E German	Sigmund Jahn	Aug	1978	Soyuz 31	Salyut 6
Bulgaria	Georgi Ivanov	Apr	1979	Soyuz 33	Salyut 6
Hungary	Bertalan Farkas	May	1980	Soyuz 36	Salyut 6
Vietnam	Pham Tuan	Jul	1980	Soyuz 37	Salyut 6
Cuba	Arnaldo Tomayo Mendez	Sep	1980	Soyuz 38	Salyut 6
Mongolia	Jugderdemidiyn Gurragcha	Mar	1981	Soyuz 39	Salyut 6
Romania	Dumitru Prunariu	May	1981	Soyuz 40	Salyut 6
France	Jean-Loup Chretien	Jun	1982	Soyuz T-6	Salyut 7
India	Rakesh Sharma	Apr	1984	Soyuz T-11	Salyut 7
Syria	Mohammed Faris	Jul	1987	Soyuz TM-3	Mir
Bulgaria	Alexander Alexandrov	Jun	1988	Soyuz TM-5	Mir
Afghan	Abdul Ahad Mohmand	Aug	1988	Soyuz TM-6	Mir
France	Jean-Loup Chretien	Nov	1988	Soyuz TM-7	Mir
Japan	Toyohiro Akiyama	Dec	1990	Soyuz TM-11	Mir
G Britain	Helen Sharman	May	1991	Soyuz TM-12	Mir
Austria	Franz Viehboeck	Oct	1991	Soyuz TM-13	Mir
Kazakhstan	Takhtar Aubakirov	Oct	1991	Soyuz TM-13	Mir
Future	**Cosmonaut**	**Flight**			
Germany	R. Ewald or K.D. Flade	1992, second half of year			
USSR	journalist to be named	1992, six training for one slot			
France	M. Tognini or J.L.Chretien	1992-93, France wants flight every two years			
Malaysia	to be named	1992-94 date to be announced			

Israeli	to be named	1993-94 date to be announced
U.S.	to be named	1993-94 date to be announced
France	J.L.Chretien or M. Tognini	1993-94 date to be announced
Canada	to be named	invited; pending decision on flight
China	to be named	pending decision to train for flight

Mir Advertising

Needing foreign-currency hard cash, the USSR has been selling advertising space on spacecraft, space stations, space suits and anything else space.

The Soviets sold commercial advertising space on the Soyuz TM-8 flight in September 1989. One stage of the launch rocket was painted orange to highlight a display ad for the Italian insurance company Generali. Beside the launch pad and in a room where reporters interviewed cosmonauts were advertisements for New Dawn perfume and a Russian electronics company.

Advertisements for Japanese products painted on the Soyuz TM-11 launch rocket in December 1990 boosted Ohtsuka Chemicals, Unicharm women's hygiene products, and Sony electronic products.

Mir station has space for two advertising billboards. Cosmonauts have space on their flight suits for publicity patches. The Bienne, Switzerland, agency Punto was arranging for USSR space advertising.

Mir's Future

What is to become of the orbiting Mir space station in the new USSR after government restructuring brought about by the 1991 coup d'etat? Some new operating conditions seem in effect, according to Moscow TV in the fall of 1991.

The State Commission for Manned Space Flight was not changed. Government organizations controlling Soviet spaceflight would continue their work. Budgets were frozen, but international agreements continued in force. Guest cosmonauts would be allowed to fly as planned. An Austrian cosmonaut flew in October 1991. Spacewalk plans were reduced.

Selling Mir? Rumors surfaced in September 1991 that Mir would be leased or sold to a foreign company. But Yuri Semenov, general designer for Energia Scientific and Production Amalgamation collective, denied it. He told Tass News Agency the collective would own Mir. Energia Scientific and Production Amalgamation designed and developed Mir and operates it in space.

Energia Scientific and Production Amalgamation is successor to the Sergei Korolev Design Bureau. Korolev, an academician, was the first chief designer of USSR spacecraft in the 1950's.

The Soviet Union's commercial space agency, Glavcosmos, has no control over Mir, Semenov said. Foreign contracts for Mir operations are Energia's prerogative.

Glavcosmos markets Soviet rocket launches to orbit and other space services to foreign customers. A private American firm, Space Commerce Corp., Houston, Texas, is affiliated with Glavcosmos.

An idea of the U.S. buying Mir appeared in a London newspaper in September 1991. U.S. government officials denied such a plan existed. U.S. and Soviet space equipment generally are not directly compatible. The superpowers developed different technology designs during the Cold War.

Also, a 1955 U.S. law bars imports of Soviet military equipment including surplus space materiel. The White House would have to approve loosening restrictions on Soviet-built imports.

Buran shuttle. The USSR was building a second Buran space shuttle with plans to fly it to Mir space station in 1992-93. Buran No. 2 would have a docking collar and manipulator arm for use in linking with the orbiting space station. The 1992-93 unmanned flight would test a cosmonaut rescue system. Buran No. 1 made an unmanned test flight November 15, 1988, and later was displayed at the Paris Air Show.

Space Colonies

The ambitious Russian space exploration program probably will lead to the first colony in space by the year 2000, according to a long-time researcher into the effects of spaceflight, Oleg G. Gazenko, head of the USSR Academy of Sciences physiology department.

Gazenko had worked on Russian manned flights from their start in the 1960's. He predicted in 1987 the USSR will have several space stations in Earth orbit at the same time, exploring and using space, leading to the first colonies on the Moon and Mars.

Noting that colonies will be possible if they can produce food, water and oxygen, Gazenko said experiments aboard the orbiting Mir space station show water and air can be regenerated through solar power. Mir cosmonauts are able to produce fifteen percent of their food today.

Mars. The first space colony likely will lead to the first manned mission to Mars. After orbiting Earth only 237 days in 1987 in Mir, Yuri Romanenko had traveled the equivalent of half the distance to Mars. He stayed in space 326 days. Later, Musa Manarov and Vladimir Titov stayed in space 366 days on one trip.

Romanenko, Titov and Manarov all approached the time in space needed for a flight to Mars—probably nine months to a year out to Mars and twice as long to return. Including the time needed to explore the planet and its two moons, a Mars visit would take three years.

Romanenko, Titov and Manarov demonstrated humans can sustain zero gravity for a long time. If they could not, revolving spacecraft would be needed to create artificial gravity. USSR flights have become longer as researchers try to determine how long the human body can sustain weightlessness. USSR officials say stays of at least six months at space stations have become the norm.

Cosmonauts Leonid Kizim, Vladimir Solovyev and Oleg Atkov set the 237-days-in-space record in 1984. Romanenko stayed in space 326 days in 1987. Titov and Manarov stayed 366 days, through 1988.

Astronauts Gerald Carr, Edward Gibson and William Pogue, set the U.S. endurance record, staying 84 days at the American space station Skylab from November 16, 1973, to February 8, 1974. NASA is hoping to launch construction materials for its own second-generation space station before the year 2000.

USSR Will Build Space Cities
by Cosmonaut Vitali Sevastyanov

The USSR took up permanent occupancy in its third-generation Mir space station in Earth orbit in 1987. This report on plans for building space cities was written in Summer 1987 by USSR cosmonaut and space pilot Vitali Sevastyanov, in response to a request from the editors for information on long range plans.

When Yuri Gagarin, a 27-year-old Russian made his historic flight around our planet on April 12, 1961, he proved that the human race would not be bound to the Earth forever. His flight was but the first step. "But even a road thousands of kilometers long begins with a first step," the cosmonaut's mother said.

Yuri Gagarin was born in a small village near Smolensk. His father was a joiner on a collective farm. At 15 Yuri enrolled at a technical school to learn the trade of a steel maker. He went in for sport and read a lot. Later, Gagarin said that it was at that time that he developed an insatiable thirst for flying. His passion for space came from reading books by Herbert Wells, Jules Verne and later Konstantin Tsiolkovsky, the Russian space theorist.

Academician Sergei Korolev, the famous designer of space ships and a man who knew Gagarin well, said about him, "Yuri epitomizes the eternal youthfulness of the Russian people. He happily combines natural courage, an analytical mind and exceptional industriousness." To this must be added that Gagarin was a very amiable and cheerful person and had a good sense of humor. All that combined with his excellent skills played the decisive role in his becoming the first Soviet cosmonaut.

200 have been there. As many as 200 spacemen have been in orbit to date, with many having flown several missions. Soviet spacecraft also lifted cosmonaut researchers from socialist countries, that are members of the Intercosmos program, and from India and France.

A total of 16 days was spent by Soviet cosmonauts in the first single-seater ships of the Vostok type. On board the Salyut 6 station, its crews worked in space for about 700 days and on board Salyut 7 for more than 800 days. An equivalent shuttle program, with its weekly missions, would have required more than 200 launchings to achieve the same duration.

Salyut 1, the first Soviet orbital spacecraft, was provided only with one docking unit. The station could be used either in the unmanned mode or in the manned one, with a crew deliverable by a Soyuz ship that docked with the Salyut.

In 1977 Salyut 6, a station of the second generation, was launched. It was fitted with two docking units, which changed radically the research program of Soviet orbital stations. Now, if one of the docking units was occupied by a Soyuz that brought up the main crew, the other docking unit made it possible to receive another Soyuz with a visiting crew, or an automatic Progress ferry ship. This advantage of Salyut 6 was used extensively: in four years of orbital service it accommodated 33 people.

Besides, the existence of two docking ports enabled large-sized "space trains" to be formed, consisting of craft that in size and mass were similar to Salyut 6. If these craft, in their turn, had two docking ends, multi-sectional orbital stations of any length could be built up in that way.

Three craft docked. In 1983, a research complex made up of Soyuz T-9, Salyut 7 and Cosmos 1443 operated in orbit for six weeks, with a total mass of 47 tons (94,000 lbs.) and a length of almost 35 meters (115 feet). It was a major step in orbital docking of craft having large masses and volumes.

By that time, Salyut stations had been perfected to such an extent that they became real testing laboratories, but one step before a permanent orbital station with rotating crews, a station that could be supplied with all it needed and could function normally for an indefinite period of time.

The extended exploitation of Salyuts provided Soviet men of science with many insights into the behavior of the human body in space. Since man's adaptability to space and his possible return to normal terrestrial existence without bad effects was one of the unknowns of space flight, this information was especially valuable.

There were, of course, problems in the early stages. Some of the cosmonauts, upon returning to Earth from their orbital missions, experienced difficulties on the firm ground. For example, in walking. To preclude this from recurring, new programs of work, physical training and recreation in orbit were drawn up. These programs owe their origins to Salyut flights.

Under a space-research agreement between the USSR and the U.S., the Soviet Union shares with the United States data on how extended zero-gravity affects the human body. As American scientists pointed out on many occasions, they are especially interested in these Soviet experiments and their results, since they are unable to obtain such information from their national program of studies.

The next generation. The experience gained by seven Salyut orbital stations has allowed Soviet scientists to prepare and make a fundamentally new step in the development of the space technology. I refer to the launching in February 1986 of the Mir orbital station. As distinct from its predecessors, it has six, not two, docking ports. Four are on the periphery and two at the ends along the central axis. The result is that different configurations can be effected and flexible technology can be adopted more widely in orbit. This is the basic idea contained in the conception of a multi-purpose permanently-manned complex with specialized modules concerned with nature studies and technological, astrophysical, biological and other research. The six docking ports mean, above all, the new quality of the station, the possibility of using its costly scientific and technical equipment in full. The cargo bridge between Earth and orbit can operate very efficiently.

The Mir has twice as much (electrical) power as the Salyut and this is very important in conducting power-demanding experiments. For example, in space technology and materials science. This station is expected to do a particularly large volume of applications research.

A permanent space station in near-Earth orbit offers vast benefits. With the help of such a station, everything—from the manufacture of perfect crystals at zero gravity to the organization of expeditions to other planets—can be translated from the stage of costly experiments into daily reality.

Humans have been directly exploring space for little more than 25 years. Before that, we felt cramped in the space we lived in. But, having started space exploration, we were surprised and then pleased to discover that there were new areas for human endeavor. By exploring and using outer space, we are possibly making immense discoveries, determining our place in the Universe, first judging our terrestrial "frailty" and glimpsing the future which will from now on be never connected entirely with the Earth.

Mir Spacewalks

A spacewalk, extra-vehicular activity or EVA is a trip outside an orbiting space station or other spacecraft. Cosmonauts frequently take trips outside the Mir space station to place equipment, to place and retrieve experiments and to make repairs.

Cosmonauts exit and re-enter Mir through airlocks. Each of the large modules composing Mir has an airlock. All spacewalks are hard work and dangerous.

Cosmonauts and astronauts find it difficult moving around in their life-supporting spacesuits. Trying to do real work in a spacesuit can be difficult. Cosmonauts always are in danger while in space. Outside the space station, they risk being struck by a micrometeorite which could punch through a spacesuit. Extreme heat and cold alternate relentlessly. A misstep could send a cosmonaut flying away from the station.

Spacewalking cosmonauts most often wear a tether which ties them to Mir. In 1989, Mir cosmonauts received what the USSR calls a space motorcycle—a manned maneuvering unit, or MMU, a rocket-powered backpack for spacewalks. Below are reports on selected unusual EVAs.

Spacewalk Saves Kvant-1

Cosmonauts took a dramatic walk in "'raw space" from the Mir orbiting station April 11, 1987, to remove a small white cloth bag which blocked the Kvant-1 astrophysics module from docking.

Soviet television, broadcasting from mission control outside Moscow, showed the first live broadcast of a Soviet space walk as Yuri V. Romanenko and Alexander I. Laveikin left Mir and floated to the docking port attached to the astrophysics module Kvant-1.

Launched March 31, 1987, Kvant-1 is a biotechnology and astrophysics space laboratory. It has an observatory, life-support systems, radio, telephone, telegraph and television equipment.

April 5. Kvant-1 was moved within 200 yards of Mir station but could not dock due to a mysterious control-system problem.

April 9. Another docking attempt was made. Kvant-1 touched Mir but the two spacecraft were not firmly linked. The joining was not airtight. The Mir crew could not walk through a tunnel in the docking port to the Kvant-1 module and "get down to astrophysical research," as Radio Moscow put it.

April 11. Romanenko and Laveikin went into what Radio Moscow called "raw space" to find the problem and fix it. They were successful. The airtight linkage permitted the cosmonauts to move freely between Mir and the 11-ton Kvant-1.

To clear the problem, Romanenko and Laveikin spent 3 hours 40 minutes outside Mir, floating in space, removing what they called an "alien object."

Laveikin spotted the small white cloth bag stuck between the craft. He radioed mission control to move Kvant-1 back a bit. Ground controllers ordered small thrusters in Kvant-1 to fire, backing it off a few inches from the Mir dock so Laveikin could grab the bag. Romanenko reported the bag was removed and the joint between the two craft was clear for a solid docking. He asked mission control to pull Kvant-1 up to the dock.

April 13. Romanenko and Laveikin opened the hatch from Mir to inspect Kvant-1 and remove grease safety coatings from its research equipment so they could conduct experiments in the module.

USSR newspapers carried detailed accounts of the spacewalk and Radio Moscow broadcast the conversation between cosmonauts and ground control. Romanenko and Laveikin had blasted off for space February 6, 1987.

Kvant-1. The Kvant-1 astrophysics module is 19 ft. long and 13.6 ft. in diameter. It has been attached permanently to one of Mir's docks, increasing the working room for cosmonauts.

The addition was an astronomer's delight—an X-ray observatory including electronic equipment from Western Europe's space agency. The Soviet Academy of Sciences, the European Space Agency, and West Germany each supplied special spectrometers. Great Britain and the Netherlands supplied a telescope for electronic images. An observatory above Earth's clouds and dusty atmosphere sees the Moon, Sun, planets, Milky Way galaxy and deep space much more clearly.

Cosmonauts Install Panels

Two Soviet cosmonauts walked in space twice in June 1987 to install more electricity-generating solar panels outside their orbiting Mir space station complex.

Working during spacewalks June 12 and June 17, Yuri V. Romanenko and Alexander I. Laveikin put together solar panels to generate additional electricity for use in Mir.

Each day, Romanenko and Laveikin opened an exit hatch and pulled themselves out to float alongside the Mir laboratory. They bolted a hinged truss, designed to carry two sunlight-collecting panels, to the outside of Mir on June 12. They attached two wings of the new solar array to Mir during that spacewalk and a second set of panels five days later during the June 17 walk. They unfolded all four wings so the solar panels could start gathering sunlight and generating electricity.

Power hungry. The solar wings are the space lab's third set. New astronomy equipment sent to the station in April 1987 in the Kvant-1 astrophysics module required more electricity than the previous solar panels could provide. The new panels boost the quantity of sunlight collected and converted to electricity. That electricity is stored in batteries and used to power science experiments aboard Mir.

USSR television broadcast live pictures to the public of Romanenko and Laveikin assembling pieces of the solar cell apparatus outside their orbiting complex.

The solar panels are composed of hundreds of semiconductor voltage-generating cells converting sunlight into electricity. They add up to a total of 26 square yards of new electricity-generating cells mounted on the station's port and starboard sides. The wings are equipped with solar gauges and drive motors to aim them at the Sun. The new third power source increases the amount of electricity available for science research in the growing Mir complex.

By erecting the 35-ft. panels, Romanenko, Laveikin and Russian ground crews learned about construction techniques for large objects in orbit. The cosmonauts also installed cassettes with samples of various structural and heat-shield materials on Mir's surface to see how they will be affected by the space environment.

June 12 walk. At 9 p.m. Moscow time June 12, mission control told the cosmonauts, "We allow opening of the exit hatch." From space, Romanenko replied, "Getting to work."

Soviet television broadcast live pictures of the two cosmonauts moving slowly as they mounted a T-shaped truss and waffle-like panels to the outside of Mir. After Romanenko reported the mounting was complete, the cosmonauts pushed themselves back into the orbiting laboratory.

June 17 walk. The crew worked more than three hours to lug the second part of the solar array outside, connect it to the first pieces they had installed June 12 and plug in the electric connectors. The cosmonauts later drew power from the solar-cell wings.

Mir was sent to orbit February 20, 1986, atop a Proton rocket. At the time of the spacewalk, Romanenko and Laveikin had been living in the station since February 6, 1987. Kvant-1, an add-on module containing sophisticated astronomy telescopes, was ferried to orbit March 31, 1987. It was docked to Mir on April 11 when Romanenko and Laveikin last had walked in space.

Spacewalk To Attach Solar Panel

USSR cosmonauts Vladimir Titov and Musa Manarov went outside Mir station February 26, 1988, to attach a new solar-panel for generating additional electricity.

During the 4 hour 25 minute spacewalk, they replaced the lower half of the array atop Mir with a new set of solar cells providing more electrical power.

It was the first spacewalk since Titov and Manarov took over from Yuri Romanenko as permanent crew in charge of the Mir space station.

Broken Wrench Ends Spacewalk

Cosmonauts Vladimir Titov and Musa Manarov went outside the orbiting space station Mir for five hours June 30, 1988, in an unsuccessful attempt to repair a faulty detector on an X-ray telescope attached to the Kvant-1 astrophysics module.

During the extravehicular activity (EVA) a wrench broke, ending their effort to repair the 88-lb. detector on the upper part of a telescope built by British and Dutch scientists.

The telescope was pointed away from Mir toward deep space during the spacewalk. The detector was attached with clips and small bolts. Not designed for replacement in space, the detector was covered with 20 layers of thermal blankets.

Brass clamp. The cosmonauts were able to peel back the blankets, snap off the bolts and undo a brass clamp. But the wrench broke as they struggled with the brass clamp. They ran out of time and were unable to install the new unit, reconnect its cable, and cover it with insulation.

Working with small bolts through their pressurized gloves was laborious, necessitating several rest breaks. The cosmonauts took turns holding each other in place so one could work with both hands.

Telescope Repair In New Spacesuits

Two USSR cosmonauts spent four hours twelve minutes outside the Mir space station October 20, 1988, replacing a melted lens on an X-ray telescope, a task they previously had to abort when a wrench broke during a five-hour spacewalk June 30, 1988.

Using new tools sent to the station in September in the Progress 38 cargo ship, Vladimir Titov and Musa Manarov removed an 88-lb. block of electronic detectors, developed by British and Dutch scientists, from a telescope on the Kvant-1 observatory module attached to Mir.

Components in the detector had been melted from exposure to cosmic radiation. Titov and Manarov replaced the part with an improved detector, prolonging the life of the telescope.

New spacesuits. During the October spacewalk, Titov and Manarov wore new spacesuits designed for more freedom from tethers to their spacecraft. The suits are equipped with new life support systems. Wearers communicate by radio. However, they did remain tied to Mir during the spacewalk since they had to repair the telescope before concentrating on testing the suits.

While outside Mir, Titov and Manarov installed, near the airlock, a spacesuit fastener used later to keep cosmonauts tethered to the station during a Soviet-French spacewalk in December 1988.

Ham radio antenna. During the October 1988 spacewalk, Titov and Manarov also fastened a 19-in. amateur radio antenna to the outer surface of the space station.

The antenna and a portable, battery-powered, handheld amateur-radio transceiver, used for recreation, were capable of transmitting a two-watt FM signal in the two-meter amateur band of frequencies from 144-146 MHz. Amateur radio space-to-ground chats by Mir cosmonauts with hams on the ground started in November 1988.

Working In Open Space

An additional special report on the October 1988 spacewalk by cosmonauts at the orbiting Mir space station, by Mikhail Chernyshov, space writer, Novosti Press Agency

On October 20, 1988, the Mir crew carried out a number of intricate operations in open space. The job was done by Vladimir Titov and Musa Manarov, while Dr. Valery Polyakov, the third member of the crew, was on standby in the descent module of the Soyuz TM-6 craft docked with the orbital complex.

"The spacewalk pursued several aims," Victor Blagov, deputy flight director said. "The first two were to test new spacesuits and to complete repairs on the Dutch TTM telescope on the external wall of the Kvant-1 astrophysical module, which forms part of the orbital complex.

"Additional operations were also planned: to assemble an aerial for radio communication with amateurs, to mount a special anchor for a Soviet-French crew that plans a spacewalk late in December of this year, and to remove dust from portholes and a TV camera on the Kvant-1's outer wall."

As for the spacesuits, their tests were planned in advance, before the start of the space expedition last December. However, it so happened later that the lens of the Kvant-1 telescope—a block of electronic detectors—deteriorated and had to be repaired.

The repairs were begun last June but were not finished because the key for unlocking the block ring was broken. Replacement of the lens was postponed. Now it was decided to combine the two tasks during one spacewalk. If everything went smoothly and the cosmonauts had time to spare, they would perform other jobs.

"The key, broken when unlocking the fastening ring, developed into quite a problem," according to Oleg Tsygankov, a space repairs specialist.

"We had to devise three methods of separating or breaking up the ring embedded in the telescope's tube. A set of seven tools—including an abrasive electric cutter, drills, removers and special nippers—were developed.

"In fact, about three-quarters of the work to get to the ring was already done by the cosmonauts during their previous spacewalk in June. If everything were to go smoothly, Titov and Manarov would only need ten minutes to open the ring and install a new detector block.

"The rest of the time could be spent on erecting an amateur radio aerial, on cleaning up the portholes and TV camera lens, and on installing an anchor. The anchor saved the coming Soviet and French cosmonauts some 30 minutes during their spacewalk December 9, 1988, for testing solar panels mounted on an intricate structure which opens up only in space."

"As for the spacesuits," says Mikhail Balashov, one of the suit's developers, "the cosmonauts first tested them in the station, working on the exact copies of the fastening ring sent by the Dutch. The new spacesuits do not differ from the old ones in outward appearance. It is the sleeves that have been modified mostly. They now can be adjusted for height and size.

"Normally, one spacesuit is used many times. The leg and arm sleeves, being most vulnerable, are sometimes damaged," Balashov explained. "It is understandable that to glue together or sew up a torn sleeve on board the spacecraft is not the best way of repairing them. Sleeves of the new spacesuit—whether for legs or arms—can be replaced totally.

"Some modifications have been made in other parts of the suit. The gloves now are more flexible. Life-support systems last longer," Balashov said.

A spacesuit for extravehicular activities does not look too complex—but that is so only at first sight. Experts once counted that a spacesuit contains more elements and systems than a car.

Indeed, it has the heat regulation system, a medical monitoring block, and radio communications facilities...to name a few. In fact, a spacesuit is a real spacecraft in miniature.

Experts stress that the new spacesuit has its own power sources and can operate independent of the station. True, the lifeline remains, for there are no means at the Mir orbital station as yet to get back a cosmonaut who has broken free.

Oxygen supplies in the main and standby cylinders make up, all in all, 1,500 liters. This allows a cosmonaut to work in open space for up to eight and a half hours—but that is the limit. That time includes all preparatory operations so actual time in space doing most work rarely exceeds six hours.

For the October spacewalk, it was early morning when the Mir crew began final preparations for their job outside. At 10 a.m., Titov and Manarov opened the outer hatch and emerged from the station. Difficulties began at once...first with the umbilical cord which got tangled. But, somehow, the cosmonauts coped with all problems as they arose. In the end, the planned operations were completed an hour ahead of schedule.

Shake-The-Web Spacewalk

French Cosmonaut Jean-Loup Chretien and USSR cosmonaut Alexander Volkov overcame problems outside the orbiting Mir space station December 9, 1988, to unfurl a carbon-plastic web and fasten it to a platform they assembled on the station's outer shell. They were scheduled to complete their work in four hours 20 minutes, but problems stretched the spacewalk to six hours.

The test of space construction techniques took longer than expected, especially laying a cable linking the plastic platform to a control panel inside Mir. Then the web refused to unfold on electronic command from cosmonaut Sergei Krikalev inside Mir.

While scientists at Mission Control Center in Kaliningrad near Moscow considered the problem, the cosmonauts were out of radio contact. When contact was restored, the cosmonauts reported, "It's opened." Chretien and Volkov had given the web a good shake to open it.

The platform will hold antennas and telescope mirrors.

Coating test. Also during the spacewalk, Chretien and Volkov attached a panel with samples of materials and coatings to the outside of Mir. The panel will be removed after six months and flown to Earth to see how materials and coatings withstood solar radiation and sharp temperature fluctuations.

Live Moscow TV coverage of the spacewalk showed Chretien and Volkov in white spacesuits, floating outside Mir as they worked on the experiments. At one point, they stopped work for a time as they passed into a shadow.

Along with flight engineer Krikalev, cosmonauts Vladimir Titov, Musa Manarov and Valery Polyakov remained inside Mir during the spacewalk.

Cosmonauts In Spacewalk Danger

The USSR daily newspaper Izvestia headlined a Midnight Emergency in Outer Space. The Moscow television evening news program Vremya said, "Was there a threat to the crew's lives? Yes, there was." Anatoly Solovov, commander of the space station crew said they needed "road signs" along the outside of the orbiting Mir space station.

The flurry of commentary followed a six-hour spacewalk July 17, 1990, in which USSR cosmonauts Anatoly Solovov, 42, and Alexander N. Balandin, 36, almost didn't get back inside Mir.

They had gone outside to a station dock to reattach a thermal blanket on their Soyuz TM-9 flyer. Half of the blanket had been hanging loose—one three-ft. piece was attached only on one side—probably torn away when a ground-rescue structure accidentally scratched the thermal blanket during launch of TM-9 from Baikonur Cosmodrome February 11.

Solovov and Balandin, planning to fly down to Earth in TM-9 on August 9, feared the dangling cover might disrupt a safe trip.

The pair had been scheduled to fly down July 30, but the trip was postponed a fortnight to give the cosmonauts more time for repairs. They were in residence aboard Mir six months when they flew home August 9.

Dangling blanket. Solovov and Balandin had boarded the station February 13 to relieve Alexander S. Viktorenko and Alexander A. Serebrov who were at Mir from September 6, 1989, to February 19, 1990. As Viktorenko and Serebrov flew home February 19, Serebrov photographed the TM-9 damage. His pictures showed large heat-shield blankets on the outside of TM-9 ripped away and blocking sensors needed to line up the capsule for reentry. The metal skin of TM-9 was exposed, releasing heat to space. To keep TM-9 warm, the 109-ft.-long Mir was turned twice a day to keep TM-9 in sunlight. Hot air was pumped into TM-9 from Mir to prevent a moisture buildup.

Five of eight thermal blankets apparently were not damaged. The three ripped from some of their attachment points were floating out at a sharp angle from the capsule. Loss of the blankets allowed TM-9 to cool, leading to a buildup of water condensation inside. Engineers feared electrical shorts through the moisture.

Partial repair. Materials to repair TM-9 were sent up in the unmanned Progress 42-M freighter which docked at Mir May 8. The USSR's Kristall Technological Module, which docked at Mir June 9, carried a 20-ft. ladder to reach the capsule. Controllers at the mission control center in Kaliningrad outside Moscow knew reattaching the loose blankets would be the more difficult job. On the other hand, cutting away the damaged blankets would clear the sensors, but do nothing for temperature-control problems.

Outside Mir on July 17, Solovov and Balandin carried oversized shears and other tools to repair or remove the three loose blankets. Working without tethers or hoses connected to the station, they crawled along the station's outer surface, using fasteners like mountain climbers, to reach the dock and TM-9.

Propping the 20-ft. ladder between Mir and TM-9, Solovov and Balandin walked across to the capsule three hours after they had left the relative safety inside Mir. They videotaped the insulation problem and then managed to fold back and clamp two of the three damaged insulation blankets, but one damaged piece just wouldn't budge.

Hatch door. The incident which Moscow TV called life-threatening occurred suddenly at the end of the long spacewalk as Solovov and Balandin tried to crawl back through the hatch in the Kvant-2 section of the Mir complex to the interior of the space station. Somehow, they couldn't get the outer door of the airlock to shut completely and it did not form a good hermetical seal.

The insides of their spacesuits were soaked with perspiration. Their six-hour supply of air was so low the cosmonauts had to hook their packs to an emergency oxygen-supply tube on the outside of the station for a refill. They left the Kvant-2 hatch and climbed across the skin of the space station to another hatch in the expansion modules, Kvant-2 and Kristall, recently added to Mir. The outer hatch of that airlock sealed properly so they took off their spacesuits and entered the station.

They had gone outside at 5:06 p.m. Moscow time and returned inside just after midnight, one of the longest space walks ever. Altogether, they had spent about seven hours outside Mir, two hours longer than planned.

Another spacewalk. Solovov and Balandin gave themselves medical examinations July 20, standard procedure following a spacewalk, including tests of cardiovascular response to exercise.

The cosmonauts made another extra-vehicular activity July 26 to close the slightly-ajar Kvant-2 hatch. Also, while rushing to get back inside Mir July 17, the cosmonauts had left behind the ladder used as a platform for working on TM-9. The ladder could damage the capsule as it separates from Mir so the cosmonauts removed and stowed the ladder and fixed the one remaining loose thermal blanket. The EVA lasted 3 hours 31 minutes. Mission control said their flight home would go ahead as planned on August 9.

Crew. Despite the danger, the cosmonauts joked about it afterward. Solovov and Balandin were the USSR's sixth crew-in-residence at the station, and the 22nd and 23rd cosmonauts to go to Mir since it was launched February 20, 1986. They were the second crew to occupy Mir since it was reopened in September 1989 after a four-month vacancy.

Anatoly Solovov flew to space once before, in Soyuz TM-5 which left June 7, 1988, for ten days in space and a visit to Mir. When he returned August 9, he had spent a total of 189 days in space.

Solovov and Balandin, a rookie on his first spaceflight, boarded the station February 13 to relieve Alexander S. Viktorenko and Alexander A. Serebrov who were in space 166 days, from September 6, 1989, to February 19, 1990. When they returned home August 9, Solovov and Balandin had spent 179 days at the station.

Two Spacewalks In One Week

Cosmonauts Anatoly Artsebarsky and Sergei Krikalev took two long spacewalks outside Mir during the week of June 25-29, 1991.

On June 25, they went outside at 12:12 a.m. Moscow time to replace a Kurs antenna on the Kvant-1 astronomy module. The antenna, which guides spacecraft to dock at the station, had been broken by an unmanned Progress freighter. A new antenna had been sent up in the Progress M-8 cargo which had arrived June 1.

While outside for four hours and 58 minutes, Artsebarsky and Krikalev also installed an experimental girder on the space station's Kvant-2 science-lab module to practice erecting large structures in open space.

At 10:02 p.m. Moscow time June 28, the cosmonauts went through the airlock again, this time to mount two science-experiment arrays on the outside of the station. The experiments, called Track, devised by Soviet and American researchers, were part of a study of the distribution of super-heavy nuclei throughout the Milky Way galaxy.

Artsebarsky and Krikalev also installed a set of detectors to measure streams and spectra of charged particles arriving at Earth from deep space. Then the cosmonauts mounted an extra TV camera on a solar battery and tested it. They deployed a cargo boom.

They stayed outside three hours and 24 minutes, returning through the airlock into Mir early in the morning hours of June 29. Artsebarsky and Krikalev had arrived at Mir station on May 20.

USSR Space Tracking Ships

The Soviet Academy of Sciences sends nearly a dozen civilian research ships around the globe to track manned space flights. The USSR Navy also assigns a pair of ships to help with tracking the orbiting Mir space station and flights of the shuttle Buran.

Tracking ships remain at sea up to ten months at a time. There are five classes of civilian space tracking ships: Gagarin, Komarov, Korolev, Belyayev and Borovichi.

Soyuz, Mir, Buran. Cosmonauts fly to and from the Mir station in Soyuz capsules. The Soyuz retrofire positions over the South Atlantic are the most important tracking locations. They are the last places from which the USSR can track a Soyuz manned spacecraft during re-entry, prior to landing in Soviet central Asia.

South Atlantic retrofire tracking ships are stationed so any emergency requiring cosmonauts to abandon Mir and return to Earth in a Soyuz capsule can be accomplished quickly. Mir is tracked and monitored even when the tracking vessels are in port for refueling.

Tracking ships relay telemetry, TV and voice communications received from Mir and Soyuz via communications satellite to the Kaliningrad flight control center.

Busy schedule. Soviet tracking ships also monitor other launches from time to time. And sometimes they have a very full schedule:

★During the maiden flight of the USSR's shuttle Buran November 15, 1988, tracking ships were sent to the North Atlantic, South Atlantic, Caribbean Sea and the South Pacific.

The tracking ships Kosmonaut Vladimir Komarov and Akademik Sergei Korolev were in the North Atlantic. The Kosmonaut Pavel Belyayev was off Cape Verde Islands. The Kosmonaut Vladislav Volkov was off Togo. The Marshal Nedelin was in the South Pacific off the West Coast of South America and the Kosmonaut Georgi Dobrovolsky was 1,500 miles south of Easter Island.

★The tracking ship Borovichi sailed from its usual position in the Gulf of Mexico to a station between Jamaica and Haiti to track the November 28, 1988, docking of the manned Soyuz TM-7 at the Mir station.

★A week later, for the December 2 launch of U.S. shuttle Atlantis, the Borovichi sailed east to a point near Antigua where NASA has a downrange tracking station. Probably to the disappointment of the Borovichi crew, the space shuttle didn't take off eastbound. Atlantis flew northeast along the Atlantic Coast of the U.S. Of course, there also were USSR tracking ships to the northeast in the North Atlantic.

★Progress 39 arrived December 27, 1988, and boosted Mir. Down on Earth's surface, the tracking ships at sea have to be repositioned each time the altitude of Mir is raised.

The orbits of spacecraft decay. That is, an orbiting craft slowly is pulled by gravity toward Earth. Satellites designed to stay in space for a long time, such as the Mir space station, need to be boosted up in altitude from time to time to prevent their orbits from decaying too much and letting the craft slide down into the atmosphere.

The USSR often uses the rocket engine in a Progress cargo freighter to boost Mir back up to its desired altitude near 350 miles above Earth. Progress freighters arrive every eight weeks or so and, when empty, have their engines fired to push Mir higher in altitude.

★Progress 40 boosted Mir February 24, 1989. Travelling at speeds of 12 to 18 knots, tracking ships cross the ocean in straight lines either across or parallel to the space station's ground track. As the altitude of Mir is raised, the time of the space station's arrival over a tracking ship's horizon is delayed, usually by half an hour. That makes it necessary to update orbit predictions.

The Soviet Academy of Sciences also operates a larger fleet of oceanographic and weather research ships. The list below includes class of ship, names of individual ships, radio callsigns, ship descriptions and stations on the world's oceans:

GAGARIN class
Ship: Kosmonaut Yuri Gagarin
Namesake: The first cosmonaut, who died in 1968.
Description: World's largest research ship.
          Largest vessel using turbo-electric propulsion.
Ocean station: Usually North Atlantic, Bermuda to Cape Race, Newfoundland, late Feb to late Aug.

Built: Baltic Shipyard, Leningrad	In use since: 1971
Home port: Odessa	Range: 24,000 nautical mi.
Length: 773 ft.	Speed: 17.7 knots
Crew: 160	Scientists: 180
Radio callsign: UKFI	

KOMAROV class
Ship: Kosmonaut Vladimir Komarov
Namesake: The first cosmonaut killed in spaceflight.
Description: Converted diesel-powered dry cargo ship. Flag ship of the civilian fleet.
Ocean station: Operates in North Atlantic from coast of Portugal south to Canary Islands, late Oct to
          early July. Port calls at Las Palmas, Canary Islands and Ceuta, Spanish enclave in North
          Africa across from Gibraltar. In eastern Caribbean for Dec 1988 launch of U.S. shuttle
          Atlantis. Port calls: Veracruz, Mexico; Willemstad, Curacao.

In use since: 1967	
Home port: Odessa	Range: 16,500 nautical mi.
Length: 510 ft.	Speed: 17.5 knots
Displacement: 17,500 tons	
Crew: 115	Scientists: 125
Radio callsign: UUVO	

KOROLEV class
Ship: Akademik Sergei Korolev
Namesake: Father of USSR space program.
Description: Diesel-powered tracking ship.
Ocean station: North Atlantic, Cape Race, Newfoundland to Bermuda, early Aug to late Feb.

Built: Black Sea Shipyard, Nikolajev	In use since: 1970
Home port: Odessa	Range: 22,500 nautical mi.
Speed: 16 knots	
Length: 596 ft.	Displacement: 21,500 tons
Crew: 190	Scientists: 170
Radio callsign: UISZ	

BELYAYEV class

Ship: Kosmonaut Pavel Belyayev
Namesake: Belyayev flew to orbit in Voskhod 2 in 1965.
Description: Vytegrales-class freighters first converted for missile test firings
into Indian and Pacific Oceans.
Ocean station: Belyayev-class tracking ships sometimes go to Atlantic Ocean, Gulf of Guinea first
for months, then to other Atlantic and Pacific Ocean tracking positions. At sea up to 10
months. Belyayev itself usually at sea from late Mar to early Dec. During Soyuz flights
to Mir space station and flight of Buran shuttle, Belyayev-class tracking ships were
beneath four orbit-changing retrofire locations to monitor changes and reentry.
NORTH ATLANTIC: One ship operates out of Lome, Togo; Abidjan, Ivory Coast; and Las
Palmas, Canary Islands, centered on Gulf of Guinea, ranging from Cape Verde Islands to
Angola, directing activities of South Atlantic and South Pacific tracking ships.
SOUTH ATLANTIC: One ship, refueling at Buenos Aires and Montevideo, ranges from
Recife, Brazil, to Montevideo, Uruguay, usually 1,200 miles east of Porto Alegre, Brazil.
SOUTH PACIFIC: A ship usually is off New Zealand, 1,500 miles south of Easter Island,
one half orbit, or 45 mins, down-range from Baikonur Cosmodrome at a point where
rockets may be fired to make an orbit circular.
Built: Belyayev-class ships were freighters converted at Zhdanov Shipyard, Leningrad.
In use since: 1977-78

Home port: Tallinn, on the Baltic Sea	Range: 7,500 nautical mi. with 300-350 tons diesel fuel.
Speed: 15 knots	
Length: 400 ft.	Displacement: 5,790 tons
Crew: 100 including scientists	Radio callsign: UTDX

Ship: Kosmonaut Georgi Dobrovolsky
Namesake: one of three cosmonauts killed in 1971 landing of Soyuz-11 capsule.
Ocean station: Belyayev-class ship Dobrovolsky usually at sea June to Jan. During 1988 flight of
shuttle Buran, Dobrovolsky was west of Chile 1,500 miles south of Easter Is.
Radio callsign: UZZV

Ship: Kosmonaut Vladislav Volkov
Namesake: one of three cosmonauts killed in 1971 landing of Soyuz-11 capsule.
Ocean station: Belyayev-class ship Volkov usually at sea Oct to Apr.
Radio callsign: UIVZ

Ship: Kosmonaut Viktor Patsayev
Namesake: one of three cosmonauts killed in 1971 landing of Soyuz-11 capsule.
Ocean station: Belyayev-class ship Patsayev usually at sea late Feb to Sept.
Radio callsign: UZYY

BOROVICHI class

Ship: Borovichi
Description: Diesel power, back-up boilers, last of original Vytegrales-class timber carriers
converted in 1960's. (See Belyayev-class ships above.) Most others were refitted to
support spaceflight with helicopter landing pads for ocean recoveries.
Ocean station: Borovichi-class ships are assigned to the South Atlantic off Argentina and to the
Caribbean Sea and Gulf of Mexico. They leave the Baltic in pairs. Morzhovets and
Kegostrov at sea from late December to late July. Nevel and Borovichi operate late July
to late December. One of each pair goes to the Caribbean, the other off South America.
The Caribbean ship monitors U.S. shuttle launches near a NASA tracking station north
of Antigua Island, just outside the Caribbean. Borovichi class ships make port calls in
South Atlantic at Buenos Aires and Montevideo. In the Caribbean and the Gulf of
Mexico, port calls are at Havana, Cuba, Veracruz, Mexico, Willemstad, Curacao,
Paramaribo, Suriname.
Built: Converted at Zhdanov Shipyard, Leningrad.
In use since: Borovichi-class ships were timber carriers converted in 1960's for tracking.
Home port: Tallinn, on the Baltic Sea.

Length: 400 ft.	Displacement: 7,600 ton
Crew: 80 including scientists	Radio callsign: UVAU

Ship: Kegostrov
Ocean station: Usually at sea from late Dec to late July.
Radio callsign: UKBH

Ship: Morzhovets
Ocean station: Usually at sea from late Dec to late July.
Radio callsign: UUYG

Ship: Nevel
Ocean station: Usually at sea from late July to late Dec.
Radio callsign: UUYZ

USSR Navy Space Tracking Ships
Ship: Marshal Nedelin
Namesake: Field Marshal Mitrofan I. Nedelin, early chief of military rockets.
Ocean station: Navy has two ships out of Vladivostok, Marshal Nedelin and Marshal Krylov,
        tracking Mir space station and shuttle Buran, based in Western Hemisphere at Santiago,
        Cuba. During 1988 flight of Buran, Nedelin was in South Pacific off Chile. Other Navy
        ships track ICBM tests in Sea of Okhotsk and Kuril Island test ranges.
Built: Admiralty Yards, Leningrad.   In use since: 1983
Home port: Vladivostok, assigned to the Pacific Ocean fleet.
Speed: 20 knots
Length: 702 ft.   Displacement: 25,000 tons
Crew: 450, including scientists
Radio callsign: RMLP during Buran flight but change often.

Ship: Marshal Krylov
Description: sister ship to Marshal Nedelin.
Radio callsign: callsigns change often.

Skylab: America's First Station

America's first and only space station was Skylab, built from modified Apollo moonflight hardware and the empty third stage of a Saturn V Moon rocket. The rocket was dry of fuel and converted into two stories, living quarters in one, lab in the other.

The space station was 84.2 feet long. It weighed 84 tons, triple the size of the USSR's Salyut space station. The station was launched from Cape Canaveral to a low Earth orbit 275 miles above Earth on the last Saturn V Moon rocket May 14, 1973.

Despite extraordinary launch vibrations which shook loose a micrometeorite shield, tearing off one solar wing and jamming the other, the house-trailer-sized station reached orbit and was visited successfully by three sets of American astronauts later that year.

The micrometeorite shield was supposed to shade the cabin-in-the-sky from the Sun's heat. Breaking the shield threatened to scuttle the project as inside temperatures reached 190 degrees. But, NASA ground controllers moved the giant satellite so the temperature inside settled down at 110 degrees. And they did it without pointing the operating solar panel away from the Sun which would have cut electrical power to the space station.

Over nine months after Skylab was sent to orbit, three teams of astronauts flew to the station in Apollo command modules (CM) launched on smaller Saturn 1B rockets. America could afford only to have three teams of astronauts visit Skylab.

Astronauts. The nine astronauts who visited Skylab were Charles P. Conrad Jr., Joseph P. Kerwin, Paul J. Weitz. Alan L. Bean, Owen K. Garriott, Jack R. Lousma, Gerald P. Carr, Edward G. Gibson, and William R. Pogue.

There were three Skylab launches in 1973, to ferry crews to work in the station. The astronauts flew in Apollo capsules on smaller Saturn 1B rockets. The total time spent by nine astronauts at Skylab: 171 days 13 hours. The astronauts brought back from Skylab important information in astrophysics, biomedicine, engineering, materials processing, solar physics, and other technology.

Out of orbit. Skylab was used by three teams of astronauts for a total of only nine months, to February 1974, then the station went unused until July 11, 1979, when it was allowed to drop into Earth's atmosphere to burn.

After five years of no visits, technical problems caused the orbit of Skylab to deteriorate. NASA made a weak attempt to keep it in space by remote control from the ground, but no astronauts were sent up for repairs. NASA's hopes failed July 11, 1979, when Skylab re-entered the atmosphere and burned.

SKYLAB 1
Rocket: Saturn V   Launch: May 14, 1973
Events: Unmanned launch of space station to 250-mi.-high orbit. Station 84.2 feet long, 84 tons. Skylab fell July 11, 1979.

SKYLAB 2
Astronauts: cmdr Charles P. Conrad Jr., scientist-pilot Joseph P. Kerwin, pilot Paul J. Weitz
Rocket: Saturn 1-B   Launch: May 25, 1973
Spacewalks: 3, totaling 5 hrs 51 mins   Crew landed: June 22, 1973
Duration: 28 days 49 min (672 hrs 49 mins 49 secs)
Event: First manned flight in Apollo capsule to space station. Astronauts had to overcome problems which jeopardized the Skylab project. During launch to orbit, a meteorite shield and heat shield were ripped off the station, striking solar panels so they would not unfold correctly. Controllers changed the orbit and built a shield the first crew members took to space 11 days later. Conrad, Weitz and Kerwin, first to visit Skylab, opened the station and made repairs. Conrad, Kerwin did three EVAs. First a stand-up-in-the-hatch exposure for 37 mins May 25. Second a 3 hr 30 min spacewalk June 7. Third a 1 hr 44 min spacewalk June 19. Total EVA time was 5 hrs 51 mins. They salvaged the mission with a spacewalk to adjust the solar panels and erect a makeshift parasol Sun shade, dropping the inside temperature to a comfortable shirt-sleeve level. Set spacewalk duration record. First pictures of solar flares shot above the atmosphere. Flew to Earth June 22.

SKYLAB 3
Astronauts: cmdr Alan L. Bean, scientist-pilot Owen K. Garriott, pilot Jack R. Lousma
Rocket: Saturn 1-B   Launch: July 28, 1973
Spacewalk total: 13 hrs 44 mins   Crew landed: September 25, 1973
Duration: 59 days 11 hrs (1427 hrs 9 mins 4 secs)
Events: Second manned flight to station. After docking, 2 of 4 capsule thrusters leaked, had to be shut down. NASA considered aborting, began planning rescue, but decided breakdowns had known causes and Apollo could fly on 2 thrusters. Astronauts were allowed to go ahead. Garriott, Lousma did 3 EVAs. First was 6 hr 29 min spacewalk Aug 6. Second was 4 hr 30 min spacewalk Aug 24. Third was 2 hr 45 min spacewalk Sept 22. Total EVA time was 13 hrs 44 mins. They placed a new Sun shield over the Skylab 2 makeshift parasol, replaced film in cameras, installed panels to measure micrometeorites. Skylab was moved so sensors pointed to most of the U.S. and parts of 33 countries. Ecological, agricultural, forestry, mapping, geological, water resource, fishing, oceanography, mineral prospecting, and meteorological data collected. Sun was very active so more solar observations were made. Spider Arabella was able to spin her web in orbit.

SKYLAB 4
Astronauts: cmdr Gerald P. Carr, scientist-pilot Edward G. Gibson, pilot William R. Pogue
Rocket: Saturn 1-B   Launch: November 16, 1973
Spacewalk total: 22 hrs 21 mins   Single spacewalk record: 7 hrs 1 min
Crew landed: February 8, 1974   Duration: 84 days 1 hr (2017 hrs 16 mins 30 secs)
Events: Third manned flight in Apollo capsule to the space station. Record U.S. man-in-space duration of 84 days. The astronauts had problems, at first, getting experiments started. Morale suffered as they fell behind on work. The schedule was shuffled for more time off. Things improved. A malfunctioning radio antenna was repaired during a Thanksgiving Day spacewalk. Other equipment was repaired. But, then, the star tracker failed completely and had to be shut down. The crew spotted and photographed Comet Kohoutek. A solar flare was photographed for the first time from space. By coincidence, in space at the same time using an ultraviolet camera to photograph solar X-rays and stars were USSR Soyuz 13 cosmonauts Pyotr Klimuk, Valentin Lebedev, for a total of five persons in orbit at one time. Gibson, Pogue did four EVAs. The first was 6 hours 33 minutes November 22. The second was 7 hrs 1 min December 25. The third was 3 hours 28 minutes December 29. The fourth was 5 hours 19 minutes February 8. Total EVA time 22 hours 21 minutes. Carr, Gibson and Pogue flew home Feb. 8.

Space Station Freedom

U.S.-international space station Freedom would be a long bridge-like truss orbiting 200 miles above Earth by the year 2000. Four habitable modules attached to the truss would be U.S. living quarters, a U.S. laboratory, Canada's mobile maintenance equipment, a Japanese lab known as JEM, and European Space Agency's lab Columbus.

Feeling almost no effect of gravity, space station astronauts would have an excellent 250-mi.-high vantage point for observing stars above and our planet below. They would study effects of prolonged weightlessness on humans and create a life support system for long voyages, while manufacturing new materials in space.

Problems. U.S. President Ronald Reagan announced plans for a permanently-manned space station in his 1984 State of the Union address, but Congress hasn't shown much interest in the project. Faced with budget problems, intense public debate, technical issues and schedule slips, NASA has had to scale down plans for the future space station.

The station is referred to as U.S.-international space station Freedom because the Canadian Space Agency, European Space Agency and Japanese National Space Development Agency are international partners of the U.S. in the project.

A scaled-down Freedom still would be permanently manned, but the number of U.S. astronauts in a standard space station crew has been cut from eight to four. Full-time operation of the station has been delayed until at least 1997. The crew might be increased to eight later, after Japanese and European modules are installed.

The reduced station would have four habitable modules attached to a long truss. They would include a U.S. laboratory module, a U.S. living quarters module, Canada's mobile maintenance equipment, Japan's JEM module, and the European Space Agency's Columbus module.

New dimensions. Length of the space station was cut from 493 feet to 353 feet. The long truss, the backbone of the station holding all of the modules, was to have been assembled like a jumbo erector set in orbit, with astronauts performing numerous spacewalks. Now, truss segments will be assembled on the ground, cutting in half the number of spacewalks. The previously-larger station would have required 3,800 hours of spacewalks a year—six a week—to replace batteries and do other maintenance. The smaller station would require only three spacewalks a week.

NASA's Attached Payload Accommodations Equipment for large external payloads has been dropped, but ports for small external payloads will be situated along the truss.

Smaller labs. Freedom's mission has been scaled back. The orbiting lab now will do only biological and material-sciences experiments. The redesigned U.S. lab and living-quarters modules were cut from 44 feet to 27 feet long and 14.5 feet in diameter. Now, the modules can be outfitted with lab gear on the ground before launch.

The U.S. lab will have 24 eight-ft. racks, including 15 for science work. Later, when permanently manned, 28 experiment racks would be available—twelve in the U.S. lab, eleven in the ESA lab and five in the Japanese lab.

Electricity. Science experiments conducted at the space station need electricity. NASA once planned on solar panels generating 75 kilowatts of electricity, but now is planning on only 65 kilowatts. Only three of four planned solar arrays will be attached to the truss by the year 2000.

Timetable. Plans call for assembly of the station Freedom in a low Earth orbit between 1995 and 1999. Eighteen U.S. shuttle flights will ferry station construction materials to orbit. In addition, a series of eight station-utilization flights by the end of the century will start in 1997. The timetable would require 26 flights in four years, or a shuttle flight about every two months.

The station would have a man-tended capability by the end of 1996, with permanent occupancy by a four-person crew in mid-1997. Individual crew members would be relieved every three to six months. Elements of the station would include the long truss to hold laboratory and astronaut living quarters modules, solar panels to generate electrical power, an air lock for spacewalks, remote manipulator arm, cupola, and Japan's JEM and European Space Agency's Columbus.

Man tended. In the man-tended period of operations, astronauts would be ferried in shuttles to Freedom to work in the U.S. laboratory for two-week periods. They would fly back down to Earth in shuttles. One set of Freedom's solar arrays would generate 22 kilowatts of power with 11 kilowatts available for use by astronauts doing science experiments.

Permanent manning. Freedom would achieve permanently-manned status by the year 2000. The U.S. laboratory and living quarters, the European and Japanese laboratory modules, and the Canadian mobile servicing system would be in place.

Accommodations would be ready for a live-in crew of four. Three sets of solar arrays would supply 65 kilowatts of electrical power, with 30 kilowatts available for science experiments and the rest for station housekeeping chores.

Life boat. Before permanently occupying the station, NASA's Assured Crew Return Vehicle (ACRV) would have to be ready and on hand to fly crew members down to Earth in an emergency.

Costs. Recent estimates find the completed scaled-down station would cost more than $20 billion. NASA has been budgeted at about $2 billion a year for the project in recent times. The 512,000-lb. space station would come to $39,000 per lb.

National Research Council and other space station critics have claimed the science results possible from the smaller space station would not justify its cost. Altogether, preparations for space station Freedom would employ 20,000 persons directly and 100,000 indirectly for the rest of the decade.

Future expansion. NASA would make Freedom expandable. Additional solar arrays could bring power up to 75 kilowatts. Four crew members could be added, bringing the regular crew to eight. Labs and other modules could be added, after the year 2000.

Processing facility. A 457,000-sq.-ft. space station processing facility (SSPF) is under construction at NASA's Kennedy Space Center, Florida, to be completed in 1994. The three-story building will be used for preflight checkout of space station elements. It will have communications and electrical control areas, laboratories, logistics staging areas, operational control rooms, and 63,000 square feet of payload processing space, including both high bay and intermediate bay.

A 5,000-sq.-ft. airlock will connect the primary processing area. Airlock and processing area will be 100,000 parts-per-million clean rooms. A visitor viewing window will allow NASA tourists to see space station preflight operations.

Cupola. The world's best view undoubtedly will be found in Freedom's small module with two octagonal glass-window rooms attached forward on the truss.

NASA calls the rooms cupolas. One cupola will face Earth and one will look out into space. Each room will hold two astronauts at a time and have two computer work stations. The symmetrical rooms, encircled with windows, will be used:

★ to provide a clear viewing for astronauts involved in command and control operations such as docking of a space shuttle or operating outside robots,

★ to watch astronauts during assembly and maintenance work outside the station,

★ for general inspection of the exterior of the station,

★ for leisure time star viewing.

A metal, wood and glass mock-up of a space station cupola was delivered by manufacturer McDonnell Douglas to NASA's Johnson Space Center, Houston, Texas, in 1989. Personnel at JSC use the mock-up to simulate astronauts working in a cupola, evaluating size, shape, interior design and work stations.

International. Freedom will be the largest international civilian venture in space to date. The day shuttle Discovery returned America to space on September 29, 1988,

U.S. Secretary of State George P. Shultz and representatives of 11 other countries signed an agreement to go ahead with construction of the orbiting station.

The Freedom partners are Canada, Belgium, Denmark, France, Great Britain, Italy, Japan, the Netherlands, Norway, Spain, the United States and Germany. The permanent station will be designed, developed, operated and used by all of the partners.

The European countries, all members of the European Space Agency, together will put more than $4.2 billion into the project. Japan will spend more than $2 billion, Canada more than $1 billion and the U.S. more than $20 billion. Heinz Riesenhuber, German chairman of the European Space Agency, said the station will be used for peaceful purposes. The U.S. State Department said a manned station is needed to lay groundwork for colonies on the Moon and Mars, stepping-stones for human exploration of the Solar System.

The U.S. will provide the basic truss structure for the manned station, plus electrical power and living quarters. The U.S. also will construct a pressurized lab where astronauts will work in a shirt-sleeve environment.

Canada will send hardware for assembling the station in orbit and for maintenance and servicing the station. Japan will build JEM, a pressurized lab with shirt-sleeve environment. Europe will add Columbus, a pressurized shirt-sleeves lab.

Japan's JEM. Japan is building the Japanese Experimental Module (JEM) for Freedom. To be permanently attached to the station truss, JEM itself will be composed of three enclosed modules plus an exposed facility. One will be a pressurized laboratory module, two will be experiment logistics modules and the exposed facility will allow experiments to be open to the environment of space. Astronauts would process materials and do life sciences research in the lab. A logistics module would ferry materials between the space station and Earth. Another would store experiment specimens, gases and consumables. Japanese astronauts would work in Freedom.

Europe's Columbus. Columbus is Europe's large manned scientific module to be attached to the U.S.-international space station Freedom. European Space Agency is building Columbus as well as Hermes, a small manned space shuttle to carry men and supplies to Columbus. Besides being a pressurized module attached to space station Freedom, Columbus also could be used as an unmanned platform, a stand-alone manned module, or part of an independent European space station. European astronauts would travel to Freedom for science experiments.

Canada's robot. Canada's contribution to the space station effort will be a roving service robot to act as hands and arms of the station. The computerized machine will have sight, touch and reasoning powers and be able to unload cargo, repair satellites, service spacecraft visiting the station, and maybe even fly away from the station's normal orbit path to retrieve and service satellites. Canadian astronauts would go to the station three months a year for experiments.

Robot farmers. Tracy and Harvey will pick the lettuce and potatoes astronauts eat in the future U.S.-international space station Freedom. Tracy, a tray-handling robot, and Harvey, a harvesting robot, will be in a farm in the station after 2000. Viruses and funguses carried by astronauts could wipe out crops, leaving nothing to eat. The sanitary robots would do the chores and protect crops. The compact farm would be as big as a school bus, powered by solar cells. Fiber optics would spray sunlight directly on plants flowering in Tracy's trays of water. Harvey, equipped with trowel and gardening tools, would roll along the farm on a 45-ft. track. They would call for help only if something were to break. Humans wouldn't be allowed near crops, except in emergencies.

Second station. America's first space station was Skylab, launched in 1973. Skylab was used only three times, in 1973 and early 1984. It fell from orbit in 1979.

America once had wanted to have the new space station in use in 1992, the 500th anniversary of Columbus' discovery of America, but now Freedom is to be inhabited permanently by the year 2000. Freedom is expected to have a useful life of 30 years.

Freedom Construction Schedule

Plans call for assembly of the U.S.-international space station Freedom in a low Earth orbit between 1995 and 1999. Eighteen U.S. shuttle flights will ferry station construction materials to orbit. In addition, a series of eight station-utilization flights by the end of the century will start in 1997.

NASA shuttles will start trucking station parts to space with a flight code-named SSF/MB-01(FEL) in November 1995. It will take 18 flights, through SSF/MB-18 in December 1999, to get everything up there and bolted together. That's 26 flights in four years, or a shuttle flight about every two months.

Construction of Freedom would be sufficiently far along as of flight SSF/MB-06 in December 1996 that the station would have what NASA calls a man-tended capability.

The station then would have a permanent two-person or four-person crew as of the eighth shuttle flight in the series, SSF/UF-01 in May 1997. The size of the permanent crew could expand as the station grows. Individual crew members would be relieved every three to six months.

Elements of the station requiring shipment to space probably would include a long truss to connect various modules, including work spaces, laboratories and astronaut living quarters. At least one large solar panel would generate 15 to 20 kilowatts of electrical power.

In order of construction, the truss probably would go to space first, followed by solar panel, air lock, mobile servicing center, temperature control system, remote manipulator arm, laboratory module, lab equipment, crew living quarters, module connectors, cupola, crew supplies, additional power system equipment, Japan's JEM module, and European Space Agency's Columbus module.

Below is NASA's most-recent tentative construction schedule. Vehicle is launch vehicle, in all cases a space shuttle. Flight is NASA's designated number. SSF is space station Freedom. MB is manned base. UF is utilization flight. FEL is first equipment launch. MTC is man-tended capability.

Launch		Vehicle	Flight	Effort
1995	Nov	shuttle	SSF/MB-01	FEL, first assembly launch of module
1995	Dec	shuttle	SSF/MB-02	second assembly launch of module
1996	Mar	shuttle	SSF/MB-03	third assembly launch of module
1996	Jun	shuttle	SSF/MB-04	fourth assembly launch of module
1996	Sep	shuttle	SSF/MB-05	fifth assembly launch of module
1996	Dec	shuttle	SSF/MB-06	MTC, sixth assembly launch of module
1997	Mar	shuttle	SSF/MB-07	MTC, seventh assembly launch of module
1997	May	shuttle	SSF/UF-01	MTC, first use of the space station
1997	Jun	shuttle	SSF/MB-08	MTC, eighth assembly launch of module
1997	Aug	shuttle	SSF/UF-02	MTC, second use of the space station
1997	Sep	shuttle	SSF/MB-09	MTC, ninth assembly launch of module
1997	Nov	shuttle	SSF/UF-03	MTC, third use of the space station
1997	Dec	shuttle	SSF/MB-10	MTC, tenth assembly launch of module
1998	Mar	shuttle	SSF/MB-11	MTC, eleventh assembly launch of module
1998	May	shuttle	SSF/UF-04	MTC, fourth use of the space station
1998	Jun	shuttle	SSF/MB-12	MTC, twelfth assembly launch of module
1998	Aug	shuttle	SSF/UF-05	MTC, fifth use of the space station
1998	Sep	shuttle	SSF/MB-13	MTC, thirteenth assembly launch of module
1998	Nov	shuttle	SSF/UF-06	MTC, sixth use of the space station
1998	Dec	shuttle	SSF/MB-14	MTC, fourteenth assembly launch of module
1999	Mar	shuttle	SSF/MB-15	MTC, fifteenth assembly launch of module
1999	May	shuttle	SSF/UF-07	MTC, seventh use of the space station
1999	Jun	shuttle	SSF/MB-16	MTC, sixteenth assembly launch of module

1999	Aug	shuttle	SSF/UF-08	MTC, eighth use of the space station
1999	Sep	shuttle	SSF/MB-17	MTC, seventeenth module assembly launch
1999	Dec	shuttle	SSF/MB-18	MTC, eighteenth assembly launch of module

Other Nations' Space Stations

Europe, Japan and China could be in a position to orbit their own space stations around the turn of the century.

China. The People's Republic of China plans to build its own station in Earth orbit. The government announced the independent station, planned for completion by 2000, at a Beijing meeting of space authorities from Pacific Ocean countries June 8, 1987.

Europe. Columbus is Europe's large manned scientific module to be attached to the U.S.-international space station Freedom. European Space Agency is building Columbus as well as Hermes, a small manned space shuttle to carry men and supplies to Columbus.

Besides being a pressurized module attached to space station Freedom, Columbus also could be used as part of an independent European space station.

Japan. A small space shuttle, called Hope, is being planned. It could carry astronauts to a Japanese space station after the year 2000. Japan is interested enough in space science to be building probes to send to the Moon and Venus around 1995-96.

The Mir Space Station Post Office

Yes, the USSR's orbiting space station Mir does have its own post office and postmark cancellation stamper. Thousands of letters are said to have been processed through the Mir post office.

Shrewd dealers, working the dark side of capitalism in the then-communist state, lightened the wallets of stamp collectors around the world by more than $1 million early in 1988, selling philatelic covers which supposedly had been flown in the Progress 33 supply ship to the USSR's Mir space station and back to Earth in a Soyuz manned capsule.

Some 1,000 small envelopes—each with a Russian 10-kopeck stamp and a half-dozen rubber-stamp cancellation marks—were arranged by an American stamp dealer. He sold 500 to a West German dealer, 400 to an Italian dealer and 100 to a Japanese dealer. They, in turn, sold them to collectors for $1,500 per envelope. A few of the envelopes were wholesaled to other dealers at $1,200 who sold them to collectors at $1,500.

Chronology. The envelopes were made in New York in 1987 and flown by airplane to Moscow where they were hand addressed to the Mir crew and the postage stamps were pasted on. The 10-kopeck stamp commemorated the 30th anniversary of Sputnik 1. The stamps were cancelled October 4 at the Moscow post office.

The envelopes were sent on to Baikonur Cosmodrome where they were cancelled again by the local post office November 21 and placed aboard the space-bound cargo freighter, Progress 33, right before blast-off.

In space. After Progress 33 docked at Mir, cosmonauts Yuri Romanenko and Alexander Alexandrov again cancelled the envelopes with a hand stamp and autographed each cover. According to reports in the philatelic press, the cancelling in space was tedious. The envelopes and canceller kept floating away during the autograph session.

Back on Earth. The covers (envelopes) returned to Earth with cosmonauts in Soyuz TM 3 on December 29. They were rushed to the post office nearest the landing site, at the Arkalyk airport, where the envelopes again were cancelled. Four rubber-stamp cancels appear on the face of each envelope.

Later, the covers were numbered and signed on the back by the head of Glavkosmos, the USSR commercial space agency. For $1,500, each collector received with his philatelic cover an advertising videotape of Russian space pictures, pushing the Glavkosmos commercial space launch services. All 1,000 sold out quickly to collectors.

Second helping. The Soviets re-opened their space postoffice in 1988. Mir Post Office re-opened for business November 29 with its own rubber stamp.

Vladimir Titov and Musa Manarov were in residence at Mir when the cosmonaut team of Alexander A. Volkov, Sergei M. Krikalev, and Frenchman Jean-Loup Chretien blasted off from Baikonur Cosmodrome November 26, 1988, in Soyuz TM-7, bound for the station with a new postoffice rubber cancellation stamp. They arrived November 28 and turned over postal implements and philatelic covers to mission commander Titov who was named postmaster, letting the cosmonauts postmark their own mail for delivery home. The Titov-Manarov covers sold out just as quickly as the Romanenko covers.

Meanwhile, amateur radio operators around the globe, who had been in contact with the cosmonauts aboard Mir at the end of 1988, were wondering if their QSL (confirmation) cards would be postmarked at the Mir Post Office. They weren't.

Space Station Glossary

AAP. Apollo Applications Program featuring the wet orbital workshop space station which later was changed to the dry orbital workshop space station Skylab.

Apollo. America's third manned spacecraft series, with three-seat capsules.

Atlantis. The fourth U.S. space shuttle to fly to orbit.

Boosters. Helper rockets strapped to a main rocket engine.

Buran. The first orbiter of the USSR's space shuttle fleet. Buran is Russian for snowstorm or blizzard. Buran No. 1 flew to Earth orbit in 1988. Buran No. 2 may fly to space station Mir in 1992-93.

Challenger. The second U.S. space shuttle, destroyed in 1986 launch explosion.

Columbia. The first U.S. space shuttle to fly to orbit.

Columbus. Europe's module in development for space station Freedom.

Command service module. CSM, the crew compartment of U.S. Apollo Moon spacecraft.

Cosmos. A generic name for many USSR space satellites.

Cruise engine. Rocket engines used to maneuver the Mir station around space.

Discovery. The third U.S. space shuttle to fly to orbit.

Ehricke, Krafft. An associate of German-American space scientist Wernher von Braun in the 1950's, Ehricke proposed blasting an Atlas rocket to Earth orbit, then furnishing it as a space station. His idea was boosted later by Douglas Aircraft Corp. in its Manned Orbiting Laboratory. See also: Manned Orbiting Laboratory, MOL.

Endeavour. The fifth U.S. shuttle to fly to orbit, it replaced shuttle Challenger.

Energia. The newest USSR super-powerful space-launch rocket.

EVA. Extra-vehicular activity. See also: Spacewalk.

Freedom. U.S. President Ronald Reagan proposed in 1984 building a second-generation U.S. space station—an $8-billion permanently-manned space station—within 10 years. The station would include modules built by Europe, Japan, Canada and others. It would be named Freedom. Some politicians and scientists objected to the expenditure, saying a station was not needed. NASA's Johnson Space Center was assigned to lead the project. European and Japanese space agencies began developing modules to be attached to the station. If funded, U.S.-international space station Freedom may be in operation by the year 2000. See also: Skylab.

Gyrodyne. The flywheel in the USSR space station used to turn Mir in space.

Hermes. Europe's space shuttle under development.

Hope. Japan's space shuttle under development.

JEM. Japan's module to be attached to U.S.-international space station Freedom.

Kristall. A large expansion "microfactory in orbit" added to Mir in 1990. The third expansion module added to Mir. Kristall is Russian for Crystal, reflecting the equipment installed in the Kristall module—furnaces to produce semiconductor crystals in the weightless environment of space.

Kvant-1. A small astronomy and science research module added to the Mir space station in 1987. Kvant is Russian for Quantum.

Kvant-2. A large expansion module added to the Mir space station in 1989.

Liquid fuel. Frozen gases used as fuel in space rockets.

Liquid rocket. Space rockets using frozen-gases for liquid fuels.

Manarov, Musa. A cosmonaut. Musa Manarov has the most time accumulated in space. Manarov and Vladimir Titov made the longest one-trip stay in space.

Manipulator. A mechanical arm to lift cargo from the shuttle bay of a U.S. shuttle. Also a mechanical arm to move modules between the USSR's Mir space station ports.

Manned Orbiting Laboratory. MOL was a space station plan advanced by Douglas Aircraft Corp. in the U.S. in the 1960's. MOL would have been the Saturn S-IV second stage of a Saturn I rocket launched to Earth orbit. The spent rocket body would have been outfitted as a space station. MOL was not built, but the U.S. Apollo program adopted some of its proposals. In the Apollo Applications Program, a rocket previously filled with fuel was described as "wet" and a space station built in such a spent rocket would be a "wet orbital workshop." The Apollo Applications Program proposal would have used a Saturn S-IV-B stage of the Saturn I-B rocket. Astronauts would ride a different rocket to space to build the station in the spent Saturn S-IV-B. Budget problems and doubts about an astronaut's ability to do such work in a weightless environment prompted program officials to opt instead for a "dry orbital workshop" with a telescope mount attached instead of launched separately. Skylab was the result. See also: Skylab.

Mir. The USSR's third-generation space station, still in use. See also: Salyut.

MOL. See: Manned Orbiting Laboratory.

Noordung, Hermann. Noordung was a pen name used by Captain Potocnik of the Austrian imperial army in 1929 to write The Problems of Space Flight, describing a toroid space station, rotating to generate artificial gravity through centrifugal force.

Oberth, Hermann. The father of modern astronautics, Oberth was first to discuss problems of building space stations.

O'Neill, Gerard K. Princeton University physicist proposing large colonies in space housing thousands in an Earth-style environment, built from materials mined from the moons of Earth and Mars and from asteroids.

Progress. The USSR's series of cargo-transporting spacecraft. Unmanned capsules ferrying food, fuel, supplies and science equipment to the USSR's Mir space station.

Progress M. A redesigned Progress. M for modified.

Proton. USSR's series of heavy-lifting space rockets.

Romanenko, Yuri. A cosmonaut, the person with the second-longest time accumulated in space. See also: Manarov.

Salyut. The series of seven first-generation and second-generation USSR space stations, launched to Earth orbit in the 1970's and 1980's. Cosmonauts lived for record periods of time in space aboard Salyut stations. Salyut is Russian for Salute, a name given to the first USSR space station in 1971 honoring the tenth anniversary of the flight of the first man in space, Yuri Gagarin.

Saturn V. A series of heavy-lifting U.S. rockets for Apollo capsules.

Shuttle. A partly-reusable spacecraft which can land on a runway.

Skylab. The first and only U.S. space station launched to date was Skylab, fired to Earth orbit in 1973 on a two-stage Saturn V rocket. Astronaut crews flew separately to orbit in Apollo command service modules (CSMs) launched on Saturn I-B rockets. Skylab was manned for a total of 172 days by three three man crews in 1973-74. Skylab was damaged during launch in 1973 and was repaired in orbit by the first crew, indicating astronauts actually could do real work in the weightless environment of outer space. NASA planners wanted a larger station—33 feet wide compared with Skylab's 22 feet. It would have been launched on a Saturn two-stage rocket. Appropriations were not forthcoming and some studies showed the large station would be impractical without a reusable space transportation system (STS) to re-supply it. Use of Skylab was abandoned and limited funds were switched to building only the reusable space transportation system now known as the space shuttle program. See also: Freedom, MOL.

Solid fuel. A chemical rocket fuel with a rubbery consistency, not gas or liquid.

Solid rocket. A space rocket using a rubbery, so-called solid, fuel.

Soyuz. The long series of USSR cosmonaut-transporting spacecraft. USSR capsules ferrying cosmonauts to and from space. Soyuz is Russian for Union, symbolizing the spacecraft's rendezvous and docking capability.

Soyuz T. A redesigned Soyuz. The T symbolizes: (1) transport, (2) troika indicating a space transport capable of carrying three cosmonauts to the Mir station, and again (3) troika indicating the third generation of the Soyuz design.

Soyuz TM. A redesign of Soyuz T. The M is for modified and for Mir, its main destination. Soyuz TM is a third generation modified Soyuz.

Space station. A base in space. A structure with quarters for long-duration living and working in orbit around a planet or moon. Since the first modern rocket experiments in the 1920's, future space stations have been depicted as science research stations, astronomy outposts, manufacturing facilities, and inhabited colonies. Serious proposals for such stations emerged in the 1920's.

Spacewalk. Extra-vehicular activity, a trip outside a shuttle or space station.

Titov, Vladimir. Cosmonauts Vladimir Titov and Musa Manarov had the longest one-trip stay in space.

von Braun, Wernher. A German-American space scientist, Wernher von Braun popularized the concept of a wheel-shaped space station in 1951 in Across the Space Frontier. See also: Krafft Ehricke.

Wet space station. A space station built in a used rocket in space. Astronauts would clean out left-over fuel inside the rocket and outfit the interior for living and working quarters.

Space Shuttles

★★★★★★★★★★★★★★★★★★★★★

History Of Space Shuttles

The U.S. Air Force came up with the first concept for a winged spacecraft—a space shuttle—in 1962. The Air Force had its B-52 bomber and an experimental rocket plane known as X-15. The idea was to bolt an unmanned X-15 to the top of a B-52 and fly it as high as the plane could go. From that altitude, near space, the pilot would launch the unmanned rocket plane. The X-15 would continue on to space, drop off a satellite in orbit and fly back down to Earth. While the X-15 didn't fly to space, didn't drop a satellite, and the Air Force shelved the plan in 1965, the idea of a reusable space shuttle hung around.

Forerunners. A number of experimental prototypes were flown before the American shuttle fleet was built, including the X-15 which was tested from 1959-1968. X-15 was a 50-ft. rocket aircraft launched 199 times from under the wing of a converted B-52. It flew at speeds up to Mach 6.7, reaching an altitude of 67 miles in 1963. Fastest was 4,534 mph in 1967. Among a dozen X-15 pilots was Neil Armstrong who later became the first man on the Moon and Joe Engle who flew an actual space shuttle.

The X-20, flown from 1960-1963, was a design for a small military suborbital and orbital shuttlecraft called Dynasoar. It was to be launched by Titan 1 rocket, but never was flown. It was cancelled in 1963 as too expensive.

Asset was the name for half a dozen launches in 1963-1965 of small Dynasoar models testing aerodynamics and thermal protection. M2 was a lifting body testing shuttle handling at trans-sonic speeds from 1966-1967. It flew in 1966, crashed in 1967, and was rebuilt for more tests. Prime was the name for three launches of maneuverable lifting bodies in 1966-1967 to near-orbital speeds. HL-10 was a manned lifting body testing shuttle handling at trans-sonic speeds in the 1960's.

X-24a was a manned lifting body testing shuttle handling at trans-sonic speeds in the 1970's. X-24b was manned lifting body tests of handling at trans-sonic speeds. It flew 36 times in 1973-1975 with some landings simulating shuttle landings on concrete runways at Edwards Air Force Base.

Apollo. Meanwhile, NASA had enjoyed large budgets during the 1960's space race and wanted to continue into the 1970's, expanding Project Apollo into an Apollo Applications Program. AAP would have built a space station in Earth orbit as a pit stop for astronauts on their way to the Moon and Mars.

But, after six spectacular manned Apollo landings on the Moon, much of NASA's money dried up. Many politicians said space spectaculars were unnecessary since the U.S. had won the race.

Forced to choose, NASA cancelled three Moon landing flights and built one space station. The name AAP was changed to Skylab and the station was launched to Earth orbit in 1973. Astronauts were shipped to the station separately in 1973, in three groups of three each in left-over Apollo Moon capsules.

NASA drew up plans for a larger space station, but sharply reduced space budgets again forced tough decisions. The first Skylab was not used after just three visits and plans for the second Skylab were dumped.

Shuttle. NASA decided a reusable space transportation system was needed to carry men and equipment to orbit, so the limited amount of money available was switched to designing and building space shuttles. Eventually, six shuttles were built, starting with Enterprise in 1974, and dozens of week-long spaceflights have been made since 1981 (all flights are listed below).

NASA never again enjoyed the powerful budgets it commanded in the 1960's. In 1984, U.S. President Ronald Reagan announced a NASA project to build a permanently-manned space station in Earth orbit within 10 years. The station was to be named Freedom and Canada, Japan and the European Space Agency were to build major parts of the project. But, like AAP back around 1970, Freedom station was opposed vigorously by some scientists and politicians.

After the 1986 Challenger disaster killed seven astronauts on their way to orbit, an outpouring of national emotion supported the allocation of federal funds to build one replacement orbiter. The shuttle Endeavour was constructed and was expected to make its maiden voyage in 1992.

Some scientists and politicians continue today to argue against manned spaceflights, forcing NASA to choose between shuttles and space stations vs. unmanned interplanetary probes and orbiting astronomy observatories.

Space station Freedom configurations and construction dates repeatedly have grown smaller and less elaborate. NASA still hopes to construct Freedom station from 1995-2000, but federal funding has not been certain. If another shuttle orbiter were to go out of service, federal funding for a replacement might not be assured.

Enterprise. A U.S. space shuttle is a winged spaceplane with wheeled landing gear and large cargo-carrying capacity. It is launched to Earth orbit by rocket, then leaves orbit to glide to a runway landing.

Space shuttle Enterprise was the first orbiter, but it never orbited, never shuttled, never went to space.

Enterprise was known as OV-101 when work was started on it in June 1974. After 100,000 Star Trek TV fans wrote in, former U.S. President Gerald Ford named the orbiter Enterprise on September 8, 1976. NASA didn't like the name, wanting to call the shuttle Constitution.

Enterprise had no engines. The 130-ton orbiter only flew glide tests, bolted to the back of a Boeing 747 jet. The airplane-shuttle combo did not go to space, but was to drop Enterprise to test its gliding ability. The combo took off briefly February 18, 1977, to see if they could fly together. For the first free flight August 12, 1977, astronauts Fred Haise Jr and Charles Gordon Fullerton were at the controls when the 747 dropped Enterprise from an altitude of 4.5 miles over California. They glided 5.5 minutes to land on a strip at Edwards Air Force Base.

Five test flights were made over two years. Haise and Fullerton piloted flights one, three and five. Astronauts Joe Engle and Richard Truly piloted flights two and four. Engle and Truly later flew shuttle Columbia on the second actual spaceflight. Truly went on to head the space agency from 1989 as NASA Administrator.

Challenger. Originally, after the glide tests, Enterprise was to go back to the workshop for engines. But, it became cheaper to convert a structure-test orbiter into a ready-to-fly orbiter, so the plan to fly Enterprise was abandoned. The structure-test orbiter, numbered OV-099, became Challenger, the shuttle which exploded in 1986 after nine successful spaceflights.

Smithsonian. Enterprise also was used for so-called fit tests to get launch pads ready for real shuttles. Then the vehicle was flown city to city, piggyback on its 747 jumbo jet, as a NASA publicity stunt—a big hit with the public. After the tour, Enterprise was stripped of instruments and other gear and displayed at Kennedy Space Center for a time, then flown to Washington, D.C. for the Smithsonian Institution to build a museum around it. The museum has not been built and, today, Enterprise sits among weeds in the outback of Washington's Dulles International Airport, paint peeling, out of sight, out of mind.

Columbia. Another orbiter was numbered OV-101, later to be named Columbia. It was the first shuttle to fly in space. That first spaceflight was in 1981. Challenger's first flight was in 1983.

Discovery. Vehicle number OV-102 was named Discovery and made its first spaceflight in 1984.

Atlantis. Vehicle number OV-103 became Atlantis and went to space in 1984. NASA named its orbiters for historic sea vessels used in research and exploration.

Buran. The USSR participated in the Space Race in the 1960's, including developing capsules to carry a man to the Moon and a space station. While the U.S. won the Moon race in 1969, the USSR won the competition for first space station when Salyut 1 was launched in 1971. Replacement space stations were launched in 1973, 1974, 1976, 1977, 1982 and 1986.

Space station activities have given the USSR the lead in man-hours in orbit. Cosmonauts have been launched in small Soyuz capsules on scores of flights to space stations. They return to Earth in the capsules which parachute to land. Supplies have been sent to the stations in unmanned Progress cargo-freighter capsules which then are jettisoned to burn in the atmosphere.

As the U.S. aggressively pursued development of reusable space shuttles in the 1970's, the USSR designed its own shuttle. The first was named Buran—Russian for snowstorm or blizzard. The first unmanned flight test of Buran No. 1 was a success November 15, 1988. Buran No. 2 was expected to make an unmanned flight to Mir space station in 1992-93.

Space station visits by USSR shuttles carrying men and supplies may increase in the 1990's, although it probably will remain less expensive to ferry men and supplies in time-tested Soyuz and Progress capsules.

Challenger. In nine distinguished spaceflights, the old U.S. orbiter Challenger compiled an extraordinary record: first American women in space Sally Ride, first black man in space Guion S. Bluford Jr., first shuttle spacewalk, first American female spacewalk by Kathryn Sullivan, first untethered spacewalk using a manned maneuvering unit (MMU), first satellite repair in orbit on Solar Max, even the first Coke and Pepsi in orbit.

Challenger's tenth flight, on January 28, 1986, set a tragic record: first Americans to die during a spaceflight, the first persons to die enroute to space.

The seven U.S. astronauts aboard Challenger shuttle flight STS-51L, later renumbered STS-25, were Francis R. "Dick" Scobee, Michael J. Smith, Judith A. Resnik, Ellison S. Onizuka, Ronald E. McNair, Gregory B. Jarvis and Concord, New Hampshire, high school social studies teacher Sharon Christa McAuliffe.

They were killed when a solid-fuel booster rocket leak led to an explosion during lift off from a Cape Canaveral launch pad. People around the world watched the accident on NASA videotape as mission control announced at one minute thirteen seconds into lift off, "Obviously a major malfunction. We have no downlink. The vehicle has exploded."

Millions of American students were watching television in classrooms as Challenger exploded. They had been planning to participate in lessons teacher Christa McAuliffe was to have taught by TV from orbit.

After Challenger. Work was started immediately on an orbiter to replace Challenger. Designated OV-105, the space agency needed a name for the new shuttle. The name Challenger was retired in honor of the seven dead astronauts. NASA wanted to continue the tradition of naming orbiters for famous explorers' ships.

Directed by Congress to do so, the space agency let American school students choose the name of the new shuttle. Students in public and private schools in the United States and territories, Department of Defense overseas dependents schools and Bureau of Indian Affairs schools entered the national competition in two divisions—kindergarten through the 6th grade, and 7th through 12th grades. Elementary and secondary students formed teams to research names. Each team prepared a classroom project to justify its name. More than 6,100 teams, including 71,650 students, completed research projects justifying their name choices. Each state, territory and agency announced a winner in each division. NASA decided the final winner in each division and announced the name of the new shuttle in 1989.

Endeavour. Sailing ships of Capt. James Cook, the 18th-century British explorer, turned out to be most popular among American children. Thirty-one of 111 state winners wanted the new shuttle to be named Endeavour, although some used the American spelling Endeavor.

Cook had chosen the name Endeavour for a 98-ft. ship in which he explored the South Pacific from 1768 to 1772. Cook circumnavigated the globe in Endeavour by sailing southwest from Plymouth, England, around South America, exploring coasts of New Zealand, Australia and New Guinea, and returning to England by way of Africa's Cape of Good Hope.

Twelve students chose the name Resolution. Cook had used the name Resolution for a ship for his second and third voyages exploring the Pacific Ocean. Cook sailed Resolution north to the Arctic Ocean and south to the Antarctic Circle. During those voyages, he found and named Sandwich Islands—now known as Hawaii. Cook was killed in the Sandwich Islands in 1779.

Eleven students selected Victoria, the name of the ship used by Ferdinand Magellan, a 16th century Portuguese navigator. Magellan sailed Victoria from 1519 to 1522. Victoria was the first to sail around the world. Unfortunately, Magellan was killed in the Philippines before the voyage was completed.

The American students also nominated the names Adventure, Blake, Calypso, Deepstar, Desire, Dove, Eagle, Endurance, Godspeed, Griffin, H.M.S. Chatham, Hokule'a, Horizon, Investigato, Meteor, Nautilus, North Star, Pathfinder, Phoenix, Polar Star, Rising Star, Royal Tern, Trieste and Victory. Discovery, already the name of a U.S. shuttle, was the second ship in Cook's company on his final sail.

The orbiter built to replace Challenger, first designated OV-105, officially was named Endeavour in 1989. Its maiden spaceflight was to be in 1992.

Flight numbers. NASA shuttle flight numbers can be confusing. All flights are given STS (space transportation system) numbers. The first nine flights were easy, numbered STS-1 through STS-9. Then the flight numbering system was changed. STS-9 also was known as STS-41A. What would have been STS-10 became STS 41B.

The new numbers were supposed to convey more information: The first numeral was the last digit of the government's fiscal year in which the launch was to take place. For example, the 4 in STS-41B stood for the 4 in 1984. The second numeral indicated the launch site. The number 1 meant Kennedy Space Center, Florida. Number 2 was Vandenberg Air Force Base, California. No shuttles ever were launched from Vandenberg. The letter represented the order of launch assignment within a fiscal year. For instance, the letter B in STS-41B meant it was the second launch scheduled for fiscal year 1984.

Things changed after the 1986 Challenger tragedy. NASA returned to the original easy-to-understand numbering system based on sequential flight numbers. All launches planned from Vandenberg were switched to the Cape.

The disastrous Challenger flight, previously known as STS-51L, actually was the 25th U.S. shuttle flight. It was renamed STS-25. The first mission when flights resumed after Challenger was the 26th shuttle so it was numbered STS-26.

Unfortunately, even the simple scheme became confused when NASA was forced to move flights around in the schedule. With STS numbers assigned 19 months in advance, the agency decided each flight would keep its number through all schedule changes. Thus, STS-36 flew before STS-31 and STS-41 flew before STS-38. U.S. shuttle space transportation system (STS) flights:

STS	Launch	Lands	Orbiter	Highlight
1	1981 Apr 12	1981 Apr 14	Columbia	maiden spaceflight
2	1981 Nov 12	1981 Nov 14	Columbia	remote arm test
3	1982 Mar 22	1982 Mar 30	Columbia	thermal tests
4	1982 Jun 27	1982 Jul 4	Columbia	9 Utah GAScans
5	1982 Nov 11	1982 Nov 15	Columbia	EVA canceled
6	1983 Apr 4	1983 Apr 9	Challenger	maiden flt & TDRS-A
7	1983 Jun 18	1983 Jun 24	Challenger	Sally Ride
8	1983 Aug 30	1983 Sep 5	Challenger	night launch/landing
9	1983 Nov 28	1983 Dec 8	Columbia	SpaceLab 1
10	1984 Feb 3	1984 Feb 11	Challenger	Palapa & Westar

11	1984 Apr 6	1984 Apr 13	Challenger	LDEF & Solar Max
12	1984 Aug 30	1984 Sep 5	Discovery	maiden flight
13	1984 Oct 5	1984 Oct 13	Challenger	female spacewalk
14	1984 Nov 8	1984 Nov 16	Discovery	2 sats retrieved
15	1985 Jan 24	1985 Jan 27	Discovery	first military flight
16	1985 Apr 12	1985 Apr 19	Discovery	flyswatter & Sen.Garn
17	1985 Apr 29	1985 May 6	Challenger	SpaceLab
18	1985 Jun 17	1985 Jun 24	Discovery	3 sats
19	1985 Jul 29	1985 Aug 6	Challenger	abort to orbit
20	1985 Aug 27	1985 Sep 3	Discovery	sat repair
21	1985 Oct 3	1985 Oct 7	Atlantis	maiden flt/military
22	1985 Oct 30	1985 Nov 6	Challenger	8 in crew/German
23	1985 Nov 26	1985 Dec 3	Atlantis	erector set
24	1986 Jan 12	1986 Jan 18	Columbia	U.S. Rep. Nelson
25	1986 Jan 28	1986 Jan 28	Challenger	7 die in explosion
26	1988 Sep 29	1988 Oct 3	Discovery	return to space
27	1988 Dec 2	1988 Dec 6	Atlantis	military spysat
29	1989 Mar 13	1989 Mar 18	Discovery	TDRS-D
30	1989 May 4	1989 May 8	Atlantis	Magellan
28	1989 Aug 8	1989 Aug 13	Columbia	military
34	1989 Oct 18	1989 Oct 25	Atlantis	Galileo
33	1989 Nov 22	1989 Nov 27	Discovery	military
32	1990 Jan 9	1990 Jan 20	Columbia	LDEF retrieval
36	1990 Feb 28	1990 Mar 5	Atlantis	military
31	1990 Apr 24	1990 Apr 29	Discovery	Hubble Telescope
41	1990 Oct 6	1990 Oct 10	Discovery	Ulysses
38	1990 Nov 15	1990 Nov 20	Atlantis	military
35	1990 Dec 2	1990 Dec 11	Columbia	Astro-1
37	1991 Apr 5	1991 Apr 11	Atlantis	Gamma Ray Obsrvtry
39	1991 Apr 28	1991 May 6	Discovery	military
40	1991 Jun 5	1991 Jun 14	Columbia	SLS-1
43	1991 Aug 2	1991 Aug 11	Atlantis	TDRS-E
48	1991 Sep 12	1991 Sep 18	Discovery	UARS
44	scheduled 1991 Dec		Atlantis	military
42	scheduled 1992 Feb		Discovery	IML-1
45	scheduled 1992 May		Atlantis	Atlas-1
49	scheduled 1992 May		Endeavour	Intelsat retrieval
50	scheduled 1992 Jun		Columbia	USML-1
46	scheduled 1992 Sep		Atlantis	tethered sat
47	scheduled 1992 Sep		Endeavour	Spacelab Japan
52	scheduled 1992 Nov		Columbia	USMP-2
53	scheduled 1992 Dec		Discovery	military
54	scheduled 1993 Jan		Endeavour	military
55	scheduled 1993 Mar		Columbia	Spacelab Germany
51	scheduled 1993 Apr		Discovery	ACTS
56	scheduled 1993 May		Endeavour	Atlas-2
57	scheduled 1993 Jul		Atlantis	Spacehab-1
58	scheduled 1993 Jul		Columbia	SLS-2
59	scheduled 1993 Aug		Discovery	to be determined
60	scheduled 1993 Oct		Endeavour	SRL-1
61	scheduled 1993 Dec		Atlantis	Spacehab-2
62	scheduled 1994 Jan		Columbia	USMP-2
63	scheduled 1994 Feb		Discovery	Hubble repair
64	scheduled 1994 Apr		Endeavour	Atlas-3
65	scheduled 1994 May		Atlantis	military
66	scheduled 1994 May		Columbia	IML-2
67	scheduled 1994 Jun		Discovery	Spacehab-3
68	scheduled 1994 Aug		Atlantis	to be determined
69	scheduled 1994 Sep		Columbia	Astro-2

70	scheduled 1995 first quarter	undetermined	Japan sat retrieval
71	scheduled 1995 first quarter	undetermined	SRL-2
72	scheduled 1995 second quarter	undetermined	Spacehab-4
73	scheduled 1995 second quarter	undetermined	to be determined
74	scheduled 1995 third quarter	undetermined	USMP-3
75	scheduled 1995 third quarter	undetermined	Spacehab-5
76	scheduled 1995 third quarter	undetermined	USML-2
77	scheduled 1995 fourth quarter	undetermined	Spacelab German
78	scheduled 1995 fourth quarter	undetermined	technology
79	scheduled 1995 fourth quarter	undetermined	Spacehab-6
80	scheduled 1996 first quarter	undetermined	Atlas-4
81	scheduled 1996 second quarter	undetermined	1st space station part
82	scheduled 1996 second quarter	undetermined	SLS-3
83	scheduled 1996 second quarter	undetermined	2nd space station part
84	scheduled 1996 third quarter	undetermined	EUVE retrieval
85	scheduled 1996 third quarter	undetermined	3rd space station part
86	scheduled 1996 third quarter	undetermined	Spacehab-7
87	scheduled 1996 fourth quarter	undetermined	SRL-3
88	scheduled 1996 fourth quarter	undetermined	4th space station part
89	scheduled 1996 fourth quarter	undetermined	USMP-4
90	scheduled 1997 first quarter	undetermined	technology
91	scheduled 1997 second quarter	undetermined	5th space station part
92	scheduled 1997 second quarter	undetermined	Hubble repair
93	scheduled 1997 third quarter	undetermined	6th space station part
94	scheduled 1997 third quarter	undetermined	Spacehab-8
95	scheduled 1997 third quarter	undetermined	7th space station part
96	scheduled 1997 third quarter	undetermined	Atlas-5
97	scheduled 1997 fourth quarter	undetermined	1st space station use
98	scheduled 1997 fourth quarter	undetermined	Spacelab Europe
99	scheduled 1997 fourth quarter	undetermined	8th space station part

Buran	Launch	Lands	Orbiter	Highlight
1	1988 Nov 15	1988 Nov 15	Buran No. 1	maiden flight
2	1992-93	1992-93	Buran No. 2	to Mir space station

U.S. Shuttle Launch Schedule

Date	Flight	Shuttle	Payload
1992 Feb	STS-42	Discovery	International Microgravity Lab (IML-1)
1992 May	STS-45	Atlantis	Atmospheric Lab for Appl'ns and Science (Atlas-1)
1992 May	STS-49	Endeavour	retrieval, repair and release of Intelsat VI comsat
1992 Jun	STS-50	Columbia	U.S. Microgravity Laboratory (USML-1)
1992 Sep	STS-46	Atlantis	European Retrievable Carrier, Tethered Satellite
1992 Sep	STS-47	Endeavour	Spacelab-J (Japan)
1992 Nov	STS-52	Columbia	U.S. Microgravity Payload (USMP-1)
1992 Dec	STS-53	Discovery	Dept. of Defense military flight
1993 Jan	STS-54	Endeavour	Tracking and Data Relay Satellite (TDRS-6)
1993 Mar	STS-55	Columbia	Spacelab D-2 (Germany)
1993 Apr	STS-51	Discovery	Advanced Communications Technology Sat
1993 May	STS-56	Endeavour	Atmospheric Lab for Appl'ns and Science (Atlas-2)
1993 Jul	STS-57	Atlantis	Spacehab-1 private module
1993 Jul	STS-58	Columbia	Spacelab Life Sciences (SLS-2)
1993 Aug	STS-59	Discovery	an available flight with payload to be determined
1993 Oct	STS-60	Endeavour	Space Radar Laboratory (SRL-1)
1993 Dec	STS-61	Atlantis	Spacehab-2 private module

1994 Jan	STS-62	Columbia	U.S. Microgravity Payload (USMP-2)
1994 Feb	STS-63	Discovery	Hubble Space Telescope (HST) retrieval and repair
1994 Apr	STS-64	Endeavour	Atmospheric Lab for Appl'ns and Science (Atlas-3)
1994 May	STS-65	Atlantis	SPAS-3 (Shuttle Pallet Satellite)
1994 May	STS-66	Columbia	International Microgravity Laboratory (IML-2)
1994 Jun	STS-67	Discovery	Spacehab-3 private module
1994 Aug	STS-68	Atlantis	an available flight with payload to be determined
1994 Sep	STS-69	Columbia	Astro-2

NASA's Shuttle Numbering System

NASA adopted a cumbersome numbering system for space shuttle flights 10 through 25. Before STS-41B, flights were numbered simply STS-1 through STS-9. After Challenger, simple numbers were returned.

STS stands for space transportation system. All shuttle flights have been given STS numbers. The first nine were easy, then the flight numbering system was changed. STS-9 also was known as STS-41A. STS-10 became STS 41B. The new numbers were supposed to convey more information:

★The first numeral was the last digit of the government's fiscal year in which the launch was to take place. For example, the 4 in STS-41B stood for the 4 in 1984.

★The second numeral indicated the launch site. The number 1 meant Kennedy Space Center, Florida. Number 2 was Vandenberg Air Force Base, California. Of course, no shuttles ever were launched from Vandenberg.

★The letter represented the order of launch assignment within a fiscal year. For instance, the letter B in STS-41B meant it was the second launch scheduled for fiscal year 1984.

After the 1986 Challenger tragedy, NASA returned to the original easy-to-understand numbering system based on sequential flight numbers. All scheduled flights from Vandenberg were cancelled.

The disastrous Challenger flight, known as STS-51L, actually was the 25th U.S. shuttle flight—so NASA renamed it STS-25. The first mission when flights resumed after Challenger was the 26th shuttle flight so it was numbered STS-26.

Unfortunately, even the simple scheme became confused. When NASA was forced to move flights around in the schedule, the space agency decided each flight would keep its STS number assigned 19 months before flight.

The Challenger Disaster

It was NASA shuttle flight STS-51L, the 25th shuttle blast-off, the first firing from Launch Pad B at Kennedy Space Center's Launch Complex 39. The flight had been scheduled six times before, but delayed for bad weather and technical problems.

Finally, at 11:38 a.m. Eastern Standard Time January 28, 1986, Challenger was on its way to space with a TDRS tracking and data relay satellite and a Spartan free-flying astronomy module to study Comet Halley. One minute thirteen seconds after liftoff, a defective seal on one of two solid-fuel booster rockets spit flame, igniting an explosion which devastated the orbiter, the big fuel tank and the two boosters. All seven crew members were killed:

★Francis R. Scobee, commander,

★Michael J. Smith, pilot,

★Judith A. Resnik, mission specialist,

★Ellison S. Onizuka, mission specialist,

★Ronald E. McNair, mission specialist,

★Gregory B. Jarvis, payload specialist,

★S. Christa McAuliffe, a New Hampshire school teacher and the first space shuttle passenger in NASA's Teacher in Space program. McAuliffe had planned to teach lessons to school children on Earth during live TV transmissions from space.

After the accident, three orbiters remained in the fleet: Columbia, Discovery and Atlantis. They were grounded. Work was started on a replacement shuttle, later named Endeavour. A Presidential Commission investigated. Booster rockets were changed, as was NASA itself. Discovery took America back to space September 29, 1988. Endeavour was to make its maiden voyage in May 1992.

Cosmonauts Carry Pictures. USSR cosmonauts Vladimir Solovyev and Leonid Kizim took a photograph of the seven U.S. shuttle Challenger astronauts up to the USSR's orbiting Mir space station in 1986.

Solovyev and Kizim flew from Baikonur Cosmodrome March 15, 1986, in Soyuz T-15, just six weeks after Challenger exploded in January. The cosmonauts said they "wanted those seven brave astronauts to go to outer space."

While visiting the U.S. after returning from Mir, Kizim and Solovyev said they had paid tribute in space to the dead U.S. astronauts.

Kizim and Solovyev accompanied a teenage Young Cosmonauts group from the USSR. They toured the U.S. launch facility at Cape Canaveral, as well as Washington, Disney World, and NASA space centers at Huntsville, Alabama, and Houston, Texas. Earlier, Young Astronaut teens from the USA had toured the USSR.

Challenger Point. A 14,081-ft. mountain peak near Kit Carson Peak in the Sangre de Cristo range in south-central Colorado was named Challenger Point to honor the seven astronauts. The U.S. Board on Geographical Names voted for the name suggested by Colorado Springs electrical engineer Dennis Williams who said mountain peaks symbolize the spirit of adventure.

Some south-central Colorado residents opposed the name. County commissioners voted against it. The sheriff said residents were afraid it would encourage people to climb and fall off. In fact, statistics showed ten persons had died in eleven years from climbing accidents and small plane crashes near the point. Supporters of the commemorative name included a former state governor and the Colorado Mountain Club.

Challenger 7 Mural At Capitol. A colorful painting of the seven lost astronauts was placed in a corridor of the U.S. Senate in Washington, DC. The Challenger mural was painted by Temple University professor Charles Schmidt.

The mural is in the Capitol's Brumidi corridor, named for an Italian immigrant who decorated the Senate in a florid 19th Century style. Blank ovals, left to record future U.S. history, line the brightly-painted corridor. Across the corridor from the Challenger mural is a 1975 painting reflecting the landing of the Apollo moonship in 1969.

No Centaurs In Shuttles. A powerful rocket fueled with liquid hydrogen and liquid oxygen was ruled out of bounds on U.S. space shuttles.

NASA decided the Centaur rocket was too dangerous to carry into space aboard a shuttle. Centaur's fuel is the same combination which destroyed shuttle Challenger.

The shuttle version of Centaur is an adaptation of a rocket which has been used for years as a final stage atop unmanned rockets. It sent Viking to Mars. Centaur was to have blasted off from outside an orbiting shuttle in May 1986, sending the Galileo probe to Jupiter. Another planned shuttle Centaur would have sent the Magellan probe to radar-map Venus. Others would have carried secret military heavyweight satellites away from shuttles. The various interplanetary missions went forward, but not with Centaur firing from a shuttle orbit. Centaur was returned for use atop unmanned rockets. The U.S. Air Force ordered unmanned Titan rockets which could carry Centaurs.

Wreckage Buried In Silos. A year after Challenger exploded on its way to orbit, workers at Cape Canaveral buried wreckage from the space shuttle Challenger in old abandoned Minuteman ICBM missile silos and underground buildings there.

The orbiter and its payload are kept in four underground buildings. Pieces of the external fuel tank and solid-fuel booster rockets are in two 78-foot-deep silos. Giant concrete caps were lowered onto the abandoned missile silos in February 1987.

Parts of the right-side booster which ruptured in the explosion were kept elsewhere for study by engineers.

The wreckage had been held in a Kennedy Space Center hangar before burial. The Smithsonian Institution's National Air and Space Museum wanted to display pieces of wreckage in Washington, but NASA didn't think it proper.

To the ocean. Did the crew aboard Challenger, after it exploded 73 seconds into flight, survive the two-and-a-half minutes it took the shuttle to crash into the ocean?

A Florida newspaper in November 1988 quoted anonymous NASA investigators who said yes, the crew survived the blast to die when the pressurized cabin slammed into the Atlantic Ocean at 200 miles per hour.

The Miami Herald newspaper's Sunday magazine, Tropic, said NASA found no evidence the crew cabin lost pressure. If pressure did drop, it was slow, without a sudden loss of air that would have knocked the astronauts out within seconds. Thus, the astronauts might have remained conscious long enough to catch a glimpse of the ocean rushing toward them. The magazine said, if Challenger had landed softly, they could have swum home.

Air pack. The wife of one of the Challenger astronauts said she wasn't surprised by the speculation. She said she had thought her husband might have been alive during the 2.5 minutes because his PEAP (personal emergency air pack) was found still attached to his seat. Three-quarters of the air in it had been breathed. She said that amount would match the amount of time to impact. The PEAP, used only on demand, was found turned on, indicating the astronaut was breathing through it to receive his air.

NASA said, in its opinion, the astronauts died the instant the spacecraft exploded. The space agency said, if the crew cabin had survived the blast intact, it would have been ripped apart by aerodynamic forces while falling.

Death certificates. The remains of seven astronauts were found entangled in the crushed cabin on the Atlantic Ocean floor six weeks after the disaster and taken to the Air Force Institute of Pathology. NASA's official report claimed no cause of death could be determined from the remains. Death certificates issued from the Johnson Space Center, Houston, Texas, on January 30, 1986, two days after the tragedy, certified the astronauts "died when the shuttle spacecraft Challenger...exploded."

Cape Canaveral and the Kennedy Space Center are located in Brevard County, Florida. The chief assistant to the medical examiner in Brevard County told Tropic magazine the Johnson Space Center death certificates were invalid because of the date, and because they were not signed by Florida authorities. He said the public may never know exactly how the astronauts died since the remains were not made available to Florida pathologists, whose records would be public.

Photos. NASA decided not to release photos of the Challenger crew cabin. Officials said it would be an invasion of privacy of the dead crew and their families.

Photos and videotapes reportedly included shots of astronaut personal effects strewn across the ocean floor. They will be shown only to NASA workers.

Photos of other parts of the wrecked shuttle are available and one photo of crew cabin debris was released in 1986 before the policy changed.

Tapes. Saying release would invade the privacy of astronaut families, NASA tried to hold back audio recordings of the crew cabin during launch, but a federal appellate court panel ruled the space agency had to release the tapes. The ruling by the U.S. Court of Appeals affirmed a lower court decision that the tapes be made public under the Freedom of Information Act.

NASA After Challenger. To improve safety and performance, NASA officials revamped the space agency in 1986 and 1987 during 975 painful and fearful days after the

Challenger disaster froze the U.S. man-in-space program. Some of the post-Challenger changes were:

★redesigned boosters for return to flight
★improved shuttle main engines
★checked all critical hardware and maintenance
★established a quality control and safety team
★improved weather forecasting for landing sites
★improved safety at landing sites
★added an escape hatch for a disabled shuttle
★overhauled top NASA management
★new administrator
★new deputy administrator
★new management running the shuttle program
★new directors controlling the space station program
★all three major shuttle field centers received new leadership
★several astronauts assigned to top positions
★shuttle management concentrated in Washington
★space station authority centered in Washington
★new associate administrator posts created.

Endeavour Space Shuttle

Sailing ships of Capt. James Cook, the 18th-century British explorer, turned out to be most popular among American children in a contest to name a new space shuttle.

Thirty-one of 111 state winners wanted the new shuttle to be named Endeavour. Many used the American spelling Endeavor. Twelve chose Resolution. Eleven selected Victoria, the name of the ship used by Ferdinand Magellan, a 16th century Portuguese navigator, NASA revealed March 21, 1989.

Challenger. Millions of American students were watching television in classrooms January 28, 1986, when the shuttle Challenger exploded 73 seconds after liftoff, killing teacher Christa McAuliffe and six other astronauts. Many pupils had planned to participate in lessons McAuliffe was to have taught from orbit.

A replacement shuttle being built was scheduled for its maiden flight in 1992. The replacement orbiter had the temporary name of OV-105.

NASA's orbiters still in use—Discovery, Atlantis and Columbia—were named after sea vessels used in research and exploration. NASA wanted to continue that tradition. The name Challenger was retired in honor of the seven dead astronauts. NASA, directed by Congress to do so, let American school students choose the name of the new shuttle. The new shuttle was completed in 1991.

Contest. Students in public and private schools in the United States and territories, Department of Defense overseas dependents schools and Bureau of Indian Affairs schools entered the national competition in two divisions—kindergarten through the 6th grade, and 7th through 12th grades. Elementary and secondary students formed teams to research names. Each team prepared a classroom project to justify its name. More than 6,100 teams, including 71,650 students, completed research projects justifying their name choices. Each state, territory and agency announced a winner in each division. NASA chose the final winner in each division and announced the name of the new shuttle in 1989.

Endeavour. Cook chose the name Endeavour for a 98-foot ship in which he explored the South Pacific from 1768 to 1772. Cook circumnavigated the globe in Endeavour by sailing southwest from Plymouth, England, around South America, exploring coasts of New Zealand, Australia and New Guinea, and returning to England by way of Africa's Cape of Good Hope.

Resolution. Cook chose the name Resolution for a ship for his second and third voyages exploring the Pacific Ocean. Cook sailed Resolution north to the Arctic Ocean and south to the Antarctic Circle. During those voyages, he found and named Sandwich Islands—now known as Hawaii. Cook was killed in the Sandwich Islands in 1779.

Discovery. Discovery was the second ship in Cook's company on his final sail. Discovery already is the name of a U.S. space shuttle.

Victoria. Ferdinand Magellan, a Portuguese explorer, sailed in the ship Victoria from 1519 to 1522. The Victoria was the first to sail around the world. Unfortunately, Magellan was killed in the Philippines before the voyage was completed.

Other names. The students also nominated these names: Adventure, Blake, Calypso, Deepstar, Desire, Dove, Eagle, Endurance, Godspeed, Griffin, H.M.S. Chatham, Hokule'a, Horizon, Investigato, Meteor, Nautilus, North Star, Pathfinder, Phoenix, Polar Star, Rising Star, Royal Tern, Trieste and Victory.

Runaway Shuttle Safety Net

A nylon safety net to stop a space shuttle from rolling off a runway during an emergency landing was tested in 1987 at an airport in Virginia.

A shuttle lands at about 200 miles per hour. During a real emergency, the net would stop a fully-loaded shuttle rolling at 100 miles per hour within 1,000 feet.

The safety net would be suspended 25 feet above a runway surface on a break-away cable. The corners of the net would be attached by cable to energy-absorber drums anchored in concrete at the edges of the runway to provide the braking force for a rolling shuttle.

As a shuttle slammed into it, the net would break loose and blanket the vehicle. As the shuttle continued down the runway, it would unwind nylon tapes from the energy absorber drums and stop rolling in 1,000 feet.

The test. The old dummy *Enterprise* shuttle was rolled slowly through a remote field into the web of hundreds of nylon straps to see which parts of the shuttle came in contact with the net and to find what stress the net had to endure.

The 130-ton spaceship did not hurtle down a runway into the net. Rather, the non-flying *Enterprise* test craft, retired on a Washington Dulles International Airport back lot for years, was winched a few inches at a time to see how the net draped over it. NASA didn't want to catch and break the nosewheel, the most vulnerable part of the orbiter.

Bad brakes. The shuttle fleet is notorious for bad brakes. If one had to land on a short runway, the giant net could be erected to halt its roll, preventing the shuttle from overrunning the landing strip. NASA calls it a Shuttle Orbiter Arresting System and the space agency needs it on short runways at remote emergency landing sites in Dakar, Senegal, and Zaragoza and Moron, Spain.

Similar systems for airplanes are on many airport runways around the world. The Naval Air Engineering Center at Lakehurst, New Jersey, helped NASA with the test since the Navy had experience snagging planes landing on aircraft carrier decks. The shuttle net is similar, but bigger.

Shuttle Launch Weather Rules

NASA has had weather problems over the years. The Cape Canaveral area of Florida is a thunderstorm-prone area. NASA rules for launching in wind, rain, clouds and lightning:

★WIND—a launch will be put off if wind gusts peak at 32 knots or if wind is steady at 22 knots.

★RAIN—a launch will be postponed if there is any precipitation in a rocket's flight path.

★CLOUDS—if the edge of a storm comes within 10 miles of the pad, launch will be held.

★LIGHTNING—if lightning is spotted within 10 miles of the pad, as much as 30 minutes before launch, firing will be postponed.

Shuttle Emergency Landing Strips

NASA has a worldwide maze of shuttle emergency landing strips. The network of fields strewn across the globe means a shuttle in distress would find a runway within three revolutions around Earth. One orbit takes about 90 minutes.

North Pacific landing strips are at Hickam Air Force Base, Hawaii, and Andersen Field, Guam. A South Pacific emergency field is on Easter Island. African strips are at Ben Guerir, Morocco; Banjul, The Gambia; and Dakar, Senegal. Strips in southern Europe are at Zaragoza Air Base, Spain, and Moron Air Base, Spain.

Continental U.S. landing fields are Kennedy Space Center, Florida, Edwards Air Force Base, California, and White Sands, New Mexico.

Runways at Hickam and Andersen have giant nets to catch skidding spaceplanes, high-tech landing aids and brilliant Xenon lights for night landings. Like giant tennis nets, the barriers known to NASA as Shuttle Orbiter Arresting System, would catch a shuttle with bad brakes. Some emergency landing sites have very long runways and do not require nets. The sites are manned by 30 to 40 persons during a shuttle flight.

Regular airports. NASA also lists 20 airports around the world, such as Orlando, Florida, as backup emergency landing strips. They do not have the large shuttle nets.

All shuttles take off from Kennedy Space Center. Kennedy and Ben Guerir are launch abort sites for a shuttle with engine trouble moments after liftoff. Kennedy now is the normally-scheduled landing site for U.S. shuttles, with Edwards and White Sands secondary.

A major shuttle launching facility was built in the 1980's at Vandenberg Air Force Base, California, but mothballed after the 1986 Challenger disaster. Should that facility ever be re-opened, Easter Island would be an emergency landing strip for shuttles taking off from Vandenberg.

Bad brakes. The space agency needs the Shuttle Orbiter Arresting System on short runways at emergency landing sites because the American shuttle fleet previously had weak brakes. If one had to land on a short runway, the giant net would be erected to halt its roll, preventing the shuttle from overrunning the strip.

Similar systems for airplanes are in use on many airport runways around the world. The Naval Air Engineering Center at Lakehurst, New Jersey, helped NASA with the safety net since the Navy had experience snagging planes landing on aircraft carrier decks. A shuttle net is bigger.

The net. A shuttle lands at 200 mph. During a real emergency, the net would stop a fully-loaded shuttle rolling at 100 miles per hour within 1,000 feet.

The nylon safety net would be suspended 25 feet above a runway surface on a break-away cable. The corners of the net would be attached by cable to energy-absorber drums anchored in concrete at the edges of the runway to provide the braking force for a rolling shuttle.

As a shuttle slammed into it, the net would break loose and blanket the vehicle. As the shuttle continued down the runway, it would unwind nylon tapes from the energy absorber drums and stop rolling in 1,000 feet.

Net tests. The safety net was tested in 1987 at an airport in Virginia. The old dummy shuttle Enterprise was rolled slowly through a remote field into the web of hundreds of nylon straps to see which parts of the shuttle came in contact with the net and to find what stress the net had to endure.

The 130-ton spaceship did not hurtle down a runway into the net. Rather, the non-flying Enterprise test craft, retired for years on a back lot at the Washington, D.C., Dulles International Airport, was winched a few inches at a time to see how the net draped over it. NASA didn't want to catch and break the nosewheel, the most vulnerable part of the orbiter.

African strips. NASA constructed emergency landing sites for shuttles at an abandoned Strategic Air Command base near Ben Guerir, Morocco, and at an airport in the tiny country of The Gambia in northwest Africa.

The former SAC base at Ben Guerir, 40 miles north of Marrakech, abandoned in 1963, has a 14,000 foot runway. It was resurfaced to make it usable by a shuttle which would fail during launch to reach the speed necessary to get to orbit.

The airport at Banjul, The Gambia, was upgraded to accommodate shuttles. Banjul, with an 11,800-foot runway, serves The Gambia as its international airport. Banjul airport is about 13 degrees north of the Equator on a flat plain seven miles inland from the Atlantic Ocean. The airport lies almost directly beneath the normal flight path of a U.S. shuttle launched into a standard 28.5 degree orbit.

Foundations for shuttle approach and landing aids were laid at the African sites. Operations and storage buildings were constructed. Each location received equipment, tugged across the Atlantic Ocean in an ocean-going barge from Kennedy Space Center, including ten Jeeps, a microwave landing system, lights, generators for power, portable satellite communications systems, automated weather stations and tools. Ben Guerir received fire-fighting equipment.

Easter Island. Chile owns Easter Island, 50 sq. mi. of land off the western coast of South America, 2,270 miles due west of Chile in the Pacific Ocean. Easter Island also is known as Isla de Pascua.

Easter Island, a Polynesian mote formed by volcanic action, is famous for its ancient tall carved-stone busts known as moais. The Easter Island Family Chiefs Council, composed of descendants of original Easter Island tribal chiefs, opposed the shuttle emergency landing strip project, saying it did archaeological damage to their island and made Easter Island a likely military target in a U.S. war.

Chile would like an astronaut to be selected from its country for flight in a U.S. shuttle, so a 9,612-ft. landing strip on Easter Island was lengthened to 11,000 ft. and paved. Approach lights were installed and a maintenance office built with construction materials ferried across the ocean in a barge from Chile.

After Challenger, NASA decided to launch all shuttles from Florida. Shuttles are not expected to take-off from California until 1994 at the earliest. The Easter Island strip has been used by NASA for airplane flights to 70,000 ft. for studies of the ozone hole over Antarctica.

Spacelab

NASA and the European Space Agency agreed in 1973 that ESA would build a pressurized science laboratory to be carried to space occasionally in the cargo bay of U.S. space shuttles. The Europeans spent $1.2 billion to build Spacelab modules and the U.S. gives free shuttle rides.

A Spacelab is a set of modules 8.9 feet long and 13 feet in diameter, placed in a shuttle's cargo bay and connected to the shuttle airlock by a tunnel.

Astronaut-scientists float back and forth through the tube between the lab and crew quarters in the shuttle. Experiments are mounted in the lab in racks.

Spacelab modules can be joined together to compose a double-size unit, carrying 4,600 lbs. of research equipment.

Experiments needing exposure to space are mounted on pallets 9.5 feet long and 13 feet wide. If a Spacelab flight does not require pressurization, then five pallets can be laid

SPACE SHUTTLES 209

along the length of the cargo bay.

SPACELAB 1
Launch: November 28, 1983   Shuttle: Columbia
Flight: STS-9   Duration: 247:47:41 (hours:minutes:seconds)
Crew: John W. Young, Brewster H. Shaw Jr., Owen K. Garriott, A. Robert Parker,
    Ulf Merbold, Byron K. Lichtenberg
Events: First test of Spacelab concept. Carried 71 experiments in astronomy, solar physics, life sciences, atmospheric physics, Earth observation, plasma physics, and materials processing technology. Columbia was modified for the lab with new engines, a bigger antenna to transmit data, new galley forward of the hatch. Astronauts in pressurized module operated 38 instruments: 20 in lab, 16 on pallets, 2 shared. X-ray, ultraviolet, solar telescopes. Ten days in orbit. Merbold from West Germany was the first non American to go to space in a U.S. craft. He and Lichtenberg were the first payload specialists. Merbold, Lichtenberg and Garriott worked in the lab.

SPACELAB 3
Launch: April 29, 1985   Shuttle: Challenger
Flight: STS-51B   Duration: 168:07:55
Crew: Robert F. Overmyer, Frederick D. Gregory, Don L. Lind, Norman E. Thagard,
    William E. Thorton, Lodewijk van den Berg, Taylor G. Wang
Events: Pressurized module like S/L 1. Biological response experiments and processing of materials. Liquid droplets levitated with sound waves. Grew bright-red mercuric-oxide crystals for 104 hours, larger and purer than possible on Earth. Carried two monkeys, 24 rats, in cages. When food and droppings floated away, astronauts had to put on surgical masks while vacuuming the air. Lodewijk van den Berg, of the European Space Agency, was the first Dutch astronaut.

SPACELAB 2
Launch: July 29, 1985   Shuttle: Challenger
Flight: STS-51F   Duration: 190:45:35
Crew: Charles Gordon Fullerton, Roy D. Bridges Jr., Loren W. Acton, F. S. "Story" Musgrave,
    Anthony W. "Tony" England, Karl G. Henize, John-David F. Bartoe
Events: Mostly astronomy. Telescopes on three pallets. No pressure module. Crew of seven cramped in Challenger cabin a week. First engine failure during shuttle flight to space. Henize was the oldest person to fly in space, age 58.

SPACELAB D-1
Launch: October 30, 1985   Shuttle: Challenger
Flight: STS-61A   Duration: 168:45:00
Crew: Henry W. Hartsfield Jr., Steven R. Nagel, James F. Buchli, Guion S. Bluford Jr.,
    Bonnie J. Dunbar, Reinhard Furrer, Ernst Messerschmid, Wubbo J. Ockels
Events: First German flight. Did 75 experiments, mostly materials processing in zero-gravity. Acceleration tests on track zooming length of lab. Record 8-man crew with Furrer and Messerschmid from West Germany, Ockels from Holland. Mission control shared with West German Space Operations Center at Oberpfaffenhofen near Munich. Largest crew: 8.

SPACELAB ASTRO-1
Launch: December 2, 1990   Shuttle: Columbia
Flight: STS-35   Duration: 8 days 23 hrs 5 mins 8 secs
Crew: commander Vance D. Brand, pilot Guy S. Gardner,
    mission specialists Jeffrey A. Hoffman, John M. "Mike" Lounge,
    A. Robert Parker, payload specialists civilian astronomers Samuel T. Durrance,
    Ronald A. Parise
Events: The first shuttle flight in five years dedicated entirely to science. Four of the seven astronauts were astronomers. Used Astro-1 ultraviolet light telescopes and the Broad Band X-Ray Telescope (BBXRT-1) to study quasars, binary stars, pulsars, black holes, galaxies and high-energy stars. Specific targets included Supernova 1987a, the nearby supergiant star Betelgeuse, radio-quiet quasar Q1821, spiral-poor galaxy cluster Abell 2256, NGC 1633, NGC 1399 in the constellation Fornax, Q1821+64. Despite early telescope control problems in orbit, the Spacelab astronomy instruments in the shuttle cargo bay ended up examining 135 deep space targets during 394 observations. The observations included Hopkins Ultraviolet Telescope, 75 targets, 101 observations; Ultraviolet Imaging Telescope, 64 targets, 89 observations; Wisconsin Ultraviolet Photo Polarimeter Experiment, 70 sources, 88 observations; and Broad Band X-ray Telescope, 76 targets, 116 observations. It was the 15th longest U.S. space mission and the third longest shuttle

flight. Parise used Shuttle Amateur Radio Experiment (SAREX) equipment to make ham radio contact with amateur operators on Earth; his callsign WA4SIR. Hoffman wore the first dress shirt and tie in space when he dressed up to teach elementary pupils on the ground by TV from the shuttle Dec 7. NASA's idea of a Space Classroom became a reality as Durrance, Hoffman, Parise and Parker conducted space-to-ground TV lessons on the invisible and visible Universe, electromagnetic spectrum and how telescopes work in space. The astronauts answered questions from pupils on the ground. Parise and Durrance were the first non-NASA astronauts to fly in a U.S. spacecraft in five years, since the Challenger disaster.

SPACELAB SLS-1

Launch: June 5, 1991   Shuttle: Columbia
Flight: STS-40   Duration: 9 days
Crew: commander Bryan D. O'Connor, pilot Sidney M. Gutierrez,
       mission specialists James P. Bagian, Tamara E. "Tami" Jernigan, Margaret Rhea Seddon,
       payload specialists Francis Andrew "Drew" Gaffney, Millie Hughes-Fulford
Events: Spacelab Life Sciences (SLS-1) explored how heart, blood vessels, lungs, kidneys and hormone-secreting glands respond to microgravity, the causes of space sickness and changes in muscles, bones and cells during space flight, and readjustment to gravity upon returning to Earth, the most detailed physiological measurements in space since Skylab in 1973-74. Spacelab is a removeable pressurized science lab carried to space and back in Columbia's cargo bay. Astronauts float through a tube into Spacelab. The crew shepherded 30 rats plus 2,478 tiny jellyfish in plastic bags and bottles—the first jellyfish in orbit—for Eastern Virginia Medical School biologist Dorothy Spangenberg who wanted to see how Aurelia aurita (moon jellies) grow in weightlessness. Since moon jellies on Earth mature to free swimming in six days if iodine is added to their water, they were ideal research organisms for Columbia's nine-day spaceflight. In orbit, Jernigan injected iodine into the water of a third of them and videotaped their behavior. The experiments were duplicated the same day in a Norfolk laboratory. After flight, Spangenberg examined them to see how much calcium they lost. Crew sleep was interrupted to fix Spacelab cooling equipment. First, a backup cooling pump had to be switched on for the Research Animal Holding Facility. Then, the Spacelab refrigerator-freezer broke down. A shortage nearly required Spacelab power to be shut down for a day to conserve liquid hydrogen and oxygen for fuel cells. It was the first shuttle with three woman in one crew. Bagian, Gaffney and Seddon were MDs. Jernigan, an astronomer. Hughes-Fulford, a biological chemist. STS-40 landed June 14.

SPACELAB IML-1

Launch: scheduled February 1992   Shuttle: Discovery
Flight: STS-42   Duration: 7 days
Crew: commander Ronald J. Grabe, pilot Stephen S. Oswald,
       payload commander Norman E. Thagard, mission specialists David C. Hilmers,
       William F. Readdy, payload specialists Roberta L. Bondar, Ulf D. Merbold
Events: International Microgravity Laboratory (IML-1), first microgravity research in a long shuttle Spacelab module, studying effects on material processes and living organisms of very low gravity in space near Earth. With NASA in the project are 5 international organizations: European Space Agency (ESA), Canadian Space Agency (CSA), French National Center for Space Studies (CNES), West German Research and Development Institute for Air and Spacecraft (DLR), and National Space Development Agency of Japan (NASDA). IML Spacelabs are to fly at 17-mo. to 25-mo. intervals; separation allowing scientists to study results and apply them to new research. They will carry plants and animals, and protein crystal, vapor crystal and mercury iodide crystal growth experiments. Discovery to orbit with its tail pointing down toward Earth in the "gravity gradient" position offering the least gravity during flight. Merbold from Germany. Roberta L. Bondar MD is to be the first Canadian woman in space; first non-U.S. astronaut in a U.S. spaceflight since 1985.

SPACELAB ATLAS-1

Launch: scheduled May 1992   Shuttle: Atlantis
Flight: STS-45   Duration: 8 days
Crew: commander Charles F. Bolden, Jr., pilot Brian Duffy,
       payload commander Kathryn D. Sullivan, mission specialists C. Michael Foale,
       David C. Leestma, payload specialists Dirk D. Frimout, Byron K. Lichtenberg
Events: Sullivan to be the first female payload commander, in charge of Atmospheric Laboratory for Applications and Science (Atlas-1) to study atmospheric phenomena, energy from the Sun and changes in the solar spectrum from a Spacelab aboard Atlantis. Atlas-1 is to be the first of several Atlas missions scheduled for orbit over a decade to study interaction of Earth's atmosphere with the

Sun. Atlas is part of NASA's Mission to Planet Earth program. STS-45 astronauts also will use the Shuttle Solar Backscatter Ultraviolet Instrument (SSBUV-A1) to measure ozone in Earth's atmosphere. SSBUVs were carried in U.S. shuttles before: SSBUV-1 in STS-34, SSBUV-2 STS-41, SSBUV-3 STS-43.

SPACELAB USML-1

Launch: scheduled June 1992   Shuttle: Columbia
Flight: STS-50   Duration: 13 days
Crew: commander Richard N. "Dick" Richards, pilot John H. Casper,
       payload commander Bonnie J. Dunbar, mission specialists Kenneth D. Bowersox,
       Carl J. Meade, payload specialists Lawrence J. DeLucas, Eugene H. Trinh
Events: U.S. Microgravity Laboratory (USML-1), materials-processing experiments in a Spacelab aboard the first extended-duration mission of the refurbished Columbia, similar to earlier International Microgravity Laboratory IML-1 aboard STS-42. Columbia's 13-day flight to be the longest shuttle trip to date. Trinh and DeLucas to conduct 30 technology and science experiments in materials, fluids, biological processes. Trinh to be first Vietnamese in a U.S. spacecraft.

SPACELAB-J

Launch: scheduled September 1992   Shuttle: Endeavour
Flight: STS-47   Duration: 7 days
Crew: commander, pilot, payload commander Mark C. Lee, mission specialists N. Jan Davis,
       Mae C. Jemison, Jerome "Jay" Apt, payload specialist Mamoru Mohri
Events: Spacelab-J joint science research project of NASA and the Japanese National Space Development Agency (NASDA). Mohri will be the second native of Japan in orbit and the first Japanese to fly in a U.S. spacecraft. Lee and Davis the first married couple to travel in space. Jemison the first black woman in orbit.

SPACELAB D-2

Launch: scheduled March 1993   Shuttle: Columbia
Flight: STS-55   Duration: 9 days
Crew: commander, pilot, payload commander Jerry L. Ross, two mission specialists,
       two payload specialists
Events: Spacelab-D2. Two payload specialists to be selected from four Germans: female meteorologist Renate Luise Brummer; and physicists Hans-Wilhelm Schlegel; Gerhard Thiele; and Ulrich Walter.

SPACELAB ATLAS-2

Launch: scheduled May 1993   Shuttle: Endeavour
Flight: STS-56   Duration: 9 days
Crew: commander, pilot, payload commander, mission and payload specialist
Events: To carry Atmospheric Laboratory for Applications and Science (Atlas-2) to study atmospheric phenomena, energy from the Sun and changes in the solar spectrum from a Spacelab carried in shuttle Endeavour's cargo bay. Atlas-2 is to be the second of several Atlas missions scheduled for orbit over a decade to study interaction of Earth's atmosphere with the Sun. Atlas-1 was carried to the same altitude and inclination in flight STS-45. Atlas is part of NASA's Mission to Planet Earth program. STS-56 astronauts also will use the Shuttle Solar Backscatter Ultraviolet Instrument (SSBUV-A2) to measure ozone in Earth's atmosphere. SSBUVs were carried in U.S. shuttles before: SSBUV-1 in STS-34, SSBUV-2 STS-41, SSBUV-3 STS-43, SSBUV-A1 STS-45. Also to be aboard STS-56 is Spartan, the Shuttle Pointed Autonomous Research Tool for Astronomy (Sptn-201-1). Spartan is a small, retrievable, free-flying satellite for X-ray astronomy.

SPACELAB SLS-2

Launch: scheduled July 1993   Shuttle: Columbia
Flight: STS-58   Duration: 13 days
Crew: commander, pilot, payload commander, mission specialists, payload specialist,
       possible USSR cosmonaut
Events: The second extended-duration flight, carrying Spacelab Life Sciences (SLS-2) exploring how the heart, blood vessels, lungs, kidneys and hormone-secreting glands respond to microgravity, the causes of space sickness and changes in muscles, bones and cells during space flight and in the readjustment to gravity upon returning to Earth. The first Spacelab Life Sciences test (SLS-1) was aboard Columbia STS-40 in 1991. Columbia's earlier 13-day flight was STS-50 in 1992. At a 1991 summit meeting, U.S. President George H.W. Bush and USSR President Michael Gorbachev agreed on a first-time exchange: a Soviet cosmonaut would fly aboard a U.S. shuttle and a NASA astronaut would travel to the USSR's Mir space station. The agreement called for medical

research on how humans adapt to weightlessness preparing for lengthy flights to Mars after the year 2000. The cosmonaut might fly 13 days in STS-58.

SPACELAB ATLAS-3

Launch: scheduled April 1994   Shuttle: Endeavour
Flight: STS-64   Duration: 9 days
Crew: commander, pilot, payload commander, mission specialists, payload specialist
Events: To carry Atmospheric Laboratory for Applications and Science (Atlas-3) to study atmospheric phenomena, energy from the Sun and changes in the solar spectrum from a Spacelab carried in shuttle Endeavour's cargo bay. Atlas-3 is to be the third of several Atlas missions scheduled for orbit over a decade to study interaction of Earth's atmosphere with the Sun. Atlas-1 and Atlas-2 were carried in flights STS-45 and STS-56. Atlas is part of NASA's Mission to Planet Earth program. STS-64 astronauts also will use the Shuttle Solar Backscatter Ultraviolet Instrument (SSBUV-A3) to measure ozone in Earth's atmosphere. SSBUVs were carried in U.S. shuttles before: SSBUV-1 in STS-34, SSBUV-2 STS-41, SSBUV-3 STS-43, SSBUV-A1 STS-45 and SSBUV-A2 STS-56. Also aboard STS-64 will be the Cryogenic Infrared Spectrometer Telescope for Atmosphere (Crista-Spas), a small U.S.-German aeronomy satellite to measure changes in Earth's atmosphere and complement the UARS satellite launched by STS-48. SPAS (Shuttle Pallet Satellite) is a German-designed payload carrier outfitted with science instruments, carried into space by shuttle, and deployed for several days in its own independent orbit to gather data. Before landing, the shuttle returns to SPAS, retrieves it, and returns it to Earth for another mission.

SPACELAB IML-2

Launch: scheduled May 1994   Shuttle: Columbia
Flight: STS-66   Duration: 13 days
Crew: commander, pilot, payload commander, mission specialists, payload specialist
Events: International Microgravity Laboratory (IML-2), second microgravity research in a long shuttle Spacelab module, studying effects on material processes and living organisms of very low gravity in space near Earth. With NASA in the project are European Space Agency (ESA), Canadian Space Agency (CSA), French National Center for Space Studies (CNES), West German Research and Development Institute for Air and Spacecraft (DLR), and National Space Development Agency of Japan (NASDA). IML Spacelabs are to fly at 17-mo. to 25-mo. intervals; separation allowing scientists to study results and apply them to new research. They will work with plants and animals, and protein, vapor and mercury iodide crystal growth experiments. Columbia to orbit with tail toward Earth in "gravity gradient" position offering least gravity during flight.

Spacehab

Spacehab is a private commercial pressurized man-tended spacecraft module to be flown to orbit in U.S. space shuttles. American spacecraft manufacturer McDonnell Douglas Corp. and Italy's Alitalia airline in 1984 set up a company, Spacehab Inc., to build a laboratory for experiments aboard U.S. shuttles.

Two Japanese companies, Mitsubishi Corp. and Mitsubishi Heavy Industries, bought ten percent of Spacehab Inc. in 1990. The Japanese firms Mitsubishi Trust and Banking Corp. and Shimizu Corp. were considering buying in. The Mitsubishi firms would focus Spacehab research on experiments for Japanese companies. Mitsubishi Heavy Industries would supply equipment for experiments.

Shuttle Atlantis flight STS-57 is scheduled to carry the first Spacehab to Earth orbit for eight days 160 miles above the equator in July 1993. The second Spacehab will fly for seven days at the same altitude in Atlantis STS-61 in December 1993. The third Spacehab will fly in Discovery STS-67 to the same altitude for seven days in June 1994.

Date	Module	Shuttle	Altitude	Inclination	Duration
1993 Jul	Spacehab 1	Atlantis	160 miles	28.5 degrees	8 days
1993 Dec	Spacehab 2	Atlantis	160 miles	28.5 degrees	7 days
1994 Jun	Spacehab 3	Discovery	160 miles	28.5 degrees	7 days
1994 Aug	Spacehab 4	tba	160 miles	28.5 degrees	7 days
1995 Mar	Spacehab 5	tba	160 miles	28.5 degrees	7 days
1995 Oct	Spacehab 6	tba	160 miles	28.5 degrees	7 days

| 1996 Apr | Spacehab 7 | tba | 160 miles | 28.5 degrees | 7 days |
| 1996 Nov | Spacehab 8 | tba | 160 miles | 28.5 degrees | 7 days |

U.S. Shuttle Spacewalks

A spacewalk, extravehicular activity or EVA, is a trip outside a space shuttle or other spacecraft. Below is a list of trips outside by astronauts during U.S. space shuttle flights.

Flight	Date	Spacewalkers	Reason
STS-6	1983 Apr	Peterson, Musgrave	spacewalk tests
STS-41B	1984 Feb	McCandless, Stewart	free-flight tests
STS-41C	1984 Apr	Van Hoften, Nelson	satellite repair
STS-41G	1984 Oct	Sullivan, Leetsma	refuel satellite
STS-51A	1984 Nov	Allen, Gardner	recover satellites
STS-51D	1985 Apr	Hoffman, Griggs	satellite rescue
STS-51I	1985 Aug	Van Hoften, Fisher	satellite repair
STS-61B	1985 Dec	Ross, Spring	construction
STS-37	1991 Apr	Ross, Apt	release GRO antenna
STS-49	scheduled 1992 May		repair Intelsat

Extraordinary Shuttle Landings

All U.S. shuttle flights have been launched from Kennedy Space Center, Florida. Before 1992, most landings were at Edwards Air Force Base, California, except for one at White Sands, New Mexico, and a handful at Kennedy Space Center, Florida.

Prior to flight STS-43 in August 1991, Edwards Air Force Base, California, had been NASA's first choice for shuttle landings. With STS-43, NASA switched Kennedy to primary landing site and Edwards to back up. White Sands also is backup.

These primary and back up landing sites are for flights which make it to orbit. NASA has emergency landing sites for flights which must abort during or shortly after blast off from Kennedy Space Center. Those are Kennedy Space Center; Edwards Air Force Base; White Sands; Hickam Air Force Base, Hawaii; Andersen Field, Guam; Easter Island; Ben Guerir, Morocco; Banjul, The Gambia; Dakar, Senegal; Zaragoza Air Base, Spain, and Moron Air Base, Spain.

First return to launch site. Challenger flight STS-7 was scheduled to make the first-ever shuttle landing at Kennedy Space Center, but bad weather forced a change to Edwards where it landed June 24, 1983. The first landing at Kennedy turned out to be Challenger, but flight STS-10 which landed February 11, 1984. Challenger STS-10 was the first-ever landing of a spacecraft at its launch site.

Columbia flight STS-24 was difficult to bring back to Earth. Bad weather at Edwards stopped a landing attempt on January 16, 1986. More bad weather at Edwards forced another wave-off January 17. The flight was extended a day so Columbia could land January 18 at Kennedy in Florida. However, bad weather at Kennedy brought another wave-off. The shuttle finally landed at Edwards January 18.

The list below shows locations and causes of unusual U.S. shuttle landings. KSC is Kennedy Space Center, Florida. Edwards is Edwards Air Force Base, California. White Sands is in New Mexico.

Flight	Shuttle	Landed	Location	Reason
STS-3	Columbia	1982 Mar 30	White Sands	California weather
STS-7	Challenger	1983 Jun 24	Edwards	Florida weather
STS-10	Challenger	1984 Feb 11	KSC	planned
STS-13	Challenger	1984 Oct 13	KSC	planned
STS-14	Discovery	1984 Nov 16	KSC	planned
STS-15	Discovery	1985 Jan 27	KSC	planned

STS-16	Discovery	April 19, 1985	KSC	planned
STS-24	Columbia	1986 Jan 18	Edwards	weather
STS-38	Atlantis	1990 Nov 20	KSC	California weather
STS-39	Discovery	1991 May 6	KSC	California weather
STS-43	Atlantis	1991 Aug 11	KSC	planned
STS-48	Discovery	1991 Sep 18	Edwards	Florida weather

SAREX: Shuttle Amateur Radio

Amateur radio is a great hobby for men and women in space. Some American astronauts, who also are licensed amateur radio operators, have taken their ham gear along on U.S. space shuttle flights. In fact, the entire five-person crew of shuttle Atlantis flight STS-37 in 1991 was licensed and on the air from outer space.

Most cosmonauts in recent years at the USSR's orbiting Mir space station have whiled away countless hours during six-month assignments in space by chatting with fellow amateurs on the ground via ham radio. American radio amateurs are planning a ham shack for recreation aboard the U.S.-international space station Freedom after 1997.

SAREX. In U.S. space shuttle lingo, ham radio is the Shuttle Amateur Radio Experiment (SAREX)—a joint effort of NASA, the American Radio Relay League (ARRL), and the Radio Amateurs Satellite Corporation (AMSAT).

Amateur radio clubs and individuals around the globe help out, especially the local ham groups at NASA's Johnson Space Center, Houston, Texas, and Goddard Space Flight Center, Greenbelt, Maryland. Cost of designing and building SAREX equipment for space shuttles, as well as amateur radio communications satellites, is borne by hams.

STS-9. U.S. astronaut Owen K. Garriott, whose amateur radio callsign is W5LFL, was the first to chat with his fellow hams on Earth while orbiting the globe. He flew to space aboard Columbia in flight STS-9 on November 28, 1983.

Garriott used a small handheld voice transceiver—a transmitter and receiver in one package. Transmissions to and from the spacecraft were in the FM (frequency modulation) mode on a frequency near 145 MHz in the two-meter ham band. The experiment showed that ham signals wouldn't disrupt other shuttle business.

Owen had been in space before. The first time was in 1973, when he spent two months aboard America's Skylab space station in Earth orbit. While at the space station, he took a six-hour spacewalk outside and watched the spiders Anita and Arabella spin webs in zero gravity inside. Unfortunately, he didn't have amateur equipment in Skylab.

Ten years later, STS-9 in 1983 was his second and final flight. Garriott spent 10 days in orbit in Columbia. Spacelab-1, in the cargo bay, was a removeable pressurized science lab carried to space and back in a shuttle's cargo hold. Astronauts floated through a tube between the crew cabin and Spacelab. Today, Garriott is retired from NASA.

STS-51F. The second ham in space was astronaut Anthony W. "Tony" England whose callsign is WØORE. (The number Ø is pronounced "zero.")

England made his only trip to space July 29, 1985, in shuttle Challenger flight STS-51F. He not only had amateur radio voice capability, but also took along ham TV gear to transmit pictures down from the spacecraft. The pictures were sent down as single frames, in a technique known as slow-scan television (SSTV).

More than 6,000 young people, taking part in England's SAREX through school and scouting clubs, were thrilled to see the first-ever amateur pictures beamed down from space. SSTV signals beamed up to WØORE from the Johnson Space Center ham station, W5RRR, were the first live TV pictures ever received aboard a shuttle.

Challenger's July 1985 ride to orbit was notable for the first engine failure during flight to space. Spacelab 2 was carried in the shuttle cargo bay, outfitted for astronomy research. England spent eight days in space. Challenger was the shuttle which exploded during liftoff just 183 days later.

Delays in NASA's shuttle and space station programs led Tony England to retire from the space agency in 1988 to teach electrical engineering at the University of Michigan, Ann Arbor. While with NASA in Houston, he had taught at Rice University. A geophysicist from North Dakota, England also did research at Michigan on satellite remote-sensing technology. England is known as a leading authority on remote sensing, having worked on the shuttle imaging radar (SIR) experiment.

STS-35. The 1986 shuttle Challenger disaster delayed plans for a third SAREX until Columbia flight STS-35 launched December 2, 1990. Payload specialist and civilian astronomer Ronald A. Parise, callsign WA4SIR, used upgraded SAREX-2 FM-voice and packet-radio gear on the two-meter band to contact amateurs on the ground.

Parise added the new packet radio technology to SAREX. The advanced digital communications technique known as packet radio is the latest computer-to-computer method employed by amateur radio operators. Parise also had ham TV gear, upgraded from SSTV to fast-scan television (FSTV). It could receive pictures from Earth, the first live fast-scan pictures received by a shuttle.

Parise, of Silver Spring, Maryland, and the flight's other payload specialist, civilian astronomer Samuel T. Durrance, were the first non-NASA astronauts to fly in a U.S. spacecraft in five years, since the Challenger disaster. With them in space were crew commander Vance D. Brand, pilot Guy S. Gardner, and mission specialists Jeffrey A. Hoffman, John M. "Mike" Lounge and A. Robert Parker.

Flying at an altitude of 190 nautical miles and an inclination of 28.5 degrees, the nine-day mission was the first shuttle flight in five years dedicated entirely to science. Four of the seven astronauts were astronomers. They used the Astro-1 collection of ultraviolet-light telescopes and the Broad Band X-Ray Telescope (BBXRT-1) to study quasars, binary stars, pulsars, black holes, galaxies and high-energy stars. NASA's Space Classroom became reality as Parise, Durrance, Hoffman and Parker conducted space-to-ground TV lessons and answered questions from pupils on the ground.

Parise plans to return to space to use SAREX gear again, in Columbia flight STS-69 in September 1994.

STS-37. Atlantis flight STS-37, launched April 5, 1991, was the first shuttle voyage in which the entire crew was composed entirely of licensed amateur radio operators. The five astronauts in space together for six days were crew commander Steven R. Nagel, pilot Kenneth D. Cameron, and mission specialists Jerome "Jay" Apt, Linda M. Godwin and Jerry L. Ross. Their ham radio call letters were: Cameron KB5AWP, Apt N5QWL, Godwin N5RAX, Nagel N5RAW, Ross N5SCW.

Cameron had upgraded the SAREX-2 radio gear to include fast-scan amateur television (ATV) on a ham radio ultra-high frequency (UHF) band, as well as FM voice, packet radio and SSTV on 2 meters. He reported good UHF fast-scan ATV reception from the ground, recording the TV pictures on a VCR in the shuttle while busy with other work. Most were black-and-white, but a few were color pictures. It was the first full-motion video ever received by any orbiting manned spacecraft. Tony England in 1985 in STS-51F had received slow-scan ATV from the ground. After STS-37 NASA was looking into fast-scan ATV as a cheap way to send information to orbiting astronauts.

While the astronaut slept, the packet radio amateur equipment was left on, to operate itself, making robot contacts with live hams on the ground.

As Atlantis revolved around the globe 243 nautical miles overhead, at an inclination of 28.5 degrees, the STS-37 astronauts used SAREX equipment to contact amateur operators on the ground and to chat from space with school children at nine sites in seven states with pupils listening in from 20 other schools.

As Atlantis passed overhead, students had 20 minutes to ask questions: what is blast off like, how hard is it to eat up there, can you see Oklahoma, and what do you think when you look down on Earth? "When you float up here above the Earth in a little spacecraft and look down at the big spacecraft that we all fly on, it makes us all think about what our priorities are," Cameron replied, while Nagel told the students to "take school seriously. Worry about your grades right now no matter how young you are. Study hard and get good grades, go to college, study science or engineering, then get into the space program after that."

Mir contact. The USSR's Mir space station was orbiting nearby in space and the STS-37 crew heard cosmonaut-hams on the air. The Mir amateur radio operators called across space to Atlantis, but when the astronauts were unable to be heard in reply, two-way contact was not established.

STS-37 was the first shuttle flight of 1991. It ferried to orbit Gamma Ray Observatory (GRO), the second of NASA's four Great Observatories. The 17-ton spacecraft was the largest science satellite ever carried by a shuttle. The 31-ft. GRO filled half of Atlantis' cargo bay. On April 7, Linda Godwin reached into the cargo bay with the 50-foot robot arm, grasped GRO, lifted the observatory out of the hold and dropped it overboard into its own orbit 280 statute miles above Earth. Ross and Apt stood by in spacesuits, prepared to spacewalk to unfold GRO's high-gain antenna or solar panels if needed—and an antenna boom did jam. A thermal blanket on the boom had become hung up. Apt and Ross spacewalked in the cargo bay April 8 to shake loose a jammed 16.5-ft. antenna boom. Later, they spacewalked again to test space station construction gear in the cargo hold. While outside STS-37, Apt punctured an outer layer of his spacesuit glove and scraped the skin of his hand until it bled lightly, but was able to return inside safely.

STS-45. Pilot Brian Duffy and mission specialist David C. Leestma were to use SAREX equipment to make amateur radio contact with ham operators on Earth during shuttle Atlantis flight STS-45 in May 1992. Busy Spacelab operations on the shuttle mission allows little extra room or electrical power, so Duffy and Leestma planned to use a battery-powered FM voice radio on the amateur two-meter band. Atlantis was to travel the same high-inclination orbit as those flown by Garriott and England, so Duffy and Leestma would pass above most of the world's population. They were to issue the general ham radio call of CQ to contact as many amateurs on the ground as possible. They also were to contact some students in schools.

Flying at an altitude of 160 nautical miles and an inclination of 57.0 degrees, the eight-day Spacelab mission is known as Atmospheric Laboratory for Applications and Science (Atlas-1). It will study atmospheric phenomena, energy from the Sun and changes in the solar spectrum. Part of NASA's Mission to Planet Earth program, Atlas-1 would be the first of several Atlas missions scheduled for orbit over a decade to study interaction between Earth's atmosphere and the Sun.

STS-45 astronauts also will use the Shuttle Solar Backscatter Ultraviolet Instrument (SSBUV-A1) to measure ozone in Earth's atmosphere.

Seven astronauts will be on the flight. Kathryn D. Sullivan is to be the first female payload commander. Crew commander will be Charles F. Bolden, Jr. Another mission specialist will be C. Michael Foale. Payload specialists will be Dirk D. Frimout and Byron K. Lichtenberg.

STS-47. Mission specialist Jerome "Jay" Apt, N5QWL, would use SAREX equipment to make ham radio contact with amateur operators on Earth during shuttle Endeavour flight STS-47 in September 1992.

Flying at an altitude of 160 nautical miles and an inclination of 57.0 degrees, the seven-day Spacelab-J mission will be a joint science research project of NASA and the Japanese National Space Development Agency (NASDA). Seven astronauts will be on the flight. Payload specialist Mamoru Mohri will be the first Japanese to fly in a U.S. spacecraft. Payload commander Mark C. Lee and mission specialist N. Jan Davis will be the first married couple to travel in space. Jemison will be the first black woman in orbit.

STS-55. Payload commander Jerry L. Ross, N5SCW, would use SAREX equipment to make ham radio contact with amateur operators on Earth during shuttle Columbia flight STS-55 in March 1993.

Flying at an altitude of 160 nautical miles and an inclination of 28.5 degrees, the nine-day Spacelab-D2 mission will be a joint science research project of NASA and Germany. Seven astronauts will be on the flight. Two payload specialists would be selected for flight STS-55 from among four Germans in training: female meteorologist Renate Luise Brummer; and physicists Hans-Wilhelm Schlegel; Gerhard Thiele; and Ulrich Walter.

STS-60. Payload commander Linda M. Godwin, N5RAX, would use SAREX to make contact with amateur operators on Earth during shuttle Endeavour flight STS-60 in October 1993.

Flying at an altitude of 130 nautical miles and an inclination of 57.0 degrees, the nine-day Space Radar Laboratory (SRL-1) mission would record radar images of Earth's surface, to be used to make maps, interpret geological features and locate natural resources. Six astronauts will be on the flight.

STS-69. Payload commander and civilian astronomer Ronald A. Parise, WA4SIR, is to return to space in September 1994. He again would use SAREX equipment to make ham radio contact with amateur operators on Earth during Columbia flight STS-69.

Flying at an altitude of 190 nautical miles and an inclination of 28.5 degrees, the nine-day Columbia mission is to carry Astro-2 telescopes to study quasars, pulsars, black holes, galaxies, high-energy stars. Parise used Astro-1 at same altitude and inclination in STS-35 in 1990. Seven astronauts will be on the flight.

A small satellite referred to as a free-flying platform and known as Office of Aeronautics, Exploration and Technology-Flyer (OAET-Flyer), to be deployed from Columbia STS-69, will contain technology experiments.

Ham radio. Amateur radio is the fraternity of licensed non-commercial communicators advancing the radio art and promoting experimentation. Hams are well known for their networks of stations which handle messages in disasters, emergencies and public-service events.

Amateurs have been pioneers in space communications over the last three decades with two dozen amateur radio satellites working in orbit, as well as years of other exotic communications such as the Earth-to-Moon-to-Earth signals known as Moonbounce, meteor scatter propagation, and studies of this planet's ionosphere.

Just four years after Sputnik launched the Space Age, a 10-lb. Orbital Satellite Carrying Amateur Radio, or OSCAR for short, was launched December 12, 1961, as ballast on a Thor-Agena rocket which carried a government satellite to orbit. The first satellite ever built by amateur radio operators launched ham radio into space.

Over the decades, thousands have used OSCAR's two dozen successors built by hams in Australia, Canada, France, Great Britain, Japan, the U.S., the USSR, Germany and elsewhere. Russian ham satellites have been known as RS for Radiosputnik.

Information on SAREX, ham satellites and amateur radio is available from the American Radio Relay League Inc., 225 Main Street, Newington, Connecticut 06111.

SAREX Amateur Radio Operators

Date	Flight	Shuttle	Astronaut	Role	Callsign
1983 Nov 28	STS-9	Columbia	Owen K. Garriott	mission specialist	W5LFL
1985 Jul 29	STS-51F	Challenger	A.W. "Tony" England	mission specialist	WØORE
1990 Dec 2	STS-35	Columbia	Ronald A. Parise	payload specialist	WA4SIR
1991 Apr 5	STS-37	Atlantis	Steven R. Nagel	commander	N5RAW
1991 Apr 5	STS-37	Atlantis	Kenneth D. Cameron	pilot	KB5AWP
1991 Apr 5	STS-37	Atlantis	Jerome "Jay" Apt	mission specialist	N5QWL
1991 Apr 5	STS-37	Atlantis	Linda M. Godwin	mission specialist	N5RAX
1991 Apr 5	STS-37	Atlantis	Jerry L. Ross	mission specialist	N5SCW
1992 May	STS-45	Atlantis	Brian Duffy	pilot	tba
1992 May	STS-45	Atlantis	David C. Leestma	mission specialist	tba

1992 Sep	STS-47	Endeavour	Jerome "Jay" Apt	mission specialist	N5QWL
1993 Mar	STS-55	Columbia	Jerry L. Ross	payload commander	N5SCW
1993 Oct	STS-60	Endeavour	Linda M. Godwin	payload commander	N5RAX
1994 Sep	STS-69	Columbia	Ronald A. Parise	payload commander	WA4SIR

Most Shuttles Are Delayed

Only nine U.S. shuttle launches have blasted off on time. The others were delayed.

False starts. Columbia flight STS-61C holds the record for number of false starts with six postponements for mechanical and weather problems starting December 18, 1985. It finally got off the ground on the seventh try January 12, 1986.

Other flights with many postponements included Challenger STS-51L and Atlantis STS-36 each with five delays; Discovery STS-41D and Columbia STS-32 with four; and Discovery STS-51D with three.

Time delays. Long times between scheduled date and actual launch have included the maiden flight of Challenger flight STS-6 delayed 74 days from January 20, 1983, to April 4, 1983, by an engine problem and a communications satellite payload damaged by a storm. Also, 66 days for Discovery STS-41D, 59 days Columbia STS-9, 34 days Columbia STS-2, 25 days Columbia STS-61C, 24 days Discovery STS-51D, 23 days Discovery STS-29, 22 days Columbia STS-32, and 17 days Challenger STS-51F. Massive schedule changes after the 1986 Challenger disaster forced longer delays on several flights.

Between flights. The longest time between flights was 975 days from the ill-fated Challenger flight STS-51L on January 28, 1986, to Discovery flight STS-26 which took off on September 29, 1988. Challenger STS-51L had suffered five postponements with mechanical, weather and scheduling problems. As it finally lifted off, a solid-fuel booster rocket exploded, killing all seven astronauts in the orbiter and sending Challenger to the bottom of the Atlantic Ocean.

On schedule. The nine U.S. shuttle flights which blasted off on schedule were:

Date	Shuttle	Flight
1982 Jun 27	Columbia	STS-4
1982 Nov 11	Columbia	STS-5
1983 Jun 18	Challenger	STS-7
1984 Apr 6	Challenger	STS-41C
1984 Oct 5	Challenger	STS-41G
1985 Jun 17	Discovery	STS-51G
1985 Oct 30	Challenger	STS-61A
1985 Nov 26	Atlantis	STS-61B
1989 Aug 8	Columbia	STS-28

The scheduled and actual launch dates, and days of delay, of 25 postponed shuttles through STS-36 were:

Year	Sched	Actual	Delay	Shuttle	Flight	Problem
1981	Apr 10	Apr 12	2 days	Columbia	STS-1	computer
1981	Oct 9	Nov 12	34 days	Columbia	STS-2	mechanical
1982	Mar 22	Mar 22	1 hour	Columbia	STS-3	fuel line heater
1983	Jan 20	Apr 4	74 days	Challenger	STS-6	maiden flight, engine problem then comm. sat payload damaged in storm
1983	Aug 30	Aug 30	17 min	Challenger	STS-8	weather
1983	Sep 30	Nov 28	59 days	Columbia	STS-9	solid booster nozzle
1984	Jan 29	Feb 3	5 days	Challenger	STS-41B	auxiliary power units
1984	Jun 25	Aug 30	66 days	Discovery	STS-41D	3 different mechanical delays then a fourth

1984	Nov 7	Nov 8	1 day	Discovery	STS-51A	launch site wind shear
1985	Jan 23	Jan 24	1 day	Discovery	STS-51C	weather
1985	Mar 19	Apr 12	24 days	Discovery	STS-51D	cargo bay doors, payload changes and a ship sailed into a restricted area of the Atlantic Ocean where solid-fuel boosters were to fall
1985	Apr 29	Apr 29	2 min	Challenger	STS-51B	launch-processing system
1985	Jul 12	Jul 29	17 days	Challenger	STS-51F	engine; also engine failure in flight to space caused an abort to a lower orbit
1985	Aug 24	Aug 27	3 days	Discovery	STS-51I	equipment and weather
1985	Oct 3	Oct 3	22 min	Atlantis	STS-51J	main engines
1985	Dec 18 '85	Jan 12 '86	25 days	Columbia	STS-61C	six delays; mechanical and weather
1986	Jan 22	Jan 28	6 days	Challenger	STS-51L	5 delays; mechanical, scheduling and and weather; exploded during liftoff
1988	Sep 29	Sep 29	1 hour	Discovery	STS-26	975 days after Challenger; then 1 hour delay to repair 2 space suits & high winds
1988	Dec 1	Dec 2	1 day	Atlantis	STS-27	winds and clouds
1989	Feb 18	Mar 13	23 days	Discovery	STS-29	2 delays; mechanical, then 1 hour for launch day fog
1989	Apr 28	May 4	6 days	Atlantis	STS-30	fuel pump
1989	Oct 12	Oct 18	6 days	Atlantis	STS-34	2 delays; engine and bad weather
1989	Nov 20	Nov 22	2 days	Discovery	STS-33	hydraulics, electrical wiring
1989	Dec 18 '89	Jan 9 '90	22 days	Columbia	STS-32	4 delays; launch pad, holidays, weather
1990	Feb 22	Feb 28	6 days	Atlantis	STS-36	5 delays; sick astronaut, tracking computer, clouds and winds

Build Another Space Shuttle?

Now that Endeavour has replaced Challenger, cash-starved NASA is not likely to build another large STS space shuttle orbiter unless another emergency arises.

The aging Columbia, Discovery and Atlantis, and the new Endeavour, comprise a fleet of four reusable shuttles. Rather than build a fifth, NASA is looking into a new class of large unmanned rockets to ferry heavy payloads to Earth orbit.

There had been claims in 1989 of a 50-50 chance of a second space shuttle accident if NASA pursued an aggressive launch schedule. That led one congressional report by the Office of Technology Assessment to call for construction of another orbiter to augment the U.S. fleet. But NASA has slowed the launch schedule and taken pains to prevent destruction of another orbiter.

Office of Technology Assessment said shuttle reliability is 97 to 99 percent. At 98 percent, there would be a 50 percent probability of losing one orbiter every three years assuming a launch rate of 11 per year. NASA, on the other hand, said the odds of losing another shuttle were between 1-in-80 and 1-in-800.

Columbia is the oldest orbiter, having made its maiden voyage April 12, 1981. Endeavour was expected to make its maiden flight in May 1992. Endeavour was built with lower weight, improved design and increased safety, including an advanced escape system and better protective heat-shield tiles. An orbiter takes four to five years to build.

HL-20: Small U.S. Space Shuttle

NASA is considering a new small space shuttle to rotate crews in its new space station around the turn of the century. Like a commuter car-pool van, the HL-20 Personnel Launch System (PLS) would carry a crew of two, eight passengers and a small amount of priority cargo to and from low Earth orbit.

NASA would fly a fleet of three to four of the small astrobuses in and out of Kennedy Space Center, Florida. Besides ferrying crew members to and from the U.S.-international space station Freedom, HL-20s also could rescue astronauts and cosmonauts stranded in space. They could fly up to inspect, service, re-fuel and repair satellites in orbit. And they could provide an express delivery service for high-priority cargo.

Ready for flight. Before flight, a horizontal HL-20 would be processed in a vehicle processing facility at Kennedy Space Center. Its expendable rocket would be processed vertically in a separate facility. Later, at the launch pad, the HL-20 would be mounted on the rocket.

The spaceplane would be launched as the orbiting space station passed overhead. The spaceplane would fly up to orbit at an altitude of 100 nautical miles. From there, it would chase the space station and fly on up to the station's altitude of 220 nautical miles.

The HL-20 would rendezvous with the space station and enter a dock. Fresh astronauts would go from the HL-20 into Freedom station. Retiring astronauts would enter the HL-20 and immediately fly down to Earth where the HL-20 would land like an airplane on a runway. Turn-around time for the HL-20 up-and-down flight would be less than 72 hours.

Size. An HL-20 would be only 29 feet long, with a wingspan of 24 feet. That's about one-quarter the size of a big U.S. shuttle orbiter. It would be a lightweight people mover, weighing little more than ten percent of an STS orbiter. HL-20 would weigh 22,000 lbs. without crew; an STS shuttle orbiter weighs 185,000 lbs. empty.

An HL-20, with wings folded, could be carried inside the payload bay of an STS shuttle. Crew and passenger room in an HL-20 would be less than in an STS shuttle, but more than in a small corporate business jet.

Shape. Like large STS shuttles, the HL-20 would be a lifting body. Before NASA designed its big space shuttles, several lifting-body research craft were flown by test pilots in 1966-1975, including M2-F2, M2-F3, HL-10, X-24A, and X-24B. HL-20 would be an evolution from those early shapes.

A lifting-body spacecraft has advantages over the bell-shaped and round capsules used for early manned flights. The greater lift while flying down through the atmosphere allows a lifting body to reach more land area, increasing its landing opportunities. Gravity loads are reduced to about 1.5g during entry, protecting sick or injured space station crews returning to Earth. Landings on wheels on a runway offer more flexibility than capsules which float down, uncontrolled, on parachutes. Airplane-style landings are possible many places around the globe, including the Kennedy Space Center launch site.

Launcher. HL-20 would have small thrusters for maneuvering in space, but no main engines. Instead, it would be boosted to orbit by a large expendable rocket, or carried to space inside an STS shuttle cargo hold.

America's existing Titan IV could be modified to loft HL-20s. The big Advanced Launch System (ALS) rocket under study by NASA and the Air Force could lob HL-20s to space. Or new liquid-fuel booster rockets, being considered to replace the STS shuttle's solid-fuel boosters, might be rigged to shoot HL-20s to orbit.

One launch method under consideration would use a large, supersonic airplane to ferry an HL-20 part way to space. At an altitude around 40 miles, the plane would drop the HL-20 and a small expendable booster would rocket the spaceplane on up to orbit.

Mock up. A full-size HL-20 engineering mock-up has been constructed by students and faculty of North Carolina A&T University and North Carolina State University. The model is under study at NASA's Langley Research Center, Hampton, Virginia, and Johnson Space Center, Houston, Texas.

If NASA receives a go-ahead, a test dummy HL-20 could be launched on a Titan-3 rocket by 1995 and manned HL-20s could be zipping to and from space by the end of the 1990's. HL-20 feasibility is being examined at the famed Skunk Works—the Lockheed Advanced Development Company, Burbank, California.

HL-20 would be the first new U.S. manned spacecraft since the Space Transportation System (STS) shuttles designed in the 1970's. HL-20 would complement, but not compete with larger STS space shuttles.

Tit for tat. In an odd twist, the HL-20 looks like a small spaceplane tested by the USSR in 1982-84. After chuckling about similarities between the new USSR space shuttle Buran and older U.S. STS shuttles, NASA now seems to have borrowed from the Soviet small shuttle design.

Shuttle-C For Cargo

Taking a leaf from the Russian notebook, NASA has considered bolting an unmanned cargo pod to the fuel tank and solid rocket motors of the space shuttle system to lift 75 tons to orbit.

The USSR's huge Energia space rocket comes with interchangeable cargo pod and manned shuttle. An Energia with cargo pod flew to space in 1987. An Energia with shuttle flew to space in 1988.

On the launch pad, Shuttle-C would look like the familiar manned shuttles, but without wings. Close up, Shuttle-C would have no crew cabin. For some flights, it would have two, rather than three, main engines.

NASA's copy. A plan studied by NASA would remove the manned orbiter and attach a cargo pod. The booster would be called Shuttle-C. Such an unmanned Shuttle-C could carry 50 to 85 tons of cargo to low-Earth orbit.

That would compare with the 110 tons of cargo Energia can ferry to orbit. Energia carries about three times the payload of a manned U.S. shuttle.

The U.S. Saturn V rocket, which launched men to the Moon in 1969 and hasn't been used since it orbited Skylab in 1973, was able to carry 100 tons. Saturn V boosters no longer are built.

Shuttle-C could carry payloads ranging from 100,000 lbs. using two main shuttle engines to 170,000 lbs. using three main engines and a new advanced solid rocket motor (ASRM) NASA is developing.

Existing manned space shuttles are heavy-lift launch vehicles carrying more than 230,000 lbs. to low Earth orbit. However, most of the weight is crew compartment, wings, heat shield and other components to return the crew and cargo. The actual payload a shuttle can deliver to orbit is 39,500 lbs. to orbit 220 nautical miles high. Proponents of Shuttle-C say many shuttle payloads don't require a crew.

Mix 'n match. Shuttle-C would use engines, solid-fuel booster rockets, external fuel tank and launch facilities of the manned shuttle. The manned orbiter, usually atop the fuel tank sandwiched between solid-fuel boosters, would be replaced by a cargo pod.

Until the 1986 Challenger disaster, NASA had intended to use only manned shuttles to carry space station parts to orbit. Since Challenger, money for station, rockets, shuttles and flights has been in short supply.

NASA needs to find inexpensive ways to get pieces of its space station to orbit. Shuttle-C could carry station parts to space for assembly. If Shuttle-C were developed, NASA still could fly manned shuttles.

It was reported in 1990 a RAND Corporation study for the U.S. Air Force found the unmanned Shuttle-C heavy-lift launcher too expensive for routine transportation.

Big expendable booster. A non-reusable rocket capable of carrying heavy loads—the Advanced Launch System (ALS)—also has being designed by the Air Force and NASA.

Reusing Shuttle External Tanks

NASA would like to find some use for those jumbo fuel tanks which fly to space with a space shuttle, then are jettisoned to burn in the atmosphere.

The so-called external tank is a 33-ton hollow aluminum bullet, as tall as a 15-story building, which carries the liquid fuel burned in the shuttle's main engines. The tank is destroyed during re-entry.

It would not be hard to keep an emptied external tank in orbit after a flight, but what to do with it? NASA contracted in 1990 with Global Outpost, Inc., Alexandria, Virginia, to figure out an answer.

Once an external fuel tank arrived in a low Earth orbit, Global Outpost would use it as a platform for commercial processing and manufacturing experiments in the extremely low gravity of space. Growing semiconductor crystals, manufacturing pharmaceutical drugs and other microgravity work could be done. Rather than clean out a depleted fuel tank in orbit, experiments would be attached in space to the outer surface of the tank.

How To See Shuttles Land

NASA said in 1991 most shuttles in the future would land at Kennedy Space Center, Florida. Some shuttle flights, however, will end as most did before 1991—at Edwards Air Force Base, California, where NASA has its Ames-Dryden Flight Research Facility.

The public can see landings easily from Edwards' East Shore Viewing Site which offers an unobstructed view of landings. The site opens 24 hours prior to a scheduled landing. Vehicle passes are not required.

Parking is on unprepared surfaces, but water, restrooms, food vendors and souvenir stands are available. Listen for local news reports about shuttle flights to determine exactly when a shuttle is expected to land. Landing updates also are available by telephone at (805) 258-3520.

Access to the site, by way of secondary roads which can be congested, is cut off one hour before a landing. There are two routes to the East Shore Viewing Site:

★From Los Angeles—travel north on Highway 14, the Antelope Valley Freeway. Turn right, which is east, on the Avenue F off-ramp. Turn left and travel north on Sierra Highway to Avenue E. Turn right and go east on Avenue E to 140th Street. Turn left and travel north on 140th Street to Avenue B. Turn right and follow Avenue B east as it curves into Mercury Blvd which leads to the viewing area.

★From Highway 58—follow the Rocket Site Road off-ramp to Mercury Blvd. which leads to the viewing area.

NASA Acronyms Defined

The National Aeronautics and Space Administration (NASA) uses hundreds of acronyms to refer to shuttle flights, expendable launch vehicle (ELV) unmanned rocket flights and payloads.

2 PHASE FLOW A demonstration of a High Efficiency Thermal Interface (HETI) in an Integrated Thermal Control System.

AAFE Aeroassist Flight Experiment. An experimental vehicle that simulates the atmospheric flight phase of an Aeroassisted Orbital Transfer Vehicle (OATV) returning from geosynchronous orbit. Provides environmental and design data for an AOTV.

AC Atlas-Centaur. An intermediate Class Expendable Launch Vehicle.

ACE Advanced Composition Explorer. A free flying scientific spacecraft that may be solar, celestial or Earth pointing.

ACES Acoustic Containerless Experiment System. A technical demonstration to obtain early microgravity tests of gas transport phenomena in a 3-axis levitation furnace.

ACTS Advanced Communications Technology Satellite. Flight verification of high risk communications technology to support future satellite communications systems.

AD Animal Development-Genetics. A series of experiments to determine effects of weightlessness on animal genetics.

ADSF Automatic Directional Solidification Furnace. A technology demonstration of directional solidification of magnetic materials, immiscibles, and IR detection materials.

AF Polar Bear Air Force Polar Bear. A satellite in polar orbit to study atmospheric effects on electromagnetic propagation.

AFE American Flight Echocardiograph. Collects quantitative in-flight data on cardiovascular changes in the crew.

AFITV Air Force Instrumented Test Vehicle. An anti-satellite target vehicle.

AFP-675 Air Force Program-675. Collects infrared data to support Strategic Defense Initiative program. Formerly, Cryogenic Infrared Radiance Instrument for Shuttle (CIRRIS).

ALT Altitude. Orbit altitude in nautical miles.

AM Auxiliary Module. Provides consumables resupply, payload changeout and additional on-orbit volume for the ISF Facility Module (FM).

AMOS Air Force Maui Optical Station. Technology development/geophysical environment study. Calibrate AMOS ground-based electro-optical sensors and study on-orbit plume phenomenology using the Shuttle as a test object.

AMPTE Active Magnetosphere Particle Tracer Experiment. Satellite to study transfer of mass from the solar wind to the magnetosphere.

ANS Astronomical Netherlands Satellite. Study the sky in ultraviolet and X-ray from above the atmosphere.

APCG Advanced Protein Crystal Growth. Second generation flight system for protein crystal growth in a microgravity environment.

APE Aurora Photography Experiment. Enhance understanding of the geographic extent and dynamics of the aurora.

APM Ascent Particle Monitor. Collects particulate materials from the Orbiter during ascent, using an automated mechanical/electrical assembly.

ARABSAT Arab Satellite. Communications satellite of the Arab Satellite Communications Organization.

ARC Aggregation of Red Cells. Studies aggregation of red cells and blood viscosity under low-g conditions.

ARF Aquatic Research Facility. Houses a variety of small aquatic specimens for research on microgravity adaptation.

ASC American Satellite Company. Satellite to provide commercial communication service to continental United States, Hawaii, Alaska, and Puerto Rico.

ASEM Assembly of Station by Extravehicular Activity Methods. One of a series of experiments designed to support SSF development by demonstrating strut handling and EVA translation techniques.

ASP Attitude Sensor Package. Foreign Reimbursable Hitchhiker-G payload.

ASTRO Astronomy. Program designed to obtain ultraviolet (UV) data on astronomical objects using a UV telescope.

ASTRO-SPAS Astronomy experiment on SPAS (Shuttle Pallet Satellite), the European cross-bay carrier.

ATDRS Advanced Tracking and Data Relay Satellite. Next generation of NASA tracking, data and communications satellites.

ATLAS Atmospheric Laboratory for Applications and Science. Measures long term variability in the total energy radiated by the Sun and determines the variability in the solar spectrum.

ATLAS I Commercial and DOD intermediate class expendable launch vehicles.

ATLAS II Commercial and DOD intermediate class expendable launch vehicles.

ATLAS-E DOD medium class expendable launch vehicle.

AUSSAT Australian Communication Satellite. Direct broadcast communication satellite which provides services to continental Australia and offshore territories.

AXAF A major free flying X-Ray observatory using a high resolution telescope. Designed to operate in orbit for 15 years. Advanced X-Ray Astrophysics Facility.

AXAF-R Advanced X-Ray Astrophysics Facility-Retrieval. Retrieval of AXAF observatory.

B/U Back-up.

BATTERY Characterization of Sodium Sulfur battery performance in microgravity. Sodium Sulfur Battery Flight Experiment.

BBXRT Broad Band X-Ray Telescope. Provides high resolution X-ray spectra for both point and extended sources, including stellar coronae, X-ray, binaries, active agalactic nuclei, and clusters of galaxies.

BIMDA Bioprocessing With the Materials Dispersion Apparatus. A wide range of tests focused on the assembly of macromolecules. Uses a middeck thermal enclosure.

C360 Cinema 360. A 35mm motion picture camera for the purpose of photographing crew and mission activities.

CANEX Canadian Experiments. A group of Canadian experiments conducted aboard STS-13 (STS-41G) by a Canadian Payload Specialist.

CANEX-2 Canadian Experiments-2. Tests of Canadian developed real-time machine vision system (SVS) using an RMS deployed target (CTA), experiments in material exposure, spacecraft glow, phase partitioning, metal diffusion and space adaptation tests.

CAPL Capillary Pump Loop Experiment. Experiment to quantify behavior of a full-scale capillary pumped loop heat transfer system in microgravity.

CASSINI Saturn Orbiter and Titan Probe complements CRAF mission. Primary objective is comprehensive study of Saturn, its rings and moons.

CBDE Carbonated Beverage Dispenser Evaluation. Pepsico, Inc. experiment to evaluate packaging and dispensing techniques for space flight consumption of carbonated beverages.

CCAP Commercial Complex Autonomous Payload. Commercial secondary payload utilizing small self-contained payload hardware.

CDR Commander. Member of the Shuttle flight crew in command of the flight.

CENTAUR Upper stage system for Atlas and Titan ELVs.

CETA Crew and Equipment Translation Aid. Multi-purpose crew system that provides rapid return to the Shuttle airlock in case of emergency, allows efficient translation, and carries equipment.

CFES Continuous Flow Electrophoresis System. Demonstrates the technology of pharmaceutical processing in space.

CGBA Commercial Generic Bioprocessing Apparatus. Develops advanced systems for and investigations in bioprocessing of materials.

CHAMP Comet Halley Active Monitoring Program. Observes Comet Halley on STS flights.

CHROMEX Chromosomes Experiment. Investigation of the effects of space flight on plant tissue growth.

CLOUDS Cloud Logic to Optimize Use of Defense Systems. Hand-held 35 mm photography for observations of cloud formation, dissipation, and opaqueness.

CM-X Commercial Middeck Payload. Commercial development middeck payload (X denotes approximate number of lockers).

CMSE Extended Duration Space Environment Candidate Materials Exposure. Evaluation of candidate composite materials for space structures for degradation due to exposure in low Earth orbit (passive/active systems).

CMSE/E Candidate Materials Space Exposure (CMSE) Evaluation of Oxygen Interaction With Materials-III (EOIM-03). Evaluation of candidate composite materials for space structures for degradation due to exposure in low Earth orbit with EOIM-III for baseline data correlation.

CNCR Characterization of Neurospora Circadian Rhythms in Space. Microgravity effects on circadian rhythms of neurospora.

COBE Cosmic Background Explorer. An Earth satellite to determine the spectrum anistropy of cosmic microwave background.

COMSTAR Communications satellite for COMSAT.

CONCAP Consortium for Materials Development in Space (Complex Autonomous Payload). Investigates reactions occurring on the surface of materials when exposed to the atomic oxygen flow in Earth orbit on high temperature super-conducting films and on materials degredation/reaction samples.

CONE Cryogenic Orbital Nitrogen Experiment. A collection of cryogenic fluid technology experiments using nitrogen as the cryogen.

CRAF Comet Rendevous Asteroid Fly-by. An interplanetary spacecraft to explore an asteroid and a comet to gather new information on the origin and evolution of the solar system, prebiotic chemical evolution and the origin of life, and astrophysical plasma dynamics and processes.

CREAM Cosmic Radiation Effects and Activation Monitor. Uses an active cosmic ray monitor and seven passive packages to record on-orbit cosmic ray environments.

CRISTA-SPAS Cryogenic Infrared Spectrometer Telescope for Atmosphere. A U.S./German Joint Aeronomy Payload intended to explore the variability of the atmosphere and to provide measurements that will complement those provided by UARS. On SPAS (Shuttle Pallet Satellite), the European cross-bay carrier.

CRUX Cosmic Rays Upset Experiment. Studies on-orbit cosmic ray environments and monitors upsets on microcircuit devices.

CRW Crew. The Shuttle flight crew for a particular mission.

CRYO-HP Cryogenic Heat Pipe. GAS canister payload using liquid oxygen as the heat pipe working fluid and may be flown as a Hitchhiker.

CRYSP Crystal Sample Package. A series of experiments to determine the effects of the complex radiation environment of space on the performance characteristics of advanced materials.

CSA Canadian Space Agency.

CTA Canadian Target Assembly. A deployable target used for test of Canadian experimental space vision system (SVS) in CANEX-2.

CVTE Crystals By Vapor Transport Experiment. Investigate application of chemical vapor transport crystal growth process to materials of practical value in semiconductor and electro-optical device.

CXM Commercial Cross-bay Carrier (MSL). Commercial development cross-cargo bay payload using Materials Science Laboratory (MSL) class systems or equivalents.

DAD Dual Air Density. Measures global density of upper atmosphere and lower exosphere.

DCWS Debris Collision Warning System. Provides the capability for sensing space debris in the 1 to 10 mm size and determines albedo and spectral characteristics of a large sample of low Earth orbit debris.

DEE Dexterous End Effector. Demonstrates a sensor for the Shuttle RMS which will allow for more precise control.

DELTA Medium class expendable launch vehicle.

DFI PLT Development Flight Instrumentation Pallet. A pallet used to accommodate the DFI used on the first four Shuttle flights.

DLR Deutsche Forschungsanstalt fur Luft-und Raumfahrt. Federal German aerospace research establishment.

DMOS Diffusive Mixing of Organic Solutions. Grow crystals of organic compounds for research programs for the 3M Corporation's Science Research Laboratory.

DOD Department of Defense.

DOD M88-01 Department of Defense M88-01. Evaluates the capability of man in space to enhance air, naval, and ground force operations and assesses the feasibility of observations of space debris while in orbit.

DS Docking System. Docking system for use in assembly and servicing of the ISF.

DSCT Directional Solidification of Cadmium Telluride. Cadmium telluride will be grown using the directional solidification technique.

DSP Defense Support Program. Geosynchronous DOD satellite.

DUR Duration. Mission duration of a Shuttle flight.

DXS Diffuse X-ray Spectrometer. Shuttle experiment to conduct spectral observations of the diffuse galactic soft X-ray background to determine the ionic, elemental abundances and the plasma temperature of the hot phase of the interstellar medium.

EASE/ACCESS Experimental Assembly of Structures in EVA/Assembly Concept for Construction of Erectable Space Structures. Obtains human factors data during assembly of structures in space during Extra Vehicular Activity.

EDO Extended Duration Orbiter. Kit added to Orbiter to extend energy resources to support mission duration up to sixteen days.

EEVT Electrophoresis Equipment Verification Test. Technology demonstration of apparatus to evaluate the effects of electrophoresis on biological cells in zero-g.

ELECTROLYSIS Electrolysis Performance Improvement Experiment. Investigation of techniques for improving electrolysis technology in microgravity.

ELRAD Earth-Limb Radiance Equipment. Obtain measurements of Earth-limb radiance for various positions of the Sun from near limb up to 9 degrees below Earth horizon.

ELV Expendable Launch Vehicle.

EO Escape Orbit.

EOIM Evaluation of Oxygen Interaction with Materials. Determines effects of atomic oxygen degradation on 1100 candidate materials.

EOS Earth Observing System. A complement of polar orbiting satellites conducting Earth science observations.

EP Electric Propulsion. Evaluates performance of an arc jet electric thruster.

EQUATOR-S German cooperative, part of the International Solar Terrestrial Physics (ISTP) program, will study the ring current and Near Earth Plasma Sheet in the equational region.

ERBS Earth Radiation Budget Satellite. Collects global Earth radiation budget data.

ESA European Space Agency.

ESA-x European Space Agency-x. Foreign Reimbursable Hitchhiker-G payload sponsored by the ESA.

ESMC Eastern Space and Missile Center. USAF organization headquartered at Patrick AFB, Florida.

EURECA European Retrievable Carrier. Platform placed in orbit for six months offering conventional services to experimenters.

EUVE Extreme Ultraviolet Explorer. Produces definitive sky map and catalog of extreme ultraviolet portion of electromagnetic spectrum (100-1000 angstroms).

EUVE RETR Extreme Ultraviolet Explorer-Retrieval. RETR Mission to perform on-orbit retrieval of the EUVE payload from the Explorer platform.

EXOSAT ESA X-Ray Satellite. Provides continuous observations of X-ray sources.

FARE Fluid Acquisition and Resupply Experiment. Obtains data to evaluate fluid dynamics associated with capillary liquid acquisition devices.

FAST Fast Auroral Snapshot Explorer. Spacecraft to investigate the processes operating within the auroral region.

FDE Fluid Dynamics Experiment. A package of six experiments flown in the middeck that involve simulating the behavior of liquid propellants in low gravity.

FEA Fluids Experiment Assembly. Investigate floating zone crystal growth processing investigations on selected semi-conductor materials.

FEE French Echocardiograph Equipment. Obtains on-orbit cardiovascular system data.

FEL First Element Launch. Initial launch of components for the space station Freedom manned base (SSF/MB).

FLOATZONE Float Zone Crystal of Cadmium Telluride. Microgravity materials processing experiment to demonstrate feasibility of producing large, high quality, single-crystal specimens of cadmium telluride.

FLT Flight. The flight sequence number for Shuttle missions.

FLT OPPTY Flight Opportunity. A planned Shuttle flight without assigned payloads.

FLTSATCOM Fleet Communication Satellite. U.S. Navy communications satellite.

FM Facility Module. A man-tended module in support of ISF providing space for middeck locker inserts and common racks for payload accommodations.

FPE French Postural Experiment. Studies sensory-motor adaptations in weightlessness.

FROZEPIPE Frozen Startup of a Heat Pipe in Microgravity. Examines the thermal behavior of a high capacity heat pipe system in microgravity.

FSA FSS Servicing Aid. Electronics module to support on-orbit servicing.

FSS Flight Support System. Support system used for revisit missions.

FTS-DTF Flight Telerobotic Servicer Demonstration Test Flight. Technology development test flight of a telerobotic manipulator system that will provide a new capability and flight qualified hardware to support NASA's future missions.

FUSE Far Ultraviolet Spectroscopy Explorer. An ultraviolet astronomy satellite

GALAXY Hughes communications satellite.

GALILEO Investigates the chemical composition and physical state of Jupiter's atmosphere and satellites.

GAS Get Away Special. Alternate name for the Small Self-contained Payload (SSCP) program, providing standard canisters (GAS CANs) to accommodate low-cost space experimentation.

GAS BRIDGE Getaway Special Bridge. A structure in the payload bay that can hold up to twelve GAS canisters.

GAS CAN GAS Canister.

GAS TEST Test instrumentation to verify ability of the GAS hardware to function properly in flight.

GE General Electric American Communications, Inc.

GEOTAIL NASA-Japan cooperative mission to explore Geotail of the Earth Plasma Physics.

GHCD Growth Hormone Concentration & Distribution in Plants. Microgravity effects on growth hormone distribution of various plant life.

GLO Shuttle Glow. A Hitchhiker payload to measure optical emissions observed on the surface of spacecraft and Shuttle.

GLOMR Global Low Orbit Message Relay. A packet radio data relay satellite.

GLOW Experimental Investigation of Spacecraft GLOW. Determination of the spectral content of luminosity near Shuttle surfaces, to assess influence on optical experiments.

GOES Geostationary Operational Environmental Satellite. NOAA weather satellites.

GOSAMR Gelation of Sols: Applied Microgravity Research. Investigates gelation of multicomponent colloidal solutions and suspensions (SOL).

GP Gravity Probe. Scientific probe to test Einstein's Theory of Relativity.

GRO Gamma Ray Observatory. Investigates extraterrestrial gamma-ray sources.

GSO Geosynchronous Orbit.

GTO Geosynchronous Transfer Orbit.

HCMM Heat Capacity Mapping Mission. Produces thermal maps for discriminating of rock types, mineral resources, plant temperatures, soil moisture, snow fields, and water runoff.

HE High Eccentricity Orbit.

HEAO High Energy Astronomical Observatory. Satellite to study energetic radiation from space.

HEE Human Energy Expenditure. Human experiment to measure energy expenditure in space.

HEEO Highly Elliptical Equatorial Orbit.

HELIO Heliocentric.

HERCULES This experiment upgrades/expands the Latitude/Longitude Locator (L3) experiment using a charge coupled device with inertial reference gyros. The objective is to locate Earth surface sites within 1 nautical mile.

HETE High Energy Transient Experiment. Spacecraft to study gamma ray burst sources and source locations, and X-ray burst sources and source locations.

HH-G Hitchhiker-Goddard. Shuttle cargo bay sidewall mounted carrier for small experiments.

HH-G1 Hitchhiker-Goddard. Demonstration flight of Hitchhiker-G hardware.

HH-M Hitchhiker-Marshall. Shuttle cargo bay across-bay carrier for small experiments.

HILAT High Latitude. Evaluate propagation effects of disturbed plasmas on radar and communications systems.

HME Handheld Microgravity Experiment. Provides for middeck experiments of limited scope in order to allow for low-cost, timely testing of concepts or procedures, or the early acquisition of data.

HPCG Handheld Protein Crystal Growth. Develops techniques to produce in microgravity protein crystals of sufficient size and quality to permit molecular analysis by diffraction techniques.

HPP Heat Pipe Performance & Working Fluid Behavior in Micro-gravity. Environment experiment to study the microgravity effects of working fluids used in heat pipes.

HPTE High Precision Tracking Experiment. Demonstrates ability to propagate a low power laser beam through the atmosphere.

HRSGS-A High Resolution Shuttle Glow Spectroscopy-A. Obtains high resolution spectra, in the visible and near visible wavelength range of the Shuttle surface glow as observed on the vertical tail of the Orbiter in LEO.

HS-376 RET HS-376 Retrieval. Salvage of HS-376 communication satellites launched on the tenth Shuttle mission.

HST Hubble Space Telescope. Observes the universe to gain information about its origin, evolution and disposition of stars, galaxies, etc.

HST REV Hubble Space Telescope Revisit. Revisit mission to the Hubble Space Telescope to replace science instruments or other orbital replacement units (ORU's).

IBIS Instrument for Biological Investigations in Space. Cell and tissue culture system to investigate effects of micro-g on the function of a variety of cells.

IBSE Initial Blood Storage Equipment. Evaluates changes in blood tissue during various storage conditions.

IBSS Infrared Background Signature Survey. Obtains infrared measurements on rocket plumes, shortwave infrared Earth-limb, Shuttle environment, and chemical release from the payload bay while detached in proximity to the Orbiter.

ICBC IMAX Cargo Bay Camera.

IECM Induced Environment Contamination Monitor. A package of ten instruments designed to fly in the Orbiter payload bay on a special pallet to check for contamination in and around the Orbiter. It also has the capability to be operated on the end of the RMS outside of the payload bay.

IEF Isoelectric Focussing Experiment. Gathers experimental data on the extent of electroosmosis in space.

IEH International Extreme-UV Far-UV Hitchhiker. Hitchhiker experiment to study ultraviolet emissions.

IG Igloo. Structure which provides a pressurized and thermally controlled environment for Spacelab pallet subsystems.

IMAX IMAX Systems Corp., Toronto, Ontario, Canada. A large screen motion picture format used by the NASA/Smithsonian project documenting significant space activities.

IML International Microgravity Laboratory. Series of microgravity missions devoted to material and life sciences studies.

INCL Inclination. Orbit inclination in degrees.

INSAT Indian Satellite. Communication and meteorological satellite for the government of India.

INTELSAT-VI-R INTELSAT-VI-Reboost. The retrieval, repair and deployment of a communications satellite for the International Telecommunication Satellite organization.

IOCM Interim Operational Contamination Monitor. Measures molecular and particulate contamination in the cargo bay from prelaunch to post-landing.

IPMP Investigation into Polymer Membranes Processing. Investigate low-G environment effects on industrial processing techniques for developing polymer membranes.

IR-IE Infrared Imaging Equipment. Infrared video camera used to measure temperature gradients on the Orbiter surface.

IRAS Infrared Astronomical Satellite. All sky survey for objects that emit infrared radiation.

IRCFE Infrared Communications Flight Experiment. Demonstrates the feasibility of using diffuse infrared light as a carrier for Shuttle crew communications.

IRIS Italian Research Interim Stage. Italian upper stage for use on the Shuttle.

IRT Integrated Rendezvous Radar Target. A target for testing of Shuttle Orbiter rendezvous techniques and capabilities in orbit.

ISAC INTELSAT Solar Array Coupons. Studies atomic oxygen effects on materials (silver and zinc sulphite).

ISAIAH Israeli Space Agency Investigation About Hornets. Gravity perceptions by hornets and their reactions to changes in gravity.

ISAS Institute of Space and Astronautical Science. The Institute of Space and Astronautical Science of Japan.

ISC International Space Corporation. Commercial joint endeavor activity.

ISEM ITA Standardized Experiment. Cross-bay structure for accommodating multiple material processing experiments and other investigations.

ISF Industrial Space Facility. Commercially-owned, man-tended orbiting facility for research and manufacturing activities.

ISPM International Solar Polar Mission (now known as ULYSSES).

ITV Instrumented Test Vehicle. A target for Anti Satellite.

IUS Inertial Upper Stage. Upper stage system for Shuttle and Titan.

IUTE Industry University Technology Experiment. Series of Space Technology experiments for U.S. industry and universities to be flown on the Shuttle or ELV.

JOINT DAMPING Measurement of the damping behavior of liquids in a variety of rotating tanks.

L3 Latitude/Longitude Locator. Tests the capability of a space sextant/camera system to locate Earth surface targets within 10 nautical miles.

LAGEOS Laser Geodynamics Satellite. Spherical passive satellite covered with retroflectors which are illuminated by ground-based lasers to determine precise measurements of the Earth's crustal movements.

LDCE Limited Duration Space Environment Candidate Materials Exposure. Evaluation of candidate space structure composite materials for degradation due to exposure in LEO (passive systems).

LDEF Long Duration Exposure Facility. Free-flying satellite providing accommodations for experiments requiring long term exposure to the space environment.

LEO Low Earth Orbit.

LFC Large Format Camera. Acquire synoptic, high-resolution images of the Earth's surface.

LIFESAT Life Sciences Satellite. Life science missions with micro-gravity experimentation as the primary objective.

LIQUID MOTION Liquid Motion in a Rotating Tank. Investigation of the behavior of liquids in a variety of rotating tanks.

LITE Lidar In-Space Technology Experiment. Project to measure the atmospheric parameters from a space platform utilizing laser sensors.

LM Long Module. A spacelab crew module.

LO Lunar Observer. Geological, elemental, gravity, and magnetic field mapping of moon.

MACE Middeck Active Control Experiment. Validation of controls/structures interaction technologies in zero gravity.

MAGELLAN Spacecraft designed to globally map the surface of Venus.

MAGSAT Magnetic Field Satellite. Map the magnetic field of the Earth.

MAR Middeck Accommodations Rack. An experiment integration facility installed in the middeck of the Shuttle with stowage volume equivalent to five middeck lockers. Power distribution and active thermal control options are available.

MARS OBSERVER Spacecraft to study the surface, climate, gravitational, and magnetic fields of the planet Mars.

MBB Messerschmitt-Boelkow-Blohm. A German industrial aerospace organization.

MD Middeck. Lower deck of the Shuttle crew compartment.

MEMBRANE Permeable Membrane for Plant Nutrient Delivery System. Verification of membrane transport performance in low gravity.

MGM Mechanics of Granular Materials. Microgravity experiment to study the effects of heat and near-zero gravity on the physical properties associated with various materials.

MICROWAVE Microwave Power Transmission. Demonstrates the ability to transmit microwave power in space and to evaluate the efficiencies and performance of key hardware elements in the environment of space.

MIS I Drug Microencapsulation in Microgravity. Evaluates the effects of microgravity on methods used to encapsulate drugs within biodegradable polymers. Combines materials science with biomedical product development and results in the production of a pharmaceutical product in space.

MLE Mesoscale Lightning Experiment. Record and observe the visual characteristics of large scale lightning as seen from space using onboard television cameras.

MLR Monodisperse Latex Reactor. Produces monodisperse latex particles in the two to forty micron range.

MODE Middeck 0-g Dynamics Experiment. To study the dynamics of liquids and skewed space structures in the microgravity environment.

MORELOS Mexican communication satellite system.

MPEC Multi-Purpose Experiment Canister. An extended Hitchhiker-G. GAS canister capable of deploying an internally stowed payload.

MPESS Mission Peculiar Equipment Support Structure. A cross-bay Shuttle payload carrier and support system for payloads weighing up to 3000 pounds. Managed by Goddard Space Flight Center (see SL-MPESS).

MPSE Mexican Payload Specialist Experiment. Experiment performed by a Mexican payload specialist on the Shuttle flight which deployed the MORELOS satellite.

MS Mission Specialist. A member of Shuttle flight crew primarily responsible for Orbiter subsystem and payload activities.

MSACP Microgravity Science and Applications Cooperative Program.

MSL Materials Science Laboratory. A payload which remains attached to the Shuttle to perform materials processing experiments in low-g.

MTC Man-tended Capability. Ability to perform laboratory operations on the SSF when the Shuttle is present.

N/A Not Applicable.

NASDA National Space Development Agency of Japan.

NATO North Atlantic Treaty Organization. A communications satellite for NATO.

NOAA National Oceanic and Atmospheric Administration. Series of operational environmental satellites in polar orbit.

NOSL Night/Day Optical Survey of Lightning. Optical survey of lightning.

NOVA Advanced Navy Navigation Satellite.

NTE NASA Technology Experiments. A series of technology experiments sponsored by the Office of Aeronautics and Space Technology and flown on the Shuttle.

OA NASA Office of the Administrator.

OAET NASA Office of Aeronautics, Exploration and Technology.

OAET-1 Office of Aeronautics, Exploration and Technology-1. Several advanced space technology experiments utilizing a common data system and mounted on a platform in the Shuttle bay, previously called OAST-2.

OAET-FLYER Office of Aeronautics, Exploration and Technology-Flyer. Free flyer deployed from the Shuttle containing several space technology experiments.

OASIS OEX Autonomous Supporting Instrumentation System. Collects environmental data in the Orbiter during dynamic STS flight phases.

OAST NASA Office of Aeronautics and Space Technology.

OAST-1 Office of Aeronautics and Space Technology-1. A payload which remained attached to the Shuttle to demonstrate a large light-weight solar array capable of being restowed in flight.

OCP NASA Office of Commercial Programs.

OCTW Optical Communication Thru the Shuttle Window Flight Demonstration. Demonstrates system to allow Shuttle crews to interface with payloads without depending on Orbiter communication systems.

OEX Orbiter Experiments. Series of engineering experiments on the Orbiter.

OIM Oxygen Interaction with Materials. Tests which obtained quantitative rates of oxygen interaction with materials used on the Orbiter and advanced payloads.

OPA NASA Office of Public Affairs.

OPM Optical Properties Monitor. An experiment to determine the effects of the space environment on critical spacecraft and optical materials by evaluating optical properties over time on a deployable carrier.

ORFEUS-SPAS Orbiting and Retrievable Far and Extreme Ultraviolet Spectrometer. A German developed payload to explore the distribution and character of radiation absorbing material in the solar system and to perform direct ultraviolet observations of the direct interstellor component.

ORS Orbiter Refueling System. An experiment to demonstrate the ability of the STS to perform on-orbit satellite refueling.

OSCAR Orbiting Satellite Carrying Amateur Radio. Amateur radio communication satellite.

OSF NASA Office of Space Flight.

OSL Orbiting Solar Laboratory. Provide detailed data on the Sun, to augment our studies of distant stars and cosmic processes.

OSO NASA Office of Space Operations.

OSS-1 Office of Space Science-1. Single Pallet carrying eight experiments to demonstrate the use of the Shuttle for investigations in space plasma physics, solar physics, astronomy, etc. and to characterize the Orbiter and payload bay environment.

OSSA NASA Office of Space Science and Applications.

OSTA Office of Space and Terrestial Applications (currently OAET).

OSTA-1 Shuttle attached payload using the Shuttle Imaging Radar (SIR-A) to obtain high resolution images of Earth.

OSTA-2 Microgravity experiments.

OSTA-3 Acquire photographic and radar images of the Earth's surface.

P-CENT Gravitropic Responses of Plant Seedlings. Quantitative characterization of plant cell growth from gravitropic plant seeds without guidance from a significant gravity force.

PAL Pallet. Spacelab Pallet structure.

PALL Pallet. Spacelab Pallet structure.

PALAPA Synchronous satellite communication system for the Republic of Indonesia.

PAM Payload Assist Module. An upper stage system used on the Shuttle and the Delta ELV.

PARE Physiological & Anatomical Rodent Experiment. Study the physiological and anatomical changes that occur in mammals under weightless space flight conditions.

PBE Pool Boiling Experiment. Study fundamental mechanisms that constitute pool boiling.

PCG-II Protein Crystal Growth-II. PCG activity in controlled temperature module.

PCG-III Protein Crystal Growth-III. Obtain high quality protein crystals to facilitate analysis of structures.

PCGPROTO Protein Crystal Growth Prototype. Development of rapid response equipment to facilitate investigation of new/special PCG techniques or samples.

PDRS/PFTA Payload Deployment and Retrieval System/Payload Flight Test Article. First object to be unberthed and reberthed by the remote manipulator system, used to test the performance of the RMS in handling a massive object.

PHCF Pituitary Growth Hormone Cell Function. Microgravity induced effects on pituitary (active growth) hormones in various types of living cells.

PIONEER VENUS A probe for remote sensing and direct measurements of Venus and its surrounding environment.

PL OPPTY Payload Opportunity.

PLAN Planetary Trajectory. High Energy Trajectory to Outer Planets.

PLC Payload Commander. A member of the Shuttle crew having overall crew responsibility for planning, integration, and on-orbit coordination of payload mission activities.

PLT Pilot. A member of the Shuttle crew whose primary responsibility is to pilot the Orbiter.

PLUM Polymerization With Light Under Microgravity. Perform ultra-violet light induced polymerization of organic polymers.

PM Polymer Morphology. Determines effects of weightlessness on morphological formation of polymers as they undergo physical transition.

PMC Permanently Manned Capability. Ability for a four person crew to occupy the space station Freedom Manned Base on a permanent basis with periodic crew rotation.

PMG Plasma Motor Generator. ELV secondary payload experiment to verify ability of plasma sources to couple electric current along a wire.

PMZF Programible Multi-Zone Furnace. Materials processing apparatus located in the middeck accommodations rack.

POLAR Polar Auroral Plasma Physics spacecraft.

PPE Phase Partitioning Experiment. Study separation behavior of two phase system generated by the mixture in water of polyglucose and polyethylene glycol.

PS Payload Specialist. A member of the Shuttle crew, who is not a NASA astronaut, whose presence is required to perform specialized functions with respect to one or more payloads or other mission unique activities.

PSAS Phenytoin for Space Adaptation Syndrome. Determines the efficacy of Phenytoin for the treatment of Space Adaptation Syndrome (SAS) and explore etiology of SAS as related to partial seizures.

PSE Physiological Systems Experiment. Examines effects of hormone therapy on changes in organic systems during spaceflight.

PVTOS Physical Vapor Transport of Organic Solids. Grow crystalline films on selected substrates of organic solids

RADARSAT Radar Satellite. Remote free flyer sensing satellite that will monitor land, sea and ice for five years over the poles (U.S./Canadian).

REQ Request.

REX Radiation Experiment. Research effects of electron density irregularities on transionosphere radio signals.

RME Radiation Monitoring Equipment. Measures gamma radiation levels in the Shuttle environment.

RMS Remote Manipulator System. A Canadian developed, remotely controlled (from the Orbiter crew cabin) arm for deployment and/or retrieval of payloads from the Orbiter payload bay.

ROMPS Robotic Materials Processing System. Investigates zero gravity anealing of semiconductor thin film and investigates robot handling of thin film samples.

S Scout. Small Class Expendable Launch Vehicle.

SAC-B Satellite de Aplicaciones Cientificas-B. Argentine spacecraft carrying Hard X-Ray Spectrometer to investigate solar flares and cosmic transient X-ray emissions.

SAGE Strategic Aerosol and Gas Experiment. Map vertical profiles of the ozone, aerosol, nitrogen Rayleigh molecular extinction around the globe.

SAM Shuttle Activation Monitor. Collects gamma and X-ray data as a function of geomagnetic location from spacecraft materials.

SAMPEX Solar, Anomalous, and Magnetospheric Particle Explorer. A spacecraft to study solar energetic particles, anomalous cosmic rays, galactic cosmic rays, and magnetospheric electrons.

SAMS Space Acceleration Measurement System. Provides Orbiter acceleration measurements in support of microgravity experiments.

SAREX Shuttle Amateur Radio Experiment. Low cost space to ground voice and slow scan television experiment.

SAS Space Adaptation Syndrome. Physiological changes which occur when adapting to microgravity.

SATCOM RCA communications satellite.

SBS Satellite Business Systems. All digital domestic communication system servicing large industry, the government, etc.

SE Student Experiment. Experiments sponsored by the Shuttle Student Involvement Program (SSIP).

SEDS Small Expendable Deployer System. Experimental tether deployment device.

SFMD Storable Fluid Management Demonstration. Demonstrates transfer of room-temperature fluids in zero-g using various transfer techniques.

SFP Space Flight Participant. A Shuttle crew member whose presence is not required for operation of payloads or mission unique activities, but is determined by the NASA Administrator to contribute to other approved NASA objectives or to be in the national interest.

SFU-RETR Space Flyer Unit Retrieval. A reusable, retrievable unmanned free flyer to be launched on the Japanese H-II rocket and retrieved by Shuttle.

SHARE Space Station Heat Pipe Advanced Radiator Element. Demonstrates and quantifies the thermal performance of a high capacity, 50 foot, space constructible, heat pipe radiator element.

SHOOT Super Fluid Helium On Orbit Transfer Demonstration. Demonstrates the feasibility of on-orbit transfer of superfluid helium using thermomechanical techniques.

SII Space Industries, Inc. A U.S. company providing commercially-owned Industrial Space Facility (ISF).

SIR Shuttle Imaging Radar. Series of synthetic aperture radar experiments.

SIRTF Space Infrared Telescope Facility. Will span the infrared part of the spectrum with a thousand-fold increase in sensitivity.

SKYNET United Kingdom military communication satellite.

SL-D1 Spacelab D1. First dedicated German Spacelab mission.

SL-D2 Spacelab D2. Second in a series of German Spacelab Missions. Objectives include microgravity research and technology preparation for space station use.

SL-D3 Spacelab D3. Third in a series of German Spacelab Missions. Objectives include microgravity research and technology preparation for space station use.

SL-E2 Spacelab E2. ESA sponsored science mission directed toward multidiscipline research in material science, fluid science, life science, space science, observation and technology.

SL-J Spacelab J. Combined NASDA/NASA Spacelab mission. Objectives include life sciences, microgravity, and technology research.

SL-MPESS Spacelab MPESS. MPESS Carrier managed by Marshall Space Flight Center

SLS Space Life Sciences Laboratory. Investigates the effects of weightlessness exposure using both man and animal specimens.

SLSTP Space Life Sciences Training Program. Series of payloads to support a broad range of life science studies.

SMEX Small Explorer. Payloads being designed to fly on Small Class ELV.

SMR San Marco Range. Italian small class ELV launch range off Kenya coast.

SMRM Solar Maximum Repair Mission. Conducted a technology demonstration of the STS capability to rendezvous, service, checkout and deploy.

SOHO Solar Heliospheric Observatory. Provides optical measurements as well as plasma field and energetic particle observations of the Sun system for studies of the solar interior, atmosphere and solar wind.

SOLAR PROBE Study unexplored region of the solar atmosphere, measure electromagnetic fields and study the particle populations close to the Sun.

SOOS Stacked OSCAR on Scout. Two Navy OSCAR navigation satellites. Not amateur radio communications satellites.

SPACEHAB U.S. company providing commercially-owned pressurized module for conducting experiments in a man-tended environment. Also a series of payloads to be flown on the Space Shuttle.

SPACELAB 1 Demonstrated Spacelab's capabilities for multidisciplinary research.

SPACELAB 2 Demonstrated Spacelab's capabilities for multidisciplinary research and verified system performance.

SPACELAB 3 Dedicated materials processing mission emphasizing research in microgravity conditions.

SPADVOS Spaceborne Direct Viewing Optical System. Evaluates the crew's ability to utilize direct viewing system to allow realtime detection of ground and airborne targets.

SPAS Shuttle Pallet Satellite. Payload Carrier developed by MBB of W. Germany.

SPAS-01/01A German Shuttle Pallet Satellite. Demonstrates the utilization of the MBB platform and systems as a carrier for science experiments.

SPAS-III Shuttle Pallet Satellite III. A reflight of the Infrared Background Survey (IBSS) mission.

SPTN Shuttle Pointed Autonomous Research Tool for Astronomy. X-ray astronomy, medium energy survey mission, using retrievable free flyer.

SPTN-HALLEY SPARTAN-HALLEY. Search for molecules containing nitrogen, carbon or sulfur and observes the UV spectrum between 2100 and 3400A.

SRL Space Radar Laboratory. Series of flights to acquire radar images of the Earth's surface. The images will be used for making maps, interpreting geological features, and conducting resource studies.

SS Sun Synchronous. Sun-synchronous polar orbit.

SSBUV Shuttle Solar Backscatter Ultra-Violet Instrument. Series of flights to measure ozone characteristics of the atmosphere.

SSCE Solid Surface Combustion Experiment. Determines the gas-phase flamespread over solid fuel surfaces in microgravity.

SSF Space Station Freedom. Orbiting space station.

SSF/MB Space Station Freedom Manned Base. Assembly launches of modules for the SSF Manned Base.

SSF/UF Space Station Freedom/Utilization Flight.

SSIP Shuttle Student Involvement Program. Competitions held between 1981-1985 in which the winning High School students proposed experiments which were accepted for Shuttle flights.

STEX Sensor Technology Experiment. Demonstrates radiation measurement technology.

STL Space Tissue Loss. An experiment to validate or confirm model of skeletal and cardiac muscle atrophy, collect data on catabolic pathway and control mechanisms, and test candidate pharmaceuticals for efficacy.

STP-XX Space Test Program-XX. A series of payloads which include DOD STP secondary experiments.

STS Space Transportation System.

STTP Life Sciences Space Technology Training Program. Activity to develop and encourage interest on the part of college students in space biology and medicine.

SWAS Submillimeter Wave Astronomy Satellite. Spacecraft to study how molecular clouds collapse to form stars and planetary systems.

SYNCOM Hughes Geosynchronous Communication Satellite. Provides communication services from geosynchronous orbit principally to the U.S. government.

TANK VENT Tank Venting Experiment. Investigation of concepts to provide tank fill-while-venting to 90 percent full capacity.

TAPS Two Axis Pointing System. An instrument support system which allows pitch, roll, or combinations thereof to precisely point instruments at different targets.

TBA To Be Announced.

TBD To Be Determined.

TDRS Tracking and Data Relay Satellite. Series of NASA tracking, data and communications satellites to replace the NASA ground based network.

TELESAT Canadian Telecommunication Satellite. Communication satellite built for Telesat Canada to provide voice and TV coverage to trans-Canada network of Earth stations.

TELSTAR AT&T Communications Satellite. AT&T COMSTAR replacement provides communication services to the continental U.S., Alaska, Hawaii, and Puerto Rico.

TEMP Two Phase Mounting Plate Experiment. Operates a mechanically pumped two phase head acquisition transport and rejection system in microgravity.

TERRA SCOUT Evaluates the ability of an expert imagery analyst to conduct realtime analysis from low Earth orbit.

TIP Transit Improvement Program. Improved configuration Transit Navigation Satellite.

TIS Teacher in Space.

TITAN II DOD medium class expendable launch vehicle.

TITAN III Commercial intermediate class expendable launch vehicle.

TITAN IV DOD large class expendable launch vehicle.

TLD Thermoluminescent Dosimeter. Gamma ray measurements of the Shuttle environment.

TOMS Total Ozone Mapping Spectrometer. Study of Stratospheric ozone.

TOS Transfer Orbit Stage. Upper stage system for Shuttle and Titan.

TPCE Tank Pressure Control Experiment. A study to determine the effects of microgravity on the thermal stratification of fluids and to validate the effects of jet induced mixing.

TPITS Two Phase Integrated Thermal System. Evaluation and demonstration of on-orbit thermal performance of prototypical SSF 2 phase thermal bus control system.

TSS Tethered Satellite System. System capable of deploying and retrieving satellite attached by a wire tether from distances up to 100 km from the Orbiter.

U.S. United States.

UARS Upper Atmosphere Research Satellite. Satellite to study chemical processes acting within and upon the stratosphere, mesosphere, and lower thermosphere.

ULYSSES Formerly ISPM (International Solar Polar Mission). Investigates the properties of the heliosphere (Sun and its environment).

USML United States Microgravity Laboratory. Series of flights of a microgravity materials processing laboratory attached to the Shuttle.

USMP United States Microgravity Payload. Conduct materials processing experiments in the microgravity environment available in the Orbiter cargo bay while in low Earth orbit.

USS Unique Support Structure.

UV Ultraviolet.

UVPI Ultraviolet Plume Imager. Free-flying satellite observation of Orbiter OMS burns.

VCS Voice Controlled System. Evaluates effectiveness of voice controlled system on the Shuttle cargo bay closed circuit television.

VFT Visual Function Test in Space. A biomedical study to determine effects of microgravity on human visual performance.

VIPOR Visual Investigation Program on Orbiter Operations. A series of experiments to study elements that can affect and degrade the performance of any optical (photo, visual, or video) system.

WESTAR Western Union Telegraph Communication Satellite. A C-band satellite to replenish and expand the Westar system (Western Union domestic communication system).

WFF Wallops Flight Facility.

WIND Satellite to measure solar wind input to magnetosphere.

WINDEX Window Experiment. To obtain calibrated measurements of environmentally induced optical emissions.

WISP Waves in Space Plasma. Active experiments using sensors on a free-flyer to measure space plasma excitation by radio transmitters in the Shuttle payload bay.

WOSE Weather Officer in Space Experiment. Assesses feasibility of expert weather observations from space to see, photograph, videotape atmosphere and ionosphere phenomena.

WSF Wake Shield Facility. Molecular and chemical beam epitaxy growth of compound semiconductors, high temperature superconductors, and other materials using techniques requiring ultra-high vacuum, high pumping speeds, and relatively large working volumes.

WSMC Western Space and Missle Center. A USAF organization with Headquarters at Vandenberg Air Force Base, California.

XTE X-Ray Timing Explorer. A payload to be used in Earth orbit to investigate the physical nature of compact X-Ray sources by studying fluctuations in X-Ray brightness over time-scales ranging from microseconds to years.

USSR Space Shuttle Buran

The sound of mighty rocket engines thundered through the autumn steppe of Kazakhstan. A monumental yellow flame sent huge clouds of white steam billowing away from the Baikonur Cosmodrome launch pad November 15, 1988, as the world's mightiest space rocket, Energia, shot through a cold rain into the sky, ferrying the brand-new Soviet space shuttle Buran successfully to two orbits of Earth and a picture-perfect automated landing on the center line of its runway.

Threatening weather. Buran, Russian for snowstorm, almost got caught in one. Liquid fueling of the towering Energia booster started 14 hours before flight. Then Moscow Weather Service predicted rain at Baikonur three hours before the unmanned launch. The temperature was 39 degrees.

Soviet television said the launch might be in jeopardy. Launch would be held if winds rose to a squall or shuttle and booster become encrusted with ice, officials said, as technicians continued pouring 2,000 tons of liquid hydrogen, oxygen and kerosene fuel into the 198-ft. Energia under the glare of 670 floodlights.

Liftoff. Squall and ice crusts didn't develop. Energia's four first-stage rockets lifted the reusable shuttle slowly off the pad, lighting the dark early-morning sky on schedule at 6 a.m. Moscow time. The rocket and shuttle disappeared into the sky. Strap-on boosters were dropped in pairs at 2.75 minutes into the flight as fuel ran out at 38 miles altitude. TV showed controllers applauding as green computer screens tracked the flight.

Buran separated properly from the liquid-fuel Energia eight minutes after liftoff from Baikonur at 99 miles altitude. The shuttle's engines fired 2.5 minutes later, for 67 seconds, finishing the flight to space. The shuttle was in a 155-mi.-high orbit. At 47 minutes into the flight, over the Pacific Ocean, the shuttle's engines fired again for 42 seconds, circularizing the orbit.

Twice around. In space, Buran was controlled entirely by computers to test launching and landing equipment and procedures. Engineers on the ground were in radio control of the craft via ground stations, four satellites in orbit and four ships in the Atlantic and Pacific oceans.

Buran carried a cylindrical dummy satellite loaded with test instruments in the cargo bay. As it passed over the Pacific for the second time after two 155-mi.-high circular orbits of the globe on fully-automatic control with its left wing facing Earth, Buran automatically started to re-enter the atmosphere, turning its tail forward and switching on a retrofire engine at 8:20 a.m. After retrofire, the orbiter turned again, to face the direction of re-entry. Buran entered the top of the atmosphere at an altitude just above 75 miles. During the hottest part of re-entry, there was an expected 20-min. radio blackout. When it dropped to an altitude of 25 miles, Buran rotated to land like an airplane at Baikonur.

TV from a MiG-25 chase jet pictured the dark delta-winged silhouette against a pale blue-gray sky, approaching the landing strip just eight miles from its original Energia launch pad at the space center in the Kazakhstan Republic of Soviet Central Asia.

Robot landing. A 34 mph cross wind blew against Buran as it touched down at 9:25 A.M. Ground TV showed the fully automated approach and touchdown of the black-and-white shuttle, landing at 204 mph like an airplane.

Dust puffed out behind the rear wheels as Buran touched down on a 2.8-mi. concrete runway on the barren, brown plain. The main landing gear was just five feet from the center line painted on the runway.

Three parachutes blossomed from the aft end as the shuttle rolled to a halt, ending a precision unmanned 3-hour 25-minute maiden flight.

The Soviets said the 205-min. flight went as planned. All on-board tests were completed. Only five of 39,000 heat-shield tiles broke off during flight.

An earlier delay. The flight had been scheduled for October 29, 1988, but that countdown had been stopped at T minus 51 seconds by a watchdog computer checking 140 elements of the launch system. A crew escape arm had failed to pull away fast enough from the rocket. Analysis during the postponement showed it would take 38 seconds, not the three seconds originally programmed in computers, to retract the access arm.

The words "automatic termination of preparations" had flashed on computer screens at Baikonur October 29 as dozens of scientists and engineers watched the abort. The sudden halt recalled a similar last-minute drama during liftoff of the world's first shuttle. The U.S. shuttle Columbia had been delayed two days in April 1981 with computer software problems.

The USSR Defense Ministry's chief specialist on multiple-use space transport systems told Baikonur workers at an outdoor ceremony for the Buran maiden flight, "Truly this is an historic task. Today's launch can compare to the launch of the first artificial satellite." The USSR began the space age in 1957 with the launch of Sputnik.

Paris Air Show. After its successful first flight in November 1988, Buran was flown piggyback on a six-jet Antonov 225, the world's largest aircraft, from Baikonur to Le Bourget, France, airport for display at the Paris Air Show.

American look-alike. The size of a passenger airliner, Buran is 119 feet long and 18.5 feet in diameter with a 79-ft. wingspan. It has a railroad-car-sized cargo hold. It will carry 10 cosmonauts—four crew and six passengers—for a month in space.

The black-and-white Buran looks very much like an American shuttle. The two aren't the same, however. The U.S. shuttle is a system, including the familiar rust-colored jumbo fuel tank, the white orbiter spacecraft and two solid-fuel booster rockets.

By contrast, the large core of the Russian system is an Energia rocket, not just a fuel tank. Either a shuttle or an unmanned cargo pod can be strapped to an Energia rocket.

The Russians claimed Buran would be safer, more flexible and more powerful than U.S. shuttles. The USSR shuttle was said to be superior to the American orbiter because of greater payload capacity and ability to fly automatically so cosmonauts can pay more attention to science experiments.

Additional Buran landing strips are under construction, near Simferopol along the Crimean Black Sea coast and in the USSR Far East.

The name Buran. Buran, or Snowstorm in Russian, has become a generic name for USSR space shuttles. The orbiter which flew the maiden voyage of all USSR shuttles in November 1988 now is called Buran No. 1 and now is retired from spaceflights.

The new orbiter to be flown to Mir in 1992 is Buran No. 2. Later, when manned shuttle flights occur, each manned Buran will be given its own nickname by the team of cosmonauts who fly it.

Next Buran flight. The USSR has built another Buran shuttle to fly unmanned to the orbiting Mir space station. It may be launched on an Energia rocket toward the end of 1992. After Buran arrives at a special dock on Mir's Kristall expansion module, the cosmonauts in Mir will take a spacewalk to enter the shuttle and check it out. The shuttle then would leave the dock and fly to Earth. A manned Buran flight might occur in 1993.

Pilotless Buran. Since the 100-ton Buran can be controlled remotely from the ground, it requires no cosmonauts to get to space and back. Cosmonauts would travel in Buran only to accomplish science research or manufacturing tasks requiring their presence. While two to four cosmonauts might be a nominal flight crew, up to ten researchers could go to space for flights of seven days to a month in Buran.

Buran pilots. The chief Buran test pilot is cosmonaut Igor Volk who gained spaceflight experience by flying to Salyut 7 station in 1984 in Soyuz T-12.

Another Buran pilot was Ukrainian cosmonaut Anatoly Levchenko. He gained experience in an eight-day spaceflight to Mir space station in 1987. Levchenko, 47, died of a brain tumor in 1988, eight months after returning to Earth. He had been a cosmonaut since 1981, but still a rookie pilot when he blasted off in Soyuz TM-4. The tumor turned up after he returned home from Mir. An operation by Moscow doctors in

August 1988 was unsuccessful.

Changing government. What is to become of shuttle Buran and the orbiting Mir space station in the new USSR after government restructuring brought about by the 1991 coup d'etat? The State Commission for Manned Space Flight was not changed after the coup. Government organizations controlling Soviet spaceflight continued their work. Budgets were frozen, but international agreements continued in force. Guest cosmonauts were to be allowed to fly as planned. An Austrian cosmonaut flew in October 1991. Space station spacewalk plans were reduced.

What's it for? Firing the Russian shuttle to space is very expensive. There have been pressures to find something appropriate for it to do. Of course, the USSR already has time-tested, relatively-inexpensive Soyuz capsules to fly cosmonauts and Progress unmanned cargo freighters to ferry materials to space.

At the same time, cosmonauts reportedly have pressured for manned launches, rather than more unmanned test flights.

The American shuttle was designed to ferry satellites to orbit. The head of the USSR space agency Glavkosmos said in 1988 shuttles are not as economical as single-use rockets for firing satellites to orbit. He said shuttles are valuable, however, for servicing space stations and satellites, and recovering objects from orbit.

The Soviets hold the record for launching satellites on single-use boosters, sending up more than 150 a year, more than any other country. The USSR also uses expendable rockets to maintain the permanent manned presence in the Mir space station.

The USSR probably will continue to use less-expensive expendable rockets until the late-1990's. Then the USSR fleet of two to four shuttles may be ready to fly two to four times a year.

Europe's Space Shuttle Hermes

The European Space Agency is forging ahead with a $17.2 billion plan to put men in space aboard a space shuttle named Hermes by 1999, despite British refusal to join the "frolic." The European Space Agency project will build a small shuttle, Hermes, and upgrade the veteran Ariane rocket to launch it. The improved rocket will be Ariane 5.

Frolic. Great Britain called the project a "hugely expensive industrial frolic" and "rather too much smacking of me-tooism." In fact, half of the 12-year $31 billion ESA budget from 1988 to 1999 was dedicated to man-in-space.

France, with Europe's largest space industry, was pitted against cost-conscious Great Britain on the Hermes question. British failure to agree with the 12 other ESA nations in the project was taken at the time as a sign of Prime Minister Margaret Thatcher's reluctance to commit Britain to European integration. France's support for the shuttle, on the other hand, attempted to lessen European dependence on the U.S., the only Western nation to launch manned flights.

Shuttle partners. Involved in the Hermes shuttle project are Austria, Belgium, Canada, Denmark, France, Ireland, Italy, the Netherlands, Norway, Spain, Sweden, Switzerland and Germany. The earliest manned Hermes flight would be in 1999.

European Space Agency (ESA) is based in Paris. The project did not require unanimous adoption. Britain was the only nation to turn it down.

The shuttlecraft. Hermes is to be a delta-wing shuttlecraft. It will be 59 feet long, half the size of an American space shuttle, with a wingspan of 33 feet. Hermes will carry 10,000 pounds of payload in a 1,236 cubic-ft. cargo bay about ten feet across.

Plans call for Hermes to be shot to orbit atop an Ariane 5 rocket. For crew safety, it will have ejection seats. Hermes will glide back to Earth, without power, to a landing strip in southern Spain. French Guiana's main airport at Cayenne may be extended for use as an emergency landing strip.

Design work is underway on propulsion, electronics, fuel cells, guidance control, a manipulator arm, temperature and environment control, life support, computers, communications, and an airlock for spacewalks.

Plans call for a 725-ton Ariane 5 rocket to ferry European astronauts aboard the Hermes spaceplane to Earth orbit before the end of the century.

Station tender. Hermes will carry two pilots and two to four other crew members to service and maintain unmanned platforms in space, manned scientific modules in space, and space stations. It will be able to replace crew members on any space station and restock stations with food and other consumable supplies.

Europeans expect the use of large unmanned space platforms to increase. Hermes can carry maintenance men to those platforms to restock fuel and return materials produced on the platforms to Earth.

To Mir. CNES, the French national space agency, will build Hermes so it can fly to, and dock with, both the USSR's Mir space station and the U.S.-international space station Freedom.

Mir already is in low orbit about 200 miles above Earth. Construction on the U.S.-international space station Freedom is expected to get underway by 1997. Hermes may get to space as early as 1996, if the European nations continue to give their space agency sufficient money.

Selling shuttle services. ESA already has an agreement with the USSR to send Hermes to the Mir space station and would like to sell regular shuttle flights to the Russians as a freighter service. Hermes is expected to be able to ferry three tons of men and cargo to space.

The U.S. and USSR manned space shuttles are much larger, like trucks to carry heavy cargo to space. The USSR has been using unmanned Progress cargo ships to carry food, water and supplies to Mir. The Europeans plan for Hermes to carry more than a Progress can in one trip, and be manned.

Rescue craft. Hermes is too small to make much profit launching commercial satellites, so its French manufacturer, Aerospatiale, also foresees the Hermes shuttle being used as a crew rescue vehicle for the space station Freedom.

To Columbus. Hermes could carry men and supplies to Columbus, Europe's large manned science module to be attached to space station Freedom in the late 1990's. Columbus, which can be used as an unmanned platform, a stand-alone manned module, or as a pressurized module attached to a space station, also could become part of an independent European station after the year 2000.

Launch site. ESA is building mammoth new facilities to launch men into space from French territory in South America at the ESA space center at Kourou, French Guiana.

It is the biggest European building site after the Channel Tunnel construction project between Britain and France. Thousands of tons of concrete are being poured into a vast clearing of the Amazon jungle. The new launch facilities will dwarf launchpads nearby from which four generations of unmanned Ariane rockets have blasted off since 1979.

City of buildings. Kourou is an isolated outpost 4,000 miles from Paris. The new Ariane 5 facility stretches over a 20 square mile area hacked from jungle and savannah along the Atlantic Ocean. Four million cubic meters of earth already have been removed and one-half million cubic meters of granite will be dynamited.

Engineers are building what ESA calls the world's largest mixing bowl, to concoct solid fuel for the Ariane 5 from two highly volatile 100-ton blocks of chemicals.

A spacious test bed, 200 feet deep and 655 feet long, has been gouged out of bedrock. Thousand-ton concrete blocks will hold down the Ariane 5 during tests. Surrounding granite will vitrify during two-minute burns. Ariane 5 will match large USSR and U.S. satellite launchers.

ESA decided in 1987 to spend $1.8 billion dollars to build the factories, test beds, launch pad and control center at Kourou. France is the biggest investor with 43.5 per cent of the project. Great Britain doubted the program and put in no money.

Japan's Space Shuttle Hope

Japan has a space shuttle on the drawing boards for a first flight sometime between 1997-1999. The orbiter has been dubbed Hope, for H-2 Orbiting Plane, after the powerful new H-2 rocket Japan will use to launch the shuttle.

Unlike, U.S., Soviet and European shuttles and plans, Japan's shuttle would be unmanned at first. The computerized robot shuttlecraft would be 60 feet long with 40-ft. wingspan, capable of ferrying three tons of payload to a low orbit.

Japan's National Space Development Agency (NASDA) plans to fly Hope to the planned U.S.-international space station Freedom. Japan is a partner with the U.S., Europe and Canada in the Freedom project.

H-2 rocket. The powerful H-2 is to be the first space rocket designed entirely in Japan. Besides shuttle flights to low orbits, NASDA also wants to use the H-2 to launch two-ton payloads to stationary orbit 22,300 miles above Earth.

That would allow Japan to compete commercially with rockets currently made in Europe, the USSR and the U.S.

The H-2 also would be used to launch probes to the Moon and Venus, as well as Hope space shuttle missions in low Earth orbit.

Space station Freedom will orbit about 280 miles above Earth. A Japanese astronaut will ride a U.S. shuttle to a tour of duty at Freedom station. The first Japanese spacewalk will be made at Freedom during attachment of Japan's JEM module to the station.

Hope vs. Hermes. Hope will be similar in size to the European Space Agency's Hermes manned shuttle, which is under development for first flights in the same time period around the end of the 1990's. Hope and Hermes will be much smaller than U.S. and USSR manned space shuttles.

Spaceport. Hope will have the shape of an airplane, with wings to land on a runway. Pacific Spaceport Group, an organization formed by Japan's seven largest companies, wants to build a six-mile-long runway, launch towers and control buildings for the coming Hope spaceplane.

Australia's proposed spaceport at Weipa in the northern state of Queensland probably will be the chosen site. Australia has organized a Cape York Space Agency to develop the Queensland launch facility.

Problems. Originally planned for a 1991 maiden flight, H-2 has been delayed to 1993 or later by technical difficulties which cropped up in the high-performance first-stage rocket engine known as LE-7. The LE-7 uses new combustion technology for a high-power mix of liquid hydrogen and oxygen fuels. High stress in the engine sparked fires during tests.

In addition, NASDA engineers were shocked when they discovered the LE-7 design would not generate enough thrust to heft a two-ton payload to orbit. The Japanese newspaper Yomiuri Shimbun reported the LE-7 design would be changed to increase its boosting power by ten tons. Design changes were likely to delay further the first flight.

H-1 rocket. Previously, Japan used the H-1 rocket modified from an old Thor-Delta rocket used in the U.S. from the mid-1960's.

With the H-1 and smaller space rockets, Japan has been building a convincing space-launch record. In 1990, it became only the third nation to send a probe to orbit the Moon.

Technology. Japan has been a late entry in the aerospace field because of prohibitions imposed by the United States at the end of World War II. Today, Japan spends billions each year across several space programs.

The Hope construction project will require basic research on robots capable of assembling and repairing satellites and for developing the H-2 rocket, with a two-ton capacity. The Japanese plan to spend billions during the next 15 years to build Hope and other rockets, satellites, interplanetary probes and ground facilities.

Space Shuttle Glossary

Atlantis. The fourth U.S. space shuttle to fly to orbit.

Abort. To end the launch or flight of a space rocket.

Atlas. The name of one kind of unmanned U.S. space launch rocket.

Boosters. Helper rockets strapped to a main rocket engine.

Buran. The first orbiter of the USSR's space shuttle fleet. Buran is Russian for snowstorm or blizzard. Buran No. 1 flew to Earth orbit in 1988. Buran No. 2 may fly to space station Mir in 1992-93. Buran is launched on an Energia rocket.

Challenger. The second U.S. space shuttle, destroyed in 1986 launch explosion.

Columbia. The first U.S. space shuttle to fly to orbit.

Columbus. Europe's module in development for space station *Freedom*.

Delta. The name of one kind of unmanned U.S. space launch rocket.

Discovery. The third U.S. space shuttle to fly to orbit.

Endeavour. The fifth U.S. shuttle to fly to orbit, it replaced shuttle Challenger.

Energia. The newest USSR super-powerful space-launch rocket. It is used to launch the shuttle Buran to Earth orbit.

Enterprise. The 1970's prototype U.S. shuttle, it was not flown to space.

EVA. Extra-vehicular activity, a spacewalk.

External tank. Large liquid-fuel tank attached to U.S. shuttle orbiter at launch.

Freedom. The U.S.-international space station in development.

Hermes. Europe's space shuttle in development.

Hope. Japan's space shuttle in development.

JEM. Japan's module in development to be attached to station Freedom.

Liquid fuel. Frozen gases used as fuel in space rockets.

Liquid rocket. Space rockets using frozen-gases for liquid fuels.

Manipulator. A mechanical arm used to move freight from a shuttle's cargo hold.

Mir. The USSR's third-generation space station.

Progress. An unmanned capsule ferrying food, fuel, supplies and science equipment to the Mir space station.

Progress M. A redesigned Progress. M for modified.

Shuttle. A partly-reusable spacecraft which can land on a runway.

Shuttle-C. A proposed unmanned cargo version of a U.S. shuttle.

Solid fuel. A chemical rocket fuel with a rubbery consistency, not gas or liquid.

Solid rocket. A space rocket using a rubbery, so-called solid, fuel.

Soyuz. The long series of USSR cosmonaut-transporting spacecraft. USSR capsules ferrying cosmonauts to and from space. Soyuz is Russian for Union, symbolizing the spacecraft's rendezvous and docking capability.

Soyuz T. A redesigned Soyuz. The T symbolizes: (1) transport, (2) troika indicating a space transport capable of carrying three cosmonauts to the Mir station, and again (3) troika indicating the third generation of the Soyuz design.

Soyuz TM. A redesign of Soyuz T. The M is for modified and for Mir, its main destination. Soyuz TM is a third generation modified Soyuz.

Space station. Structure in orbit with long-duration living and working quarters.

Spacewalk. Extra-vehicular activity, a trip outside a shuttle or space station.

Titan. The name of one kind of unmanned U.S. space launch rocket.

Voskhod. A redesigned Vostok. Voskhod is Russian for Sunrise, following USSR Premier Khrushchev's observation that "the Sun rises in the East."

Vostok. The first-ever manned spacecraft. Vostok is Russian for East, symbolizing the international-relations East-West competition.

Space Rockets
★★★★★★★★★★★★★★★★★★★★★

Space Rockets Table Of Contents Continues On Page 242

Space Rockets
★★★★★★★★★★★★★★★★★★★★★

History Of Space Rockets

A cold rain pelting Baikonur Cosmodrome couldn't quench the monstrous belch of yellow flame, but fed the billowing clouds of white steam roiling off the launch pad as Energia, the world's mightiest space rocket shot up through the November sky, ferrying the brand-new space shuttle Buran to Earth orbit.

That 1988 rocket success was even more important than its payload, the USSR's first shuttle. Energia proved it not only had enough power to ferry shuttles to orbit, but also could carry 110-ton satellites to orbit, space stations large enough to hold 12 cosmonauts, or manned missions to the Moon and Mars.

Energia's main core—197 feet tall, 26 feet in diameter—had four large rocket engines fueled with liquid oxygen and liquid hydrogen. Four booster rockets, 13 feet in diameter, fueled with kerosene and liquid oxygen, were strapped around the core. The 4.4 million-lb. Energia delivered 6.6 million lbs. of thrust. Energia is able to fire 36,000-lb. satellites to stationary orbit, 64,000-lb. probes to the Moon or 56,000-lb. spacecraft to Mars.

China. We've come a long way in the 2,300 years since Chinese religious mandarins first tossed ceremonial bamboo tubes packed with gunpowder into festival fires to drive off evil spirits.

Sometime between 300 B.C. to A.D. 1000, "fire arrows" were used in China, but historians aren't sure if those were rockets or merely conventional arrows burning. Either way, when firecrackers finally did turn into rockets, the sulphur, saltpeter and charcoal in gunpowder formed the earliest solid fuel for rockets. By A.D. 1045, gunpowder rockets were important weapons in in China's military arsenal.

The Sung Dynasty improved gunpowder projectiles in the 13th century with new explosive grenades and cannon to hold off growing Mongolian hordes. Fire arrows repelled Mongol invaders at the battle of Kai-fung-fu in A.D. 1232.

Old records show the huge Chinese gunpowder rockets carried iron shrapnel and incendiary material, and may have had the first combustion-chamber "iron pots" to direct thrust. Blast off of a fire arrow was heard for 15 miles and its impact demolished everything within half a mile.

Europe. Mongols brought gunpowder rockets to Europe by 1241, firing them against Magyar forces at Sejo before capturing Budapest that year.

Arabs. After Mongols used gunpowder rockets to invade and capture Baghdad in 1258, the Arabs added them to their arsenals. Arab rockets were used in 1268 against King Louis IX's French army of the Seventh Crusade.

France. With the largest army in Europe, France used rockets in 1429 at the siege of Orleans in the Hundred Years War with England.

Italy, Germany. Italians had them in the 1500's when they rediscovered the use of rockets for festive fireworks displays, as the Chinese had 1,700 years earlier.

England. The book Art of Gunnery, published in London in 1647, included 43 pages on gunpowder rockets.

The Netherlands. The Dutch were using gunpowder rockets by 1650.

Germany. Germans experimented with gunpowder rockets in 1668. Field artillery colonel Christoph Fredrich von Geissler manufactured 55-lb. to 120-lb. rockets in 1730.

India. The British not only fought the French in the 18th century for the riches of India, but also the Mogol forces of Tippoo Sultan of Mysore. Tippoo Sultan's father, Hyder Ally, had 1,200 rocketeers in his army in 1788. Tippoo Sultan increased the rocket force to 5,000 men, about one-seventh of his Army. They fired gunpowder rockets in 1792 and 1799 against the British in two battles of Seringapatam. Today, the British have a Tippoo Sultan rocket in the Royal Ordnance Museum, Woolwich Arsenal, London.

Congreve. Sir William Congreve developed a series of 18-lb. to 300-lb. British barrage rockets which were used against Napoleon.

Napoleon. Surprisingly, Napoleon was an artillery officer who didn't use rockets in the French Army, but he wasn't able to see the advantage over cannons.

Denmark. Some 25,000 Congrieve rockets were fired in an 1807 barrage on the Danes at Copenhagen. Many houses and warehouses burned. Later, an official rocket brigade was added to the British Army in 1818.

America. The British 85th Light Infantry fired rockets in 1814 against an American rifle battalion led by U.S. Attorney General William Pickney during the Battle of Bladensburg in the War of 1812. "Never did men with arms in their hands make better use of their legs," British Lieutenant George R. Gleig wrote later.

Mexico. A brigade of rocketeers accompanied Maj. Gen. Winfield Scott's 1846 expedition into Mexico. The U.S. Army's first battalion of rocketeers—150 men with 50 rockets—was led by First Lieutenant George H. Talcott.

The rocket battery was used in the siege of Veracruz March 24, 1847, against Mexican forces. On April 8, the rocketeers moved inland, placed in their firing position by Captain Robert E. Lee. They fired 30 rockets in the battle for Telegraph Hill, then used their rockets in the capture of the Chapultepec fortress, forcing the surrender of Mexico City.

Civil War. The rocket battalion was disbanded after the Mexican War and the remaining rockets were stored for 13 years By the time they were hauled out in 1861 for the Civil War, the rockets had deteriorated. Replacements were required.

Maj. Gen. J.E.B. Stuart's Confederate cavalry fired the first rockets of the Civil War at Maj. Gen. George B. McClellan's Union troops at Harrison's Landing, Virginia, July 3, 1862.

The Union Army's New York Rocket Battalion was 160 men under the command of British-born Major Thomas W. Lion. When it tried in 1862 to fire rockets against Confederates defending Richmond and Yorktown, the missiles skittered wildly across the ground, between the legs of mules. One detonated harmlessly under a mule, lifting the animal several feet off the ground.

Maj. Gen. Alexander Schimmelfennig's Union troops used rockets at night to chase Confederate picket boats away from Charleston, South Carolina, in 1864.

First satellite? In his book, Our Incredible Civil War, Burke Davis recounted how a Confederate 12-ft. solid-fuel ballistic missile was fired at Washington from outside Richmond. The rocket had 10 lbs. of gunpowder in a brass warhead engraved with the letters C.S.A. for Confederate States of America. Jefferson Davis and others who witnessed the event, saw the rocket ignite and roar up and out of sight. Since no one saw the rocket land, some have speculated whether a satellite engraved with C.S.A. might have been launched into orbit 93 years before Sputnik.

Whaling. The international whaling industry developed explosive-tipped harpoon rockets to kill the mammals in oceans around the world.

Tsiolkovsky. Russian theorist and school teacher Konstantin Eduardovich Tsiolkovsky was born in 1857, the son of a Polish forester who had moved to Russia. Childhood scarlet fever left Tsiolkovsky nearly deaf, but he taught himself mathematics and became a high school teacher in the small town of Kaluga, 90 miles south of Moscow. Like Transylvanian physicist Hermann Oberth, Tsiolkovsky was captivated at an early age by Jules Verne's books. Verne's novels inspired him to design airships, planes, and rockets. He was encouraged by chemist Dmitry Mendeleyev and foreign astronautics pioneers such as Oberth.

Tsiolkovsky wrote science-fiction space adventures describing artificial satellites, spacesuits, space colonies, and asteroid mining. The titles included On the Moon published in 1895, Dreams of the Earth and Sky in 1895, and Beyond the Earth in 1920.

From 1896 to 1913, Tsiolkovsky's space science articles appeared infrequently in Russian journals, but were not noticed abroad. Best known was his description of the motion of a rocket in weightlessness and in a vacuum in an article, Research into Interplanetary Space by Means of Rocket Power, in 1903.

Goddard. Robert Hutchings Goddard was born in 1882. In 1898, the 16-year-old Goddard was excited by War of the Worlds, a new book by English novelist H.G. Wells. Three decades later he would fire the first liquid-fuel rocket.

Wright brothers. In 1903, Wilbur and Orville Wright were the first men to fly, making the first airplane flight at Kill Devil Hill near the village of Kitty Hawk, North Carolina. Even then, Robert Goddard was scheming over rockets to explore the upper atmosphere and probe space.

World War I. In France, Navy lieutenant Y.P.G. LePrieur was designing aerial rockets. Despite objections from pilots who didn't like shooting fire arrows from biplanes covered with flammable cloth and varnished wings, the aerial rockets were used in World War I to shoot down enemy hydrogen observation balloons.

Goddard. Robert Goddard formulated rocket plans while working for a doctorate at Clark University, Worcester, Massachusetts. His paper, A Method of Reaching Extreme Altitudes, was published in 1919. In it, he explained rocketry, even suggesting a demonstration rocket be flown to the Moon. That led many to think of Goddard as eccentric, which turned him against publicity.

Tsiolkovsky. Konstantin Tsiolkovsky retired in 1920 because of ill health and received a government pension. When Oberth's works were published in Europe in 1923, Tsiolkovsky's earlier astronautics articles were expanded and republished. They were popular in Russia and finally brought him international recognition.

Tsiolkovsky wrote about multistage rockets in 1929. He never built rockets, but encouraged young engineers who did, including Sergei Korolev who became chief designer of USSR spacecraft in the 1950's.

von Braun. Wernher von Braun was born in 1912. While Goddard's Moon rocket plan made him look like an oddball, German teenager Wernher von Braun's bent for explosives almost earned him the label of juvenile delinquent. Two things sparked the idea of delinquency:

His father couldn't understand why a 13-year-old liked to play with fireworks. Would Wernher grow up to become a safecracker?

Then the teenage von Braun lit six skyrockets tied to his wagon. In a cloud of smoke, the little red wagon whizzed five blocks into the center of town. The rockets exploded, charring the wagon. Wernher was dragged home by a policeman.

Not dissuaded of his interest in rocketry, Wernher von Braun had earned a doctorate in physics by age 22.

Oberth. Rumanian writer Hermann Oberth's paper, The Rocket into Interplanetary Space, was published in 1923 and inspired future rocket designers and spacecraft planners. Wernher von Braun read it in 1925 and joined Oberth in rocket experiments in 1930.

Goddard. On March 16, 1926, a year after von Braun had burned up his red wagon, Robert Goddard launched the first liquid-fuel rocket ever to fly.

He ignited the rocket in a snowy pasture on his Aunt Effie's farm at Auburn, Massachusetts. It flew up to altitude of 184 feet and landed in a cabbage patch 152 feet from the launch site—about the same distance as the Wright Brothers flight.

On December 30, 1930, Robert Goddard fired a liquid-fuel rocket to 2,000 feet altitude from Roswell, New Mexico.

He was approached in 1930 by the American Interplanetary Society but refused the publicity. Goddard earned 214 rocket patents by the time he died in 1945. NASA named its Goddard Space Flight Center, Greenbelt, Maryland, to honor Robert Goddard.

Tsiolkovsky. Like Goddard, Konstantin Tsiolkovsky had expected rockets to use liquid fuels because solid propellants of the time were not sufficiently powerful. One difference: Goddard went beyond theory to build rockets, while Tsiolkovsky wrote about astronautics and designed liquid-fuel rockets from 1903-1926, but didn't build any before he died in 1935. Later, to honor Tsiolkovsky, a museum was built at Kaluga and a crater on the back of the Moon was named for him.

Germany. Verein fur Raumschiffahrt, the Society For Space Travel, was founded in 1927 at Breslau, Germany. Wernher von Braun was an early member.

Winkler. Johannes Winkler, a member of Verein fur Raumschiffahrt launched Europe's first liquid-fuel rocket on February 21, 1931.

Ley. American Interplanetary Society founders Edward and Lee Pendray visited Prof. Willy Ley of the German Rocket Society in 1931 for a tour of his rocket launch site in the suburbs of Berlin.

Dornberger. The next year, in April 1932, the German army assigned Capt. Walter Dornberger to start work on a liquid-fuel military rocket. Dornberger's command was to build small rockets with the help of Verein fur Raumschiffahrt. Dornberger got to know Von Braun when the society test-fired a Mirak rocket for the army in July 1932. That October, von Braun went to work for Dornberger on liquid-fuel rockets. The society disbanded when most of the club followed von Braun into national service.

Pendray. Back in the U.S., Edward and Lee Pendray persuaded the American Interplanetary Society to build a rocket. The first test in November 1932 fizzled.

German A2. By December 1934, von Braun and the German army scored their first success with an A2 rocket fueled with ethanol and liquid oxygen, fired from their testing ground at Kummersdorf on the outskirts of Berlin.

Cal Tech. In the U.S. October 31, 1936, California Institute of Technology students and friends blasted off a small liquid-fuel rocket in the dry Arroyo Seco riverbed, near where today's Jet Propulsion Laboratory stands outside Pasadena.

Peenemuende. Rocket research outgrew Kummersdorf and was removed in 1936 to the remote island, Peenemuende, on Germany's Baltic Sea coast.

Peenemuende today is a village on a small island at the mouth of the Peene River, in what until 1990 was East Germany.

The infamous Peenemuende Proving Ground and its secret industrial town were created in 1936 by Major General Dornberger, who became director, and von Braun, now a rocket scientist. The remote proving ground and village, four hours north of Berlin near the post-war border with Poland, housed an enormous missile factory, a liquid-oxygen rocket fuel plant, bunkers, underground tunnels, and a rail network to move the V-2 rockets.

The Dornberger-von Braun team of 100 engineers grew into the secret industrial town of Peenemuende with a population of 17,000. Dornberger later wrote that the telephonists were physicists, the drivers were engineers and the kitchen help aerodynamics experts. World War II prisoners were used as forced labor.

A4. The German army had completed development of its A3 rocket by 1936. In October of that year, German engineers started designing a new, very much larger, liquid-fuel rocket they labeled A-4.

Dornberger's team launched 70 A3 and A5 rockets into the Baltic between 1937 and 1941, testing parts of the future A4. After years of break-neck development by the Third Reich, Dornberger and von Braun finally tested an A-4 from Peenemuende in June 1942. It splashed into the Baltic Sea one-half mile from the launch site. In a second launch in August, an A4 climbed seven miles high, then exploded.

While the earlier rockets had been used for research, von Braun and Adolf Hitler's engineers saw the A4 as a final product, a practical military weapon.

Vengeance. By its third flight, the A-4 had been mutated into the World War II terror weapon Vergeltungswaffe Zwei (Vengeance No. 2). That flight was launched successfully October 3, 1942, on a perfect ballistic flight 120 miles downrange from Peenemuende Proving Ground test pad number 7 into the Baltic.

The infamous Vengeance No. 2 rocket came to be known as V-2. Prototypes were painted dark green, leading to their nickname "cucumber." Later production models were painted green and brown camouflage colors.

JPL. In the U.S., the Jet Propulsion Laboratory was founded November 20, 1943, to expand on the rocket research work done by California Institute of Technology students from 1936 in the dry Arroyo Seco riverbed outside Pasadena.

A9. In Germany, for the fuhrer, the army rocket engineers wanted to develop a far more powerful A-9 rocket they called Amerika, capable of reaching the United States.

Hitler saw the advantage of having atomic bombs in his arsenal, but apparently was slow to see strategic value in the big rockets. The massive A-9 was not built.

Dornberger said after the war crucial time between 1939 and 1942 had been squandered when Hitler didn't grasp the potential and refused to give priority to A-4, V-2 and A-9.

V-2 production. In 1943, the Nazi high command did see a Peenemuende demonstration of a 40-ft. V-2 carrying a one-ton conventional warhead. When the high command decided to speed up the program at that late hour, the V-2 became the first mass-produced long range rocket.

Despite a severe pounding of Peenemuende in August 1943 by British bombers, the first V-2s were delivered in September 1944. They were fired immediately against London.

Hitler demanded 2,000 V-2s per month, which was four times what the factory could produce. The laborers actually were able to roll 500 V-2s off the production line each month onto trains to launch pads along the occupied English Channel coast. But, the V-2s arrived too late to affect the outcome of the war.

Red hot darts. For the victims, V-2s appeared suddenly, at the last minute, as glowing red darts in the sky. Tens of thousands of persons were killed and even more buildings were destroyed.

"For the first time, a machine built by man hit the Earth with a force equivalent to that of fifty 100-ton express locomotives travelling at 60 mph," Dornberger wrote after the war.

Some 6,000 were made during World War II with 3,225 blasted across the Channel at England, at Antwerp, Belgium, and on France. A thousand were fired against the Russians on Germany's Eastern Front.

Surrender. The German Army was in full retreat everywhere by April 1945. Hitler had committed suicide in his Berlin bunker. Soviet troops captured Peenemuende that month.

Hitler had ordered his SS to kill all German rocket engineers to prevent their capture. In early May, Wernher von Braun and more than 100 rocket engineers were hiding out at Haus Ingeburg, an inn near Oberjoch. Magnus von Braun, Wernher's brother, slipped into American lines to arrange for their safety. On May 2, the day the Soviet army captured Berlin, Wernher von Braun led his scientists and engineers behind American lines. World War II ended May 8, 1945.

The U.S. Army thoroughly debriefed the von Braun team, keeping them away from the Russians. Assembled V-2s and parts for other V-2s were collected and shipped to the U.S.

Clarke. Arthur C. Clarke wrote of communication satellites in 1945, suggesting the possibility of a global communication system using satellites in stationary orbit. He assumed it would involve manned space stations.

Korolev. Gen. Dwight D. Eisenhower permitted his Allied troops one last series of V-2 launches in Europe, from the North Sea port of Cuxhaven at the mouth of the Elbe River. Cuxhaven is in Lower Saxony in what after World War II became West Germany. Three V-2s were fired from Cuxhaven in June 1945.

Most of von Braun's team were not on hand for the Cuxhaven launches. However, on the reviewing stand for the Cuxhaven blast offs was Russian Army Col. Sergei Korolev.

Ten years later, Korolev would be chief designer of USSR spacecraft, responsible for the Vostok, Voskhod and Soyuz capsules which have carried all cosmonauts orbited since 1961.

Coming to America. Some 65 captured V-2s, along with spare parts and data files, were removed in 1945 by the U.S. to the New Mexico desert near Las Cruces. The hardware and files filled 300 railroad freight cars.

As the last V-2s were blasting off from Cruxhaven in June 1945, most of the von Braun team of rocket experts were at Fort Bliss, Texas, near El Paso. In 1946, they moved on to nearby White Sands Proving Ground, New Mexico. By February 1946, the entire Peenemuende team had reorganized at White Sands.

NRL. The U.S. Navy, Bureau of Aeronautics, started in October 1945 planning to build a liquid-fuel space rocket to launch a man-made artificial Earth satellite. The Naval Research Laboratory started planning sounding rockets to send TV pictures from high altitudes.

Corporal. JPL was developing three military missiles for the U.S. Army: Private, Corporal and Sergeant. The version of the Corporal rocket used for science research was called WAC—for Women's Army Corps—because it was thought of as the military Corporal's little sister.

Jet Propulsion Laboratory's first sounding rocket was a WAC-Corporal launched October 11, 1945, from the new U.S. Army site at White Sands, New Mexico. Reaching 33 miles altitude, the WAC-Corporal was the highest flier to that time.

Van Allen. In 1946 in the U.S., James A. Van Allen, a cosmic ray physicist at Applied Physics Lab, Johns Hopkins University, fired rockoons. They were small sounding rockets carried to high altitudes by balloons, then blasted above the atmosphere to measure ozone and cosmic rays.

V-2 Panel. A U.S. group known as the V-2 Panel, including Harvard astronomer Fred Whipple, Johns Hopkins' James A. Van Allen, Princeton's M.H. Nichols, Michigan's W.G. Dow, was formed February 27, 1946, to design science gear to ride in empty nosecones of V-2 test flights, to study cosmic rays, solar energy and the atmosphere.

V-2 launches. The U.S. space program began when Wernher von Braun launched a captured V-2 for the U.S. Army at White Sands, New Mexico, on April 16, 1946. That first V-2 flew up about 3 miles, then a fin fell off.

A month later, on May 10, von Braun successfully launched a V-2 from White Sands to an altitude of 71 miles. From then to September 1952, 64 V-2s were launched from White Sands. All carried science instruments, none had warheads.

Life forms. The first primates in space, the monkeys Albert 1 and Albert 2, died in nose cones during early V-2 test flights.

Aerobee. The U.S. Navy, Bureau of Ordinance, in 1946 designed the successful Aerobee high-altitude sounding rocket.

V-2 Aerobee. The U.S. Air Force fired V-2 Aerobee high-altitude research rockets from Wallops Island, Virginia, from 1946-1950. V-2 Aerobee added an Aerobee upper stage to a V-2 rocket.

Viking. The U.S. Naval Research Laboratory in 1946 designed the Viking high-altitude sounding rocket.

Satellites. The U.S. Navy started a space satellite program in 1946.

Juarez V-2. A gyroscope broke during a V-2 test flight from White Sands on May 29, 1947. The missile streaked over El Paso, Texas, crossed the Rio Grande River, smashing into a hill beyond Juarez, Mexico.

Midway. The U.S. Navy launched a V-2 from the carrier Midway in September 1947.

Bumper-WAC. Jet Propulsion Laboratory had been developing the Corporal missile for the Army. The WAC-Corporal for science research was JPL's first sounding rocket. A WAC-Corporal rocket was bolted on top of a V-2 rocket to gain altitude. The two-stage combination was called Bumper-WAC.

First in outer space. On May 13, 1948, the first of the two-stage Bumper-WAC rockets was fired from White Sands. The first four Bumper-WAC launches failed.

The fifth Bumper-WAC flight was a success on February 24, 1949, reaching a record altitude of 244 miles. It was the first rocket in outer space.

Cape Canaveral. A need for more launch-pad acreage led to establishment of the Joint Long Range Proving Ground in 1949 at the deserted, remote Cape Canaveral, Florida.

Bumper-WAC. A two-stage Bumper-WAC was the first rocket launched from Cape Canaveral, July 24, 1950.

Redstone Arsenal. In 1950, Werhner von Braun and his U.S. Army rocket team moved to Redstone Arsenal, Huntsville, Alabama. Wernher von Braun died in 1977.

Atlas. The U.S. Air Force started designing the Atlas Intercontinental Ballistic Missile (ICBM) in 1951. Atlas-D, a conversion of the ICBM rocket, would carry the first American to orbit in 1962—astronaut John Glenn in Mercury-Atlas 6. Mercury capsule weight required the heavy-lifting ability of Atlas-D, the only U.S. rocket in the 1960's with sufficient power to lift a heavy satellite to 100 miles.

Life in space. A monkey and mice were shot to the edge of space on a U.S. Aerobee rocket early in 1951. They died when their parachute failed to open.

In the first successful spaceflight for living creatures, a monkey and 11 mice were shot to the edge of space on an Aerobee rocket September 20, 1951, and were recovered alive.

Redstone rocket. Wernher von Braun expanded the V-2 rocket design and named the new booster after his new post, Redstone Arsenal at Huntsville, Alabama. The first test launch of a Redstone rocket was August 20, 1953.

Eight years later, a Redstone would carry the first American to space in 1961, astronaut Alan Shepard in Mercury-Redstone 3. Redstones were used to launch America's first two manned capsules which reached suborbital speed and altitude.

Titan. The U.S. Air Force started designing its Titan ICBM in 1955. Today, the Titan-4 rocket is America's most powerful launcher for unmanned payloads.

Jupiter. Wernher von Braun set up the U.S. Army Ballistic Missile Agency at Redstone Arsenal in 1956 to develop Jupiter, an intermediate range ballistic missile (IRBM), from the Redstone rocket.

Jupiter-C. The U.S. Army launched its first Jupiter-C version of the Redstone rocket on September 20, 1956. Sixteen months later, a Jupiter-C would carry America's first satellite to space in 1958.

USSR space rockets. Sergei Korolev, who witnessed the 1945 Cuxhaven launches, returned to the USSR after the war. His Korolev Design Bureau developed the Soviet Union's first intercontinental ballistic missile (ICBM), known as R-7.

The first flight of the R-7 was August 3, 1957. Just three months later, a version of the R-7 would orbit the first Russian Sputnik satellite.

Two confusing numbering systems for USSR space rockets were added in the U.S. The Congressional Research Service listed rockets as A, B, C, etc. The Department of Defense used SL, for space launch, as in SL-1 or SL-2. ICBM versions of the rockets were given SS numbers, as in SS-4 or SS-5. The A or SL-1 rocket lobbed Sputniks 1, 2 and 3 to orbit. Its ICBM version was called SS-6.

Thor. The U.S. launched its first Thor rocket September 20, 1957.

Sputnik. The Space Age started October 4, 1957, when the USSR launched Sputnik, the world's first man-made artificial satellite to orbit Earth.

Vanguard. A Vanguard rocket failed December 6, 1957, in the Navy's first attempt to beat the Army in launching the first U.S. satellite. In case it were to fail, the Navy had referred to the rocket before launch as TV-3 or Test Vehicle No. 3. With worldwide media tuned in, Vanguard lost thrust two seconds after launch, just four feet off the pad. It fell back and exploded. The six-in.-diameter satellite popped out of the flames and rolled away, transmitting its radio signal on the ground. Newspapers called it names such as Kaputnik and Stayputnik.

IGY. During the 1957-58 International Geophysical Year (IGY), dozens of sounding rockets were fired toward space to measure the atmosphere and collect ultraviolet and X-ray energy from the Sun and solar flares.

The V-2 Panel, established in 1946, was renamed the Rocket and Satellite Research Panel. It conducted the IGY experiments.

The sounders included many U.S. Navy Aerobee and U.S. Army Nike-Cajun rockets. Nike-Cajuns were Nike-Ajax air defense missiles with the Ajax stage replaced by a Cajun stage to lift science instruments to an altitude of 145 miles.

Vanguard. The U.S. Navy tried again to beat the Army to space with America's first satellite on January 25, 1958, but the Vanguard rocket fizzled within 14 seconds of ignition.

Explorer. The U.S. Army successfully launched the first U.S. satellite to orbit from Cape Canaveral on a Jupiter-C rocket on January 31, 1958, making the U.S. the second nation with a space rocket. The Explorer 1 satellite had been designed by Jet Propulsion Laboratory.

Vanguard. The U.S. Navy finally launched successfully its Vanguard satellite from Cape Canaveral March 17, 1958.

Atlas. The U.S. successfully launched its first Atlas rocket August 2, 1958.

NASA. The National Aeronautics and Space Act of 1958 legislation, enacted July 29, 1958, enabled the National Aeronautics and Space Administration (NASA) which then was born October 1. The Department of Defense had fought the Space Act legislation, wanting to keep space research entirely military. NASA incorporated the former National Advisory Committee for Aeronautics (NACA).

USSR A-1. The USSR added an upper stage to its A rocket in 1959 for extra power needed to lift cosmonauts to orbit in Vostok capsules and for firing probes to the Moon. Cosmonaut Yuri Gagarin was the first human in space, making one orbit of Earth in Vostok 1 on April 12, 1961. A-1 also is known as SL-3.

Redstone Arsenal. The Army Ballistic Missile Agency and the von Braun team were transferred to NASA in 1960. The Army Missile Command owns Redstone Arsenal.

Saturn. In 1961, President John F. Kennedy committed the U.S. to sending astronauts to the Moon by the end of the 1960's. New rockets were needed, each powerful enough to carry three astronauts and equipment to the Moon. The resulting Saturn V, the most powerful rocket to that time, carried teams of astronauts to land on the Moon from 1969-72. The first manned Saturn V flight was December 21, 1968. Another Saturn V launched the U.S. Skylab space station to Earth orbit May 14, 1973. The final Saturn V orbited America's Apollo capsule in 1975's joint U.S.-USSR Apollo-Soyuz Test Project.

USSR A-2. The USSR added an even more powerful upper rocket stage to the A rocket in 1963 to launch even heavier payloads, including the three-man Voskhod capsules. Voskhod 1 was launched October 12, 1964. The A-2 also has been used since 1967 to launch all Soyuz manned capsules. The first Soyuz was launched April 24, 1967. A-2 also is known as SL-4.

USSR A-2e. In 1961, a version of the A rocket, referred to in the U.S. as A-2e, had a third stage added to carry satellites very high to stationary orbits. A-2e also is known as SL-6.

USSR B-1. A version of the SS-4 medium-range ballistic missile (MRBM), known as B-1 or SL-7, was used to orbit small payloads from 1962-77.

USSR C-1. Since 1964, the C-1 or SL-8 version of the SS-5 intermediate-range ballistic missile (IRBM), has been used to orbit medium-weight payloads and for multiple-payload flights.

USSR D-1. The D-1 or SL-9, known as Proton and used since its first flight on July 16, 1965, is not derived from an ICBM. The USSR uses it to orbit large payloads such as Salyut and Mir space stations.

Asterix. France was the third nation able to fly a satellite to Earth orbit. The 19-lb. Asterix 1 was lobbed to space November 26, 1965, on a Diamant rocket. The radio beacon in Asterix lasted two days. The satellite fell from orbit to burn up in the atmosphere March 25, 1967.

USSR F-1r and F-1m. The F-1r and the F-1m have been used since 1966. F-1r may be an A rocket with re-entry stage added. F-1m may be an A rocket with maneuvering stage added. They also are known as SL-11.

USSR D-1e. The D-1e or SL-12, used since 1967 for interplanetary flights, has an extra Earth-escape stage added. The upper stage of the rocket goes into Earth orbit, then the escape stage fires, sending a probe away from Earth to a distant Solar System body.

Japan. Japan launched its first satellite, Ohsumi 1, to orbit February 11, 1970, making it the fourth nation with a space rocket.

USSR D-1m. The D-1m, was used in 1970 to launch the highly-maneuverable Cosmos 382. The satellite may have been a maneuvering third stage added to the rocket.

Long March. China's first satellite, named Mao 1, was launched to Earth orbit April 24, 1970, on a Long March 1 rocket.

Black Arrow. Great Britain's first satellite, Black Knight 1, was launched on a Black Arrow rocket from Woomera, Australia, October 28, 1971.

USSR F-2. The D-1 or SL-14 has been used since 1977, replacing the A-1 for launches to polar orbit since 1985.

Ariane. The first Ariane rocket launched by the European Space Agency from Kourou, French Guiana, carried ESA's first satellite, referred to as CAT, to orbit December 24, 1979.

SLV-1. India became the seventh nation to demonstrate a space rocket when it launched its first satellite, Rohini 1, on a Satellite Launch Vehicle (SLV-1) rocket July 18, 1980.

Space shuttle. NASA needed powerful new rockets to lift heavy space shuttles to Earth orbit. Liquid-fuel main engines and solid-fuel boosters were designed. The three main engines were built to be reused to 7.5 hours. Columbia flew in 1981. Challenger, in 1983. Discovery, in 1984. Atlantis, in 1985. Challenger exploded during lift off in 1986. Its replacement, shuttle Endeavour, was expected to fly in 1992.

Shavit. Israel orbited its first satellite, Horizon 1 or Ofek 1, firing it to space on a three-stage Shavit rocket September 19, 1988, from a military launch pad in the Negev Desert. Shavit is Hebrew for comet. Horizon 1 fell from orbit after 4 months.

Energia. The world's most powerful rocket is Energia, or G-1 in the U.S. numbering system. It was launched unmanned May 15, 1987. Energia later carried the USSR's first shuttle, Buran, unmanned, from Baikonur Cosmodrome to orbit 155 miles above Earth on November 15, 1988. Energia has greater lifting power than the mammoth Saturn V rocket used by the U.S. to send astronauts to the Moon in 1969-72.

Titan 4. Currently the most powerful U.S. unmanned rocket, Titan 4 thundered to space from Cape Canaveral in its maiden voyage June 14, 1989.

Iraq. Iraq launched a 48-ton three-stage rocket into orbit December 5, 1989, from the Al-Anbar Space Research Center 50 miles west of Baghdad, a six-orbit flight.

Shavit. Israel launched a second satellite, Horizon 2, to Earth orbit on a Shavit rocket on April 3, 1990.

Pegasus. Two days later, the first all-new unmanned U.S. space rocket since the 1960's was launched April 5. Pegasus can orbit satellites up to 600 lbs. The winged, three-stage, solid fuel Pegasus is strapped under the wing of a B-52 bomber, ferried seven miles above Earth, dropped free and blasted to orbit.

The Race To The Moon: 1961-1969

If Sputnik slapped America awake in 1957, the spaceflight of Yuri Gagarin in 1961 was the final prod to action. Just six weeks after Gagarin flew Vostok in Earth orbit, U.S. President John F. Kennedy told the world of America's great challenge—to land a man on the Moon by the end of the 1960's.

The United States seemed short on everything, except determination. Only three weeks earlier Alan Shepard had flown 15 minutes along a suborbital path. The U.S. had yet to send a man to orbit. There was no rocket big enough to send men and gear to the Moon, no planned route to the Moon, not even knowledge of whether the Moon's surface could hold up a man if he could land there.

It took 2,978 days and billions of dollars, but an American did land on the Moon, within the decade, on July 20, 1969. Here are some Moon Race mileposts:

The prod. Before Kennedy made his announcement May 25, 1961, the USSR posted several important space firsts which prodded America into action:

Year	First Achievement
1957	Space Age started when USSR launched the first artificial satellite, Sputnik 1.
1957	USSR sent the first higher life form to orbit, the dog Laika in Sputnik 2.
1959	USSR sent Luna 1, the first craft to fly by the Moon, first to leave Earth's gravity.
1959	USSR's Luna 2 was the first craft to impact on the Moon.
1959	The solar-powered Luna 3 circled Moon and recorded first photos of the dark side.
1961	Venera 1 was first to fly by Venus, first interplanetary probe to escape Earth orbit.
1961	Yuri Gagarin became the first man in Earth orbit, in Vostok 1 April 12.

How to start? The United States needed to know what the Moon really was like, as well as how to get there. Unmanned explorers had to be built and sent, including probes named Ranger and Surveyor.

Ranger. The first NASA program to investigate another Solar System body—the Moon—was a series of probes named Ranger in the early 1960's.

Rangers were supposed to bounce onto the lunar surface carrying a seismometer, but the first six failed. The first five, launched between August 1961 and October 1962, failed. After that, they were redesigned for photography. Ranger 6, launched January 30, 1964, reached its target, but its cameras broke down.

Ranger again was redesigned. Pictures were to be snapped as a craft approached its bulls-eye on the Moon. Thus, the last three Rangers became successful. They took the first close-ups of the Moon surface, showing boulders and three-foot craters.

Ranger 7, the first success, was launched July 28, 1964. It impacted in an unnamed mare area later named Mare Cognitum, which meant "the sea which has become known." The level mare without craters was chosen because it might make a good landing spot for later Apollo manned flights. More than 400 photos from six sequential TV cameras were transmitted to Earth. The widest angle covered a square mile and showed 30-ft. craters. The most detailed pictures revealed 3-ft. craters in an area 100 ft. by 160 ft. Ranger 7 sent 4,300 photos.

Ranger 8 was fired February 17, 1965. Its bulls-eye was in the southwest part of Mare Tranquillitatis. Ranger 8 transmitted 7,100 photos.

Ranger 9, which blasted-off March 21, 1965, was targeted on the floor of the Alphonsus Crater. It landed there, between the central peaks and rilles near the crater's east wall. Ranger 9 returned 5,800 photos.

The Rangers showed that the lunar mare were free of small craters, cracks and rock fields, and would support the weight of a manned spacecraft. They accurately measured the Moon's radius and mass.

Surveyor. A series of seven U.S. unmanned TV and trenching probes named Surveyor, designed to land softly on the Moon, were launched between May 1966 and

January 1968. The seven Surveyors were part of America's search for Apollo moonship landing sites. Five were successful in soft landings.

Surveyors had a moveable TV camera powered by solar cells and a trench digger along with the capability to retrieve and analyze a soil sample.

In the spring of 1966, Surveyor 1 made the first soft landing on the Moon for man's first close-up look at the surface. The three-legged spacecraft flew 240,000 miles to land within nine miles of target. Over eight months it transmitted 11,150 pictures, including panoramas and close-ups.

Surveyors found lunar maria to be basaltic and the lunar highlands rich in aluminum and calcium.

Within two years, four other Surveyors landed to study Moon chemistry and physics. Information retrieved by the five Surveyors was useful in the manned Apollo flights from 1969-72.

Probe	Launch	Destination
Surveyor 1	1966 May 30	Oceanus Procellarum, soft landing; 11,150 photos sent to Earth.
Surveyor 2	1966 Sep 20	Oceanus Procellarum, crash landing.
Surveyor 3	1967 Apr 17	Oceanus Procellarum, soft landing, tested soil, sent 6,315 photos back to Earth, Apollo 12 visited later.
Surveyor 4	1967 Jul 14	Sinus Medii, soft landing, radio contact lost.
Surveyor 5	1967 Sep 8	Mare Tranquillitatis, soft landing, analyzed soil, 18,006 photos.
Surveyor 6	1967 Nov 7	Sinus Medii, soft landing, analyzed soil, 30,000 photos.
Surveyor 7	1968 Jan 7	near Tycho, soft landing, analyzed soil, 21,000 photos.

Man in space. Capsules had to be built to carry first one, then two and finally three men to Earth orbit and then on to the Moon. Those were Mercury, Gemini and Apollo.

Mercury. America's first men in space went there in Mercury capsules. There were a total of 23 launches from Cape Canaveral in the first U.S. manned spaceflight program— Project Mercury.

Of the 23 launches in the Mercury series, only about a quarter carried men. Of the 23, thirteen carried no crew, four carried animals and six carried astronauts. Sixteen of the flights were successful. There were seven launch failures. There were no flight failures during manned launches.

Various rockets were used for the flights, including seven Little Joe boosters, three Atlas, six Atlas D, one Atlas Big Joe, five Redstone and one Blue Scout. The time between launches ranged from three days to 224 days, an average of 61 days between tests.

Year	Mercury Flight
1959	Mercury Big Joe, unmanned suborbital, Atlas rocket.
1959	Mercury Little Joe 1, unmanned suborbital, Little Joe rocket.
1959	Mercury Little Joe 2, unmanned suborbital, Little Joe rocket failed.
1959	Mercury Little Joe 3, suborbital, Little Joe rocket, carried the monkey Sam, biomedical and escape system test.
1960	Mercury Little Joe 4, suborbital, Little Joe rocket, carried the monkey Miss Sam, biomedical and escape system test.
1960	Mercury-Atlas 1, unmanned suborbital, Atlas rocket failed.
1960	Mercury Little Joe 5, unmanned suborbital, Little Joe rocket, escape rocket fired prematurely.
1960	Mercury-Redstone 1, unmanned suborbital, Redstone rocket.
1961	Mercury-Redstone 2, suborbital, Redstone rocket, carried Chimp Ham on 16-min. flight.
1961	Mercury-Atlas 2, unmanned suborbital, Atlas rocket.
1961	Mercury Little Joe 5A, unmanned suborbital, Little Joe rocket, escape system failed.
1961	Mercury-Redstone BD, unmanned suborbital, Redstone rocket.
1961	Mercury-Atlas 3, unmanned suborbital, Atlas rocket failed.
1961	Mercury Little Joe 5B, unmanned suborbital, Little Joe rocket failed.

1961   Mercury-Redstone 3, May 5 suborbital, Redstone rocket, astronaut Alan B. Shepard Jr. in
        Freedom 7 capsule, time 15 mins 22 secs, first American manned spaceflight.
1961   Mercury-Redstone 4, July 21, suborbital, Redstone rocket, astronaut Virgil I. "Gus"
        Grissom in Liberty Bell 7 capsule, second American manned spaceflight,
        capsule sank to Atlantic Ocean floor, astronaut rescued.
1961   Mercury-Atlas 4, rocket Atlas D, first orbital flight to orbit.
1961   Mercury-Scout 1, unmanned, Blue Scout rocket.
1961   Mercury-Atlas 5, rocket Atlas D, carried chimp Enos, two orbits of Earth, November 29.
1962   Mercury-Atlas 6, rocket Atlas D, February 20, astronaut John H. Glenn Jr., Friendship 7
        capsule, first U.S. manned orbital flight, first American in orbit,
        third person to orbit, three revolutions, 4 hours 55 minutes 23 seconds.
1962   Mercury-Atlas 7, rocket Atlas D, May 24, astronaut M. Scott Carpenter in Aurora 7
        capsule, 2nd U.S. manned orbital flight, 3 revolutions, 4 hrs 56 mins 5 secs.
1962   Mercury-Atlas 8, October 3, rocket Atlas D, astronaut Walter M. "Wally" Schirra Jr. in
        Sigma 7 capsule, third U.S. manned orbital flight, 6 revs, 9 hrs 13 min 11 sec.
1963   Mercury-Atlas 9, May 15, rocket Atlas D, astronaut Leroy Gordon "Gordo" Cooper Jr. in
        Faith 7 capsule, fourth U.S. manned orbital flight, first U.S. astronaut to stay
        a day in space, final Mercury flight, first live TV from U.S. craft,
        22 revs in 34 hrs 19 min 49 sec.

Gemini. America's second man-in-space program featured two-man capsules and
was called Project Gemini. There were 10 launches of 20 astronauts in 1965 and 1966, all
successful. Each Gemini capsule was lofted by a Titan 2 rocket from Cape Canaveral.

Year	Gemini Flight
1965	Gemini 3, March 23, capsule nicknamed Molly Brown, the first Gemini capsule to orbit, astronauts Virgil I. "Gus" Grissom and John W. Young, the first manned craft to change orbit, the first corned beef sandwich in orbit. Flight duration (hours:minutes:seconds): 4:53:00 for 3 orbits.
1965	Gemini 4, June 3, the second Gemini capsule to orbit, astronauts James A. McDivitt and Edward H. White 2nd. White took the first American spacewalk for 20 minutes. Flight duration: 97:56:11 for 62 orbits.
1965	Gemini 5, August 21, third Gemini to orbit, first use of fuel cells for electricity, astronauts L. Gordon "Gordo" Cooper Jr. and Charles "Pete" Conrad Jr. Time: 190:55:14, 120 orbits.
1965	Gemini 6, December 15, astronauts Walter M. "Wally" Schirra Jr. and Thomas P. Stafford, completed the first space rendezvous with Gemini 7. The dual flights of Gemini 6 and 7 were called "The Spirit of Gemini 76." Duration: 25:51:24 for 16 orbits.
1965	Gemini 7, December 4, astronauts Frank Borman and James A. Lovell Jr., completed the first space rendezvous with Gemini 6, the dual flights of Gemini 6 and 7 called "The Spirit of Gemini 76." Gemini 7 was longest Gemini flight. Duration: 333:35:31 for 206 orbits.
1966	Gemini 8, March 16, astronauts Neil A. Armstrong and David R. Scott, first space docking, with Agena rocket. A broken control ended the flight. Duration: 10:41:26 for 6.5 orbits
1966	Gemini 9, June 3, astronauts Thomas P. Stafford and Eugene A. Cernan, rendezvous with Agena rocket but docking didn't work, Cernan took a spacewalk of two hours, his facemask fogged from heavy breathing. Flight duration: 72:21:00 for 44 orbits. The originally scheduled crew, Elliot See and Charles A. Bassett II had been killed February 28 in a crash of their T-38 training aircraft at Lambert Field, St. Louis, Missouri.
1966	Gemini 10, July 18, astronauts John W. Young and Michael Collins, docked with an Agena rocket. Collins made two spacewalks, one to retrieve a detector from the Agena. Flight duration: 70:46:39 for 43 orbits.
1966	Gemini 11, September 12, astronauts Charles "Pete" Conrad Jr. and Richard F. Gordon Jr., photographed stars and galaxies. They docked with an Agena target rocket. Gordon made a spacewalk, during which he tied Gemini and Agena together with a 100-ft. nylon rope. As the two craft circled each other in orbit, artificial gravity was created for the first time in space. Gemini 11 set the

Gemini altitude record of 739.2 miles. Flight duration: 71:17:08 for 44 orbits.
1966   Gemini 12, November 11, last Gemini flight, astronauts James A. Lovell Jr. and Edwin
E. "Buzz" Aldrin Jr., visual docking with an Agena rocket, then turned on
Agena rocket to fly the docked pair to a higher orbit. Aldrin made three
spacewalks for a record 5.5 hours of extravehicular activity (EVA) time. Flight
duration: 94:34:31 for 59 orbits.

Saturn V. A primary question in the early 1960's was, what path should be
followed to the Moon? Should a way-station be built in Earth orbit from which men could
journey on to the Moon? Should men fly to lunar orbit and descend from there to the
surface? Or should they leave Earth in a powerful direct flight to a Moon landing, with
the ability to lift off the Moon in a direct flight back to a landing on Earth?

A station in Earth orbit would have been too slow in building and too costly. The
direct surface-to-surface flight required too large a rocket and was too expensive. The
route to lunar orbit was chosen. Thus, the lunar excursion module (LEM) was added to the
command and service modules (CSM) and the mighty Saturn V rocket was built.

Saturn V was the world's most powerful rocket, able to carry 200,000 lbs. to Earth
orbit. Altogether, 13 Saturn Vs blasted off from Cape Canaveral between 1967 through
1973, including four to Earth orbit before manned Moon landings, eight to the Moon
and the last to carry America's Skylab space station to Earth orbit. The U.S. abandoned
Saturn V in 1973. The most powerful rocket today is the USSR's Energia which can carry
220,000 lbs. to orbit.

Apollo. America's third man-in-space program featured the three-man capsules
which would go to the Moon. In the program called Project Apollo, there were 12
launches carrying 36 astronauts between 1968 and 1972, all relatively successful.
Apollo capsules were lofted by Saturn 1B and Saturn V rockets from Cape Canaveral. The
first trip manned around the Moon was Apollo 8. The first manned Moon landing was
Apollo 11. The last Apollo Moon flight was Apollo 17. Apollo 18 was the joint U.S.-
USSR Apollo-Soyuz flight in Earth orbit in 1975.

Year	Apollo Flight
1967	Apollo 1, January 27, astronauts Virgil I. "Gus" Grissom, Edward H. White 2nd and Roger B. Chaffee. Not a launch, the crew was killed in launch-pad fire during ground test.
1967	Apollo 4, November 9, first unmanned flight of the mighty Saturn V rocket.
1968	Apollo 5, January 22, Saturn 1B rocket, first unmanned test of lunar excursion module (LEM) in Earth orbit.
1968	Apollo 6, April 4, second unmanned Saturn V launch.
1968	Apollo 7, October 11, Saturn 1B rocket, astronauts Walter M. "Wally" Schirra Jr., Donn F. Eisele, R. Walter Cunningham, first manned flight of Command Service Module (CSM), first colds caught in space. Duration (hr:min:sec): 260:09:03 for 163 orbits of Earth.
1968	Apollo 8, December 21, Saturn V rocket, astronauts Frank Borman, James A. Lovell Jr., William A. Anders, first manned flight around Moon, sent TV pictures of Moon. Duration: 147:00:42, including 10 orbits of the Moon in CSM.
1969	Apollo 9, March 3, Saturn V rocket, astronauts James A. McDivitt, David R. Scott, Russell L. Schweickart, CSM Gumdrop, LEM Spider, first manned flight of Lunar Excursion Module (LEM). Duration: 241:00:54, 151 orbits of Earth.
1969	Apollo 10, May 18, Saturn V rocket, astronauts Thomas P. Stafford, John W. Young, Eugene A. Cernan, full dress rehearsal for landing with CSM Charlie Brown and LEM Snoopy in Moon orbit. Duration: 192:03:23 including 31 orbits of the Moon.
1969	Apollo 11, July 16, Saturn V rocket, astronauts Neil A. Armstrong, Michael Collins, Edwin E. "Buzz" Aldrin Jr., first landing of men on Moon July 20. LEM Eagle landed at Mare Tranquillitatis. Armstrong first outside, Aldrin second. Armstrong said, "That's one small step for man...one giant leap for mankind."

Collins stayed in Moon orbit in CSM Columbia. Armstrong and Aldrin collected 48.5 lbs. of rocks and soil during 21 hrs 36 mins 21 secs on surface. Flight duration: 195:18:35 including 30 CSM orbits of the Moon.

1969   Apollo 12, November 14, astronauts Charles P. "Pete" Conrad Jr., Richard F. Gordon Jr., Alan L. Bean, lightning struck the Saturn V rocket twice during liftoff. Conrad & Bean in LEM Intrepid made man's second Moon landing, at Oceanus Procellarum (Ocean of Storms), a very large young area of the Moon previously visited by unmanned Luna 9, Luna 13, Surveyor 1 and Surveyor 3. Astronauts collected 74.7 lbs. of rocks & soil, 31 hrs 31 mins on surface. Gordon stayed in CSM Yankee Clipper. Duration: 244:36:25 including 45 orbits of the Moon by the CSM.

1970   Apollo 13, April 11, Saturn V rocket, astronauts James A. Lovell Jr., John L. Swigart Jr., Fred W. Haise Jr., CSM Odyssey oxygen tank ruptured in an onboard explosion causing the Moon landing to be aborted. Crew returned to Earth safely using LEM Aquarius as a lifeboat for oxygen and power. Duration: 142:54:41.

1971   Apollo 14, January 31, Saturn V rocket, astronauts Alan B. Shepard Jr., Stuart A. Roosa, Edgar D. Mitchell, man's third Moon landing by Shepard and Mitchell in LEM Antares at Fra Mauro formation cone crater, collecting 96 lbs. of rock, soil samples in 33 hrs 31 mins on surface. First 6-iron golf shot on Moon. Roosa stayed in CSM Kitty Hawk. Duration: 216:01:57 including 34 orbits of the Moon by the CSM.

1971   Apollo 15, July 26, Saturn V rocket, astronauts David R. Scott, Alfred M. Worden, James B. Irwin, man's fourth Moon landing by Scott and Irwin in LEM Falcon at Hadley Rille, first use of the lunar rover, they collected 170 lbs. of samples, 66 hrs 55 mins on surface. Worden in CSM Endeavor. First deep-space walk during flight. Duration: 295:11:53 including 74 orbits of the Moon by CSM.

1972   Apollo 16, April 16, Saturn V rocket, astronauts John W. Young, Thomas K. Mattingly 2nd, Charles M. Duke Jr., man's fifth landing by Young and Duke in LEM Orion at Cayley Descartes formation in the lunar highlands, collected 213 lbs. of samples in 71 hrs 2 mins on the surface. Mattingly in CSM. Duration: 265:51:05 including 64 orbits of the Moon by the CSM Casper.

1972   Apollo 17, December 7, astronauts Eugene A. Cernan, Ronald E. Evans, Harrison H. Schmitt. For this last manned Moon flight, the Saturn V rocket made the first U.S. manned launch at night. Man's sixth Moon landing by Cernan and Schmitt in LEM Challenger at Taurus-Littrow, a highland area on the border of Mare Serenitatis, collecting 243 lbs. of samples in a record 75 hours on the surface. A geologist, Schmitt was the first scientist to land on the Moon. Evans stayed in the CSM America. Cernan was the last human being to walk on the Moon. Duration: 301:51:59 including 75 orbits of the Moon by CSM.

1975   Apollo 18 also known as Apollo-Soyuz, July 15, Saturn 1B rocket, astronauts Thomas P. Stafford, Vance D. Brand, Donald K. "Deke" Slayton. Soyuz 19 also known as Soyuz Apollo, July 15, A-2 rocket, cosmonauts Alexei Leonov, Valeri Kubasov. The first US/USSR joint space mission. Not a Moon flight, but a show of international cooperation in Earth orbit. First simultaneous American and Russian space flights. The American crew rendezvoused and docked the Apollo 18 capsule with the Soyuz 19 capsule. The five spacemen did experiments, shared meals, held a news conference from space. Slayton was the first person over 50 in space. U.S. flight duration 217:28:23 including 136 orbits of Earth. USSR duration 143:31 including 96 orbits of Earth.

No let-up. How were the Russians performing in the Moon Race in the 1960's? The USSR continued to display a string of space firsts, even after the Moon Race began May 25, 1961:

Year   First Achievement

1963   Valentina Tereshkova, first woman in space, flew in USSR's Vostok 6 capsule.

1964   Konstantin P. Feoktistov, Vladimir M. Komarov and Boris B. Yegorov, first three-man crew in space, in Voskhod 1 capsule, first live TV, first without spacesuits.

1965   Alexei A. Leonov made the first spacewalk, 10 minutes on a 10-ft. tether, while Pavel I.

1965   Belyayev accompanied him in Voskhod 2.

1965   USSR fired powerful Proton rocket, capable of carrying 50,000 lbs. to orbit.

1965   Venera 3, the first spacecraft to impact on Venus, the first human artifact to reach the surface of another planet.

1969   Soyuz 4, Soyuz 5 completed first linkup of two craft in orbit with crew transfer.

1970   Luna 17 ferried Lunokhod 1 to the Moon, the first robot Moon rover.

1971   The first space station, Salyut 1, was launched to Earth orbit.

Russian Moon Probes. Thirty USSR Luna and Zond spacecraft explored the Moon between 1959 and 1976. The 24 craft in the Luna, or Lunik, series were unmanned probes returning photos and information on Earth's natural satellite.

The Luna program included many firsts: Luna 1 in 1959 was the first-ever Moon probe from Earth. It missed the Moon by only 3,700 miles. Luna 2 in 1959 was the first Earth probe to impact on the Moon. And Luna 3 in 1959 made the first photographs of the Moon's farside. Luna 9 in 1966 made the first soft landing on the Moon. Luna 10 in 1966 was the first Moon orbiter. Luna 16 in 1970 returned a soil sample to Earth. The Lunokhod-1 rover vehicle flew to the Moon in Luna 17 in 1970.

Eight Zond launches from 1964 to 1970 included one to Venus, two to Mars and five flights preparing for manned Moon trips, including three flights around the Moon which returned to land on Earth. Zond 1 flew to Venus and Zonds 2 and 3 were Mars flights. Of the eight Zond shots, Zonds 3 through 8 explored the Moon.

Moscow announced that Zonds 4 through 6 in 1968 were preparations for a manned flight to the Moon. However, problems developing more-powerful rockets in that era prevented the final step of sending men around the Moon after Zond 8. The U.S. landed men on the Moon in 1969.

Probe	Launch	Results
Luna 1	1959 Jan 2	missed Moon by 3,700 miles; first of 24 Luna (Lunik) flights.
Luna 2	1959 Sep 12	the first Earth probe to impact on the Moon.
Luna 3	1959 Oct 4	sent the first photographs of the Moon's farside.
Luna 4	1963 Apr 2	missed the Moon by 5,300 miles.
Luna 5	1965 May 9	failed soft landing, impacted on the Moon.
Luna 6	1965 Jun 8	missed the Moon by 100,000 miles.
Zond 3	1965 Jul 18	This Mars flight flew within 5,717 miles of Moon, sent 25 photos of the far side of Moon. This was the third of 8 Zond launches between 1964-70 including 1 Venus, 2 Mars and 5 flights preparing for manned Moon trips including 3 around Moon which returned to land on Earth. Zond 1 launched April 2, 1964, had flown within 60,000 mi. of Venus. Zond 2 launched November 30, 1964, flew within 1,000 miles of Mars.
Luna 7	1965 Oct 4	failed soft landing, impacted on the Moon.
Luna 8	1965 Dec 3	impacted on the Moon, failed in soft landing attempt.
Luna 9	1966 Jan 31	first soft landing on the Moon, sent back 27 photos.
Luna 10	1966 Mar 31	first Moon orbiter, sent data for two months.
Luna 11	1966 Aug 24	transmitted data to Earth from Moon orbit.
Luna 12	1966 Oct 22	In Moon orbit, sent back high-resolution photos.
Luna 13	1966 Dec 21	soft landing on the Moon, sent photos, tested soil sample.
Zond 4	1968 Mar 2	test launch away from the Moon, preparing for later manned Moon trips.
Luna 14	1968 Apr 7	measured gravity from Moon orbit.
Zond 5	1968 Sep 14	a biosat with turtles, worms, flies, bacteria, spiderwort plant and seeds of wheat, pine and barley; around Moon at 1,212 mi., far side photos; night return splash down in Indian Ocean, first USSR water landing.
Zond 6	1968 Nov 10	biosat with micrometeorite and cosmic ray detectors, flew around Moon at a distance of 1,398 miles, sent back stereo photos of surface; skip re entry, landed in USSR.
Luna 15	1969 Jul 13	orbited the Moon, impacted on the surface of the Moon.
Zond 7	1969 Aug 7	like Zond 6, color photos of Moon & Earth.
Luna 16	1970 Sep 12	soft landing on the Moon, lifted off to return soil sample to Earth.
Zond 8	1970 Oct 20	last Zond flight circled Moon at 684 miles; color photos; first TV of

		Earth from 40,000 miles; North Pole re-entry, no skip, Indian Ocean night splash down.
Luna 17	1970 Nov 10	soft landing on the Moon, Lunokhod-1 rover vehicle travelled surface.
Luna 18	1971 Sep 2	orbited the Moon, impacted on the Moon.
Luna 19	1971 Sep 28	orbited the Moon and transmitted photos to Earth.
Luna 20	1972 Feb 14	soft landing on the Moon, lifted off to return soil sample to Earth.
Luna 21	1973 Jan 8	soft landing on the Moon, Lunokhod-2 rover vehicle travelled surface.
Luna 22	1974 May 29	orbited the Moon, transmitted photos to Earth.
Luna 23	1974 Oct 28	soft landing on the Moon, sample return failed.
Luna 24	1976 Aug 9	soft landing on the Moon, lifted off to return soil sample to Earth.

USSR Moon Craft. Westerners were shown the actual spacecraft, built to land Russians on the Moon in 1968, for the first time in 1989. A touring group of U.S. college professors in November 1989 were allowed to examine the landing craft and return-to-Earth module stored in a back room of the Moscow Aviation Institute.

Existence of the never-used lunar spacecraft confirmed the USSR in the 1960's was in a race with the U.S. to put a man on the Moon.

It was only in October 1989, in a Pravda interview with chief spacecraft designer Vasily P. Mishin, that the USSR had finally acknowledged they once had a manned lunar program. Previously they would not admit they were trying to get to the Moon in the 1960's.

The American professors, from Massachusetts Institute of Technology and California Institute of Technology, were in the USSR in an exchange program with the Moscow Aviation Institute. Professor Oleg Alifanov of the Institute told the visiting Americans the Moon ship had been ready to fly in 1968, a year before the U.S. Apollo 11 capsule reached the Moon.

Soviet engineers were pressured by preliminary U.S. Apollo successes, but their flight never got off the ground because of difficulty perfecting the huge N-1 rocket designed to carry the equipment to Earth orbit.

Korolev. Sergei P. Korolev had been imprisoned by Soviet dictator Josef Stalin, but had kept on working in the 1950's in a laboratory built especially for him behind bars. He emerged to fire the Soviets into space first with Sputnik.

Korolev was the Soviet Union's chief rocket designer and head of the space program, though his name never was published by official Soviet media during his lifetime. He was known to the Soviet public only as "the chief designer."

Flushed with the Sputnik victory, Korolev began designing a lunar orbiter in 1957. In 1961, he sent Yuri Gagarin into space on the first manned flight. That same year, he took control of a powerful Proton rocket design of Vladimir Chelomei, and started working on a lunar module to land a cosmonaut on the Moon.

Moon cosmonaut. Valeri F. Bykovsky was known to Soviet citizenry as Cosmonaut No. 5 after he made the Soviet Union's fifth manned flight June 14, 1963. It was a team flight with Valentina Tereshkova, the first women to fly in space. He stayed in space five days, the longest-ever solo space flight.

After proving he could stay in space that long, he was on his way to being the first man to walk on the Moon—if only the mighty N-1 rocket would fly. He was first to begin training to fly the new, secret lunar orbiter code-named Zond—probe in Russian.

As Korolev's rocket engineers rushed development of the Proton booster, Bykovsky accelerated training in the Zond. Proton had enough thrust to carry the Zond to lunar orbit. Orbital rendezvous was tested. Alexei Leonov made the first spacewalk.

Premier Nikita Khrushchev wanted his man on the Moon in 1967, for the 50th anniversary of the Bolshevik Revolution. When Korolev said it couldn't be done, the Moon landing was scheduled for 1968.

That would have been a year before America's Neil Armstrong walked on the lunar surface. Soviet officials were saying publicly they considered sending men to land on the Moon too risky, too wasteful, and not as productive as sending robot probes.

Sudden death. Bykovsky was training intensely for 1968. Preparations for a Moon shot and lunar landing were well underway when the project was disrupted abruptly by the sudden death of the chief designer.

Korolev, 60, died on the operating table January 14, 1966, after a botched operation for hemorrhoids, according to one-time dissident scientist Zhores Medvedev, now in England. An obituary in the official newspaper Pravda said Korolev died of "cardiac insufficiency" after a cancer operation. Whatever happened, his death dispirited Soviet space scientists and beheaded the program.

Moon rocket. Proton was transformed into a huge Moon rocket, called N-1. Its four stages had 43 engines. The American Saturn 5 was a three-stage rocket with 11 engines.

Soviet engineers were not able to keep thrust stable as N-1's 30 first-stage engines fired. All four tries to launch the massive N-1 to Earth orbit failed. One N-1 did make it as high as 70,000 feet before exploding.

One on the Moon. The Soviet spacecraft would have carried one person to the surface of the Moon. The lander and return capsule resembled the U.S. command service module and lunar excursion module, except the USSR required two rockets, rather than one, to transport it all to Earth orbit.

One MIT professor said the Soviets showed a mixture of pride that they had built the Mooncraft, but sorrow that they had not used it and concern that no one had heard about it. The antique flight equipment is used today to teach Moscow Aviation Institute students who crawl around in it and make drawings of it.

Giving up. There seem to have been no flights of the Moon capsule. Even though the manned equipment was ready for a trip to the Moon, the rocket couldn't be perfected in time. Having given it a valiant effort, the Soviets lost the Moon race.

The Russians turned their attention to space stations, launching the world's first, Salyut 1, into Earth orbit in April 1971. The Moon program was put on hold in 1972 and canceled in 1974 as the USSR perfected Earth-orbiting space stations.

Finish line. The U.S. won the Moon race because it was able to perfect the huge Saturn V rockets which lifted the Apollo command service modules and lunar excursion modules to Earth orbit, and then on to the Moon.

American astronauts Neil A. Armstrong and Edwin E. "Buzz" Aldrin Jr. became the first men to step on the Moon when they flew there in Apollo 11 in 1969.

The Moon roundtrip, for Armstrong, Aldrin and Michael Collins, lasted 195 hours 18 minutes 35 seconds and included 30 orbits of the Moon by the command service module Columbia.

Armstrong and Aldrin flew the lunar excursion module Eagle down to the surface July 20, 1969, landing at Mare Tranquillitatis.

Armstrong, the first outside, said, "That's one small step for man...one giant leap for mankind." Armstrong and Aldrin collected 48.5 lbs. of rocks and soil during 21 hours 36 minutes 21 seconds on the surface. Collins stayed in Moon orbit in Columbia.

Old film clips. The first era of men from Earth on the Moon ended three years later when astronauts in Apollo 17 made the sixth and final lunar landing in December 1972.

The USSR wouldn't broadcast film clips of Armstrong on the Moon until the 20th anniversary in 1989. The man who had wanted to beat Neil Armstrong to the moon flew again—in Earth orbit only—in Soyuz capsules in 1976 and 1978.

During the 1978 trip, he visited the Salyut 6 space station. Altogether, he spent 21 days in space in three trips. Later he headed a Soviet engineering research center in East Germany.

Another Moon race? There won't be another Moon Race. After the year 2000, U.S. astronauts and USSR cosmonauts are likely to share the same new spacecraft. Their rocket may be old, however. Reports hold that plans for the old Saturn Moon rocket will be dusted off and boosters constructed for new trips to the Moon and Mars.

American Spaceports

The United States fires rockets and shuttles toward space from four well-known spaceports: Kennedy Space Center, Florida; Vandenberg Air Force Base, California; Wallops Flight Facility, Virginia; and White Sands Space Harbor, New Mexico; and it has other launch sites.

The first U.S. sounding rockets were fired toward space from White Sands and Wallops in the late 1940's after World War II. The first launch from Cape Canaveral was in 1950. Other spaceports have been built since then.

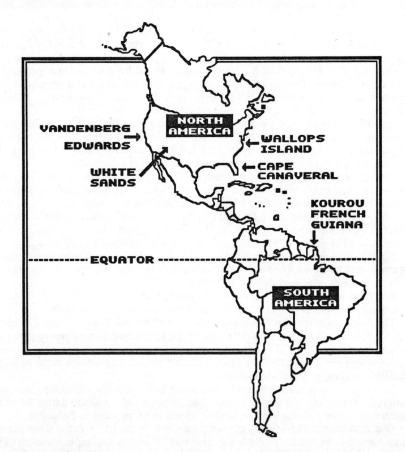

Cape Canaveral. A need for more space led to establishment of the Joint Long Range Proving Ground in 1949 at deserted, remote Cape Canaveral, Florida. The U.S. Army's two-stage Bumper-WAC was the first rocket launched there, on July 24, 1950.

Today, Kennedy Space Center is NASA's preeminent spaceport, located on 138,000 acres of Merritt Island north of Cape Canaveral, Florida. Cape Canaveral Air Force Station is nearby.

In 1961, launch pads 39A and 39B were constructed for the mighty Saturn V moon rockets. The spaceport was upgraded in the 1970's for space shuttle launches and landings. All U.S. space shuttles have been launched from Cape Canaveral. In 1991, it also became the landing site of choice for U.S. shuttles.

On November 28, 1963, after U.S. President John F. Kennedy was assassinated November 22, Cape Canaveral was renamed Cape Kennedy in honor of his memory. Later, local residents voted to change back to the original Cape Canaveral. However, NASA kept the name for its spaceport as the John F. Kennedy Space Center.

Apparently wasps swarm each fall around the tops of towers on the launch pads at the Kennedy Space Center. NASA called in the U.S. Agriculture Dept. in 1990 to trap them with a very special bait: the sexually-attractive scents known as pheromones.

Wallops Island. Wallops Flight Facility at Wallops Island on the east coast of Virginia, is part of NASA's Goddard Space Flight Center. Its pads are used for launching smaller suborbital rockets east over the Atlantic Ocean. NASA's use of America's new air-launched commercial space rocket, Pegasus, is controlled from Wallops. Goddard Space Flight Center is located inland, west of the Chesapeake Bay, at Greenbelt, Maryland.

Along with full-size government rockets, scale models of famous rockets built by hobbyists blast off regularly from Wallops Island Flight Center. The models of past and present rockets, such as Mercury-Redstone and space shuttles, are one to three feet tall. Some reach altitudes of 1,000 feet. The general public observes the 45-minute events. Many bring their own rockets.

NASA's Wallops Island visitor center is located on Route 175 six miles from Route 13 and five miles from Chincoteague, Virginia. It is open from 10 a.m. to 4 p.m. Thursday through Monday. Admission is free, but insect repellant is advised. Additional information may be obtained by telephoning (804) 824-1344 or (804) 824-2298.

Wallops lightning. Bolts of lightning caused three solid-fuel rockets to blast off unexpectedly from Wallops launch pads on the night of June 9, 1987. The same lightning put the United States' prime weather satellites out of business for several hours by disabling their ground receiving station.

The lightning around Wallops Island turned on igniters and launched one Orion and two Pacer rockets. The Pacers, used to align tracking equipment, flew three miles high and 2.5 miles down range to ditch in the Atlantic Ocean. They had been pointed up at a 75 degree angle. Although not tracked, they probably flew to 15,000 feet altitude.

The Orion had been on the pad in a horizontal position. It flew straight ahead about 300 feet and dumped in the water. Ironically, it had been scheduled to test lightning's effect on the ionosphere.

Four rockets had been on the launch pads when the bolts struck. The Orion and two Pacers which launched had been fitted with fuel igniters. The fourth rocket had not been fitted with an igniter. The launch pads, 150 ft. apart, shared a common electrical ground system. Electrical energy from the lightning might have raced through the ground system to each pad. More than 13,000 rockets had been fired from Wallops Island up to that time. None previously had been launched by a lightning strike. A launch crew was in the blockhouse and the pad was clear when the bolts arrived. No one was hurt.

Film at eleven. Television viewers across the country were deprived of their nightly-news weather pictures from space. The National Oceanic and Atmospheric Administration (NOAA)—the weather bureau—said lightning struck the agency's satellite ground station at Wallops three times. The electrical energy in the bolts overwhelmed lightning rods, a grounding system and surge suppressors.

NOAA's Wallops Command and Data Acquisition Center was the only ground station receiving weather pictures from space. Damage was extensive. NOAA workers required nine hours to restore service. The strike was the worst in the station's 18 years of operation. Previous bolts had caused the loss of less than an hour of pictures. Lightning never had hit multiple antennas at Wallops before. The storm in Virginia and Maryland coastal areas made a direct hit on an antenna and hit power lines feeding the system.

More lightning. Mighty Mouse injured a French electrical engineer at Kennedy Space Center August 27, 1991, but was unable to duplicate Ben Franklin's feat by lifting a wire in a lightning storm. It wasn't the cartoon character, but one of NASA's tiny 4-ft.-long Mighty Mouse rockets being used in the experimental Rocket Triggered Lightning project. The 2.75-in. diameter rocket was set to lift electric field sensor wires used in lightning research when Might Mouse inadvertently was touched off at 1:15 pm. The rocket flew up a hundred feet, then dropped on its usual target in the Mosquito Lagoon.

The engineer, with minor burns to his left leg, was taken to the launch complex 39 dispensary, then to the Parrish Medical Center, Titusville, Florida, where he was treated and released. The rocket had been unarmed and safety procedures had been observed, NASA said. An inquiry board has been convened.

Vandenberg. America's West Coast spaceport is Vandenberg Air Force Base, a coastal military complex sprawling across 98,400 acres near Lompoc, California, 140 miles northwest of Los Angeles. The Air Force launches space rockets and test-fires intercontinental ballistic missiles from Vandenberg.

Rockets fired from Vandenberg place their satellite payloads into polar orbit which can't be reached from Cape Canaveral, Florida, because they would fly over populated areas. Many spy satellites fly in polar orbits so they can see all of the Earth below.

Top-secret payloads are common in flights from Vandenberg, where a mammoth space rocket's roar often lasts five minutes, even after the 20 seconds or so it takes many boosters to disappear into cloud banks overhanging Vandenberg. Space rockets and military missiles often fade into a fog bank over Vandenberg, blasting out deep-throated roars, rattling windows in the town of Lompoc.

End of an era. Vandenberg has a never-used space shuttle launch pad which cost $3.3 billion, but was mothballed after the 1986 Challenger disaster. Now the Air Force is converting it into a launch pad for unmanned Titan-4/Centaur rockets.

Space Launch Complex 6 at Vandenberg has been under construction since it first was started in the 1960's, yet nothing has been launched from it. Originally, it was to have been a launch pad for a lunar spacecraft that was canceled. The Air Force spent $3.3 billion rebuilding it as a base for shuttles flying to polar orbits. It was mothballed following the Challenger disaster, then the Air Force couldn't depend on shuttles to meet its busy launch schedule. Heavy payloads were switched to new Titan 4 rockets, which became America's most powerful unmanned rocket with its maiden flight in 1989.

Titan 4-Centaur. The most powerful version of Titan 4 would be one with a Centaur rocket added as an upper stage. Converting Space Launch Complex 6 for Titan 4-Centaur will put Vandenberg and California out of the space shuttle launch business.

The Air Force says it will cost $500 million to convert the pad for Titan 4-Centaur, and be cheaper and environmentally safer than building a new complex for $800 million.

The jumbo military satellites which require the heavy-lifting Titan 4-Centaur would include Milstar communications satellites and Space-Based Wide-Area Surveillance System (SBWASS) radar satellites which track planes, ships and submarines.

The first launch of a Titan 4 rocket augmented with a Centaur upper-stage booster, from Space Launch Complex 6 is expected in 1996—if Congress funds the larger military surveillance and communications satellites suited for Titan-Centaurs.

The Air Force wants the pad in case an accident would prevent use of Space Launch Complex 4 East, which already is available for Titan 4 launches. A spysat was expected to be launched there on a Titan 4. The first two Titan 4s launched by the Air Force flew from Cape Canaveral. The Pentagon wants to be launching ten Titan 4s a year by 1995.

Edwards. Part of a Titan 4 solid-fuel booster went off accidentally during a test at Edwards Air Force Base, California, in September 1990, killing one person and injuring two. The accident spewed a toxic cloud over the Mojave Desert after a bottom section of the booster ignited when it was dropped while being moved at the Air Force Astronautics Laboratory.

NASA has its Ames-Dryden Flight Research Facility at Edwards Air Force Base, 70 miles north of Los Angeles. Ground firing tests sometimes are held at Edwards. During one recent June test, thunderstorms and winds over the base forced tests to be scrubbed several times. Officials feared winds would blow rocket exhaust fumes over a residential area near the Mojave Desert research center. The Air Force worried about lightning and electrical interference around the test firing stand anchored to the ground.

Shuttle landings. Some space shuttle landings can be see from Edwards' East Shore Viewing Site. The site opens 24 hours prior to a scheduled landing. Vehicle passes are not required. Parking is on unprepared surfaces, but water, restrooms, food vendors and souvenir stands are available. Local radio stations report when a shuttle is expected to land. Landing updates are available by telephone at (805) 258-3520.

Access to the site, by way of secondary roads which can be congested, is cut off one hour before a landing. There are two routes to the East Shore Viewing Site:

From Los Angeles, travel north on Highway 14, the Antelope Valley Freeway. Turn right, which is east, on the Avenue F off-ramp. Turn left and travel north on Sierra Highway to Avenue E. Turn right and go east on Avenue E to 140th Street. Turn left and travel north on 140th Street to Avenue B. Turn right and follow Avenue B east as it curves into Mercury Blvd which leads to the viewing area.

From Highway 58, follow the Rocket Site Road off-ramp to Mercury Blvd. which leads to the viewing area.

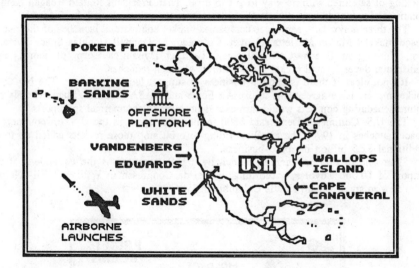

Poker Flats, Barking Sands. NASA also launches small sounding rockets on suborbital flights from White Sands Missile Range, New Mexico; Poker Flats, Alaska; and Barking Sands, Hawaii, as well as from foreign locations such as Italy's launch site in the Indian Ocean off the coast of Kenya, Africa. NASA's Poker Flats Research Range is a few miles outside of Fairbanks at Chatanika, Alaska.

Airborne. The U.S. has a new space rocket which can be launched anywhere in the world from an airplane. Pegasus was the first all-new unmanned American launch vehicle since the 1960's. To get to space, Pegasus is attached to a B-52 bomber, ferried more than seven miles above Earth, dropped free and blasted to orbit. Pegasus can carry 600-lb. satellites to orbit at a fraction of the cost of conventional launches. It can carry twice the payload of a comparable ground launcher because of its take-off altitude, the ability to fly the B-52 to any latitude , and starting three-fourths of the way to space.

Offshore. NASA and the U.S. Air Force have plans to use an enormous large oil-drilling platform off the California coast as a launch pad for powerful Advanced Launch System (ALS) rockets in the late 1990's. To conserve land at Vandenberg, the Air Force Space and Missile Test Organization and NASA have considered launching from a remote Pacific island, from a ship, from a man-made island or from one of the Channel Islands.

In one plan, the Air Force would tow a large drilling platform into Vandenberg harbor, load a rocket aboard and tow it several miles offshore. Out in the Pacific, legs on the platform would be lowered so it could stand on the ocean floor. The ALS rocket then would be fueled and launched by remote control cable from Vandenberg.

NASA and the Air Force have been considering up to 50 ALS launches a year, beginning in 1996. Operators of the U.S. Strategic Defense Initiative and the U.S.-international space station Freedom would be major users of heavy-lifting ALS rockets.

Local spaceports. Eight states are promoting their commercial interests in space as private commercial launches become more popular in the 1990's: Alabama, California, Colorado, Florida, Hawaii, Texas, Utah and Virginia. Florida, Hawaii and Virginia actually have plans to build commercial spaceports.

The U.S. Transportation Department encouraged officials in Florida, Hawaii and Virginia to pursue commercial spaceports. The states began thinking of commercial spaceports after the 1986 Challenger disaster grounded the U.S. shuttle fleet, creating a backlog of satellites with no way to get to orbit. Then President Ronald Reagan banned commercial payloads from shuttle flights to encourage private launch business.

The three major U.S. aerospace companies market commercial launches of their large space rockets: Martin Marietta, Denver, Colorado; McDonnell Douglas Space Systems Co., Huntington Beach, California; and General Dynamics Corp. of San Diego, California. Several small rocket manufacturers also sell launches.

It's not clear if there is enough business to support local spaceports. The big three rocket makers have agreements to use Air Force and NASA facilities. State officials say future scheduling conflicts would turn excess business to commercial spaceports.

The U.S. Commerce Dept. said $600 million was spent in the U.S. on commercial space launches in 1990. The satellites riding to orbit atop those rockets added up to an additional $2.5 billion in space business.

"The sale of one commercial launch by a U.S. company is the equivalent of the import of 10,000 Toyotas," according to Florida Congressman William "Bill" Nelson who flew to space aboard shuttle Columbia in 1986.

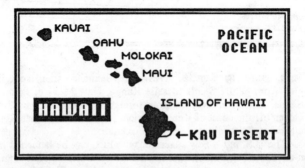

Spaceport Hawaii. Hawaii wants America's first exclusively-commercial space rocket launching pad. State engineers have been studying Kau, a thinly populated southeast coast area of the island of Hawaii. Rockets could be lofted from Kau into either equatorial or polar orbits without flying above populated areas. The U.S. government launches satellites to equatorial orbits from Florida and to polar orbits from California.

Spaceport Florida. Florida, out to get a piece of the $3 billion space satellite launch business, wants to restrict public access to a wildlife refuge and the Canaveral National Seashore parkland, as well as endanger the area's sensitive environment with toxic rocket fuels, to build the nation's first commercial spaceport in the northern reaches of Kennedy Space Center.

The state wants to complete a $58 million commercial Spaceport Florida in the mid-1990's, either at Kennedy Space Center or the adjacent Cape Canaveral Air Force Station.

State officials prefer to pave over the space center's remote Shiloh corner which is managed as a park known as the Merritt Island National Wildlife Refuge. The state would build one launch pad on 50 acres at Shiloh. Later, three more pads would be constructed.

Building at Shiloh would increase construction costs by 20 percent, however, and three launch pads abandoned by the military at Cape Canaveral could be renovated for use. But, fearing restrictions on money and access posed by a military installation, state officials preferred Shiloh, according to the director of the Florida Department of Commerce space programs office.

Concerned naturalists have been fighting the spaceport, according to activist Charles Lee, senior vice president of the Florida Audubon Society. More than a dozen environmental and civic groups drew up battle plans to oppose the Shiloh location, saying that public officials giving away chunks of parkland for commercial ventures would set a bad precedent. Most environmentalists support building the spaceport on already developed areas of the Canaveral Air Force Station. The state also wanted to build launch pads for suborbital sounding rockets at a new base on Cape San Blas in the Florida panhandle.

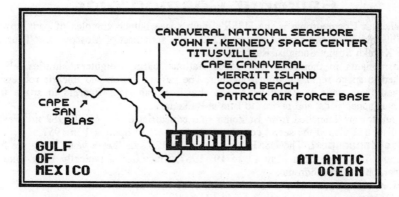

Spaceport Florida would be run by a Florida Spaceport Authority, which would develop a commercial launch complex like an airport. It was estimated Spaceport Florida might see one commercial flight per month. At $45 million each, commercial launches would inject $540 million per year into Florida's economy. Such a spaceport also might attract related private industries worth $60 billion by the end of the century.

After Florida Gov. Bob Martinez signed the Spaceport Florida Authority Act into law, he said, "It won't be long before cargo that is launched into space is stamped with the words 'Shipped from Florida'."

A seven-member board of commissioners, appointed by the governor and approved by the Florida Senate, oversees Spaceport Florida Authority activities, much like an airport authority. Florida Spaceport Authority is based near Kennedy Space Center and the adjacent Cape Canaveral Air Force Station.

Spaceport Authority is negotiating with the Air Force for use of abandoned launch pads at the Air Force station and for use of Eglin Air Force Base's Test Site D-3A facility at Cape San Blas in Gulf County.

Florida began plugging its non-military, non-NASA commercial Spaceport Florida launch facility at Cape Canaveral in 1988, to take advantage of facilities and equipment already in place at Cape Canaveral Air Force Station and NASA's Kennedy Space Center.

NASA and the U.S. Air Force have launch pad priority at existing Cape Canaveral facilities. Spaceport Florida probably could not guarantee launch dates unless the state does build its own expensive pad, control center and tracking facilities.

Russians at Cape Canaveral. Would you believe...the Russians want to use Cape Canaveral launch pads to blast off their big Proton space rockets?

Space Commerce Corp., the Houston, Texas, marketer selling launches on USSR rockets to U.S. satellite owners, has asked permission to ship rockets to Florida and launch them from the Cape.

It would seem an expensive proposition since the USSR would have to build new launch pads and support facilities in Florida, but Cape Canaveral is closer to the equator than sites in the USSR, requiring less rocket fuel to get a payload to orbit. And launching from the Cape might bypass U.S. laws banning launches in the Soviet Union of American satellites to prevent transfer of technology to the USSR.

The Orlando, Florida, Sentinel newspaper said the request was in a letter from Space Commerce Corp. to the Spaceport Florida Authority at Cocoa Beach—a state agency in charge of private commercial space development. It is not known how the U.S. Air Force would react to Russian launches at or near the Cape Canaveral Air Force Station.

Baikonur Cosmodrome

Baikonur Cosmodrome is the USSR's main space launch complex of pads, towers, assembly buildings and control centers 1,200 miles southeast of Moscow near Tyuratam in the Kazakh Soviet Socialist Republic.

Cosmonauts in their capsules, space stations, cargo freighters, shuttles, all are launched to space regularly from Baikonur. The mighty Proton and Energia rockets are launched from pads at Baikonur. Space shuttle Buran rides Energia to space from Baikonur launch pads, and returns to land at Baikonur.

Satellites are launched from Baikonur into equatorial orbits. Until 1989 no Western journalists had visited the secret complex, which launched Sputnik 1 in 1957.

Busy spaceport. The USSR launches 80 to 120 satellites a year. Many are from Baikonur. The days from May 11 to 19, 1987, illustrated a typically busy time for engineers at the cosmodrome:

★Launch operators fired a communications satellite to orbit May 11.

★A Cosmos spy satellite was launched May 13.

★Just 15 minutes later on May 13, a different Cosmos spy satellite blasted off from a nearby launch pad.

★Energia, the world's most powerful rocket, roared into the sky on its maiden flight May 15. USSR General Secretary Mikhail S. Gorbachev was on hand for ceremonies.

★The supply ship Progress 30 was launched to the Mir space station May 19.

And the Soviets still had unused capacity. Today, the Proton booster is being sold around the world as a means of launching commercial satellites to orbit. For $30 million, Proton can carry a 4,000-lb. satellite to stationary orbit 22,300 miles high.

Eight on one rocket. Of course, the yearly total increases when more than one satellite is ferried to space on one rocket. The Russians accomplish multiple launches from time to time. They frequently loft two or more satellites at a time and have carried half a dozen and more on one rocket more than once.

For instance, the USSR launched eight Cosmos satellites to orbit on one rocket

March 25, 1989. The satellites were said to carry science equipment.

The Soviet Union used a single booster to launch six satellites to orbits 870 miles above Earth on September 8, 1987. Cosmos 1875, 1876, 1877, 1878, 1879 and 1880, were said to be carrying science equipment, but they probably were military satellites.

Roll-out tradition. The 35-year-old, 600-sq.-mi., Baikonur Cosmodrome has six main launch pads. By tradition, USSR space engineers roll a space rocket out of its assembly building to the launch pad at 7 a.m., commemorating the hour when the Vostok rocket was rolled out for Yuri Gagarin's first manned space flight April 12, 1961.

Soyuz rockets are rolled out in a horizontal position, on a rail line which treks across the steppe to the old Gagarin launch pad No. 1. Enormous flood lights on trucks illuminate the scene. Usually, by 9 a.m., Soviet engineers wearing orange hard hats and black weather gear, are ready to use a hydraulic lift to crank the booster upright on the pad. The gantry clamps the rocket, holding it upright for a day or two until launch.

Kazakhstan. The Kazakh Soviet Socialist Republic is one republic in the Union of Soviet Socialist Republics. Kazakhstan is in the northern plains of central Asia, northeast of the Caspian Sea. It is a vast territory, second in area only to the Russian Soviet Federative Socialist Republic. Kazakhstan is 2,000 miles east to west and 1,000 miles north to south, bordering Siberia in the north and three central Asian republics in the south. The Kazakh capital, Alma-Ata, once a dusty nowhere, now is home to a million persons. Yurts no longer are used as portable houses, but Kazakh folklore still is practiced by akyns acting out improvised verse.

President Nursultan Nazarbayev established Kazakhstan's own space program in September 1991. The Kazakh Institute of Space Research would coordinate spaceflights with Moscow. Wildcatters from the Alma-Ata Geological Institute in 1991 discovered an oil field around Baikonur. The semi-arid desert land has no rivers and lakes, but is rich in gas and oil. Drawn up from a mile under the plain, oil from the South Turgai field will be used for automobile and aircraft fuel. Kazakhstan is the second largest oil producer in the USSR, after Russia.

Other cosmodromes. The USSR also makes space launches from its Northern Cosmodrome near Plesetsk northeast of Moscow, and the Volgograd Cosmodrome near Kapustin Yar southwest of Moscow.

USSR vs. Russia. The Union of Soviet Socialist Republics (USSR) is composed of several republics. Russia is one of those republics: the Russian Soviet Federated Socialist Republic has slightly more than half of the population of the USSR and three-fourths of the territory.

Russia has 700 years of history while the USSR (Russia and 14 other Soviet Socialist Republics or SSRs) has only about 75 years of history.

The USSR covers one-sixth of the face of the Earth, about 2.5 times the size of the U.S. It stretches 6,000 miles from eastern Europe, across northern Asia, to the Pacific Ocean. Terrain, weather, agricultural pursuits, industrial development, urbanization, all vary widely. Population is growing from 280 million.

The Russian language is the strongest uniting bond. But, in the Asian SSRs, only about one-third of the people ever have spoken Russian. Even so, the USSR's two prime TV channels broadcast in Russian. Most magazines are in Russian. Pravda and Izvestia, the two official newspapers, are in Russian. Most of the circulation of other USSR newspapers are in Russian.

There are a dozen religions—including Islam which itself is a way of life and a tradition. Islam is strong in southern regions of the USSR giving that area the USSR's greatest growth spurt. Moslems could outnumber all of the rest of the USSR population together by 2150.

With such cultural and social differences, no wonder perestroika (restructuring) and glasnost (openness) bring considerable smyateniye (turmoil).

Northern Cosmodrome: Plesetsk

The world's busiest spaceport is set among peat bogs, wild birches and charming lakes in ancient Russia. The Northern Cosmodrome near Plesetsk has fired more than 1,200 space satellites to orbit Earth's poles since 1957.

Week after week, as rockets ferrying satellites for communications, weather observation and other uses rise above the great fir and birch forests surrounding Lake Plesetskaya, heads turn in Mirny, the bedroom community next door to the launch center, south of Archangel, 500 miles north of Moscow.

Secret spaceport. Sputnik 1 blasted off in 1957 from a different launch pad, the Baikonur Cosmodrome on the Asian plain 1,200 miles southeast of Moscow in the Kazakh Soviet Socialist Republic—2,000 miles from Plesetsk.

That same year, the Northern Cosmodrome was carved from those northern forests and launch pads, towers, assembly buildings and control centers were built. Since then, it has been the busiest spaceport on Earth. Satellites launched from Plesetsk go into north-south polar orbits, while spacecraft launched from the southern Baikonur Cosmodrome orbit east-west along the equator.

Scrap metal. Over 34 years of blasting giant rockets toward space, the Arctic region around Archangel, down range from the Northern Cosmodrome, was polluted with vast fields of scrap metal. Workers from Plesetsk in September 1991 hunted down used booster rockets, engines, nosecones and other things which had fallen from the sky onto two large areas—the tundra of the Nenets Autonomous Region and the Mezen district. Altogether, helicopters carried away 245 tons of scrap metal.

The clean-up was a mixed blessing. Local residents actually had come to like what they called "gifts from the sky." Some didn't want to part with their "celestial souvenirs," as the Tass News Agency called them. The local reindeer herders had been in the habit of using nosecones as all-around containers for storing and carrying goods. Officials decided to clean up the valuable scrap metal every year anyway.

Fuel explosion. An accident at Plesetsk in 1973 killed nine workers. The fuel tank of a Soyuz rocket being filled with liquid oxygen and kerosene exploded.

Seven years later, a terrible disaster at Plesetsk in 1980 killed 50 workers. A Soyuz rocket was being fueled with liquid oxygen and kerosene when an explosion demolished the launch pad. Forty-five were killed immediately; five died later from burns.

Today, a memorial for the fallen technicians rests in Mirny's main square. The monument's red granite slabs, displaying pictures of the dead, topped by a carved granite rocket, overlooks Lake Plesetskaya and the birch and fir forest.

Baikonur. The USSR launches 80 to 120 satellites a year—two or three a week, mostly from the Northern Cosmodrome. Even so, until a 1989 media event at Plesetsk, Baikonur Cosmodrome had been the USSR's better known space launch complex. The base at Plesetsk was locked up tight and little was known about it.

All cosmonauts and space stations have been launched from Baikonur. The mighty Proton and Energia rockets are launched from pads at Baikonur. The Soviet shuttle fleet rides Energia to space from there, and returns to land at Baikonur. Baikonur also is a busy spaceport.

In 1989, the USSR took reporters to the Plesetsk to see a double-header. A workhorse Soyuz rocket blasted a Molniya TV satellite to orbit, followed moments later by the launch of another Soyuz rocket carrying a pair of satellites to investigate Earth's ionosphere. Nine minutes after launch, Molniya was in orbit ranging from 2,485 miles to 24,233 miles altitude, ready to broadcast TV programs to remote areas. The pair of satellites included a large mother ship and a smaller satellite.

Volgograd. The USSR also makes launches from its Volgograd Cosmodrome near Kapustin Yar southwest of Moscow.

Europe's Jungle Spaceport

Europe launches its Ariane rockets from pads at the Guiana Space Center near Kourou, French Guiana, on the northeast coast of South America.

Guiana Space Center is an 11-mile-long strip of land on the Atlantic Coast between Kourou and Sinnamary. It is close to the equator at latitude 5.23 degrees north. Its launch window over the Atlantic stretches from north (minus 10.5 degrees) to east (93.5 degrees), making it ideal for launching satellites toward stationary orbit.

The Kourou complex was set up by the French government and carved from French jungle territory by the French space agency—CNES—starting in April 1964. Kourou, an isolated outpost 4,000 miles from Paris, went into operation in April 1968 with the launch of a Veronique sounding rocket.

Diamant and Europa rockets were launched from Kourou before the Ariane program. The first Ariane 1 blasted off from the jungle December 24, 1979.

ESA owns the launch pads which are operated by Arianespace. Pads in use include ELA 1, used since 1979, and ELA 2, used since 1986. Arianes 1, 2 and 3 have blasted off from ELA 1 with two month intervals between launches. Arianes 2, 3 and 4 have been fired from ELA 2, where preparation buildings and launch pad are connected by a half-mile railway. The time interval between launches from ELA 2 is one month.

Ariane 5 site. ESA is building mammoth new facilities at Kourou to launch men into space. It is the biggest European construction project after the Channel Tunnel between Britain and France. Thousands of tons of concrete are being poured into a vast clearing of the Amazon jungle. The new launch facilities will dwarf launchpads nearby from which four generations of unmanned Ariane rockets have blasted off since 1979.

The new Ariane 5 facility stretches over a 20 square mile area hacked from jungle and savannah along the Atlantic Ocean. Four million cubic meters of earth already have been removed and one-half million cubic meters of granite will be dynamited.

Engineers are building what ESA calls the world's largest mixing bowl, to concoct solid fuel for the Ariane 5 from two highly volatile 100-ton blocks of chemicals. Manual construction work is done by Brazilian laborers.

A spacious test bed, 200 feet deep and 655 feet long, has been gouged out of bedrock. Thousand-ton concrete blocks will hold down the Ariane 5 during tests. Surrounding granite will vitrify during two-minute rocket burns.

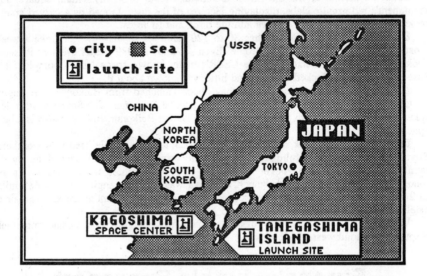

Japan's Tanegashima Spaceport

Japanese space rockets are launched by the National Space Development Agency of Japan from its launch site on small Tanegashima Island off Japan's southern main island of Kyushu, part of the Kagoshima Space Center.

Japan has made more than two dozen launches since 1975 from the island at the southern tip of Japan. The country has enjoyed a very high success rate with its N-series and H-series rockets.

Japan got a late start in space work under restrictions at the end of World War II. So far it has used technology from the U.S., but the agency is developing its own powerful all-Japanese H-2 space rocket. An H-2 is expected lift two tons 22,300 miles to stationary orbit.

The 1,210-lb. CS-3a and CS-3b communications satellites, launched in 1988, were the largest satellites Japan's National Space Development Agency had sent to space to date. Signals from communications satellites are received by Japan at its Masuda, Katsuura and Okinawa tracking stations.

China's Spaceports

The world's most populous nation launches its Long March rockets, carrying satellites to Earth orbit, from four major space complexes:

★Jiuquan launch site in the northern Gobi Desert in far-western China, not far from Mongolia,

★Taiyuan launch site in Shanxi Province, 60 miles southwest of Beijing, in north-central China,

★Xichang launch site in an isolated corner of Sichuan province in southern China, not far from Burma, Laos and Vietnam.

★Hainan Island off the southern coast of China, separating the South China Sea from the Gulf of Tongking. Only 19 degrees north of the equator, it is southeast of Hanoi, Vietnam, across the Gulf of Tongking.

All of the sites are remote and usually closed to foreigners, although the press was taken to the Xichang Satellite Center in 1988 as China was pressing the U.S. for approval of its launch business.

The site, 40 miles north of Xichang, is a small town accessible from the provincial capital of Chengdu by a 12-hour train ride and airplane flights twice-a-week.

Dusty backroad. Along the road to the Xichang Satellite Center, surrounded by green mountains in a remote corner of Sichuan province, reporters saw turbaned peasants lugging firewood, barefoot children scampering around mud-baked houses, women washing clothes in streams, water buffalo pulling wood plows through rice paddies and farmers spreading grain on the road so truck wheels would grind it.

China sent its first satellite, Mao 1, to space in 1970. Since then it has launched more than two dozen and retrieved a dozen. China built the remote launch site near Xichang in the 1970's because it had clear winter days for launches, was sparsely populated, and was easy to defend.

The handful of western reporters who visited in 1988 said the site looked almost abandoned between launches. Run by the People's Liberation Army, the Xichang launch complex employs 1,200 persons. Cavernous buildings house satellites and 141-ft. Long March rockets being readied for launch. The reporters noted windows in an 11-story gantry, blown out during a March 1988 launch, had not been replaced. Two engineers were at monitors in a gymnasium-sized control center four miles from the launch pad. A smaller control center was built in a mountain at the site.

Commercial boost. As China competes with other nations in the lucrative international satellite-launching market, Xichang Satellite Center is the takeoff point. China offered commercial launches in 1986 to satellite owners while U.S. shuttles and European rockets were grounded. Contracts were signed.

Great Wall Industry Corp. is the government body which contracts for commercial space flights from the People's Republic.

China is expecting to do a lot of satellite launch business in future years. They recently built a second pad and a building where satellites can be checked and repaired at Xichang. They also built a new hotel for visiting businessmen.

Simplicity of operation and a high success rate allow China to offer prices lower than the U.S. and the European Space Agency. Rocketry in China is run by the Ministry of Astronautics, a government agency.

Long March. China's space boosters are Long March 1, 2, 3 and 4. The latest, most powerful version is Long March 4. In 1988, China used the first Long March 4 to launch a heavy weather satellite to stationary orbit from Taiyuan. Great Wall Industrial Corp. said it hoped the bigger rocket would draw foreign customers for satellite launches.

Man in space. China may be preparing to launch its own astronauts to orbit in the 1990's from one of its four launch sites. The first flight is "not far off," the official newspaper People's Daily said in 1986. Life-support systems and a crew cabin reportedly were ready to fly atop a modified Long March.

China has been bringing satellites back to Earth successfully since the 1970's. Spacesuits and food have been in the works since 1980. Astronauts training in a mock spaceship first showed up in Science Life, a Chinese magazine, in 1980. A first-flight crew was being selected, according to the People's Daily newspaper. A Chinese shuttle also may be on drawing boards.

Thieves delay launch. It was the launch of the made-in-American satellite ASIAsat 1, shown live on Chinese Central Television and repeated on TV in North America and elsewhere—the first time since China started launching satellites in 1970 live pictures of a blast off were telecast. With all the world watching April 7, 1990, they were attempting to launch a Long March 3 rocket carrying ASIAsat when a hotline linking Xichang launch pad in southwestern China with a spaceflight command center at Beijing broke down. Someone had chopped yards of copper out of a transmission cable.

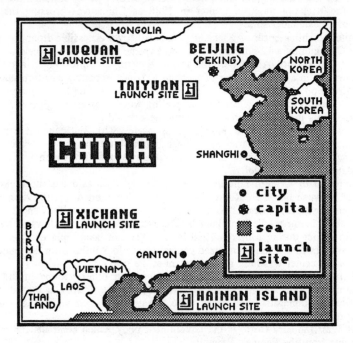

As ASIAsat rested atop the glistening white three-stage rocket, branded with a red flag and gold star of the People's Republic, communications between Xichang and Beijing were disrupted for hours, delaying the launch. Engineers bypassed a cut cable, but thick clouds over the launch pad in a hilly farm area delayed the flight another hour.

When the blast-off moment finally arrived at 9:30 p.m. local time and the countdown reached zero, there was an unexplained delay of many seconds. The Chinese-to-English telecast translator first cried, "Blast off!" Then after a pause of several seconds with no rocket action, he again cried, "Blast off!" Again nothing happened for several seconds. Then the translator said, "There seems to be a problem. They're going to try again." After a few more long moments while engineers switched cables to bypass the cut line, flames appeared under the rocket and it started its slow-motion leap into the sky. Sure that it was on its way this time, the translator again cried, "Blast off!"

From then, everything went smoothly. Launch control engineers broke into applause. The rocket's orange tail flame dwindled as it rose into the cloud cover over Xichang. Half an hour later, control center observers applauded again as the ASIAsat payload separated from the rocket and entered orbit.

Second chance. The 2,750-lb. communications satellite already had been flown to space in 1984 as Westar 6 in U.S. shuttle Challenger, but had failed. The satellite was captured by shuttle Discovery later in 1984, returned to Earth and refurbished. A Hong Kong company bought it from the insurance company which paid for its loss. Westar, renamed ASIAsat, became the first completely-foreign-built satellite launched by China.

China uses most of the satellite's 24 transponders for domestic telephone, television and data communications. The rest are rented to nearby countries. Burma, Hong Kong,

Nepal, Thailand, South Korea, Pakistan and others use ASIAsat during its ten-year life in orbit. The satellite is capable of reaching 30 Asian nations with 2.5 billion persons.

What of the copper thieves? Local Chinese radio later reported two men were arrested for cutting the lines between the launch site and a monitoring base 220 miles away in the provincial capital, Chengdu. They did the job sometime between November 1988 and January 1990. It wasn't clear why the missing link didn't show up until the April launch night. The official newspaper Legal Daily said the thieves would be executed.

India's Sriharikota Spaceport

India fires its space rockets from the Sriharikota Space Center in the southeastern state of Tamil Nadu. Rockets which fail during launch fall safely into the Bay of Bengal.

India Space Research Organization builds the Satellite Launch Vehicle (SLV), Augmented Satellite Launch Vehicle (ASLV) and Polar Satellite Launch Vehicle (PSLV).

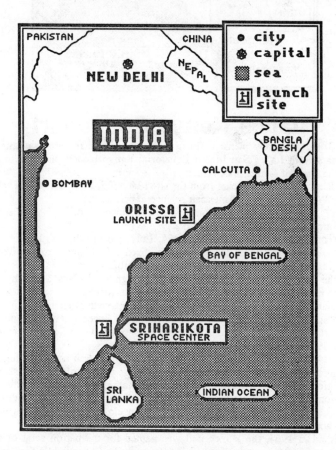

India launched its first satellite, Rohini 1, to orbit using the SLV rocket on July 18, 1980. India was the seventh nation with a space rocket. The SLV successfully placed three 80-lb. satellites in orbit between 1980-1983. The ASLV is designed to carry 330-lb. satellites. An ASLV has an SLV at its core, four solid-fuel stages and two strap-on solid-fuel boosters. PSLV will carry 4,400-lb. satellites to polar orbit in the mid-1990's. India also is building a military rocket range in the eastern state of Orissa. Other India satellites in orbit were launched by the U.S., USSR and European Space Agency.

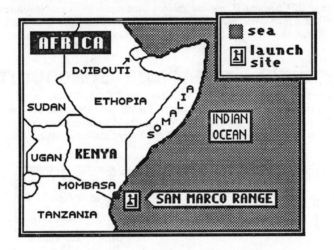

Italy's Kenya Spaceport

The U.S. space agency plans to fire a series of science research satellites over the next ten years from Italy's San Marco Equatorial Range launch platform in the Indian Ocean off the coast of Kenya, East Africa.

NASA launched a Scout rocket from the site March 25, 1988, carrying a satellite with five Italian, West German and American science experiments.

The 1988 launch was the first from San Marco since 1974. NASA has scheduled its next San Marco Scout launch for May 1992.

Historic. Back on December 15, 1964, Italy became the third nation to send a satellite to space when an Italian launch team fired a U.S. Scout rocket from the San Marco Range carrying the San Marco 1 satellite to orbit.

It was the first launch of a NASA rocket by foreign technicians. Since that time, Italy has been the most-active foreign user of NASA's Scout rocket.

The Italians have big future plans for their Indian Ocean launch site, too. They want to take two solid-fuel booster rockets from a European Space Agency Ariane rocket and strap them around a four-stage Scout. The new booster, known as Scout 2, could be flying early in the 1990's with double the capacity of Scout 1.

The platforms. The launch site is a pair of platforms anchored together three miles offshore in Kenya's Ngwana Bay, similar to rigs used in offshore oil drilling.

The platforms, operated by the University of Rome, are 90 miles north of Kenya's port city Mombasa. They have been the site of nine launches from 1964 through 1988, all successful, all U.S. Scout rockets. The location close to the equator gives satellite controllers access to an equatorial orbit.

Established in 1964, the launch pad was named for the patron saint of navigators. Launches from San Marco have been so popular with the Italian public that Pope Paul VI blessed them.

The Scout. Scout has been one of NASA's most reliable boosters. The 1988 San Marco launch was the 95th success in 109 attempts since 1960. Scout can lift a 475-lb. payload to a 300-mi.-high orbit.

The 10:55 p.m. launch March 25, 1988, came 12 hours after a daylight launch was postponed. Weather forecasters had predicted a thunderstorm which did not materialize. The four-stage solid-fuel Scout rose on a plume of smoke and fire into a cloudless sky beneath a quarter moon. The 522-lb. satellite was monitored from a Rome control room while it orbited 100 miles above Earth for a year.

Townsend. Marjorie Townsend, the first woman to launch a spacecraft, sent the U.S. satellite Explorer 42 (SAS-1) to Earth orbit from the San Marco platform on December 12, 1970. Townsend named the satellite Uhuru, a Swahili word for "freedom," in honor of Kenya's independence day December 12. Uhuru detected a black hole in the constellation Cygnus on January 30, 1971, receiving X-rays and radio waves from a faint blue star with a tiny, unseen, dark companion, more massive than a neutron star.

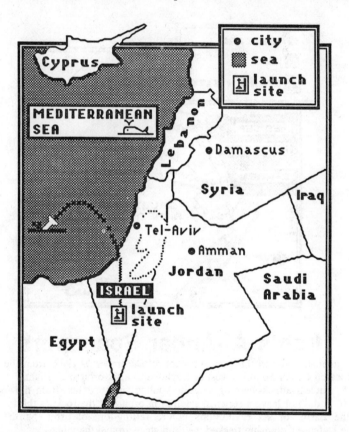

Israel's Negev Desert Spaceport

Israel has made several launches of its three-stage Shavit space rocket from a new military launch pad south of Tel Aviv and Jerusalem in the Negev Desert, including sending two satellites to Earth orbit. Shavit, a converted Jericho II medium-range ballistic missile, was test fired in May 1988 on a short suborbital flight to a splashdown in the Mediterranean Sea. Shavit is Hebrew for comet.

Israel became the eighth country to demonstrate a rocket powerful enough to fire a satellite to orbit when it launched its first satellite, Ofek 1 or Horizon 1, on September

19, 1988, on a Shavit fired over the Mediterranean Sea.

The three-stage rocket flew from the Negev launch pad, carrying a small satellite to a low elliptical orbit. Onlookers saw Shavit arch through the noon Middle Eastern sky where stages separated and fell into the sea.

The satellite launch brought Israel into the international club of spacefaring nations with China, France, Great Britain, India, Japan, USSR and the United States. Horizon 1 fell from orbit after 4 months.

On April 3, 1990, Israel launched another satellite, Horizon 2 or Ofek 2, on a Shavit from the Negev Desert to an elliptical orbit ranging in altitude from 125 to 923 miles. Horizon 2 fell into the atmosphere July 9, 1990.

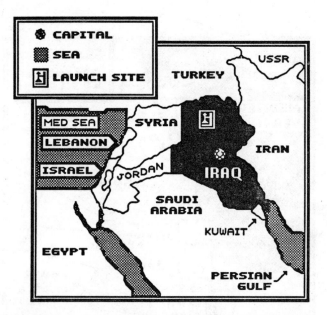

Iraq's Al-Anbar Spaceport

Iraq launched a 48-ton three-stage rocket on December 5, 1989, from the Al-Anbar Space Research Center 50 miles west of Baghdad to a six-orbit spaceflight.

A U.S. nuclear-attack-warning satellite spotted fiery exhaust from the rocket as it blasted off. It did not leave a separate satellite in orbit, but the rocket's third stage swept around the Earth for six revolutions before falling out of orbit. North American Aerospace Defense Command tracked the third stage around the globe.

The launch made Iraq the ninth nation with a rocket powerful enough for space launches. It was the first time Iraq had exposed its space research.

The 75-ft., three-stage rocket probably was similar to a U.S. Scout rocket used to send small satellites to low orbits. The booster may have been a modified version of Argentina's Condor ballistic missile. Such a missile could carry a nuclear warhead 1,240 miles. Iraq had a 600-mi. missile built around the USSR's Scud.

Scuds were launched by Iraq against Iran during their eight-year war in the 1980's. Iraq again launched Scuds, against Israel and Saudi Arabia in the 1991 Persian Gulf War. Iraqi launch sites, including Al-Anbar Space Research Center, were damaged in the 1991 war, casting doubt on Iraq's ability to continue in the Middle East space race with Israel.

Australian Spaceport Weipa

If you'll be needing a spaceport in the next decade, you might want to consider Weipa, Australia. The Queensland state government has been putting a lot of energy into trading the excellent climate in the remote, northern, sparsely-populated 4,000-square-mile Cape York Peninsula—including its overpopulation of crocodiles—for a slice of the worldwide spaceflight launching pie in the 1990's.

It's certainly a low-overhead area at the moment. And it probably can guarantee longer life in space for satellites since it's only about 15 degrees below the equator. An orbiting stationary satellite, which had been launched from Cape York Peninsula, would need 120 lbs. less fuel to maintain its orbit. That would mean a couple of years more working life in space.

Weipa. The premier of Queensland thinks the weather at Weipa is as near perfect as one could want. That might be attractive to governments and private companies who could replace Mother Nature with steel towers, concrete runways, and hotels.

Australia has organized a Cape York Space Agency to develop a Queensland launch facility. A Queensland University scientist says spaceport operators would earn millions of dollars in profits, even after the expense of moving the crocodiles elsewhere.

Meeting resistance. Thousands of Aborigines who call tropical Cape York Peninsula home disagree with the premier of Queensland. They are campaigning against the proposed launch site, calling it a "second invasion" of their country.

The spaceport is backed by the government of Australia and the state of Queensland, but the Wuthathi and Kuku Yau communities say it was planned in Tokyo, Moscow and Washington, without consideration for local land rights.

The local Cape York Aboriginal Land Council tried legal means of stopping the development. Wuthathi people are traditional Aborigine owners of the outlying land on the east coast of Cape York Peninsula.

Japanese spaceport. Japan would like to land its Hope space shuttle at Weipa. Japan has ambitious plans for its shuttle to carry freight to space station Freedom.

The name Hope is short for H-2 Orbiting Plane. H-2 is a powerful space rocket Japan is building for use in the 1990's. Hope shuttle will be shot to orbit atop an H-2.

Pacific Spaceport Group, an organization formed by Japan's seven largest companies, wants to build a six-mile-long runway, plus launch towers and control buildings for the Hope shuttle, at the Weipa spaceport.

The Japanese government plans to spend billions through the 1990's to build rockets and ground facilities.

Russian spaceport. The USSR would like to ferry U.S. satellites to orbit from Weipa on Zenit rockets. The Zenit medium-lift booster would carry U.S. satellites starting in 1995—if Aborigines permit United Technologies Corp. of Hartford, Connecticut, to build the Cape York spaceport on their lands.

Construction. The U.S. government informally approved start of the construction project, after questions of transferring U.S. technology to the USSR had held up the project a year.

Some 70 USBI workers would design the spaceport for the Cape York Space Agency (CYSA), then 300 would supervise construction of the base in northern Queensland and 200 would manage the launch site on Australia's northernmost tip after 1995.

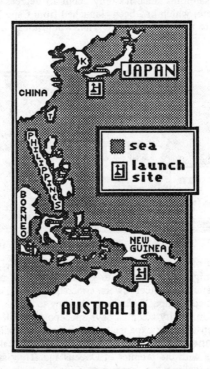

Zenit. The USSR commercial space agency, Glavkosmos, would sell rockets to CYSA and supply engineers to handle launches, but would not share in ownership of the spaceport. Zenits would ferry satellites to space from Cape York for customers from around the world, including the United States.

The $500 million Cape York venture has been backed by an Australian real estate developer and would not receive government money. It would increase competition for American rocket makers, already competing with Europe and China for launch business.

Secret Space Base. Across the continent from Weipa, the United States and Australia operate a secret satellite communications and control base at an old British-Australian rocket testing range near Nurrungar in the state of South Australia.

Great Britain and Australia tested rockets on the range at Woomera from the end of World War II through the 1970's.

Word of the secret space center slipped out in 1988 while the Australian Broadcasting Corp. was making a documentary on the town of Woomera.

Australia's Defense Department then publicized photographs of the space-tracking center, showing a close-up of an antenna at the station inside a protective radome, an outside shot of a radome, and a wide angle view of three radomes and buildings.

A radome is a protective covering over a radar antenna which does not interfere with signals to and from an antenna inside.

Three sites. Nurrungar is the Joint Defense Space Communications Station, one of three shared bases in Australia. The others are Pine Gap, near Alice Springs, Northern Territory, and North-West Cape, Western Australia.

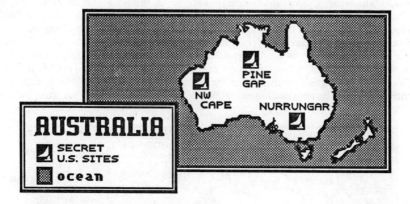

News reports indicated Nurrungar controls U.S. infrared early-warning and arms control verification satellites which observe Soviet missile launches.

During the 1991 Persian Gulf War, Australian prime minister Bob Hawke acknowledged Nurrangar gives early warning of ballistic missile launches and nuclear detonations, while the satellite ground station at Pine Gap receives intelligence data.

Nurrungar. Infrared sensors in a U.S. Defense Support Program (DSP) satellite over the Indian Ocean scan Europe, Africa, the Middle East and Southwest Asia, Asia and Southeast Asia, and Australia.

The heat from the exhaust of any rocket launch is spotted within seconds by the sensors, triggering DSP to send down an alert to Nurrungar, 300 miles northwest of Adelaide.

Within a minute of blast off, 500 Australian and U.S. technicians pinpoint a rocket's origin within three nautical miles, calculate its trajectory and determine its target.

That information is rushed to commanders, providing time for air-raid warnings. Location of a launch site is used for search-and-destroy missions.

Nurrungar is a relic of the Cold War in which it served as an early-warning system against a massive Soviet nuclear attack. It was a key Allied warning link in the Persian Gulf War, detecting Scud missiles as Iraq lobbed them at Israel and Saudi Arabia.

Pine Gap. Pine Gap, in central Australia, uses an extraordinarily-sensitive eavesdropping satellite to monitor two-way radio and other communications.

The satellite's receivers are so sensitive, it is said they can hear low-power handheld portable radios during Soviet military exercises.

In the Persian Gulf War, the base reportedly was tuned to Iraqi military signals to see which stations had been knocked out.

Australia, one of 28 nations in the coalition against Iraq, sent three ships to the Gulf. The secret satellite ground stations are under U.S. command.

Other U.S. satellite receiving stations dot the globe, including secret listening posts at Fort Meade, Maryland; Bad Aibling, West Germany; and Menwith Hill, Great Britain.

Brazil's Atlantic Spaceports

Brazil's Ministry of Aeronautics has space launch and tracking facilities in northeast Brazil on the country's 4,603-mi.-long Atlantic Ocean coast near Alcantara and Natal, not far south of the equator. The European Space Agency space launch site is nearby on the Atlantic Coast at Kourou, French Guiana, north of the equator.

The Brazilian Complete Space Mission (MECB), has a control and tracking station at Cuiaba, Mato Grosso. Satellites transmit data to an analysis station at Cuiaba for distribution across the country. Mission control is in Cachoeira Paulista in Sao Paulo. Development is at Instituto de Pesquisas Espacias (INPE) at San Jose dos Campos.

Brazilian scientists have launched hundreds of sounding rockets since the mid-1960's. Alcantara, known as CLA, has a launch pad for Sonde sounding rockets, VLS space satellite launchers, a launch control center and blockhouse. The People's Republic of China, looking for a launch site on the equator, may ship Long March rockets to Alcantara. Brazilsat communications satellites have been launched by ESA from Kourou.

South African Spaceport?

Israel's Shavit space booster could orbit another satellite in the 1990's, this time for the Republic of South Africa. The South Africans have tested the Israeli Jericho 2 intermediate-range ballistic missile which converts to the Shavit space rocket.

South Africa could launch a satellite to space, firing it from the nation's southern area near Cape Town into the sky over either the Atlantic or Indian Ocean.

The South African government says it has the science and technology to launch a space satellite, but is not interested because of high costs. However, Pretoria has kept the pot boiling by setting up a committee to keep abreast of space technology and coordinate development. The committee includes the Department of Posts and Telecommunications, the Weather Bureau, the Council for Scientific and Industrial Research (CSIR), the Industrial Development Corporation (IDC), and other agencies, according to the Minister for Economic Affairs and Technology.

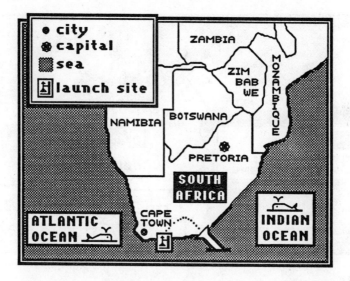

Israel created a space agency in 1983 and started work on space rockets and spy satellites. A joint South African-Israeli rocket project is said to have been under way since 1987. Israel first test-fired a Shavit space rocket, converted from a Jericho 2 medium-range ballistic missile, in May 1988 on a sub-orbital flight into the Mediterranean. In 1988, Israel became the eighth country to fire a satellite to orbit. Israel sent a second satellite to orbit in 1990.

As a medium-range ballistic-missile weapon, Jericho 2 can carry a nuclear warhead 900 miles to a target on Earth. As a Shavit space rocket, it can lift a small spacecraft such as a spy satellite to Earth orbit. Shavit is Hebrew for comet.

Mueller. The Republic of South Africa should have its own space program, according to pioneering U.S. space engineer George Mueller who directed Gemini and Apollo man-in-space programs in the 1960's.

He said investment in space would bring a technological fallout, improving lives of everyone in South Africa. Mueller said South Africa could have its own weather and television satellites.

Hamsat. He spoke to students, teachers and businessmen at Rand Afrikaans University, Johannesburg, and at University of Stellenbosch, east of Cape Town, where students and teachers are working on an amateur radio satellite to be fired to Earth orbit on an American or European launcher.

Ham radio operators in South Africa are building a satellite for the University of Stellenbosch. The small communications satellite, to be launched in the 1990's, would have a radio signal repeater and house student science experiments. The university recently built a new technology park for its Bureau of Systems Engineering which is in charge of the satellite project.

The first nine spacefaring nations, in order of appearance, were USSR, U.S., France, Japan, China, Great Britain, India, Israel and Iraq.

Canada's Churchill Spaceport

Canadian Space Agency operates the Churchill Rocket Research Range in the northeastern part of the country on Hudson Bay. NASA has launched small sounding rockets—Nike-Orion and Black Brant—from Churchill.

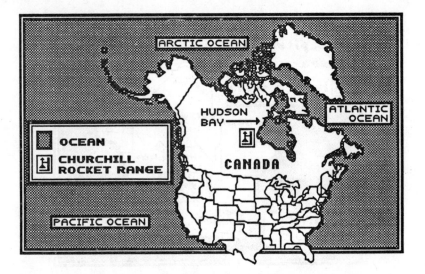

Pakistan's Space Launches

China launched Pakistan's first satellite, Badr-A, to orbit July 16, 1990, because Pakistan has no spaceport. However, Pakistan does have a space program.

Scientists and engineers at Pakistan's Space and Upper Atmosphere Research Commission (SUPARCO) built the tiny radio repeater placed by Long March 2E rocket in a 375-mi.-high circular orbit. Badr means "new moon" in the Urdu language.

Badr-A once was to have been launched by a U.S. space shuttle, but the plan changed after the 1986 Challenger disaster. The spacecraft then was shipped to Xichang Launch Center in southwestern China for the first flight for China's new Long March 2E.

SUPARCO has two sites—at Karachi, a major seaport in southern Pakistan on the Arabian Sea, and at the University of the Punjab at Lahore, a prominent border city in eastern Pakistan not far from Delhi, India. Satellite ground stations have been built at Karachi and Lahore where amateur radio operators took part in a digital communications experiment with a British hamsat. Badr-A used digital store-and-forward techniques.

Pakistan also has a national communications satellite project to build two large satellites for stationary orbit at altitudes near 22,300 miles. One would be placed over the equator at 38 degrees east and the other at 41 degrees east. SUPARCO has built antennas and receivers for direct-to-home TV reception from the two big satellites. SUPARCO also is building small terminals for rural telephone connections via satellite. A large ground station has been built by SUPARCO at the nation's capital, Islamabad.

Pakistan may consider launching its own rockets in the future, possibly using a southern site on the Arabian Sea, away from the seaport. India fires its space rockets from a southeastern state so launchers which fail can fall safely into the Bay of Bengal.

Space Rocket Names

There have been many names used for space rockets. Here's a sampler:

★The United States has used Vanguard, Jupiter, Bumper-WAC, Aerobee, Viking, Redstone, Thor, Scout, Delta, Atlas, Centaur, Titan, Pegasus, Conestoga, Starfire, Industrial Launch Vehicle, Mighty Mouse, Black Brandt, and more, plus shuttles Columbia, Discovery, Atlantis and Endeavour.

★European Space Agency uses the name Ariane.

★The USSR uses names like Vertikal, Zenit, Proton, Energia and shuttle Buran.

★Japan has H-1, H-2, N-1, N-2 and Mu-3S-2 and shuttle Hope.

★Israel has Shavit or Comet.

★India uses Satellite Launch Vehicle (SLV), Augmented Satellite Launch Vehicle (ASLV) and Polar Satellite Launch Vehicle (PSLV).

★People's Republic of China has Weaver Girl and Changzheng or Long March.

★Brazil is building a rocket known as Veiculo Lancador de Satelites (VLS).

Additional rockets are being tested by various nations and their names will be heard in the future.

U.S. Rockets: Unmanned Flights

NASA, the U.S. space agency, plans to use expendable launch vehicles (ELVs) to carry satellites to orbit in the 1990's, in addition to ferrying payloads to space in reusable space shuttles.

In NASA's unmanned rocket launch schedule below: DATE is year and month. CL is class of rocket, in which I is intermediate, L is large, M is medium, and S is small.

TBD is to be determined. RKT is rocket name. INC is inclination in degrees. ORB is payload orbit, in which GSO is geosynchronous orbit, GTO is geosynchronous transfer orbit, HE is high-eccentricity orbit, LEO is low Earth orbit, EO is escape orbit, PLAN is high-energy trajectory to outer planets, and SS is Sun synchronous.

LS is launch site, in which ESMC is Eastern Space and Missile Center, Kennedy Space Center, Florida, and WSMC is Western Space and Missile Center, Vandenberg Air Force Base, California. PAY is satellite payload.

DATE	CL	RKT	INC	ORB	LS	PAY
1992 Jun	S	Scout	TBD	TBD	WSMC	SAMPEX
1992 Jul	M	Delta 2	28.7	HE	ESMC	GEOTAIL
1992 Sep	I	Titan 3	28.5	EO	ESMC	MARS OBSERVER
1992 Dec	I	Atlas 1	28.5	GSO	ESMC	GOES-I
1992 Dec	M	Delta 2	28.7	HE	ESMC	WIND
1993 Jun	M	Delta 2	90.0	HE	WSMC	POLAR
1993 Aug	I	Atlas 1	28.5	GSO	ESMC	GOES-J
1993 Sep	S	Pegasus	TBD	LEO	WFF	TOMS-1
1993 Dec	M	Atlas E	98.7	SS	WSMC	NOAA-J
1993 Dec	TBD	TBD	TBD	HEEO	ESMC	EQUATOR-S
1994 May	M	Delta 2	TBD	LEO	WSMC	LAGEOS III
1994 Jul	S	Pegasus	TBD	LEO	WFF	SAC-B/HETE
1994 Jul	M	Titan 2	98.7	SS	WSMC	NOAA-K
1994 Sep	S	Pegasus	TBD	LEO	WFF	FAST
1994 Dec	M	Delta 2	98.6	LEO	WSMC	RADARSAT
1995 Apr	I	TBD	28.5	GSO	ESMC	TDRS-7
1995 Jun	S	Pegasus	TBD	LEO	WFF	SWAS
1995 Jul	I	Atlas 1	28.5	GSO	ESMC	GOES-K
1995 Jul	I	Atlas 2AS	28.5	HE	ESMC	SOHO
1995 Dec	L	Titan 4/ Centaur	TBD	PLAN	ESMC	CASSINI

1996 Jan	S	Pegasus	TBD	TBD	WFF	SMEX-04
1996 Feb	L	Titan 4/	TBD	PLAN	ESMC	CRAF
		Centaur				
1996 Jun	M	Delta 2	TBD	TBD	WSMC	LIFESAT-01
1996 Jul	M	Titan 2	98.7	S S	WSMC	NOAA-L
1997 Feb	M	Titan 2	98.7	S S	WSMC	NOAA-M
1997 Feb	I	Atlas 1	TBD	GSO	ESMC	GOES-L
1997 Apr	I	TBD	TBD	GSO	ESMC	ATDRS-1
1997 Jun	S	Pegasus	TBD	TBD	WFF	SMEX-05
1997 Jun	M	Delta 2	TBD	TBD	WSMC	LIFESAT-02
1997 Jun	S	Pegasus	TBD	LEO	WFF	TOMS-2
1997 Aug	M	Delta 2	TBD	HE	ESMC	ACE

U.S. Rockets: The Titan Family

America's most-powerful unmanned space launcher today is the Titan 4 rocket, the latest evolution of a family of non-reusable expendable rockets which started more than three decades ago when Titan 1 blasted off a launch pad at Cape Canaveral, Florida, February 6, 1959, on its maiden flight.

Compared with today's Titan 4, the Titan 1 was puny. In its initial launch, only the first stage of the U.S. Air Force rocket had fuel. The second stage fuel tanks were filled with water for weight. Titan 1 traveled only 300 miles down the Atlantic Tracking Range.

Titan 1. Soon after the maiden voyage, both stages of the rocket came into use and Titan 1 was lobbing dummy warheads 5,000 miles down range at 17,000 mph. It became America's first two-stage intercontinental ballistic missile (ICBM). Titan 1, the giant of the U.S. launch vehicle family, was the most powerful bomb delivery vehicle in the American nuclear arsenal.

Titan 2. It didn't take long for the Air Force to visualize an improved ICBM, making Titan 1 obsolete. It was replaced in 1962 by a larger and more powerful Titan 2.

Gemini-Titan. Modified Titan 2 rockets were used by NASA in 1965 and 1966 to loft 10 pairs of Gemini astronauts to orbit, preparing for later three-man Apollo astronauts to the moon.

Gemini 3 carried two astronauts to orbit March 23, 1965. Gemini 4, June 3, 1965. Gemini 5, August 21, 1965. Gemini 6, December 15, 1965. Gemini 7, December 4, 1965. Gemini 8, March 16, 1966. Gemini 9, June 3, 1966. Gemini 10, July 18, 1966. Gemini 11, September 12, 1966. Gemini 12, November 11, 1966.

Titan-Centaur. NASA added a Centaur upper stage to a few Titan rockets to launch interplanetary and deep-space probes such as Voyager and Viking payloads to explore the planets and the Helios deep-space probe.

The Helios spacecraft were two 815-lb. interplanetary probes sent to observe the Sun and its solar wind. They were constructed in West Germany and launched by the U.S. from Cape Canaveral on Titan-Centaur rockets in December 1974 and January 1976. Their orbits took them within 28 million miles of the Sun, closer than previous probes.

Underground ICBM. As an ICBM, Titan 2 and its fuel could be stored for years in underground silos, ready to be fired on a moment's notice across the North Pole to the USSR. Fifty-four of the 103-ft., liquid-fuel Titan 2 rockets stood on alert for 20 years in 165-foot silos in Arizona, Arkansas and Kansas. Each Titan 2 ICBM carried a nine-megaton hydrogen bomb. Over the years, fuel corroded the missiles and sometimes they exploded. The nuclear warheads did not explode in those mishaps. Old age brought removal of the obsolete missiles from the silos. The last one was deactivated in 1982.

They were replaced as ICBM's by Minuteman and submarine-launched Trident missiles. Minuteman and Trident used easier-to-handle solid fuel.

Converted for orbit. The Air Force reconditioned thirteen of the old Titan 2 rockets as space boosters at a cost of $528.9 million, because it was cheaper than building new expendable rockets.

The first Titan 2 converted to a space rocket roared into the sky from Vandenberg Air Force Base, California, September 5, 1988, carrying four Navy White Cloud ocean-spying satellites to eavesdrop on radar and radio signals from the Russian fleet. The missile faded into a fog bank with a deep-throated roar that lasted five minutes, rattling windows in the nearby town of Lompoc.

Titan 2 is useful for a variety of payloads including Defense Meteorological Support Program (DMSP) weather satellites. NASA plans to use Titan 2s to launch weather satellites to polar orbit from Vandenberg—NOAA-K in July 1994, NOAA-L July 1996, and NOAA-M February 1997. The Air Force may recycle more Titan 2 rockets if most of the first thirteen work well as space boosters.

Titan 3. Titan 2 rocket motors were improved, fuel tanks were enlarged, solid-fuel booster rockets were strapped around the liquid-fuel core and a solid-fuel rocket was tacked on top as a new upper stage. Titan 3 was created.

The improved rocket comes in versions offering different weight-lifting capabilities, including Titan 3, Titan 3B and Titan 34-D.

Most Titan 3 rockets have been used by the Air Force to launch communications, navigation and reconnaissance satellites. NASA plans to use a Titan 3 at Cape Canaveral in September 1992 to launch the Mars Observer interplanetary probe on its way to the Red Planet.

Titan 3B. Titan 3B was one of the larger rockets available to the U.S. military for years. It was stronger than Titan 3, but less-powerful than Titan 34-D. Titan 3B can carry 7,000 lbs. to a 100-mi.-high orbit.

A Titan 3B may have boosted a U.S. Air Force ear-in-the-sky spy Jumpseat satellite to orbit 100 miles above Earth on February 11, 1987. Such a secret satellite would have been able to pick up radio signals from foreign countries. The launch from Vandenberg Air Force Base left the payload in a polar orbit, which took the satellite over the USSR. An additional Agena rocket attached to the satellite pushed the communications-reconnaissance spysat on out to an orbit ranging from 1,000 miles to 22,000 miles above Earth.

Titan 34-D. A version of Titan 3, known as Titan 34-D, was America's most powerful space launcher in operation at the end of 1988. A Titan 34-D is composed of two 96-ft.-tall solid-fuel booster rockets strapped around a 161-ft.-tall two-stage liquid-fuel Titan 3B rocket. Titan 34-D was sent to space three times in 1987 and 1988.

A Titan 34-D exploded August 28, 1985, at Vandenberg when fuel leaked and a fuel pump didn't work during launch. One of its liquid-fuel engines shut down prematurely. Its spy satellite cargo was destroyed.

Seven months later at Vandenberg, on April 18, 1986, flame burned through insulation on a solid-fuel booster strapped to a Titan 34-D rocket, igniting an impressive fireball above the launch pad. The booster exploded nine seconds after launch, destroying the Titan 34-D and its spy-satellite payload. The KeyHole KH-11 satellite had been meant to replace one deactivated in 1985.

The Titan 34-D fleet was grounded 18 months until launch pad damage and rocket problems could be fixed. Titan 34-D was sent to space three times in 1987 and 1988.

The Titan failures were among a series of setbacks to the American space program in the 1985-1987 period, including the January 28, 1986, explosion which destroyed shuttle Challenger, killing seven astronauts.

Confident in February 1987 that its Titan 34-D problem was with the solid-fuel booster, not the Titan 3-B core rocket, the Air Force successfully launched a Titan 3-B from Vandenberg carrying a radio-eavesdropping satellite. Then a two-minute ground-test firing of one of the 96-ft. solid-fuel boosters from a Titan 34-D was successful June 15, 1987, at Edwards Air Force Base, California, 70 miles north of Los Angeles.

Thunderstorms and winds over the base forced tests to be scrubbed several times that June. Officials feared winds would blow rocket exhaust fumes over a residential area near the Mojave Desert research center. The Air Force also had been worried about lightning and electrical interference around the test stand on which the rocket was ignited while anchored to the ground.

Eye in the sky. Finally, Titan 34-D made its return to space October 26, 1987, carrying a spysat to orbit in the first successful blast-off since the April 1986 explosion. Titan could be seen ascending from Vandenberg Air Force Base, California, 170 miles northwest of Los Angeles, for 20 seconds before it disappeared with its top-secret payload into the clouds.

The Air Force wasn't saying, but the October 1987 payload may have added to U.S. capability for monitoring Soviet compliance with arms treaties. It could have been a KeyHole KH-11 photo-reconnaissance satellite or a new model of the Jumpseat radio-eavesdropping satellite.

After August 1985, the U.S. had been in the ticklish position of being down to its last working photo-reconnaissance satellite. Military planners worried the last satellite might fail, leaving the U.S. with no eyes in space. Of course, the military wouldn't be deaf since radio-eavesdropping satellites were in orbit. The Titan 34-D success allowed the Air Force to plan regular launches of reconnaissance satellites, such as models KH-9, KH-11 and KH-12.

Two in a row. Titan 34-D made it two in a row with the successful launch of another spy satellite November 28, 1987. The Titan 34-D blasted off from the Air Force's Eastern Space and Missile Center at Cape Canaveral toward a stationary orbit 22,300 miles high. The payload was one of a series intended to provide 30 minutes warning of nuclear attack.

The success continued America's recovery from 1986 failures and improved launch chances for the backlog at that time of 30 military satellites. It was the second straight jackpot for Titan 34-D after the April 1986 grounding.

Cape Canaveral was busy. Another U.S. rocket, Delta, grounded by a 1986 failure, had resumed flights.

September failure. A setback for Titan 34-D came with an orbital failure September 2, 1988, after what at first looked like a successful launch from Cape Canaveral. The unmanned Titan 34-D thundered into the sky to a preliminary low orbit. But then the upper stage of the core Titan 3-B failed to ignite. The upper stage was needed to push its spysat payload out to a planned 22,300-mi.-high stationary orbit. The rocket's nose cone covering may have hit the satellite. The satellite was rendered useless, in a sharply elliptical orbit ranging from 100 to 22,300 miles above Earth.

The electronic-ear satellite, with a folded antenna as large as a baseball diamond, had been designed to eavesdrop on missile tests and radar, radio, telephone and electronic diplomatic and military communications.

The September failure, first launch of a Titan 34-D since the 1987 flights, had been delayed several months by technical problems. The 1987 flights had been the only heavy military payloads launched successfully in nearly three years. Only three Titan 34-D rockets remained in inventory after the September failure.

November success. The picture brightened when a Titan 34-D thundered up into heavy morning cloud cover and away from Vandenberg's Space Launch Complex 4 on November 6, 1988, ferrying military communications satellites to orbit.

The payload probably was a pair of Satellite Data System (SDS) communications satellites to be used if the U.S. had to launch nuclear attack. However, the payload might have been one KeyHole KH-9 or KeyHole KH-11 spysat.

The rocket was to be the last Titan 34-D fired from Vandenberg. Two were left for launch from Cape Canaveral in 1989. America's most powerful booster was being replaced by an even more powerful Titan 4, due for maiden launch from the Cape in 1989.

When manned space shuttles began flying, the Pentagon and NASA decided to phase out unmanned rockets. But, even before Challenger exploded, the Air Force needed unmanned expendable rockets as backup. It started to develop Titan 4, a more powerful version of Titan 34-D. After Challenger, the Air Force ordered additional Titan 4's, totaling 27. And it asked for money for 25 more. The Air Force also decided to refurbish the 13 obsolete Titan 2 intercontinental ballistic missiles to serve as space boosters.

Thunder in the ground. The towering Titan 34-D, ferried what seemed to be a pair of communications satellites to orbit on a secret military flight May 10, 1989.

The ground around launch complex 40 at Cape Canaveral Air Force Station shook from the thunderous blast off as the Titan 34-D punched upward through a partly-cloudy sky, the Defense Satellite Communications System (DSCS-3) payload shrouded by its nosecone. Like all secret military launches, hundreds of civilian spectators watched, turned out by Coast Guard warnings to boaters of an offshore danger zone where debris from the rocket's first stage was to fall.

A launch attempt May 9 had been scrubbed three minutes before blast off. On May 10, visible from the Kennedy Space Center for miles along east-coast Florida beaches, the 1.5-million-pound Titan curved eastward over the Atlantic Ocean on a fiery cloud of sooty exhaust from two 90-ft., strap-on, solid-fuel boosters. Two minutes into flight, the 10-ft.-wide boosters jettisoned and the central two-stage liquid-fuel Titan rocket powered on, disappearing from sight on its way to space. The rocket launched May 10 was one of only two Titan 34-Ds left in the Air Force inventory.

DSCS sats. The payload may have been a pair of DSCS-3 communications satellites to provide cloaked communications between the White House and commanders of U.S. fighting forces flung around the globe.

DSCS satellites require an inertial upper stage (IUS) booster for the final drive to stationary orbit 22,300 miles above the equator. It was not clear whether the satellite carried by the Titan 34-D had an IUS attached or a less-powerful trans-stage booster used for spy satellites sent into lower orbits. The Air Force wasn't confirming anything, so the May 10 payload might have been a Vortex or Chalet radio-eavesdropping spysat, augmented by a liquid-fuel Transtage booster.

Orbiting 22,300 miles above the equator, a satellite completes one revolution around our planet every 24 hours. Thus, a satellite seems to hang at one point in the sky, providing uninterrupted radio or photographic coverage of the surface below.

A 2,143-lb. DSCS-3 satellite is 9-ft. by 6-ft. with solar-panel wings spreading 38 feet. It is said to be jam-proof, hardened against attack and maneuverable in orbit. It has a narrow transmitter beam to restrict eavesdropping by an enemy. A DSCS-3 is expected to work ten years in Earth orbit. Three DSCS-3 satellites were launched before the 1986 Challenger disaster. If one were carried in the May 10 launch, a total of five would have been in stationary positions over the Atlantic Ocean, east Pacific Ocean, west Pacific Ocean, Indian Ocean and a spare on standby in orbit.

Last Titan 34-D. The Air Force blasted its last Titan 34-D rocket to space September 4, 1989, ferrying a secret satellite from Cape Canaveral. It was the fifteenth and final flight for the workhorse booster, about to be replaced by Titan 4.

The 1.5 million-lb. rocket shook the beaches near Kennedy Space Center as it shot east across the Atlantic Ocean, blazing a trail of orange flame from two 90-ft. solid-fuel boosters, lighting the 2 a.m. darkness. Two minutes out of launch complex 40, the two 10-ft.-wide strap-on boosters were dropped and the two-stage liquid-fuel Titan core continued toward space.

It was the sixth Titan 34-D launch since failures in 1985 and 1986. As usual, the payload carried by the Titan 34-D September 4 was not revealed. It could have been the two Defense Satellite Communications System (DSCS) communications satellites or spy satellites sent into lower orbits.

Powerhouse. Titan 34-D, built by Martin Marietta Astronautics Group, Denver,

Colorado, was capable of carrying 31,650-lb. payloads to low orbit and 4,000 lbs. to stationary orbit. The rocket's first flight was October 30, 1982.

Three of 15 Titan 34-D rocket launches failed, but Martin Marietta considered the Titan 34-D a member of the larger Titan 3 rocket family which had 135 successes in 141 launches. Test flights before the rocket became operational were not in the total.

Titan 4. America's newest and currently-most-powerful unmanned space rocket, Titan 4, thundered to space June 14, 1989, from launch complex 41 at Cape Canaveral Air Force Station in a proud maiden voyage, carrying an early warning satellite to orbit.

The first Titan 4 blast off had been scheduled for October 1988, but problems with the nose cone, and a shortage of IUS boosters, delayed the debut to June 14.

The launch provided a close call. It was reported that one of two engines in the second stage swiveled four degrees out of position late in the ascent. That forced the other engine to swivel the other way to compensate. The rocket stayed on course and delivered the spysat payload to its planned orbit.

A 174-ft. Titan 4 with two solid-fuel boosters can ferry 39,000-pound payloads to an orbit 150 to 300 miles high. Or it can lift a 10,000-lb. package to stationary orbit 23,000 miles high. Titan 4 replaced Titan 34-D.

Returning ELVs. The Titan 4 flight was a milestone in recovering from the 1986 Challenger disaster as well as failures of Titan 34-Ds in 1985 and 1986.

As shuttle Challenger blasted off on its fatal final voyage, unmanned expendable launch vehicles (ELVs) were being phased out, the result of political decisions in the 1970's to spend less money on space, forcing NASA's manned shuttle to become America's only vehicle for carrying large payloads to space.

The Air Force was uncomfortable relying on a fleet of complex shuttles, thinking an accident could ground high-priority military satellites. Even before Challenger, the Air Force was given money to construct ten Titan 4 rockets big enough to carry shuttle-ready payloads. Following Challenger, during the three-year delay in shuttle flights, the Air Force obtained more money to build 13 additional Titan 4s, contracts worth $5.1 billion. In addition, the Air Force plans to buy another 20 rockets with options for six more.

Grounded shuttle flights left 40 spy satellites and other payloads in storage awaiting launch vehicles. That led the Defense Department to more than double the number of Titan 4s on order while proceeding to develop two new smaller rockets, the Delta 2 and Atlas 2.

Between 1983 and 1988, the Air Force budgeted $685 million for ELVs. Between 1988 and 1994, the expense is expected to be at least $14.1 billion.

The June 14 launch was the first of at least 23 Titan 4s to be built by Martin Marietta under the $5.1 billion contract. Titan 4s average $217.8 million per rocket.

It was the most expensive unmanned space flight for the United States to that time—including the $217 million rocket, the $180 million satellite, a $45 million payload booster stage (IUS) and other costs pushing the total close to half a billion dollars.

Titan 4 delays. The first of the new Titan 4s was rolled out to the launch pad on May 15, 1988, but it had to wait more than a year to blast off.

Actually, it was supposed to lift off in October 1988, but various technical problems caused delays stretching over many months, costing extra millions of dollars. Trouble areas included the rocket's 56-foot nosecone and the satellite's inertial upper stage (IUS) booster.

Finally, on June 14, 1989, it was smooth sailing all the way as the ten-story-tall boosters roared to life, shattering morning tranquility with a tower of fire and a sizzling roar, shaking the ground for miles around. The 20-story-tall Titan 4, focal point of the $14 billion military space buildup, bellowed away from launch complex 41, punching into the eastern sky over the Atlantic Ocean in front of a broiling trail of dingy smoke.

Two minutes later, the 1.9-million-lb. Titan 4's liquid-fuel first-stage fired. Boosters fell away, their 591,000-lbs. of fuel used up. The remaining two Titan stages rushed on upward, fading into the translucent blue sky. After 8 minutes, the rocket's flight was over. Its payload was dropped off in a planned preliminary orbit.

Mystery payload. Shrouded in secrecy, only the military knows for sure what the payload was, how fast the rocket travelled or what altitude it reached.

Reports indicated the Titan 4 carried the first of a series of Defense Support Program (DSP) satellites using a sensitive infrared telescope to detect missile launches. Reportedly, at least nine of the 33-ft.-long, 2.5-ton DSP early-warning satellites will be launched to warn of Soviet land or sea missile attack.

Most powerful? Titan 4 is capable of carrying payloads as heavy as those carried by space shuttles. The Titan family of rockets, however, would be no match for the extraordinary Saturn V rockets which sent Americans to the Moon in 1969-1972 and which lifted America's first-generation Skylab space station to orbit in 1973. A Saturn V could lift 200,000 lbs. to low orbit. The U.S. abandoned Saturn V in 1973.

The USSR's flight-tested Energia rocket is even more powerful than Saturn V, carrying 220,000 lbs. to low orbit. Energia, with an unmanned cargo pod, triples the payload capability of a U.S. space shuttle. Energia can lift payloads almost five times heavier than Proton, previously the largest Russian rocket. Energia also can lift five times the payload weight of Titan 4.

American and Russian space shuttles fly only to low orbits, roughly 100 to 400 miles above Earth. A USSR shuttle can carry 67,000 lbs. to low orbit, about 10,000 lbs. more than an American shuttle.

Commercial Titan. While Americans were warming up for some spectacular end-of-the-decade New Year's Eve parties, a Titan 3 rocket rang in the new decade of the 1990's with a thunderous launch from Cape Canaveral launch complex 40 at 7:07 p.m. EST December 31, 1989. The first launch of what Martin Marietta called Commercial Titan turned night into day as it ferried two satellites to Earth orbit.

The launch came on the tenth try. It had been scheduled to debut December 7, 1989, but inaccurate computer programming forced a 24-hour delay. Then eight attempts ended in bad weather, mostly high winds 25,000 to 35,000 feet above the launch pad.

Riding atop an updated Titan 3 were the British military communications satellite, Skynet 4A, and the Japanese communications satellite JCsat 2.

The solar-powered JCsat 2 is a high-powered KU-band TV satellite, partially owned by its builder, Hughes Communications, a subsidiary of the U.S. firm, Hughes Aircraft Co. The primary owner is the Japan Communications Satellite Co., made up of C. Itoh & Co. Ltd. and Mitsui & Co. Ltd. JCsat 2 has 32 transponders, each carrying one television channel, 250 telephone circuits or 45 million bytes of data per second to private companies across Japan.

Skynet 4A was built by British Aerospace Space Systems for the United Kingdom's Ministry of Defense. The first Skynet was lobbed to orbit in 1988 on a European Space Agency Ariane rocket.

NASA's Mars Observer was to be launched toward the Red Planet on one of the modernized Titan 3s in September 1992.

The 155-ft. Commercial Titan was composed of a Titan 3 enhanced two 90-ft. booster rockets. It delivered both payloads to low-Earth orbits, from where their onboard kick motors pushed them on to stationary orbit 22,300 miles above Earth.

The launch was sold by Commercial Titan, Inc., a Martin Marietta subsidiary established to sell Titan space rocket launches. Martin Marietta in 1986 had been the first American company to enter the commercial launch vehicle business. Martin Marietta leased the launch pad from the Air Force. The company said Titan 3 had a 96 percent success record in 142 launches over 25 years.

Commercial satellites were banned from U.S. shuttles after the 1986 Challenger disaster. That led three major American rocket companies into commercial sales of satellite launches: McDonnell Douglas Space Systems Co. and its Delta rockets, General Dynamics with Atlas-Centaur boosters, and Martin Marietta. Other competitors in the commercial launch industry included the USSR, China and the European Space Agency.

U.S. Rockets: The Atlas Family

Atlas first was an American intercontinental ballistic missile (ICBM). Later, it became the veteran workhorse space rocket it is today. Coupled with other rockets, Atlas has boosted to space scores of satellites, unmanned Moon probes and most of the nation's planetary probes.

The first Atlas ICBM was launched June 11, 1957. The Atlas-Centaur space rocket made its maiden flight on May 8, 1962, from Cape Canaveral. Ironically, both flights ended in explosions.

Atlas went on to become America's first operational long range military rocket. Later, it served as the first stage for two powerful space boosters, the Atlas-Agena and Atlas-Centaur. A Mercury capsule on an Atlas rocket formed MA-6, which carried the first American to orbit. Atlas-Centaur boosted satellites to Earth orbit, and Moon and planet probes.

Mercury-Atlas. Size and weight of the Project Mercury capsules required the lifting capability of the powerful Atlas-D rockets used to launch the first four manned U.S. flights to orbit. Converted from an Atlas ICBM, Atlas-D was the only U.S. rocket in the 1960 arsenal with sufficient reliability and power to loft a heavy manned satellite to a 100-mile-high orbit.

John H. Glenn Jr. rode to orbit February 20, 1962, in the Mercury-Atlas 6 capsule he nicknamed Friendship 7. He revolved around Earth three times in 4 hours 55 minutes 23 seconds, then splashed down in the Atlantic Ocean. It was the third U.S. manned spaceflight and the first U.S. manned orbital flight. Glenn was the first American in orbit, but the third person to go to orbit. He was the third American in space.

Following a pair of manned suborbital test flights and the Mercury-Atlas 5 flight of Enos the chimp, an Atlas-D rocket carried Glenn in the Friendship 7 from Cape Canaveral to space.

Atlas-Centaur. The powerful Atlas-Centaur, also known as Atlas 1, was a two-stage workhorse space rocket from the earliest days of the U.S. space program. The Centaur stage was on top with the Atlas underneath.

From the 1960's, Atlas-Centaurs ferried a variety of military and commercial satellites and science probes to space. Built by General Dynamics Corp., San Diego, California, Atlas-Centaur launches were 95 percent successful.

Atlas-Centaurs rocketed some famous names to outer space: the Pioneer spacecraft which flew by Jupiter and Saturn and now are off to interstellar space; Mariner probes which made close-up inspections of Venus, Mercury and Mars; and even the series of seven Surveyors which soft-landed on the Moon, dug trenches and sent back TV pictures from 1966-1968. Today's Atlas-Centaur—renamed Atlas 2—is sold for commercial satellite launches by General Dynamics.

Atlas-Centaur stats. The 73-ft. first stage weighs 320,701 lbs. at launch. It is an upgraded version of the original rocket launched by NASA in 1959.

The Atlas first stage has a Rocketdyne MA-5 engine system with two nozzles and a smaller sustainer engine with a single nozzle between the two main engine nozzles. Two small vernier engines burn at liftoff, fine tuning the rocket's trajectory. The first stage has a total thrust of 438,922 lbs.

The Atlas first stage actually is a stage and a half. About 2.5 minutes into flight, the main engine cuts off. The nozzles and other systems are blown away to reduce weight. The vernier engines and the sustainer continue to burn until fuel is exhausted 4.5 minutes after liftoff. When the first stage shuts down, the rocket is travelling at 9,486 mph.

The 30-foot-long Centaur D-1A second stage weighs 38,800 lbs. at launch. It consumes supercold liquid hydrogen and liquid oxygen to produce 32,815 lbs. of thrust. Centaur is the most powerful rocket, for its size, of any stage yet built.

The main propulsion system is two Pratt & Whitney RL-10A-3-3A engines which can be stopped and restarted. The Centaur stage does not ignite on the ground at launch.

Centaur usually ignites four minutes and 44 seconds after launch and burns 2.2 minutes, then shuts down. During a long coast period, twelve small thrusters keep the rocket on course. They also serve to keep Centaur fuel settled in the bottom of their tanks. At 20 minutes into flight, Centaur re-ignites for 1.5 minutes. When Centaur finally shuts down, the rocket is flying at 22,513 mph.

The whole operation—Atlas and Centaur—is controlled by a digital computer unit (DCU) built into Centaur.

The most-common use for rockets today is launching satellites to Earth orbit. Some remain in low orbits 100-500 miles above Earth. Others fly on out to stationary orbit 22,300 miles above Earth. Workhorse launchers such as Atlas-Centaur, usually don't have a third stage attached for flights to Earth orbit. Instead, most commercial satellites are equipped with on-board rockets that take the place of the third stage.

Lightning bolts. A four-bolt lightning blast punched a hole in an Atlas-Centaur nosecone as the rocket rose through a cloud bank above Cape Canaveral March 26, 1987. The force of the punch drove the huge vehicle into a sharp right turn, forcing NASA ground controllers to explode it.

Four lightning strikes were recorded on the ground near the rocket launch pad less than a minute after the rocket lifted off and seconds before it broke apart. A 2-ft. charred segment of fiberglass and cork nosecone with a pinhole burned in it was recovered. The hole in the fiberglass was as large as a quarter.

The rocket had been hidden in clouds as it left the pad. NASA photographs later showed a bolt of lightning streaking from the rocket to the ground. About that time, the rocket veered off course and began to break apart. Control room telemetry registered electrical failures in both the Atlas and Centaur stages at the time of the lightning strike.

The powerful rocket blastoff itself triggered lightning strokes. A voltage surge from the lightning had found its way into the rocket's digital computer where it altered a single word in the computer's memory. That told the Centaur engines to swivel, changing the rocket's flight path from vertical to horizontal just 54 seconds after liftoff. The $161 million vehicle and cargo then broke apart.

NASA, criticized for blasting off in severe weather, said the launch team didn't violate rules, including one regulation that liftoff not occur if lightning is within five miles of the pad. A lightning hit had been recorded 16 minutes before liftoff, but 16 miles from the pad. About 48 seconds after launch, four lightning strikes were recorded within 1.9 miles of the launch area. More than 2,000 lightning flashes were striking Florida when the rocket was exploded.

Investigators. A NASA investigating board blamed a bad call by Air Force weathermen who allowed the rocket and its satellite cargo to be launched into an electrical storm. Investigators found weather gauges near the launch pad had indicated dangerous lightning at the time of liftoff, but that data was ignored. Strange as it may seem, NASA had paid attention to those gauges only during manned shuttle launches, not during unmanned rocket firings. A shuttle would not be launched if the gauges measured electrical charges as low as minus 1,000 volts per meter. Readings at the time of the Atlas-Centaur launch were as low as minus 7,360 volts per meter.

Air Force weathermen gave a go-ahead for launch four minutes before the engines were turned on. Neither the Air Force nor NASA argued against blastoff because they had launched in inclement weather before. The inquiry board recommended future rockets be hardened against lightning strikes. It suggested the flight path of a rocket must not be through any electrical clouds. Use of a weather-observing aircraft should be mandatory unless there is a cloud-free line of sight in the vehicle's flight path, the board instructed.

Nose cone. The rocket went out of control 54 seconds into the flight at an altitude of 14,250 ft., only one-half mile downrange from launch complex 36B, as it passed through the rain clouds. NASA destroyed the 137-ft. vehicle. Debris fell on parts of Cape Canaveral Air Force Station and in the ocean. More than 100 pieces were recovered.

The $83 million communications satellite was inside a nosecone shroud atop the Centaur stage. The nosecone was in place to protect the communications satellite and would have been opened like a clamshell at high altitude. The composite material in the 30-ft.-long, 10-ft.-diameter nosecone had been known to build up static electricity and produce sparks when flying at high speed through rain. NASA recovered the unfired explosive bolts which would have blown away the nosecone in space.

Previous bolts. A NASA videotape showed a bolt of lightning darting from where the rocket was hidden in clouds and striking near the launch pad. Critics recalled that bad weather at launch time also contributed to the 1986 Challenger disaster. NASA knew machines flying through storms could create lightning. An F-106 jet airplane used by NASA for tests was hit 700 times with lightning between 1980 and 1986. A researcher said that lightning actually was triggered by the presence of the plane.

However, in 2,000 launches from the Cape over 30 years, only one other rocket had been known to have been hit by lightning. That was a mammoth Saturn V rocket bound for the Moon in 1969 with three men in the Apollo 12 capsule. It took a bolt which disrupted power. The astronauts fixed the problem and went on their way.

Last Atlas-Centaur. NASA's last Atlas-Centaur rocket carried Fltsatcom, a military communications satellite, to orbit September 25, 1989, from Cape Canaveral.

Clouds had delayed blast off from launch pad 36B for 44 minutes when the 5 a.m. darkness finally was illuminated by the 137-ft. liquid-fuel Atlas-Centaur blasting skyward over the Atlantic Ocean on the final major unmanned launch for the time being by American's civilian space agency.

In the Atlas-Centaur nosecone was a $125 million, 5,075-lb., solar-powered Navy fleet-communications satellite Fltsatcom equipped with a modern UHF (ultra high frequency) radio for routine communications and use by the president in a crisis.

Half an hour after launch, the Atlas-Centaur dropped off the satellite in an oval orbit ranging from 104 miles altitude to 22,300 miles. Later, a rocket attached to the satellite fired to make the orbit circular at a "stationary" altitude of 22,300 miles. There, the satellite takes 24 hours to revolve around the Earth, making it seem to hang stationary over one spot on the surface.

Two-year delay. The $78 million General Dynamics Atlas-Centaur was supposed to have been launched in 1987, but July 13, 1987, a launch pad work platform fell onto the rocket, destroying a $4 million liquid-hydrogen fuel tank in the Centaur second stage. That stage had to be rebuilt. Repairs cost $15 million.

A Fltsatcom had been the payload in the failed March 26, 1987, launch. Its Atlas-Centaur was struck by lightning a minute after liftoff in a Cape Canaveral thunderstorm.

Last launch? News media made much of the September 25 launch being NASA's last. The launch was the last unmanned flight to be conducted by NASA, but not NASA's last unmanned rocket launch.

After the 1986 Challenger disaster, then-President Ronald Reagan established a new policy banning commercial payloads from manned space shuttles. To start an American commercial launch industry, Reagan ordered NASA to hire private companies to conduct unmanned launches. General Dynamics and other rocket companies started selling launches. That did not mean NASA would no longer have unmanned rocket launches, only that NASA would pay someone else to run the launches.

Upgraded Atlas-Centaurs and other unmanned boosters will be launched in the future. The flights will be conducted by the rocket builders, not NASA.

Fltsatcom. A Fltsatcom (pronounced "fleet-sat-com") satellite has 23 UHF radio channels and an experimental EHF (extremely high frequency) channel said to be less susceptible to interference and jamming.

Ten UHF channels are used by the Navy to communicate with aircraft, submarines and ships. Twelve channels are used by the Air Force to command nuclear forces around the world. One channel is reserved for the U.S. president.

Of seven Fltsatcom launched to date, five still work in orbit. A Fltsatcom launched in August 1981 was damaged during the climb to orbit while the satellite in the March 1987 launch was destroyed.

NASA plans. NASA has scheduled seven Atlas launches from 1992-1997. An Atlas-E was to carry the NOAA-I weather satellite to polar orbit from Vandenberg Air Force Base, California, by the first quarter of 1992. An Atlas-1 was to carry the GOES-I weather satellite to orbit from Cape Canaveral in December 1992. Another Atlas-1 was to carry the GOES-J weather satellite to orbit from Cape Canaveral in August 1993.

Another Atlas-E was to carry the NOAA-J weather satellite to polar orbit from Vandenberg Air Force Base, California, in December 1993. Another Atlas-1 was to carry the GOES-K weather satellite to orbit from Cape Canaveral in July 1995. An Atlas-2AS was to carry the SOHO satellite to orbit from Cape Canaveral in July 1995. Another Atlas-1 was to carry the GOES-L weather satellite to orbit from Cape Canaveral in February 1997.

Atlas 2. Atlas 2 is an improved version of the much-used Atlas-Centaur rocket built by General Dynamics Corp. It is considered a "medium" rocket as it carries 6,000-lb. payloads to orbit. The Pentagon had been referring to the booster as the Gregory B. Jarvis Medium Launch Vehicle II, until General Dynamics received a contract from the Air Force in 1988 for a medium-lift rocket to ferry communication satellites to orbit. Gregory Jarvis was one of the seven astronauts killed in the 1986 Challenger disaster.

The contract paid Atlas 2 research and development expenses, plus construction and launch of the first two boosters. Altogether, the Air Force planned to purchase eleven Atlas 2 boosters at about $40 million each over several years.

Atlas 2 was part of an acceleration of unmanned-rocket construction started by the Pentagon after the 1986 loss of shuttle Challenger. The Air Force selected General Dynamics over McDonnell Douglas Corp. and Martin Marietta Corp. for the Gregory B. Jarvis Medium Launch Vehicle II.

Besides military payloads, two Intelsat 7 civilian commercial communications satellites were to be launched on Atlas 2 in 1993. The NASA/European Space Agency Solar Heliospheric Observatory (SOHO) is to be orbited on an Atlas 2 in July 1995.

Atlas-E. Three small satellites were lobbed to polar orbits April 11, 1990, when the U.S. Air Force fired a 29-year-old Atlas-E rocket from Vandenberg Air Force Base. The Atlas-E had a Scout-Altair rocket attached as an upper stage for extra power.

The research satellites, circling the globe at an altitude of 400 miles, are measuring the planet's magnetic field, looking into interfering radio static in the upper atmosphere and testing computer memories.

Commercial launch. General Dynamics' first commercial launch was July 25, 1990. A commercial Atlas-Centaur boosted the U.S. Combined Release/Radiation Effects Satellite (CRRES) from Cape Canaveral launch complex 36B to orbit to study Earth's trapped radiation belts.

The 3,758-lb. CRRES was ferried to an elliptical orbit, ranging from 217 miles to 22,236 miles altitude, where the satellite studied the effects of space radiation on microelectronic parts and how the ionosphere and magnetosphere interact.

The CRRES launch contract with NASA was an unusual barter arrangement. General Dynamics gave the launch to the government at no cost in exchange for title to $64-million-worth of tools and parts from NASA's completed Atlas-Centaur and Shuttle-Centaur programs.

CRRES, built by Ball Aerospace Systems Group, Denver, Colorado, was to have been carried to space in a U.S. shuttle in 1987, but the 1986 Challenger disaster delayed the flight three years. The satellite was refitted for launch by unmanned Atlas rocket. A launch attempt had been postponed at the last minute July 20, 1990, when a leak turned up in an Atlas-Centaur liquid helium line. The rocket finally blasted off July 25.

CRRES background. Combined Release and Radiation Effects Satellite (CRRES) illuminated Earth's invisible magnetic field in sensational displays resembling Northern Lights. CRRES was a joint NASA–U.S. Air Force mission to study the effects of space radiation on electronic equipment as well as Earth's magnetic field.

CRRES was loaded with 24 canisters of lithium, barium, calcium and strontium chemicals to be ejected at various altitudes over a year. Nine canisters were released into the upper ionosphere 250 miles above Earth, the rest at lower altitudes.

One or two canisters were ejected at a time. Each had a 25-minute timer to ensure the satellite got away to a safe distance before chemical vapors were released. The chemicals painted the sky with glowing, 60-mi. clouds as they were ionized by ultraviolet light from the Sun. The aurora-like clouds were visible to the naked eye as they flowed along the lines of Earth's magnetic field. Scientists observed the temporarily-visible magnetic field lines through telescopes on the ground and from specially-outfitted airplanes.

Here's how the man-made aurora worked: ultraviolet light arriving from the Sun ionized the chemicals, changing negatively-charged electrons in each atom to give the gas an electrical charge. The resulting plasma cloud traveled along Earth's magnetic field lines—like iron filings line up around a bar magnet to make its field visible.

Colorful clouds. NASA controllers radioed orders to CRRES to release the first canisters in September, 1990, over American Samoa in the South Pacific. The first display to be seen over North America was in January and February 1991. The entire continent saw the barium clouds, looking like Northern Lights or aurora borealis. Seven high-altitude chemical releases, visible across the entire Western Hemisphere, were made in wintertime. Others were released over the Caribbean in the summer of 1991. Florida and southern Alabama and Georgia saw those. Scientists looked for interaction between the clouds and the normal invisible charged particles in the ionosphere and magnetosphere. Scientists from the USSR helped NASA observe the Caribbean releases.

In other experiments carried out by the solar-powered CRRES, experimental microelectronic parts were exposed to space radiation to see if they would fail. And instruments in the satellite probed the Van Allen radiation belts which girdle Earth. CRRES traveled through the inner and outer Van Allen radiation belts, intense radiation zones which can harm spacecraft electronics. Altogether, CRRES had five major experiments using 55 instruments. Late in its planned three-year life, CRRES became part of NASA's constellation of Global Geospace Science satellites.

The Pegsat satellite, carried to space April 5,1990, by the new Pegasus rocket, also released two barium canisters over northern Canada in April. Similar canisters also were ejected from ten sounding rockets launched from Kwajalein Atoll in the Republic of the Marshall Islands in the South Pacific and from Puerto Rico in 1991.

Second commercial. The second Atlas-Centaur commercial launch failed spectacularly. One of two second-stage engines failed to ignite in an Atlas-Centaur rocket April 18, 1991, pitching the big commercial booster off course. The 144-ft. liquid-fuel rocket's three first-stage engines had blasted off 16 minutes late because of high winds. Atop the rocket was Japan Broadcasting Corp.'s BS-3H direct-to-home TV broadcast satellite.

The booster lifted smoothly away from Cape Canaveral launch complex 17. The first stage burned out as planned and separated 4.5 minutes into flight. Two Centaur second-stage engines then were supposed to fire, but only one did. The rocket tumbled out of control. At 4 minutes 45 seconds into flight, the rocket exploded, destroying the Japanese satellite.

A two-month investigation determined the accident probably happened because a piece of debris had been sucked into a pump, triggering the engine failure April 18.

In devastating irony, the 1,620-lb., solar-powered BS-3H satellite had been intended to replace one that was destroyed in a February 1990 Ariane failure. BS-3H had three 200-watt TV transmitters to broadcast programs directly to Japanese homes. It had been built early in the 1980's for another company, then mothballed, but later refurbished and sold to Nippon Hoso Kyokai (NHK) the Japan Broadcasting Corp. for the April 1991 launch.

Vandenberg. General Dynamics has launched Atlas-E rockets and operated at Vandenberg since 1958. Besides the commercial Atlas, General Dynamics assembles Atlas boosters as medium-launch vehicles for the U.S. Air Force and Navy.

The company built ten Atlas rockets for the U.S. Navy to lift a new generation of Hughes ultra-high frequency communications satellites to Earth orbit throughout the 1990's, starting in 1992.

Atlas commercial launch services are marketed by General Dynamics Commercial Launch Services Inc.

Atlas Ferries Weather Sat. An Atlas-E rocket on December 1, 1990, lofted a U.S. military weather satellite to space from Vandenberg Air Force Base. The Defense Meteorological Satellite Program (DMSP) satellite went into polar orbit 500 miles above Earth. Weather patterns recorded by optical scanners aboard the satellite are reported to all branches of the military, allowing forecasters to track conditions in remote areas. The information also is released to civilian agencies such as the weather bureau—the National Oceanic and Atmospheric Administration (NOAA).

U.S. Rockets: The Delta Family

America's first Delta rocketed to space in 1960, then went on to become one of America's workhorse space boosters, carrying more than 175 satellites to orbit for NASA and the Defense Department.

Delta flights carry satellites to low Earth orbits, high stationary orbits, highly elliptical orbits and polar orbits. There were 205 Delta launches from 1960 to 1991. Of those, 193 were successful. That's a 94 percent success rate. All 20 launches from August 1989 through May 1991 were successful.

Delta's biggest assignment in the last four years has been ferrying 21 military navigation satellites to Earth orbit. More than half of the Navstar global positioning system (GPS) constellation was in space by the end of 1991.

Delta has carried some important science and astronomy satellites to orbit in recent years, including Infrared Astronomy Satellite (IRAS) in 1983, Cosmic Background Explorer (COBE) in November 1989 and the Roentgen Satellite (Rosat) in June 1990. It recently carried GOES weather satellites, and communications satellites for NATO, Indonesia and the International Maritime Organization (Inmarsat).

NASA has purchased Delta launches to ferry satellites to Earth orbit, including Geotail and Wind in 1992, Polar in 1993, Lageos and Radarsat in 1994, Lifesat in 1996 and 1997, and ACE in 1997.

Commercial Delta. There were two successful commercial Delta space shots in less than a month in 1987. GOES-7, a U.S. East Coast weather satellite, was sent to orbit from Cape Canaveral February 26 aboard a Delta 1. Then the last Delta 1 was used for a commercial flight March 20, launching Indonesia's Palapa B2-P to orbit from the Cape.

Palapa B2-P, second of three spacecraft in the Palapa B series of communications satellites, is a radio station in space capable of relaying 24,000 simultaneous telephone calls to 165 million Indonesians on that nation's 3,100-mile-long archipelago of 13,677 islands, and to other southwest Pacific nations.

Palapa B2-P added telephone, television, telegraph and data communications channels for Indonesia and the Association of South East Asian Nations (ASEAN).

ASEAN includes the Philippines, Thailand, Malaysia, Singapore and Papua, New Guinea. Communications with many islands is possible only via satellites.

Palapa B2-P, in stationary orbit 22,300 miles over the southwest Pacific Ocean, once was to have been ferried to space aboard an American shuttle in 1986, but that plan was dropped after the Challenger disaster. It was the first payload switched from shuttle to non-reusable rocket after Challenger.

The change was made after Indonesia's foreign minister made a personal request during a visit of then-President Ronald Reagan to Indonesia in May 1986. The foreign minister told President Reagan that Indonesia wanted to replace an aging satellite already in orbit before elections. The U.S. state department persuaded the Pentagon to delay a Star Wars test flight so the Indonesians could use the Delta rocket. The satellite was launched March 20 and on April 23, 1987, Indonesians elected members of their national House of Representatives. The U.S. Stars Wars launch was rescheduled for November.

Palapas B1 & B2. An earlier satellite in the series, Palapa B1, had been ferried to orbit by an American shuttle in June 1983. Later, Palapa B2 was carried to orbit, along with Western Union's Westar 6 satellite, by shuttle Challenger in February 1984. They were dropped off in orbits lower than planned so both the Palapa B2 and Westar 6 satellites failed. The pair were recaptured by spacewalking astronauts from shuttle Discovery in November 1984, returned to Earth in the shuttle and refurbished.

Today, Palapa B2 is known as Palapa B2-R. After B2 was brought back to Earth, an Indonesian company bought the satellite from the insurance company which had paid off after the failed launch. The U.S. Dept. of Transportation licensed McDonnell Douglas, manufacturer of the Delta rocket, to launch Indonesia's Palapa B2-R on a Delta 2 rocket from Cape Canaveral in 1990. Back in space, Palapa B2-R will be a backup for Palapa B2-P when Palapa B1 stops working.

The retrieved Westar 6 was renamed AsiaSat and launched by China using a Long March rocket, the first American satellite sent to orbit by a non-Western country.

Star Wars Delta. After the last commercial launch, three Delta 1 rockets remained in the American inventory. They were set aside for military Strategic Defense Initiative (SDI) Star Wars flights.

The first of the three was used for a secret flight February 8, 1988. It whisked a cluster of key American military research satellites to orbit, simulating 15 Russian nuclear missiles to be tracked through space. The two-stage Delta, clearing the Cape Canaveral launch pad after a secret countdown, shot upward through low hanging clouds. The second stage fired as planned four minutes after launch. The test was to clarify whether a split-second response by a space-based missile defense system would work. The Delta flew for the U.S. government's Strategic Defense Initiative (SDI) Organization.

In orbit, the Delta 1 second stage—itself a 6,000-lb. satellite—released 15 small satellites over four hours. Four of the little payloads fired rocket motors to simulate flame plumes from USSR rockets hurtling from launch pads. Eleven simulated Soviet missiles coasting through space on their way to drop swarms of nuclear warheads.

Seven sensors on the Delta second stage, and sensors at 250 ground stations, shadowed the midget satellites as they wheeled through space for eight hours. Laser beams, radar scanners and optical telescopes traced the payloads against the background of space, land, sea and the atmosphere just above the ground known as Earth's limb. Infrared and ultraviolet telescopes locked onto the firing rockets. So much data was collected it took the main satellite 10 days to transmit it to ground stations.

The February 8 Delta flight followed Star Wars experiments in a September 5, 1986, Delta flight in which two satellites tailed each other around the globe, charted the launch of another rocket on the ground, and finally demolished each other in a premeditated collision.

A Delta-Star missile-hunting satellite thundered into orbit March 24, 1989, on a Delta rocket for a six-months test of Star Wars defenses against nuclear rockets.

Challenger. A U.S. government decision in the 1970's made NASA's manned space shuttle the nation's main launcher for civilian and military payloads. Prior to the 1986 shuttle Challenger disaster, unmanned expendable rockets—such as Delta, Titan and Atlas-Centaur—were being phased out by the military and NASA.

Even before Challenger, the Defense Department had become concerned about relying on one launcher for all high-priority payloads. The Pentagon had ordered ten Titan 4 rockets as shuttle backups.

Then the 1986 loss of Challenger grounded the shuttle fleet, leaving the U.S. with no way to send payloads to orbit. The Pentagon rushed into an unprecedented $14 billion military space buildup, including purchase of 20 new $43 million, upgraded Delta 2 rockets from manufacturer McDonnell Douglas, to carry the GPS Navstar navigation satellites that had been scheduled for launch by the shuttle fleet.

The Pentagon also ordered 11 Atlas-Centaur and 13 more Titan 4 rockets. Titan 4, now America's most powerful unmanned launcher, made its maiden flight in 1989 and the Atlas 2 flew in 1990. Atlas is built by General Dynamics. Titans are manufactured by Martin Marietta. Delta is made by McDonnell Douglas Space Systems Co.

Delta 2. Delta 1 was upgraded to Delta 2, a medium-lift, three-stage rocket capable of carrying middle-weight satellites to orbit. The slender, 8-ft.-diameter, 126-ft., blue-and-white Delta 2, with strap-on solid-fuel boosters, is 14 percent more powerful than Delta 1. Delta 2s are launched from an old Delta 1 launch pad at Cape Canaveral, renovated to handle the larger rocket.

Delta 2 is capable of delivering 8,780-lb. payloads to low orbit or 3,190 lbs. to a 22,300-mi.-high stationary orbit used for communications and some weather satellites.

By comparison, Titan 4 is a heavy-lifter, while Delta 2 is a medium-lifter. Titan 4 can haul 32,000 pounds to a 100-mi.-high orbit. A shuttle can carry 55,000-60,000 lbs. to a similar low orbit. The USSR's Energia rocket can carry 220,000 lbs. to low orbit.

Boosters and third stage. The Delta 2 first stage was extended 12 feet over earlier versions of the rocket to allow more fuel to be carried. It uses an RS-27B main engine, built by Rocketdyne, to blast out 231,700 lbs. of thrust. The main engine is fueld with liquid oxygen and RP-1 kerosene.

Nine solid-fuel boosters are strapped around the first stage. They burn a more powerful solid propellant than earlier models, to generate 97,100 lbs. of thrust. The boosters are made by Morton Thiokol Inc.

Delta 2's second stage is a 9,645-lb.-thrust engine built by Aerojet Tech Systems. Like all high-tech modern rockets, it can start and stop repeatedly . The second stage is fueled with nitrogen tetroxide and Aerozine 50.

A third stage depends on payload and required orbit. For instance, to launch military Navstar GPS satellites, Delta 2 has a third-stage solid-fuel rocket motor, Morton Thiokol Star 48B. It generates 15,000 lbs. of thrust. The rocket also has a new 9.5-foot-wide nosecone to house the GPS satellites.

First flight. America's first upgraded Delta 2 rocket rode a tongue of flame off launch pad 17A at Cape Canaveral into a clear blue sky February 14, 1989, on its maiden spaceflight, carrying the first of 21 Navstar Global Positioning System (GPS) satellites.

Trailing a billowing cloud of white exhaust, visible for miles along the Florida coast, Delta 2 powered into the sky for 25 minutes, then dropped the 1,860-lb. Navstar into a highly-elliptical orbit ranging from 100 miles to 12,500 miles above Earth.

Two days after launch, a ground station sent a signal to fire a PAM-D rocket motor attached to the $65 million navigation satellite to move Navstar to a 12,500-mi.-high circular orbit. The 59-second burn went perfectly, correcting the Navstar orbit to a low of 12,335 miles and a high of 12,636 miles. The slightly-imperfect circle was corrected later with small maneuvering jets.

Delta 1 had a record of 12 failures in 183 launches, with the last on May 3, 1986. Delta 2 has been successful to date.

U.S. Rockets: Scout

America's reliable space launcher for small payloads has been Scout, with more than 100 blast-off successes since the first in 1960.

Scout can lift a 475-lb. payload to a 300-mi.-high orbit. It has been fired over the Atlantic Ocean from Wallops Island on the Virginia coast, over the Pacific Ocean from Vandenberg Air Force Base on the California coast, and over the Indian Ocean from a platform in the water off the East African coast of Kenya.

Over the years, payloads carried aloft by Scout have included communications, biology, astronomy, navigation, meteorology, geodesy, meteoroid-detection, Earth-sensing, re-entry test materials and atmosphere-sensing satellites. Scout has launched 26 of the Navy's widely-used Transit navigation satellites to 600-mi.-high orbits.

Bad start. Scout seems reliable today, but it wasn't always so. The first launch of the four-stage Scout rocket was July 1, 1960. It started off well, but an improperly calibrated radar tracker made it seem to move off course. The range safety officer destroyed it in flight. NASA still counts that one a success since there really was nothing wrong with the rocket.

The first four Scouts were built by engineers at NASA's Langley Research Center in Virginia. To save money and time, they cobbled together Scouts from off-the-shelf components. The first-stage rocket, known as Algol, was scavenged from a Navy Polaris missile. The second stage, Castor, was borrowed from an Army Sergeant missile. The third and fourth rocket stages, known as Antares and Altair, were swiped from the Navy's Vanguard. After the first four, Scout rockets were constructed by a company then known as Chance Vought Aircraft, known today as LTV.

Unfortunately, ten of Scout's first 23 flights blew up on the way to space. The second launch was good, but the third failed. The fourth was good, but the fifth and sixth failed. The program almost was cancelled. NASA paused, rebuilt 27 Scouts on hand, standardized procedures for quality control, and started over.

After a science satellite made it successfully to orbit from Vandenberg in December 1963, Scout went into a magnificent three-decade string of successes. The earlier 57 percent success rate faded into the later 96 percent success story.

Spacefaring nations. Besides NASA and the U.S. Defense Department, Scout has carried payloads to space for France, Great Britain, Italy, The Netherlands, West Germany and even the European Space Agency.

Italy has been the most-active foreign user of NASA's Scout rockets. In fact, Scout gave Italy the prestige of being the third country to send a satellite to space. The USSR had been first in space with Sputnik in 1957. The U.S. had been second with Explorer in 1958. On December 15, 1964, an Italian launch team fired a U.S. Scout rocket from a platform in the Indian Ocean off Kenya, East Africa, carrying Italy's San Marco 1 satellite to orbit. It was the first firing of a NASA rocket by foreign technicians.

Later, the first woman ever to launch a spacecraft, Marjorie Townsend, sent the U.S. satellite Explorer 42 to Earth orbit from the San Marco platform on December 12, 1970. Townsend named the satellite Uhuru, a Swahili word for "freedom," in honor of Kenya's independence day December 12.

Another 1988 launch from San Marco Range was the 95th success in 109 attempts since 1960. NASA launched a Scout there March 25, 1988, carrying a satellite with five Italian, West German and American science experiments—the first Scout launch from San Marco since 1974.

Today's reliability. At first, a Scout could carry a 131-lb. payload to a 300-mi.-high orbit. Over the years, capacity was raised. Today, the four-stage solid-fuel Scout can lift a 475-lb. payload to a 300-mi.-high orbit. Payload space inside the nose-cone is six times greater today than in 1960.

A Scout launch is cheap by today's standards. It costs $10 million versus $45 million for a Delta or $100 million for a Titan 3 launch.

In the late 1970's, when NASA imagined it could accomplish 50 shuttle flights a year, it didn't want to use expendable rockets so the last Scout blast-off went on the calendar for 1983. Things didn't work out, however, and nowadays Scout launches are back in demand.

114th Scout. NASA launched the 114th Scout rocket June 29, 1991, from Vandenberg Air Force Base, California. It carried the U.S. Air Force's 188-lb. Radiation Experiment satellite to a 450 nautical mile polar orbit.

The Air Force satellite tested sophisticated military communications in a high radiation environment, information needed for development of equipment to withstand the harsh environment of space.

Scout 2. The Italians have considered modifying Scouts for future launches from their Indian Ocean launch site. They would adapt two solid-fuel boosters from a European Space Agency Ariane rocket, strapping them around a four-stage Scout. Known as Scout 2, the high-power booster would double the weight-lifting capability of Scout 1.

U.S. Rockets: Pegasus

Pegasus, the first all-new unmanned American launch vehicle to appear since the 1960's, was launched to space April 5, 1990. It was launched from an airplane and carried the lightweight Pegsat to a low orbit.

To get to space, the white 49-ft. Pegasus first was slung under the right wing of an eight-engine B-52 bomber, then ferried more than seven miles above Earth, dropped free and blasted to orbit.

Pegasus carried to polar orbit a payload known as Pegsat containing a small Navy communications satellite, a package of instruments and two barium canisters.

Maiden voyage. On April 5, retired astronaut Gordon Fullerton piloted the big warplane, the same B-52 used back in the 1960's to launch more than 100 X-15 rocket-plane flights. He was cruising due south over the Pacific Ocean, 60 miles southwest of Monterey, California, when the 41,000-lb. Pegasus dropped away. Fullerton immediately banked the lumbering jet to the left to get out of the way.

After a short five seconds, solid fuel ignited in the first stage of the falling rocket. Flames burst from its stern as the 49-ft. booster shot ahead of the B-52. Chase aircraft fell behind as the main computer signaled steering fins at the base of the first stage to direct Pegasus upward in a steep climb. The rocket's 22-ft. delta-shaped wing was severely stressed by 100,000 lbs. of aerodynamic force as the slender cylinder rushed out into space on a nine-minute 37-second flight to orbit.

Some 82 seconds into the flight, the exhausted first stage dropped away. The second and third stage engines then smoothly popped the 450-lb. Pegsat into polar orbit 370 miles above earth.

Cheap ride. The winged, three-stage, solid-fuel Pegasus is 50 feet long and 50 inches in diameter. That's about the size of the X-15, the first aircraft to reach the fringes of space. Pegasus was launched from the same B-52 which had carried the X-15 on its first flight in 1962. Pegasus was released over the ocean at 40,000 feet.

Orbital Sciences Corp., a small aerospace company in Virginia, builds Pegasus. Using Pegasus, OSC can launch satellites up to 600 pounds for $10 million, a fraction of the cost of conventional launches. Pegasus can carry twice the payload of a comparable ground launcher because of its take-off altitude, the ability to fly the B-52 to whatever latitude is chosen, and being three-fourths of the way through the atmosphere.

OSC expects the new booster to fire an interest in small payloads. Pegasus will enable a new group of customers to get their satellites to orbit, including small satellites for communications, data collection, search and rescue communications, tracking of ships and cargoes, and remote sensing. A Pegasus can be ready for launch in just seven days. That could be reduced to a day or two if there were a rush order, such as a scientific phenomenon demanding immediate research.

OSC. OSC, a private company formed in 1962, previously developed a space transfer vehicle to carry satellites to high orbits after they are launched from the space shuttle or by a ground-based rocket. Pegasus is OSC's second product. No government money was spent developing Pegasus. OSC's partner in the Pegasus venture is Hercules Aerospace Co., Wilmington, Delaware, an old-line rocket maker that built parts of Polaris, Trident, Poseidon, Minuteman, Pershing, and MX nuclear-warhead missiles.

Easter barium. Folks going out to sunrise services on Easter, April 15, 1990, may have thought the Easter bunny had spilled some of his egg colors across the sky.

Pegsat emptied one of its two canisters of barium at 1:36 a.m. EDT 362 miles above Churchill and Yellowknife in northern Canada. The chemical spread a yellowish cloud, changed to green and white, then faded, while purple streaks grew up across the sky.

Vaporized barium struck by sunlight glows with electrical charges. Residents of central Canada and the north central United States may have been able to see the rainbow of colors in the pre-dawn sky. Scientists watched for interaction of barium with Earth's magnetic and electrical fields from Stony Rapids, Saskatchewan; Churchill Research Range, Manitoba; Lynn Lake, Manitoba; Fort Smith, Northwest Territories; Boston, Massachusetts; west Texas; New Mexico; California; Washington; and Puerto Rico.

STEPsat. The U.S. Air Force is developing a lightweight satellite, known as Space Test Experiments Platform (STEP), especially for launch by Pegasus.

The under-1,000-lb. STEPsat is expected to accommodate a variety of space and technology experiments. TRW Space & Technology Group, Redondo Beach, California, may build as many as a dozen STEPsats for the Air Force Space Systems Division by 1997. The first STEPsat was to be launched in 1992, housing four experiments in an elliptical polar orbit ranging from 300 miles to 1,000 miles above Earth.

Pegasus rocket vs. satellite. In 1965, the U.S. used Saturn 1 rockets to fire three large satellites to Earth orbit: Pegasus 1 on February 16, Pegasus 2 on May 25 and Pegasus 3 on July 30. A dummy Apollo capsule flew to space with each Pegasus. Pegasus satellites sprouted 2,300 sq. ft. panels which extended 96 ft. to measure the number of penetrations by meteoroids. Watchers with binoculars on the ground could see the satellites at night. Scientists discovered that interplanetary dust particles were considerably fewer in number than had been expected. That news allowed engineers to cut 1,000 lbs. of protective weight from each Apollo capsule.

NASA launches. NASA planned to use Pegasus rockets to launch satellites in September 1993, July 1994, June 1995, January 1996, and two in June 1997.

U.S. Rockets: Sounders

Sounding rockets are small launchers, capable of carrying lightweight payloads on suborbital flights up to the edge of space, but not all the way into orbit.

NASA's Black Brant is a well-known sounding rocket. Black Brant is a small two-stage launcher, 44 ft. long, 17 in. in diameter. Black Brants often are fired from NASA's Goddard Space Flight Center launch site at Wallops Island, Virginia, but not always. For instance, a Black Brant was fired from White Sands Missile Range in New Mexico in 1987, carrying a 685-lb. package of science instruments to the edge of space.

Wallops Island, in recent years, had been overseeing 40-45 Black Brant firings a year from several pads around the world. NASA recently ordered 50 more Black Brant V rocket motors. Built in Canada by Bristol Aerospace Ltd., the motors actually are used in 13 different rocket configurations in NASA's family of suborbital launchers.

Solar X-rays. NASA launched a Nike Black Brant two-stage rocket and a Taurus Nike Tomahawk three-stage rocket 40 seconds apart July 27, 1988, from the Atlantic

Coast site at Wallops Island. The suborbital flights carried instruments to record lightning in the upper atmosphere.

Both rockets used solid fuel and were 48 feet long. Nike was 17 inches in diameter and Taurus was 23 inches in diameter. The smaller rocket probed the upper atmosphere while the larger went into the ionosphere. Earlier rocket research had found lightning bolts penetrate to outer space above the atmosphere.

Sun corona. A June 24, 1988, launch of a Black Brant IX rocket from White Sands Missile Range, New Mexico, carried an 870-lb. payload, including a telescope to film X-rays from the Sun's corona, to an altitude of 163 miles.

The outer corona of the Sun is so hot, nearly 4 million degrees Fahrenheit, it emits X-rays as well as visible light.

Black Brant IX, also known as Terrier Black Brant, was 52 feet long and 18 inches in diameter. The payload was recovered by parachute.

Alaska explosion. A Black Brant 10 sounding rocket failed seconds after launch from NASA's Poker Flats, Alaska, launch site in March 1987. The lost payload was a plasma physics experiment to be carried out in the upper reaches of Earth's atmosphere on the fringe of space.

Black Brant is a solid-fuel rocket. Its second stage failed about ten seconds after ignition and the rocket exploded at 20 seconds into the flight.

Noctilucent clouds. The NASA sounding rockets Black Brant VB, Super Arcas, Nike-Orion and ViperDart were used in July and August 1991 in an international project examining the highest, coldest clouds around Earth.

They were launched from the Swedish Space Corporation's Esrange launch pads at Kiruna, Sweden, above the Arctic Circle, to study noctilucent clouds.

Noctilucent clouds occur only during the summer at 52 miles altitude in northern polar latitudes. U.S. and European scientists also flew in aircraft and turned ground radars upward to study the clouds.

Artificial comet. NASA fired two cannisters of barium near space on a three-stage, solid-fuel Black Brant X sounding rocket from Wallops Island May 3, 1989.

The experiment was designed to check a theory on the formation of the Solar System by creating an artificial comet 300 miles above the Atlantic Ocean and visible along the East Coast. Hannes Alfven in 1954 had theorized why planets have different chemical compositions and how they were formed.

The 1989 launch had been delayed several times by bad weather from April 23. When finally launched May 3, the rocket ejected the chemical barium, which produced colored clouds observed by radar antennas and low-light-level TV cameras.

Six minutes after liftoff, a canister of barium was ejected from the rocket nosecone. Ninety seconds later the second canister was ejected. Both were timed to explode, the first while the nose cone rose and the second as it began to fall back to Earth, creating a greenish-blue cloud visible for 20 minutes from the U.S. East Coast west to Ohio.

The rocket failed to go as high as planned so the Sun's illumination of the clouds was reduced, making the clouds dimly visible from the ground. Even so, the cloud looked blue with rose streaks through it, according to reports from Franklin Lakes, N.J., 20 miles east of New York City. The cloud was large, about three times the size of the Moon.

U.S. National Launch System

More-powerful U.S. rockets will be needed if the nation wants to send astronauts to Mars or deploy heavy Strategic Defense Initiative (SDI) Star Wars satellites.

NASA and the U.S. Dept. of Defense recently have been studying a new rocket known as the National Launch System (NLS). Actually, NLS would be a family of launch vehicle modules to be used for the next 20 years by the Air Force and NASA.

National Launch System is an outgrowth of earlier joint studies of Advanced Launch

System (ALS) rockets and Shuttle-C. Since 1988, NASA and the Air Force have been studying various launcher and propulsion system proposals to augment the shuttle fleet.

NLS would have a core rocket which could be upgraded with strap-on boosters and various payload nosecones, depending upon requirements for civilian or military flights.

The new rocket is said to be necessary because current U.S. rockets are descended from 1950's ballistic missiles and 1970's shuttle designs. NLS would replace rockets such as Titan 4, Atlas-Centaur and Delta. If full development of NLS starts in the mid-1990's, its first flight could be made in the year 2000.

The Office of Technology Assessment (OTA), a research agency for Congress, said in 1988 the best buy in future space launchers would be found in improving the shuttle fleet and existing expendable rockets such as Titan, Atlas-Centaur and Delta. That would require less money and technical risk, OTA found. OTA suggested modifying existing rockets to carry more cargo, building more rockets and launch sites, developing the Shuttle-C cargo freighter from present technology, and constructing an all-new rocket using current technology. Besides NLS, new technologies studied recently include the Air Force's proposed Advanced Launch System (ALS) expendable rocket, a second-generation space shuttle, and a National Aerospace Plane hypersonic aircraft, known as Orient Express, capable of flying from a runway to orbit and back.

U.S. Space Commercialization

Commercialization of space. Private businesses launching private satellites. The satellite launch business is booming around the world—no pun intended!

For years, NASA sought business for its fleet of shuttle "space trucks" to haul cargo to orbit. The European Space Agency and its marketing arm, Arianespace, scoured the world for launch business for Ariane rockets. After U.S. shuttles and European Ariane rockets were grounded back in 1986, others were tempted. The USSR started advertising Proton rockets. The Chinese rushed for the same market with their Long March rockets.

Small start-ups. Several small, private, U.S. space launch firms started up and three major American rocket manufacturers sold launches. The states of Florida, Hawaii and Virginia started planning their own spaceports. Some of the small companies:

★E-Prime Aerospace Co. of Titusville, Florida, blasted-off a 10-ft. LOFT-1 rocket from Cape Canaveral, carrying four commercial payloads on a five-minute suborbital flight in 1988—the first private, commercial "space" flight. The little rocket lofted its payload only three miles high, making it debatable as a space flight.

It was the first time a privately-owned commercial rocket had been launched from either Air Force or NASA facilities at Cape Canaveral. After spending $200,000 on the launch, E-Prime said customers paid for the ride, but not enough to make it profitable.

★Space Services Inc. (SSI), a small commercial space-launch firm at Houston, Texas, successfully fired its romantically-named Conestoga 1 rocket with a 490-lb. dummy payload from Matagorda Island, Texas, September 9, 1982. The payload splashed down 300 miles east of the launch site in the Gulf of Mexico.

That suborbital flight from a private launch site made SSI the first private U.S. company to do such a thing with private funds. It demonstrated an ability, not only to build a rocket, but to build a launch site and control launch operations.

On March 29, 1989, SSI gave its first paying customer a ride near space when it blasted the Consort 1 payload on a Starfire 1 rocket to a suborbital altitude of 198 miles for the University of Alabama.

The Starfire 1 flight was the first private commercial space rocket to be licensed by the U.S. Transportation Department's Office of Commercial Space Transportation. The 625-lb. payload, known as Consort 1, made a 15-minute suborbital flight from White Sands Missile Range, New Mexico.

The brief trip through space cost the payload customer $1 million, or $1,600 per pound of payload. The project cost SSI $2 million, including paying White Sands for facilities and personnel. The $1 million launch contract was paid by a NASA grant.

Houston Industries Inc., a local power company, sunk millions into SSI in the hope that private launches of commercial satellites would catch on. Back in 1982 there was no precedent for a private rocket flight so Space Services Inc. had to arrange with 18 agencies, including the Federal Aviation Administration, Federal Communications Commission, Defense Department, and even the local sheriff's department, for permission to fire Conestoga 1.

After that, the U.S. Department of Transportation started licensing private launches. Starfire 1 was the first commercial flight licensed by the department. The Transportation Department may process 10 private commercial rocket launch licenses in 1989. Of course, LOFT and Consort didn't go to orbit. The small companies have yet to launch a satellite to orbit.

Who needs 'em? Who needs the services of small launch companies? Nobody seems to have a small satellite needing a lift to orbit right now, but the government could be a future customer. News organizations would like to have their own spy satellites overhead. Oil companies and users of natural resources need remote sensing satellites. The U.S. Defense Dept. or the CIA could become customers for launches of lightweight satellites. Maybe even NASA.

However, there's strong new competition from the winged, air-launched Pegasus which already has flown successfully to orbit. It is designed to carry payloads up to 600 lbs. And the big rocket makers have been signing up customers for their Delta, Atlas and Titan rockets.

USSR Rockets: Space Glasnost

The USSR in recent years has been very candid about the occasional problems met by its cosmonauts flying to and from the orbiting Mir space station. In addition, the Soviets have revealed 228 deaths in launch pad explosions and space flights.

Some 165 workers were killed in a 1960 explosion on a USSR launch pad. The first cosmonaut to die in flight was killed in 1967. The second cosmonaut in-flight tragedy killed three in 1971. At least nine were killed in a USSR launch-pad rocket explosion in 1973. At least 50 were killed in a 1980 rocket explosion on a USSR launch pad.

The openness with which the Soviets have been talking about space problems highlights how much the space program has changed under President Mikhail S. Gorbachev. It's a long way from the shrouded missions after the October 4, 1957, launch of the world's first artificial satellite, Sputnik 1.

Since Yuri Gagarin blasted off in Vostok 1 in 1961 as the first person to go to space, manned launches have become routine in the USSR, but the first live TV coverage of a Soviet manned space launch didn't come until the joint U.S.-USSR Apollo-Soyuz mission in 1975.

Even then, the first launch of an exclusively-USSR crew to be shown live didn't come until 1984. Nowadays, the USSR invites foreigners to watch blast-offs. It shows launches and landings live on television.

While America coped with the loss of seven astronauts in the 1986 Challenger disaster, the Soviets looked back 17 years for its last fatal in-flight disaster. That was June 30, 1971, when three cosmonauts, returning home from the Salyut 1 space station, suffocated as air leaked out of their capsule during re-entry.

The other flight fatality in Russian space history came when the Soyuz 1 capsule crash landed April 24, 1967, on its way home from space. Cosmonaut Vladimir Komarov was killed.

USSR Rockets: Designations

Two numbering systems are used in America to label Russian rockets: a simple alphanumeric system designed by Charles Sheldon of the Congressional Research Service and the Pentagon's space launch (SL) numbers. The list below shows both numbers and the intermediate range ballistic missile (IRBM) or intercontinental ballistic missile (ICBM) from which a space rocket was derived. Soviet rockets often are known by a popular name taken from their major payloads such as Soyuz and Proton.

A

Popular name:	Sputnik
Alpha number:	A
SL number:	SL-1, SL-2
ICBM:	SS-6
Length:	104 ft.
Diameter:	33 ft.
Lifts:	4,400 lbs. to low Earth orbit (LEO)
First used:	1957
Payloads:	Sputniks 1, 2 and 3

A-1

Popular name:	Vostok
Alpha number:	A-1
SL number:	SL-3
ICBM:	SS-6
Length:	110 ft.
Diameter:	33 ft.
Lifts:	11,000 lbs. to LEO or 970 lbs. to the Moon
First used:	1959
Payloads:	Vostok, Korabl Sputnik, Luna 1, 2 and 3, Elektron, Meteor, Cosmos weather, Cosmos photo spysat, Cosmos electronic spysat

A-2

Popular name:	Voskhod and Soyuz
Alpha number:	A-2
SL number:	SL-4
ICBM:	SS-6
Length:	140 ft.
Diameter:	33 ft.
Lifts:	16,500 lbs. to LEO
First used:	1963
Payloads:	Voskhod, Soyuz, Soyuz T, Soyuz TM, Progress, Cosmos science sats, Cosmos photo spysat

A-2e

Popular name:	Molniya
Alpha number:	A-2e
SL number:	SL-5, SL-6
ICBM:	SS-6
Length:	140 ft.
Diameter:	33 ft.
Lifts:	4,600-lb. to Molniya elliptical orbit, 4,000 lbs. to the Moon, 2,650 lbs. to Mars and Venus
First used:	Molniya: 1960, Moon-Mars-Venus: 1962
Payloads:	Molniya, Venera 1 to 8, Zond 1 to 3, Mars 1, Luna 4 to 14, Cosmos early warning, Prognoz

B-1

Popular name:	Intercosmos
Alpha number:	B-1
SL number:	SL-7
ICBM:	SS-4

Length:	105 ft.
Diameter:	6 ft.
Lifts:	1,000 lbs. to LEO
First used:	1962
Payloads:	Intercosmos 1 to 9, Cosmos science, Cosmos radar calibration

C-1

Popular name:	Vertical, Cosmos
Alpha number:	C-1
SL number:	SL-8
ICBM:	SS-5
Length:	103 ft.
Diameter:	8 ft.
Lifts:	3,750 lbs. to LEO
First used:	1964
Payloads:	Oreol, Intercosmos 10 to 20, Ariasat, Bhaskara, Sneg, Cosmos science, Cosmos radar calibration, Cosmos navigation, Cosmos electronic spysat, Cosmos military tactical communications, Cosmos anti-satellite ASAT target

D-1

Popular name:	Proton
Alpha number:	D-1
SL number:	SL-9
Length:	318 ft.
Diameter:	42 ft.
Lifts:	D-1: 27,000 lbs. to LEO, D-1h: 44,000 lbs. to LEO, D-1h Mir: 47,000 lbs. to LEO
First used:	D-1: 1965, D-1h: 1970, D-1h Mir: 1986
Payloads:	Proton, Salyut space stations, Star space station modules, Mir space station, Cosmos 557 space station, Kvant, Kristall

D-1e

Popular name:	Gorizont
Alpha number:	D-1e
SL number:	SL-12, SL-15
Length:	312 ft.
Diameter:	42 ft.
Lifts:	3.300 and 4,400 lbs. to stationary orbit 22,300 mi. altitude 14,000 to Moon, 11,000 to Mars and Venus
First used:	1967: stationary/Moon, 1971: Mars, 1975: Venus, 1986: SL-15 4th stage
Payloads:	Gorizont, Ekran, Molniya, Raduga, Zond 4 to 8, Luna 15 to 24, Mars 2 to 7, Venera 9 to 12, Astron, Cosmos communication, Cosmos navigation, Cosmos electronic spysat, Cosmos early warning

F

Popular name:	Cosmos
Alpha number:	F-1m, F-1r, F-1s
SL number:	SL-11
ICBM:	SS-9
Length:	148 ft.
Diameter:	10 ft.
Lifts:	11,000 lbs. to LEO
First used:	1966: Rorsat, 1974: Eorsat
Payloads:	Cosmos Rorsat, Eorsat, Cosmos anti- satellite interceptor, Cosmos, Fobs

F-2

Popular name:	Cyclone (Tsyklon) or Meteor
Alpha number:	F-2
SL number:	SL-14
ICBM:	SS-9
Length:	158 ft.
Diameter:	10 ft.
Lifts:	12,000 lbs. to LEO
First used:	1977

Payloads: Meteor, Cyclone, Cosmos geodetic, Cosmos communications, Cosmos earth resources, Cosmos Elint electronic spysat

G-1
Popular name: N-1e
Alpha number: G-1
SL number: SL-15
Lifts: 297,000 lbs. to LEO
First used: late 1960's, early-1970's launch attempts failed
Payloads: Moon rocket abandoned in 1974

G-1e
Popular name: N-1m
Alpha number: G-1
SL number: SL-15
Lifts: 110,000 lbs. to Moon
First used: late 1960's, early-1970's launch attempts failed
Payloads: Moon rocket abandoned in 1974

J-1
Popular name: . Zenith
Alpha number: G-1
SL number: SL-16+
Lifts: 33,000 lbs. to LEO
First used: 1982
Payloads: Zenith, small spaceplane shuttle, Cosmos satellites

K-1
Popular name: Energia
Alpha number: G-2
SL number: SL-17
Lifts: 220,000 lbs. to LEO, 40,000 lbs. to stationary orbit, 70,000 lbs. to Moon, 44,000 lbs. to Mars, 73,000 lbs. shuttle cargo to LEO
First used: 1987: cargo pod, 1988: shuttle Buran
Payloads: Enrgia, Mir-2 large space stations, Buran large manned shuttles, unmanned cargo pods, large satellites to stationary orbit, manned interplanetary spacecraft to Moon, Mars

USSR Rockets: Energia

A column of yellow flame sent billowing clouds of white steam rolling away from the Baikonur Cosmodrome launch pad November 15, 1988, as the world's mightiest space rocket, Energia, shot through a cold rain into the sky, ferrying the brand-new Russian space shuttle Buran successfully to orbit.

The launch of a shuttle was an important step for USSR space planners, but the successful flight of the mammoth Energia rocket probably was even more important.

As the world's largest operational launcher, Energia not only can ferry Russian shuttles to orbit, but it also can carry 110-ton satellites to orbit, or a larger fourth-generation space station module expanded to hold 12 cosmonauts, or manned missions to the Moon and Mars. Energia is designed to carry voluminous unmanned cargo pods to space with the same ease it lifted Buran.

Earlier test. The November 1988 shuttle maiden flight was the second launch of an Energia rocket. The Soviet Union first successfully launched Energia May 15, 1987. The unmanned cargo pod carried by the rocket in the 1987 test flight was a 150-ft.-long, 10-ft.-diameter black cylinder with nose cone. The dummy test payload inside probably weighed 100,000 to 200,000 lbs.

Not reflecting on Energia capabilities, the dummy satellite in the cargo pod apparently malfunctioned in the 1987 test flight. A U.S. Air Force missile-warning satellite watched the fireball as the satellite fell into the Pacific Ocean.

The USSR reported the first and second stages of Energia worked well, but, when it came time for the payload to separate, a circuit error in one of the dummy satellite's onboard instruments prevented it from reaching the speed necessary to get to orbit after separation.

More powerful than Saturn. Energia is the world's most powerful rocket. A 4.4 million-lb. Energia develops 6.6 million lbs. of thrust. It can carry 220,000 lbs. to orbit. That's 10 tons more than the U.S. Saturn V which sent men to the Moon in 1969-72 and lofted the American space station Skylab in 1973. The U.S. abandoned Saturn V in 1973.

The 110 tons Energia can carry to orbit triples the lifting ability of a U.S. space shuttle. Energia can lift payloads five times heavier than payloads carried by Proton, the USSR's second most powerful rocket.

Core and boosters. Energia has four rocket engines inside its main core, fueled with liquid oxygen and liquid hydrogen, plus four booster rockets strapped around the core, fueled with kerosene and liquid oxygen. Either a shuttle or cargo pod can be strapped onto the outside of Energia, between the two pairs of boosters. The 4.4 million-lb. Energia delivers 6.6 million lbs. of thrust.

The Energia core is 197 feet tall with a diameter of 26 feet. Each of the strap-on boosters are 13 feet in diameter. Altogether, Energia's eight cryogenic engines develop ten times the power of the rocket used by the USSR to launch cosmonauts to space in Soyuz capsules.

V. I. Gubanov, chief designer of the Energia rocket, wrote in Pravda in 1988 that Energia is able to fire 36,000-lb. satellites to stationary orbit as well as 64,000-lb. probes to the Moon and 56,000-lb. probes to Mars.

He said the boosters on Energia have their own control system and may be used as a third stage. They also can carry payloads within themselves.

Using Energia, the USSR will be able to boost giant payloads with a single rocket. The United States cannot do that and has no plans to build a rocket as large as Energia. The most powerful unmanned rocket currently in the U.S. inventory—Titan 4—can lift about one-third the payload of Energia.

The U.S. once had a launcher almost as powerful as Energia in the Saturn V series of rockets used to send astronauts to the Moon in 1969-72. After taking men to the Moon, Saturn V was used to launch America's first-generation space station, Skylab, in 1973. But the Saturn V boosters were abandoned by NASA after 1973 as the space shuttle project was started.

Energia can lift ten tons more than the extinct Saturn V could. A large new American rocket now on the drawing boards for the year 2000—the National Launch System (NLS) space rocket—will lift 100,000 to 150,000 lbs. to orbit. Energia can lift 220,000 lbs.

In regular service, Energia is expected to carry Russian shuttles to orbit, 110-ton satellites to orbit, a fourth-generation space station expanded to hold 12 cosmonauts in the early 1990's, and even manned missions to the Moon and Mars.

Different shuttles. A U.S. shuttle's jumbo fuel tank feeds main engines in the aft end of the orbiter. But, for the USSR shuttle, the main engines and boosters are those in and on Energia. The shuttle vehicle itself does not contain its own main engines as in the U.S. shuttle. The Russian shuttle, in effect, is a special cargo pod carrying cosmonauts to orbit.

Energia engines develop more thrust than the 400,000 lbs. from the U.S. shuttle main engines. Energia's strap-on boosters are similar to engines used in the USSR's SL-16 medium-lift booster used to send Cosmos satellites to orbit.

The USSR has two Energia launch pads at the Baikonur Cosmodrome in central Asia. Shuttles fly back to land at that vast spaceport. Additional shuttle landing strips are under construction, near Simferopol along the Crimean Black Sea coast and in the USSR Far East.

Using it over. The Russians made a major advance when they designed Energia to be reusable. The 198-ft. rocket has a three-part core body with four booster rockets strapped around the sides. After using all its fuel on a flight to space, the rocket will separate into seven sections, each of which will drop back to Earth by parachute. The main rocket body, housing the first-stage and second-stage engines, will break into three parts. Those three sections and the four booster rockets will float down separately, all to be cleaned and used again.

After using its fuel, the first-stage rocket parachutes to solid ground. The second stage, reaching sub-orbital speed, separates and splashes down at a predetermined spot in the Pacific Ocean, avoiding cluttering of space with big fragments of rocket debris.

After the second stage separates, acceleration to orbital speed comes from a rocket built into the payload. In effect, a third-stage rocket is in the payload, whether unmanned cargo pod or manned shuttle. For instance, the shuttle Buran had an engine which pushed it the final distance to orbit.

Energia is an expensive state-of-the-art system so using one rocket over is desirable. Reusing the rocket parts should reduce significantly the cost of a heavy-lift launch. The USSR plans to improve unmanned flight to and from space and make it routine. It already has robot landings.

The need for cheaper rides to space pushed development of the Energia super rocket. The greater the payload carried in one launch, the cheaper the cost per pound of weight ferried to space.

Earlier powerhouses. The USSR is said to have tried unsuccessfully to launch secretly Saturn V-class rockets, known as N-1, between 1969 and 1972.

The first attempt reportedly was in 1969, the year a Saturn V launched U.S. astronauts to land on the Moon. The USSR rocket is said to have exploded on the launch pad. Effects of the blast, which covered many acres, reportedly still can be seen in satellite photos.

The second attempt reportedly was in 1971. A third was in 1972. In each, the booster reportedly exploded at low altitudes.

The USSR has been trying to sell its previously strongest booster, Proton, around the world as a means of launching commercial satellites to orbit. For $30 million, Proton can carry a 4,000-lb. satellite to stationary orbit 22,000 miles high.

Tech specs. Energia is a two-stage launcher with four first-stage rockets arranged around a central core which holds the second stage. The payload weight is distributed asymmetrically. The starting weight of the rocket could be as much as 2,400 tons.

Each first stage section is equipped with four-chambered liquid-propellant rocket engines (LPRE), using liquid oxygen and hydrocarbon fuel. The thrust of the first-stage engine is 740 tons at Earth's surface and 806 tons in a vacuum.

The second stage uses oxygen and hydrogen fuel and has four single-chambered LPREs, each with a thrust of 148 tons at Earth's surface and 200 in a vacuum.

Start-up of the first and second stages takes place almost simultaneously at launch. The overall thrust at the beginning of the flight is about 3,600 tons. Firing up both stages while on the ground frees engineers from the problem of starting engines in weightless conditions, making success more likely.

Combo launchers. Energia is a flexible launcher. Its first-stage rocket, second-stage rocket and boosters can be used together in various combinations to create heavy launchers and super-heavy launchers with different amounts of booster power. The amount of weight any one combination could lift would depend on the number of modules involved. Such a modular system is under study in the U.S. for the proposed National Launch System (NLS) to fly in the year 2000.

Cold fuel. The high-energy fuel used by Energia includes liquid oxygen cooled to a temperature of –366 degrees Fahrenheit as an oxidizing agent. The propellant for the second stage, liquid hydrogen, is cooled to a temperature of –491 degrees Fahrenheit.

Special materials. Designing Energia required new piping, tanks, hydraulic parts and construction materials of sufficient strength for cryogenic temperatures. New heat-resistant coverings and insulation were created. A series of new types of high strength steels were developed with aluminum and titanium alloys. Overall, 70 per cent of the weight of Energia is made from new materials.

Rocket engines. Over the years, the USSR has done much research and development into liquid-fuel rockets. Energia engineers faced complex problems building reliable first-stage and second-stage rockets. The standard rocket engine designed for the first stage of the new generation of launchers is known as the RD-170.

The development of heavy thrust sustainer engines with a long life for the second stage of the Energia launcher was a significant achievement, minimizing gas dynamic losses, regenerative cooling and minimizing resistance of materials in liquid hydrogen.

To ensure reliability and viability, backup has been built in for all vital systems and parts. The robot control complex has backups for every part and circuit.

The rocket's emergency systems were designed from scratch to diagnose the condition of the engines of both stages, ordering prompt shutdown should trouble develop. The rocket has fire and explosion prevention systems. If a problem should arise on the way to space, an Energia could continue its controlled flight to a low circular orbit with only one engine working on the first or second stage.

USSR Rockets: Glavkosmos

Glavkosmos is the USSR's commercial space agency, established by the Kremlin in 1985 to take the USSR into the worldwide satellite launch business and run the civilian space program.

Glavkosmos competes with European Space Agency's Arianespace, China's Great Wall Industry and several individual American companies for private commercial launch contracts. Launch sales have been slow due to Western restrictions on high technology exports.

Glavkosmos offers satellite boosts to space on a variety of rockets including the powerful Proton. China offers various models of its Long March while ESA sells launches on Ariane models. The U.S. companies offer Atlas, Delta, Titan plus a variety of smaller rockets.

Cheap lifts. Following the 1986 U.S. shuttle Challenger disaster, the USSR announced in 1987 it would make inexpensive space flights available to other countries needing satellites ferried to orbit. To make it easy for other nations to use the USSR's new profit-making service, the Russians exempted payloads from customs inspection when entering the USSR. Payloads were to be permitted to be shipped in sealed trucks to Baikonur Cosmodrome in the Central Asian desert.

Three USSR boosters were to be used for commercial launches—the heavy-lifting Proton which then had flown successfully on 90 out of 97 tries since 1970; the SL-4 which the USSR used to launch its own Soyuz manned spacecraft and Progress cargo ships; and a smaller rocket known as Vertical which had been used for atmospheric soundings with recovery capsules. Vertical could provide commercial space research service short of carrying large cargos to be dropped off in orbit. Subsequently, the Zenit launcher was added.

The USSR provided insurance coverage from its own government insurance company. Flights on behalf of developing countries in Asia, Africa and Latin America were to be carried at price discounts.

Channels for rent. To profit from past space efforts, the Russians also started renting out channels in their Gorizont communications satellites. And they planned to sell data about resources on the planet collected by various Earth-observation satellites they have in orbit. They offered to develop custom experiments and manufacturing systems for use aboard USSR spacecraft on behalf of commercial firms around the world.

Proton rocket. The superstar of their commercial launch business is the Proton rocket. It can carry 20 tons of goods to orbit 125 miles above Earth. Or it can ferry two tons of payload to stationary orbit some 22,300 miles out in space.

Main competitors for Proton are the European Space Agency's Ariane rockets, the Chinese Long March launchers and the American Titan, Atlas and Delta launchers. The U.S. space shuttle travels only to low Earth orbit and then only with medium-weight payloads.

The USSR has said it would send 20 tons of payload, aboard Proton, to stationary orbit for $43 million payable in Swiss francs.

Proton has three rocket stages when flying to low orbit. At the bottom of Proton, the first stage is a central fuel tank with six booster rockets strapped to its sides. A second stage is bolted above the first stage and a third stage is mounted atop the second stage. A fourth stage is added when a payload must be carried 22,300 miles out to stationary orbit, or on an interplanetary flight.

Unlike U.S. technique, USSR rockets are assembled while resting horizontally on their sides. They are transported by railway in that position from assembly building to launch pad where they are erected to a vertical position for blast off.

European Space Agency

European Space Agency is a 13-nation international space research and technology organization. Member countries are Austria, Belgium, Denmark, France, Germany, Great Britain, Ireland, Italy, the Netherlands, Norway, Spain and Switzerland. Canada is an observer member. Poland and Hungary have applied to become the first East European members of ESA.

The European Space Research Organization was established in 1962, then the European Launcher Development Organization was founded in 1964. The two were merged in 1975 to form ESA.

Most ESA satellites are launched on Ariane rockets from Kourou, French Guiana. The first Ariane was blasted to space from Kourou December 24, 1979. In orbit, satellites are controlled by the European Space Operations Center at Darmstadt, Germany. Tracking stations are in Belgium, Germany, Italy, and Spain. European Space Range sounding-rocket launch stations are located in Norway and Sweden.

ESA is headquartered in Paris. Its European Space Research and Technology Center is located at Noordwijk in the Netherlands. The European Space Research Institute is at Frascati, Italy, as is the Space Documentation Service. The Meteorological Program Office is in France at Toulon.

ESA members control their own annual financial contributions, deciding which projects they will support. For instance, the British have not supported the Hermes space shuttle project. No ESA member may start a space project without inviting the others. ESA issues contracts on a pro rata basis.

ESA owns and operates Spacelab, a science lab flown to space and back in U.S. space shuttles. The agency is designing Hermes, a small space shuttle to fly at the end of the 1990's, and Columbs, a lab module for the U.S.-international space station Freedom. ESA also builds communications, geophysics, and astronomy satellites.

ESA's Giotto spacecraft was launched July 2, 1985. It made the closest approach of any probe to Comet Halley, passing within 372 miles of the nucleus March 13, 1986.

Giotto survived a battering by high-speed particles in the comet's tail. Its radio fell silent, but returned after 20 minutes. A month later, European Space Operations Center, Darmstadt, put Giotto in hibernation, rounding the Sun every 10 months. Giotto was parked in that orbit from April 1986 to February 1989.

To use minimum electrical power, radio transmissions shut down. Giotto's camera had been broken during the Halley flyby, but European scientists said it still could relay valuable information. ESA decided to send Giotto to a July 10, 1992, rendezvous with Comet Grigg Skjellrup. Giotto swung by Earth July 2, 1990, for a gravity boost to catch its second comet.

European Rockets: Arianespace

Arianespace, Europe's commercial space transportation company, is a consortium of 50 European companies responsible for commercial exploitation of Europe's Ariane rockets. Arianespace is a private French industrial, commercial company selling Ariane rocket launches for the 13 nation-members of ESA. ESA and Arianespace launches are from the Guiana Space Center, Kourou, French Guiana, South America.

Ariane rockets are developed by the European Space Agency (ESA). Models have included Ariane 1, Ariane 2, Ariane 3 and Ariane 4. A more powerful Ariane 5, to fly after 1995, is slated to lob Europe's Hermes space shuttle to orbit.

The last Ariane 1 flight was in 1986. The last Ariane 2 launch was in 1989. The last Ariane 3 rocket flew in 1989. Today, all launches are on Ariane 4 rockets, until the bigger Ariane 5 starts to fly after 1995.

50 more rockets. Dozens of European companies manufacture and assemble Ariane parts, giving each an economic benefit matching its investment share in the cost of research and development.

Arianespace ordered 50 more Ariane 4 space rockets in 1989 for $3 billion. The new rockets will be fired to space through the 1990's. French companies benefiting from the big order were France's state-owned Aerospatiale SA (20.4 percent share), France's state-controlled Societe Europeenne de Propulsion (21.1) and Matra SA (8.7). West German companies sharing in the big order include MAN (9.3) and ERNO (5.1).

Contractors with smaller shares were Belgium's Fabrique Nationale, France's Air Liquide, Great Britain's British Aerospace, Great Britain's Ferranti, Italy's SNIA, Italy's Aeritalia, Spain's Construcciones Aeronauticas SA, Sweden's Volvo AB, and The Netherlands' Fokker.

Arianespace reduced costs 20 percent by ordering 50 rockets at once, cutting the cost of an Ariane 4 launch to $90 million to $100 million. An Ariane 4 can carry a 4.6-ton payload to orbit. Arianespace plans to launch as many as 100 Ariane 4 and Ariane 5 rockets by the year 2000.

Chinese Long March rockets had provided tough competition for Ariane, with Chinese launches selling for $20 million to $25 million.

Commercial space. Europe launched its newest commercial launch workhorse, Ariane 4, for the first time June 15, 1988. Ariane 4 can haul 9,200 lbs. or 4.6 tons to orbit. Ariane 3, the previous generation, could carry 5,900 lbs. or 2.9 tons.

Ariane 4 technicians strap on different numbers of solid and liquid-fuel booster rockets and adapt the upper portion of the launcher to the payload, tailoring the vehicle to a client's budget and needs. The version used for the June 15 flight included two liquid and two solid-fuel boosters.

Developed over six years at a cost of $575 million, Ariane 4 competes in the global launch business with Russian Proton, Chinese Long March, and American Titan 4 by Martin Marietta Corp., Delta 2 by McDonnell Douglas and an improved Atlas-Centaur known as Atlas 2 by General Dynamics. ESA's Arianespace was hauling half the world's commercial payloads.

Arianespace pictures Titan as its toughest competitor in a worldwide market for some 25 launches a year. Titan is the only U.S. rocket able to lob satellites weighing more than 5,400 lbs. or 2.7 tons to space. However, the new Delta 2 is bigger and more powerful than the older Delta 1 model.

Ariane 4 was developed by 11 countries—Belgium, Denmark, France, Great Britain, Ireland, Italy, The Netherlands, Spain, Sweden, Switzerland and Germany. The French space agency, Centre National D'Etudes Spatiales (CNES) was prime contractor for the launch project. Some 500 companies built parts of the rocket.

European Rockets: The Ariane Family

Ten European nations got together in the early 1970's to develop a new rocket for launching satellites to Earth orbit. The European Space Agency started developing its Ariane space rocket in 1973.

Decisions were dominated by France and, to a lesser extent, West Germany, after the French kicked in 64 percent of the budget and Germany 20 percent. The other eight nations together contributed a total of 16 percent. Centre National d'Etudes Spatiales, the French space agency CNES, was prime Ariane development contractor.

Ariane was designed to carry large spacecraft—such as communications and weather satellites—to the 100-mi.-high Earth orbit from which they can blast themselves on out to a so-called stationary orbit at 22,300 miles above Earth.

Ariane rockets could be used for less dramatic flights, of course, as well as for sending spacecraft away from Earth on deep-space missions. None has been used for out-of-the-ordinary purposes to date.

Going commercial. After four test flights, the Ariane rocket system became operational in 1981. Four free promotional launches were made, then Ariane rockets went commercial. Arianespace, created by ESA to sell launches, was the first space transportation company to operate commercially.

Ariane is a three-stage rocket. The first two stages are fueled with liquid hydrazine and nitrogen tetroxide. The upper third stage is said to be "cryogenic" because it uses super-cold liquid hydrogen and liquid oxygen as fuel.

Over the years, ESA has improved on the original Ariane 1 design to make the rocket capable of lifting heavier loads to orbit. The improvements flown to date have been known as Ariane 2, Ariane 3 and Ariane 4.

Ariane types. Ariane 3 rockets added two booster rockets strapped to the outside of the three-stage core. Ariane 4 went beyond that with various combinations of more strap-on boosters. An even more powerful version, Ariane 5, is under development.

Rocket	Years Used	Payload
Ariane 1	1979-1986	4,080 lbs.
Ariane 2	1984-1989	4,850 lbs.
Ariane 3	1984-1989	5,950 lbs.
Ariane 4	1988-1999	9,260 lbs.
Ariane 5	1994-2010+	42,000 lbs.

All Ariane flights have lifted off from launch pads cut out of the jungle at Kourou, French Guiana, a department of France on the northeastern coast of South America. The Guiana Space Center is a mere five degrees north of the equator. Arianespace would like to average about eight launches per year.

Ariane 4. Ariane 4 is the most powerful launcher marketed by the 13-nation European consortium Arianespace in competition with American space boosters Delta, Atlas and Titan. It is built by companies in 11 European countries.

Ariane 4 first flew June 15, 1988. It can carry around 10,000 lbs. of payload to high orbits required by communications satellites. Only the U.S. Titan 4 rocket, U.S. space shuttles, and the USSR's Proton and Energia rockets are more powerful.

Ariane 4's 77.6-ft.-tall first stage has two identical cylindrical 24.3-ft.-long, 12.5-ft.-diameter fuel tanks. The propulsion system is four Viking 5 engines putting out 607,000 lbs. of thrust. The first stage fires 3.5 minutes burning 485,000 lbs. of fuel.

There are six versions of an Ariane 4 rocket, each with a different arrangement of strap-on boosters to provide extra power. Some versions have solid-fuel boosters or combinations of both.

The Ariane 4 second stage is 37.5 feet long, weighs 7,936.7 pounds empty and is 8.5 feet in diameter. It has one Viking 4 engine generating 176,700 lbs. of thrust, burning a hydrazine and nitrogen tetroxide mixture, the same fuel used in the first-stage. The second stage fires for two minutes and then falls away.

The Ariane 4 third stage is an ultra-modern rocket concept burning more-powerful cryogenic fuels—super-cold liquid hydrogen and liquid oxygen. Those are the fuels used by the U.S. space shuttles. The third stage weighs 2,646 lbs. empty, is 8.5 feet in diameter and is 32.5 feet tall. The third stage's single HM7B rocket engine blasts for 12 minutes, turning out 14,000 pounds of thrust while burning 23,590 pounds of propellant.

Failures. ESA has had five launch failures with its powerful Ariane space rocket since it started launching the big boosters December 24, 1979. Ignition failures in the cryogenic third stage caused three of the five failures—L-5, V-15 and V-18—but apparently not the LO-2 and V-36 failures.

LO-2. The first failure was the second Ariane 1 launch. The experimental mission had to be aborted when one of the Viking 5 engines in the Ariane malfunctioned May 23, 1980. Known as flight LO-2, the Ariane 1 carried AMSAT's Phase 3-A amateur radio satellite and a satellite known as Firewheel from West Germany's Max Planck Institute. They sank to the bottom of the Atlantic and flights were suspended to June 1981.

L-5. The second failure of an Ariane 1 was experimental flight L-5 on September 10, 1982. The International Maritime Organization's Inmarsat Marecs-B and ESA's Sirio 2 satellites were destroyed. Flights were grounded to June 1983 when an Ariane 1 successfully lofted AMSAT's Phase 3-B amateur radio satellite and the Eutelsat ECS-1 satellite to orbit. Experimental Ariane launches ended and commercial launches began with a flight labeled V-1 in May 1984.

V-15. The third failure was an Ariane 3 on flight V-15 September 12, 1985. GTE's Spacenet 3 and Eutelsat's ECS-3 communications satellites were lost. Further flights were postponed to February 1986.

V-18. The fourth failure was an Ariane 2 on flight V-18 May 31, 1986. The Intelsat V (F14) satellite was lost. Launches were suspended from May 1986 to September 1987.

V-36. ESA had 17 straight launch successes after recovering from the V-18 explosion. Then, on February 22, 1990, a cloth stuck in a pipe caused an Ariane 4 carrying two Japanese communications satellites to loose power and explode during lift-off. The Ariane fleet was grounded five months. Flight V-36 was the first loss of an Ariane 4, the fifth loss in 36 launches since Arianes started to fly in 1979 and the third since commercial launches began in 1984.

The French magazine Express in a sensational report May 3, 1990, claimed saboteurs may have been responsible for Ariane failures. France plays a leading role in the European Space Agency (ESA) which builds the Ariane family of space rockets. The magazine suggested sabotage might have been possible in three of five failures, particularly the V-36 explosion February 22, 1990, which destroyed two Japanese communications satellites during lift off from ESA's launch site at Kourou. ESA set up an inquiry board to probe the V-36 explosion. It found that something had clogged a water line. Flights resumed with V-37 on July 24, 1990.

European Rockets: Future Flights

European Space Agency planned nine unmanned Ariane 4 rocket launches in 1992. ESA and its commercial launch-sales arm, Arianespace, has been blasting unmanned Arianes to space at a rate of about one satellite per month.

Date	Flight	Rocket	Satellite Payload
1992 Jan	V-49	Ariane 44L	Superbird-B and either Inmarsat 2 (F4) or Arabsat 1C
1992 Feb	V-50	Ariane 44L	Eutelsat 2 (F4) and Insat 2A
1992 Mar	V-51	Ariane 44L	Telecom 2B and either Arabsat 1C or Inmarsat 2 (F4)
1992 Jun	V-52	Ariane 42P	Topex-Poseidon and Asap #3 carrying Kitsat-A and S 80/T
1992 Jul	V-53	Ariane 44L	Hispasat 1A and Satcom C4
1992 Aug	V-54	Ariane 42P	Galaxy 7
1992 Oct	V-55	Ariane 44L	Eutelsat 2 (F5) and Insat 2B
1992 Nov	V-56	Ariane 44L	Hispasat 1B or Superbird-A and an open flight opportunity
1992 Dec	V-57	Ariane 42P	Galaxy 4

1992 Payloads	Origin
Arabsat 1C	Arab League
Asap #3, Kitsat-A, S 80/T	ESA experimental
Eutelsat 2 (F4)	European Organization for Telecommunications by Satellite
Eutelsat 2 (F5)	European Organization for Telecommunications by Satellite
Galaxy 4	U.S., Hughes
Galaxy 7	U.S., Hughes
Hispasat 1A	Spain
Hispasat 1B	Spain
Inmarsat 2 (F4)	International Maritime Organization
Insat 2A	India
Insat 2B	India
Satcom C4	U.S., General Electric
Superbird-A	Japan
Superbird-B	Japan
Telecom 2B	France
Topex-Poseidon	France-U.S.

1993 Payloads	Origin
Astra 1C	Luxembourg
Astra 1D	Luxembourg
Brasilsat B1	Brazil
Brasilsat B2	Brazil
Helios	Europe
Intelsat VII (F1)	International Telecommunications Satellite Organization
Intelsat VII (F4)	International Telecommunications Satellite Organization
Intelsat VII (F5)	International Telecommunications Satellite Organization
ISO	Europe
Mop-3	Europe, Meteosat weather satellite
MSat1	Canada
Solidaridad 1	Mexico
Solidaridad 2	Mexico
Spot-3	France
Turksat 1	Turkey
Turksat 2	Turkey

European Rockets: Past Flights

Ariane space rocket launches by the European Space Agency from December 1979 to October 1991. Two dots after Ariane model number indicates flight failure.

No.	Date	Ariane	Satellite
LO1	1979 Dec 24	1	Cat
LO2	1980 May 23	1••	Cat, Firewheel, Amsat
LO3	1981 Jun 19	1	Cat, Apple, Meteosat
LO4	1981 Oct 20	1	Marecs A
L5	1982 Sep 10	1••	Marecs B, Sirio 2
L6	1983 Jun 16	1	ECS 1, Oscar 10

L7	1983 Oct 19	1	Intelsat V (F7)
L8	1984 Mar 5	1	Intelsat V (F8)
V9	1984 May 22	1	Spacenet 1
V10	1984 Aug 4	3	ECS 2, Telecom 1A
V11	1984 Nov 10	3	Spacenet 2, Marecs B2
V12	1985 Feb 8	3	Arabsat 1, Brazilsat S1
V13	1985 May 8	3	G-Star 1, Telecom 1B
V14	1985 Jul 2	1	Giotto
V15	1985 Sep 12	3••	Spacenet 3, ECS 3
V16	1986 Feb 22	1	Spot 1, Viking
V17	1986 Mar 28	3	G-Star 2, Brazilsat S2
V18	1986 May 31	2••	Intelsat V (F14)
V19	1987 Sep 16	3	Aussat K3, ECS 4
V20	1987 Nov 21	2	TVsat 1
V21	1988 Mar 11	3	Spacenet 3R, Telecom 1C
V23	1988 May 17	2	Intelsat V (F13)
V22	1988 Jun 15	4	Meteosat P2, Oscar 13, PanAmSat-1
V24	1988 Jul 21	3	Insat 1C, ECS 5
V25	1988 Sep 8	2	G-Star 3, SBS 5
V26	1988 Oct 27	2	TDF 1
V27	1988 Dec 11	4	Skynet 4B, Astra 1A
V28	1989 Jan 26	2	Intelsat V (F15)
V29	1989 Mar 6	4	JCsat-1, Meteosat (MOP 1)
V30	1989 Apr 1	2	Tele-X
V31	1989 Jun 5	4	Superbird A, DFS-Kopernikus 1
V32	1989 Jul 12	3	Olympus 1
V33	1989 Aug 8	4	Hipparcos, TVsat 2
V34	1989 Oct 27	4	Intelsat VI (F2)
V35	1990 Jan 22	4	Spot 2, microsats OSCARs 14-19
V36	1990 Feb 22	4••	Superbird B, BS-2x
V37	1990 Jul 24	4	TDF-2, DFS-Kopernikus 2
V38	1990 Aug 30	4	Skynet-4C, Eutelsat 2-F1
V39	1990 Oct 12	4	Galaxy VI, SBS-6
V40	1990 Nov 20	4	Satcom C-1, GStar IV
V41	1991 Jan 15•	4	Eutelsat 2-F2, Italsat-1
V42	1991 Feb 21•	4	Astra-1B, MOP-2
V43	1991Apr 4	4	Anik-E2
V44	1991 Jul 17	4	ERS-1, microsats Orbcom-X, Sara, TUBsat, UoSAT-F
V45	1991 Aug 14	4	Intelsat VI (F5)
V46	1991 Sep 26•	4	Anik-E1
V47	1991 Oct 29•	4	Intelsat VI (F1)

European Rockets: Ariane V-19

Europe got space launches back on track September 16, 1987, when its Ariane 3 rocket blasted off from the jungle space center at Kourou, French Guiana, lifting two communications satellites to orbit.

Carrying Europe's first payload in 16 months, Ariane lit the night with a blaze of fire, touching off applause from ground controllers at the space center on South America's northeast coast.

The European communications satellite ECS-4 and the Australian Aussat K3 satellite were carried to orbit without difficulty.

Western Europe's space program had been troubled by technical failures. The September 1987 liftoff of the 160-ft. rocket not only put two satellites in orbit, but gave the European Space Agency (ESA) and its commercial arm, Arianespace, a lead for a time in the battle to provide launch services for the West.

As the 240-ton Ariane successfully completed ESA's flight V-19, Arianespace became the non-communist world's leader in launch services. At the time, Arianespace had 46 satellites waiting to be launched on Ariane rockets.

Months of delay. The European Space Agency had grounded its Ariane launcher fleet in 1986 and 1987 for investigation of a blast-off failure. Three rocket models, Ariane 2, 3 and 4, were affected. An Ariane 2 failed in May 1986 triggering the investigation. An Ariane 3 failed in September 1985. The new Ariane 4 was untried. The U.S. shuttle disaster and failure of two other major U.S. launchers in 1986 had persuaded some satellite owners to use Ariane to get to orbit. When Ariane also was grounded, some asked China and other spacefaring nations for reliable launching rockets.

ESA. The European Space Agency is a joint project of Austria, Belgium, Denmark, France, Great Britain, Ireland, Italy, the Netherlands, Norway, Spain, Sweden, Switzerland and West Germany. Finland is an associate member and Canada has an agreement for close cooperation.

Four of the previous 18 Ariane shots had failed, three because of problems in a rocket's third stage. ESA's rockets were grounded 16 months after the third-stage ignition system malfunctioned May 31, 1986, on an Ariane 2, forcing ground technicians to destroy the rocket, and its $55 million communications satellite payload, during flight to orbit.

The rocket's third stage caused the 1986 accident when it shut off early. ESA spent $83 million to redesign it. Power of the ignition system was tripled. The third stage, known as HM7B, is manufactured at Vernon, France, where six spies were arrested by French police. The spies wanted information on the HM7B, one of the most advanced rockets in the world despite its difficulties.

In the September 1987 launch, the third stage operated correctly, turning on at four minutes and 36 seconds into the launch and continuing for a perfect burn.

Ariane 3 is an improved model of Europe's original Ariane 1 which first placed satellites in orbit early in 1981. Two solid-fuel boosters strapped on give Ariane 3 the ability to lift more weight to orbit than Ariane 1. Ariane 3 can place 5,950 lbs. in a low-Earth geosynchronous transfer orbit compared to a total payload of only 4,080 lbs. for the older Ariane 1.

Ariane rockets are fired from Kourou to get a helping hand from Mother Nature. Since the site is almost on the equator, where Earth rotates fastest, less booster power is needed to push a satellite to Earth orbit. Satellites are launched north and east from Kourou. The tropical site has not had a launch delayed by bad weather.

Stationary orbits. From the point of view of a person standing on the surface of Earth, some man-made satellites appear to whiz around the globe, passing overhead every hour or two. They usually are in low orbits, near Earth, from 100 to 1000 miles out in space.

Others seem to stick at one point in the sky overhead. They are some 22,300 miles out from Earth's surface. At that distance, the satellite's speed circling the globe is the same as the speed at which Earth itself spins. Such matching speeds gives the appearance the satellite is stationary in the sky over one spot on Earth's surface. Engineers refer to a stationary satellite as geosynchronous.

Communications satellites, such as the pair lifted by Ariane 3, are built for a stationary position overhead. To get to their assigned spots in the sky, they are ferried by rocket to a low transfer orbit, known as a geosynchronous transfer orbit (GTO).

Back on track. Two launch windows were available to ESA on September 16, 1987. Controllers missed the first when the countdown was halted six minutes before liftoff due to problems with a pressure sensor measuring leaks in the rocket's third stage. Blast-off came during the second period of time suitable for launch that day, within two minutes of the end of that window.

Ariane's liquid-fuel rockets fired. A second later solid-fuel boosters ignited, providing a brilliant flash, illuminating the dark jungle around the coastal launch site.

Old-timers said later they were reminded of the relatively slow, deliberate climb of the old Ariane 1 rocket from the launch pad. By contrast, the Ariane 3 seemed to catapult into the cloud cover hanging over the launch complex. Ariane 1 had only a liquid-fuel first stage while Ariane 3 had solid-fuel boosters pushing it upward rapidly, as if fired from some giant slingshot.

The Ariane 3 rocket worked smoothly, placing Europe's ECS-4 and the Australian Aussat K3 satellite into transfer orbits from where they would be sent, two days later, to geostationary orbit 22,300 miles over the Earth.

The 1,430-lb. Aussat K3 arrived at its proper transfer orbit right on schedule, just 18 minutes 27 seconds after liftoff from the Kourou pad. Then the 1,540-lb. ECS-4 popped out of its carrier into orbit at 22 minutes 2 seconds. The entire mission, from liftoff to depositing the satellites in orbit, took less than 25 minutes.

Extraordinary accuracy. The extraordinary accuracy of the Ariane 3 launch and flight was said to be remarkable for a new launch system. French President Francois Mitterrand at Paris congratulated ESA on the launch, pointing out the success gave Europe's space program a new beginning.

ECS-4 was left in its transfer orbit 125 miles out in space. From there, a small rocket attached to the satellite—an apogee boost motor—was fired to push the satellite 22,000 miles farther away from Earth. The satellite was transferred from a low orbit to a high orbit. The transfer was made, about 37 hours after launch from French Guiana, on September 17 when the apogee boost motor attached to ECS-4 was fired successfully to send the satellite out to a stationary orbit.

Huge solar panels on ECS-4 capture sunlight and turn it into electricity to power computers, transmitters and receivers in the satellite. Once in stationary orbit, the ECS-4's solar wings were extended. Radio telemetry signals confirmed that 1,280 watts of electricity were being generated for use in ECS-4.

From arrival at stationary orbit over the equator September 17 until mid-October, the satellite was allowed to drift east across the Atlantic Ocean to a point over West Africa. From there, ECS-4 looks down on a vast landmass including all of Europe. The satellite was tested for two months, then put into regular commercial service in January 1988, retransmitting telephone, television and computer data signals received from ground stations.

Eutelsat. ECS-4 was fourth in a series of five telecommunications satellites designed and built entirely in Europe. ESA prepared and launched the satellite. For several weeks of testing, ESA controlled the satellite in orbit from its European Space Operations Center at Darmstadt, West Germany.

After testing, control of ECS-4 was turned over to the European Telecommunications Satellite Organization, or Eutelsat, for regular service. The 26 countries belonging to Eutelsat control ECS-4.

The satellite was renamed Eutelsat 1-F4 with ground control at an operations center at Redu, Belgium. Older satellites ECS-1 and ECS-2, still working in orbit, also are controlled from Redu.

ECS-4 has 12 radio-repeating transponders which can handle 12,000 telephone calls or 12 television programs. It also has two transponders for business data communications. The satellite is expected to work about seven years, until 1994, in orbit.

Eutelsat uses the satellite primarily to send TV broadcasts to and from distant locations. Eight of the 12 transponders distribute TV programs. Others provide backup for transponders in ECS-1 and ECS-2, handling mostly telephone calls and business data communications.

ECS-4 also carried to space a message from European youth members of the Red Cross about the universal ideals of Red Cross and Red Crescent.

Future shots. Riding on the success of the September 1987 V-19 flight were

$2.45 billion in contracts for Arianespace. Some 46 satellites were waiting at the time to be launched, including 11 for customers who had signed up after the 1986 Ariane failures and Challenger disaster. The ESA launch schedule assumed a blast-off about every 6 weeks.

The final version of the European communications satellite, ECS-5, was launched on an Ariane 3 July 21, 1988, the 24th Ariane blast-off.

Launch biz. ESA has control of the launch services market in the West for the time being. Even after the U.S. shuttle fleet returned to space in 1988, NASA gave priority to military and science payloads. Most commercial contracts were cancelled.

Meanwhile, Ariane 2, Ariane 3 and Ariane 4 rockets were scheduled to ferry 50 commercial, scientific, weather, communications, military and amateur radio satellites to space. Included in the manifest were satellites for India, Australia, Canada, Italy, France, United States, Latin America, Europe and international groups.

Besides Arianespace, Western satellite owners can book rides on U.S. Titan, Delta and Atlas-Centaur and lesser-known rockets offered for private commercial flights by American companies. The USSR and China also offer commercial launch services.

European Rockets: Ariane V-20

A West German television satellite failed to unfold one of two solar panels in space November 21, 1987, just hours after being placed in orbit in a perfect launch of an Ariane 2 rocket from Kourou, French Guiana.

It was flight V-20 for the European Space Agency, the second successful Ariane launch following 16 months in which all Ariane rockets were grounded due to technical problems. The return to space had been successful September 16, 1987.

French Premier Jacques Chirac praised the twentieth Ariane flight while a West German official claimed the solar-panel problem on the TVsat-1 satellite was not dramatic.

There were no countdown holds before launch of V-20. The critical Ariane third-stage engine separated properly four minutes and 34 seconds after liftoff. The third-stage engine ignited correctly two seconds later. Third-stage ignition problems had been behind the failure of the eighteenth launch which forced the European Space Agency's rocket program into a 16-month shutdown. Three of four Ariane launch failures involved third-stage problems.

TVsat-1. TVsat-1 reached a 125-mile-high transfer orbit over Africa November 20, just 19 minutes after the late night liftoff of the Ariane 2 rocket from ESA's base in the South American jungle. The launch made Ariane the West's only satellite launcher currently active at the time in the profitable business of carrying satellites to space.

TVsat-1 was a joint project of West Germany and France. West German technicians in Kourou did not appear upset by their satellite's performance while attempting maneuvers in space to get the panel to open. A similar problem some years before was straightened out in space. But, this time, they were unable to unfold the wing of electricity-generating solar cells. The satellite could function only at half power, broadcasting on half of its TV channels.

TVsat-1 was to be a direct-to-home broadcast satellite, letting consumers tune in directly to television broadcasts using only three-ft.-diameter, or smaller, dish antennas. At full power, TVsat-1 was to have transmitted four television channels with the highest quality image available to 300 million viewers around Europe. The $108 million TVsat-1 cost $97 million to launch. It had been designed to work 10 years in space.

TVsat-2. TVsat-1 primarily was to serve Germany. TVsat-2, to cover France, was scheduled to be launched in 1990. The date was moved forward to August 1989.

The TVsat's were to be among the largest, most powerful commercial satellites built.

Each weighed 2.2 tons and had solar arrays generating 3.2 kilowatts of electrical power. Such energy was needed to run a powerful transmitter with spot beam pattern so home users would need only small dish antennas on the ground.

Broken sat. West Germany eventually had to abandon efforts to salvage the crippled TVsat-1. The West German Space Center at Oberpfaffenhofen in southern West Germany tried repeatedly to free the stuck solar panel blocking a TV receiving antenna on the orbiting $105-million TVsat-1.

The satellite could receive commands from Earth but the stuck panel blocked television relay. TVsat-1 was designed to provide four television channels and 16 digital radio frequencies.

Failure of TVsat-1, after a year of launch delays, was a technical and financial setback for West German efforts to expand television programming. TVsat-2, a twin to TVsat-1, was launched on an Ariane 4 rocket in 1989.

West Germany signed a contract with Arianespace, the European Space Agency's commercial arm, to rush the launch of a second TV satellite. The deal was cut just one month after efforts were abandoned to salvage West Germany's crippled TVsat-1.

TVsat-1 and TVsat-2 were to have operated in tandem. To make up for the loss, Germany leased channels in a French satellite, TDF-1, launched on Ariane 2 flight V-26 October 27, 1988.

European Rockets: Ariane V-21

America's Spacenet IIIR-Geostar and France's Telecom 1C communications satellites flew to space successfully March 11, 1988, atop a European Space Agency Ariane 3 rocket in launch V-21 from the South American jungle pad at Kourou, French Guiana.

The Ariane 3 ignited properly four minutes after blast-off. Engineers cheered, recalling the ignition had failed in September 1982, September 1985 and May 1986 launches. The 1986 failure, V-18, grounded the rocket.

GTE Spacenet, Europe's first American client, has used Ariane to launch five satellites. One was lost in the September 1985 failure. Spacenet IIIR was the first U.S. communications satellite to be launched in two years. Telecom 1C, replacing one that quit working January 15, 1988, is used for telephone, radio, television and military communications. Spacenet IIIR and Telecom 1C are in stationary orbit at an altitude of 22,554 miles.

Grounded. Ariane rockets had been grounded 16 months after flight V-18 failed. A third-stage ignition failure in flight V-18 had forced the European Space Agency's rocket program into the 16-month shutdown. Three of four Ariane launch failures had involved third-stage problems.

The program restarted successfully with launch of V-19 September 16, 1987. The last launch was an Ariane 2 on flight V-20 November 21, 1987, the second successful Ariane launch following the 16 months in which all were grounded.

European Rockets: Ariane V-22

European Space Agency launched a new, more-powerful generation of Ariane rocket June 15, 1988, lifting three satellites to orbit and showing future customers the vehicle was ready for sale.

The three satellites orbited in flight V-22 were Europe's Meteosat, the American Panamsat and the AMSAT Phase 3C amateur radio satellite renamed OSCAR 13 in orbit.

Near the last moment, the Ariane 4 was delayed a few days by mechanical and computer problems. Wiring between payload and rocket had to be changed and a new on-board computer had to be fitted in the vehicle equipment bay and tested. Then Ariane 4 blasted off without a hitch June 15 from the second launch pad, known as ELA-2, at the ESA space center, Kourou, French Guiana. Less than 13 minutes after blast-off, Ariane reached orbit with its international payload.

Lancements doubles. Ariane 4 was able to carry all three satellites because it used a new SPELDA designed to carry and release two satellites. SPELDA is an acronym from the French name Structure Proteuse Externe pour Lancements Doubles Ariane. In English, External Support Structure For Ariane Dual Launches. SPELDA was built by British Aerospace.

The SPELDA was inside a payload cover which opened in space. Meteosat was mounted on a metal cylinder enclosing OSCAR 13, both in the top half of the SPELDA. Panamsat was in the bottom half. About 20 minutes after launch, Meteosat popped off the top of the exposed payload stack, into its own separate-but-close path. Four seconds after Meteosat was deployed, the cylinder with OSCAR 13 inside was separated from Meteosat. OSCAR 13 would remain inside the metal cocoon for another hour.

One minute after OSCAR 13 separated from Meteosat, the top part of the SPELDA was ejected, exposing Panamsat. Two minutes later, Panamsat popped from its SPELDA section. Some 80 minutes after launch, explosive bolts sheared off a clamp, freeing OSCAR 13 from its cylinder. Three springs, which had been compressed, popped out, pushing OSCAR 13 away from the cocoon.

The spent rocket, SPELDA parts, and other bits and pieces became space junk flying in formation with the three satellites.

GTO. Their orbit was a near-Earth path, about 100 miles high, known as a Geosynchronous Transfer Orbit or GTO. From that low point, small rockets in each satellite were fired, kicking each toward a much higher orbit. Meteosat and Panamsat are in stationary orbit. OSCAR 13 is in a highly elliptical orbit, swinging out as far as stationary satellites and then dropping back in near Earth every 12 hours or so.

A stationary or geosynchronous orbit is 22,300 miles above Earth where a satellite revolves around the globe at the same speed as Earth rotates. The synchronized speeds make a satellite seem to remain stationary over one spot on the planet.

On June 16, ground controllers fired the small rockets—called kick motors—in Meteosat and Panamsat to shove them farther out, toward stationary orbit. AMSAT controllers later fired a kick motor in OSCAR 13 to push it toward a final orbit.

Satellite trio. The OSCAR 13 path is known as a Molniya orbit, after a family of Russian communications satellites following the same kind of elliptical path. Three Molniya satellites provide the Russian half of the White House-Kremlin emergency hotline.

OSCAR. AMSAT's OSCAR 13 is a 330-lb. worldwide communications satellite built and used by amateur radio operators. It has four radio relay systems which receive signals from ham stations on Earth and retransmit them on frequencies ranging from 145 to 2400 MHz. OSCAR 13 has radio beacons which can be heard with communications receivers on frequencies of 145.812 MHz, 435.651 MHz and 2400.325 MHz.

AMSAT is the Radio Amateur Satellite Corp., an international club of hams building satellites and using them in space. Two dozen amateur radio satellites have been launched over the past three decades on free rides to space as ballast or demonstrators.

Meteosat. Meteosat P2 transmits weather data to 16 European countries. The 1,559-lb. satellite monitors weather over most of Europe, Africa, the Middle East and eastern South America. Meteosat P2 replaced Meteosat-F2, launched in 1981, which until recently sent back pictures every half hour.

Panamsat. Panamsat or PAS-1, owned by the American telecommunications company Pan American Satellite, provides low-cost communications between Latin America, Europe and the United States. It gives Latin American countries the ability to create national television networks.

Arianespace used the June 15 flight to show potential customers Ariane 4—Europe's biggest, most-sophisticated rocket to date—is ready for operational flights. The three satellites got a free ride as part of the demonstration. Paid Ariane 4 flights started with launch V-27 on December 11, 1988.

Radio listeners around the world, and television viewers in some areas of Europe and North America, received a live broadcast of the June 15 countdown and blast-off. In the U.S., the C-SPAN public-service cable TV network carried the launch live.

Ham radio operators in the U.S. linked with others in England, Argentina, Japan, New Zealand and South Africa to transmit news of the Ariane 4 takeoff worldwide. Even an older ham radio satellite, OSCAR 10, broadcast the event from orbit to the Western Hemisphere and Europe.

Commercial space. Europe's newest commercial launch workhorse, Ariane 4 can haul 9,200 lbs. or 4.6 tons to orbit. Ariane 3, the previous generation, could carry 5,900 lbs. or 2.9 tons.

Ariane 4 technicians strap on different numbers of solid and liquid-fuel booster rockets and adapt the upper portion of the launcher to the payload, tailoring the vehicle to a client's budget and needs. The version used for the June 15 flight included two liquid and two solid-fuel boosters.

Arianespace plans to launch 100 or more Ariane 4 and Ariane 5 rockets by the year 2000. The larger Ariane 5 is to start flying in 1995. The last Ariane 1 flight was V-16 on February 22, 1986. The last Ariane 2 was flight V-30 on April 1, 1989. The last Ariane 3 was flight V-32 in July 1989.

Developed over six years at a cost of $575 million, Ariane 4 competes in the global launch business with Russian Proton, Chinese Long March, and American Titan 4 by Martin Marietta Corp., Delta 2 by McDonnell Douglas and an improved Atlas-Centaur known as Atlas 2 by General Dynamics. ESA's Arianespace, already hauling half the world's commercial payloads, had 67 launch contracts totaling $3.4 billion at the time of flight V-22.

Arianespace pictures Titan as its toughest competitor in a worldwide market for some 25 launches a year. Titan is the only U.S. rocket able to lob satellites weighing more than 5,400 lbs. or 2.7 tons to space. However, the new Delta 2 is bigger and more powerful than the older Delta 1 model.

Ariane 4 was developed by 11 countries—Belgium, Denmark, France, Great Britain, Ireland, Italy, The Netherlands, Spain, Sweden, Switzerland and West Germany. The French space agency, Centre National D'Etudes Spatiales (CNES) was prime contractor for the launch project. Some 500 companies built parts of the rocket.

European Rockets: Ariane V-23

European Space Agency, firing an Intelsat V communications satellite to orbit in a near-flawless operation May 17, 1988, scored its 18th success in 22 launches of Ariane rockets.

Officials at ESA's jungle launch pad on the northeastern shoulder of South America at Kourou, French Guiana, called the Ariane 2 flight a complete success, their second success of 1988.

Despite rain and moderate winds, normal for the season in the French territory on the north coast of South America, the only incident was a 15-minute countdown halt as a storm cloud passed six minutes before liftoff of flight number V-23.

Sixteen minutes forty seconds after the evening launch, the 4,374-lb. satellite separated smoothly into a low orbit in preparation for transit to a higher orbit.

Batting over 800. Four of 21 previous Ariane shots were not successful. Three of four failures were caused by problems with the rocket third stage. A failure in flight V-18 in May 1986 grounded the program 16 months.

The program had restarted successfully with launch of V-19, an Ariane 3, September

16, 1987; V-20, an Ariane 2, November 21, 1987; V-21, an Ariane 3, March 11, 1988; and then V-23.

Intelsat V. The Intelsat V (F13) satellite, which later boosted itself into a stationary orbit 22,500 miles above Earth, is one in a series of 15, each providing two television channels and 15,000 telephone circuits.

The International Telecommunications Satellite Organization (ITSO) operates Intelsat satellites.

Intelsat V (F14) was launched on Ariane in May 1986. Intelsat V (F15) was launched by Ariane in January 1989. Intelsat V has 15,000 telephone circuits and two color TV channels.

A new generation of larger Intelsat communications satellites went to space with the launch of Intelsat VI (F2) on an Ariane in 1989. An Intelsat VI contains 40,000 telephone circuits.

European Rockets: Ariane V-24

A European Space Agency Ariane 3 rocket carrying satellites for India and Europe blasted off with only one brief hitch July 21, 1988, from ESA's launch site in the jungle at Kourou, French Guiana.

Cloudy weather above the pad caused a six-minute delay in the countdown. At lift-off, brilliant orange flame from the Ariane 3 blossomed, then flickered and disappeared into low clouds moments after takeoff.

The Ariane 3 was carrying the ECS-5 communications satellite for Eutelsat, the European communications satellite organization, and Insat 1-C for India. Insat 1-C is a dual-purpose satellite to collect weather data as well as expand television and long-distance telephone service in India.

Seventeen minutes after takeoff the rocket attained orbit and began dropping off its payloads.

The evening liftoff from the northeast coast of South America was the 24th for Arianespace, the company responsible for commercial flights of European rockets. At the time, Arianespace had put 11 satellites in orbit in 11 months.

Workhorse. An Ariane 3 rocket can lift 5,900 lbs. to low orbit. At launch, the workhorse Ariane 3 had been used seven times since first launched in August 1984. ESA's bigger booster, Ariane 4, fired for the first time June 15, 1988, can lob 9,200 lbs. to space.

ECS-5. The 2,607-lb. ECS-5 satellite was meant to upgrade long distance telephone service across Europe—from Portugal to Finland and Turkey to Ireland—during a seven-year life in space. ECS stands for European Communications Satellite.

Eutelsat. ECS satellites are renamed Eutelsat in orbit. Launch of the first European Communications Satellite ECS-1 was June 16, 1983. ECS-1 became Eutelsat 1 F-1. The F-1 was short for flight one. In space, ECS-5 was renamed Eutelsat 1 F-5.

By the time of the ECS-5 launch in 1988, Eutelsat 1 was beyond its design life, but still working in space. Its transmitters had operated more than 400,000 hours.

Each Eutelsat has nine relay channels for television and radio transmissions to cable and satellite viewers across Europe. Transmitters have power output strong enough to permit reception on the ground with dish antennas less than three feet in diameter.

Eutelsat 1 F-5 was the final satellite in the Eutelsat 1 series. The satellites are managed by a 26-nation European Telecommunications Satellite Organization.

Insat 1-C. The 2,618-lb. Insat 1-C, expected to have a life of 10 years, was to expand television and long-distance telephone service while collecting meteorological data. Insat stands for India Satellite. Both satellites are kept in stationary orbit 22,500 miles above the equator.

The Indian satellite originally was to have been ferried to space in a U.S. shuttle. After the 1986 Challenger disaster, India switched to Ariane. Meanwhile, Eutelsat had reserved a place on an American Atlas Centaur rocket in case of an Ariane setback. U.S. space shuttles were grounded nearly three years after seven astronauts were killed January 28, 1986, during liftoff of shuttle Challenger.

Setback for India. Unfortunately, part of the electrical power system in India's Insat 1-C communications satellite broke down in space July 29, just a week after riding Ariane 3 to orbit.

Insat 1-C seemed to function normally until July 29, according to the Indian Space Research Organization (ISRO). Then a part in one of two solar-powered electricity-generating systems malfunctioned. Half of the satellite was connected to a power system that was operating properly while engineers struggled to fix its faulty counterpart.

Engineers from the satellite manufacturer, Ford Aerospace and Communications Corporation of Palo Alto, California, joined ISRO scientists working on the problem.

The power supply problem was the second space setback in 1988 for India. On July 13, ISRO's solid-fuel Augmented Satellite Launch Vehicle multistage rocket fell into the Bay of Bengal shortly after lift-off from the Sriharikota Space Center in southeast India.

Insat 1-C was the tenth Indian payload to reach orbit. Most had been launched by ESA, the USSR and the U.S. Before Insat 1-C, the last previous launch was India Remote Sensing (IRS) satellite fired to space March 17, 1988, from the USSR. India's first space rocket, the Satellite Launch Vehicle, carried three 80-lb. satellites to orbit between 1980 and 1983.

European Rockets: Ariane V-25

Europe's Ariane space booster completed another successful launch September 8, 1988, from the northeastern coast of South America, but one of two American satellites it ferried to low orbit failed to travel on correctly to higher orbit.

The Ariane 3 rocket lifted off in an evening launch from a European Space Agency pad at the edge of the jungle near Kourou, French Guiana. After 17 minutes of flight, the huge booster reached a proper low orbit 100 miles high and ejected both of the satellites into their own low orbits. From there, both were to fly on their own to stationary orbits 22,300 miles high.

The difficulty. The two large communications satellites, G-Star 3 and SBS-5, aboard the Ariane 3 rocket on flight V-25, were designed to expand and improve television, telephone and data communications across the 50 states.

Each satellite had a built-in rocket, or kick motor, to push it out from the low 100-mi. orbit to an altitude of 22,300 miles. The 2,799-lb. G-Star 3 had a kick motor failure and did not fly on to proper stationary orbit 22,300 miles above Earth.

G-Star 3 was to have completed GTE's network of three Spacenet satellites, all built by General Electric's Astro Space Division and launched on European Space Agency Ariane rockets.

The kick motor failed when GTE controllers tried to boost G-Star 3 on to stationary orbit. The satellite did reach 22,300-mile altitude and its kick motor did fire in an attempt to push it into a circular orbit. But the kick motor did not fire long enough, so the satellite moved into an elliptical egg-shaped orbit.

Engineers wasted valuable fuel firing small thrusters, normally used to keep the satellite from drifting out of position, trying to nudge G-Star 3 into a path where its speed matched the rotation of Earth. A similar maneuver was performed on a NASA TDRS tracking and data relay satellite in 1983. G-Star 3 cost $85 million to build and launch. It was insured for $60 million.

Good news. Flight V-25 was the 25th Ariane launch. The 2,735-lb. SBS-5, constructed by Hughes Aircraft Corp., flew to proper stationary orbit where it is expected to work 10 years. A similar satellite was launched in 1984. A third in the SBS series is under construction.

European Rockets: Ariane V-26

The European satellite-TV war heated up October 27, 1988, with a successful firing by the European Space Agency of an Ariane 2 rocket carrying the powerful French telecasting satellite TDF-1 to space from a jungle launch pad at Kourou, French Guiana.

Ariane flight V-26 rose smoothly into clear skies above the ESA pad on the northeastern coast of South America. It was the 13th successful blastoff in 13 months. Sixteen minutes later, the rocket reached its low transfer orbit and popped TDF-1 into space.

About nine minutes later, solar panels opened on the satellite, bringing cheers from technicians who recalled failure in 1987 of the West German TVsat-1, a sister to the French satellite. TVsat-1 solar panels failed to open. Both satellites were built under a 1977 French-German agreement.

TDF-1. The five-channel French satellite beams television shows directly to homes across Europe, from Portugal to the Ural Mountains of the Soviet Union and from Britain to North Africa. Previously, satellite signals were picked up only by large dish antennas owned by cable companies and networks which relayed programs to viewers.

TDF-1 will operate eight years from a stationary orbit 22,300 miles above the Atlantic Ocean, bringing viewers who have proper receiving equipment a better TV picture with stereo sound.

A first step toward high-definition television for Europe, TDF-1 was designed by the French broadcasting authority to increase the number of channels for 66 million French-speaking viewers in France, Belgium, Switzerland, Luxembourg and Monaco, and 400 million viewers in North Africa.

Times are changing. TDF-1 was a hot idea back in the 1970's when French-speaking viewers had only three stations to choose from. But, in the early 1980's, deregulation opened up European airwaves, doubling the number of French-language channels. After the successful TDF-1 launch, critics called the $250 million satellite outmoded even as a member of Europe's parliament said the satellite "turns a new page in the history of television in Europe."

Recent technology had put TDF-1 at a disadvantage against five competing satellites scheduled. At blastoff, four of the five TV channels in TDF-1 had not been rented. Only one channel had been leased—the French cultural network La Sept signed on for arts programming.

Only 50,000 satellite dishes existed in Europe at the time of TDF-1 launch. Also, equipment for receiving signals from TDF-1 is larger and more expensive than the competition's gear. TDF-2, a second French satellite, was launched in July 1990 to backup TDF-1.

Stiff competition. Newer technology has made smaller satellites and smaller receiving antennas. TDF-1 found competition quickly as the the medium-power 16-channel Astra 1-A satellite was launched by ESA on December 11, 1988, aboard an Ariane 4. Astra 1-A, started by Luxembourg in 1983, already had most of its channels rented when it was launched.

TDF-1 also faces competition from the satellites Eutelsat 2-A, Eutelsat 2-B and Eutelsat 2-C, owned by a group of European telecommunications governments. They were launched in 1990 and 1991.

To answer the competition, the French government decided to subsidize the cost of leasing TDF-1 channels. Instead of $20 million to $22 million per channel, the satellite will cost about the same as Astra 1-A, $6 million to $7 million per channel.

The money's there. Sponsors' wallets will be ready if satellite-to-home broadcasting becomes popular with viewers on the ground. In 1992, trade barriers will be dropped across Europe by international agreement. Satellites then could give advertisers access to 350 million people in one giant European audience.

Luxembourg is a mini-country, surrounded by Belgium, France and West Germany. Not to be topped by its powerful neighbor, the tiny European nation of Luxembourg thinks it has changed the TV picture for Western Europe with launch of its Astra 1-A satellite, ferried to space December 11 on an Ariane 4.

Astra is a project of Luxembourg's Societe Europeenne des Satellites. Millions of European viewers now need only small satellite dish antennas, the size of an umbrella, to receive TV broadcasts without paying for the service.

More channels. While TDF-1 has only five channels, Astra 1-A is the first satellite capable of beaming 16 channels directly to homes in the most densely populated areas of Europe. Programs are paid for by advertising.

By comparison, low-power satellite signals received in North American homes require large, expensive dishes. Such signals usually are received and distributed through cable networks which charge viewers. On the other hand, the signals beamed down by the higher-power Astra 1-A require only a small antenna costing about the same as a video recorder.

The French and West Germans started planning the TDF-1 style of satellites in 1977. In the early 1980's, Luxembourg saw broadcast monopolies of European governments beginning to fade with the advent of cable TV and the prospect of continent-wide satellite broadcasts. Luxembourg took a financial gamble to push through innovations, starting Astra in 1983. Astra is U.S.-made, manufactured by RCA.

European Rockets: Ariane V-27

The number of players in space, battling for the attention of 400 million European TV viewers, increased December 11, 1988, as the European Space Agency's most powerful rocket, the new Ariane 4, blasted off from South America on its first commercial flight.

The countdown for flight V-27 had been halted minutes before liftoff December 9 when a storm passed and a sensor broke on the rocket's third stage. Twenty-four hours later, a last-minute hitch developed when a meter reported a bad valve in the rocket third stage. The countdown was stopped at T minus 26 seconds. Technicians checked the valve and found it working. The countdown resumed after a 30-minute delay.

Carrying two satellites, the rocket blasted off successfully from launch pad ELA-2 at the European space complex in the jungle at Kourou, French Guiana on the Atlantic Coast of northern South America. Thirty minutes after blastoff, the rocket popped two satellites into low 100-mi.-high transfer orbits from where they later boosted themselves to 22,300-mi.-high stationary orbits.

Astra 1-A. One of the satellites was Astra 1-A, a 3,894-lb. control-room-in-space to broadcast TV directly to European homes. Astra 1-A, owned by a Luxembourg company, broadcasts 16 channels of TV programs to 2-ft.-diameter home dish antennas. Its signal covers 85 percent of West Europe homes.

Eleven channels started broadcasting immediately, including all-news, all-sports, all-movies and other programs. Luxembourg will reach 55 million homes directly and 25 million more through cable systems. A month before Astra 1-A launched, an Ariane had boosted a competing French direct-to-home TV-broadcasting satellite, TDF-1, to space.

One tiny state. Luxembourg is a European mini-country, surrounded by Belgium, France and West Germany. Astra 1-A is Luxembourg's medium-power 16-channel satellite. New technology has made possible smaller TV satellites and home receiving antennas. The tiny nation plans to change the TV picture for Western Europe as

Luxembourg's Astra TV satellites carry TV pictures directly into homes of viewers across Europe and North Africa. Hundreds of millions of European viewers will need only small satellite dish antennas, the size of an umbrella, to receive TV broadcasts without paying for the service.

Astra 1-A is the first capable of beaming 16 channels directly to homes in the most densely populated areas of Europe. Programs are paid for by advertising. By comparison, low-power satellite signals received in North American homes require large, expensive dishes. The U.S.-made Astra 1-A was manufactured by RCA.

The Astra satellite control center is in Betzdorf, 13 miles east of the capital of Luxembourg.

The satellite's markets, where viewers need only small antennas, are Britain, West Germany, France, Austria, Switzerland, northern Italy, Denmark, the Netherlands, Belgium and Luxembourg. Viewers in other nations will need larger dish antennas.

Skynet 4-B. The other satellite carried by Ariane 4 on December 11, 1988, was Skynet 4-B, a 2,800-lb. British military communications satellite, the first of four for submarines, surface ships and ground stations.

Ariane 4. Ariane 4 is an expendable, non-reusable, unmanned launcher with two liquid-fuel and two solid-fuel strap-on boosters. Ariane 4 made a non-commercial demonstration flight in June 1988, carrying three satellites to orbit.

Flight V-27 was the seventh launch in 1988. By December 1988, ESA had fired 16 satellites to orbit in 15 months using Ariane 2, Ariane 3 and Ariane 4 rockets.

Arianespace is a private French industrial, commercial company producing Ariane rockets for 13 ESA countries. Dozens of European companies manufacture and assemble Ariane parts, giving each an economic benefit matching its investment share in the cost of research and development. Arianespace, selling launches and handling firings from the Guiana Space Center, had a launch backlog of three dozen satellites.

European Rockets: Ariane V-28

An Ariane 2 rocket rose into the night sky January 26, 1989, carrying the international communications satellite Intelsat V (F-15) to orbit above the European Space Agency launch pad at Kourou, French Guiana on the Atlantic coast of South America.

The late evening blast-off of flight V-28 followed a 53-minute hold triggered by an automatic warning light which reported a malfunctioning hydrogen reservoir valve and stopped the countdown at T minus 18 seconds. Technicians traced the problem quickly, determined the valve was okay and the indicator was faulty, and flashed thumbs up for blast-off.

Geosynchronous. The Ariane 2 rocket ferried the 4,400-lb. satellite to a 100-mi.-high orbit. The satellite boosted itself from there to a stationary orbit 22,300 miles above Earth. Ground controllers adjusted its position to spot it over the Indian Ocean.

Intelsat V (F-15), like 13 other series-5 Intelsat satellites in orbit, has 36 transponders to carry 15,000 telephone circuits and two channels of color television programs.

The satellite network, operated by a non-profit cooperative of 115 countries, the International Telecommunications Satellite Organization (ITSO), carries most trans-oceanic television transmissions and more than half of all international telephone calls.

Last of its kind. Intelsat V (F-15) was the last of the series-5 Intelsat communications satellites. Intelsat V (F-14) was launched on Ariane in May 1986. Intelsat V (F-13) was launched by Ariane in May 1988. (The F stands for flight.)

A new generation of larger Intelsat VI communications satellites went to space with the launch of Intelsat VI (F-1) on an Ariane rocket in October 1989. Intelsat VI (F5) was launched in August 1991 and Intelsat VI (F2) was launched in October 1991. Intelsat VI contains 40,000 telephone circuits.

Ariane rockets have been developed by the European Space Agency (ESA). The last previous Ariane launch was flight V-27 in December, the seventh launch in 1988.

The January 26 blast-off of flight V-28 was the first launch of 1989 for ESA and Arianespace, Europe's commercial space transportation company, the consortium of 50 European companies responsible for commercial exploitation of Europe's Ariane rockets.

European Rockets: Ariane V-29

The 29th Ariane rocket lifted off March 6, 1989, from European Space Agency's South American launch center at Kourou, French Guiana, ferrying Japanese and European satellites to orbit.

Blast off was to have been February 28, but was stopped when technicians went on strike. After the strike was settled, launch was reset for March 4, but had to be called off when high winds unplugged two umbilical cables, ventilating a weather satellite and purging oxygen reservoirs, as the gantry support tower was being pulled away from the side of the rocket three hours before launch. A new purge connector was flown to the South American jungle launch pad by jet from France and launch was rescheduled for March 6.

Flight V-29 finally got off the ground March 6. During 17 minutes 41 seconds of flight, the satellites were carried to a 100-mi.-high orbit. After three minutes in orbit, the Ariane 4 nosecone opened and JCsat-1, the first of its payload of two satellites, popped out. Four minutes later, Meteosat MOP-1, the second satellite, popped from the nosecone into its own orbit. From there, liquid-fuel rockets attached to the satellites pushed them higher to stationary orbit 22,300 miles above Earth.

The strike. Some 230 workers of the French company Thomson CSF at the Kourou space center went on strike February 26 for a $200-a-month pay raise, more insurance and improved social benefits. It was the first time a worker strike had led to postponement of an Ariane flight.

Many of the striking Thomson employees were technicians manning radar tracking gear, operating computers and handling other important launch jobs.

The satellites. JCsat-1 is a Japanese commercial communications satellite and Meteosat MOP-1, a European weather satellite.

The American-made JCsat-1 was built by Hughes Aircraft Co., Los Angeles, for the Japanese Communication Satellite Co. Japan has other communications satellites in orbit, but JCsat-1 was the Asian nation's first commercial communications satellite.

JCsat-1 will be above the equator at 150 degrees west longitude. The satellite's speed, as it whisks around the globe, will be the same as the speed at which Earth rotates. Thus, it will seem to hang stationary in the sky, providing uninterrupted communications relay service.

HS-393. JCsat-1 was the first of a new generation of radio relay satellites from Hughes known as HS-393. The spin-stabilized HS-393 satellite is equipped with 32 high-frequency KU-band transponders, each capable of carrying one high-quality television channel, a high-speed data channel or 250 telephone calls.

JCsat-2, another HS-393, was the second built by Hughes Aircraft Co. for JCSAT of Japan. It was launched from the Cape December 31, 1989, on the maiden flight of Commercial Titan, a Titan 3 built by Martin Marietta Astronautics Group, Denver.

Meteosat. MOP-1, ESA's fourth meteorological satellite, sends weather pictures to ground stations every half hour. MOP-1 was the first of three upgraded meteorological satellites operated by ESA's 16-nation organization, Eumetsat.

The 695-lb. MOP-1 will be over the equator at zero degrees longitude. The weather satellites MOP-1 and MOP-2—and MOP-3 to be launched in 1993—provide television pictures for weather forecasting like the satellite pictures seen on American TV.

On a roll. Flight V-29 was the ESA's second 1989 launch, the 11th straight Ariane success. The last previous launch was January 26, when an Ariane 2 put a communication satellite into orbit.

The March 6 launch was the third flight for the 195-ft. Ariane 4, Europe's most powerful space booster. With four strap-on boosters for extra boost, Ariane 4 is one of a family of boosters marketed by ESA's Arianespace in competition with U.S., Russian and Chinese rocket makers.

Arianespace has contracts to launch more than two dozen satellites, including several owned by American companies. An Ariane 4 rocket can carry payloads ranging from 4,190 lbs. to 9,260 lbs. to orbit. The U.S. Titan 4 unmanned rocket and the manned shuttle fleet can carry heavier loads.

The first Ariane rocket was launched December 24, 1979. Since then, the agency suffered only five flight failures in 47 launches. The 1986 failure resulted in an 18-month delay in the launch schedule. Blast-offs restarted in September 1987. ESA in 1989 ordered 50 more Ariane 4 rockets to be delivered over eight years.

European Rockets: Ariane V-30

A European Space Agency Ariane 2 rocket carried the Scandinavian communications satellite Tele-X to orbit April 1, 1989, from Kourou, French Guiana.

ESA flight V-30, the 30th Ariane launch and the final use of the older Ariane 2 launcher, lofted the telephone and television satellite built by Sweden, Norway, Denmark and Finland.

Scandinavian squabble. The project was started by the four nations in 1979. Nordic Satellite Co. was set up by the four to own the satellite. Some $228 million was spent.

Squabbling over programming led Denmark to pull out, leaving Sweden with 85 percent of Nordic Satellite Co. and Norway with 15 percent. Finland, in a separate agreement with Sweden, controlled a token three percent of Sweden's interest.

On the eve of launch, Norway backed out, giving its 15 percent share in Nordic Satellite Co. to Sweden. In exchange, Norway got control of one of three television channels in the satellite.

The last-minute deal gave Sweden 100 percent of Nordic Satellite Co. Finland retained its token agreement with Sweden.

Unfortunately, the satellite will remain unused in space while the Scandinavians reach agreement on a joint television program to be broadcast to northern European homes via the satellite.

The launch. Blast-off of the Ariane 2 went smoothly 24 hours after a helium leak in ground equipment and problems with the guidance system had led to postponement. With only 7 minutes to go in the first try, final countdown for the 162-ft. Ariane 2 had been halted. The older Ariane 2 was small compared with powerful Ariane 4 rockets fired recently from Kourou.

It was European Space Agency's third flight of seven 1989 flights from its jungle launch complex on the northeast coast of South America. The flight was completed only 26 days after a successful launch of an Ariane 4 March 6. The Ariane record went to 26 successes and four failures. Flight V-30 was the 12th good shot since the last failure in May 1986. After the launch, Arianespace had $2.2 billion in contracts to launch 34 satellites, five owned by American companies.

The satellite. Solar-powered Tele-X, owned by Nordic Satellite Co. and operated by Swedish Space Corporation, was built to provide direct-to-home television broadcasts to Sweden, Norway, Denmark and Finland. Built by the French firm Aerospatiale, it also was designed to relay telephone calls and data for businesses.

The 2,800-lb. Tele-X was ejected from the Ariane 2 nosecone into a 100-mi.-high elliptical transfer orbit 19 minutes after blastoff. Later, three firings of a small "kick motor" rocket attached to the satellite would push it into a circular orbit. Eventually, the satellite would be in a stationary orbit at 22,300 miles altitude.

A satellite's orbiting speed depends on its altitude. At 22,300 miles altitude, it takes 24 hours to go around Earth one time. That makes the satellite seem to be stationary at one point in the sky. TV satellites usually are in such geosynchronous orbits.

European Rockets: Ariane V-31

An upgraded Ariane 4, the most powerful rocket Europe has ever built, blasted off June 5, 1989, from European Space Agency's site at Kourou, French Guiana. It ferried to orbit two expensive communications satellites—Superbird-A, a $100 million Japanese communications satellite, and DFS Kopernikus-1, owned by the West German post office to provide television communications service for West Germany.

Known as flight V-31, it was the fourth Ariane 4 blast off, but the first using four strap-on liquid-fuel boosters. The satellites had a combined weight of 8,613 lbs. Climbing 25 minutes from the jungle launch complex on the northern coast of South America, the 192-ft. three-stage rocket successfully placed the pair of satellites in temporary elliptical orbits, ranging from 115 miles to 22,349 miles altitude, as the spent final stage soared over Zaire, Tanzania and central Africa. Later firings of on-board rockets pushed the satellites into circular orbits 22,300 miles over the equator.

Superbird-A, built in America by Ford Aerospace Corp., is owned by the Space Communications Corp., a Japanese common carrier formed by Mitsubishi Corp., Mitsubishi Electric Corp. and 27 other Mitsubishi Group companies. The 5,482-lb. solar-powered satellite is the first of two for telephone, television and data communications for the Japanese islands and Okinawa. Superbird-B failed to launch in February 1990 and was rescheduled for January 1992.

DFS Kopernikus-1, a 3,115-lb. radio relay built in Europe by Consortium R-DFS for Deutsche Bundepost, provides coverage of West Germany and West Berlin.

There had been 27 Ariane successes in 31 launches since 1979, 13 in a row since a 1986 failure. The most recent previous flight was April 1 when the Scandinavian satellite Tele-X was launched on a less-powerful Ariane 2.

The June 5 launch originally was to have lifted off May 26, but was delayed as engineers fixed pressure in the rocket's third-stage liquid-hydrogen fuel tank. The problem was in insulation which let room-temperature helium come in contact with the super-cold liquid-hydrogen tank's forward bulkhead, raising the hydrogen temperature and thus the pressure in the tank.

V-31 was the fourth of seven Ariane flights in 1989. The next was V-32 on July 12. It was an Ariane 3 lofting the Olympus 1 communications satellite to space. The satellite, once known as L-Sat, had a number of experimental communications channels plus two direct-to-home TV transmitters, one used by the Italian Television Agency RAI.

European Rockets: Ariane V-32

The Olympus 1 experimental high-power communications satellite bounded to space July 12, 1989, aboard Europe's last Ariane 3 rocket. The 161-ft. liquid-fuel rocket, with a pair of solid-fuel boosters strapped on, was 11 days late leaving launch complex ELA 1 at Kourou, French Guiana. Even so, it rushed flawlessly to Earth orbit, chalking up the fourteenth straight success for the European Space Agency.

The satellite. Olympus 1, built by British Aerospace, portends a new generation of large European communications satellites in Earth orbit. The 6,900-lb. sophisticated radio relay station will test state-of-the-art communications technology from a position over the equator above central Africa at 19 degrees west longitude. From that vantage point, Olympus 1 will look down on most of Europe.

A joint venture of eight countries—Austria, Belgium, Canada, Denmark, Italy, the Netherlands, Spain and the United Kingdom—Olympus 1 has four main communications channels:

★Two direct-to-home television broadcasting channels, one covering southern Europe dedicated to use in Italy by the broadcaster RAI and the other to be used by the British Broadcasting Corporation (BBC) for coverage of northern Europe. There will be some so-called "distant learning experiments" during off-peak hours on the northern space-to-ground television beam;

★ A third channel will carry special communications for various business applications using small receiving stations on the ground. Users of this service will include postal telephone and telegraph (PTT) administrations of governments of various European countries as well as universities and science and technology centers. The satellite will link computer networks for data transmission, demonstrate high-speed data networking among several countries and provide experimental business communications among small Earth stations;

★ A fourth channel will be used for advanced communications experiments at extraordinarily high frequencies not previously used. Applications for use at the end of the century will be tested, including adding thousands of telephone lines where hundreds have been possible in the past. The satellite will relay data to and from Europe's Eureca science satellite which was scheduled to be launched by U.S. space shuttle in 1992. Before that, Olympus was to be used by more than 150 experimenters.

Canada is able to take part in the experiments since Olympus has moveable antennas for what is called "steerable beams."

The launch. The flight, designated V-32, was delayed a week when two mechanical arms carrying liquid hydrogen and liquid oxygen fuel to the Ariane rocket did not pull back properly and the countdown had to be stopped four seconds before ignition July 5.

On July 12, the arms as well as the Ariane 3 rocket's three stages worked well. Twenty minutes after liftoff from launch pad ELA 1, Olympus 1 was dropped off in space in an elliptical "transfer" orbit ranging from 124 miles to 22,420 miles altitude.

About 36 hours later, a small booster rocket built into the satellite pushed it, or transferred it, to a circular stationary orbit 22,300 miles high. In such a stationary or geosynchronous orbit, the satellite circles Earth at a speed that keeps it over a single point on the equator. It seems to hang stationary in the sky as Earth turns, eliminating the need for expensive moveable or steerable receiving antennas on the ground. Olympus is positioned above the equator at 19 degrees west over Africa.

Flight V-32 was the last time Kourou's ELA 1 launch complex will be used. A total of 25 launches have left from ELA 1 since Ariane flight V-1 in December 1979. The launch pad has been superseded by the more modern and flexible ELA 2, first used in March 1986.

The successes. The V-32 blast off had been scheduled for June 30, but a launch pad fuel line feeding liquid hydrogen to the rocket's third stage failed to pull back from the vehicle four seconds before lift off.

The July 12 firing was the fifth of seven Ariane launches in 1989. The last previous flight had been June 5 when a two communications satellites were ferried to orbit by an Ariane 4. Ariane's record after V-32 stood at 28 successes in 32 launches since 1979. The next Ariane launch, flight V-33, carried Hipparcos, an astronomy satellite, and TVsat-2, a West German direct-to-home TV broadcasting satellite, to orbit on August 8.

The Arianes. Ariane 3 and Ariane 4 are enhanced versions of the European Space Agency's original Ariane 1 rocket.

★Ariane 1, able to carry 4,080-lb. payloads to Earth orbit, was used from 1979 to 1986.

★Ariane 2, carrying 4,850 lbs, was used from 1984 to 1989.

★Ariane 3, carrying 5,950 lbs. to orbit, was used 1984-1989.

★Ariane 4, the most powerful European rocket to date, carrying 9,260 lbs., has been in use since 1988 and will continue in service until about 1999. It compares with America's Titan rocket.

★Ariane 5, to carry 42,000 lbs., will go into service about 1995 and may be used until the year 2010 or later.

European Rockets: Ariane V-33

An unmanned Ariane 4 rocket blasted off August 8, 1989, from a European Space Agency launch pad at Kourou, French Guiana, ferrying the Hipparcos astronomy satellite and the TVsat 2 television broadcasting satellite to Earth orbit.

It was the 33rd Ariane flight. The TVsat-2 communications satellite was designed to modernize West German television. Hipparcos was set to expand man's knowledge of the stars. The launch to low Earth orbit by the 194-ft., three-stage, Ariane 4 was successful, but then Hipparcos failed in orbit. Two days after launch, on August 10, an on-board kick motor designed to push Europe's extraordinary star-cataloging satellite to a stationary orbit over Africa failed, stranding Hipparcos in an elliptical orbit.

The launch. Engineers at the Guiana Space Center on the northern coast of South America had to launch the rocket by 7:32 p.m. August 8 or Earth would move out of proper position for the satellites. Missing the deadline would have necessitated postponing launch to another day.

The countdown had reached seven seconds before ignition when a fail-safe computer stopped the blast off. A glitch in a mechanical-controlling computer had prevented a fuel line attached to the rocket from retracting. Technicians scrambled to fix the problem before 7:32 p.m. Countdown was re-set to six minutes.

After a delay of 39 minutes, the Ariane lifted off with just seven minutes left in the launch window. The muggy tropical scene was called "lovely" as blast off of Europe's most powerful launcher lit up the nighttime sky.

Just 10 hours after the U.S. shuttle Columbia had lifted off from Florida on a secret military mission, Ariane flight V-33 streaked away from launch complex ELA 2. The liquid-fuel Ariane blasted skyward for 21 minutes, ferrying TVsat-2 and Hipparcos to a low orbit.

Transfer orbit. TVsat-2 was ejected first from the Ariane nosecone into an elliptical "transfer" orbit ranging from 125 miles to 22,304 miles altitude. Two minutes later, Hipparcos was dropped off in a similar orbit. Some 37 hours later, small solid-fuel rockets, known as kick motors or apogee boost motors, attached to the two satellites were to be fired to push them into individual circular orbits 22,300 miles above Earth's equator. In such a stationary or geosynchronous orbit, the satellites speed around Earth at a velocity which makes each seem to stay over one point on the equator. They appear to hang stationary in the sky as Earth turns, eliminating the need for moveable receiving antennas on the ground.

The problem. The kick motor aboard TVsat-2 apparently worked, while the similar rocket in Hipparcos did not. From August 10 to 17, repeated radio commands failed to induce the Hipparcos kick motor to ignite as the satellite sailed through the high point of its orbit. ESA engineers, who didn't know why the engine wouldn't turn on, said the deep-space star-mapping satellite was in a safe orbit, meaning it would stay in space for a time. It was not clear whether the satellite could carry out any of its precision astronomy functions in an elliptical orbit.

Hipparcos. Equipped with a reflecting telescope, Hipparcos was designed to measure accurately the positions of 120,000 stars. Astronomers would use this so-called astrometry information to calculate distances to the stars and create highly-detailed and precise new star maps by 1995.

Finding the distance to a star is an astrometric measurement. Astronomers the find distance from Earth to a star by measuring the star's position position from each side of Earth's orbit. The star has an apparent motion. It seems to move slightly—like an object held at arm's length seems to jump back and forth as you look with one eye and then the other.

The diameter of Earth's orbit around the Sun is 184 million miles. The apparent motion of a star is small, but sufficient for astronomers to sketch a triangle with a 184 million mile base. Trigonometry is used to calculate the length of the other two sides of the triangle.

From orbit above Earth's murky atmosphere, the 2,491-lb. solar-powered Hipparcos, built by Matra for the European Space Agency (ESA), could improve astrometric measurements as it spins slowly, sweeping the whole sky every two hours.

Hipparcos was supposed to be positioned over the equator above Africa at 12 degrees west longitude, but its kick motor wouldn't work. After considerable study, European engineers decided nearly all of Hipparcos' mission could be accomplished from the long elliptical orbit in which it was stranded. Small control thrusters were fired to push the satellite into a slightly higher orbit where it would work for a few years. Solar panels were unfolded and the satellite was put to work.

Long-term project. Hipparcos is Europe's first astrometry satellite. It was proposed in the 1960's and approved by ESA in 1980. Construction began in 1984. Hipparcos scientists will draw up a celestial catalog, using the satellite to map 120,000 stars with an accuracy of up to 0.002 seconds of arc. That precision is more than 10,000 times better the naked eye.

As it spins, Hipparcos will sample five or six stars at a time. During its 2.5 year life, Hipparcos will look at the entire celestial sphere, transmitting 24,000 bits of data each second to researchers on the ground at ESA's European Space Operations Centre (ESOC) at Darmstadt, near Frankfurt, in West Germany. Data from the satellite will be dispatched across Europe to two separate groups of scientists who will double-check each other and coordinate mapping.

An improved full-sky catalog of star positions will be available in the 1990's. By 1995, the complete new Hipparcos Star Catalog should be available. The unprecedented accuracy of that catalog should have a major effect on astronomy.

For instance, Hipparcos measurements will provide a basic yardstick for determining the size of the Universe. It will create new basic knowledge of the positions and motion in space of many tens of thousands of stars, leading to a deeper understanding of the evolution of stars and even our own Milky Way galaxy.

TVsat-2. The other satellite in the Ariane payload—West Germany's TVsat-2, replaced TVsat-1 which failed in orbit shortly after an Ariane launch in November 1987. TVsat-1 was rendered useless when one of its solar electricity-generating panels failed to unfold in space following launch.

TVsat-2 is positioned over the equator above central Africa at 19 degrees west. From there, the television relay station will be look down on most of Europe. The 4,696-lb. TVsat-2, built by Eurosatellite GmbH, has five high-power direct-to-home TV broadcast channels operated by the West German Bundespost telecommunications authority.

TVsat-2 will make more TV channels available to West German viewers. It also will be able to telecast the so-called D2-MAC compatible high-definition TV coming to Europe in the 1990's.

Success string. ESA seemed to have better luck with rockets than with satellites. While some satellites such as TVsat-1 and Hipparcos fail in orbit, the Ariane family of rockets had been enjoying a string of successes.

After flight V-33, ESA had launched 24 satellites in 23 months with 15 Ariane launches. Flight V-33 was the fifteenth success in a row, the twenty ninth success in 33 missions since 1979. V-33 was the sixth Ariane launch of 1989. The last previous Ariane flight, designated V-32, came just 27 days before V-33 on July 12, 1989, when a less-powerful Ariane 3 ferried the experimental Olympus 1 communications satellite to space.

Ariane 4 is the most powerful launcher in the European family of rockets. The first Ariane 4 flew June 15, 1988. Ariane 4 rockets have six different configurations, tailored to individual payloads, capable of carrying 4,190 to 9,260 lbs. to orbit. The August 8 launch carried 7,186 lbs. Ariane 4 is comparable to a U.S. Titan rocket.

ESA represents eleven European countries. Arianespace, the launch-sales arm of ESA, plans monthly Ariane 4 launches, half for European clients and half from outside Europe. After V-33, Arianespace still had 33 satellites booked for launches worth $21.4 billion.

European Rockets: Ariane V-34

The largest commercial communications satellite ever orbited, capable of handling 120,000 telephone calls at once, was blasted to space October 27,1989, on a European Space Agency Ariane 4 rocket from Kourou, French Guiana, South America.

The $131 million, 5,645-lb. radio relay station, known as Intelsat VI (F2), owned by the International Telecommunications Satellite Organization, was manufactured by nine international companies under the direction of the Hughes Aircraft Co., El Segundo, California. F2 is the first of a new generation of five Intelsat VI satellites planned by Washington-based Intelsat, a nonprofit cooperative of 117 nations. Intelsat, with a network of 13 satellites in orbit, paid Hughes, a unit of General Motors, $785 million for five Intelsat VI satellites and associated equipment. The Intelsat system currently handles most global telephone traffic and most worldwide television transmissions.

World's largest. Intelsat VI, tall as a four-story building in orbit, measures 39 feet by 12 feet, with 24,000 simultaneous two-way telephone circuits and three color TV channels, will relay telephone, television, fax, telex and data.

Using state-of-the-art digital switching techniques, Intelsat VI can handle 120,000 telephone calls through its 24,000 circuits—triple the capacity of the latest trans-Atlantic fiber optic telephone cable. The capacity of an Intelsat VI satellite is double that of an Intelsat V satellite. One Intelsat VI can handle 500 times as much traffic as Intelsat I, the historic Early Bird satellite of the 1960's.

Spinning constantly in orbit for gyroscopic stability, the cylindrical drum-shaped Intelsat VI has 38 C-band radio transponders and 10 K-band transponders. Six antennas can cover an entire hemisphere or target specific regions.

The solar-powered Intelsat 6 (F2) will be in a stationary orbit, positioned over the Atlantic Ocean at 335.5 degrees East longitude to provide communications between the eastern United States, South America, Europe and parts of Africa.

Launch. The blast off was the 34th ESA Ariane launch since the first in 1979. It was the sixth Ariane 4, comparable in power to America's heavy-lift Titan rocket. Before the October 27 launch, ESA had four Ariane failures and 29 successes, including 15 successes in a row since a third-stage failure destroyed an Intelsat V satellite May 31, 1986.

The 192-ft. four-stage Ariane had been scheduled to lift off on flight V-34 October 5, but problems turned up in electrical components of the command system.

Streaking upwards from launch pad ELA-2 at ESA's jungle complex on the northern coast of South America October 27, the Ariane 4 rocket needed its four liquid-fuel strap-on boosters to push the heavy satellite to an elliptical orbit from 125 to 22,350 miles.

Clarke Belt. A small rocket attached to the satellite, known as an apogee kick motor, boosted the satellite into a circular stationary orbit at 22,350 miles altitude.

Revolving around the globe at the same speed as Earth rotates, Intelsat VI (F2) will appear stationary.

The region of space girdling the globe over the equator at an altitude of about 22,300 miles sometimes is referred to as the Clarke Belt, after science-fiction writer Arthur C. Clarke who first proposed placing communications satellites in such stationary orbits back in 1945.

ESA and Intelsat. After on orbit tests, Intelsat placed the satellite in commercial operation over the Atlantic Ocean.

ESA had launched five earlier Intelsat V communications satellites, the last in January 1989. Five more Intelsat VI satellites are scheduled for launch on Ariane 4 rockets in the next few years. Intelsat VI (F5) was launched in August 1991 and Intelsat VI (F1) in October 1991.

ESA's launch company, Arianespace, a consortium of 11 European countries, has 32 satellites from various nations and companies booked for launches in the 1990's. Six were launched in 1990.

Intelsat VI (F2) is the first of five built under a $700 million contract awarded in 1982—the largest contract ever for a commercial satellite program. In 1986, Intelsat gave Hughes an $85 million contract to increase the transmitter power of several channels on the Intelsat VI satellites. The satellites are to work 13 years in space.

European Rockets: Ariane V-35

Six amateur radio communications satellites were launched successfully to Earth orbit January 22, 1990, on one European Space Agency Ariane 4 rocket.

Flight V-35 from Kourou, French Guiana, carried OSCAR 14, OSCAR 15, OSCAR 16, OSCAR 17, OSCAR 18 and OSCAR 19, as well as France's Earth-monitoring photography satellite Spot-2.

Launch of the Ariane 4 had been delayed several times, needing a gimbal inertial platform fixed in its guidance system and replacement of a magnetic recorder used by Spot-2 to store photographs in orbit.

Microsats. The six amateur radio satellites, described as microsats, shared a new style of compact satellite framework. Some were as light as 27 lbs.

All six were in about the same polar orbit. They passed over the Northern Hemisphere two or three times each evening between 8 p.m. and midnight, then two or three times each morning between 9 a.m. and Noon.

Amateur radio operators on the ground used the satellites to relay various kinds of voice and digital data communications around the globe. OSCAR stands for Orbital Satellite Carrying Amateur Radio and is not the same as the U.S. Navy series of Oscar navigation satellites.

Spot 2. Launched to sun-synchronous orbit with the six microsats was Spot-2, France's second photography satellite. It was a commercial observation satellite like Spot 1 which had made some of the most detailed snapshots of secret Warsaw Pact military installations available to the public.

Spot, for Systeme Probatoire d'Observation de la Terre, was conceived and designed by the French space agency Centre National d'Etudes Spatiales (CNES). France operates Spot satellites with Belgium and Sweden. Spot customers can order pictures of any place on Earth's surface. Spot 1 had been launched to a 500-mi.-high orbit February 22, 1986. Both Spot satellites revolved around Earth every 101 minutes.

Tuning in. Radio signals from some of the OSCAR satellites are very easy for a novice to hear. OSCAR 17 has been especially easy to hear, transmitting to Earth on a frequency of 145.825 MHz. Listeners can use readily-available radios to receive OSCAR 17 as it passes overhead. The satellite has a relatively-high transmitter power of two watts. In some modes, OSCAR 17 transmits two minutes, then stops 30 seconds.

To hear OSCAR 17, a listener needs a simple receiver, such as a police scanner or monitor or ham two-meter rig, with an antenna. An outdoor antenna works better than one indoors, although there have been numerous reports of signals received from OSCAR 17 using the typical flexible "rubber duck" style of antenna on a handheld portable ham receiver, or the typical pull-up whip antenna which come with many scanners.

Sometimes OSCAR 17 transmits messages in synthesized voice. At other times, it sends bursts of data signals.

European Rockets: Ariane V-36

A cloth stuck in a pipe grounded Europe's space program for months when it caused an Ariane 4 rocket carrying two Japanese communications satellites to loose power and explode during lift-off February 22, 1990.

The 192-ft. three-stage rocket was model 44L, the most powerful Ariane 4 version with four Viking 5 first-stage rocket engines at its core with four Viking 6 boosters strapped on. As the launch began at 8:17 p.m. local time, combustion pressure dropped in one of the four core rockets forcing the other three to compensate. With one engine giving only half power, a guidance computer swiveled two first-stage nozzles to adjust for unbalanced thrust. It worked only briefly as the Ariane veered close to the gantry tower. Drifting sideways, the rocket came within seven feet of the tower as it blasted up and away from the pad. Exhaust flames scorched the top of the tower as flight V-36 rose into the sky.

After blasting away for 90 seconds, the three good core rockets weren't able to provide sufficient compensation. By the time the rocket had flown eight miles from the pad to an altitude of almost six miles, it had tipped over so far it was starting to fly sideways. The flight control system couldn't compensate for the unbalanced thrust. Controllers were stunned as they realized the structure couldn't withstand such powerful force in the wrong places and they watched the Ariane 4 explode. The vehicle was torn apart by excessive aerodynamic pressure 95 seconds into flight. A safety engineer radioed a command to the rocket to explode anything not already blown up.

The $430 million fireball created a three-mile-thick cloud of dangerous gas drifting some six miles above the European Space Agency's launch site and nearby Kourou, French Guiana. The 12,000 residents of that South American town and ESA workers fled inside as the noxious chemical cloud drifted for three hours to a distance of 50 miles out over the Atlantic Ocean. Small pieces of metal and insulation showered Devil's Island, a former prison camp.

The next day, French military helicopters searched for wreckage. Rocket parts found in the dense jungle and impenetrable mangrove swamps along the coastline had to be winched up to the helicopters and airlifted out. Pieces of the two Japanese satellites washed ashore from the shallow Atlantic waters off the coast onto beaches.

Past failures. ESA has developed ever more powerful rockets since Ariane 1 flew for the first time in December 1979. Arianes are launched from Kourou. Subsequent models were known as Ariane 2, Ariane 3 and Ariane 4. An even more powerful Ariane 5 is being designed to fly around 1995. Ariane blast offs had seemed routine with successful flights about once a month. Flight V-36 was the fifth loss since ESA started launching Ariane rockets December 24, 1979. That success rate had been 86 percent.

Two satellites were destroyed in the V-36 explosion. At $430 million, including the satellites and the rocket, it was the worst commercial rocket industry catastrophe since the 1986 Challenger disaster.

V-36 satellites. Space Communications Corp. (SCC) of Japan lost a 5,500-lb.

solar-powered Superbird B communications satellite. Nippon Hoso Kyokai (NHK) lost its 2,750-lb. BS-2x direct-to-home broadcasting satellite (dbs).

BS-2x, with three K-band transponders to serve the Japanese archipelago, was built by General Electric's Astro Space Division of Princeton, New Jersey, for $100 million. BS-2x was a backup for the state television network NHK's satellite channels. NHK operates two satellite channels broadcasting directly to dish antennas on home roofs and balconies.

Superbird B was second in a series of high-power Japanese communications satellites. Superbird B, with 31 radio transponders, was to have worked 10 years in space providing telephone, telegraph and telex communications. Superbird A was launched on Ariane 4 flight V-31 on June 5, 1989. Superbird A and Superbird B were built by Ford Aerospace Corp., Palo Alto, California, for $600 million.

SCC part of Japan's industrial giant Mitsubishi, wanted the second satellite in orbit to expand coverage of the Japanese islands. TV crews transmit news coverage to studios via satellite. The loss may delay expansion for three years of Japan-wide satellite news-gathering services to television stations. SCC's main rival, Japan Communications Satellites, part of the industrial firm C. Itoh, already is in operation. Superbird-B was rescheduled for flight V-49 in January 1992.

There are six versions of an Ariane 4 rocket, each with a different arrangement of strap-on boosters to provide extra power. The V-36 rocket had four liquid-fuel boosters, each a single Viking 6. Total thrust during first-stage flight was to have been 1.2 million pounds.

A commission of enquiry said in April 1990 the February 22 launch failed because a rag blocked a water pipe to one of the engines. Various parts for the water cooling system for that rocket came from the Aerospatiale Mureaux factory near Paris and the Societe Europeenne de Propulsion at Vernon west of Paris. Some persons supposedly are under surveillance at those plants, according to the magazine.

The magazine alleged that a leading space inspector, someone named Jean Gruau, wrote a report in 1982 addressed to four top bodies, including the president of France. The magazine described the report as alleging the unsuccessful launch of an Ariane rocket, on May 23, 1980, could have been caused by sabotage, for instance "a minuscule capsule of explosive...introduced into the combustion chamber" of one of the rocket motors. None of these extraordinary claims have been substantiated.

Fine business. When the European Space Agency's Ariane 4 flight V-36 blew up February 22 during lift-off, observers thought the setback might affect orders for more launches. But shortly after the disaster, the ESA marketing arm Arianespace signed three new satellite launch contracts. ESA restarted flights in July 1990.

Galaxy. An Ariane 4 rocket launched two new U.S. communications satellites, for Hughes Communications Inc.—Galaxy VI in 1990 and Galaxy VII scheduled for 1992.

The contract with the American satellite builder was announced just a week after the February explosion. The agreement included an option for launch of two more of the same satellites. Hughes calls itself the world's largest private satellite operator.

Galaxy VI and Galaxy VII, each weighing 5,665 lbs., were from Hughes' HS-601 series of communications satellites, expected to operate 12 years in Earth orbit, relaying television and private telephones and data.

Telecom. An Ariane 4 was contracted to carry a second-generation telecommunications satellite, Telecom 2-B, for France. The 4,840-lb. satellite would relay civilian telephone, television and data signals as well as military communications. The Telecom 2-B contract was the 80th commercial agreement for ESA in 10 years, third since the February failure. It was scheduled for March 1992.

Inquiry. An ESA accident inquiry board found the V-36 explosion apparently was caused by a piece of cloth lodged in a water pipe. The obstruction in a water line cut the flow to a pump in the fourth first-stage rocket known as motor D.

In an apparently unrelated failure in the same launch, at 2.4 seconds after firing of the strapped-on Viking 6 liquid-fuel booster rockets, a fuel leak caused a fire. ESA continued to manufacture Ariane 4 rockets during the inquiry.

The inquiry board blamed the explosion on a rag mislaid in the cooling system, concluding Ariane 4 launchers did not need to be redesigned. The panel did make nine major and 35 minor recommendations for future flights, including reinforcement of procedures and verifications. The board suggested double-checking the air-tightness of fuel lines in the first and second stages of the liquid-fuel strap-on boosters. The next flight would be V-37 on July 24, 1990.

European Rockets: Ariane V-37

After the V-36 explosion, Ariane 4 flights resumed with V-37 blasting off from ESA's South American spaceport at Kourou, French Guiana, in a colorful evening launch on July 24, 1990.

A threatened storm didn't materialize and, by twenty-five minutes after blast off, the Ariane rocket had dropped the two satellites off in space.

It was the 37th flight for the European Space Agency's series of Ariane space rockets. The 190-ft.-tall Ariane 4 rocket had four boosters strapped on, fueled by liquid oxygen and liquid hydrogen. It ferried the French TDF-2 and West German DFS-2 Kopernikus communications satellites to a low geostationary transfer orbit (GTO). Later the satellites boosted themselves on out to stationary orbit.

DFS-2 Kopernikus, owned by Deutsche Bundespost Telekom, was Germany's second 11-channel commercial communications satellite. DFS-2 weighed 3,126 lbs. DFS-1 already was in orbit.

TDF-2 was France's second five-channel, direct-to-home television satellite. The 4,621-lb. satellite broadcasts five channels. Its twin, TDF-1, already was in orbit. France estimated 400 million viewers would be able to see TDF telecasts.

Commercial flight V-37 had been arranged by Arianespace, a consortium of electronics, banking and aerospace companies in eleven West European countries: Belgium, Denmark, France, Germany, Great Britain, Ireland, Italy, the Netherlands, Spain, Switzerland and Sweden. Ariane rockets already had carried to orbit 45 commercial satellites from Canada, France, West Germany, Britain, India, Italy, Spain, the U.S., and the international organizations Intelsat and Eutelsat.

European Rockets: Ariane V-38

An unmanned Ariane 4 rocket lobbed two satellites to Earth orbit August 30, 1990, from a European Space Agency launch pad in the jungle outside Kourou, French Guiana, on the northeastern shoulder of South America.

Europe's 38th commercial launch carried the Eutelsat 2-F1 and Skynet 4C communications satellites to a low Earth orbit. Later, the satellites boosted themselves on out to stationary orbit 22,300 miles above Earth.

Skynet 4C channels military communications for the British government. Built by British Aerospace for the Ministry of Defence, the 2,117-lb., solar-powered Skynet 4C was outfitted with three radio transponders. It was stationed over the equator off the west coast of Africa.

Eutelsat 2-F1 carries telephone calls, telex messages and TV broadcasts to 600 million persons in Europe, North Africa and the Middle East. The 2,500-lb., solar-powered Eutelsat 2, built by Aerospatiale, was sent to a post over the equator above central Africa. It had 16 transponders to transmit radio, TV and data across Europe.

The three-stage launcher was an Ariane model 44-LP, a powerful version of the Ariane

4 series. It was augmented by two liquid-fuel and two solid-fuel boosters. It dropped Skynet off in space 100 miles above Earth just 20 minutes after blast off. Four minutes later, it ejected Eutelsat into a similar orbit.

European Rockets: Ariane V-39

Hurricane Lili was passing through the Caribbean to the north, but weather was good at Kourou with only slight winds as an Ariane 4 rocket carried two U.S.-made communications satellites to Earth orbit October 12, 1990.

It was an evening launch from the European Space Agency center in French Guiana. The rocket carried two satellites—Galaxy VI and SBS-6—built by Hughes Aircraft Co., Pasadena, California, for Hughes Communications Inc. Galaxy 6 is used for retransmission of network and cable TV programs. SBS-6 is for video communications.

The countdown was held briefly before launch so workers could top off the gas tank. Extra liquid hydrogen and oxygen fuel were needed so the Ariane 4 would have the power needed to boost the hefty pair of satellites. Together, they weighed nearly 8,100 lbs.

Europe's 39th commercial launch carried the Galaxy and SBS satellites to a low Earth orbit. The rocket dropped the satellites off in space less than half an hour after blast off. Later, they boosted themselves on out to stationary orbit 22,300 miles above Earth, to be stationed over the Galapagos Islands west of Ecuador.

The 192-ft., three-stage, unmanned launcher was a powerful Ariane model 44-L. Four liquid-fuel boosters were strapped to the Ariane 4. It was the eleventh Ariane 4 shot to space and the fifth time the 44-L had blasted off.

European Rockets: Ariane V-40

An unmanned Ariane 4 rocket blasted off November 20, 1990, from the European Space Agency's jungle launch pad at Kourou, French Guiana, ferrying the U.S. communications satellites Satcom C-1 and GStar IV to Earth orbit.

It was the 40th Ariane flight. GStar IV was GTE-Spacenet's seventh communications satellite in space. SatCom C-1, owned by GE Americom, replaced SatCom 1R, the industry's first in-orbit back-up satellite.

Satcom C-1 and GStar IV were carried to a low Earth orbit by the 192-ft., three-stage, Ariane 4. Later, the satellites boosted themselves on out to stationary orbit 22,300 miles above Earth.

European Rockets: Ariane V-41

An unmanned Ariane 4 rocket blasted off January 15, 1991, from Kourou, French Guiana, ferrying the Eutelsat 2-F2 and Italsat-1 communications satellites to Earth orbit. It was the 41st Ariane flight. The satellites were carried to a low Earth orbit by the 192-ft., three-stage, Ariane 4.

European Rockets: Ariane V-42

An unmanned Ariane 4 rocket blasted off February 21, 1991, from a European Space Agency launch pad at Kourou, French Guiana, ferrying the television broadcast satellite Astra-1B and the weather satellite MOP-2 to Earth orbit. It was the 42nd Ariane flight.

The satellites were carried to a low orbit by the 192-ft., three-stage, Ariane 4. Later, the satellites boosted themselves on out to stationary orbit 22,300 miles above Earth.

European Rockets: Ariane V-43

Thick clouds parted over the European Space Agency's Kourou, French Guiana, launch pad April 4, 1991, allowing an Ariane 4 rocket to heft the first of a new series of advanced Canadian communications satellites.

Europe's 43rd commercial launch carried the three-ton Anik-E2 communications satellite to a low Earth orbit. It was the first Canadian communications satellite launched by Ariane. Later, the satellite boosted itself on out to stationary orbit 22,300 miles above Earth.

Anik-E2 works in conjunction with a twin satellite, Anik E-1, which was launched later, in September 1991. Anik is the Eskimo word for brother.

The 192-ft., three-stage, unmanned launcher was an Ariane model 44-P, a powerful version of the Ariane series. Four solid-fuel boosters were attached to the Ariane 4. The rocket dropped Anik off in space 100 miles above Earth just 20 minutes after blast off. With the V-43 flight, Ariane had a record of 38 successful launches and five failures.

The solar-powered Anik-E2, first in a new generation of Canadian radio repeaters in the sky, sported a total of 24 C-band voice transponders, 16 KU-band high-speed data transponders and 32 television channels.

It was built in Canada by Spar Aerospace and is owned by Canada's national telecommunications company Telesat Canada. Anik-E2 channels communications across Canada's sprawling territory and northern parts of the U.S. The satellite was expected to work 13.5 years in orbit, stationed above the equator at 107.3 degrees west longitude.

Ariane launches are sold for ESA by Arianespace, an 11-nation European consortium. Ariane 4 rockets are marketed in configurations of boosters, depending on how many satellites are carried to space and their weight. Currently, the most powerful Ariane 4 can lift 10,000-pound satellites to orbit, comparable to a U.S. Titan rocket.

European Rockets: Ariane V-44

An Ariane 4 rocket blasted off July 17, 1991, from the European Space Agency's South American launch pad at Kourou, French Guiana, ferrying the satellites ERS-1 and the microsats Orbcom-X, Sara, TUBsat and UoSAT-F to Earth orbit.

ERS-1 and the microsats were carried by an unmanned, 192-ft., three-stage Ariane 4. It was the 44th Ariane flight.

European Rockets: Ariane V-45

An unmanned Ariane 4 rocket lobbed a huge communications satellite to Earth orbit August 14, 1991, from European Space Agency launch pad ELA-2 in the jungle outside Kourou, French Guiana, on the northeastern corner of South America.

Europe's 45th commercial launch carried the Intelsat VI (F5) satellite to a low Earth orbit. Later, the satellite boosted itself on out to stationary orbit 22,300 miles above Earth. Intelsat VI is the largest civilian communications satellite on the market.

The three-stage launcher was an Ariane model 44-L, a powerful version of the Ariane series. It was augmented by four liquid-fuel boosters. The extra-high-power Ariane was needed because of the 9,471-lb. heavyweight satellite. The rocket dropped Intelsat off in space 100 miles above Earth just 25 minutes after blast off.

Intelsat VI (F5) was the second of three to be launched by the European Space Agency for the International Telecommunications Satellite Organization. It was one of five sixth-generation Intelsats built by the U.S. firm Hughes Aircraft Company.

Intelsat VI, with 24,000 telephone circuits and three television channels, is in service over the east Atlantic Ocean.

European Rockets: Ariane V-46

An Ariane 4 rocket hauled the second of a new series of advanced Canadian communications satellites to Earth orbit from the European Space Agency's Kourou, French Guiana, launch pad September 26, 1991.

Europe's 46th commercial launch carried the three-ton Anik-E1 communications satellite to a low Earth orbit. Later, the satellite boosted itself on out to stationary orbit 22,300 miles above Earth. It was the second Canadian communications satellite launched by Ariane. Its twin, Anik-E2. had been launched earlier, in April 1991.

Anik-E1 works in conjunction with Anik-E2. Anik is the Eskimo word for brother.

The 192-ft., three-stage, unmanned launcher was an Ariane model 44-P, a powerful version of the Ariane series. Four solid-fuel boosters were attached to the Ariane 4. The rocket dropped Anik off in space 100 miles above Earth just 20 minutes after blast off.

The solar-powered Anik-E1, second in a new generation of Canadian radio repeaters in the sky, sported a total of 24 C-band voice transponders, 16 KU-band high-speed data transponders and 32 television channels.

It was built in Canada by Spar Aerospace and is owned by Canada's national telecommunications company Telesat Canada. Anik-E1 transmits communications across Canada's sprawling territory and northern parts of the U.S. The satellite was expected to work 13.5 years in orbit.

European Rockets: Ariane V-47

An unmanned Ariane 4 rocket lobbed a huge communications satellite to Earth orbit October 29, 1991, from European Space Agency launch pad ELA-2 in the jungle outside Kourou, French Guiana, on the northeastern corner of South America.

Europe's 47th commercial launch carried the Intelsat VI (F1) satellite to a low Earth orbit. Later, the satellite boosted itself on out to stationary orbit 22,300 miles above Earth. Intelsat VI is the largest civilian communications satellite on the market.

The three-stage launcher was an Ariane model 44-L, a powerful version of the Ariane series. It was augmented by four liquid-fuel boosters. The extra-high-power Ariane was needed because of the 9,471-lb. heavyweight satellite. The rocket dropped Intelsat off in space 100 miles above Earth just 25 minutes after blast off.

Intelsat VI (F1) was the third of three to be launched by the European Space Agency for the International Telecommunications Satellite Organization. It was one of five sixth-generation Intelsats built by the U.S. firm Hughes Aircraft Company.

Intelsat VI, with 24,000 telephone circuits and three television channels, is in service over the east Atlantic Ocean.

German Rocket: Sanger

German engineers have been designing a two-stage space rocket called Sanger. The first stage would have an air-breathing propulsion system and wings for an airplane-style of take off. The Sanger rocket would accelerate to speeds near Mach 6. The second stage then would separate and its own rocket would blast it to orbit.

British-USSR Rockets: Hotol

British Aerospace and the Soviet Ministry of Aviation Industry are studying a new low-cost way of launching satellites. Their unmanned reusable satellite launch vehicle would be sent to orbit from one of the USSR's Antonov AN-225 super-transports, one of the world's largest airplanes. Hotol, for Horizontal Take-Off and Landing, is a project which British Aerospace dropped in 1987 when the government withdrew funding. The new British-Soviet idea was described as an interim-Hotol, a half-way stage to the Hotol.

The Antonov AN-225, flying at an altitude of 29,500 feet, would drop a rocket on a parachute. The rocket would ignite, blasting itself up to space where a satellite would be deposited in Earth orbit.

They say such a system might be able to drop off payloads in low Earth orbit at about one-fifth the cost of a U.S. space shuttle. British Aerospace said it might take ten years and $4.35 billion dollars to get such a launcher in production. Then the cost of launching a satellite to orbit would be a low $15 million. U.S. space shuttle prices have been near $60 million.

Japanese Rockets: N-1, N-2, H-1, H-2

Japan was a late bloomer in space work because of restrictions imposed at the end of World War II. Then, on February 11, 1970, thirteen years after Sputnik, Japan launched its first satellite, Ohsumi, to orbit, making it the fourth nation with a space rocket, after the USSR, U.S. and France. The Japanese have made two dozen launches since, enjoying a nearly perfect success rate with their series of N-1, N-2, and H-1 rockets.

N-1/N-2 rockets. Japan's first workhorse space rockets were the two-stage N-1 and N-2. The last N-2, number 8 in the series built by Mitsubishi, was used to launch the Maritime Observation Satellite MOS-1, the nation's first Earth-observation satellite, to a 564-mi.-high orbit February 19, 1987.

In orbit, MOS-1 was renamed Momo for Peach Blossom. MOS-1 or Peach Blossom had three sensors to study vegetation, land use and marine phenomena.

The N-2, which launched Peach Blossom from the National Space Development Agency of Japan's Osaki Launch Site at Tanegashima Space Center, was similar to a U.S. Delta 1 space rocket. MOS-1 was the final N-1 flight. It was replaced by the more-powerful H-1 rocket. The Japanese had improved their lift-to-orbit capability by turning the N-2 into a two-stage rocket.

H-1 rocket. The H-1 rocket became Japan's principal space launcher, replacing earlier-model N-1 and N-2 rockets. The H-1 was the first Japanese booster to incorporate a third stage in its design.

The first H-1 was fired to space August 13, 1986. It included only two rocket stages in a test of the inertial guidance system and a powerful liquid oxygen second-stage engine, the LE-5, both developed in Japan.

The H-1 third stage was all-new, Japanese-developed and used solid fuel. The August 1987 three-stage H-1 lifted off in the early evening hours from the Tanegashima Space Center on a small island at the southern tip of Japan, carrying the 1,210-lb. ETS-V communications satellite to orbit. The two-stage N-1 and N-2 rockets had been able to loft only smaller payloads of 330 and 770 pounds.

Japan launched another big satellite, the 1,210-lb. CS-3a communication satellite February 19, 1988, after delays including one caused by tests of an electronic part from a U.S. company. The CS-3a satellite was the largest satellite Japan's National Space Development Agency ever sent to space.

The 132-ft. H-1 rocket carried the satellite in an evening blast off from Tanegashima, Japan's southern island launch site.

The CS-3a launch had been scheduled for February 1, but technicians didn't like test results from U.S.-made semiconductors in the rocket's electronic circuitry. The delay was controversial as it raised questions about American quality and use of American products in Japanese spacecraft. The launch also was delayed three more times by weather.

The H-1 rocket was about 80 percent made in Japan. The payload cover and other parts came from the U.S.

H-2 rocket. Japan had been using older technology borrowed from the U.S., but now the National Space Development Agency is preparing for a next-generation space booster which it hopes will propel two-ton payloads into space.

Moon orbiters. Sun-survey satellites. Venus probes. Heavy-lift rockets. Free-flying station platforms in space. Space tugs. Manned stations. Even a small space shuttle named Hope.

Japan has ambitious plans for a wide range of space launches, satellites and interplanetary probes through the 1990's, even including the small space shuttle Hope to carry freight to the U.S.-international space station Freedom.

All-new technology H-2 is a powerful new space-launch rocket Japan is developing entirely by itself for use by the mid-1990's.

Technology in Japan's older H-1 rocket had been borrowed from the United States. Agreements between the two countries prevented Japan from seeking commercial launch contracts for H-1 flights. The H-2 will not be so encumbered and Japan is expected to be competitive in the space-launch business when the 147-ft. H-2 rocket is ready.

Mitsubishi Heavy Industries Ltd, Ishikawajima-Harima Heavy Industries Co., Kawasaki Heavy Industries Ltd., NEC Corp., Nissan Motor Co. and Mitsubishi Corp. and 69 other Japanese industrial companies, banks, trading houses and electronic equipment makers formed Rocket Systems Inc. in July 1990 to sell rockets and launch services to foreign satellite owners once the H-2 is ready to fly. Rocket Systems Inc. will place a $3 million order for the first H-2 to be launched in 1993.

The Japanese government plans to spend billions through the 1990's and beyond to build Hope, rockets, satellites, interplanetary probes and ground facilities.

LE-7 troubles. The advanced LE-7 engine is to power the first stage of Japan's big new H-2 rocket, which will free Japan from restrictions on using U.S technology. The LE-7 rocket engine, fueled with liquid oxygen and liquid hydrogen, similar to U.S. space shuttles, will be in the first stage of the two-stage H-2 rocket and will have a thrust of 110 tons, rivalling a shuttle engine.

H-2 engineers have had their share of problems getting the new rocket ready for spaceflight, particularly in testing the LE-7 which must survive extremely high stresses in the engine during firing.

Three explosions in four years rocked the lab where the H-2 and its LE-7 turbopump are tested at a site in Kakuda, a city on the northeast Pacific coast of Japan, 180 miles northeast of Tokyo.

In September 1987, the roof of the building was blown off by a liquid hydrogen leak. A blast caused by a similar leak did minor damage in November 1987.

Engine builders Mitsubishi Heavy Industries and Ishikawajima-Harima Heavy Industries had trouble developing pressure-resistant fuel pumps for the LE-7 in Spring 1989. Later that year, in September and November 1989, gases escaped an LE-7 during firing tests and ignited, damaging the engine and the test stand on Japan's southern Tanegashima Island.

The turbines of the turbopump in the LE-7 engine broke down during a propulsion test in 1989. The LE-7 caught fire four times during engine tests.

Japan was able to fire the powerful LE-7 successfully in a four-second ground test in March 1990. Then fire engulfed a test stand being used to fire the H-2 rocket's LE-7 engine in July 1990.

An explosion in May 1991 again blew the roof off the remote-controlled testing laboratory. A broken pipe had leaked highly-volatile liquid hydrogen as engineers fed the fuel to an LE-7 in a turbopump test. No one was injured in any of the explosions, but the structure's frame was warped. Damage was less than the National Space Development Agency of Japan had expected because there was no fire. Windows were broken in five homes within a mile of the facility.

An engineer was killed in August 1991 in an explosion during pressure testing of an

H-2 engine part at the Mitsubishi Heavy Industries aeronautics factory in Komaki near Nagoya in central Japan. When air pressure was raised to 143 atmospheres, a manifold in the main injector of the first-stage LE-7 engine burst, blasting off the door of the test chamber. Unfortunately, the door hit engineer Arihiro Kanatani, 23, in the head, breaking his skull.

The laundry list of development difficulties has pushed the first H-2 flight back to no earlier than the first quarter of 1993.

TR-1. Japan conducted three test fights in 1988 and 1989 of one-quarter scale models of the H-2. Labeled TR-1, the experimental single-stage solid-fuel one-quarter scale model rockets were launched from the Tanegashima Island space-launch center off southwestern Japan near the main southern island of Kyushu.

Each single-stage TR-1 rocket was 47 feet long and four feet wide. Each made a one-minute suborbital flight and splashed down in the Pacific Ocean.

Two dummy solid-fuel booster rockets attached to each TR-1 main rocket body separated about 50 seconds after lift-off. The rockets continued on up to around 50 miles altitude. Nose cones on the dummy boosters, loaded with instruments for gathering flight technical data on heat, sound and vibration, were recovered from the ocean by a NASDA ship 60 miles south of Tanegashima.

Besides data recorders, the TR-1 rockets carried materials processing experiment similar to equipment planned for space station Freedom in the late 1990's.

Hope. The shuttle name Hope is short for H-2 Orbiting Plane. The spaceplane will be fired to orbit atop an H-2 rocket, carrying three tons of baggage to a low orbit. Pacific Spaceport Group, an organization formed by Japan's seven largest companies, plans to construct control building, launch tower and six-mile runway for the Hope shuttle at an Australian spaceport at Weipa, Queensland. Australia has organized a Cape York Space Agency to develop the Queensland facility.

A shuttle larger than Hope, to carry men as well as freight, could be built later and launched on an H-2 with extra booster rockets strapped to its sides.

Here are some of Japan's plans for becoming a major spacefaring nation:

1993 H-2 rocket goes into service
1993 GMS-5 next-generation stationary weather satellite
1994 ADEOS advanced Earth observation satellite
1995 stationary orbit platform
1996 Planet-B probe to Venus
1997 lunar surveyor probe
1998 manned platform
1999 manned research lab module to space station Freedom
2000 space shuttle Hope
2005 larger space shuttle
2010 independent Japanese space station
2015 participation with the U.S., USSR and Europe in manned flights to Mars
2020 participation with the U.S., USSR and Europe in Moon and Mars colonies

Chinese Rockets: Long March Family

China's first long-range military ballistic missiles, and eventually space rockets, were developed by Tsein Weichang, an immigrant to the U.S. from China, educated in Canada, who worked for the U.S. government at the Jet Propulsion Laboratory, Pasadena, California, only to be forced back to his homeland in 1949 in America's spasm of anti-communist witch hunting.

The Peoples Republic of China opened its first Missile and Rocket Research Institution October 8, 1956, while still on friendly terms with the USSR. Hindered by what China today recalls as "technical blockades put in by the imperialist countries," nothing much happened until the 1960's when experiments with liquid-fuel rockets picked up steam. Satellites, and space rockets to carry them to orbit, were designed.

Mao 1. The PRC launched its first satellite—known as China 1 or Mao 1—on its own Long March space rocket in 1970. That 390-lb. electronic ball floated around the Earth blaring the patriotic The East Is Red song.

China made a total of 32 successful satellite launches from 1970 through the end of 1991. The 32 included remote sensing, communications and weather satellites for both civilian and military use. The PRC sent a dozen to orbit with packages to be retrieved. They recovered all successfully. Two satellites were launched for foreign owners.

Homing satellite. In November 1975, the first Long March 2 rocket carried China's first "homing satellite" to orbit. That made China the third nation capable of retrieving a satellite. The pace of China's space industry picked up in the 1980's. In September 1981, the PRC successfully launched three satellites to orbit with one rocket.

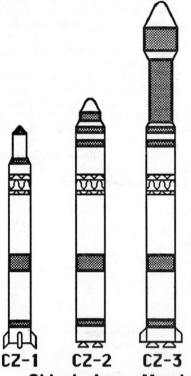

CHINESE ROCKET sizes compared in artist's conception of selected boosters including Long March 1 (CZ-1), Long March 2 (CZ-2) and Long March 3 (CZ-3). All use liquid fuel. The three stage Long March 1, in service since 1970, is 97 ft. tall, 7 ft. diameter. It can carry 661 lbs. to a circular 273 mile high orbit. The newer two-stage Long March 2, used since 1975, is 107 ft. tall, 11 ft. diameter. The big three-stage Long March 3, in use since 1983, is 142 ft. tall, 11 ft. diameter. It can lift a 3,086-lb. payload to a 120-mi. geosynchronous transfer orbit. The powerful Long March 4, first used in 1988, is not shown. The People's Republic of China made 32 successful satellite launches from 1970 through the end of 1991. The 32 included remote sensing, communications and weather satellites for both civilian and military use. Among them, the PRC sent a dozen to orbit with packages to be retrieved from space. They recovered all successfully. Two satellites were launched for foreign owners. China's commercial space launch firm is the Great Wall Industry Corp.

CZ-1   CZ-2   CZ-3

China's Long March 1, 2 And 3 Space Rockets

For a time after the 1986 Challenger disaster in the U.S., there was a worldwide shortage of launch vehicles. China began offering commercial launch services to foreign countries in 1987. Great Wall Industrial Corp. sells launch services. Simplicity of operation and a high success rate have allowed China to offer prices lower than U.S. commercial launch firms and the European Space Agency.

In April 1990, the U.S.-made ASIAsat 1 commercial communications satellite was launched on the first commercial flight of the Long March 3 rocket. The PRC has commercial launch contracts with other nations as well. Xichang Satellite Center is China's takeoff point in the international satellite-launch market.

Astronautics. Rocketry in China is run by the Ministry of Astronautics, a government agency. Lin Zongtang is minister of China's aerospace industry.

China may launch a second ASIAsat in the 1990's. Sweden is discussing launch of a weather satellite by China in the 1990's. An Australian company was to pay the Chinese to launch two brand-new Hughes communications satellites, Aussat 1 and Aussat 2, from Xichang in 1992, with approval from the U.S. government.

CAST. The state-owned Chinese Academy of Space Technology (CAST) is China's only designer and manufacturer of space satellites. The state-run Overseas Telecommunications Commission (OTC), Australia's international telecommunications arm, signed a cooperation agreement with CAST in 1991. In a satellite joint venture, CAST would be responsible for launch facilities, while OTC would manage the launch.

Man-in-space. Manned flights, in two-place spacecraft similar to the U.S. Gemini capsules of the 1960's, are being prepared. The Chinese also want to launch a 24-ft. 22-ton manned space station to low Earth orbit by 1998.

The world's most populous nation launches its Long March rockets, carrying satellites to Earth orbit, from four major space complexes:

★Jiuquan launch site in the northern Gobi Desert in far-western China, not far from Mongolia,

★Taiyuan launch site in Shanxi Province, 60 miles southwest of Beijing, in north-central China,

★Xichang launch site in an isolated corner of Sichuan province in southern China, not far from Burma, Laos and Vietnam, and

★Hainan Island off the southern coast of China, separating the South China Sea from the Gulf of Tongking. Only 19 degrees north of the equator, it is southeast of Hanoi, Vietnam, across the Gulf of Tongking.

All of the sites are remote and usually closed to foreigners, although the press was taken to the Xichang Satellite Center in 1988 as China was pressing the U.S. for approval of its launch business.

Long March. Chinese space rockets are known as Changzheng, or Long March. CZ is short for Changzheng.

★Long March 1 (CZ-1), first sent to space April 24, 1970;

★Long March 2 (CZ-2), first fired July 26, 1975;

★Long March 3 (CZ-3) first used August 19, 1983;

★Long March 4 (CZ-4) used September 6, 1988.

The Chinese may have sustained a launch failure in 1969, three failures in 1974 and another in 1979.

Long March 4. The latest, most-powerful version of China's workhorse series of Long March space rockets is the Long March 4. It can lift 11,000 lbs. to a low orbit, using the same first and second stages as a Long March 3. Using a different third stage for shots to low and medium altitude orbits, it can carry more than 8,000 lbs.

With 600,000-lbs. thrust, Long March 4 will lift 5,000 lbs. to stationary orbit. That's nearly double the previously-most-powerful Long March 3, in use since 1983. Long March 3 can lift 2,800 lbs. to stationary orbit or 6,000 lbs. to a low orbit. A Long March 2 can ferry a satellite of less than 2.5 tons to a low-Earth orbit.

China first used a Long March 4 to launch a heavy weather satellite to stationary orbit from Taiyuan in 1988. Great Wall Industrial Corp., which sells launch services, hopes the bigger rocket will draw more foreign customers for commercial satellite launches.

Space triple. The PRC hit a space triple September 6, 1988:

★Inaugural use of a third space launch site, in north-central China at Taiyuan, Shanxi Province, south of Beijing,

★Maiden voyage of a new heavy-lifting Long March 4 space booster rocket,

★First launch of a 1,650-lb. Feng Yun 1, or Wind and Cloud No. 1, weather satellite to a 560-mi.-high polar orbit.

Wind and Cloud No. 1 has infrared and visible light sensors. It transmits pictures to Earth on a frequency of 137.78 MHz. Wind and Cloud No. 1 is similar to U.S. NOAA weather satellites. Wind and Cloud No. 1 was designed to radio data on clouds, ocean surface temperatures, marine water color, Earth's surface, vegetation growth, and ice and snow cover to ground stations around the world.

Unfortunately, Feng Yun 1 seemed to have problems. Possibly tumbling out of control, it may have been shut down for a time.

Weaver Girl. China hit a space double header December 19, 1988, when it launched a new space rocket from a new space center on southern Hainan Island.

The new rocket, known as Weaver Girl 1, named after a Chinese legend, ferried a recoverable satellite to space. The payload remained in space 2.5 hours, then returned to Earth 40 miles from the launch site. The blast off was the first time Chinese scientists had researched Earth's atmosphere from a low-latitude equatorial launch site.

Weaver Girl I was developed by the Chinese Academy of Sciences and the University of Defense Science and Technology.

Hainan. Hainan Island is off the southern coast of China, separating the South China Sea from the Gulf of Tongking. It is southeast of Hanoi, Vietnam, across the Gulf of Tongking. The Hainan space base, only 19 degrees north of the equator, is used for low-latitude, low-altitude space research launches. Initially, rockets blasting off from Hainan fly only to a height of about 74 miles above Earth.

Long March 2e. China launched Pakistan's first satellite, Badr-A, to a 375-mi.-high circular orbit July 16, 1990, using the new Long March 2E rocket for the first time. The launch was from Xichang Launch Center in a hilly farm area in an isolated corner of Sichuan province in southwestern China.

Long March 2E was designed to lift 15,000 lbs. to a low elliptical orbit ranging from 250 to 500 miles above Earth. The rocket, called Cluster Carrier, blasted off from a new pad built to launch bigger boosters. Long March 2E, with four boosters strapped on, carried a large Australian dummy satellite and the 150-lb. Badr-A.

India's Rockets: The SLV Family

India launched its first satellite, Rohini 1, to orbit on a Satellite Launch Vehicle (SLV) rocket July 18, 1980. India was the seventh nation with a space rocket. The SLV successfully placed three 80-lb. satellites in orbit between 1980-1983.

The nation's Augmented Satellite Launch Vehicle (ASLV) is designed to carry 330-lb. satellites to Earth orbit. A future Polar Satellite Launch Vehicle (PSLV) is to carry 4,400-lb. satellites in the mid-1990's. ASLV failures could mean PSLV won't fly until later.

A satellite launch attempt by India failed in July 1988 when an ASLV solid-fuel rocket fell into the Bay of Bengal moments after liftoff from the Sriharikota Space Center in the southeastern state of Tamil Nadu.

It was the second failure in a series of four ASLV test flights. The first, March 24, 1987, ended as the ASLV fell into the Bay of Bengal one minute after liftoff. The five-stage 40-ton 77-ft. rocket was carrying a 300-lb. weather and communications satellite.

India Space Research Organization builds ASLV space boosters from the technology used in India's older liquid-fuel Satellite Launch Vehicle (SLV). An ASLV has an SLV at its core, four solid-fuel stages and two strap-on solid-fuel boosters.

India had eight satellites in orbit at the end of 1991. Four satellites had descended earlier. A giant Insat communications satellite, designed in India and launched from the U.S., provides a national TV link, long-distance telephones, and meteorological reports for the country which is two-thirds agricultural and dependent on weather patterns.

Insats have been ferried to space by U.S., European and Soviet rockets. The USSR also launched the India Remote Sensing (IRS) satellite in 1988. India is building a military rocket launch site in the eastern state of Orissa.

Brazilian Rockets: VLS

With two Brazilsat telephone satellites, launched by others, already relaying calls from orbit, Brazil is working overtime to build its own launch pads, rockets and spacecraft to boost the South American country into the club of spacefaring nations.

Through the 1980's, a program called MECB, short for The Brazilian Complete Space Mission, has grown toward a self-contained space capability for the Portuguese-speaking nation, largest on the continent of South America. For a decade-long expenditure of $1 billion, Brazil's spacemen have been building:

★a powerful new space rocket,
★launch pads and tracking stations,
★fast space computers for ground stations,
★four satellites.

The Ministry of Aeronautics is building launch and tracking facilities in northeast Brazil on the country's 4,603-mi.-long Atlantic Ocean coast near Alcantara and Natal, not far south of the equator.

The European Space Agency launch site is nearby on the Atlantic Coast of northeast South America at Kourou, French Guiana, north of the equator.

Brazil is building a tracking station and data gathering center at Cuiaba, Mato Grosso. The MECB satellite control center is at Cuiaba. Mission control is in Cachoeira Paulista in Sao Paulo.

Development work is in the university-campus-style collection of research labs of the Instituto de Pesquisas Espacias (INPE) at Sao Jose dos Campos. INPE employs 1,550 including 1,000 scientists and engineers.

First satellites. Rockets are being tested with an eye toward launching the first of four satellites in the 1990's. It might be fired in 1989 if work goes ahead of schedule.

The 250-lb. octagonal satellite will collect environmental data and transmit the information to ground stations. It will work in a 466-mi.-high orbit for at least six months, passing over Brazil eight times a day.

A receiver in the satellite will listen for signals from on-the-ground robot environmental data collection platforms strewn across Brazil's 3.29 million sq. mi. countryside. That's an area larger than the contiguous 48 United States.

The robot platforms are small, automatic, unattended Earth stations which record weather and other environmental data for transmission to the satellite. The satellite collects these short bursts of radio data and retransmits them down to a central analysis station at Cuiaba for distribution to users across the country. The aluminum satellite would be powered by solar cells generating 35 watts of electricity.

Satellites to be launched during the 1990's would include two weather and environment data collection satellites and two remote-sensing satellites searching for natural resources.

VLS. rocket. The space launcher being developed is a four-stage solid-fuel rocket known as Veiculo Lancador de Satelites (VLS). It will be capable of lifting 250 lbs. to a 466-mi. orbit. Or 350 lbs. to 400-mi. altitude. The MECB satellites will be in circular orbits at those heights.

The first stage of VLS has four rocket motors. The second stage is one rocket engine, identical to the first-stage rockets, nested inside the first-stage motors. The third and fourth stages each have one rocket engine.

The first-stage will be test fired three times before the first satellite will be launched.

Past experience. Brazilian scientists have launched hundreds of sounding rockets in the last 25 years, some as high as 375 mi.

More than 200 two-stage Sonde 1 rockets have gone aloft in search of weather data, measuring winds and temperatures. Sonde 1 can carry 10 lbs. to a 45-mi. altitude.

The single-stage Sonde 2 can ferry 65 lbs. to a height of 65 miles. Sixty of these have blasted experiments to the upper atmosphere over the years.

Sonde 3, a two-stage launcher capable of lifting 135 lbs. 375 miles high, is made up of a pair of Sonde 2 rockets, one atop the other. Two Sonde 3 rockets have been fired.

Two Sonde 4 vehicles also have blasted off. The Sonde 4 can lift 1,100 lbs. to 375 miles. The second stage of Sonde 4 is the same as the first stage from Sonde 3. The first stage of Sonde 4 will be used as the third stage in the VLS satellite-orbiting rocket.

The main space launch site at Alcantara, known as CLA, has a launch pad for Sonde 3 and Sonde 4 as well as VLS satellite launchers. It has a control center and blockhouse.

A deal with China. People's Republic of China also is aggressively pursuing a future in space with its Long March series of inexpensive rockets. Looking for a launch site on the equator, China may ship Long March rockets to Alcantara for blast off.

Besides earning money or goods in trade for the services of Alcantara, Brazil would gain launch experience for its engineers by firing Chinese rockets to space.

The Brazilsat satellites were launched by the European Space Agency at Kourou, French Guiana. Two more are scheduled for Ariane launches. INPE wants its own launch capability to send satellites to space to photograph land use, geological formations, forestry, agriculture, water, pollution, urban sprawl and oceanography from space. Brazil is home to 135 million persons.

1990's slowdown. In the 1990's, budget cuts slowed work on Brazilian satellites, rockets and construction of the Alcantara launch site in the state of Maranhon. The National Space Research Institute told Congress in 1990 that Brazil's Satellite Launch Vehicle could not be launched before 1993 or 1994.

Brazil's first natural resources satellite originally was to be launched by a state-run space firm. Since the 250-lb. satellite was ready before the Satellite Launch Vehicle rocket, a launch by one of the European Space Agency's Ariane rockets was purchased to carry the satellite to space in 1992.

Brazil also is considering launching other satellites on rockets made in the U.S., USSR and China. Space budget cuts were forced by an economic crisis in Brazil. Restrictions on military technology transfers to Brazil also slowed the rocket program. However, in exchange for two launch contracts for other Brazilian satellites, Europe's Arianespace is furnishing Brazilian technicians with technology from France's Viking rocket and gyroscope guidance systems.

Manned vs. Unmanned Rockets

The United States government said in July 1991 it will shift emphasis in satellite launches from the reusable space shuttle fleet back to traditional disposable rockets.

U.S. Vice President J. Danforth Quayle was at America's major West Coast spaceport, Vandenberg Air Force Base, California, when he said space shuttles are too valuable for jobs which don't need their unique capabilities.

Many expendable rockets have been launched from Vandenberg, while no shuttles have been launched there. Some shuttle critics have calculated that single-use rockets actually are more cost-effective than reusable vehicles for carrying payloads to space.

"In all probability, we have purchased the last space shuttle," Quayle said about the new shuttle Endeavour which replaced Challenger. Quayle called for conservative use of space shuttles in the future, to make them last longer.

The U.S. Advisory Committee on the Future of the U.S. Space Program, agreed, telling the space agency to "proceed expeditiously" on unmanned rockets as an alternative to manned shuttles to head off possible future disruptions from the loss of another shuttle. The committee expressed concern over the "instability" of NASA's budget. Despite already tight budgets, proposals to cut funds for the U.S.-international space station Freedom and other NASA projects had surfaced in Congress.

Astrodynamics

The physical laws which control spacecraft motion are based on Isaac Newton's laws of motion and Johannes Kepler's three laws of planetary motion.

Laws of motion. Newton's laws of motion control the motion of a spacecraft.

(1) Unless acted upon by an outside force, an object at rest or in motion will remain at rest or in motion with the same speed and direction.

(2) The change in motion of an object acted upon by an outside force is directly proportional to the force and inversely proportional to the object's mass.

(3) Every action force has a reaction force equal in magnitude and opposite in direction. Usually stated as, to every action there is an equal and opposite reaction.

Law of gravitational attraction. The force of attraction between any two particles of matter is directly proportional to the product of their masses and inversely proportional to the square of the distance between their centers of mass.

Planetary motion. Kepler found the empirical laws of planetary motion.

(1) The orbit of each planet is an ellipse with the Sun at one focus.

(2) Each planet revolves so an imaginary line between it and the Sun sweeps over equal areas of space in equal intervals of time.

(3) The square of the planetary periods is proportional to the cube of the mean distance to the Sun.

Rockets and spacecraft either can be in an elliptical or a circular orbit, or be on a parabolic or hyperbolic trajectory. Circular and elliptical orbits are closed, periodic flight paths. Any Earth satellite is in either an elliptical or circular orbit.

A parabolic trajectory is an escape path. Escape velocity is the speed along a parabolic trajectory. A hyperbolic trajectory always has a velocity greater than escape velocity, leaving a spacecraft with extra speed after it escapes from a planet. Interplanetary spacecraft leave Earth on hyperbolic trajectories.

Rockets Glossary

A. The USSR's rocket which carried to Earth orbit the first space satellite, Sputnik I, in October 1957. It was a central core rocket engine with four strap-on booster engines. The core was fueled with kerosene and liquid oxygen. All engines were ignited at lift off. The boosters were dropped off along the way and the central core continued on with its payload to orbit. In orbit, Sputniks 1 and 3 separated from their spent rockets, while Sputnik 2 remained attached to its spent rocket in orbit. The A rocket had 20 main nozzles and 12 steering rockets. It could lift two metric tons to a low Earth orbit. The A space rocket was a variation of the SS-6 intercontinental ballistic missile (ICBM).

A-1. A variation of the USSR space rocket A used to launch Luna 1-3, Korabl Sputnik, Vostok, Elektron, Meteor, and various Cosmos spy and weather satellites.

A-2. A variation of the USSR space rocket A used to launch Voskhod, Soyuz, Soyuz T, Progress and various Cosmos satellites.

A-2e. A variation of the USSR space rocket A used to launch Venera 1-8, Zond 1-3, Mars 1, Molniya, Luna 4-14 and various Cosmos spy satellites.

Abort. To end the launch or flight of a space rocket.

Atlas. A class of unmanned U.S. space launch rockets used for decades to lift medium-weight to heavy-weight payloads to Earth orbit.

B-1. A USSR space rocket, derived from the SS-4 medium-range ballistic missile (MRBM). The lower engine of the two-stage rocket was fueled by kerosene and nitric acid. The upper stage burned liquid oxygen and methyl hydrazine. The B-1 rocket was used between 1962 and 1977 to carry small Cosmos satellite payloads to Earth orbit.

Big Joe. A rocket used in combination with an Atlas rocket in a launch in the Project Mercury series. The rocket was known as Atlas Big Joe. See also: Mercury.

Blue Scout. A rocket used in a launch in the Project Mercury series. See also: Mercury.

Boosters. Helper rockets strapped to a main rocket engine. See also: Separation.

Burn. Firing a rocket engine. See also: Rocket Engine.

C-1. A USSR space rocket, derived from the SS-5 intermediate range ballistic missile (IRBM). The C-1 rocket has been used since 1964 to carry medium-weight mostly-military Cosmos payloads, including multiple satellites, to Earth orbit.

Capsule. A small pressurized compartment or spacecraft, atop a rocket, in which astronauts travel.

Cosmos. The generic name given to most of the USSR's unmanned satellites.

Countdown. The high number to low number counting of final seconds leading to a rocket launch. Ten, nine, eight, seven...

Cruise engine. Rocket engines to maneuver the Mir space station around space.

Cutoff. Stopping combustion to end thrust by shutting off the flow of fuel in a rocket engine. See also: Rocket Engine, Thrust.

D. A class of USSR heavy-lifting space rocket, derived from the original A rocket with two core engines and six strap-on boosters instead of four. Known also by the name Proton, the D rocket has been used since July 1965 to carry heavy payloads to Earth orbit. The USSR has tried to sell commercial space launch services using the D rocket. See also: D-1, D-1e, D-1m, Proton.

D-1. D rocket with special upper stage to lift Salyut and Mir space stations to Earth orbit. See also: D, Proton.

D-1e. D rocket with escape-velocity upper stage to boost a spacecraft, such as Zond 4-8, Luna 15-24, Mars 2-7, and Venera 9-12, on interplanetary flights. It also has been used to launch to Earth orbit Molniya, Raduga, Gorizont, Ekran, Astron and various Cosmos spy, communications and navigation satellites. See also: D, Proton.

D-1m. D rocket with special upper stage to lift and maneuver in Earth orbit the very heavy Cosmos 382 satellite in 1970. See also: D, Proton.

Delta. A class of unmanned U.S. space launch rockets used for decades to lift medium-weight payloads to Earth orbit.

Ehricke, Krafft. An associate of German-American space scientist Wernher von Braun in the 1950's, Ehricke proposed blasting an Atlas rocket to Earth orbit, then furnishing it as a space station. His idea was boosted later by Douglas Aircraft Corp. in its Manned Orbiting Laboratory. See also: Manned Orbiting Laboratory, MOL.

Energia. The newest USSR super-powerful space-launch rocket. The most powerful space rocket today. Energia is used to carry the space shuttle Buran to Earth orbit.

Engine. A rocket. See also: Motor, Rocket.

Escape velocity. The speed required for a rocket to leave a planet or celestial body without falling into orbit around it. See also: Orbital Velocity.

External tank. Large liquid-fuel tank attached to U.S. shuttle orbiter at launch.

F. A class of USSR heavy-lifting space rocket, derived from the SS-9 intercontinental ballistic missile (ICBM). Variations have included F-1r and F-1m which lifted various spy satellites to Earth orbit, and F-2 which carried Meteor and various Cosmos communications, scientific and military satellites.

Fuel. Liquid or solid propellant used in an internal reaction to provide thrust for a space rocket.

G. A USSR very-heavy-lifting space rocket.

Geostationary. An orbit 22,300 miles above Earth where satellites revolve around the planet at the same speed at which Earth rotates. Satellites in this orbit seem to hang stationary over one spot on Earth's surface. Also known as Geosynchronous. A geostationary transfer orbit, or GTO, is a much lower orbit, around 100 miles above Earth's surface, from where a satellite may be boosted on out to geosynchronous orbit.

Geosynchronous. See: Geostationary.

Gimbal. Swiveling of a rocket exhaust nozzle to control direction of flight.

Ground test. Testing a rocket, tied to the ground, without a launch.

GTO. See: Geostationary.

Jupiter. The three-stage rocket which carried to Earth orbit the first U.S. space satellite, Explorer I, in January 1958.

K. A huge USSR very-heavy-lifting space rocket. See: Energia.

Launch. The lift off toward space of a rocket. The start of space flight with the take off of a rocket.

Launch pad. The platform, tower or tube from which a rocket blasts off for space.

Launch window. A time interval during which a spacecraft must be launched to achieve a desired orbit or flight path away from Earth.

Liquid fuel. Frozen gases used as fuel in space rockets.

Liquid rocket. Space rockets using frozen-gases for liquid fuels.

Little Joe. Class of rocket in seven launches in the Project Mercury series.

Mercury. America's first men in space went there in Mercury capsules. There were a total of 23 launches in the first U.S. manned spaceflight program—Project Mercury. Of the 23, only about a quarter carried men. Of the 23, thirteen carried no crew, four carried animals and six carried astronauts. Sixteen of the flights were successful. There were seven launch failures, but no flight failures during manned launches.

Motor. A rocket. See also: Engine, Rocket.

Orbit. The closed elliptical path travelled through space by an object around another object, following Newton's laws of gravitation and motion. For instance, satellites orbit planets. In a satellite's orbit around Earth, perigee is closest to the planet and apogee is farthest away. The sidereal period is the time taken by a body in completing an orbit around another object. The synodic period is the time between oppositions of a satellite. A revolution is completed when a satellite in orbit passes over the longitude of its launch site. Period is the time interval of one orbit.

Orbital velocity. The speed required for a rocket to go into orbit around a planet or celestial body. To reach Earth orbit from the surface, a velocity of five miles per second is required. See also: Escape Velocity.

Pegasus. A class of unmanned U.S. space launch rockets used to lift lightweight payloads to Earth orbit. The rocket is carried by airplane to high altitude from where it blasts on up to Earth orbit. Pegasus can orbit small satellites.

Propellant. A liquid or solid fuel used in an internal reaction to provide thrust for a space rocket.

Proton. The USSR's series of heavy-lifting space rockets. See also: D.

Redstone. The class of rocket used in five launches in the Project Mercury series. See also: Mercury.

Retrograde orbit. The orbit of a satellite in a backward or westerly direction, opposite the easterly motion of an observer on Earth. See also: Orbit.

Rocket. A transportation device containing at least a rocket engine, fuel and a payload. See also: Engine, Motor, Rocket Engine.

Rocket engine. A propulsion motor to drive a rocket. See also: Engine, Motor, Rocket.

Satellite. A small body orbiting a larger object. A moon in orbit around a planet. Natural satellites are called moons. Earth has one Moon; Mars has two moons. Man-made artificial moons, usually referred to as satellites, are machines in orbit.

Saturn V. A series of heavy-lifting U.S. rockets for orbiting Apollo capsules and Mon flights in the 1960's and 1970's.

Scout. A class of unmanned U.S. space launch rockets used for decades to lift lightweight payloads to Earth orbit.

Separation. The release and falling away of a space rocket's boosters after fuel for the boosters has been exhausted. See also: Booster.

Shuttle. A partly-reusable manned spacecraft which can land on a runway.

Shuttle-C. A proposed unmanned cargo version of a U.S. shuttle.

Solid fuel. A chemical rocket fuel with a rubbery consistency, not gas or liquid.

Solid rocket. A space rocket using a rubbery, so-called solid, fuel.

Soyuz. A name for the rocket which carries to Earth orbit the series of USSR cosmonaut-transporting spacecraft known as Soyuz.

Specific impulse. The ratio of pounds of thrust to pounds of fuel in a rocket.

Telemetry. Data transmitted from a rocket or spacecraft.

Thermal protection system. A heat shield on a rocket or manned spacecraft.

Thrust. The propulsion force generated by a rocket during firing. A rocket's thrust is used to carry a satellite to orbit.

Titan. A class of unmanned U.S. space launch rockets used for decades to lift heavyweight payloads to Earth orbit. A Titan 4 rocket currently is America's most powerful space booster.

Umbilical. A bundle of wires and pipes carrying electricity and fluids between a tower on a launch pad and an upright rocket before launch. An umbilical also is a life support line between an astronaut outside a shuttle or space station on a spacewalk and the spacecraft.

V-2. A German rocket carrying a bomb used in World War II.

von Braun, Wernher. A German-American space scientist, Wernher von Braun popularized the concept of a wheel-shaped space station in 1951 in Across the Space Frontier. See also: Krafft Ehricke.

Vostok. A name for the rocket which carried to Earth orbit the first-ever manned spacecraft, known as Vostok.

Space Satellites

★★★★★★★★★★★★★★★★★★★★★★

The Space Age

The Space Age started on a Friday morning in 1957 as a thermometer, radio transmitter and battery in a 23-in. aluminum ball called Sputnik 1 left the surface of Earth at 6 a.m. Moscow time October 4 on Russia's Old Number Seven rocket.

For three weeks, spinning around the world every 96 minutes in a globe-girdling orbit 560 miles above our heads, the 184-lb. Sputnik 1 beep-beeped its message of a future above the ocean of air. On January 4, 1958, Sputnik 1 burned as it fell from orbit.

The rocket. Early in the 1950's, the U.S. and the USSR built hydrogen bombs. The U.S. wanted to use airplanes to carry H-bombs, while the USSR decided to develop a new means of transporting them to targets—intercontinental ballistic-missile rockets.

Russian engineers, headed by Sergei Korolev, designed the first intercontinental ballistic missile (ICBM) in the mid-1950's.

The ICBM rocket was known officially as R-7, but referred to informally as Old Number Seven. West of the Iron Curtain, the North Atlantic Treaty Alliance (NATO) called it SS-6 and Sapwood. When R-7 was used later as a space launcher, the U.S. Air Force labeled it Space Launcher-1 (SL-1). The U.S. Library of Congress catalogued it as an A-class booster.

Tests of parts of the ICBM started in 1956. The first long-range flight-test rocket exploded on the launch pad in May 1957. Eight test firings were made that year, including two rockets which flew 4,000 miles that August.

Space race. The idea had dawned on Soviet engineers that Old Number Seven not only could blast a warhead thousands of miles, but it also could carry a payload to such speed and altitude it would be above the atmosphere orbiting Earth. An artificial Earth satellite could be created.

The first satellite was completed in June 1957 to be launched on the 100th anniversary of the birth of Konstantin Tsiolkolvski, known as the father of cosmonautics. Soviet Premier Nikita Khrushchev gave the okay to launch it toward space August 27, just 24 days after the first successful 4,000-mi. ICBM test. However, technical problems pushed the blast-off back to October 4 when the beach-ball-sized Sputnik 1 rode Old Number Seven into history.

Those times. It was an intense period in the Cold War, less than a year after Khrushchev had boasted, "We will bury you," and just six weeks after the USSR had demonstrated its intercontinental ballistic missile.

Sputnik horrified the American ego, sparking a comeback rally in U.S. schools and colleges. Teachers and scientists fondly recall the end of the 1950's as a golden era with U.S. President Dwight D. "Ike" Eisenhower marshalling the intellectual elite to resuscitate math and science education, but others remember America was mortified by what a writer called that "humiliating beep-beep in the high heavens."

Each beep was an "outer-space raspberry to a decade of American pretensions that the American way of life was a gilt-edged guarantee of our national superiority," according to Claire Booth Luce, a member of Congress and wife of the publisher of Time magazine.

Sen. Henry M. Jackson of Washington reeled under the "devastating blow to the prestige of the United States as the leader in the scientific and technical world." Labor leader Walter Reuther declared Sputnik a "bloodless Pearl Harbor."

When reporters at an October 9 press conference questioned the President closely about Sputnik, Eisenhower called it "one small ball in the air. I wouldn't believe that at this moment you have to fear the intelligence aspects of this." Admitting Sputnik's 184-lb. weight "astonished our scientists," Eisenhower conceded the USSR had "a great psychological advantage throughout the world."

More sputniks. Not only was the USSR's first satellite launch attempt successful, but a second try just 30 days later was too. The 1,119-lb. Sputnik 2 was launched November 3 carrying the live dog Laika on a life-support system.

The capsule remained attached to the converted intercontinental ballistic missile. The dog captured hearts around the world as life slipped away from Laika a few days into her journey. Later, Sputnik burned in the atmosphere April 14, 1958.

Sputnik 2 forced the U.S. administration to action. On Nov. 7, the President made a radio-TV broadcast, naming the first White House science adviser. Later, Eisenhower approved $1 billion for the first direct federal aid to education "to meet the pressing demands of national security in the years ahead."

Vanguard. The U.S. government wanted to launch a satellite. The Navy tried to beat the Army in launching the first U.S. satellite, but achieved only a spectacular failure of its Vanguard rocket December 6, 1957. Worrying in advance that it might fail, the Navy had referred to the rocket before launch as TV-3 or Test Vehicle No. 3.

With world-wide media tuned in, Vanguard lost thrust two seconds after launch, just four feet off the pad. It fell back and exploded. The tiny six-in. satellite popped out of the flames and rolled away, transmitting its radio signal on the ground. Newspapers called it Kaputnik and Stayputnik.

The Navy tried again to beat the Army on January 25, 1958, but the Vanguard rocket fizzled within 14 seconds of ignition.

Explorer 1. The Army won the race January 31, 1958, launching the Explorer 1 satellite on a Jupiter-C rocket from Cape Canaveral. The 31-lb. satellite was designed in California by the Jet Propulsion Laboratory.

The Army's Explorer 2 failed to reach orbit March 5 when the rocket fourth stage didn't ignite. The Navy finally reached launch success when it sent the tiny 3-lb. satellite Vanguard 1 to orbit from the Cape March 17, 1958. The Army's Explorer 3 made it to orbit March 26 carrying instruments to measure cosmic rays, meteorites and temperature.

Sputnik 3. Meanwhile, the third time was not a charm for the USSR. A rocket failed to boost a large geophysical observatory to orbit February 3, 1958. But, the fourth try, carrying another geophysical observatory, was successful May 15. Sputnik 3 was solar powered and weighed 2,925 lbs.

Explorer 4. The U.S. Army launched Explorer 4 from the Cape July 26, 1958. James Van Allen, analyzing data from Explorers 1, 3 and 4, discovered belts of radiation trapped in the magnetic field surrounding planet Earth.

NASA. Despite resistance from the Department of Defense which wanted to keep space research entirely military, legislation was enacted in the U.S. on July 29, 1958—the National Aeronautics and Space Act of 1958—enabling the birth of the civilian National Aeronautics and Space Administration (NASA). On October 1, NASA was founded. It included the old National Advisory Committee for Aeronautics (NACA).

A crash program—the National Defense Education Act (NDEA)—started that September, granted money to graduate students and helped local school districts pay for math and science teachers, school equipment and buildings.

The National Science Foundation budget nearly tripled from 1958 to 1959, then doubled again by 1962. NSF persuaded scientists to write new chemistry and physics books for high schools, and held summer sessions for high school science teachers.

Explorer-6. The earliest TV pictures of Earth's cloud cover were sent down by Explorer 6 after it was launched August 7, 1959.

Tiros-1. The prototype weather satellite, Television Infra-Red Orbital Satellite 1 (Tiros 1), was launched April 4, 1960. It sent down 22,952 cloud photos in 77 days.

Echo-1. A 100-ft. aluminum-coated balloon was launched August 12, 1960. Radio and TV signals from New Jersey were bounced off the balloon to France.

Man in space. In the 1960's, U.S. and USSR launches were highlighted by manned flights to Earth orbit in small capsules, manned flights to the Moon, and unmanned probes to Venus, Mars and the Moon.

Yuri Gagarin was launched to orbit in the USSR's Vostok-1 April 12, 1961, followed by Alan Shepard on a U.S. suborbital trip to space in Mercury-3 on May 5, 1961.

The Vostok man-in-space program became a pet project for Khrushchev who mined political propaganda from spaceflight. His push for spectaculars gave the space race a solid start before he was overthrown in 1964.

Today, after decades of the Space Age, each anniversary of the Sputnik launch brings recollections of $1 movies, $1.50 restaurant dinners, $2,000 automobiles, Vanguard, Redstone, Explorer, Discoverer, the 100-ft. balloon-in-space Echo, Vostok, Mercury, Gemini, Soyuz, Apollo, Salyut, Skylab, Mir, Gagarin, Shepard, Glenn, Grissom, Armstrong, Romanenko, Titov, Manarov, Atlas, Saturn, Delta, Titan, Energia, Challenger, Buran, Cape Canaveral and Baikonur Cosmodrome.

Satellite Scoreboard

Satellites of 27 nations and international organizations were in orbit June 30, 1991. The tally below shows total objects ever in orbit, including useless debris and valuable payloads. The 7,025 in orbit include 1,989 payloads and 5,036 chunks of debris. Since 1957, all nations have sent 21,549 payloads and pieces of debris to space. A total of 14,524 objects no longer are in orbit, including 2,189 payloads and 12,335 pieces of space junk. Most no longer in orbit burned as they descended. Space junk debris in this table is large enough to be tracked by ground radar.

Payloads And Debris In Orbit
as of June 30, 1991

Country	OBJECTS IN ORBIT			NO LONGER IN ORBIT		
	Payload	Debris	Total	Payload	Debris	Total
Argentina	1	0	1	0	0	0
Australia	4	0	4	1	0	1
Brazil	3	0	3	0	0	0
Canada	15	0	15	0	0	0
China	10	79	89	20	57	77
Czechoslovakia	0	0	0	1	0	1
ESA	21	140	161	3	426	429
ESRO	0	0	0	7	3	10
France	15	20	35	7	54	61
France/FRG	2	0	2	0	0	0
Germany	10	1	11	4	5	9
India	8	0	8	4	7	11
Indonesia	5	0	5	1	1	2
Israel	0	0	0	2	2	4
ITSO	39	0	39	1	0	1
Italy	2	0	2	5	0	5
Japan	44	46	90	7	61	68
Luxembourg	2	0	2	0	0	0
Mexico	2	0	2	0	0	0
NATO	7	2	9	0	0	0
The Netherlands	0	0	0	1	3	4
Pakistan	0	0	0	1	0	1
Saudi Arabia	2	0	2	0	0	0
Spain	1	0	1	0	0	0
Sweden	2	0	2	0	0	0
UK	16	2	18	8	3	11
USA	580	2634	3214	607	2673	3280
USSR	1198	2112	3310	1509	9040	10549
Totals	1989	5036	7025	2189	12335	14524
Grand Total						21549

ESA   European Space Agency
ESRO   European Space Research Organization

FRG   Federal Republic of Germany (former West Germany)
ITSO   International Telecommunications Satellite Organization (Intelsat)
NATO   North Atlantic Treaty Organization
UK   United Kingdom of Great Britain and Northern Ireland
USA   United States of America
USSR   Union of Soviet Socialist Republics

Spacefaring Nations

The first nine nations to launch satellites to orbit were the USSR, U.S., France, Japan, China, Great Britain, India, Israel and Iraq. A satellite in orbit showed each nation had a rocket powerful enough for space launches. Since 1957, more than 4,100 satellites have been launched.

The USSR's famous Sputnik 1 started the Space Age when it was launched October 4, 1957. Since then, the USSR has launched some 2,800 satellites.

The U.S. Explorer 1 satellite followed on January 31, 1958. The U.S. since has launched some 1,200 satellites.

French space age. Asterix 1 was the 19-lb. satellite launched from North Africa which made France the third nation on Earth in space.

The launch pad was outside of the adobe village of Hammaguir, 900 miles into the Sahara Desert from Algiers. Camels, goats and sheep foraging among brittle weeds were startled November 26, 1965, when French technicians wearing burnooses blew charges, exploding bolts holding down the silvery, needle-like Diamant single-stage rocket.

Exhaust pouring from the rocket's one large nozzle pushed the small satellite to space. France's network of receiving stations tracked Asterix 1, named for a red-whiskered Celtic barbarian in French comics.

Old-fashioned chemical batteries powering the radio beacon in Asterix lasted only two days. The satellite fell from orbit to burn up in the atmosphere March 25, 1967.

Shortly after the Asterix launch, another French satellite, known as FR 1, rode a U.S. Scout rocket to orbit December 6 from Cape Canaveral. Since then France has orbited more than 50 satellites, as a key partner in the European Space Agency and on its own.

Asia. Japan launched its first satellite, Ohsumi 1, to orbit February 11, 1970, making it the fourth nation with a space rocket. China's first satellite, named Mao 1, was launched to Earth orbit April 24, 1970, on a Long March rocket.

Europe. Great Britain's first satellite, Black Knight 1, was launched on a Black Arrow rocket from Woomera, Australia, October 28, 1971. The first Ariane rocket launched by the European Space Agency from Kourou, French Guiana, carried ESA's first satellite, referred to as CAT, to orbit December 24, 1979.

South Asia. India became the seventh nation to demonstrate a space rocket when it launched its first satellite, Rohini 1, on a Satellite Launch Vehicle (SLV) rocket July 18, 1980.

Middle East. For a time, Israel and Iraq seemed to be pitted against each other in a Middle Eastern space race. However, the 1991 Persian Gulf War may have ended Iraq's ability to participate for many years.

Israel orbited its first satellite, Horizon 1 or Ofek 1, firing it to space on a three-stage Shavit rocket September 19, 1988, from a military launch pad in the Negev Desert. Shavit is Hebrew for comet. Horizon 1 fell from orbit after 4 months. Another satellite, Horizon 2, was launched April 3, 1990.

Iraq launched a 48-ton three-stage rocket December 5, 1989, from the Al-Anbar Space Research Center 50 miles west of Baghdad on a six-orbit flight. It was not a separate satellite, but the third stage of a rocket which swept around the Earth for 6 revolutions before falling out of orbit.

A U.S. nuclear-attack-warning satellite spotted the fiery Iraqi rocket exhaust as it blasted off. North American Aerospace Defense Command tracked the third stage around the globe. The 75-ft. rocket was similar to a U.S. Scout rocket used to send small satellites to low orbits. The Iraqi booster may have been a modified version of Argentina's Condor ballistic missile. The launch was the first time Iraq exposed its secret space research. Such a ballistic missile could carry a nuclear warhead 1,240 miles. Iraq already had a 600-mi. ballistic missile built around the USSR Scud. Scuds were launched against Iranian cities in the eight-year Persian Gulf war in the 1980's and Israel and Saudi Arabia in the 1991 Persian Gulf War.

Nation	Launch	Satellite	Rocket	Launch Site
USSR	1957 Oct 4	Sputnik 1	A	Baikonur Cosmodrome
USA	1958 Jan 31	Explorer 1	Jupiter-C	Cape Canaveral
France	1965 Nov 26	Asterix 1	Diamant	Algeria
Japan	1970 Feb 11	Ohsumi	Lambda 4S-5	Kagoshima
China	1970 Apr 24	Mao 1	Long March 1	Inner Mongolia
Great Britain	1971 Oct 28	Black Knight 1	Black Arrow	Woomera Australia
Europe	1979 Dec 24	CAT	Ariane 1	Kourou, French Guiana
India	1980 Jul 18	Rohini 1	Satellite Launch Vehicle	Sriharikota Island
Israel	1988 Sep 19	Horizon 1	Shavit	Negev Desert
Iraq	1988 Sep 19	rocket 3rd stage	3-stage rocket	Al-Anbar

Satellite Orbits

Most people think of satellites as flying east-west, circling the globe above the equator. Most do fly in such low equatorial orbits, but polar-orbiting satellites travel a different path, north-south across the Earth's poles several times a day. Stationary satellites actually are in equatorial orbits so far above the planet's surface they seem to stand still in the sky.

Equatorial and polar satellites fly low, near Earth, generally orbiting at altitudes between 100 and 1,000 miles. Stationary satellites, on the other hand, orbit at altitudes around 22,300 miles. They travel around the world at the same speed the globe is spinning, which makes them seem to hang stationary over one spot on Earth's surface.

Polar orbits. Polar-orbit satellites are designed to look down on the entire surface of the Earth every day. Satellites in polar orbit circle the Earth, passing above the North and South poles several times a day. As the satellite loops around the globe, the Earth seems to rotate under the orbit. The satellite passes over the entire surface daily.

Observation of the ground is improved if the surface always is illuminated at about the same Sun angle when viewed from the satellite. Most weather, Earth-resource and reconnaissance satellites are in so-called Sun-synchronous polar orbits. Examples include Nimbus, Tiros, Landsat, Geos, NOAA and Navy Oscar satellites. Geodetic-survey and navigation satellites usually are in almost-perfectly-circular polar orbits. By comparison, TV satellites and GOES weather satellites are in stationary orbits.

The first polar-orbiting U.S. weather satellite was Nimbus 1, launched in 1964, which automatically sent down high-resolution TV pictures of cloud cover in visible and infrared light. America's first polar-orbiting NOAA weather satellite was launched in 1970. It had a camera for continuous day and night pictures.

The polar-orbiting NOAA-11 weather satellite doubles as a search and rescue (SARSAT) satellite. It listens for distress signals from ships and downed planes from 500 miles high, circling Earth every 102 minutes. As the Earth rotates beneath NOAA 11, the satellite scans the entire globe every 12 hours. If it hears an emergency, NOAA 11 relays the distress signal to a rescue service on the ground.

Landsat-1, the first U.S. Earth-resources observation satellite, was sent to polar orbit in 1972. It shot photos and scanned the planet's surface 14 times a day, recording the environment of forests, oceans, deserts, croplands and urban areas.

Launch sites. Rockets can be fired in any direction from a launch pad. Neighboring human populations usually dictate the direction of launch.

From NASA's Cape Canaveral, Florida, launch pads, rockets mostly are launched east over the Atlantic Ocean, but sometimes northeast along the Atlantic Coast. From Vandenberg Air Force Base, California, rockets usually are fired west and north over the Pacific Ocean into polar orbits.

Rockets launched from Baikonur Cosmodrome in the southern USSR carry satellites to equatorial orbits, while rockets from the USSR's Northern Cosmodrome at Plesetsk carry satellites to polar orbits.

Rockets from European Space Agency's Kourou, French Guiana, launch site near the equator in South America go east or north over the Atlantic Ocean to equatorial orbits.

Air launches. Satellites don't have to be launched from the ground to reach polar orbit. In 1990, the U.S. fired Pegsat to polar orbit on a 50-ft. Pegasus rocket carried aloft by an Air Force B-52 bomber. The winged rocket was strapped under the airplane's wing, ferried seven miles above Earth and dropped. Pegasus ignited and blasted on up to a 350-mi.-high polar orbit. Pegsat contained a communications satellite and experiments.

Transfer orbit. A satellite can be fired from Earth into a special equatorial orbit known as a geostationary transfer orbit (GTO). A transfer orbit is highly elliptical—the satellite swings out as far as 22,300 miles and back in to an altitude of 100 miles above Earth. At an assigned time and place along the transfer orbit, a "kick motor" rocket attached to the satellite pushes it on out to a circular orbit at the stationary altitude of 22,300 miles.

Stationary orbit. A stationary orbit also is known as synchronous, geostationary or geosynchronous orbit. A satellite which whirls around the globe at the same speed as Earth rotates is in stationary orbit. From the ground, the satellite at 22,300 miles altitude seems to remain over the same point on Earth's surface.

Such a synchronous orbit provides a high, fixed vantage point for continuously viewing a large portion of Earth. For instance, stationary satellites are ideal for monitoring continent-wide weather patterns and environmental conditions.

Good orbits. Stationary orbits are popular for communications satellites which relay signals between points on Earth's surface. Well-known stationary communications satellites have included Syncom, Intelsat and TDRS. The common TV satellites, sending programs to backyard dish antennas, are in stationary orbits. American stationary weather satellites are known as GOES.

The U.S. communications satellite Syncom 2 was the first satellite launched successfully to stationary orbit. Relay 1 was a TV satellite launched in 1962 to an elliptical polar orbit ranging from 800 to 4,500 miles above Earth.

Satellite history was made in 1964 when Syncom 2 relayed TV pictures across the Pacific Ocean from the Olympic games in Japan. From America, the pictures were bounced on across the Atlantic Ocean to a European audience via Relay 1.

Today, dozens of TV, weather and other kinds of satellites are operated in geosynchronous orbit by many nations and international organizations.

Shuttles and stations. Shuttles flying in space, and space stations, are satellites of Earth in low equatorial orbits. U.S. space shuttles have been launched from Cape Canaveral to equatorial orbits ranging from 100 to 380 miles altitude. USSR space stations have been launched from Baikonur Cosmodrome to orbit at 200-mi. altitudes.

For example, as shuttle Discovery flew west to east 350 miles above the U.S. in September 1991, the orbit transported the shuttle as far north as Hudson Bay, Canada, in the Western Hemisphere and Moscow in the Eastern Hemisphere, and as far south as the southern tip of South America.

The crew openly relished the view. "Just about the whole United States is visible from window to window. We can see the lights from all the major cities right on down the coast from Maine all the way down almost to Florida. We're coming south right over the city of New York. We can see Long Island clearly illuminated and looking down the coast we can follow I-95 right on through Philadelphia, Baltimore, Washington, Richmond, right on into Norfolk and Virginia Beach, all the way down the Florida peninsula as far as Miami. It's hard to believe, but you can see the individual casinos in Atlantic City from here," pilot Kenneth Reightler radioed down to Houston.

The Clarke Belt

The region of space girdling the globe above the equator at an altitude around 22,300 miles sometimes is referred to as the Clarke Belt, after the science-fiction writer Arthur C. Clarke who first proposed the possibility of placing communications satellites in such stationary orbits back in 1945.

Satellites in the Clarke Belt, travelling around Earth at the same speed as Earth turns, appear to hang stationary in the sky over one spot on the ground.

Many communications, weather, science and other satellites occupy positions in the Clarke Belt today. For instance, revolving around the globe at the same speed as Earth rotates, the giant commercial communications satellite Intelsat VI (F2) appears to occupy a stationary position over central Africa.

At the time the largest commercial communications satellite ever orbited, Intelsat VI (F2) was blasted to space October 27, 1989, on a European Space Agency Ariane 4 rocket from launch pad ELA-2 at Kourou, French Guiana, South America. The 9,300-lb. satellite, known as Intelsat VI (F2), is owned by the International Telecommunications Satellite Organization.

The Ariane 4 rocket, needing four liquid-fuel strap-on boosters, pushed the satellite to a low elliptical orbit ranging from 125 miles to 22,350 miles. About five hours later, a small rocket, or apogee kick motor, attached to the satellite pushed it into a circular stationary orbit at 22,350 miles altitude.

Satellite Uses

Satellites orbiting Earth either look outward to the Sun, planets and stars, or are focused back down on Earth. They either are in low orbits circling the world every hour or two, or in high stationary orbits where they stay above one spot on Earth's surface.

Many satellites which look out from Earth have telescopes which stare at stars, galaxies and other astronomical objects, looking for visible, infrared and ultraviolet light, and other energy from deep space, and transmitting data about what they see to eager astronomers on Earth.

For example, NASA launched eight Orbiting Solar Observatories from 1962-75. Solar Max was sent to Earth orbit in 1980 to study the Sun. To look deeper into space, beyond our Solar System, NASA launched two Orbiting Astronomical Observatories in 1968-72, three Small Astronomical Satellites in 1970-75, and three High Energy Astrophysical Observatories in 1977-79.

In a joint project with the European Space Agency (ESA) and Great Britain, NASA launched IUE, the International Ultraviolet Explorer, in 1978. In a project with The Netherlands, the Infrared Astronomical Satellite (IRAS), was launched to orbit in 1982.

Using space shuttles, NASA ferried its large Hubble Space Telescope to Earth orbit in 1990 and the Gamma Ray Observatory in 1991.

Looking down. Many satellites have gone to orbit for science research and specific work applications. They have studied Earth, relayed communications, guided navigation, reported weather conditions and done other jobs.

Earth-studying satellites have included NASA's six Orbiting Geophysical Observatories launched from 1964-69, NASA's many Explorer satellites, the USSR's

Cosmos satellites, and others from several countries.

The USSR and the United States also have used various Earth-oriented satellites, such as Cosmos, Samos, Lacrosse and KeyHole-11, for military reconnaissance since the early 1960's.

Accurate navigation positions come from tracking the U.S. Navy's Transit, Oscar and Navstar satellites and Glonass satellites. Very accurate geodetic data is obtained from pulsed laser beams. Lageos, the Laser Geodynamic Satellite, orbited in 1976, measured movements of Earth's crust to within 0.8 inch, important in predicting earthquakes.

Some government, commercial and amateur communications satellites are in low 100 to 1,000-mi.-high orbits. Other communications satellites, such as the USSR's Molniya series, are in long elliptical orbits which repeatedly take them out beyond 22,000 miles before they swing back to within 1,000 miles of Earth's surface. Stationary communications satellites, orbiting at 22,300 miles altitude, can spritz radio signals over an entire hemisphere.

So-called lightsats, smallsats, cheapsats and lowsats are inexpensive, lightweight, single-purpose satellites fired on small rockets to low orbits. They have short lives in space, falling back into the atmosphere to burn up after only a few months or years. Higher satellites can take centuries to slip down to burn in the atmosphere.

Beyond Earth. Science spacecraft which leave Earth and fly on to the Moon, planets or Sun are satellites. The power of their launching rocket boosted them to such a high speed, they were able to break away from the pull of Earth's gravity. Having rocketed beyond Earth orbit, they fall into orbit around some other body.

If the destination is a planet, the spacecraft may become a satellite of that planet. For instance, the interplanetary probe Magellan flew away from Earth in 1989 and into orbit around Venus in 1990.

Japan's satellite Hiten, launched in January 1990 was orbiting Earth on such a long elliptical path, it was able in March to drop off a smaller satellite, Hagoromo, in orbit around the Moon.

Planet probes which miss their target fall into orbit around the Sun. For example, the USSR's probe Mars 4, launched in 1973, now is orbiting the Sun after a braking rocket failed and the spacecraft overshot the Red Planet in 1974.

The attraction of the Sun's gravity is so strong, few spacecraft have received sufficient boost to leave the Solar System. One which did was Pioneer 10, launched in 1972. The first spacecraft to cross the Asteroid Belt, it flew by Jupiter in 1973. Ten years later, it crossed an invisible boundary in 1983, departing the known Solar System and entering interstellar space.

Weather Satellites

Television news programs often show pictures of cloud patterns snapped from overhead only an hour or so before air time. A weatherman will point out differences between places with clear skies and swirling storm systems. Sometimes a series of time lapse photos made over a 24-hour period show movement of a storm. As familiar as they are today, such weather satellite photos were not possible before 1960.

Harry Wexler, director of research for the U.S. government Weather Bureau, was a leading proponent of weather satellites. He had imagined that cameras in space satellites, orbiting above the clouds, would see thick bands of clouds swarming along weather fronts, popcorn fair weather clouds, and enormous, revolving hurricanes over the oceans.

Tiros. Tiros 1, the first weather satellite, was launched April 1, 1960. Tiros stood for Television and Infra-Red Observation Satellite. Wexler was proven correct when, within hours of launch, government scientists were showing U.S. President Dwight D. Eisenhower photos of cloud patterns from space.

The first weather satellite picture was of clouds over the Gulf of St. Lawrence. The 270-lb. Tiros 1 worked only 78 days, but sent down 23,000 photos.

Tiros 2 was launched in November 1960. Tiros 3 was launched in July 1961. The last in the series named Tiros was Tiros 10 launched in 1965. The Tiros satellites flew at altitudes around 400 miles above Earth.

Nimbus. Tiros satellites were considered experimental. The second generation of experimental weather satellites was called Nimbus. Seven Nimbus satellites were sent to Earth orbit between 1964 and 1978. They had better still-photo cameras, television cameras for mapping clouds, and infrared radiometers for night photography. Nimbus satellites flew at 600 miles altitude.

ESSA. Tiros and Nimbus were followed by nine civilian weather satellites in the ESSA series launched from February 1966 to February 1969. ESSA photos were recorded and transmitted automatically. Anyone with proper equipment could receive them. Other nations installed gear and, today, more than 120 nations get their weather pictures from U.S. satellites. The ESSA satellites flew at altitudes around 900 miles above Earth.

NOAA. The NOAA series of weather satellites are Tiros satellites which started work December 11, 1970. NOAA-1 to NOAA-5 were like ESSA satellites, but launched into 900-mi.-high polar orbits so they would see the entire surface of Earth each day. Nighttime cloud photos were possible with infra-red sensors aboard the NOAA satellites.

Tiros-N was a research prototype for the following operational series which were called NOAA-A to NOAA-M before launch. Tiros-N was launched October 13, 1978. Starting with NOAA 6, they were placed in 800-mi.-high orbits.

In 1990, the original Trios-N cast doubt on the existence of a "greenhouse effect" around planet Earth. Scientists at NASA's Marshall Space Flight Center and University of Alabama, Huntsville, reported that analysis of data recorded over twelve years by Tiros-N showed no conclusive evidence of global warming from a greenhouse effect.

NOAA-A was launched June 27, 1979, renamed NOAA-6 and deactivated March 31, 1987. NOAA-B was launched May 29, 1980, but failed to reach its proper orbit. NOAA-C was launched June 23, 1981, renamed NOAA-7 and deactivated in June 1986 after its electrical power system failed.

NOAA-D was dropped out of sequence in favor of NOAA-E, which was a longer spacecraft and could hold more equipment, including a search and rescue radio receiver. NOAA-E was launched March 28, 1983, and renamed NOAA-8. Its clock and electrical power system broke down and it stopped working December 29, 1985.

NOAA-F was launched December 12, 1984, and renamed NOAA-9. In 1992, it was on standby in orbit with some data still being processed.

NOAA-G was launched September 17, 1986, and designated NOAA-10. It was working well in 1992, except for its Earth Radiation Budget Experiment (ERBE) scanner. The scanner would stick or hang-up sometimes.

NOAA-H was launched September 24, 1988, and renamed NOAA-11. Its instruments were working well in 1992, however its attitude control system had lost two of four gyros and was being controlled by special software.

NOAA-D was designated NOAA-12 after it reached orbit May 14, 1991, on an Atlas rocket from Vandenberg Air Force Base, California. It had five primary instruments, including the Advanced Very High Resolution Radiometer. The satellite was placed in a 522-mi.-high polar orbit inclined 98.7 degrees to the equator.

NOAA satellites are modernized Tiros satellites, part of a cooperative program between Canada, France, Great Britain, NASA and the U.S. National Oceanic and Atmospheric Administration (NOAA). NOAA-12, like the others, is a Tiros-N class satellite built by General Electric to provide day and night global environmental data.

GOES. After 1970, meteorologists realized they needed pictures shot more frequently than once a day. They needed a satellite parked overhead, continuously taking pictures. GOES 1 was launched in May 1974.

GOES stands for Geostationary Operational Environmental Satellite. At 22,300 miles above Earth, the GOES satellites are much higher in altitude. At that distance, they seem to hang stationary above assigned places on Earth's surface. The cloud pictures usually seen by the public on television today are from the series of GOES satellites. Compared with the small Tiros, today's weather satellites weigh more than 2,000 lbs.

NOAA satellites are lower and see more detail over a smaller area, while the much higher GOES satellites see an entire hemisphere. GOES can photograph one third of the entire planet every 30 minutes, while satellites in lower polar orbits might pass over a developing hurricane, for example, only once a day.

There are supposed to be two GOES in orbit watching North America—from offshore over the Atlantic and Pacific oceans—collecting and transmitting visible and infrared images every half hour. Unfortunately, the Pacific GOES failed in 1989 after six years of service. Since then, NOAA forecasters have been relying on the single remaining satellite, NOAA-7, which was launched in 1988. NOAA-7 was moved over the center of the U.S., although NOAA lets it slip toward the East Coast during hurricane season.

NOAA-7 was working well, but was running low on fuel at the end of 1991. It was supposed to run out of fuel in 1992, but NASA said conservative use might leave enough fuel to keep NOAA-7 in place until 1995. The fuel is burned in small thruster rockets fired to keep the satellite at its assigned position in stationary orbit. After NOAA-7 runs out of fuel, it will drift away from its position over the equator.

GOES-NEXT. The current GOES series is to be replaced by an improved series called GOES-NEXT. A replacement GOES-NEXT satellite was to have been launched in 1989, but remained on the ground at the end of 1991 with technical problems. Mirror flaws, wiring problems and infrared detector defects had been found.

Computer models and hardware tests found temperature extremes—such as exposure to raw sunlight in space—could warp the surface of the satellite's mirror. Five GOES-NEXT satellites are being built with mirrors to reflect images to various sensors. Unlike the flawed Hubble Space Telescope mirror which is concave, the GOES-NEXT mirrors are flat. Engineers coated the mirrors to reduce susceptibility to temperature extremes.

The U.S. declared a weather satellite emergency in 1991. Lease of European and Japanese weather satellites was considered. European Space Agency moved a weather satellite farther out over the Atlantic to give the U.S. weather bureau additional forecasting capability. If the last GOES were to run out of fuel before the first GOES-NEXT is launched, the U.S. would have to rely on pictures from the polar-orbiting NOAA-9, NOAA-10 and NOAA-11 weather satellites, as well as European, Japanese and U.S. military meteorological satellites.

Military satellites. Many U.S. military weather satellites have been developed secretly and launched over the years after 1960. A rebuilt Atlas ICBM ferried an 1,815-lb. Defense Meteorological Satellite from Vandenberg Air Force Base, California, to a polar orbit in 1988. It was dropped off in an orbit 527 miles above Earth, replacing one launched from Vandenberg in 1983. The Air Force previously had launched a Defense Meteorological Satellite in 1987 from Vandenberg.

Two such weather satellites usually are in orbit at a time, circling the globe in a north-south flight path every 12 hours. Tracking storms, they scan 1,600-mi.-wide swaths of Earth's surface, feeding cloud cover and temperature information to the military. NOAA also uses the information for civilian weather forecasts.

Meteor. The first USSR weather satellite was Cosmos 4 launched April 26, 1962. Today's Soviet weather satellites are known as Meteor. Many have been launched to polar orbit in two series. For instance, the USSR sent its fifteenth Meteor 2 weather satellite to a 600-mile-high orbit from the Northern Cosmodrome near Plesetsk on January 5, 1987, despite subzero temperatures which disrupted Europe and Asia that day.

Unofficial listeners in England reported radio signals from the new Meteor were interfering with transmissions from an older Meteor weather satellite in orbit and a secret military satellite known as Cosmos 1766 which used radar for ocean surveillance.

All three satellites apparently were transmitting on the same frequency. On January 14, 1987, the USSR changed the new satellite frequency, but the older Meteor apparently was stuck on the old frequency where it continued interfering with the military satellite.

Europe. The European Space Agency (ESA) also has launched a series of weather satellites called Meteosat. They are controlled from ESA's Operations Centre at Darmstadt, West Germany. Weather data is fed to national weather bureaus in Belgium, Denmark, West Germany, Finland, France, Greece, Ireland, Italy, Norway, Portugal, Spain, Sweden, Switzerland, The Netherlands, Turkey and Great Britain.

Meteosat is a combined weather observer and radio relay in the sky. Its main payload is a sharp-eyed radiometer which takes infrared, water vapor and visible light pictures of the Earth. In infrared, it can resolve objects down to three miles. Visible light photos show objects down to 1.5 miles.

Meteosat takes pictures every half hour and transmits them to Darmstadt for processing. The processed images then are sent by radio back up to the satellite for relay down to the 16 countries using the weather data.

The Meteosat series started with a first satellite placed in orbit in 1977 and a second in 1981. The 1981 satellite was expected to work three years, but still is working in stationary orbit 22,245 miles above England. The 705 lb. cylinder is seven feet in diameter and ten feet tall.

New weather satellites have been launched to replace the 1977 and 1981 satellites. The new satellites have the same capabilities plus the added ability to relay weather charts and written weather reports to all 16 countries.

Japan. Japan's weather satellites are called Himawari or Sunflower. Four have been launched to orbit. The 715-lb. weather satellite Sunflower No. 4 was blasted off the Tanegashima Island launch pad September 6, 1989, on a three-stage H-1 rocket.

Sunflower No. 4 was sent to a stationary orbit 22,370 miles above New Guinea. The 1989 flight was the fourth successful launch in the Himawari series started in 1986 by Japan's National Space Development Agency (NASDA).

The $36.4 million satellite Sunflower No. 4 replaced the aging Sunflower No. 3 already in orbit sending weather pictures to ground stations. Sunflower No. 3 continued snapping cloud photos and transmitting them to Earth for another year.

NASDA operates the Tanegashima Island Space Center 616 miles southwest of Tokyo at the southern tip of Japan off Kagoshima Island on the southern tip of Kyushu, Japan's southernmost main island.

China. China launched its first weather satellite in 1988. Its first two weather satellites have been called Feng Yun No. 1 and Feng Yun No. 2, or Wind and Cloud No. 1 and No. 2. They were sent to polar orbits on Long March 4 rockets from Taiyuan space center in China's northern Shanxi province.

India. India has built and launched its own space satellites, and has had others build and launch satellites. The series called Insat, manufactured in the U.S., have been dual-purpose satellites to collect weather data as well as relay television and long-distance telephone service across India. Insat stands for India Satellite.

India's remote-sensing and Insat weather satellites have been launched by the U.S., Europe and the USSR. For instance, the Indian satellite Insat-1C was to have been ferried to space in a U.S. shuttle, but was switched to Ariane after the 1986 Challenger disaster. A European Space Agency Ariane 3 rocket ferried Insat-1C to space July 22, 1988.

The 2,618-lb. satellite was expected to have a working life of 10 years in a stationary orbit 22,500 miles above the equator. Unfortunately, part of the electrical power system in Insat 1-C broke down in space July 29, a week after riding Ariane 3 to orbit, according to the Indian Space Research Organization (ISRO). Then a part in one of two solar-powered electricity-generating systems malfunctioned. Only half of the satellite was connected to a power system that was operating properly.

Insat-2A and Insat-2B were scheduled for launch on Ariane in 1992. The India Remote Sensing (IRS) satellite was launched by the USSR March 17, 1988.

Latest technology. The latest weather satellites record atmosphere and ocean temperatures at various altitudes and depths, gauge rainfall for forecasting of droughts and harvests, survey chlorophyll content for crop health studies, spot forest fires, map ocean currents, see volcanic eruptions, and chart ice in shipping lanes. In fact, they monitor the total global environment. Some also have search-and-rescue capability. They carry radio equipment listening for faint emergency signals from ships and aircraft.

NASA recently devised a satellite thermometer 100 times more sensitive than earlier equipment. The new infrared radiometer measures ocean surface temperatures from space.

It can tell the difference between the surface of the ocean and the atmosphere immediately above it—a major breakthrough. It had been difficult to draw accurate climate maps before, because sensors couldn't make the distinction between surface and atmosphere. To test it, NASA put the new gadget aboard the Goodyear blimp Columbia and flew it over the Pacific Ocean between Los Angeles and Santa Catalina Island. The radiometer gauged sea temperatures by receiving and analyzing infrared light in sunlight reflected naturally from the ocean surface.

Meteorologists say they need the better temperature and wind maps from the radiometer to draw more-detailed pictures showing the interaction of worldwide weather patterns. For example, they hope to learn how storms in the tropics spread to other latitudes. Eventually, they want to understand powerful weather such as El Nino, the unusually-warm water in the eastern Pacific Ocean which can damage fishing industries.

The new infrared radiometer is passive—it only receives light in the infrared part of the energy spectrum. Weather radar, on the other hand, is an active device because it both transmits and receives radio signals to monitor weather patterns.

Tiros	Launch	
Tiros 1	Apr 1	1960
Tiros 2	Nov 23	1960
Tiros 3	Jul 12	1961
Tiros 4	Feb 8	1962
Tiros 5	Jun 19	1962
Tiros 6	Sep 18	1962
Tiros 7	Jun 19	1963
Tiros 8	Dec 21	1963
Tiros 9	Jan 22	1965
Tiros 10	Jul 2	1965
Tiros N	Oct 13	1978

Nimbus	Launch	
Nimbus 1	Aug 28	1964
Nimbus 2	May 15	1966
Nimbus 3	Apr 14	1969
Nimbus 4	Apr 8	1970
Nimbus 5	Dec 11	1972
Nimbus 6	Jun 12	1975
Nimbus 7	Oct 24	1978

ESSA	Launch	
ESSA 1	Feb 3	1966
ESSA 2	Feb 28	1966
ESSA 3	Oct 2	1966
ESSA 4	Jan 26	1967
ESSA 5	Apr 20	1967
ESSA 6	Nov 10	1967
ESSA 7	Aug 16	1968
ESSA 8	Dec 15	1968
ESSA 9	Feb 26	1969

NOAA	Launch	
NOAA 1	Dec 11	1970
NOAA 2	Oct 15	1972
NOAA 3	Nov 6	1973
NOAA 4	Nov 15	1974
NOAA 5	Jul 29	1976
NOAA 6	Jun 27	1979
NOAA 7	Jun 23	1981
NOAA 8	Mar 28	1983
NOAA 9	Dec 12	1984
NOAA 10	Sep 17	1986
NOAA 11	Sep 24	1988
NOAA 12	May 24	1991
NOAA I	*Dec	1991
NOAA J	*Dec	1993
NOAA K	*Jul	1994
NOAA L	*Jul	1996
NOAA M	*Feb	1997

GOES	Launch	
GOES 1	Oct 16	1975
GOES 2	Jun 16	1977
GOES 3	Jun 16	1978
GOES 4	Sep 9	1980
GOES 5	May 22	1981
GOES 6	Apr 28	1983
GOES 7	Feb 26	1987
GOES I	*Dec	1992
GOES J	*Aug	1993
GOES K	*Jul	1995
GOES L	*Feb	1997
GOES M	*Jul	2000

*scheduled to be launched

Communication Satellites

Communications satellites are radio relay stations in the sky. Signals are transmitted to a satellite. The satellite receives them and retransmits them down to Earth. Below are selected highlights from the early history of these orbiting radio relays:

1945. Author Arthur C. Clarke outlined the possibility of a global communication system using satellites in stationary orbit. He assumed it would involve manned space stations.

1955. April, John R. Pierce of Bell Laboratories published a paper laying out technical requirements for a satellite communication system.

1957. October 4, the USSR launched Sputnik 1, the first artificial satellite in orbit. Sputnik 2 was launched November 3, 1957.

1958. January 31, the U.S. launched Explorer 1, America's first satellite in orbit.

1958. December 18, the Score satellite was launched to broadcast a pre-recorded message, taped by President Dwight D. Eisenhower.

1960. April 4, the U.S. Tiros 1 satellite was the first weather satellite in orbit.

1960. May 13, NASA failed in an attempt to launch Echo A-10, a 100-ft. Mylar balloon to be a passive communications satellite.

1960. August 12, Echo 1, a 100-foot Mylar balloon satellite was placed in orbit as a passive communications satellite.

1960. October 4, Courier 1B was launched to become the first successful active communications satellite. It had the first store-and-forward message bulletin board equipment in space.

1961. April 12, Yuri Gagarin became the first man in space, flying the USSR's Vostok 1 capsule.

1961. May 5, Alan Shepard in his Mercury 3 capsule became the first American in space, a suborbital flight.

1961. October 21, the satellite Midas 4 was launched with the West Ford passive dipoles which failed to disperse after release in space. Radio signals were to have been bounced off the metal needles, reflecting signals back to Earth. A second try in 1963 was successful.

1961. December 12, OSCAR 1, the first amateur radio satellite was launched.

1962. February 20, John Glenn in the Mercury 6 capsule was the first American in orbit.

1962. April 26, Cosmos 4 was the USSR's first weather satellite and its first photo-reconnaissance spysat.

1962. July 10, NASA launched Telstar 1, owned by AT&T Bell Laboratories, the first active real-time communications satellite, it received, amplified and retransmitted signals from Earth.

1962. August 31, President Kennedy signed legislation creating the Communications Satellite Corp., with a mission to establish a worldwide satellite network to promote world peace and understanding.

1962. December 13, Relay 1, an active real-time communications sat by RCA was launched. It had 12 telephone channels, 1 TV channel.

1963. February 14, electronics failed in the Syncom 1 launch, an attempt to place the first satellite in stationary orbit.

1963. May 7, Telstar 2 was launched.

1963. May 9, the West Ford metal-needle dipoles dispersed successfully in space as passive communications reflectors.

1963. July 26, Syncom 2 became the first communications satellite launched successfully to stationary orbit; TV pictures for broadcast in U.S. received from 1964 Olympics in Japan.

1964. January 21, Relay 2 was launched.

1964. January 25, Echo 2, another 100-ft. Mylar balloon, was launched as a passive radio reflector in the first joint space project with the USSR.

1964. August 19, Syncom 3 was launched, the second satellite in stationary orbit.

1965. February 11, the active real-time communications satellite Les 1 was launched.

1965. March 9, OSCAR 3 was launched, the first active real-time amateur radio communications satellite.

1965. April 6, Early Bird was launched as the first commercial communications satellite. It also was known as Intelsat 1.

1965. April 23, Molniya 1A was launched to an elliptical orbit, the USSR's first active real-time communications satellite.

1965. May 6, the communications satellite Les 2 was launched.

1965. October 14, Molniya 1B was launched to become the USSR's second communications satellite.

1965. December 21, OSCAR 4 was launched to be the first high-altitude amateur radio communications satellite.

1967. The first USSR communications satellites were orbited for clandestine spy missions.

1967. June 25, "Our World," a TV program using five satellites, brought together artists such as the Beatles in London, Marc Chagall and Joan Miro in Paris, and Van Cliburn and Leonard Bernstein in New York.

1974. April, Western Union launched Westar I, the first satellite for communications within the United States.

1975. December, the USSR's Raduga stationary communications satellites became operational.

1976. October 26, the USSR's first Ekran stationary communications satellite launched.

1976. September, Taylor Howard of San Andreas, California, became the first American to receive satellite TV signals on a home system.

1978. October 26, the first two USSR amateur radio communications satellites, RS-1 and RS-2, were launched.

1978. December 19, the USSR's first Gorizont stationary communications satellite launched.

1979. December, Neiman Marcus featured a home satellite TV system on the cover of its Christmas catalog for about $36,000.

1984. October, President Reagan signed a law legalizing private reception of unscrambled satellite TV programming.

1986. January 15, Home Box Office Inc. announced it would be the first programmer to scramble its satellite TV signal full time.

1986. January 28, shuttle Challenger exploded, killing seven crew members. The loss left the United States without a launch vehicle for satellites for nearly three years.

1986. April 27, John MacDougall of Ocala, Florida—also known as Captain Midnight—illegally interrupted a Home Box Office (HBO) broadcast for more than four minutes to display a message vowing not to pay for the movie service.

1987. June 6, Pope John Paul II led Roman Catholics on five continents in prayers for peace in a live broadcast involving a record 23 communications satellites and more than 35 languages.

1988. September 29, With the launch of shuttle Discovery, America regained its satellite-launching capability.

Echo Satellites

Many times during the decades of the Space Age which started in 1957, America seemed to be playing catch-up with the Russians.

For instance, entering the 1960's, America was looking for ways to recover prestige lost to the USSR following successful unmanned Sputniks and the Korabl Sputnik carrying two live dogs which parachuted back to Earth.

The world was anticipating a manned flight and, as we came to know, Vostok would carry Yuri Gagarin on April 12, 1961, as the first man in space.

Meanwhile, before the Gagarin flight, U.S. President Dwight D. Eisenhower in 1960 itched for some swift, spectacular, low-cost space achievements. One was to be Echo.

Shiny balloon. Echo A-10, a giant Mylar balloon to be inflated in orbit, was the first passive communications satellite, but NASA failed to get it to orbit in a May 13, 1960, launch attempt.

Three months later, Echo 1, the 49th satellite ever sent successfully from Earth to orbit by any country, was launched August 12, 1960. It, too, was a Mylar balloon inflated in orbit to a diameter near 100 feet.

It was the world's first successful passive communications satellite, or comsat. Radio signals beamed toward the satellite from the ground bounced off the highly-reflective exterior of Echo 1 to other listening posts on the ground.

Visually, Echo turned out to be brighter than anything else in the sky, except for the Sun and Moon. Sighting it by naked eye was popular around the world for several weeks. Echo 1 descended to burn up in the atmosphere May 24, 1968.

First cooperation. Echo 2, the 740th satellite ever orbited, went up January 25, 1964, and came down June 7, 1969. Another 100-ft. Mylar balloon, Echo 2 was the first ever space-cooperation project with the USSR.

Needle dipoles. Bouncing radio signals off of space objects was popular. A cluster of metal needles were fired to orbit by the U.S. and released in space May 9, 1963, in a project called West Ford. The needles were cut to appropriate lengths to act as radio dipole antennas. They reflected signals sent up from Earth. An earlier October 21, 1961, launch of West Ford dipoles from the satellite Midas 4 had failed when the needles didn't disperse.

Moonbounce. The Moon also has been used as a passive radio signal reflector over the years. In a technique known as EME for Earth-Moon-Earth, signals of very high power are transmitted from antennas pointed at the Moon. The radio waves bounce off the Moon, reflecting back to Earth to be received by listeners. U.S. Army Signal Corps was first to bounce radar off the Moon on January 10, 1946.

Early Communications Satellites

Early communications satellites were real-time repeaters, except for Score, Echo, Courier 1B and West Ford. Score had a pre-recorded taped message broadcast from space. Echo satellites were large passive Mylar balloons. Courier 1B was the first successful active communications satellite with the first store-and-forward message bulletin board equipment. West Ford was a cluster of metal needles cut to an appropriate length to become dipole radio antennas reflecting signals back to Earth.

Below, PER/AP is perigee (lowest altitude) and apogee (highest altitude) miles:

Satellite	Launch	Per/Ap	Details
Score	12/18/58	115/914	broadcast a taped message
Echo A-10	5/13/60	---	failed to orbit, NASA
Echo 1	8/12/60	941/1052	first successful passive
Courier 1B	10/4/60	586/767	first successful active
Midas 4	10/21/61	2058/2324	West Ford dipoles failed to disperse
Telstar 1	7/10/62	593/3503	first active real-time, AT&T
Relay 1	12/13/62	819/4612	active real-time, RCA
Syncom 1	2/14/63	21195/22953	electronics failed, NASA
Telstar 2	5/7/63	604/6713	active real-time
West Ford	5/9/63	2249/2290	dipoles successful
Syncom 2	7/26/63	22062/22750	1st successful stationary orbit
Relay 2	1/21/64	1298/4606	active real-time
Echo 2	1/25/64	642/816	last passive, first joint space project with USSR
Syncom 3	8/19/64	22164/22312	2nd in stationary orbit
Les 1	2/11/65	1726/1744	active real-time
OSCAR 3	3/9/65	565/585	1st active real-time hamsat
Early Bird	4/6/65	21748/22733	first commercial comsat, was known also as Intelsat 1
Molniya 1A	4/23/65	309/24470	1st USSR active real-time
Les 2	5/6/65	1757/9384	active real-time
Molniya 1B	10/14/65	311/24855	2nd USSR active real-time
OSCAR 4	12/21/65	101/20847	first high-altitude hamsat

NASA TDRS Communications Sats

TDRS tracking and data relay satellites are NASA's global communications link, relaying voices of astronauts laboring in space shuttles amidst millions of data bits from orbiting observatories.

NASA has launched five of the 2.5-ton satellites. Four are in orbit. Just one TDRS could relay all of the data in a 24-volume encyclopedia in five seconds.

The first satellite in the TDRS series was hauled to orbit April 4, 1983, in the maiden voyage of shuttle Challenger. The second TDRS was destroyed in Challenger's last flight, January 28, 1986, as the shuttle exploded during lift off.

Like the thrown rider climbing back on his horse, NASA returned to space after Challenger with a TDRS in Discovery's cargo hold September 29, 1988. Six months later, Discovery carried another of the 50-ft. TDRS communications satellites on March 13, 1989. The fifth TDRS was ferried to space in shuttle Atlantis August 2, 1991. NASA plans to send shuttle Endeavour to space with TDRS-6 in January 1993.

Each TDRS has two arrays of solar cells, converting sunlight to electricity, and seven antennas ready to transmit more than 300 million bits of information per second per radio channel to TDRS ground control at White Sands, New Mexico.

Besides relaying astronaut chatter data from space shuttles in orbit, TDRS satellites retransmit science data to White Sands from unmanned science satellites and observatories such as Hubble Space Telescope, Cosmic Background Explorer, Compton Observatory and the Upper Atmosphere Research Satellite.

TDRS-1 is stationed over the equator at 171 degrees west. TDRS-2 was lost with Challenger. TDRS-5 replaced TDRS-3 which is standing by as emergency backup, at 62 degrees west longitude. TDRS-4 is positioned at 41 degrees west. TDRS-5 operates at 174 degrees west longitude over the Gilbert Islands, south of Hawaii.

Prior to TDRS, NASA relied on a network of more than 20 ground stations dotted around the globe. Controllers could communicate with a shuttle or satellite during only about 15 percent of its orbit. With TDRS, controllers could communicate with a satellite during 85 to 100 percent of its orbit. The TDRS network is managed by NASA's Goddard Space Flight Center, Greenbelt, Maryland.

Amateur Radio Satellites

Ham radio is flying high these days! Thirty-nine amateur communications satellites have been launched since the first in 1961.

Private groups of amateur radio operators around the world have built and sent those hamsats to orbit since the first—OSCAR 1—on December 12, 1961. The satellites are financed by hams completely through donations of time, hardware and cash. They receive free rides to space as ballast on various government rockets carrying other commercial and government satellites to orbit. U.S., USSR, European and Japanese governments have provided the free rides.

The number of satellites has been growing rapidly in recent years. Only four amateur satellites were orbited in the 1960's. Six went to space in the 1970's. Seventeen were launched in the 1980's. In the first two years of the 1990's alone, twelve have been launched. Record launch years were 1981 and 1990, with seven each. Close behind was 1991 with five hamsats launched. Altogether the total is 39 amateur satellites sent to Earth orbit in 30 years. Most hamsats remain in orbit today and many still are in use.

Most hamsats have been called OSCAR or RS. OSCAR stands for Orbital Satellite Carrying Amateur Radio. The OSCAR amateur radio satellites are not the same as U.S. Navy Oscar navigation satellites. RS stands for Radiosputnik, the name of most USSR amateur radio satellites. A complete list is below.

How they work. Most amateur radio satellites are in low orbits near Earth. They circle the globe, coming within range of a ham radio station on the ground every hour or so. On the other hand, a few such as OSCAR 10 and OSCAR 13 are in long elliptical orbits which keep them in view of many ground stations for hours at a time.

Most hamsats have been radio repeaters in the sky, with transponders which act as radio-signal relays. A ham operator on the ground would transmit his voice or Morse code signal up to a satellite which would repeat it down over a wide area.

Some ham satellites have had other kinds of experimental gear for digital computer-to-computer communication, television pictures of Earth from space, ionosphere research, radio propagation tests and radio astronomy.

OSCARs. American amateur radio operators have had international help from hams in West Germany, Japan, Canada, Australia, Great Britain, and elsewhere during design and construction of the satellites in the OSCAR series.

The beeping space beacon OSCAR 1 was the first satellite ever built by amateur radio operators anywhere. The 10-lb. satellite was launched December 12, 1961, as ballast on the Thor-Agena rocket which carried Discoverer 36 to space.

OSCAR 1 transmitted a weak radio signal from orbit on a frequency of 145 MHz. Its power was only 140 milliwatts, yet that was 14 times the power of the 10-milliwatt radio transmitter in America's first government satellite, Explorer 1, in January 1958. It did not carry a two-way repeater or transponder, but OSCAR 1 transmitted the greeting HI in International Morse code. The speed of the message was controlled by the temperature inside the satellite. Hundreds of amateurs around the globe mailed in reception reports. OSCAR was at such a low altitude it remained in orbit only 50 days before descending into the atmosphere and burning up January 31, 1962.

OSCAR 2, very similar to the first hamsat, was launched June 2, 1962, only six months after OSCAR 1. Transmitter power was lowered to 100 milliwatts to make batteries last longer. Another OSCAR was built about the same time, with a 250-milliwatt transmitter, but was not launched.

OSCAR 3 was the first hamsat to have a two-way signal-repeating transponder, so it could act as radio relay in the sky. It was launched March 9, 1965. It had a receiver for 146 MHz signals and a 1-watt transmitter at 144 MHz. Along with the transponder, the satellite also had two radio beacons. One sent a continuous signal for tracking and propagation studies. The other sent telemetry data about temperatures and battery voltages. A few solar cells were included to back-up the battery powering the beacons, thus the first use of solar power on an amateur spacecraft. But, the satellite was not fully solar powered. The solar cells allowed the beacons to work months longer than the transponder. More than 1,000 hams from 22 nations talked through OSCAR 3 during its 18 days of operation. Exciting first contacts included New Jersey to Spain, Massachusetts to Germany and New York to Alaska.

OSCAR 4 was the first amateur satellite to have a partial launch failure. It blasted off December 21, 1965, aboard a Titan 3-C rocket, but the upper rocket stage failed and the satellite did not make it to its intended 21,000-mi.-high circular orbit. The satellite ended up in an orbit ranging from 100 miles out to 21,000 miles. OSCAR 10 had a similar problem 18 years later. It was the first satellite to be powered fully by solar cells generating electricity. The transponder received on 144 MHz and transmitted on 432 MHz with a power of 3 watts. No telemetry beacon was included, so when the radio failed after a few weeks, hams were unable to know why. The battery may have overheated or radiation may have knocked out the solar cells. The first U.S.-USSR satellite contact was made through OSCAR 4.

The first four OSCAR projects had been managed by hams in a U.S. West Coast group known as Project Oscar. AMSAT, the Radio Amateur Satellite Corp., was formed in 1969 as a U.S. East Coast group to build and fly hamsats.

OSCAR 5, also known as Australis-OSCAR 5, was designed and built by students in the Astronautical Society and Radio Club at the University of Melbourne, Australia. AMSAT managed the U.S. launch of OSCAR 5 which flew to space aboard a NASA rocket January 23, 1970. The first amateur satellite which could be controlled from the ground, OSCAR 5 had a command receiver allowing ground stations to turn the 29.450 MHz beacon transmitter on and off. OSCAR 5 had telemetry beacons transmitting at 144.050 MHz and 29.450 MHz. It did not use solar cells and had no transponder. It did have a magnetic attitude-stabilizing system.

The first generation of hamsats ended with OSCAR 5. OSCAR 6 was the beginning of the second generation of amateur radio satellites, or Phase 2 hamsats. OSCARs 7, 8, 9, 11 and 12 and others have been Phase 2 satellites. The six microsats launched in 1990 were Phase 2 satellites.

Third-generation, or Phase 3, satellites came with Phase 3A which was lost in a 1980 launch failure. Phase 3B was orbited in 1983 and named OSCAR 10. Phase 3C became OSCAR 13 in 1988.

OSCARs 9, 11, 14, 15 and 22 also are known as UoSAT for University of Surrey satellite. They were built by students at that university in England.

Other students, at Weber State College, Utah, built NUsat which stands for Northern Utah Satellite. It was carried to space in shuttle Challenger and dropped overboard. Later, the students built WEBERsat/OSCAR 18. It has a TV camera taking still photos of Earth.

Fuji-OSCAR 12 and Fuji-OSCAR 20, built by Japanese amateur radio operators and launched on a Japanese rocket from Japan, were known as JAS-1a and JAS-1b before launch. JAS stood for Japan Amateur Satellite.

The Brazilian hamsat OSCAR 17, known as DOVE for Digital Orbiting Voice Encoder, was like NUsat and WEBERsat in that it was designed to educate the public about space. DOVE has a recorder and voice-synthesizer attached to its transmitter. The satellite rebroadcasts peace messages written and spoken in various languages by school children around the world and sent up to space by hams in Brazil.

OSCAR 17 messages can be heard by anyone with a VHF FM receiver or scanner of the kind used to monitor police radio. A school teacher's guide to using the satellite with classroom exercises and experiments is available from AMSAT Science Education Advisor Rich Ensign, 421 N. Military, Dearborn, Michigan 48124 USA.

OSCAR 23 is not a typical amateur radio satellite used for two-way communications among hams on the ground. Rather, it is a first-of-its-kind amateur radio astronomy telescope in orbit. Known as SARA before launch, the satellite was built by ESIEESPACE, an amateur astronomy club at France's Ecole Superieure d'Ingenieurs en Electrotechnique et Electronique (ESIEE). SARA listens for high-frequency radio signals from Jupiter and relays information about what it hears to the ground. SARA, built with off-the-shelf consumer equipment, identifies itself with the amateur call FXØSAT.

Electromagnetic energy from Jupiter is very strong. SARA uses three pairs of 15-ft. antennas to snare the naturally-generated radio waves from the planet. SARA's receiver listens to eight 100-KHz-wide channels between 2 MHz and 15 MHz.

Radiosputniks. A dozen Russian hamsats have been called Radiosputnik or RS for short. There also were two USSR hamsats known as Iskra, spark in Russian.

The first Russian amateur radio satellites were RS-1 and RS-2. Later, students built Iskra 2 and Iskra 3 which were launched by hand from the Salyut 7 space station.

The largest clutch of satellites orbited at one time was six Radiosputniks—RS-3 through RS-8—in 1981. That launch was matched in 1990 by six OSCARs launched on a European Space Agency Ariane rocket.

RS-10 and RS-11 are two sets of gear aboard one large satellite in Earth orbit. RS-12 and RS-13 are similar Radiosputniks riding piggyback on one larger satellite.

More Info. Information on amateur radio satellites is available from the American Radio Relay League (ARRL), 225 Main Street, Newington, Connecticut 06111 or from AMSAT, P.O. Box 27, Washington, D.C. 20044.

ARRL is the primary national fraternity of amateur radio hobbyists in the U.S. AMSAT is the Radio Amateur Satellite Corporation, a worldwide organization of amateur radio operators, and other persons interested in space communications and research, building and using space satellites. AMSAT has local affiliates in Germany, Great Britain, the USSR, France, Japan and many other nations around the globe.

In this list of amateur radio satellites sent to space over the years, Ap is apogee, the satellite's maximum altitude in miles. Cntry is the country of origin. Life is the satellite's operational life in orbit—down is the date when the satellite descended into the atmosphere and burned up. A satellite still is in orbit if it has no down date, although it may be dead in orbit. TBA is to be announced.

Launch	Hamsat	Ap	Cntry	Life
1961 Dec 12	OSCAR 1	290	USA	operated 21 days, down 1/31/62
1962 Jun 2	OSCAR 2	240	USA	operated 19 days, down 6/21/62
1965 Mar 9	OSCAR 3	590	USA	transponder 18 days, beacon months
1965 Dec 21	OSCAR 4	21,000	USA	bad orbit, oper'd 85 days, down 4/12/76
1970 Jan 23	OSCAR 5	925	Australia	operated 52 days
1972 Oct 15	OSCAR 6	910	USA	operated 4.5 years
1974 Nov 15	OSCAR 7	910	USA	operated 6.5 years
1978 Mar 5	OSCAR 8	570	USA	operated 5.3 years
1978 Oct 26	RS 1	1,050	USSR	lasted several months
1978 Oct 26	RS 2	1,050	USSR	lasted several months
1980 May 23	Phase 3A	0	USA	launch failed
1981 Oct 6	OSCAR 9	338	UK	UoSAT operated 8 yrs
1981 Dec 17	RS 3	1,030	USSR	operated for years
1981 Dec 17	RS 4	1,030	USSR	operated for years
1981 Dec 17	RS 5	1,030	USSR	lasted until 1988
1981 Dec 17	RS 6	1,030	USSR	operated for years
1981 Dec 17	RS 7	1,030	USSR	lasted until 1988
1981 Dec 17	RS 8	1,030	USSR	operated for years
1982 May 17	Iskra 2	210	USSR	operated 53 days. down 7/9/82
1982 Nov 18	Iskra 3	210	USSR	operated 37 days, down 12/16/82
1983 Jun 16	OSCAR 10	22,060	USA	improper orbit, but still operating
1984 Mar 2	OSCAR 11	431	UK	UoSAT still operating
1985 Apr 29	NUsat 1	250	USA	down 12/15/86
1986 Aug 12	OSCAR 12	900	Japan	JAS-1a Fuji pacsat turned off in orbit
1987 Jun 23	RS 10	620	USSR	still operating
1987 Jun 23	RS 11	620	USSR	still operating
1988 Jun 15	OSCAR 13	23,400	USA	still operating
1990 Jan 22	OSCAR 14	500	UK	UoSAT-D, operating
1990 Jan 22	OSCAR 15	500	UK	UoSAT-E, failed in orbit
1990 Jan 22	OSCAR 16	500	USA	Pacsat, operating
1990 Jan 22	OSCAR 17	500	Brazil	Dove, operating
1990 Jan 22	OSCAR 18	500	USA	WEBERsat, operating
1990 Jan 22	OSCAR 19	500	Argentina	LUsat, operating
1990 Feb 7	OSCAR 20	1100	Japan	JAS-1b Fuji pacsat replaced OSCAR 12
1991 Jan 29	RS 14/OSCAR 21	625	USSR	Radio-M1, Rudak-2 pacsat
1991 Feb 5	RS 12	600	USSR	like RS 10 & RS 11
1991 Feb 5	RS 13	600	USSR	like RS 10 & RS 11
1991 Jul 17	OSCAR 22	500	UK	UoSAT-F, operating
1991 Jul 17	OSCAR 23	500	France	SARA, radioastronomy Jupiter observer

Possible Future Hamsat Launches

1992	Arsene	France	to be launched
1992-93	Itamsat-1	Italy	to be launched
1992-93	KITsat	South Korea	to be launched
1992-93	tba	Great Britain	to be tossed out of Mir station
1993-95	SEDsat	U.S.	science research and communications
1992-93	SUNsat	South Africa	to be launched
1992-93	TECHsat	Israel	to be launched
1994-95	Phase-3D	German/U.S.	to be launched

Soviet-American Biosatellites

The U.S. and USSR have cooperated since 1971 in launching satellites—known as biosats—for biological studies in the weightlessness of outer space. A U.S.-USSR Space agreement was signed in 1971, setting up a joint working group on space biology and medicine between the USSR Academy of Sciences and NASA. In 1974, the USSR asked the U.S. to join in actual spaceflights. The five biosats flown between 1975-1985 were known as Cosmos 782, Cosmos 936, Cosmos 1129, Cosmos 1514, and Cosmos 1667.

A new agreement called Cooperation in the Exploration and Use of Outer Space for Peaceful Purposes was signed in 1987. Again, the U.S. participated in a USSR biosat flight, Cosmos 1887. For the 1987 flight, the U.S. provided 26 major life-sciences experiments from more than 50 scientists at U.S. universities and NASA's Ames Research Center. In addition, the U.S. placed eight radiation detector packages inside and outside the Cosmos 1887 biosat to measure dosages in space that might be harmful to astronauts. Eight American scientists and engineers travelled to the USSR in October 1987 for the Cosmos landing. The USSR invited the U.S. to participate again, in Cosmos biosat flights in 1989 and 1991.

Cosmos 1887. After five boring days in an orbiting space capsule, Yerosha, one of two rhesus monkeys launched September 29, 1987, by the USSR in Cosmos 1887 decided to have fun for the rest of his jaunt.

Yerosha and his companion, Dryoma, had been strapped into little chairs in a hermetically-sealed biological research spacecraft for a two-week flight. But, as his ground controllers noted, Yerosha was an intelligent, lively passenger. In fact, Yerosha was such an interesting space flier, the monkey may be sent back to orbit.

Yerosha grabbed the attention of ground controllers on his fifth day in space when he appeared on TV screens without his metal name tag. The tag had been pinned on his hat which somehow had become askew. Yerosha had managed to free his left front paw from a restraining cuff to play with the hat—and everything else he could reach.

Yerosha, whose name means troublemaker in Russian, was smaller and livelier than Dryoma. Ground control considered cutting the flight short to see what Yerosha had done with his name tag. They set up a mock capsule at Moscow, with a similar monkey and the same temptations, to see what it might do with a free hand. When nothing much happened, they decided Yerosha couldn't get into trouble and let the flight go on as planned.

Dryoma, the name of Yerosha's companion, means drowsy or dreamer. In Russian, the name Yerosha is associated with trouble-making, mischievous animals from Russian folk tales.

Drowsy. Scientists had found out just how excitable Yerosha was when his pulse shot up to 200 beats per minute during liftoff. On the other hand, Dryoma reacted quietly to launch and weightlessness.

In space, they followed commands sent by radio and displayed as light signals, then were fed. Yerosha apparently was so busy playing he forgot to do some activities to get all his food.

Equipped with telemetry to transmit data on the animals to Earth, Cosmos 1887 was the third USSR biosatellite to go to orbit with monkeys. Along with Yerosha and Dryoma, Cosmos 1887 housed 10 white Czechoslovakian rats, newts, fruit flies, fish, fish eggs, plants and organisms as small as paramecia, for research into space sickness.

The launch was eighth in a series testing weightlessness on animals. Space sickness is a problem faced by men in space. Researchers know little about its cause. Fifteen sensors, implanted in the heads of the two monkeys, detected changes in electrical impulses as they adapted to zero gravity.

More juice. Soviet TV broadcast pictures of Yerosha flexing his free paw and sucking one of his food tubes. In case a tube was blocked, ground control increased his supply of concentrated juices. Russian media reported Yerosha "studies with interest the design of his space suit, but evidently treats with caution the helmet from which wires stretch to the information systems."

Radio data indicated Yerosha's health was good. He was isolated in a sealed chamber.

Other than Yerosha's fun, the orbiting menagerie flew smoothly.

Recovery. The Vostok-style craft had lifted off September 29, 1987, from the Northern Cosmodrome at Plesetsk, in the western USSR, to a 140 to 250 mi. high orbit. It returned to Earth 13 days later on schedule October 12, 1987, only to land in the wrong place.

Problems with the craft's braking rocket led the capsule to land more than 1,000 miles off course.

Arrival of the capsule surprised residents near Mirnyy, a city in the diamond-rich Yakutia area of central Siberia. The monkeys were reported okay despite a frosty early-morning landing in a pine-tree taiga, or forest, 25 miles from Mirnyy, 2,000 miles east of the Plesetsk cosmodrome.

Helicopters flew from Mirnyy airport. Spotting the capsule in a forest, biologists erected a tent over the capsule to raise the temperature inside, protecting the warm-weather animals against minus 5 degrees Fahrenheit.

All okay. It took scientists some time to open the capsule and the monkey containers to confirm their condition. The Soviet Institute of Biomedical Problems confirmed the monkeys survived.

Biologists said Dryoma felt fine but Yerosha seemed weak as the menagerie from the capsule flew by airliner 2,500 miles west to Moscow.

Yerosha, didn't withstand stress of the flight to Earth as well as Dryoma. At the landing site, recovery workers had to use intensive methods to resuscitate Yerosha. But, he and Dryoma survived the harrowing re-entry, and post-flight examinations, to recover fully.

Back to space. Eugene Ilyin, director of the USSR's Cosmos program, visiting NASA's Ames Research Center, Mountain View, California, in 1989, noted that Yerosha and Dryoma were not sacrificed for science after the space trip.

"Now this monkey is in excellent shape and we are discussing the possibility of launching him again just to see how he can tolerate repeated spaceflight," Ilyin said.

Cooperative project. Cosmos 1887 was the first joint U.S.-USSR project under an April 1987 treaty which had restarted space cooperation after a five-year hiatus.

More than fifty scientists from American colleges and NASA centers were involved in experiments aboard Cosmos 1887. The cooperative effort with the Moscow Institute for Biomedical Problems was one of 16, in the agreement signed by the U.S. and USSR in April, 1987, to investigate space effects on bones, muscles, nerves, heart, liver, glands and blood. Following the flight, tissue from five rats was flown to the United States for examination. Pituitary cells show growth hormones. Spleen and bone marrow cells reveal effects of low gravity on the immune system.

Penn State. A brigade of biologists welcomed the returning biosat passengers. Among the scientists, Pennsylvania State University biochemist Wesley Hymer and research technician Kim Motter were among eight U.S. scientists in the USSR for the landing of Cosmos 1887. Hymer was to examine pituitary glands from five of the rats, extracting growth hormone producing cells. He previously had studied pituitary tissue from rats flown to space in 1985.

Two squirrel monkeys and 24 rats went to space April 29, 1985, for seven days in the European Space Agency's Spacelab module tucked into the cargo compartment of U.S. shuttle Challenger. The flight became memorable when seals on the $10 million cages broke. Food and animal droppings floated from Spacelab into Challenger's cockpit.

Hymer found the Challenger 24 rats, while in space, suffered a 50 percent reduction in the release of growth hormones. They lost muscle and bone strength. Cosmos 1887 was to tell Hymer the effect of twice as much time in space.

In adult humans, hormones affect growth and repair of tissue, muscle and bone, as well as the immune system and red blood cell production. The growth hormone has been diminished in American astronauts returning from short flights. Understanding hormones is critical to planners of Mars flights and Moon settlements.

Astronauts experience weakness after space flights. Cosmonauts in orbit for extended periods have had trouble walking after landing.

France, Hungary, East Germany, Poland, Rumania and European Space Agency also had experiments aboard Cosmos 1887. The space achievement came at the 30th anniversary of the October 4, 1957, launch of the world's first satellite Sputnik 1.

Remote-Sensing Satellites

One of the most useful capabilities gained in the Space Age has been the ability to step back and look at Earth as a whole.

Satellites away from the surface of the planet use remote sensing to observe wide areas of Earth, both land and ocean, to spot weather patterns and other environmental conditions, and to explore for natural resources.

Earth-observation satellites in orbit above the planet include the U.S. NOAA weather satellites and Landsat photo satellites, France's SPOT photo satellites, Japan's Maritime Observation Satellites (MOS), European Space Agency's ERS satellite and numerous others launched by the USSR, Europe and the U.S.

Remote-sensing satellites take pictures in visible light and infrared light of Earth's surface from hundreds of miles above. Sometimes images of Earth are made with radar beams. America invented remote sensing in the 1960's, selling satellite photos of the Earth to governments, geologists, foresters, oil men and others interested in resources.

The series of U.S. remote sensing satellites known as Landsat have provided photos for mineral exploration, checking soil conditions and vegetation, urban and rural planning, monitoring volcanos, and tracing ocean currents.

Some pictures have become famous. Landsat 5 revealed to the world the Chernobyl nuclear plant meltdown in the USSR in 1986. Two years later, the U.S. Forest Service looked to Landsat photos for help battling forest fires in Yellowstone National Park. Landsat's infrared camera in 1989 photographed the Valdez oil spill in Alaska.

Several generations of government-owned Landsats were sent to orbit. Since 1985, Landsat has been a commercial venture known as the Earth Observation Satellite Co.

Spot. In February 1986, the French launched Spot-1, followed by Spot-2 in 1990. News media around the globe became interested in Spot pictures from space which showed airfields and a chemical weapons plant in Libya. Reporters also broadcast satellite photos of the melted Chernobyl nuclear reactor and a USSR naval base.

Resolution. Spot photos show more detail than Landsat, with 30-ft. resolution versus Landsat's 90-ft. A 30-ft. resolution from a camera in space will show an area as small as half a tennis court. The USSR beats both, selling photos with 18-ft. resolution.

Photo satellites are in demand. Several have been launched to orbit recently and more are planned for the 1990's. India sent a remote-sensing satellite to space from the USSR in 1987. Later, India sent its India Remote Sensing (IRS) satellite to space March 17, 1988, on a Soviet rocket from the USSR.

Japan added an observation satellite in orbit in February 1988. European Space Agency launched one in 1991. Canada wants to launch a radar remote-sensing satellite in 1994. China has a remote-sensing satellite and Brazil has one on the drawing board.

Earthquake faults. Until earthquake fault lines turned up in Landsat photos of California, the U.S. government had been ready to get out of the space-photography-of-Earth's-resources business entirely.

Scientists at NASA's Jet Propulsion Laboratory (JPL), Pasadena, California, discovered several previously-unknown active earthquake fault lines in California's Mojave Desert in Landsat photos. The fault lines appeared in data from a thematic mapping instrument aboard Landsat 5. The photos showed the earthquake faults in the central and eastern parts of the Mojave Desert. They were discovered by studying a series of optical and infrared photos.

The faults, some of which are inactive while others may be active, are part of a complex fault region between Death Valley and the San Andreas fault region. Without Landsat images, the faults would not have been discovered, according to scientists at JPL and Louisiana State University.

Rift Valley fever. Mosquitos carried a virus from the Rift Valley on the eastern side of Africa across the continent to Egypt in 1977 killing 2,000 people with its deadly fever. A Maryland research lab found a way to track the path of the disease by satellite.

The Rift Valley stretches from Mozambique to the Red Sea. The fever was first diagnosed there in 1968. The fever killed most of the sheep in Egypt. Some 200,000 humans caught the fever while slaughtering sheep.

Researchers today use satellite photos of land conditions across Africa to predict spread of the fever virus. They map vegetation density which gives clues to mosquito movements. The 1977 outbreak was stemmed by animal vaccine from South Africa and human vaccine from the U.S. Army's Fort Detrick at Frederick, Maryland.

Monitoring the oceans. Japan's first remote-sensing satellite, MOS-1 was built by Japan's National Space Development Agency (NASDA) and blasted to space February 18, 1987, from Tanegashima Space Center. The 1,628-lb. MOS-1 went to a 564-mi.-high orbit, providing ocean data to 60 Japanese organizations as well as 16 other countries. MOS-1 focuses on the oceans, noting color and temperature. Of course, it also looks at land surfaces when over them, checking on agriculture, fishing, forestry and environmental problems.

MOS-1 has three sensors, known as MESSR, VTIR and MSR. All are passive collectors of sunlight reflected from the surface of Earth or electromagnetic energy radiated from the planet's surface. MESSR monitors the colors of the surface of the oceans. It looks at both visible and infrared light reflected from land surfaces. VTIR looks at wide swaths of Earth in visible and infrared light, measuring water vapor in the atmosphere and surface temperature. MSR monitors microwave radiation from the surface to measure quantities of water vapor, snow and ice.

The satellite is in a polar orbit floating over the entire Earth every 17 days. It takes only 103 minutes to go around Earth one time, of course, completing 14 trips a day. But due to Earth's rotation, its path moves slowly to the west along the equator. Every 237 orbits, or 17 days, it has covered the entire face of the planet and is back where it started.

Japan later orbited MOS-2. Both satellites are controlled by NASDA's Tsukuba Space Center. Signals also are received by tracking stations at Katsuura, Masuda and Okinawa. NASDA's Earth Observation Center, established in 1978, is at Hatoyama-machi, Saitama prefecture, 30 miles northwest of downtown Tokyo.

Pollution hunter. Garbage washing ashore along New Jersey's coast in the summer of 1987 was so bad federal officials looked to the skies for help. They requested a satellite as spy-eye-in-the-sky against polluters. Believing that ships illegally dump trash in the oceans, New Jersey's U.S. Sen. Bill Bradley called for a satellite to snap photos of the ocean surface regularly so the federal inspector could see the bad guys dumping junk. To clamp down on illegal ocean polluters, U.S. Deputy Commerce Secretary Clarence Brown agreed in December to study putting a satellite on the lookout for ocean dumpers. Bradley thought up the idea after becoming "revolted and angered" by the floating filth which tormented his state.

Brown and Bradley released photographs from weather satellites mapping the coast in detail. With resolution down to 90 feet, smaller than a football field, the photos showed a ship clearly visible along the New York-New Jersey coast, apparently snaking a trail of trash. Bradley said a watchdog satellite would deter would-be polluters who face fines and jail for illegal dumping. Enforcement has been weak, but this may be a space program the law-and-order politicians finally can sink their teeth into.

Spring bloom. A weather satellite peering down at the ocean since 1978 has charted all of the plankton near the surface of the North Atlantic.

Plants in the ocean take carbon dioxide, from the atmosphere and the oceans, and change it to plant tissue in the important process known as photosynthesis.

NASA's unmanned Nimbus 7 weather satellite, in orbit since October 1978, stared at sunlight reflecting from a 500-mile-wide swath of ocean every time it flew overhead, noting the color of the surface of the sea.

Information was sent by radio to researchers on the ground. Many such images were combined by the scientists to make an ocean-wide color picture of how the microscopic plant life, phytoplankton, looks to the satellite.

The most striking feature is spring bloom, a big concentration of plankton extending all the way across the North Atlantic into the North Sea.

Chlorophyll is a pigment giving plants such as phytoplankton a photosynthesis ability. Researchers put the satellite data through mathematical equations to compute how much chlorophyll is in the water.

Nimbus 7, built to scan coastal waters, saw five colors of light. Its pictures revealed land and ocean plants on the edges of continents bordering the North Atlantic.

Land vegetation also was recorded by a sharp-eyed NOAA 6 satellite, in orbit since June 1979. The Sahara Desert, tropical rainforests of South America and the spring greening of North American fields and forests were major features captured by NOAA 6.

The Nimbus ocean scanner died in 1986 as NASA planned an improved land and ocean eye for science aboard the U.S. space station in the 1990's.

Ocean floor map. Without wind or waves, Earth's oceans would seem calm and flat. But, the reality is, oceans have hills and valleys—similar to the hills and valleys on land. Ocean surfaces are like blankets dropped over rugged terrain. At the bottom of the oceans, gravity pulls water down into valleys and pushes it up around mountains. The result: gentle hills and valleys on an ocean surface.

Ocean surfaces are so uneven, for example, looking north from Puerto Rico toward Bermuda, the ocean surface drops 60 feet over 30 miles.

NASA drew maps using data transmitted to Earth from altimeters aboard the Geodynamic Experimental Ocean Satellite (GEOS-3) and the Sea Satellite (SEAsat). The maps depicted the hills and valleys of ocean surfaces more accurately and with greater detail than before. GEOS-3 and SEAsat measured the distance from space to ocean surface with radar. A computer used the radar information to draw a map of the surfaces of the world's oceans. Altimeters in the satellites are so accurate, scientists can detect small surface features only three to six feet in height.

NASA assembled satellite-altimeter profiles of the ocean into a map showing the shape of the surface—wave crests and valleys—and the shape of the underlying ocean floor. The map helped scientists study earthquakes and volcanos. The measurements by GEOS-3 and SEAsat also helped researchers understand how oceans circulate and how currents flow in the oceans.

The success of the experiments with GEOS-3 and SEAsat led to plans for the Ocean Topography Experiment Satellite, also known as Topex/Poseidon. NASA and the French space agency CNES plan to launch Topex/Poseidon in 1992-1993 for ocean studies.

The altimeters aboard future oceanographic satellites, like Topex/Poseidon, will be so accurate, scientists will be able to see most of Earth's ocean-surface currents.

Scientists at NASA's Goddard Space Flight Center, Greenbelt, Maryland, say the ocean transports heat around the globe. Topex/Poseidon will help geophysicists study the circulation of ocean currents and how they influence worldwide weather patterns.

NASA/Goddard will build the satellite in Maryland. CNES, the Centre National d'Etude Spatiale, will launch Topex/Poseidon to orbit aboard a European Space Agency Ariane rocket from South America. To map surface features of oceans around the globe, NASA and CNES both will provide sensors for the satellite. They will share data received from it. Topex/Poseidon will measure the sea level of Earth's oceans at least until 1995. Scientists expect to determine the general circulation pattern of the oceans, see surface features, and find a better understanding of how oceans affect climates.

Volcano watching. It wasn't a rain cloud. It wasn't ozone. It was a large blob of sulphur dioxide from a Mexican volcano which showed up in weather satellite pictures recently. Scientists used to look at conventional satellite cloud pictures to guess which shapes might be plumes of sulphur dioxide from volcanoes. Then they discovered the ozone-mapping spectrometer aboard the weather satellite Nimbus 7 sees these plumes, highlighting them on maps. Nimbus 7 has checked sixteen erupting volcanoes recently.

Mt. Pinatubo. An ozone mapper aboard the Nimbus 7 satellite watched sulfur dioxide gas from the eruption of Mt. Pinatubo in the Philippines spread across the tropical Northern Hemisphere. The sulfur dioxide—a toxic air pollutant—spread 4,800 miles after an eruption June 16, 1991.

Rosy sunsets months later were blamed on suspended aerosol particles showered into the atmosphere by the volcano. The Mt. Pinatubo eruptions blasted twice as much sulfur dioxide into the atmosphere as did the eruption of El Chichon in 1982.

Lost Arabian Society. Radar satellites staring down from space at the local topography along remote reaches of the globe have penetrated 600-ft. mountains of windswept sand to make a startling find on the fringe of the Arabian Desert.

A faint shadow of the lost civilization of Ad has turned up like a ghost in a computer-enhanced image of ancient ground under the Rub al Khali desert in the sultanate of Oman. The timeworn network of roads under the dunes seem to point through the desert to the burial place of the legendary Ad society, believed to be the bustling hub of the world's frankincense trade 5,000 years ago.

Ad is referred to in the Koran, in the tales of The Arabian Nights, and in the Holy Bible. Some biblical archaeologists suggest wise men traded there for frankincense they bore as gifts for the infant Jesus.

The grinding sand of the Rub al Khali had defeated water-short British explorer Bertram Thomas looking for Ad's ancient trade routes in the 1930's. Today, unable to search the entire perilous desert environment, modern archaeologists have worked in their laboratories, feeding data from satellite radar pictures to computers searching for long-lost clues.

In the satellite photos, Charles Elachi and Ronald Blom of NASA's Jet Propulsion Laboratory, Pasadena, California, were able to see a 100-yard-wide hoof-trodden path hidden under tons of sand in giant dunes. A 1990 scouting expedition tracked that trail which may have been formed by frankincense traders riding camels.

Atlantis of the sands. Frankincense is an aromatic resin from the sap of Middle Eastern and East African trees. Ages ago, it was an incense used by everyone, monarchs and common folk alike, in cremations, religious rituals, ceremonies and imperial processions.

Backed by wealthy businessman Armand Hammer, the Ad expedition included Blom, Elachi, British explorer Sir Ranulph Fiennes, Los Angeles attorney and part-time explorer George Hedges, and an archaeologist, geologist, computer scientist and documentary filmmaker.

Following their satellite-drawn compass, a scouting expedition looked for geological evidence of a trail through the now-barren land to the once-thriving city of Ubar, once the main frankincense shipping center of Ad. The adventurer T.E. Lawrence once described Ubar as "the Atlantis of the sands." The city was a starting point for worldwide shipments—to markets as far away as China and Rome—of frankincense, an important commodity in the ancient world before the rise of Christianity.

Ad society lasted from 3000 B.C. to the 1st century A.D. In the end, it was victimized by politics, economics and climate after a drop in demand for the frankincense fragrance as Christianity preached burying bodies instead of burning them. The abandoned villages of Ad eventually were inundated by tides of shifting sands. Today those dunes reach heights of 200 to 600 feet.

Subtle signs. There have not been many scholarly writings about Ad. Because of a lack of ruins to study, many archaeologists have said the Ad civilization was mythical. The JPL scientists found both modern tracks and ancient ones in their satellite maps. The new ones go around the dunes while the old ones go underneath. The main 100-yard-wide path seen in satellite photos is very subtle with the ground worn slightly into the desert floor.

The explorers had been preparing for six years to start excavating in January 1991 when they stumbled upon Ad artifacts—900 pottery shards and flint pieces—on the trade route in the sprawling Rub al Khali desert during a three-week scouting expedition in July 1990. High winds drove the team away, leaving the artifacts in the hands of Oman's Department of National Heritage until the expedition returns. The main expedition in 1991 will try to prove the Ad people existed.

The winds of war in the Middle East delayed the main expedition in 1991. Oman, once known as Muscat and Oman, is a sultanate on the southeast side of the Arabian Peninsula. It is bounded on the north by the Gulf of Oman and on the east and south by the Arabian Sea. To the southwest is Yemen. To the west is Saudi Arabia. On the northwest border is the United Arab Emirates. The Rub al Khali desert extends into the western area of Oman, but is mostly in Saudi Arabia.

About one million people live in Oman. Besides commercial quantities of oil found in 1964, they export dates, limes, cereals and fish.

Oman was ruled for centuries by emirs controlled by a caliphate at Baghdad. Later it was controlled by Portugal followed by the British government of India. Today, the ruling sultan has close ties with great Britain.

Earth Observing System

Astronaut Sally K. Ride, the first American woman to go to space, as she was leaving NASA in 1987 called for a Mission to Planet Earth to preserve our planet's environment. NASA's Earth Observing System is the key element in Mission to Planet Earth.

EOS would observe land, sea and atmosphere from high above Earth to predict changes in the environment. NASA has proposed a handful of very large "platforms" in space—unmanned satellites crammed with sensing and measuring equipment staring down as they revolve around the globe on a north-south path.

EOS platforms. EOS platforms will study Earth as a total system—atmosphere, oceans, biological cycle, chemistry and the "greenhouse" effect. The platforms would not be space stations. General Electric Astro Space, Princeton, New Jersey, is building the first platform for launch in 1998 on a Titan 4 rocket from Vandenberg Air Force Base, California. Altogether, six would be launched on Titan 4s from 1998 through the year 2011. Each platform would be expected to last at least five years in space.

NASA's Goddard Space Flight Center, Greenbelt, Maryland, manages EOS. Federal budget problems have led some in Congress to suggest a constellation of small EOS satellites, rather than a few large spacecraft. The platforms suggested by NASA and their tentative launch dates:

Launch	Platform
1998 Dec	EOS-A1
2001 Jun	EOS-B1
2003 Dec	EOS-A2
2006 Jun	EOS-B2
2008 Dec	EOS-A3
2011 Jun	EOS-B3

UARS. An early EOS platform actually to go to orbit was the Upper Atmosphere Research Satellite (UARS) on September 12, 1991. Astronauts in U.S. space shuttle Discovery ferried UARS to orbit. From 372 miles above Earth, UARS is able to study the upper atmosphere on a global scale.

Nine sensors in the satellite explore winds, chemical composition and energy of the stratosphere, providing details on how natural processes and man-made pollutants destroy Earth's ozone layer.

Sounder. NASA's Jet Propulsion Laboratory issued a contract in 1991 for an Atmospheric Infrared Sounder instrument for the first true EOS platform in 1998. The infrared sounder will be a global thermometer taking profiles of atmospheric temperatures with an accuracy to one degree Celsius. Like the entire EOS platform, Atmospheric Infrared Sounder will work continuously for five years in orbit.

Safire. Other EOS instruments with odd acronyms will include Spectroscopy of the Atmosphere Using Far Infrared Emission (SAFIRE), Tropospheric Radiometer for Atmospheric Chemistry and Environmental Research (TRACER), and the Stratospheric Aerosol and Gas Experiment (SAGE III).

SAFIRE will do something science hasn't done before—provide molecular-level chemistry profiles of the Earth's ozone layer to help scientists understand what breaks it down and enhances it. SAFIRE will involve science teams from the U.S., England, Italy and France.

Tracer. TRACER will measure distribution around the globe of carbon monoxide at various levels in the troposphere. It will map worldwide motion of the atmosphere and look at chemical changes brought on by motion of the atmosphere.

TRACER will be so sensitive it will detect variations of carbon monoxide in different parts of the world. From part measurements, scientists have learned that carbon monoxide released by burning of tropical vegetation rivals that created by transportation and industry in the Northern Hemisphere.

Sage. SAGE III actually will be three science instruments on the EOS platform. They will draw vertical profiles of aerosols and other gases in the atmosphere from near the surface up through cloud tops into the stratosphere and mesosphere.

Navstar Global Positioning Satellites

The TV picture tube sunk in your automobile dashboard displays a colored map of city streets surrounded by country roads. As you steer your car along a thoroughfare, a radio receiver behind the dash hears and understands signals from a Navstar satellite orbiting 12,532 miles overhead. A small car-shaped cursor rolls along the TV map. A trail of dashed lines reveals where you've been.

The U.S. Air Force is creating the constellation of 21 Navstar Global Positioning Satellites (GPS) in Earth orbit, circling the planet twice a day. By 1991, full coverage of North America was achieved. By the mid-1990's, four of the space beacons will be in view from any spot on Earth at any time.

You won't notice the Navstar radio receiving antenna molded into your automobile roof. It will feed received signals to a GPS receiver lost in the jumble of computers and other electronic innards cluttering the hidden interior of your dashboard.

With four Navstar satellites looking down on your car, three-dimensional navigation is possible. The receiver tuned to the time signals broadcast by the satellites can determine its own longitude, latitude and altitude. You'll know when you're up on Pike's Peak and when your down in Death Valley. Of course, the GPS system can be used by planes and ships, as well as land vehicles.

Atomic clock. The heart of Navstar is its accurate clock. Each satellite has an atomic oscillator with rubidium cell and cesium beam. Clock accuracy is within one second in 300,000 years!

The space beacons beep away cheerily, 24 hours a day, with never-ending messages telling the time. Navstar Control Center at Colorado Springs, Colorado, checks the satellite clocks regularly to make sure they are beeping in sync.

The receiver in your car checks its own clock, too, comparing it with time signals from whichever satellites happen to be overhead at the moment. After checking four satellites, the computer in your receiver computes how far away the satellites are, then displays your location on the dashboard map screen.

When can you have it? Handheld and mobile GPS receivers, which read out your location in latitude and longitude, already are being sold. For your car, you need a receiver set up to drive a dashboard map.

Manufacturers have demonstrated such automobile navigation systems based on Navstar. The most innovative combine a safer "heads-up display" projected on the windshield with a computer which understands spoken words. The driver speaks street names out loud. The Navstar receiver responds by showing how to get there.

Less sophisticated automobile receivers are about to be sold—at steep prices, which will drop eventually toward the mass consumer market—below $500 by 1995.

Walk-arounds. Receivers already are so small and inexpensive that one-pound handheld personal navigators are being lugged by backpackers into the wilderness. Receivers in small private planes are permitting in-the-dark landings within ten feet of intended touchdown. Police and fire vehicles soon will be tracked continuously by dispatchers. Nationwide tracking of refrigerator railcars and uranium shipments is about to start. Truck, train and ship terminal bottlenecks might even be relieved if dispatchers know ahead of time who is coming and when. Engineers are predicting all ships, airplanes and most automobiles will have Navstar receivers by 1999.

In the future, ships at sea and on inland waterways will avoid collisions with a warning device built around Navstar. Maybe even robot crop dusting aircraft will be run entirely by autopilot using location information from Navstar. Similarly, road grading could be done by a robot equipment operator receiving Navstar location information.

Downed aircraft or boats lost at sea could be located readily if they transmitted exact positions based on Navstar receiver calculations. Robots could do land surveys. Ocean buoys could become self-monitoring. Air traffic control might become automatic. Map making and wild-animal tracking could be completed instantly by machine.

Moving landmarks. The navigation satellites already have been moving American landmarks. Of course, the landmarks haven't actually moved physically. Rather, cartographers moved them on maps after Navstar pinpointed their true positions.

The biggest leap: the flagpole in front of the judiciary building in Honolulu, Hawaii, hopped 1,480.8 feet to the southeast.

Some other examples: the Bismarck, North Dakota, water tank had to be relocated 101 feet west. The Empire State Building in New York City was moved 120.5 feet to the northwest.

The District of Columbia's Washington Monument now turns up 94.8 feet northeast. The head of the statue on the dome of the U.S. Capitol at Washington, D.C., was moved 94.8 feet to the northeast.

Nearby, in Annapolis, the Maryland state capitol dome moved 98.5 feet northeast. In Richmond, Virginia, the war memorial moved 99.8 feet northeast. The border between the United States and Canada moved as much as 66 feet on maps.

Latitude and longitude form an imaginary grid of lines covering the world. The basic reference points were measured last in 1927 with portable towers, plumb lines and sighting telescopes.

Now, in a 12-year project, some 250,000 of those basic reference points have been recalculated for engineers, military, surveyors, highway builders, utilities, regional planners, even geologists studying the movements of continents.

This time, Navstar satellites orbiting Earth radioed data to 5,000 GPS receivers which calculated latitude and longitude from the signals. Those readings were used to recalculate the other 245,000 points. The new readings were 100 times more accurate than the 1927 findings, with an error of only one inch in six miles. Homeowner property lines weren't changed, but state and national boundaries moved a few feet.

Gulf War. Military forces in the Persian Gulf War made extensive use of Navstars. American and allied military forces on land, sea and in the air received global, three-dimensional, speed and position information.

Navstar receivers also are in use by civilians for land surveying, exploring off-shore for oil, oceanography and guiding planes, ships, trucks, buses and private automobiles.

New Navstars. The first of a new generation of Navstar Global Positioning System navigation satellites was blasted to orbit February 14, 1989, on a new-generation Delta 2 rocket from Cape Canaveral.

The U.S. Air Force launches Navstars on Delta 2 rockets from Cape Canaveral Air Force Station, adjacent to the Kennedy Space Center in Florida. Delta 2s ferry the 1,860-lb. Navstars to highly-elliptical orbits in which they range from 100 miles to 12,500 miles above Earth. A ground station radios a signal to fire a PAM-D rocket attached to the $65 million satellite, moving Navstar to a 12,500-mi.-high circular orbit.

Challenger. The Delta launches are a result of the 1986 loss of shuttle Challenger which grounded the shuttle fleet almost three years, leaving the U.S. with no way to send payloads to orbit. The Pentagon rushed to buy 21 Delta 2 rockets to carry the Navstar GPS navigation satellites which once were to have been launched by shuttle.

Delta 2. The second-generation Delta 2 is an upgraded Delta 1, one of America's workhorse boosters for decades. Delta 2 was modified specifically to lift Navstars. Each of the blue-and-white, 126-ft. Delta 2 rockets cost $30 million and has nine solid-fuel boosters strapped on. Each Delta carries a solar-powered GPS on a 25-minute flight to a low, preliminary, elliptical orbit of about 100 miles altitude. Later, a "kick motor" rocket built into the satellite pushes it into a circular, 12-hour, 12,500-mi.-high orbit.

The tenth Delta 2-Navstar launch, in 1990, was the first flight with more-powerful solid-fuel boosters strapped around the slender rocket. In other launches, the souped up boosters would allow the rocket to lift payloads up to 4,010 lbs. to the low preliminary orbits used by communications satellites. Delta 2s are made by McDonnell Douglas Space Systems, Huntington Beach, California.

Navstar GPS Block 2 Launches On Delta 2 Rockets

1	1989 Feb 14		7	1990 Mar 25
2	1989 Jun 10		8	1990 Aug 2
3	1989 Aug 18		9	1990 Oct 1
4	1989 Oct 21		10	1990 Nov 26
5	1989 Dec 12		11	1991 Jul 3
6	1990 Jan 24			

Block 2. The GPS program began in 1973 with so-called Block 1 first-generation satellites. Ten were launched, half still work. The second-generation Block 2 solar-powered Navstar launched in February 1989 joined seven first-generation Navstar satellites still working in orbit at that time.

Stored in orbit. Navstars are built by Rockwell International's Satellite Systems Division in Seal Beach, California. The U.S. Air Force's Navstar Control Center is at the U.S. Air Force space command, Falcon Air Force Base, Colorado Springs, Colorado.

Rockwell received a $1.2 billion contract in 1983 to build 28 GPS satellites—eighteen to be used immediately in space, three to be sent to space to be stored in orbit as back-up spares, and seven replenishment spares available on the ground for launch as needed. The Air Force would like to launch a Navstar every 60 to 90 days.

The Navstars already in orbit by 1991 provided full 24-hour GPS coverage of North

America. By the mid-1990's, the eighteen-satellite constellation should be in place, providing worldwide navigation

Global positioning. Navstars continuously broadcast their own positions and atomic-clock time signals. Receivers on Earth have clocks. Computers in the receivers calculate how long it takes a signal to arrive from a satellite. Then, knowing the speed at which radio waves travel, the distance to the satellite is calculated.

Signals are received from three or four satellites, letting the receiver on the ground know the range, or distance, to three or four points in the sky. The receiver's computer then uses trigonometry to find its own position from the known points. A user can pinpoint his location and altitude to within 53 feet anywhere in any weather.

Originally intended for U.S. and allied military use, signals from the six new Block 2 space beacons are being heard and used by civilians. The USSR also has its own global network of similar navigation satellites. Some receivers can be tuned to both systems.

Shafting civilians. Navstar is a U.S. Defense Department project. The Pentagon plans to reduce GPS accuracy for civilians after the entire Navstar flotilla is in orbit.

Each GPS satellite broadcasts two channels of data. One channel, giving an accuracy to 330 feet, broadcasts clear, non-coded signals for use by civilians. The more-accurate 53-ft. position transmissions are available now for all to use, but will be secretly encoded so only military units can use it. The less-precise 330-ft. civilian fixes use a publicized code known as C/A.

Silent rendezvous. A GPS satellite transmits a stream of information continuously, telling where it is at a given moment and what time it is at that moment. Computers in receivers on the ground—even small hand-sized portable radios—analyze time and location signals beamed down from the four nearest satellites. Receivers use the signals to compute location and altitude. The receivers can compute velocity to better than one foot per second. The system is valuable to military forces needing to rendezvous without using radio transmitters or radar. For example, aircraft are able to link up for mid-air refueling without using voice communications.

Radio frequency. Each satellite transmits its stream of C/A-coded information on a frequency of 1575.42 MHz (megahertz). Along with its own time and location information, a Navstar also sends out information about other satellites nearby. In effect, it is transmitting a periodically-updated celestial almanac and the user's receiver on the ground is holding that data in memory.

How fast is it? A so-called two-channel receiver calculates faster since it hears both kinds of information at the same time—time-location data and celestial almanac data. A single-channel receiver takes in one at a time, leading to slower computations. A typical consumer single-channel receiver fixes its position GPS every 10 to 15 seconds. A two-channel receiver might be updated every one-half second.

Professional sets for surveyors are five-channel receivers. They have four channels listening for time-and-place data from four satellites plus a fifth channel listening to celestial almanac data from one satellite. The position fixes from such five-channel receivers are extraordinarily accurate and updated many times each second.

Searching for satellites. When a user turns his receiver on, it starts searching the airwaves for signals from the closest satellite. Finding a satellite, the receiver accepts almanac data and records it in memory. Next the receiver listens to three nearby satellites for their positions and time signals. A fourth satellite in view is received and used for a time reference time.

The receiver checks its own clock to see when it received a time signal. It compares its time with the time reported in a satellite transmission to see how long it took the radio signal to get from the satellite in orbit to the receiver on the ground. Using the known speed of radio signals, the receiver calculates the distance to the satellite. Using its calculated distance to three satellites, the receiver can triangulate or fix its position with great accuracy.

Three dimensions. Boats on water and other slow-moving vehicles not requiring altitude fixes can triangulate with only three satellites. With three satellites, the fix is two-dimensional. Altitude or height above Earth's surface is the missing third dimension. For altitude, data must be received from a fourth satellite.

Transit satellites. Navstar GPS gives more-precise pinpoint location fixes than the older Loran non-satellite navigation system or Transit satellite navigation system.

So-called "satnav" receivers, using the older Transit satellite system, rely on a satellite passing overhead every 90 minutes for a new position fix. By comparison, with eighteen or more Navstar GPS satellites in six polar orbits, any point on the globe always will have four satellites in view all the time. And a receiver on the ground has to hear from only three satellites to find its own latitude and longitude location. The fourth satellite lets the receiver calculate its own altitude.

Some have said GPS stands for "great pie in the sky" and the "gold-plated specter" and even "government project stalled." But, now that the worldwide navigation system is working well, one magazine has said GPS stands for "good product, sport."

Wildlife Tracking By Satellite

A satellite looks down on a wide area of the globe from its high vantage point. That can make it a useful tool for biologists, naturalists and conservationists working with animals, birds and fish in their natural environments.

They have attached tiny radio transmitters to wild animals for years, tracking the signals with radio receivers nearby on the ground. Now they can track wildlife ranging over wider areas with receivers high above the ground. A receiver in an orbiting space satellite can hear a transmitter attached to an animal. Even if the animal is out of sight over the horizon from a ground tracker, a satellite in space can hear the transmitter and repeat its signal down to trackers on the ground.

Whale tracking. A three-month tracking by satellite in the summer of 1987 of three young male pilot whales that beached themselves on Cape Cod in December 1986 showed the giant mammals can be nursed back to health and returned to the wild sea.

More than 70 whales had died, running aground on Cape Cod in 1986. Four out of the previous five winters, whales had stranded themselves on Cape beaches. Others were pushed out to sea by concerned humans.

Tag, Notch and Baby, the young pilot whales, were too ill to survive without treatment for shock. After treatment, radio transmitters were attached to their dorsal fins. Tag carried a transmitter capable of reaching a space satellite in low Earth orbit.

100 days. Named by the New England Aquarium at Boston, the whales were returned to the ocean June 29, 1987, the first time scientists had released more than one whale at a time. Baby lost his radio as he entered the water. Batteries for two other transmitters failed after a month. But scientists were able to track the animals until October 7, 1987, when batteries quit in the satellite transmitter attached to Tag.

Tracking didn't reveal why whales beach themselves, but gave new knowledge of whale diving and migration. Radio signals emitted by Tag's transmitter were received by a satellite overhead in Earth orbit. The signals were relayed by the satellite to scientists at a research station on land.

Tag carried the satellite transmitter on more than 200,000 dives in 95 days, staying underwater for an average of 34 seconds and as long as 7 3/4 minutes during the longest tracking ever of a whale or dolphin. He swam north to Maine and south to Delaware.

Pod of pilots. Last signals before the transmitter failed indicated Tag was 20 miles off Cape Cod, the closest to shore since being released. Aquarium scientists decided the whales, heading south, were not in danger of beaching.

Tag, Notch and Baby had to join a pod of pilot whales for survival. Tag was seen in August 1987 with a pod. And researchers think Notch and Baby remained with Tag.

The Cape strandings occurred among whales southbound to warmer waters. Scientists speculated the enclosed bay and unusual weather, or maybe the whale's herd instinct, caused mass beachings. Researchers now know they can rehabilitate and return a whale. The scientists planned to tag the next beached whale needing treatment with a transmitter with longer-life battery.

Swan Tracking. The Wild Bird Society of Japan wanted to know exactly where swans go when they migrate to the Arctic each summer, so they used an Earth-orbiting satellite to track signals from tiny 1.4-oz. radio transmitters attached to the 12-lb. birds.

Japan's phone company, the giant telecommunications utility Nippon Telegraph and Telephone Corp., made the miniature transmitters for the birders.

Larger transmitters used recently to track migrations of dolphins and seals were too heavy for swans to carry in flight.

The first radios were attached during one recent Spring to four swans on Hokkaido, Japan's northernmost island. Signals beaming upward from the flying birds were received by Argos, an orbiting U.S.-French environmental satellite. Information relayed by the satellite to scientists on the ground in Japan revealed the swans' flight path to their Arctic summer home.

Power Station In Space

The USSR has been planning a large Earth-orbiting satellite for a power station in space, converting sunlight into energy for use on Earth. The station also would reflect sunlight to Earth for night lighting.

The Soviet Union's powerful Energia space rocket can lift 110 tons to low orbit in one flight. It flew successfully first in 1987 and 1988. It could lift heavy power-generating station components to orbit. An unmanned power station in orbit could convert sunlight to useful electricity to operate many different projects in space. It could electrify manned space stations or even a large space city composed of many station modules joined together in orbit.

Beaming down. The power station would collect solar energy and beam it down to receiving stations on Earth's surface where it would be converted to electricity to meet ground power needs. Separate large reflectors would light cities on Earth at night.

Building a power station in Earth orbit could take several years. First, one or more solar reflectors would be assembled in space to bounce sunlight to Earth for night lighting. Later, a large power-generating station could be built in modules on the ground and sent to space to be assembled in orbit to capture sunlight and turn it into some form of energy for transmission to Earth. Antennas would have to be constructed on the ground to receive the energy from space and convert it to useful electricity.

Man-tended. Men shuttling back and forth from the USSR's orbiting Mir space station, or other manned stations, could tend the power station. The USSR's Buran shuttle could take maintenance cosmonauts to the power station.

After the May 15, 1987, Energia launch, Guri I. Marchuk, head of the Soviet Academy of Sciences, mentioned Energia had the booster power needed to ferry experimental solar power plants to orbit.

Spy Satellites

Just as civilians can spy on the weather from space or ferret out natural resources by satellite, so military units can spy on each other.

Most military satellites are for communications, meteorology or reconnaissance. The U.S. Dept. of Defense uses satellites in various orbits above Earth to relay communications, help pinpoint locations for navigation, report weather and environmental conditions, scan land and sea for enemy deployments, monitor arms-control agreements and give early warning in the event of nuclear attack.

Military satellites already in orbit spy on facilities and forces, scan radio waves as electronic ferrets, watch for nuclear explosions and missile launches, map Earth's surface and relay communications. Weather forecasting is essential in battle. The U.S., the USSR and other nations have advanced weather satellites to aid low-level bombers, detect solar flares and find holes in cloud cover for reconnaissance work. The latest navigation satellites provide the pinpoint accuracy needed by the giant MX nuclear missile. And they allow troops to synchronize watches to billionths of a second.

Some satellites are in low orbit, 100 to 300 miles above the surface of our planet. Others are at intermediate altitudes from 500 to 1,000 miles high.

Some orbit the equator while others fly over the poles. Some pace the whirling Earth from 22,300 miles altitude, making them seem stationary in the sky.

Peeping Tom photo-satellites, radio-eavesdroppers and radar imagers are among the latest rage in spy satellites. The biggest U.S. military satellite in orbit is said to be Lacrosse, a radar-beam eye-in-the-sky which scans the planet surface over which it passes, radioing pictures of what it sees to headquarters. Lacrosse was ferried to space in December 1988 in a secret flight of shuttle Atlantis.

Ears in the sky are called ELINT or electronic intelligence satellites. They are radio receivers in the sky used by ground controllers to tune in telephone conversations elsewhere on the ground. They also receive radar beams, missile telemetry, and other signals. One such radio-eavesdropping satellite has been known as Jumpseat.

Key Hole. The worst-kept secret, but best eye-in-the-sky, was the humorously-named series of U.S. photo-reconnaissance spysats known as Key Hole. Recent models have been KH-9 and KH-11. A bigger, heavier and more versatile KH-12 may be in use.

A KH-9 Big Bird photo-reconnaissance satellite ejects actual camera-made photos in a re-entry capsule which drops from space and floats by parachute to an ocean surface where it is picked up by ship. The cargo of the Titan 34-D rocket which exploded in 1985 may have been a KH-9 Big Bird.

KH-11, on the other hand, is a television camera peering down from space. It converts TV pictures to digital radio signals and beams them to Earth.

A KH-11 may have a relatively-short working life in space of about three years. A worn out KH-11 was turned off in orbit in August 1985. The 1986 Titan 34-D rocket explosion probably destroyed a KH-11 meant to replace the one deactivated in 1985.

The U.S. was not left completely blind in space. One last KH-11 was working in orbit to watch USSR military events. Unfortunately, that satellite was near the end of its three-year life. After August 1985, the U.S. was in the ticklish position of being down to its last working sky-eye spy sat. Military planners worried that satellite might fail, leaving the U.S. with no eyes in space. Of course, the military wouldn't have been deaf since radio-eavesdropping satellites remained in orbit.

Undoubtedly, at least one of the successful Titan 34-D launches in 1987 and 1988 carried a new KH-11 to orbit. The Titan 34-D launched October 26, 1987, probably carried a KH-11. However, there is the possibility it might have carried a new model of the Jumpseat radio-eavesdropping satellite if the Air Force chose not to risk the last available KH-11. If it did carry a KeyHole, it would have been the last KH-11, an engineering test model dusted off and refurbished for flight.

The KH-12 has more capability than a KH-11, with a TV camera plus radar to take pictures through clouds and bad weather. However, a KH-12 is very heavy. Only a shuttle or Titan 4 unmanned rocket can carry it to space.

Miscellaneous others. One recent payload carried to space was one of a series of satellites intended to provide 30 minutes warning of nuclear attack. Another spysat now in orbit is monitoring Russian compliance with arms treaties.

In 1988, a Titan 2 rocket carried four U.S. Navy White Cloud ocean spysats to eavesdrop on radar and radio signals from the Russian fleet.

A Titan 2 also could carry a Defense Meteorological Support Program (DMSP) weather satellite. One of those was launched on a different kind of rocket from Vandenberg in 1988.

Delta 2 rockets launch a flotilla of 21 Navstar global-positioning satellites so ships and planes can navigate accurately and aim weapons precisely. GPS receivers were very popular with allied troops in the desert in the 1991 Persian Gulf War.

One spysat payload in a 22,300-mi.-high stationary orbit is an electronic-ear satellite, with an antenna as large as a baseball diamond to eavesdrop on missile tests and radar, radio, telephone and electronic diplomatic and military communications.

A 1988 launch ferried a pair of military communications satellites to orbit. They were Satellite Data System (SDS) communications satellites to be used if the U.S. had to launch nuclear attack.

Other spysats include an early-warning satellite to detect a Soviet nuclear missile attack, an eavesdropping satellite called Vortex or Chalet, undoubtedly other supersecret Key Hole photo-reconnaissance satellites, and other communications, spy, navigation, reconnaissance, weather, surveillance and early-warning satellites.

Lacrosse. The American space shuttle Atlantis made a secret flight to space December 2, 1988, from launch pad 39B at Cape Canaveral, carrying Lacrosse, a top-secret military payload reported to be an all-weather radar spysat.

NASA and the Pentagon refused to disclose details about the payload in flight STS-27, but the 18-ton, 150-ft.-wide, $500 million, intelligence-gathering satellite was said to be in an orbit which passes over 80 percent of the Soviet Union.

Lacrosse. The spysat is said to be crammed with advanced-technology gear. Lacrosse has an imaging radar as well as optical reconnaissance with digital imaging. That means the spysat snaps radar and optical photos through clouds of the ground and oceans. Its 10,000-watt radar signal penetrates cloud cover for both ocean and land surveillance. Photos are sent to Earth as digital radio signals.

The Atlantis five-man all-military crew picked up Lacrosse from the cargo bay with the shuttle's 50-ft. robot arm and dropped it overboard into its own orbit. Lacrosse was designed to be refueled by astronauts in a later shuttle flight or even retrieved and returned to Earth for improvements.

Atlantis. NASA had been preparing Atlantis for the December 1988 flight since March 1987. Some 200 modifications improved safety and performance. It was the shuttle's third flight, the 27th in shuttle history, and the second flight of the fleet since the 1986 Challenger disaster. Discovery had flown successfully September 29, 1988.

The Atlantis flight was commanded by Navy Cmdr. Robert "Hoot" Gibson. The pilot was Air Force Lt. Col. Guy Gardner. Mission specialists were Air Force Lt. Col. Jerry Ross, Air Force Col. Richard "Mike" Mullane and Navy Cmdr. William Shepherd.

Top secret. The astronauts, not allowed to discuss the mission, poked fun at the secrecy by donning Lone Ranger masks before launch day for reporters. Gibson told reporters he only could answer, "Yes. No. I don't know. I can't tell you. I can tell you but I'd have to kill you."

The exact Atlantis launch time, between 6:32 a.m. and 9:32 a.m., wasn't publicized until minutes before actual lift off at 9:30:34 a.m. Eastern Standard Time, less than a minute before the launch window was to close December 2.

The public was allowed to see Atlantis liftoff, straight up on a northward track from Kennedy Space Center, into a windy Florida sky.

There was a news blackout while Atlantis was in space. Brief still-up-there announcements were made every 24 hours. The actual landing, at Edwards Air Force Base, California, at 3:35 p.m. Pacific Standard Time December 6, was announced only hours in advance.

The Defense Department cloaked details of Air Force plans for other experiments on

Atlantis as well as future flights to see if astronauts are useful for reconnaissance, surveillance of naval forces and military operations. Such tests would be coordinated with troops on the ground, ships at sea and missile launches. The Russians were said to have orbited Cosmos 1870, a satellite similar to Lacrosse, 18 months earlier.

Unarmed satellites. Like the 200 or more other military satellites currently wheeling through space, sent there by the United States and the Soviet Union, Lacrosse was not armed. One of the main assignments for the big satellite was to find targets for America's new stealth bomber and to look for long-range targets for other missiles and bombers deep behind the lines of Warsaw Pact countries. Critics described such satellites as attractive targets for destruction in some future space warfare. Of course, the Warsaw Pact later was disbanded. The U.S. and USSR subsequently became friends, although American officials said some military threat still existed.

Europe's Helios. American spy satellites are so busy Western European nations say they have to build their own to complete their intelligence-gathering missions. Since 1986, France, Italy and Spain have been building the Helios espionage satellite to be launched on an Ariane to a polar orbit in 1993 at a cost of more than $1 billion. Ground receiving stations in France, Italy and Spain will process images from the 5,500-lb. Helios satellite.

The Netherlands and other NATO-member governments plan to join France, Italy and Spain in a project to build an even bigger spysat for launch around the year 2000.

Most NATO allies are dependent on whatever American spysat photos the U.S. government allows to circulate. But the load on American espionage satellites increased sharply with verification work from the INF treaty and START agreement.

The U.S.-USSR Intermediate-Range Nuclear Force (INF) treaty eliminates all medium- and shorter-range nuclear missiles. The U.S. and USSR negotiated reductions of strategic nuclear arsenals in START.

Fortress Europe. Spysats apparently still are thought necessary to strengthen Europe within NATO. The Netherlands, for instance, was pushing for more military cooperation in Western Europe, even as economic cooperation was building across the continent. One significant result would be more involvement by France in the defense of Western Europe. France is a member of the Atlantic Alliance, but does not take part in NATO's integrated military command.

The Dutch have favored closer European military cooperation. In 1987, the Netherlands promoted formation of a Western European naval presence in the Persian Gulf, arguing that Europe had a vital interest in keeping oil shipping lanes open. Of course, that was before the 1991 Gulf War.

The Hague would like Europe to be less dependent on America for intelligence-gathering from space. The Dutch see a French lead in space expertise.

Moscow-Washington Hotline

The Moscow-Washington hotline, hooked up June 20, 1963, after the 1962 Cuban missile crisis brought the U.S. and USSR to the brink of nuclear war, helped cool tensions during the Cold War years.

Believe it or not, the hotline is not a red telephone. The White House and Kremlin communicate via teleprinter. The only telephone is one technicians use to keep the line in repair.

Teleprinter messages from the White House travel 40 miles north to Frederick, Maryland. There, at the Army's Fort Detrick satellite Earth station, messages are transmitted to a Russian Molniya communications satellite in Earth orbit. The Molniya satellite retransmits the messages down to a USSR satellite Earth station south of Moscow. From there, the messages are sent to the Kremlin. To make sure it's open and working, communications technicians at Fort Detrick check the line every hour. Another ground station on the American end is at Etam, West Virginia.

Crisis management. Discovering in 1962 the USSR had placed missiles in Cuba capable of reaching U.S. cities, then-President John F. Kennedy set up a blockade, ordering Navy ships to turn away vessels delivering missiles to Cuba. War seemed at hand for a week until the Soviets removed the missiles.

Jess Gorkin, senior editor for Parade, wrote an open letter in his magazine, asking the superpower leaders to communicate directly. Kennedy liked the idea and a hotline to Moscow was set up in 1963. Satellite links through Fort Detrick started in 1978.

American presidents have other ways to communicate with the USSR. The hotline is a direct link between the two heads of state, used when either is planning military action in some part of the world. The leaders can bypass diplomatic channels.

Not used often. The Hot Line has been used fewer than twenty times over the years. It was used during the Arab-Israeli Six-Day War in 1967. Then-President Ronald Reagan also used the hotline as he sent troops to invade Grenada.

Fax was added in 1986 so leaders can supplement verbal messages with maps and charts. The link between Soviet and American presidents was modernized in 1990.

Each Hot Line message actually takes three routes to the other nation. The written messages are sent simultaneously via land cable and through a U.S. satellite and a USSR satellite. Duplication is a guarantee the messages actually arrive in a crisis.

The satellite connection wasn't always there. The first Hot Line communications went by land cable only—until a Finnish farmer plowed through the cable one day.

Search And Rescue Satellites

A dozen countries around the globe support a network of fulltime Sarsat/Cospas search-and-rescue satellites in orbit to find ships lost at seas and downed aircraft.

Sarsat is shorthand for Search and Rescue Satellite-Aided Tracking System. Cospas is a Russian acronym for Space System for Search of Vehicles in Distress.

The Sarsat/Cospas network is built around American and Russian spacecraft. The U.S., USSR, France and Canada founded the system. Today more than a dozen nations participate. Authorities credit Sarsat/Cospas, since 1982, with having saved more than 1,150 persons around the world from shipwrecks, plane crashes and even a dogsled race which went awry.

Ham pioneers. OSCAR 7, a private amateur satellite built by ham radio operators in the 1970's, led to today's governmental Sarsat/Cospas network. Under the direction of the Radio Amateur Satellite Corporation (AMSAT), OSCAR 7 in orbit in December 1975 proved Sarsat/Cospas would work when it repeated low-power radio signals, received from the ground, down to Goddard Space Flight Center in Maryland, proving a weak uplink could provide accurate tracking to within two to four miles of an emergency site.

Officially, Sarsat/Cospas started in 1982. Since then, it is said 1,000 lives of air and sea accident victims—mostly Americans—have been saved by rescue teams on the ground alerted by Sarsat/Cospas satellites. Today, there are more than half a dozen government Sarsat/Cospas satellites listening in space.

ELTs. The Sarsat/Cospas satellites, while carrying out assigned primary tasks in orbit, also continuously monitor international radio distress frequencies for signals from emergency locater transmitters (ELTs). By analyzing the Doppler shift of an emergency radio signal heard and repeated by a Sarsat/Cospas satellite, ground crews can "fix" or pinpoint the ELT location to within two to four miles.

One of the latest Sarsat/Cospas satellites to go to orbit was the USSR's Cosmos 1861 on June 23, 1987. Cosmos 1861, stationed some 600 miles above Earth, also houses the amateur radio communications gear known as RS-10 and RS-11.

An American Sarsat/Cospas satellite is NOAA 11 launched September 24, 1988. Like other weather satellites—NOAA 9, NOAA 10, and GOES-7—NOAA 11 carries a radio which receives emergency signals from lost and disabled ships and downed planes.

The satellite circles Earth every 102 minutes in a 540-mi.-high orbit. As the Earth rotates beneath it, the polar-orbiting satellite scans the entire globe every 12 hours. If it hears an emergency beacon, NOAA 11 relays the distress signal to a rescue service on the ground.

NOAA 9 was launched December 12, 1984. It had to be relieved as some of its temperature and humidity sensors failed. NOAA 10 was launched September 17, 1986. NOAA 11 replaced NOAA 9, joining NOAA 10 in service in orbit.

The U.S. National Oceanic and Atmospheric Administration—the National Weather Service—operates NOAA satellites. NOAA satellite headquarters is in Camp Springs, Maryland.

Another NOAA Sarsat/Cospas satellite is GOES-7, launched February 26, 1987. GOES-7 is one of NOAA's stationary weather satellites, the tenth meteorological satellite to be launched to 22,300-mile-high stationary orbit by NASA since 1974 and the eighth in the GOES series. It transmits cloud pictures, atmosphere temperatures and data on the space environment, while on the lookout for airplanes and ships in distress.

New frequency. GOES-7 uses a new Sarsat/Cospas equipment frequency of 406 MHz. Several USSR and U.S. Sarsat/Cospas satellites are closer to Earth in polar orbits with receivers tuned to the worldwide aircraft emergency beacon frequency of 121.5 MHz. The higher altitude of GOES-7 and the new frequency used lets it hear a distress call first and signal the first alert if an emergency is occurring. Then, the polar satellites in lower orbits pinpoint the location of an emergency beacon signal. GOES-7, at launch in 1987, was expected to have a useful life of seven years in orbit, or until about 1994.

Sailors. Two Maryland men sailing a 35-ft. boat from Massachusetts to Bermuda owe their lives to an international Sarsat/Cospas satellite which relayed their radio distress signals in a storm.

The boat was tacking 140 miles southeast of Nantucket December 17, 1987, when heavy wind and waves broke the mast. When the small craft took on more water than bilge pumps could handle, Chris Burtis of Annapolis and Troy Wilson of Bowie switched on their distress transmitter.

An overhead Sarsat/Cospas search and rescue satellite, picked up the signal and relayed it to Scott Air Force Base in Illinois. The Air Force used the distress signal to determine the location of the sailboat. That information was sent to a Coast Guard air station on Cape Cod. The Coast Guard rescued Burtis and Wilson.

Sarsat/Cospas satellites are navigation, communications, weather and research satellites which have extra radio gear stowed aboard to receive and relay distress messages from emergency locator beacons carried by ships and planes. The global Sarsat/Cospas program is operated jointly by Canada, France, the USSR and the United States.

Ontario plane crash. A brand-new NOAA 10 weather satellite, unlimbering during its first day of operation in space, saved the lives of Rory Johnston and three other Canadians when it found their small plane where it had crashed in a remote area of Ontario.

Johnston's small Cessna airplane lost power during takeoff in poor weather, forcing an emergency landing on a lake. The plane sank, nose down, in eight feet of water.

With face cuts, bruised shoulder and dislocated wrist, Johnston swam 200 yards to shore where he found a canoe. He paddled back out to his sunken craft and brought his injured passengers ashore.

Meanwhile, radios in NOAA 10, launched on an Atlas E rocket from Vandenberg Air Force Base in California on September 17, 1986, had just been activated during its 76th orbit around the Earth. The same day, during its 90th revolution, a special Sarsat/Cospas search-and-rescue radio on board NOAA 10 heard Johnston's distress signal.

Sarsat. A satellite equipped with Sarsat/Cospas search and rescue equipment picks up distress signals from ships and planes in trouble and relays them to rescue crews on the ground. NOAA 10's Sarsat/Cospas receiver heard Johnston's emergency beacon while flying over Canada. It relayed the distress signal to Canadian rescue forces at Trenton.

At the same time, a Russian satellite equipped for Sarsat/Cospas also heard Johnston's beacon coming up from Ontario. NOAA 10 picked up the signal again on its 91st orbit. A pilot of a private plane also heard and reported Johnston's emergency beacon to authorities.

The combination of reports from Soviet and American Sarsat/Cospas satellites and the private pilot alerted rescue teams in Edmonton, Alberta. A four-engine C-130 Hercules plane with paramedics aboard took to the air in search of Johnston. Poor weather prevented the C-130 crew from spotting Johnston's plane that night. The next morning, when fog lifted, they found the Cessna and two medical technicians parachuted in to provide first aid.

ELT. NOAA 10's report had been so accurate the C-130 crew reported hearing the distress beacon exactly where the Sarsat/Cospas satellite said it was. Such airplane distress signals are sent by an emergency locater transmitter (ELT) activated automatically upon impact.

After receiving first aid, Johnston and his three passengers were flown to an airstrip at Sachigo Lake, transferred to another plane and taken to Winnipeg, Manitoba, where they were hospitalized.

Nadezhda Hope. The USSR launched another Sarsat/Cospas on July 4, 1989, aboard a civilian navigation satellite named Nadezhda, or Hope. Hope was orbiting at an altitude around 625 miles.

Space Junk: Orbiting Danger

When the U.S. Space Command spotted the derelict hulk of a burned out Russian rocket closing with shuttle Discovery on September 16, 1991, commander John "J.O." Creighton and pilot Kenneth S. Reightler Jr. fired the shuttle's maneuvering thrusters to swerve to a lower orbit.

The uncomfortably-close USSR booster, cast aside after boosting Cosmos 955 to orbit 14 years earlier on September 20, 1977, had been on track to pass a mere 1,150 feet below, 3,000 feet north and 1.2 miles ahead of 1991 shuttle flight STS-48. Discovery's dive kept the vagrant rocket beyond 10 miles.

Paint flecks. In recent years, the North American Air/Space Defense Command and NASA have been tracking thousands of Earth-orbiting chunks of debris the size of a softball or larger. Amidst the worn out satellites and spent booster rockets are millions of pieces of loose junk too small to track by radar.

Larger stuff includes camera lenses, panels from rockets, snippets of wire, fragments from satellites destroyed on purpose or accidentally, nuts and bolts, even depleted nuclear reactors.

In the vacuum of space, things stay in orbit a long time, and pieces accumulate. Altogether, as many as 70,000 half-inch and larger chunks are orbiting Earth. Ninety percent of them are too small to be tracked by ground radar. Zipping along at 25,000 mph, they collide with each other and break up into smaller hard-to-see debris. Flakes of paint, particles of exhaust and countless other grains, smidgens, beads, specks and pellets may add up to 3.5 million or more bits of debris. Even such tiny objects can crack a spacecraft by crashing into it at high speeds.

Sources. About half the debris in the vast space junk yard came from explosions. Spy satellites often are terminated deliberately to keep them from falling into enemy hands. Other satellites malfunction and blow up.

Delta. Plenty of debris came from explosions in the 1970's of the second stages of seven U.S. Delta rockets after long periods in orbit. Delta was modified later.

Ariane. The unexpected explosion of a spent Ariane booster in orbit in 1986 spewed out 400 fragments of debris. The rocket had been launched February 22 ferrying a Spot photography satellite to orbit. Nine months later, the spent rocket third stage suddenly blew apart into a cloud of debris. The harsh space environment had ignited small quantities of left-over liquid oxygen and liquid hydrogen fuel.

Ground radar showed 400 pieces of one-half-inch diameter or larger. Thousands of other particles, too small for radar to see, were floating in the cloud. Space engineers around the globe were concerned for the many satellites in orbits passing through the area of space where the Ariane debris was located.

If a particle crashed into a satellite, shuttle or space station, with or without warning, the spacecraft could be damaged or destroyed. All space junk, no matter what size, threatens satellites and astronauts.

Shuttle banging. The windshield of shuttle Challenger was dinged in orbit in 1983 by a minute particle and had to be replaced after the flight. Chemical analysis found the window had been hit by a fleck of paint no larger than eight one-thousandths of an inch in diameter. The tiny paint chip probably could have punctured the spacesuit of an astronaut on a spacewalk.

In fact, NASA engineers find dents and holes in most returning spacecraft, although shuttle collisions are rare because they fly in low orbits and don't stay long in space. Longer stays require additional protection.

Freedom. U.S.-international space station Freedom, planned for launch to space in the late 1990's, would be used 30 years. Engineers hope shields will protect its crew from debris the size of a marble—one centimeter in diameter. Such a shield adds 2,000 lbs. to the weight of each module. Shuttles will have to carry that extra ton of weight to orbit for each module used.

Pollution. U.S. Vice President J. Danforth Quayle said in 1989 most space junk cluttering orbital highways is from USSR satellites. Warning the environment in outer space is being polluted like the environment on Earth, Quayle said, "One of the problems that we have right now, some people laugh at it but it's a problem, is the debris. The Soviets do a lot differently than we do. Everything they take up there, it's like an ocean—they throw it out. These things start whirling around, it's quite dangerous. Plus every time a satellite breaks up, you've got these things flying all over. It's really quite a problem."

The U.S. congressional Office of Technology Assessment, indicating the U.S. and USSR have been the largest contributors to the space junk problem, said "An international treaty or agreement specifically devoted to orbital debris may be necessary." OTA is the analytical agency advising Congress on technical matters.

OTA called on spacefaring countries to reduce debris, saying, "Unless nations reduce the amount of orbital debris they produce each year, future space activities could suffer loss of capability, loss of income and even loss of life as a result of collisions between spacecraft and debris."

American newspaper columnist Jack Anderson reported in 1990 NASA was holding junk-reduction talks with the USSR.

U.S. Sen. Ernest Hollings, D-S.C., chairman of the Senate Committee on Commerce, Science and Transportation, said, "We can no longer be so cavalier about what we leave behind in space."

What to do? Ideas for removing debris from outer space have included large balloons to sweep up debris and shuttles to capture inactive satellites, but engineers say it would be too expensive to build a Vacuum Cleaner Satellite to clean house.

However, the OTA report found continued littering of space could make the most-used low-Earth orbits too risky for use by 2000 or 2010. The growing collection of space junk above Earth led the British science magazine New Scientist to predict astronauts could be stuck on Earth within 30 years.

Rather than cleaning up outer space, engineers suggest adding shielding to satellites and around astronauts. In addition, satellites could be brought down to low altitudes and destroyed so their debris would descend quickly to burn in the atmosphere.

Future boosters might be designed to fall back toward Earth, to burn in the atmosphere, rather than staying in orbit. Redesigning rockets to do that might increase their cost by 10 to 20 percent, however, since retro-rockets would be needed.

OTA suggested countries and international organizations design rockets and satellites with little likelihood of breaking up, protect batteries from electrical shorts which cause explosions, and sending spent rockets to lower altitudes where the atmosphere would drag them down.

Kessler Syndrome. New Scientist magazine reported on a so-called Kessler Syndrome theory in which so much debris would accumulate above Earth that random secondary collisions would cascade. The resulting vast belt of orbiting junk would stop spaceflight for centuries.

Space Junk: Falling Danger

Pieces of space junk frequently re-enter Earth's atmosphere and burn. Infrequently, a chunk makes it all the way to the surface of our planet, usually falling onto the 70 percent of the surface covered by ocean. There have been some spectacular examples of large satellite scraps making it down to the surface.

Transit. Satellite nuclear accidents have caused worldwide concern since a U.S. Transit-5BN-3 navigation satellite failed to achieve orbit April 21, 1964. Its nuclear-powered electricity generator disintegrated in the atmosphere, spewing radioactive plutonium-238 around the globe.

Cosmos 954. A rorsat is a radar ocean reconnaissance satellite used mostly to track warships. Nuclear reactors are compact sources of large quantities of electrical power used in some science and military spacecraft. Rorsats use nuclear reactors to operate powerful radar to detect warships. Cosmos is a generic name used by the USSR to identify satellites without revealing their purposes.

Several USSR rorsats have fallen to Earth. Cosmos 954, powered by 110 lbs. of uranium, came down in January 1978. It re-entered the atmosphere over North America, raining radioactive debris across 20,000 square miles of western Canadian wilderness January 24, 1978. The USSR repaid Canada half of $3.5 million spent to clean up.

After the accident, Cosmos 954-style rorsats were redesigned to allow the reactor to be jettisoned so it would not have the protection of the satellite body as it re-entered the atmosphere, making it more likely to burn up. The powerplant would eject its fuel rods as the satellite re-entered the atmosphere, theoretically burning up the radioactive waste.

Skylab. The danger of being hit by chunks of a falling satellite worried many persons as much as radioactivity. Concern stirred before pieces of the U.S. space station Skylab came down in remote western Australia July 11, 1979. The 84-ft., 84-ton satellite had been launched May 14, 1973, and occupied nine months in 1973-74.

Cosmos 1402. One of the new-design rorsats, Cosmos 1402, dropped from orbit in two parts on January 25, 1983, over the Indian Ocean. No radioactive debris was found on land. The four-ton main body fell into the Indian Ocean in January and its 110-lb. nuclear-reactor electricity generator dropped into the South Atlantic in February.

Cosmos 1813. Cosmos 1813, a Russian spy-photo satellite, was launched January 15, 1987. It was parked for a short time in an orbit ranging from 130 to 240 miles out. Later it was moved to a 250-mile-high orbit for a better view of the United States and Western Europe.

The military reconnaissance satellite flew two weeks and was supposed to drop its film to a Russian recovery team, but it did not respond to re-entry commands.

USSR controllers exploded it in space to prevent danger from an uncontrolled impact of the recovery capsule which weighed several tons. The explosion blew the spacecraft into more than 100 pieces of debris out to 350 miles and prevented recovery by others.

Cosmos 1817. Possibly a communications satellite on its way to stationary orbit 22,300 miles out in space, Cosmos 1817 was launched on a Proton rocket January 30, 1987. It was destroyed when the Proton fourth stage failed to restart in the low 150-mi.-high transfer orbit.

Cosmos 1818. To avoid the threat of a nuclear accident, the USSR experimented with solar panels to generate electricity in the rorsat Cosmos 1818.

Cosmos 1824. Nuclear rorsat Cosmos 1824 was launched June 18, 1987.

Cosmos 1871. A 10-ton Russian spysat reached only an insufficient 80 to 90 miles altitude after blast-off August 1, 1987, on an SL-16 rocket. It fell harmlessly into the South Pacific 1,500 miles east of New Zealand near Antarctica August 10.

Cosmos 1871, which would have been the heaviest satellite ever sent to polar orbit, was the first time the USSR announced beforehand one of its satellites was dropping. The last time before 1987 the Soviets had made any announcement about a premature return was in January 1983 after Cosmos 1402 burned over the Indian Ocean.

Cosmos 1932. The nuclear-powered rorsat design was changed, adding an automatic action sequence to separate the reactor core and blast it up to a graveyard orbit 500 miles above Earth. There, radioactive material would vaporize and remain in orbit. Nuclear rorsat Cosmos 1932 was launched March 14, 1988.

Gorizont-16. A workhorse SL-12 rocket was launched by the USSR August 18, 1988, carrying the Gorizont-16 communications satellite to orbit. After the satellite separated in orbit, the spent SL-12 fell back into the atmosphere August 21, lighting up the U.S. sky from New Mexico to Michigan as it crashed into northeastern Canada.

The burning spectacle was not the anticipated fall of Cosmos 1900. Police switchboards were flooded with calls from alarmed observers who thought Cosmos 1900 was falling on the U.S. No damage was reported after the rocket plummeted to Earth.

The falling rocket was tracked by the U.S. Space Surveillance Center, Cheyenne Mountain, Colorado, coming down where the Space Surveillance Center had predicted.

It's not unusual for a rocket body to leave orbit. There's nothing to keep it there. An average of one space vehicle reenters Earth's atmosphere each day. U.S. officials don't try to recover rocket body parts as they are 95 percent burned in reentry.

Cosmos 1900. By 1987, the USSR had sent dozens of nuclear-powered satellites into space. Cosmos 1900, the third to fall to Earth in a decade, was launched December 12, 1987. The newspaper Pravda said Cosmos 1900 was a research satellite studying ocean surfaces, but U.S. experts called it a rorsat tracking U.S. warships.

Cosmos 1900 weighed 2,200 lbs. and was 20 ft. long and 7 ft. wide. Its nuclear reactor, weighing about 220 lbs., was fueled by 110 lbs. of uranium U-235.

Previous rorsats had been observed firing rockets twice a week to correct their orbits. Western spacewatchers noticed in April 1988 that Cosmos 1900's rockets were not firing. Soviet ground controllers were forced to acknowledge they had lost radio contact. A worldwide alert for radioactive debris raining on Earth was sounded.

The alert was called off October 1 after Cosmos 1900 properly separated its nuclear power supply from its instrument section, shooting the reactor to a higher, safer graveyard orbit. The satellite had been designed to boost its powerplant to a 500-mi.-high parking orbit to delay re-entry while radioactive elements decayed to safe levels. The reactor successfully rocketed up to a 476-mi. orbit.

Normal use of the powerplant in orbit had consumed most of the fuel, reducing it to radioactive isotopes. From the 500-mi.-high graveyard altitude, the isotopes eventually would disintegrate to dust suspended in the upper atmosphere. Some, no larger than snowflakes, might settle slowly to ground, offering little danger.

One large, non-nuclear piece of Cosmos 1900 housing instruments splashed into the Atlantic Ocean off Africa's northwest coast. Another non-radioactive piece, orbiting for a time at 91 miles, burned up as it slid down into denser layers of the atmosphere.

The possibility Cosmos 1900 could rain radioactive debris had frightened some Europeans. Paris sounded the alert when observers thought the radioactive satellite had fallen there September 30, 1988. But, a smoking shiny sphere that closed the Paris-Lille superhighway for two hours turned out to be a 6-ft. aluminum ball which had fallen from a carnival truck after dark, igniting internal electrical circuits.

French space specialists had given Cosmos 1900 only one chance in a thousand of falling in France. But, about 9 p.m., a motorist near Peronne, 85 miles north of Paris, saw smoking, shiny debris and called police. Gendarmes diverted traffic and set up a security perimeter. The superhighway was reopened by 11 p.m.

Nuclear ban. A top Soviet space official and a private American scientist in 1988 called for a ban on nuclear reactors in orbit. Roald Sagdeev, former director of the Soviet Space Research Institute and advisor to President Michael S. Gorbachev, and Frank von Hippel of the Federation of American Scientists wanted an international agreement to ban nuclear reactors from near-Earth orbit.

Such a ban would end Soviet use of nuclear-powered rorsats. It also would end the American SP-100 nuclear space project. SP-100 is a project for orbiting nuclear reactors to power Star Wars weapons.

Development of its new solar-powered rorsat may have led to the Soviet decision to call for a ban on reactors in orbit. The proposed ban would not apply to nuclear reactors aboard missions in deep space, such as a manned mission to Mars.

French satellite. A large section of a fallen French satellite was found June 24, 1989, on a farm in a remote section of northeastern Australia 1,100 miles northwest of Brisbane. A resident of the nearby town of Conclurry described the debris, probably from Operation Globus, as very large with two cylinders attached and two parachutes. The hulk was carted away by helicopter.

Solar Max. On December 2, 1989, Solar Max fell to burn in the atmosphere over the Indian Ocean near Sri Lanka. The large Solar Maximum Mission satellite had been an observatory with instruments to study the Sun from Earth orbit. It had been launched February 14, 1980, then repaired in orbit in 1984 by astronauts spacewalking from U.S. shuttle Challenger.

AFP-731. U.S. shuttle Atlantis carried the military reconnaissance satellite AFP-731 to orbit February 28, 1990. The astronauts dropped the spysat into its own orbit March 1. Atlantis returned to land March 4.

Strangely, the AFP-731 left in space must have malfunctioned shortly thereafter. By March 7, the USSR was reporting the spysat was disintegrating with four large remnants about to plunge into the atmosphere.

The first chunk burned March 19 over the Pacific Ocean 900 miles north of Midway Islands. The second disintegrated March 20. The two remaining disintegrated by mid-May in the atmosphere without reaching ground.

Satellites Glossary

Angular momentum. A quantity of motion caused by rotation.

Apogee. The orbit location farthest from the body being orbited. For instance, the point where a satellite orbiting Earth is farthest from Earth. See also: Orbit.

Azimuth. Direction to the horizon under an object in space. See also: Elevation.

Celestial sphere, equator, poles. The celestial sphere is the imaginary inverted bowl of stars surrounding Earth. The celestial equator is a projection of Earth's equator onto the celestial sphere, equidistant from the celestial poles, dividing the celestial sphere into two hemispheres. The celestial poles are points on the celestial sphere above Earth's North Pole and South Pole. The celestial sphere seems to rotate around the celestial poles. The meridian is an imaginary line crossing the celestial sphere, passing through both Poles and the Zenith. The prime meridian is the meridian passing through the vernal equinox. Zenith is the point in the sky directly above the observer. Nadir is the point on the celestial sphere opposite Zenith. Precession is the shift of celestial poles due to the gravitational tug of the Sun and the Moon on Earth's equatorial bulge. Right ascension is the eastward angle between an object and the vernal equinox, expressed in hours-minutes-seconds. Declination is the angle between a celestial object and the celestial equator. Inclination is the angle between the orbital plane of a satellite and the plane of the ecliptic.

Communications satellite. An unmanned satellite to relay radio and television signals.

COMSAT. The trade name for one specific communications satellite operating company. Also, slang for any communications satellite.

Cosmos. The generic name given to most of the USSR's unmanned satellites.

Decay. The loss of altitude by a satellite, resulting from a reduction in kinetic energy caused by atmospheric friction.

Declination. Angle between satellite and celestial equator. See: Celestial Sphere.

Docking. Mechanical linking of two spacecraft.

Downlink. A radio or television transmission from a spacecraft.

Drag. The resistance of air in the upper atmosphere to the motion of a satellite.

Elevation. Angle of an object in space above the horizon. See also: Azimuth.

Elliptical. Orbits of planets around the Sun are ellipses.

Escape velocity. The speed required for a spacecraft to leave a planet or celestial body without falling into orbit around it. See also: Orbital Velocity.

Geostationary. An orbit 22,300 miles above Earth where satellites revolve around the planet at the same speed at which Earth rotates. Satellites in this orbit seem to hang stationary over one spot on Earth's surface. Also known as Geosynchronous. A geostationary transfer orbit, or GTO, is a much lower orbit, around 100 miles above Earth's surface, from where a satellite may be boosted on out to geosynchronous orbit.

Geosynchronous. See: Geostationary.

Gimbal. Swiveling of a rocket exhaust nozzle to control direction of flight.

Gravitation. The force of attraction between two bodies.

Greenwich Mean Time. See: Universal Time.

GTO. See: Geostationary.

Inclination. Angle between the orbital plane of a satellite and the plane of the ecliptic. See: Celestial Sphere.

Interplanetary. Within the Solar System but outside the atmosphere of any planet or the Sun. Between the planets. For example, a probe which flies away from Earth, beyond the Moon, to or near planets and other bodies in our Solar System is an interplanetary spacecraft.

Lagrangian point. The place in space where gravity from two large bodies is neutralized.

Launch window. A time interval during which a spacecraft must be launched to achieve a desired orbit or flight path away from Earth.

Microgravity. The space science term for the nearly-zero gravity of Earth orbit. See also: Zero-G.

Node. The point in a satellite's orbit around Earth where it crosses the equator.

Orbit. The closed elliptical path travelled through space by an object around another object, following Newton's laws of gravitation and motion. For instance, satellites orbit planets. In a satellite's orbit around Earth, perigee is closest to the planet and apogee is farthest away. The sidereal period is the time taken by a body in completing an orbit around another object. The synodic period is the time between oppositions of a satellite. A revolution is completed when a satellite in orbit passes over the longitude of its launch site. Period is the time interval of one orbit.

Orbital velocity. The speed required for a spacecraft to go into orbit around a planet or celestial body. To reach Earth orbit from the surface, a velocity of five miles per second is required. See also: Escape Velocity.

Outer space. The domain outside Earth's atmosphere.

Perigee. The orbit location closest to the body being orbited. For instance, the point where a satellite orbiting Earth is closest to Earth. See also: Orbit.

Period. See: orbit.

Photovoltaic cell. See: Solar Cell.

Precession. See: Celestial Sphere.

Probe. An unmanned interplanetary spacecraft which flies away from Earth, beyond the Moon, to or near planets and other bodies in our Solar System.

Remote sensing. Obtaining data about Earth's biosphere by a non-contact method, usually by remote-sensing satellites.

Retrograde orbit. The orbit of a satellite in a backward or westerly direction, opposite the easterly motion of an observer on Earth. See also: Orbit.

Right ascension. See: Celestial Sphere.

Satellite. A small body orbiting a larger object. A moon in orbit around a planet. Natural satellites are called moons. Earth has one Moon; Mars has two moons. Man-made artificial moons, usually referred to as satellites, are machines in orbit.

Satellite dish. A round concave antenna for transmitting signals to, or receiving signals from, a satellite in orbit.

Sensor. Device to receive energy.

Sidereal period. Time required for a body to complete an orbit around another object. See also: Orbit, Synodic Period.

Solar cell. A wafer of material, usually made of silicon, to convert sunlight into direct-current electricity. A photovoltaic cell.

Space science. The study of aerospace science, aerospace engineering, rocketry, and astronomy.

Specific impulse. The ratio of pounds of thrust to pounds of fuel in a rocket.

Sputnik. The first artificial satellite, launched by the USSR October 4, 1957. It initiated the era of space exploration known as the Space Age.

Synodic period. The time between oppositions of a planet or satellite. See also: Orbit, Sidereal Period.

Synthetic aperture radar. SAR is a satellite radar-imaging system in which the antenna size is, in effect, increased by the motion through space of the satellite.

Telemetry. Data transmitted from a spacecraft.

Terminator. A line dividing dark and light hemispheres of a planet or satellite. The line between day and night seen from a spacecraft orbiting Earth.

Thermal protection system. TPS is a heat shield on a manned spacecraft.

Transponder. A transceiver which receives radio signals on one frequency and retransmits the signals on a different frequency.

Universal time. UTC, GMT, Greenwich Mean Time. Mean solar time at the zero-degree meridian at Greenwich, England.

Uplink. A radio or television transmission to a spacecraft.

Weather satellite. An unmanned spacecraft photographing clouds and weather patterns from above Earth and transmitting meteorological data to Earth.

Zero-G. The space science term for the nearly-zero gravity of Earth orbit. See also: Microgravity.

Solar System
****

Solar System Table Of Contents Continues On Page 400

Solar System
****

The Solar System

The corner of the Universe we call home—our Solar System—formed 4.6 billion years ago as the newborn Sun spun slowly in the nebulous disk of leftover matter from which it had grown to life.

The slowly-rotating disk of nebula material was extraordinarily broad, stretching well beyond the edges of today's Solar System.

Today we find nine major planets with at least 54 natural satellites or moons, plus thousands of minor planets or asteroids, more than 1,000 observed comets and plenty of meteoroids and other debris leftover from that original stellar formation time. The planet nearest the Sun is Mercury. The farthest is Pluto. Is there another beyond Pluto? Is there a vast cloud of debris out there beyond Pluto, too, from which comets occasionally rip into the inner Solar System?

The original disk. There was a sizeable amount of rocky metallic matter left in the dusty chaos after the Sun formed. And tiny particles of frozen water and frozen carbon dioxide. All were suffused through a vast cloud of hydrogen and helium.

The Sun's nuclear burning threw off vast quantities of heat. Nearby in the boiling soup, metallic grains and chunks of rock formed. Iron and stone floating in a vast cloud of dust. Over time, some collided and stuck together. As that accretion continued over million of years, large rocky bodies known as planetesimals were formed—tiny worlds, maybe as large as 60 miles or so in diameter.

After some rocks collided, they broke apart, blasting more chunks through the nebula to crash into more rocks, continuing the process.

After 75 million years or so of such chaos, what we know today as the major and minor inner planets were growing recognizable. The bigger they became, the more their gravity was strong enough to attract smaller chunks, adding to their size.

There were still plenty of planetesimals flying around and, over a few hundred million years, they crashed one after another into various major planets. Some would have been deflected, of course, even out of the Solar System, as the system began to clean itself up.

Inner planets. Mercury, Venus, Earth and Mars are the rocky, metallic, major planets which formed near the Sun. Today we refer to them as terrestrial or planets since they are much like Earth.

The Asteroid Belt, probably composed of smashed planetesimals, is in an orbit farther from the Sun than Mars, but not as far as Jupiter, the first of the outer planets.

Outer planets. The Sun's heat was not as strong farther out in the nebula. Cooler temperatures allowed chunks of ice, floating among the rocks in the hydrogen and helium cloud, to grow.

The four planets formed in these cooler regions—Jupiter, Saturn, Uranus and Neptune—were larger worlds of mostly ice and gas. Referred to as Jovian planets after Jupiter, they have very little rock content.

Did the Jovian planets form from accretion or merely coalesce? After all, the Sun coalesced from gases. It's hard to say today, but future explorers may find out.

Was it unusual? It may be that planets don't always form out of a nebula disk of matter around a new star. It seems that many times two or a cluster of stars form as the nebula coalesces. Does that leave additional debris to form planets?

Already unmanned interplanetary spacecraft from Earth have visited the Sun, the Moon, the Asteroid Belt and seven of the eight other planets in our Solar System. Space scientists would like to send a probe to Pluto and astronomers have begun an earnest search outside the Solar System for other stars with planets.

What is it today? The Solar System is a collection of bodies held near each other in space by the gravitational attraction of a star we call the Sun. The bodies orbit the Sun, which in turn is one of a hundred billion stars orbiting the center of our galaxy.

Even today, billions of years after the major planets formed from the disk of matter around the Sun, all of the bodies of the Solar System swim in a thin soup of rocky dust particles. The rocks and dust may be left over from the original formation accretion process, or ejected by comets as they pass through the inner Solar System or they may be debris from collisions between minor planets.

The Sun is the only star we know for a fact to be accompanied by such an extensive swarm of objects, although some astronomers guess there may be other stars with planets nearby in our own Milky Way galaxy. Could it be that other solar systems exist and have technological civilizations?

The Sun

Blemishes were breaking out all over our star as activity in and on the Sun reached a peak at the beginning of the 1990's.

Solar conditions vary on a regular schedule covering eleven years on man's calendar. That Sunspot Cycle is best known for its affects on radio, television and even household electricity on Earth.

The Sun. Our Sun is a large, hot, spherical star nearly a million miles across. That's 109 times bigger than planet Earth. The Sun is our prime source of light, heat and energy. Earth and other planets in our Solar System revolve around the Sun.

Astronomers looking at the Sun discovered a long time ago that sometimes its face is pockmarked. The blemishes are sunspots.

Sunspots are dark, cool spots on the Sun's surface where magnetism inside prevents heat from bubbling out to the surface.

A sunspot looks like a patch on the surface of the Sun, a small light ring with a dark center. As the Sun rotates, a sunspot will seem to move across the face and disappear behind the Sun. If it has not abated in the meantime, a sunspot will reappear 27 days later when the Sun completes one rotation.

Seeing spots. Sunspots appear and disappear. Their number reaches a maximum every eleven years. Magnetic storms and disruptions of radio communications on Earth are associated with large numbers of sunspots. More sunspots mean more solar activity going on. More activity in the Sun means improved worldwide radio transmission on Earth.

Many groups of sunspots were seen by naked eye on the Sun in 1989-91. The Sunspot Cycle turned out to be the second most intense since Galileo first saw solar eruptions in 1610.

NASA scientists had said the so-called solar maximum, or peak of the Sunspot Cycle, would be the biggest since 1610 when Italian astronomer-physicist Galileo Galilei first observed sunspots. But, the cycle apparently did not turn out to be as intense as the 1957-59 solar maximum.

A Sunspot Cycle lasts 11 years. The last previous maximum was late in 1979. The last minimum was September 1986.

Second most active. Sunspot cycle 22, which started at a low count of 12 blemishes on the face of Ol' Sol in September 1986, peaked with a rash of blots in 1990.

The count now is dropping to a new bottom. The end of cycle 22, should be reached in 1994 or 1995. After, that the number will begin to climb again as cycle 23 starts.

250 years' observations. Sunspot-count records have been kept since the mid-1700's. Previously, the three most energetic were Cycle 19 with a peak sunspot number of 201 in 1957-59, Cycle 21 with a maximum of 165 in 1979-80, and Cycle 3 with 159 in 1778.

Cycle 22 probably will have beaten all but Cycle 19. The Royal Observatory of Belgium is the official keeper of sunspot records. Its astronomers keep daily counts and average them across a month to give a "smoothed sunspot number." It already had recorded a monthly mean sunspot number of 173 by November 1989. The Solar-Terrestrial Physics lab of the U.S. National Geophysical Data Center, Boulder, Colorado, figured the peak was in February 1990 at 189. Canada's National Research Council, Ottawa, expected a peak in March or April 1990 while the Rutherford Appleton Laboratory in Great Britain expected it in April 1990.

Dark spots. Sunspots are cooler dark eruptions, some 1,200 to 60,000 miles in diameter, on the surface of the Sun. Each spot appears for three to seven weeks where the solar magnetic field is about 1,000 times stronger than in other locations. As many as 30 spots appear at one time.

Popping up where the solar magnetic field is most intense, they explode into solar flares which send electrons, protons, X-rays, and other radiation streaming out from the Sun. Some of that energy showers Earth, causing magnetic storms by disrupting our planet's magnetic field.

The stronger-than-usual radiation from the Sun might knock satellites off course, cause scattered electric-light blackouts and disrupt telephone calls and radio broadcasts. The effects of a major magnetic storm are subtle, however. The average person might note flickering lights, poor connections for satellite-relayed phone calls, noise on long-distance lines, fading radio stations and interrupted television shows transmitted through orbiting satellites.

The Space Environment Services Center in Boulder, Colorado, the government forecasting agency, says magnetic storms are not dangerous to people on Earth. The planet's atmosphere protects us.

Lethal stream. However, the stream of radiation carries a lethal proton hazard for astronauts on spacewalks and for electronic equipment inside satellites in orbit.

How can it make a satellite fall from orbit? Excessive ultraviolet light from solar flares heats Earth's atmosphere, making it expand. As it expands out around low satellites, it produces drag on those satellites in low orbits. That makes the satellites fall into the atmosphere sooner than predicted and burn up prematurely. The U.S. first-generation space station, Skylab, fell after a big solar flare in 1979.

Of course, there's a positive result. The extra drag works on the thousands of pieces of space debris orbiting Earth. They too are dragged down to burn up in the atmosphere. Getting rid of a lot of space junk makes it safer for men and machines in orbit. And it removes some of the light-scattering debris above Earth which can interfere with astronomers' telescopes.

Northern lights, or aurora borealis, may be seen farther south than usual during a magnetic storm. To see northern lights in northern latitudes, go to a dark site and face north when interference from a bright Moon isn't too great.

Gas jets. Solar jet streams are east-west currents of gas parallel to the Sun's equator. The new observations of solar jet streams support a theory proposed in 1987 at California Institute of Technology. It suggested the 11-year cycle—during which the number of sunspots reaches a maximum, then a minimum and then another maximum—is merely part of a little-known 18-year to 22-year cycle producing visible sunspots only during its latter 11 years.

If the new idea is correct, jet stream gas currents appear during the earliest stages of the longer cycle near the Sun's poles—like smoke rings which grow and roll down the Sun toward the equator. When the jet streams reach a solar latitude of 30 or 35 degrees they produce sunspots.

In 1988, the jet streams were detected at a latitude of 70 degrees on the Sun. They eventually will migrate into the region where sunspots form, probably about 1997.

The jet streams could explain why the number of sunspots changes. Astrophysicists have been seeing solar jet streams closer to the Sun's poles in more detail this time than most other cycles. They have been measuring the speed of the gases moving across the

Sun's surface 20 times a day using a 150-ft.-tall solar observatory and a 75-ft.-deep spectrograph on Mount Wilson in the San Gabriel Mountains northeast of Los Angeles.

Sunlight arriving at a lens near the top of the observatory tower is focused on a hole letting light enter the spectrograph. The spectrograph breaks sunlight into a rainbow of colored light, allowing astronomers to measure the speed of solar gases and detect jet streams the same way a police radar gun measures the speed of a car.

Old sunspots. The famous 11-year Sunspot Cycle has been going on for millions of years.

Huge chunks of ancient rock, imprinted with 100-foot-long dark lines every half inch, have been dug from the bottom of an Ice Age lake in Australia.

Over millions of years, ice in glaciers melted and the water running off carried silt to the lake bed. The marks were written on the rocks when layers of silt were laid down each spring at the bottom of the lake bed. The warmer a spring, the greater the runoff and the more silt deposited on the lake bed.

In prehistoric times, as now, the changing pattern of spots on the face of the Sun led to varied weather on Earth, changing the amount of spring runoff from land with glaciers.

Geologists have correlated the lines on the Australian rocks with modern sunspot cycle records back to 1800.

Warning. It's best not to try to look at sunspots. But, if you must try to look for spots on the Sun, be VERY CAREFUL. Looking directly at the Sun can damage your eyes almost immediately. To look at sunspots, look through a safe solar filter such as number 14 welder's glass.

1991 flares. Solar flares are extra-hot spots erupting outward on the Sun's surface. A week-long series of intense solar flares in June 1991 triggered a severe geomagnetic storm on Earth. The solar disturbances resulted in a "major proton event," according to the National Oceanic and Atmospheric Administration (NOAA) Environment Services Center, Boulder, Colorado. The proton event caused severe interference with high-frequency radio communications on Earth.

The Sun produced several of the most intense solar flares that can be recorded, resulting in a continuous geomagnetic storm on Earth for a week. Flares were at the X-12 level, intense enough to saturate detectors in monitoring equipment on Earth.

Earth was awash in high-energy particles. Managers of electrical power distribution networks, pipe lines and satellites took steps to avoid damage to operations during the storm. The storm resulted in spectacular northern lights displays far south into the U.S.

Giant flares. One of the biggest solar flares ever recorded exploded on the surface of the Sun March 6, 1989. The flare-up erupted near the eastern edge of the Sun so most of the material expelled by the gargantuan explosion belched across the Solar System away from Earth, sparing our planet the communications black-outs and other effects sometimes associated with solar flares.

Even so, the flare was so powerful it saturated detectors on NOAA's GOES satellite which routinely measures strength of solar flares and which recorded the event, according to the U.S. Space Environment Services Center which monitors solar activity.

Astronomers report only a dozen or so flares of the powerful X-5 strength each 11-year solar cycle. The March 6 flare was so extraordinary it was estimated to have strength at the X-15 level. Only two other flares in the 1980's were so big. The flare, which overwhelmed the satellite's instruments, was the most powerful to erupt on the face of the Sun since April 1984.

Northern lights. Police telephones rang off their hooks across the country as the explosive solar activity touched off rare sightings of northern lights as far south as Texas and Florida.

Aurora borealis, or northern lights, shine as tiny charged particles flowing from the Sun crash into Earth's upper atmosphere hours or days later after a solar flare.

Worried callers thought the odd lights might be wildfires or debris from the shuttle Discovery. Early risers saw a shifting curtain of purple and red glow spotted with vertical white streaks. The curtain seemed to move northeast to northwest and back.

The multicolored lights in the nighttime sky usually are seen only near the Arctic Circle. But, they were reported March 12-13, 1989, in Florida, Texas, Utah, Pennsylvania, New York, Detroit, Colorado, Delaware, Indiana and Georgia.

In Utah, the northern light display covered 40 percent of the sky. National Weather Service's Salt Lake City office chief meteorologist said it was the most brilliant display of northern lights in 30 years.

Unfortunately, the show in skies over most of the rest of the nation was obscured by clouds or haze.

Australian balloon. An unmanned balloon on a leisurely float from Alice Springs, outback in Australia, caught some rays for a couple of weeks in February 1987 as it floated around the globe at 69 miles per hour.

NASA's balloon, cruising above 100,000 miles in altitude, carried instruments to see small hard X-ray flares and super-hot flare plasma from the Sun. When it slowed to 10 mph over Brazil and started down, NASA ended the flight February 21.

An earlier balloon had been blown down by a thunderstorm over Paraguay February 6. When that balloon wandered into cool air, it dropped below 60,000 feet. A pressure sensor pushed the payload overboard for a parachute ride to the ground.

The later balloon had been launched February 9, 1987, to circumnavigate the globe. Neither balloon made the complete trip.

Longer cycle? An Air Force scientist at the National Optical Astronomy Observatories (NOAO) has found an overlapping cycle, beginning every 11 years and lasting 17 to 19 years.

According to a solar astronomer from the Air Force Geophysics Laboratory working at the National Solar Observatory, Sacramento Peak, New Mexico, the overlapping cycle begins at high latitudes on the Sun's globe a few years after the 11-year Sunspot Cycle minimum. The overlapping cycle migrates slowly toward the Sun's equator where it blends into the start of the 11-year cycle.

The overlapping-cycle idea came from studies of activity in the solar corona. The solar corona is the tenuous upper atmosphere of the Sun, which can be observed to high latitudes.

The solar corona was observed with a Photoelectric Coronal Photometer at Sacramento Peak in 1973. The photometer was built to observe low-latitude activity on the Sun and coronal holes. Coronal holes are giant rips in the corona through which the Sun spews streams of energetic particles.

The corona is hard to see because the center of the Sun is a million times brighter than the corona. And the sky next to the Sun is very bright—ten times to hundreds of times brighter than the corona. The photometer was able to subtract the sky background from the sunlight, revealing the relatively weak light from the corona.

The photometer made daily observations of a green emission line of so-called 13-times-ionized iron. The astronomer averaged his iron-line observations starting in July 1975 at Sacramento Peak with 1973 and 1974 data from America's Skylab space station to make the overlapping cycle of activity show up in the faint coronal light.

The Sun. Our star is 864,948.7 miles in diameter. Compare that to Earth's diameter of 7926.7 miles. The surface of the Sun is about 6,000 degrees Celsius. Inside it increases to 13 million degrees.

That internal temperature is hot enough for nuclear fusion to take place. Hydrogen is converted to helium and tremendous amounts of energy are released as electromagnetic radiation of all wavelengths, including radio and TV frequencies.

Brief energy outbursts are seen on the Sun's surface as bright solar flares. Solar flares give off extra strong radio waves as well as protons and electrons.

The Sun radiates in all directions and much energy arrives at Earth, only 93 million miles distant. Most of the energy here is absorbed by our atmosphere.

The Ionosphere. In the highest reaches of Earth's atmosphere, near space, beyond our surface weather, is a layer filled with clouds of electrically-charged particles called ions. These floating areas of ions form the ionosphere. Radiated energy from the Sun charges the particles in the ionosphere. More sunlight means more ions and a more thoroughly reflecting mirror for radio signals.

A radio signal moves around the world bouncing between the ionosphere and Earth's surface. Originating at a transmitter, the radio signal travels up to the ionosphere. The ionosphere acts like a mirror to the radio wave, reflecting it back down to Earth. The surface bounces it up again and the ionosphere reflects it back down. The radio signal seems to skip around the globe. Listeners with receivers along the path can detect the signal even if they are on the far side of the Earth.

Yesterday & tomorrow. Casual radio and television users may have noticed sunspot peaks in the past. Distant television stations could have interrupted local programs. CB radios may have heard stations 1,000 miles away better than transmitters across town. Shortwave listeners would have found it easier to tune in the BBC or Radio Moscow. Police dispatchers sometimes had local calls interrupted by signals from cities hundreds of miles away. Ham radio operators found it easier to talk with amateurs in more than 200 countries around the world. Brief electrical outages on Earth have been noted at times of worst solar disturbances.

It's not known why more sunspots appear during some cycles and fewer at other times. Flat cycles with little enhancement to radio transmissions also have been recorded. Scientists still can't predict for sure which type we will experience each 11 years, but they want to learn how.

Dust rings. The Sun is circled by huge rings of interplanetary dust. One ring of dust, between Jupiter and Mars, has been known for a long time. It causes what astronomers call Zodiacal Light.

Another ring was discovered among the asteroids between Mars and Jupiter by the Infrared Astronomy Satellite (IRAS) in 1983.

Japanese and Indonesian astronomers found a third ring in 1983 near the Sun. The dust in that ring looks like it has been spiralling slowly inward from the outer Solar System. The Sun vaporizes it and the gas is blown back outward from the Sun by the pressure of solar radiation.

Less bright. Our star lost about one-tenth of one percent of its brightness during the 1980's, but scientists say there's nothing to worry about. Although the dwindling brightness could mean a drop of temperature by one-fifth to two-fifths of a degree on Earth, that seems to have been offset by the greenhouse effect around this planet.

Solar Max—the Solar Maximum Satellite—has been reporting on the Sun's brightness since 1980. Solar Max found about one-thousandth of the light radiated from the Sun was missing by 1985. If solar output continued to dwindle, life on Earth would be threatened. But, solar astronomers at NASA's Jet Propulsion Laboratory, Pasadena, California, think there is no chance of the phenomenon creating a new Ice Age on Earth. Mostly because they haven't seen any further decrease since mid-1985.

The scientists haven't been able to agree on whether or not the dip in Sun brightness relates to the 11-year sunspot cycle. The last sunspot cycle peak was in 1980, after a Nimbus 7 satellite first reported the diminishing brightness in 1978. Some solar physicists think the 1980's loss of brightness may be related more to increased solar flare activity than to sunspots.

The amount of light radiated from the Sun started to increase slightly in 1987 as the sunspot cycle started to build again toward a peak.

Frigid history. There was a period of very low sunspot activity between the years 1650 and 1725. For about 300 years, centered on the late 1600's, Earth shivered through what has come to be known as the Little Ice Age.

Farming was hurt and glaciers grew southward when temperatures dropped two to three degrees Fahrenheit around the world.

Greenhouse effect. If the 1980's dimming were to continue for another 50 years, another Little Ice Age could result. However, the 1980's loss appears to have been offset by the greenhouse effect. Environmental pollution gases from our industrial society warm the Earth by trapping heat, like glass panes in a greenhouse.

Little torches. The news from Solar Max disturbed scientists who had been laboring under the belief the Sun would become more dim as more cool sunspots appeared on its face. Now astrophysicists say the Sun may be brighter during sunspot peaks because the spots are accompanied by bright hot spots called faculae or "little torches."

Downturns and upturns in solar output, as measured by satellites like Solar Max in orbit around Earth, seem tied to decreases and increases in hot spot activity on the Sun.

Sun WIMPs. A Swiss scientist is blaming WIMPs for a shortage of neutrinos coming from the Sun. Neutrinos are tiny parts of atoms which overflow from the Sun and stream outward in all directions. Some of these particles, landing on Earth, are found by physicists who think there should be three times as many arriving here as are counted. To explain the mysterious shortage of neutrinos from the Sun, a California astrophysicist designed the WIMP theory.

A WIMP is a Weakly Interacting Massive Particle found in his theory at the center of our star. It increases the density of matter in the Sun and increases the gas pressure by accounting for extra hydrogen in the Sun's heart. Without substituting big particles like WIMPs for the missing small neutrino particles, scientists would be forced to lower their estimates of the temperature in the middle of the Sun.

A lower temperature, however, would mean lower gas pressure inside the Sun. It wouldn't be able to hold up its outer layers and would collapse.

A Swiss scientist at the World Radiation Center has found what he says may be proof that there really are WIMPs in the Sun. Without WIMPs the density of stuff in the Sun's core is about 166 grams of matter per cubic centimeter. If there are Weakly Interacting Massive Particles, density could be as high as 196 grams per cc. That's about 196 times as dense as water on Earth.

Global network. The Sun never will set on GONG—a global chain of solar monitors giving astronomers an uninterrupted series of observations of our star's activity. GONG stands for the Global Oscillations Network Group of the U.S. National Solar Observatory.

From 1985 to 1988, scientists from the U.S. National Optical Astronomy Observatories (NOAO) Solar Observatory installed solar-radiation monitors at ten sites: Yuma, Arizona; Big Bear Solar Observatory, California; Learmonth Solar Observatory, Australia; Mount Wilson Observatory, California; Haleakala, Hawaii; Mauna Kea, Hawaii; Izana, Canary Islands; Cerro Tololo Inter-American Observatory, Chile; Las Campanas Observatory, Chile; and Udaipur, India.

NOAO's Cerro Tololo Inter-American Observatory and the Carnegie Institution of Washington's Las Campanas Observatory are in the Southern Hemisphere, on the foothills of the Andes mountains, 50 miles from La Serena, Chile, near the southern edge of the dry Atacama Desert.

Data from the ten sites was examined for the smallest number of gaps in solar observations caused by clouds, fog, rain, smog or other atmospheric disturbances blocking sunlight. Scientists selected the six best sites to place solar telescopes.

Through lengthy uninterrupted observations of the Sun's surface vibrations, using new techniques of helioseismology, the six telescopes will tell astronomers about the interior structure and dynamics of the Sun. Analyzing vibrations caused by internal pressure waves, astronomers should be able to unlock secrets of the Sun's interior the same way seismologists study earthquake waves to learn about the Earth's interior.

GONG astronomers wonder about the Sun's internal rotation and the size of its core. Answers may explain how the Sun affects Earth's climate, what causes sunspots to appear and disappear, what pushes the 11-year sunspot cycle from minimum to maximum.

The instruments at the six sites are pyroheliometers. They follow the Sun on a clock-driven mount and measure solar radiation with an accuracy within one percent. Data from the pyroheliometers is transmitted through a cable to a data logger—a digital voltmeter, a calculator and a digital micro-cassette drive.

A GONG television camera at each site, stabilized by laser, takes a Sun picture every 75 seconds to record ripples.

NOAO. The National Optical Astronomy Observatories are three well-known facilities operated by the Association of Universities for Research in Astronomy (AURA), Inc., contract to the National Science Foundation: Kitt Peak National Observatory, Arizona; Cerro Tololo Inter-American Observatory, Chile; and National Solar Observatory, Sacramento Peak, New Mexico.

Viewing time of astronomers from around the country is scheduled on the relative merit of research proposals. There also are resident astronomers doing research. They are designing the proposed 15-meter National New Technology Telescope.

Balloon sextant. Astronomers have been using a gargantuan high altitude balloon to measure the diameter of the Sun to see how it affects Earth's climate. The vast balloon, as big as a football field, carries a solar-disk sextant to measure minute changes in the shape and size of the Sun.

The Sun is the easiest star for astronomers to observe. They try to understand its behavior to understand stars in general. The sextant was designed by Yale astronomy professor Sabatino Sofia and researchers at NASA's Goddard Space Flight Center, Greenbelt, Maryland.

Sunquakes. The device also senses fluctuations resounding on the Sun's surface from sound and gravity waves deep in its core, something like the waves sent through our planet by an earthquake. The mammoth balloon carries the unmanned sextant to the top of Earth's atmosphere, an altitude of 120,000 feet. From that vantage point, the sextant measures the Sun's diameter, more precisely than possible from the ground.

The atmospheric distortion around the balloon at that altitude of almost 23 miles was much less than on the ground. Less distortion made it easier for the equipment to measure gradual changes in energy coming from the Sun and its impact on Earth's climate.

As the Sun fluctuates in size, its energy output changes. In the past, the only way to measure that was timing total eclipses of the Sun. Previous weather research focused on how the Sun's energy penetrated Earth's atmosphere, not on the Sun's energy output.

Future flights. The sextant project began at the end of the 1970's. Sofia, head of the university's Center for Solar and Space Research, joined Yale after leaving Goddard in 1985. He and the Goddard researchers plan to launch the sextant farther into space on a small rocket in 1992.

From there, the sextant would provide continual solar measurements with no atmospheric distortions. By 1997, they plan to install it in an unmanned satellite in orbit. It would orbit 15 years, recording the Sun ten to fifteen minutes every day.

The Nine Planets

The nine known planets of the Solar System are Mercury, Venus, Earth, Mars, Jupiter, Saturn, Uranus, Neptune and Pluto.

In the table below: Diameter is planet diameter in miles. Sun is average distance in miles from planet to Sun. Earth is average distance in miles from planet to Earth. Day is length of day in Earth days (d), hours (h) and minutes (m). Year is length of year in Earth days (d) or years (y). D Temp is surface temperature in degrees C at noon or on the light side of planet. N Temp is surface temperature in degrees C at night. C Temp is temperature in degrees C at cloud surface. Axial rota is equatorial axial rotation period. Axial tilt is in degrees, minutes, seconds. Inclin is inclination of orbit to ecliptic in degrees, minutes, seconds. Density is mean density in grams per cubic centimeter. Mass assumes Earth equals one. Magnitude is at brightest. Escape is escape velocity in kilometers per second.

Planet	Color	Diameter	Sun	Earth	Moons
Mercury	orange	3031	36.04	50-136	0
Venus	yellow	7521	67.11	25-161	0
Earth	green	7926	92.99		1
Mars	red	4217	141.61	35-248	2
Jupiter	yellow	88734	483.66	367-600	16
Saturn	yellow	74566	886.72	744-1028	21
Uranus	green	31566	1783.38	1606-1960	15
Neptune	yellow	30758	2794.38	2677-2910	3
Pluto	yellow	1429	3666.19	2670-4700	1

Planet	Day	Year	D Temp	N Temp	C Temp
Mercury	176 d	87.97 d	350	-200	
Venus	2760 d	224.70 d	500		
Earth	23 h 56 m	365.26 d	av 0.22		
Mars	24 h 37 m	686.98 d	-20	-80	
Jupiter	9 h 50 m	11.86 y			-150
Saturn	10 h 14 m	29.46 y			-180
Uranus	24 h	84.01 y			-21
Neptune	22 h	164.79 y			-220
Pluto	6 d 9 h	248.00 y	-230		

Planet	Axial rota	Axial tilt	Inclin	Density
Mercury	58d 15h 30m	0º	7º00'26"	5.42
Venus	243d 24m 29s	178º18'	3º23'40"	5.25
Earth	23h 56m 4.07s	23º24'		5.52
Mars	24h 37m 26s	25º12'	1º51'9"	3.94
Jupiter	9h 50m 33s	3º06'	1º18'29"	1.31
Saturn	10h 39m 22s	26º42'	2º29'	0.69
Uranus	17h 14m	97º54'	0º48'26"	1.30
Neptune	18h 26m	29º36'	1º46'27"	1.66
Pluto	6d 9h 17m	94º	17º9'3"	1.80

Planet	Mass	Magnitude	Escape
Mercury	0.056	−1.9	4.3
Venus	0.815	−4.4	10.3
Earth	1.000		11.2
Mars	0.107	−2.8	5.0
Jupiter	317.830	−2.6	61.0
Saturn	95.160	−0.3	35.6
Uranus	14.500	+5.6	21.2
Neptune	17.200	+7.7	23.6
Pluto	0.0003	+13.6	1.0

Mercury

Its heavily-cratered globe is only about 3,000 miles in diameter, making Mercury the second smallest planet in the Solar System, after Pluto. Mercury has no natural satellite.

Mercury orbits the Sun every 88 days at an average distance of about 36 million miles. It is the closest planet to the Sun. Earth, the third planet, is three times as far from the Sun as Mercury.

The U.S. probe Mariner 10 flew by Mercury three times in 1974 and 1975, measuring a daytime temperature on the surface of 374 degrees Fahrenheit. And that might reach as high as 842 degrees when Mercury is closest to the Sun. At night, temperatures drop to minus-292 degrees.

In photos, Mercury resembles Earth's Moon. Its surface is heavily cratered. The planet's mass is only five percent of Earth's and half of the mass of Mars. It probably has a large nickel-iron core.

Mercury has only a trace of atmosphere, mostly composed of sodium, helium and hydrogen. There appears to be more sodium than anything else in the thin atmosphere.

Mercury has an odd swing, showing first one face and then the other during close approaches to the Sun.

The nine known major planets of the Solar System usually are divided into two groups: the inner terrestrial planets and the outer Jovian planets. The inner planets—comparable in size, density, and other characteristics to the Earth—are Mercury, Venus, Earth, and Mars. Earth is largest of the four.

Mercury, closest of all to the Sun, has been studied the least. Scientists don't know much about what's deep inside the planet, or even just below the surface. Only 40 percent of the planet's surface has been photographed. The rest is unmapped. Despite being relatively close to Earth, Mercury is among the least understood of the planets.

Radio Photos. The Solar System's innermost planet, too close to the Sun to see, has a surface too hot to touch in the daytime while too cold to touch at night, and an unbreathable atmosphere.

Scientists at NASA's Goddard Space Flight Center are trying to go beyond data sent back from the planet Mercury 13 years ago by the Mariner 10 spacecraft. They are using giant dish antennas in New Mexico to shoot long-distance radio-wave photographs of the intriguing planet.

VLA. One new remote-sensing tool available to astrophysicists is a radiotelescope known as the Very Large Array (VLA) at Socorro, New Mexico. That set of sensitive radio antennas was pointed at Mercury on July 6, 1986, to make the first-ever radio snapshots of the planet. The antennas are so sensitive they even allowed scientists to peer three feet beneath the Hermean surface.

Scientists refer to Mercury's parts as Hermean, rather than Mercurian. In ancient Earth mythology, Hermes was the Greek-god counterpart of the Roman god Mercury. Mercury was messenger to the gods as well as god of commerce, travel and thievery. Hermes was messenger to the gods while god of commerce, invention, cunning and theft.

Examining soil just below the surface of planet Mercury allowed scientists to measure the Sun's long-term heating effects on Mercury. The radio-wavelength pictures confirmed a theory that Mercury has a hot pole—like a north or south pole always pointed toward the Sun—caused by a unique spinning in orbit as the planet is domineered by the Sun. Scientists now are able to predict the electrical properties of the surface.

Venus

Venus, a burning-desert world hidden under bitter clouds of sulfuric acid and carbon dioxide, may once have been awash with oceans of near-boiling water.

For hundreds of millions of years, most of the water on Venus was liquid near the boiling point. But the water finally dried up, according to three scientists at NASA's Ames Research Center in Mountain View, California, south of San Francisco.

In what they call a moist-greenhouse or wet-greenhouse effect, the Ames team says the water on Venus was able to remain liquid for a few hundred million years because planet temperatures actually were cooler than Earth scientists once thought, and because of the proportion of carbon dioxide to water vapor in the atmosphere of Venus.

Venusian seas dried up slowly as hydrogen eventually escaped into space. Oxygen formed compounds with other elements and was incorporated into the planet's crust.

Scientists say liquid water seems essential to life. The Ames researchers said the oceans of Venus might still have been liquid at 200 to 300 degrees Fahrenheit because of intense pressure. But that may have been too hot for life to begin. If not, it probably did not last long enough for life to begin.

Venus, nearly Earth's twin in size, has atmospheric pressure on the surface 90 times that of Earth. The temperature on the surface may be as high as 900 degrees Fahrenheit. Venus is the second planet from the Sun, after Mercury. Earth is third. Mars is fourth. There are nine known major planets.

Old theories. Scientists already had a runaway-greenhouse explanation in which water existed on Venus, but only as vapor.

In a greenhouse effect, Sun heat is trapped by gases in a planet's dense atmosphere. The wet-greenhouse and the runaway-greenhouse theories have a greenhouse effect in common.

In the runaway-greenhouse explanation, Venus was said to be so hot that its water existed only as vapor and had no chance to condense to liquid on the surface. Water vapor rose into the atmosphere, where radiation from the Sun cracked it into separate oxygen and hydrogen atoms. The hydrogen escaped into space and water couldn't form.

But the Ames researchers, looking at different climates on Venus, Earth and Mars, didn't like the runaway-greenhouse explanation. The old theory forgot that the Sun was 25 to 30 percent cooler 4.5 billion years ago, they noted. It also did not account for the water loss.

Earth. The wet-greenhouse theory might be a model for Earth. Scientists are concerned that burning fossil fuels puts too much carbon dioxide into Earth's air, causing the atmosphere to trap heat from the Sun. That could raise air temperatures gradually, increasing sea levels as polar ice melts.

However, Earth is farther from the Sun and has a different atmosphere which is expected to prevent the climate catastrophe that turned Venus into a desert.

Fool's gold. NASA scientists have bounced radar waves off cloud-shrouded Venus and detected colossal highly-reflective blobs and rings that may be lava flows or volcano crater rims rich in fool's gold.

The radar survey by researchers from NASA's Jet Propulsion Laboratory (JPL), Pasadena, California, turned up more evidence that volcanoes have erupted on Venus within the last few million years, maybe even within recent centuries.

JPL radar scientists and geologists transmitted radio signals from a giant antenna at Goldstone, in California's Mojave Desert, toward the equator of Venus. Half the radio waves bounced back to three 210-foot-diameter, dish-shaped receiving antennas at the desert facility.

The picture made by bouncing a radar beam off the cloud-shrouded planet covers 620 miles across Venus's prime meridian along the equator and includes a feature called Heng-o Chasma.

In the radar image, Heng-o Chasma appears to be an egg-shaped oval structure known as a corona. At 744 miles in diameter, it is the largest one seen so far on Venus.

The fuzzy radar pictures portrayed bright blobs as well as highly-reflective rings on the surface. The largest blob, 180 miles wide and 350 miles long, may have lava flowing from a jumbo mound—possibly a volcano—at its west end.

The largest ring, 60 miles across, looks like the rim of a crater caused by meteorite impact. Other rings, 18 miles across, may be rims of large volcano craters known as calderas.

Rings and blobs. The rings and blobs could cause bright radar reflections if they are lava flows and crater rims rich in iron pyrite—known as fool's gold because it looks like gold—or other metallic minerals blasted to the surface of the planet by volcanic eruptions and meteorite impacts.

Fool's gold or metallic minerals should decompose, losing reflectivity over time. Their detection might be evidence of recent volcano activity. Old craters don't have bright rings.

Recent activity might be within the last few million years, or even within the last few centuries. Planetary geologists don't know yet how quickly rocks break down on Venus.

Lightning. New research indicates Venus may have lightning similar to Earth. The lightning may be related to volcanic activity on the surface of the planet.

However, radio signals seem to indicate it is an activity that occurs during the afternoons in a heavy layer of clouds 35 miles above the surface, not from volcanos.

Magellan. The interplanetary space probe Magellan left Earth in 1989 and fell into orbit around Venus August 10, 1990. Since then, it has sent back spectacular radar images for a new, more-detailed map of Earth's cloud-shrouded sister planet.

NASA had trouble holding a steady radio contact with the spacecraft for a time as it whirled around Venus. Even so, spectacular images of rugged terrain were received.

Magellan completed one complete mapping of 90 percent of the surface of Venus with radar able to peer through the blanket of thick clouds which block visible light for optical cameras, then went into a second mapping cycle.

Magellan's Venus pictures reminded geologists of California earthquake faults, Hawaiian volcanoes, the rift valleys of East Africa and Europe's Rhine Valley. The strips of photos covering territory 1,000 miles long by 15 miles wide, called "noodle" pictures, depicted a violent Venus-scape sculpted with long, parallel valleys and ridges— like those between the Rocky Mountains and the Sierra Nevada range—as well as deep impact craters, jagged quake faults and expansive lava flows similar to those on Hawaii and in the Snake River plains of Idaho.

Overlapping lava flows six to ten miles wide, of different ages, were bright and dark splotches in the photos. Many Venus-quake faults and fractures suggested movement of the planet's crust had shaped the landscape. Parallel sets of elongated valleys and ridges looked like the basin-and-range area of the intermountain region of Utah and Nevada, or the Great Rift Valley in eastern Africa.

Apparently, part of Venus' crust had been stretched apart. Zigzag lines across the surface were tinsel fractures caused by pulling the crust apart, like the the rift valleys in East Africa or the Salt Trough earthquake fault under California. A pit crater about a mile wide in the photo, like depressions on the slopes of Hawaiian volcano Mauna Lea, may have been volcanic in origin, not the result of a meteor impact. The Devana Chasm, a giant valley, had fault patterns like those in the Rhine Graven region of Germany.

Pioneer-Venus 1. NASA's Pioneer-Venus probe is a small satellite still orbiting Venus. The U.S. sent two interplanetary Pioneer probes to Venus in May and August 1978, one to orbit the planet and the other to deliver four smaller probes to Venus.

Pioneer 12, also known as Pioneer-Venus 1, went into orbit around the planet December 4, 1978. Its highly-elliptical path around Venus brought it to within 100 miles of the surface to do radar mapping, cloud studies and magnetometer readings.

Pioneer-Venus 1 still is in orbit around Venus, using radar to map the planet surface and send back data about the solar wind. The spacecraft intercepts particles in the solar wind as they pass Venus, flying out from the Sun during the 11-year sunspot cycle. Pioneer-Venus 1 reportedly has less than five lbs. of fuel left but that amount should keep it in proper orbit around Venus until at least 1992.

Radar maps suggested plateaus, volcanoes, and valleys on Venus larger than similar features found on Earth. Previous radar pictures made from Earth, and from Soviet spacecraft at Venus in 1983, displayed lava flows, volcanic craters and craters from meteorite impacts.

Pioneer-Venus 2. Pioneer 13, or Pioneer-Venus 2, was one big probe carrying four smaller probes to be dropped into the atmosphere of Venus. The mothership fired one large 160-lb. probe toward the surface of the planet November 15, 1978. Later, on November 19, it sent three smaller 50-lb. probes.

The probes penetrated the atmosphere of Venus at different locations on December 9, 1978, gathering data and relaying it back to Earth as they descended and hit the planet. One small sounder probe actually survived the hard landing and transmitted data for 68 minutes from the surface. The main Pioneer-Venus 2 mothership also arrived at Venus December 9, entering the upper atmosphere as a probe. It burned in the atmosphere.

Mars

Today it's a very dry, very cold place, but Mars once may have had lakes and seas of water. Ancient features seeming to resemble islands, sandbars and shorelines on Earth can be seen in photos of Martian terrain taken from 1976 to 1981 by the Viking spacecraft in orbit around the Red Planet.

Mars, the Red Planet, the fourth planet from the Sun is named for the Roman god of war. Seen as a red dot in Earth's night sky, it is 1.5 times farther from the Sun than the Earth is. The globe of Mars is only half as large as the Earth.

Mars is more like Earth than any other planet. Its period of rotation and the inclination of its axis are similar. Its density indicates it's made of rocky materials, although with less iron and more lightweight elements and volatiles than the Earth. Its atmosphere is thin enough to allow us to see the surface.

Water. Lakes or seas might have formed in Mars' northern lowlands as a result of massive floods which created the famous large Martian channels some 2 billion to 3 billion years ago, according to scientists at NASA's Jet Propulsion Laboratory (JPL), Pasadena, California.

Most of the old Viking photos, however, are too fuzzy for detailed study of ancient Martian beaches and shorelines. New close-up photos are needed.

There is disagreement among scientists over Mars' geological past. The JPL scientists called the lakes and seas a hypothesis that won't be settled until future camera-carrying spacecraft send back new photos and data.

Some scientists have suggested the large channels on Mars could have been formed by volcanic lava, ice or wind. Most agree they probably were created by groundwater released when overlying rock suddenly collapsed.

Most of the channels seem to empty into large, seemingly featureless plains. As the JPL scientists looked at Viking photos of these channel mouths, the question was what happened to water as it reached the ends of the channels. If floodwaters carried large quantities of sediment and soaked into the ground, that should have left deposits in enormous alluvial fans like those on Earth. But, no channeling like alluvial fans on Earth can be seen in Viking photos.

If water had formed a sea by collecting in Martian lowlands, it should have created deltas at the channel mouths. The delta deposits should be thin and widespread, because of the nature of the flooding and lower gravity. Mars gravity is about one-third Earth gravity.

Looking to see if landforms at the edges of plains resembled features along the shores of lakes and oceans on Earth, the scientists came up with several sites that seemed to support their hypothesis.

One was in Cydonia Mensae, an area in Mars' northern hemisphere marked by clusters of eroded mountains and mesas connected by curved ridges. The site looked a lot like groups of islands on Earth linked by coastal barriers and sandbars.

The rim of an eroded crater in Acidalia Planitia, a region west of the Cydonia Mensae site, had been eroded into a loose, broken chain of mountain-like knobs, unique to the Martian lowland plains. The shape of the knobs might have been produced by wave action.

A third place looked like an actual shoreline, in Deuteronilus Mensae, another area of eroded mesas along the lowland margin in Mars' northern hemisphere.

Photographs from Mars Observer, an American probe to be launched in 1992, and pictures from several Russian spacecraft going to Mars in coming years, should give a better look at surface features.

Moons. Only Mars and Earth among the terrestrial planets in the inner Solar System have natural satellites. Venus and Mercury do not. The two tiny moons orbiting Mars, Phobos and Deimos, were discovered in 1877 by Asaph Hall.

The Movie. Planet scientists at NASA's Jet Propulsion Laboratory, Pasadena, California, let a supercomputer sift for 37 days through pictures of Mars and elevation data from the U.S. Viking probes to create a five-minute movie of a flying tour around the Red Planet.

The supercomputer needed time equivalent to 24 hours a day for 37 days to generate 3,500 frames that make up the three-dimensional computer graphic simulation of a small area of the Martian surface. For a touch of class, the scientists set the tour to music—Gustav Holst's The Planets.

1988 Dust Storms. Mammoth dust storms flared up on the Red Planet in 1988, first in June, then in November.

Fortunately, the June storm subsided and Mars features were very visible from Earth as the planet passed close by in September. When the Red Planet closed with our green planet on September 21, it was just over 36 million miles away. That close, it rivaled Jupiter as the brightest object in the sky after the Sun, Moon and Venus.

Later Mars faded some as it moved farther away from Earth, but remained interesting to look at through a telescope.

June. The gargantuan dust storm which erupted on Mars in June 1988—4,800 miles long and 2,500 miles wide—was seen by amateur astronomers around the world.

The storm may have started in an 1,100-mi.-wide, 2.5-mi.-deep basin in Mars' southern hemisphere known as Hellas. To Earth observers, Hellas looks like a bright area on the surface of the Red Planet. It may be an old impact crater. The U.S. Mariner 9 spacecraft in 1971 and the Viking spacecraft orbiting Mars in 1976 photographed the round canyon, finding it to be the largest of Mars' impact basins.

The towering yellow dust cloud, 60 million miles from Earth, was spotted from Arkansas June 13 and confirmed from Florida June 14. By June 15 the cloud, as seen from Earth, had grown to a vast dumbbell shape with two bright centers. By June 16, the dust was throwing a shadow across Cimmerium Mare, Tyrrhenium Mare and Hesperia. The cloud shined yellow along the entire edge of the planet in late June as the planet revolved.

Close encounter. Dust settled from the June storm and the Red Planet made its nearest approach to Earth in 17 years September 21, 1988, providing some of the best views of the Martian surface since 1875. During the close encounter, the two planets were 36.54 million miles apart. Mars won't be so close again until August 2003. After 2003, the next close encounter will be in the year 2025.

Orange UFO. Mars did not disappoint Earth-bound viewers, as had the ballyhooed comets Halley and Kohoutek. Being a planet, the brightness of Mars was more predictable than light reflected from those comets. The planet was seen easily around the globe, bringing on a spate of UFO sightings by folks not used to finding an unblinking orange-red light in their night sky.

The cold and arid red desert planet has a gigantic 2,500-mile-long super-version of the Grand Canyon. And an 18-mi.-high mountain, three times higher than Mount Everest, capped by a crater the size of the state of Georgia.

Backyard astronomers said the view from Earth was good in September with the close approach sandwiched between June and November dust storms. Views of Mars' puffy white clouds, towering volcanos and shrinking southern ice cap continued in October.

November. At its peak, the November storm made much of the southern hemisphere of Earth's neighbor-planet appear featureless to astronomers as yellow dust clouds blocked their view of mountains, valleys and plains. The familiar Eye of Mars was nearly wiped from view. The storm subsided in December.

Because Mars appeared much farther north in Earth's skies than usual during such encounters, views for Northern Hemisphere observers were better than at any time since 1875 or until 2025.

Many viewing parties were held across the country, from Oregon to Alabama. About 100 of 500 amateur astronomers participating in a worldwide Mars Watch '88 observation project sent 1,500 photographs and sketches of the planet to a group named the International Mars Patrol.

It was summer in Mars' southern hemisphere. The southern ice cap extended only to 81 degrees south latitude, much smaller than in winter when it can stretch almost halfway to the equator. The largest volcanos—especially Olympus Mons, the biggest at three times taller than Mount Everest—were seen by observers using large telescopes. Viewers also spotted round and oval white puffs of mountainous clouds around the volcanos.

Mars changes continuously, according to astronomers, who say there's always something to discover. Mars could be seen with the naked eye, but surface features weren't visible. To Northern Hemisphere watchers, it was as a reddish-orange spot, the brightest object in evening skies.

On August 12, 1988, the planet was at perihelion or closest approach to the Sun. At 11:25 p.m. EDT September 20, 1988, it reached opposition, meaning it was on the opposite side of Earth from the Sun. The combination of the two events—perihelion and opposition—led to Mars' close approach to Earth.

Again in 1990, backyard astronomers have spotted Martian winds kicking up dust. A small cloud, which appeared October 4, 1990, in the Chryse desert, expanded in one day to cover Aurorae Sinus, then spread over the northern areas of Mare Erythreaum, Aurora Sinus and Eos. Dust was seen in Valles Marineris canyons. Some usually-visible Martian features faded beneath the yellow dust cloud.

Is There Life On Mars? The U.S. fired two unmanned Viking spacecraft toward Mars in 1975. Each carried a lander to detach itself from the main spacecraft and drop down to land on Mars.

Viking 1 was launched September 9, 1975, arrived in Mars orbit June 19, 1976. Its lander touched down on the surface July 20, 1976. Data was sent back for 6.5 years.

Viking 2 was launched August 20, 1975, arrived in Mars orbit August 7, 1976, and its lander touched down September 3, 1976. Data was sent back 3.5 years.

Life on Mars? Earth wanted to know if there ever had been life on Mars. The biology labs in the landers came up with a "not sure." There wasn't a flat-out "no," but there wasn't a dramatic "yes" either. Two tests for the existence of life—detecting gases produced in metabolism—showed vigorous activity suggesting life processes like Earth. But, a third test—to detect organic materials—was negative.

Some scientists, looking for explanations for the puzzling results, supposed they were seeing some exotic chemistry, maybe an oxidizing Martian soil unlike Earth soil because of heavy bombardment of Mars by ultraviolet radiation from the Sun and electrified atomic particles in the solar wind. It was just a guess.

Of course, there was more to Viking than the search for life. Viking conducted the most detailed science investigation of a planet ever made by an unmanned craft. It found:

Nitrogen. The atmosphere of Mars had all the elements needed to support life. Viking 1 found nitrogen, which is essential to life as we know it on Earth. Nitrogen amounted to two or three percent of the Martian atmosphere.

Carbon dioxide. The landers found an atmosphere 96.2 percent carbon dioxide.

Oxygen. The atmosphere was 0.1 to 0.4 percent oxygen.

Water. The atmosphere was one to two percent argon-40. The ratio of argon-36 to argon-40 in the atmosphere told scientists the atmospheric pressure near the surface once may have been 10 to 100 times higher than now. That would have allowed liquid water. Viking photographs from the main spacecraft high overhead in Mars orbit confirmed considerable water once flowed over the surface of Mars. However, the atmosphere on Mars today is only one percent as dense as that on Earth so liquid water would vaporize instantly.

Rare gases. Traces of the rare gases krypton and xenon were detected in the atmosphere by the Viking 2 lander.

Bright sky. The atmosphere was thin so the brightness of the Martian sky surprised some scientists.

Pink light. The pink color the scientists saw in photos was a surprise. Scientists guessed the color came from the scattering and reflection of sunlight from reddish particles adrift in the lower atmosphere. The surface of Mars was covered with orange-red matter which itself seemed to cover darker bedrock. The redness on the surface may have been due to oxidation of iron-rich soil.

Inorganic elements. Aluminum, calcium, chlorine, iron, silicon, sulfur and titanium were among inorganic elements detected in the soil by the Viking landers.

Desert soil. The soil of Mars seemed a lot like deserts on Earth.

Iron. Like only the most iron-rich areas on Earth, 16 percent of the Martian soil was iron.

Salts. Sulfur and chlorine detected by Viking 1 indicated that at one time there may have been sulfide and chloride salts, left behind when permafrost below the surface of Mars melted and evaporated quickly in the thin atmosphere.

Trace elements. Rubidium, strontium and zirconium are trace elements commonly found in Earth samples. They either were not found or were found only in very low concentrations in the Viking soil samples. A lack of trace elements dated the rocks around where the landers touched down as more primitive than rocks on Earth.

Windmills On Mars. Mars colonists after the year 2010 may heat their houses and operate scientific equipment with electricity from windmills.

The two U.S. Viking spacecraft, landing on Mars in 1976, found 15 mile per hour winds on the surface. Hills and valleys, to channel the wind naturally to man-made windmills, can be seen from satellites orbiting Mars.

Windmills were used first on Earth 1,400 years ago in the 7th Century. In the 21st Century, men on Mars may hook turbines to modern windmills, called giromills, to generate electricity for water and oxygen generating stations.

Viking photos showed sand dunes and wind erosion, suggesting winds on Mars even stronger than 15 miles per hour. The probe Phobos 2 from the USSR in 1989 found water vapor in the Martian atmosphere.

Mars rocks. How did eight rocks get here from Mars? Geologists, picking through 10,000 meteorites found on Earth, have wondered for years how eight of those rocks came all the way here from Mars. The most recent answer: an asteroid did it.

A meteorite is a chunk of rock now on Earth which came from somewhere else in our Solar System or beyond. The eight mystery meteorites display a chemistry similar to rocks analyzed back in 1976 when two Viking spacecraft landed on Mars and ran chemical tests. The eight found on Earth show a chemical makeup different from other Earth rocks.

Scientists estimate a rock on Mars would have to accelerate up to at least 11,300 miles per hour to escape the planet's gravity and travel on out into space. What would cause a rock to move so fast?

Suppose an asteroid of, say, 500 to 5000-ft. diameter were floating past Mars at, say, 16,000 miles per hour. Suppose it struck the planet a glancing blow at an angle of, say, between 25 and 60 degrees. It would be like a golfer making a long drive.

The crashing asteroid would kick up dust and gas travelling at 50,000 miles per hour.

That vapor would be powerful enough to lift rocks and send them flying out and away from Mars at 11,300 mph or faster.

Sailing along through interplanetary space, some would have come close enough to be attracted down onto the face of Earth by our planet's gravity, according to astrogeologists at the California Institute of Technology.

Ancient Martian Carved Face? What is that mile-long rock shaped like a human face on Mars? A handful of scientists have speculated it may have been carved from a mountain by a long lost civilization.

Around the face in the Cydonia region of Mars is a complex of objects which the speculating scientists say look like they were built by intelligent design—a city, a five-sided mountain that looks like a pyramid, a group of rocks that could be a fortress and a cliff that could have been an astronomical marker.

One researcher claimed a line drawn from the city across the carved face to the cliff would point to the Sun at the moment of the planet's summer solstice 500,000 years ago.

Light magic. Most scientists, however, found the face merely light and shadow, just a trick of nature. NASA noticed the formation in the Viking photographs in the 1970's, but dismissed it.

NASA has not said the face is either science or nonsense, according to the chief of space and Earth sciences at Goddard Space Center, Greenbelt, Maryland, who recalled that some of the world's leading geologists examined the photos.

The speculating scientists replied there is enough uncertainty about the origin of the mysterious outcroppings they should be a target for future spacecraft with high-resolution cameras. The Mars Observer unmanned probe will have a camera which might settle the question if pointed toward the face.

Cosmonauts Flash Thumbs Up For Mars. Cosmonauts Alexander S. Viktorenko and Alexander A. Serebrov, one of the crews aboard the USSR's orbiting Mir space station, gave a television interviewer two thumbs up recently when asked about a joint U.S.-USSR manned mission to Mars in the 21st century.

Mars Visit. President George Bush marked the 20th anniversary of man's first landing on the Moon by asking the National Space Council and NASA to plan Moon and Mars flights. NASA turned in its ambitious plans to U.S. Vice President J. Danforth Quayle, chairman of the Bush Administration's National Space Council in November 1989.

The U.S. space agency visualizes astronauts returning to land on Earth's only natural satellite starting in 2001. They would stay 30 days. An aggressive flight schedule would lead to permanent occupancy of the Moon in 2002. Nuclear-generation of electrical power would start in 2003.

"Just a three-day journey from Earth, the Moon provides the ideal location to develop the systems and experience to prepare for the next step of the initiative: an outpost on Mars," NASA said.

After several years of work on the Moon, including garnering experience for long stays on Mars and extracting oxygen and other resources from the Moon, astronauts would head on out to the Red Planet as early as 2011.

Mars "exploration focuses on studying past and present geologic and climatic environments, including the search for past and present life and water environments," NASA said. A 600-day visit to Mars would follow, leading to a permanent presence.

Before manned flights. To get things started, an unmanned Lunar Polar Observer (LPO) satellite might be sent to scout territory in the mid-1990's.

After LPO, an unmanned rover vehicle could be sent to land on the Moon, grab soil samples and return them to Earth, after the mid-1990's. Bigger rovers then could be sent on more complex missions in 1997-1999.

Freedom station. NASA is on the way to building the U.S.-international space station Freedom in orbit about 200 miles above Earth from 1996-2000. In use by 1998, the station will be key to the Moon and Mars flights.

First, a moonship and spider-like lunar lander larger than the Apollo lunar modules would be assembled at the space station to ferry crews in the year 2001 from Freedom across some 240,000 miles of space to the Moon. About 32 years after men first went there, U.S. astronauts would return to the Moon.

Outpost on the Moon. An unmanned flight would drop off needed hardware and supplies on the cratered body in the year 2000. In 2001, the first manned flight would leave Freedom for the Moon.

NASA engineers at Johnson Space Center, Houston, have proposed that astronauts would dig in on the Moon, building houses buried under lunar soil.

By the year 2003, the first tiny 21st century Moon base would have a population of only four to eight persons at a time. Two four-person crews would stay on the Moon 12 months at a time. Their base would have living quarters, of course, plus a science lab, telescopes for astronomy, and landing and launching pads.

Up to two lunar landings per year might be made from 2001-2005 as a town site was prepared. By 2005, a Moon refinery would be turning out sufficient oxygen to fuel flights from there. Construction of a Moon town would be completed between 2005 and 2010. Four manned flights would be made per year during the years 2005-2010.

On the far side of the Moon, astronomers and astrophysicists would build a huge platform to study deep space, free from Earth's murky atmosphere. Scientists say a major observatory on the dark side of the Moon would be away from Earth's radio and light pollution.

On to Mars. The NASA work force would grow to manage the big Mars project. A new fleet of bigger spacecraft would be built. The space agency would need a new heavy-lifting rocket to carry Mars-ship parts from Earth's surface up to Freedom. New interplanetary propulsion and life support systems would be necessary for a three-year roundtrip flight to Mars.

The first manned Mars ship and lander would be assembled at Freedom. Astronauts would leave on an eight-million-mile sail to the Red Planet about 2011.

A series of Mars flights would take place. On the first, four astronauts would land, set up living quarters with airlock and utility systems, do science experiments within a few miles of their camp and send out remote-control roving vehicles to explore up to 40 miles from camp.

A Mars ship could be fueled in space with lunar oxygen, making it lighter to lift from Earth. Moon oxygen could power flights to build a Mars colony during the 10 years after 2010. Later, oxygen mined from lunar material could fuel ships bound for Mars, Venus, the Asteroids, Jupiter and Saturn.

The forerunners. American astronauts Neil A. Armstrong and Edwin E. "Buzz" Aldrin Jr. were the first men to step on the Moon when they flew there in Apollo 11 in 1969. The Moon roundtrip, for Armstrong, Aldrin and Michael Collins, lasted 195 hours 18 minutes 35 seconds and included 30 orbits of the Moon by the command service module Columbia. Armstrong and Aldrin flew the lunar excursion module Eagle down to the surface July 20, 1969, landing at Mare Tranquillitatis. Armstrong, the first outside, said, "That's one small step for man...one giant leap for mankind." Armstrong and Aldrin collected 48.5 lbs. of rocks and soil during 21 hours 36 minutes 21 seconds on the surface. Collins stayed in Moon orbit in Columbia.

Avoid Dead End. Scientist Carl Sagan told the U.S. House of Representatives in 1989 the Moon is a dead end and Mars should be the next American target for exploration. The Moon is not a way station between the future U.S.-international space station Freedom and Mars, he pointed out. Going back to the Moon before Mars would be a waste of time and resources, he said.

A manned flight to Mars might cost NASA $60 billion over 20 years. Other scientists have said that cost could be shared by a number of nations.

Phobos is cheaper. Landing astronauts on one of the Martian moons, instead of the surface of the planet, would be the fastest and cheapest way to explore the Red Planet, according to astronomer Fred Singer who is the U.S. Department of Transportation's chief scientist.

Astronauts sitting on either Phobos or Deimos could send down robot rovers to explore and collect samples from the planet's surface, he said.

Singer said exploring Mars, which may once have been warmer and had water, is important for the survival of Earth because it would tell us how planetary climates change. Today, Mars is dry, barren and sand-blasted regularly by dust storms.

Singer noted three ways to explore Mars: unmanned robots controlled from Earth, landing men and women on the surface, or landing astronauts on a moon.

Robots from Earth are frustrating because Mars is so far away it takes an hour for radio signals to make a round trip, he said. Landing on the surface would be very expensive because it would necessitate building a new spacecraft which could both land and take off from Mars. Fuel for the blast off toward Earth would add extra weight.

It would take less fuel to fly home from Phobos or Deimos because the tiny moons have almost no gravity. Radio signals to control robots exploring the planet surface would take only a few seconds to make a roundtrip between Mars and one of its moons.

USSR Mars plans. An astronomer who worked in the USSR's 1988-89 effort to study Mars and Phobos said a manned flight to Mars could be made in the year 2005.

Vasily Moroz of Moscow's Space Research Institute said at Baltimore, Maryland, in August 1988 that a six-person crew of U.S. and USSR astronauts/cosmonauts could leave in 2005 for Mars, land for a week of exploration, then fly back home. The whole trip would take two to three years.

Two unmanned Phobos probes were launched from the USSR July 7 and 12, 1988, as a first step toward putting men on Mars. In 1994, the Soviets plan to send a Mars rover, with artificial intelligence software, to pick up and analyze Martian soil samples for fossils. Moroz said he didn't think it necessary to send cosmonauts to the Moon in preparation for a Mars flight.

He said the USSR needs to develop a new nuclear-electrical rocket propulsion system to carry the extra weight needed for a Mars flight. They also want to improve bio-detective technology for the search for life on Mars.

The U.S. was expecting to launch in 1992 an unmanned Mars Observer to orbit the Red Planet in 1993. Mars Observer would be the first U.S. flight to Mars since the 1976 Viking I and Viking II probes.

Mars wanderer. A non-profit group headquartered at Pasadena, California, is helping out with research leading to a 1996 USSR Mars probe with a package of science experiments to roam around the Red Planet by balloon.

The Planetary Society is doing preliminary research on a thermal balloon filled with helium or hydrogen. During a Martian day, sunlight would heat the balloon, letting it rise three miles and float across the planet's surface. The balloon would settle back down to the surface during cooler night temperatures. Such a rover couldn't destroy itself by driving off a cliff. And it could cover great distances if there are sufficient Martian winds.

The paper-thin balloon will have to lug a 60-lb. payload while withstanding temperatures down to 200 degrees below zero Fahrenheit. The payload, shaped like a flexible snake and dangling on a 100-ft. fishing line, would be dragged along by the balloon. Its shape could prevent it from becoming stuck in a crater or snagged on a rock.

In the daytime, the balloon might drift thousands of miles across the surface. Pictures from balloon altitudes of three miles would resolve objects as small as four inches.

French and Soviet scientists are cooperating on the Mars probe to be fired to Mars on a Russian rocket. A version of the balloon and payload already have been tested in a Lithuanian desert and in California's Mojave desert. The Planetary Society is helping out with engineering problems and testing. The society was trying to raise $500,000 from private donations to finance its research work.

Japan Mars Camera. In the first cooperative space project between Japan and the USSR, Japan is making a high-performance camera for a USSR Mars probe to be launched in 1994, according to a Tokai University professor. Others contributing science instruments to the 1994 mission include the U.S., France and Germany.

German Mars Camera. German space agency DARA is contributing a high-resolution stereo camera so the 1994 Soviet Mars orbiter can take three-dimensional photographs of the planet.

Another German camera, the wide-angle optoelectronic stereo scanner for the 1996 lander, will enable panoramic views of Mars and be used in the study of the Martian atmosphere and environment. Germany is assisting with development of a land rover to be sent down to the surface in a landing capsule, as well as orbiting balloons and monitoring stations.

Latvian scanner. A compact radio scanner for the 1994 Soviet Mars probe has been built by the Institute of Engineers for Civil Aviation at Riga, Latvia, and the Institute of Aeronomy at Verrieres, France. The scanner will be in a balloon instrument package dropped to the martian surface. The radio antenna doubles as balloon ballast.

Latvian News Agency Leta said the radio will transmit 300 MHz electromagnetic pulses to probe Martian rock up to a mile beneath the surface. Signals reflected back up to the surface may reveal something of the rock's structure, composition and density. They might signal the presence of ice or water in the Martian soil.

Jupiter

Jupiter, fifth planet from the Sun, may itself have been a prototype Sun which just wasn't big enough to become a star. As a planet, however, Jupiter is by far the largest of nine planets circling the Sun.

Jupiter alone is more than two-thirds of the total mass of all planets in our Solar System. It's about 318 times the mass of Earth. Although a thousand times smaller than the Sun, Jupiter is a big planet—as big as 1,317 Earths.

Still, Jupiter wasn't big enough. If the planet had been several times more massive, it might have become a star as the pressure and temperature at its core triggered nuclear fusion. Jupiter is a big gas bag, with a relatively low density.

Jupiter is more than five times farther from the Sun than the Earth, at a distance of 483.6 million miles. It orbits the Sun every 11.9 years.

Jupiter spins rapidly—once every 9 hours 55.5 minutes. That causes it to be slightly flat. Its diameter at the equator is 88,700 miles, but the distance from its north pole to south pole is only 83,000 miles.

A sun. The Sun was formed billions of years ago when matter in this area of the galaxy coalesced into a massive clump. The clump collapsed into itself, exploding the matter into the ball of fire we know as our Sun.

Jupiter might have formed the same way. It, too, might have been a major clump of coalesced matter, but it didn't have the mass needed to fire a nuclear engine. Instead, it fizzled into the lukewarm gas bag we see today.

Of course, there's the possibility the primeval nebula was less extensive, only sufficiently massive to build one large body which became the Sun.

Jupiter. Could it be that particles of dust, floating in the nebula after the Sun exploded into life, collided and coalesced? One clump became large enough for its gravity to pull in an envelope of gas from the nebula. The clump became the rocky core which might be at Jupiter's heart. It would have a mass several times the entire Earth.

On the other hand, the primeval nebula in which the Sun, Jupiter and other planets formed was a soup of hydrogen and helium. Jupiter's temperature today is warm enough to suggest there is nothing solid under the gas in the planet's atmosphere—no solid ground, no hard surface, only a gradual change from gas to liquid.

Below one-quarter of the way from the outer cloud tops down into the planet, pressure and temperature must be so high liquid would be metallic. Molecules would have been stripped of their outer electrons.

There is more than just hydrogen and helium in Jupiter's atmosphere today—trace amounts of water, ammonia, methane, and organic carbon compounds.

Cloud layers. Astronomers figure there are three cloud layers in Jupiter's atmosphere, each separated by 19 miles. The top clouds probably are ammonia ice. Middle-level clouds may be ammonia-hydrogen sulfide crystals. The lowest layer probably is water ice or water droplets.

Of clouds actually seen by astronomers, blue ones are warmer indicating they are lowest in altitude. Brown, white, and red clouds, in that order, are higher.

The color shades tip off scientists to out of balance chemicals with sulfur, phosphorus, and organic compounds coloring the clouds. The condition may have been forced by lightning or impact of charged particles or even a rapid vertical swirling of chemicals through the temperature levels.

Auroras. In 1979, Voyager heard lightning and saw auroras of charged particles on Jupiter's night side.

Jupiter's wind jets move parallel to the equator. Some are eastward, some westward, varying with latitude, and distance from the equator. Winds blow hundreds of miles per hour in jet streams near colored bands of orange brown and white clouds. Those colors may change as gas rises in some bands and descends in others.

Jupiter's weather seems to include eddies and storms. Some last only a few days, others much longer. Some get caught between regions of different east-west wind speeds and are sheared apart. Larger eddies, such as long-lived white spots and the Earth-sized Great Red Spot, seem to be able to survive by rolling like ball bearings between zones.

Whirlpools of metallic hydrogen inside Jupiter generate a magnetic field, just as Earth's molten iron core does. Jupiter's magnetic field, however, is 4,000 times stronger than Earth's.

Radio signals. The big planet's magnetic field is like a bar magnet with its centerline off of the center of the planet by about 6,200 miles and tipped 11 degrees.

As Jupiter spins, the magnetic field wobbles up and down with electrically charged particles trapped in it. The wobble generates a natural radio signal. Scientists, tuning in to those natural outbursts, keep track of how often they appear, then calculate how long it takes the planet to rotate—the length of Jupiter's day.

Plasma, a gas of charged particles, is locked to the magnetic field so that it rotates with it. This is Jupiter's magnetosphere. It forms an extremely large, intense radiation region around the planet in which some particles are accelerated to speeds of tens of thousands of kilometers per second. The natural satellite Io and, to a lesser degree, other natural satellites sweep up these energetic electrons. Electric currents probably generated in Io may be responsible for long-puzzling radio bursts received on Earth from the direction of Jupiter.

Those energetic particles striking Io probably also help remove atoms and ions of sodium, sulfur, and other elements from Io and pull them into a doughnut-shaped cloud surrounding Jupiter along Io's orbit. Of course, some material in the cloud may have been ejected by volcanos on Io. High-energy particles from the plasma around Io probably spiral along magnetic-field lines to Jupiter's atmosphere where they create the aurora Voyager saw.

Electrons Shower Earth. Jupiter sprays 216 trillion watts of electricity across 480 million miles onto Earth every 27 days, an alternative explanation for the hole in the atmosphere's ozone layer over Antarctica.

Scientists at Los Alamos National Laboratory, New Mexico, found the electron showers, which last about 2.5 days and seem to come from Jupiter, the Sun or both.

Detected by an instrument aboard a satellite, the electron showers dump about a billion watts of energy a second into Earth's atmosphere. The electrons spiral toward Earth's north and south poles along the planet's magnetic field lines.

Ozone screen. At an altitude of 15 miles, Earth's ozone screen keeps out 99 percent of ultraviolet light from the Sun. A thinning of the ozone layer, popularly pictured as a widening hole in the layer, has been observed over Antarctica each year. Diminishing the ozone layer could change climates worldwide, lead to crop failures, and increase human skin cancers. Antarctic ozone drops 60 percent each October.

High-energy electrons from Jupiter could mix with gases in Earth's atmosphere, producing nitrogen and hydrogen compounds that reduce ozone. Stable air circulation over Antarctica concentrates the compounds produced by electron showers, creating the conditions there for thinning of the ozone layer. The north pole has an unstable air pattern which might lessen ozone depletion.

Many scientists, concerned about the ozone layer since the 1970's, had blamed chemical pollutants, chlorofluorocarbons, used in plastic foam softdrink cups, coolants in refrigerators and air conditioners, propellants in aerosol sprays and solvents in manufacturing of computer chips. Work is underway to see if the electron showers relate to chlorofluorocarbons in the ozone layer.

Changing Face. For a hundred years, Jupiter has been famous for its Great Red Spot, a voluminous hurricane in the atmosphere of the gaseous giant planet. Now Saturn, the gas planet just beyond Jupiter, has its own hurricane known as the Great White Spot. It's even larger than the Great Red Spot.

Both seem to intensify and fade on some natural cycle. By odd coincidence, Jupiter's spot has faded recently while Saturn's intensified. Usually, Jupiter's Great Red Spot is so dark it can be seen easily from Earth with a small telescope. At times, though, it has looked more like a thin colorless streak.

When the Great Red Spot was hot in 1988, Jupiter's South Equatorial Belt vanished. After the South Equatorial Belt began darkening, the Great Red Spot started fading.

Astronomers have been admiring the oval spot for at least 100 years. In fact, it may be the same as a similar feature seen on Jupiter in the 17th century.

In the 20th century, the Great Red Spot has run 9,000 miles north-south and 15,000 to 25,000 miles east-west, varying in color from brick to pink.

Astronomers once thought it was an island floating in the planet's atmosphere, or maybe a cloud over a mountain. But the Pioneer 10 and Pioneer 11 probes flown out from Earth found the spot to be a giant cyclone—a whirling vortex of cold turbulent clouds rotating counter-clockwise. Voyager spacecraft found that the Great Red Spot was not unique in the Solar System.

Interestingly, the spot is only half as large at the end of the 20th century as it was at the end of the 19th century. Could it disappear entirely?

Volcanos On Io. Io, nearest to the giant planet, is the smallest of Jupiter's moons. It has a remarkable range of volcano landforms and processes, according to a scientist at NASA's Jet Propulsion Laboratory (JPL), Pasadena, California.

JPL found eruptive plumes, large calderas, and abundant flows indicating Io could be the most geologically-active body in the Solar System.

A series of infrared remote-sensing observations from Earth have determined that Io's heat flow is 30 times that of Earth, on average, and 100 times that of Earth's Moon, a body of equal size.

Such intense heat flows are characteristic of geothermal areas. For instance, the heat flow averaged over the geothermal area at Wairakei, New Zealand, is slightly less than that of Io.

The source of the heat energy being dissipated by Io? The moon may be heated by tidal interaction with Jupiter and other moons.

Saturn

Saturn is big—the second largest planet in the Solar System, with huge globe-girdling rings and 18 moons.

Saturn's largest moon, Titan, is the second-largest moon anywhere in the Solar System. The other known moons are Rhea, Iapetus, Dione, Tethys, Enceladus, Mimas, Hyperion, Prometheus, Pandora, Phoebe, Janus, Epimetheus, Helene, Telesto, Calypso, Atlas, and Pan, plus some small unseen shepherd moons. Pan is a newly-discovered 12-mi.-wide moon found in 1990 by a NASA astronomer.

Dry Land? Radar signals bounced back to Earth from Titan have contradicted what the interplanetary spacecraft Voyager 1 reported from the scene as it passed Titan in 1980. Titan may have continents of dry land as well as oceans of liquid natural gas.

Methane and ethane compose what we call natural gas. The touring probe Voyager 1 in 1980 radioed back data revealing that methane and ethane condense in Titan's mostly nitrogen atmosphere and fall to the surface as rain. Some planetary scientists suggested then that Titan might be covered completely by a mile-deep ocean of liquid natural gas. That is, an ocean of liquid ethane and methane. But, scientists who bounced radar off its surface in 1989 say it may have some dry land or even an icy surface.

If there is an ethane-methane ocean, it would make Titan have a low reflectivity, creating weak radar echoes. To test the question, planetary scientist at the California Institute of Technology fired 360,000-watt radar signals at Titan during three 5.5-hour periods on June 3, 4 and 5, 1989. Instead of weak radar echoes, the scientists saw strong radar echoes coming back to Earth during one of the three observation sessions. As Titan rotated on its axis, weak echoes were received during the other two observation periods. That suggested to planetary scientists that Titan has oceans in some places and solid surface in others.

Cassini. NASA and the European Space Agency (ESA) plan to fire a probe in 1996 to Titan. NASA will launch the American-made mothership toward Saturn. It will fly by an asteroid and the planet Jupiter along the way. The craft will be named Cassini after the French-Italian astronomer who discovered several of Saturn's moons and features of the planet's rings in the 17th Century. It will fly by the asteroid 66 Maja in 1997 and the planet Jupiter in 1999 on its way to an October 2002 rendezvous with the planet Saturn.

Arriving at Saturn, Cassini will drop an ESA probe, named after Christian Huygens, the Dutch astronomer and physicist who discovered Titan and the rings of Saturn in 1656, into Titan's atmosphere. A large airbrake cone will slow Huygens to 600 miles per hour. At an altitude of 112 miles above Titan's surface, parachutes will deploy. Huygens then will drop slowly for two to three hours toward a soft landing on Titan's surface. During the descent, Cassini will remain in orbit as a radio relay station retransmitting to Earth data received from the descending Huygens probe. Huygens will be travelling only about 11 miles per hour when it bumps down on the surface. Scientists hope that impact will be soft enough to allow analysis of soil or whatever's on the surface before the probe dies.

After the end of the Huygens part of the mission, Cassini will spend four years circling Saturn 30 times, sometimes as low as 600 miles above the colossal planet's surface.

Saturn. The sixth major planet from the Sun, Saturn is second largest, after the planet Jupiter, with a diameter of 75,000 miles. That compares with Earth's diameter under 8,000 miles.

Saturn is nine to ten times as far from the Sun as Earth, taking 29.5 years to complete one orbit of our star. Saturn's average distance from the Sun is 893 million miles. The planet's density is only 0.7 that of water, lowest of any planet. Like Jupiter, Saturn seems to have yellowish dark and light cloud bands parallel to its equator. These belts, zones and spots aren't as prominent as those on Jupiter.

The planet is mostly hydrogen. The upper atmosphere of Saturn is thought to be hydrogen, methane and ethane with floating clouds of ammonia crystals. At its heart, Saturn may have an iron core as big as Earth. Strangely, Saturn, like Jupiter, radiates more energy than it absorbs from the Sun. It probably has an internal primordial energy reservoir.

Probes from Earth already have revealed that at least 22 moons surround the jumbo planet. Saturn and its system of rings and satellites, including Titan, were probed by U.S. Pioneer 11 in 1979 and U.S. Voyager 1 and Voyager 2 in 1980 and 1981.

Through a telescope, Saturn is stunning, with prominent rings girdling its equator. The rings, 375,000 miles across, make the planet shine brightly for astronomers. Galileo saw the rings in 1610. Huygens recognized them as a system in 1656.

Saturn's moon Titan, with an organic chemistry in its atmosphere which may resemble the primitive Earth before the dawn of life, is the second largest moon in our Solar System and the only moon known to have a thick, organic-rich nitrogen atmosphere. Titan's surface temperature is minus-290 degrees Fahrenheit.

Uranus

The planet Uranus, its moons, its rings, orbiting as they do at the far reaches of our Solar System, seem even stranger as scientists analyze data received by radio from Voyager 2 in January 1986.

Uranus is the third most distant planet from the Sun. The length of a Uranus day is 17.24 hours. Uranus is four times the size of Earth. The planet does have a magnetic field, but it's tilted 60 degrees away from the axis upon which Uranus rotates making it unreadable from Earth. That tilt may mean the magnetic field is undergoing a planet-wide polarity shift.

The 15 moons of Uranus are eye-openers. They have statuesque mountains towering more than ten miles high. Incredibly deep valleys. Vast plains, some with a mysterious dark surface. Moon diameters range from a bit fatter than 25 miles up to about 1,100 miles, circling Uranus from every eight hours to every 14 days.

While looking at the rings around Uranus, Voyager uncovered strange incomplete-circle partial formations now called ring arcs. The many rings and arcs varied in depth and density.

After passing Uranus, Voyager 2 sailed on toward a rendezvous within 3,000 miles of the planet Neptune in August 1989.

Lightning On Uranus. French scientists listening to radio signals recorded in 1986 at Uranus by the U.S. Voyager 2 spacecraft heard lightning bolts.

The same kind of static crashes heard in an AM radio during an electrical storm on Earth were found among data transmitted back to Earth from Uranus.

Apparently the lightning strikes on Uranus are more powerful than those on Earth, but nowhere near as strong as the magnificent electrical discharges on Saturn. The bolts are giant sparks of static electricity generated by turbulence in the Uranian atmosphere.

Radio waves. When electricity flows, it generates electromagnetic radiation. We call such energy a radio wave. A radio wave can be controlled by man using a transmitter to generate a signal. The signal can be modulated with sounds or made to carry other forms of information. On the other hand, when electromagnetic energy is not controlled, we call it noise or static. Lightning is merely a very large, uncontrolled burst of static.

The Voyager spacecraft's receiver heard strong static as it sailed past Uranus. The chaotic signals received had the same characteristics as lightning noise.

Voyager. The interplanetary probe Voyager 2 flew by within 50,600 miles of Uranus' cloud tops on January 24, 1986, discovering a magnetic field and much ultraviolet light, known as dayglow.

Voyager 2 found 10 previously-unknown moons, the largest only 90 miles in diameter, bringing the total to 15. Astronomers previously had seen five large ice-and-rock moons. The innermost, Miranda, was found by Voyager to be one of the strangest bodies in the Solar System, with fault canyons up to 12 miles deep, terraced layers and mixed young and old surfaces.

Titania was marked by huge faults and deep canyons. Ariel had the brightest and youngest surface of Uranian moons, with many deep valleys and broad ice flows. Dark-surfaced Umbriel and Oberon looked older.

Shakespeare. Nine of the new moons have been named for Shakespearean characters: Bianca, Cordelia, Cressida, Desdemona, Juliet, Ophelia, Portia, Puck and Rosalind. The tenth was named for Belinda whose lock of hair was stolen in The Rape of the Lock.

Three of five previously-known Uranus moons—Oberon, Titania and Miranda—have Shakespearean names. Umbriel is from Pope's poem. Ariel is a name in both The Rape of the Lock and Shakespeare's The Tempest.

Moons of other planets mostly have been named for characters from Roman mythology. Use of literary names for moons of Uranus was started by English brewer and astronomer, William Lassell. He discovered Ariel, Oberon, Titania and Umbriel in 1851.

Voyager discovered the moons shortly before the January 28, 1986, shuttle Challenger disaster. Two U.S. congressmen had proposed naming the 10 moons for dead Challenger and Apollo 1 crew members. They prompted thousands of children, citizens and officials to write letters and sign petitions favoring naming moons for astronauts.

The International Astronomical Union (IAU), a worldwide association of 6,000 professional astronomers headquartered in Paris, has been responsible for naming planetary bodies since IAU was formed in 1919.

Harold Masursky, a planetary geologist with the U.S. Geological Survey in Flagstaff, Arizona, heads the IAU planetary system nomenclature working group.

Astronomers say, even though the Uranus moons were discovered by the United States, the natural satellites are international. Naming features or moons is not the property of one country. Going along with tradition received international acceptance.

The IAU did name craters on Earth's Moon for the seven crew members—Francis Scobee, Michael J. Smith, Judith Resnik, Ronald McNair, Ellison Onizuka, Gregory Jarvis and teacher-astronaut Sharon Christa McAuliffe—killed in the 1986 shuttle Challenger launch explosion. The craters, 20 to 60 miles wide, are in a vast crater named Apollo on the far side of the Moon.

Also, seven asteroids were named in 1986 by the IAU for the Challenger astronauts, two months after the shuttle exploded.

The IAU previously had named craters on Earth's Moon for nine American astronauts and nine Soviet cosmonauts—most living, some dead—including Roger Chaffee, Gus Grissom and Ed White, killed when a launch pad fire swept the Apollo 1 spacecraft in 1967.

Voyager found nine rings around Uranus, but they were quite different from Jupiter and Saturn rings. Uranus' rings may be young remnants of a shattered moon.

40 Moons? British astronomers said Uranus may have up to 25 more natural satellites than previously seen.

Two moons, Cordelia and Ophelia, are shepherds, keeping eleven large rings around Uranus from spreading into space. But, astronomers at London University say the two shepherd moons are not enough to explain the thin sharp edges on the rings. They figure something like 25 more small moons must be there, their gravity pulling the rings into shape. The small shepherd moonlets may range from one to 12 miles in diameter.

Gaps in some Voyager 2 photos, combined with a mysterious wave in one ring, may indicate two of the unseen moons could be 11 miles in diameter. They could be orbiting at 28,200 and 28,800 miles from the center of the planet.

Miranda: The Movie. Twelve years. Two billion miles. You breathe a sigh of relief as your personal interplanetary spaceflyer eases slightly to the left, pointing toward a sparkling jewel hanging in space just ahead.

You're two years beyond Saturn. For months you have been watching the jumbo gasball planet, Uranus, grow bigger and brighter as you approached. Rings and arcs girdling the enormous 32,000-mi.-wide globe have become visible in recent weeks, along with some of Uranus' 15 moons.

Your radio has been crackling with static from monster lightning bolts flashing in the planet's atmosphere. Now you're ready to face your goal—Miranda, the small 300-mi.-wide moon with the strange face.

Roll the film. The flyer goes in low and fast, holding altitude at nine miles above the strangely cracked and broken surface of the bizarre icy moon.

You sit in stunned silence, engrossed, as you flit around statuesque mountains towering miles above the surface, down incredibly steep canyons and across mysteriously dark plains.

Miranda's craters, strange grooves, scarps, and fractures sweep by your window as you speed across the hostile terrain for 60 seconds.

Denouement. For now, it's all a dream. Unable to go there in person, NASA's Jet Propulsion Laboratory (JPL) image processing engineers created a 3-D movie simulating flight over the odd little moon Miranda orbiting Uranus.

A one-minute trip over bizarrely-contoured terrain, the film was created from actual photos of Miranda sent back in January 1986 by the American Voyager 2 spacecraft. "Miranda: The Movie" was produced by JPL's Digital Image Animation Laboratory.

Not much was known about Miranda, except its size and position in orbit around Uranus—until the Voyager flyby. The spacecraft from Earth flew within 18,000 miles of Miranda's surface, shooting some of the clearest close-ups ever snapped of another body in the Solar System.

Scientists were startled to find such a variety of landforms on such a small moon—only 300 miles in diameter.

Imaging engineers took the Voyager photos and combined them with elevation data from the U.S. Geological Survey to create the animated tour of Miranda's most striking features. They were able to visualize what it would look like to fly nine miles above the surface of Miranda even though they never had been there.

There were nine original two-dimensional Voyager photos of Miranda, each with 64,000 pixels, or picture element points. The nine pictures had a total of 576,000 pixels. The photos covered areas on Miranda as small as an acre.

Depth was added to the photos using elevation data from the U.S. Geological Survey, Flagstaff, Arizona. While adding the third dimension, the image processors exaggerated the vertical scale by a factor of three. Mountains in the movie look three times as tall as they would in person.

Planetary scientists are using the movie to study Miranda's surface in three dimensions, as well as in more understandable detail.

Neptune

Trying to explain slight changes in the orbit of the planet Uranus, John Couch Adams in 1845 in England and Urbain Jean Joseph Leverrier in 1846 in France independently calculated where an eighth planet should be found. Johann Gottfried Galle and Heinrich Louis d'Arrest looked at the predicted location in the heavens above Berlin, Germany, on the night of September 23, 1846, and spotted the new planet.

Named for the Roman god of the sea, the brother of Jupiter, Neptune is the most remote of the jumbo gas planets in the Solar System. It is the eighth planet out from the

Sun among nine known planets. The nine, in order out from the Sun, are Mercury, Venus, Earth, Mars, Jupiter, Saturn, Uranus, Neptune and Pluto.

Big planet. Neptune is 2,794,000,000 miles, on average, from the Sun. The length of its year is about 165 Earth years for one trip around the Sun. Neptune's orbit is even more nearly circular than Earth's. It has an 18 hour day plus-or-minus 24 minutes.

Neptune's diameter at the equator is about 31,000 miles. Neptune's mass is about 17 times Earth, slightly smaller and heavier than Uranus, with an internal structure similar to Uranus. The planet's atmosphere is mostly hydrogen, methane, and ammonia, similar to Uranus, except for a warmer stratosphere. The planet's density is 1.67 (water is 1).

For Earth astronomers, Neptune reaches a brightness of magnitude 7.8, five times too faint to see with the naked eye. It looks like a small blue-green disk when seen through a large telescope.

Triton. William Lassell saw Neptune's large moon Triton less than three weeks after the discovery of Neptune in 1846. His find was confirmed in 1847. Triton is in an unusual retrograde circular orbit about 221,000 miles from Neptune.

Nereid. Gerard Kuiper saw Nereid in 1949. It's about 3.5 million miles from Neptune.

Voyager. On August 24, 1989, the U.S. interplanetary spacecraft Voyager 2 flew within 3,000 miles of the large gas planet Neptune. The spectacular science results have been hailed as one man's greatest achievements.

Voyager 2 had been launched from Cape Canaveral in 1977. Originally the outer Solar System probe was to fly by Jupiter and Saturn. It worked so well its job was extended after Saturn to Uranus and Neptune. It was called a Grand Tour.

Moons. Voyager 2 flew by Jupiter in 1979, Saturn in 1981, Uranus in 1986, then Neptune in 1989. At Neptune, Voyager found six previously-unknown moons, bringing the total to eight. Astronomers on Earth already had seen Triton and Nereid. The found moons ranged in diameter from 33 miles to 250 miles. In 1991, they were named Naiad, Thalassa, Despina, Galatea, Larissa, Proteus. The names Galatea and Larissa were controversial since asteroids previously had been given those names. Names are assigned by the nomenclature committee of the International Astronomical Union (IAU).

Spots. Voyager also found large dark spots on Neptune, like Jupiter's Great Red Spot hurricane. The largest was named the Great Dark Spot. A small irregular cloud flitting across cloud tops was named Scooter.

The strongest winds of any planet were measured on Neptune, with winds near the Great Dark Spot blowing at 1,200 mph. The probe found an odd magnetic field, as it had at Uranus.

Rings. Voyager found four complete rings around Neptune and no partial ring arcs. The fine material in the rings is so diffuse they cannot be seen from Earth. In 1991 the three newly-discovered rings were named Galle, Leverrier and Adams, in order outward from the planet.

Triton up close. Neptune's largest moon, Triton, turned out to be one of the most intriguing satellites in the Solar System. Voyager sent back photos of erupting geysers spewing invisible nitrogen gas and dark dust particles high in the extremely thin atmosphere which extends 500 miles above Triton's surface.

Triton may not always have been a moon of Neptune. If so, tidal heating might have melted Triton, leaving the moon liquid for a billion years after its capture by Neptune.

Today it may have thin clouds of nitrogen snowflakes a few miles above its firm surface. With a surface temperature of −391 degrees Fahrenheit, Triton is the coldest body known in the Solar System. Triton's atmospheric pressure was 1/70,000th the surface pressure on Earth.

Voyager 2's pass came at a time when Neptune was the most distant in the Solar System. Now Voyager 2 is headed out of the Solar System to interstellar space. Pluto once again will become the most distant planet in 1999.

Pluto

A young Kansas farmer, working a telescope by feel in the dark of a wooden dome on a frozen Arizona mesa in 1930, discovered Pluto, the ninth planet of our Solar System.

Clyde W. Tombaugh, who had taught himself to be an astronomer, became the first 20th century man to discover a planet when he found Pluto that February 18. Pluto was one of the top 10 news stories of 1930.

Later, Tombaugh taught Navy and Marine officer candidates in World War II, developed optical tracking systems for Army missiles, surveyed paths to the Moon for NASA and started the astronomy department at New Mexico State University, Las Cruces.

No respect. From the first, some astronomers heaped scorn on Pluto's status as a planet. Recently, the director of a Houston planetarium said, "It seems likely that Pluto will be demoted" to asteroid. An asteroid is a minor planet.

Pluto's moon Charon was discovered in 1978. NASA scientists at the Jet Propulsion Laboratory, Pasadena, California, reported in 1987 observations showed Pluto has a substantial atmosphere. School children wrote in, saying they didn't want Pluto demoted.

Planets and most moons have gassy atmospheres while asteroids don't have atmospheres or moons. Pluto has both a moon and an atmosphere.

"This latest discovery greatly enhances the stature of Pluto," NASA said. Tombaugh said in 1988, if Pluto looks like a planet, feels like a planet and smells like a planet, it must be a planet. Tombaugh, in his eighties, still used a homemade telescope on a platform in his yard to gaze at the desert night sky.

Most Distant, Least Understood. Pluto—most distant, smallest and least understood of the Sun's known planets—recently has been the closest that mysterious body has come to Earth since 1740. As it swung within 2.8 billion miles, astronomers tried their best to unlock some of its secrets.

The nine known planets of the Solar System, in order out from the Sun, are Mercury, Venus, Earth, Mars, Jupiter, Saturn, Uranus, Neptune and Pluto. Pluto is 3.6 billion miles from the Sun.

With a diameter of only 1,370 miles, Pluto is the smallest planet, smaller even than Earth's airless Moon. Pluto is so far away and so small that little was learned about it for decades. Observations intensified as its 248-year orbit around the Sun brought it closer to Earth.

Observing Pluto's close approach, especially during eclipses, astronomers have tried to recalculate the size of Pluto and its moon Charon, examined their composition and looked for surface features. The planet and Charon eclipsed each other while astronomers watched.

Pluto's discovery February 18, 1930, by Clyde W. Tombaugh at the Lowell Observatory was based on mathematical predictions made earlier by Percival Lowell who had seen unexplained motions in the orbits of the planets Uranus and Neptune.

Close-up. If you were out there, Pluto would seem eerie and dark, with thin air above an an icy landscape and temperatures so cold a warm day would be –350 degrees Fahrenheit. The Sun, if you noticed it, would be only a small luminous disc. The sky wouldn't be all black, however, as sunlight would filter through the haze of methane and other gases. Pluto is 40 times farther from the Sun than the Earth is. Sunlight takes six hours to reach the planet.

Charon. Pluto has a moon, Charon, which orbits the planet every 6.4 Earth days. Astronomers didn't discover Charon until June 1978 when James W. Christy at the U.S. Naval Observatory, Flagstaff, Arizona, found a strange bump in photos of Pluto, 48 years after Tombaugh discovered the planet.

Shocked scientists could hardly believe their calculations which showed Pluto has only 1/400th the mass of Earth. And Charon has only one-tenth the mass of Pluto. These are really small bodies for a planet and a moon. Pluto is only four times larger than the biggest asteroid, Ceres. Pluto's diameter is 1,000 miles less than the diameter of Mercury, the second smallest planet. Charon's diameter is about half that of Pluto.

Pluto and Charon are quite different. Pluto is a huge iceberg with a reddish tint, mostly water ice with a methane ice crust. Its only moon Charon is a much darker body.

Pluto is the only planet cold enough to keep methane solid. Buried in methane snow, frozen by summertime temperatures of minus 378 degrees, the surface of Pluto wouldn't be strong enough to support a mountain. The methane snow seemed to melt and evaporate just slightly as the planet approached its closest point to the Sun. The extensive methane atmosphere expanded a bit.

Pluto is so small its gravity is not strong enough to hold the methane vapor on the planet's surface. Molecules of the gas may be shooting out into space. Some of the gas probably is smashing into Charon at 700 mph.

Astronomers continue to look closely for other moons orbiting Pluto. Most planets in the outer Solar System have lots of moons. Uranus, for example, has a dozen and Jupiter has several dozen. Telescopes in Texas and Colorado have been used to search for other moons of Pluto.

Charon orbits 12,000 miles above Pluto. That compares with the distance of 240,000 miles separating Earth from its Moon. Pluto is only twice as large as its moon. The combined total mass of Pluto and Charon comes to only 20 percent of the mass of Earth's Moon.

Some astronomers have suggested Pluto and Charon might even be a double-planet system with the atmosphere of the two bodies interacting with each another.

Eighth Or Ninth? Pluto is referred to as the ninth planet but, in recent years, Pluto has not been as far from the Sun as the planet Neptune. In that sense, Pluto temporarily is number 8. Neptune is 9.

Although it usually is the ninth and farthest planet from the Sun, Pluto's highly-inclined elliptical orbit brought it in 1980 closer in to the Sun than Neptune. Pluto will swing farther out than Neptune in 1999 to regain its position as the true ninth planet.

Pluto is a long way from Earth, of course—about 3.98 lighthours as astronomers reckon distances. Light takes 3.98 hours to travel from Pluto to Earth at its closest approach when Pluto is within 2.8 billion miles of Earth. Visually, the planet is a dim magnitude 13.7. You need a very large telescope of at least 200mm aperture to see Pluto.

Pluto made its closest approach to Earth on May 7, 1990. It was a mere 2.67 billion miles (28.69 astronomical units) away. Most backyard astronomers didn't see the 14th magnitude Pluto, unless they had a very large telescope and a very dark observing site.

Trying to decide. NASA astronomers have rescued some of the reputation of Pluto, recently bad-mouthed as a second-class planet or less because of small size, odd orbit and features more like an asteroid than a planet.

A team found evidence that Pluto, like all real planets, has a significant atmosphere. Earlier studies suggested Pluto had only a thin methane gas atmosphere. More recent studies showed the methane to be far thicker than previously calculated. The evidence is in Pluto's temperature. Measurements show Pluto to be too warm for a body insulated with as little methane as thought earlier.

Pluto has the only solid surface among planets in the outer reaches of the Solar System. Planets five, six, seven and eight are the giant gas planets Jupiter, Saturn, Uranus and Neptune.

Some astronomers had been thinking of Pluto only as a wayward asteroid because its orbit is highly elliptical, sharply tilted from the plane of the other planets.

The latest Pluto data was collected over several recent years by a NASA Jet Propulsion Laboratory (JPL) team using ground-based telescopes, an airborne observatory, and IRAS, the Infrared Astronomical Satellite in Earth orbit.

Atmosphere verified. Astronomers flying over the southern Pacific Ocean in

June 1988 directly observed the atmosphere of Pluto for the first time during a temporary disappearance of a star behind the planet. They found the atmosphere surrounding Pluto, but were unable to measure Pluto's diameter satisfactorily.

Astronomers had been pretty sure about the atmosphere before they looked through a 36-in. telescope aboard NASA's Kuiper Airborne Observatory (KAO) to get the first direct look at it.

The same telescope, plane and technique had been used to discover rings around Uranus in 1977. NASA had used KAO to watch Comet Halley and Supernova 1987a from above Earth's clouds.

Flying KAO at an altitude of 41,000 feet over the Pacific, Massachusetts Institute of Technology (MIT) researchers used a solid-state video camera attached to the telescope to record light from a faint 12th-magnitude star. They watched as tiny Pluto moved across their field of view in front of the 12th-magnitude star.

The edge of the planet moved toward and in front of the star. A decreasing intensity of light from the star as Pluto approached confirmed an atmosphere surrounding the planet. Visible light from the star declined gradually as the planet approached.

As Pluto passed in front of the star, the planet's atmosphere acted like a lens and the two bodies appeared like a sunset, astronomers explained.

Measuring the amount of light with a telescope, the planetary scientists calculated the average surface temperature of Pluto at minus 360 degrees. The temperature of the atmosphere 40 miles above the surface is minus 279 degrees.

The star, of course, was not near Pluto except in the field of view through the telescope. Pluto was 2.5 billion miles from Earth while the star was hundreds of trillions of miles farther away in the background.

The effect, which lasted just over a minute shortly after midnight, was visible only from Earth's Southern Hemisphere where Pluto cast an almost-imperceptible 1,200-mi.-wide shadow across Earth.

The scientists had hoped the airplane flight through Pluto's shadow also would result in a precise measurement of the planet's diameter when data was combined with observations made on the ground at Lowell Observatory at Flagstaff, Arizona, and Hobart Observatory, Australia.

The KAO airborne-observatory flew from Hawaii to Pago Pago, American Samoa. The observations, made 500 miles south of Pago Pago, did not resolve immediately the debate over how extensive Pluto's atmosphere really is.

Carbon Monoxide, Nitrogen, Argon. Pluto's atmosphere contains large amounts of carbon monoxide and some nitrogen and argon, in addition to the methane already seen there, according to University of Arizona lunar and planetary laboratory scientists in 1989.

NASA JPL scientists in Pasadena, California, had reported in 1987 Pluto has a substantial atmosphere of methane, or natural gas.

They claimed 1983 space observations of Pluto by IRAS showed the planet's temperature was uniform over its entire surface. They said that could happen only if a significant atmosphere were present.

However, astronomers at University of Arizona and the Planetary Science Institute, both in Tucson, contradicted the finding in 1987 when they reported evidence from IRAS suggesting Pluto's atmosphere is very thin, at least 900 times less dense than Earth's.

Pluto's erratic wobbling made the exact location of the shadow on Earth June 9 difficult to predict. The planet's position was observed for several weeks prior to the flight. The track was predicted from photos made at the U.S. Naval Observatory, Flagstaff.

Portrait. Astronomers used the Hubble Space Telescope's Faint Object Camera in 1991 to photograph the disks of Pluto and its moon Charon. The picture was taken with Charon at it maximum separation from Pluto—about 13,000 miles.

Is There A Tenth Planet?

A tenth planet may exist beyond the nine known in our Solar System.

If it does, it must travel in an orbit at nearly right angles to the orbits of the nine known planets. The tenth-planet's orbit might be 10 to 20 billion miles across. The orbit would be so elongated that the tenth planet would come near the Sun and the other known planets only every 700 to 1,000 years.

The nine known planets of the Solar System in order out from the Sun are Mercury, Venus, Earth, Mars, Jupiter, Saturn, Uranus, Neptune and Pluto. Pluto is 3.6 billion miles from the Sun. Mercury, Venus, Earth and Mars compose what astronomers call the inner Solar System. Jupiter, Saturn, Uranus, Neptune and Pluto are the outer Solar System.

NASA scientists examined astronomical measurements made over decades, comparing the data with the recently observed lack of tenth-planet gravity effects on the Pioneer 10 and Pioneer 11 spacecraft.

Pioneer 10, launched in 1972, and Pioneer 11, launched in 1973, are in the far outer reaches of the known Solar System on their way to deep space.

200 years' data. Over the past 200 years, astronomers have suspected there might be another large planet, or other natural space object, beyond the orbits of Neptune and Pluto. The suspected planet may be 10 to 20 billion miles from the Sun.

When astronomers have calculated the orbits of planets, gravity from something seems to have been tugging at the outer planets. Until 1978, they thought the extra gravity pull was coming from Pluto. That theory was upset when an astronomer at the U.S. Naval Observatory discovered a moon orbiting Pluto. He calculated that neither Pluto nor its moon, Charon, were big enough to cause the waverings in their orbits of the outer planets observed over many decades.

In fact, many astronomers are ready to demote Pluto from rank as a planet. They used to think Pluto was 88,000 miles in diameter but more precise modern measurements have whittled it down to 1,375 miles, about the size of Earth's moon. They don't think it's a comet which wandered close enough to be seen since Pluto appears to be a rocky ball with a metal core. Comets usually are dirty snowballs spattered with chunks of rock.

1810 to 1910. Over the years from 1810 to 1910, techniques for measuring the orbits of planets were about as good as those in use today. During that time, strong evidence was found for an additional large Solar System body. However, the gravity seen back then appears not to be tugging now at the Pioneer 10 and Pioneer 11 spacecraft.

Examining data over 200 years, NASA scientists found, since 1910, measurement techniques have failed to show any unexplained outer planet variations in orbits. Most orbit experts had been assuming the effects had been continuing through this century.

Long periods of time are required to calculate very small changes in orbits of planets because the outer planets take a long time to circle the Sun. Uranus goes around once every 84 years. Neptune, every 165 years. And Pluto takes 248 years to orbit the Sun. Slight drifting of planets in their orbital paths takes decades to be seen and measured.

The new interpretation seems to mean something was there from about 1810 to 1910, but hasn't been there since. Scientists maintain that the best explanation, there 100 years and then seeming to disappear, would be a planet on an elongated path around the Sun.

Comet cataclysms. If true, the tenth planet might explain why comets sometimes shoot into the inner Solar System. Some comets, over the millions of years of Earth history, may have come close or even struck our planet. In that sense, the tenth planet might have caused cataclysms which some scientists say brought mass extinctions of life on Earth from time to time. Dinosaurs, for instance, may have been wiped out that way.

If not a tenth planet, maybe a brown-dwarf star, companion to our own Sun, could be the giant body affecting the planets. But scientists says a tenth planet, rather than a star, fits the data better.

No brown dwarf has ever been discovered, but astronomers theorize they are large star-like clumps of gas which are not quite heavy enough for nuclear reactions to turn on in their cores. Such a brown dwarf would be somewhat larger than the planet Jupiter, but smaller than the Sun. Astrophysicists think a gas ball has to be at least eight percent the size of our Sun to become a glowing star.

Tracking Pioneers. Today both Pioneers are in the far outer reaches of the Solar System on their way to deep space. Together they compose a unique, extraordinarily-sensitive instrument for measuring gravity and its effects. The Pioneers are managed by NASA's Ames Research Center, Mountain View, California.

Tracking them, NASA keeps an eye out for undiscovered heavenly bodies and gravity waves. The Pioneers are good indicators of the gravity pull of celestial objects because the two spacecraft generate almost no forces of their own affecting their trajectories. They are like miniature planets floating free in the gravity fields of all of the other bodies in the Solar System.

The Pioneers are stabilized by their own spin, rather than the thrust of control jets. An unexpected change in their speed would show the presence of an uncharted star or planet. They are precise gravity sensors because they are on paths moving rapidly outward from the Sun and radio tracking is exact.

NASA's Deep Space Network telescopes at Madrid, Spain, Canberra, Australia, and Goldstone, California, can transmit a signal to the Pioneer craft and measure the so-called Doppler shift, change in frequency, of signals coming back from the craft. The faster a Pioneer pulls away from the Sun, the longer the wavelength received by the Deep Space Network antennas. Such two-way Doppler tracking gives an accuracy of one millimeter per second in speed measurements.

Recent years. Over several recent years of such precise measuring, NASA found no strange gravity effect on Pioneer 10 or Pioneer 11. All effects measured could be explained by the nine known planets. Reviewing older data turned up no tenth-planet effects over the past 75 years.

But, scientists say data collected from the 17th Century to early in the 20th Century may be valid. That data showed irregularities, probably caused by some unknown large body. The current negative data, then, could mean that whatever disturbed the outer planets back then is either a huge distance from the Sun now or is orbiting on the far side of the Sun opposite Uranus and Neptune.

Pioneer 10, the first probe from Earth to fly through the Asteroid Belt, made the first fly-by of Jupiter in 1973. It sent back pictures of the planet and three of its natural satellites, Europa, Ganymede and Callisto.

Pioneer 11, sometimes called Pioneer-Saturn, flew by Jupiter in 1974 and Saturn in 1979. It sent back pictures from inside Saturn's rings.

Calculation errors. Celestial mechanics is the field of study in which scientists measure the orbits of planets. Skeptics say the abnormal orbits of Uranus and Neptune come from calculation errors. But, a model of the Solar System used by NASA's Jet Propulsion Laboratory was good enough to fly Voyager 2 past Uranus and Neptune, but now appears inconsistent with recent observations on the position of Uranus. For a time, the JPL calculations seemed okay, but new observations show Uranus is not exactly on the expected path.

Hot pursuit. Several ten-planet Solar System schematics were drawn up in 1990 by a computer at the U.S. Naval Observatory main office at Washington, D.C. The models were inspected to see what effect Planet X might have on the orbits of Uranus and Neptune. Some small areas of the sky, which looked promising in the models, were searched for the phantom giant using an astrographic telescope at Black Birch Astrometric Observatory, Blenheim, New Zealand. Black Birch took pictures several nights in a row and the photos will be analyzed at Washington.

Tenth Planet. Planet X has inspired stargazers for years. Speculation started after calculations showed something—maybe a titanic object orbiting the Sun at the outer edge of the Solar System—seemed to be delivering a gravity nudge, disrupting orbits predicted for the planets Uranus and Neptune.

When astronomers have calculated the orbits of those big outer planets, gravity from something seems to have been tugging at them. Since the astronomers still can't predict exactly the locations of Uranus and Neptune, they think something is amiss along the far rim of the Solar System.

Pluto isn't enough. Early this century, astronomers conceived of additional planets beyond Neptune after Uranus failed to travel as predicted along its path around the Sun. A search led to the discovery in 1930 of the outermost and ninth planet, Pluto. Astronomers said the mysterious gravity pull was coming from Pluto. Additional searching turned up, in 1978, a moon orbiting Pluto.

But, new calculations showed the combined mass of Pluto and Charon are one thousand times too small to explain irregularities in the orbits of Neptune and Uranus.

Moons Of The Solar System

There are at least 61 natural satellites of planets in the Solar System. Each moon is much smaller than its planet.

Earth's Moon is one of the larger natural satellites with a diameter of 2,160 miles. It is the only moon close enough to us that details of its surface can be seen through a telescope from Earth.

Largest. The largest moon is Ganymede with a diameter of 3,280 miles, even larger than either of the planets Mercury and Pluto. Saturn's Titan is the second largest moon in the Solar System with a diameter of 3,200 miles, twice Earth's Moon.

Smallest. The smallest moon is Deimos, only seven miles in diameter, except for possibly-smaller shepherd moons yet to be counted and named in the rings around Saturn and other giant gas planets in the outer Solar System.

Charon is the moon closest in size to its planet, Pluto. Earth's Moon is second in that comparison.

Neptune. The interplanetary probe Voyager 2 in 1989 found six previously-unknown moons orbiting Neptune. They ranged in diameter from 33 miles to 250 miles. In 1991, they were named Naiad, Thalassa, Despina, Galatea, Larissa, Proteus.

The names Galatea and Larissa were controversial since asteroids previously had been given those names. Names are assigned by the nomenclature committee of the International Astronomical Union (IAU).

Triton. Before Voyager 2, astronomers knew Neptune had two moons, Triton and Nereid. With a surface temperature of –391 degrees, Triton was found by Voyager 2 to have a thin veneer of methane and nitrogen on top of water ice on its surface.

Triton had been thought to have a diameter of 2,361 miles, close in size to Earth's Moon, but turned out to be smaller, around 1,690 miles. Nereid is 210 miles in diameter.

Pan. A new moon only 12 miles in diameter was discovered in 1990 circling the planet Saturn. It is Saturn's 18th and most distant moon. Photos of the planet, moons and rings, left over from the 1981 Saturn flyby by Voyager 2, had been filed away at NASA's Ames Research Center in Mountain View, California, for a decade. A planetary scientist was tipped off to Pan's presence by a disturbance in a 200-mi.-wide gap—Encke's Gap—in Saturn's outermost A ring. Pan's gravity moved particles in the large A ring, creating a gap with waves along the edges like the wake of a motorboat.

Checking the wavy edges, he calculated the probable position and mass of the moon and compared them with Voyager 2 positions and camera angles in 1980-81. When finally uncovered, the moon stood out as a small bright spot in 11 pictures among

30,000 photos scanned by a computer. In 1991, the moon officially was named Pan.

It was the second time since Neptune was discovered in 1894 that a gravity disturbance had been used to pinpoint a previously-unknown Solar System body. Saturn has more moons than any other planet in the Solar System. Jupiter has 16 moons and Uranus has 15 moons. There may be more.

Solar System. The inner Solar System includes the planets Mercury, Venus, Earth and Mars. Of 61 moons, only three are in the inner Solar System. Mars has two. Earth has one. Mercury and Venus are the only planets without moons. The outer Solar System includes Jupiter, Saturn, Uranus, Neptune and Pluto—and the other 58 moons.

Atmospheres. Most moons are airless, but Jupiter's Io, Saturn's Titan and Neptune's Triton seem to have atmospheres. Titan may be flooded with an ocean of liquid ethane. Triton may be covered by an ocean of liquid nitrogen. Io seems to have a thin sulphur dioxide atmosphere from volcanos.

Titan. Titan appears to have an organic chemistry in its atmosphere which may resemble the primitive Earth before the dawn of life. It is the only moon known to have a thick, organic-rich nitrogen atmosphere. Surface temperature is around −290 degrees Fahrenheit. Solar System's moons are listed below:

Planet	Moons	Names
Mercury	none	
Venus	none	
Earth	1	Moon
Mars	2	Phobos, Deimos
Jupiter	16	Ganymede, Callisto, Io, Europa, Amalthea, Himalia, Elara, Thebe, Metis, Adrastea, Pasiphae, Carme, Sinope, Lysithea, Ananke, Leda
Saturn	18	Titan, Rhea, Iapetus, Dione, Tethys, Enceladus, Mimas, Hyperion, Prometheus, Pandora, Phoebe, Janus, Epimetheus, Helene, Telesto, Calypso, Atlas, Pan, plus half-a-dozen small shepherd moons
Uranus	15	Titania, Oberon, Umbriel, Ariel, Miranda, Puck, Bianca, Cressida, Desdemona, Juliet, Portia, Rosalind, Belinda, Cordelia, Ophelia
Neptune	8	Triton, Nereid, Naiad, Thalassa, Despina, Galatea, Larissa, Proteus.
Pluto	1	Charon

The Moon

Earth's only natural satellite, The Moon, may have resulted from a mid-space collision between planet Earth and some other huge planetary body eons ago when our Solar System was forming.

The mystery planet, about one-seventh the size of Earth, approached Earth [1] and they collided [2]. Heat from the explosive impact melted the two planets' iron cores together and threw chunks of mantle from both bodies into space [3]. The iron core remained as Earth while the weight of the spun-off debris led its gravity to clump it into a Moon-shaped ball [4]. The new theory from New Mexico researchers explains differences in amounts of chemicals found in the separate Moon [5] and why the Moon has only a tiny iron core.

Double planet. Earth and the Moon could be called a double-planet system. The Moon is 2,160 miles in diameter, nearly one-quarter the size of Earth. Earth is 81 times as massive as the Moon. There are larger and heavier moons in the Solar System, but only Pluto's moon Charon is so similar to its planet. Most moons are far smaller than the planets they orbit.

The Moon moves around the Earth in an elliptical orbit, while Earth revolves around the Sun. The Earth-to-Moon distance varies slightly during a month, ranging from 221,000 to 253,000 miles. It averages 238,900 miles. Even at their closest, Venus and Mars are 100 times farther away than the Moon.

Moon mining. Earth engineers have the mining technology to extract valuable minerals from the Moon, according to a 1989 report to the American Mining Congress.

As Earth's human population continues to deplete natural resources, it seems likely to become necessary to find other sources of minerals and metals. Lunar soil is 30 percent metal, including titanium, iron and aluminum.

Many Moon materials could be useful for industry, construction and fuels. Technology could develop large scale lunar mining.

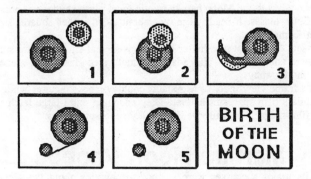

Helium dust. It turns out that an odd kind of helium dust, which NASA had thought was a useless find on the Moon, may make nuclear power plants on Earth safer.

Geologists think there may be one million tons of the rare Helium-3 in moon dust. That would supply the world's needs for hundreds, even thousands of years.

Electricity has been produced by nuclear fission in which atoms are split. Now some scientists want to switch to nuclear fusion where atoms are joined together.

Fusion plants would be safer. If they used Helium-3 for fuel they would be even safer since waste from the plant would be so low in radioactivity it could be buried near Earth's surface.

To get the fuel, a mining station would have to be placed in space to mine the Moon's surface, refine the Helium-3, and ship it back to Earth. Proponents say all that effort would still be ten times as efficient as present uranium energy yields.

Scientists had known about the fusion possibilities for Helium-3. They had discounted it as impractical since Helium-3 is scarce on Earth. Apollo Moon missions uncovered the rich deposits.

Iron dirt. Scientists think they can make iron on the Moon from lunar ilmenite. The iron, which would be extracted from the mineral in a process using hydrogen to remove oxygen, then could be made into building blocks for houses, according to a May 1990 report from the Space Studies Institute.

Moon food. Will wheat and soybeans grown in dirt on the Moon? Minnesota scientists have powdered about 200 lbs. of rock quarried near Duluth. They say it's just like lunar soil. They even put in some glass.

The light porous glass in Moon dirt comes from the heat of countless meteor impacts over eons so they used a plasma-arc furnace to make some. The furnace also will boil off moisture in the Earth rock making it more like lunar soil.

The stuff was trucked to Florida's Epcot Center where wheat and soybean seeds were planted in it. Moon residents 20 years from now may need only a greenhouse with atmosphere to feed themselves.

Top ten. The U.S. Apollo Moon landing was one of America's ten greatest engineering feats, according to the National Academy of Engineering which named the ten greatest engineering achievements from 1964-1989. America's Apollo manned landing on the Moon in 1969 topped the list. The other nine top engineering feats were applications satellites, microprocessors, computer-aided design (CAD) and computer-aided manufacturing (CAM), CAT scanners, composite materials, jumbo jets, lasers, fiber-optics and genetic engineering. Academy members include 1,500 persons and professional societies.

Moon camera. The first snapshot camera carried by American astronauts on the Moon was European, originally designed in 1948 in Goteborg, Sweden, by Hasselblad.

The first one taken to the Moon had an erratic focal plane shutter and some American-made Kodak Ektar lenses. It made negatives about 2-1/4 inches square, larger than the popular 35mm format.

The astronauts didn't like the shutter and lenses so leaf shutters were substituted and Zeiss lenses were added. NASA also adapted a motorized Hasselblad model 500EL for lunar surface photography.

They built a frame viewfinder so astronauts could compose pictures while encumbered in the unwieldiness of a spacesuit helmet. They added an extra-large latch so astronauts could change film magazines while wearing bulky spacesuit gloves.

Martian Moon Phobos

Although its major successes were science discoveries about the planet Mars, the USSR's 1989-90 flight to the Red Planet riveted everyone's attention on the small Martian moon Phobos, not on the planet itself.

Scientists had hoped the miniature natural satellite Phobos would reveal clues to how the Solar System came to exist billions of years ago, present more accurate distance measurements for astronomers, and generally add to man's picture of the Red Planet and its moons. The spacecraft Phobos didn't send that information back.

Earth is the third planet out from the Sun, at a distance of about 93 million miles. Mars is fourth, at a distance averaging about 142 million miles from the Sun.

Mars, Earth, Venus and Mercury are said to be in the inner Solar System. Planets farther from the Sun, beyond the Asteroid Belt, are in the outer Solar System. They are Jupiter, Saturn, Uranus, Neptune and Pluto.

Long haul. The distance from Earth to Mars ranges from 35 to 248 million miles, depending upon where each planet is in its orbit around the Sun. Mars came within 36 million miles of Earth in September 1988. Phobos 2, however, had to travel 111 million miles on its journey to the Red Planet from July 1988 to January 1989.

When radio contact with Phobos 2 was lost March 27, 1989, Mars was 166 million miles away from Earth. Radio signals travelling at the speed of light took 15 minutes to get to Mars. It took 30 minutes for command signals from Moscow mission control to get to Phobos 2 and an acknowledgement to return to Earth.

Discovery. Astronomers say that after Venus, Mars is the planet most like Earth. Venus has no moons, Earth has one and Mars has two. The American astronomer Asaph Hall discovered two tiny natural Martian satellites in 1877 using a 26-inch telescope at the U.S. Naval Observatory. He called the moons Phobos and Deimos.

For almost a century, astronomy couldn't do much more than stare across millions of miles at the moons. Telescopes were not powerful enough to determine size, weight or other characteristics.

Interplanetary probes. Eventually, engineers became able to build and send interplanetary science probes. Data radioed back from various Mars probes included startling information on Phobos and Deimos.

USSR Mars 1, launched in November 1962, was the first attempt to probe Mars. Unfortunately, contact was lost only 60 million miles along its route to Mars.

U.S. Mariner 4 launched in November 1964 was the first successful probe to reach Mars. It sent back 22 photos of the planet as it flew by in July 1965.

Mariner 9 brought the first close-ups of Phobos and Deimos. Launched toward Mars in May 1971, Mariner 9 arrived in a 12-hour orbit around the Red Planet in November 1971, the first man-made satellite to orbit a planet other than Earth. Mariner 9 had two TV cameras which sent back 7,329 photos.

Later, Viking 1 and Viking 2 carried the American flag across millions of miles to photograph Mars, Phobos and Deimos. Viking 1 was launched September 9, 1975. It arrived at Mars June 19, 1976. Viking 2 was launched August 20, 1975, and arrived at Mars August 7, 1976.

The small moon. Phobos, innermost of Mars' two natural satellites, measures only 17 miles by 9 miles. It is shaped like a potato. A Phobos day lasts only eight hours. Its gravity has only one-thousandth the strength of Earth gavity.

Deimos, Mars' other moon, is the smallest natural satellite in the Solar System at about seven miles diameter. By comparison, Earth's Moon is one of the largest natural satellites in the Solar System at 2,160 miles diameter.

Tantalizingly, Viking showed the density of rocks on Phobos' pitted surface differs from Earth rocks. But, Viking photos could say nothing definite about the composition of the tiny moon.

Gravity pulls Phobos gradually closer to Mars. The moon will fall on the planet sometime in the next 30 to 70 million years.

In July 1988, the USSR sent two Phobos probes to Mars with experiments from 14 countries. Phobos 2 arrived at Mars in January 1989. Phobos 2 shot close-up photos of Phobos, one of the Solar System's smallest moons. The probe also found water vapor in the atmosphere of the planet Mars. Radio contact was lost with Phobos in March 1989. Phobos 1 failed along the way to Mars.

Life On Earth

How did life get started on Earth? Where did it come from? Looking out at other planets may give clues to what might have happened once-upon-a-time on Earth.

When Solar System scientists got together recently to discuss how Jupiter's mammoth atmospheric storms, Martian soils and evidence of lakes on Venus relate to the search for life's origins, they decided they have only glimpsed what may be Out There. They wanted to do more.

Scientists have discovered that Mars, Earth's nearest planet neighbor, once had more atmosphere. Martian soils show that lakes and deep rivers once covered the Red Planet. Evidence of lakes on Venus suggests in ancient times the Solar System's inner planets had an abundant water supply—probably a necessary condition for life.

Large storms seen on Jupiter suggest a natural laboratory for chemical evolution. Scientists also have found oceans of something on Jupiter's satellite Europa.

Such discoveries help trace the origins of Earth, the Solar System and life itself. Interplanetary probes now at Venus and on the way to the Sun, Mars, Jupiter, Saturn and its moon Titan, may help explain the origin of life.

Life from cosmic dust? An astronomer at University of Leiden, The Netherlands, says cosmic dust particles, comets and other matter from space may have showered ancient Earth with the basic organic chemicals from which life could have evolved.

Laboratory studies of cosmic dust and space studies of Comet Halley returned strong evidence that comets easily could have supplied Earth with raw materials for evolution of life, the Dutch astronomer reported.

Some other astronomers agree that comets may have been the main source of organic molecules, the first building blocks of life. The chemicals got in the comets through billions of years in space from cosmic dust particles, according to the Dutch researcher.

Cosmic dust, so small that 100 billion particles would make a ball the size of a pea, are spewed out of exploding stars. The particles cool and form vast deep-space clouds.

While in clouds, the particles become coated with ice, methane and ammonia. Over billions of years, heat and radiation from stars cook the molecules in the coating into complex organic chemicals.

The particles clump together, join with other clumps and eventually are big enough to be recognized as comets and asteroids.

Comet Halley looked like it was made of ice and dust particles. Comets have a low density and are almost fluffy.

If such comets struck the atmosphere of a primitive Earth, the Dutch astronomer said, they would break apart. Part would fall all the way to Earth's surface unchanged.

For instance, all of the water in the oceans may have been supplied by comets. Many large complex organic molecules may have arrived in that water, creating a primordial soup from which life could have developed.

If things really happened that way, they might have happened elsewhere too. Life may have developed on other planets and near other stars. If a star has a solar system and a cloud of comets, it may have its own supply of organic chemicals, the Dutch scientist said.

Did life start later? Apparently a thick cloud of fine debris surrounded our planet for the first few hundred million years of Earth history. If so, could life get a foothold without sunlight?

If life couldn't get going until sunlight reached the surface of primitive Earth, then scientists will have to recalculate the time required for the origin of life. If life on Earth had to wait for sunlight to get started, then several hundred million years less time would have been available than had been thought for the origin of life.

In other words, life on Earth may have come about faster than scientists once assumed—because comets and meteorites struck the young planet, shrouding it in dust, preventing a more leisurely evolution from starting earlier, according to University of Arizona planetary scientist David Grinspoon and Cornell University astronomer Carl Sagan.

Scientists generally believe Earth and the eight other major planets were formed when material orbiting the newborn Sun collided, clumped and coalesced. After planets formed, there was plenty of debris left—comets, meteorites and asteroids, called planetesimals. The debris kept crashing into the planets, as seen in the craters on the Moon and Mars. Erosion later erased most craters from Earth's face.

At least three chunks of debris, each body larger than two miles in diameter, seem to have struck Earth *every year* for the planet's first several hundred million years, according to Grinspoon and Sagan. That's based on the calculated rate of lunar crater impacts and of the number of planetesimal collisions thought to be needed to form Earth.

Earth formed 4.6 billion years ago. Computer calculations indicate life started about 4 billion years ago. The earliest known life forms on Earth were highly-evolved bacteria that existed at least 3.5 billion years ago, according to study of fossils. The time period for the origin of life is squeezed from both ends. The geological record pushs it earlier, even as calculations of comet impacts push it later.

Was there a dust shroud? Some planetary scientists figure there may not have been enough objects hitting Earth to maintain a continuing cloud. But, even if there were, biologists have said life could have started on a dark, cold, dust-shrouded Earth. It's impossible to know for sure how often comets and meteorites struck primitive Earth.

The quantity of dust spouted into the sky by nuclear bomb blasts on the ground in the 1950's was used by Grinspoon and Sagan to estimate how much dust would be shot into the air by three comet or meteorite strikes per year.

A paleobiologist at the University of California, Los Angeles, said the first organisms on Earth probably did not use light. Even with a dust cloud, life could have evolved. A fossil expert at the University of California at Berkeley said surface temperatures didn't matter because life was starting in the oceans.

The Grinspoon and Sagan theory recalls a controversy about whether dinosaurs became extinct after a comet or asteroid struck Earth more recently—only 65 million years ago.

What of Mars? Old channels seen on the Red Planet make it appear running water might have been there at the same time life was starting on Earth. Maybe there was time for life to start on Mars.

Ocean warmth. A runaway greenhouse effect on the early Earth may have driven daily temperatures on the surface to 190 degrees Fahrenheit, according to a 1990 NASA report. Through hundreds of million of Earth's earliest years, enormous quantitites of carbon dioxide and carbon monoxide spewed from erupting volcanos into the primitive atmosphere. More of the gases were injected by meteorite and comet impacts.

Heat radiating from Earth's surface was trapped by the greenhouse gas carbon dioxide, blasting temperatures to 190 degrees Fahrenheit, according to the NASA scientists at Ames Research Center, Mountain View, California. That temperature is just 22 degrees below the boiling point of water.

If the searing temperatures destroyed the large organic molecules which are building blocks of life, where did life originate? The NASA group suggested mid-ocean ridges— mile-deep chains of underwater mountains. Heavy pressure at that depth could have kept the large biological molecules from breaking down in intense heat.

Primitive bacteria, which did not live by photosynthesis, were discovered in 1982 circulating around volcanic vents along the mid-ocean ridges. The bacteria not only thrived in hot temperatures, but metabolized sulfur. They were able to live in an acid.

If large amounts of carbon dioxide from the atmosphere were dissolved in the ocean, it would be acidic.

Differing scientists argue the planet's surface would not have been as hot during Earth's early history 3.9 billion years ago. A less-bright and younger Sun would have delivered 25 percent less heat to Earth's surface, they say. Most scientists find plentiful carbon dioxide then would have kept the temperatures up to what it is today. The Ames group suggests there was far more carbon dioxide present.

A University of Michigan geochemist found early Earth's atmosphere was 80 percent carbon dioxide and had pressures twelve times those of today. Early Earth may have poured three times more carbon dioxide than today into the atmosphere.

Other scientists suggest complex molecules of life may have arrived on comets, meteorites and asteroids. Collisions were common in the early Solar System. Scientists have found amino acids, the basic ingredients of proteins, in meteorites from the asteroid belt. Some scientists believe all the water and organic carbon now in the Earth's biosphere could have arrived via comets and asteroid fragments.

A warm period would have been required for the river valleys and other marks of flowing and standing water seen on Mars. A greenhouse of dense carbon dioxide could have warmed the frozen Red Planet to make water liquid. Mars is 48 million miles farther from the Sun than Earth is. It receives less than half as much solar heat.

Water On Earth

How did water come to be on Earth? Did it boil up from underground or was it brought here from space? An Iowa astrotheorist speculates that space-traveling slush balls brought most of Earth's water.

Scientists had thought that water boiled up from inside our planet four billion years ago as crusty layers of rock cooled. Now it looks like ice comets have been splashing into our atmosphere over billions of years and vaporizing into clouds of water.

A glow of ultraviolet light shines from Earth's upper atmosphere. Satellite pictures show dark spots in that glowing area.

Researchers think water clouds from melted comet snow and ice block ultraviolet light and make the dark shadows 35 to 85 miles across in the pictures. Sometimes 10 spots appear per minute, stay for a few minutes and disappear.

Black snowballs. Blackened snowball comets falling by the millions from space onto Earth—building oceans, lakes and icecaps with their melted water—have caused a lot of discomfort for an Iowa scientist.

But, recent findings around the globe are beginning to ease the professional pain felt by University of Iowa physicist Louis A. Frank who published the theory in 1986.

Research in North America, Europe and Canada is adding scientific weight to a concept that once threatened to undermine Frank's career. And no one has found data showing black snowball comets don't exist.

The Iowa scientist had looked at data gathered between 1981 and 1986 by the Dynamics Explorer I satellite in Earth orbit. In satellite photos made to study air glow around Earth, Frank found strange holes which seemed to have been punched through the upper atmosphere. Puzzled for months, he discarded dozens of answers before deciding that only 100-ton ice comets covered with a black hydrocarbon could have made the holes. He figured 10 million of the black snowballs were falling to Earth each year.

30-ft. iceballs. The 30-ft. snowball comets, actually more like ice balls, would break apart as they approached Earth and flash-melt to vapor in the atmosphere. Eventually, that vapor would fall as rain, joining water already on Earth.

The jumbo snowballs may have added the equivalent of one inch of water all over the Earth's surface every 10,000 years. It doesn't sound like much, but since it started 4.5 billion years ago when the Earth was formed, black snowballs could have poured enough water on our planet to fill the oceans and lakes and build the polar ice packs. If snowball comets continue for billions more years, they will make big changes in Earth's climate.

Initially, Frank's idea was met with ridicule. He feared for his career. But, some took it seriously and tested the concept. Now support is growing. Papers in support have been presented to the American Geophysical Union.

A physicist at NASA's Jet Propulsion Laboratory, Pasadena, California, said he was skeptical at first. But then he used a powerful space search telescope at Kitt Peak Observatory, Arizona, to scan the skies for snowballs. He did capture photo views of the speeding comets approaching Earth, results agreeing exactly with predictions.

Hydrogen atoms. A space physicist at University of Michigan thought Frank's theory could be disproved by measuring the amount of hydrogen atoms released by water vapor in space above Earth. He turned to ultraviolet photos from Voyager 2, expecting to find few hydrogen atoms. Instead, data showed hydrogen was much denser than expected, enough to support Frank's theory.

Canadian scientists looked at images from a camera aboard a Viking spacecraft and discovered that it, too, detected the atmospheric holes Frank found in the Explorer satellite data.

A Pennsylvania State University researcher used microwave detectors to confirm the presence of water at high altitudes. Similar results were reported by West German scientists using rockets to study the atmosphere 60 to 80 miles above Earth.

Comet Halley was one of history's famous space snowballs. Scientists looking at it found a deuterium-to-hydrogen ratio the same as in Earth's oceans. That might indicate comets contributed water to the planet.

The new-found support helps, but Frank says, "My life would certainly have been easier" if the theory had never been published.

Clouds from space. Those colorful wispy 50-mile-high clouds you see after sunset in the summer are water and dust from meteors vaporized as they splash into the top of our atmosphere.

The thin shining colored clouds, called noctilucent as they appear before sunrise and after sunset, can't be water vapor climbing up there from the surface of our Earth. They're just too high. We've known how high they are for the last hundred years but we haven't been able to figure how they get there.

Now a NASA physicist says the atmosphere at that altitude is cold enough for the clouds to form. Meteors pass Earth all the time. Some plunge into the atmosphere and are vaporized into water and dust.

The water and dust condense into clouds. If enough meteors splash into the warm area of the atmosphere above the equator, they can briefly cool it and permit clouds to form.

Over the polar ice caps, where temperatures drop to minus 260 degrees Fahrenheit, it's plenty cold enough for clouds to form. Between the equator and polar ice caps, the upper atmosphere is cold enough for clouds of meteor water and dust to condense.

Earth's Lumpy Core

Where on Earth can you find mountains several miles high? Or valleys six times deeper than the Grand Canyon?

No, not on the hard outer crust we call land. Not at the bottom of the oceans. Maybe we should have said, where in Earth? That lumpy blob of molten iron and nickel at the very core of our planet, is the answer.

Geologists have been listening to the way surface shock waves from earthquakes rattle down to the center of the Earth, bounce off the core and return to the surface. Just like a CAT-scan doctors use to make pictures inside the human body, the geologists read the waves and drew pictures of the shape of Earth's core.

It had been thought the core would be a smooth sphere of red-hot metal. It actually is a rough ball. The geologists used seismic tomography in which a computer draws a cross-section of the Earth. The technique is similar to computer-assisted tomography (CAT) scanners used by doctors to make cross-section pictures of bodies. Seismic tomography analyzes all of the shock waves from thousands of earthquakes, large and small, over ten years.

The core is 4,000 miles in diameter. The Earth's crust is about 25 miles thick. The mantle is composed of viscous molten rock about 2,000 miles deep from the bottom of the crust to the top of the core.

The core's deep valleys are found under the parts of the crust we know as Europe, Mexico, the East Indies and the Philippine Sea. The canyon under the Philippine Sea looked to be more than six miles deep.

The mountains on the core are under what we know as central North America, the northeast Pacific Ocean near Alaska, Central America, southcentral Asia and eastern Australia. The mountain under the ocean near Alaska may be more than six miles high.

Geologists think the mountains may arise when currents flow up and down within the mantle. Hot parts of the mantle would be less dense. They would float up, away from the core like hot air from a radiator. At the top of the mantle, near the crust, some of the material may cool and sink. These up and down convection currents would pull parts of the core up into mountains and push other parts down into valleys.

Can Earth Be Slowing Down?

Yes, it appears to be true. Earth rotates on its axis. Scientists at NASA's Jet Propulsion Laboratory at Pasadena, California, have been looking at ancient Chinese records of solar eclipses to calculate just how much that rotation may be slowing. They say, nowadays, one day is seven hundredths of a second longer than a day 4,000 years ago in the year 1876 B.C.

It seems Earth's rotation slows as tidal interactions make the Moon orbit Earth more quickly—sort of the way a spinning ice skater slows down by extending his arms. The Moon becomes more distant from the planet. Calculations show, around four billion years ago, the Moon was one-third as far from Earth as the 240,000 miles it is today.

It's mind-boggling to contemplate but, four billion years ago, a day was only eight hours long. By 1876 B.C., the day had lengthened and was only 70 thousandths of second shorter than a day today. By 899 B.C., a day was only 42 thousandths of second shorter than a day today. By 532 A.D., a day was only 22 thousandths of second shorter than a day today.

Earth's rotation does vary slightly today as our oceans and atmosphere produce drag. Molten rock inside our planet sloshes against solid rock to produce more drag.

Ancient Chinese annals report thousands of eclipses. Researchers found eclipses at sunrise or sunset most useful. Knowing times they were visible in China, scientists are able to calculate how much Earth's rotation has slowed since the eclipses. For instance, had the day always been 24 hours long, the 899 B.C. eclipse would have been seen in the Middle East instead of China. The oldest records from Arabia and Babylonia of solar eclipses date to about 700 B.C.

Ancient Chinese chronicles, known as the Bamboo Annals, mentioned a time when "the day dawned twice." With a computer simulating the history of Earth's rotation around the Sun, scientists determined the phrase meant that the Moon eclipsed the Sun just after dawn on April 21, 899 B.C.

That let them figure out the Earth's rotation rate at the time and, also, place a date on the reign of King Yi of the Western Zhou dynasty.

Ozone Hole

A hole opens in the ozone layer of Earth's atmosphere over the North and South Poles each Fall. The holes grow wider each year and, in 1991, the Antarctic hole opened earlier than ever.

NASA's Upper Atmosphere Research Satellite (UARS) was launched in September 1991 as an ozone watchdog. UARS looks down on the ozone layer with unprecedented clarity as scientists try to gauge the health of the atmosphere and how human's damage it. UARS was to make a three-dimensional map of the ozone layer over 18 months.

The ozone layer shields Earth from cancer-causing ultraviolet (UV) light from the Sun. Industrial chlorine, such as chlorofluorocarbons (CFCs) destroys ozone, allowing UV to get through to Earth's surface. CFCs are used as air-conditioner refrigerants and to make cleaning solvents and plastic foams. Without the ozone shield, humans couldn't survive, according to a researcher in NASA's Earth science and applications division. Ultraviolet light would destroy many of Earth's life forms.

Without the ozone layer, rainfall patterns and other global weather patterns would change, affecting crops and aquatic life, according to scientists monitoring ozone depletion. The Antarctic ozone hole was seen first in 1985.

Researchers don't know why the hole has started forming earlier, according to the National Oceanic and Atmospheric Administration's climate monitoring laboratory at Boulder, Colorado.

Cosmoid snowballs. The ozone hole over Antarctica may be caused by cosmoids—the giant space snowballs known as cosmic meteoroids, according to two scientists at NASA's Goddard Space Flight Center at Greenbelt, Maryland.

Like comets, cosmoids circle the Sun on long elliptical orbits. Some encounter Earth, falling heavily onto Antarctica which faces away from the Sun and toward incoming cosmic meteoroids during polar winters.

Cosmoids are zapped into snowflakes before they enter Earth's atmosphere. That is, electrical forces break them into microscopic, fiber-like, icy particles, which mostly vaporize in sunlight but accumulate in Antarctica's stratosphere during that continent's cold, dark winters. Most ozone is lost over Antarctica in September and October as the Sun rises after the winter-long polar night.

The Goddard scientists proposing the cosmoid theory say thousands of space snowballs, much colder than snowballs on Earth, ranging from inches to 30 feet in diameter, plunge into Earth's atmosphere daily. They disintegrate into clouds of icy particles that destroy ozone.

A temporary annual decrease of up to 50 percent in Antarctic ozone concentrations has come to be called the ozone hole.

The other view. Critics admit high-altitude clouds may help cause the ozone hole that appears for a couple of months each year in the ozone layer over Antarctica, but say the clouds probably are formed by water from Earth's lower atmosphere, not by ice from outer space. An atmospheric scientist at NASA's Ames Research Center in Mountain View, California, called the cosmoid snowball theory "a ridiculous idea."

A physicist at NASA's Jet Propulsion Laboratory in Pasadena, California, says the theory is "really off the wall." On the other hand, the atmospheric chemistry director at the National Center for Atmospheric Research in Boulder, Colorado, said, "I'm going to be hard to convince, but it's an interesting thing to follow up."

Ozone at ground level is a toxic air pollutant, but the ozone layer 12 to 30 miles above Earth's surface blocks harmful solar ultraviolet rays. Scientists say depletion of that high-altitude layer may cause crop failures, climate changes and an increase in human skin cancers.

Plastic foam. Most scientists speculate that man-made chemicals are damaging Earth's ozone shield. Chlorine compounds called chlorofluorocarbons (CFC's) react in clouds to destroy ozone.

CFC's are used as coolants in refrigerators and air conditioners, in making plastic foams, and as spray-can propellants. Their use as propellants was banned in the U.S. in 1978. Other researchers have suggested wind patterns, sunspots, electrically charged particles called electrons and nitrogen oxide pollutants are possible causes of ozone destruction.

In the controversial cosmoid theory, ozone-destroying chemical reactions in clouds involve a hydroxyl which forms when sunlight hits the icy particles. The particles attract ozone through a known chemical reaction. Sunlight hitting them triggers another reaction that produces hydroxyl, which destroys ozone more rapidly than CFC's do.

One problem with the theory: the ozone hole has appeared annually only since the mid-1970's, yet should have existed earlier if the cosmoid theory is correct. If the new theory is correct, the number of incoming cosmoids varies annually, and a smaller influx before 1975 would explain a lack of an ozone hole before then.

The new theory recalls a projection a year earlier by a University of Iowa physicist who suggested Earth's oceans were slowly filled by water from billions of comets much larger and more numerous than cosmoids.

Arctic ozone hole. An ozone hole is developing in the stratosphere near the North Pole each winter, similar to the ozone hole over the South Pole.

A scientist for Canada's Department of Environment, who sent instruments aloft in balloons, found a vast crater of depleted ozone forming near the North Pole during winter months. The crater is a big "sink drain" for ozone from throughout the Northern Hemisphere.

Shaped like a crater with a deep center and high sides, the hole was discovered by balloons first launched from Alert, a Canadian town near the North Pole, in 1986. Instruments carried by the balloons measured ozone in the upper reaches of Earth's atmosphere, radioing data to the ground.

Why there is a loss of ozone from the atmosphere is not clear, but many scientists have been blaming it on chemical reactions started by chlorofluorocarbons (CFC's), chemicals used in refrigeration fluids, computer chip cleaners, blowing agents in making plastic foam and as power for aerosol sprays.

Ozone forms a natural shield in the atmosphere against ultraviolet rays from the Sun. Overexposure to ultraviolet causes skin cancer. The U.S. Environmental Protection Agency (EPA) estimated a one percent thinning of Earth's ozone shield would cause a five to six percent increase in skin cancers.

Normally, different weather patterns leave 15 percent more ozone in the atmosphere over the Arctic than over the Antarctic.

So far, the Arctic ozone loss is smaller. A hole above the North Pole probably would reduce ozone protection of the Northern Hemisphere just as the hole above the South Pole drops protection over the Southern Hemisphere.

Even if ozone depletion is less in the North, the effect could be more important because the North has most of the human population and most of the land mass.

Meanwhile, the protective ozone layer over the United States has shrunk 2.3 percent since 1969, according to NASA.

Annual ozone depletion over the mid-latitudes of the Northern Hemisphere is at a much higher rate. The latitudes, 30 to 64 degrees, include all of the United States except southern Texas, Louisiana and Florida and extend halfway up Canada's Yukon Territory.

If the use of CFC's were to continue increasing at 2.5 percent per year, EPA said, 20 percent of the ozone barrier could be gone in the next century.

Ozone is a pollutant on the ground. CFC's don't break up at ground level because not enough ultraviolet light gets through the atmosphere. On the other hand, ozone is considered good when it is 15 to 25 miles high in the atmosphere where it filters out the Sun's ultraviolet rays, making life on Earth possible.

The high-altitude problem starts when the Sun's ultraviolet rays break up CFC compounds in the atmosphere, freeing chlorine, which in turn destroys ozone.

When CFCs were banned as aerosol propellants in the United States in 1978, worldwide emissions declined.

The National Academy of Sciences even went so far as to say in 1983 that future ozone decline would be only two to four percent. But CFC use started growing again, partly because one chlorofluorocarbon compound was good for cleaning computer chips and electronic parts.

Rapid Ozone Loss. Satellites looking down from space show less ozone in the upper atmosphere than expected as the near-space layer shrinks rapidly.

Added crop damage and skin cancers are resulting from Earth's shrinking ozone filter. Ozone protects us from the Sun's deadly ultraviolet radiation. A one percent loss of ozone from the layer brings 15,000 new skin cancer cases each year.

Fast-food retailers in the United States are being pressured to stop using foam cups, plates and boxes. So far this decade, the ozone layer has shrunk almost three percent. It had been thought that much loss would take 40 years.

Melting glaciers. A pair of British scientists at Edinburgh University say global warming from an atmospheric greenhouse might change sea levels. They studied ice sheets in Greenland and easaten Antarctica.

The Geenland glacier might melt at the edges, but grow at the center, raising sea level a bit. The Antarctic sheet, on the other hand, might expand slightly, lowering sea level. The net result would be only a tiny increase in worldwide sea level over centuries.

Alps Ice. Something is slowly melting the sparkling blue-green sheets of ice which form huge glaciers in the Swiss Alps, according to annual figures published by the Swiss glacier commission. Scientists there are concerned.

Alps ice has been observed and measured for 110 years. Recently, it has been receding ten feet per year. Glaciologists say it's shrinking more on the southern side of the Alps. Milder winters and hot summers may have caused most of the melting.

Wobbly Earth

Earth has a quiver like the wobble of a clothes washer with an off-center load. Shifting masses of air in the atmosphere make our planet flutter like an unbalanced laundry.

Through the centuries, we've thought of Earth as terra firma. Now scientists report a shifting atmosphere makes the planet actually wobble slightly as it rotates. The wobble is only two feet wide at the poles, but it's useful to scientists who use it as a means of studying the atmosphere.

Over three years, a researcher at the U.S. Naval Observatory, Washington, D.C., and scientists at NASA's Jet Propulsion Laboratory, Pasadena, California, and Atmospheric and Environmental Research Inc., of Cambridge, Massachusetts, found the wobble related to changes in air pressure around the globe.

Air pressure changes when the atmosphere shifts. Fluctuations as slight as one-tenth percent to three-tenths percent of normal atmospheric pressure were found to contribute to the wobble.

Scientists knew Earth's axis regularly pulsates, with oscillations lasting about 433 days and about a year. The longer cycle is linked to the atmosphere. The recently discovered fluctuations have cycles of only two weeks to several months.

Earth bulges at the equator, so the weight of the atmosphere does not push exactly toward the center of our planet, but a bit off-center. It tends to nudge Earth off its rotational axis.

For instance, if pressure is high over North America, low over South America, high over Australia and low over Siberia, it would tend to rotate the globe as if a huge hand on North America were pushing south.

What happens when air pushes on oceans isn't understood as well as air pushing on land. Ocean water seems to respond to pressure over long periods by flowing off to the side, the way a waterbed responds if one slowly sinks a foot into it. But, the atmosphere still may exert some force through oceans. Oceans may respond more like a waterbed being stomped on quickly, with a complex sloshing back and forth that could transmit pressure to Earth at the ocean floor.

Northern Lights, Southern Lights

Northern Lights are known to scientists as Aurora Borealis. A similar South Pole phenomenon is called Aurora Australis. Astronomers know that flares on the Sun blast energy into space. That energy creates a magnetic storm as it engulfs Earth, charging the upper atmosphere and creating Northern Lights and Southern Lights.

Now, two NASA scientists say they have discovered how the lights can persist after a magnetic storm subsides: intense interplanetary electric fields lasting up to three hours.

Northern and Southern Lights last for days or weeks after a magnetic storm and are not just the storm decay. Satellite data convinced the NASA scientists that the lights also are caused by so-called Alfven Waves.

Alfven Waves. To find out what causes long-lasting auroras, the researchers turned to data from 500 days of solar observations by the International Sun Earth Explorer-3 (ISEE-3) satellite. Operated by NASA's Goddard Space Flight Center, ISEE-3 worked in an orbit 850,000 miles from Earth in 1978 and 1979.

They don't know what causes Alfven Waves, but the pair of scientists, from NASA's Jet Propulsion Laboratory, Pasadena, California, and from the Instituto de Pesquisas Espaciais in Sao Jose dos Campos, Brazil, found the waves originate near the Sun and move through interplanetary space from Sun to Earth.

The waves cause directional changes in the interplanetary magnetic field. When the field becomes opposite to the north direction of Earth's magnetic field, an interaction couples the two fields together.

Particle slingshot. The interplanetary magnetic field, starting at the Sun, connects with Earth's magnetic field and pulls the field downstream away from the Sun. The field reconnects behind Earth and causes charged atomic particles to slingshot toward the night side of Earth. When they get close to the planet, an instability occurs and scatters the particles.

Ions and electrons race along the magnetic field lines to the surface of Earth at both poles. Along the way, they crash into atoms in the upper atmosphere, exciting them to high energy. The excited atoms radiate the excess energy by giving off light—the wavy red and green light of auroras.

Small magnetic storms occur every 3 hours. Large storms come once every month or two. Very large storms are rare, coming only about once every three or four years.

Solar wind. The solar wind is a stream of particles emitted constantly by the Sun. The solar wind may be accelerated by Alfven Waves near the Sun.

Northern Lights are remnants of the accelerated solar wind striking Earth's magnetic field, injecting solar wind energy into Earth's atmosphere.

ISEE-3 was renamed International Cometary Explorer (ICE) in 1985 when it was sent for a close-up look at Comet Giacobini-Zinner.

See the lights. To see Northern Lights, check the sky every clear night:

★go outside after dark;

★wait a few moments for your eyes to adjust to the darkness;

★look into the northern half of the sky;

★expect clouds to fool you from time to time.

A real aurora will be a pale green curtain or streaks, slowly changing shape as you watch for several seconds. Sometimes an aurora can look like patches of clouds which fade away and then return, growing brighter by the minute.

Shoot the lights. It's easy to photograph Northern Lights. Here's how:

★use a camera offering time exposures;

★load slide or print film rated at ISO 400 or higher, although slower film will work;

★support the camera on a tripod;

★set the lens focus at infinity;

★open the lens as far as it will, that is set the f/stop at the lowest number on the aperture ring;

★set the shutter for a time exposure;

★point the camera at the aurora;

★ensure success by bracketing your exposures. Make at least three exposures if the aurora is bright, one each at five, ten and thirty seconds. Also make one-minute and two-minute exposures if the aurora is dim.

Winter Solstice

Winter solstice, coming as it does at the annual time of least sunlight, has been celebrated for centuries for the triumph of the Sun, unconquered by autumn. Strangely, rather than celebrating the winter solstice today, we dread it—as the start of winter.

Winter solstice once was the most popular pagan celebration. Today, it's overshadowed by the religious holidays and the New Year. Winter solstice comes at the low point of the year for the Sun's apparent travel. After the solstice, the Sun sends its warming rays farther and farther north until it peaks in June at the summer solstice.

Primitive peoples worried about the Sun's departure. Their folk tales about seasons commemorated the winter solstice as a time of rebirth, foretelling the return of summer.

In ancient Rome the season was Saturnalia, celebrated with the noisy merrymaking seen today on New Year's eve. Since ancient times, hilltop bonfires have been lit to persuade the Sun to return. The German commemoration is the Yule feast, burning of a Yule log, with holly, ivy and mistletoe as magic decorations, and wassail.

When does winter start? In medieval times, November 23 was considered the season's beginning. Even though there really aren't any official definitions of seasons, today's calendars list the winter solstice date as the start of winter.

Meteorologists, on the other hand, consider winter beginning December 1. Most people, however, judge winter's arrival by how cold they feel when they go outdoors.

The modern bad feeling about winter solstice stems from the months afterward, the coldest time of the year. Although sunlight is increasing gradually, it takes months to overcome the cooling of autumn when sunlight was declining.

Spring arrives, of course. In medieval tradition it came February 22. Meteorologists today say it begins March 1. Calendar makers print it as March 20. For most people, it's the first balmy day.

Fall Equinox

The autumnal equinox, when daylight hours nearly equal the hours of darkness, is the moment the Sun appears to cross the Equator on its annual journey south.

The equinox will make its earliest appearances this century when it comes at 18:44 Coordinated Universal Time (UTC) on September 22, 1992, and also at 18:01 UTC on September 22, 1996, according to the Nautical Almanac Office at the U.S. Naval Observatory.

In the year 2000, the equinox will edge all the way back to 17:29 UTC on September 22.

The previous earliest appearance this century was in 1988. Not since 1897 had the equinox come so early.

Coordinated Universal Time (UTC) is also known as Greenwich Mean Time (GMT). UTC is five hours ahead of Eastern Standard Time (EST), four hours ahead of Eastern Daylight Time (EDT). For instance, 3:29 p.m. EDT, the same as 1529 on a 24-hour clock, equals 1929 UTC.

The tendency of the equinox to occur earlier will continue, at least until the year 2100 when the absence of a leap year will shift it to later dates again.

Calendar makers and others often list the autumnal equinox as the start of fall, and the spring equinox as the start of spring.

However, there are no official definitions for those seasons. Inserting an extra day in leap years every four years tends to cause back-and-forth shifts in calendar dates for natural phenomena such as equinoxes and solstices.

Moving around the exact equinox times comes about because a true Earth year isn't exactly an even number of days.

The Gregorian calendar we use is 365 days long most years, and 366 days in leap years. That leap day added every four years makes a calendar year average out to 365 days, 6 hours.

The problem is it actually takes 365 days, 5 hours 48 minutes 46 seconds to go from one equinox to the next. The difference is 11 minutes 14 seconds—not much, but it adds up over centuries.

To compensate, the calendar skips leap days in most century years. The year 1900, for example, was not a leap year.

There is an exception. Century years that can be divided by the number 400 are leap years. Therefore, the year 2000 will be a leap year, getting an extra day.

That means the calendar year will be a bit longer, causing the equinox to be a little earlier that year.

The year 2100 won't be a leap year since it can't be divided by 400. So, the equinox won't be able to head back to later dates until after the year 2100.

Solar Eclipses

The Sun is eclipsed when the Moon passes directly in front of it. The Moon casts a shadow, less than 300 miles wide, on Earth's surface. As the Moon travels its orbital path around Earth, the shadow moves across Earth's surface.

The Moon orbits Earth every 29.5 days. Earth orbits the Sun once a year. Daily rising and setting of Sun and Moon are caused by rotation of the planet, not by movements along orbital paths. Earth's daily rotation does not cause eclipses.

Since the paths of the Sun and Moon are at small angles from each other, eclipses happen only when the orbits of the Sun and Moon cross—about every six months.

A 1991 solar eclipse was seen by millions as it blocked the Sun on July 11 over Hawaii, North Pacific Ocean, Mexico, Central America and South America.

On rare occasion, if the first solar eclipse of a year is early in January, three may be squeezed into one year. There were four solar and three lunar eclipses in 1982.

In a lunar eclipse, the Moon passes through the shadow of the Earth. Usually a lunar eclipse occurs within 14 days of a solar eclipse. Sometimes, three eclipses occur within one month.

Solar eclipses can be total, partial and annular. If Sun, Moon and Earth are in a straight line, a total eclipse blocks our view of the Sun completely.

If the Moon is slightly off a line to the Sun, a partial eclipse is seen. If the Moon is too far from Earth to completely fill the Sun's face, but blocks most of the Sun, it is an annular eclipse.

Never look directly at the Sun with the naked eye, nor with binoculars, telescope or camera. The eye can be damaged seriously, even though no pain is felt. Instead, look at an image of the Sun projected through a pinhole in a cardboard onto white paper. Here is a schedule of total solar eclipses from 1992 through 2017:

Year	Date	Area on Earth seeing total solar eclipse
1992	Jun 30	South America, South Atlantic Ocean
1994	Nov 3	South Pacific Ocean, South America, South Atlantic, Indian Ocean
1995	Oct 24	Middle East, India, Southeast Asia, South Pacific Ocean
1997	Mar 9	Siberia, Arctic Ocean
1998	Feb 26	Pacific Ocean, South America, North Atlantic Ocean
1999	Aug 11	North Atlantic Ocean, Europe, Middle East, India
2001	Jun 21	South Atlantic Ocean, Southern Africa, Madagascar, Indian Ocean
2002	Dec 4	South Atlantic Ocean, Southern Africa, Indian Ocean, Australia
2006	Mar 29	Atlantic Ocean, Northern Africa, Middle East, USSR
2008	Aug 1	Northern Canada, Greenland, North Atlantic Ocean, USSR, China
2009	Jul 22	India, China, South Pacific Ocean
2010	Jul 11	South Pacific Ocean, Southern South America
2012	Nov 13	Australia, South Pacific Ocean
2015	Mar 20	North Atlantic Ocean
2016	Mar 9	Indonesia, Borneo, Pacific Ocean
2017	Aug 21	North Pacific Ocean, USA, North Atlantic Ocean

Lunar Eclipses

Sometimes a full Moon passes through Earth's shadow, creating the exciting spectacle of an eclipse.

There may be two or three partial lunar eclipses in any year but the more spectacular total eclipses happen less frequently.

They can last up to an 1 hour 40 minutes. Some sunlight passing through Earth's atmosphere is refracted to fall on the Moon, giving the Moon a copper hue.

Lunar eclipses are much more pleasant to observe than solar eclipses. Here is a list of expected total lunar eclipses:

Year	Date		Year	Date		Year	Date
1992	June 15		1995	April 15		1997	September 16
1992	December 9		1996	April 4		1999	July 28
1993	June 4		1996	September 2		2000	January 21
1993	November 29		1997	March 24		2000	July 16
1994	May 25						

In a total solar eclipse, the disk of the Sun is completely hidden briefly by the Moon. Solar eclipses find the Moon directly in line between the Sun and Earth. It's a partial eclipse when the Moon isn't exactly in line.

Caution: protect your eyes. Never look directly at the Sun. You must use screens and filters to avoid eye damage.

Solar Eclipses happen two to five times a year. The so-called annular eclipse finds the Moon lined up, but so far away the dark center of its shadow doesn't reach Earth. Most people don't see a solar eclipse, anyway, because the shadow travels a narrow path across the face of the Earth and lasts just over seven minutes at totality.

If Sunlight sparkles through mountains on the edge of the Moon during a total eclipse, the phenomena is called Baily's Beads. A flash of light at the start or end of totality is known as the diamond ring effect.

Cleaning Earth's Atmosphere

A summary of the major provisions of the 1990 U.S. Clean Air Act:

Ozone-depleting chemicals. Chlorofluorocarbons (CFCs) are used in refrigerants, as cleaning solvents for computer chips and during production of plastic foam products. The 1990 U.S. Clean Air Act phases out chlorofluorocarbon production by the year 2000.

The year 2030 is the end-of-production deadline for hydrochlorofluorocarbons, a less-ozone-depleting substitute for CFCs.

The year 2000 is the end-of-production deadline for halons used in the cleaning solvent carbon tetrachloride and in fire extinguishers.

The end-of-production deadline is the year 2002 for the solvent methyl chloroform.

Automobile pollution. Vehicle tailpipe emissions cause urban smog. The Clean Air Act requires automobile manufacturers to cut tailpipe emissions in half by the mid-1990s. After the year 2000, the Environmental Protection Agency (EPA) could order even more cuts if feasible technologically.

Lower carbon monoxide emissions from vehicles are required by the Act.

The Act requires industry to develop new low-polluting vehicles to be sold in a pilot program in California. That specification might be met with gasoline-powered cars, so automobile manufacturers would be able to avoid having to mass produce cars using cleaner-burning alternative fuels such as natural gas or wood alcohol. Operators of commercial fleets of ten or more vehicles might have to switch to non-gasoline fuels.

Alternative fuels. When added to gasoline, the corn derivative, ethanol, has fewer contaminants and burns more cleanly. The Clean Air Act requires the oil industry to produce gasoline containing ethanol.

The reformulated gasoline must be available in cities with excess carbon monoxide pollution by 1994. The new blend must be sold in the nine smoggiest cities by 1995. The gasoline must have 15 percent fewer smog-forming compounds by 1995 and 20 percent fewer by the year 2000.

Urban smog. The Clean Air Act tightens controls on volatile organic compounds and nitrogen oxides from factories and commercial operations in 96 cities where smog exceeds federal health limits. In the smoggiest, the bill would cover small facilities such as gasoline stations not subject to emission limits under current law.

Carbon monoxide exceeds federal health standards in 41 cities. Soot is at unhealthful levels. The Act includes steps to reduce these.

The Act sets an air clean-up schedule for cities. Most must bring themselves into compliance within three, six, or nine years. The dirtiest cities would get the most time:

With severe smog, Chicago, Philadelphia, Milwaukee, San Diego, and Muskegon, Michigan, would have until 2005.

With even worse smog, New York, Baltimore and Houston would have until 2007. America's dirtiest city, Los Angeles, would have until 2010.

Emission allowance trading system. The Clean Air Act sets up an emission allowance trading system to ensure acid rain emissions will not increase after the year 2000.

Plants which exceed pollution limits would be subject to penalties of $2,000 per ton of emissions.

The Act also would lessen the pollution control cost burden for utilities facing major reductions.

Each plant would be permitted to release a fixed amount of pollution. Those plants which reduce emissions below assigned levels would free up emission allowances which then could be sold to other utilities seeking to expand coal-burning capacity.

Toxic air pollution. The Clean Air Act requires approximately 250 different industries to cut by at least 75 percent within ten years their emissions of 189 types of toxic and cancer-causing pollutants.

A wide range of facilities, from oil refineries and chemical plants to dry cleaners and bakeries in polluted cities, would be affected. The law also will apply to hospital, industrial and municipal garbage incinerators.

EPA would require an initial round of cuts based on the maximum achievable emission control technology available for each industry. EPA would review remaining emissions and decide whether a second round of cuts was needed to make sure the remaining risk to persons living near a plant would not be excessive.

EPA is required to assure the most exposed individual has an ample margin of safety. The courts have interpreted that as a cancer risk no greater than one in 10,000.

Acid rain. Sulfur dioxide and nitrogen oxide are the primary precursor pollutants linked to acid rain which has damaged lakes, streams, forests and buildings in the northeast United States and Canada.

The Clean Air Act requires 111 big coal-burning power plants, mostly in the Midwest, Southeast and Appalachia, to lower acid rain emissions by ten million tons below 1980 levels within ten years. That is about a 50 percent cut.

The Act also requires the same plants to cut nitrogen oxide emissions by approximately two million tons.

Worker assistance. The Clean Air Act authorized a five-year $250 million program to help workers forced out of work by pollution reduction requirements. It will let displaced workers continue receiving unemployment compensation beyond their normal six-month allotment if they enter a retraining program within the first 13 weeks of receiving benefits. Extended benefits will last as long as a worker is in retraining.

Asteroids Near Earth

An asteroid could threaten life on Earth, according to planetary scientist Eleanor Helin of NASA's Jet Propulsion Laboratory, Pasadena, California.

A relatively-small mile-wide asteroid could ruin climates around the globe, end most farming, spreading starvation, death and destruction, even wiping out mankind, according to David Morrison, space science chief at NASA's Ames Research Center, Mountain View, California.

The NASA researchers were among 160 planetary scientists, astronomers and engineers at an International Conference on Near-Earth Asteroids held by NASA and the Planetary Society in June 1991. The Planetary Society is a non-profit group with 120,000 members worldwide.

Cosmic shooting gallery. Morrison described Earth as in a cosmic shooting gallery, suggesting higher priority should be placed on searching out asteroids which might collide with our planet. The scientists discussed ways to avoid a collision, including placing a rocket on an asteroid to drive it away or bumping an asteroid with a nuclear bomb to nudge it off course.

Earth has the technology to change the orbit of an asteroid, according to the American Institute of Aeronautics and Astronautics. The scientists at the 1991 conference urged appropriation of extra money right away to build telescopes around the globe to search for asteroids approaching Earth.

Rocky debris. Asteroids are giant chunks of rock left over from the time the Solar System was formed billions of years ago. They orbit the Sun, sometimes crossing Earth's path around the Sun. Each year, some asteroids whiz by within 10 million miles of Earth. Small asteroids smash into Earth every 100 years or so.

An asteroid is a tiny extraterrestrial Solar System body—a large chunk of rock, a minor planet, a planetoid. Nine out of ten known asteroids are orbiting the Sun in the so-called Asteroid Belt between Mars and Jupiter.

Thousands of asteroids, composing the Belt, are in a circular orbit some 200 to 300 million miles from the Sun. It usually takes any one asteroid three to six years to circle the Sun.

As with Earth's Moon, the minor planets shine by reflecting sunlight. Only one asteroid, Vesta, can be seen with the naked eye. About 3,000 asteroids have been named or numbered. Astronomers think 100,000 could be seen and photographed.

Asteroids range up to 600 miles in diameter which is the size of Ceres, the largest known minor planet. Most, however, are much smaller, although 200 are over 50 miles in diameter with ten nearly 200 miles in diameter.

Asteroids may not be remains of an old exploded planet since the weight of all the asteroids in the Belt totals less than one tenth of one percent of the weight of the Earth. More likely, asteroids simply are debris from collisions between a few small objects which condensed between Mars and Jupiter when the Solar System was forming. Meteorites arriving at Earth probably are fragments from these minor planets.

About one percent of asteroids may be on paths which cross the orbits of planets. Asteroids crossing planets actually collide with a planet about once in 10 million years.

American Institute of Aeronautics and Astronautics said such an asteroid hitting Earth would create a blast 100,000 times as powerful as the atom bomb dropped on Hiroshima.

Some scientists believe a six-mi.-wide asteroid plunged to Earth 65 million years ago, killing two-thirds of all living species, including dinosaurs. They suggest such a large-scale collision happens every 50 to 100 million years. Smaller asteroids about one mile wide strike Earth every 300,000 to 1 million years.

If one hit Earth. On average, an asteroid larger than one-quarter-mile diameter hits Earth every 50,000 years. The last crater formed by an asteroid crash was 50,000 years ago in what today we call Arizona.

Hundreds or thousands of asteroids are believed to pass relatively near Earth over periods of many years. The question for some astronomers is when, not if, one will hit our planet. Astronomers calculate the chances of an asteroid hitting Earth by computing each known asteroid's orbit. Today they do not know of any asteroids which might hit Earth, but they say people would be "totally unprepared" if an asteroid were to hit.

A hit would trigger earthquakes, force volcanic eruptions, spawn giant tidal waves and wreak other natural disasters around the globe. Local effects would be like nuclear war without radiation. Buildings would be leveled for hundreds of miles around a point of impact. A Florida astronomer said, if an asteroid fell into the ocean off the coast of

Bermuda, Florida would be submerged by a 200-foot tidal wave. Cape Canaveral, Fort Lauderdale, Jacksonville and other coastal cities would be wiped out almost instantly.

The father of the hydrogen bomb, Edward Teller, and a few other scientists think such a catastrophe could be averted be firing a missile at an approaching asteroid. A nuclear bomb, exploding as the missile hit the asteroid, would divert the asteroid, they say.

The odds of an asteroid whacking Earth are extremely small. However, so many persons would be killed in a comet disaster there's more chance of being killed by an asteroid than of dying in an airplane crash.

Novosibirsk. The USSR has proposed the United Nations set up a full-time comet-watch operation at the abandoned Novosibirsk radar station in Kazakhstan. Astronomers at Alma-Ata and Novosibirsk in July 1991 suggested transferring the military radar from the USSR Defense Ministry to a scientific organization. The equipment would be used continuously to scan the skies for doomsday bodies.

Craters. The scientists had been studying what they called "circular structures," ancient craters in the ground. Several craters have been found in Kazakhstan, including the 13-mi.-diameter Baltatarak Crater and the 39-mi.-wide Araganaty Crater. The youngest meteorite crater in Kazakhstan is the 700,000-year-old Zhamanshin Crater.

USSR territory has been struck at least twice this century. In 1908, the Tungus meteorite kicked up a blast equal to a 12-20 megaton bomb. In 1947, the Sikhote-Alin meteorite shower did great damage.

The Soviet scientists suggested a combined U.S.-USSR intercontinental ballistic missile (ICBM) with nuclear warhead be sent to divert any comet threatening Earth.

Hunting asteroids. To find new asteroids, JPL astronomers photograph large areas of the sky each month. The pictures are inspected immediately and follow-up observations are made right away if an asteroid or comet is found. The 18-inch Palomar Schmidt telescope has turned out to be just right for asteroid searches.

Searches for asteroids near Earth using telescopes with wide-field cameras turned up asteroids 1982db, and 1982xb, with orbits so close to Earth they are favored for future interplanetary-probe rendezvous flights.

Outside the Belt. There are a few groups of minor planets with orbits different from the Asteroid Belt. For example, those bodies named Trojan, Apollo and Amor. Achilles is in the Trojan group near Jupiter.

The group named after the minor planet Amor, discovered in 1932, travels closer to the Sun than Mars but not as close as Earth.

The Apollo group comes within Earth's orbit around the Sun. Hermes and Icarus are minor planets in the group named after the half-mile-wide Apollo, discovered when it came within seven million miles of Earth in 1932. 1989pb is an Apollo asteroid.

No. 2060. Chiron, first noticed in 1977, is a unique minor planet. With most of its orbit out beyond Saturn, the 112-mile-wide asteroid takes 50 years to circle the Sun. Chiron could be alone or it could be the brightest of an unseen swarm of asteroids in the far reaches of the Solar System. Or maybe it's not an asteroid at all, but a comet.

The 112-mi.-diameter Chiron seemed to have a faint coma in 1989 observations through the 3.6-meter Canada-France-Hawaii optical and infrared telescope atop Mauna Kea mountain in Hawaii. Comets have comas.

Chiron was discovered in 1977 by U.S. astronomer Charles T. Kowal. It was designated officially an asteroid, or minor planet, number 2060. Astronomers later found Chiron in photos dating to 1895. Whatever it is, the dark-surfaced Chiron is in an orbit different from any other known asteroid. Some say it may pass very close to Saturn in about 10,000 years.

If a comet, Chiron must be bigger than others we know of. It's about 250 times as bright as Comet Halley was at the same distance. Chiron and the small major planet Pluto once may have been moons of the larger major planet Neptune, wrenched away by the cataclysmic passing of some unknown massive object in the ancient past.

1991ba. Earth's closest call ever recorded was a 30-ft.-wide asteroid, which astronomers labeled 1991ba. It sped by within 106,000 miles of Earth January 18, 1991. That's less than half the distance to the Moon. When an astronomer at Arizona's Kitt Peak Observatory found the fast-moving 1991ba, it was 500,000 miles from Earth. Just 12 hours later, it zipped by at 106,000 miles distance. Roughly the size of a small house, 1991ba was the smallest asteroid ever seen.

1991rc. The known asteroid second closest to the Sun, 1991rc, was spotted crossing near Earth on September 3, 1991, by an Australian astronomer. The asteroid skims within 17 million miles (0.185 astronomical unit) of the Sun, half the distance from the Sun out to the first planet, Mercury. Asteroid 1991rc takes 13.5 months to fly around the Sun on a long path stretching beyond the orbit of Mars.

1991oa. Asteroid 1991oa was discovered in July 1991 only five million miles from Earth. An eight-inch telescope was needed to see it.

1991jx. The Apollo asteroid 1991jx was discovered by Eleanor Helin in May 1991. As it passed within a few million miles of Earth, radar signals from Arecibo, Puerto Rico, were bounced off of 1991jx. Again an eight-inch telescope was needed to see it. Asteroid 1991jx takes 3.47 years to circle the Sun.

1991jw. The asteroid 1991jw also was discovered near Earth in Spring 1991. It takes only 13 months to orbit the Sun.

1991ju. The Atens asteroid 1991ju was discovered near Earth in Spring 1991 when it was 62 million miles from the Sun. It requires 336 days to orbit the Sun.

1991jr. The Apollo asteroid 1991jr was discovered near Earth in Spring 1991. It requires 1.7 years to trek around the Sun.

1991jg1. The Apollo asteroid 1991jg1 was discovered by Eleanor Helin as it was near Earth May 9, 1991. Asteroid 1991jg1 approaches only as close as 104 million miles (1.12 astronomical units) from the Sun. It orbits the Sun once every 19 months.

1991da. Australian astronomers found the unusual asteroid 1991da near Earth February 18, 1991. It requires 27.5 years to trek around the Sun in an odd orbit ranging from the orbit of Mars to beyond the orbit of Saturn.

1990mu. Australian astronomers on August 17, 1990, found a quarter-mile-long asteroid after it had crossed Earth's orbital path around the Sun in June. Known as 1990mu, the asteroid came within 1.2 million miles of Earth. It was discovered by astronomers at Siding Spring Observatory near Coonabarabran, New South Wales. They say 1990mu may cross Earth's path again in two years, possibly closer to our planet. But the chances of 1990mu hitting Earth within the next million years are remote. About 70 asteroids have been found crossing Earth's path. Astronomers have calculated 1,000 asteroids exist in the Solar System, mostly between the orbits of Mars and Jupiter.

1990mf. Planet Earth had another close call with an asteroid. This one streaked by at the relatively-short distance of three million miles July 10, 1990. Known as 1990mf, the asteroid was a chunk of rock between 300 to 1,000 feet in diameter. It flashed past Earth at 22,000 mph. Asteroid 1990mf was eight million miles away and inbound toward Earth at 12,500 mph on June 26 when it was discovered by a team lead by NASA-JPL planetary scientist Eleanor Helin working at Palomar Observatory on Palomar Mountain, California. The object, from the Apollo group of asteroids, probably takes two years four months to make one complete trip around its 161-million-mile orbit. The asteroid received a gravity assist from our planet, speeding it up as it passed. Astronomers said 1990mf had a weak magnitude-15 gleam. That's 140,000 times fainter than the north star, Polaris. The smaller the magnitude number, the brighter the object. On a clear night, we see objects as faint as magnitude 5 or 6 with the naked eye. With binoculars we see objects as weak as magnitude 8 or 9. A six-inch backyard telescope will show objects to magnitude 13. The rocky body crossing Earth's orbit was tracked by Arecibo Observatory in Puerto Rico; NASA's Deep Space Tracking Network at Goldstone, California; Siding Spring, Australia; Mount John Observatory, New Zealand;

the Infrared Telescope Facility, Mauna Kea, Hawaii; and Harvard observing station, Oak Ridge, Massachusetts. Helin's JPL sky survey team has found 39 asteroids near Earth since 1973. Asteroid 1990mf was one of the closest in 50 years.

1990mb. An astronomer using an 18-inch Schmidt telescope on Palomar Mountain, California, found the Amor asteroid 1990mb on June 20, 1990. Amor asteroids cross the orbit of Mars, but not of Earth.

1989pb. Astronomers, judging from radar reflections made with the 1,000-ft. Arecibo, Puerto Rico, radiotelescope antenna, say the large asteroid 1989pb which just missed Earth by only three million miles August 24, 1989, has a shape like a dumbbell. The mile-wide mountain of rock floating through space has two big ends, each half a mile across, connected by a narrow neck. Asteroid 1989pb was one of the closest objects to pass Earth this century. One of the ten all-time closest approaches by an asteroid or comet, the August visit was the third such close encounter in 1989. The so-called Apollo asteroid or minor planet was visible in small telescopes as it swung around the Sun on October 27 at a distance of 51 million miles. Astronomers previously only had guessed that asteroids could be shaped like dumbbells. Asteroid 1989pb passed at 10 times the distance between the Earth and the Moon. Sailing through space at many thousands of miles per hour, the big rock presented no danger to Earth.

The asteroid was spotted first on August 9 by planetary scientist Eleanor Helin, of NASA's Jet Propulsion Laboratory, Pasadena, California, in a photography made early that morning with an 18-inch Schmidt telescope at Palomar Observatory northeast of San Diego. Helin told reporters so many asteroids near Earth is ominous, suggesting a large population of asteroids could be devastating. Robert Staehle, president of the World Space Foundation, South Pasadena, California, said an asteroid hitting Earth could be as destructive as a nuclear weapon. Astronomer Brian Marsden, Cambridge, Massachusetts, director of the Central Bureau for Astronomical Telegrams, an astronomy reporting agency of the International Astronomical Union's said he didn't want to cause worry, but anything crossing Earth's orbit could hit our planet, but that likely would be millions of years in the future.

1989ja. A two-mile-wide asteroid passed within eight million miles of Earth June 3, 1989, just two months after a closer call with a different asteroid which came within 450,000 miles of our planet March 23. The June asteroid, known as 1989ja, was between one and two miles in diameter. Speeding by at 20 miles per second, 1989ja was visible through powerful telescopes. The object looked like a faint star in the constellation Virgo. Astronomers had been watching 1989ja for two months as it approached Earth. There was no danger to Earth, according to a University of Florida, Gainesville, astronomer.

1989fc. Earth had a close call with a passing asteroid in March 1989. No one was aware of the minor planet 1989fc as it rushed by Earth at a distance of 465,000 miles March 23. Only twice as far away as the Moon, it may have been the closest approach by any comet or asteroid in 300 years. Six hours after it went by, Earth was where the asteroid had been. The last time any space object came so close was in 1937 when the asteroid Hermes rushed past Earth. Comet Lexell came within 1.5 million miles of Earth in 1770 and Comet IRAS-Araki-Alcock came within 3 million miles in 1983. After it passed Earth, the faint 17th magnitude chunk of rock was spotted by astronomers N. G. Thomas and H. E. Holt using the 18-in. Schmidt telescope on Palomar Mountain, California, March 31. It may have been slightly brighter, about 12th magnitude, as it swept past Earth eight days before.

3200 Phaethon. Asteroid 3200-Phaethon will not hit Earth in the year 2115. The asteroid, discovered in 1983 by IRAS, the Infrared Astronomy Satellite, is only a few miles in diameter. Its orbit is about the same as the Geminid Meteors which come by every year.

Geminid Meteors probably are dust from a vanished comet so Phaethon may be the naked nucleus left after the comet lost all of its outer shell. However, Phaethon has a rocky surface, not the crust of dust astronomers think a dead comet should have.

Phaethon will be unusually close to Earth that year as it swings along its orbital path. It may even provide some exciting sky watching. But it won't hit Earth. The world will not come to an end in 2115.

Asteroids And Dinosaurs

A pinprick of night light grew quickly into a fiery ball and, finally, executioner of the dinosaurs some 65 million years ago.

A five-mile-wide asteroid or comet chanced across Earth's orbit around the Sun, smashing into our planet. The impact blew vast clouds of dust and debris into the atmosphere. Sunlight was blocked from Earth's surface by the spreading cloud.

Overheated, starving dinosaurs gradually became extinct after the asteroid disaster turned on a planetary greenhouse effect.

Probably three-fourths of the world's animal species at that time died. Some scientists think the scenario could have happened, but where exactly did the asteroid hit?

A fat crater underneath Manson, Iowa, has been suggested. A big crater in the southern USSR may have resulted from the crash. A University of Capetown geologist has found a giant bowl-shaped depression at the bottom of the Indian Ocean east of Africa known as the Amirante Basin.

The Indian Ocean crater is exciting those geologists who say an asteroid did splash down in the Indian Ocean, because they found a thick coating of clay had blanketed Earth about 65 million years ago. The clay contains an unusual amount of iridium, a metal rare on planet Earth, but plentiful on meteors and asteroids.

Some scientists say they think the object from outer space dunked itself somewhere in an ocean because the clay is similar to ocean crust. Others think it smashed into land because the clay contains tiny bits of quartz and feldspar. These bits are largest in the clay in western United States and Canada so some geologists think the Iowa site most likely. A few suggest more than one impact.

Greenhouse effect. The asteroid started a devastating greenhouse effect in Earth's atmosphere, according to the theory of a geophysicist and a planetary scientist at California Institute of Technology whose studies indicate an asteroid crashed into a limestone layer, liberating a vast cloud of carbon dioxide gas from the smashed rocks into the atmosphere.

Carbon dioxide traps solar heat in Earth's atmosphere, like glass holding heat in a greenhouse. Warming of the atmosphere by increasing the concentration of carbon dioxide is known as the greenhouse effect.

A whopping increase in carbon dioxide 65 million years ago might have boosted temperatures dramatically across the whole Earth, with a catastrophic aftermath. Plants and animals dinosaurs ate might have been killed. Dinosaurs and many other species might have been killed slowly through dehydration and starvation.

The Cal Tech scientists shot steel bullets into rocks at 4,500 mph. Then they checked the amount of carbon dioxide released from the smashed rocks. Their calculations showed that, if an asteroid six miles in diameter hit limestone, it would release enough carbon dioxide to increase the amount of that gas in Earth's atmosphere by two to five times. That would warm our planet from 9 to 36 degrees Fahrenheit over 10,000 years after the asteroid struck.

Leafy plants which dinosaurs ate might have been destroyed selectively by the heat, while some plants could have survived. Competition for available food would have increased. Gradually, the supply of food would run out and some species, including dinosaurs, might become extinct. Such an extinction would not be sudden.

This theory of an asteroid whack leading to a carbon dioxide greenhouse effect differs from an earlier theory in which scientists said a smashing asteroid might have caused extinctions by kicking up enough dust to block sunlight. Dust would have blanketed Earth with a freezing darkness that might have killed food plants.

A University of California, Los Angeles, geochemist said the greenhouse theory probably is sound. But, a paleontologist at University of California, Berkeley, was skeptical of the greenhouse theory, calculating that dinosaurs started to die earlier than 65 million years ago. However, he called the study a good step because it attempted to explain why only some animals became extinct and why extinctions were gradual, not sudden.

Fossils found in recent centuries seem to indicate dinosaurs and some other creatures became extinct gradually, while many species survived. Many paleontologists and biologists have suggested an extinction 65 million years ago might have been caused by a gradual climate change, not an asteroid or big volcano eruption.

Other scientists like the theory that a greenhouse effect caused extinctions, but they say the excess carbon dioxide was generated by acid rain that dissolved carbon-rich rocks, or by changes in ocean circulation. A Dartmouth College geophysicist said geologic evidence shows volcanos spewed enough ash skyward to block sunlight and cause extinctions.

Will another big asteroid crash? Will natural changes in Earth's environment cause new extinctions? Scientists don't know, but they are excited by the prospect of finding out what killed off dinosaurs.

Many believe rising levels of carbon dioxide and other greenhouse gases will warm Earth by 3 to 5 degrees by the year 2050. That could bring droughts, threaten crops, and melt some of the polar ice caps to raise sea levels and threaten coastal cities.

Acid rain. Some scientists suggest acid rain wiped out dinosaurs and other animals after a giant meteorite struck Earth 65 million years ago.

The researchers, after spotting strontium in fossils, say the severe acid rain was as corrosive as battery acid. Such rainfall over two years might have led to extinctions, according to a Scripps Institution of Oceanography geochemist.

Meteorite shower. Scientists have been arguing about what caused mass extinctions. Space scientists and geologists say either one big meteorite or a shower of medium-size meteorites struck the Earth, kicking up a dust storm that blocked sunlight, turning the planet cold and dark, killing food plants. If not that, then intense heat from the meteorite plunging through the atmosphere led oxygen and nitrogen to react, forming nitrogen oxides. Nitrogen oxides may have mixed with water to make acid rain that killed many animals and destroyed the food of others.

Others like the blocked sunlight theory, but say gas and dust from a colossal volcano eruption did it.

Then there are biologists and fossil experts who look back and see a gradual evolutionary and climate change, not a sudden catastrophe.

The Scripps geochemist looked at the proportion of strontium-87 in fossilized shells of forams, tiny ocean-floating animals. Strontium-87 was highest during mass extinctions 65 million, 94 million and 225 million years ago, supporting the acid-rain theory. Strong acid rain would have eroded rocks rich in strontium-87. The erosion would have washed out to sea where it would have ended up in foram shells.

If acid rain scorched Earth, how did birds, snakes, fish, turtles, lizards, amphibians, even crocodiles come through unscathed? Maybe eggs in underground burrows were shielded from the corrosive downpour.

Of course, a meteorite striking Earth could have caused everything—cold, darkness, acid rain, warming of the atmosphere. Even if dinosaurs had started to dwindle and Earth was in a cooling period before a monster meteorite impact 65 million years ago, the collision still might have blasted up a spout of dust, spurring on the smog, acid rain and colder weather that could have been a final blow to some animals.

Fulgurite. A giant fulgurite found in Michigan is leading some researchers to challenge the idea that dinosaurs were killed when a colossal meteorite collided with Earth. Meteorites are rich in iridium which is rare on Earth. When geologists found layers of iridium on Earth, they took that as a sign that a large meteorite carrying iridium had smashed into Earth. Supposedly the meteorite crash threw up dust, clouding the atmosphere, making life impossible for many animals and plants.

But now they have found a really big fulgurite—usually a small glass tube formed when lightning strikes sand, rock or oil—or when a meteorite strikes Earth. Looking through a bunch of fulgurites in southeastern Michigan, geologists uncovered the biggest yet. It's one foot in diameter and sixteen feet long.

Analyzing it, scientists found unknown minerals and odd metals. They say it appears that hits by tiny meteorites can produce unusual metals.

They speculate the rich deposits of iridium previously attributed to one big meteor might have been created by an endless string of small meteorites falling on Earth over many eons. If they are right, a new cause will have to be deduced for the extinction of dinosaurs.

Mini-dinos. Miniature dinosaurs and other little animals may have flourished on Earth, while their more husky brethren were dying out in an asteroid disaster.

Scientists have unearthed fossils of an unknown, two-ft., 20-lb., grass-eating dinosaur which thrived in Wyoming 130 million years ago. The long-tailed dinosaur, nicknamed Little Big Foot, was formally labeled Drinker Nisti by a University of Colorado paleontologist who found it and three other unknown animals in the so-called Breakfast Bench fossil area.

The other petrified bones came from a neurologically-sophisticated turtle and two ferocious rodents weighing less than a pound each. They had lived there when Breakfast Bench was a swamp 130 million to 65 million years ago. There may have been several mass extinctions in ancient history in which many families of huge dinosaurs died out while small animals flourished, they say.

Astroblemes

An astrobleme is an ancient meteorite scar in Earth's crust. Known astroblemes are younger than ten million years.

One big astrobleme in South Africa is 25 miles across. Germany has one 15 miles in diameter. Others include a nine-miler in Saskatchewan, Canada; a six-miler in Ghana; and one four miles across at Serpent Mound, Ohio.

A big meteorite, weighing thousands of tons, isn't slowed much by Earth's atmosphere as it enters. It hits the surface at tens of thousands of miles per hour and explodes, dishing out a deep crater 20 to 60 times the size of the meteorite. Only a tiny fragment of the original rock remains buried in the crater wall.

Comets

A cloud of gas and dust at the edge of the Milky Way galaxy collapsed to form our Solar System 4.6 billion years ago.

One big clump of matter in that cloud was massive enough to start nuclear burning on its own, and now is our Sun at the center of our Solar System.

Other large clumps, not massive enough to collapse into stars, are the nine major planets: Mercury, Venus, Earth, Mars, Jupiter, Saturn, Uranus, Neptune and Pluto. Miscellaneous smaller clumps now are known as moons, asteroids and comets.

Today, we see comets as giant snowballs of dirty slush which scientists think have collected at the far outer edge of the Solar System. A ring of comets out there, known as the Oort Cloud, is 30,000 to 100,000 times farther away from the Sun than Earth.

Sometimes, gravity from a star in the Sun's neighborhood will nudge a comet, bumping it out of the Oort Cloud so that it falls inward toward the Sun at the center of the Solar System.

Most comets take a long period of time—up to thousands of years—to fly in and out of the Solar System. But, not too often, a comet zipping in toward the Sun from the Oort Cloud will pass close to a planet. Gravity from the planet may affect the comet's orbit. The comet might become trapped, at least temporarily, in the inner Solar System, reappearing in view of Earth in shorter periods of time.

Comet Tempel 2 is an example of such a short-period comet, orbiting the Sun every 5.5 years. Comet Tempel 2 travels from just inside the orbit of Jupiter to perihelion (its closest point to the Sun) near the orbit of Mars, 140 million miles from the Sun.

By comparison, the famous Comet Halley is a long-period comet which requires 76 years to swing out as far as Neptune and back in between Mercury and Venus.

Orbits of comets. A comet orbits the Sun just as planets do, but the comet's path is highly elliptical while planet orbits are more nearly circular.

We only notice a comet when it is on the portion of its orbital path near the Sun and within sight of Earth. At other times it is out of sight and may be farther from the Sun than any known planet.

A comet probably is a lightweight clump of slush, maybe as large as 1.5 million miles in diameter, with a tiny relatively-solid nucleus surrounded by a big cloud, a nebulous coma.

A long curved plume of vapor trails the coma when the comet is close to the Sun. The whole dirty snowball probably is water, carbon dioxide, methane and ammonia.

Weighing comet dust. The world's most sensitive scale will measure the weight of one particle of dust when NASA sends it out after a comet's tail in the early 1990's.

An instrument maker made the space scale now for NASA. It consists of a hollow tube which vibrates like a tuning fork. When a dust particle from a comet's tail lands on a weighing platform attached to the tube, the particle's weight slows down the vibrations. A computer uses the change in vibration to calculate the weight of the particle.

The scale, in the weightlessness of deep space, will measure what dust particles in a comet's tail would weigh if they were on Earth—to one-trillionth of a gram.

Astrophysicists think they will find clues to the origin of our Solar System in the weight of dust and ice particles shed by comets.

The instrument maker was tiny Rupprecht & Patashnick Co. in Albany, NY.

Are comets tainted? New comets may not be the virgin snow astronomers thought, according to scientists at the University of Colorado, Boulder.

Scientists have said for years that comets are dirty snowballs formed at the birth of the solar system. Astronomers study them for clues to construction of the Sun and planets.

Now the Colorado researchers say comet properties may have been refashioned over 4.5 billion years by heating from bright stars and supernova star explosions. A pristine comet, of the material that existed at the birth of the Solar System, may not have survived.

Trillions of comets probably orbit the Sun, but at distances 10,000 to 100,000 times as far away from the Sun as the Earth. Most have not been seen, but Earth observers sometimes do spot comets that have left the far-distant orbit to swing in toward the Sun.

While still in their distant orbits, comets may have been heated to boiling, erasing the original structure and chemical composition of their surfaces, according to the new speculation.

Skeptics point out that comets which regularly return near Earth have been heated and modified by trips near the Sun, but comets making their first trips to the inner Solar System have seemed unspoiled. The new conjecture is not likely to be tested until future flights out to comets measure their composition and ice structure directly.

Earth water from comets? Ultraviolet photographs of Earth shot by NASA's Dynamics Explorer 1 spacecraft from 1981 to 1986 turned up countless dark spots that looked like tiny holes in Earth's atmosphere.

A University of Iowa physicist, Louis Frank, in 1986 calculated that 10 million comets—mostly fluffy dark-coated 30-foot snowballs—have been plunging into Earth's atmosphere each year, forming holes before falling apart and vaporizing. In his theory, millions of small, dark, fast-moving, water-bearing comets strike Earth's atmosphere every year. He visualized an early-Earth history in which billions of the tiny comets provided enough water to fill all the oceans.

Other scientists have been skeptical, but now NASA physicist Clayne Yeates at Jet Propulsion Laboratory, Pasadena, California, says he has uncovered evidence supporting Frank. Yeates was deputy project scientist for the launch of Galileo to Jupiter.

Yeates used the 36-inch Spacewatch Telescope at Kitt Peak National Observatory in Arizona to convince himself a million or more tiny, dark comets speeding along at 22,000 mph hit Earth's atmosphere each year. The comets appear as white streaks on Yeates' enhanced telescope images.

Sweden's Viking spacecraft recently spotted the holes in the top of the atmosphere but some scientists remain disbelievers. They say something is there, but they may not be 30-ft. snowballs.

The comets orbit the Sun in the same direction as Earth, in Frank's view. They are very dark, absorbing 98 percent of the sunlight striking them and reflecting only two percent of the sunlight that hits them.

University of Michigan planetary scientist Thomas Donahue checked 1977 data from the U.S. Voyager 2 interplanetary probe and found Earth's atmosphere contained extra hydrogen atoms from water vapor. That could agree with Frank, but also could be explained if hydrogen were freed from many fewer comets that were smaller but faster and more active. Fewer comets wouldn't carry enough water to fill oceans.

★Why haven't astronomers on Earth's surface seen the comets? Only two special telescopes in the U.S., including Spacewatch, are capable of observing the comets.

★Why haven't the comets hit artificial satellites orbiting Earth? Despite the large number of comets, there is so much space around Earth that the odds indicate a comet would strike any single man-made spacecraft only once every 100 million years.

★The Moon has little water vapor. Why don't the comets also strike the Moon, creating a lot of water there? Supporters of the theory haven't explained that one yet.

Comet watching. Finding a never-before-seen comet can be the highlight of an astronomer's life. Six or eight new comets are discovered in an average year. Some watchers hunt a lifetime without a discovery. In 1987, a record 16 previously-unknown comets were discovered, mostly by amateur astronomers.

Patience is required during a lot of hours looking at background star patterns, watching for one object to move oddly. That's how an Australian had found 13 new comets by 1987. William Bradfield made his thirteenth discovery of a comet on August 11, 1987. The very-faint magnitude-9 comet was dubbed by astronomers as 1987s or Comet Bradfield.

The smaller the magnitude number, the brighter the object. Objects as weak as magnitude-5 sometimes are seen through binoculars or even by naked eye. The few bright naked-eye stars seen from Earth are around magnitude 1.

William Bradfield's 13 comets means he has more than any other living comet hunter.

Other comet hunters. On June 14, 1991, David Levy, a Tucson, Arizona, astronomer discovered his seventh comet in seven years after going 19 years without spotting one. Because of persistence, Levy has more new comet sightings than anyone.

Levy, who spends one to four hours each clear night scanning the sky with one of his several telescopes, has been searching for 26 years. After 19 years without a find, he discovered Comet Levy-Rudenko in 1984. In January 1987, he spotted the first new find of that year, 1987a. Levy discovered his second comet of 1987 in October. Identified as 1987y, the comet had no tail and apparently was travelling away from the Sun.

Levy built a 6 ft. x 8 ft. wooden platform atop his home for a better view of the horizon. Three days after building the platform, he spotted his third comet. He found his fourth just above the horizon an hour before dawn in March 1988 using a 16-inch telescope.

An American amateur astronomer spotted a previously unknown comet in 1987. Michael Rudenko, using a six-in. reflector telescope in Massachusetts, found 1987u or Comet Rudenko, about magnitude 9 and 10.

Halley's Comet

Comet Halley visits the inner Solar System on its way around the Sun every 76 years. The last time it was in here with Earth was in 1985-86. Various nations sent deep-space probes for a close-at-hand look at Halley.

Probes. In March 1986 six spacecraft from Earth—Giotto, Vega 1, Vega 2, Suisei, Sakigake and ICE—encountered Halley at various fly-by distances on different days.

Giotto, sent by the European Space Agency (ESA), was named after Italian painter Giotto Di Bondone who depicted Halley's Comet in A.D. 1304 as the Star of Bethlehem in the Scrovegni Chapel in Padua.

Vega 1 and Vega 2 were sent by the USSR's Intercosmos agency in Moscow. Suisei and Sakigake were from the Japanese Institute of Space and Astronautical Science (ISAS). The International Cometary Explorer (ICE) was from NASA.

Earthbound observations by 1,000 professional astronomers and 1,200 amateur astronomers were coordinated by the International Halley Watch (IHW) in eight observing networks. IHW started collecting data in 1983 when Halley's Comet was found again, and continued until 1989.

Chunk explodes. A chunk of rocky slush weighing 20 million pounds was blown away from Halley's Comet March 18, 1986, in a mysterious explosion.

Strong energy from the nearby Sun, rare in the comet's 76-year elliptical swing around the Sun, probably heated carbon monoxide and carbon dioxide in Halley's core. Both kinds of particles are more volatile than water and can explode if heated suddenly.

The comet was traveling away from the Sun at a distance of about 90 million miles when the blast occurred.

Although ten thousand tons of matter isn't much from a comet weighing 100 billion tons, it may have left a scar. Astronomers will be on the lookout for the mark when Halley's Comet returns to the inner area of the Solar System, near the Sun and Earth, in 2061.

End Of The Trail? After another 2,000 years or so, when it has passed near Earth about 30 more times, Comet Halley will have melted away to almost nothing.

Some of it melts away on each trip through the inner Solar System. By the year 4000, all that will be left of the comet will be a dark inactive nucleus, according to USSR astronomers reading data collected by ultraviolet telescopes aboard the Astron satellite.

Ultraviolet light data collected and sent down by Astron was combined with information sent back to Earth from the probes Vega and Giotto to give an estimate of the amount of matter lost by the comet in its most recent turn around the Sun.

1991 sighting. Professional astronomers with powerful telescopes tracked Comet Halley long after most Earth observers had given up. In 1991, the comet was floating along, two billion miles away from the Sun, beyond the orbit of Saturn on its way out to or beyond the orbit of Uranus. Something caused Halley to flare in brightness briefly in February 1991, which let astronomers at the European Southern Observatory in Chile spot it. It had been magnitude 25, but brightened temporarily to magnitude 22. The comet's coma was more than 100,000 miles long.

Halley's formaldehyde. Could Halley's Comet be older than the Solar System? The gas halo around Halley's Comet contains tiny chains of formaldehyde molecules that may be older than our Solar System, according to evidence gathered by Europe's Giotto spacecraft as it flew near the comet in March 1986.

Scientists had thought comets were remnants of the original gas and dust cloud, parts of which compacted to become our Sun and its planets. But the chains of formaldehyde molecules in Halley suggest the comet formed before the Sun appeared.

That was big news to the researchers who analyzed data Giotto collected on the mass of ions, or charged particles, in the gas around the hard core of the comet. The pattern they saw suggested the presence of formaldehyde chains called polymers.

Formaldehyde is made of oxygen, hydrogen and carbon. Those basic elements are among the most abundant in space.

Formaldehyde wasn't totally unexpected. Some astronomers earlier had uncovered evidence of formaldehyde chains in deep-space clouds.

Did the Sun have a companion? An unusual concentration of chemicals noticed in Halley's Comet as it passed Earth in 1986 has suggested a companion star may have exploded near the Sun during formation of our Solar System 4.6 billion years ago.

Spectrum analysis of comet elements revealed the proportion of various carbon isotopes was unlike anything recorded before in the Solar System. That suggested an ancient star near the Sun might have exploded into a supernova, contaminating the cloud of gas and dust which later condensed into planets around the Sun.

Star cluster? Such a supernova could mean the Sun once was not alone, but part of a pair or cluster of three or more stars, according to astronomers at Arizona State University. Many distant stars across our Milky Way galaxy formed in pairs and clusters.

If a star exploded near the Sun, new chemical elements would have been formed in the nuclear fireball and spewed out into space to contaminate the cloud of dust and gas from which Earth and other Solar System bodies formed, leaving traces of isotopes from the supernova to be found in Solar System objects.

Comet Halley might have escaped the contamination if it formed in a faraway region now called the Kuiper Belt, 100 times farther from the Sun than the outermost planet Pluto.

65:1 ratio. The idea of a supernova during Solar System birth is based on a difference astronomers found in the proportion of the isotope carbon 12 to the isotope carbon 13. They previously had measured 89 parts of carbon 12 to one part carbon 13 in Earth, Moon samples, meteorites and atmospheres of Jupiter and Saturn. By comparison, the Comet Halley ratio was 65 parts of carbon 12 to one part carbon 13.

The 65-parts measurement has only ever been found in Comet Halley, while the 89-parts ratio is found in all other objects examined. No other comets have been checked.

Other than a companion star exploding, what could have caused the difference in Comet Halley? The difference might be explained if Comet Halley formed outside the Solar System and later became trapped in an orbit around the Sun through some gravity disturbance. Astronomers don't like that argument because of vast distances between the Sun and other stars. The Arizona theory was to be tested in Summer 1989 as Comet Brorsen-Metcalf passed near by the Sun.

Chiron: Asteroid Or Comet?

Chiron is a small but significant body orbiting the Sun at a great distance, beyond Saturn, but not as far as Uranus. Astronomers list it as an asteroid, but is it a comet?

About 112 miles in diameter, Chiron has been looking more like a comet since

1988. In observations through the 3.6-meter Canada-France-Hawaii optical and infrared telescope atop Mauna Kea mountain in Hawaii December 27, 1989, Chiron seemed to have a faint coma.

Asteroids. Minor planets are asteroids and planetoids, small rough rocky bodies orbiting the Sun. Some 95 percent of all known minor planets are in an orbit known as the Asteroid Belt between the orbits of the major planets Mars and Jupiter.

There may be more than 100,000 asteroids, but only 3,000 have been catalogued. Ceres is the largest. They probably are not the remains of an exploded planet as, together, they total only about four-ten-thousandths the mass of Earth. More likely, they are debris from collisions of several small bodies between Mars and Jupiter at the time the Solar System was forming around 4.6 billion years ago.

Astronomers looking through telescopes see the asteroids shine as they reflect sunlight. Vesta is the only asteroid bright enough to be seen by naked eye.

No. 2060. Chiron was discovered in 1977 by U.S. astronomer Charles T. Kowal. It was designated officially minor planet number 2060. After its discovery, astronomers looked back and found Chiron in photos dating to 1895.

Whatever it is, the dark-surfaced Chiron is in an orbit different from any other known asteroid. It takes around 50 years to circle the Sun. Some say it may pass very close to Saturn in about 10,000 years.

Chiron may be the brightest member of an otherwise-unseen swarm of minor planets...or is it a comet? If a comet, it must be bigger than any previously known. It's about 250 times as bright as Comet Halley was at the same distance. It even may be that both Chiron and the major planet Pluto once were moons of the major planet Neptune, wrenched away by the cataclysmic passing of some massive object in the ancient past.

Largest Solar System Structure

A 20-million-mile-long hazy cloud of sodium atoms, stretching 20 million miles away from the planet Jupiter and its moon Io, is the largest permanently-visible structure in the Solar System. The sodium atoms probably are blasted outward from volcanos erupting on Io. They collide with electrically-charged particles in Jupiter's powerful magnetic field, which squirts them in a fast narrow jet into space.

Described as a "magneto-nebula" by the team of Boston University astronomers who found it, the cloud is 400 times the radius of the planet Jupiter. They studied and photographed the nebula from November 1989 through February 1990 using a telescope at McDonald Observatory, Texas.

Astronomers have known since 1975 that Jupiter and one of its moons, Io, are enveloped by a cloud of sodium. The size of the cloud had not been measured and it was thought to be close to the giant gas-bag planet. Boston astronomers Michael Mendillo and the Jeffrey Baumgardner were startled to detect the sodium atoms far from Jupiter.

Neon In Space

Scientists were excited in 1973 when parts of helium, nitrogen and oxygen atoms were discovered by the Interplanetary Monitoring Platform orbiting Earth, and by the interplanetary Pioneer 10. The Voyager twins in 1985 and 1986 counted hydrogen, helium, neon, nitrogen, and oxygen atomic particles, even a few particles of carbon and argon. Carbon is abundant in nature so researchers were puzzled about why Voyagers found only a few carbon particles. Then they realized the Sun's magnetic field was repelling carbon ions. Only after Voyagers leave the Sun behind will they count more carbon particles. Voyagers are rushing out of the Solar System to encounter whatever is between stars in interstellar space.

Interplanetary Spacecraft

If we can't yet travel far out of our own Solar System, we certainly can explore where we are. Earthmen since the late 1950's have hurled more than 100 automated spacecraft into the vast emptiness between planets of our Solar System. These sophisticated robots have been the first to explore the Moon, the planets, and the Sun.

The 1990's are even more exciting for space watchers with space stations growing in Earth orbit and exploring machines on the way to the Sun, Mars, the Asteroid Belt, Jupiter, Saturn and comets. Here's where we're going:

The Sun. The Ulysses spacecraft, designed by the European Space Agency to study the Sun at close range, flying over the Sun's poles in 1994-95.

Venus. The Magellan spacecraft is in orbit around Venus, using radar to create maps of the planet surface.

Japan plans to launch the Planet B spacecraft toward Venus in March 1996 atop an improved Japanese solid-fuel space rocket. Planet B will carry out scientific studies.

The U.S. Pioneer 12, launched in 1978, may continue to send back radar images of Venus beyond 1992.

The Moon. Japan sent the Muses-A science explorer, later renamed Hiten, around the Moon in 1990, dropping off the miniature lunar satellite Hagoromo as it swung by. The dual satellite was practice for future interplanetary spaceflights. Japan plans to send a large satellite to orbit the Moon around 1996.

The USSR, the U.S. and other countries, may send cosmonauts and astronauts to land on the Moon and establish a base camp by the end of the 1990's.

Mars. The last unmanned landings on Mars were made in 1976 by the U.S. Viking spacecraft. The U.S. Mars Observer was to be launched in 1992 to orbit the Red Planet 1993-94, sending back photos from 227 miles above the surface. Unlike the earlier Vikings, Mars Observer will not have a lander.

The USSR will send a Mars orbiter with small landing probes in 1994. Then, it will send a large Mars lander, with rover vehicle and roving balloon, in 1996. Rock-sample return flights could be made between 1998 and 2000.

The U.S. may send a Mars Observer 2 with a lander after 1998. It might land, and pick up and return rock samples to Earth.

All of the unmanned Mars orbiters and landers from the USSR and the U.S. would be in preparation for a manned flight from Earth to Mars and back about the year 2005.

Asteroids. All NASA flights to the outer Solar System check out the Asteroid Belt along the way. The Mariner Mark II spacecraft on the Comet Rendezvous Asteroid Flyby (CRAF) mission would have a camera and science instruments if it leaves Earth in the 1990's in search of clues to the origins of our Universe. CRAF would go beyond the orbit of Mars to that dangerous area of big rock chunks known as the Asteroid Belt. Asteroids are rocks, boulders and tiny planets travelling together in orbit around the Sun, filling a ring between the orbits of Mars and Jupiter. The Belt includes about 1500 rocky objects ranging in size from boulders up to minor planets of about 300 miles diameter. Under control from Earth, the CRAF spacecraft would fly by an asteroid for a close look. Then it would fly on to probe a comet.

The Galileo and Cassini spacecraft will fly by asteroids on their way to Jupiter and Saturn. USSR Mars probes after 1998 could visit the Red Planet, then fly on for closer examination of asteroids after the year 2000.

Jupiter. Galileo is on a six-year course to Jupiter. It will orbit the planet in 1995, dropping a science probe on a parachute into the colossal planet's stormy atmosphere.

The Ulysses spacecraft, designed by the European Space Agency to study the Sun at close range and launched by U.S. shuttle in 1990, will whip by Jupiter on its way to polar orbit above the Sun.

The Cassini spacecraft, launched by the U.S. and the European Space Agency in

1996, also will fly by Jupiter in 1999 on its way to Saturn.

Saturn. NASA and the European Space Agency plan to fire the interplanetary probe Cassini in 1996 toward Saturn and its natural satellite Titan, the solar system's biggest moon and the only one with an atmosphere much like ancient Earth.

Cassini will fly by the asteroid 66 Maja in 1997 and the planet Jupiter in 1999 on its way to an October 2002 rendezvous with the planet Saturn. At Saturn, Cassini will drop an ESA probe, named after Christian Huygens, the Dutch astronomer who discovered Titan and the rings of Saturn, into Titan's atmosphere.

Neptune. No new Neptune probes are planned, following Voyager 2's spectacular success there August 24, 1989. Voyager 2 now is headed out of the Solar System.

Pluto. There are no plans to send a spacecraft to Pluto, which never has been visited by an interplanetary probe from Earth.

Planet X. Some astronomers suspect beyond Pluto there may be another planet yet to be discovered. If so, that far-away body would be a tenth major planet in our Solar System. There are no plans to send a spacecraft to search for the so-called Planet X, however Pioneers 10 and 11 and Voyagers 1 and 2 are on their way out of the Solar System headed for interstellar space. Their instruments might reveal a Planet X.

Comets. After passing through the Asteroid Belt, the probe CRAF would travel on out to rendezvous with a comet. It would slow down and fly alongside the comet for three years. Monitors in the spacecraft would analyze dust, gas and plasma around the comet as it nears the Sun and its tail grows. Information would be sent by radio to Earth. About midway through their tandem travel, CRAF would be commanded to fire a probe filled with scientific instruments three feet deep into the nucleus of the comet.

Space telescope. Two of NASA's four Great Observatories already are in orbit: the 2.4-meter Hubble Space Telescope and Gamma Ray Observatory (GRO). Major astronomy findings are likely in the 1990's.

Space shuttle. The U.S. will make 50-100 shuttle flights to Earth orbit through the 1990's. Twenty after mid-decade will carry parts for the U.S.-international space station Freedom to a low Earth orbit.

The USSR will fly its shuttles Buran. The Europeans and Japanese will build their own shuttles.

Space station. In the 1990's, the U.S. and other nations are likely to join the USSR as countries with space stations orbiting Earth. The USSR has had space stations since 1971. Its third-generation Mir station will continue in use.

First Explorations Of The Planets

Unmanned interplanetary spacecraft from Earth have explored the Sun, the Moon, the Asteroid Belt and seven of the eight planets in our Solar System since 1959. The very first explorations were fly-bys. Spacecraft orbiting the planets and landings on the planets came later.

Flew by	Year	From	Craft
Sun	1960	USA	Pioneer 5
Mercury	1974	USA	Mariner 10
Venus	1961	USSR	Venera 1
The Moon	1959	USSR	Luna 1
Mars	1965	USA	Mariner 4
Asteroid Belt	1973	USA	Pioneer 10
Jupiter	1973	USA	Pioneer 10
Saturn	1979	USA	Pioneer 11
Uranus	1986	USA	Voyager 2
Neptune	1989	USA	Voyager 2
Pluto	has not been explored		

Orbited	Year	From	Craft
Sun	1960	USA	Pioneer 5
Mercury	an orbiter has not been sent		
Venus	1967	USSR	Venera 4
The Moon	1966	USSR	Luna 10
Mars	1971	USA	Mariner 9
Asteroid Belt	an orbiter has not been sent		
Jupiter	an orbiter has been sent, due 1995		
Saturn	an orbiter is to be sent in 1996 to arrive 2002		
Uranus	an orbiter has not been sent		
Neptune	an orbiter has not been sent		
Pluto	an orbiter has not been sent		

Landed on	Year	From	Craft
Sun	a lander has not been sent		
Mercury	a lander has not been sent		
Venus	1965	USSR	Venera 3
The Moon	1959	USSR	Luna 2
Mars	1976	USA	Viking 1 & 2
Asteroid Belt	a lander has not been sent		
Jupiter	a lander has been sent, due 1995		
Saturn	a lander is to be sent in 1996 to arrive 2002		
Uranus	a lander has not been sent		
Neptune	a lander has not been sent		
Pluto	a lander has not been sent		

Moon Probes: USSR Luna & Zond

Thirty USSR Luna and Zond spacecraft explored the Moon between 1959 and 1976.

Luna. The 24 craft in the Luna, or Lunik, series were unmanned probes returning photos and information on Earth's natural satellite.

The Luna program included many firsts: Luna 1 in 1959 was the first-ever Moon probe from Earth. It missed the Moon by only 3,700 miles. Luna 2 in 1959 was the first Earth probe to impact on the Moon. The USSR intentionally crashed Luna 2 into the Moon to gain information on Earth's natural satellite. It was the first time Man had landed anything on the Moon. The wreckage still is there.

Luna 3 in 1959 made the first photographs of the Moon's farside. Luna 9 in 1966 made the first soft landing on the Moon. Luna 10 in 1966 was the first Moon orbiter. Luna 16 in 1970 returned a soil sample to Earth. The Lunokhod-1 rover vehicle flew to the Moon in Luna 17 in 1970.

Zond. Eight Zond launches from 1964 to 1970 included one to Venus, two to Mars and five flights preparing for manned Moon trips, including three flights around the Moon which returned to land on Earth. Zond 1 flew to Venus and Zonds 2 and 3 were Mars flights. Of the eight Zond shots, Zonds 3 through 8 explored the Moon. Moscow announced that Zonds 4 through 6 in 1968 were preparations for a manned flight to the Moon. However, problems developing more-powerful rockets in that era prevented the final step of sending men around the Moon after Zond 8. The U.S. landed the first men on the Moon in 1969.

Probe	Launch	Results
Luna 1	1959 Jan 2	missed Moon by only 3,700 miles; first of 24 Luna or Lunik flights to the Moon.
Luna 2	1959 Sep 12	the first Earth probe to impact on the Moon.
Luna 3	1959 Oct 4	the first photographs of the Moon's farside.
Luna 4	1963 Apr 2	missed the Moon by 5,300 miles.
Zond 1	1964 Apr 2	first of 8 Zond launches 1964-70 including 1 Venus, 2 Mars and 5 flights preparing for manned Moon trips including 3 around Moon

		which returned to land on Earth; Zond 1 was not a Moon flight; flew within 60,000 mi. of planet Venus.
Zond 2	1964 Nov 30	flight within 1,000 miles of planet Mars.
Luna 5	1965 May 9	failed soft landing, impacted on the Moon.
Luna 6	1965 Jun 8	missed the Moon by 100,000 miles.
Zond 3	1965 Jul 18	Mars flight flew by Moon enroute at 5,717 miles, sent 25 photos of far side of Moon.
Luna 7	1965 Oct 4	failed soft landing, impacted on the Moon.
Luna 8	1965 Dec 3	impacted, failed soft landing.
Luna 9	1966 Jan 31	first soft landing on the Moon, 27 photos.
Luna 10	1966 Mar 31	first Moon orbiter, sent data two months.
Luna 11	1966 Aug 24	transmitted data to Earth from Moon orbit.
Luna 12	1966 Oct 22	Moon orbit, sent high-resolution photos.
Luna 13	1966 Dec 21	soft landing, sent photos, tested soil sample.
Zond 4	1968 Mar 2	test launch away from Moon.
Luna 14	1968 Apr 7	measured gravity from Moon orbit.
Zond 5	1968 Sep 14	a biosat with turtles, worms, flies, bacteria, spiderwort plant and seeds of wheat, pine and barley; around Moon at 1,212 mi., far side photos; night return splash down in Indian Ocean, first USSR water landing.
Zond 6	1968 Nov 10	biosat with micrometeorite and cosmic ray detectors, around Moon at 1,398 miles, stereo photos of surface; skip re-entry, landed in USSR.
Luna 15	1969 Jul 13	orbited, impacted.
Zond 7	1969 Aug 7	like Zond 6, color photos of Moon & Earth.
Luna 16	1970 Sep 12	soft landing, returned soil sample to Earth.
Zond 8	1970 Oct 20	last Zond flight circled Moon at 684 miles; color photos; first TV of Earth from 40,000 miles; North Pole re-entry, no skip, Indian Ocean night splash down.
Luna 17	1970 Nov 10	soft landing, Lunokhod-1 rover vehicle.
Luna 18	1971 Sep 2	orbited, impacted.
Luna 19	1971 Sep 28	orbited, transmitted photos to Earth.
Luna 20	1972 Feb 14	soft landing, returned soil sample to Earth.
Luna 21	1973 Jan 8	soft landing, Lunokhod-2 rover vehicle.
Luna 22	1974 May 29	orbited, transmitted photos to Earth.
Luna 23	1974 Oct 28	soft landing, sample return failed.
Luna 24	1976 Aug 9	soft landing, returned soil sample to Earth.

Moon Probes: U.S. Ranger

Ranger, the first NASA space program to investigate another Solar System body—the Moon—was a series of probes in the early 1960's.

At first, Rangers were to bounce onto the lunar surface with a seismometer, but the first six failed. After that, they were redesigned for photography. The first five, launched between August 1961 and October 1962, failed. Ranger 6, launched January 30, 1964, reached its target, but its cameras broke down.

Ranger was redesigned completely. Pictures were to be snapped as a craft approached its bulls-eye on the Moon. The last three Rangers, thus, became successful. They took the first close-ups of the Moon surface, showing boulders and three-foot craters.

Ranger 7. Ranger 7, the first success, was launched July 28, 1964. It impacted in an unnamed mare area later named Mare Cognitum, which meant "the sea which has become known." The level mare without craters was chosen because it might make a good landing spot for later Apollo manned flights. More than 400 photos from six sequential TV cameras were transmitted to Earth. The widest angle covered a square mile and showed 30-ft. craters. The most detailed pictures revealed 3-ft. craters in an area 100 ft. by 160 ft. Ranger 7 sent 4,300 photos.

Ranger 8. Ranger 8 was fired February 17, 1965. Its bulls-eye was in the southwest part of Mare Tranquillitatis. Ranger 8 transmitted 7,100 photos.

Ranger 9. Ranger 9, which blasted-off March 21, 1965, was targeted on the floor of the Alphonsus Crater. It landed there, between the central peaks and rilles near the crater's east wall. Ranger 9 returned 5,800 photos.

The Rangers showed that mare were free of small craters, cracks and rock fields, and would support the weight of a manned spacecraft. They accurately measured the Moon's radius and mass.

Moon Probes: U.S. Lunar Orbiter

Lunar Orbiter was a U.S. series of five satellites launched between Aug 1966 and Aug 1967 to orbit and map the Moon. Lunar Orbiters 1, 2 and 3 sought Apollo landing sites, while Lunar Orbiters 4 and 5 did global mapping of the lunar landscape. The Lunar Orbiter series discovered excess concentrations of mass under the maria, known as mascons, and photographed 99 percent of the lunar surface.

Year	Date	Spacecraft	Description
1966	Aug 10	Lunar Orbiter 1	First of 5 satellites launched to orbit and map the Moon in 1966-67. L.O. 1 was the first U.S. craft to circle the Moon. After sending 211 high resolution photos, L.O. 1 was crashed into surface to avoid conflict with L.O. 2.
1966	Nov 6	Lunar Orbiter 2	Second. Sent photos of 13 possible Apollo landing sites. L.O. 2 was crashed into surface to avoid L.O. 3.
1967	Feb 4	Lunar Orbiter 3	Third. After sending Apollo landing sites photos, L.O. 3 was crashed into surface to avoid L.O. 4.
1967	May 4	Lunar Orbiter 4	Fourth. Made first photos of Moon's south pole. Measured gravity, radiation. After sending Apollo landing sites photos, L.O. 4 was crashed into surface to avoid L.O. 5.
1967	Aug 1	Lunar Orbiter 5	Fifth. First photos of Moon's south pole. Photos and measurements of both sides of the Moon. Sent 212 photos, then crashed into surface.

Moon Probes: U.S. Surveyor

Surveyor was a series of seven U.S. unmanned TV and trenching probes designed to land softly on the Moon, launched between May 1966 and January 1968.

The seven Surveyors were part of America's search for Apollo moonship landing sites. Five were successful in soft landings.

Surveyors had a moveable TV camera powered by solar cells and a trench digger along with the capability to retrieve and analyze a soil sample.

In the spring of 1966, Surveyor 1 made the first soft landing on the Moon for man's first close-up look at the surface. The three-legged spacecraft flew 240,000 miles to land within nine miles of target. Over eight months it transmitted 11,150 pictures, including panoramas and close-ups.

Surveyors found lunar maria to be basaltic and the lunar highlands rich in aluminum and calcium.

Within two years, four other Surveyors landed to study Moon chemistry and physics. Information retrieved by the five Surveyors was useful in the manned Apollo flights from 1969-72.

Probe	Launch		Destination
Surveyor 1	1966	May 30	Oceanus Procellarum, soft landing, 11,150 photos transmitted to Earth.
Surveyor 2	1966	Sep 20	Oceanus Procellarum, crash landing.

Surveyor 3	1967	Apr 17	Oceanus Procellarum, soft landing, tested soil, 6,315 photos sent to Earth, Apollo 12 visited later.
Surveyor 4	1967	Jul 14	Sinus Medii, soft landing, radio contact lost.
Surveyor 5	1967	Sep 8	Mare Tranquillitatis, soft landing, analyzed soil, 18,006 photos.
Surveyor 6	1967	Nov 7	Sinus Medii, soft landing, analyzed soil, 30,000 photos.
Surveyor 7	1968	Jan 7	near Tycho, soft landing, analyzed soil, 21,000 photos.

Moon Probes: Japan's Hiten

Japan, the fourth nation ever to send a satellite to Earth orbit, became the third nation ever to send a spacecraft to the Moon when its Muses-A probe blasted off from Japan's Kagoshima Space Center on the southern island of Kyushu January 24, 1990.

After the spectacular night launch, the robot science explorer was renamed Hiten. It looped out from Earth and around the Moon. The 430-lb. mothership dropped off a miniature 26-lb. lunar satellite, Hagoromo, as it swung by the Moon in March 1990.

Japan fired the Muses-A dual satellite to practice for future interplanetary spaceflights including a bigger Moon orbiter and a Venus probe in 1996.

Hiten. The main spacecraft Hiten was a 430-lb., 5-ft.-diameter, 3-ft.-tall cylinder. Detachable from one end was the 26-lb. 1-ft.-tall miniature satellite Hagoromo.

Aboard the mothership was a West German micrometeorite counter from the Munich Technical University which recorded the weight, speed and direction of dust particles striking the craft. Both the large and small orbiters were built by NEC Corp., a major Japanese computer company, with government funding.

MU-3S-2 rocket. Muses-A rode to space atop an MU-3S-2 rocket built with government funds by car-maker Nissan Motor Co., Japan's second-largest automobile maker. Nissan's aerospace division dates to 1953 and has had an important role in developing Japan's solid-fuel rocket engines. Development of the 92 ft. long, 5.5 ft. wide MU-3S-2 rocket began in 1980. It first flew in 1985. The three-stage, 62-ton MU-3S-2, not a particularly powerful rocket, was just strong enough to lift the 430-pound Muses-A payload.

Launch. There wasn't much publicity in Japan—the seemingly-momentous countdown wasn't live on television—however reporters flocked to a hillside some miles from the oceanside launch site to watch the blast off from Kagoshima Space Center.

Unfortunately, the countdown stopped at T-minus-18 seconds when an electrical switching problem cut off power to a hydraulic pump aiming the nozzle of an auxiliary booster rocket. It was the first time in five launches of the slender solid-fuel MU-3S-2 a countdown had to be stopped in the last 60 seconds.

Winds had blown volcanic dust from the 3,668-ft. mountain, On-take, a nearby live volcano also known as Sakurajima. The grit blanketed much of the launch area, but engineers said the dust didn't cause the electrical problem.

Technicians, rushing that night to fix the red-and-silver solid-fuel space booster on Kagoshima's only launch pad, shivered as heaters in sheds and blockhouses were turned off to save electricity for launch equipment. Volunteers from nearby towns stood watch in red-and-black firefighter uniforms.

Overcast skies blanketed the remote site and snow fell nearby as the rocket was ready to fly again. Amidst billowing clouds of smoke, it bounded into the night sky over the Pacific from its oceanside pad nestled between mountains. Minutes later, the satellite separated from the rocket and was in a 250-mi.-high Earth orbit on an eight-week rendezvous with the Moon. After two circuits around the globe, solid-fuel rockets pushed the renamed Hiten out into a long oval orbit. NASA's Deep Space Network antennas at

Tidbinbilla, Australia, and Goldstone, California helped the Japanese track Muses-A. They showed the satellite on its first day in space orbiting Earth out to a distance of 186,000 miles.

The Moon orbits Earth every 27.322 days at an average distance of 238,861 miles. To navigate its course, the spacecraft looked at bright stars and the Moon's edge.

Hiten was in an elliptical orbit swinging farther out from Earth, coming closer to the Moon as time passed. The mothership arrived at the Moon on schedule March 19, 1990, initiating Japan into the exclusive club of nations having spacecraft circling Earth's natural satellite. Only the U.S. and the USSR have done it before.

Traveling at 2,237 mph on an oblong path around Earth, Hiten reached two goals: it did a swing-by of the Moon, using lunar gravity to boost its speed and enlarge its long elliptical orbit around Earth. And it disgorged the smaller basketball-sized satellite, Hagoromo, into orbit around the Moon.

A swing-by uses gravity to accelerate or decelerate a spacecraft's speed. The familiar technique has been employed by a number of American deep space probes.

Release of the 26-lb. satellite into orbit around the Moon showed great accuracy on the part of Japan's flight engineers. Kuninori Uesugi, chief scientist at Japan's Institute of Space and Astronautical Science, put it in baseball terms when he told reporters the success was like hitting the eyeball of a bug in the outfield from home base.

Hiten came within 9,100 miles of the Moon that day. It then continued to loop around the Moon and Earth eight times, four to speed up and four to slow down. Hiten means space flyer. The small satellite will remain in orbit around the Moon. Neither Hiten nor Hagoromo were meant to land on the Moon.

The polyhedral-shaped tiny lunar orbiter is covered with solar cells to gather energy from the Sun and generate electricity. A tiny cross-shaped antenna allows direct radio contact with controllers on Earth. The orbiter carried 11 of its 26 lbs. of weight in the "retro" motor it fired to move into orbit around the Moon. The small lunar satellite was designed to record temperatures and electrical fields around the Moon and radio the data to the mothership for relay to Earth.

Dead radio. A broken transistor radio caused Japan's space scientists to lose track of Hagoromo March 19. Apparently Hagoromo's tiny rocket motor had fired on schedule to pull away from Hiten while the mothership was 12,500 miles from the Moon that day, but the small satellite's tracking transmitter failed immediately, leaving a record of rocket firing but no signal to track the location of the satellite. Fortunately, Japanese astronomers using large optical telescopes later were able to see Hagoromo orbiting the Moon.

Kuninori Uesugi, chief scientist at Japan's Institute of Space and Astronautical Science (ISAS), said a transistor in the satellite's radio must have failed.

Later on. Hiten continued on its long oval orbit, encompassing Earth and the Moon, collecting data on space dust and other phenomena.

By its fifth fly-by of the Moon on August 6, 1990, Hiten was travelling out more than 600,000 miles from Earth.

Japanese scientists expected to maintain that 600,000-mi. apogee distance for the rest of the spacecraft's life. Among other jobs, Hiten was a test flight for a future mission called Geotail, part of an international Sun-Earth physics research project. Geotail will fly a long elliptical orbit.

ISAS. The Muses-A/Hiten/Hagoromo project was sponsored by Japan's government-funded Institute of Space and Astronautical Science (ISAS). Affiliated with the Education Ministry, ISAS focuses on science research in space.

Japan's other space agency, the National Space Development Agency (NASDA), focuses on applied uses of space. NASDA is developing a small space shuttle and a powerful all-Japanese H-2 rocket to lift payloads comparable to those carried by U.S., Soviet and European rockets. The H-2 is expected to lift two tons to orbit later in the

1990's. Japan's Science and Technology Agency, part of NASDA, develops communications and weather satellites.

The 1,210-lb. CS-3a and CS-3b communications satellites, launched in 1988, are the largest satellites NASDA has sent to space so far.

Both ISAS and NASDA are preparing for future manned space flights and interplanetary travel. The Muses-A mission will give Japan valuable experience in targeting orbits and in the use of swing-bys to guide future spacecraft traveling to distant planets.

Japan got a late start in space work under restrictions at the end of World War II and has been using technology from the U.S. It has had low space budgets, spending only $1.06 billion for space projects last year.

Building a respectable launch record, Japan enjoyed a 100 percent success rate with its smaller N-series rockets and nearly 100 percent success with its larger H-1 rockets. NASDA had its first failure in 20 launches in August 1989 when the first stage of an H-1 rocket failed to ignite.

Launch sites. The country also has a shortage of good launch sites. Kagoshima Space Center is at the southern tip of Japan. Rockets are fired toward space from a pad on small Tanegashima Island offshore. Japan has made two dozen launches since 1975 from Tanegashima Island.

The Uchinoura pad at Kagoshima Space Center and the pad on Tanagashima Island, Japan's two launch sites, are very small when compared with space centers in other nations. Because tuna fishermen near the center complain that blastoffs are dangerous, launches are permitted in a 90-day period each year—45 days in January-February and 45 days in August.

Kagoshima. Kyushu, a mountainous island with famous peaks rising to 6,500 ft., is southernmost of the four main islands of Japan. One prefecture in Kyushu is Kagoshima whose two million people mostly are foresters and coastal fishermen.

The ancient seaport city of Kagoshima is a well-protected harbor on the the west coast of Kagoshima Bay, a deep inlet on the southern coast of Kyushu. For centuries, Kagoshima was the castle city of a powerful daimyo of the Satsuma clan. The city was bombarded by British warships in 1863, destroyed by fire in 1877, damaged by eruption of the volcano On-take on an island in the bay in 1914, and bombed from June to August 1945 by planes of the Allies in World War II. Today the city, with a population of half a million, is best known for its Satsuma ware.

Ohsumi. Japan's first satellite was Ohsumi, launched February 11, 1970. Launch of Ohsumi made Japan the fourth nation capable of launching satellites to Earth orbit, after the USSR, the U.S. and France.

Since then, Japan's space program has relied more on resourceful engineering than large budgets. Muses-A mission chief Hiroki Matsuo told reporters, "This time we are going to the Moon, but our objective is not the Moon itself. Our institute is getting into interplanetary missions in the 1990's and for that we need to refine our technology." Japan has launched four dozen spacecraft since 1970 with a third still in operation today.

Venus. Japan plans to launch a spacecraft called Planet B toward Venus in March 1996 atop an improved version of the MU-3S-2 rocket.

The current MU-3S-2 version is 91 feet long and five feet in diameter. It will be lengthened to 98 feet and fattened to seven feet diameter for the interplanetary shot. Planet B will carry out scientific studies at Venus.

Competition. The likelihood of Japan challenging other nations for commercial space business is strong. Last May, U.S. President George Bush charged that space satellites are one of three areas in which Japan indulges in unfair trade practices. Wood products and supercomputers were the other two.

The U.S. trade representative claimed Japan favors buying satellites from its own companies, despite the availability of quality U.S. satellites. The trade representative said the idea is to enhance Japanese satellite builders at the expense of U.S. industry.

Swing-by. A swing-by is a familiar technique used to boost a satellite's speed by taking advantage of the force of gravity from planets and the Sun. The U.S. used it first when Mariner 10 flew by Venus in 1973 on its way to Mercury in 1974.

Launched successfully November 3, 1973, Mariner 10 was the first craft from Earth to visit two planets. It flew by Venus February 5, 1974, and went on to fly by Mercury three times. The first Mercury fly-by was March 29, 1974. The second, September 21, 1974. The third, March 16, 1975.

Mariner 10 contained two TV cameras which transmitted the first close-ups of Mercury and Venus. It also had an infrared radiometer to sense the temperature of Mercury's surface, ultraviolet spectrometers to find atmosphere air glow and measure how the Sun's ultraviolet rays were absorbed, and magnetometers to sense changes in the magnetic field as Mariner 10 flew its course.

The craft flew by Venus at a distance of 3,579 miles, sending back 3,500 photos. Pictures in ultraviolet light showed cloud cover and circulation of the atmosphere clearly. The gravity of Venus then was used to slingshot Mariner 10 toward Mercury.

The first pass by Mercury was at a distance of 460 miles above the baked planet's surface. The second was at a distance of 29,827 miles. The third, at 205 miles. Altogether 10,000 photos of Mercury were transmitted to Earth. The planet's magnetic field was discovered. Mariner 10 now is in orbit around the Sun.

First since '76. Muses-A/Hiten added prestige to Japan's space program as it closed a 14-year era with no Moon missions from anywhere on Earth. Previously, only the U.S. and USSR had sent craft to the Moon. Most recent was the USSR's Luna 24 with an unmanned rover which landed on the Moon in 1976.

From the earliest days of their space programs in the 1950's, the U.S. and the USSR have flung rockets at the Moon. Most memorable from those early days may have been the USSR's robot probe Luna 3 which radioed the first pictures of the dark side of the Moon in October 1959. Later unmanned American probes, including a successful 1960's series of Surveyor automatic landers, led to six manned Apollo landings on the lunar surface from 1969 to 1972. Thirty USSR Luna and Zond spacecraft explored the Moon between 1959 and 1976 while Ranger, Surveyor and Lunar Orbiter probes explored the Moon for the U.S. in the 1960's.

Moon Probes: Japanese Lander

A Japanese government panel on space development policy decided in 1991 to launch a large Moon probe within five years, using Japan's next-generation M-5 rocket, being developed by the Institute of Space and Aeronautical Science.

The first M-5 rocket would be test-launched between April 1994 and March 1995. Then the lunar orbiter would be launched sometime between March 1995 and March 1996. The lunar probe would carry moonquake measuring equipment and three instruments to pick up soil samples to help study the moon's origin.

Manned Moon Flights

Altogether, there were three American astronaut flights around the Moon and six flights to the Moon with landings.

Apollo	Astronauts	Landing Site	Days At Moon	Lbs. of Rocks	Other
Apollo 8 Dec 21-27 1968	Borman Lovell Anders	orbited Moon only			1st manned circumnavigation

Apollo 10 May 18-26 1969	Stafford Young Cernan	orbited Moon only			test of complete spacecraft 2nd manned circumnavigation
Apollo 11 Jul 16-24 1969	Armstrong Aldrin Collins	Mare Tranquillitatis	0.9	22	first manned landing
Apollo 12 Nov 14-24 1969	Conrad Bean Gordon	Oceanus Procellarum	1.3	34	landed near Surveyor 3
Apollo 13 Apr 11-17 1970	Lovell Swigart Haise	no landing			aborted mission circumnavigation only
Apollo 14 Jan 31-2/9 1971	Shepard Mitchell Roosa	Fra Mauro	1.4	44	highlands landing; pushcart
Apollo 15 Jul 26-8/7 1971	Scott Irwin Worden	Hadley Rille Apennine Mountains	2.8	77	first Moon buggy
Apollo 16 Apr 16-27 1972	Young Duke Mattingly	Descartes, Cayley Plains	3.0	97	another Moon buggy
Apollo 17 Dec 7-19 1972	Cernan Schmitt Evans	Taurus Littrow	3.1	110	last manned landing

Moon Town

The idea of building a town on the Moon after the turn of the century is catching on with some space scientists. Engineers at Johnson Space Center, Houston, have proposed a Moon camp. Astronauts would dig in on the Moon, building housing buried under lunar soil. Exploration alone would be a benefit to mankind, some say, but a station on the Moon also would be a testbed for new technologies and even help U.S. national security.

On the far side. Astronomy and astrophysics would find the Moon a huge platform in space from which to study the far reaches of the Universe, free from Earth's murky atmosphere. Scientists envision a major observatory on the dark side of the Moon, away from Earth's radio and light pollution.

Oxygen mines. Oxygen mined from lunar material could fuel interplanetary ships bound for Venus, Mars, the Asteroids, Jupiter and Saturn.

LPOsat. To get things started, an unmanned Lunar Polar Observer (LPO) satellite might be sent to scout territory in the late-1990's. After LPO, an unmanned rover vehicle could be sent to land on the Moon, grab soil samples and return them to Earth. Bigger rovers then could be sent on more complex missions.

Freedom station. Two or more lunar landings per year might be made from 2000-2005 as a town site was prepared. Those Moon flights would leave from the U.S.-international space station Freedom or the USSR's Mir station.

By 2005, a Moon plant could be turning out sufficient oxygen to fuel flights away from there. Construction of a Moon town would be completed between 2005 and 2010. Four manned flights might be made per year during those years.

For those who see the Moon as a jumping off point for Mars trips, lunar oxygen could power flights to build a Mars colony during the 10 years after 2010. A Mars ship could be fueled in space with lunar oxygen, making it lighter to lift from Earth.

Makin' Do On The Moon

MAKE DO WITH WHAT'S HERE might well be the motto of 21st century residents of a Moon town.

Early European settlers going to the New World didn't transport everything needed to sustain themselves across the Atlantic Ocean. Instead, they carried tools to America and constructed what they needed for shelter and living from local materials.

The same settler's rule of thumb will apply to colonists building a new world on the Moon after the year 2005, or on Mars. Tools and knowledge from Earth, and resources found near their new homes, will give them what they need.

Don't think of space as a vacuum, devoid of everything. Instead, think of it as an ocean, as early settlers had to view the Atlantic. It's a hostile environment, but one with many resources including readily-available solar power and the promise of surface mines on the Moon, Mars and asteroids.

Take it or make it? The dilemma of space flight is that everything carried away from Earth must be powered to sufficient speed to escape Earth's gravity. In the past, men have manufactured things on Earth and ferried them to space with rockets. But the cost-per-pound of boosting anything to orbit is huge. Lack of money for space projects always is a political problem.

If they don't carry all supplies and materials from Earth to space, where will future astronauts and cosmonauts get them? The Moon.

Earth's natural satellite is a world of useful materials. And, even better news, it is 20 times easier to lift a payload away from the lunar surface than it is to accelerate an identical weight away from Earth's surface.

Also, the Moon has no atmosphere. That would permit use of so-called electromagnetic accelerators, or mass drivers, to launch payloads from the Moon surface.

Lunar soil. Raw Moon dirt will be used as shielding, not just down on the lunar surface, but also out in space for orbiting stations.

Moon soil will be broken down into its basic chemical elements. For example, oxygen is the biggest part of water and of rocket fuel. Lunar soil is rich in oxygen which could be cooked out of the dirt.

Most of the equipment to convert soil to oxygen is low-tech hardware so an extraction plant itself could be manufactured right on the Moon.

Lunar soil could be made easily into concrete for houses, buildings and all sorts of other structures.

Glass made from lunar soil could be used in construction projects on the Moon as well as in space. Crude but effective techniques are available. Brittleness could be overcome by glass composites.

Moon-made glass could be turned into glass-glass composite beams which could be used on the surface as well in orbit for structural elements of a solar power satellite.

The USSR, Western Europe and Japan are interested in solar power satellites. The Soviets want to build one in Earth orbit by the year 2000, using materials launched from Earth.

More than 90 percent of such a satellite power station could be made of materials from the Moon.

Mining. Mines on the Moon will be different from those on Earth as there is no water available.

And asteroids might be mined using mass-driver engines. A mining spacecraft would capture a small asteroid or a piece of a large asteroid and extract minerals and other useful materials.

Slag or other waste products might even be pulverized and used by a mass driver to propel a captured asteroid back to an Earth orbit.

Materials mined from moons, asteroids and planets could be used in the construction of entire free-floating space colonies.

Machinery. As men build towns on the Moon, they probably will do surface mining as well as go below the surface for materials. Earthbound mining machines are too heavy to lift to space so new techniques will be found. One problem: rock requires force to move and break up.

The brute force method used on Earth won't be used on the Moon or asteroids due to low gravity. Surface mines might be a slusher-scoop. It scoops up material in a bucket dragged across the surface by cables and winch. A laser cutter might be used, if sufficient electrical power can be generated on the Moon.

Fusion pellets, which create shock waves by impact, might be used to break up rocks. Nuclear charges probably won't be practical as they would throw material too far in low gravity.

Venus Probes From Earth

Not long after their 1957 launch of Sputnik, the first artificial Earth satellite, space engineers in the Soviet Union fired the first rocket toward another celestial body. Venera 1, in 1961, was the first spacecraft sent beyond Earth.

The Magellan radar-mapping interplanetary probe fired toward Venus from the U.S. shuttle Atlantis May 4, 1989, was the sixth American spacecraft and the 30th from Earth built to explore Venus.

One of the six American flights failed, while nine of 24 Russian missions failed. Successful USSR and U.S. encounters with Venus were:

1961 February 12, USSR Venera 1 probe, second launch of unmanned Venera to explore planet Venus, first interplanetary probe to escape Earth orbit, flew within 62,000 miles of Venus May 19. Radio failed 14 million miles from Venus, but probe was tracked by radar.

1962 December 14, U.S. Mariner 2 flew past Venus at a distance of 21,300 miles, the first successful U.S. planetary fly-by.

1965 November 16, the USSR's Venera 3 probe was first human artifact to reach the surface of another planet, first spacecraft to crash on Venus, 280 miles from planet center.

1967 October 18, the USSR's Venera 4 probe parachuted into the atmosphere of Venus and radioed data to Earth about pressure, density and chemical composition.

1967 October 19, U.S. Mariner 5 flew past Venus at 2,500 miles, sending back surface temperature and magnetic field data.

1969 May 16, USSR Venera 5 radioed data 53 minutes before being crushed by increasing pressure as it descended through the atmosphere of Venus.

1969 May 17, USSR Venera 6 transmitted 51 minutes before being crushed as it floated down.

1970 December 15, USSR Venera 7 landed on Venus and radio 23 minutes of data.

1972 July 22, USSR Venera 8 landed on the surface of Venus and transmitted data for an hour.

1974 February 5, U.S. Mariner 10 spacecraft flew past Venus at 3,500 miles, then flew on to Mercury as the first two-planet probe.

1975 October 22, USSR Venera 9 was the first to transmit a picture from the surface of another planet. Data was radioed to Earth for three hours.

1975 October 25, USSR Venera 10 transmitted pictures from the surface.

1978 December 4, U.S. Pioneer-Venus 1 radar mapper entered Venus orbit, mapping the surface with a resolution of 50 miles.

1978 December 9, U.S. Pioneer-Venus 2 arrived at Venus, releasing probes into the atmosphere.

1978 December 21, USSR Venera 12 flew past Venus, dropping probes to the surface. TV failed.

1978 December 25, USSR Venera 11 flew past Venus, dropping probes to the surface. TV failed.

1982 March 1, USSR Venera 13 landed on the surface, conducting the first soil analysis.

1982 March 5, USSR Venera 14 landed on the surface, conducting soil analysis.

1983 October 10, USSR Venera 15 orbited Venus, mapping the surface with low-resolution radar.

1983 October 16, USSR Venera 16 orbited Venus, mapping the surface with low-resolution radar.

1985 June 11, USSR Vega 1, bound for Halley's Comet, dropped probe to the surface. Probe releases instrumented balloon to float through the atmosphere.

1985 June 15, USSR Vega 2, headed for Halley's Comet, dropped probe to the surface. Probe releases instrumented balloon to float through the atmosphere.
1990 August 10, U.S. Magellan, entered orbit around Venus for radar mapping.

Venus Probes: USSR Venera & Vega

Probe	Launched	Result
Venera	1961 Feb 4	called publicly Tyazheily Sputnik 4, A-1 rocket only reached Earth orbit.
Venera 1	1961 Feb 12	A-1 rocket, first interplanetary probe to escape Earth orbit, within 60,000 miles of Venus 5/19/61.
Venera	1962 Aug 25	unannounced launch, new A-2e rocket only reached Earth orbit, fell from Earth orbit 8/28/62.
Venera	1962 Sep 1	unannounced, A-2e only Earth orbit, fell from orbit 9/6/62.
Venera	1962 Sep 12	unannounced, A-2e only Earth orbit, fell from orbit 9/17/62.
Cosmos 27	1964 Mar 27	announced launch under generic name, A-2e rocket only reached Earth orbit, fell from Earth orbit 3/29/64.
Zond 1	1964 Apr 2	Spacecraft passed within 60,000 mi. of Venus 7/19/64.
Venera 2	1965 Nov 12	A-2e carried larger probe, passed within 15,000 miles of Venus 2/27/66.
Venera 3	1965 Nov 16	A-2e rocket, larger probe, first human artifact to reach the surface of another planet 2/27/66, first spacecraft to crash on Venus, 280 miles from planet center.
Cosmos 96	1965 Nov 23	announced launch under generic name, A-2e rocket only reached Earth orbit, fell from Earth orbit 12/9/65.
Venera 4	1967 Jun 12	A-2e carried heavier spacecraft which dropped 844-lb. probe 10/18/67 to parachute on 96-min. descent to Venus surface with gas analyzers, thermometers, barometer, radio altimeter, atmospheric densitometer; transmitted down to altitude of 15 mi.; main craft had hydrogen & oxygen sensors, magnetometer, charged particle trap, cosmic ray counter; altogether returned important data on atmosphere, pressure, temperature, magnetic field, hydrogen corona.
Cosmos 167	1967 Jun 17	announced launch under generic name, A-2e rocket only reached Earth orbit, fell from Earth orbit 6/25/67.
Venera 5	1969 Jan 5	Like Venera 4 with added heat and pressure shield on probe for smaller parachute, faster fall, sent data 53 min., landed on night side 5/16/69.
Venera 6	1969 Jan 10	Like Venera 5, landed day after Venera 5, sent data 51 min. 5/17/69.
Venera 7	1970 Aug 17	Like Venera 6, stronger probe to land on surface; sent data 25 min. during descent and 23 min. from night side surface 12/15/70; first ever probe soft landing; Venus surface temperature 887º, pressure 90 times Earth.
Cosmos 359	1970 Aug 22	announced launch under generic name, A-2e rocket only reached Earth orbit, fell from Earth orbit 11/6/70.
Venera 8	1972 Mar 27	Like Venera 7, added soil analyzer, sent data 50 min. after landing on day side 7/22/72; soil like granite with density of 1.5 grams/cc.
Cosmos 482	1972 Mar 31	announced launch under generic name, A-2e rocket only reached Earth orbit.
Venera 9	1975 Jun 8	New Proton rocket, heavier spacecraft was orbiter with lander to drop; orbiter studied cloud tops, upper atmosphere, solar wind with 4 spectrometers, 2 photometers, photopolarimeter, infrared radiometer, magnetometer, panoramic camera, charged particle traps; 6-ft. lander, cooled for long surface life, had panoramic TV camera with signals sent to orbiter for

		relay to Earth, 3 photometers, 2 spectrometers, accelerometers, anemometer, radiation sensors, densitometer; TV 115 minutes after landing 10/22/75 showed young active planet, bright surface, floodlights not needed.
Venera 10	1975 Jun 14	Like Venera 9, landed 10/25/75 in a congealed lava field 1,400 mi. from Venera 9, transmitted 65 min. from surface, data on temperature, rock density, wind velocity.
Venera 11	1978 Sep 9	Like Venera 10 but lighter spacecraft, main craft intentionally flew by within 22,000 mi. rather than into orbit around Venus, lander on surface 12/25/78 500 mi. from where Venera 12 had landed; fly-by craft monitored high-energy particles, solar wind, plasma and carried French SMEG gamma ray detector; lander recorded electrical activity, chemicals of atmosphere during descent; relayed temperature, pressure data to Earth from surface 110 min. via main craft; TV failed.
Venera 12	1978 Sep 14	Like Venera 11, fly-by plus lander on surface 12/21/78, 500 mi. from where Venera 11 would touch down 4 days later; transmitted 95 min. from surface; TV failed.
Venera 13	1981 Oct 30	Like Venera 12, fly-by plus lander; carried Austrian magnetometer; during descent 3/1/82 lander used 3 spectrometers, nephelometer, gas chromatograph, hydrometer to measure atmosphere; sent color TV of land and sky from surface, collected and analyzed soil samples, took seismic readings; landing sites in volcanic area chosen from U.S. Pioneer-Venus 1 radar map.
Venera 14	1981 Nov 4	Like Venera 13, lander 3/5/82 sent data 57 min. for relay by fly-by craft.
Venera 15	1983 Jun 2	Venera 9-13 style of craft modified to orbit only; collected cosmic rays, charged particles enroute; entered orbit 10/10/83; side-looking Polyus V radar with 18-ft. antenna mapped 75 million sq. mi. of Northern Hemisphere from orbit with 1-2 km resolution through 3/85; orbiter carried East German spectrometer, interferometer to measure atmosphere temperature.
Venera 16	1983 Jun 6	Venus orbiter like Venera 15, arrived 10/14/83, radar worked through 1986; Venera 15 & 16 compiled temperature map of Venus Northern Hemisphere.
Vega 1	1984 Dec 15	Vega is acronym for Venus-Gallei meaning Venus-Halley. The probe flew by Venus 6/11/85, dropping off lander, on way to Comet Halley; the Venus lander released at 33 mi. altitude French atmosphere balloons with gondola of instruments which radioed data 46 hours tracked by 20 Earth stations in Australia, Great Britain, Sweden, U.S., USSR; lander had TV, soil analysis, aerosol analyzer, mass/gamma ray/ultraviolet spectrometers, hydrometer, gas chromatograph; lander transmitted 21 min. from surface; soil sample failed; the 265 lb. instrument probe whipped by Venus gravity toward Comet Halley, armored against speeding dust particles, passed Halley 3/9/86 at 6,000 mi., one of five probes from various Earth nations to Halley; countries with experiments aboard both Vega 1 & 2 were Austria, Bulgaria, Czechoslovakia, East Germany, France, Hungary, Poland, U.S., USSR and West Germany.
Vega 2	1984 Dec 21	The Vega 2 spacecraft was like Vega 1. It flew by Venus, dropped lander 6/15/85, whipped probe on to Comet Halley; Venus lander soil analysis found dirt like Moon highlands analyzed earlier by USSR Luna craft; Vega 2 comet probe passed within about 4,000 miles of Comet Halley's nucleus three days after Vega 1 comet probe.

Venus Probe: Pioneer 12

Pioneer 12 had been designed to orbit Venus only 243 days when it left Earth in 1978, but it has flown through far more than a decade of solid science work, a compliment to the engineers who kept the probe working by remote control.

The interplanetary spacecraft mapped 93 percent of the broiled surface of Venus, providing new understanding of volcanic action and land formation on Earth.

By 1988, a shopworn but serviceable Pioneer 12 had transmitted back to scientists on Earth more than 10 trillion bits of data, including a description of the extreme greenhouse effect which has trapped the Sun's heat to drive the planet's surface temperature to 900 degrees Fahrenheit.

Venus is the second planet from the Sun, after Mercury. Earth is third. Mars is fourth. There are nine known major planets.

Yet so different. Venus once was thought of as a twin to Earth. The planet is similar in size and composition, with continents, mountains and canyons. But, there are some big differences.

The atmospheric pressure on the surface of Venus is 90 times that of Earth. The temperature on the surface may be as high as 900 degrees Fahrenheit.

The U.S. sent two exploration spacecraft in 1978, Pioneer 12 and Pioneer 13, to investigate the planet. Pioneer 12, also known as Pioneer-Venus 1, still is on station, in orbit around Venus, using radar to map the planet surface and sending back data about the solar wind. The spacecraft intercepts particles in the solar wind as they pass Venus, flying out from the Sun during the 11-year sunspot cycle.

Pioneer 12, responding to the effect of solar gravity on Venus, is in a variable orbit which swings the probe as close as 93 miles above the surface and as distant as 1,426 miles. Pioneer 12 has less than five lbs. of fuel left but that amount should keep it in proper orbit around Venus until at least 1992.

Greenhouse effect. The greenhouse effect, in which clouds or pollution trap the Sun's heat near the planet surface, is more developed than on Earth, accounting for some of the differences between Venus and Earth.

Venus is a stark example of a polluted atmosphere. Chlorine in the atmosphere of Venus is 100 to 1,000 times the concentration found in air above Earth. Such extreme stratospheric pollution on Venus may explain some of the extraordinary conditions.

Pollution on Venus is natural, not man-made, but the chemistry is the same as on Earth. Venus provides a warning about the future of Earth under the greenhouse effect.

Pioneer 13 also was sent in 1978 to Venus. It was a spacecraft bus carrying five Venus landers. After arriving at the planet, the five landers were dropped successfully through the atmosphere toward the surface, sending data back to Earth on their way down. One sounder actually landed on Venus and sent back data from the surface for 68 minutes. The main bus burned up in Venus' atmosphere in 1978.

Comet Wilson. Pioneer 12 recently turned its attention temporarily to Comet Wilson, discovered in 1986 by a Cal Tech graduate student. The reliable old satellite, which already had studied three other comets while on duty near Venus, spent April 1987 measuring the rate of water evaporation from Comet Wilson, named after Christine Wilson, the California Institute of Technology Ph.D. candidate who discovered it in August 1986.

Pioneer looked to see how much carbon and oxygen the comet gave off as it swung close to the heat of the Sun.

Wilson was streaking past Venus in an arc around the Sun. Its closest approach, called perihelion, was April 20. Pioneer 12 spent 20 hours a day studying Wilson while it was nearby.

Pioneer 12 was situated well to study Comet Wilson. Ultraviolet measurements cannot be made from Earth since the ozone layer in our atmosphere blocks most ultraviolet light. Pioneer 12 was ordered to turn 10 degrees so it could see the comet.

Other spacecraft in orbit around Earth helped in the study of Wilson. International Ultraviolet Explorer (IUE) satellite in Earth orbit scanned the comet, looking for other chemicals. IUE's prime mission was looking at distant stars and galaxies.

Venus Probe Magellan

Zipping along at 85,000 miles per hour, the unmanned interplanetary spacecraft Magellan from Earth dropped into orbit around the planet Venus August 10, 1990.

Magellan had been carried to space from the U.S. spaceport at Cape Canaveral, Florida, May 4, 1989, in U.S. shuttle Atlantis. Astronauts dropped it overboard in low Earth orbit from where its booster rocket blasted it on a 463-day, 948-million-mi. trip to Earth's cloud-shrouded sister planet. Magellan was sent to map 90 percent of the planet's surface with radar which was able to peer through the blanket of thick clouds which block visible light for optical cameras.

Early in the flight, Magellan's solid-fuel rocket motor and two equipment compartments were hotter than expected. Engineers used the high gain antenna to shade them from the Sun. Problems with gyroscopes and radar data also were solved.

A star scanner used to find the position of the spacecraft by looking at pairs of reference stars went through a period when it saw unexpected glints of light. Engineers decided it either was a bombardment of protons during large solar flares or else small particles flaked off the craft's cover as the star scanner went from shade to sunlight. Magellan had sailed through three big solar outbursts in its first six months. The engineers radioed software changes to narrow the star fix and ensure the scanner remained in shade before and during calibrations.

At Venus. While Magellan was out of sight of Earth behind Venus on the morning of August 10, a computer aboard the spacecraft fired Magellan's solid-fuel rocket motor as planned to slow the probe so it would fall into an elliptical polar orbit around Venus. Then the rocket motor was jettisoned.

The American space agency, agonizing over shuttle fuel leaks and a bad mirror in the Hubble Space Telescope, enjoyed the euphoria of success for six days. Then harsh reality returned. NASA lost radio contact with Magellan August 16.

It took 15 hours for NASA controllers at the Jet Propulsion Laboratory, Pasadena, California, using emergency systems built into Magellan, to find the silent spacecraft. It was found on August 17, but another eight hours was needed to stabilize communications with the spacecraft. Finally, Magellan heard one of a series of commands transmitted from Earth which stopped its spinning and pointed one of its antennas at Earth. Signals from the probe were lost again August 21 and found again August 22.

Saying Magellan was healthy, NASA still had not fingered the communications culprit five days later when it showed off exciting test pictures sent back by Magellan on August 21.

First pictures. Magellan's first Venus pictures reminded geologists of California earthquake faults, Hawaiian volcanoes, the rift valleys of East Africa and Europe's Rhine Valley.

JPL called the strips of photos covering territory 1,000 miles long by 15 miles wide "noodle" pictures. The images depicted a violent Venus-scape sculpted with long, parallel valleys and ridges—like those between the Rocky Mountains and the Sierra Nevada range—as well as deep impact craters, jagged quake faults and expansive lava flows similar to those on Hawaii and in the Snake River plains of Idaho.

Overlapping lava flows six to ten miles wide, of different ages, were bright and dark splotches in the photos. Many Venus-quake faults and fractures suggested movement of the planet's crust had shaped the landscape. Parallel sets of elongated valleys and ridges looked like the basin-and-range area of the intermountain region of Utah and Nevada, or the Great Rift Valley in eastern Africa. Apparently, part of Venus' crust had been stretched apart.

The photos were of an area of Venus known as Beta Region Volcanic Highlands. Zigzag lines across the surface were tinsel fractures caused by pulling the crust apart, like the rift valleys in East Africa or the Salt Trough earthquake fault under California.

A pit crater about a mile wide in the photo, like depressions on the slopes of Hawaiian volcano Mauna Lea, may have been volcanic in origin, not the result of a meteor impact. The Devana Chasm, a giant valley, had fault patterns like those in the Rhine Graven region of Germany.

The test pictures were made August 16 as Magellan bounced its radar beam off Venus for the first time. They did not show any of the big volcano mountains seen by previous robot explorers from Earth. The noodle pictures also did not show whether any volcanoes are active nor if the planet's crust is broken into drifting plates like those carrying continents across the face of the Earth.

Lost again. But again there was harsh reality. The $744 million mapper's second major shock to the space agency in five days appeared just eight hours after NASA's picture show August 21 when radio signals from Magellan stopped arriving at NASA's big deep-space dish antenna near Canberra, Australia.

This time it took 17 hours for JPL controllers to find the spacecraft. It was found on August 22 and commands were sent to stop Magellan's spinning and point one of its antennas at Earth. NASA continued to say Magellan was healthy, but still was unable to identify the cause of the communications breakdown.

What went wrong? Magellan looks at stars regularly to make calibrations and keep its computer up-to-date on where the probe is in space. Contact was lost August 16 as the spacecraft routinely tried to find and lock onto a guide star. Magellan went into a "safe mode" when a computer controlling the probe's attitude suddenly stopped sending timing signals to the master computer.

Unable to find that heartbeat from the atitude-control computer, the master computer activated emergency programs which put Magellan into a first-stage safe mode called "RAM safing." RAM is a computer's random access memory.

Magellan stopped what it was doing, faced its solar panels broadside to the Sun and began looking around for the familiar star Sirius. Knowing where it was in relation to the bright Sirius and to the Sun, Magellan should have been able to point an antenna at Earth to receive further instructions by radio. But, then another problem turned up.

What did it see? Magellan locked onto something, but it was not the star Sirius. Was the craft's star sensor confused by a dust speck, by Venus, by another planet? NASA doesn't know what the probe saw, but when Magellan started looking for Earth, it couldn't find the home planet. That triggered a second-stage emergency. Magellan's master computer started running diagnostic programs to uncover what had gone wrong. That operation alerted a third stage emergency, causing Magellan to begin searching harder for Earth by rotating about its vertical axis, sweeping the Solar System with its radio beam. Eventually the beam swept across Earth and was received by NASA.

The signal arriving at Earth held telemetry data sent very slowly, indicating the probe was in hibernation—the safe mode protecting the craft from problems its computer couldn't handle. Both high-data-rate radio transmitters are switched off in safe mode.

Earlier problem. Although the Magellan trip to Venus was relatively quiet, a scanner used to find the position of the spacecraft by looking at pairs of reference stars did go through a period when it saw unexpected glints of light. Such "spurious interrupts" affected daily calibrations. Engineers decided it either was a bombardment of protons during large solar flares—Magellan sailed through three big solar outbursts in its first six months—or else small particles flaked off the craft's cover as the star scanner went from shade to sunlight. They radioed software changes to narrow the star fix and ensure the scanner remained in shade before and during calibrations.

Why now? What caused Magellan to break off radio contact August 16 and 21? A dust speck on a lens? A high-energy particle from the Sun blitzing computer memory? A previously-undetected electrical field around Venus affecting electronics? JPL engineers don't know if Magellan has a hardware or software problem.

NASA planners had expected cosmic ray hits to ping on Magellan's electronic gear about once a year. Similar events have been known to bother the Pioneer-Venus 1 orbiter known as Pioneer 12, launched in 1978, as well as NASA's Tracking and Data Relay Satellites (TDRS) and other Earth-orbiting satellites. NASA recalled radio contact had been lost for eight days in 1977 with a Voyager interplanetary probe launched that year.

Mapping. Magellan eventually took the planned 243 days to complete one cycle of mapping Venus during one complete rotation of the planet. On its path around Venus, Magellan swings down to a low of 155 miles above the surface and up to a high of 1,180 miles. Each trip around the planet, taking 3.1 days, gives a 40-minute window when the craft is near the surface.

Magellan has a special synthetic-aperture radar to map the planet. During each 40-minute swoop near the surface, radar peers through Venus' billowy clouds, shooting pictures of the surface for relay to Earth. The radar maps a swath 15 miles wide and 10,000 miles long, each orbit. It took 1,853 such fragments over 243 days to piece together a chart of the entire planet. The first map was complete in April 1991, after which Magellan started a second 243-day sweep.

Side-by-side, the noodle photos together provided a picture of the entire planet with ten times better resolution than pictures taken by the 24 U.S. and Soviet spacecraft which have investigated Venus previously. The images show surface details the size of two football fields. By the end of 1991, more than 90 percent of Venus had been mapped by Magellan.

Rough landscape. Two NASA pictures in September 1991 revealed a Venus even more torn by tectonic forces than had been suspected before Magellan. Earth's nearest neighbor-planet is dynamic with searing winds, a violently-deformed crust and giant volcanic eruptions.

One high-resolution radar image showed Akna Montes, mountains on the west side of the elevated smooth plateau Lakshmi Planum. Surrounded by mountains, the high plains were formed by extensive volcanic eruptions. The meteorite impact crater Wanda was in the picture. Wanda looked like it had be gouged out after the Akna mountains were formed. The western ridge of the crater had collapsed onto its floor.

A second high-resolution radar image depicted the Danu mountains south of Lakshmi Planum and a round volcanic dome 12 miles in diameter. The southern part of the dome probably had been deformed when the mountains were created. The mountains seemed to have been formed by uplift resulting from compression. The high plains surface was bent and broken at the base of the mountains.

Venus topography. A cover of thick clouds obscures Venus so astronomers know a lot more about the topography of the Moon, Mars, Mercury and Earth. Venus is Earth's closest neighbor in our Solar System, but the two planets are quite different. The temperature is 900 degrees Fahrenheit on the surface of Venus. The atmosphere is almost entirely carbon dioxide. Atmospheric pressure is 90 times Earth's.

Scientists want to know why the two planets evolved so differently. Magellan's imaging radar is uncovering more of the face of Venus, giving scientists their best detailed pictures yet of craters, landslides, volcanoes and surface formations changed by wind and flowing liquid.

Longest river. NASA scientists were puzzled n 1991 when Magellan discovered the longest channel known in the Solar System. The channel crosses the plains of Venus for 4,200 miles, longer than the Nile River, which is the longest river on Earth.

Earlier probes. Over the years since 1961, the U.S. has sent five probes to Venus while the USSR has sent 19:

★Between 1961 and 1985, the USSR sent 16 Venera, one Zond and two Vega probes to fly-by, orbit and land on Venus. Venera 1, in 1961, was the first interplanetary probe to escape Earth orbit and the first to fly by Venus. Venera 3, in 1966, was the first spacecraft to impact on Venus, the first human artifact to reach the surface of another planet.

★The U.S. sent three Mariner probes to Venus, in 1962, 1967 and 1974. The Mariners were fly-bys. Mariner 2, in 1962, completed the first successful American fly-by of any planet. Mariner 10, in 1974, also went on to visit the planet Mercury.

★Pioneer 12 and 13, also known as Pioneer-Venus 1 and Pioneer-Venus 2, were sent in 1978. Pioneer-Venus 1 was an orbiting radar mapper. It mapped 93 percent of the planet's surface after it arrived there in 1978. It remains in orbit around Venus today with enough fuel to keep working until around 1992. Pioneer-Venus 2 dropped probes to ground through the planet's atmosphere.

Venus Probe: Japan's Planet B

Japan plans to launch a spacecraft called Planet B toward Venus in March 1996 atop an improved version of the Japanese MU-3S-2 solid-fuel space rocket. Planet B will carry out scientific studies at Venus.

The current MU-3S-2 version is 91 feet long and five feet in diameter. It will be lengthened to 98 feet and fattened to seven feet diameter for the interplanetary shot.

Mars Probes From Earth

By the 1960's, space engineers were able to build and send interplanetary science probes. A total of 19 unmanned Mars explorers have been fired into interplanetary space from the U.S. and the USSR to look at the Red Planet and its moons Phobos and Deimos.

The USSR's Mars 1, launched in November 1962, was the first attempt to probe Mars. Unfortunately, contact was lost with the spacecraft only 60 million miles along its route to the Red Planet.

U.S. Mariner 4 launched in November 1964 was the first successful probe to reach Mars, sending back 22 photos as it flew by in July 1965. Mariner 9 brought the first close-ups of the moons Phobos and Deimos. Launched toward Mars in May 1971, Mariner 9 arrived in a 12-hour orbit around the Red Planet in November 1971, the first man-made satellite to orbit a planet other than Earth. Mariner 9 had two TV cameras which sent back 7,329 photos.

Later, Viking 1 and Viking 2 carried the American flag across millions of miles of interplanetary space to photograph Mars, Phobos and Deimos. Viking 1 was launched September 9, 1975. It arrived at Mars June 19, 1976, and landed. Viking 2 was launched August 20, 1975, and arrived at Mars August 7, 1976, and landed.

Phobos 2 carried the USSR flag 111 million miles to Mars orbit January 29, 1989. It detected water vapor in the Martian atmosphere.

Launch	From	Probe	Result
1962	USSR	Mars 1	radio contact lost, 60 million miles.
1964	USA	Mariner 3	failed to achieve Mars trajectory.
1964	USA	Mariner 4	passed within 6,200 miles on 7/14/65, 22 photos.
1964	USSR	Zond 2	passed Mars 1965, sent no data.
1965	USSR	Zond 3	Mars flight, enroute flew by Moon at 5,717 miles, sent 25 photos of far side of Moon.
1969	USA	Mariner 6	passed Mars by 2,100 mi. 7/31/69 photoed equator, sent

			100 photos, measured surface temperature, atmosphere pressure, atmosphere composition.
1969	USA	Mariner 7	passed Mars at 2,200 mi. 8/5/69 photographed southern hemisphere and polar cap, 100 photos, measured surface temperature, atmosphere pressure and composition.
1971	USA	Mariner 8	launch failure.
1971	USSR	Mars 2	entered Mars orbit 11/71 studied surface and atmosphere, landing capsule crashed.
1971	USSR	Mars 3	entered Mars orbit 12/71, studied surface and atmosphere, lander stopped transmitting in 2 min.
1971	USA	Mariner 9	orbited Mars 11/13/71, two TV cameras, sent 7329 photos, entire surface mapped, photos of Phobos and Deimos, studied atmosphere and surface temperature, saw violent planet-wide dust storm 9/71, craft was turned off 10/72.
1973	USSR	Mars 4	braking rocket failed craft overshot Mars 2/74.
1973	USSR	Mars 5	entered Mars orbit 2/74, snapped photos, quit working after few days.
1973	USSR	Mars 6	flew past Mars 3/74 dropped lander which crashed.
1973	USSR	Mars 7	flew past Mars 3/74 dropped lander which missed planet.
1975	USA	Viking 1	orbited Mars 1976, two TV cameras, 26,000 photos, lander parachuted successfully to surface 7/20/76 with weather station, seismometer and soil analyzer, seismometer failed, TV showed red rocky surface, dusty pink sky, sand dunes, no large life forms, soil mostly silicon and iron, temps 20º to -120º, winds 30 mph lander worked 6.5 years on surface.
1975	USA	Viking 2	orbited Mars 1976, two TV cameras, 26,000 photos, lander parachuted successfully to surface 9/3/76 with weather station, seismometer and soil analyzer, found wind and minor marsquakes, red rocky surface, dusty pink sky, sand dunes, no large life forms, soil mostly silicon and iron, temps -20º to -120º, 30 mph winds lander worked 3.5 years on surface.
1988	USSR	Phobos 1	left Earth 7/7/88, travelled 12 million miles of 111 million-mile route to Mars, accidentally turned off by ground controller error 8/29/88, now aimless in solar orbit.
1988	USSR	Phobos 2	left Earth 7/12/88, arrived Mars 1/29/89, mapped planet, found water vapor in atmosphere, photos of moon Phobos, radio contact lost 3/27/89, unable to drop hopping lander on Phobos 4/89.

Mars Probes From The USSR

Since 1962, eleven unmanned spacecraft have been sent by the USSR to study Mars.

PROBE	LAUNCH	RESULT
Mars 1	1962 Nov 1	radio contact lost 3/21/63 after 60 million miles. Flew by 6/19/63 within 120,000 of Mars. Now orbiting Sun. Venera-style capsule.
Zond 2	1964 Nov 30	passed within 1,000 miles of Mars in 1965, sent no data.
Zond 3	1965 Jul 18	Mars flight, enroute flew by Moon at distance of 5,717 miles, sending back 25 photos of far side of Moon as communications test. Eight Zond launches 1964-70 included 1 to Venus, 2 to Mars and 5 unmanned flights preparing for manned Moon trips including 3 unmanned around Moon which returned to land on Earth.
Cosmos 419	1971 May 10	Mars attempt, first use of big Proton D-1e rocket in interplanetary flight, failed to go beyond low Earth orbit. Fell 5/12/71.
Mars 2	1971 May 19	orbiter and lander, entered Mars orbit 11/27/71, studied surface and

		atmosphere, landing capsule touched down 11/27, may have crashed. Data transmitted by orbiter until 3/72.
Mars 3	1971 May 28	orbiter-lander identical to Mars 2, entered Mars orbit 12/2/71, studied surface and atmosphere, lander touched down 12/2, stopped transmitting in less than 2 min. on surface, maybe from dust storm. Data transmitted by orbiter until 3/72.
Mars 4	1973 Jul 21	Mars orbiter, braking rocket failed, craft overshot Mars, fly by 2/10/74 within 1,400 miles of Mars. Photos of Mars transmitted as craft flew by. Now in solar orbit.
Mars 5	1973 Jul 25	orbiter identical to Mars 4, entered Mars orbit 2/2/74, made photos, quit working after few days. Only complete success in USSR 1973-74 Mars flights. Carried TV camera, radio telescope, spectrometer, photometers, polarimeters. Mars 4 & 5 photos later combined with pix from U.S. Mariner 9 to compose Martian surface map.
Mars 6	1973 Aug 5	flew past Mars 3/12/74, dropped lander on way by. Lander touched down 3/12. Data from lander stopped 148 seconds into parachute descent. Mars 6 now orbiting Sun.
Mars 7	1973 Aug 9	lander like Mars 6, flew past Mars 3/9/74, dropped lander. Lander braking rocket failed. Lander missed planet 3/9 by 800 miles.
Phobos 1	1988 Jul 17	left Earth 7/7/88, travelled 12 million miles of 111-million-mile route to Mars, accidentally turned off by ground controller error 8/29/88, now aimless in solar orbit.
Phobos 2	1988 Jul 12	left Earth 7/12/88, arrived Mars 1/29/89, mapped planet, found water vapor in atmosphere, photos of moon Phobos, radio contact lost 3/2789, unable to drop hopping lander on Phobos 4/89.

Mars Probes: U.S. Viking

Carrying the U.S. bicentennial flag across millions of miles of space, the American orbiters and landers photographed Mars, and the Martian moons Phobos and Deimos.

Viking	Launch	at Mars	on Mars	Worked
1	9/9/75	6/19/76	7/20/76	6.5 yrs
2	8/20/75	8/7/76	9/3/76	3.5 yrs

Mars Probes: USSR Phobos Twins

The USSR's unmanned Phobos 2 science probe went into orbit around Mars January 29, 1989, completing a six-month, 111-million-mile passage from Earth—the first successful USSR Mars probe in 15 years and the first spacecraft from Earth to reach Mars in 13 years.

Many photos of the Red Planet and its potato-shaped moon Phobos were sent back along with data on the Martian atmosphere. The first water vapor in that dry atmosphere was recorded.

Unfortunately, Moscow mission control lost radio contact March 27, 1989, with the Phobos 2 probe while it was in a circular orbit around the planet preparing to buzz Phobos and drop landers on the tiny moon.

Only one brief signal from the spacecraft was received as a top-level government commission tried to figure out what went wrong. Engineers wanted to know if a transmitter had failed or an error had appeared in navigation calculations.

By early April 1989, it became clear to mission control that sufficient contact could not be restored to permit dropping two landers onto Phobos, larger of the planet's two natural satellites. The landings had been scheduled for the first half of April. Attempts to regain contact were abandoned.

Undaunted, USSR space engineers immediately started dusting off an unused ground-test spare spacecraft, studying charts for a Mars launch opportunity in 1992.

The Phobos 1/Phobos 2 expedition had been the most ambitious of 19 Soviet and American unmanned expeditions to Mars since the USSR sent Mars 1 in 1962. Scheduled to work 460 days around Mars, Phobos 2 lasted only 57 days in Mars orbit.

Like Venus, Mars is much like Earth. Some scientists believe life may once have existed on the Red Planet. If the little Martian moon Phobos, described as looking like a "bitten apple," is an asteroid grasped and held by Mars' gravity, as some scientists suggest, the Phobos probes were to have studied matter unchanged since the infancy of the Solar System. The soil might have revealed clues about how the Solar System formed 4.6 billion years ago.

First there were two. Phobos 1 and Phobos 2 had been hurled on their 201-day flights to Mars July 7 and July 12, 1988.

Delegations from around the world gathered each time at Baikonur Cosmodrome to see the white probes atop their silver rockets, standing out against the flat brown plain. Massive clouds of gray and white smoke pierced by spurts of orange and blue flames split the Soviet Central Asia night sky twice.

Each time the Soviet Union fired one of the pair of science probes on its odyssey to explore the mysteries of Mars in preparation for a manned flight at the turn of the century, a powerful 198-ft.-tall Proton rocket roared to life, lifting its solar-powered probe off the steppe.

Hours after each launch, the on-board navigation system in the craft was switched on. Around the globe, giant dish antennas swiveled to track Man's latest outbound marvel. Even NASA's Deep Space Network prepared to help out with radio tracking data so the Russian probes could locate the midget Martian moon.

Asleep at the switch. Phobos 1 was more than 12 million miles along its 111-million-mile route on the night of August 29-30 when a ground control operator at the USSR mission control center near Moscow transmitted a deadly radio command. A watchdog computer, failing to detect the human error, did not flash a "bad command" alert. In violation of control center rules, a backup human operator was not on duty to check the work of the controller who made the mistake.

The erroneous signal ordered Phobos 1, in effect, to commit suicide by turning its vital energy-absorbing solar panels away from the Sun, causing the spacecraft to loose orientation. Within 48 hours, the probe's systems were frozen. The mistake wasn't discovered for three days.

The probe was rendered useless, falling into what one Soviet space official called a "deep lethargic sleep." Ground controllers kept on transmitting signals at Phobos 1, but with no luck. The probe was rushing through space, away from Earth, its orientation and guidance systems turned off. Unable to communicate with the spacecraft, operators no longer could control the flight path of Phobos 1 and its batteries died. A month later, Moscow mission control gave up. Phobos 1 was lost.

Phobos 1 and Phobos 2 carried almost identical equipment, but the lost probe also carried instruments to register neutron radiation from Phobos' surface and to study Mars' magnetic field.

As Phobos 2 rushed on toward a rendezvous with the Red Planet, Phobos 1 sailed by, thousands of miles from Mars, on its way into orbit around the Sun.

Then there was one. Meanwhile, Phobos 2 continued its cruise toward the Red Planet, sending back data on the solar wind encountered along the way. In the first two months after blast-off, Russian scientists communicated via radio with the spacecraft 75 times. They were able to measure trajectory and monitor operation of on-board systems and scientific equipment. It's traditional for interplanetary craft to study the Sun and interplanetary space while enroute. Phobos 2 was no exception on its journey to Mars.

In December, the official Tass News Agency said some instruments aboard Phobos 2, including three TV cameras and a spectrometer, had failed enroute, but most had been fixed although one of ten particle stream instruments remained broken.

The Phobos 2 probe also may have suffered problems with its communications gear as it approached Mars. Reports indicated trouble with the probe's main 50-watt radio transmitter, used to send data to Earth at a high rate. Phobos 2 still could use its 5-watt low-data-rate back-up transmitter.

The high-data-rate system could send 4,000 computer bits per second. The low-power system could send at about 10 percent of that rate. Phobos 2 had internal computer memory storage capacity of 30 megabytes.

At Mars. As 1989 began, Phobos 2 began to register the gravitational pull of Mars.

Giant antennas across the Soviet Union tracked the spacecraft. The U.S. space agency's three Deep Space Network antennas were turned toward Phobos 2 ten times in January 1989 as the probe neared Mars. Nine of those times NASA focused two of the antennas on a deep-space quasar of known location to calibrate tracking before switching to Phobos 2.

NASA's Jet Propulsion Laboratory, Pasadena, California, used the reception to compute precise locations for Phobos 2 as it trekked toward the Red Planet. Moscow control also did computer calculations from its own deep-space antennas. The U.S. and USSR data was compared to estimate the exact locations of the probe.

Braking rockets. Moscow control ordered Phobos 2 to fire course-correction rockets for several seconds January 23. Then, on January 29, breaking rockets fired for 201 seconds as Phobos 2 arrived at its rendezvous point. The probe arrived in equatorial orbit around Mars January 29, 1989.

The last previous spacecraft from Earth to arrive at Mars had been the American Viking landers which had touched down on Mars in July and September 1976. The last USSR spacecraft to reach Mars had been Mars 7 launched in 1973. It had flown past Mars in March 1974 and had dropped a lander which missed the planet.

Science work starts. Phobos 2 was in an elliptical path around Mars, at altitudes ranging from 530 miles by 49,500 miles above the planet. On February 1, magnetometers and plasma-wave instruments in Phobos 2 began collecting and sending to Earth data on the planet's atmosphere and structure. Infrared light sensors and gamma ray meters were turned on.

Phobos 2 took infrared light and gamma ray readings while recording daily and seasonal temperature changes on the planet, making a temperature map of the surface, and identifying areas where the soil was permanently frozen. Data on the planet's minerals and atmosphere also was collected.

Rockets were fired February 18, 1989, to make the orbit more circular and just 217 miles farther from Mars than the orbit of the tiny moon Phobos. The principal ballistic goals of the flight were met with the very close approach to Mars and Phobos.

Phobos photos. On February 21, Phobos 2 transmitted its first images of the 17-mi.-wide natural moon circling the planet Mars. Nine high-quality television images of Phobos from various angles were recorded and transmitted to Moscow mission control.

Mission control communicated numerous times with the spacecraft through February and March, with no problems reported.

On March 25, mission control had regular communications with Phobos 2, taking pictures of Phobos. No problems were encountered. Another photo opportunity was scheduled two days later.

Mars was about 166 million miles away as March 27 dawned on Earth. Radio signals travelling at the speed of light took 15 minutes to get to Mars. The USSR's probe orbiting the Red Planet was so far away it took 30 minutes for command signals from Moscow mission control to get to Phobos 2 and an acknowledgement to return to Earth.

Final days. Telemetry radio signals were received normally in the morning of March 27, indicating all systems were working well. Phobos 2 was in its circular orbit.

During the afternoon photo session at 4 p.m. Moscow time, the probe was to take pictures of the moon from a distance of 133 to 231 miles.

Flightpath engineers wanted the photos to see exactly where Phobos 2 was in relation to the moon. They would use that information to calculate the exact time and position for the coming low-level flight over the surface of Phobos.

To make its low-level pass days later, early in April, Phobos 2 rockets would have to have been fired to change the orbit to make Phobos 2 swoop low over the pitted surface of the moon Phobos. Phobos 2 was to have drifted across the surface of the moon for 20 minutes at an altitude of 100 to 260 feet at a speed of 4 to 11 miles per hour. During the 20-minute fly-over, an amazing amount of scientific detection work was to have been underway and two landers were to have been dropped.

Contact lost. Instead, radio contact was lost March 27. The loss occurred after the probe was ordered at 3:59 p.m. Moscow time by its on-board computer to turn away from Earth, take pictures of the Martian moon Phobos and then point its antenna again toward Earth to send back picture data.

Phobos 2 must have made the swing away, but failed to swing its antenna back toward Earth after the photo session between 6:59 and 7:05 p.m. Mission control was unable to re-establish stable radio contact with the probe as scheduled. At 8:50 p.m. Moscow time, the probe's signal was received for 13 minutes but then disappeared again.

Phobos 2 had been finding its own position in space by looking at the Sun and the star Canopus. Engineers weren't sure why it couldn't regain its lock onto those two stars and correct its position. The brief 13-minute signal received at 8:50 p.m. seemed to sweep across Earth, as if the spacecraft had not been able to regain its stable Mars obit and was spinning aimlessly.

What Novosti Press Agency described as "12 task forces of specialists" immediately started looking for a solution. Their task was complicated by the probe being hidden behind Mars for an hour and a half each day.

Some good things. Phobos 1 and 2 were a $480 million USSR-international project which included technology and experiments from more than 13 countries and the European Space Agency. By the time its signal was lost, Phobos 2 already had transmitted pictures of the surface of Mars, the moon Phobos and much other science data.

It reportedly had registered the presence of water vapor in the atmosphere of Mars for the first time. Evidence was received from Phobos 2 that Mars has radiation belts.

The Moscow TV evening news program Vremya showed Mars surface maps compiled from photos sent back by Phobos 2.

Early preparations. Soviet space officials had said the Phobos 1 and 2 mission would help them to scout out landing sites for a manned expedition to Mars about the year 2005.

Phobos 2 was a duplicate of Phobos 1, except for a device to measure X-rays from the Sun. Both were to have soft-landed science outposts on Phobos to analyze soil samples and both carried plaques commemorating American astronomer Asaph Hall who discovered the tiny moon Phobos and the other Martian moon, Deimos, in 1887.

The Phobos landers. Phobos 1 and Phobos 2 each had two moon landers. When radio contact was lost, Phobos 2 was within a few days of its low-level flight over the moon Phobos and the dropping of its pair of landers. Some of the now-useless experiments aboard the Phobos spacecraft:

10 megawatt laser. To study moon soil, a 10 megawatt laser beam from Phobos 2 was to be focused on a minute 10 square feet section of real estate on the surface of Phobos as the mother ship drifted across the surface. Nearly 150 laser pulses were to be beamed down to evaporate particles of Phobos soil.

Vapors of evaporated material were to blast upward in a plasma cloud which would have surrounded the Phobos 2 probe 100 feet above the surface. Traps mounted on

Phobos 2 were to capture and analyze rising particles. On board Phobos 2 was a mass-spectrometer to analyze that plasma cloud. It would have transmitted to Earth its findings on the composition of bedrock on the moon Phobos.

Ion gun. Meanwhile, an ion gun aboard Phobos 2 was to be fired at the surface of the moon to knock dust loose. The debris would have been collected by the probe as it sailed overhead and analyzed to see what atoms compose the dust covering the surface of Phobos.

Beams shot by the ion gun as the spacecraft drifted across the moon were to knock off secondary ions from the surface. These ions would have been analyzed by the probe's onboard mass-spectrometer.

Gamma ray meter. A gamma ray recorder was to sniff Phobos' rocks for radiation from naturally-radioactive elements and from irradiation of the soil by cosmic rays. Each basic element which can form rocks—iron, magnesium, silicon, aluminum, potassium, calcium, sodium and oxygen—emits its own gamma energy. Measuring the level of energy, Phobos 2 would have discovered which basic elements are in the moon's surface rocks.

Television. Phobos 2 was to turn on its videospectrometric complex as it approached the moon. As it closed with the moon, at 164 feet above the surface, it was to start sending live TV pictures. The pictures would have been so sensitive that topographic features and objects as small "as a spectacle lens" were to be discernible.

The Phobos television camera and picture memory were to transmit 1,100 color pictures to Earth. As the probe swooped across the moon, television cameras were to make panoramic surveys of Phobos. TV transmissions were to continue for a time as Phobos 2 flew away from the moon.

Hopping around Phobos. In addition to remote analysis from 100 feet above Phobos, direct contact with the surface was planned. The most spectacular event of the 20-minute fly-over was to have been ejection of two 120-lb. landers—a small hopping rover and a long-duration stationary TV lab—to soft-land on the moon's surface. Each probe was 10 feet across with solar-panel wings for power.

The hopper was to be ejected from Phobos 2 to bounce onto the surface of the moon. It was to use built-in "whiskers" to right itself to a working position. The hopper would have studied physical characteristics of the moon, the chemical composition of surface soil, and the magnetic field of Phobos.

To do that, the rover would have used rotating legs to hop a distance of 165 to 330 feet, analyze the soil and transmit information. Then it would have hopped again, analyzed again and transmitted new information. At least 10 hops had been expected. The transmitted information was to have been relayed to Earth by the main Phobos 2 spacecraft, by flying away from the moon into Mars orbit.

The lander had an accelerometer, an X-ray fluorescence spectrometer, and other science instruments. The small moon is less than 17 miles in diameter. Its gravity is one-thousandth of Earth's.

Moon TV. The Phobos 2 mother ship was to drop another 120-lb. lander. It would have glided to Phobos' surface, fired small rockets skyward to force itself down to the ground, then sunk harpoons with ropes. It would have reeled itself in with the ropes and affixed itself to the tiny moon.

The lander had a miniature rotating TV camera, a transceiver for maintaining contact with Earth, a solar panel and scientific instruments.

The TV lander was designed to last a year, shooting panorama photos of the landing site. It also was to study the physical and mechanical properties of the soil and its structure. The lander was to measure temperature, surface composition and moonquakes. Its radio beacon to Earth was to have been used to precisely measure the motion of the moon Phobos orbiting Mars.

The lander had a seismic sensor and a device that would have dropped tiny balls inside the probe to calculate Phobos' gravity and predict its orbital position in detail.

Phobos falling. One of the most important objectives was determining the degree of the acceleration of the motion of the moon Phobos in orbit around Mars. The fact that the little moon has been spiraling inward toward Mars faster and faster has been known for some time. It is being drawn ever nearer by the planet's gravity.

Astronomers calculate the moon Phobos will fall on the planet in 30 to 70 million years. A more accurate calculation of the acceleration would let researchers trace the history of the natural satellite's orbit to gain insight into its origin.

International project. NASA's Deep Space Network system of giant dish antennas operated by the Jet Propulsion Laboratory in California, Spain and Australia had been providing USSR scientists with precise tracking information. Also tracking Phobos 2 was the European Space Agency's operations center (ESOC) and the French CNES space agency tracking station at Toulouse, France.

NASA scientists had planned to measure the wobble of the moon Phobos by tracking radio transmissions from one of the landing craft, to gain insights into the nature of gravity, the internal structure of Phobos and the weight of asteroids passing near Mars.

Phobos 2 carried research instruments and science experiments from more than 13 nations to Mars. The 13 countries were Austria, Bulgaria, Czechoslovakia, East Germany, Finland, France, Hungary, Ireland, Poland, Sweden, Switzerland, USSR, West Germany and the European Space Agency. The United States contributed ground support only. The USSR Academy of Sciences, Institute of Space Research, was in charge of the project. The USSR said the Phobos mission was first of a series of expeditions to be completed by the landing of human beings on Mars about the year 2005.

Hall memorial. It was an indicator of coming U.S.-Soviet cooperation in space, when the USSR's Phobos probe carried a plaque commemorating that natural satellite's discovery in 1877 by American astronomer Asaph Hall. NASA said the Russians agreed to the idea in a "continuing spirit of cooperation in space."

The plaque. The two moons of Mars—Phobos and Deimos—were discovered in August 1877 by American astronomer Asaph Hall using a 26-inch telescope at the U.S. Naval Observatory in Washington, D.C.

In 1988, NASA gave the Soviet Academy of Sciences a photographic transfer on aluminum of Hall's August 17, 1877, telescope logbook page. The USSR placed the aluminum plaque on the spacecraft.

The plaque, in addition to reproducing the original Hall notes, carried two phrases in both English and Russian: "U.S.S.R. 'Phobos' Mission 1988" and "Discovery of Phobos—Asaph Hall—U.S. Naval Observatory—August 17, 1877."

The discovery. Hall was looking for satellites around Mars when he detected what he logged as a Mars star moving against the background of stars on August 11. The log for August 17 recorded his discovery of a second object moving with Mars. He recorded position angles and separations of Phobos and Deimos, noting Phobos was "faint and difficult to observe." George Anderson, Hall's assistant, also saw the satellites.

The Naval Observatory announced the discovery of the natural Martian satellites, naming them for the horses of Mars, the Roman god of war.

Since 1877, the Naval Observatory has continued to observe Mars, Phobos and Deimos using the same telescope. The Russian probes followed data based on U.S. Naval Observatory studies dating back over the past 100 years.

Asaph Hall's great-great grandson, Andrew Hyde, fortuitously worked for a U.S. Senator and was able to persuade NASA to ask the USSR to commemorate the discovery.

"This is something my great, great grandfather, in his wildest imagination, never could have imagined," Hyde said.

What next? Phobos 1 and 2 were the first of a new series of probes designed by the Babakin Engineering Research Center, replacing the earlier Venera-style spacecraft.

The USSR plans to use the Phobos-style probe again for several Mars flights in the 1990's. An unmanned spacecraft will be sent in 1994 to orbit the planet and drop small landers. Another unmanned spacecraft will fly from the USSR to Mars in 1996, landing there with a rover vehicle and a roving balloon.

Meanwhile, the multi-billion-dollar Phobos 2 package continues to circle Mars. Someday explorers will recover it and see why it lost its lock on Canopus and the Sun.

Mars Probes: USSR Phobos Twins
Well, What Is Phobos?

When the USSR fired two interplanetary probes—Phobos 1 and Phobos 2—toward Mars in 1988, the editors of SPACE TODAY asked for information on the spacecraft. This disptach was written by Mikhail Chernyshov, science correspondent for Novosti Press Agency.

A powerful Proton rocket, on the night of July 7, lifted off and put on course toward Mars the first automatic station, Phobos 1. On July 12, Phobos 2 blasted off following the same route.

Priority in this unprecedented international project is given not to Mars, but to its small natural satellite—Phobos. Why? Specialists believe that the material of which the Solar System was once made has been preserved better on Phobos. The project is expected to clear up many cardinal aspects of our knowledge of the surrounding world. This is why the small celestial body is now riveting the attention of practically all world science.

In 1877, the American astronomer Asaph Hall discovered two tiny Martian satellites, which he named Phobos and Deimos. But, for a long time, things did not go beyond a mere acknowledgement of the existence of the satellites. Instruments then available were not powerful enough to determine their size, weight and other characteristics. It was not until a century later that, with the aid of interplanetary probes, we gained closer knowledge of Phobos and Deimos.

It emerged that Phobos was shaped like a potato measuring 27 by 15 kilometers [17 by 9 miles]. After estimating the density of Phobos rocks, scientists came to the conclusion that they differed from basic terrestrial rocks. But the measurements were indirect and nothing definite can as yet be said about the composition of Martian satellite substance.

The Phobos stations are equipped with perhaps all the latest research tools: 14 project countries and some bodies of the European Space Agency have pooled their scientific and technical capabilities. But, perhaps the main point is that those probes are to be dropped from the station directly on the surface of Phobos.

One of the [lander] probes is of a stationary type. It is a kind of research platform that will anchor itself to the satellite's surface by means of a harpoon.

The platform has its own power station—solar panels—to energize its equipment and to send radio signals to Earth. The panels will not unfold at once upon landing. If there is dust on Phobos, it might settle on the panels and render them inoperative. So the unfolding has some delay built into it. Then, the three panel petals will open up and, like a sunflower, will seek the Sun and turn to face it. Mars is at a greater distance from the Sun than the Earth and gets less solar energy. For this reason, the [turning toward the Sun] step is doubly important: panels will be getting more energy and the receiving and transmitting aerials, connected with [the] orientation, will be trained on Earth more accurately.

The platform will have most of its operations performed automatically. It takes a radio signal several [nearly 20] minutes to travel from Earth to Mars and back again to Earth, while the entire procedure of the [lander] probe separating itself from the [orbiting] station, landing and being geared up for operation will take much less time. [The radio signals will travel about 186,000 miles per second.]

A Phobos day lasts eight hours. This will be the frequency with which the probe will come in contact with the ground-based network of radio telescopes [on Earth] and send all that it has been able to gather over the eight hours. The platform is fitted with instruments for direct analysis of soil. It has a photo-television device that can scan not only the surrounding terrain, but also photograph the microstructure of rock.

Phobos, due to gravitational pull, gradually approaches Mars and will fall on the planet within the next 30 to 70 million years. The motion characteristics of both celestial bodies are generally known to scientists, but still there is something peculiar about their ballistics, separating Mars and Earth. All large radio telescopes of the USSR, the U.S. and Western Europe will be taking part in the effort. Measurements will also help to specify the so-called astronomical unit [AU]—the mean distance between the Earth and the Sun. It is, of course, known and equals 150 million kilometers [about 93 million miles]. But its present value, registered in astronomical books with many decimal places after the point, no longer suits ballistic experts. They want a better approximation of its absolute value.

The second landing probe is a kind of jumper. It is a semi-spherical apparatus which, by means of special legs, will make jumps of 100 to 300 meters [roughly 300 to 1000 feet] over the surface of Phobos. The service life of the jumper is less than that of the stationary platform, but it can explore a much greater part of the Phobos surface.

It will take the interplanetary stations more than six months to reach Mars. The main events will take place in 1989.

Mars Probes: USSR Phobos Twins
200 Days of Flight Lie Behind

When the USSR interplanetary probe Phobos 2 reached Mars January 29, 1989, the editors of SPACE TODAY asked for information on the arrival. This dispatch was written by Oleg Borisov of the Novosti Press Agency.

The Phobos spacecraft entered an intermediate elliptical orbit of Mars as an artificial satellite, completing the first stage of the project which started last July.

Fourteen countries—Austria, Bulgaria, Hungary, the German Democratic Republic, Ireland, Poland, the Soviet Union and the United States (ground-based support of the mission), Finland, France, West Germany, Czechoslovakia, Switzerland, Sweden—and the European Space Agency take part in the Phobos project. The USSR Academy of Sciences, Institute of Space Research, was the head organization for preparing and implementing the project.

Circular orbit. After entry into the Mars orbit, it was necessary within 30 days to correct the orbit by a self-contained power-plant, turning the elliptical orbit into a circular one which will exceed the distance from moon Phobos to Mars by 350 kilometers.

The spacecraft stayed in the circular orbit around Mars for a period of one to four months during which the characteristics of moon Phobos' motion (ephemerides) were calculated. Then the rendezvous with Mars' moon Phobos was started.

The space probe entered the orbit of Phobos and, having reached the target, drifted over the surface of Phobos at an altitude of 30 to 80 meters (98 to 262 feet) at a speed of 2 to 5 meters per second (6.5 to 16 feet per second or 4.47 to 11.18 miles per hour).

The rendezvous will last not more than 20 minutes during which it is planned to carry out studies including seven independent experiments.

10 megawatt laser. To study soil, a laser beam whose energy density will exceed 10 megawatts will be focused on a tiny one-square-meter section (10.76 square feet) of the surface of Phobos. The upper layer of substance will immediately evaporate and, in the form of a plasma cloud, will reach the onboard mass-spectrometer. The analysis will show the composition of bedrock of this celestial body.

One of the instruments is designed for recording gamma radiation of Phobos' rocks. This radiation is emitted by natural radioactive elements and due to the irradiation of the surface by cosmic rays. Each of the basic rock-forming elements—iron, magnesium, silicon, aluminum, potassium, calcium, sodium and oxygen—emits gamma-quanta of a definite energy typical of the element. Measurements of energy levels will make it possible to determine the elemental composition of surface rocks.

Ion gun. Another method of determining the chemical composition is to record neutrons coming from the satellite due to the interaction of cosmic rays and solar radiation with its surface. An ion gun will be used to study the atomic composition of dust which covers Phobos. Beams shot by the ion gun at the stage of the spacecraft's drift will knock off secondary ions from the surface. These ions will be analyzed by the onboard mass-spectrometer.

The videospectrometric complex carried by the spacecraft will transmit television pictures from a height of 50 meters (164 feet) above Mars' moon. From these pictures, one will be able to discern topographic features as small as a spectacle lens.

The complex will be switched on at a close distance to Phobos, and will operate while at the minimum distance and while flying away from the Martian moon. During the flight at a minimum height, television cameras will carry out plan and panoramic surveys of Phobos.

Landing on Phobos. In addition to remote studies of Phobos, direct contacts with its surface are envisaged. Landing sondes will be dropped on Phobos. One of the sondes (landers) has a miniature TV camera, a transceiver for maintaining contact with Earth, a solar panel, scientific instruments and a device for a reliable hold on the surface since the gravitational force on Phobos is 1,000 times weaker than Earth's gravitational force.

Using this lander, whose active service life is designed for about a year, it is intended to obtain a panorama of the landing site, to study the physical and mechanical properties of the soil and its structure.

But the main objective of the module is to determine the degree of the acceleration of Phobos' motion in orbit around Mars—this acceleration has long been noticed by astronomers—and the inexorable drawing nearer to Mars. Scientists believe that Phobos will fall on the planet in several tens of millions of years.

A more precise determination of the acceleration value will enable scientists to trace the history of the satellite's orbit and hence to get an insight into the mystery of its origin.

Hopping around Phobos. The other lander is a hopper. It will be delivered to the surface of Phobos where it will assume a working position by means of special "whiskers." It will study the chemical composition of the soil surface layer, its physical characteristics and the magnetic field of Phobos.

The hopping rover will transmit information to ground-based antennas and will hop a distance of 20 meters (66 feet). Then it will begin everything again. There will be about ten such cycles.

The overall duration of the Mars-Phobos expedition will be 460 days. The rendezvous with Phobos and the landing of the descent modules on its surface are scheduled for April 1989.

However, experiments were started on the flight path to Mars: the Sun and interplanetary space became the first objects of investigations.

While these objects are traditional, and are studied during every interplanetary flight, the study of Phobos at a close range is a new task and experiments to be conducted there will signify a new step in studying small bodies of the Solar System. The Mars part of the program will usher in a new stage of intensive studies of the Red Planet. This is a prelude to a series of expeditions which, as many scientists believe, will be completed by the landing of human beings on Mars.

Mars Probe: U.S. Mars Observer

America planned to launch its Mars Observer interplanetary spacecraft on a Titan 3 rocket from Cape Canaveral in September 1992.

Mars Observer is the only currently-scheduled U.S. entry in the race to expand knowledge of the Red Planet, while the USSR is sending at least two large spacecraft to Mars—in 1994 and 1996. The U.S. last studied Mars with a pair of Viking orbiters and landers in 1976.

The unmanned Mars Observer won't land on the Red Planet. Rather, its computer is programmed to maintain a low, circular, polar orbit 227 miles above the surface of Mars, photographing, collecting and transmitting information for two years. Mars Observer will fly closer to the planet's surface than did the ill-fated Phobos 2 probe from the USSR in 1989.

Earlier delay. NASA received a go-ahead in 1984 to blast Mars Observer away from Earth in 1990 from a shuttle in low orbit. However, after the 1986 Challenger disaster, the take off was delayed and it became necessary to bolt Observer atop an unmanned Titan rocket for a ride to space.

Key science questions about Mars remain, but federal budgets for space research have been tight. To save money, Mars Observer was designed after a successful communications satellite with some weather satellite parts added on. It is small and lightweight with limited data handling capability, yet its few sophisticated instruments are to add much to man's knowledge of Mars.

Mars Observer has a moving-mirror infrared eye to see what's in the dust in Mars' atmosphere as well as on its surface. It also has a sharp camera for the best pictures ever taken of Mars from orbit. The camera includes a compact telescope. Pictures are expected to show two-mile squares of the Martian landscape.

Mars Observer has a radar altimeter to map hills and valleys on the surface and electronics to read Mars' magnetic field. A gamma-ray scanner will look for uranium, iron, potassium, calcium, magnesium and other key elements.

In orbit around the dry Red Planet, Mars Observer is to catalog the Martian atmosphere, especially its famous dust. And the satellite will spot different kinds of rocks on the surface.

French balloons. The Russians asked the U.S. to modify Mars Observer so it can act as a radio relay to Earth, boosting along data from small Soviet landers in a 1994 flight and relaying Martian snapshots taken by a Soviet-French roving balloon floating along the surface in a 1996 flight. The change cost little and was significant cooperation as it is likely to double or triple the amount of data transmitted by the Russian landers from the surface of Mars.

Second Observer? NASA is considering another unmanned probe to follow Mars Observer. That flight might be launched in 1998, land a robot rover on the surface in 1999, gather samples of soil, rock and atmosphere, and return to Earth by 2001.

Experts in both the U.S. and the USSR say unmanned landings are necessary before manned flights. Unmanned probes are complex, stretching today's automation technology and robotics engineering capabilities. A Martian rover, for instance, would have to avoid hazards on the planet surface when it lands, negotiate terrain, rendezvous with its launcher for a flight back to Mars orbit, and dock in Mars orbit.

NASA's Johnson Space Center (JSC), Houston, Texas, and Jet Propulsion Laboratory (JPL), Pasadena, California, were working out a scenario for a robot spacecraft to leave the U.S. for Mars in 1998 or later, and return home in 2001 or later with rock samples.

The spacecraft would land on Mars, rove around for sightseeing, gather 11 lbs. of soil, rock and atmosphere samples and bring them back to Earth. Such an unmanned flight would be a forerunner to later manned visits to Mars.

JPL designed a Mars-orbiting craft and a surface rover equipped with scoops and tongs. Both would rely on their own computer programs for most work since remote-control radio signals would take up to a half hour to get from Earth to Mars. JPL managed the Viking lander flights to Mars in 1975 and 1976 as well as most other American interplanetary spaceflights.

JSC designed a lander capable of flying down into the Martian atmosphere and slowing down for a gentle landing. The Martian atmosphere is only one one-hundredth as dense as Earth's atmosphere. JSC also is designing the lift-off rocket and other equipment needed to blast the lander back to Mars orbit for rendezvous with the orbiter before return to Earth.

Returned Martian samples would be dropped off at the U.S.-international space station Freedom in Earth orbit for lab analysis, similar to the chemical tests applied to Moon rocks nearly 20 years ago, and then carried on down to Earth by shuttle.

The USSR has said it plans to send men to Mars in the years 2005 to 2010. The U.S., Europe and Japan are likely to cooperate in such a manned flight, and send astronauts.

Mars Probes: USSR 1994 & 1996

The Russians plan to use an atomic rocket to send a six-member crew to Mars between the years 2005-2010 in a two-stage spacecraft assembled in Earth orbit, according to Russian space expert Vasiliy I. Moroz, a scientist in the Soviet Space Research Institute.

The USSR wants American scientists and astronauts to take part in the flight, but even without U.S. participation the Soviets are determined to explore the red planet, Moroz told the International Astronomical Union in 1988.

To get to Mars, USSR engineers are developing an nuclear-electric propulsion rocket for the manned voyage.

Water and life. Mars has only been partially explored by the American Viking probes, he pointed out, saying the USSR thinks it is time to return to Mars. Soviet scientists want to search for liquid water on the Martian surface. They hold out hope there may be life, even though Viking found none.

An atomic-electric rocket engine would be used to accelerate an explorer craft on an 18-month trip to Mars. The nuclear-powered rocket would accelerate a manned craft toward Mars after that craft had been boosted to Earth orbit by a chemical rocket, according to Moroz who said a Mars mission would require the new type of propulsion.

The current Russian space fleet includes the world's most powerful rocket, Energia, but Moroz said a chemical rocket couldn't accelerate a manned craft to Mars. He said the USSR has a high priority on exploration of Mars. Their timetable:

★In 1994, the USSR plans to send an unmanned craft to Mars carrying small Martian landers and surface penetrators. Most data from the 1994 flight will come from the orbiter observing Mars.

The U.S. Mars Observer should be in orbit over Mars when the USSR orbiter arrives. Mars Observer will help by receiving weak signals from the USSR landers and relaying them to Earth. The Soviet orbiter will relay signals to Earth also.

★In 1996, the USSR plans to send an unmanned craft to Mars carrying a rover vehicle and a roving balloon carrying science instruments. Most data from the 1996 flight will come from the rover vehicle and roving balloon.

The robot rover, exploring hundreds of miles of the planet's surface, would drill for rock samples and measure the atmosphere. The U.S. Mars Observer may still be on hand, in Mars orbit, to send relay data to Earth.

★A manned Martian flight might leave Earth as early as 2005 or 2010. A six-member crew would fly a two-stage ship from Earth orbit.

The Mars mission might use artificial gravity to keep crew members healthy. USSR cosmonauts have experience in weightlessness, however Russian and American studies have shown that long exposure to the weightlessness of space causes the heart to weaken and the body to lose bone mass.

Nuclear-electric power. An nuclear-electric rocket uses charged particles as the propulsion medium. Such rockets give only small thrusts, but can keep pushing for a long period of time since they are powered by atomic energy. Such a ship would accelerate slowly, but achieve a high speed eventually. It would be able to get by with less weight in fuel than chemical rockets for the same energy.

Such a nuclear spacecraft drive is about twice as efficient as chemical rocket systems now in use. American space scientists have talked for years of a need to design propulsion systems that do not depend upon chemical reaction of fuel and oxidizer. NASA looked into electric and ion drive rockets, but received no money to build a rocket.

Mars Probes: U.S.-USSR Together

NASA is helping the USSR pick landing sites on Mars for robot spacecraft to be launched in 1994 and 1996. The 1994 spacecraft will drop small landers. The 1996 probe will land a rover vehicle and balloon rover.

From its highly successful Viking landers, NASA accumulated photos and maps of Mars between 1975 and 1982. Two Viking craft orbited Mars and two went down to the surface of the Red Planet. Now, on the detailed American maps of the Martian surface, the Russians have found four possible landing sites, all within 30 degrees of the equator.

Swapping instruments. Russian-American cooperation will extend to using NASA's Deep Space Network to help track and communicate with the Soviet Mars probes; exchange of scientists for work on specific projects; exchange of data on Venus, a planet explored by both countries; exchange of Apollo lunar material for meteorites collected by the USSR; and exchange of science instruments for use on planetary probes.

USSR scientists planning launch of a robot rover to explore Mars may use data sent back by the U.S. Mars Observer to be launched in 1992. The unmanned Mars Observer will orbit the Red Planet two years, photographing the Martian surface. While doing that, it will relay to Earth signals from the Soviets' small 1994 landers.

Red And Black Mars Balloons. A red heart-shaped toy balloon tied to a large black plastic balloon floating across the California sky over NASA's Jet Propulsion Laboratory, Pasadena, was a prototype for a probe that will float a gondola of experiments across a Martian landscape in 1996.

NASA scientists were searching for a good method to explore Mars' surface terrain and composition in the 1990's. The idea was for a balloon inflated by Sun heat to float a basket of experiments around Mars to sample new territory everyday.

Now the balloon idea will become reality, with a balloon rover flying in a USSR spacecraft to Mars in 1996.

The technique uses a helium-filled or hydrogen-filled balloon attached to a solar-heated hot-air balloon. The two-balloon combo would support a gondola filled with instruments hardy enough to survive repeated landings and draggings across the ground.

Overnight. At night, the hot-air balloon would deflate. The gas balloon would just barely be able to hold it upright. The entire contraption, then, would stand still with the instrument payload sitting on the Martian terrain sampling the local environment.

Morning Sun. In the morning, as the Sun rises, the black hot-air balloon would absorb heat from the Sun, inflate with warm air, rise above the surface of Mars and carry the instrument gondola to a new location.

French Centre National d'Etudes Spatiales—CNES, France's space agency—chief scientist Jacques Blamont thought up the concept of exploring Mars with a solar-heated balloon while visiting NASA's Jet Propulsion Lab, Pasadena, California.

The JPL experiments with toy balloons were followed by tests with a 30-ft. hot-air balloon at NASA's Dryden Flight Research Facility, Edwards Air Force Base, California. The non-profit Planetary Society also backed the project.

Mars ballooning will allow scientists to study different locales from Martian polar caps to volcanic terrain. Landing sites could not be controlled easily, but knowledge of Martian wind patterns would permit a rough plan.

Mars Flights: Selecting Astronauts

NASA will have to do deep psychological tests, some lasting years, before it will be able to select crews for deep space flights such as a voyage to Mars, according to an advisory panel.

The committee said NASA will have to learn much more about the effects of weightlessness and forced confinement on the human body before the space agency can hope to attempt long-term space flight aboard a space station or a voyage to Mars.

The only U.S. experience in long-term space flight is 84 days three astronauts spent in the space station Skylab in 1973-74. By comparison, USSR cosmonauts Vladimir Titov and Musa Manarov hold the single-trip space record of 366 days. Cosmonaut Manarov also has the most accumulated time in space, 541 days.

The committee found NASA doesn't have enough information on effects of long flights to start missions lasting years, such as a voyage to Mars. The U.S. hopes the Soviet Union will share physiological data from Mir missions.

Weightlessness damage. The low gravity of space causes muscles to atrophy, the cardiovascular system to lose conditioning and bones to lose minerals and become thinner. The destructive reaction, known as the space adaptation syndrome, must be understood thoroughly before long-term space flight is possible, the panel said.

Former astronaut Gerald Carr commanded the 84-day Skylab 3 flight. He helped prepare the panel's report, saying NASA must study the psychological aspects of long-term space flight. The space agency's methods are not adequate for choosing crew members for flights that last years, Carr said.

He said studies have not been made of the effects of years' confinement of small groups in hazardous situations, such as space flight. Carr said past criteria must be changed because future astronauts will face far more complex stresses than those in Apollo and Skylab.

600-day lock-down. Former NASA executive William C. Schneider, who headed the Skylab program and was on the advisory panel, saw it "quite likely" NASA would have to conduct 500-day to 600-day tests to determine criteria for crew of long-duration space flights. Nobel laureate Frederick C. Robbins, committee chairman, said the space agency should strengthen research into how to send people into space for long periods, explore the Universe for life forms and study Earth, planet home of the human species.

In the study called Exploring the Living Universe, the panel also said NASA should:

★do intense life support experiments in orbit, especially to develop self-contained oxygen and water recirculation systems,

★increase work on the effects of gravity and weightlessness on various Earth life forms, including plants,

★increase studies of the possible existence of life forms beyond the Earth and the evolution of life.

Mars Flights: Nuclear Rockets

The USSR is designing a two-stage nuclear-fuel rocket to send cosmonauts to Mars about 2005, according to a Russian space expert.

Vasiliy I. Moroz, a scientist in the Soviet Space Research Institute, said a Mars mission would require a new type of propulsion. The current Russian space fleet includes the world's most powerful rocket, Energia, but he said a chemical rocket couldn't accelerate a manned craft to Mars.

A nuclear-powered rocket would be boosted to Earth orbit by a chemical rocket. Then the atomic-electric rocket engine would accelerate the manned ship on the 18-month trip to the Red Planet. After 1994, the USSR plans to test a new nuclear-electric rocket.

An atomic-electric rocket uses charged particles as the propulsion medium. Such rockets give only small thrusts, but can keep pushing for a long period of time since they are powered by atomic energy. Such a ship would accelerate slowly, but achieve a high speed eventually. It would be able to get by with less weight in fuel than chemical rockets for the same energy.

Rover. American space scientists have described a need to design propulsion systems that do not depend upon chemical reaction of fuel and oxidizer. NASA looked into electric and ion drive rockets, but received no money to build a rocket.

The U.S. government spent $1.4 billion between 1955 and 1973 on a nuclear propulsion system called Rover, mainly to develop a solid-core heat exchanger.

Such a nuclear drive is about twice as efficient as chemical rocket systems now in use. But the politics of sending nuclear reactors to Earth orbit overwhelmed the project.

Recently, the U.S. Air Force said there is a need for efficient propulsion systems late in the 1990's and after the turn of the century. If political problems could be resolved— and the public convinced nuclear powerplants wouldn't shower Earth with dangerous radioactivity—nuclear propulsion might be used to power a manned U.S. mission to Mars. The Galileo Jupiter probe has a plutonium-fuel electric generator.

The Los Alamos National Laboratory in New Mexico has spent a few million dollars recently to pick up the Rover project where it left off in 1973. The scientists hope it will benefit civilian, as well as military, space programs.

Sun Probes: Helios

Helios were two 815-lb. interplanetary probes sent to observe the Sun and its solar wind. They were constructed in West Germany and launched by the U.S. from Cape Canaveral on Titan-Centaur rockets in December 1974 and January 1976. Their orbits took them within 28 million miles of the Sun, closer than previous probes.

Sun Probe: Ulysses

The European-American solar probe Ulysses, once known as International Solar Polar Mission (ISPM), was ferried 160 miles above Earth in U.S. shuttle Discovery October 6, 1990. Its built-in 17-ft. inertial upper stage (IUS) rocket fired it away from Earth orbit toward Jupiter on a four year voyage across the Sun's north and south poles.

Ulysses was to swing around Jupiter February 8, 1992. Gravity from the planet was to fling the probe above the plane in which planets of the Solar System orbit the Sun, boosting it over the Sun's south pole in 1994, then over the Sun's north pole in 1995.

Ulysses is designed to measure magnetic field, plasma, electrons, protons, ions, interstellar neutral gas, gravity waves, solar particles, cosmic rays, plasma waves, radio signals, X-rays, gamma rays, cosmic dust. It will be the first probe to explore the Solar System's so-called "third dimension."

The probe's radio and plasma-wave receiver has two boom antennas, one 24.3 ft. long, the other 238 ft. During deployment of the 24.3-ft. boom November 4, 1990, a wobble appeared in the spacecraft's spin, but the shake seemed gone by 1991. One radio stopped working for a time in June 1991.

As Ulysses passed within 280,000 miles of Jupiter in February 1992, it swept down the huge planet's northern hemisphere and across its southern hemisphere, measuring the planet's magnetic field and its interaction with the solar wind for two weeks.

Heliosphere. The 11-ft., 121-lb. Ulysses is a joint American-European project to explore the heliosphere—the Sun and its environment. Ulysses had to use Jupiter's gravity to boost itself above the plane of the Solar System from where it will dive back toward the Sun. There were no rockets with sufficient power to send Ulysses directly.

The IUS rocket has a good track record, having been used several times in shuttle flights. It has launched from shuttles NASA Tracking and Data Relay Satellites (TDRS) and secret military satellites. An IUS booster and a TDRS were in Challenger when that shuttle exploded in January 1986.

Ulysses was to have been launched in May 1986 with a Centaur rocket fired from outside the shuttle Challenger cargo hold. Launch was delayed by the 1986 Challenger disaster. NASA banned Centaur upper-stage rockets from shuttle flights. The new launch date was selected to bring maximum return of data from the important science flight. Some 120 scientists from 44 labs in 12 countries are involved in the experiments.

Jupiter Probe: Galileo

The unmanned interplanetary spacecraft Galileo will flash past its home planet December 8, 1992, in the last bounce of a three-bank celestial billiard shot driving the probe toward Jupiter.

Galileo is on a roundabout 2.4 billion-mi. course to Jupiter which, in 1990, made it the first interplanetary craft from Earth to approach Earth. Astronomers got their first glimpse of our globe as they have seen other planets.

Galileo measured greenhouse warming of Earth's atmosphere as well as water vapor at high altitudes affecting the ozone layer. It measured the planet's magnetic field and solar plasma flows, while snapping pictures of the crescent Moon. The far side of the Moon always faces away from Earth. Galileo's route let lunar astronomers explore the far side for the first time with modern instruments.

Asteroid Belt. The spacecraft grazed the Asteroid Belt, snapping photos and taking measurements as it passed within 600 miles of the 20-mi.-wide asteroid Gaspra on October 29, 1991. Then it fell back toward the Sun and the inner Solar System, again streaking by Earth for a speed boost.

It will zip past our globe, 185 miles above South Africa, on December 8, 1992. Earth's gravity will give it one final boost in speed by 8,280 mph as Galileo heads out to Jupiter. It will be travelling at 87,000 mph relative to the Sun.

Dust collector. The spacecraft collects cosmic dust particles and ultraviolet light, as well as readings on electrical and magnetic fields in the deep space environment. Information collected enroute is radioed to Earth every four or five days.

The ultraviolet instruments, magnetometer, heavy ion counter and dust detector have been on and taking in science data during the cruise from Venus to Earth. Power was turned on for the energetic particle detector, the plasma wave and plasma science instruments. The photo-polarimeter-radiometer was turned on.

Galileo, called the most sophisticated interplanetary spacecraft ever built, is managed by NASA's Jet Propulsion Laboratory (JPL), Pasadena, California.

Venus pix. As Galileo approached Earth in November 1990, NASA scientists received 80 photographs and data recorded on a digital magnetic tape recorder as Galileo sailed within 10,000 miles of Venus February 9, 1990.

Galileo's main umbrella-shaped antenna, which was furled to protect it from sunlight in the vicinity of Venus, was to be opened after the craft passed Earth in 1990. Unfortunately, it stuck partially closed. One side of the antenna seemed hampered by something. NASA engineers sent a number of commands to loosen it, with no luck. Galileo is able to send data and photos, but in smaller quantities limited by Galileo's secondary antenna.

Dropping the probe. Jupiter, the Solar System's largest planet and the fifth planet from the Sun, is 480 million miles from the Sun. Galileo will study Earth, its Moon and one or two asteroids over a period of years before separating into two parts in July 1995. About 150 days before the orbiter reaches Jupiter, a small atmosphere probe will detach and start a long fall into the planet's atmosphere of swirling gases.

Arriving at Jupiter. On December 7, 1995, both orbiter and probe will arrive at Jupiter, completing the 2.4-billion-mile trip. The probe—a 760-lb., 4-ft.-diameter, teardrop-shaped instrument package—will plummet into Jupiter's atmosphere at 115,000 mph for the first-ever direct measurements. A parachute on the small probe will open as it plunges 370 miles through the planet's atmosphere, collecting and measuring the cloud environment of the Jovian atmosphere for 75 minutes as it sinks.

Data will be radioed up from the probe to a 900-megabyte digital magnetic tape recorder aboard the Galileo mother ship orbiting above the huge planet's atmosphere. As the probe falls, increasing atmospheric pressure and temperature eventually will crush and vaporize the probe.

The probe's demise won't be the end of the Galileo mission. The mothership will transmit the recorded probe data to Earth, then dance for 22 months through ever-changing orbits, using its electronic eyes and ears to scout the Solar System's largest planet, its rings, ten of its sixteen known moons, its magnetosphere and environment of electrically-charged particles.

The orbiter will sail a changing elliptical orbit past the planet's four major moons—Io, Ganymede, Callisto and Europa. The orbiter will take gravity assists from those big moons to set up low-altitude flybys of other moons as it sends TV postcards home to Earth—close-up snapshots of Jupiter and the moons. Galileo's on-board photography system provides image resolution ten-times better than the spectacular images sent home by the two Voyager spacecraft.

1989 launch. Galileo was launched to Earth orbit October 18, 1989, from Cape Canaveral in the U.S. shuttle Atlantis. The spacecraft's inertial upper stage (IUS) rocket fired the 5,000-lb. Galileo away from the shuttle that day on a looping six-year trip to Jupiter. The flight plan required Venus and Earth to provide gravity boosts, like slingshots hurling the craft on to Jupiter.

Galileo, a joint U.S.-German project named after the 16th century Italian astronomer who discovered Jupiter's moons, was to have been launched from the shuttle Atlantis payload bay in Earth orbit in May 1986. A hydrogen-fueled Centaur rocket would have driven Galileo quickly to Jupiter. The direct trip out to Jupiter would have taken only two years, but the voyage was delayed four years by the Challenger disaster.

After Challenger exploded January 28, 1986, killing seven crew members, the shuttle fleet was grounded until 1988. NASA decided it would be too dangerous to carry liquid-fuel Centaurs aboard a shuttle. So, Galileo's route to Jupiter was redesigned for a weaker rocket, making the flight last six years and cover 2.4 billion miles. The trip also was lengthened because Earth and Jupiter no longer were conveniently aligned as in 1986.

Changing the flight plan required new insulation, heaters and sun shades to protect the spacecraft from thermal stress as it flew by Venus. New radio antennas, computer software and special instruments were needed.

Nuclear power. The outer planets of the Solar System—Jupiter, Saturn, Uranus, Neptune, Pluto—are so far from the Sun that electricity-generating solar panels are not feasible. All deep space probes which have traversed the Asteroid Belt, including Voyager 2 which beamed back spectacular views of Neptune in 1989, have had nuclear power supplies. Such nuclear power packs also were used in the six manned Apollo flights to the Moon.

Galileo is powered by a pair of 122-lb. radioisotope thermoelectric generators

(RTGs) which convert heat generated by the decay of plutonium 238 into electricity. Each RTG started with 24 pounds of plutonium 238 dioxide fuel.

RTG power packs work by converting heat, released in the radioactive decay of plutonium-238 dioxide, to electricity. The RTG casings heat up to 350 degrees as they generate 572 watts of electricity.

Galileo's mileposts:

October 18, 1989. Galileo rode an afternoon launch to Earth orbit in U.S. shuttle Atlantis.

October 18, 1989. After six hours in orbit, Galileo will be blasted away from Earth by its Inertial Upper Stage (IUS) booster.

February 9, 1990. Flying around Venus at less than 10,000 miles above the planet, Galileo stared down at the cloudy planet, looking for lightning flashes and listening for radio whistlers which could indicate lightning. Ultraviolet photos of sunlight reflecting from the cloudtops were made. Infrared images of the planet's edges were snapped. To protect its sensitive antennas from the great heat near the Sun at Venus, Galileo's antenna was covered with a heat shield and pointed away from the Sun. Data collected was stored for transmission after Galileo left Venus.

December 8, 1990. In man's first incoming-from-deep-space encounter, Galileo flew directly along the tail of Earth's magnetic field, counting electrically-charged particles trapped there. The spacecraft looked at our planet's atmosphere for evidence of global warming. A spectrometer looked at near-infrared light, mapping the spread of chlorofluorocarbons (CFCs) and other gases contributing to a suspected greenhouse effect. A spectrometer looking at ultraviolet light observed the ozone hole over the South Pole while recording Earth's airglow of ultraviolet fluorescence. The craft looked for very high clouds pushed up into the mesosphere by thunderstorms. Scientists followed Galileo's radio signal closely, using tiny changes in the spacecraft's trajectory to calculate more precisely Earth's mass. Readings by a spectrograph as Galileo passed the Moon showed differences between minerals on the near and far sides of Earth's natural satellite. There is an unmapped area on the near side of the Moon, south of the vast Orientale Basin. Galileo photographed that strip of lunar territory.

October 29, 1991. Nine months after the end of its first encounter with Earth, Galileo was zipping through interplanetary space at 18,000 mph when it happened past the tiny asteroid Gaspra. Galileo came within 600 miles of that 9-mi.-wide rock. The spacecraft made a few quick snapshots and some basic data for transmission to Earth.

December 8, 1992. Galileo's second and final return home will see it pass very close to Earth—within about 186 miles. Lots of Earth-orbiting satellites fly higher than that! The flyby will fling the spacecraft into a larger six-year Sun orbit, wide enough to encompass Jupiter. Galileo will fly over the Moon's north pole where scientists want to look for ice in shaded crater sides. If past comet impacts on the Moon have left water, now ice, future manned lunar bases could put it to use.

August 28, 1993. After picking up speed along the way to 28,000 mph, Galileo will find itself sailing briefly alongside the asteroid Ida. At 18 miles diameter, this rocky body is about twice the size of Gaspra. Galileo will try for a few snapshots and some data about Ida to radio to Earth.

July 7, 1995. Two years after the Ida encounter, the Galileo spacecraft mothership will be homing in on colossal Jupiter when it kicks a small probe ahead into a trajectory directly down into the planet's tempestuous atmosphere.

December 7, 1995. The small Galileo probe will plunge head-on into Jupiter's stormy atmosphere, furiously reading everything it can about the environment, sending data up to the mothership entering an orbit above the planet. After falling inward for 75 minutes, the small probe will be crushed by the increasingly dense Jovian atmosphere.

December 7, 1995. The main Galileo spacecraft, reinforced years before on Earth with radiation-hardened electronics, will plunge through Jupiter's powerful radiation belts for a close flyby of Io and some close-up photos of volcanos on that Jupiter moon.

December 7, 1995. Galileo then will become a Jupiter orbiter, spending at least two years looking over the jumbo planet and its moons Europa, Ganymede and Callisto and sending back photos and data.

Saturn Probe: Cassini

Another grand tour of the outer Solar System is about to be launched. This time, NASA and the European Space Agency (ESA) plan to fire the interplanetary probe Cassini in April 1996 toward Saturn and its natural satellite, Titan, the solar system's biggest moon and the only one with an atmosphere much like ancient Earth.

Cassini is scheduled to be be launched on a Titan 4-Centaur space rocket, America's most-powerful booster. The mission would last at least eleven years, including seven just to get there. Along the way, Cassini would visit one or more asteroids, then fly by Jupiter, arriving at Saturn in the year 2002.

Cassini is named after French-Italian astronomer Jean Dominique Cassini who discovered several Saturn moons and features of the planet's rings in the 17th Century.

Spacecraft Cassini will fly by the asteroid 66 Maja in 1997 and the planet Jupiter in 1999. Cassini will pick up speed with a gravity boost from Jupiter, while taking a close-up look at Jupiter and its moons, then be on its way to an October 2002 rendezvous with the planet Saturn.

At Saturn, the planet-orbiter built by NASA and the Titan moon-lander built by ESA will take a long, close-up look at the giant gas planet, its moons, rings, atmosphere and magnetosphere.

Huygens. Titan is an immense, mysterious, gas-shrouded satellite of Saturn. Arriving at Saturn, Cassini will drop a small ESA probe, named after Christian Huygens, the 17th century Dutch astronomer and physicist who discovered Titan and the rings of Saturn in 1656, into Titan's atmosphere.

The Huygens probe will carry eight science instruments on its three-hour plunge through Titan's atmosphere to the surface. A large airbrake cone will slow Huygens to 600 miles per hour. Huygens will examine Titan's atmosphere and clouds.

During the descent, Cassini will remain in orbit as a radio relay station retransmitting to Earth data received from the descending Huygens probe.

At an altitude of 112 miles above Titan's surface, parachutes will deploy. Huygens then will drop slowly for two to three hours toward a soft landing on Titan's surface.

Huygens will be travelling only about 11 mph when it bumps down on the surface. Scientists hope that impact will be soft enough to allow analysis of soil or whatever's on the surface before the probe dies. If it survives landing on Titan, Huygens will send back panoramic TV pictures of the surface.

Scientists wonder if Titan is covered with lakes or oceans of ethane or methane. Liquid hydrocarbons might result from photochemical processes in Titan's upper atmosphere.

Huygens also will look for interactions between Saturn's magnetosphere and dust and moonlets in the rings, and with Titan's atmosphere. That may help scientists understand how plasma, dust and radiation interact—as in the formation of planets in the early Solar System.

Cassini. After the Huygens part of the mission, Cassini will spend four years circling Saturn 30 times, sometimes as low as 600 miles above the colossal planet's surface. It will be the first spacecraft from Earth to visit Saturn since the 1981 flyby by Voyager 2.

Cassini's onboard cameras will snap detailed pictures of the cratered, icy moons, map their topography with high-resolution radar and determine surface composition with spectrometers. Radar will map much of Titan's cloud-shrouded surface, similar to the way Magellan mapped Venus in 1990-91. Instruments will probe chemical processes which produced Titan's unique atmosphere. Some scientists say those processes may be something like the primitive chemical evolution on Earth.

Cassini science instruments will include cameras, a Titan radar mapper, ion and neutral mass spectrometer, cosmic dust analyzer, magnetospheric camera, ultraviolet imaging spectrometer, magnetometer and a visual and infrared mapping spectrometer.

Titan. The moon, with an organic chemistry in its atmosphere which may resemble the primitive Earth before the dawn of life, is largest in our Solar System and the only moon known to have a thick, organic-rich nitrogen atmosphere. Titan's surface temperature is –290 degrees Fahrenheit.

Saturn's 17 other moons are Rhea, Iapetus, Dione, Tethys, Enceladus, Mimas, Hyperion, Prometheus, Pandora, Phoebe, Janus, Epimetheus, Helene, Telesto, Calypso, Atlas, and Pan. There also may be half-a-dozen small unseen shepherd moons.

Saturn. The sixth major planet from the Sun, Saturn is second largest, after the planet Jupiter, with a diameter of 75,000 miles. That compares with Earth's diameter under 8,000 miles.

Saturn is nine to ten times as far from the Sun as Earth, taking 29.5 years to complete one orbit of our star. The planet's density is only 0.7 that of water, lowest of any planet. Like Jupiter, Saturn seems to have yellowish dark and light cloud bands parallel to its equator. These belts, zones and spots aren't as prominent as those on Jupiter.

The planet is mostly hydrogen. The upper atmosphere of Saturn is thought to be hydrogen, methane and ethane with floating clouds of ammonia crystals. At its heart, Saturn may have an iron core as big as Earth. Strangely, Saturn, like Jupiter, radiates more energy than it absorbs from the Sun. It probably has an internal primordial energy reservoir.

Ringed planet. Through a telescope, Saturn is stunning, with prominent rings girdling its equator. The rings, 375,000 miles across, make the planet shine brightly for astronomers. Galileo saw the rings in 1610. Huygens recognized them as a system in 1656.

The rings range from 150 feet to 300 feet thick. A 35,000-mi.-thick blanket of hydrogen extends above and below the rings.

Rings actually are thousands of tiny ringlets made up of countless small particles. The particles mostly are an inch or so in diameter. However, in the so-called B ring, chunks of matter may be up to a few yards in diameter.

B, the brightest ring, seems to have spokes and red particles, and is punctuated by lightning flashes. C ring, which looks grooved and has many individual ringlets, fills the gap between B and D. Ring D, closest to Saturn, seems to go right down to the cloud tops.

The so-called Cassini Division, one of the prominent features of the ring system, is a 3,000-mi. gap between ring A and ring B. Gravity interaction between Saturn's natural satellite Mimas and the particles in the B ring may cause the gap.

Ring A, actually many ringlets and small gaps, is not as bright as ring B. Ring A has a sharp outer edge, indicating a tiny shepherd satellite nearby.

There are minor rings scattered beyond the A ring. For instance, the F ring seems braided and is not circular. Two tiny shepherd satellites bracket ring F. Even more tenuous is ring G, farther out, between the orbits of Mimas and two small satellites. Most distant is the E ring which extends nearly to the orbit of Saturn's moon Enceladus.

Probes from Earth already have revealed that at least 22 moons surround the jumbo planet. Saturn and its system of rings and satellites, including Titan, were probed by U.S. Pioneer 11 in 1979 and U.S. Voyager 1 and Voyager 2 in 1980 and 1981.

Scientists. NASA in 1990 appointed the main Cassini scientists. They come from eleven U.S. universities, three U.S. laboratories, thirteen foreign countries and NASA's Ames Research Center, Goddard Space Flight Center, and Jet Propulsion Laboratory. They will design 62 science experiments for Cassini and Huygens. The experimenters will study the structure and composition of Saturn's atmosphere, and the physical properties of ring particles, moons and moonlets within the rings.

Comet Halley Explorers

The list includes spacecraft name, country or group of nations which launched it, launch date, date and universal time (with small Z) when it was closest to Comet Halley, distance from the Sun in millions of miles, distance between spacecraft and comet in miles, and the speed in miles per hour at which the spacecraft flew by the comet.

Craft	Vega 1
From	USSR
Launch date	12/15/84
Fly-by date	3/6/86 0720z
To Sun	73.47 miles
To comet	5523.990 miles
Speed	177,165.4 mph

Craft	Vega 2
From	USSR
Launch date	12/21/84
Fly-by date	3/9/86 0720z
To Sun	77.19 miles
To comet	4989.611 miles
Speed	171,796.7 mph

Craft	Sakigake
From	Japan
Launch date	1/8/85
Fly-by date	3/11/86 0418z
To Sun	79.98 miles
To comet	3.73 million miles
Speed	168,441.3 mph

Craft	Giotto
From	Europe
Launch date	7/2/85
Fly-by date	3/14/86 0003z
To Sun	82.77 miles
To comet	372.8227 miles
Speed	153,006.4 mph

Craft	Suisei
From	Japan
Launch date	8/18/85
Fly-by date	3/8/86 1306z
To Sun	76.26 miles
To comet	93,827.05 miles
Speed	163,296.3 mph

Craft	ICE
From	USA
Launch date	8/12/78
Fly-by date	3/25/86 1031z
To Sun	99.51 miles
To comet	17.46 million miles
Speed	145,177.2 mph

Comet Halley Probes: Vega 1 & 2

Half An Hour In The Comet's Coma

In 1986, various countries and international organizations sent science probes for a close-up look at Comet Halley as it made its closest approach to Earth and the Sun in 76 years. The USSR sent two Vega spacecraft. When the editors of SPACE TODAY asked the USSR for information on Russian deep-space science, the response was this story of how the USSR's Vega probe inspected Comet Halley as it passed near Earth in 1986 written by USSR Academician Roald Z. Sagdeyev and Dr. Leonid V. Ksanfomality, a USSR physicist, mathematician and scientist.

In March 1986, two Vega spacecraft for the first time transmitted the image of Comet Halley's nucleus and performed numerous scientific measurements. The results of research performed by Vega, Sakigake and Giotto spacecraft became a subject of many scientific articles and conferences.

Today is about the 30th anniversary of the first spacecraft launch to another celestial body. Nowadays, spacecraft are able to perform much more complicated multipurpose programs than 30 years ago. Vega was precisely this: a complicated, multipurpose program.

In Russian, the name Halley is pronounced Galley. Thus, the Vega spacecraft's name came from the abbreviated form of Venus-Halley (Venus-Galley) or Ve-Ga. The abbreviation reflected the sequence of goals. That is, exploration of Venus in the first part of the program, then passage to Comet Halley and the encounter with the comet's nucleus.

Before this direct exploration started, information about the comet nuclei was indirect. It was obtained by ground observations of a very extensive head, or coma, of the comet. It was rather difficult to explore Comet Halley due to many factors, some of which we will note here:

★unknown physical phenomena to be faced by the spacecraft near the nucleus,

★the fantastic speed of the mutual approach between the spacecraft and the comet,

★bombardment of the spacecraft by dust particles,

★and the not-well-known motion of the comet which is subjected to slightly irregular alterations.

Extremely complicated celestial-mechanics calculations were required to prove the feasibility of the encounter with Comet Halley's nucleus. But even more unprecedented technical problems were waiting ahead to be solved.

Preparations for the Comet Halley explorations drew mass-media attention as well. Having interviewed an engineer or scientist, the correspondents presented material to their readers in their own way, depending on the extent of their understanding. For instance, having interviewed Professor A. Galeev, one correspondent reported about a new "Galeev Comet." One correspondent, with an idea to interview an elderly scientist, found one who remembered well the appearance of Comet Halley in 1910. It turned out the elderly scientist was born in 1912.

There were a lot of funny situations but the development and creation of the Vega spacecraft were time and labor-consuming processes. But, finally, all the trials came to an end. The day of launch arrived. Then the day of approach to Venus and, at last, the day of encounter with Halley's Comet.

The half hour arrives. Early in the morning of March 9, 1986, there was a comet session with Vega 2 at the Space Research Institute of the USSR Academy of Sciences. The spacecraft was plunging quickly into the head (coma) of the comet, approaching its mysterious nucleus. Radio signals were traveling 160 million kilometers (almost 100 million miles) from the spacecraft. The path from the coma was nearly 100,000 kilometers (62,000 miles) long. Even so, the speed of approach was so great—75 kilometers per second or 168,000 miles per hour—that it would take only half an hour to pass through the coma.

All the people sitting in a brightly illuminated room full of numerous monitors and keyboards, pointers and plotters, were united by expectation. The businesslike atmosphere was supplemented with festive elation shared by the experimenters with the "outsiders" though they were not numerous.

Strictly speaking, unauthorized visitors were not to be let in. But several journalists still managed to get through the obstruction. Now they were rather high-handedly driven away from the tables piled with calibration tables and diagrams, and illuminated by colored flashes of characters and figures on the displays.

But even the experimenters themselves were eager to know what was going on at their neighbors' tables in other experiments. Only by comparing all the data obtained could one understand the complicated phenomena occurring near the comet nucleus not yet seen by anybody.

Sun melt. The nucleus of Comet Halley is a giant block of muddy ice, of dimensions believed earlier to be two to five kilometers (1-1/4 to 3 miles). But further processing of the TV survey images indicated much greater dimensions. It was a body of irregular shape, looking like an old shoe worn down at the heel. The longest axis of this celestial iceberg was about 16 kilometers (10 miles).

Once every 76 years, Halley's Comet passes relatively close to the Sun, and then goes away for a long time, somewhere beyond the orbit of Neptune. When the comet approaches the Sun, its surface is heated. The evaporation increases drastically. The flows of evaporating gases—first and foremost the water vapor—carry the finest solid particles away. Near the Sun, the general amount of ejected matter reaches tens of tons per second. Due to this fact, when the number of passes by the Sun is enough, the comet's nucleus will be destroyed almost completely.

The main mass of the jets falls on gas, and the amount of dust for Halley's Comet is about one-fifth of it. The products of evaporation into vacuum are accelerated up to considerable rates—about one kilometer per second or over 2,000 miles per hour.

Ultraviolet radiation from the Sun destroys the molecules of the primary gases. From the other side, the density of the gas diminishes quickly as it leaves the nucleus.

The ejected flows are affected by at least four factors: gravity, the pressure of sunlight, interaction with the solar-wind plasma, and electromagnetic fields interacting with any charged particles in the jet. These affects are displayed in the formation of the coma and different types of comet tails.

Some interesting peculiarities of the dust particles had been predicted by theory. Their trajectories look like a parabola, depending in a certain way on the mass of the particle. But the final answer was to be given by the Vega spacecraft.

Watching dust flow. Aboard each spacecraft there were several instruments especially for the exploration of the dust flows: to define the mass of particles, their number, composition and trajectory. Scientists from many countries participated in the preparation and the performance of the experiments aboard the new spacecraft.

Let me tell you about an experiment called DUCMA. It was to register dust particles with sizes starting at half of one-thousandth of a millimeter. This experiment was scientific cooperation between the Space Research Institute in Moscow and the E. Fermi Institute of the University of Chicago.

Professor John A. Simpson, the scientific supervisor of the experiment from the University of Chicago side, described figuratively the sensitivity of DUCMA this way: The finest comet dust particles recorded, as compared to cigarette smoke, are ten times as little.

The particles registered by the instrument were graded by mass. Dust particles with a mass exceeding the lowest by 1,000 times were recorded by cumulative number. The experiment enabled us to determine how the particles were distributed over the distance from the nucleus of the comet.

Americans became involved. That morning of March 9, 1986, Professor

Simpson and his American and Soviet colleagues watched the readings of the instruments with emotion. Although according to celestial measures the spacecraft was not far from the nucleus of the comet, the instruments still were recording few dust projectiles. However, the scientists were ready for that because three days earlier, on March 6, the main events had started aboard Vega 1 only some ten minutes before the closest approach to the nucleus.

So far there were few dust particles. We had enough time to review how it all had started:

In late 1983, when preparation of the Vega project was in full swing and the engineering model of the spacecraft was ready, one of the leading U.S. physicists, John A. Simpson, applied to the Space Research Institute of the USSR Academy of Sciences, through West German participants in the project, with a proposal to supplement the completed list of Vega scientific payloads with a new instrument aimed at registering dust particles ejected by the comet nucleus.

Comet Halley Glossary	
Coma	comet's large bright gassy head
Ducma	spacecraft dust particle counter
Galley	Russian version of Halley
Giotto	Europe's craft exploring Halley
Halley	large well-known comet
Nucleus	hard core inside comet's coma
Perihelion	closest approach to Sun
Sakigake	Japan's craft exploring Halley
Vega 1	USSR's first craft exploring Halley
Vega 2	USSR's 2nd craft exploring Halley
Venus	second planet from the Sun

Professor Simpson heads the prominent school in the Space Research Laboratory at the Fermi Institute, University of Chicago. His works on cosmic rays, anti-protons, etc., are widely known. About 40 space experiments have been carried out under his guidance. The name of Professor Simpson is known to non-physicists, as well, through his participation in the antiwar Bulletin of Atomic Scientists publication.

The new instrument was developed under the guidance of Professor Simpson by his pupil and colleague, Dr. A. Tuzzolino. The idea was to use extremely thin film of the polarized polymer, polyvinyldene fluoride, as a detector for the instrument. Materials of this type are called electrets. Once applied, this material can preserve a strong electric field. Electrets are similar to permanent magnets. The difference is they have a static electric field and not a magnetic field.

The new material turned out to be very sensitive, both to mechanical effects and to heat (pyroeffect). The comet dust projectiles cut into the film and burn out a certain microscopic volume of it. The film reacts by a voltage inrush on its conductive coating.

By the way, that new material has found many new applications in various fields of technology. But let's come back to our topic.

Breaking the ice. Everything was okay with the physics but Professor Simpson had to labor much in order to get official permission from the U.S. government to mount the new space instrument of the University of Chicago aboard the Soviet spacecraft. Then, in February 1984, a scientific meeting was held in Budapest to specify the scientific goals of the experiment and the technical measures necessary to "catch up with the train." That is, keep up with the spacecraft development program.

It was interesting to watch how, an hour after the meeting, the ice of formality and distrust started melting among the participants who were meeting for the first time. Let me note that American scientists and engineers always got every kind of assistance. They were treated with favor and friendliness by their Soviet colleagues during their further work at the Moscow Space Research Institute through 1984. The Americans also did their utmost best.

At the Budapest meeting, it was decided that scientists and engineers of both countries would join the efforts in the preparation of the experiment. Invaluable help was given by engineers from Hungary and West Germany who also became participants in the experiment.

Preparation of the experiment also required assistance of another kind. The abbreviation DUCMA stands for Dust Particle Counter and Mass Analyzer. DUCMA could become an actual analyzer only after calibration on a special dust-particle accelerator— the best of which is in Heidelberg, West Germany. That accelerator is a unique and extremely complicated instrument. The scientists and engineers of the Heidelberg Institute rendered assistance. The DUCMA instruments were calibrated for each Vega individually.

Of course, first of all there was the "General's question" to be settled concerning the experiment. Having weighed the pros and cons, the scientific management of Vega took an official decision to include DUCMA in the scientific payload (within the framework of cooperation between the Space Research Institute of the USSR Academy of Sciences and the University of Chicago, USA).

So, the University of Chicago group joined scientists from nine participating countries. That day, March 9, 1986, they were standing by the display exchanging comments on the rapid growth of the dust particle flow.

Professor Simpson managed to record the most interesting comments on a portable dictaphone. He explained that he wanted to supplement his phonogram collection with Vega.

Preparing DUCMA. The preparation of DUCMA had not been easy. One of the reasons was that the instrument had to be developed in the shortest possible time. That could be done only by real enthusiasts like our guests A. Tuzzolino, M. Perkins and G. Lentz.

A lot of special problems sprang up. For example, the number of electrical connectors in the spacecraft. Commands and telemetry channels were strictly calculated and fixed, and should not be changed. Then there was no free space aboard. But, anyway, we managed to build the instrument into the package of scientific payloads.

With regard to radio telemetry, sometimes there were blank spaces—windows—in the telemetry data flow. DUCMA intercepted a part of the flow, searched for such windows, and filled them with its own data. Great assistance was rendered by other experimenters and technical managers of Vega as well.

Flight to Halley. By the end of 1984, all the stages of preparation were over. That December, both spacecraft, Vega 1 and Vega 2, were launched.

While passing Venus, the sounders were released, exposing the DUCMA detectors which started recording sparse interplanetary dust particles. About ten particles per month were counted. That number changed only slightly until the spacecraft met the comet.

Now, five minutes before the closest approach, the instruments were registering several dust projectiles every second. Then 20, 30 50, etc. The instruments showed the peak, then the particle flow went down smoothly.

That was very unlike Vega 1 data, obtained on the 6th of March. Then, after the approach, the number of particles had jumped and increased by 30 times, reaching 1,000 per second. Evidently, the Vega 1 spacecraft had crossed the jet ejected from the active side of the nucleus. Before the closest approach of Vega 1 to the comet, the particle flow started to pulsate. Periodically the spacecraft was passing through areas of increased density of the dust. The distance between these areas of increased density was 300 to 500 kilometers (about 200-300 miles). It was assumed that the pulsations were connected with the jet.

Now, with Vega 2, there were no pulsations. During the three days between Vega 1 and Vega 2, the nucleus had made almost a revolution and a half and, now, it turned its passive side to the spacecraft (and to the Sun). Such was its location when it was approached by Giotto (European Space Agency's probe of Halley's Comet) on March 14, 1986.

But the structure of the dust flows became not the only scientific prey of DUCMA. Despite, the theoretical predictions, the finest particles occurring very far from the nucleus were the most numerous ones. On the other hand, some theoretical conclusions were immediately confirmed by the experiment.

Work on the far side. After the encounter with Comet Halley, the DUCMA instruments and the two Vega spacecraft were still active. During the first days of 1987, both spacecraft passed through a part of Comet Halley's tail which was very far from the nucleus. Again the dust particles were recorded, but this time their total number was less than ten.

Now, years after the encounter, the scientists have been able to fathom the processes occurring in the comet's nucleus better. The results of different experiments promoted the better understanding, not only when in agreement with each other, but the contradictory ones as well.

For example, as it has been mentioned, DUCMA noted a very intense jet with a sharp front, but meanwhile other dust experiments showed much weaker growth of the particle-count rate. One of the most probable reasons for this discrepancy is the physics of the comet and it is defined by the rather high resolution time of DUCMA.

Snowflakes. It turned out that the comet dust particles have a rather friable (easily pulverized) structure similar to that of a snowflake. In such a "shaggy" structure, the separate crystals and needles of the solid matter are weakly fixed with each other and occupy only a small portion of its volume.

The density of such a particle is only 0.1 to 0.4 gram per cubic centimeter. Its single elements may be fixed with each other by, for example, water hoar-frost. Under the action of the Sun, the particles soon disintegrate. But the disintegrated microparticles, even if at a considerable distance from the nucleus, continue moving in a tight group. That was discovered among the DUCMA results.

In many cases, the dust particle impacts were recorded, not as those of separate particles, but as a cluster consisting of six to eight—sometimes 30—particles. The distance between these clusters was only several dozen centimeters (a few feet or less).

More than ten publications have been done particularly on DUCMA results. Many articles are devoted to other experiments. For example, to the TV that obtained the unique images of the Halley Comet nucleus. Now we know that the mean (average) density of this celestial body is less than a unit. Its mass is close to 200 billion tons. Near the perihelion (closest approach to the Sun), it loses up to 45 tons every second. Of that, six to ten tons falls to this special dust I've described. But, then, Halley's Comet is so "extravagant" only once in 76 years. That will provide it with almost 50,000 years of survival.

Comet Probe: Giotto

The successful European Space Agency's interplanetary spacecraft Giotto, which brushed Halley's Comet in 1986, was assigned a second tour of duty—a two year trip to rendezvous with and explore Comet Grigg-Skjellrup on July 10, 1992.

The probe's first tour took it for a close-up look at Comet Halley in March 1986, after being launched July 2, 1985. After visiting the famous comet, ESA's probe was supposed to be exhausted, crippled and heading for oblivion. In fact, some of its instruments were damaged and knocked out on the approach to the comet.

Damaged, not dead. On March 14, 1986, Giotto was hustling along at 153,000 miles per hour 83 million miles from the Sun when it brushed Comet Halley at the extraordinarily close distance of 373 miles. As it passed Halley, Giotto's radio fell silent for 20 minutes—but it started up again to the applause of mission controllers at ESA's European Space Operations Center, Darmstadt, West Germany. Floating through the comet's tail, the 1,300-lb. probe survived a fierce battering by high-speed particles.

A month later, European controllers radioed orders putting Giotto in a state of electronic hibernation, rounding the Sun every 10 months. From April 1986 until February 1989, Giotto was in that parking orbit, running on low power with radio transmissions shut down.

Giotto's camera was broken during the Halley fly-by, but European scientists said it still could send back valuable information. Studying half a dozen options—including letting the battered old probe die in space from lack of funds—ESA decided to send Giotto for a 1992 rendezvous with Comet Grigg-Skjellrup.

Wake-up call. ESA sent a wake-up call February 19, 1989, to the hibernating Giotto. The probe then was 62 million miles from Earth and approaching the point of its solar orbit closest to our planet.

For the wake-up call, ESA fed 100,000 watts of transmitter power to a 230-ft. dish antenna at NASA's Deep Space Network tracking station at Madrid, Spain, to radio a command ordering Giotto to power up. They had to wait 2 hours 21 minutes for a reply from the probe, a whisper across 62 million miles of space.

That signal told Darmstadt technicians Giotto was okay. After the probe's solar-panel wings resupplied Giotto's batteries with electricity, ESA mission control ordered the spacecraft's high-gain antenna to point toward Earth so better signal strength could be received in Spain. That improved reception of telemetry data. The color television camera and science instruments still worked, so engineers radioed a signal July 2, 1989, giving Giotto its new itinerary.

Flashing by within 13,600 miles of Earth July 2, 1990, Giotto used terrestrial gravity as a slingshot to speed up to catch Grigg-Skjellrup—the comet first discovered by New Zealand amateur astronomer John Grigg in 1902 and rediscovered by astronomer J. Skjellrup in 1922. Comet Grigg-Skjellrup circles the Sun every 5.09 years. Only one other comet, Encke, has a shorter solar orbit than Comet Grigg-Skjellrup. Encke takes only about 3.3 years to circle the Sun. Encke has been observed more than any other comet. It was found in 1786, recovered in 1822 and 54 more times in the last 168 years. Encke has been proposed as a target for a future deep-space probe from Earth.

Comet Halley. Human science excelled in March 1986 as a flotilla of six deep-space probes from Earth—Giotto, Vega 1, Vega 2, Suisei, Sakigake and ICE—flashed past Comet Halley, sending home a wealth of information about that sooty, potato-shaped rock as big as a mountain.

Giotto was named after Italian painter Giotto Di Bondone who depicted the vagabond Comet Halley in A.D. 1304 as the Star of Bethlehem in the Scrovegni Chapel in Padua.

Vega 1 and Vega 2 were sent by the USSR's Intercosmos agency in Moscow. In Russian, the name Halley is pronounced Galley. The name Vega came from the abbreviated form of Venus-Halley (Venus-Galley) or Ve-Ga, reflecting the mission's sequence of goals. Vega 1 and Vega 2 flew by Venus in 1985, dropping off landers, enroute to Comet Halley in 1986.

Suisei and Sakigake were from the Japanese Institute of Space and Astronautical Science. The International Cometary Explorer (ICE) was a NASA probe, renamed and redirected from an earlier mission.

Comet Encke. Only one other comet, Encke, has a shorter solar orbit than Comet Grigg-Skjellrup. Encke takes only about 3.3 years to circle the Sun. Encke has been observed more than any other comet. It was found in 1786, recovered in 1822 and 54 more times in the last 168 years. Encke has been proposed as a target for a future deep-space probe from Earth.

Comet & Asteroid Probe: CRAF

NASA hopes to send CRAF—the Comet Rendezvous Asteroid Flyby—spacecraft for a close-up inspection of a large comet and a flyspeck asteroid in the late 1990's.

Specific small Solar System bodies to be visited would depend on when CRAF gets off the ground. Budget pressures have raised and lowered CRAF's priority in Congress, making its future unclear. Bodies mentioned for the project have included Comet Kopff, Comet Tempel 2 and asteroid 46 Hestia. Tiny 46 Hestia is only 82 miles in diameter.

Mariner Mark II. To cut costs, increase efficiency and avoid duplication, NASA has designed CRAF and the Saturn probe Cassini to share one standard interplanetary spacecraft design known as Mariner Mark II. Most of the CRAF and Cassini flight hardware would be identical, except where minor modifications may be required for unique science investigations.

A generic Mariner Mark II spacecraft, built from off-the-shelf hardware from past spacecraft and a few new designs, would be relatively inexpensive. The CRAF Mariner Mark II would be launched on a Titan 4-Centaur rocket.

Asteroid flyby. Whichever bodies CRAF would visit, the probe would fly by the asteroid first, on its way to a comet approaching the Sun. It would fly alongside the comet for three years, between the orbits of Mars and Jupiter, measuring the composition of comet dust and ice. CRAF's work would complement science investigations by Cassini at Saturn and the Galileo spacecraft at Jupiter.

Flying formation. It probably would take two years to get to the asteroid. Photos would be snapped of the asteroid and science data collected as the probe passed by. Then it might take another year to reach the comet. CRAF might be able to fly in formation as the comet swings around the Sun and heads back to the far reaches of the Solar System.

The flight would be the first time a spacecraft from Earth and a comet would have flown in formation. ICE, the International Comet Explorer, inspected Comet Giacobini-Zinner in 1985. Russian, Japanese and European probes flew by Comet Halley in 1986. The European probe Giotto was to visit Comet Grigg-Skjellrup on July 10, 1992.

Flying alongside the comet, CRAF would look for original, unchanged matter left behind when a cloud of gas and dust at the edge of the Milky Way galaxy collapsed to form our Solar System 4.6 billion years ago.

Slush balls. Comets are giant balls of dirty slush which scientists think have collected at the far outer edge of the Solar System. The ring of comets out there, known as the Oort Cloud, is 30,000 to 100,000 times farther away from the Sun than Earth. Sometimes, gravity from stars in the Sun's neighborhood nudge a comet out of the Oort Cloud so that it falls inward toward the Sun at the center of the Solar System.

Most comets take a long period of time—up to thousands of years—to fly in and out of the Solar System. Not too often, a comet zipping in toward the Sun from the Oort Cloud will pass close to a planet. Gravity from the planet may affect the comet's orbit. The comet may become trapped, at least temporarily, in the inner Solar System, reappearing in view of Earth in shorter periods of time.

Streaming tail. As a comet approaches the Sun's heat, a ton of gas and dust boils off every second into a cloud, called coma. A long tail streams away from the coma as solar wind and light pressure from the Sun blow the material away.

At first, before the coma and tail begin to build, CRAF could fly extremely close, within a few miles of the nucleus. Later, the craft would back off a few thousand miles to avoid damage from particles and to watch coma and tail develop. From time to time, it would make brief excursions into the dust cloud to collect samples and then move off again to a safe distance. CRAF's science payload might include:

★the Cometary Ice and Dust Experiment (CIDEX), to make two chemical analyses: X-ray fluorescence spectrometry and gas chromatography to determine the elemental and chemical composition of the comet's dust and ice particles.

★cameras to photograph the nucleus, coma and tail, and changes that occur as the comet orbits the Sun. Photos would help scientists determine the comet's size, structure, location of its poles, rotation rate and geology structure.

★surface harpoon—a penetrator lander—to be fired into the nucleus of the comet. Instruments in the harpoon would search for 20 chemical elements. A gamma-ray spectrometer in the harpoon's tip would be buried three feet below the comet's surface. The harpoon would have thermometers to measure temperatures beneath the surface, and an accelerometer to sense the comet's surface strength and resistance to puncture. The harpoon would radio its findings to CRAF which would relay them to Earth.

★mass spectrometers to study gases released by the nucleus and the plasma cloud of ionized gas around the nucleus.

★dust collectors, counters and analyzers would sample the comet's dust and study it inside CRAF, helping scientists learn which chemical elements are in the ice and dust. Weight, shape, size and composition of individual dust grains would be measured.

★an infrared-light and visible-light mapping spectrometer to watch changes in chemical composition of the coma and the surface of the nucleus.

★a magnetometer and plasma-wave analyzer would see how energy particles from the Sun affect the coma. The magnetometer also would measure the comet's magnetic field, if it has one.

Interplanetary Probes: The Mariners

Mariner was a series of U.S. interplanetary science probes to the Solar System's inner planets in the 1960's and early 1970's. Then Mariner 11 and Mariner 12 became Voyager 1 and Voyager 2 which went on grand tours of the Solar System's outer planets.

Mariner 1. Launch toward Venus failed.

Mariner 2. Launched successfully toward Venus August 14, 1962, the 445-lb. Mariner 2 flew by within 21,642 miles of Venus December 14, 1962. Mariner 2 was the second planet fly-by, the first successful planet fly-by and the first American planet fly-by. The probe confirmed the existence of a solar wind. It measured the temperature of Venus but did not find a magnetic field. Today Mariner 2 is in orbit around the Sun.

U.S. space researcher James A. Van Allen, who earlier had discovered the radiation belts surrounding Earth, made the first instruments ever carried to another planet, those taken by Mariner 2 to Venus.

Mariner 3. Launched toward Mars November 5, 1964, failed to follow the proper trajectory and now is in orbit around the Sun.

Mariner 4. Launched successfully toward Mars November 28, 1964, the 575-lb. Mariner 4 flew by Mars July 14, 1965. Mariner 4 was the first successful probe to Mars, passing the Red Planet at a distance of 6,117 miles. It sent back 22 photographs showing the planet's arid, cratered surface. Today it is in orbit around the Sun.

Mariner 5. Launched successfully toward Venus June 14, 1967, the 540-lb. Mariner 5 flew by Venus October 19, 1967. It passed Venus at a distance of 2,479 miles, measuring the planet's temperature and collecting a profile of its atmosphere. Mariner 5 accurately measured the mass and diameter of Venus, but didn't find a magnetic field. Today it is in orbit around the Sun.

Mariner 6. Mariner 6 and Mariner 7 were 910-lb. twins. Launched successfully toward Mars February 24, 1969, Mariner 6 flew by Mars July 31, 1969. Mariner 6 passed Mars at a distance of 2,120 miles. It sent back photos of the region around Mars' equator.

It measured surface temperature, atmospheric pressure, composition and diameter of Mars. Today Mariner 6 is in orbit around the Sun.

Mariner 7. Mariner 6 and Mariner 7 were 910-lb. twins. Launched successfully toward Mars March 27, 1969, Mariner 7 flew by Mars August 5, 1969. It passed at a distance of 2,128 miles. Mariner 7 photographed the southern hemisphere of Mars and the Martian South Pole ice cap. Mariners 6 and 7 together transmitted to Earth a total of 201 photographs of the Red Planet. They measured surface temperature, atmospheric pressure, composition and diameter of Mars. Today Mariner 7 is in orbit around the Sun.

Mariner 8. Failed during launch to Mars.

Mariner 9. Launched successfully toward Mars May 30, 1971, the 2,273-lb. Mariner 9 arrived in a 12-hour orbit around Mars November 13, 1971. Mariner 9 was the first man-made satellite to orbit a planet other than Earth. The orbit ranged from 1,025 to 10,626 miles above the planet. Mariner 9 had two TV cameras which sent back 7,329 photos.

A violent dust storm on the planet's surface in September 1971 threatened the photography. Nearly all surface features were obscured by dust as Mariner 9 closed with Mars. Four blurry spots in first photos turned out to be peaks of huge volcanos on Mars' Tharsis Ridge. But, the dust settled over a few weeks and photos showed canyons and channels which suggested water once flowed on the surface. Seasonal changes in polar ice caps were noticeable. The entire surface of the planet was mapped. Close-up shots were made of Phobos and Deimos, Mars' two natural moons.

Mariner 9 had an infrared spectrometer and an ultraviolet spectrometer to study the planet's atmosphere. An infrared radiometer measured surface temperature. Today Mariner 9 is dead in orbit around Mars.

Mariner 10. Launched successfully November 3, 1973, the 1,109-lb. Mariner 10 was the first spacecraft from Earth to visit two planets. It flew by Venus February 5, 1974, and went on to fly by Mercury three times. The first Mercury fly-by was March 29, 1974. The second, September 21, 1974. The third, March 16, 1975.

Mariner 10 contained two TV cameras which transmitted the first close-ups of Mercury and Venus. It also had an infrared radiometer to sense the temperature of Mercury's surface, ultraviolet spectrometers to find atmosphere air glow and measure how the Sun's ultraviolet rays were absorbed, and magnetometers to sense changes in the magnetic field as Mariner 10 flew its course.

The craft flew by Venus at a distance of 3,579 miles, sending back 3,500 photos. Pictures in ultraviolet light showed cloud cover and circulation of the atmosphere clearly. The gravity of Venus then was used to slingshot Mariner 10 toward Mercury.

The first pass by Mercury was at a distance of 460 miles above the baked planet's surface. The second was at a distance of 29,827 miles. The third, at 205 miles. Altogether 10,000 photos of Mercury were transmitted to Earth. The planet's magnetic field was discovered. Mariner 10 now is in orbit around the Sun.

Mariner 11. The spacecraft was renamed Voyager 1 and launched September 5, 1977. It flew by Jupiter March 5, 1979, Saturn November 12, 1980, and now is leaving the Solar System.

Mariner 12. The spacecraft was renamed Voyager 2 and launched August 20, 1977. It flew by Jupiter July 9, 1979, Saturn August 25, 1981, Uranus January 24, 1986, Neptune August 24, 1989, on its way out of the Solar System.

Probe	Launched	Destination
Mariner 1	Jul 22 1962	Veered off course and was destroyed.
Mariner 2	Aug 14 1962	flew within 21,598 mi. of Venus Dec 14 1962, now orbiting Sun.
Mariner 3	Nov 5 1964	Launched to Mars, but rocket burned incorrectly, now orbiting Sun.
Mariner 4	Nov 28 1964	flew by within 6,115 mi. of Mars Jul 14 1965, now orbiting Sun.
Mariner 5	Jun 14 1967	flew by within 2,479 mi. of Venus Oct 19 1967, now orbiting Sun.
Mariner 6	Feb 24 1969	flew by within 2,120 mi. of Mars Jul 31 1969, now orbiting Sun.

Mariner 7	Mar 27 1969	flew by within 2,190 mi. of Mars Aug 5 1969, now orbiting Sun.
Mariner 8	May 8 1971	Launched failed, rocket fell into Atlantic Ocean.
Mariner 9	May 30 1971	arrived in orbit around Mars Nov 13 1971, the first spacecraft to orbit another planet. Mariner now is dead in orbit.
Mariner 10	Nov 3 1973	flew by within 3,585 mi. of Venus Feb 5 1974, flew by within 169 mi. of Mercury Mar 29 1974, 2nd fly-by within 29,827 mi. of Mercury Sep 21 1974, 3rd fly-by within 203 mi. of Mercury Mar 16 1975, now orbitng the Sun.
Mariner 11	Sep 5 1977	renamed Voyager 1, crossed the Asteroid Belt to fly by Jupiter Mar 5 1979, flew by Saturn Nov 12 1980, moved above the Solar System, photographed sweeping portrait of the entire Solar System Feb 13 1990 at 3.7 billion miles from Earth, leaving Solar System.
Mariner 12	Aug 20 1977	renamed Voyager 2, crossed the Asteroid Belt to fly by Jupiter Jul 9 1979, flew by Saturn Aug 25 1981, flew by Uranus Jan 24 1986, Neptune fly-by Aug 24 1989, now leaving the Solar System.

Interplanetary Probes: The Pioneers

The Pioneer series of spacecraft were American interplanetary probes to the Moon, Sun, Jupiter and Venus. A total of 14 launches were attempted between 1958-78.

Pioneers 1, 2 and 3 were aimed at the Moon in 1958 but failed. Pioneer 4 passed within 37,280 miles of the Moon in 1959.

Pioneers 5, 6, 7, 8 and 9 went to solar orbit to study the Sun's flares, solar wind and magnetism. Pioneer 5 worked only three months.

Pioneers 6, 7, 8 and 9 all were built in the 1960's to study solar flares and related phenomena. Each of the four spacecraft measures 32 inches long and 37 inches in diameter. The four probes also have provided valuable information on Earth's magnetic field.

The oldest spacecraft still working in interplanetary space is Pioneer 6, first of the series.

The 140-lb. Pioneer 8, launched from Cape Canaveral in 1967, was supposed to last six months but NASA still is receiving signals decades later. Pioneer 8 was equipped with eight science instruments. All detectors but one have quit as sensors failed and solar cells deteriorated. The many solar flares Pioneer 8 has been exposed to, and its close exposure to the Sun's ultraviolet radiation, damaged the probe. Pioneer 8 did verify Earth has a magnetic tail like a comet.

NASA had turned off the electric-field detector aboard Pioneer 8 from 1971 to 1984 to save power but lost contact with Pioneer 9 in 1984 just when Pioneer 8 was about to pass through Earth's tail. Radio commands were sent to Pioneer 8 to turn the electric-field detector back on. It responded immediately after 13 years hibernation.

Over 20 years, Pioneer 8 has transmitted 11 billion bytes of information. It orbits the Sun every 388 days ranging from two to 186 million miles from Earth. NASA declared Pioneer 9, launched in 1968, dead in 1987.

After the launch of Pioneer 9, the 10th launch in the Pioneer series was attempted in 1969. It failed. That probe was to have been Pioneer 10 but became known as Pioneer E.

Later, a new Pioneer 10 and Pioneer 11 were sent to the planet Jupiter. The flight following E—11th in the series of Pioneer launches—was successful in 1972. It became known as Pioneer 10. It passed within 82,000 miles of Jupiter in 1973. Today it is on its way out of the Solar System toward interstellar space.

Pioneer 11, then, was the 12th launch in the series, in 1973. It passed within 26,600 miles of Jupiter in 1974 and then within 13,000 miles of Saturn in 1979. It too is on its way out of the Solar System.

Pioneers 12 and 13 were sent in 1978 to investigate the planet Venus.

Pioneer 12, sometimes called Pioneer-Venus, still is on station, in orbit around

Venus, using radar to map the planet surface and sending back data about the solar wind. The spacecraft intercepts particles in the solar wind as they pass Venus, flying out from the Sun during the 11-year sunspot cycle. Pioneer-Venus reportedly has less than five lbs. of fuel left but that amount should keep it in proper orbit around Venus until at least 1992.

Pioneer 13 was sent in 1978 as a spacecraft bus carrying five Venus landers. After arriving at the planet, the five landers were dropped successfully, sending data back to Earth on their way down. One sounder actually landed on Venus and sent back data from the surface for 68 minutes. The main bus burned up in Venus' atmosphere in 1978.

A total of 14 U.S. probes called Pioneer have been launched to study the Sun, Moon and planets:

Pioneer	Launch	Goal	Results
1	10/11/58	Moon	failed Moon; studied Earth radiation
2	11/8/58	Moon	failed, rocket 3rd stage didn't ignite
3	12/6/58	Moon	failed Moon; studied Earth radiation
4	3/3/59	Moon	passed 37,280 mi. of Moon
5	3/11/60	Sun	studied flares, solar wind, to 6/26/60
6	12/16/65	Sun	with Pioneers 7, 8, 9 studied Sun
7	8/17/66	Sun	measured Earth's magnetic tail
8	12/13/67	Sun	studied magnetic fields, cosmic rays
9	11/8/68	Sun	studied Sun atmosphere, solar orbit
E	8/27/69	Sun	failed, was to have been Pioneer 10
10	3/3/72	Jupiter	passed 82,000 mi. of Jupiter 12/3/73, now leaving Solar System
11	4/5/73	Jupiter	passed 26,600 mi. of Jupiter 12/2/74 then 13,000 mi. of Saturn 9/1/79, now on way out of Solar System
12	5/20/78	Venus	Venus 1, orbiting radar mapper
13	8/8/78	Venus	Venus 2, main craft, five landers. Main burned in atmosphere 12/9/78, landers sent data on way to impact. One transmitted 68 min. after landing.

Interplanetary Probes: Pioneer 10, 11

The radio signal from Pioneer 11, America's venerable robot exploring interstellar space beyond the Solar System, has become wobbly.

Meanwhile, telemetry data from America's vintage Pioneer 10 interplanetary probe shows a slight decline in gas pressure in the thrusters used to perform alignment maneuvers. The thrusters keep the probe's high gain antenna pointing toward Earth and are not used to stabilize the spacecraft. Pioneer is stabilized by its own spin, rather than the thrust of control jets.

Engineers at NASA's Ames Research Center, Mountain View, California, control Pioneer operations. So far, they have found no problem resulting from Pioneer 10's low gas pressure. Pioneer 10 should have enough pressure for several years of operations.

Pioneer 10. Pioneer 10 was eleventh in a series of 14 Pioneer launches when it blasted off March 2, 1972, from Cape Canaveral, Florida. The little 570-lb. probe, sent out only for the first ever close look at the Asteroid Belt and the planet Jupiter, has gone far beyond its original assignment.

Arching along a 620 million mile route from Earth, Pioneer 10 sailed through the Asteroid Belt to approach within 82,000 miles of the jumbo planet on December 3, 1973. The probe sent home spectacular pictures as well as data about the planet's atmosphere, magnetic field, interior structure and moons. It was first to chart the giant planet's intense radiation belts, locate Jupiter's magnetic field, show that the planet is mostly liquid, and measure the mass of Jupiter's four planet-size moons—Io, Europa, Ganymede and Callisto.

Beyond Jupiter, Pioneer 10 raced outward from the Solar System toward interstellar space. Officially departing the known Solar System on June 13, 1983, it has been one of America's most successful space missions. Since it was launched from Cape Canaveral in 1972, Pioneer 10 has travelled farther than any other human-made object.

Pioneer 10 was the first spacecraft from Earth to leave the Solar System. It passed an important milestone September 22, 1990, when it was 50 times farther from the Sun than our planet Earth is from the Sun. Earth is at an average distance from the Sun of 93 million miles. Astronomers refer to that distance as an astronomical unit, or AU. One AU equals 93 million miles. Pioneer 10 was 50 AU from the Sun—that's 4.65 billion miles. Pioneer 10 had travelled farther than any other human artifact.

Pioneer 11. Pioneer 11, sometimes called Pioneer-Saturn, is a twin to Pioneer 10. It was the twelfth Pioneer launch when it went to space April 5, 1973. Pioneer 11 shot across the Asteroid Belt and barreled through Jupiter's deadly radiation belts at 107,373 mph, the fastest speed ever traveled by a man-made object. It passed within 26,600 miles of the planet December 2, 1974.

A big boost from Jupiter's gravity shot Pioneer 11 high above the plane of the planets and 1.5 billion miles across the Solar System to Saturn. On September 1, 1979, Pioneer 11 was the first unmanned robot probe to explore Saturn, passing within 13,000 miles of the ringed planet. Pioneer 11, sometimes called Pioneer-Saturn, was the first to send back pictures from inside Saturn's rings.

Pioneer 11 crossed the orbit of Neptune 2.8 billion miles from Earth February 3, 1990, leaving behind forever the known parts of the Solar System.

It joined three other U.S. spacecraft in the unexplored wide open spaces beyond the farthest planet in our Solar System. Already out there were Pioneer 11's sister spacecraft Pioneer 10, the first to leave, as well as Voyager 1 and Voyager 2, the second and third to fly beyond the planets.

Neptune's orbit at the time was the outer limit of the known Solar System because a slight irregularity in the orbit of Pluto took that usually-most-distant planet inside Neptune's path until the turn of the century. Pioneer 11 was 2,807,364,100 miles from the Sun February 23, 1990.

Rickety radio. Pioneer 11's radio signal has become variable, possibly indicating a failure in a timeworn driver amplifier feeding the transmitter traveling wave tube. Ames engineers say they don't think it is in the transmitter tube itself.

NASA's Deep Space Network radiotelescopes at Madrid, Spain, Canberra, Australia, and Goldstone, California, transmit and receive radio signals to and from the Pioneer spacecraft.

The Pioneers are equipped with low-power radios transmitting data with eight-watt signals. By the time data from one of the Pioneer probes arrives at Earth, signal strength has dropped to four billionths of a trillionth of a watt. NASA needs the huge antennas of its Deep Space Network to pick up such weak signals.

Travelling at 186,000 miles per second—the speed of light—signals take half a day to travel from NASA to Pioneer and back to the 230-ft. deep-space antenna on Earth. NASA expects to be able to communicate with both Pioneer 10 and Pioneer 11 until the year 2000. It will be difficult to operate the craft's radio transmitter and scientific instruments at the same time, but Pioneer 11's life may extend to 2000.

Weak signals. Pioneer 10 radio signals now are returning to Earth from 50 times the distance between Earth and the Sun. Energy in the signals picked up from Pioneer 10 by a 230-ft. deep-space antenna on Earth is so weak it would have to be continuously captured and stored for 11 billion years to accumulate enough to light a 7-1/2 watt nightlight for one thousandth of a second. Still, those signals, from both Pioneer 10 and Pioneer 11, are strong enough for scientists to read out high-quality data every day.

Spinning away. Pioneer spacecraft are stabilized by their own spin, rather than the thrust of control jets. The momentum of each is driving it out of the Solar System.

Today, NASA tracks the Pioneer 10 and Pioneer 11, looking for undiscovered celestial bodies and gravity waves. The Pioneers are good indicators of the gravity pull of celestial objects because the two spacecraft generate almost no forces of their own affecting their trajectories. They are tiny bodies floating free in the gravity fields of all of the other bodies in the Solar System.

Stabilized by their own spin rather than rocket thrust, any unexpected change in their speeds would reveal the tug of an uncharted star or planet. The Pioneers are precise gravity sensors because they are on paths moving rapidly outward from the Sun and radio tracking is exact.

Doppler shift. NASA's Deep Space Network radiotelescopes at Madrid, Spain, Canberra, Australia, and Goldstone, California, can transmit a signal to the Pioneer craft and measure the so-called Doppler shift or change in frequency of signals coming back from the craft. The faster a Pioneer pulls away from the Sun, the longer the wavelength received by the Deep Space Network antennas. Such two-way Doppler tracking gives an accuracy of one millimeter per second in speed measurements.

Over several recent years of such precise measuring, NASA found no strange gravity effect on Pioneer 10 or Pioneer 11. All effects measured could be explained by the nine known planets. Yet the seventh and eighth planets, Uranus and Neptune, seem to be disturbed by something sometimes, maybe a previously unknown planet.

Helping Voyager. Pioneer 11 was a traveler's aide to Voyager 2 during the younger probe's close encounter with Neptune. Although Pioneer 11, viewed as one of the elder statesmen of spacecraft, couldn't communicate directly with Voyager 2, it sent data to NASA. The data from Pioneer 11's magnetometer and plasma analyzer was used to supplement information from Voyager instruments and predict when Voyager would encounter Neptune's bow shock—the solar wind striking the planet's plasma or sea of charged particles.

It was the second time the 570-lb.Pioneer 11 has assisted the 1,800-lb. Voyager. Pinoeer provided similar data to NASA as Voyager passed Uranus.

Planet X. Pioneer 10 remains the most distant spacecraft, but Pioneer 11 has since been passed by both Voyager 1 and Voyager 2.

Nowadays, the Pioneers are looking for signs of a mysterious, unknown tenth planet at the far edge of the Solar System, the so-called Planet X. Pioneer 10 was relaying evidence suggesting either one mysterious Planet X or maybe even two small planets exist somewhere beyond Pluto, the ninth planet out from the Sun.

Heliosphere. Our Sun is a star, surrounded by planets and other smaller bodies in what we call the Solar System. The extent of the Sun's atmosphere is the heliosphere, the area directly influenced by the Sun's electromagnetic energy. The outer edge of the Solar System where the heliosphere ends and interstellar space begins—where the Sun's influence gives way to the Milky Way galaxy's influence—is known as the heliopause.

From its location almost a billion miles above the plane of the planets, Pioneer 11 is helping astronomers chart the heliosphere and looking for the heliopause. Recently the probe sent back data showing most cosmic rays in the heliosphere originate in gas in the interstellar space between stars.

Another decade. The tiny craft should continue to provide clues to some of the puzzles of the Universe for many years. They will transmit information until their power sources diminish, sometime after the year 2000.

The spacecraft may give scientists clues to the shape of the heliosphere, the boundary of the Sun's influence, by sensing cosmic rays and the flow of the solar wind at heretofore unexplored distances from the Sun. Astronomers hope Pioneer 10 finds the outer boundary of the heliosphere as it flies on into nearby interstellar space. Pioneer 10 also will keep searching for those long gravity waves proposed by Albert Einstein.

As they continue out of the Solar System and into interstellar space, both Pioneer 10 and Pioneer 11 should send back information about the heliosphere.

Nametag. Even after their radio signals disappear after the turn of the century, the Pioneer twins will survive for millions of years in the vacuum of space, perhaps as long as the Earth itself. The spacecraft carry a plaque for communication with any intelligent species which might come across the probe in a future millennium. The plaque shows the location of the Sun and its Solar System within the Milky Way galaxy, a map of the Solar System locating Earth, and a man and a woman.

Interplanetary Probes: The Voyagers

The nuclear-powered, grand-touring Voyager twins were nuclear-powered, automated spacecraft launched from Cape Canaveral, Florida, in 1977 to fly out through the Asteroid Belt to explore Jupiter and Saturn.

Voyager 2 blasted off August 20 on a Titan-Centaur rocket. Voyager 1, launched on a Titan-Centaur September 5, soon overtook its twin on a shorter path.

The probes crossed the Asteroid Belt in 1978 and 1979. Voyager 1 swept past Jupiter March 5, 1979, using the giant planet's gravity for a slingshot boost toward the ringed planet, Saturn. Voyager 2 sped by Jupiter on July 9, 1979.

Voyager 1 flew past Saturn November 12, 1980, using the planet's gravity to fling it up out of the plane of the Solar System for an encounter with Saturn's moon Titan. Voyager 2 went by Saturn August 25, 1981.

Voyager 1 pulled above the swirling disk of planets surrounding our Sun, rushing for the edge of the Solar System and beyond on an endless space odyssey. Today, it is more than four billion miles from Earth and two billion miles above the plane of the Solar System.

The Jupiter and Saturn results were so electrifying, Voyager 2's assignment was extended to Uranus and Neptune—the Grand Tour. The small spacecraft called on the big outer planets for repeated gravity assists as it cruised on to encounter Uranus on January 24, 1986, and Neptune on August 24, 1989. Today, Voyager 2 also is cruising out of the Solar System, on its way toward interstellar space.

Control signals. The Voyager twins are controlled by NASA's Jet Propulsion Laboratory (JPL), Pasadena, California. Photos and data sent back are analyzed at JPL.

Radio signals to and from both Voyagers are transmitted and received via the huge dish-shaped antennas of NASA's Deep Space Network near Goldstone, California, Madrid, Spain, and Tidbinbilla, Australia. Even though radio signals travel at the speed of light, the Voyagers are so far from Earth that signals take hours to reach home.

Discoveries. Equipped with TV cameras transmitting extraordinary close-ups of uncharted worlds, the Voyagers discovered and photographed:

★ active volcanos spewing sulfur on Jupiter's moon Io, nitrogen-ice on Neptune's moon Triton, and icy glacier flows on some moons of Saturn and Uranus;

★ dozens of monumental hurricanes, including the jumbo planet Jupiter's 16,000-mi.-wide Great Red Spot and Neptune's 6,200-mi.-wide Great Dark Spot;

★ orange smog and methane-ethane rain on Saturn's moon Titan;

★ two dozen previously unknown moons, including three in orbit around Jupiter, four circling Saturn, ten orbiting Uranus, and at least half a dozen around Neptune;

★ new rings encircling Jupiter, Saturn, Uranus and Neptune, incredible complexity in rings encircling Saturn, Uranus and Neptune, a thin ring around Jupiter and ring arcs at Neptune;

★ complex geology on the Uranian moon Miranda and Neptune's moon Triton, including an ice cliff on Miranda twice the height of Mount Everest, and smokestack plumes of dark nitrogen ice on Triton;

★ low temperatures on Triton, the coldest object in the Solar System with an atmosphere four degrees above absolute zero or minus 273.15 degrees centigrade, the point at which liquid hydrogen freezes;

★vast belts of windblown clouds like a collar around Neptune's south pole;

★spectacular northern lights around Neptune and its moon Triton, produced by an enormous magnetic field surrounding the planet.

The recording. Flying out of the Solar System, both Voyagers are sniffing for electrically-charged particles and magnetic fields as they go. They should be able to send back some measurements for up to 30 more years.

Each one-ton probe, the size of a compact car, has a 12-in. copper disk with a diagram showing how to spin out two-hours of sound from our civilization by rotating the record while touching its surface with a needle. The disk has a human hello in 60 languages, whale songs, sounds of rain, thunder, a railroad train, heartbeats, a crying baby, a hyena's laugh, the soft sound of a kiss, Mozart's Magic Flute and rock-'n-roller Chuck Berry doing Johnny B. Goode.

A message written June 16, 1977, by then-U.S. President Jimmy Carter is encoded on the disk, "This is a present from a small distant world, a token of our sounds, our science, our images, our music and our feelings...this record represents our hope and our determination, and our good will in a vast and awesome universe."

Interplanetary Probe: Voyager 1

The U.S. outer Solar System probe Voyager 1 was launched September 5, 1977. It flew across the Asteroid Belt and by Jupiter on March 5, 1979.

Jupiter. Passing 216,865 miles above the giant planet, Voyager 1 snapped color photos of clouds and rings, and the moons Io, Ganymede and Callisto. Io surprised astronomers with volcanos, making it the first volcanic body found beyond Earth. Ganymede and Callisto were scarred by meteorite craters. Ganymede seemed to be a mix of rock and ice.

From Jupiter, Voyager 1 moved on to inspect Saturn, arriving at the ringed planet November 12, 1980.

Saturn. Passing 78,300 mi. above Saturn's cloud tops, it snapped 17,500 color photos revealing six additional moons, raising the total to eighteen. Voyager 1 confirmed the Pioneer 11 finding that Saturn radiates more heat than it gets from the Sun.

Clouds revealed 1,100 mph winds sweeping the planet. Saturn's major A, B and C rings were found to have ringlets. Shepherding moons were found in the F ring. The moons Mimas, Tethys, Dione and Rhea turned out to be heavily cratered. The largest moon, Titan, displayed a nitrogen atmosphere 1.5 times denser than Earth's atmosphere.

Solar System portrait. Today, Voyager 1 is zipping along, making its way out of the Solar System toward interstellar space. In 1990, the spacecraft turned briefly to snap the first-ever portrait of the Solar System.

Floating through space 3.7 billion miles from Earth, Voyager 1 looked back toward home February 13, 1990, and snapped its two aging cameras one last time to make a sweeping portrait of the planets of the Solar System and the Sun.

The photos were be transmitted to NASA's Deep Space Network tracking antennas which were standing by to catch the extraordinarily weak radio signals. Computers processed the pictures, allowing astronomers at NASA's Jet Propulsion Laboratory (JPL), Pasadena, California, to assemble them into a mosaic.

In the Solar System portrait, Mercury was hidden in the Sun's glare and Pluto was too distant to appear. But the major planets Venus, Earth, Mars, Jupiter, Saturn, Uranus and Neptune materialized as small but important dots against a dark background of the constellation Eridanus.

Photo op. It was the first opportunity scientists had to take a picture of the planets from outside the Solar System. There probably won't be another opportunity for many years. When it made the snapshot, Voyager 1 was looking down from a position 32 degrees above the plane of the Solar System. There are no flights planned to send a craft so high above the ecliptic plane.

Voyager 1 was 12 years out from Earth when its two 1970's television cameras took the 64 photographs of the Solar System over a four-hour period. The Voyager 1 picture session began with three narrow-angle shots of Neptune, the giant blue planet scouted by Voyager 2 in 1989. Then Voyager 1's wide-angle camera exposed a set of overlapping pictures covering space from Neptune to Uranus. Three narrow-angle views of Uranus were snapped next as the cameras started to track inward toward Saturn for more narrow angle views. The cameras then made a series of shots of the inner Solar System, followed by photos of the Sun, Venus, Earth and Mars, and finally one last picture of Jupiter.

Photographing the Sun probably warped shutter blades on Voyager 1's wide-angle camera, but NASA doesn't plans to use it again and Voyager 1 was in better position to make the portrait than Voyager 2. The images were recorded on magnetic tape and radioed to scientists on Earth. Even though they traveled at 186,000 miles per second, it took the radio signals 5.5 hours to reach Earth.

JPL engineers combined narrow shots of the inner planets with wide views to produce a mosaic of the inner Solar System against the star-speckled backdrop of the Universe. Earth, Venus and Mars were bright star-like points of light near the larger Sun. Among the nine planets, only Jupiter was big enough to show up as a disk.

Interplanetary Probe: Voyager 2

The extraordinary Grand Tour of the Solar System started August 20, 1977, with the lift off of the U.S. interplanetary spacecraft Voyager 2 from Cape Canaveral. The probe was assigned to fly through the Asteroid Belt into the outer Solar System, visiting Jupiter and Saturn. It did so well, the mission was extended after Saturn to include Uranus and Neptune—the Grand Tour.

Jupiter. Voyager 2 flew by Jupiter on July 9, 1979. At 404,000 miles above Jupiter's clouds, the probe photographed cloud weather patterns. Turning to look at the moons Europa, Callisto, Ganymede and Io, Voyager 2 photographed a massive volcano erupting on Io and a large impact basin on Callisto. Voyager 2 found the planet's rings were 4,040 miles wide with particles right down to cloud tops.

Saturn. Voyager 2 arrived at Saturn on August 25, 1981. At 62,760 miles above Saturn's cloud tops, the probe measured the thickness of Saturn's rings and found spokes in the rings which may be clouds of micrometeoroids.

Photos radioed to Earth showed the moons Mimas, Tethys, Dione and Rhea were heavily cratered. Voyager found evidence of motion of the crust of the moon Enceladus.

Titan. Saturn's largest moon, Titan, had a nitrogen atmosphere 1.6 times denser than Earth's atmosphere. Methane was in abundance. Titan's surface temperature of –292 deg. F., density twice water.

Uranus. The probe made its first solo planet flyby, passing within 50,600 miles of Uranus' cloud tops on January 24, 1986. Voyager 2 discovered the planet has a magnetic field. The planet's average temperature was –350 degrees Fahrenheit. Uranus radiated a great deal of ultraviolet light, known as dayglow.

Voyager 2 found 10 previously-unknown moons, the largest only 90 miles in diameter, bringing the total for Uranus to 15. Astronomers looking out from Earth previously had seen only five large ice-and-rock moons. The innermost moon, Miranda, was found to be one of the strangest bodies in the Solar System. Voyager 2 photos showed fault canyons up to 12 miles deep, terraced layers and mixed young and old surfaces. The moon Titania also was marked by huge faults and deep canyons.

The moon Ariel had the brightest and youngest surface, with many deep valleys and broad ice flows. The dark-surfaced moons Umbriel and Oberon looked older.

Voyager found nine rings around Uranus. They were quite different from rings around Jupiter and Saturn. Uranus' rings may be young remnants of a shattered moon.

Neptune. In a spectacular completion of its assignment, Voyager 2 swept by Neptune on August 24, 1989, while the planet was the most distant in the Solar System. Pluto will regain its usual rank as most distant planet in 1999.

The probe flew within 3,000 miles of the planet, finding large dark spots on Neptune like Jupiter's Great Red Spot hurricane. The largest was named the Great Dark Spot. A small irregular cloud seen flitting across the cloud tops was named Scooter.

The strongest winds of any planet were measured on Neptune, with those near the Great Dark Spot blowing at 1,200 mph. The probe also found an odd magnetic field, as it had at Uranus.

Voyager 2 found there are four complete rings around Neptune and no partial ring arcs. The fine material in the rings is so diffuse the rings cannot be seen from Earth. The rings discovered by Voyager 2 were named Galle, Leverrier and Adams, in order outward from the planet.

The spacecraft found six moons, bringing the total for Neptune to eight. The newly-discovered moons were named Naiad, Thalassa, Despina, Galatea, Larissa, and Proteus. The large moons Triton and Nereid already were known to astronomers.

The largest moon, Triton, turned out to be one of the most intriguing satellites in the Solar System. It had erupting geysers spewing invisible nitrogen gas and dark dust particles high in an extremely thin atmosphere which extended 500 miles above Triton's surface.

Triton may not always have been a moon of Neptune. If so, tidal heating might have melted Triton, leaving the moon liquid for a billion years after its capture by Neptune. Today it may have thin clouds of nitrogen snowflakes a few miles above its firm surface.

With a surface temperature of –391 deg. Fahrenheit, Triton is the coldest body known in the Solar System. Triton's atmospheric pressure was 1/70,000th the surface pressure on Earth.

Leaving Neptune behind, Voyager 2 now is headed out of the Solar System toward interstellar space.

Mission To Planet Earth

Mission to Planet Earth is a NASA plan to map global weather and environmental conditions at the end of the 1990's. A plethora of sensors would be located in satellites in polar orbit and on the future U.S.-international space station Freedom, complemented by more than 1,000 probes in and on the land and oceans around the world

Mission to Planet Earth was proposed by retired astronaut Sally K. Ride, America's first woman in space. She told the U.S. Senate in 1989 the project would add to human understanding of Earth as an ecological system. With Mission to Planet Earth, she said, man's ability to spot unusual rainfall patterns, drought and concentrations of fish would improve, bringing scientific and financial benefits.

Earth Observing Satellites. For Mission to Planet Earth, NASA wants to build a network of six large Earth-observing satellites (EOS) to monitor our planet's atmosphere, especially ominous changes in the ozone layer. The polar-orbiting laboratories also would monitor the greenhouse effect in which the Earth's temperature is gradually rising because of increased carbon dioxide in the atmosphere.

Each satellite would have synthetic-aperture radar for radio-wave snapshots of areas hidden by vegetation and clouds, and laser ranging gear to measure slight movements of geological features such as California's San Andreas Fault.

The six would include one satellite each from Japan and Europe, and four from the U.S. The USSR could participate. NASA has received more than 1,000 letters from scientists proposing experiments for the orbiting labs.

UARS. NASA's Upper Atmosphere Research Satellite (UARS) launched in 1991 is the first in a series of Mission to Planet Earth spacecraft.

Mission to Planet Earth might cost $30 billion dollars over 20 years. Each big satellite would cost $500 million. Federal budget restrictions have led critics to suggest NASA substitute a large number of smaller satellites for the six large spacecraft.

Contemporary Planet Visits

Current and planned interplanetary spacecraft travels to planets and other major Solar System bodies. Chronologically by launch date. Dot (•) indicates tentative launch date.

Probe	Origin	Launched	Target	Arrival
Pioneer 10	USA	1972 Mar 3	interstellar	2000+
Pioneer 11	USA	1973 Apr 5	interstellar	2000+
Voyager 2	USA	1977 Aug 20	interstellar	2000+
Voyager 1	USA	1977 Sep 5	interstellar	2000+
Pioneer 12	USA	1978 May 20	Venus	1978 Dec 4
Pioneer 13	USA	1978 Aug 8	Venus	1978 Dec 9
Magellan	USA/Europe	1989 Apr 28	Venus	1990 Aug
Galileo	USA/Europe	1989 Oct 12	Jupiter	1995
Ulysses	USA/Europe	1990 Oct 5	Sun	1994
Mars Observer I	USA	1992 Sep	Mars	1993
Mars 94	USSR	1994 •	Mars	1995
Hagoromo 2	Japan	1995	Moon	1995
Cassini	USA/Europe	1996 •	Saturn	2002
Planet B	Japan	1996 •	Venus	1996
Mars 96	USSR	1996 •	Mars	1997
CRAF	USA	1996 •	comet/asteroid	1997

Minor Planets

Minor planets are asteroids and planetoids, small roughly-shaped rocky bodies orbiting the Sun. Some 95 percent of all known minor planets are in an orbit known as the Asteroid Belt between the orbits of the major planets Mars and Jupiter.

Astronomers looking through telescopes see asteroids shine as they reflect sunlight. Vesta is the only asteroid bright enough to be seen by naked eye.

There may be more than 100,000, but only 3,000 have been catalogued. Ceres is the largest. They probably are not the remains of an exploded planet as, together, they total only about four-ten thousandths of the mass of Earth. More likely, they are debris from collisions of several small bodies between Mars and Jupiter at the time the Solar System was forming around 4.6 billion years ago.

Poet's planet. An body in the Asteroid Belt between Mars and Jupiter has been named after one of Russia's favorite 20th Century poets, Anna Akhmatova, even though her work was barred from publication during most of her life.

Soviet astronomers named the asteroid in honor of the poet's 100th birthday. Poetry is popular in the USSR today and her works are in demand.

Akhmatova was born in 1889. Inspired by the style of Alexander Pushkin, her works were famous before the 1917 Bolshevik Revolution. Akhmatova's husband, poet Nicholas Gumilev, was executed by the Bolsheviks in 1921 and the government discouraged publishing and reading of her works until 1940 when she was brought back for the duration of World War II. Returned to official disfavor in 1946, she was expelled from the Soviet Writers Union for "bourgeois decadence" until 1959 when she was reinstated. She died in 1966.

Star's asteroid. Planet Samantha Smith, an asteroid between Mars and Jupiter, was named by the USSR astronomer who discovered it after the American teenager who visited Russia, became a television star and died in a 1985 plane crash in Maine.

Astronomer Lyudmila Chernykh discovered the space object in 1986 and decided to name it after Samantha.

The late Russian leader Yuri Andropov had invited Samantha to visit Russia in 1983 after she wrote him a letter in a personal campaign for better U.S.-USSR relations. Later she starred in an American television program until she died in the 1985 air mishap.

Samantha, who was afraid of nuclear war in 1982, wrote a letter from Maine to Andropov about her fear. He replied with a letter saying Russians wanted peace and inviting her to visit the Soviet Union. Her July 1983 trip to Leningrad and to the Artek Pioneer Youth Camp near the Black Sea made her world famous as a peace symbol, boosting her acting career. Tragically, Samantha, her father Arthur, and six others died in an August 1985 plane crash in Maine after filming her TV show in London.

Before the minor planet was named for her, the USSR already had honored Samantha's memory with a special commemorative postage stamp, had given her name to a ship, named a new species of flower after her, and named a 12,000-foot Caucasus Mountain peak for her.

In the United States, a statue of Samantha was erected near the Maine State House. A Samantha Smith Foundation was started in Hallowell, Maine. Maine Governor John R. McKernan Jr. signed a bill designating the first Monday of every June as Samantha Smith Day. Schools across Maine remember the day with discussions and videotapes of Samantha, remembering her as someone who did something for world peace.

Some of the best known asteroids or minor planets are:

No.	Name	Discovered	Diameter
1	Ceres	1801	623 miles
2	Pallas	1802	336 miles
3	Juno	1804	153 miles
4	Vesta	1807	349 miles
5	Astraea	1845	110 miles
6	Hebe	1847	125 miles
7	Iris	1847	130 miles
10	Hygiea	1849	280 miles
15	Eunomia	1851	169 miles
31	Euphrosyne	1854	230 miles
433	Eros	1898	14 miles
588	Achilles	1906	31 miles
944	Hidalgo	1920	9 miles
1221	Amor	1932	550 yards
1566	Icarus	1949	1,100 yards
1862	Apollo	1932	1,100 yards
2060	Chiron	1977	186 miles
2062	Aten	1976	1,100 yards
--	Adonis	1936	350 yards
--	Hermes	1937	550 yards

Solar System Glossary

Accretion. There was a large amount of rocky metallic matter left in the hot dusty chaos after the Sun formed, along with tiny particles of frozen water and frozen carbon dioxide, all suffused through a vast cloud of hydrogen and helium. As the bits and pieces collided, some stuck together, forming chunks of rock in a process called accretion. Eventually, through collisions, small chunks became big rocks. Big rocks smashing into each other led to planetesimals which, through accretion, became minor and major planets.

Achondrite. A stony meteorite.

Anorthosite. An aluminum-rich silicate rock on the Moon.

Aphelion. The orbital point farthest from the Sun for any body in the Solar System. See also: Apogee, Perigee, Perihelion, Orbit, Solar System.

Apogee. The orbit location farthest from the body being orbited. For instance, the point where a satellite orbiting Earth is farthest from Earth.

Apollo group. A collection of asteroids whose orbits cross Earth's path.

Ashen light. A mysterious dim glow on the dark side of Venus when the planet is seen as a thin crescent.

Asteroid. An asteroid, also known as a minor planet or planetoid, is a rock of debris orbiting the Sun. There are thousands of asteroids, mostly in an orbit, known as the Asteroid Belt, between the orbits of Mars and Jupiter. See also: Planetesimal.

Astronomical unit. AU is a unit of measurement. The average distance between Earth and Sun is about 92.96 million miles. Astronomers refer to that distance as one astronomical unit (AU). The outer edge of the Solar System is estimated to be 75 to 100 times the distance from the Earth to the Sun. That is 75 to 100 AU. When it passed Neptune in 1989, the probe Voyager 2 was 47 astronomical units from the Sun.

AU. See: Astronomical Unit.

Aurora. Energy from the Sun strikes particles in Earth's upper atmosphere, causing a glow over polar territory.

Autumnal equinox. See: Equinox.

Basin. A large multi-ringed lunar crater, produced by asteroid impact, later filled with lava to form mare. See also: Moon.

Bode's Law. A theory in which the planets range outward from the Sun in an orderly sequence of astronomical-unit distances.

Bowshock. As the Sun moves through space, at about 60,000 mph, orbiting the center of the Milky Way galaxy, a large bubble of energy surrounding the Sun is deformed to a teardrop shape, compressed from the front with a long tail streaming behind in its wake. Ahead of the bubble is a bowshock, a disturbance something like a wave breaking on the bow of a ship moving through water. Similarly, as Earth sweeps through space along its orbital path around the Sun, a shock wave is created where the solar wind strikes the planet's magnetic field. See also: Magnetopause.

Caldera. A large bowl-shaped crater left by the explosion and collapse of the cone of a volcano. See also: Moon.

Carbonaceous chondrites. A type of meteorite containing carbon compounds which may be the most primitive form of matter in Solar System. See also: Meteor.

Cassini's division. A gap between the planet Saturn's A and B rings.

Celestial sphere, equator, poles. The celestial sphere is the imaginary inverted bowl of stars surrounding Earth. The celestial equator is a projection of Earth's equator onto the celestial sphere, equidistant from the celestial poles, dividing the celestial sphere into two hemispheres. The celestial poles are points on the celestial sphere above Earth's North Pole and South Pole. The celestial sphere seems to rotate around the celestial poles. The meridian is an imaginary great circle line crossing the celestial sphere, passing through both Poles and the Zenith. The prime meridian is the meridian passing through the vernal equinox. Zenith is the point in the sky directly above the observer. Nadir is the point on the celestial sphere opposite Zenith. Precession is the shift of celestial poles and equinox due to the gravitational tug of the Sun and the Moon on Earth's equatorial bulge. Right ascension is the eastward angle between an object and the vernal equinox, expressed in hours-minutes-seconds. Solstice is the sky positions where the Sun is at maximum angle or declination from the celestial equator. Declination is the angle between a celestial object and the celestial equator. A node is a point in orbit where an ascending or descending object crosses the ecliptic.

Chondrite. Meteorite composed of small spherical objects known as chondrules.

Chromosphere. The outer atmosphere layer of the Sun above the photosphere.

Cislunar space. The region between Earth and Moon.

Coma. The fuzzy head of a comet.

Comet. Comets, described as "dirty snowballs," are clumps of gas, dust, dirt, rock and ice in long elliptical orbits around the Sun. More than 1,000 comets have been seen. The best known is Comet Halley. See also: Oort Cloud, below.

Conjunction. Visual alignment of Earth, the Sun and a planet. As seen from Earth, objects appear to be in line with each other. At superior conjunction, a planet is on the opposite side of the Sun from Earth. The planet is between the Sun and Earth at inferior conjunction.

Copernican System. The first heliocentric, or Sun-centered, theory of the Solar System, published by Nicolaus Copernicus in 1543, holding that planets revolve around the Sun. See also: Solar System.

Core. The central mass and region of highest density in the Sun or a planet.

Cosmic rays. Energetic particles, or ions, travelling through space after being emitted by the Sun or other stars.

Crater. A cone or bowl-shaped depression or rocky formation on the surface of the Moon, or other Solar System body, gouged out by a meteoritic impact. Also, a round pit on some bodies, possibly at the summit of a volcano. See also: Moon.

Crepe ring. The inner ring around the planet Saturn.

Crust. The outer layer of a planet or moon. Diastrophism is the process in which crust is deformed by internal forces.

Declination. See: Celestial Sphere.

Deep space. The realm beyond the Earth and the Moon.

Earth. The fifth largest planet of the Solar System and third from Sun.

Eclipse. When one celestial body obscures another. For instance, the Moon obscures the Sun in a solar eclipse.

Ecliptic. The Sun's path across the sky, through a band of constellations known as the Zodiac.

Electrons. Parts of atoms. The solar wind is a stream of negatively-charged electrons and positively-charged protons moving away from the Sun. See also: Protons.

Elliptical. Orbits of planets around the Sun are ellipses.

Encke's Division. A gap in ring A around planet Saturn.

Ephemeris. Daily position tables for planets, comets and other celestial bodies.

Equinox. The autumnal and vernal equinoxes are the points where the ecliptic crosses the celestial equator. The autumnal equinox occurs as the Sun's apparent north-to-south path crosses the celestial equator. The vernal equinox occurs as the Sun's apparent south-to-north path crosses the celestial equator. At equinox, the Sun crosses the equator and day and night are equal in length. Precession is the shift of the equinox due to the gravitational tug of the Sun and the Moon on Earth's equatorial bulge.

Escape velocity. The speed required for a spacecraft to leave a planet or celestial body without falling into orbit around it. See also: Orbital Velocity.

Ether. An imaginary substance once said to fill outer space.

Faculae. Bright spots on the Sun's surface.

Fireball. A bright meteor or shooting star seen in Earth's sky.

Fly-by. Exploring planet, moon, asteroid, comet, Sun or other body by a passing, remote-controlled, unmanned reconnaissance spacecraft.

Flying saucer. An imaginary flying machine said to be from a planet other than Earth. See also: UFO.

Fraunhofer lines. The most prominent absorption lines researchers find in the spectrum of light from the Sun.

Galilean satellites. The four largest moons of Jupiter, first seen by Galileo Galilei and named Callisto, Europa, Ganymede, and Io. See also: Jupiter.

Gas giant. A large low-density planet. One of the major Solar System planets. Of the nine major planets of the Solar System, the four largest planets, Jupiter, Saturn, Uranus, and Neptune, sometimes referred to as the Major Planets, are gas giants. See also: Major Planets.

GDS-2. See: Great Dark Spot No. 2.

Gegenschein. A faint glow in the sky opposite the Sun, caused by light reflecting off cosmic dust.

Geomagnetic. An effect of the planet Earth's magnetic field.

Geomagnetic tail. Earth's geomagnetic field blown away from the Sun side of the planet by the solar wind.

Grand Tour. A popular name for the interplanetary spacecraft Voyager 2's trip past four outer Solar System planets: Jupiter, Saturn, Uranus and Neptune.

Granules. The smallest things visible on the surface of the Sun.

Gravitation. The force of attraction between two bodies.

Great Dark Spot. Voyager 2 photos showed a huge storm swirling in Neptune's southern hemisphere. Larger in diameter than planet Earth, the hurricane was named Great Dark Spot after the Great Red Spot storm in Jupiter's atmosphere.

Great Dark Spot No. 2. Smaller cloud formations in Neptune's atmosphere have been called Great Dark Spot No. 2, Great Dark Spot No. 3, etc.

Greenhouse effect. The heating of a planet's atmosphere by trapping infrared radiation. For example, the hot atmosphere of the planet Venus.

Greenwich Mean Time. See: Universal Time.

Heliopause. The boundary between the heliosphere and the interstellar medium.

Heliosphere. The Sun's magnetic field dominates the local space environment, while the solar wind is a cloud of electrons and protons streaming at supersonic speeds away from the Sun. A large bubble of this energy surrounds the Sun as it moves through space at about 60,000 mph. The front of the bubble is compressed, while a long tail extends behind in the bubble's wake. This invisible teardrop-shaped bubble of energy, with the Sun at its core, is the heliosphere.

Inferior planet. A planet orbiting the Sun inside Earth's orbit around the Sun.

Infrared telescope. Equipment on Earth, aboard satellites and in interplanetary probes to sense light at infrared wavelengths.

Interplanetary. Within the Solar System but outside the atmosphere of any planet or the Sun. Between the planets. For example, a probe which flies away from Earth, beyond the Moon, to or near planets and other bodies in our Solar System is an interplanetary spacecraft.

Interplanetary travel. Hypothetical manned flights to other planets in the Solar System.

Io flux tube. A powerful electrical current flowing between the ionosphere of the planet Jupiter and Jupiter's moon Io.

Ion drive. An electric-drive space rocket which expels ions to create thrust.

Ion. A charged nuclear particle having either an electron missing or added.

Ionization. Electrically charging an atom by removing an electron.

Ionosphere. A layer in Earth's upper atmosphere where atoms have been ionized by solar radiation.

Jovian. Refers to Jupiter, as in Jovian moons.

Jupiter. The largest planet of the Solar System and fifth from the Sun. Seen through binoculars or telescope, yellowish-white Jupiter is brighter than any star. Four of Jupiter's moons, looking like four points of light on a straight line through the planet, are seen easily with binoculars. A telescope will show faint bands of colored clouds blanketing the planet. Jupiter is famous for its Great Red Spot.

Libration. A slight wobbling or rocking of a celestial body such as the Moon.

Lightyear. The distance light travels at 186,000 miles per second through space in one year, about 5.9 trillion miles. See also: Parsec.

Limb. The edge of the visible disk of the Sun, Moon, planet or other object.

Lunar. Referring to the Moon.

Magellan. An unmanned interplanetary spacecraft launched from a U.S. space shuttle to orbit Venus and map the planet. See also: Venus.

Magnetic field. The Sun generates a powerful magnetic field which dominates the Solar System environment. Planets have magnetic fields. See also: Heliosphere.

Magnetic storm. A heavy flow of energetic particles from the Sun which disrupts Earth's magnetic field.

Magnetometer. An astronomer's device to measure strength of a magnetic field.

Magnetopause. The Earth-space boundary where the solar wind strikes Earth's magnetic field. See also: Bow Shock.

Magnetosphere. The upper region surrounding a planet where the planet's magnetic field is stronger than the interplanetary magnetic field.

Major planets. Of the nine major planets of the Solar System, the four largest planets—Jupiter, Saturn, Uranus, and Neptune—sometimes are referred to as the Major Planets. See also: Gas Giants.

Mantle. The part of a planet's surface between crust and core.

Mare. A dark, dry, sea-like basin on the Moon's surface. See also: Moon, Terrae.

Mariner. A series of NASA interplanetary spacecraft used for planet flyby explorations. See also: Voyager.

Mars. The seventh largest planet of the Solar System and fourth from the Sun. It is referred to as the Red Planet—seen through binoculars or telescope, Mars looks orange-red. It's faint, but when closest to Earth can be as bright as Jupiter. Sky watchers often confuse it with red stars.

Mascon. A mass concentration in the Moon's crust.

Mean solar day. The time between Sun crossings of the meridian. The 24 hour day on Earth.

Mercury. The eighth largest planet of the Solar System and the planet nearest to the Sun. Seen through binoculars or telescope, Mercury is yellowish and brighter than a star, but not as bright as Venus or Jupiter.

Meridian. See: Celestial Sphere.

Meteor, meteorite, meteoroid, micrometeoroid. Meteors are streaks of light associated with the burning of small chunks of rock or interplanetary debris, known as meteoroids, often the size of a grain of sand or smaller, which arrive regularly in Earth's atmosphere from space. Very fine grains of space dust are micrometeoroids. Friction heats the rock plunging through Earth's atmosphere and makes the meteoroid glow in the air, causing the streak of light also known as a shooting star or fireball. On a moonless, dark night, a dozen meteors per hour may be seen all over the sky. During meteor showers, seen at certain times of the year, as many as 30 to 100 meteors per hour may be seen. A meteorite is a meteoroid large enough to survive the fall to Earth's surface.

Minor planet. See: Planet, Planetesimal.

Mohorovicic discontinuity. The boundary between Earth's crust and mantle; also referred to as MOHO.

Moon. A natural satellite. A body in orbit around a planet. Natural satellites of planets are moons. Earth has one Moon, Mars has two moons. Artificial moons are man-made satellites. There are at least 54 moons in the Solar System. The Moon is the natural satellite of Earth, the third planet from the Sun. American astronauts flew in Apollo spacecraft to visit the surface of the Moon six times from 1969 to 1972. Features seen on the Moon include craters, rays, maria and mountains. Craters are circular, ranging from as small as one inch in diameter to 100 miles across. They were formed when meteoroids struck the lunar surface at thousands of miles per hour. Such a collision scoops out a large hole and raises a mountain of rock around the edge. The craters easiest to see are Tycho, Copernicus and Kepler. Some craters have mountain peaks in their centers. Rays are the thin lines of rock splashed out of a crater hole by the meteor impact. Some rays from Tycho extend outward for a thousand miles. Maria is Latin for seas. These large, dark features on the lunar surface may have formed several billion years ago as the Moon was being created and it still may have had a thin crust. Meteoroids may have crashed through the crust, allowing molten lava to bubble up and fill shallow basins on the surface. That lava may have cooled into large flat maria. The maria are darker than surrounding territory because the rock is volcanic basalt which is darker than the material in the surrounding mountains. Mountains which circle the edges of maria may have been formed four billion years ago, after the crust formed but the Moon still was hot and molten. The outer layer buckled as the Moon cooled and shrank slightly. Rocks were pushed up to form mountains. There's no atmosphere on the Moon so there is no erosion. Most features have changed little over the billions of years, unlike Earth's mountains which have been eroded by water and wind.

Nadir. See: Celestial Sphere.

Neptune. The fourth largest planet in the Solar System, usually the eighth planet from the Sun, but sometimes farther from the Sun than ninth-planet Pluto.

Noctilucent clouds. Faint, luminous, high-altitude clouds seen at night in Earth's upper atmosphere.

Node. See: Celestial Sphere.

Nutation. A nodding motion of a body such as the Moon.

Occultation. One celestial object covering another in the view from Earth.

Olympus Mons. The largest volcano on the planet Mars, one of the largest in the Solar System. See also: Mars.

Oort Cloud. Astronomers suggest a swarm of comets is at the outer edge of the Solar System in what is called the Oort Cloud. See also: Comet.

Opposition. When a superior planet, or satellite, is opposite the Sun from Earth.

Orbit. The path travelled through space by an object, often a circle or ellipse. The closed path of one body around another. Satellites orbit planets. Planets, comets, asteroids and satellites orbit the Sun. The Sun and its Solar System orbit the center of the Milky Way galaxy. In a satellite's orbit around Earth, perigee is closest to the planet and apogee is farthest away. Perihelion is the point of a planet's orbit closest to the Sun. The farthest point from the Sun is aphelion. The sidereal period is the time taken by a body in completing an orbit around another object. The synodic period is the time between oppositions of a planet or satellite.

Orbital velocity. The speed required for a spacecraft to go into orbit around a planet or celestial body. To reach Earth orbit from the surface, a velocity of five miles per second is required. See also: Escape Velocity.

Outer space. The domain outside Earth's atmosphere.

Parallax. The apparent shift of position of a nearby object against a distant background when viewed from two places. A seeming change in position of celestial bodies with change in viewing angle. The apparent shift in position or direction of a celestial body when viewed from Earth at opposing points of its orbit around the Sun. A slight difference in a camera's field of view between lens and viewfinder. The slight difference between the image in a camera's viewfinder and the actual image as seen through the camera's lens. Spectroscopic parallax is the distance to a star as based on its absolute magnitude calculated from the relative intensities of selected spectrum lines.

Parsec. Parallax second, astronomy unit of measurement equal to 3.26 lightyears, the distance required to produce a parallax angle of one arc second. See also: Lightyear.

Penumbra. The partial shadow around the main shadow of the Moon during a solar eclipse or the Earth during a lunar eclipse. See also: Sunspots.

Perigee. The orbit location closest to the body being orbited. For instance, the point where a satellite orbiting Earth is closest to Earth.

Perihelion. The orbital point closest to the Sun for any body in the Solar System. See also: Aphelion, Apogee, Perigee, Orbit, Solar System.

Photometer. A device to measure light.

Photon. A unit of light.

Pioneer. A NASA series of unmanned interplanetary probes exploring distant reaches of the Solar System.

Planet. A non-luminous body in orbit around a star such as our Sun. The known planets are nine massive bodies in our Solar System revolving around the Sun and reflecting its light. The nine planets circling the Sun: Mercury, Venus, Earth, Mars, Jupiter, Saturn, Uranus, Neptune, Pluto. Astronomers say there may be a tenth planet in our system plus planets orbiting other stars. A minor planet is an asteroid. There are thousands of minor planets, mostly in an orbit known as the Asteroid Belt between the orbits of Mars and Jupiter.

Planetarium. A theater with a domed ceiling on which images of the Solar System and stars are projected.

Planetesimal. A clump of matter in space. Large rocky planetesimals were created as the new Sun's heat acted on nearby metal grains and chunks of rock as the Solar System was forming. The planetesimals were tiny worlds, up to 60 miles in diameter. Over time, some collided and stuck together. After 75 million years of chaos, what we know today as the major and minor inner planets were growing. Over a few hundred million years, the planetesimals smashed into various major planets. Others would have been deflected, even out of the Solar System. See also: Asteroid, Planet.

Planetoid. See: Asteroid, Planet, Planetesimal.

Plasma. Stream of ionized gas flowing from the Sun.

Pluto. The smallest planet of the Solar System, usually ninth and farthest from the Sun, but with a highly eccentric orbit which sometimes brings it closer to the Sun than eighth-planet Neptune. See also: Neptune, Planet, Solar System.

Plutonium 238 dioxide. The interplanetary probes Voyager 1, Voyager 2, Galileo and other spacecraft have been outfitted with long-life electricity-generating radioisotope thermoelectric generators (RTGs) powered by the decay of radioactive plutonium 238 dioxide.

Polarimeter. A device to measure the polarization of a ray of light.

Precession. The turning of Earth on its axis like a spinning top, caused by the pull of the Moon's gravity on Earth's equatorial bulge. See also: Celestial Sphere.

Prime meridian. See: Celestial Sphere.

Probe. An interplanetary spacecraft which flies away from Earth, beyond the Moon, to or near planets and other bodies in our Solar System.

Prominence. A tongue of flaming gas licking outward from the corona on the edge of the Sun.

Protons. Parts of atoms. Solar wind is a stream of positively-charged protons and negatively-charged electrons in the solar wind. See also: Electrons.

Pyroheliometer. Device to measure energy radiating from the Sun.

Quadrature. A superior planet 90 degrees east or west of the Sun.

Radar astronomy. Radar signals from orbiting probes have been used to measure and map Venus, Mercury and other Solar System bodies.

Radioisotope thermoelectric generators. The interplanetary probes Voyager 1, Voyager 2, Galileo and other spacecraft have been outfitted with long-life electricity-generating radioisotope thermoelectric generators (RTGs) powered by the decay of radioactive plutonium 238 dioxide.

Ranger. NASA series of unmanned lunar photography probes. See also: Surveyor.

Red Spot. A reddish hurricane seen in the planet Jupiter's southern hemisphere.

Regolith. A clumpy topsoil powder on the surface of the Moon.

Retrograde motion. The passage of a planet through a star field in a backward or westerly direction, opposite the easterly motion of an observer on Earth.

Right ascension. See: Celestial Sphere.

Rille. Canyon or narrow valley on the surface of the Moon.

Ring plain. A large-diameter lunar crater with relatively smooth interior. See also: Moon.

RTG. See: Plutonium 238 Dioxide, Radioisotope Thermoelectric Generators.

Satellite. A small body orbiting a larger object. A moon in orbit around a planet. Natural satellites are called moons. Earth has one Moon; Mars has two moons. Man-made artificial moons, usually referred to as satellites, are machines in orbit.

Saturn. The second largest planet of the Solar System and sixth from the Sun. Seen through binoculars or telescope, yellowish Saturn is fainter than the brightest stars. Saturn is famous for its rings which can be seen with a telescope, but usually not with binoculars.

Scooters. Nickname for fast-moving storm clouds photographed by Voyager 2 in Neptune's atmosphere. The clouds rotate around Neptune faster than the planet spins on its axis. Such nicknames are informal; formal names for Solar System bodies and their features must be approved by the International Astronomical Union.

Seismometer. Device to measure impacts and quakes, such as Moonquakes.

Sensor. Device to receive energy.

Shield volcano. A volcano with layers of lava on its sides.

Shooting star. A bright meteor or fireball seen in Earth's sky. See also: Meteor.

Sidereal period. Time required for a body to complete an orbit around another object. See also: Orbit, Synodic Period.

Siderite. A nickel-iron meteorite. See also: Meteor.

Solar Constant. The quantity of energy from the Sun radiating onto the top of Earth's atmosphere—1.94 calories per minute per square centimeter.

Solar flare. A sudden brightening of a spot on the surface of the Sun from which a high-speed jet of hot plasma gas erupts.

Solar System. The Solar System is the Sun, the surrounding bubble of energy it generates, and the nine known major planets, asteroids, comets and other matter. A solar system is a group of planets orbiting a star. The Solar System is a collection of bodies held near each other in space by the gravitational attraction of a star we call the Sun. The Solar System bodies, including major planets, their moons, minor planets, comets, asteroids, and debris, all orbit the Sun. Billions of years ago the major planets formed from a disk of matter which surrounded the Sun. Now all of the bodies of the Solar System swim in a thin soup of rocky dust particles. The rocks and dust may be left over from the original accretion formation process, or ejected by comets as they pass through the inner Solar System, or they may be debris from collisions between minor planets. The nine major planets in order from the Sun are Mercury, Venus, Earth, Mars, Jupiter, Saturn, Uranus, Neptune and Pluto. Some astronomers speculate there may be a tenth planet in our system. The Solar System also has moons, asteroids, minor planets, planetoids, comets and other objects. See also: Planet.

Solar wind. A supersonic stream of energized particles, negatively-charged electrons and positively-charged protons, flowing away from the Sun through interplanetary space. A stream of ionized hydrogen and helium.

Solstice. The sky positions where the Sun is at maximum angle or declination from the celestial equator. See also: Celestial Sphere.

Spectral bands. Lines in the spectrum of light from the Sun, caused by absorption or emission by molecules.

Spectroscopic parallax. See: Parallax.

Spokes. Radial structures in the rings of the planet Saturn. See also: Saturn.

Sun. A star is a sun. The star we call The Sun is one of at least a hundred billion stars orbiting the center of our Milky Way galaxy—a blazing pinwheel of stars 100,000 lightyears in diameter. So far, the Sun is the only star known to be surrounded by an extensive solar system, although many astronomers suggest other stars with planets likely will be found nearby in the Milky Way. Stars are fireballs of bubbling-hot gas with surface temperatures ranging from 3,000 to 100,000 degrees Fahrenheit—hot enough to melt almost anything on Earth. The surface temperature of the Sun is about 11,000 degrees Fahrenheit. A star's color tells how hot it is. A cool star, at about 3,600 degrees, is red. A hot star, at about 90,000 degrees, is blue. The Sun is yellow and provides light and warmth to the Earth The Sun isn't a very large star, being only 865,400 miles in diameter. It is, however, about 100 times the size of Earth and the heart of a Solar System with nine orbiting planets, comets and asteroids. The Sun is a yellow dwarf star composed mostly of hydrogen and helium, with about one percent of its mass in heavier elements. See also: Planet, Solar System.

Sunspots. Large storms in the Sun's atmosphere are called sunspots. The storms generate massive explosions from the surface, firing high-speed atomic particles outward. The spots range from specks to 90,000 miles across. They usually are noticed by observers on Earth when they grow as large in diameter as Earth. The largest sunspots can be seen with the naked eye. However, never look directly at the Sun with the naked eye, nor with binoculars, telescope or camera. The eye can be damaged seriously, even though no pain is felt. Each sunspot has a dark core, known as umbra, and an outer gray band known as penumbra. The spots look dark because they are cooler than the surrounding surface. The cool spots are 7,000 degrees Fahrenheit, while surrounding areas are 11,000 degrees. The spots change in size and position every day as the rotation of the Sun makes them seem to travel across the surface in 14 days. The number of sunspots at one time varies with a maximum about every 11 years. Increased spots bring increased auroras on Earth as well as disruptions of radio, television, telephone and electrical systems. See also: Sun, Solar Flare.

Superior planet. A planet orbiting the Sun outside Earth's orbit around the Sun. At opposition, a superior planet is opposite the Sun from Earth. See also: Planet.

Surveyor. A NASA series of unmanned photography and soil-testing soft-landing lunar probes. See also: Moon, Ranger.

Synodic period. The time between oppositions of a planet or satellite. See also: Orbit, Sidereal Period.

Tenth Planet. Planet X is a suspected, but yet to be discovered, most-distant major planet from the Sun.

Terminator. A line dividing dark and light hemispheres of a planet or satellite. The line between day and night seen from a spacecraft orbiting Earth.

Terrae. Light colored areas on the surface of the Moon, higher than the dark mare basins. Observers once saw mare as seas and terrae as land masses. See also: Mare.

Terrestrial planets. The four inner planets of the Solar System: Mercury, Venus, Earth, and Mars. See also: Planet, Solar System.

Titan. The largest satellite of Saturn and the largest satellite in the Solar System. See also: Planet, Saturn, Solar System.

Transit. The passage of an object across an observer's meridian or an object across the face of another object. See also: Celestial Sphere.

Triton. The largest satellite of the planet Neptune and the only body in the Solar System with active ice volcanism. See also: Neptune, Planet, Solar System.

Tropopause. The part of the atmosphere between troposphere and stratosphere.

UFO. Unidentified Flying Object; a hypothetical spacecraft from another Solar System. See also: Flying Saucer.

Ultraviolet telescope. Equipment on Earth, aboard satellites and in interplanetary probes to sense light at ultraviolet wavelengths.

Umbra. The main cone of shadow cast by the Moon during a solar eclipse or by Earth during a lunar eclipse. See also: Sunspots.

Universal time. UTC, GMT, Greenwich Mean Time. Mean solar time at the zero-degree meridian at Greenwich, England.

Uranus. The third largest planet of the Solar System and seventh planet from the Sun. See also: Solar System.

Van Allen Belts. Zones of electrified particles trapped at high altitudes by Earth's magnetic field.

Venera. Series of USSR spacecraft exploring the planet Venus. See also: Venus.

Venus. The sixth largest and brightest planet of the Solar System and second from the Sun. Seen through binoculars or telescope, Venus is yellowish-white due to being completely covered with clouds. Venus is brighter than a star or any other planet. Because it orbits the Sun inside Earth's orbit, Venus usually is seen as a crescent.

Vernal equinox. see: Equinox.

Vesta. A minor planet in our Solar System. See also: Asteroid, Solar System.

Viking. A NASA series of unmanned soft-landing Mars probes. See also: Mars.

Visible-light telescope. Television camera on Earth, aboard satellites and in interplanetary probes to sense light at wavelengths visible to the human eye.

Voyager. A NASA series of interplanetary probes, based on the Mariner series, used to fly by and explore the outer planets of the Solar System, including Jupiter, Saturn, Uranus, and Neptune. See also: Mariner.

Zenith. See: Celestial Sphere.

Zodiac. Band of 12 constellations on the celestial sphere with the ecliptic as its middle line, including the paths of all the major planets except Pluto. The zodiac is a band of the celestial sphere, 8 degrees wide on each side of the ecliptic. It is divided into 12 equal zones of 30 degrees each, and used in astrology. See also: Celestial Sphere.

Zodiacal band. A vague band of light along the ecliptic, joining the zodiacal light and gegenschein, caused by the scattering of sunlight by interplanetary particles. See also: Ecliptic, Gegenschein, Zodiac, Zodiacal Light.

Zodiacal light. A subtle cone of light reaching up from the horizon, caused by the scattering of sunlight by interplanetary particles. A faint glow in the East just before dawn and in the West just after sunset.

Deep Space
★★★★★★★★★★★★★★★★★★

Deep Space Table Of Contents Continues On Page 532

Deep Space
★★★★★★★★★★★★★★★★★★

Deep Space Table Of Contents Continues On Page 533

Deep Space

★★★★★★★★★★★★★★★★★★

Grains Of Sand On A Vast Beach

There are untold trillions of galaxies—clouds of stars, planets, moons, comets, asteroids, gas, dust—strewn across the Universe like grains of sand on a vast beach. From our earliest times we've looked up and wondered what's out there.

Stars. Our Sun is a star. A star is a sphere of hot glowing gas, a single ball of fire in space, often a million of miles or more in diameter. The stuff of some stars is 10,000 times as thin as Earth's air at sea level. Other stars are so dense a cupful of material would weigh tons on Earth.

Inside, stars have temperatures measured in millions of degrees. At their surfaces, temperatures up to 55,000 degrees are common.

Stars are a long way off, yet all stars visible from Earth, even with telescopes, are within our own Milky Way galaxy. Proxima Centauri, the star nearest Earth, is so far away it's only a pinpoint of light in the largest telescopes on Earth.

Many other stars may be ringed by planets and other small bodies, just as our Sun has its Solar System of planets, moons, comets and asteroids. Planets reflect starlight, while stars shine with their own light. Stars differ in size, brightness and color.

From our perspective on Earth, stars seem to hold the same positions in the sky year after year. Over many centuries they do appear to move, however, as our Solar System slowly circles the core of our own Milky Way galaxy.

Neighborhood galaxies. A galaxy is a vast island of stars floating through the Universe—millions or billions of stars attracted to each other and held close by their gravity. Astronomers have seen galaxies of many different shapes and sizes. Most common are flat pinwheel spirals, egg-shaped ellipticals, and oddly-shaped irregulars. The galaxies most widely known are Andromeda Galaxy, Large Magellanic Cloud galaxy and our own Milky Way galaxy

★MILKY WAY. Our Earth circles the Sun which is one star among 100 billion in the Milky Way galaxy. Viewed from afar, the Milky Way would look like a pinwheel, a spiral, a flat disk 100,000 lightyears in diameter with a bulge at the center. A lightyear is the distance light travels through space in one year, about 5.9 trillion miles.

★ANDROMEDA. The Andromeda Galaxy, also known as M-31, is the spiral galaxy nearest to our Milky Way and the brightest galaxy in Earth's sky. Bigger than the Milky Way, it contains 300 billion individual stars in a disk 180,000 lightyears in diameter. Andromeda Galaxy is 2.2 million lightyears from Earth.

★MAGELLANIC CLOUDS. In 1987, the world focused on an irregular galaxy, the Large Magellanic Cloud, when a star in that galaxy exploded into view as the supernova closest to Earth in 400 years. The outburst was labeled Supernova 1987a.

The Large Magellanic Cloud is a galaxy neighbor of our Milky Way galaxy—only 163,000 lightyears from Earth. Light from Supernova 1987a left the Large Magellanic Cloud 163,000 years ago traveling at 186,200 miles per second. The Large Magellanic Cloud is only 39,000 lightyears in diameter, less than half as wide as the Milky Way.

The Large Magellanic Cloud has a smaller companion galaxy known as the Small Magellanic Cloud. The Small Magellanic Cloud is under 20,000 lightyears in diameter and is 196,000 lightyears from Earth.

Some astronomers have wondered if the Small Magellanic Cloud they see isn't two small galaxies lined up in their telescopes. If so, astronomers would call the third galaxy the Mini Magellanic Cloud. It might be some 230,000 lightyears from Earth.

At any rate, astronomers think the Large Magellanic Cloud may have been closer to Earth once upon a time as it drifted past the Milky Way a billion years ago. A trail of gas and dust, which astronomers call the Magellanic Stream, extends across our sky. The Magellanic Stream probably is matter ripped away from the Large Magellanic Cloud by Milky Way gravity. It's not clear whether the Magellanic Clouds are satellite galaxies of the Milky Way, forever circling us, or simply passers-by on a one-time chance encounter as they make their way through the Universe.

While astronomers ponder clusters and super-clusters of galaxies, even stranger objects wash up from time to time on the vast beach. Supernovas, neutron stars, pulsars, black holes, quasars and more...

Supernovas. Astronomers were riveted when the previously-unremarkable star known as Sanduleak 69.202 exploded in a fiery death as Supernova 1987a. After it blinked into view as the supernova closest to Earth in 400 years, Europeans and Americans checked satellite views of the event. Russians at the orbiting Mir space station photographed it. U.S. and Australian scientists fired rockets, floated instrumented balloons and flew an observatory airplane, all to get as high as possible for a better view of the brilliant Supernova 1987a.

Neutron stars. A supernova is the extraordinarily powerful death explosion of a massive star. The star brightens for a short time, then usually reduces to a small remnant—a neutron star. Astronomers wonder if a neutron star, maybe even a pulsar, hasn't formed at the heart of Supernova 1987a.

Pulsars. During a supernova explosion, the star collapses inward, explodes, spews out gas and dust, and collapses again into a small heavy core, a neutron star. If the neutron star spins and radiates energy toward Earth in blips—like a lighthouse beacon— it is referred to as a pulsar. A pulsar is so crushed by the weight of its own gravity, a thimbleful would weigh a million tons.

Astronomers have been watching one pulsar, 12,000 lightyears from Earth, which has such a regular spin it could replace atomic clocks as our most reliable timepiece. Known by its sky-map location, 1937+21, the pulsar spins at a steady 642 twirls a second. The official U.S. clock in a laboratory at Boulder, Colorado measures natural vibrations in cesium atoms to mark time. However, they do fluctuate. Pulsar 1937+21, on the other hand, apparently hasn't deviated at all since 1982.

Black holes. At the end of a total collapse of a star or several stars, matter is crushed into a black hole, a very small and extremely massive body, even more densely packed than a neutron star. It is so dense, its gravity so intense, light can't escape—it's invisible. However, a black hole's presence can be seen in its interaction with normal space around it. If a black hole sucks matter away from a nearby star, a fountain of X-rays might spew out from the event horizon, the boundary of a black hole.

Black holes once seemed to be only faraway objects, but in recent years astronomers have begun to pick up clues that one may be lurking at the heart of our Milky Way galaxy. It looks like a gargantuan black hole near the core of the Milky Way may be sucking in matter from magnificent dust and gas clouds which obscure our view of the center. A river of gas, 90 trillion miles long, has been found streaming into the center of the Milky Way. It could be fueling a prodigious black hole at the very heart of the galaxy, according to reports from astronomers who also suspect mini black holes and super-massive black holes are strewn across the Universe.

Quasars. A quasar is a quasi-stellar object, a powerful source of light and natural radio energy. Many have been seen. They may be highly-luminous galaxies, extremely remote from Earth at the far distant edge of the known Universe, probably undergoing violent explosions.

They may be the most distant objects in the Universe. Quasars are billions of lightyears from Earth, racing farther away at 150,000 miles per second. In their wake, they leave huge quantities of light and other radiation. Even though they seem smaller than galaxies, some emit 100 times as much energy as common galaxies. They may be violently exploding galaxies or the cores of very active galaxies.

Observers at the Royal Astronomical Society, London, recently found a cluster of quasars which may be the largest object ever known in the Universe. The elongated group of 10 to 13 quasars was spotted at a distance of 6.5 billion lightyears from Earth. The cluster measures 650 million lightyears long by 100 million lightyears wide.

British astronomers also have seen what may be the most luminous object ever seen, a massive quasar hiding in a dust cloud at the far edge of the Universe 16 billion lightyears from Earth. The mysterious body pumps out huge amounts of energy—300 trillion times more than the Sun and 30,000 times more than our Milky Way galaxy.

Some astronomers suggest the exploding giant originally may have formed into a galaxy shortly after the beginning of time in a Big Bang, itself an explosion which some say created the Universe. Some say the Universe is only 15-20 billion years old. At 16 billion lightyears from Earth, the puzzling object would be 80 percent of the way back to the Big Bang. Finding the oldest galaxy has become the Holy Grail of astronomy, one astronomer has said.

The Lightyear

A lightyear is a measure of distance. Just as you would say your cat and dog are fifteen feet apart, or two towns are forty miles apart, objects in space—such as planets, stars, galaxies, pulsars, quasars—can be some number of lightyears apart.

One lightyear is the distance one ray of light would travel in one year in a vacuum. That's about 5.9 trillion miles or 9.5 trillion kilometers.

Space is not 100 percent vacuum, of course, but it's close enough, despite all of those beautiful clouds of dust and gas strewn through the Universe.

The distances are very great. Here are some examples:

★Our Solar System is about 30,000 lightyears from the center of our own Milky Way galaxy.

★From one side of the Milky Way galaxy to the other is about 100,000 lightyears.

★The star nearest to our Sun is Proxima Centauri, about 4.5 lightyears away. That's about 26 trillion miles.

Magnitude

The magnitude we assign to a planet or star depends on the amount of light it gives off and its distance from Earth. A star farther from Earth sending out a lot of light may appear brighter here than a closer star not giving off much light. A large dull star might seem weaker to us than a small bright star.

The lower the magnitude number, the brighter the object. Thus a first-magnitude, or magnitude 1, star such as Aldebaran (magnitude +0.9) seems brighter to us than Castor (magnitude +2.0).

As the numbers approach and pass zero, they become negative numbers on the magnitude scale. Thus the brightest star we see is Sirius (magnitude –1.4), only nine lightyears away from Earth.

Sometimes, magnitudes are rounded off. For instance, magnitudes from -1.5 to -0.6 often are considered magnitude -1.0. Magnitudes from -0.5 to +0.4 often are reported as magnitude 0.0 and magnitudes from +0.5 to +1.4 sometimes are referred to as magnitude +1.0.

Brightest. The very-brightest objects in our sky are the Sun at magnitude –27 and the Moon at magnitude –13.

On a clear night, we can see objects as faint as magnitude 5 or 6 with the naked eye. With binoculars we see stars as weak as magnitude 8 or 9. A six-inch telescope will show stars down to magnitude 13. The faintest objects photographed so far by astronomers using the largest telescopes are about magnitude 26.

Stars in the constellation we call the Little Dipper range in magnitude from 2 to 5. Some planets can be seen with the naked eye. Mars, for example, varies in magnitude from –2.8 to +1.6 depending on distance from Earth. Jupiter is –2.5 to –1.4 magnitude.

Galaxies Of Stars

A galaxy is a vast cloud of millions or billions of stars attracted to each other and held close by their gravity while floating as a group through the Universe. There seem to be more galaxies strewn across the Universe than grains of sand on a beach.

Galaxies have different shapes and sizes. Most common are flat-disk pinwheel spirals, egg-shaped ellipticals, and oddly-shaped irregulars. Best-known galaxies are our Milky Way and the nearby Large Magellanic Cloud and Andromeda.

The Sun is one star among 100 billion in the Milky Way which is a flat disk-shaped spiral galaxy 100,000 lightyears in diameter with a bulge at the center, like a huge blazing pinwheel floating through space. A clutch of stars 15 billion lightyears away, known as 4C41.17, is one of the most distant galaxies. A lightyear is approximately 5.9 trillion miles.

Messier Catalogue Numbers

An astronomical catalog is a table of data on celestial objects with stars numbered. Hipparchus of Rhodes compiled the first such reference with 850 stars about 150 B.C. Ptolemy of Alexandria tabulated 1,022 stars, with position and brightness, about A.D. 127. Moslem astronomers composed numerous catalogs during the Middle Ages.

French astronomer Charles Messier in 1784 printed a catalog of what, at the time, he said were the brightest "nebulae." Galaxies looked like fuzzy extended objects in his telescope so Messier thought they were clouds in deep space. He wanted to avoid confusing them with comets.

Messier listed 103 objects in his catalog in 1784 and six more in 1786. Later astronomers found the objects actually were not nebulae, or clouds of dust and gas, but galaxies and star clusters.

Today's astronomers refer to the stars from the old-time catalog of celestial objects by their Messier number, such as M-42 or M-66.

NGC: New General Catalogue

NGC stands for New General Catalogue of Nebulae and Clusters of Stars. Compiled by Danish astronomer J.L.E. Dreyer at Armagh Observatory, Ireland, and published in 1888, it listed 7,840 known clusters, galaxies and nebulae.

The brightest objects in the NGC list previously had appeared in the Messier Catalog. Objects in the new catalog are known by their NGC numbers, such as NGC 3628 near the galaxy M-66.

Dreyer added to his catalog with supplements he called the first and second Index Catalogues (IC). They were published in 1895 and 1908. Today all three—NGC and two IC's— are printed as one volume covering the entire sky and covering 13,000 objects.

Andromeda Galaxy

Andromeda is said to be the best-known galaxy, after our own Milky Way. Andromeda is the nearest spiral galaxy beyond the Milky Way, at a distance of 2.2 million lightyears. Known to astronomers as M-31 or NGC-224, Andromeda also is the most distant object visible to the naked eye.

The galaxy is huge. Astronomers estimate its mass at 300 billion times our Sun. That's triple the size of our Milky Way galaxy. Andromeda's diameter is 130,000 lightyears. That's 1.3 times our galaxy. Andromeda has globular clusters, open clusters and nebulas of gas and dust. The Milky Way has them also. In fact, astronomers first thought it was a nebula in the Milky Way. They named it Andromeda Nebula. A great debate from the late 1800's through the 1920's centered on Andromeda, until Edwin Hubble's work confirmed it was something new to astronomers—an independent galaxy outside the Milky Way.

The name Andromeda comes from a mythical Greek princess.

Constellation Andromeda. Ancient astronomers named a constellation, prominent during autumn in the heavens over the Northern Hemisphere, Andromeda.

Andromeda contains the Andromeda Galaxy and the planetary nebula NGC-7662 within our Milky Way galaxy at a distance of 5,000 lightyears from Earth. The brightest star in the constellation is Alpheratz.

Black holes nearby. Gigantic black holes, like cancers eating their victims from the inside out, are hiding in the hearts of two neighboring galaxies, not far from Earth and our Milky Way galaxy.

Some astronomers think those massive energy-stealing black holes could kill their galaxies in the distant future.

Andromeda Galaxy has dark matter about 70 million times the mass of our Sun at its core. And a galaxy satellite of Andromeda, known as M-32, one-tenth the size of Andromeda and revolving around Andromeda, has about one tenth as much dark matter concentrated at its center.

The only deep-space objects with so much unseen non-luminous matter are black holes. A black hole is a concentrated ball of matter so dense after collapsing in on itself that nothing—not even light—can escape the grip of its intense gravity.

M-31 and M-32 are a long way from Earth, yet they are neighbors to us in the vast expanse of the Universe. The pair of galaxies are two million lightyears away from our Solar System. A lightyear is the distance light travels through space in a year—about six trillion miles. Andromeda is visible to the naked eye on Earth as a faint glow high in the summer sky. M-32 requires a telescope to be seen from Earth.

Black holes. Astronomers had speculated for some time that black holes, or some other dense bodies, were responsible for the huge reservoirs of energy radiating from the centers of the great star clusters known as galaxies. Until now, those suspicions were based on circumstantial evidence. The new conclusions are based upon fresh telescope observations and computer calculations of weights and speeds within the galaxies.

No one knows for sure, but astronomers recently stepped up from only a "possible" explanation to what now is the "likely" explanation by finding the first evidence of gravitational pull of the black holes on the surrounding matter of their galaxies. Black holes seem certainly to be active powerhouses in the center of galaxies, pulling in gas and dust from space between stars, and from stars themselves, for fuel to burn.

Many millions of years from now, the black holes may cause the collapse and death of the M-31 and M-32 galaxies.

As a black hole at the heart of a galaxy feeds and grows larger, it may eventually suck in and compress enough outside matter to give rise to a quasar. A quasar is an intensely powerful engine that radiates extraordinary amounts of energy. Quasars, dotted about the Universe, seem like beacons in the heavens to observers on Earth.

Fresh observations. A Carnegie Institution astronomer used the 200-inch Hale telescope at the Palomar Observatory in California to measure the motion of stars near the centers of the M-31 and M-32 galaxies.

Change in frequency of radio waves coming from the stars told him the speed with which the stars are moving about, their distances from the center of the galaxies, and how massive they are.

A University of Michigan astronomer used advanced math to explain patterns of orbits of stars in the galaxies. He figured that what we see isn't enough to explain the way the stars are moving. Only the presence of a giant unseen object in each galaxy, with mass 10 million to 100 million times greater than that of our Sun, could force the stars to move as they do. Large black holes seemed the only plausible explanation.

M-31 and M-32 are ordinary spiral-shaped pinwheel galaxies. A black hole in each suggests that such big dark objects may be common in other galaxies—maybe even our own spiral-shaped Milky Way galaxy.

Globular Clusters

Globulars are clusters of hundreds of thousands of stars which, through binoculars, look like small fuzzy patches of light deep in space. Through small telescopes, they look like fuzzy blobs.

A backyard astronomer can see 200 to 300 individual stars in a cluster. In the late 1700's, William Herschel studying M13, the Great Cluster in Hercules, counted 30,000 individual stars. Since then, clusters have been estimated at 100,000 to 1,000,000 stars.

Globulars are not distributed evenly across the sky. Most are in one hemisphere of Earth's sky. One third are seen in the constellation Sagittarius. That constellation covers only two percent of the sky. Globular clusters compose a large sphere centered in Sagittarius.

The clusters orbit the center of our Milky Way galaxy just as the planets of our Solar System surround our Sun. Globular clusters form a large 100,000-lightyear-diameter sphere around the heart of the Milky Way. Some 131 globulars have been counted. Knowledge of globular clusters made it clear that our Solar System is toward the edge of the Milky Way rather than at the center as previously thought.

A parsec is a measure of distance, about 19 trillion miles. Globular clusters are far away, up to 60,000 parsecs from our Sun. The stars in the clusters are billions of years older than the stars of the Milky Way galaxy. They condensed at a time when the gas of our galaxy was too turbulent to condense as the Milky Way galaxy.

Cluster stars are mostly helium and hydrogen. Metals became available for star formation at a later date via supernova explosions. Milky Way stars formed more recently contain metals.

Stars are two lightyears apart in the outer parts of a cluster but only a fraction of a lightyear apart at the center of a cluster. A lightyear is about six trillion miles. Stars at the center of a cluster are so tightly packed, most probably orbit each other. Some globulars may have black holes at their centers.

Globular clusters circle the galaxy as the entire system travels the Universe. Orbiting the center of our galaxy in elliptical paths, clusters pass through the disc of the galaxy every ten billion years and are torn apart or distorted by gravity, losing some stars. Thus, globular clusters close to the galactic core are smaller than those in outer areas.

As Comet Halley passed near Earth, rounding our Sun in 1986, some observers thought they had seen a second comet in their binoculars. Astronomers with larger telescopes quickly resolved the hazy patch into thousands of stars. It was the bright globular cluster Omega Centauri.

Globules are not globular clusters. A globule is a dense, round, compact cloud of space dust, probably precursor to birth of a star.

Superclusters Of Galaxies

Imagine one grain of sand among billions on an endless beach stretching out of sight, to a distance where you know there are more beaches than you can see.

Such a mental snapshot may help you grasp the astronomer's view of the relatively insignificant place of our vast Milky Way galaxy in an immense Universe. Each grain of sand on the beach is a galaxy of stars.

A University of Hawaii astronomer describes the Milky Way galaxy as just one part of a "supercluster complex" of galaxies 100 times larger than any previously identified interacting collection of stars and galaxies in the Universe.

He sees our Milky Way galaxy as near one end of a supercluster containing millions of galaxies that stretch across ten percent of the observable Universe.

Light travels nearly six trillion miles through space in one year. The supercluster complex described by the Hawaii astronomer is so large, light leaving the Sun takes a billion years to reach the other end of the structure. The supercluster complex has a flat, oblong shape, 150 million lightyears across.

The researcher collected data gathered over the years by other astronomers and sent it through his computer to find the size and shape of the supercluster complex. At one billion lightyears in length and 150 million lightyears in width, the supercluster complex is about seven times as long as it is wide.

Since the University of Hawaii researcher is the first to map the Universe on such a large scale, other astronomers are examining his chart. If he turns out to be right, basic understanding of how galaxies are strewn through the Universe will change.

Islands of stars. Space watchers once thought galaxies were islands of stars spread evenly through the Universe. But then there were reports of galaxies found clustered together in long filaments, like strings of pearls 10 million to 100 million lightyears in length. A lightyear is about 5.9 trillion miles. Smaller strings-of-pearls clusters of galaxies themselves are arranged in a pattern to form a supercluster structure that is 100 times larger.

Current astronomy textbooks describe our Milky Way galaxy as one whirlpool of stars among a group of galaxies called the Local Supercluster. While mapping the Local Supercluster, the University of Hawaii astronomer noticed it was larger than previously thought.

Glittering clusters. As he sifted the data, rich clusters of galaxies hundreds of millions of lightyears away, turned up across ten percent of Earth's sky. Each rich cluster glittered with thousands of galaxies, each galaxy with at least 100 billion stars.

Putting it altogether, his computer depicted 60 of the rich clusters in a much larger pattern. He labeled it the Pisces-Cetus Supercluster Complex after the constellation in which it is found. Constellations are arbitrary star patterns imagined by astronomers years ago to portray areas of Earth's night sky.

But, the Pisces-Cetus Supercluster Complex won't be the end of the story. Finding something to peer at in the future, the astronomer spotted what looked like parts of four other superclusters that intrude into the area of the sky he surveyed. Of course, they may not be mapped soon since they extend beyond the limits of Earth's current astronomy technology.

Mystery tugs. American and British astronomers in 1989 turned up startling new clues about a mysterious, but magnificent, flow of stars across the Universe. Clouds of galaxies in almost-inconceivably-vast spaces are being pulled at millions of miles per hour through space toward some unseen attraction.

Even our own Milky Way galaxy is streaming through space, caught up in the herd of galaxies sailing toward the tremendously-powerful force. Where are all of the trillions upon trillions of stars headed? Astronomers say in the direction of a constellation sketched on their night-sky charts as the Southern Cross.

Studying 400 elliptical galaxies over six years, researchers were struck by the massive concentrations of galaxies they refer to as the Virgo Supercluster and the Hydra-Centaurus Supercluster.

Large as it seems, our galaxy of 100 billion stars is but a tiny speck drifting on one side of the Virgo Supercluster. The numbers of stars accumulated in those sprawling clouds of galaxies surely is more than all of the grains of sand on all the beaches known to man...

Cosmic Picket Fence

U.S. and British astronomers in 1990 reported finding a "cosmic picket fence" composed of a dozen or more clusters of galaxies posted at surprisingly even intervals throughout part of the Universe.

The unexpected pattern again challenged accepted notions about how galaxies formed billions of years ago. The theory had been that the Universe is a smooth, uniform place.

Other astronomers previously had reported what they called a "Great Wall" of galaxy clusters spread across space in the largest structure ever found.

The cosmic picket fence made the Great Wall seem like just one of many great walls flung across the Universe. The series of great walls might be spaced at 400-lightyear intervals across the Universe, like a picket fence. Astronomers said there was relatively little matter visible in the voids between the superclusters. A lightyear is about 5.9 trillion miles. Astronomers don't know how extensive the picket fence is as they have reviewed only a small fraction of the Universe.

The picket-fence data was collected by researchers from the University of California-Santa Cruz, Johns Hopkins, the University of Chicago at the Kitt Peak National Observatory in Arizona, and the University of Durham in Great Britain using the Anglo-Australian Observatory in Australia.

Most Luminous Object In Deep Space

British and American astronomers have discovered a very distant object which they describe as the brightest, most luminous object ever seen in the Universe. The massive cloud of dust is emitting 300 trillion times as much energy as the Sun.

Sifting through data from the Infrared Astronomy Satellite (IRAS), the scientists discovered a massive dust cloud in the constellation Ursa Major, 16 billion lightyears from Earth. That makes it one of the most distant objects ever observed.

The cloud appeared to be radiating 99 percent of its energy as infrared light. The astronomers wondered if the dust cloud was being heated by a billion extremely hot, young stars formed in the early stages of the birth of a galaxy. Or if the cloud was being heated by a more powerful quasar than had been seen before.

If it is a galaxy in the process of formation, with dust illuminated by a billion hot, young stars, it would be the first time astronomers have witnessed the birth of a galaxy.

IRAS was a joint U.S., British and Dutch project which performed an all-sky infrared survey during 1983. The dust-cloud team in 1991 included researchers from Queen Mary and Westfield College, London; the Infrared Processing and Analysis Center at NASA's Jet Propulsion Laboratory, Pasadena, California, and California Institute of Technology; and the National Radio Astronomy Observatory, Charlottesville, Virginia.

Largest Galaxy Ever Found

The largest galaxy ever seen was found lurking at the heart of a mammoth cluster of galaxies known as Abell 2029, a billion lightyears from Earth, in 1990.

Astronomers said one galaxy at the very center of the cluster has a diameter of six million lightyears—that's five times wider than the galaxy previously thought to be largest and sixty times larger than our own Milky Way galaxy which has a diameter of about 100,000 lightyears.

The jumbo galaxy also is one of the most luminous ever observed, shining with two trillion times the light produced by Earth's Sun. More than a quarter of all the light radiated by the huge cluster of galaxies is contributed by that one galaxy.

The mammoth galaxy must have existed for a very long time since it would have taken billions of years to accumulate all of the matter it contains and organize its network of stars. It took light from the galaxy a billion years to reach Earth.

A lightyear is the distance light travels at 186,000 miles per second through space in one year, about 5.9 trillion miles. Light from the newly-discovered gargantuan galaxy took one billion years at a speed of 186,000 miles per second to reach Earth where a team of astronomers used a 36-in. telescope at Kitt Peak National Observatory, Arizona, to see it.

The researchers—from Haverford College, Haverford Pennsylvania; Michigan State University, East Lansing, Michigan; and the National Radio Astronomy Observatory (NRAO), Socorro, New Mexico—used a CCD (charge-coupled device) camera to make sixteen photographs of the galaxy. A CCD is a light-sensitive electronic integrated-circuit (IC) chip.

Glowing halo. A mosaic of the sixteen images showed a halo of diffuse light glowing around an enormous object at the center of galaxy cluster Abell 2029. The halo was so uniform, the astronomers interpreted it as a single galaxy. The light spread smoothly from the central galaxy and had the elliptical shape of the central galaxy.

A star is a single ball of fire in space, a sphere of hot glowing gas. Inside, stars have temperatures measured in millions of degrees. At their surfaces, temperatures up to 55,000 degrees are common.

Many stars are millions of miles in diameter. Some have atmospheres 10,000 times as thin as Earth's air at sea level, while some stars are so dense a cupful of material would weigh tons on Earth.

Galaxies of stars. A galaxy is a vast cloud of billions of stars attracted to each other and held close by their gravity, floating through the Universe together.

Earth is one planet circling one star—the Sun. In turn, the Sun is one star among 100 billion stars collected in the Milky Way galaxy. The Milky Way, shaped like a pinwheel spiral, is a flat disk 100,000 lightyears in diameter with a bulge at the center.

Distances across the galaxy are great. All the stars visible from Earth, even with small telescopes, are near us, inside our own Milky Way galaxy.

Even as its stars revolve around its center, the Milky Way galaxy itself is traveling through the Universe. Galaxies sometimes are thought of as islands of stars, or clouds of stars. There are untold trillions of galaxies strewn across the Universe.

Astronomers have seen galaxies of many different shapes and sizes. The most common are flat pinwheel spirals, egg-shaped ellipticals, and oddly-shaped irregulars. The best-known galaxies probably are our own and those nearby: the Milky Way, the Andromeda Galaxy, and the Large Magellanic Cloud.

Clusters of galaxies. Just as stars are parts of galaxies, most individual galaxies are parts of giant clusters of galaxies. Galaxy clusters can contain millions or billions of galaxies, all revolving around the center of the cluster.

The Milky Way galaxy, the Andromeda Galaxy, the Magellanic Clouds and other nearby galaxies form our local cluster in the Universe.

The Coma Cluster is a large collection of galaxies located in the sky in the constellation Coma Berenices. The Virgo Cluster is a large cluster of galaxies seen in the constellation Virgo.

Faraway Galaxy Ages Universe

Astronomers are continuing to find some of the oldest, most-distant galaxies ever seen across the reaches of the Universe.

The latest find, something ten times the size of our Milky Way galaxy, probably coalesced from deep-space gas and dust when the Universe was less than 2 billion years old. An astronomer at the University of Hawaii, Manoa, came across the object while identifying naturally-generated radio waves coming from a distant deep-space source recorded in 1981. He used Great Britain's Infrared Telescope and the Canada-France-Hawaii optical telescope, both atop Hawaii's Mauna Kea, to pick up infrared light and visible light from the precise point where the radio emission had been detected.

His discovery and others like it could force astronomers to change their conception of the Universe.

Natural radio signals. Cataclysmic explosions cause faraway galaxies and other natural deep-space bodies to emit radio signals as well as visible light, infrared light, ultraviolet light, X-rays and other radiation.

The University of Hawaii researcher identified the radio source as a very old, mature galaxy. The Universe was only a few billion years old at the time the light he was seeing left there. But, the galaxy itself already may have been one billion or two billion years old when the light left. That pushed back the calculation on when galaxies first formed in the early Universe.

Dark matter. If more such faraway objects are found, the cold dark matter scenario favored by astronomers will be in trouble. That scenario describes most of the Universe as full of cold, dark stuff we can't see. The theory predicts massive galaxies, like the one discovered, should not have existed so early in the history of the Universe. Now that one so old has been found, the theory is in doubt.

Other astronomers, who have looked at far distant galaxies, say the new find completes a substantial jump back in time to an important result. Astronomers really don't know how or when galaxies form. The new find will help just a bit in clearing up the picture of the early Universe.

Competing ancients. University of Arizona astronomers earlier had reported even older and more distant galaxies, but their results have yet to be confirmed.

The University of Hawaii astronomer can claim the most-distant galaxy until the University of Arizona galaxies are confirmed. Since the Hawaii galaxy implies the Arizona galaxies may exist, the Arizona and Hawaii discoveries actually mesh.

To determine a deep-space object's age and distance, astronomers observe the so-called redshift of light coming from the object. The greater the redshift, the older and more distant the object.

The Hawaii galaxy seemed to have a redshift of 3.4. The oldest and most distant galaxy observed before had a redshift of 1.8.

The age of the Universe hasn't been determined precisely. Best estimates put it between 10 billion and 20 billion years. If 15 billion years ago is chosen arbitrarily for the birth of the Universe, then the redshift of the Hawaii galaxy means the light that researcher saw left the faraway galaxy when the Universe was only 3 billion years old. Observations of the galaxy itself indicate it was then 1 billion or 2 billion years old. That means the galaxy was formed when the Universe was 1 to 2 billion years old.

Another Most-Distant Galaxy. A cluster of stars 15 billion lightyears away, accumulated eons before our own Milky Way, Sun and Earth, is the most distant galaxy known, according to astronomers.

The distant galaxy, known as 4C41.17, is too faint to be seen by naked eye. Astronomers found the galaxy when they encountered its naturally-occurring radio signals. Those signals are a billion times more powerful than emissions from our Sun.

Galaxy 4C41.17, one of the most powerful radio galaxies ever spotted, is thought to be primitive, in an early stage of formation.

The galaxy's distance from Earth was calculated from the kind of optical light it sends out. Galaxy 4C41.17 transmits light characteristic of hydrogen and carbon. The wavelength of the light, however, is stretched by movement away from observers on Earth. Such a stretch causes the color of the light to shift toward red.

By measuring this so-called red shift, astronomers calculated the distance to the galaxy. The red shift of 3.8 found for galaxy 4C41.17 converts to 15 billion lightyears away. At 15 billion lightyears, the galaxy is 90 percent of the distance to the edge of the universe, astronomers computed.

A window back in time. The light we see coming from galaxy 4C41.17 left there 15 billion years ago, so viewing the galaxy is like looking back in time. Light arriving at Earth today from galaxy 4C41.17 started on its journey only a few billion years after the Big Bang. By comparison, the Sun and Earth probably formed as recently as only 4.6 billion years ago.

Yet another most-distant quasar. Late 20th century astronomy tools have grown very sophisticated, triggering an avalanche of extraordinary deep-space finds. As the ninth decade of the century ends, astronomers continue to sight stunning objects in deep space. This time, an extraordinarily-old quasar has been discovered at the astounding distance of 14 billion lightyears from Earth—probably the most distant object seen to date in the far reaches of the Universe.

Light races through space at a speed of 186,000 miles per second. A lightyear is the distance light travels in one year—about 5.9 trillion miles or 9.5 trillion kilometers. Several quasars at lesser distances of around 12 billion lightyears have been seen.

Quasar stands for quasi-stellar object. They are the brightest objects known in the Universe. A quasar may be small—as tiny as the diameter of our Solar System—but it can radiate more light than 1,000 galaxies each with 100 billion stars like our Sun. Astronomers suppose that light is generated in the gargantuan explosion of energy as matter is sucked into the maw of a black hole at the heart of an old galaxy.

Astronomers using the Hale Telescope at California Institute of Technology's Palomar Observatory in 1989 found the newly-discovered quasar, known as PC 1158-4635, in the sky constellation Ursa Major below the bowl of the Big Dipper. Light from PC 1158-4635 took 14 billion years to arrive at Earth. Seeing light from the quasar is like looking back in time. The quasar may have exhausted all fuel available around it and burned out long before the light we see finally reached Earth in 1989.

Quasar PC 1158-4635, announced in November 1989, was found by Maarten Schmidt and Francis Mosely, California Institute of Technology; Donald Schneider and Eugene Higgins, Princeton; and James Gunn, designer of the 4-Shooter.

Our Sun and the Earth are relatively young, probably having formed as recently as 4.6 billion years ago. Quasar PC 1158-4635 is slightly older than a then-most-distant quasar reported in 1988. It was only 13.8 billion lightyears away.

The quasar cannot be seen by naked eye—it is 400,000 times too faint for that. Astronomers found it using a camera with four detectors extremely sensitive to light. The camera, known as a 4-Shooter, recorded data as numbers on computer tape, rather than spots of light on photographic film.

Big Bang. Many scientists like to think the Universe began at a certain time, 10 to 20 billion years ago, when a Big Bang explosion in a small super-dense lump of matter sent bits of that matter flying outward from a point. According to the Big Bang theory, the Universe has been expanding ever since.

Quasar PC 1158-4635 contradicts the Big Bang theory about the beginning of the Universe. Its existence when the Universe was only seven percent of its current age, and 15 percent of its current size, suggests galaxies formed much faster and earlier than astrophysicists had imagined.

Quasar PC 1158-4635 pushes the beginning of the era of quasars back to 14 billion years ago. Since galaxies seem necessary for quasars, galaxies must have formed as early as one billion years after the Universe began...if it started at one point in a Big Bang.

Vast expanses. Distances across the Universe are very great. For example, from the Sun out to its most distant planet, Pluto, is about three-quarters of a lightyear. The diameter of our Solar System is only around 1.5 lightyears—and it is only one of 100 billion star systems in the Milky Way galaxy.

We are 30,000 lightyears from the center of the Milky Way galaxy. The star nearest us in the galaxy is Proxima Centauri, 4.5 lightyears away. From one side of the Milky Way to the other is 100,000 lightyears.

In 1987, a star was seen exploding as Supernova 1987a in a nearby galaxy known as the Large Magellanic Cloud. That galaxy, at 39,000 lightyears diameter, is less than half as wide as the Milky Way. It is 163,000 lightyears from Earth.

In our neighborhood of the Universe is the well-known Andromeda, the spiral galaxy nearest us. Andromeda is bigger than the Milky Way with 300 billion stars in a disk 180,000 lightyears in diameter. Andromeda Galaxy is 2.2 million lightyears from Earth.

By comparison, the newly-discovered quasar shines at a distance of 14 billion lightyears. A billion is a thousand million so quasar PC 1158-4635 is more than 6,000 times as far away as Andromeda.

Clouds of stars. A galaxy is a vast cloud of millions or billions of stars attracted to each other and held close by their gravity while floating through the Universe. There seem to be more of these unfathomed flocks strewn across the Universe than grains of sand on a beach. Many galaxies are in the intervening space between our Milky Way galaxy and the newly-found 14-billion-year-old quasar.

Black hole. A black hole may be the end result of a total collapse of a star or several stars. The matter in a black hole is crushed into something even more densely packed than a neutron star. It is so dense that even light cannot escape, making a black hole invisible. However, its presence can be felt from its interaction with more normal space around it. For instance, if near a visible star, a black hole might suck matter away from the star. Vast quantities of X-rays and other energy might be given off by the matter as it crossed over onto the black hole.

Quasar birth. A great deal of gas and dust may pour in toward the center of a galaxy, fueling a quasar being born. As unusually large quantities of dust and gas fall onto a black hole, tremendous energy may be generated at the edge of a black hole, possibly producing the intense light we see as a quasar.

Birth explosions of ten quasars in the far reaches of the Universe were under study recently by California Institute of Technology astrophysicists. Infrared light from the ten galaxies was studied in photos by the Infrared Astronomical Satellite (IRAS) in Earth orbit in 1983. The galaxies ranged in distance from 200 million to one billion lightyears from Earth.

13.8 billion years. Since the first quasar was detected in 1963, several thousand have been discovered. Better astronomy equipment, and improved deep-space searching, sensing and seeing techniques, have brought a spate of discoveries recently of very-distant quasars, including half a dozen uncovered in the last three years.

A quasar 13.8 billion lightyears, or about 81 billion-trillion miles, from Earth was found in the sky constellation Sculptor by astronomers at the Kitt Peak National Observatory, Tucson, Arizona, and the Institute of Astronomy, Cambridge, England, using a telescope in Australia.

How old really? Analysis of lightwaves seems to reveal the distance to the quasar, but calculating the distance depends on assumptions about the age of the Universe. If the Universe is 15 billion years old, as some say, the quasar is 13.8 billion lightyears away. But various researchers calculate differing Universe ages, ranging from 10 billion to 20 billion years.

An astronomer using Great Britain's 12-ft.-diameter Schmidt telescope at Siding Spring in New South Wales, Australia, discovered the quasar 13.8 billion lightyears away. Scientists at Sydney, Australia, from the Anglo-Australian Observatory, said the faraway body turned up in 1988 in an old photograph taken in 1983. The quasar was discovered by a British astronomer, now at the University of Pittsburgh.

Even at the enormous speed at which light travels through space—186,000 miles per second or 5.9 trillion miles in a year—energy from the latest find took 13.8 billion years to travel to the observatory telescope in Australia. A lightyear is the distance light travels in a year, roughly 5.9 trillion miles.

The farther, the older. A quasar is a distant object in space, extraordinarily far from Earth, sending out a massive stream of energy. Some liken a quasar to a beacon near the edge of the Universe.

Farther away quasars are older than closer quasars. Scientists know the faster a quasar recedes from Earth, the farther away it is. They measured the lengthening of the newly-discovered quasar's lightwaves as it moved away from Earth. The lightwaves indicated the object was as far or farther away than any other known quasar.

Astronomers like to find quasars because they divulge clues about how stars and galaxies formed in the early Universe. However, the finding of such a distant, and therefore very old, quasar is confusing to some astronomers whose theories suggest galaxies formed later in the evolution of the Universe.

Astronomers were excited at the time because the quasar was much brighter than two previously known quasars with a redshift of more than four.

13-Billion-Year-Old Quasar

An astronomer, using a 12-ft.-diameter telescope at Siding Spring in New South Wales, Australia, has discovered a quasar at least 13 billion lightyears away, one of the most distant deep-space energy sources ever seen.

Light seen today on Earth left the quasar 13 billion years ago and travelled through space at a speed of 186,000 miles per second. Seeing light from the quasar is like looking back in time. Astronomers expect the 13-billion-year-old quasar to reveal something of the structure and composition of the Universe at an early time, when it was less than than one-fifth its present age.

Many astronomers think the early Universe came to life and started expanding outward with a massive explosion they call the Big Bang, maybe 20 billion years ago. If there was such a universal start, light arriving at Earth today from the quasar started on its journey only a few billion years after the Big Bang.

Spring chickens. The Sun and Earth are relatively young, probably having formed as recently as only 4.6 billion years ago.

Earth is one of nine planets in a Solar System circling a star we call the Sun. Our star is one of 100 billion stars bunched together in our Milky Way galaxy.

The Milky Way galaxy is about 100,000 lightyears in diameter. Our Solar System is about 30,000 lightyears from the center of the Milky Way galaxy. The star nearest our Sun in the Milky Way galaxy is Proxima Centauri, about 4.5 lightyears away, or only about 26 trillion miles. A lightyear is the distance a ray of light travels through space in one year's time—about 5.9 trillion miles.

Many other galaxies are in the intervening space between our galaxy and the 13-billion-year-old quasar.

Brightest quasar. Scientists at Sydney, Australia, from the Anglo-Australian Observatory, said the faraway, ancient, body is significant because it is the brightest found so far.

Quasars are extremely distant celestial objects that emit huge quantities of light and powerful radio waves. They are so far away they can't be seen by naked eye or even small telescopes. Each is a million times fainter to the naked eye than an ordinary star. In very large telescopes, they appear in the sky as tiny points of light.

But, even though they are bright in visible light, most of their energy is released as infrared light and X-rays. Scientists think one quasar may give off as much light as our entire Milky Way galaxy.

Since the first quasar was detected in 1963, several thousand quasars have been discovered. The word quasar is astronomer shorthand for quasi-stellar object.

The newly-reported quasar actually turned up in 1988 in an old photograph taken in 1983 with Great Britain's Schmidt telescope at Siding Spring, Australia. The 12-ft. Schmidt telescope is one of the largest in the Southern Hemisphere.

Black holes. Quasars probably are born when stars are sucked into black holes in the centers of galaxies. Many star systems feeding a black hole would create a stupendous light and energy display as they pass over the edge into the black hole. Astronomers think a quasar light show could be equivalent to 100 trillion blazing Suns.

Astronomers have been excited because the quasar is much brighter than two previously known quasars with a redshift of more than four.

Redshift. The wavelength of light is stretched when a body in space, such as the quasar, moves away from observers on Earth. Such a stretch causes the color of the light to shift toward red.

Redshift is a measure of the speed at which the quasar is going away from Earth. Thus, it is a measure of distance to the quasar in the expanding Universe. A quasar with a redshift of four is farther away than one with a redshift of three.

By measuring redshift, astronomers have calculated the distance to the quasar.

Some of the numbers astronomers plug into their formulas for calculating redshift are controversial. Thus, there sometimes are discrepancies in reports of how far away these extraordinarily-distant objects really are. And the age of the Universe could fall between 15 and 20 billion years.

For example, a red shift of 3.8 was computed for the faraway galaxy 4C41.17 calculated by its discoverers to be 15 billion lightyears away. At 15 billion lightyears, the astronomers figured that galaxy to be 90 percent of the distance to the edge of the universe.

Of course, unless a scientist is intent on precisely dating the Universe, what's a few million lightyears among friends?

Vast Blue Arcs Are Mirages

Glowing blue lights arching across trillions of miles of uncharted space are merely illusions, according to astronomers who have been staring at them since the vast energy arcs were discovered in 1987.

However, the fact there are such mirages suggests the Universe may end eventually in an inward collapse—a big crunch.

Three mysterious, vast blue arcs of light in deep space were revealed in 1987 by Arizona and California astronomers who later had "nightmares" trying to explain them.

The astronomers at Kitt Peak National Observatory, Tucson, and Stanford University, California, pictured one arc stretching 1.9 million-trillion miles through a cluster of galaxies named Abell 370.

One galaxy, such as our own Milky Way, is a collection of 100 billion or more stars like our Sun. The galaxy cluster Abell 370 is about seven billion lightyears from Earth. That's about 41 billion-trillion miles. A lightyear is the distance light travels through space in a year, just some 5.9 trillion miles.

Another blue arc of light seemed to be in a galaxy cluster known as 2244-02. And there were faint signs of a third arc in a cluster of galaxies labelled Abell 2218.

Cosmic mirage. Analyzing light from the arc in the Abell 370 cluster led the astronomers finally to view the arc as a "cosmic mirage."

The mirage was created when a massive galaxy in the foreground bent light from a galaxy in the background. The background galaxy was directly behind the foreground galaxy and twice as far away. The astronomers suspect the other arcs also are illusions.

Light moving toward Earth from a faraway galaxy spreads out. Along the way, gravity from an intervening galaxy in Abell 370 acted as a "gravitational lens," bending light like a glass lens. The bent light looked like an arc to observers on Earth.

The arc illusion was like a mirage on Earth which lets you look across flat land and see mountains hidden over the horizon. Light bouncing from mountains is bent by an inversion layer of cold air over hot air. The image of mountains bends down toward you.

Interestingly, the light-bending galaxies contain billions of stars but not enough stars with sufficient brightness to be counted from Earth. The stars exist in those distant galaxies, giving off great quantities of energy, as do all stars. But the energy they radiate as visible light doesn't amount to much.

Cataclysmic explosions cause distant galaxies and other natural deep-space bodies to emit radio signals, visible light, infrared light, ultraviolet light, X-rays and other radiation.

Dark matter. Since they are massive enough to bend light from another galaxy, they must contain considerable "dark matter" invisible to astronomers on Earth.

Many astronomers think the Universe started expanding outward with a massive "Big Bang," 15 to 20 billion years ago. The theory suggests the Universe may eventually stop expanding outward and start shrinking back in on itself.

Something, of course, would be needed to stop the expansion so astronomers came up with the idea of dark matter. It might make the difference, they say, supplying enough gravity to slow the outbound expansion of the Universe.

Thus, astronomers have been searching for dark matter lately, thinking the amount of bright visible matter seen in the Universe is not enough to exert the gravitational force needed to stop the Universe from expanding forever.

Big Crunch. If all galaxy clusters sprinkled through the Universe contain as much dark matter as those which apparently formed the blue-arc mirages, the expansion eventually would grind to a halt. The Universe would reverse direction and collapse inward in a "Big Crunch," advocates of the theory say.

The National Optical Astronomy Observatories (NOAO) operates the Kitt Peak Observatory. NOAO referred to the blue arcs in 1987 as the largest visible structures in the Universe. However, since then, astronomers have discovered a supercluster complex of galaxies 500 times larger.

Other known illusions have been blamed on gravitational lenses created by massive objects in space. For instance, a double image of a single quasar has been seen. A quasar is a quasi-stellar object. Quasars are the brightest objects in the Universe.

Doppler effect. The astronomers found the arc in Abell 370 to be an illusion by measuring the so-called redshift or Doppler effect on light from the arc.

The Doppler effect makes a train whistle seem to increase in pitch as a train moves toward you. It causes an apparent decrease in pitch after the train passes and is moving away from you.

Higher pitch is a shorter wavelength. A lower pitch is a longer wavelength. If the Universe is expanding, distant galaxies are moving away from Earth. Their light increases in wavelength, shifting toward the red end of the light spectrum. The farther a galaxy is from Earth, the faster it and the Earth are moving apart and the more its light is shifted toward red wavelengths.

The redshift in the arc in Abell 370 was about twice as great as the red shift of light from stars in that cluster. That indicates the light in the arc comes from a source twice as far away as the cluster.

Mystery Tugs At Galaxies

American and British astronomers recently turned up startling new clues about a mysterious, but magnificent, flow of stars across the Universe. Clouds of galaxies in almost-inconceivably-vast spaces are being pulled at millions of miles per hour through space toward some unseen attraction.

Even our own Milky Way galaxy is streaming through space, caught up in the herd of galaxies sailing toward the tremendously-powerful force. Where are all of the trillions upon trillions of stars headed? Astronomers say in the direction of a constellation sketched on their night-sky charts known as the Southern Cross.

Studying 400 elliptical galaxies over six years, the researchers were struck by the massive concentrations of galaxies they refer to as the Virgo Supercluster and the Hydra-Centaurus Supercluster. Small as it seems, our galaxy of 100 billion stars is but a tiny speck drifting on one side of the Virgo Supercluster. The numbers of stars accumulated in those sprawling clouds of galaxies surely is more than all of the grains of sand on all the beaches known to man.

Black Hole Feeds Quasar

Vast amounts of gas and dust concentrated near the open maw of a hungry black hole may be the fuel for a new quasar as two spiral galaxies collide.

A black hole is a small dead star so extraordinarily dense that its powerful gravity won't even let light escape. It sucks in loose gas and dust in space and strips matter from stars which come too close as they float through space. Black holes may be the generators powering the extremely bright deep-space objects known as quasars.

Some 3,600 quasars have been seen since the first was identified in 1963. Most are among the objects farthest away from Earth. That makes them the oldest since light we see today would have taken longer to travel the greater distance to Earth. The most powerful quasars are the brightest objects astronomers see in the Universe.

The core of a quasar is at least as wide as our entire Solar System—a distance as great as if the Sun had ballooned out to a diameter large enough to encompass all of the known planets. A quasar core gives off tremendous quantities of ultraviolet light as well as a great deal of visible light. One quasar apparently is brighter than an entire galaxy of billions of stars.

Ultra-luminous galaxies. Astronomers decided some time ago that quasars are formed when galaxies collide. A billion suns smashing into and becoming enmeshed with a billion other suns forms a new very-luminous infrared star system. These ultra-luminous infrared galaxies are the intermediate step as a quasar is formed.

The infrared galaxy gives off lots of energy in the infrared part of the spectrum but little visible light. In fact, after the collision, the new galaxy gives off ten times more heat than light. The new infrared galaxy has a distorted central disk of stars as the two sets merge. Some stars are stripped off each of the parent galaxies and are thrown out into long tails streaming away from the new galaxy.

Tremendous energy. A great deal of gas and dust is left in the center of the new galaxy. This fuels the quasar being born. When such unusually large quantities of dust and gas fall onto a black hole, tremendous energy is generated at the edge of the black hole producing the intense light we see from a quasar.

A small amount of visible light can be seen coming from the intermediate-stage infrared galaxy. That visible light looks more like light from a quasar than light from a single spiral galaxy or a galaxy merely forming new stars.

Astronomers see more infrared from the intermediate-stage galaxy than visible light because the dust and gas filling the new galaxy block their view of the core. The dust and gas absorb ultraviolet and visible light being given off by the quasar forming at the core and are heated by that light. Eventually most dust and gas is blown away and a mature quasar can be seen giving off plenty of visible light.

Birth explosions of ten quasars in the far reaches of the Universe were under study by California Institute of Technology astrophysicists. The galaxies were photographed by the Infrared Astronomical Satellite in Earth orbit in 1983. The ten ranged in distance from 200 million to one billion lightyears from Earth.

Cygnus X-1 Black Hole Gamma Rays

Powerful gamma rays spewing out of a strange object 7,500 lightyears across space from Earth have convinced astrophysicists at NASA's Jet Propulsion Laboratory that a black hole exists.

The astronomy satellite HEAO-3 orbited Earth, collecting energy from deep space in in 1979-1980. The researchers in 1987 found gamma rays from Cygnus X-1 in data collected by JPL's Gamma Ray Spectroscopy equipment aboard HEAO-3.

Cygnus X-1, in our own galaxy, had been considered a good bet for a black hole, which probably formed after a massive star exhausted its fuel supply and collapsed.

The tremendous gravity of Cygnus X-1 indicates a mass about 10 times that of our Sun had been squeezed into a small ball, a sphere 40 miles across.

Cygnus X-1 has a supergiant companion star, with a mass 30 times that of the Sun. The black hole's gravity is so intense, it sucks matter away from the supergiant companion. As that drawn-off matter spins around the black hole, it forms a hot accretion disk of very-high-energy plasma which shoots X-rays across space.

Neutron stars. X-ray astronomers knew of neutron stars, a type of collapsed star whose strong gravitational field can attract and heat gas from companion stars.

That generates X-rays with many thousands of electron-volts energy. But those neutron stars usually don't produce gamma rays like those observed coming from Cygnus X-1. Usually the neutron stars have strong magnetic fields, which force matter to rotate with the star. Friction of spinning gases is reduced so temperatures are not high enough to generate gamma rays. Strong magnetic fields are opaque to gamma rays.

Black holes should not have magnetic fields. Finding gamma rays able to escape from Cygnus X-1 labeled it as a black hole. In fact, gamma rays seem to be a useful test for distinguishing black holes from neutron stars.

Astronomers say Cygnus X-1 is similar to the unknown powerful object emitting energy at the center of our Milky Way galaxy.

Positrons. The gamma rays from Cygnus X-1 were evidence for the creation of anti-matter in the form of positrons, or anti-electrons which have a positive charge, in the hot innermost region of the accretion disk, possibly in a spherical volume 30 miles across surrounding the black hole. That almost-unimaginable area must be highly packed with positrons—100 million billion per cubic centimeter—and heated to a temperature in the billions of degrees, a million times hotter than the surface of our Sun.

The penetrating gamma rays leaving Cygnus X-1 had an energy of one million electron volts, similar to those used on Earth for the treatment of cancer with radioactive cobalt. The gamma rays were so powerful they could travel across the entire galaxy without being deflected.

Violent Hidden Galaxy

A galaxy satellite of our own Milky Way blocks most of their view, but scientists still think they have glimpsed what may be one of the most violent energy sources in nature—a mammoth clump of stars throwing off the heat of 650 billion Suns.

Astronomers from California Institute of Technology, Kitt Peak National Observatory in Arizona and Cerro Tololo Inter-American Observatory in Chile, were using an optical telescope in Chile, trying to see visible light at the same points in the sky where the Infrared Astronomical Satellite (IRAS) had seen stars sending out infrared light. Suddenly they came upon what looked like an unusually warm galaxy located behind the Small Magellanic Cloud, a satellite galaxy of the Milky Way, visible only in the southern sky.

Galaxies. A galaxy is a clump of stars, gas and dust rotating around a central point. Most of the matter we see in the Universe is collected in these giant star systems.

The gravity of its stars holds a galaxy together. The gravity of a galaxy, as a whole, affects other galaxies so the entire Universe might be seen as a system of galaxies.

Astronomers list three kinds of galaxies by shape: spiral, elliptical and irregular. Spirals are flat disks with arms wrapped around a swollen core, like a giant pinwheel revolving in deep space. An elliptical galaxy looks like a sphere or ball of stars with no pinwheel arms. Irregulars are just that, blobs of stars with no particular pinwheel or spherical shape. Three-quarters of all galaxies are spirals. One in five is elliptical. The rest are irregular.

Our own Milky Way galaxy is composed of 100 billion stars like our Sun. The Universe is populated by millions or billions of galaxies. Just as stars most often are found in pairs or groups, galaxies frequently are seen in pairs. In fact, immense clusters of galaxies have been seen. Beyond that, clusters of galaxy clusters—called superclusters—have been reported.

Local Group. The small cluster of 30 to 40 galaxies in this neighborhood of the Universe, of which our Milky Way galaxy is a part, is called the Local Group. Some of the bigger and better-known galaxies in the Local Group, along with the Milky Way, are Andromeda Galaxy and the Large Magellanic Cloud.

Our Local Group cluster, along with a dozen other such clusters, forms a local supercluster—the Virgo Supercluster—about 300-million lightyears in diameter in our region of the Universe. A lightyear, about six trillion miles, is the distance light can travel through space in a year. Our supercluster is named after the Virgo Cluster of some 2,500 galaxies near the core of the Local Supercluster. As stars rotate around a galaxy, galaxies rotate around a cluster and clusters rotate around a supercluster.

Serendipitous find. The scientists at Cerro Tololo were searching for visible-light objects to match the sources logged by IRAS. Suddenly, they realized the object IRAS 00521-7054—supposedly a star in the Small Magellanic Cloud galaxy—was in fact an optically faint galaxy much farther away. They were surprised, since galaxies usually are much cooler. They had been trying to avoid galaxies in their study.

Checking the object with 1.5-meter and 4-meter telescopes at Cerro Tololo, they discovered its infrared brightness was 650 billion times greater than the Sun, as bright as the most luminous objects previously known.

Despite the fact that galaxies include very hot stars, they still are mostly cold, empty space. Even the extraordinarily hot and bright IRAS 00521-7054 has an overall temperature of 400 degrees below zero Fahrenheit.

It also is one of the reddest galaxies ever seen, with dust temperatures about 1100 degrees Fahrenheit. The redness may be caused as starlight passes through thick dust, like Earth's dusty atmosphere reddens at sunset. The dust cloud has a mass a million times that of the Sun, so thick that little radiation leaks out.

The dust might be in a compact cloud, close to the galaxy's nucleus, heated by something hidden inside. Inside might be the violently active nucleus of a Seyfert type galaxy, the accretion disk of a black hole, or the extremely compact, inexplicably powerful core of a quasar.

IRAS 00521-7054 does not look like two galaxies merging, dragging no tail of stars. The astronomers wondered if they had stumbled across a previously-unknown kind of galaxy in IRAS 00521-7054.

Galaxy Birth In A Heavenly Backwater

Astronomers trying to tune up their radiotelescope by pointing it where nothing was supposed to be have stumbled across a magnificent new protogalaxy—an extraordinarily vast cloud of hydrogen ready to collapse with a bang.

The unborn galaxy, ten times larger than the Milky Way, appeared unexpectedly while scientists were adjusting the big Arecibo, Puerto Rico, radiotelescope.

The hydrogen cloud was reported in August 1989 by Cornell University astronomy professor Riccardo Giovanelli, head of radio astronomy at the National Science Foundation's National Astronomy and Ionosphere Center, Arecibo, Puerto Rico, and Cornell associate professor of astronomy Martha Haynes.

The cloud is 65 million lightyears from Earth. One lightyear is 5.9 trillion miles. Actually, 65 million lightyears is not too far away as distances go across the Universe.

Armada. A galaxy is a mighty armada of stars—billions of stars attracted to each other by gravity—sailing the Universe. Typical galaxies hold 10 billion to 100 billion stars. Our Sun is a star, one of 100 billion in our own Milky Way galaxy.

The hydrogen atom—a proton and an electron—is the simplest element in the Universe. Astronomers think galaxies are created when vast clouds of hydrogen condense over eons to form stars.

The newly-seen pre-galaxy is an oval cloud of hydrogen in the area of sky south of what astronomers call the Virgo Cluster of galaxies. Such a pre-galaxy, or protogalaxy, may indicate galaxies still are being formed. Some astronomers had speculated that all such star systems were formed billions of years ago, while others have supposed some galaxies might still be forming.

"Before stumbling into this thing, neither I nor Martha were great believers in the existence of such objects. But this cloud indicates that the discs of galaxies can form slowly throughout the history of the Universe and not just be something that happened during some magic period in the distant past," Giovanelli said.

The cloud seems to be two large clumps of gas that look as if they are merging. Astronomers aren't sure the cloud will collapse into a star system, but it seems organized to evolve into a galaxy.

Radio signals. The hydrogen cloud is much larger in diameter than the Milky Way, but has only one-tenth the mass. The cloud has not condensed into stars so it does not give off visible light. It does, however, generate very strong natural radio signals.

The lack of visible light coming from the cloud explains why it hadn't been seen before by astronomers using ordinary optical telescopes. The astronomers were calibrating receivers wired to the radiotelescope at Arecibo when they found the cloud. They had pointed the 1,000-ft. dish at a region of the sky thought to be empty when the characteristic radio signal of hydrogen suddenly appeared in their receivers.

Double-checking ruled out interference, spurious signals and noise generated inside the detectors. The cloud really had formed independently, in a remote corner of space isolated from other galaxies. Despite the strong radio signal, they found no stars in the cloud. Prior to the discovery, other large clouds found in space had been aligned with existing galaxies. The discovery disputed the theory that all galaxies formed within the first few million years after a theoretical Big Bang explosion sent hydrogen flying throughout the Universe 10 to 20 billion years ago. Previous searches for hydrogen-cloud protogalaxies had turned up nothing, despite strong clues that the clouds existed.

Quasars, bright objects more than 10 billion lightyears from Earth, had provided clues. Part of the light from quasars arriving at Earth was missing. The frequency of the missing light was the same as frequencies which would have been absorbed by hydrogen if it had been blocking the light's path to Earth.

240 Miles-Per-Second Galaxy Wind

Scientists, scratching their heads over an odd light from deep space, think they have found a gassy wind whooshing at speeds up to 240 miles a second out of a far-away galaxy known as M82.

A galaxy is a cluster of stars, gas and dust. Earth is one planet of a Solar System surrounding one star—the Sun—in the Milky Way galaxy. There are countless galaxies.

The M82 gas wind seems to blow out from opposite sides of the galaxy. It may be driven by continuing supernova star explosions inside the galaxy, astrophysicists suggest.

M82 is visible from Earth with a small telescope. It is 60 trillion miles away, in the direction of the constellation Ursa Major. A constellation is a segment of Earth's sky arbitrarily drawn by astronomers over the centuries.

Observations by earlier astronomers had suggested the idea of a wind from M82. No other galaxy is known to have such a wind.

The wind, mostly hydrogen gas, may have begun blowing at least 3 million years ago, if not tens of millions of years ago. It appears to be going faster the farther it gets from the galaxy.

BL Lacerta Object Found

Astronomers from the Royal Astronomical Society, London, in 1991 proved the existence of a BL Lacerta object in a galaxy dominated by a large flattened disk. Previously, the strange BL Lac objects had been assumed to exist only in the center of elliptical galaxies.

BL Lac objects fire high-speed jets from the active nuclei of elliptical galaxies. X-rays, radio waves and visible light steam out of the explosions.

Forty BL Lac objects have been at various locations in deep space. Now that the Royal Astronomical Society has uncovered such an object in a non-elliptical galaxy, previous theories about the objects have been thrown into doubt.

The discovery was made by astronomers from Southampton University, Oxford University, Massachusetts Institute of Technology and the European Southern Observatory, using the 165-in. Herschel Telescope in the Canary Islands.

Seyfert Galaxies

In 1943, Carl Seyfert realized some galaxies have extremely bright, starlike, nuclei. Later, astronomers using data received at the end of the 1970's from the orbiting HEAO-2 satellite, known as the Einstein Observatory, learned there are two types of Seyfert galaxies—those which give off strong X-rays and others which release weak X-rays.

About two percent of all galaxies are Seyferts. Most are seemingly-normal spirals. More than 150 have been discovered. Radio waves come from compact centers of Seyfert galaxies. They are brighter at radio frequencies than common spiral galaxies, but less luminous than radio galaxies.

Some astronomers suggest Seyfert galaxies are something like quasars, but 100 times less powerful. The object cataloged as NGC 4151 is the nearest bright Seyfert. Astronomers have observed it with the International Ultraviolet Observer (IUE) satellite.

CHAMPs vs. WIMPs

Nine-tenths of the mass in the Universe is invisible to observers here on Earth. Astrophysicists feel its presence only through the effects of the gravity it generates. For some researchers, CHAMPs are competing with WIMPs for an explanation of the mass missing throughout the cavernous span of the Universe.

In recent years, scientists have visualized the invisible stuff—the so-called "dark matter"—first as flocks of small, cold, dim stars; then fogs of weakly interacting massive particles (WIMPs); and now as swarms of charged massive particles (CHAMPs).

Swamped with disagreement, researchers at Harvard University in 1989 hypothesized

CHAMPs are stable particles left over from the Big Bang. Of course, to believe in CHAMPs, you have to accept that the Universe burst to life in the explosion of all time during the first moment of time around 20 billion years ago.

CHAMPs would be big—estimated at 20,000 to a million times the mass of a single hydrogen atom. If unearthed on this planet, they might seem bizarre heavy isotopes of chemical elements. A CHAMP would attract an electron to become a neutral atom. Other than its phenomenal mass, it would react chemically like a hydrogen atom.

One place to look for CHAMPs might be on the Moon where there is no atmosphere and almost no geological tumult. A flurry of CHAMPs crashing there might settle into a greater concentration than on Earth.

Mass and gravity. Mass is a measure of the quantity of matter in a body. The international unit of mass is the kilogram. A pound is equal to 0.45 kilograms. The astronomical unit is solar mass. One solar mass is 2×10^{30} kilograms.

Mass determines the resistance of a body to change in its motion as well as the gravity field a body can generate.

Gravity, the weakest force in nature, is the ability of bodies to attract each other. First expressed by Isaac Newton in 1687, the tug of gravity is greater with more mass and less as distance between bodies increases. Despite its weakness, gravity holds together the galaxies and even our own Solar System.

Cosmic Strings

Cosmic strings may sound like Brazilian beachwear, but they're far from it. Space theorists figure they are leftovers from some Universe-wide cataclysms in past eons and that we should be able to see them.

Cosmologists try to decide how the Universe came to be. However, some think it has undergone large-scale shifts in phase from time to time. These phase shifts somehow may have altered space and time. They are on the lookout for strange objects in deep space, millions or billions of lightyears from Earth. Those objects, if they exist, would be strings—remnants of a previous space and time, sort of a ragged edge where some previous incarnation of our Universe didn't mate up exactly with the present version.

Fast-lane relics. These relics of an unknown past might be moving through our Universe at nearly the speed of light. They would shed a great deal of light on how our Universe came to be if one or more could be found.

Strings would have so much energy in our Universe they probably would give off radio signals. Looking at the skies with radio telescopes, cosmologists hope to see radiowave-emitting objects shaped like long, thin strings.

In the mid-1980's, some cosmologists could be heard saying, "Ah ha!" They were looking at a deep space radio source known as G357.7-0.1. It appeared to be string-like and some cosmologists were saying it might be a cosmic string. They still had to calculate the rate at which filaments of G357.7-0.1 were moving through space to see if they were near light speed.

Imperfect snowflakes. What exactly is a cosmic string? An imperfect snowflake comes to mind. Everybody knows each snowflake is perfect. Imagine an imperfect snowflake and you'll be thinking of something akin to a cosmic string. Of course, a cosmic string would be on an almost-unimaginably larger scale than a snowflake.

Theorists say cosmic strings are defects in the structure of our Universe. Sort of defects in the fabric of space. They are imagined to be leftover from some earlier time when the vacuum of space was somehow different from now. Cosmic strings may form immense loops which vibrate. Some scientists think they see notes from such oscillating strings when an unexpected shower of gamma-ray bursts pass by Earth.

Radio Waves Bent In Deep Space

The first Einstein Ring, a celestial illusion predicted by the famed physicist Albert Einstein a half century ago, apparently has been spotted.

A Massachusetts Institute of Technology radio astronomer reported radio waves being bent into an elliptical ring by the gravity of a distant cosmic object. The stellar mirage may be an example of a gravitational lens in which a galaxy closer to Earth acts as a lens, bending light from a more distant object. During the 1980's, several odd deep-space apparitions were spotted, usually looking like two objects where only one existed. However, the latest suggested lens was bending radio waves, rather than light waves.

Albert Einstein speculated in 1936, before radio astronomy, that such things might exist, but would be too small to see. The ring was observed in 1986 and 1987 with the National Radio Astronomy Observatory's Very Large Array (VLA) radiotelescope near Socorro, New Mexico.

Missing Gravity Source

Astronomers say they now can account for more dust, gas and gravity in space than before, but there's still something missing.

They have calculated for some time there is a lot more gravity in the Universe than could be coming from the known stars. They used to think loose gas and dust between stars generated the extra gravity. Now they think the Universe has been more efficient than they expected in making stars and galaxies out of dust and gas.

There still are huge intergalactic clouds of dust and gas, but new studies have seemed to indicate that 90 percent of known matter already has been gathered into stars. The 10 percent left over in clouds of dust and gas is not enough to account for the calculated total gravity of the Universe. Astronomers say there must be an unknown source of gravity. They will look for it somewhere other than in the gas and dust clouds. Could it be in clouds of small stars, too weak to be detected from Earth?

White Holes and Wormholes

Years ago, astronomers figured there were black holes in the Universe and that they led somewhere. They assumed a black hole would have to lead to a matching white hole—just the opposite of the black hole—in another universe.

The edge or opening or maw of a black hole was called the event horizon. Space and time were thought to be distorted at the event horizon. The connection through the event horizon to another universe was known as the Einstein-Rosen Bridge or, more colorfully, as a Wormhole.

Matter falling into a black hole in our Universe would pass through the wormhole and pour out into the other universe from a white hole.

Way back when, very bright deep-space objects such as quasars were suspected of being white holes into our own Universe. Unfortunately, contemporary calculations throw a wet blanket on that exciting idea. Links to other imagined universes and white holes probably don't exist.

Birth Of The Milky Way

Stars in the Milky Way galaxy exploded to life in a series of chaotic outbursts starting six billion years ago, rather than in an orderly progression over time, according to University of Texas astronomers and others in the American Astronomical Society.

Milky Way stars first blossomed six billion years. The Sun formed early, about 6 billion years ago. More stars came along four to five billion years ago. The final surge of star birth was only about 200 million years ago. The galaxy is at the end of the last burst of production of new stars, so star birth still is high today. Previously, astronomers said stars bloomed at a relatively-steady rate, though slowing over time.

To measure the galaxy's age, one astronomer measured the stars, finding the masses, which decline with age, were not uniformly distributed. They were in four clusters, suggesting the star formation outbursts. Another researcher estimated the ages of stars by measuring their chemical history. Stellar activity declines with age. A third astronomer measured lithium levels. Stars lose lithium at a known rate. Age can be determined by measuring the amount of lithium left in a star.

Milky Way Rifts

Anybody looking through binoculars or a telescope at our Milky Way galaxy—a beautiful snowfall of stars strewn in a band across a clear night sky—is struck by noticeable breaks or rifts in the mosaic.

Those rifts in the band of stars do not mean our galaxy isn't continuous. Rather, they are vast, dark dust and gas clouds which obscure our view of thousands of stars behind them. Astronomers using radio telescopes can see through these tiny dark nebulae to examine what's behind them.

Vast Cloud Banks In Milky Way

Astronomers have found an extraordinary number of small, lightweight gas clouds which have been floating unnoticed through our Milky Way galaxy.

The clouds are so numerous, scientists say there may be as many as 500 to 1,000 for every star in the Milky Way. With 100 billion suns blazing throughout the galaxy, that could mean more than a trillion hot, gaseous clouds, each as wide as the distance from Earth to our Sun.

Astronomers hadn't noticed them before, despite their vast number, because the clouds are less massive than previously reported cosmic clouds.

The clouds deplete some of the energy in radio signals arriving at Earth from deep-space objects, according to researchers at the Naval Research Laboratory in Washington, D.C., Mullard Radio Astronomy Observatory in Cambridge, England, and Virginia Polytechnic Institute and State University in Blacksburg, Virginia.

The astronomers, wondering why strength of incoming radio waves seemed to dip from time to time, looked at 36 quasars and other objects. They decided that small floating clouds pass between the quasars and Earth, deflecting some of the radio waves and cutting down on the intensity of signal arriving at Earth.

They checked waves coming from the 36 objects every day between 1979 and 1985. A number of quasar radio signals would jump briefly to a higher intensity, then dip sharply in signal strength for 10 weeks to three months, then jump again to higher-than-usual intensity. That made it seem that clouds were passing between the objects and Earth, scattering radio waves off to each side.

The small clouds probably are as hot as 3,000 to 4,000 degrees Fahrenheit. Such heated gas would be ionized and could deflect radio waves.

As a cloud starts to pass between a radio source and Earth, our planet gets an extra dose of waves in the shower deflected off the leading edge of the roaming cloud. Then Earth gets less signal strength as it is sheltered by the cloud. Then Earth is bombarded with an extra shower of waves deflected off the trailing edge of the cloud as it moves out of the way. The clouds are as far from Earth as 300 million times the distance between the Earth and Sun. Just one shadow cast by such a cosmic cloud is larger than our whole Solar

System. The clouds are close to Earth, however, when compared with the distance to the quasars. That explains how a small cloud can blot out radio waves just as Earth's own tiny Moon can eclipse our much larger Sun. Previously-known large cosmic clouds are at least 20,000 times wider than the newly discovered small clouds.

The Milky Way's Curiosities

For decades astronomers have sketched our Milky Way galaxy as a towering pinwheel of blazing stars sailing through space—a smooth spiral plane of stars. But now that picture is distorting a little.

The disk-shaped reservoir of stars apparently is not perfectly flat. New observations show something reminiscent of a warped 45-rpm phonograph record with a bulging spindle adaptor at the center.

Most surprisingly, the disk seems to be trimmed with a dainty scalloped fringe. Like the phonograph record had been cut with pinking shears.

Once measured as only a typical spiral galaxy, the Milky Way now seems a more titillating quantity:

★A whopping black hole, fueled by matter and energy in the bloated center of the Milky Way, may be a stupendous engine spinning the galaxy through the Universe.

★Two powerful compact objects, electrifying the galaxy's heart, might be remnants of star birth associated with the black hole.

★Apparently 90 percent of the Milky Way is invisible to us. At least, contemporary telescopes can't show Earth's astronomers the huge halo of dark matter estimated to be just beyond the outer edge of the visible spiral arms of our galaxy.

★The dark halo might be full of WIMP's—Weakly Interacting Massive Particles—remnants of the so-called Big Bang which some assume formed the known Universe.

Our Sun is a star. It's but one of a hundred billion in the Milky Way, a slim star-studded disk about 100,000 lightyears across with a bulge at the center. A lightyear is about 5.9 trillion miles. Astronomers estimate countless trillions of other galaxies are spread across the Universe.

Scalloped brim. At the outskirts of the Milky Way's visible disk of stars is a cloud of hydrogen gas stretching the galaxy's edge another 25,000 lightyears. Astronomers describe the brim of the hydrogen gas cloud as warped and scalloped.

Beyond that outermost fringe remains more than 99 percent of the mass of the Universe. Invisible, astronomers don't know what it is.

Maybe it's saturated with a trillion-trillion faint midget stars, each less than one-tenth the size of our Sun. Or, filled with black-hole relics of giant stars. Or, a more radical explanation, loaded with WIMP's.

The Milky Way's curiosities aren't only at its farther limits. The inner 10 percent of the galaxy has an extraordinary star density.

The bulging hub is an energetic fountain of radiation, beaming prodigious quantities of X-rays, infrared light, ultraviolet light, radio waves and gamma rays outward. A black hole, driving the galaxy's rotation, probably lurks at the core of the Milky Way, shrouded by an impenetrable overload of energy which swamps Earth telescopes. Future telescopes may resolve the enigma.

Star Birth In The Milky Way

An immense cloud of Milky Way gas and dust is collapsing upon itself, giving spectacular birth to a ring of giant blue stars.

The celestial cloud, known to radio astronomers as M-49a, is about 300,000 trillion miles away in a bright area of the sky. A dozen hot young stars already have blazed to life at the heart of the cloud. Light from the ring of stars, 50,000 times the mass of our Sun, is a million times brighter than the Sun.

The cloud is 100 lightyears across and 50,000 lightyears from Earth, but still within the Milky Way galaxy which includes our Sun and Solar System. A lightyear is about 5.9 trillion miles, the distance light travels in a year.

Even more stars are about to turn on as gas and dust molecules race at 45,000 miles per hour toward the center of the cloud.

Prolific clouds. M-49a reveals star birth on a scale larger than previously imagined by astronomers. Apparently, dozens of stars can form in one cloud at the same time.

Starbirth takes a very long time. Stars are born over millions of years as dust and gas molecules are pressed together by their own gravity. The molecules become so densely packed they heat up, eventually to the point of nuclear fusion.

Using radio telescopes at the Hat Creek Radio Observatory in northern California and at the Very Large Array in New Mexico, astronomers from University of California, Berkeley, Massachusetts Institute of Technology, California Institute of Technology and Rensselaer Polytechnic Institute tracked the direction of movement of carbon monoxide and hydrogen. The molecules gave off radio signals the astronomers could measure in frequency.

Blue-shifted gas. Direction of movement was determined by Doppler shift, which causes signals moving away from a receiver to have a different frequency from those moving toward a receiver. Movement away is redshift. Motion toward is blueshift.

Gas from both sides of the cloud was moving toward the center. Blue-shifted gas moving toward the radio telescopes was on the back side of the cloud. Red-shifted gas going away was on the near side of the cloud. The cloud was collapsing.

Usually, a gas cloud is so far away, with movement so slight, collapsing motion is hard to detect. But, in M-49a, astronomers were excited to find more motion than usual as molecules spiraled rapidly inward.

Milky Way Black Hole Clues

It's beginning to look more and more like there is a gargantuan black hole near the core of the Milky Way, sucking in matter from the magnificent dust and gas clouds which obscure the view of the center of our galaxy.

A river of gas, 90 trillion miles long, has been found streaming into the center of the Milky Way. It could be fueling a prodigious black hole at the very heart of the galaxy, according to new reports from astronomers.

Black holes. Stars like our Sun are born over millions of years in clouds of gas and dust floating in space. The dust and gas molecules are pressed together by their own gravity, becoming so densely packed they heat up, eventually to the point of nuclear fusion.

A new star is a hot bubble of basic elements held together by gravity, burning like a continuous nuclear explosion. Matter in the star is fuel for its nuclear burning.

As millions of years pass, the fuel is used up and the star collapses. If a star were small to start with, it cools to a black-dwarf cinder. A larger star collapses in a supernova explosion with the remnant a neutron star. Shock waves from supernova blasts compress gas and dust, creating new stars.

The largest stars collapse inward to something smaller than a neutron star, yet with extraordinary mass. These stars are so massive when they finally collapse, the remnant is a single infinitely-dense speck. Astronomers call that a black hole.

The gravitational pull of a black hole is so strong that nothing can escape it, not even light or radio waves. Thus, black holes cannot be seen. Astronomers find clues to the existence of black holes in the gravity tug they give other bodies nearby in space.

New evidence. The new evidence is a thin stream of gas being siphoned from nearby stars by the awesome gravity of a black hole or something at the center of the Milky Way galaxy. It was found by a team of astronomers at the Harvard-Smithsonian Center for Astrophysics in Cambridge, Massachusetts, the Max Planck Institute in Munich, the University of Cologne, West Germany and the Massachusetts Institute of Technology whose radiotelescope observations seemed to support the idea of a gravity well at the heart of the galaxy, making stars drift toward the center.

Our Sun is one among 100 billion stars in the Milky Way galaxy. The galaxy is a slim star-studded disk about 100,000 lightyears across with a bulge at the center.

Astronomers estimate countless trillions of other galaxies are spread across the Universe. A lightyear is about 5.9 trillion miles, the distance light travels in a year.

The Milky Way's gravity holds together a vast structure. Astronomers sketch our galaxy as a towering pinwheel of blazing stars sailing through space.

Many stars at the center make the galaxy bulge, while many more stars are clumped in spiraling arms that stretch across 100,000 lightyears—about 588,000 trillion miles—of space. Our Solar System lies off center, on one side of the Milky Way Galaxy. The Sun is 22,000 lightyears (130,000 trillion miles) from the center of the galaxy. From our position out on one of the spiral arms, the galaxy looks like a dense white band of stars crossing Earth's night sky. Thus, the name Milky Way.

Seeing through dust. Astronomers have mapped vast regions of deep space beyond our galaxy, using powerful optical and infrared telescopes looking above or below the flat disk of the Milky Way. But those optical and infrared telescopes can't see into the heart of our own galaxy very well.

Visible light is blocked by voluminous clouds of dust clogging space between the stars of the Milky Way. Because the galaxy has hoards of infrared light sources, it's hard to distinguish individual objects seen in infrared light. If a black hole lurks at the core of the Milky Way, driving the galaxy's rotation, it is shrouded by an impenetrable overload of energy which swamps present-day Earth telescopes.

Radiotelescopes. Luckily, radio waves penetrate deep-space dust clouds. Radiotelescope surveys are able to map the Milky Way by detecting objects that emit naturally-generated radio waves. Astronomers use radiotelescopes to track the direction of movement of carbon monoxide and hydrogen molecules which give off radio signals. The astronomers measure the frequency of those signals. Direction of movement is determined by Doppler shift which causes signals moving away from a receiver to have a different frequency from those moving toward a receiver. Movement away is redshift. Motion toward is blueshift.

Radiotelescope maps of the entire Milky Way show exploding supernova stars, startlingly-dense supernova neutron-star cinders, pulsars, seemingly-endless dust-and-gas clouds where new stars are born, and maybe even that black hole.

The scientists have found the inner 10 percent of the galaxy crammed with an extraordinary number of stars. The bulging hub is an energetic fountain of radiation, beaming prodigious quantities of X-rays, infrared light, ultraviolet light, radio waves and gamma rays outward.

The colossal black hole, fueled by matter and energy in the bloated center of the Milky Way, may be a stupendous engine spinning the galaxy through the Universe. A few astronomers have asked whether there may even be two powerful compact objects, electrifying the galaxy's heart, possibly remnants of star birth around the black hole.

The edge or opening or maw of a black hole is known as the event horizon.

Sagittarius clouds. Over the centuries, Earth observers have divided the sky into arbitrary constellations, giving each a folk name such as Sagittarius. Astronomers see an immense gas cloud obscuring the center of the Milky Way. It's about 30,000 light years from Earth, in the sky constellation Sagittarius.

Astrophysicists say the gas cloud at the heart of the Milky Way may be swirling around a black hole, a million times more massive than the Sun. If so, parts of the cloud may be crossing the event horizon, tumbling continuously into the black hole. Possibly as much as the equivalent of one-thousandth of the mass of the Sun streams into the black hole each year. That much matter crossing the event horizon would release tremendous amounts of kinetic energy, the same energy released by motion.

The latest information suggests the stream of gas seen snaking across space toward the event horizon is the fuel that keeps the black hole going.

Where did this particular river of gas come from? One theory: a supernova in the Sagittarius gas cloud may have shoved some of the dust and gas toward the center of the galaxy where it could be sucked in by the gravity of the black hole. Whatever caused it, the gas is streaming into a new cloud orbiting the center of the galaxy.

Earlier evidence. In 1987, astronomers examining data from orbiting observatories reported a bright object toward the center of the galaxy appeared to have a lot of positronium.

Under unusual conditions, electrons bind together with their opposite particles, positrons, forming positronium. Their destruction unleashes a flood of gamma rays. Some new mystery object at or near the center of the Milky Way seemed to vary in the brightness of its gamma ray emissions. It looked huge, maybe more than 10 trillion miles wide. Something that big might be a hot cloud around the event horizon of a black hole.

Decidedly peculiar. American astronomers returning from orbit in 1985 reported the Spacelab X-ray telescope allowed them to detect a never-before-seen glow around the core of the galaxy. Astronaut Loren Acton was an astronomer aboard when the shuttle Challenger lifted Spacelab 2 to space July 29, 1985.

He and six crewmates used several kinds of telescopes and instruments to look at the Sun and deep space for eight days of the Spacelab 2 flight. What they described as a "decidedly peculiar" source of X-rays was found at or not far from the center of the galaxy.

Aluminum. What is the Universe made of? As far as we know, 92 elements occur naturally. Chemistry students list these in order by weight, starting with hydrogen as the lightest. As composition of atoms increase, weights become greater.

Astrophysicists think the Universe started with hydrogen. Nuclear reactions in stars create more complex, but still lightweight, elements up to carbon and oxygen. Heavier elements are formed in more dramatic events, such as the supernova explosion of a star.

Gamma ray readings taken by an Earth-orbiting satellite have uncovered the production of medium-weight elements such as aluminum near the center of our Milky Way. The energy seen represented aluminum-26, an isotope with a brief life of about a million years. The short life and strong energy indicated aluminum is being manufactured now by some unusual events involving stars at the heart of our galaxy.

Skeptics. A massive black hole at the center of the Milky Way still is a controversial idea. Many, but not all, astronomers believe a black hole is there. Skeptics say the imagined nature of black holes doesn't fit the basic rules of physics. Physicists admit they really don't know what happens when a body with a million times the mass of our Sun collapses down to a point. The laws of physics, as verified in the laboratory, don't explain such an extreme event.

Sitting For New Portrait

The most detailed snapshot ever made of the Milky Way has been started at Socorro, New Mexico, where 27 giant dish antennas of Earth's most-sensitive radiotelescope have begun a systematic scan of the entire galaxy.

Known as the Very Large Array (VLA), each of the 27 antennas is 80 feet in diameter. The radiotelescope started operations in the late 1970's. Its antennas move on 20 miles of tracks west of Socorro, detecting fainter radio signals than any other telescope.

Two years of astronomers peering through deep-space dust clouds and computers drawing maps will result in the most comprehensive portrait of our Milky Way galaxy ever made at any wavelength including visible light, radio and infrared radiation. The detailed survey will give an overall picture of Earth's galaxy. Cataloging the contents will present a more clear picture of what the galaxy is.

Cataclysmic explosions cause faraway galaxies, pulsars, quasars and other natural deep-space bodies to emit radio signals as well as visible light, infrared light, ultraviolet light, X-rays and other types of radiation.

100 billion suns. Our Sun is a star, one of billions in the Milky Way galaxy. The disk-shaped Milky Way is a flat spiral of stars, reminiscent of a pinwheel with a bulging center. It's a vast structure of 100 billion stars clumped in spiraling arms that stretch across 100,000 lightyears—about 588,000 trillion miles of space.

The Sun and its Solar System, including Earth, is 22,000 lightyears—130,000 trillion miles—from the center of the galaxy. From our position out on one of the spiral arms, the galaxy looks to human observers on Earth like a dense white band of stars crossing Earth's night sky. Thus, the name Milky Way.

Surveys have been made before of vast regions of deep space beyond our galaxy, using powerful optical and infrared telescopes looking above or below the flat disk of the Milky Way. But optical and infrared telescopes can't see our galaxy very well. Visible light is blocked by voluminous clouds of dust clogging space between the stars of the Milky Way. Because the galaxy has hoards of infrared light sources, it's hard to distinguish objects seen in infrared light.

Seeing through dust. Radio waves, on the other hand, penetrate the deep space dust clouds. The new survey will be able to map the 70 percent of the Milky Way visible from Earth's Northern Hemisphere by detecting objects that emit radio waves.

The new map will show exploding stars known as supernovas, startlingly-dense supernova cinders known as neutron stars, pulsars, the seemingly-endless dust-and-gas clouds where new stars are born, and maybe even a black hole.

Astronomers hope to find supernova explosions which otherwise are invisible to optical telescopes. Shock waves from supernova blasts may compress gas and dust, creating new stars. The survey may show if supernovas happen to be in areas of space where star formation is notable.

Some stars leave almost nothing behind after they explode. Other supernovas collapse into super-dense neutron stars, or rotating neutron stars called pulsars, or even denser objects known as black holes.

Life elsewhere. The new map won't show other life in the galaxy, but it may help astronomers improve their estimate of the odds that life exists elsewhere.

Astronomers at University of California, Davis, and Columbia University, New York City, are spending 300 hours using the VLA to view the galaxy. Computers will digest the VLA data over nearly two years, creating a map of all detected radio signals.

Gomez's Big Mac Spotted

A footprint, an egg, a red rectangle and now a hamburger in space. Those are some of the strange objects catalogued by deep-space observers at the National Optical Astronomy Observatories (NOAO).

Officially called Object Gomez, the new hamburger-shaped deep-space rarity joined Minkowski's Footprint, the Egg Nebula, and the Red Rectangle in the select company of celestial oddities in the catalog.

It was named Object Gomez in the catalog after its discoverer, Arturo Gomez, a

Chilean assistant observer at the Cerro Tololo Inter-American Observatory near La Serena, Chile, who first detected the weird object in May 1985. Astronomers unofficially dubbed it Gomez's Big Mac for its appearance.

The object offered a rare look at a fast-moving process in which dying stars seed interstellar space with material which later spins up into the denser stuff of new stars, planetary nebulae and the chemical ingredients of life.

Out of place. Arturo Gomez was looking through a 60-in. telescope at a portion of sky, very rich in stars, near the center of the Milky Way. That part of the sky is not known to have many strange objects which don't look like stars. The object he discovered appears tiny from Earth. It apparently is the only one of its kind in its galactic neighborhood of our Milky Way galaxy for it does not show on other Cerro Tololo photos of the central part of the Milky Way.

The scattered bright material looks like a smudge of light divided horizontally, like a bun. Only five similar objects are known. They have been described as resembling feet, eggs and geometric shapes. Astronomers don't know what these peculiar deep-space things are. Some see them as new stars, recently formed and still surrounded by cocoons of dust and gases. On the other hand, some think they are dying stars ejecting clouds of hot dust and gas.

Another galaxy? To the NOAO astronomers in Chile, Object Gomez first seemed to be a galaxy viewed edgewise. The obvious horizontal division, they thought, was caused by a band of interstellar dust which often can be seen surrounding a galaxy. However, galaxies aren't noticed much in that part of Earth's southern sky as they are concealed behind the dense dusty center of our own Milky Way.

After a longer look at the strange object, they changed their minds. Breaking down the light coming from the object into its full spectrum rainbow, the astronomers found it resembled light from a normal star somewhat warmer than our Sun. However, astronomers have photographed many stars of such a general type before without encountering anything like Object Gomez.

Only a star? Strangely, infrared light in the spectrum of Object Gomez didn't look anything like infrared light from a hot star. It seemed to be at least ten times too cold. The blue light in the spectrum, indicating heat, may be light scattered back toward Earth from dust near the poles of the star. So, the scientists temporarily are hanging onto a view of Object Gomez as a star with a ring of dust around its equator thick enough to hide part of the star from view, creating the odd bun-like appearance.

Object Gomez may be a star which has aged beyond typical stars, past the supergiant stage, and now is exploding rapidly into a planetary nebula. Old red giants, up to eight times the size of our Sun, explode into planetary nebulas, reducing themselves down to a white dwarf only about half of the size of the Sun. The old star returns part of itself to interstellar space to be used by succeeding generations of stars.

If the suggestion is correct, Object Gomez wouldn't be so odd. Many planetary nebulae are known. But, astronomers don't know a lot about the stages in a star's life immediately prior to planetary nebulae. As with life itself, old age happens all too fast. In the case of a star, that last phase may pass in a thousand years or less, very quickly as the Universe goes.

Clouds Of Bacteria In Space?

Is the Universe filled with germs? Interstellar dust strewn across the Universe may be teaming with bacteria and minute plants.

Astrophysicist Gurcharan Kalra of India's University, Delhi, told a July 1990 meeting of the International Astronomical Union at Sydney, Australia, samples of light from the center of our Milky Way galaxy resembled light passed through E. Coli bacteria and through tiny sea plants called diatoms. Thus, warm dusty areas of deep space may be filled with simple forms of life. Clouds of bacteria may be swirling out there.

Kalra told his fellow conferees the concept seemed "obviously absurd" at first, until statistical correlation between the two sets of light was very strong.

Clouds of bacteria in space were first suggested by British astrophysicist Fred Hoyle in 1982. He was ignored by scientists. Kalra had set out to debunk Hoyle, but found his own results hard to deny.

Both Hoyle and Kalra studied infrared-light photos of dust swirling in an area at the heart of the Milky Way known as GC-IRS 7. The galactic dust was calculated to be as hot as 630 degrees Fahrenheit. They heated E. Coli bacteria to 630 degrees, then compared the photos with graphs of the heated bacteria. The two graphs were similar.

Planetary Nebula

A planetary nebula is a vast cloud of dust spreading out from a star in a disk as wide as, or wider than, our Solar System. Some 10,000 are suspected in the Milky Way galaxy, but only about 1,000 actually have been cataloged by astronomers. Several new planetary nebula may start each year in the galaxy.

As seen by astronomers through telescopes from Earth, a planetary nebula seems to shine. The glow comes from ultraviolet light, generated by the very hot central star, which heats the gas in the nebula.

That gas usually is expanding outward from the central star at about 45,000 miles per hour. Since the star is losing material, a nebula indicates a stage in the life of a star between middle age and old age. The lifespan: a hot young blue star evolves into an older red-giant which then blows out to become a white dwarf. The red giant is so large gravity at its surface is too weak to hold matter which blows away to form the nebula. A nebula may last only 10,000 years or so, expanding while the star fades.

William Herschel first called them planetary nebula in 1785. Those seen have a variety of shapes which have led to popular names. Some examples:

NEBULA	CATALOG No.	CONSTELLATION
Butterfly	NGC 6302	Scorpius
Dumbbell	NGC 6853, M27	Vulpecula
Eskimo	NGC 2932	Gemini
Owl	NGC 3587, M97	Ursa Major
Ring	NGC 6720, M57	Lyra
Sunflower	NGC 7293	Aquarius

Galaxy Noise Heard At South Pole

Scientists at work in Antarctica near Earth's South Pole can hear galaxy noise. Australian specialists in upper-atmosphere physics have heard it at Casey Station in the Australian Antarctic Territory on the Greater Antarctica mainland at Budd Coast, Wilkes Land, on the Vincennes Bay of the Southern Ocean.

Working with the engineers is a physicist from the People's Republic of China. During the Antarctic winter of 1989, when there were few daylight hours and daytime temperatures were about 10 below zero Fahrenheit, they set up a small geophysical observatory to monitor Earth's magnetic field as well as ionized layers of this planet's atmosphere up to altitudes of 300 miles.

About twice a week, the Australians measured the magnetic field and radioed the data by space satellite to the Bureau of Mineral Resources at Canberra, Australia.

One of their instruments, known as a flux-gate magnetometer, could read very tiny changes in the magnetic field referred to as micropulsations. A coil with 250,000 turns of wire senses the tiny magnetic-field fluctuations, which are amplified and recorded on a chart. Earth's magnetic field, as well as activity on the Sun, affect the ionosphere.

The Australians had two sensitive pieces of gear monitoring the ionosphere. One, known as a rimeter, was a sensitive radio receiver tuned to a frequency of 30.1 MHz (megahertz). At that frequency, the engineers were listening for so-called galactic noise.

The noise actually is generated 50 miles above Antarctica when electrons blowing along in the solar wind arrive to excite what scientists have labeled the D-layer of the ionosphere. The D-layer absorbs the incoming energy. How much it absorbs indicates the electron density, revealing information about solar activity.

Satellites in Earth orbit are vital to people wintering over in Antarctica. The Australians, for instance, used an Intelsat satellite for telephone, fax and data links back home. They had a 21-ft. dish antenna which beamed signals up to the satellite 22,300 miles overhead and received signals coming down. For emergency backup, they had a 10,000-watt shortwave transmitter for direct contact with the rest of the world.

The researchers weren't studying the ozone layer. That work was being done elsewhere, at Australia's Macquarie Island in the South Pacific Ocean.

25 Brightest Stars

Looking up from your backyard? Here are the 25 brightest stars in the sky.

A star's magnitude depends on the amount of light it gives off and its distance from Earth. A star farther from Earth sending out a lot of light may appear brighter here than a closer star not giving off much light. A large dull star might seem weaker to us than a small bright star. The lower the magnitude number, the brighter the object. Thus a first-magnitude, or magnitude 1, star such as Aldebaran (magnitude +0.9) seems brighter to us than Castor (magnitude +2.0). As the numbers approach and pass zero, they become negative numbers on the magnitude scale. Thus the brightest star we see is Sirius (magnitude –1.4), only nine lightyears away from Earth.

The brightest objects are the Sun at magnitude –27 and the Moon at magnitude –13. On a clear night, we can see objects as faint as magnitude 5 or 6. With binoculars we see stars as weak as magnitude 8 or 9. A six-inch telescope will show stars down to magnitude 13. Some planets can be seen with the naked eye.

Here are the 25 brightest stars in the sky in order of magnitude, including their popular names, their proper designations and their distances in lightyears from Earth, according to the Royal Astronomical Society of Canada:

MAG	NAME	DESIGNATION	DIST
–1.4	Sirius	α Canis Majoris	9
–0.7	Canopus	α Carinae	98
–0.1	Arcturus	α Bootes	36
0.0	Vega	α Lyrae	26
0.0	Alpha Centauri	α Centauri	4
+0.1	Capella	α Aurigae	45
+0.1	Rigel	β Orionis	900
+0.4	Procyon	α Canis Minoris	11
+0.4	Betelgeuse	α Orionis	520
+0.5	Achernar	α Eridani	118
+0.6	Beta Centauri	β Centauri	490
+0.79	Alpha Crucis	α Crucis	370
+0.8	Altair	α Aquilae	16

+0.9	Aldebaran	α Tauri	68
+0.9	Antares	α Scorpii	520
+0.9	Spica	α Virginis	220
+1.2	Fomalhaut	α Piscis Austrini	23
+1.2	Pollux	β Geminorum	35
+1.3	Deneb	α Cygni	1600
+1.3	Beta Crucis	β Crucis	490
+1.4	Regulus	α Leonis	84
+1.5	Adhara	ε Canis Majoris	680
+1.6	Shaula	λ Scorpii	310
+1.6	Bellatrix	γ Orionis	470
+2.0	Castor	α Geminorum	45

Stars Nearest To Earth

The Sun is Earth's star, at the center of our Solar System. It seems to be a single star, probably not part of a double-star or triple-star system.

By comparison, many of the stars nearest to our Solar System are part of double and triple systems. Sometimes the companion stars are bright enough to be seen, sometimes they are unseen.

For example, Alpha Centauri is a triple-star system which includes the third brightest star in Earth's sky and the brightest star in the constellation Centaurus. One of the three stars in the trio, Proxima Centauri, is the star closest to our Sun. The two brightest stars in the trio, Alpha Centauri A and Alpha Centauri B, compose a yellow-orange binary pair circling each other every 80 years. Each is about the size and mass of the Sun.

The fourth-nearest star to our Sun is Barnard's Star, a red-dwarf star seen in the constellation Ophiuchus. It was discovered in 1916 by E.E. Barnard. Astronomers see Barnard's Star wobbling in its path as it orbits the center of the Milky Way galaxy. The wobble leads them to suspect one or more jumbo planets, the size of Jupiter, may be in orbit around the star.

Distances to stars in the list below are in parsecs. Parsec, short for parallax second, is a measurement of distance to objects outside the Solar System. One parsec is about 3.2616 lightyears or 19.24 trillion miles.

In the list, Lalande 21185 B also is known as BD 36-2147. Luyten 726-8 B also is known as UV Ceti. BD is Bonner Durchmusterung. CD is Cordoba Durchmusterung.

STAR	MAGNITUDE	DISTANCE
Proxima Centauri	15.1	1.31
Alpha Centauri A	4.4	1.31
Alpha Centauri B	5.8	1.31
Barnard's Star	13.2	1.83
Wolf 359	16.8	2.32
Lalande 21185 A	10.5	2.49
Lalande 21185 B	unseen	2.49
Sirus A	1.4	2.67
Sirus B	11.5	2.67
Luyten 726-8 A	15.4	2.74
Luyten 726-8 B	15.8	2.74
Ross 154	13.3	2.90
Ross 248	14.7	3.16
Epsilon Eridani	6.1	3.28
Luyten 789-6	14.9	3.31

Ross 128	13.5	3.32
61 Cygni A	7.5	3.43
61 Cygni B	8.3	3.43
Epsilon Indi	7.0	3.44
Procyon A	2.7	3.48
Procyon B	13.0	3.48
BD 59-1915 A	11.1	3.52
BD 59-1915 B	11.9	3.52
BD 43-44 A	10.3	3.55
BD 43-44 B	13.2	3.55
CD -36-15693	9.6	3.59
Tau Ceti A	5.7	3.67
Tau Ceti B	unseen	3.67
BD +5-1668	11.9	3.76
CD -39-14192	8.7	3.85
Kapteyn's Star	10.8	3.91
Kruger 60 A	11.8	3.94
Kruger 60 B	13.4	3.94

Greek Letters Identify Stars

The bright stars on star charts are designated by Greek letters. The brightest star in a constellation is alpha or α, the next brightest is beta or β, and so on. To identify a star, add the Greek letter to the Latin form of the constellation name. For example, the star in the constellation Orion commonly referred to as Betelgeuse is properly designated Alpha Orionis or simply α Orionis. The Greek alphabet:

α alpha	η eta	ν nu	τ tau
β beta	θ theta	ξ xi	υ upsilon
γ gamma	ι iota	o omicron	φ phi
δ delta	κ kappa	π pi	χ chi
ε epsilon	λ lambda	ρ rho	ψ psi
ζ zeta	μ mu	σ sigma	ω omega

Constellations

The 88 constellations are arbitrary groupings of stars in Earth's sky, used by astronomers down through the ages to identify particular sky locations.

Two stars seen in one constellation may not be near each other since one could be relatively close to Earth—say, within a few lightyears distance—while the other could be thousands of lightyears away. For example, the star Alpha Centauri is less than five lightyears from Earth while Beta Centauri is more than 450 lightyears away. They seem close in the sky, but are not.

The 88 constellations are not equal in size of sky area they cover. Hydra covers the largest area of the sky. Crux Australis, the smallest.

The list below shows astronomers' names for constellations, common names and first-magnitude stars, if any, in a constellation.

ASTRONOMY	COMMON	STARS
Andromeda	Andromeda	
Antlia	The Air Pump	
Apus	The Bee	
Aquarius	The Water Bearer	
Aquila	The Eagle	Altair

Ara	The Altar	
Aries	The Ram	
Auriga	The Charioteer	Capella
Bootes	The Herdsman	Arcturus
Caelum	The Graving Tool	
Camelopardalis	The Giraffe	
Cancer	The Crab	
Canes Venatici	The Hunting Dogs	
Canis Major	The Great Dog	Sirius
Canis Minor	The Little Dog	Procyon
Capricornus	The Sea Goat	
Carina	The Keel	Canopus
Cassiopeia	Cassiopeia	
Centaurus	The Centaur	Agena, Alpha Centauri
Cepheus	Cepheus	
Cetus	The Whale	
Chamaeleon	The Chameleon	
Circinus	The Compasses	
Columbia	The Dove	
Coma Berenices	Berenice's Hair	
Corona Australis	The Southern Crown	
Corona Borealis	The Northern Crown	
Corvus	The Crow	
Crater	The Cup	
Crux Australis	The Southern Cross	Acrux, Beta Crucis
Cygnus	The Swan	Deneb
Delphinus	The Dolphin	
Dorado	The Swordfish	
Draco	The Dragon	
Equuleus	The Foal	
Eridanus	The River	Achernar
Fornax	The Furnace	
Gemini	The Twins	Pollux
Grus	The Crane	
Hercules	Hercules	
Horologium	The Clock	
Hydra	The Water Snake	
Hydrus	The Little Snake	
Indus	The Indian	
Lacerta	The Lizard	
Leo	The Lion	Regulus
Leo Minor	The Little Lion	
Lepus	The Hare	
Libra	The Balance	
Lupus	The Wolf	
Lynx	The Lynx	
Lyra	The Lyre	Vega
Mensa	The Table	
Microscopium	The Microscope	
Monoceros	The Unicorn	
Musca Australis	The Southern Fly	
Norma	The Rule	
Octans	The Octant	
Ophiuchus	The Serpent Bearer	
Orion	Orion	Betelgeux, Rigel
Pavo	The Peacock	
Pegasus	The Flying Horse	
Perseus	Perseus	
Phoenix	The Phoenix	

Pictor	The Painter	
Pisces	The Fishes	
Piscis Australis	The Southern Fish	Fomalhaut
Puppis	The Poop	
Pyxis	The Compass	
Reticulum	The Net	
Sagitta	The Arrow	
Sagittarius	The Archer	
Scorpius	The Scorpion	Antares
Sculptor	The Sculptor	
Scutum	The Shield	
Serpens	The Serpent	
Sextans	The Sextant	
Taurus	The Bull	Aldebaran
Telescopium	The Telescope	
Triangulum	The Triangle	
Triangulum Australe	The Southern Triangle	
Tucana	The Toucan	
Ursa Major	The Great Bear	
Ursa Minor	The Little Bear	
Vela	The Sails	
Virgo	The Virgin	Spica
Volans	The Flying Fish	
Vulpecula	The Fox	

Search For Planets At Faraway Stars

Our star, the Sun, and nine confirmed planets form the only definitely-known Solar System. Those nine verified planets are Mercury, Venus, Earth, Mars, Jupiter, Saturn, Uranus, Neptune and Pluto. Some astronomers suspect there may be a tenth undiscovered planet—the so-called Planet X—in our own Solar System, orbiting our Sun at a great distance, even beyond Pluto.

But there are other stars in this corner of the Universe and some are not unlike the Sun. Could they, too, have planets? If so, they remain unseen. The very best telescopes on Earth and in orbit are not powerful enough to see tiny bodies orbiting distant stars.

There are other means, however. For instance, in 1987 three Canadian astronomers used the 3.6-meter Canada-France-Hawaii Telescope atop Mauna Kea, Hawaii, to measure the line-of-sight velocities of 16 nearby stars. Two of the stars definitely, and seven probably, appeared to speed up and slow down rhythmically, as if they were being tugged back and forth by unseen companions. That led to the conclusion that at least one star as large as Jupiter was accompanying each of the stars.

Jupiter. Jupiter may itself have been a prototype Sun which wasn't big enough to become a star. Jupiter is the largest of the nine planets circling our Sun. In fact, Jupiter alone is more than two-thirds of the total mass of all planets in our Solar System. It is 318 times the mass of Earth. Although a thousand times smaller than the Sun, Jupiter is a big planet—as big as 1,317 Earths. Jupiter is used as a standard of comparison by astronomers looking for planets in orbit around other stars.

Still, Jupiter wasn't big enough. If the planet had been several times more massive, it might have become a star as the pressure and temperature at its core triggered nuclear fusion. Instead, Jupiter is a big gas bag, with a relatively low density. Could there be similar large planets orbiting nearby stars?

Vega. Back in 1983, the telescope in the orbiting Infrared Astronomical Satellite (IRAS) indicated other stars could generate planets when it reported large disks of dust and gas orbiting Vega and other nearby stars. Astronomers said planets may have been able to form as lumps or patches in those disks of matter.

Van Biesbroec. Then, in 1984, University of Arizona astronomers thought they saw a planet near the star Van Biesbroec 8. However, two teams of astronomers elsewhere couldn't find the planet later. Some scientists thought the planet either had moved or a wandering dust cloud had fooled the Arizona telescope.

Brown dwarfs. Some astronomers visualize large gassy more-than-a-planet-but-not-quite-a-star objects which they call brown dwarfs. They would be something like the planet Jupiter, but larger. They would be massive clumps of gas, but not quite heavy enough to set off the nuclear fires needed to turn themselves into stars. Larger than any known planet, a brown dwarf still would be only 15 percent as big as our Sun.

Nobody knows for sure just how much mass is required to make a lump in a cloud of dust and gas ignite nuclear burning and become a glowing star. If a brown-dwarf sub-star were near a real star, it might cause the glowing star to wobble in its path. Despite searches, no brown dwarfs actually have been found so far.

Gamma Cephel, Epsilon Eridani. Canadian astronomers said in 1987 they had discovered planets circling two distant stars, Gamma Cephei and Epsilon Eridani.

Both stars are similar to our Sun. Epsilon Eridani is 11 lightyears from our own Solar System. Gamma Cephei is 48 lightyears from the Sun. A lightyear is a measure of the distance one ray of light would travel in one year, about 5.9 trillion miles.

The astronomers calculated that each of their planets would have to be from two to five times as large as Jupiter. It was not known whether small Earth-sized planets also exist in those distant solar systems.

The astronomers, at the Dominion Astrophysical Observatory, studied 16 stars within 50 lightyears of Earth for six years before compiling sufficient evidence to announce the discovery of the two giant planets. The astronomers felt they had clear evidence of planets in two stars. Five other stars revealed tentative evidence of planets.

The astronomers suggested the discovered objects were fully-formed planets, too small to be brown dwarfs. The planet they believed to be orbiting Epsilon Eridani was said to be two to five times larger than Jupiter. The planet companion to the star Gamma Cephei was something like half again as large to twice as large as Jupiter.

At the speed of light, it would take eleven years to travel to Epsilon Eridani and 48 years to get to Gamma Cephei. Mankind does not now have the ability to accelerate a spacecraft to more than ten percent of the speed of light. With today's technology, it would take more than 100 years to get to Epsilon Eridani and more than 500 years to get to Gamma Cephei.

Seeing planets. Neither of the two stars with probable planets is among the brightest in Earth's night sky. Planets don't generate light. They only reflect it. So they are almost impossible to see from Earth. To make their discovery, the Canadian astronomers watched how the stars Gamma Cephei and Epsilon Eridani wobbled slightly in their annual travels through the Milky Way galaxy.

The unusual motion they found seemed to indicate that gravity from unseen large planets probably was tugging slightly at the stars. The researchers were able to see changes in the speed of the stars down to 25 miles per hour. That's 50 to 100 times more accurate than earlier methods.

Giclas. Astronomers in California and Hawaii reported in 1987 they had evidence a large Jupiter-like planet or object was in orbit around a star not far from Earth.

The object they discovered was big, but not burning like a star. That might make the discovery a planet about the size of Jupiter, or slightly larger, with a surface temperature double the 900 degrees of Venus, according to an astronomy professor at the University of California, Los Angeles, and a University of Hawaii astronomer.

The California-Hawaii mystery object seemed to be orbiting a white-dwarf star known as Giclas 29-38 about 46 lightyears from Earth. That's close in the vast expanse of the Universe where objects can be billions of lightyears away.

The astronomers used NASA's infrared telescope on Mauna Kea, Hawaii, to find more heat coming from the vicinity of Giclas 29-38 than could come from the star itself. Heat radiated by a brown dwarf was said to be the most likely explanation. They didn't suppose the heat was given off by asteroids or dust around the star as that wouldn't agree with rules of physics, they said.

The evidence pointed to an object not burning hydrogen, not a star. It seemed big, but not nearly as big as the star it was near. It could be a gaseous body, like Jupiter, only even more massive. Yet it would have been much less massive than the white-dwarf star it orbits. Evidence of the object orbiting Giclas 29-38 was more direct, more convincing, than evidence for the Epsilon Eridani, Gamma Cephei or Van Biesbroeck 8 planet-like bodies.

Faraway civilization. One unlikely possibility, which can't be ruled out, is a civilization on a planet in a solar system of a distant star such as Giclas 29-38. A few observers suggested that might mean the heat we see is being reflected by solar panels collecting energy in orbit around a planet.

Can we go there to find out? At the speed of light, it would take 46 years to travel to Giclas 29-38. Mankind does not now have the ability to accelerate a spacecraft to more than ten percent of the speed of light. With today's technology, it would take 500 years to get to Giclas 29-38.

HD 114762. Astronomers routinely measuring standard stars in 1988 came up with what they called the first confirmed evidence of a planet orbiting a star other than our Sun. Orbiting a sun 90 lightyears from our own, the unseen planet was really big— 30,000 times larger than Earth. The astronomers said it probably was lifeless with a surface temperature hundreds of degrees hotter than an oven. The planet was so close to its star that it orbited once every 84 days. That would be similar to Mercury, the planet closest to our Sun which orbits every 87 days. Mercury has a very harsh environment.

Using a 61-inch reflector telescope at Oak Ridge Observatory in Massachusetts, the astronomers found the planet while running instrument tests on a star known only as HD 114762. The star was one of a group of so-called "candidate" standard stars. Astronomers have been measuring those for years to log characteristics of basic star types so astronomers around the world can have a standard for calibrating instruments.

Photon counter. The astronomers sent starlight through a photon counter which detected a slight wobble in the motion of the star HD 114762. The wobble could only have been caused by a body orbiting nearby, they said.

To give star HD 114762 such a wobble, the newly-discovered planet must be massive. It may be 20 times larger than Jupiter. The minimum size for a true star was said to be 100 times the size of Jupiter.

The newly-discovered planet may be jumbo, but it still was too small and too distant to be seen. Trying to see the planet would be like trying to look at a candle beside a searchlight on the Moon. Rather than seeing it directly, scientists calculate the effect of the planet's gravity on the star it orbits.

Star HD 114762, about 522 trillion miles from Earth, shines at magnitude 7, too faint for the naked eye. It is what astronomers call a spring star, seen through telescopes from March through July. Earth's tilt causes it to be blocked from view by our Sun from August through February. Detection of the planet was confirmed by a Swiss astronomer at Geneva Observatory.

At the speed of light, it would take 90 years to travel to the star HD 114762. With today's technology, it would take nearly 1,000 years to get to the star HD 114762.

Naked stars. A University of Colorado astronomer dampened enthusiasm for finding stars around other planets for a time in 1988. He said stars which had been assumed capable of producing planets actually can't. He found many young Sun-like stars lacking the surrounding clouds of gas and dust essential for formation of planets.

Stars in the planet-formation stage often are so-called T Tauri. The dust enveloping T Tauri stars disappears after about 10 million years. Then they are called post-T Tauri stars. The Colorado astronomer checked star-forming areas in the constellations Taurus, Orion and Scorpius and found many post-T Tauri stars actually were naked T Tauri stars, young stars with no dust clouds. He told the American Astronomical Society naked T Tauri stars are ten times more numerous than dusty T Tauri stars.

Nine planets. After eight years of searching nearby regions of our galaxy, and staring particularly hard at 18 Sun-like stars nearby, a Canadian astronomer said in 1989 there seem to be at least nine planets orbiting distant stars in the Milky Way.

Without claiming the nine celestial bodies he found actually are planets orbiting stars, a physicist at the University of Victoria, British Columbia, reported his evidence points to planet-sized objects outside of our Solar System. The objects range from one to ten times the size of Jupiter, the largest planet in our Solar System.

50 percent. Half of 18 stars the astronomer analyzed since 1981 showed signs of companions which might be planets. The suspected planets could be like the tip of an iceberg. Each may be the largest planet in its solar system. The higher-than-expected percentage might indicate as many as half of the 100 billion stars in our Milky Way galaxy could have planetary companions. Our Solar System has nine planets. Could their be 100 billion to 900 billion planets across our Milky Way galaxy?

Such large numbers would seem to increase the likelihood that planets like Earth, capable of supporting life, may be strewn across the galaxy. The Canadian astronomer planned to expand his search to 50 or 100 stars to see if the percentage holds up.

Even if giant planets like Jupiter are inhospitable to life as we know it, there may be smaller Earth-like planets sheltering life in those distant solar systems.

Gamma Cephei. The clearest finding among the nine located was something about the size of Jupiter orbiting the star Gamma Cephei. The other eight ranged from that size up to ten times the mass of Jupiter.

No astronomer actually has seen a planet outside of our Solar System and life has yet to be found beyond Earth. The view of all presumed distant planets is washed out by the bright light from their stars. Instead, the existence of the planets is inferred from minute changes in the paths of their stars. Finding a distant planet orbiting a star is like detecting a speck of dust next to a 1,000-watt light bulb, astronomers say.

The 18 stars studied by the Canadian astronomer were similar to each other:

★they are similar to the Sun in size and temperature,

★they are bright enough to be seen with the naked eye from Earth,

★they are within 50 lightyears of Earth, a distance of about 294 trillion miles,

★they are single stars, rather than members of close pairs of stars known as binary systems.

Color shift. The new discoveries were made by measuring the speeds at which individual stars were moving through the galaxy, as well as changes as small as 23 miles per hour in the speed of a star. The computations were 100 times more precise than previous calculations.

The astronomer analyzed the shifting color of light from the stars. The color of light from a distant object changes toward red if the source is moving away. It shifts toward blue if the source is moving toward the observer. By carefully measuring the amount of color shift, scientists can detect the wobbly movements of stars as they are tugged by the gravitational attraction of nearby bodies. The velocity of that movement is used to calculate the size of an invisible object. The calculated mass indicates whether the object might be a planet, a brown dwarf or another star.

Scutum. British radio astronomers reported a planet outside the Solar System in July 1991. They found a pulsar with an irregular rotation period in the constellation Scutum 30,000 lightyears from Earth.

Radio energy bursts from pulsars usually are very regular. They are so regular they make good cosmic clocks. But the one in the constellation Scutum runs two percent fast for three months, then slows down by two percent for three months.

The British radio astronomers suggested the pulsar is wobbling under the influence of an orbiting massive planet. The wobble Doppler-shifts radio waves from the pulsar.

They calculated the unseen planet at ten times the mass of Earth. It may circle its star as far away as Venus is from the Sun. Hardest to understand is the planet's location, since a pulsar is a remnant of a past supernova which surely would destroy a nearby planet.

Search for life. Scientists say verification of planets anywhere else in our Milky Way galaxy would shed a great deal of light on Earth's origins, electrify the search for extraterrestrial life, and spark man's desire to go there for a close look.

Four Stars Locked In Orbit

An Arizona scientist discovered one of astronomy's rarest phenomena, four stars locked in orbit around each other.

The quadruple system, about one quadrillion miles from Earth in the constellation Leo, is one giant gravity web. The ensnared four are pulling and tugging at each other, producing magnificent tidal effects, spewing out spectacular flares from hot gas eruptions on their surfaces.

The system is two pairs of stars circling each other. At the heart of the quad, two stars actually touch each other. One is larger than the other making them look like a lopsided dumbbell. They rotate every six hours around a common center of gravity.

Circling them is another pair about a billion miles away. The stars in the second pair are about 1.4 million miles apart. They rotate every 18 hours around a common center of gravity. The outer pair circles the inner pair every 20 years.

The four may have formed from one disk of hot gas and dust billions of years ago. But, because the orbit planes of the two pairs are slightly different, the two sets may have formed in separate clouds of gas and dust. Some powerful gravity source could have torn a pair from its original position, moving it near the other pair, leaving all four locked in a new gravity embrace.

There are billions of stars in the Universe, most in pairs called binary stars. About a dozen quadruples have been found. Our Sun is unusual in being a single star with no apparent companion.

Dwarfs And Novas

A star is a globe of hot gases, floating in space, giving off light and other invisible energy. Our Sun is a star. A string of sparkling stars flung across the far reaches of the Universe have fired imaginations and inspired endless nights of searching Earth's sky.

A nova is a star which has used up its hydrogen fuel and flares briefly before becoming smaller. As it flares up, the star spews out a cloud of hot gas and becomes 5,000 to 10,000 times brighter for a few days. Then its light weakens and becomes faint as seen from Earth.

Nova vs. supernova. A supernova also is a dying star. Having used up all of its hydrogen fuel, it contracts. The outer layers of the star's atmosphere fall in toward the core. But then that contraction produces extraordinarily-high temperatures, even for a star, giving rise to thermonuclear reactions which produce heavy elements by nuclear fusion. Such a supernova star, after collapsing inward, then explodes, becoming a trillion times brighter than the Sun. The residue is referred to as a white dwarf star.

Dwarf stars. Dwarf nova stars are small objects which flare up with sudden increases in brightness from 2 to 6 magnitude numbers at intervals from weeks to months. The added brightness lasts only a few days.

U Geminorum was the first dwarf nova to be found so the type of star has come to be known as U Geminorum stars. The brightest dwarf nova ever seen is SS Cygni.

Dwarf nova stars always are part of binary pairs of stars. The main star in the pair is a white dwarf. The secondary star is a cooler star. The size of the two stars usually is about the same. They spin around each other in periods ranging from three hours to 15 hours.

The lesser star usually is expanding. Hydrogen-rich gas streams from it, feeding a disk orbiting the primary star. Astronomers think the hydrogen disk is heated so it changes from opaque to transparent and back. That change may cause the brightening seen from Earth.

The brightening is not an explosion. Nothing is spewed out. However, the gas in the disk must spiral down onto the white dwarf where it eventually would cause a nova explosion.

White dwarf. A white dwarf is a small, dense, dull star giving off little energy in the twilight of its life. Because of its size it has a high surface temperature and appears white. It is so dense that a one-inch cube would weigh several tons.

White dwarf is the final stage in the evolution of a small star. As much as 90 percent of its matter has been lost into a planetary nebulae, or cloud of gas and dust spreading far beyond the star. After the nuclear fuel is exhausted, the core of the star shrinks. A white dwarf star is one which has collapsed to about one percent the diameter of our Sun.

Interestingly, the most massive white dwarfs shrink to the smallest diameters and highest densities.

The light seen from a white dwarf doesn't come from an internal nuclear reaction. Rather, it comes from the star's thin gassy atmosphere slowly leaking away, releasing the star's heat into space.

A white dwarf star is very faint with a magnitude between +10 and +15. White is a poor name as the star changes from white through yellow and red to black as it cools down. At the end, it will be called a black dwarf.

Astronomers estimate there may be billions of white dwarfs in our Milky Way galaxy alone. Most would have cooled to black dwarfs.

Sirius. The best known white dwarf probably is Sirius B, a companion to Sirius. It was discovered in 1862 after an 1844 prediction that it was causing Sirius to have an unusual motion. Sirius B is only about twice the size of planet Earth, but has a mass as great as the Sun.

Stars which originally were too large to become white dwarfs collapse to become neutron stars, pulsars and black holes.

Ultraviolet white dwarf. Scientists took a snapshot of extreme ultraviolet radiation coming from a distant white-dwarf star with a telescope and spectrometer sent to the edge of space on a Black Brant rocket in 1987.

Part of a continuing NASA study of our Milky Way galaxy, the suborbital rocket flight from White Sands Missile Range in New Mexico carried a University of California, Berkeley, telescope and a spectrometer tuned to extreme ultraviolet light rays.

Extreme ultraviolet light (EUV) falls between visible light and X-rays in the electromagnetic energy spectrum. EUV is absorbed by Earth's atmosphere so research must start above the atmosphere.

The rocket was aimed to discover the density of helium in the interstellar medium—the void between the stars. The so-called interstellar medium is extremely thin gas—mostly hydrogen and helium—which fills the space between stars.

The density of helium in space between stars in the Milky Way had not been measured before. The Berkeley experiment provided a comparison of helium density with the density of hydrogen in the same part of space.

The spectroscope looked at the very hot white-dwarf star known to astronomers as G191 B2B.

Even though the star G191 B2B is hundreds of lightyears away from Earth, scientists know how much ultraviolet light leaves the star. The trick to finding the density of the interstellar medium is to see how much ultraviolet light is absorbed by the helium and hydrogen between the star and Earth.

Black Brant is a two-stage sounding rocket, 44 feet long and 17 inches in diameter. The 685-lb. spectrometer payload was parachuted back to the desert at White Sands where it was refurbished for later use. NASA launches more than 40 sounding rockets a year from locations around the world.

Variable stars. A star which seems to change brightness and other physical characteristics over time is said to be variable. The brightness of individual stars varies over periods from minutes to years.

Of the untold billions of stars strewn across the Universe, Man has catalogued about 30,000 variable stars over the years.

There are at least three reasons why stars vary in brightness. Astronomers categorize them as pulsating, eclipsing and cataclysmic:

★The surface of a pulsating variable star expands and contracts making it seem to brighten and fade.

★An eclipsing star, on the other hand, temporarily blocks light as it passes in front of another star, making it seem to fade.

★A cataclysmic event occurs in a binary pair—where two stars orbit each other—and one draws matter from the other in an explosive flare-up.

An unusual star known as Chanal's variable, in the Orion nebula, normally shines with a brightness seen from Earth as magnitude 16 or 17. It made news in 1988 when it brightened to 14th magnitude for the first time in years.

Geminorum. A dwarf nova star known to astronomers as U Geminorum flared up December 6, 1988. Normally fainter than magnitude 13, the variable star brightened to magnitude 9 by December 8. The star U Geminorum is half of a binary pair of stars orbiting each other every 4 hour 15 minutes. Its companion is a cool red star pouring material onto the white dwarf. U Geminorum, only 800 lightyears from Earth, erupts at irregular intervals.

Ophiuchus. A Japanese astronomer in 1988 discovered a new nova star in the constellation Ophiuchus.

The nova brightened to magnitude 10 for a time, according to Tokyo Astronomical Observatory and Siding Spring Observatory in Australia.

A constellation is an arbitrary pattern of stars in the sky which appear to observers on Earth to form a shape. Stars in a constellation are not necessarily near each other when comparing their distances from Earth.

Nova Scuti. A star in the constellation Scutum exploded into a nova in August 1989 and was discovered in September by an astronomer at Bern University, Switzerland. Officially designated Nova Scuti 1989, the star had not appeared on the Palomar Sky Survey before the September discovery.

When Nova Scuti 1989 was found September 20, it had an apparent brightness of about magnitude 10. It probably had brightened by as much as 160,000 times for a gain of 13 magnitudes. Sky watchers needed only large binoculars or small telescopes to see the nova in the early evening sky. The star brightened to magnitude 9.5 early in October, but then began to fade rapidly later in October.

Sagittarii. An astronomer at the Siding Spring Observatory in Australia discovered a distant 10th-magnitude nova in the constellation Sagittarius.

Nova Sagittarii 1987, seen in May 11, 1987, photographs, was fainter than magnitude 11.5. As it was observed for two weeks, the red star's brightness increased to magnitude 10. The larger the magnitude number, the fainter a star appears to astronomers on Earth.

After the discovery, a search turned up nothing visible in old photographs of that area of space made with a British telescope.

A star is a globe of hot gases, floating in space, giving off light and other invisible energy. Our Sun is a star.

A nova is a star which has used up its hydrogen fuel and flares briefly before becoming smaller. As it flares up, the star spews out a cloud of hot gas and becomes 5,000 to 10,000 times brighter for a few days. Then its light weakens and becomes faint as seen from Earth.

Nova Sagittarii 1987 is a relatively-common nova and should not be confused with the magnificent, rare Supernova 1987a discovered in February 1987.

Vulpecula. A lot of patience and a stroke of luck let two astronomers in on that brief moment when a faraway star explodes, flickering briefly among millions of stars in Earth's sky.

Kenneth Beckmann of Louiston, Michigan, discovered an exploding star—a nova—in the constellation Vulpecula November 14, 1987. Beckmann spotted the magnitude 7 star explosion at 8 p.m.

Just two hours later the blazing star was discovered independently by Peter Collins of Phoenix, Arizona, using binoculars. By then it already had faded to magnitude 7.3. By the next day, they reported, the nova had faded to magnitude 7.4.

A nova occurs when one of a pair of stars draws sufficient hydrogen gas from its companion to ignite a nuclear reaction.

Typically, a common "main-sequence" star and a "white-dwarf" star form a binary system—locked in orbit around each other. The common star expands and a cloud of its hydrogen gas is sucked down onto the white dwarf star. After 10,000 to 100,000 years, sufficient gas accumulates on the white dwarf to kindle an enormous nuclear explosion. The binary pair—visible to Earth astronomers as one object—becomes thousands of times brighter for a few days. The pair of stars aren't changed much after the explosion fades, except that the common star has given up a percentage of its available fuel, shortening its life.

Collins and Beckmann both have discovered novas before. The 1987 find was Collins' third and Beckmann's second. In 1984, Collins found the last previous nova to be discovered visually.

Constellations are arbitrary regions of Earth's sky defined by astronomers. Vulpecula, the fox, is a constellation in the Milky Way seen from Earth's Northern Hemisphere.

Red giant. Like a jumbo smoking machine inhaling and exhaling, the red giant star Mira Ceti can puff itself up to extraordinary size and brightness. The latest remarkable brightening came in November 1989 when the star pumped itself up to magnitude 2.8.

That compares with the orange giant star Diphda, or Beta Ceti, which at magnitude 2 usually is the brightest star in the constellation Cetus. Another bright star in the same constellation with Mira and Diphda is Alpha Ceti at magnitude 2.7.

Mira means "the wonderful." Ancient astronomers knew Mira Ceti for its regular brightening and dimming—it would seem to appear and disappear from their night skies.

In 1596, it was found to be a variable star. In fact, it is the prototype for a class of numerous long-period variable stars known as Mira stars.

Mira Ceti is on a 331-day cycle in which it puffs up in diameter by 20 percent and shrinks. At maximum, it is more than 330 times the diameter of our Sun. The surface temperature of Mira Ceti varies from 2,960 degrees to 4,760 degrees Fahrenheit.

The visible brightness of Mira Ceti, as seen by astronomers on Earth, varies by about six magnitudes, averaging between magnitude 3 and 4. At minimum it is magnitude 8 to 10. In 1779 it reached magnitude 1.2 and in 1969 it reached magnitude 2.1.

Dust grains in its expanding gas shell make Mira Ceti give off infrared light. Mira Ceti has a companion star orbiting with it.

Cetus, or "the Whale," is a large constellation seen in Earth's night sky near Orion. There are 88 constellations, or arbitrary groupings of stars in Earth's sky, used by astronomers through the ages to identify particular locations overhead. The 88 constellations are not equal in size of sky area they cover.

Chanal. Chanal's Star is an unusual deep-space body in the Orion Nebula discovered varying in brightness in 1983 by French amateur astronomer Roger Chanal. Today the star is known as V1118 Orionis.

Old 20th century astronomy photos showed light was constant from Chanal's Star at magnitude 16. It brightened during 1983-84 to magnitude 13, then dropped by late 1984 to magnitude 15. It turned up brighter again in a Dallas, Texas, amateur astronomer's November 1989 photograph of Orion Nebula.

Living dead star. When stars like the Sun exhaust their fuel, they cool and shrink to become white-dwarf stars, sometimes smaller than planet Earth. Astronomers had thought, because white dwarfs were cooling, they couldn't evolve further. But now they're scratching their heads over new data from three observatories and an orbiting satellite.

The star 0950+139, 1,500 lightyears from Earth, is surrounded by a 50,000-year-old glowing cloud of gas about the size of our Solar System. It looks like, after diminishing to the white dwarf stage, the star somehow went back to the red-giant stage and shed more gas.

The discovery of new activity in a star thought to be a white dwarf stellar corpse surprised astronomers at the Space Telescope Science Institute, Baltimore, Maryland, who said a gas cloud has not been seen before surrounding such an old, cool, small star.

After a star ages and cools, strong gravitational force at the surface of the resulting white dwarf should hold the gas inside the star, preventing release of a cloud. Could hydrogen drop below the surface to mix with carbon rising from the interior? Could that mixture rekindle nuclear burning?

If the star 0950+139 ballooned back out to the red giant stage, the event must have been brief and unnoticed by astronomers. Astronomers hope the star they found in the constellation Leo may lend new insight into how stars are born, evolve and die.

Chi Cygni. The erratic variable red-supergiant star Chi Cygni has gone into an unusual fade.

The brightness of Chi Cygni usually ranges from magnitude 6.5 to magnitude 8.5. Recently, the star's brightness as seen from Earth dimmed to magnitude 9, the weakest it's been since astronomers first logged it in 1926.

Cygnus. Chi Cygni is seen from Earth in the imaginary sky pattern, or constellation, astronomers call Cygnus, or Swan.

Cygnus is a large constellation, conspicuous in the Milky Way for Northern Hemisphere observers. The five brightest stars in the Cygnus constellation are:

★Alpha Cygni, magnitude 1,
★Gamma Cygni, magnitude 2,
★Epsilon Cygni, magnitude 2,
★Beta Cygni, magnitude 3,
★Delta Cygni, magnitude 3.

Those five form a pattern well-known as the Northern Cross. Alpha Cygni is the extremely bright supergiant star known popularly as Deneb. Beta actually is a spectacular binary pair of two stars orbiting each other, known as Albireo.

Chi is an irregular variable star seen just beyond the Swan's imaginary northwest wing tip. It is referred to as a Mira type of variable star, after the first such variable. These huge stars pulsate and are not very stable. Apparently shockwaves reverberate through such a star's atmosphere, traveling outward, heating gas, producing bursts of extra energy. Such a star might have dust grains inside its atmosphere which belch out simple molecules and infrared light.

Magnitude. The magnitude we assign to a star depends on the amount of light it gives off and its distance from Earth. A star farther from Earth sending out a lot of light might appear brighter here than a closer star not giving off much light. A large dull star might seem weaker to us than a small bright star.

The smaller the number, the less magnitude. Thus a first-magnitude star such as Aldebaran at magnitude +0.9 seems brighter to us than Castor at magnitude +2.0.

As the numbers approach and pass zero, they become negative numbers on the magnitude scale. Thus the brightest star we see is Sirius (magnitude –1.4), only nine lightyears from Earth.

The brightest objects in our sky are the Sun at magnitude –27 and the Moon at magnitude –13.

On a clear dark night, we can see objects as faint as magnitude 5 or 6 by naked eye. With binoculars we see stars as weak as magnitude 8 or 9. A six-inch telescope will show stars down to magnitude 13.

Some planets can be seen with the naked eye. Mars, for example, varies in magnitude from –2.8 to +1.6 depending upon its distance from Earth. Jupiter ranges from –2.5 to – 1.4 magnitude.

Supernovas

A supernova is the death blast of a large star. It collapses inward, explodes, spews out gas and dust, and collapses again into a small heavy cinder—a neutron star.

For astronomers, supernovas blossom unexpectedly, last a few weeks and fade as a star dies. Astrophysicists say those brief moments are important because that's when heavier chemical elements are created. The supernova death of an old star gives birth to new stars and planets as the chemical elements created within the dying star are blasted into space.

The best known supernova in recent years was Supernova 1987a, the astronomical event of the 1980's and one of the top ten happenings in astronomy in the 20th century.

The supernova was designated 1987a as it was the first supernova to be found in 1987. Amateur astronomer Albert Jones in New Zealand and professional astronomer Ian Shelton of the University of Toronto, working in Chile, separately spotted the exploding star February 24, 1987, in the skies above Earth's Southern Hemisphere. The event was unfolding nearby, only 163,000 lightyears away from Earth in a galaxy known as The Large Magellanic Cloud, a close neighbor to our own Milky Way Galaxy. Light from Supernova 1987a took 163,000 years to travel the distance to Earth.

Supernova 1987a

Supernova 1987a was the astronomical event of the 1980's, one of the top ten happenings in astronomy in the 20th century. Here's what the excitement was all about:

What. A supernova is the death blast of a large star. It collapses inward, explodes, spews out gas and dust, and collapses again into a small heavy cinder—a neutron star. The supernova was designated 1987a as it was the first supernova to be found in 1987.

When. Amateur astronomer Albert Jones in New Zealand and professional astronomer Ian Shelton of University of Toronto, working in Chile, separately spotted the exploding star February 24, 1987, in the skies above Earth's Southern Hemisphere.

Where. The event continues to unfold 163,000 lightyears from Earth in a galaxy known as The Large Magellanic Cloud, a close neighbor to our own Milky Way Galaxy.

A galaxy is a collection of stars hanging together in space. A lightyear is roughly 5.9 trillion miles. Light from Supernova 1987a took 163,000 years to travel to Earth.

Before. Supernova 1987a was the first such explosion near Earth since 1604 and first since the invention of the telescope. Where astronomers now see an expanding fireball, there were a pair of red stars, a couple of almost-unseen blue stars and a prominent blue giant superstar named Sanduleak 69.202. The view of that area of the sky is obscured by bright light from the supernova, but the bet is that Sanduleak exploded.

During. The brighter an object, the smaller the magnitude number. Visible light from Sanduleak was only magnitude 12 before it exploded. A very large telescope was needed to see it. The supernova, on the other hand, could be seen by naked eye from Earth as it reached peak brightness of magnitude 2.7 in May 1987. After that it dimmed slowly. While visible light revealed the supernova, invisible ultraviolet light from it was seen by a satellite in Earth orbit and infrared light beaming out from the explosion was seen. Cosmonauts in the USSR's Mir space station, as well as an unmanned Japanese satellite, collected X-rays arriving from Supernova 1987a.

Mysteries. Pulsars are stars beaming out pulses of light and other energy. Supernova 1987a seemed to be beaming pulses for a time in January 1989, but then stopped. Later that year, astronomers at the European Southern Observatory found an unusual amount of infrared light from the supernova. If it was from dust heated by some powerful central engine, that would be further evidence of a spinning neutron star.

Hubble Space Telescope in 1991 photographed the cloud of debris blasting away from the supernova. It had mushroomed to a diameter of 0.1 lightyear. Astronomers got a look at a ghostly relic of Sanduleak, a glowing halo around the blast site. The distant ring couldn't have been produced in the supernova explosion because it was more than one lightyear across. Not enough time had passed for debris to fill a space as large as the ring. The ring must have existed before Sanduleak exploded. How did it get there?

Astronomers were puzzled again in 1991 after chemical analysis showed the supernova surrounded by a ring of bright gas, not a spherical shell. That might mean Sanduleak was spinning rapidly when it exploded, which could mean it was half of a close pair of stars. Two stars colliding or one sucking a large lump of matter from the other might explain Sanduleak's blue color and the ring surrounding the supernova.

After. The vast shell of gas and dust blowing away from the dying Sanduleak becomes ever thinner as it disperses. That veil is expected to become so thin, astronomers on Earth eventually will be able to see through it to the heart of the supernova. They expect to see a small, extremely dense, extraordinarily heavy cinder—a neutron star—left behind at the core of the exploding supernova. If it rotates, seeming to pulsate as it blinks flashes of light toward Earth, it will be a pulsar.

The meaning. Scientists report the heavy elements in our own planet, in our bodies, the stuff of life itself, came from dust created and scattered by some unknown ancient supernova explosion. The death of Sanduleak as Supernova 1987a is the birth of some life somewhere millions of years in the future.

Supernovas Make Galaxy Sparkle

Stars in the galaxy NGC 6240 are exploding at a rate 150 times the rate of our own Milky Way galaxy. As many as three supernovas appear each year in NGC 6240. That compares with two per 100 years in the Milky Way.

For some reason, NGC 6240 appears to be a super-luminous galaxy—1,000 times as luminous as the Milky Way. Through a powerful telescope, a strange litter of stars can be seen hiding amidst copious clouds of dust. Astronomers wonder if the object actually is two galaxies colliding. Whatever, NGC 6240 is resounding with shock waves and spraying immense volumes of infrared light across the Universe.

Supernova Within A Supernova

A star may have exploded 4,000 years ago inside the remnant of an ancient supernova. The older supernova, Puppis A, has strange intertwined filaments near the center of its expanding shell.

Astronomers speculate the odd filaments may be the visible remnants of a second supernova explosion. In effect, a newer supernova inside the shell of an older supernova. If so, that would be a very rare cosmic event.

The filaments look like they are made of different elements—such as sulfur and oxygen—which could have been ejected by an exploding star. In other words, the filaments look like a supernova remnant. And the speed at which they are moving is nearly the same as the speed of material ejected by any supernova.

Supernova remnants, of course, give off X-rays and radio waves. No X-rays and radio waves come from the filaments. That suggests to scientists that the filaments might be inside the shell of the older supernova. A high-pressure, low-density environment inside the shell could trap X-rays and radio waves, preventing them from escaping to be read from Earth.

What stars exploded? Could the double supernova have been the explosion of two similar stars orbiting each other.

Skeptics think the filaments came from the older supernova explosion. Filaments are an odd feature which should be explained, they say, but it's not likely two neighbor stars would explode at about the same time. No other such double supernova has been been found by astronomers on Earth, they point out.

The most recent suggestion of a supernova within a supernova was made by researchers at Middlebury College in Middlebury, Vermont, the Harvard-Smithsonian Center for Astrophysics and the Cerro Tololo Inter-American Observatory in Chile.

Naked Supernova

Three unusual exploding stars have been spotted in as many years. The uncommon "naked" supernovas seen in 1984, 1985 and 1986 let astronomers look into the hearts of dying stars.

Late in a life of billions of years, a star usually explodes in a quick bright flash of light. The outer ball of hydrogen gas surrounding the star's core blows out. It spreads in all directions from the star, preventing Earth observers from seeing what may be happening at the core.

Astronomers can study the core only sometime later when the gas shell has dissipated. First-hand viewing of crucial events in the death of the star are left to imagination.

But, fortuitously, the astronomers spotted the 1984-86 naked supernovas. The gas shell around each star had been stripped away before the final ball of fire. Astrophysicists think it may have happened in each case when two stars were orbiting each other. A companion star's gravity may have sucked hydrogen away from the supernova star before the final core collapse.

Without the mask of hydrogen gas, the natural light waves and radio waves from the dying core were able to travel unblocked to Earth. Astronomers happened to be looking at the right spot in the sky to receive an intimate view of the details of the aging star in the early moments of its cataclysmic explosion.

Seeing A Star Start To Die

Astrophysicists have seen a star start to die. A nondescript star 60 million lightyears from Earth in the galaxy known as M99 began its final moment of glory in May 1986. Buried in the Virgo cluster of galaxies, the star exploded into a supernova.

Supernovas blossom unexpectedly, last a few weeks and fade as a star dies. Those brief moments are important. That's when the heavier chemical elements are created. The supernova death of an old star gives birth to new stars and planets as the chemical elements created within the star are blasted into Space.

Scientists had come across remnants of supernovas before, but had been frustrated at not seeing one from beginning to end. University of California, Berkeley, physicist Frank Crawford, looking over pictures shot May 17, 1986, comparing them with photos made May 9, spotted the bright new light from the M99 supernova. He was on a search team using a 30-inch telescope near Berkeley to photograph 1,000 galaxies every night.

Supernova 1989b In M66

Another star has exploded into a supernova, this time much farther away in the spiral galaxy known as M66 in the constellation Leo. The last previous supernova to gain notoriety had been Supernova 1987a, only 163,000 lightyears from Earth. M66 and its nearby twin spiral galaxy M65, however, are 20 million lightyears from Earth.

Light arriving at Earth from Supernova 1989b is a faint 13th magnitude. Galaxy M66 itself is 9th magnitude. The brighter an object appears from Earth, the lower the magnitude number.

Robert Evans, New South Wales, Australia, first discovered Supernova 1989b with an optical telescope January 30, 1989. Then Federico Manzini, Novara, Italy, found it later the same night.

The International Ultraviolet Explorer (IUE) satellite in Earth orbit turned to look at the exploding star, finding it to be a so-called Type I supernova—the death blast of an old white-dwarf star. Astronomers on the ground analyzed the spectrum of light coming from the explosion, confirming the supernova is Type I.

By February 1, 1989, the supernova had brightened to 12th magnitude as its fireworks continued. Backyard astronomers on Earth needed at least a six-inch telescope to see Supernova 1989b.

Constellations. Leo is one of 88 constellations or arbitrary groupings of stars in Earth's sky, used by astronomers to mark off certain areas of the sky.

Two stars seen in one constellation may not be near each other since one could be relatively close to Earth—say, within a few lightyears' distance—while the other could be thousands of lightyears away. For example, the star Alpha Centauri is less than five lightyears from Earth while Beta Centauri is more than 450 lightyears away. They seem close in the sky, but are not. M65 and M66, however, apparently are relatively close together.

The 88 constellations are not equal in size of sky area they cover. Hydra covers the largest area of the sky. Crux Australis, the smallest. Leo, or the Lion, where M65 and M66 are seen, is a large constellation in the sky over Earth's Northern Hemisphere near the constellation Ursa Major. The brightest star in Leo is the 1st-magnitude Regulus. At 1st magnitude, Regulus is much brighter than the 9th-magnitude M66 and M65 galaxies.

Deep space catalogues. The M in M66 stands for Messier, a catalog of celestial objects prepared by French astronomer Charles Messier in 1784. He listed 103 objects in his catalog in 1784 and six more in 1786. Today's astronomers refer to the stars in the catalog by their Messier number, such as M66.

NGC stands for New General Catalogue of Nebulae and Clusters of Stars, compiled by Danish astronomer J.L.E. Dreyer at Armagh Observatory, Ireland, and published in 1888. It listed 7,840 objects. The brightest also had appeared in the Messier Catalog. Objects in the catalog are known by their NGC numbers, such as NGC 3628, seen in Leo near M66. Today, one NGC volume covers 13,000 objects.

M66 is in both catalogs so it also is known as NGC 3627. M65 is NGC 3623. Another example of a famous object being in both catalogs is the Orion nebulae, which is M42 and NGC 1976.

Supernova 1989m In M58

USSR astronomers looking at the constellation Virgo—The Virgin—on June 28, 1989, found Supernova 1989m in the spiral galaxy M58. Supernova 1989m, at magnitude 12.1, was bright enough to be seen by amateur astronomers around the globe with medium-sized backyard telescopes.

It was the second supernova in galaxy M58 in two years. The galaxy itself is a magnitude 10 barred spiral collection of stars. A year earlier, Supernova 1988a had appeared to brighten to magnitude 13.5 in January 1988.

Supernova 1989n In NGC 3646

In 1989, an engineering student at California Institute of Technology, Pasadena, found a new supernova in a spiral galaxy 137 million lightyears from Earth. Known as Supernova 1989n, it is in the large spiral galaxy known as NGC 3646 in the eastern part of the constellation known to astronomers as Leo.

Celina Mikolajczak, 19, of San Diego, was looking for comets and asteroids, in a Caltech summer program for students doing faculty-supervised research. He discovered the exploding star while examining photos made with an 18-inch Schmidt telescope at Palomar Observatory, northeast of San Diego. Supernova 1989n, the fourteenth discovered in 1989, cannot be seen by naked eye.

Mikolajczak told reporters he first thought he was seeing a dust speck but, after careful examination of the film, thought, "Wow, I think I've found something."

Usually, 15 to 20 supernovas are discovered each year. However, a total of 27 were found in 1988.

Supernova 1989n is in the constellation known to astronomers as Leo. Regulus is the best-known star in the constellation Leo—The Lion.

Supernova In NGC 4527

Amateur astronomers in Maine and California discovered a supernova in the 10th-magnitude galaxy NGC 4527 in the constellation Virgo April 14, 1991.

The explosion brightened the dying star to 13th magnitude. Backyard astronomers needed at least an eight-inch telescope to see the supernova.

Supernova In NGC 5493

Australian amateur astronomer Robert Evans discovered a supernova in June 1990. The 13.5-magnitude star exploded in galaxy NGC 5493 in the constellation Virgo.

Christmas Supernova

An exploding star, 300 million lightyears away in the distant galaxy NGC 1667, was called a Christmas supernova by scientists who found it December 22, 1986.

Colin Okada, an undergraduate astronomy student at the University of California, first spotted the supernova in the constellation Eridanus during a nightly sky sweep.

The university astronomers examine 400 galaxies by computer-controlled telescope each night, searching for previously-unseen space objects. In the first six years of looking, four supernovas were found.

A supernova is the final violent end of a large star. With hydrogen fuel running out, its center collapses. Helium and other remaining atoms under that pressure fuse together, forming heavier elements. Nuclear and gravity energy are released. The star blows itself apart. Backyard astronomers needed only a small home telescope to see the 1986 Christmas supernova. It was directly overhead at midnight, but too faint for the naked eye. The constellation Eridanus is south of the better-known Taurus constellation. The supernova was at the center of the galaxy catalogued as NGC 1667.

Black Widow Star

A black widow spider uses its companion and then destroys it. Now it seems at least one pulsar star is doing the same thing. The faraway star, devouring its companion like a colossal black widow spider, is giving astronomers clues in a long-standing celestial mystery.

It appears a high-speed pulsar known as PSR 1957+20, many lightyears from Earth, is showering a travelling-companion star with such intense heat radiation, the companion is evaporating.

The mystery: why do some of the ultra-dense stars known as pulsars, which can spin up to hundreds of revolutions per second, have companion stars while others do not?

PSR 1957-20. The pulsar, labeled PSR 1957-20 on star maps, was detected thrashing an invisible companion star in Spring 1988. Astronomers say they may have found the answer to the mystery in the black widow binary star system where a pulsar is using its companion and then destroying it.

If theories are correct, the black widow star would be a missing link, a bridge between fast-spinning stars that have companion stars and those without companions, according to an astronomer at the Carnegie Institution in Washington, Cornell University, and NASA's Goddard Space Flight Center, Greenbelt, Maryland. The black widow star would be significant in understanding life and death of stars.

The tiny energetic pulsar, orbited by its larger companion, form a binary star system. The pulsar, while drawing fuel from the companion, also is transmitting a powerful blast of energy, blowing away its companion.

What astronomers call millisecond pulsars spin 600 times or more per second, releasing extraordinary amounts of radiation onto objects nearby in space.

Tiny powerhouse. PSR 1957-20 is a high-speed pulsar seen from Earth. While PSR 1957-20 is only about fifteen miles in diameter, its mass is at least 50 percent greater than the Sun. That gives it a tremendously powerful gravity pull.

Its unseen companion star may be as large as the diameter of our Sun. Scientists deduced the presence of the companion as it regularly eclipsed the pulsar. The eclipses gave a pattern to the naturally-generated radio signals coming from the pulsar, making the blips seem to repeat every 9.2 hours.

Calculations indicate the companion is made up of various gases, but has less mass than any object ever known before outside our own Solar System. That might mean the pulsar has reduced its companion to something less than a star—a so-called brown dwarf.

The companion star probably once was larger than the Sun. It has been evaporated down to its core by heat radiating from the nearby pulsar. The companion eventually will be destroyed. Here's some background:

★The Sun is a common yellow star.

★A red giant star is a normal star, similar to our Sun, but much later in its life span. The Sun is not yet a red giant.

★Binary stars are pairs of stars. There seem to be endless numbers of stars strewn through the Universe. Most appear to be in pairs, triplets, quads, the companions circling each other. Two stars circling each other form a binary system. Such double stars are very common in our galaxy and across the Universe. A single star, alone, like our Sun, is more unusual than systems with two or more stars orbiting each other. The

PSR 1957-20 binary pair is composed of a spinning neutron-star pulsar orbited by a larger star which may have been reduced to brown-dwarf status.

★Supernovas are star explosions. When one star, at least half again as large as our Sun, reaches a stage late in its life and its fuel is used up, it collapses, then bursts into a spectacular explosion called a supernova.

★A neutron star appears in the aftermath of a supernova explosion. At the end of a supernova explosion, the remaining core collapses under its own weight, squeezing down to a very small dense cinder—a neutron star. Scientists calculate the neutron star in PSR 1957-20 may be only fifteen miles in diameter.

Like an ice skater who spins faster by pulling in legs and arms, the compressing of the star's core during collapse makes it spin faster. Atoms in the core are squeezed so tightly they break down and cease to exist as atoms.

All that remains are neutrons packed so densely a thimbleful would weigh a billion tons. Thus, a neutron star is born.

A neutron star scintillates with radiation, spritzing powerful X-rays, infrared light, ultraviolet light, visible light, radio waves and gamma rays outward.

★A pulsar is formed when the intense gravity of a neutron star captures a second star and holds it in orbit around the neutron star. As the second star circles the neutron star, it momentarily blocks our view of the naturally-occurring radio signals from the neutron star. From our observation post on Earth, the neutron star's radio waves appear to blink on and off regularly. This pulsing on and off led to the name pulsar.

In PSR 1957-20, the common star is extremely close to the pulsar—probably at a distance of no more than two million miles or so, nowhere near as far away as Mars is from Earth.

The companion regularly eclipses the spinning neutron star, circling it once every 9.2 hours, making it a pulsar. Each time it goes around, it blocks out our view of the pulsar's radio beam for about 50 minutes.

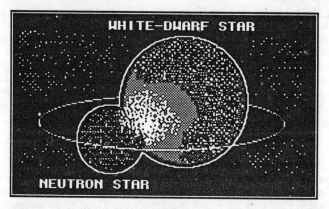

Rapid spinners. Neutron stars spin at different speeds, depending on how much fuel was available originally and how much energy is left. Most known pulsars spin about once a second and are slowing down as they lose energy.

Millisecond pulsars, which blink very rapidly, send out many bursts of energy each second. The PSR 1957-20 pulsar is a millisecond pulsar, spinning about 600 times a second. It is one of six known to spin extremely rapidly.

Pulsars usually emit no light, but broadcast twin rotating beams of radio waves, like lighthouses in the sky.

Melting the brown dwarf. Astronomers probably are seeing the terminal phase of an ancient one-sided relationship between the two celestial bodies.

The neutron star is sucking hydrogen, helium and other matter from its mate to fuel its own spin. But, as the matter streams from the companion down onto the neutron star, a magnificent explosion showers the companion with a blaze of atomic particles. The brown dwarf is being bombarded with high-energy atomic particles at least 5,000 times more intense than our Sun's radiation.

Astronomers imagine there is a fog of gas around the companion, since the radio pulse arrival times slow just before and after each eclipse.

The brown-dwarf companion is being destroyed two ways. Its matter is being sucked off by the neutron star as well as blown away by the blast from the neutron star. It would seem the companion soon will disappear, which might explain the mystery of how other pulsars turn up without companions.

Millisecond Pulsar

Large stars end up as neutron stars and some of those as pulsars. A different binary pulsar known as PSR 1855+09 was reported in 1986. Described by astronomers as a millisecond pulsar, it turns on and off 200 times per second. Its companion, a white dwarf star, is a large lightweight sphere of hot helium.

Pulsar Clock

Another pulsar, this one 12,000 lightyears from Earth, could replace Earth-bound atomic clocks as our most reliable timepiece. Known as 1937+21 for its location on astronomers' sky maps, the pulsar spins on its axis at what seems to be a very steady 642 twirls a second. That's unusual. These dense cinders left when stars burn out usually spin down over time.

Pulsar 1937+21 was discovered in 1982. It has not been seen to vary in spin rate since then. That makes it the most stable thing known to scientists.

With no detectable fluctuation in its spin rate, the pulsar is more regular than atomic clocks built on Earth. The official U.S. atomic clocks are kept in a National Bureau of Standards laboratory at Boulder, CO. They measure the natural vibrations in cesium atoms to mark time. However, they do fluctuate. For instance, two atomic clocks vibrating at nine billion cycles per second have been found to deviate by a few billionths of a second after only a few years. Pulsar 1937+21, on the other hand, apparently hasn't deviated at all since 1982.

Whirling X-Ray Stars

The pulsar PSR 1957-20 bleeps every 9.2 hours. Compare that with only 11.3 minutes for one star to whip around another in the fastest double-star system found so far. Compare it also with the month it takes Earth's Moon to go around once. And the year it takes Earth to circle the Sun.

These speed demons have been dubbed x1820-30 after their position in the sky, at least 20,000 lightyears from Earth. One of the two stars composing x1820-30 is a dying white dwarf. It's gripped firmly by a neutron star.

A lightyear is about 5.9 trillion miles, the distance light travels in a year. Astronomers used a telescope in Europe's orbiting Exosat unmanned observatory to discover the binary pair. Exosat was launched on a Delta rocket in May 1983.

Over the centuries, Earth observers have divided the sky into arbitrary constellations, giving each a folk name such as Sagittarius. The two distant stars forming the binary system x1820-30 are at the heart of a very bright source of X-rays far away, in the constellation Sagittarius. Astronomers today report an immense gas cloud obscures the center of the Milky Way galaxy, some 30,000 light years from Earth, in the sky constellation Sagittarius.

The double-star system was discovered in 1986. The outer star, a white dwarf, is only 80,000 miles from its central neutron-star companion. That's only about a third the distance from Earth to the Moon.

The neutron star in x1820-30 is a tiny stellar corpse, probably five miles across but not more than thirteen miles across. By comparison, planet Earth is 7,926 miles across. The Sun is a jumbo 865 thousand miles in diameter.

However, the neutron star is massive. It weighs a third more than our Sun. The extraordinary mass gives it even greater gravity pull than the Sun.

The white dwarf orbiting the neutron star is a small star only about the size of Earth. Its mass is less than one-tenth that of the Sun.

Passerby. How did the two stars in x1820-30 come together? Previously, astronomers would have guessed the neutron star might have captured a passing white dwarf and tugged it in. The white dwarf would have spiraled down quickly until it was destroyed by the neutron star. However, this pair doesn't seem to have come about that way.

It looks like this white dwarf was captured millions of years ago, while it still was a larger common star, older than the Sun, of a type known as red giant. A red giant is a normal star, similar to our Sun, but larger and much later in its life span. The Sun is not yet a red giant.

Astronomers suspect the red giant star was floating by when gravity from the neutron star caused tides in its surface gases and, finally, a bulge. The red giant lost momentum and moved closer, to a short distance from the neutron star. It was pulled into a highly-elliptical orbit around the neutron star.

With each pass around the neutron star, tidal bulges distorted the star. The red giant finally settled into a circular orbit around the neutron star. At that point, it might have taken ten hours for the red giant star to circle the neutron star.

Naked white dwarf. The powerful pull of the neutron star sucked matter from the surface of the red giant star leaving only a white-dwarf core. Over time, the entire outer atmosphere of the star blew away, leaving only a naked white dwarf star. After such an extreme loss of mass, the star began to spiral closer to the neutron star. Each orbit around the central star became faster.

Eleven minutes is by far the shortest orbital time period of any known binary pair.

Our Milky Way galaxy has about 50 X-ray sources similar to x1820-30. They are powered by gas falling from a common star onto a neutron star.

The gas travels quickly, at about one-third the speed of light. When it strikes the neutron star, the resulting explosion releases an enormous amount of energy from the neutron star—about 10,000 times the output of our Sun.

As the gas from the white dwarf star in x1820-30 races onto that neutron star, it becomes very hot. X-rays are produced. To the Exosat satellite, which sees X-rays, it looks very bright.

Blip every 11.3 minutes. Exosat recorded X-rays from x1820-30 sent out as a pulse or blip every 11.3 minutes. That tells astronomers the white-dwarf companion star blocks Exosat's view of the neutron star every 11.3 minutes.

Such a short orbital period indicates the former red giant now must be a tiny star. As a white dwarf, it probably is only about the size of planet Earth, and not more than three times the diameter of Earth.

That's somewhere between 8,000 and 24,000 miles in diameter, small for a star. It is a white dwarf made of pure helium.

The entire binary pair—neutron star and white dwarf star—would fit easily between Earth and the Moon. Yet, the total mass of the system is greater than our Sun. A lot of weight in a very small space!

Ancient stars. This remarkable couple lives at the center of a globular cluster called NGC 6624. A globular cluster is a dense conglomeration of very old stars which have accumulated near each other over the eons. Stars in our part of the Milky Way galaxy are far apart so there's not much chance of two stars colliding around here anytime soon. But, in a globular cluster, stars are a million times closer together.

The 9.2-hour pulsar PSR 1957-20 may be similar to the 11.3-minute pulsar but at an earlier point in its life.

Noisy Pulsar

What are you going to call a noisy neutron star when a household word like pulsar already is taken? How about noisar?

At least that's what the Los Alamos National Laboratory in New Mexico is calling Sco X-1. That deep space object is one of the most familiar X-ray sources in our Milky Way galaxy. In fact, it was the first X-ray source ever recognized. Now we see it has one of the most unusually noisy signals coming from space.

Energy radiated from pulsars and neutron stars usually has a natural rhythm to it. Noisars, on the other hand, also include a lot of seemingly random incoherent energy in what they send out.

Astrophysicists want to make sense of it. Until they do, the new class is called noisars to distinguish them from plain-jane rhythmic neutron stars.

Among the 100 billion stars in our galaxy, a mere few hundred X-ray sources have been spotted. And, of those, only a tiny handful are noisars.

Like other big signals from outer space, the pulsar Sco X-1 is a binary system. The neutron star's extraordinary gravity sucks helium and hydrogen from its companion. The blast when the gas crashes into the neutron star at high speed is such a large explosion that tremendous energy flies out into space, including X-rays.

Twinkless Stars

An Illinois engineer is de-twinkling stars so an astronomer can shoot a sharper picture. An astronomer locks his telescope onto a guide star to follow movement of an object across the sky. Now, an electrical engineer and an astronomer at the University of Illinois, Urbana, have built what amounts to an artificial guide star from a laser beam. It should help Earth-bound telescopes see as well as telescopes out in space.

The problem with Earth-bound optical telescopes: stars twinkle. That is, images of stars are blurred by turbulence in our atmosphere which makes the stars seem to twinkle. By reducing this distortion, the Illinois researchers are trying to get the clarity that astronomers formerly would have gotten only by putting a telescope above the atmosphere in space.

Previously, the best pictures of the planets Saturn and Jupiter came from the Voyager spacecraft. The Illinois researchers want that kind of clarity from down on the ground using their laser guide star. The new laser telescope may cost $3 million but it should perform many of the functions of a space telescope costing $1 billion.

The solution was a "rubber" mirror. Adjustable, thin, flexible, it forms exactly the opposite distortion pattern from what it sees, to correct the image in the telescope.

To do that, there must be a bright reference point—like a guide star—in the sky near the object being studied. But there aren't enough natural guide stars. So, the Illinois engineer created one with a laser.

The laser beams a dot of light 60 miles above the ground into a layer of sodium in Earth's atmosphere, causing the sodium layer to glow. In a test at Mauna Kea

Observatory in Hawaii, the Illinois researchers fired their laser and photographed their star, measuring its brightness. A computer connected to the telescope uses the artificial guide star to estimate atmospheric distortion. It adjusts the thin, flexible mirror 50 to 100 times a second as the atmosphere changes. That unscrambles the image of real stars, planets and galaxies letting astronomers take sharp pictures.

The next step is to find just the right laser for permanent operation. It must behave exactly as a real star. The guide star telescope may be ready for general use in five years.

Photographing The Night Sky

The sky is most mysterious, most intriguing at night. You can look up from our position at the bottom of an ocean of air anytime. But daytime, revealing little of space beyond the Sun and Moon, doesn't hold the mystique for most. Nighttime, when the heavens twinkle, captures our fancy.

We look with the naked eye. We see more with binoculars. We view the most through telescopes. At some point we want to record the excitement of what we are seeing. We want to make photos.

Astrophotography. Men have been collecting images of light from space for 100 years. Capturing starlight for our picture archives is a popular part of the hobby of astronomy. Here's how to go about it:

★Any camera may catch some light coming from space but your best bet will be one which makes negatives or slides of 35mm or larger size. The 35mm single-lens reflex (SLR) is most popular. A larger-format camera will permit bigger enlargements.

★You'll need to be able to mount your camera on a tripod to steady it. A shaking or vibrating camera will make ugly pictures. The camera should have a "bulb" or B setting for time exposures and a way to manually trigger the shutter. Some exposures require the shutter to be open for hours so beware of a battery-powered shutter. Batteries do run down.

★The lens on your camera depends on what you want to shoot. Besides the "normal" 50mm lens which came with your camera you may want to consider a wide-angle 28mm or 35mm lens. Telephoto lenses of 128mm and 200mm also will be useful.

★The film you use depends on what you wish to photograph and the format of your camera's negative. Brands to consider include Kodak Kodachrome, Ektachrome, Tri-X, Pan-X, Technical Pan and VR; Fuji Fujichrome; 3M Scotch; and Agfa XRS and RS.

More light is received from nearby planets in our Solar System so you can use the luxury of slower films which have a finer grain.

Stars, galaxies, clusters and other objects outside our Solar System are so far away that precious little light arrives at Earth. You either need extreme lengths of exposure time or a fast film. You will opt for the faster film but at the expense of pictures with more and bigger grain. Some hobbyists hypersensitize film to make it more sensitive.

Use color slide film rather than color negative film for color shots.

Kodak Tri-X 400 film is a good negative film for black-and-white photography. The grain is not fine but it sees weak light well. A good fine-grain black-and-white negative film is Kodak 2415 Technical Pan. It has an ISO rating (ASA rating) of 100. If there's enough light, Kodak 2415 gives fine detail and it can be hypersensitized.

The length of time the shutter is open depends on the type of film you use. Faster films, with higher ISO numbers, will record more light in a short time. One drawback comes when you discover that faster films also fog more quickly from background sky light. Slower films can have longer exposures but they, too, eventually will fog.

Your exposure time will depend upon how bright the sky is at your location. Shoot a series of pictures to find the right exposure. If you are using a 400-speed film and a 50mm lens try exposures of 15, 20, 25 and 30 minutes for your first experimental shots. If you use 1000-speed film with a 50mm lens open the shutter for 5, 6, 7, 8, 9 and 10 minutes. Films slower than 400 speed may be exposed for longer than 30 minutes.

Stars can look like nothing more than white specks on a frame of film. The processor may not be able to align the first frame for cutting and mounting. Either ask to receive the developed roll of film uncut and unmounted, or shoot a normal well-lit scene for the first frame on a roll to be used as a placement guide.

A telescope often is needed to record details of space objects. Prime focus, eyepiece projection and afocal photography are common ways to use a camera and telescope in astrophotography.

In prime focus the telescope is used to guide the camera and to make the photograph. The camera is attached to the telescope and you look through an eyepiece to make sure the telescope is tracking the object you want to record. Exposures of deep-sky objects often run from 30 minutes to three hours. During that time you must watch the eyepiece to make sure the object doesn't drift out of view.

Eyepiece projection is used to magnify objects in Sun, Moon and planet photos.

In afocal photography the camera sits on its tripod and is pointed into the telescope's eyepiece. Since you are adding the magnification of the camera's lens to the magnification of the eyepiece, overall magnification is greater than in simple eyepiece projection.

What to shoot. You can make pictures of the Moon, stars, meteors, comets, planets, galaxies, nebulae, deep sky objects, constellations. If it can be seen, it can be photographed. Sometimes it can be photographed even if you can't see it otherwise:

★Meteors aren't hard to record. Mount your camera on a tripod. Set the shutter for time exposures on "bulb" or B. Leave the shutter open for a period of time ranging from four minutes up to many hours depending upon the ISO (formerly called ASA) rating of the film you use. If you are using an ISO 400 film, the shutter-open period should range from four minutes to one hour.

Such long exposures will make light sources in the night sky look like streaks. The streaks or meteor trails will be straight near the North Star.

Meteor showers, which happen several times a year, are best for seeing because there are more rocks per hour.

★Stars, seen in photographs as trails or streaks, are recorded in the same manner as meteors. To go beyond pictures of trails across the sky requires a clock drive to match your camera to the rotation of the Earth. Objects in the night sky move slowly from east to west as our globe turns. A telescope with a clock drive moves in a direction opposite to this star motion. The telescope's clock drive is used by attaching your camera to the telescope.

★Constellations are recorded by pointing the camera toward the appropriate part of the sky and making exposures of several seconds each. Check your star map. If the constellation you want is above your horizon, point your tripod-mounted camera toward it. If you use a "normal" lens of about 50mm or a "wide-angle" lens of from 20 to 50mm, use fast film. Adjust the f-stop or aperture control either one or two stops below the lens' fastest setting. For example, if your lens opens up to f-1.4, set it at 1.8 or slower for the exposure. Using the B shutter setting and a cable shutter trigger, make a series of six exposures starting at 10 seconds and working up to one minute. You will have exposures of 10, 20, 30, 40, 50 and 60 seconds. Exposures longer than 60 seconds with a normal lens will cause the stars to look like streaks. With a wide-angle lens, limit exposures to under two minutes. A normal 50mm lens should be used to record Orion, the constellation with the beautiful Barnards Loop, Orion and Horsehead nebulae.

★Our Milky Way galaxy and its gigantic dust clouds are best photographed through a 28mm wide-angle lens.

★The planets of our Solar System are bright in the sky but look like star points of light because they are small and so far away. The position of a planet against the background of stars changes rapidly and can be photographed with a brief exposure

50mm lens. If you want to make detailed pictures of planets you will need a telescope to magnify the small object.

★Comets are similar to planets when it comes to photography except that their long orbital paths only bring them in close enough for us to see them for short periods from time to time.

★Asteroids are tiny chunks of rock in orbit around the Sun like planets. Since they are only a few miles across they are quite faint and only the brightest ever become visible to the eye. Like planets, they look like star points of light and they appear to move against the background of stars. A telescope must be used to make photographs of asteroids.

★Variable stars can be seen to regularly change in brightness over time. Some can be seen by eye but they should be photographed using a telescope. Try to shoot one picture when the star's brightness is near maximum and another near minimum. To start, take pictures of the same area ten days to three weeks apart.

★Binary stars are most common in the sky. These double stars actually are two or more orbiting each other. The stars in one binary are often different colors but you need a telescope to separate the single star point of light into two or more stars.

★Star clusters, beyond our Solar System and even beyond our galaxy, are deep sky objects requiring a telescope to make photographs.

★Galaxies, beyond our own Milky Way, must be photographed through a telescope to obtain detail.

★Pictures of the Moon are best during its first or third quarters. Contrast is best then so details look better. A full moon can be the worst time to shoot. Use a telescope to record craters.

★The Sun is far and away the brightest object in our sky and, therefore, the most dangerous for photographers. Never look directly at the Sun with the naked eye or through an unfiltered camera or telescope. Filters for sunlight include white-light filters which cut the amount of light by 99.9 percent and show the Sun as it looks to the naked eye. And hydrogen alpha filters which reveal only hydrogen alpha light from the Sun. White-light filters show the grainy surface of the Sun and sunspots. Hydrogen alpha filters allow you to see solar prominences and flares. A telescope is needed to shoot details of the Sun's face.

Which Camera Lens To Use

What: deep-sky objects you may want to photograph. When: the best season to make the photograph. Where: photographer's position in the Northern Hemisphere or Southern Hemisphere of Earth. Lens: 28mm, 50mm, 135mm and 200mm, for 35mm cameras.

What	When	Where	Lens
Big Dipper	Spring	North	28, 50
Andromeda galaxy	Fall	North	50, 135, 200
Sagittarius	Summer	North	28, 50
Orion	Winter	North	50
Star clouds	Summer	South	28, 50
Pleiades	Winter	North	135, 200
Scorpius	Summer	North	50
Coma star cluster	Spring	North	135
North American nebula	Summer	North	50, 135
Hyades	Winter	North	135
Summer Triangle	Summer	North	28
California nebula	Winter	North	135
Large Magellanic Cloud	Summer	South	135
Small Magellanic Cloud	Summer	South	135, 200

Rosette nebula	Winter	North	135
Eta Carina Nebula	Summer	South	135
Hyades & Pleiades	Winter	North	50

Which Film To Use

The films below are good for astrophotography. Not all films listed will fit all cameras. There are other good films available, not listed here. Film manufacturers continually offer new model numbers. Look for improved versions of the films listed here. The first and third columns show film brand names. The second and fourth columns show sky objects to photograph. Abbreviation dp spc is deep space.

Kodak	Subject	Fuji	Subject
Kodachrome 25	planets	Fujichrome 100	planets
Kodachrome 64	planets	Fujichrome 200	planets/dp spc
Ektachrome 64	planets	Fujichrome 400	planets/dp spc
Ektachrome 100	planets	Fujichrome 400D	planets/dp spc
Ektachrome 200	planets/dp spc	Fujichrome 1600D	dp spc
Ektachrome 400	planets/dp spc	**3M**	
Tri-X 400	planets/dp spc	Scotch 640	dp spc
Pan-X 100	planets	Scotch 1000	dp spc
Technical Pan 100	planets	**Agfa**	
103a-E	dp spc	XRS 100	planets
103a-O	dp spc	XRS 200	planets/dp spc
103a-F	dp spc	XRS 400	dp spc
VR 100	planets	XRS 1000	dp spc
VR 200	planets/dp spc	100 RS	planets
VR 400	dp spc	400 RS	dp spc
VR 1000	dp spc	1000 RS	dp spc

Universal Time

Although we usually deal in local time wherever we are, there is a single time usable anywhere—Coordinated Universal Time (UTC), formerly Greenwich Mean Time (GMT).

The most-frequently-used standard times in North America are Eastern (EST), Central (CST), Mountain (MST) and Pacific (PST). From April through October, standard times change to Eastern Daylight (EDT), Central Daylight (CDT), Mountain Daylight (MDT) and Pacific Daylight (PDT).

★To calculate UTC from EST add five hours; four to EDT.
★To calculate UTC from CST add six hours; five to CDT.
★To calculate UTC from MST add seven hours; six to MDT.
★To calculate UTC from PST add eight hours; seven to PDT.
For example, 1:12 p.m. PDT is the same as 4:12 pm EDT and 2012 UTC.

FOR UTC	FROM EST add 5	FROM CST add 6	FROM MST add 7	FROM PST add 8
FOR UTC	FROM EDT add 4	FROM CDT add 5	FROM MDT add 6	FROM PDT add 7

The four-digit 24-hour clock, sometimes called military, government, transportation or official time, calls time from the minute after Midnight to Noon as 0001 to 1200 and from the minute after Noon to Midnight as 1201 to 2400. The first two digits are hours, the last two are minutes. The hours after Noon are added to 12. For example, 3 p.m. is 1500 and 8:30 p.m. is 2030.

WWV Navstar. Shortwave radio time-signal stations WWV and WWVH, operated by the U.S. National Institute of Standards and Technology, transmit brief reports on the status of the Navstar global positioning system (GPS) satellites hourly.

WWV transmits continuously around the clock from Fort Collins, Colorado, on frequencies of 2.5, 5, 10, 15, and 20 MHz. WWVH transmits continuously from Kauai, Hawaii, on 2.5, 5, 10, and 15 MHz. The two stations broadcast time signals every minute, 24 hours a day, and are heard across all of North America and much of the world.

A brief description of the Navstar GPS system is broadcast by WWV at 14 minutes past each hour. Then the GPS update is broadcast by WWV at 15 minutes past each hour.

The Navstar introduction is broadcast by WWVH at 43 minutes past each hour and the update report is broadcast at 44 minutes past each hour.

These broadcasts, which can be heard with a simple shortwave receiver, are prepared by the U.S. Coast Guard's Omega Navigation Systems Center (ONSCEN) in Alexandria, Virginia, and updated once a day.

Both WWV and WWVH also transmit solar activity and geomagnetic field reports as well as Omega navigation system status reports and weather. Weather and navigation status reports are read between time signals which are heard each minute. The hourly content of WWV and WWVH is shown below. The numbers represent minutes past the hour.

WWV
:08 North Atlantic Ocean weather
:09 North Atlantic Ocean weather continued
:10 East Pacific Ocean weather
:14 GPS introduction
:15 GPS status report
:16 Omega status report
:18 Solar activity and geomagnetic field report

WWVH
:43 GPS introduction
:44 GPS status report
:45 Solar activity and geomagnetic field report
:47 Omega status report
:48 Pacific Ocean weather
:49 Pacific Ocean weather continued
:50 Pacific Ocean weather continued
:51 Pacific Ocean weather continued

Planetariums & Observatories
Museums, Planetariums and Observatories In The U.S. And Canada

ALABAMA

Birmingham	Meyer Planetarium, Birmingham So. College
Florence	University of North Ala. Planetarium and Observatory
Huntsville	Marshall Space Flight Center

ALASKA

Fairbanks	Geophysical Inst, University of Alaska–Fairbanks

ARIZONA

Amado	Multiple Mirror Telescope Observatory
Amado	Whipple Observatory
Flagstaff	Barringer Meteor Crater
Flagstaff	Campus Observatory, Northern Arizona University
Flagstaff	Lowell Observatory
Tucson	Flandrau Planetarium, University of Arizona
Tucson	Kitt Peak National Observatory
Tucson	Steward Observatory, University of Arizona

ARKANSAS

Little Rock	Planetarium, University of Arkansas-Little Rock

CALIFORNIA

Apple Valley	Apple Valley Science and Technology Center
Berkeley	Lawrence Hall of Science, University of California
Big Bear	Stony Ridge Observatory
Edwards	Dryden Flight Research Facility
Fresno	Discovery Center

Los Angeles	Astronomy Dept, University of California–Los Angeles
Los Angeles	Griffith Observatory
Moffett Field	NASA–Ames Research Center
Mt. Hamilton	Lick Observatory
Palomar Mt.	Palomar Observatory
Pasadena	NASA–Jet Propulsion Laboratory
Pleasant Hill	Planetarium, Diablo Valley College
Redding	Schreder Planetarium
Rohnert Park	Astronomy Dept, Sonoma State University
Sacramento	Scaramento Science Center
San Diego	Fleet Space Theater and Science Center
San Diego	Mt. Laguna Observatory, San Diego State University
San Francisco	Exploratorium
San Francisco	Morrison Planetarium, Calif Academy Sciences
Santa Barbara	Observatory, Santa Barbara Museum Natural Histry
Santa Barbara	Planetarium, Santa Barbara Museum Natural Histry
Santa Monica	Planetarium, Santa Monica City College
Santa Rosa	Planetarium, Santa Rosa Junior College

COLORADO

Boulder	Fiske Planetarium, University of Colorado
Colorado Springs	Black Forest Observatory
Colorado Springs	Planetarium, U.S. Air Force Academy
Denver	Chamberlin Observatory
Denver	Gates Planetarium, Denver Museum Natural History

CONNECTICUT

Danbury	Planetarium, Western Connecticut State College
New Britain	Copernican Observatory, Central Conn St University
Stamford	Stamford Museum and Nature Center
West Hartford	Grengas Planetarium, Science Museum of Conn

DELAWARE

Wilmington	Mount Cuba Astronomical Observatory

DISTRICT OF COLUMBIA

Washington	National Air and Space Museum, Smithsonian
Washington	Rock Creek Nature Center
Washington	U.S. Naval Observatory

FLORIDA

Bradenton	Bishop Planetarium
Cocoa	Astronaut Memorial Hall
Daytona Beach	Planetarium, Museum of Arts and Sciences
Fort Myers	Jr. Museum and Planetarium of Lee County
Jacksonville	Planetarium, Jacksonville Museum Sci and Hist
Kennedy Spc Center	NASA–Kennedy Space Center, Spaceport USA
Melbourne	Space Coast Science Center
Miami	Miami Space Transit Planetarium
Orlando	John Young Planetarium, Orlando Science Center
St. Petersburg	Pinellas County Science Center
Tallahassee	Planetarium, Florida State University
W. Palm Beach	Observatory, South Florida Science Museum

GEORGIA

Albany	Wetherbee Planetarium
Atlanta	Planetarium, Fernbank Science Center
Columbus	Patterson Planetarium
Macon	Mark Smith Planetarium and Observatory
Rock Spring	Planetarium, Walker County Science Center
Savannah	Planetarium, Savannah Science Museum
Valdosta	Planetarium and Observatory, Valdosta St College
Young Harris	Rollins Planetarium, Young Harris College

HAWAII

Hilo	Mauna Kea Observatory

Honolulu	Bernice P. Bishop Museum and Planetarium

IDAHO

Rexburg	Ricks College Planetarium

ILLINOIS

Champaign	Staerkel Planetarium, Parkland College
Chicago	Adler Planetarium
Chicago	Crown Space Center
Chicago	Museum of Science and Industry
Dekalb	Observatory, Northern Illinois University
Evanston	Observatory, Northwestern University
Normal	Planetarium, Illinois State University
Peoria	Planetarium, Lakeview Museum of Arts and Sci
River Grove	Cernan Earth and Space Center, Triton College
Rockford	Time Museum

INDIANA

Evansville	Koch Science Center and Planetarium
Fort Wayne	Schouweiler Planetarium, St. Francis College
Greencastle	McKim Observatory, DePauw University
Marshall	Nature Center Planetarium, Turkey Run State Park
Muncie	Observatory and Planetarium, Ball State University
Terre Haute	Observatory, Indiana State University

IOWA

Cherokee	Sanford Museum and Planetarium
Des Moines	Observatory, Drake University
Des Moines	Sargent Planetarium, Science Center of Iowa
Dubuque	Heitkamp Memorial Planetarium, Loras College
Waterloo	Planetarium, Grout Museum of Hist and Science

KANSAS

Great Bend	Planetarium, Barton County Community College
Greensburg	Pallasite Meteorite, The Big Well
Hutchinson	Kansas Cosmosphere and Space Center
Lawrence	Observatory, University of Kansas
Wichita	Lake Afton Public Observatory
Wichita	Planetarium, Omnisphere and Science Center

KENTUCKY

Berea	Weatherford Planetarium, Berea College
Georgetown	Planetarium, Georgetown College
Louisville	Museum of History and Science
Louisville	Rauch Memorial Planetarium, University of Louisville
Richmond	Hummel Planetarium, Eastern Kentucky University

LOUISIANA

Baton Rouge	Planetarium, Louisiana Arts and Science Center
Lafayette	Lafayette Natural History Museum and Planetarium
Luling	St. Charles Parish Library and Planetarium
New Orleans	NASA–Michoud Assembly Facility

MAINE

Easton	Planetarium, Francis M. Malcolm Science Center
Orono	Planetarium and Observatory, University of Maine
Portland	Southworth Planetarium, University of Southrn Maine

MARYLAND

Baltimore	NASA–Space Telescope Science Institute
Baltimore	Planetarium, Community College of Baltimore
Baltimore	Planetarium, Maryland Science Center
College Park	Observatory, University of Maryland
Frederick	Williams Observatory, Hood College
Frostburg	Planetarium, Frostburg State University
Greenbelt	NASA–Goddard Space Flight Center
Hagerstown	Washington Co, Planetarium and Space Sci Center
Westminster	Western Maryland College

MASSACHUSETTS

Boston	Observatory, Astronomy Dept., Boston University
Boston	Hayden Planetarium, Boston Museum of Science
Cambridge	Smithsonian Astrophysical Observatory
Nantucket	Loines Observatory
Springfield	Planetarium, Springfield Science Museum
Wenham	Woodside Planetarium and Observatory
Williamstown	Milham Planetarium, Williams College
Worcester	Planetarium, New England Science Center

MICHIGAN

Alpena	Jesse Besser Museum and Planetarium
Ann Arbor	Ruthven Planetarium Theater, University of Michigan
Battle Creek	Planetarium, Kingman Museum of Natural History
Bloomfield Hills	Planetarium, Cranbrook Institute of Science
Detroit	Children's Museum, Detroit Public Schools
Detroit	Planetarium, Detroit Science Center
East Lansing	Abrams Planetarium, Michigan State University
Flint	Robert T. Longway Planetarium
Grand Rapids	Chaffee Planetarium, Grand Rapids Publ Museum
Kalamazoo	Baldauf Planetarium, Kalamazoo Public Museum
Lansing	Impression Five Science Museum
Marquette	Shiras Planetarium, Marquette Senior High Schl
Mt. Pleasant	Brooks Observatory, Central Michigan University

MINNESOTA

Duluth	Alworth Planetarium, University of Minnesota–Duluth
Hibbing	Paulucci Space Theater, Arrowhead Com College
Minneapolis	Minneapolis Planetarium
Moorhead	Planetarium, Moorhead State University
New Ulm	Planetarium, Dr. Martin Luther College
St. Cloud	Observatory and Planetarium, St. Cloud St University

MISSISSIPPI

Jackson	Ronald E. McNair Space Theater
Jackson	Russell Davis Planetarium

MISSOURI

Kansas City	Planetarium, Kansas City Museum
St. Louis	Planetarium, St. Louis Science Center

MONTANA

Missoula	University of Montana

NEBRASKA

Hastings	McDonald Planetarium, Hastings Museum
Lincoln	Planetarium, University of Nebraska State Museum

NEVADA

Las Vegas	Planetarium, Clark County Community College
Reno	Fleischmann Planetarium, University of Nevada–Reno

NEW HAMPSHIRE

Hanover	Shattuck Observatory, Dartmouth College

NEW JERSEY

Mountainside	Planetarium, Trailside Nature and Science Center
Newark	Planetarium, Newark Museum
Randolph	Planetarium, Morris County College
Toms River	Novins Planetarium, Ocean County College

NEW MEXICO

Alamogordo	Planetarium, The Space Center
Albuquerque	Observatory, University of New Mexico
Socorro	National Radio Astronomy Observatory

NEW YORK

Alfred	Observatory, Alfred University
Binghamton	Link Planetarium, Kopernik Observatory
Buffalo	Ferguson Planetarium, State University NY–Buffalo

Centerport	Vanderbilt Planetarium
Fredonia	Planetarium, State University NY–Fredonia
Herkimer	Herkimer County BOCES Planetarium
Ithaca	Observatory, Cornell University
Ithaca	Spacecraft Planetary Imaging Facility, Cornell
Middle Island	Longwood District Planetarium
New York	Hayden Planetarium, American Mus'm Natural Hist
Poplar Ridge	Southern Cayuga Atmospherium-Planetarium
Potsdam	Planetarium, State University NY–Potsdam
Rochester	Planetarium, Rochester Museum and Science Center
Schenectady	Schenectady Museum and Planetarium
Staten Island	Planetarium, Wagner College
Syracuse	Discovery Center of Science and Technology
Troy	Planetarium, The Junior Museum
Yonkers	Andrus Planetarium

NORTH CAROLINA

Chapel Hill	Morehead Planetarium, University of NC–Chapel Hill
Charlotte	Planetarium, Discovery Place and Nature Museum
Greensboro	3-College Observatory, University of NC–Greensboro
Greensboro	Zane Planetarium, Greensboro Natural Sci Center
Salisbury	Woodson Planetarium, Horizons Unl Sup Ed Center
Winston-Salem	Planetarium, Nature Science Center

NORTH DAKOTA

Grand Forks	Planetarium, University of No Dak, Center Aerospc Study
Minot	Observatory, Minot State University

OHIO

Bay Village	Schuele Planetarium
Canton	Hoover-Price Planetarium, McKinley Museum
Cincinnati	Observatory, University of Cincinnati
Cincinnati	Planetarium, Cincinnati Museum Natural History
Cleveland	Mueller Planetarium, Cleveland Mus'm Natural Hist
Cleveland	NASA–Lewis Research Center
Columbus	Battelle Planetarium, Center of Science and Industry
Columbus	Ohio State University
Dayton	Planetarium, Dayton Museum of Natural History
Granville	Swasey Observatory, Denison University
Marietta	Observatory, Marietta College
Sylvania	Copernicus Planetarium, Lourdes College
Westerville	N. American Astro Observtry, Otterbein College
Youngstown	Planetarium, Youngstown State University

OKLAHOMA

Fort Sill	Fort Sill Planetarium
Oklahoma City	Kirkpatrick Planetarium

OREGON

Bend	Pine Mt. Observatory, University of Oregon
Portland	Kendall Planetarium, Oregon Mus'm Sci and Tech
Salem	Planetarium, Chemetka Community College

PENNSYLVANIA

Edinboro	Observatory, Edinboro University
Erie	Planetarium, Erie Historical Museum
Harrisburg	Planetarium, Pennsylvania State Museum
Lancaster	Planetarium, Franklin and Marshall College
Lewisburg	Observatory, Bucknell University
Lock Haven	Ulmer Planetarium, Lock Haven University
Philadelphia	Fels Planetarium, Franklin Institute
Pittsburgh	Allegheny Observatory
Pittsburgh	Planetarium, Buhl Science Center
Reading	Planetarium, Reading School District
Scranton	Planetarium, Everhart Museum

University Park	Pennsylvania State University
Villanova	Observatory and Planetarium, Villanova University
West Chester	Planetarium, West Chester State University
Williamsport	Detwiler Planetarium, Lycoming College

RHODE ISLAND

Providence	Cormack Planetarium
Providence	Ladd Observatory

SOUTH CAROLINA

Columbia	Gibbes Planetarium
Florence	Planetarium, Francis Marion College
Greenville	Daniel Observatory, Roper Mt. Science Center
Orangeburg	Planetarium, South Carolina State College
Rock Hill	Settlemyre Planetarium, York County Museum

SOUTH DAKOTA

Vermillion	University of South Dakota

TENNESSEE

Brentwood	Dyer Observatory, Vanderbilt University
Kingsport	Bays Mt. Planetarium and Nature Center
Memphis	Memphis Pink Palace Museum and Planetarium
Murfreesboro	Observatory, Middle Tennessee State University
Nashville	Stevenson Center Observatory, Vanderbilt University
Nashville	Sudekum Planetarium, Cumberland Sci Museum

TEXAS

Amarillo	Planetarium, Don Harrington Discovery Center
Corpus Christi	Planetarium, Richard King High School
Dallas	Planetarium, Richland College
Dallas	Planetarium, St. Marks School of Texas
Dallas	Science Place Planetarium
El Paso	El Paso Planetarium
Fort Davis	McDonald Observatory
Fort Worth	Noble Planetarium, Ft. Worth Mus'm Sci and Hist
Houston	NASA–Johnson Space Center
Houston	Planetarium, Houston Museum Natural Science
Lubbock	Moody Planetarium, Texas Tech University Museum
Odessa	Odessa Meteor Crater
San Angelo	Planetarium, Angelo State University
San Antonio	Planetarium, San Antonio College
Tyler	Hudnall Planetarium, Tyler Junior College

UTAH

Provo	Summerhays Planetarium, Brigham Young University
Salt Lake City	Hansen Planetarium

VERMONT

St. Johnsbury	Fairbanks Museum and Planetarium

VIRGINIA

Ashland	Keeble Observatory, Randolph Macon College
Charlottesville	McCormick Observatory, University of Virginia
Fort Union	Wicker Planetarium, Ft Union Military Academy
Hampton	NASA–Langley Research Center
Harrisonburg	Wells Planetarium, James Madison University
Newport News	Planetarium of the Virginia Living Museum
Richmond	Universityerse Space Theater, Sci Museum of Virginia
Roanoke	Planetarium, Sci Museum of Western Virginia
Virginia Beach	Planetarium, Va Beach City Public Schools
Wallops Island	NASA–Wallops Flight Facility
Williamsburg	College of William and Mary

WASHINGTON

Bellevue	Planetarium, Bellevue Community College
Goldendale	Goldendale Observatory State Park
Seattle	Planetarium, Pacific Science Center

Spokane	Planetarium, Eastern Washington Science Center

WEST VIRGINIA

Charleston	Planetarium, Sunrise Children's Museum
Davis	Blackwater Falls State park
Green Bank	National Radio Astronomy Observatory
Wheeling	Planetarium, Benedum Natural Science Center

WISCONSIN

Fall Creek	Observatory, Beaver Creek Reserve
La Crosse	Planetarium, University of Wisconsin–La Crosse
Madison	Observatory, University of Wisconsin–Madison
Milwaukee	Milwaukee Public Museum
Milwaukee	Planetarium, University of Wisconsin–Milwaukee
Stevens Point	Planetarium, University of Wisconsin–Stevens Point
Williams Bay	Yerkes Observatory

WYOMING

Casper	Casper Planetarium

CANADA

ALBERTA

Calgary	Centennial Planetarium, Alberta Science Center
Edmonton	Planetarium, Edmonton Space Sciences Centre

BRITISH COLUMBIA

Penticon	Dominion Radio Astrophysical Observatory
Vancouver	MacMillan Planetarium, Southam Observatory
Vancouver	Observatory, University of British Columbia
Victoria	Climenhaga Observatory, University of Victoria
Victoria	Dominion Radio Astrophysical Observatory

MANITOBA

Winnipeg	Lockhart Planetarium, University of Manitoba
Winnipeg	Manitoba Planetarium, Man and Nature Museum

NOVA SCOTIA

Halifax	Burke-Gaffney Planetarium, St. Mary's University

ONTARIO

Ottawa	Observatory, National Museum of Sci and Tech
Richmond Hill	David Dunlap Observatory, University of Toronto
Sudbury	Doran Planetarium, Laurentian University
Toronto	McLaughlin Planetarium, Royal Ontario Museum
Toronto	Planetarium, Ontario Science Center

QUEBEC

Montreal	Dow Planetarium

SASKATCHEWAN

Saskatoon	Observatory, University of Saskatchewan

Fifth Force In The Universe?

Most people know about two physical forces of the Universe, gravity and electromagnetism. Some also have heard of the strong force and the weak force. Gravity attracts objects to one another. Electromagnetic force is radio waves, light waves, and other radiation. The strong force binds protons and neutrons in the nucleus of an atom. The weak force makes some atoms break down in radioactive decay. Scientists now say a fifth force could counteract gravity—a feather would hit the ground before a brick.

U.S. Astronomy Clubs

Selected amateur astronomy clubs in the U.S. Others exist in particular local areas.

ALABAMA
Auburn Astronomical Society
1963 Canary Dr.
Auburn, AL 36830

Astronomical Society of U.N.A.
U.N.A. Box 5580
Florence, AL 35632

Birmingham Astronomical Soc
1425 22nd St.
Birmingham, AL 35205

Mobile Astronomical Society
P.O. Box 190042
Mobile, AL 36619-0042

Muscle Shoals Astronomical Soc
302 NE Commons
Tuscumbia, AL 35674

Von Braun Astronomical Society
P.O. Box 1142
Huntsville, AL 35807

ALASKA
Anchorage Astronomical Society
3539 Dunkirk Dr.
Anchorage, AK 99502

ARIZONA
Flagstaff Astronomical Society
25 N. San Francisco
Flagstaff, AZ 86001

Leisure World Amateur
Astronomers
1792 Leisure World
Mesa, AZ 85206

Phoenix Astronomical Society
6945 E. Gary Rd.
Scottsdale, AZ 85254

Prescott Astronomy Club
420 Canyon Springs Road
Prescott, AZ 86301

Sabino Canyon Observatory
5100 N. Sabino Foothills Dr.
Tucson, AZ 85715

Saguaro Astronomy Club
5 S. Pueblo St.
Gilbert, AZ 85234

Tucson Amateur Astronomy Assn
7222 E. Brooks Drive
Tucson, AZ 85730

ARKANSAS
Ark-La-Tex Skywatchers
Rt. 2, Box 9
Ashdown, AR 71822

Arkansas-Oklahoma
Astronomical Society
109 S. Cedar
Sallisaw, OK 74955

Astronomical Society of
Northwest AR
P.O. Box 316
Lincoln, AR 72744

MARS
711 W. 3rd Street
Little Rock, AR 72201

Mid-South Astronomical
Research Society
P.O. Box 5142
Little Rock, AR 72205

CALIFORNIA
Astronomical Assn. of N.
California
731 Camino Ricardo
Moraga, CA 94556

Amateur Astro. Projects Soc
3485 Leota 88
Simi Valley, CA 93063

Andromeda Astronomical Society
P.O. Box 2615
Palm Springs, CA 92263

Antelope Valley Astronomy Club
44-513 Sancroft
Lancaster, CA 93535

Astronomical Society of the
Desert
College of the Desert
45-300 Monterey Avenue
Palm Desert, CA 92260

Astronomers of Humboldt
c/o Bob O'Connell
College of the Redwoods
Eureka, CA 95501

Bear Valley Astronomical Society
P.O. Box 6807
Big Bear Lake, CA 92315

Celestial Observers
9534 Gierson Avenue

Chatsworth, CA 91311

Central Coast Astronomical
Society
P.O. Box 1415
San Luis Obispo, CA 93406

Central Valley Astronomers
The Discovery Center
1944 N. Winery Avenue
Fresno, CA 93703

China Lake Astronomical Society
P.O. Box 1783
Ridgecrest, CA 93555

Eastbay Astronomical Society
Chabot Science Center
4917 Mountain Boulevard
Oakland, CA 94619

Excelsior Telescope Club
265 Roswell
Long Beach, CA 90803

Fremont Peak Observatory
Association
P.O. Box 1110
San Juan Bautista, CA 95045

Idyll-Gazers Astronomy Club
P.O. Box K
Idyllwild, CA 92349

Kern Astronomical Society
2901 Renegade Avenue
Bakersfield, CA 93300

Los Angeles Astronomical
Society
Griffith Observatory
2800 East Observatory Road
Los Angeles, CA 90027

Tampalais Astronomical Society
27 Morning Sun
Mill Valley, CA 94941

MIRA Astronomy Club
2868 Forest Hill Blvd.
Pacific Grove, CA 93950

Modesto Junior College
Astronomy Club
College Avenue
Modesto, CA 93550

Modesto Society of Ast.
Observers

P.O. Box 1911
Modesto, CA 93253

Mount Diablo Astronomical
Society
1725 Sunnyvale Ave.
Walnut Creek, CA 94596

Mt. Wilson Observatory
Association
813 Santa Barbara St.
Pasadena, CA 91101

Omega One
21558 American River Dr.
Sonora, CA 95370

Orange County Astronomers
2215 Martha Ave.
Orange, CA 92667

Peninsula Astronomical Society
P.O. Box 4542
Mountain View, CA 94040

Polaris Astronomical Association
22018 Ybarra Rd.
Woodland Hills, CA 91364

Polaris Astronomical Society
228 E. Pentagon St.
Altadena, CA 91001

Pomona Valley Amateur
Astronomers
P.O. Box 309
La Verne, CA 91750

Riverside Astronomical Society
8642 Wells Avenue
Riverside, CA 92503

S.T.A.R.S., S-95
City College of San Francisco
50 Phelan Avenue
San Francisco, CA 94112

Sacramento Valley Astronomical
Society
Sacramento Science Center
3615 Auburn Boulevard
Sacramento, CA 95821

San Bernadino Valley Amateur
Astronomers
1345 Garner Avenue
San Bernadino, CA 92411

San Diego Astronomy

Association
P.O. Box 23215
San Diego, CA 92123

San Francisco Amateur
Astronomers
Morrison Planetarium
California Academy of Sciences
San Francisco, CA 94118

San Francisco Sidewalk
Astronomers
1600 Baker Street
San Francisco, CA 94115

San Jose Astronomical Society
3509 Calico Avenue
San Jose, CA 95124

San Mateo Astronomical Society
1721 Earl Avenue
San Bruno, CA 94066

Santa Barbara Astronomy Club
P.O. Box 3702
Santa Barbara, CA 93105

Stony Ridge Observatory
3019 Welsh Way
Glendale, CA 91206

Sonoma County Amateur
Astronomers
P.O. Box 183
Santa Rosa, CA 95402

Tri-Valley Stargazers
202 Kittery Place
San Ramon, CA 94583

Tulare Astronomical Association
P.O. Box 515
Tulare, CA 93275-0515

University of CA Davis
Astronomy Club
Physics Department
Davis, CA 95616

Valley of the Moon Astronomers
411 W. Napa Street
Sonoma, CA 95476

Ventura County Astronomical
Society
P.O. Box 982
Simi Valley, CA 93062

Western Observatorium

13215 E. Penn Street, Suite 411
Whittier, CA 90602

COLORADO
Denver Astronomical Society
P.O. Box 10814
Denver, CO 80210

Longmont Astronomical Society
P.O. Box 8086
Longmont, CO 80501

Southern Colorado Astronomical
Society
2801 8th Avenue
Pueblo, CO 81003

CONNECTICUT
Astronomical Society of Greater
Hartford
The Science Museum of
Connecticut
Gengras Planetarium
950 Trout Brook Dr.
Hartford, CT 06119

Astronomical Society of New
Haven
P.O. Box 5831 Yale Station
New Haven, CT 06520

Fairfield County Astronomical
Society
Stamford Museum Observatory
39 Scofieldtown Road
Stamford, CT 06903

Mattatuck Astronomical Society
Cathi Pelletier, Math/Science
Division
Mattatuck Community College
750 Chase Parkway
Waterbury, CT 06708

Thames Amateur Astronomical
Society
118 Allen St.
Groton, CT 06340

Waterbury Astronomical Society
437 Washington Avenue
Waterbury, CT 06708

Westport Astronomical Society
P.O. Box 5118
Westport, CT 06880

DELAWARE
Delaware Astronomical Society

Mt. Cuba Astro. Observatory
P.O. Box 3915
Hillside Mill Rd.
Greenville, DE 19807

DISTRICT of COLUMBIA
National Capital Astronomers
5120 Newport Ave.
Bethesda, MD 20816

FLORIDA
Ancient City Astronomy Club
P.O. Box 546
St. Augustine, FL 32084

Astro Soc of Palm Beaches
South Florida Science Museum
4801 Dreher Trail N.
West Palm Beach, FL 33405

Astronomy Club of Bay County
425 E. 19th Street, #106
Panama City, FL 32405

Brevard Astronomical Society
P.O. Box 1084
Cocoa, FL 32922

Callahan Astronomical Society
Rt. 3, Box 1062
Callahan, FL 32011

Central Florida Astronomical
Society
810 E. Rollins St.
Orlando, FL 32803

Emerald Coast Astronomy Club
26 Maple Avenue
Shalimar, FL 32579

Escambia Amateur Astronomers
Association
6235 Omie Circle
Pensacola, FL 32504

Florida Astronomical Society
1214 NW 36th Terrace
Gainesville, FL 32605

GulfStream Astronomical Society
631 Falcon Avenue
Miami Springs, FL 33166

Local Group-Deep Sky Observers
2311 23rd Avenue W.
Bradenton, FL 33505-4530

M.A.R.S. Astronomy Club of

Tampa
7407 Del Bonita Ct., Apt. #69
Tampa, FL 33617

North East Florida Astro Soc
P.O. Box 5002
Jacksonville, FL 32247-5002

St. Petersburg Astronomy Club
St. Petersburg Jr. College
Planetarium
P.O. Box 13489
St. Petersburg, FL 33733

Southern Cross Astronomical
Society
6501 S.W. 65th Street
South Miami, FL 33143

Southwest Florida Astro Soc
P.O. Box 6583
Fort Myers, FL 33911

Tallahassee Astronomical Society
325 Conradi Street
Tallahassee, FL 32304

Tampa Amateur Astronomical Soc
1915 Vandervort Road
Lutz, FL 33549

Tampa Area Astronomy Society
5405 98th Avenue
Temple Terrace, FL 33617

GEORGIA
Athens Astronomical Society
160 Plantation Drive
Athens, GA 30605

Atlanta Astronomy Club
5198 Avanti Ct.
Stone Mountain, GA 30088

Augusta Astronomy Club
P.O. Box 96
Evans, GA 30809

Middle Georgia Astronomical Soc
Museum of Arts & Sciences
4182 Forsyth Road
Macon, GA 31210

HAWAII
Hawaiian Astronomical Society
P.O. Box 17671
Honolulu, HI 96817

Maui Astronomy Club

325 Olokani St.
Makawao, Maui, HI 96768

Mauna Kea Astronomical Society
RR 1, Box 525
Capt. Cook, HI 96704

IDAHO
Idaho Falls Astronomical Society
1710 Claremont Lane
Idaho Falls, ID 83401

ILLINOIS
Amateur Astro Assn of Aurora
168 South Fork Street
Aurora, IL 60505

Argonne Astronomy Club
6 N. 106 White Oak Lane
St. Charles, IL 60174

Astro. Society @ Univ of Illinois
349 Astronomy Bldg.
1011 W. Springfield
Urbana, IL 61801

Chicago Astronomical Society
P.O. Box 48504
Chicago, IL 60648

Central Illinois Astronomical Soc
6 Forest Park East
Jacksonville, IL 62650

Fox Valley Astronomical Society
P.O. Box 508
Batavia, IL 60510

Fox Valley Skywatchers
518 Ryan Cr.
W. Dundee, IL 60118

Ill. Benedictine College Ast. Soc
c/o Fr. Robert Buday, IBC
5700 College Road
Lisle, IL 60532

Joliet Astronomical Society
P.O. Box 3893
Joliet, IL 60435

Lake County Astronomical Soc
603 Dawes
Libertyville, IL 60048

Mt. Olive Astronomical Society
705 E. Main Street
Mt. Olive, IL 62019

Naperville Astronomical
Association
525 S. Chase Ave.
Lombard, IL 60148

Northwest Suburban Astronomers
4960 Chambers Dr.
Barrington, IL 60010

Peoria Astronomical Society
P.O. Box 419
Peoria, IL 61651

Popular Astronomy Club
John Deere Planetarium
Augustana College
Rock Island, IL 61201

Quincy Astronomical Society
#22 Summer Creek
Quincy, IL 62301

Rockford Amateur Astronomers
6804 Alvina Road
Rockford, IL 61103

Sandwich Astronomical Society
168 South Fourth
Aurora, IL 60548

Skokie Valley Astronomical
Society
910 Glenwood Lane
Glenview, IL 60025

Southwest Astronomical Society
17825 67th Avenue
Tinley Park, IL 60477

Space Sciences Explorer Post
#998
910 Glenwood Lane
Glenview, IL 60025

Twin City Amateur Astronomers
P.O. Box 755
Normal, IL 61761

INDIANA
Calumet Astronomical Society
1048 Greenbriar Court
Crown Point, IN 46307

Evansville Astronomical Society
P.O. Box 3474
Evansville, IN 47733

Ft. Wayne Astronomical Society
P.O. Box 6004

Ft. Wayne, IN 46896

Indiana Astronomical Society
3 Wilson Drive
Mooresville, IN 46158

Michiana Astronomical Society
P.O. Box 262
South Bend, IN 46624

Wabash Valley Astronomical Soc
282 Littleton, #223
West Lafayette, IN 47906

Warsaw Astronomical Society
RR 8, Box 236
Warsaw, IN 46580

IOWA
Ames Area Amateur Astronomers
1208 Wilson Ave.
Ames, IA 50010

Cedar Amateur Astronomers
Ambroz Art Center
2000 Mt. Vernon Rd.
Cedar Rapids, IA 52321

Cedar Valley Astronomical Soc
Department of Earth Science
University of Northern Iowa
Cedar Falls, IA 50613

Des Moines Astronomical Soc
2307 49th Street
Des Moines, IA 50310

Quad Cities Astronomical Society
P.O. Box 3706
Davenport, IA 52808

Southeastern Iowa Astronomy
Club
610 Walnut
Burlington, IA 52601

Summit Observatory
73 Summit Ave.
Rt. 1, Box 383
Swisher, IA 52338-9759

KANSAS
Astronomical Assn of Lawrence
1082 Malott Hall
University of Kansas
Lawrence, KS 66045-2151

Kansas Astronomical Observers
Wichita Omnisphere

220 South Main
Wichita, KS 67202

Northeast Kansas Amateur
Astronomers
P.O. Box 951
Topeka, KS 66601

Wichita Astronomical Society
2100 University Avenue
Wichita, KS 67213

KENTUCKY
Blue Grass Astronomical Society
1016 Della Dr.
Lexington, KY 40504

Louisville Astronomical Society
Museum of History & Science
727 West Main St.
Louisville, KY 40402

Midwestern Astronomers
1643 Elder Court
Fort Wright, KY 41011

LOUISIANA
Astronomical Society of Acadiana
805 S. Guegnon
Abbeville, LA 70510

Astro. Society of Ama Astro
P.O. Box 66
Glenmore, LA 71433

Baton Rouge Astronomical Soc
12635 Parnell Avenue
Baton Rouge, LA 70815

Pontchartrain Astronomical Soc
1441 Avenue "A"
Marrero, LA 70131

Red River Astronomical Society
2820 Pershing Boulevard
Shreveport, LA 71109

MAINE
Astro. Soc. of N. New England
West Hill Observatory
80 Goodwin Road
Eliot, ME 03903

45th Parallel Ama Astronomers
RFD 1, Box 1220
Dexter, ME 04930

MARYLAND
Anne Arundel Ass. of Amateur

Astronomers
Anne Arundel Community
College
101 College Parkway
Arnold, MD 21012

Baltimore Astronomical Society
Davis Planetarium
Maryland Academy of Science
601 Light St.
Baltimore, MD 21230

Cumberland Astronomy Club
221 Baltimore Street
Cumberland, MD 21502

Goddard Astronomy Club (NASA)
Code 511.1, Goddard Spc Flt Ctr
Greenbelt, MD 20771

Harford County Astronomical
Society
P.O. Box 906
Bel Air, MD 21014

Tri-State Astronomers
Washington County Planetarium
 & Space Science Center
823 Commonwealth Ave.
Hagerstown, MD 21740

Westminster Astronomical
Society
3481 Salem Bottom Road
Westminster, MD 21157

MASSACHUSETTS
Amateur Telescope Makers of
Boston
60 Victoria Rd.
Sudbury, MA 01776

Amherst Amateur Astro Assn.
P.O. Box 335
North Amherst, MA 01059

Cape Cod Astronomical Society
P.O. Box 98
West Hyannisport, MA 02672

Connecticut River Valley Astro
P.O. Box 54
Monson, MA 01057

Goddard Astronomical Society
43 Regency Drive
Holliston, MA 01746

Springfield Stars Club

107 Lower Beverly Hills
West Springfield, MA 01089

MICHIGAN
Astronomy Club of Livonia
P.O. Box 121
Livonia, MI 48152

Astronomical Society of Hillsdale
3260 N. Dunes Rd.
Hillsdale, MI 49242

Capital Area Astronomy Club
Abrams Planetarium
Michigan State University
East Lansing, MI 48824

Detroit Astronomical Society
14298 Lauder
Detroit, MI 48227

Grand Rapids Ama. Astro. Assn
James C. Veen Observatory
3308 Kissing Rock Road SE
Lowell, MI 49331

Greater Muskegon Astronomy
Club
P.O. Box 363
Muskegon, MI 49443-0363

Kalamazoo Astronomical Society
5831 Downing
Portage, MI 49081

Marquette Astronomical Society
Shiras Planetarium
1201 W. Fair Avenue
Marquette, MI 49855

Michigan State University
Astronomy Club
Abrams Planetarium
East Lansing, MI 48824

Muskegon Astronomical Society
P.O. Box 363
Muskegon, MI 49443

Plutonian Society
8191 Woodland Shore 12
Brighton, MI 48116

Toledo Astronomical Association
8534 Covert Road
Petersburg, MI 49270

Warren Astronomical Society
P.O. Box 474

East Detroit, MI 48021

MINNESOTA
3-M Club Astronomical Society
14601 55th Street S.
Afton, MN 55001

Arrowhead Astronomical Society
Alworth Planetarium
University of Minnesota - Duluth
Duluth, MN 55812

Minnesota Astronomical Society
Science Museum of Minnesota
30 East 10th Street
St. Paul, MN 55101

Minnesota Valley Ama Astro
Rt. 4, Box 15A
New Ulm, MN 56073

Moorhead-Fargo Astronomy Club
P.O. Box 28
Physics Dept.
Concordia College
Moorhead, MN 56560

MISSISSIPPI
Jackson Astronomical
Association
6207 Winthrop Circle
Jackson, MS 39206

Jackson County Astronomical
Society
3803 Arlington Street
Pascagoula, MS 39567

S. Mississippi Astronomical
Association
P.O. Box 3450
Gulfport, MS 39505

MISSOURI
Astronomical Society of Kansas
City
P.O. Box 400
Blue Springs, MO 64015

Central Missouri Ama Astro
1903 S. Bus 54
Fulton, MO 65251

Eastern Missouri Dark Sky
Observers
10231 Midland Blvd.
Overland, MO 63114-1538

McDonnell-Douglas Amateur

Astronomers
3491 Heather Trail
Florissant, MO 63031

North Central Missouri
Astronomy Club
P.O. Box 772
Chillicothe, MO 64601

Rural Astronomers of Missouri
Route 1, Box 501
Winfield, MO 63389

St. Louis Astronomical Society
14055 Baywood Villages Dr.
Chesterfield, MO 63017

MONTANA
Astronomical Institute of the
Rockies
6351 Canyon Ferry Rd.
Helena, MT 59601

NEBRASKA
Omaha Astronomical Society
3390 S. 130th
Omaha, NE 68144

Prairie Astronomy Club
P.O. Box 80553
Lincoln, NE 68501

NEVADA
Astronomical Society of Nevada
825 Wilkinson Ave.
Reno, NV 89502

Carson Star Searchers
P.O. Box 1436
Carson City, NV 89701

Elko Nevada Astronomical
Society
550 S. 12th, #22
Elko, NV 89801

Las Vegas Astronomical Society
Clark Comm Coll Planetarium
3200 E. Cheyenne Avenue
Las Vegas, NV 89030

NEW HAMPSHIRE
Keene Amateur Astronomers
12 Gardner Street
Keene, NH 03431

New Hampshire Astro Society
22 Center St.
Penacook, NH 03303

NEW JERSEY

Amateur Astronomers Assn. of
Princeton
P.O. Box 2017
Princeton, NJ 08540

Amateur Astronomers Assn. of
Trenton
NJ State Museum Planetarium
CN530
205 W. State Street
Trenton, NJ 08625

Amateur Astronomers
Sperry Observatory
Union County College
1033 Springfield Ave.
Cranford, NJ 07016

Amateur Astronomy Club of
South Jersey
208 Briar Lane
Carney's Point, NJ 08069

Astronomy Club of Glassboro
Glassboro State College
Glassboro, NJ 08028

Astronomical Society of Toms
River Area
Novins Planetarium
Ocean County College
Toms River, NJ 08754-2001

Morris Museum Astronomical Soc
Columbia Tpk & Normandy Rd
Morristown, NJ 07960

New Jersey Astronomical
Association
Voorhees State Park
P.O. Box 214
High Bridge, NJ 08829

North Jersey Astronomical Group
P.O. Box 4021, Allwood Station
Clifton, NJ 07012

Sheep Hill Astro Association
P.O. Box 111
Boonton, NJ 07005

Small Scope Observer's
Association
4 Kingfisher Place
Audubon Park, NJ 08106

S.T.A.R. Astronomy Society
Monmouth County Park System

Newman Springs Road
Lincroft, NJ 07738

NEW MEXICO

Alamogordo Amateur
Astronomers
1210 Filipino
Alamogordo, NM 88310

Albuquerque Astronomers
14328 Mocho NE
Albuquerque, NM 87123

Astronomical Society of Las
Cruces
P.O. Box 921
Las Cruces, NM 88004

Clovis Astronomy Club
216 Sandzen
Clovis, NM 88101

NEW YORK

Albany Area Amateur
Astronomers
1529 Valencia Road
Schenectady, NY 12309

Amateur Astronomers
Association of NYC
1010 Park Avenue
New York, NY 10028

Amateur Observers Society of
New York
707 South 9th
Lindenhurst, NY 11757

Astronomical Society of Long Is.
1011 Howells Rd.
Bay Shore, NY 11706

Astronomical Society of NYC
HOSS Planetarium
153 Arlo Rd.
Staten Island, NY 10301

Astronomy Section
Rochester Acadamy of Science
12 Round Trail Drive
Rochester, NY 14534

Broome County Astronomical
Society
Robertson-Kopernik Observatory
Underwood Road
Vestal, NY 13850

Buffalo Astronomical Assn

Buffalo Museum of Science
Humbolt Parkway
Buffalo, NY 14211

Finger Lakes Astronomy Club
42 Colt Street
Geneva, NY 14456

Lockport Astronomy Association
3165 Roland Drive
Newfane, NY 14108

M. Martz Memorial Astro Soc
11 Nash Avenue, RD #1
Frewsberg, NY 14738

Mohawk Valley Astronomical
Society
867 Bleeker Street
Utica, NY 13501

Northern New York Astronomical
Society
P.O. Box 106
Watertown, NY 13601

Rockland Astronomy Club
110 Pascack Road
Pearl River, NY 10965

Syracuse Astronomical Society
1115 E. Colvin Street
Syracuse, NY 12310

Tri-Lakes Astronomical Society
P.O. Box 806
Saranac Lake, NY 12983

Valley Stream Astronomical Soc
P.O. Box 177
Elmont, NY 11003

Wagner College Astronomy Club
Astronomical Society of NYC
Wagner College Planetarium
Staten Island, NY 10301

NORTH CAROLINA
Astronomical Society of Rowan
Margaret C. Woodson
Planetarium
1636 Parkview Circle
Salisbury, NC 28144

Astronomy Club of Cumberland
2308 Colgate Dr.
Fayetteville, NC 28304

Blue Ridge Astronomers

Physics Dept., Appalachian St.U.
Boone, NC 28608

Cape Fear Astronomy Club
P.O. Box 704
Wrightsville Beach, NC 28480

Catawba Valley Astronomy Club
12 East Street
Granite Falls, NC 28630

Chaple Hill Astronomy Club
1036 Highland Woods Road
Chaple Hill, NC 27514

Charlotte Ama. Astronomers Club
Discovery Place
301 North Tryon Street
Charlotte, NC 28204

Cleveland County Astronomical
Society
825 Ivywood Dr.
Shelby, NC 28150

Cumberland Astronomy Club
P.O. Box 297
Stedman, NC 28391

Forsyth Astronomical Society
504 Gayron Dr.
Winston-Salem, NC 27105

Gaston Astronomy Club
408 S. Avon
Gastonia, NC 28052

Greensboro Astronomy Club
Natural Science Center
4301 Lawndale Avenue
Greensboro, NC 27408

Raleigh Astronomy Club
P.O. Box 10643
Raleigh, NC 27605

Cleveland County Astronomy Soc
285 Ivuwood Dr.
Shelby, NC 28150

NORTH DAKOTA
Dakota Astronomical Society
P.O. Box 2539
Bismark, ND 58502

OHIO
Astronomy Club of Akron
1630 Thomapple Avenue
Akron, OH 44301

Black River Astronomical Soc
11905 West Lake Road
Vermilion, OH 44089

Chagrin Valley Astronomical
Society
P.O. Box 11
Chagrin Falls, OH 44022

Cincinnati Astronomical Society
5274 Zion Rd.
Cleves, OH 45002

Cleveland Astronomical Society
Warner & Swasey Observatory
Case Western Reserve University
Cleveland, OH 44106

Columbus Astronomical Society
P.O. Box 16209
Columbus, OH 43216

Cuyahoga Astronomical
Association
P.O. Box 29089
Parma, OH 44129

Findlay Astronomy Club
c/o Physics Dept., Findlay
College
Findlay, OH 45840

Fostoria Astronomy Club
118 North Main
Bowling Green, OH 43402

Fremont Astronomical Society
518 Fourth Street
Fremont, OH 34322

Kent Quadrangle Astronomical
Society
622 Walter Street
Kent, OH 44240

Lakeland Astronomical Society
R. Schmidt
Rte. 306 & I-90
Mentor, OH 44060

LeRC Astronomy Club (NASA)
Lewis Research Lab/M.S. 142-2
21000 Brookpark Road
Cleveland, OH 44135

Lima Astronomy Club
127 E. North Street
Kenton, OH 43326

Mahoning Valley Astronomical
Society
1076 State Route 534
Newton Falls, OH 44444

Miami Valley Astronomical
Society
Dayton Museum of Natural
History
Apollo Observatory
2629 Ridge Avenue
Dayton, OH 45414

Midwestern Astronomers
5760 Richland Circle
Milford, OH 45150

Richland Astronomical Society
811 Chestnut Street
Ashland, OH 44805

Stark County Astronomical
Society
819 Dartmouth SW
Canton, OH 44710

Toledo Astronomical Association
8534 Covert Road
Petersburg, MI 49270

Tuscarawas Co. Amateur
Astronomical Soc.
RFD #1, Box 1295
New Philadelphia, OH 44663

Wayne County Astronomical
Society
218 S. Walnut Street
Wooster, OH 44691

Wilderness Center Astronomy
Club
558 Canford Avenue NW
Massilon, OH 44646

OKLAHOMA
Arbuckle Astronomical Society
P.O. Box 5516
Ardmore, OK 73403

Arkansas-Oklahoma
Astronomical Society
P.O. Box 31
Fort Smith, AR 72902

Astronomy Club of Tulsa
3828 S. Victor
Tulsa, OK 74105

Bartlesville Astronomical Soc
225 SE Fenway Place
Bartlesville, OK 74006

Central Oklahoma Astronomy
Club
P.O. Box 628
Purcell, OK 73080

Cimmarron County Star Gazers
P.O. Box 278
Boise City, OK 73933

Oklahoma City Astronomy Club
2100 NE 52
Oklahoma City, OK 73111

OREGON
Eugene Astronomical Society
Lane ES.D. Planetarium
P.O. Box 2680
Eugene, OR 97402

OMSI Astronomers
Oregon Museum of Science &
Technology
Harry C. Kendall Planetarium
4015 SW Canyon Road
Portland, OR 97221

Northwest Astronomy Group
55371 McDonald Rd.
Veronia, OR 97064

Portland Astronomical Society
2626 SW Luradel
Portland, OR 97219

PENNSYLVANIA
Amateur Astronomers Assn. of
Pittsburgh
Buhl Planetarium
Allegheny Square
Pittsburgh, PA 15212

Astronomical Society of
Harrisburg
P.O. Box 356
Harrisburg, PA 17339

Bucks-Mont Astro Assn
1335 Commonwealth Dr.
Newtown, PA 18940

Beaver Valley Astronomy Club
404 Daugherty Drive
New Brighton, PA 15066-3620

Berks Co. Amateur Astronomical

Society
Reading School District
Planetarium
1211 Parkside Dr. S
South Reading, PA 19611-1441

Bucknell Astronomy Club
Box C-2315
Lewisburg, PA 17837

Churchville Astronomy Club
5 Swallow Road
Holland, PA 18966

Delaware Valley Amateur
Astronomers
6233 Cator Ave.
Philadelphia, PA 19149

Delaware Valley Amateur
Astronomers
1115 Melrose Ave.
Melrose Park, PA 19126

Kiski Astronomers
2 Aluminum City Terrace
New Kensington, PA 15068

Lackawanna Astronomical
Society
Everhart Museum
Scranton, PA 18510

Lehigh Valley Amateur Astro.
Society
P.O. Box 368
Fogelsville, PA 18051

M-31 Astronomical Society
309 Andrews Park Blvd.
Erie, PA 16511

Rittenhouse Astronomical
Society
P.O. Box 12531
Philadelphia, PA 19131

Sir Isaac Newton Astronomical
Society
P.O. Box 591
Kane, PA 16735

Warren County Amateur
Astronomy Club
118 Falconer Street
Warren, PA 16365

RHODE ISLAND
Skyscrapers

P.O. Box 814
Greenville, RI 02828

SOUTH CAROLINA
Carolina Skygazers Astronomy
Club
Settlemyre Planetarium
Museum of York County
4621 Mt. Gallant Rd.
Rock Hill, SC 29730

Lowcountry Stargazers
58 Kiawah Loop
Charleston AFB, SC 29404

Midlands Astronomy Club
P.O. Box 477
White Rock, SC 29177

SOUTH DAKOTA
Black Hills Astronomical Society
c/o C.A. Grimm, Dept. of
Mathematics
School of Mines & Technology
Rapid City, SD 57701

TENNESSEE
Barnard Astronomical Society
P.O. Box 90042
Chattanooga, TN 37412

Bays Mountain Ama Astronomers
Bays Mountain Planetarium
Route 4
Kingsport, TN 37660

Bristol Astronomy Club
824 Hidden Valley Rd.
Kingsport, TN 37663

Memphis Astronomical Society
P.O. Box 11301
Memphis, TN 38111

Middle Tennessee Astro Soc
1305 Sycamore St.
Manchester, TN 37355

Orion
P.O. Box 3262
Oak Ridge, TN 37830-3262

Smoky Mountain Astronomical
Society
P.O. Box 6204
Knoxville, TN 37914-0204

TEXAS
Abilene Astronomical Society

1109 Highland Avenue
Abilene, TX 79605

Amarillo Astronomy Club
5303 S. Milam
Amarillo, TX 79110

Association of Amateur
Astronomers
Texas A&M University Dept. of
Physics
College Station, TX 77843

Astronomical Society of East
Texas
Rt. 14, Box 394-A
Tyler, TX 75707

Astronomy Club
College of the Mainland
Division of Math, Health, &
Science
Texas City, TX 77591

Austin Astronomical Society
P.O. Box 12831
Austin, TX 78711

Corpus Christi Astronomical
Society
3814 Marion St.
Corpus Christi, TX 78415

Denton Co. Astronomical Society
P.O. Box 2492
Denton, TX 76201

Fort Worth Astronomical Society
P.O. Box 161715
Fort Worth, TX 76161-1715

Greater Temple Astronomical
Society
P.O. Box 365
Troy, TX 76579

General Dynamics Astro Club
P.O. Box 12245
Fort Worth, TX 76116

Houston Astronomical Society
8522 Bluegate Drive
Houston, TX 77025

JSC Astronomical Society
3702 Townes Forest
Friendswood, TX 77546

San Angelo Amateur Astro Assn

212 W. 1st
San Angelo, TX 76903

San Antonio Astronomical
Association
6427 Thoreau's Way
San Antonio, TX 78239

South Plains Astronomy Club
1920 46th St.
Lubbock, TX 79412-2214

Texas Astronomical Society
P.O. Box 25162, Preston Station
Dallas, TX 75225

Texas Observers
Fort Worth Museum of Sci & Ind
Noble Planetarium
1501 Montgomery Street
Fort Worth, TX 76107

Ursa Major Astronomical Society
Baylor University
Waco, TX 76706

West Texas Astronomers
811 A Central Drive
Odessa, TX 79761

Wichita Falls Astronomy Club
c/o Museum Planetarium
Two Eureka Circle
Wichita Falls, TX 76308

UTAH
Ogden Astronomical Society
2336 West 5650 South
Roy, UT 84067

Salt Lake Astronomical Society
c/o Hansen Planetarium
15 South State Street
Salt Lake City, UT 84111

Southern Utah Astronomy Group
147 South 300 East
Cedar City, Utah 84720

Utah Celestial Observers
Association
4816 West 4895 South
Kearns, UT 84118

VERMONT
Springfield Telescope Makers
4 Russell Avenue
St. Johnsbury, VT 05819

Vermont Astronomical Society
P.O. Box 782
Williston, VT 05495

VIRGINIA
Black Bay Amateur Astronomers
2808 Flag Dr.
Chesapeake, VA 23323

Fauquier Astronomical Society
Rt. 2, Box 680
Catlett, VA 22019

No. Virginia Astronomy Club
5608 Flag Run Dr.
Springfield, VA 22151

Peninsula Nature & Science
Center
524 Clyde Morris Boulevard
Newport News, VA 23601

Richmond Astronomical Society
2218 Martin Street
Richmond, VA 23228

Roanoke Valley Astronomical
Society
Roanoke Valley Science Museum
10978 Rocky Rd.
Bent Mtn., VA 24059

Shenandoah Valley Astro. Club
James Madison University
Dept. of Physics
Harrisonburg, VA 22807

Tidewater Telescope Makers
5908 Beechwalk Drive
Virginia Beach, VA 23464

WASHINGTON
Olympic Astronomical Society
4510 NW Shelley Drive
Silverdale, WA 98383

Seattle Astronomical Society
852 NW 67th Street
Seattle, WA 98115

Southwest Washington Astro.
Society
2421 Leisure Lane
Centralia, WA 98531

Spokane Astronomical Society
P.O. Box 404
Mead, WA 99021

Tacoma Astronomical Society
7101 Topaz Dr. SW
Tacoma, WA 98498

Tri-City Astronomy Club
1907 Luther Place
Richland, WA 99352

WEST VIRGINIA
Ohio Valley Astronomical Soc
1128 S. Jefferson Drive
Huntington, WV 25701

WISCONSIN
Beloit Astronomical Society
2550 Highcrest Road
Beloit, WI 53511

La Crosse Area Astronomical
Society
P.O. Box 204
La Crosse, WI 54602

Madison Astronomical Society
205 Cameo Lane
Madison, WI 53714

Milwaukee Astronomical Society
W248 S7040 Sugar Maple Dr.
Waukesha, WI 53186

Neville Museum Astronomical
Society
Neville Public Museum
210 Museum Place
Green Bay, WI 54303

Northeast Wisconsin Stargazers
(Newstar)
109 Skyline Dr.
Appleton, WI 54915

Racine Astronomical Society
518 Emmertsen Rd.
Racine, WI 53406

Rock Valley Astronomical
Society
P.O. Box 1362
Janesville, WI 53545

Sheboygan Astronomical Society
2330 S. 11th Street
Sheboygan, WI 53081

Wehr Astronomical Society
5501 Acorn Court
Greendale, WI 53129

WYOMING
Cheyenne Astronomical Society
3409 Frontier St.
Cheyenne, WY 82001

Telescopes In Orbit

Nearly 100 satellites equipped for astronomy work have flown in Earth orbit, although most no longer function and some have fallen back to Earth. The list below shows the better-known astronomy satellites over the years. It does not include probes which have flown away from Earth to other bodies in the Solar System.

SATELLITE	LAUNCH	COUNTRY
Explorer 7	1959	USA
Vanguard 2	1959	USA
Explorer 8	1960	USA
Solrad 1	1960	USA
Explorer 11	1961	USA
Injun 1	1961	USA
Solrad 3	1961	USA
Explorer 16	1962	USA
Hitchhiker 1	1963	USA
Solrad 5b	1964	USA
Solrad 7a	1964	USA
Explorer 30	1965	USA
Explorer 31	1965	USA
OSO 2	1965	USA
Solrad 6b	1965	USA
Solrad 7b	1965	USA
Explorer 33	1966	USA
OAO 1	1966	USA
Aurora 1	1967	USA
Explorer 37	1968	USA
Explorer 38	1968	USA
OAO 2	1968	USA
OGO 5	1968	USA
OV5 5	1969	USA
OV5 6	1969	USA
OV5 9	1969	USA
Uhuru/Explorer 42	1970	USA
Explorer 45	1971	USA
Oreol 1	1971	USSR
Copernicus (OAO 3)	1972	USA
Explorer 47	1972	USA
Explorer 50	1973	USA
Oreol 2	1973	USSR
Prognoz 3	1973	USSR
Miranda	1974	UK
GOES 1	1975	USA
Prognoz 4	1975	USA
Prognoz 5	1976	USA
Solrad 11a	1976	USA
Solrad 11b	1976	USA
ESA Geos 1	1977	ESA

GOES 2	1977	USA
HEAO-1	1977	USA
Prognoz 6	1977	USA
Einstein/HEAO-2	1978	USA
ESA Geos 2	1978	ESA
GOES 3	1978	USA
IUE	1978	USA/Europe
Kyokko (Exos 1)	1978	Japan
Ariel 6 (UK 6)	1979	UK
HEAO-3	1979	USA
Intercosmos 19	1979	USSR
SAGE	1979	USA
Solwind	1979	USA
GOES 4	1980	USA
Solar Max (SMM)	1980	USA
Dynamics Explorer 1	1981	USA
Dynamics Explorer 2	1981	USA
GOES 5	1981	USA
Hinotori (Astro A)	1981	Japan
Intercosmos 22	1981	USSR
Oreol 3	1981	USSR
SME	1981	USA
Salyut 7 space station	1982	USSR
Astron	1983	USSR
GOES 6	1983	USA
IRAS	1983	USA/UK/Netherlands
CCE (AMPTE A)	1984	USA
GMS 3	1984	USA
IRM (AMPTE B)	1984	W Germany
UKS (AMPTE C)	1984	UK
Prognoz 10	1985	USSR
Mir space station	1986	USSR
Viking	1986	Sweden
Ginga (Astro C)	1987	Japan
GOES 7	1987	USA
Kvant space station module	1987	USSR
San Marcos 5	1988	Italy/USA
COBE	1989	USA
Hipparcos	1989	Europe
Kvant-2	1989	USSR
Astro-1	1990	USA
Hubble Space Telescope	1990	USA/Europe
Rosat	1990	Germany/USA
Euve	1991	USA
GRO	1991	USA
SARA	1991	France
Solar-A	1991	Japan/USA

ACRONYMS:

AMPTE	Active Magnetospheric Particle Tracer Explorers
ESA	European Space Agency
EUVE	Extreme Ultraviolet Explorer

GMS	Geostationary Meteorological Satellite
GOES	Geostationary Operational Environmental Satellite
GRO	Gamma Ray Observatory
HEAO	High Energy Astronomical Observatory
Hipparcos	High Precision Parallax Collecting Satellite
IRAS	Infrared Astronomy Satellite
IUE	International Ultraviolet Explorer
Kvant	an astrophysics module attached to the Mir space station
OAO	Orbiting Astronomical Observatory
OGO	Orbiting Geophysical Observatory
OSO	Orbiting Solar Observatory
Rosat	Roentgen Satellite
SAGE	Stratospheric Aerosol and Gas Experiment
SARA	Satellite for Amateur Radio Astronomy
SME	Solar Mesospheric Explorer
SMM	Solar Maximum Mission, or Solar Max
Solrad	Solar Radiation
Solwind	Solar Wind satellite killed in orbit by the U.S. Sept. 13, 1985, in an antisatellite weapons test.
UK	United Kingdom

Telescopes In Orbit: Hubble

Hubble Space Telescope, first of NASA's Great Observatories to go to space and the largest telescope ever sent to space, was dropped overboard from U.S. space shuttle Discovery April 24, 1990. Within weeks, it was sending down photos of deep-space objects in detail never seen before.

Unfortunately, even though the detail was better than Earth telescopes could achieve at that time, it was not as exquisite as had been predicted. The problem wasn't with faulty predictions, but faulty grinding of the huge telescope's main mirror.

For more than a decade, NASA had been touting Hubble as an astronomy satellite powerful enough to hunt for the edge of the Universe. It had a 94.5-in. mirror, five sensitive cameras and other instruments to revolutionize optical astronomy from above Earth's obscuring atmosphere.

Called the most powerful optical instrument ever built for use in space, the telescope was a joint project of the European Space Agency and NASA. They said the telescope would show astronomers far-flung objects 15 billion lightyears away from our planet.

Looking at visible light and ultraviolet light, Hubble was designed to expand the observable volume of the Universe by several hundred times. The powerful instrument was expected to relay the best pictures yet of the Universe. It would return pictures of the Universe much more distinct than those seen through telescopes on the ground.

Long time coming. NASA had started work on the project in 1977. The space agency had wanted to launch the telescope in 1985. That date slipped to the fall of 1986, then launch was delayed by the January 1986 explosion of shuttle Challenger.

After Challenger, NASA changed Hubble's flight software, upgraded the safe-mode system, and inserted a longer life electrical system with new solar arrays, longer-life components and higher-power nickel-hydrogen batteries. Originally, HST was to have cost $435 million. Delays and problems ballooned the cost to $1.5 billion.

Out of pocket by $1.5 billion, NASA was five years behind schedule launching it. The telescope was in a California warehouse generating storage and maintenance bills of $8 million a month, awaiting launch.

The agency had to take the telescope to space in a shuttle, but NASA had trouble scheduling the launch after the Challenger disaster set shuttle flights back three years. Corrections had to be made in shuttle hardware to prevent another disaster. There was a shortage of booster fuel for a time and Defense Dept. demands for shuttle rides pushed the flight date later, into a time when the Sun would be blossoming with sunspots.

The Sun has been approaching an 11-year peak in its Sunspot Cycle. Exploding gases on the surface of the star radiate energy which will heat Earth's upper atmosphere, expanding it deeper into space, causing extra drag on satellites. That might threaten to pull the 12-ton, 43-ft. telescope down to a low orbit from which it might fall to Earth like a meteor. The explosive sunspots gave NASA an exasperating choice: launch Hubble so high in the atmosphere a shuttle might not be able to reach it again for repair, or launch it so low shuttles would have to fly up every six months to boost it higher at $300 million per nudge. The experts were betting the space agency would delay the launch if sending it up on schedule meant it would have to spend extra money to reposition the telescope frequently.

Out of the warehouse. Hubble took important steps October 4, 1989, on its way to the stars. It was shipped in a special container meant for large military spy satellites aboard an Air Force C-5A cargo jet from Sunnyvale, California, to a cleanroom at Kennedy Space Center, Florida. The telescope was set up for tests in NASA's Vertical Processing Facility, a large hangar used to prepare the satellite for 15 years in space.

Hubble's wide-field planetary camera, as big as a baby grand piano, was not shipped to Florida with the telescope. One of two built-in computers to control the instrument was broken, so the camera had to be repaired at NASA's Jet Propulsion Laboratory (JPL), Pasadena, California. It was shipped to Florida in December.

The wide-field planetary camera was said to be so sensitive that, on Earth, it could see a baseball 200 miles away. In space, it would detect objects 100 times fainter than those visible from Earth's surface, with 10 times the sharpness or resolution. With the telescope, a dime could be seen 20 miles away.

Launch. After Hubble's many delays, Discovery flight STS-31 was April 24, 1990, launched to a record altitude of 380 statute miles above Earth. That was 70 miles higher than any previous shuttle.

Astronauts aboard flight STS-31 were commander Loren J. Shriver, pilot Charles F. Bolden Jr., and mission specialists Steven A. Hawley, Bruce McCandless II, and Kathryn D. Sullivan. Hubble's 15-year mission in Earth orbit started when Hawley, from inside the crew quarters, used the shuttle's outside 50-ft. remote-control mechanical arm to lift the 25,500-lb. Hubble out of the cargo bay and drop it overboard into its own orbit.

The huge satellite's solar arrays and dish antenna were extended. Sullivan and McCandless almost had to take a spacewalk when a solar panel didn't unfold until the third try. The astronauts flew shuttle Discovery down to Earth April 29.

Big Bang video. With Hubble comfortably in orbit, astronomers settled into their chairs before their computer consoles to stare back toward the dawn of the Universe on their video screens. Hubble was to be their time machine in orbit. Astronomers didn't expect to see the beginning of time, but galaxies of stars as they were 14 billion years ago. They figured that's how long light from the most distant objects would have taken to reach our planet. That 14 billion years ago might be a scant billion years or so after the so-called Big Bang, an unfathomed explosion of fiery gases some astronomers think created the Universe.

Glitches. Unfortunately, more than a dozen small and large Hubble problems heckled ground controllers. NASA wanted to take a first-light picture with the new telescope a week after Hubble went to space, but engineering problems—the space agency called them growing pains—put the meticulous work of turning on the telescope 10 days behind schedule. Most of those pains were cured right away, but the agency had three important glitches. The most difficult at that time seemed to be a vibration shuddering through the telescope when the satellite travelled into and out of sunlight.

NASA decided not to release pictures for weeks. Technicians in the Goddard control room were busy correcting small errors in the telescope pointing system which caused it to wobble slightly. Focusing was a painstaking process of finding and locking onto stars and moving the secondary mirror back and forth in minute increments until starlight is brought into focus. The secondary mirror can move only about one-tenth inch. When precisely aligned, the optical system would have ten times the clarity of observatories on the ground, NASA said.

Guide fixed. NASA cured a major Hubble headache in June 1990 when it solved a guidance problem preventing the observatory satellite from locking onto guide stars.

Hubble's cameras make time-exposures by tracking guide stars. Most of the early problems after launch centered on the telescope's inability to find guide stars. The problem solved in June 1990 was a computer program which didn't line the telescope up on correct targets. Instead of orienting itself by turning the telescope to position guide stars on the centerline in its field of view, the computer was ignoring guide stars unless they ended up by happenstance near the centerline. At least 14 hurdles disappeared after engineers discovered the centerline snag, bringing a dramatic change in Hubble's ability to lock onto guide stars.

The telescope still wobbled slightly as it passed in and out of sunlight. Temperature changes caused the large solar panels to wiggle. Those jitters were corrected somewhat by a change in computer software. Called SAGA for Solar Array Gain Augmentation, the software reduced the frequency of the vibration and cut the length of jittery time in half.

NASA soon found there were even bigger problems. The space agency was embarrassed when Hubble pictures turned out blurry. The 12-ton telescope's 94-in. mirror was curved incorrectly—it had spherical aberration.

Bad mirror. Hubble mirrors were ground early in the 1980's by Perkin-Elmer Corp. of Connecticut. The company used a measuring instrument known as a reflective null corrector to position and check its grinder, which shaped the concave Hubble mirror to reflect visible and ultraviolet light.

An investigating panel found a tiny 1.3-mm spacing error in a Perkin-Elmer grinder. The faulty measuring device contained 1.3 mm-thick spacing washers, the same size as the error in the null corrector. Considered huge for high-tech optics, the 1.3-mm error was likened to a room built 3 feet too long. The thin washers, used as spacers between the grinder and its baseplate, probably flawed the mirror.

Old records indicated a test in the early 1980's had found the error after the mirror was manufactured, but Perkin-Elmer engineers placed complete faith in their null corrector and disregarded the test results. NASA says it was not aware of the test results. Perkin-Elmer Corp. was bought in 1989 by Hughes and now is known as Hughes Danbury Optical Systems Inc. In 1991, NASA gave Hughes Danbury Optical Systems Inc. a contract to grind mirrors for another Great Observatory, the Advanced X-Ray Astrophysics Facility (AXAF) to be launched in 1998.

How fuzzy is it? Like a slightly-out-of-focus film projector. Like looking through someone else's glasses. That's how astronomers described the fuzzy pictures Hubble was capable of shooting.

Two mirrors in the satellite collect and reflect star light. The surfaces of both mirrors are perfectly smooth and curved to cause light to be magnified. But, one was built with the tiny error. Its surface curve does not align perfectly with the curve of the other mirror so light reflected to the camera is scattered slightly, making starlight look fuzzy with no crisp, sharp edges. Focus is off by only two microns—about four one-hundredths the diameter of a human hair—but enough to blur Hubble's vision.

Most affected by the spherical aberration were the Wide Field/Planetary Camera, the European Space Agency's Faint Object Camera and the High-Speed Photometer. The Goddard High Resolution Spectrograph and the Faint Object Spectrograph don't require finely focused light to complete their observations successfully.

Eye glasses. NASA came up in 1991 with a plan to fix the blurry vision—give the telescope eye glasses. Never short of pithy acronyms, the agency calls the eye glasses COSTAR for Corrective Optics Space Telescope Axial Replacement. If Hubble is the defective star on the public stage, COSTAR will be its...co-star.

COSTAR will have ten small mirrors, and mechanisms to position and support the mirrors, to correct Hubble's spherical aberration. NASA had planned to send astronauts up in shuttle Discovery in February 1994 for routine maintenance on the big satellite anyway. They will take COSTAR along and install it on Hubble.

Each of the ten mirrors is about the size of a quarter. COSTAR is to correct the aberration from the primary mirror, sending its light on to the three instruments that suffer from the aberration. Astronomers have their fingers crossed that COSTAR will restore the potential of three major instruments—the Goddard High Resolution Spectrograph, the Faint Object Spectrograph and the Faint Object Camera. COSTAR is to be built by April 1, 1993, at a cost of $30.4 million.

Fix-it flight. The bad mirror was not the end of NASA's problems with Hubble. In 1991, gyroscopes which stabilize the satellite failed and a faulty power supply cut off a key spectrograph. The spacewalking Discovery astronauts will not only fix the spherical aberration in 1994, but also fix the spectrograph power supply, replace gyroscopes and tighten loose solar panels.

The astronauts visiting Hubble will make repairs and replace batteries and fuel. Hubble's orbit will have deteriorated by then, so the shuttle will be used to boost the telescope back up to its original altitude. With such regular maintenance, Hubble is expected to be used for at least 20 years. NASA's Space Telescope Science Institute, Baltimore, Maryland, coordinates research with the telescope.

Despite the telescope's focusing problem, Hubble's imaging of bright targets still is better than the best images produced through ground-based telescopes on a clear night. However, there is one nagging concern for Hubble fans. Telescopes on the ground are getting better, by compensating for motion of the atmosphere and using optical interferometry techniques. Such improvements weren't visualized when Hubble was designed in the 1970's. Today's Earth-bound telescopes are approaching Hubble's ability to resolve very weak and distant objects.

First light. Hubble sent its first two pictures to Earth May 20. The so-called "first light" recorded by the telescope had traversed 1,260 lightyears of deep space from an open cluster of stars known to astronomers as Theta Carina.

Hubble was 381 miles above Jayapura, New Guinea, when the shutter of its electronic wide-field planetary camera snapped the first rough-focus, black-and-white picture and stored the 30-second exposure in a recorder. Two hours later, that image was transmitted in four parts to one of NASA's Tracking and Data Relay (TDRS) satellites, then to receivers on the ground at White Sands, New Mexico, then back up to another communications satellite which relayed them down to astronomers waiting at Goddard Space Flight Center, Greenbelt, Maryland.

A Goddard computer removed background noise and assembled the photo. In the first "raw" unprocessed Hubble photo, stars in the cluster looked like vague points of light against a murky field. Technicians immediately set about to improve the contrast, making the background blacker and sharpening the points of light for easier viewing.

The first picture clearly revealed two stars where astronomers on the ground previously had seen only one elongated object.

They were the first photographs ever made above Earth's clouded atmosphere by a large, complex optical telescope, made to test Hubble's initial focus. NASA said the first Hubble photo was in better focus than expected. The agency said that verified Hubble's big mirror had made it through launch in good condition. Astronomers described the digital images as three times sharper than expected.

Interestingly, most light from the stars in the first photo was in the center of the image. Astronomers said that's where they would expect it to be from a high-quality mirror slightly out of focus.

Carina. The stars in Theta Carina are a handful among the 100 billion stars in our galaxy. Astronomers arbitrarily divide the sky surrounding Earth into 88 constellations to identify particular locations. At a distance of 1,260 lightyears, the three-billion-year-old Theta Carina cluster is one source of light in a constellation known as Carina, or Ship's Keel. Astronomers on Earth see the constellation only from the Southern Hemisphere. The best-known star in Carina is Canopus. Astronomers log the three-billion-year-old Theta Carina cluster officially as NGC-3532. NGC stands for New Galactic Catalogue.

Cassegrain. Hubble is a Cassegrain telescope. Starlight enters its tube and bounces off a 94.5-inch primary mirror, back up sixteen feet to a 12-inch secondary mirror, from which it reflects back down through a hole in the primary mirror to be captured and recorded. The electronic wide-field planetary camera used for the May 20 pictures is one of five major science instruments on board the satellite. There is another camera, two light-splitting spectrographs and a photometer to study the intensity of starlight. Even with its deficiencies, Hubble probably will transform astronomy, shooting 30,000 photos during 15 years traveling around Earth.

The pictures are likely to show more details of planets, stars and galaxies then ever before. Telescopes on the ground are hampered by particles in the air which absorb and distort starlight. The space telescope is above the atmosphere in the vacuum of space where the cameras, spectrographs and photometer will see much more clearly. Astronomers still think Hubble will find clues to basic questions about the birth, age, size, shape, and even the ultimate fate of the Universe.

Star-birth photos. In August 1990, Hubble sent down its first photos showing how massive stars form in galaxies other than our own Milky Way. The test pictures snapped August 3 revealed new information about our neighboring galaxy, the Large Magellanic Cloud, and 30 Doradus, the most prolific stellar nursery in that galaxy.

A lightyear is the distance light travels at 186,000 miles per second through space in one year, about 5.9 trillion miles. The Large Magellanic Cloud and 30 Doradus are 160,000 lightyears from Earth.

The nebula 30 Doradus, seen in the constellation Dorado, the swordfish, is so bright it can be seen by naked eye from Earth's Southern Hemisphere. An American astronomer once said 30 Doradus was so bright it might cast shadows on Earth's night landscape.

The 30 Doradus region of space is smaller in diameter than the distance from Earth to the nearest star in our own galaxy—about 4 lightyears. Just ten years earlier, astronomers thought the 30 Doradus region had only one star. But now, hundreds of stars are believed to exist there and some 60 can be seen from Earth with the naked eye. Astronomers expected to count even more stars in Hubble photos.

Stellar nursery. The August 3 photo, made with Hubble's wide-field planetary camera, was used to make a finding chart to checkout another Hubble instrument, the Goddard high-resolution spectrograph. Astronomers examining the wide-field planetary camera picture saw the central stars of the R136 cluster, within the 30 Doradus nebula.

Hot stars generate more blue and ultraviolet light than cooler stars. Because Hubble took the 30 Doradus picture through a violet light filter, it highlighted the hottest, most massive stars in the picture. Several appeared to be single objects of such brightness they may be more than 100 times as massive as our Sun.

Before 1980, astronomers had thought R136 was a single star. In the 1980's, astronomers were able to resolve the single star into eight. As the 1990's opened, better Earth-bound telescopes raised the count to 27 stars in the cluster. Hubble's picture was better—60 stars could be seen.

Astronomers said future cluster pictures by Hubble would have the same resolution. When shot through different colored filters on the wide-field planetary and faint-object cameras, photographs were likely to reveal the masses of stars in the clusters. Knowing how many stars of each mass are present in a cluster, astrophysicists can calculate how stars form and how they produce the chemical elements present in space.

All massive stars probably explode into supernovas, spewing out new elements, such as iron, an essential ingredient for life as we know it on planet Earth. By determining how many such stars are present in 30 Doradus and similar clusters in other galaxies, astronomers expect to compute accurate information on the richness of chemical elements across the Universe.

Saturn's Wilber. In November, Hubble photographed Saturn's Great White Spot, a huge storm on that planet more than 50,000 miles in diameter and growing.

The storm, covering most of one hemisphere of the planet, is larger than Jupiter's famous Great Red Spot. Astronomers named Saturn's storm the Wilber Spot after amateur astronomer Stuart Wilber of Las Cruces, New Mexico, who had discovered it September 24. At that time, the storm was less than 10,000 miles in diameter. By October 2, the storm was more than 12,000 miles across, larger than planet Earth.

The mysterious hurricane further perplexed astronomers during October 1990 as new spots turned up in the main spot, while two separate round white spots formed. One side of the main spot darkened, as if Wilber were a raised feature casting a shadow, NASA said. The spot was so bright it seemed almost fluorescent in appearance.

Astronomers had been expecting a new white spot to appear on Saturn, having seen similar white spots at 27 to 33-year intervals, in 1876, 1903, 1933 and 1966. The earlier spots each lasted several weeks, spreading eastward around the planet, then breaking up into weak, tattered scraps which remained in sight for a few months. The 1990 spot was larger than the earlier spots.

Of the nine planets in the Solar System, Saturn is the sixth from the Sun. It was the outermost known planet before invention of the telescope.

Astronomers around the world were captivated by Saturn's Great White Spot which covers most of the planet's northern hemisphere and equatorial zone. Saturn spins once every 10.25 hours so, to see the storm through their telescopes, astronomers had to wait for the side of the planet with the Great White Spot to face Earth. This time, Hubble's wide-field planetary camera was on hand in Earth orbit to snap several shots of the Spot in blue light and near-infrared light each day for three days. The Great White Spot was even larger than the well-known Great Red Spot on Jupiter.

Other Hubble pix. Before the Great White Spot erupted, astronomers had used Hubble to make a photograph with engaging clarity of Saturn and its rings. Previous snapshots of Saturn sent back by visiting spacecraft may have been sensational, but the Hubble image was the best made from the vicinity of Earth.

In fact, NASA's worries about the effectiveness of Hubble lessened after astronomers used two of the telescope's cameras in the summer of 1990 to take the extraordinary pictures of Saturn and its rings, as well as a different ring around Supernova 1987a, the nucleus of Seyfert galaxy M77, the globular cluster M14, light from a quasar bent by a gravitational lens, the Orion Nebula, star cluster R136 in the Large Magellanic Cloud, a possible black hole in the crowded heart of galaxy NGC 7457, and the planet Pluto and its moon Charon.

Pluto. One of several new Hubble photos exhibited in October 1990 showed the clearest view yet of Pluto, the puzzling ninth planet from the Sun, and its moon Charon. Pluto is the only planet not to have been visited by a spacecraft from Earth.

The frigid Pluto and Charon, 3.6 billion miles from the Sun, have looked like overlapping blobs of light in previous pictures made by cameras on Earth. Hubble showed two clearly distinct spheres.

Gravitational lens. A quasar is a powerful object, very distant from Earth, exploding in extremely high amounts of energy—possibly the nuclei of ancient active galaxies. The quasar photographed by Hubble was eight billion lightyears from Earth. Its image, recorded by the European Space Agency's faint-object camera aboard Hubble, was the most detailed picture yet of a so-called gravitational lens.

The photo showed four images created as light from the object was bent by the gravity of a galaxy located in the path between the quasar and Earth. The galaxy, only 400 million light years from Earth, created a gravitational lens as light from the quasar bent around it.

A lightyear is the distance light travels at 186,000 miles per second through space in one year, about 5.9 trillion miles. Quasar is short for quasi-stellar radio source.

Supernova 1987a. A supernova is the catastrophic death of a massive old star. Supernova 1987a, a star which was first seen in February 1987 exploding 160,000 light years from Earth, has been called one of the most spectacular astronomical events of the century. On August 23, 1990, Hubble's faint-object camera was used to record a visible-light image of the famous supernova.

The 1990 photo showed the remnant of the exploded star was surrounded by an intriguing luminescent ring of hot gas and dust. The slowly-expanding ring is 1.3 lightyears in diameter. It may have formed during the million years before the star blew up and was compressed by the explosion.

Today it is left over from the enormous cloud of hydrogen gas blown off by the star—in its "red supergiant" late stage of life—in the stellar wind 10,000 years before the explosion took place. The star evolved to a "blue supergiant" stage, ejecting a high-speed stellar wind which swept up the gas and compressed into a narrow, high-density hydrogen shell.

The star apparently was most effective at compressing the gas along an equatorial plane, creating the ring around the star's equator. The circular ring is inclined along the line-of-sight so it appears elliptical to viewers on Earth.

In the first hours after the supernova blast, the ring was submerged in an overdose of ultraviolet light which heated it. The ring is still glowing years later at a temperature of more than 20,000 degrees Kelvin.

Swiftly-moving dust ejected from the supernova, travelling at one-tenth the speed of light, will catch up and overtake the ring in a few years. That shock-wave collision will heat the ring, making it glow brightly in ultraviolet light and X-rays. After a few decades, the ring will be engulfed by supernova debris. Then it will be visible for centuries as a bright supernova remnant.

The Hubble picture, taken in the green light of doubly ionized oxygen, was the first time an exploding outer envelope of a supernova had been photographed directly. Astronomers called the ring an important clue in determining the history of the star which exploded.

Globular cluster. Astronomers used Hubble to photograph a globular cluster known as M14. They thought they would find a nova, but it didn't turn up.

Black hole? At the core of a "rather boring" galaxy 40 million lightyears away, Hubble on August 17 saw something so dense it could be a long-sought "black hole."

Black holes are collapsed burned-out cinders of massive stars with gravity so intense visible light cannot escape. Gas and dust being sucked into a black hole is accelerated by the extraordinary gravity of the collapsed star, heating the particles to extremes before they cross over into the black hole, sending off showers of intense energy including X-rays, ultraviolet light, infrared light and visible light.

Many such fountains of energy sparkle across the vast reaches of the Universe, warming the extreme cold of deep space at temperatures of one million to 100 million degrees. Astronomers suspect some of these objects, many trillions of miles away from Earth, may be black holes.

On August 17, researchers were using Hubble's wide-field planetary camera to look at NGC 7457 when they realized that galaxy had what was probably the second most dense core they ever had seen.

Hubble's 800-second exposure showed the heart of galaxy NGC 7457 is composed of at least 30,000 suns, closely bunched together at 400 times the density previously seen through telescopes on Earth. Astronomers saw in the Hubble photo a hint that the mob of stars was orbiting a black hole. Since NGC 7457 is considered a typical galaxy, the tens of thousands of suns crowding its core suggests the nuclei of normal galaxies may be more densely packed with stars than thought.

The tumultuous natural environment in such galaxy centers interests astronomers who would like explanations for quasars, pulsars, cosmic jets, Seyfert galaxies and other baffling activity. NGC 7457 had been seen as a commonplace galaxy. Its pedestrian nature had led to its selection for basic Hubble tests. The galaxy's jam-packed core surprised astronomers who hadn't expected to find such an exceptionally bright kernel embedded in the sprawling galaxy.

Before Hubble, astronomers were not able to see with such high resolution a galaxy outside of the Local Group of two dozen galaxies in our neighborhood. NGC 7457 is more than half way to the great Virgo cluster of galaxies. NGC 7457 is seen in Earth's sky as an Magnitude 11 galaxy in the constellation Pegasus.

Quasar UM675. Quasars, first seen in 1963, are the brightest objects across the Universe. In October 1990, Hubble made its first observation of the chemistry of one of the faint, faraway objects. Astronomers hoped it would shed light on the early Universe.

Using the space telescope's faint object spectrograph for eight hours, astronomers from Johns Hopkins University, Baltimore, the University of California, San Diego, and NASA observed the spectrum of quasar UM675 which is 12 billion lightyears from Earth. They were looking for signs of helium in far-ultraviolet light.

Helium is the second most abundant element in the Universe, after hydrogen. There appears to be too much of it to have come only from nuclear fusion in stars. Such nuclear fusion converts hydrogen into helium.

Helium in the distant quasar writes a signature of spectral lines in the ultraviolet it radiates, but ozone in Earth's atmosphere blocks most ultraviolet light from telescopes on the ground. The space telescope, operating 400 miles above Earth's surface, is above the ozone layer and can see ultraviolet light from deep space objects.

Some astronomers suggest that most helium was made in the so-called "big bang," an explosion they say created the Universe 15 billion years ago. They hope the helium in UM675 will be sufficient to indicate the abundance of the element in the early Universe.

Astronomers are puzzled about how quasars spew out energy too intense to be produced by such small objects.

Orion. The Hubble Space Telescope also was used in October to make new photos of the Orion Nebula, finding some previously-unknown features of the nebula. Orion Nebula is one of the richest star birth regions in our local neighborhood of the Universe. It has been of interest to astronomers for centuries.

Hubble is a joint project of NASA and the European Space Agency. Hubble research projects are coordinated by the Space Telescope Science Institute at Johns Hopkins University, Baltimore, Maryland, which is operated for NASA's Goddard Space Flight Center, Greenbelt, Maryland, by the Association of Universities for Research in Astronomy, Inc. The telescope operations control center is at Goddard. NASA's Marshall Space Flight Center, Huntsville, Alabama, is in charge of maintenance of the telescope.

The other three high-tech American instruments in NASA's new generation of Great Observatories are Gamma Ray Observatory (GRO) launched in 1991, Advanced X-Ray Astrophysics Facility (AXAF) to be launched in 1998, and Space Infrared Telescope Facility (SIRTF) to be launched in 1999.

Telescopes In Orbit: GRO

The second of NASA's Great Observatories to go to space was the huge Gamma Ray Observatory (GRO) satellite launched April 5, 1991. It was carried to a 280-mi.-high Earth orbit in shuttle Atlantis and deployed April 7. The 17-ton GRO was the heaviest

NASA satellite ever launched by the space agency. Its mission is to search for highly-energetic gamma rays blasted out by some of the most violent processes in the universe.

Invisible gamma rays are powerful radiation created when the nuclei of atoms collide and matter is annihilated in the presence of antimatter. GRO senses and records high-energy gamma rays flowing from quasars, supernovas, pulsars, neutron stars, and black holes, the most violent natural processes known in the Universe.

Gear. GRO has four science instruments to investigate the invisible energy rays which may hold clues to how the Universe was formed. The 17-ton GRO's great bulk was dictated by the size and weight of the instruments needed to trap gamma rays which cannot be viewed through our planet's atmosphere.

The equipment aboard GRO includes an Oriented Scintillation Spectrometer Experiment and an Energetic Gamma Ray Experiment. The spacecraft has arrays of electricity-generating solar cells and a high-gain radio antenna.

GRO is able to keep itself up at the 280-mi. altitude, correcting its own orbit to compensate for drag. Every 14 days, GRO's view is shifted a bit so astronomers can sweep the entire Universe during the observatory's first two years in orbit.

Science. GRO science operations began May 16 when the observatory turned to point toward a pulsar in the Crab Nebula and searched for gamma rays. A full-sky sweep by GRO for gamma rays was expected to last 15 months.

Opportunity. Controllers at Goddard Space Flight Center, Greenbelt, Maryland, pointed the satellite at the Sun for GRO's first target of opportunity on June 7, 1991, to peer at two X-class solar flares.

X-class are the largest, most powerful of solar flares. Solar flares are temporary outbursts of intense solar radiation which shoot looping geysers of hot gas 430,000 miles out from the Sun into space.

Their outbursts of intense radiation have disrupted Earth's magnetic field and interfered with communications and electrical power distribution. Scientists know the composition and magnitude of solar flares, but not much about thermonuclear processes inside them.

By the end of 1991, GRO's Burst and Transient Source Experiment (BATSE) was detecting gamma ray bursts with more sensitivity than any previous receiver and in more detail than previously possible. BATSE was observing a gamma ray burst almost every day, a rate of 250 per year.

Second opportunity. The second target of opportunity, photographed August 8, 1991, was the binary star system Cygnus X-3 some 30,000 lightyears from Earth in the disk of the Milky Way. Lately, radio astronomers had found it spewing out increased radio emissions. Later, GRO photographed Vela Pulsar, Hercules X-1 and Nova Muscae in September 1991.

Strong energy. Exploring the strongest part of the energy spectrum, GRO receives a much wider range of wavelengths than earlier observatories. Astronomers hope it will return clues about the question of whether most gamma rays arriving at Earth from across the Universe come from quasars and pulsars, or are there other objects in our Universe sending out gamma rays.

Most-luminous. GRO gathers gamma rays generated in the far-flung reaches of the Universe as much as 15 billion years ago.

NASA's Energetic Gamma Ray Experiment Telescope (EGRET), one of the four major science instruments aboard GRO, found what was described as the most-distant and most-luminous gamma-ray source ever seen. It was the variable quasar 3C279, located in the constellation Virgo.

The quasar, seven billion light years from Earth, was found to be blasting out a prodigious amount of gamma rays—each gamma-ray photon has an energy greater than 100 million electron volts. The luminosity, or total energy, emitted by the quasar is ten million times the total from the Milky Way galaxy.

Quasar 3C279 is labeled variable because its intensity changes. At its present intensity, the quasar might have been spotted by two previous gamma-ray-hunting satellites—NASA's Small Astronomy Satellite in 1972-1973 and the European Celestial Observation Satellite in 1975-1982—but they didn't report it.

EGRET is the largest instrument ever assembled at Goddard Space Flight Center. The 7.3-ft. by 5.4-ft., 4,001-lb. EGRET is ten to twenty times larger and more sensitive than any high-energy gamma-ray telescope peviously sent to orbit Earth.

Fate of the Universe. Scientists will use all such information sent down from GRO in an attempt to learn more about the origin and fate of the Universe. Built to look out from Earth orbit for a minimum of two years, GRO may work as long as eight years.

NASA's other Great Observatories, planned for Earth orbits above the sight-fogging atmosphere of our planet, are Hubble Space Telescope (HST), Advanced X-Ray Astrophysics Facility (AXAF) and Space Infrared Telescope Facility (SIRTF). HST already is in Earth orbit. NASA plans to launch AXAF in 1998 and SIRTF in 1999.

Man has studied the heavens since ancient times but the view has always been made fuzzy by the ocean of air between Earth's hard surface and the near-vacuum of space. Only just now are we able to put entire astronomical observatories in space where the seeing is not blurred by an atmosphere. And now we know objects in space—stars, galaxies, pulsars, quasars—give off natural radio-wave energy at frequencies beyond visible light.

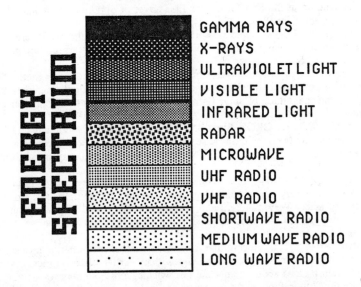

ENERGY SPECTRUM

GAMMA RAYS
X-RAYS
ULTRAVIOLET LIGHT
VISIBLE LIGHT
INFRARED LIGHT
RADAR
MICROWAVE
UHF RADIO
VHF RADIO
SHORTWAVE RADIO
MEDIUM WAVE RADIO
LONG WAVE RADIO

Energy spectrum. The electromagnetic spectrum, including radio waves and light waves, extends from very low frequencies to extremely high frequencies. Our ability to tune in those radio waves has grown in recent decades. In fact, it's only been in the 20th Century we have used any radio, starting at the long-wave end of the spectrum.

Today's AM broadcast stations transmit signals in the medium-wave portion of the spectrum. FM music stations use VHF transmitters. TV stations use VHF and UHF. Cooking ovens are at microwave frequencies, as are police-speedtrap radar transmitters and receivers. All are part of the electromagnetic, or energy spectrum. Visible light we receive with our eyes is flanked in the spectrum by infrared and ultraviolet light we can't see. We do, however, have the capability to capture infrared light, ultraviolet light and X-rays on photographic film.

Only recently have we been able to make radio receivers and sensors covering the UHF to gamma ray part of the electromagnetic spectrum, small enough to send to space. NASA is developing the set of four Great Observatories In Space to extend mankind's knowledge of astronomy and life itself. Each observatory has its own different instruments to gather data from different parts of the electromagnetic spectrum.

Telescopes In Orbit: AXAF

The Advanced X-Ray Astrophysics Facility, third of NASA's four Great Observatories in Earth orbit, is scheduled to start reading some of the most violent processes in the Universe in 1998, revealing clues to their hidden mysteries.

Earlier X-ray satellites in Earth orbit, such as Uhuru Observatory and Einstein Observatory, gave astronomers a tantalizing taste of what may lurk across the vast reaches of the Universe, hidden from view in the X-ray spectrum. So valuable were Einstein Observatory images after it was launched in 1978 that astronomers immediately called for a long-lived orbiting X-ray observatory. NASA started drawing up plans for an Advanced X-Ray Astrophysics Facility (AXAF).

AXAF is to be a 14-ft.-diameter, 45-ft.-long, 15-ton telescope satellite launched in 1998 to a circular orbit 320 miles above Earth. It would operate for at least 15 years.

AXAF. The satellite will have a high-resolution mirror and four powerful instruments: a high resolution camera built by Smithsonian Astrophysics Observatory, Cambridge, Massachusetts; a charge-coupled-device (CCD) X-ray camera from Pennsylvania State University, University Park; an X-ray spectrometer built by NASA's Goddard Space Flight Center, Greenbelt, Maryland; and a Bragg Crystal Spectrometer built by Massachusetts Institute of Technology, Cambridge.

The Charged-Coupled Device Imaging Spectrometer will be up to 1,000 times more sensitive than current X-ray telescopes and will produce X-ray images of deep-space objects 10 times as sharp. With it, astronomers will be able to measure energy levels of X-rays received from extremely faint celestial bodies. The satellite has to orbit above Earth's atmosphere because X-rays can't pass through it to the ground.

Mysteries. Many astronomers, seeing powerful outbursts of radio and light energy from the hearts of many galaxies, say those enormous pulses of radiation are prima facie evidence for the existence of black holes—the vast gravity pits from which even light cannot escape. Wanting to know more, the astronomers suggest:

★Matter falling onto a black hole probably would release strong X-rays. By creating an image from X-rays, AXAF will show details of the immediate vicinity of any suspected black hole.

★Scientists need AXAF to study quasars, probably the most powerful energy sources in the Universe.

★Astronomers want X-ray images of spinning neutron stars, composed of matter so dense a teaspoonful would weigh more than the combined weight of everyone on Earth.

★If they explore neutron stars, scientists will need X-ray images of pulsars.

★AXAF images might depict data from which astronomers could deduce the exact age of the Universe.

★Is the Universe an open or closed loop? That is, will the Universe continue to expand forever, or will it eventually retract, collapsing in on itself, triggering another Big Bang, recycling matter into a new expansion? AXAF would be able to add some clues to this ultimate mystery.

★AXAF would expand our knowledge of the so-called dark matter spread through the Universe. Is it asteroids, planets, dark stars, black holes, exotic previously-unknown particles and other unknown kinds of matter?

★AXAF might be able to find and weigh invisible matter in galactic clusters. It might highlight differences among clusters, giving more clues about dark matter.

★The value of AXAF would reach beyond astronomy as an important tool for basic research in plasma physics, the fundamental properties of matter, and laws of physics.

Great Observatories. NASA's four Great Observatories telescopes are Hubble Space Telescope (HST) already in orbit, Gamma Ray Observatory (GRO) already in observatory, Space Infrared Telescope Facility (SIRTF) to be launched in 1999, and AXAF. All would be visited from time to time by maintenance crews in U.S. space shuttles from Earth and later from the U.S.-international space station Freedom.

The four high-tech satellites with strange-sounding acronyms, above the blurring atmosphere of Earth, open previously-unseen views of the Universe. NASA says with all four in orbit, the United States would be the world focal point for astronomy again.

Uhuru. The satellite observatory Uhuru detected a black hole in the constellation Cygnus on January 30, 1971, when the small astronomy satellite received X-rays and natural radio waves from the direction of a faint blue star which had a tiny, unseen, dark companion, more massive than a neutron star.

Uhuru was the U.S. Explorer 42 satellite launched to Earth orbit from Kenya December 12, 1970. The satellite was named Uhuru, a Swahili word for "freedom," in honor of Kenya's independence day December 12.

Einstein. The Einstein Observatory received X-rays from quasars billions of lightyears away. It found X-rays from sources in deep space 1,000 times farther than previously seen. It also received X-rays from Jupiter.

Also known as High Energy Astronomical Observatory 2 (HEAO-2), the Einstein Observatory was launched by NASA November 13, 1978, into 300-mi. circular orbit.

Another satellite, High Energy Astronomical Observatory 1 (HEAO-1), already had been launched August 12, 1977, carrying telescopes to map X-ray and gamma-ray sources in deep space. The X-ray telescope found 1,500 sources and a universal background gas enveloping galaxies. The gamma-ray telescope caught the strange galaxy Centaurus-A emitting gamma rays at one million electron volts energy.

A third telescope in the series, High Energy Astronomical Observatory 3, was launched by NASA on September 20, 1979. It had two cosmic-ray telescopes.

Rosat. German scientist Wilhelm Conrad Roentgen discovered X-rays in 1895. A 5,000-lb. space telescope named Roentgen Satellite (Rosat) and containing a German X-ray telescope was ferried to a 360-mi.-Earth orbit from Cape Canaveral June 1, 1990. Rosat was the largest imaging X-ray telescope built to that time. Rosat carried out the first full-sky survey of deep-space objects emitting X-rays. Its extraordinary results pointed astronomers toward targets needing further study with the AXAF telescope.

AXAF mirror. The same company which made the Hubble Space Telescope imperfection is building mirrors for AXAF. Hughes Danbury Optical Systems Inc., formerly Perkin-Elmer, said in August 1990 it could provide good mirrors.

Hughes Danbury said the Hubble problem had nothing to do with AXAF. A 1.3-mm spacing error was found in a measuring instrument used by Perkin-Elmer to grind the Hubble mirror in the early 1980's. Hughes Danbury says it is using a different polishing method for AXAF mirrors which will be cylindrical.

The two telescopes use different optical systems to explore the vast reaches of the Universe. NASA said the AXAF mirrors will undergo rigorous testing before launch.

The first two of AXAF's six cylindrical mirrors passed NASA tests in 1991. The 48-in.-diameter mirrors were tested in a new X-ray Calibration Facility at Marshall Space Flight Center. They are the largest ever made to collect X-rays in space.

The tested pair will be outermost of six concentric mirror pairs being made for the AXAF. The entire 12-mirror set will become the High Resolution Mirror Assembly which will be tested and calibrated prior to launch.

With good mirrors, AXAF is expected to deliver X-ray images with a resolution of 0.23 arc second, 10 times better than any previous X-ray telescope.

Dewar. AXAF will have a thermally-insulated container, known as a helium dewar, to cool its X-Ray Spectrometer (XRS). The dewar would operate for five years in orbit, the longest time a stored-cryogen supercooler would ever have operated in space.

The dewar will allow AXAF to be 100 times more sensitive than previous X-ray telescopes and work over a broader energy range. Data sent down to Earth by AXAF is expected to add extensive knowledge in the fields of plasma physics, atomic and nuclear physics, general relativity, and cosmology.

Science Center. The Smithsonian Astrophysical Observatory, Cambridge, Massachusetts, was chosen by Marshall Space Flight Center in March 1991 to become the official AXAF Science Center. It would design and operate a science support center for AXAF, develop and oversee the AXAF X-ray telescope observation program, and receive and distribute the data it collects.

X-rays. Stars and galaxies shine in at many different wavelengths, depending on chemical makeup and nuclear reactions. The higher the energy, the shorter the wavelength of light and electromagnetic energy given off. X-rays fall between ultraviolet light and gamma rays on the short end of the energy spectrum.

Visible light from deep space reaches Earth's surface, but high-energy radiation such as X-rays, gamma rays and ultraviolet light mostly do not pass through the atmosphere so are invisible to observers on the ground. Rosat will look for high-energy X-rays coming from stars, black holes and those hot clouds of gas and dust which are remnants of supernovas, or exploding stars.

X-ray radiating bodies in deep space, existing at temperatures of one million to 100 million degrees, are many millions of miles away across the Universe. Such objects release enormous amounts of energy, sometimes revealing evidence of black holes and other strange objects unseen in visible light.

Black holes are collapsed burned-out cinders of massive stars with gravity so intense visible light cannot escape. Gas and dust being sucked into a black hole is accelerated by the extraordinary gravity of the collapsed star, heating the particles to extremes and sending off showers of X-rays which can be detected by Rosat in Earth orbit.

By studying X-rays arriving at Earth from deep space, astronomers better understand how stars, galaxies and the Universe itself formed.

X-rays must be seen in space because they don't penetrate Earth's atmosphere. Gamma rays and ultraviolet light from deep space also do not penetrate the atmosphere. Until recently, knowledge of the Universe was limited to objects seen in visible light, infrared light and radiowave portions of the electromagnetic spectrum which reach Earth's surface. X-ray astronomy shows a Universe quite different from what astronomers see at optical wavelengths.

Telescopes In Orbit: SIRTF

The fourth and last of NASA's Great Observatories is to be the Space Infrared Telescope Facility (SIRTF), a thousand times more sensitive than earlier infrared satellites and maybe the telescope which finally sees planets in other solar systems.

The four high-tech American instruments in NASA's Great Observatories program are Hubble Space Telescope (HST), Gamma Ray Observatory (GRO), Advanced X-Ray Astrophysics Facility (AXAF) and SIRTF. HST and GRO already are working in Earth orbit. NASA has slotted AXAF for launch in 1998 and SIRTF in 1999.

SIRTF's receiver would look out into deep space, watching the invisible infrared part of the electromagnetic spectrum, most of which does not penetrate Earth's atmosphere.

Astronomers suspect planets orbit stars not far from our Sun. SIRTF may show planets around other stars. If astronomers see planets outside our Solar System, the search for extraterrestrial intelligence (SETI) will be strengthened.

SIRTF would be the most sophisticated infrared receiver ever flown, an evolutionary leap forward from renowned, long-lived, superfluid helium-cooled systems such as the

Infrared Astronomical Satellite (IRAS) launched in 1983 and the Cosmic Background Explorer (COBE) satellite launched in 1989.

Hubble works in the visible and ultraviolet part of the spectrum. GRO measures gamma rays. AXAF will receive X-rays. SIRTF is managed by NASA's Jet Propulsion Laboratory, Pasadena, California.

Telescopes In Orbit: COBE

America's sophisticated Cosmic Background Explorer (COBE) satellite, sent to space in 1989 to find out why the Universe has a "lumpy" consistency, has not been able to uncover answers to that basic question for astronomers.

The $160 million, ultra-sensitive COBE was launched on its two-year mission November 18, 1989, on a Delta rocket from Vandenberg Air Force Base, California. COBE is in a 570-mi.-high circular orbit inclined 99 degrees to the Equator. COBE flies over Earth's poles and along the terminator line between night and day.

COBE was sent to Earth orbit to look out across the Universe, searching for explanations of why the far flung star-clouds we call galaxies are distributed so unevenly across the cosmos. The deep space observatory was to see back to a possible beginning of the Universe, to study the Big Bang, the primeval explosion that some think sent the Universe expanding outward some 15 billion years ago.

Three receivers aboard the satellite recorded weak radiation supposedly left over from the Big Bang explosion. COBE completed a scan of the entire sky for microwave radiation arriving at Earth from across the Universe. COBE has three instruments looking at the spectrum of electromagnetic energy arriving at Earth from deep space— two seeing infrared light and one seeing microwave.

Bad news? Many scientists think the Universe exploded into existence 15 billion years ago in a hot, dense fireball which instantly started expanding in all directions. That's the so-called Big Bang theory.

That would make it seem the Universe should be smooth in its expansion, with an even distribution of matter everywhere. The problem is, today the Universe looks uneven or lumpy, with great clusters of galaxies dotted across giant dark voids. COBE was supposed to find out why, but has failed to find a trace of any significant energy release after the first huge explosion.

Astronomers would like to take comfort from COBE's failure to find traces of significant energy releases after the first Big Bang explosion—because that means background radiation is smooth with the same intensity in all directions, implying the Big Bang expansion of matter across the Universe has been uniform. Unfortunately, the universe does not look uniform today.

Astronomers had hoped COBE's instruments would detect subtle variations in the intensity of the cosmic background radiation, explaining the uneven distribution of stars through space. COBE has found no such variations.

COBE sent down a huge volume of information. Puzzled astronomers on the ground need many months to sift through the data windfall. But, with no further explanation of the lumps, the popular Big Bang theory may be in deep trouble.

Good news. There was a bit of good news in April 1990 when COBE sent down the best pictures ever seen of the center of our Milky Way galaxy. NASA combined near-infrared images from the satellite to make individual pictures, clear of the dust that usually blocks the view of the heart of our galaxy, exposing millions of stars in this galaxy. The photos were made by the Diffuse Infrared Background Experiment (DIRBE).

In October 1990, COBE took its first photograph of the Milky Way galaxy at a far-infrared wavelength which revealed the dust from which planets and galaxies are formed. The photos of the Milky Way, showing radiation from cold interstellar dust, were taken by the Diffuse Infrared Background Experiment (DIRBE). The April and October images highlighted differences in the distribution of stars, and the dust and gas clouds between stars, across the galaxy.

Cold tank. Weighing 5,000 lbs., COBE is 18 feet long and eight feet in diameter on Earth. In space, the observatory expanded to 27.5 feet diameter.

The spacecraft and its instruments—FIRAS, DMR and DIRBE—were designed and built at NASA's Goddard Space Flight Center, Greenbelt, Maryland.

The instruments were in a large vacuum-insulated tank at the very cold temperature of minus-457 degrees Fahrenheit. The tank, similar to one orbited in the Infrared Astronomical Satellite (IRAS) in 1983, is kept cold by evaporation of liquid helium. IRAS made sky maps of infrared light coming from deep space for 11 months until its supply of cooling helium ran out.

Ground controllers use NASA's TDRS tracking and data relay satellite to communicate with COBE. Observatory data is fed into tape recorders in the satellite, then radioed to the ground, analyzed at Goddard, and used to draw maps of the sky.

Infrared & microwave. COBE has three instruments—two infrared and one microwave—to study the diffuse cosmic background radiation coming to Earth from all directions. Electromagnetic energy recorded by COBE is infrared light and microwave radiation. Wavelengths range from one micrometer to one centimeter.

Scientists wonder if diffuse radiation arriving at Earth is from a Big Bang explosion or from more recent events. Some researchers think, since it comes to Earth from all directions, diffuse radiation is a remnant of the start of the early Universe.

Diffuse radiation and the motion of distant galaxies are the two pieces of evidence scientists use to claim the Big Bang really happened. Distant galaxies look like they are receding from Earth at speeds proportional to their distances—as though they had been blown away by a gigantic explosion.

Astronomers can't see the edge of the Universe, however. The Universe appears the same in every direction from Earth. It extends farther than the eye can see—even when augmented by the biggest and best telescopes.

Viewed from Earth, our Milky Way galaxy seems to be in the middle of the explosion, but astronomers say the center of the explosion actually can't be located.

Astronomers don't know if the Big Bang expansion will continue forever or if it will reverse into a gigantic collapse, a Big Crunch.

FIRAS. Diffuse radiation from the Big Bang first was predicted in the late 1940's. It finally was found in 1964. Scientists say the radiation is about as bright as if it came from a perfect black object at a temperature of minus-455 degrees Fahrenheit.

The far-infrared absolute spectrophotometer (FIRAS) aboard COBE was to determine whether or not the temperature really is the same at each wavelength. It was to measure the brightness of diffuse radiation at 100 different wavelengths for 1,000 different parts of the sky. FIRAS data was to be 100 times more accurate than previous information. With it, astronomers had hoped to see events which occurred only one year after the Big Bang explosion, but the big questions remained unanswered.

DMR. Although diffuse radiation seems to be about the same from all directions, astronomers had thought there might be small differences. They said half of our sky is about one-tenth of one percent brighter than the other half because of Earth's motion.

Other small differences may show the Universe is not homogeneous, as they had thought. Maybe it's rotating, they suggested. Maybe it's not expanding uniformly. Maybe it has small seeds that form into galaxies.

The differential microwave radiometer (DMR) aboard COBE looked for those small differences in the diffuse radiation. It measured 1,000 parts of the sky, giving better data than previous instruments, but the big questions remained unanswered.

DIRBE. Diffuse cosmic radiation coming from outside our Milky Way galaxy at wavelengths from one to 300 micrometers might have held the best surprises for astronomers. At those wavelengths, astronomers had hoped to be able to see the first stars and galaxies. Those very-old deep-space objects would not have been seen before.

The diffuse infrared background experiment (DIRBE) aboard COBE measured radiation at 10 different wavelengths from one to 300 micrometers, looking at 80,000 different parts of the sky. It was to detect the polarization of the radiation from one to three micrometers wavelength, which astronomers hoped would help distinguish nearby radiation sources near Earth from distant sources, but the big questions remained unanswered.

Telescopes In Orbit: Rosat

The Roentgen Satellite (Rosat), containing a German X-ray telescope, was ferried to a 360-mi.-high orbit from Cape Canaveral on June 1, 1990, by a U.S. Delta 2 rocket.

German scientist Wilhelm Conrad Roentgen discovered X-rays in 1895. The 5,000-lb. astronomy satellite named after him was to have been carried to space in a U.S. shuttle years before, but the 1986 Challenger disaster knocked it out of the schedule.

Rosat was one thousand times more sensitive than any previous X-ray telescope. Its sensitive telescope, and wide-field camera to detect extreme-ultraviolet light, equipped the satellite to carry out the first full-sky survey of deep-space objects emitting X-rays.

Launch. The 128-ft. Delta 2 had nine solid-fuel boosters strapped on for extra power. Launch from pad 17 at Cape Canaveral Air Force Station was delayed 13 minutes when a private Beech Kingair aircraft flew through the launch zone.

Some 43 minutes after blast off, Rosat was nudged gently away from the Delta 2 as they sailed over the Indian Ocean. The satellite's solar panels unfolded so Rosat could convert sunlight to electricity. Later, the Oberpfaffenhofen ground control centre near Munich, Federal Republic of Germany, reported all Rosat systems working like clockwork in orbit. Rosat was set to work at least 18 months in orbit.

Rosat. Rosat's telescope focal length was 7.78 feet with a maximum aperture of 835 millimeters. While the satellite kept its electricity-generating solar panels pointing at the Sun, the telescope swiveled to survey the entire sky every 180 days.

The robot observatory was the most ambitious X-ray astronomy project up to that time. It was the largest imaging X-ray telescope ever built. Rosat mapped sources of X-rays arriving at Earth from deep space. Rosat focused on 1,400 targets of special interest. Previously, astronomers had logged 5,000 objects emitting X-rays. Using Rosat, they expected to find 100,000 more.

Sponsors. The telescope was manufactured for West Germany's Ministry for Research and Technology by the Dornier aerospace company. The satellite, built more than five years before launch, had been in storage at Friedrichshafen, West Germany. It was shipped to Cape Canaveral in February 1990.

Rosat was a joint project of the Federal Republic of Germany, Great Britain and the U.S. It originally had been scheduled for a manned shuttle launch, but plans changed to an unmanned rocket after the 1986 Challenger disaster.

Rosat carried a British telescope, known as the Wide Field Camera, to observe extreme ultraviolet sources from deep space.

High energy. Stars and galaxies shine at many different wavelengths, depending on chemical makeup and nuclear reactions. The higher the energy, the shorter the wavelength of light and electromagnetic energy given off. X-rays fall between ultraviolet light and gamma rays on the energy spectrum.

Visible light from deep space reaches Earth's surface, but high-energy radiation such as X-rays, gamma rays and ultraviolet light mostly do not pass through the atmosphere so are invisible to observers on the ground. Rosat will look for high-energy X-rays coming from stars, black holes and those hot clouds of gas and dust which are remnants of supernovas, or exploding stars. X-ray radiating bodies in deep space, existing at temperatures of one million to 100 million degrees, are many millions of miles away across the Universe. Such objects release enormous amounts of energy, sometimes revealing evidence of black holes and other strange objects unseen in visible light.

Black holes are collapsed burned-out cinders of massive stars with gravity so intense visible light cannot escape. Gas and dust being sucked into a black hole is accelerated by the extraordinary gravity of the collapsed star, heating the particles to extremes and sending off showers of X-rays which can be detected by Rosat in Earth orbit.

X-ray astronomy. By studying X-rays arriving at Earth from deep space, astronomers better understand how stars, galaxies and the Universe itself formed.

X-rays must be seen in space because they don't penetrate Earth's atmosphere. Gamma rays and ultraviolet light from deep space also do not penetrate the atmosphere. Until recently, knowledge of the Universe was limited to objects seen in visible light, infrared light and radio wave portions of the electromagnetic spectrum which reach Earth's surface. X-ray astronomy shows a Universe quite different from what astronomers see at optical wavelengths.

Einstein. The Roentgen Satellite was similar to the Einstein Observatory (1978-81) but with more sensitive detectors. The Einstein Observatory or High Energy Astronomical Observatory 2 (HEAO-2), was launched by NASA November 13, 1978, into a 300-mi.-high circular orbit. It carried an X-ray telescope and a TV camera.

The Einstein Observatory spotted X-rays coming from quasars billions of lightyears away. X-rays were found from sources in deep space 1,000 times farther than had previously been seen. X-rays also were found coming from Jupiter.

First light. The NASA High Resolution Imager aboard Rosat sent its first-light pictures to Earth October 3, 1990. NASA said one of the pictures, of Cygnus X-2, showed a neutron star orbiting a normal stellar companion 3,000 lightyears from Earth.

A second photograph showed a supernova remnant, Cassiopea-A, in the constellation Cassiopea. It may have resulted from a supernova explosion in the Milky Way galaxy 320 years ago. The X-rays in the images came from a hot plasma cloud of several million degrees produced when the blast wave from the supernova struck the surrounding interstellar material. The supernova remnant is 9,000-10,000 lightyears from Earth.

Galaxy clusters usually give off strong X-rays. A third X-ray photograph showed a cluster of galaxies known as Abell 2256. The X-rays in the image came from hot gas swept out of colliding galaxies and accumulating between the galaxies. Such gasses are thought to be tens of million degrees in temperature.

The HRI images represent X-rays in the energy range known to astronomers as 0.4 to 2.5 kilo electron volts.

Rosat completed its initial six-month survey of the X-ray and extreme-ultraviolet sky in the first quarter of 1991. Rosat's all-sky survey of X-ray sources had produced the most sensitive and complete X-ray source map to that time, and identified more than 50,000 objects.

Pointing. Rosat entered a new phase of science operations February 8, 1991, when it began pointing operations. It was to observe particularly interesting sources of X-ray emissions identified by the NASA-sponsored U.S. guest observer program.

Gyro. However, the satellite lost one of its four gyros on May 13, 1991, and another was erratic. Some studies of X-ray sources could continue, but the satellite was unable to observe fulltime until ground controllers reprogrammed it to use two gyros and follow the Sun and stars for orientation.

The guest scientists were interested in X-ray images of high-energy celestial objects. Rich sources of X-rays are supernova remnants, galaxy clusters, quasars and binary star systems containing neutron stars or black holes. X-rays are very high natural energy emissions—greater than one million degrees.

The U.S. Rosat Science Data Center (USRSDC) at NASA's Goddard Space Flight Center, Greenbelt, Maryland, arranged guest investigators and processed and distributed Rosat data. USRSDC was a joint project of the Goddard Laboratory for High Energy Astrophysics, the Smithsonian Astrophysical Observatory High Energy Astrophysics Division and the Goddard Space Data and Computing Division.

Telescopes In Orbit: Hipparcos

Hipparcos is a reflecting telescope in an Earth-orbiting satellite, designed in Europe to measure the positions of 120,000 stars for precise new star maps. It rode to space August 8, 1989, atop an Ariane 4 rocket thundering away from Europe's launch pad in the South American jungle. But that wasn't the beginning of Hipparcos' story.

A strange thing happened to the star-mapper on its way to space. It almost didn't make it through a meteorite bombardment in Italy—before it was shipped to Europe's space center in French Guiana.

Hipparcos had been suggested back in 1966 by Pierre Lacroute of France's Strassbourg Observatory as a way to draw up a very precise catalog of 120,000 stars visible from above Earth's atmosphere.

It took the European Space Agency some time to get the project off the ground, but the satellite finally was constructed. The Hipparcos project was approved by the European Space Agency in 1980. Construction began in 1984.

Thirty years after the original proposal, ESA finished assembly and testing of Hipparcos in its ESTEC technical center at Noordwijk, The Netherlands. The satellite was packed into a nitrogen-filled container and sent to Aeritalia in Turin, Italy, to wait for a ride to space.

The satellite was to have been launched in July 1988, but Ariane rocket troubles delayed all ESA launches. Rather than ship the satellite to its launch pad at Kourou, French Guiana, South America, ESA kept Hipparcos in the Turin warehouse.

Meteorite. That's where a really odd coincidence appeared and a very rare encounter almost happened. On May 18, 1988, a 2-lb. meteorite smacked the surface of Earth near Turin, missing the Hipparcos warehouse by less than 1,000 feet.

The rock from outer space actually landed in the Aeritalia parking lot, less than a foot from a tiny Fiat automobile parked there. Excited astronomers immediately grabbed the biggest remaining fragment of the small rock—a 1-lb. chunk—named it Aeritalia-A, and hauled it off to a laboratory for close study. Unscathed, Hipparcos was shipped to Toulouse, France, for passage to Kourou and launch to space on Ariane 4 flight V-33. But, again the story didn't end.

The launch. Two payloads had been loaded aboard the Ariane rocket—Hipparcos and the TVsat-2 broadcasting satellite. The muggy tropical scene on August 8 at the European Space Agency's Guiana Space Center jungle launch complex on the northern coast of South America was described as "lovely." Engineers had to launch their Ariane 4 rocket by 7:32 p.m. or Earth would move out of proper position. Missing the deadline would have necessitated postponing launch to another day.

The countdown was just seven seconds before ignition when a fail-safe computer stopped the blast off. A glitch in a mechanical-controlling computer had prevented retraction of a fuel line attached to the rocket. Technicians scrambled to fix the problem before 7:32 p.m. Countdown was re-set to six minutes. After a delay of 39 minutes, the Ariane lifted off with just seven minutes left in the launch window.

The 194-ft. Ariane 4 lit up the nighttime sky above Kourou launch complex ELA 2 as Europe's most powerful launcher punched through the atmosphere toward space. The liquid-fuel Ariane blasted skyward for 21 minutes.

Kickless motor. TVsat-2 was ejected first from the Ariane nosecone into an elliptical "transfer" orbit ranging from 125 miles to 22,304 miles altitude. Two minutes later, Hipparcos was dropped off in a similar orbit.

Some 37 hours later, small solid-fuel rockets—kick motors or apogee boost motors—attached to the satellites were to be fired to push them into individual circular orbits 22,300 miles above Earth's equator.

The kick motor which was supposed to push Hipparcos to a stationary orbit over Africa failed August 10, stranding the star-mapper in an elliptical orbit. TVsat's kick motor worked, while the similar rocket in Hipparcos did not.

ESA engineers on the ground were aghast. From August 10 to 17, commands were radioed to the satellite five times to induce its kick motor to ignite each time the satellite sailed through the high point of its orbit. Each time the command failed. The frustrated engineers used tiny thrusters, designed to make minor orbit-correction maneuvers, to shift the satellite a bit higher, into a safer orbit from which it would not fall for a time.

Compromise. The 2,491-lb., solar-powered satellite was built to spin slowly in orbit, changing its axis of rotation so its telescope could scan the entire sky.

ESA engineers, with no way to boost Hipparcos on up to the circular 22,300-mi. orbit, calculated the satellite could complete most of its assignment from its low elliptical orbit. From the low elliptical orbit, Hipparcos still could sweep 80 percent of the sky. The engineers salvaged the mission with a decision to put Hipparcos to work in its lower orbit. They used the small thrusters to move the satellite into an orbit ranging from 314 miles to 22,310 miles. It revolves around Earth every 10 hours 39 minutes.

Getting on with it. Hipparcos finally was able to get down to serious work, cataloging the exact positions of 80 percent of the planned 120,000 stars.

Using data from Hipparcos, astronomers are determining more precisely than ever before where stars are in space and how far they are from Earth.

Data from Hipparcos was to have been received by a ground station at Odenwald, Germany. ESA added ground stations at Perth, Australia, and Kourou, French Guiana, to compensate for the irregular orbit and help Hipparcos calculate data.

120,000 stars. Scientists hadn't been very enthusiastic in 1966 when Pierre Lacroute of France's Strassbourg Observatory proposed the satellite to measure with what then was an unbelievable precision.

Lacroute thought a precision of one-hundredth of an arc second was possible. His fellow astronomers doubted it could be accomplished as they seldom had been able to achieve a precision better than one-tenth of an arc second. Data from Southern and Northern Hemisphere observatories then usually differed by up to several arc seconds on any one star.

Lacroute won the argument and Hipparcos was developed at a cost of $360 million. The satellite seemed expensive since it had only one small telescope, but that telescope was very sophisticated.

Even better than expected. The Hipparcos telescope ended up with a precision five to ten times better than even Lacroute had imagined. It would be able to measure the locations of 120,000 stars with a precision of one to two-thousands of an arc second—more than 10,000 times better the naked eye.

To do that, Hipparcos needed to be very stable in orbit. Matra, the French company which had designed the Faint Object Camera in the Hubble Space Telescope, designed the Hipparcos stabilization system. The telescope has two viewports, separated by an angle of 58 degrees. Pictures from the two openings are bent and brought together by a mirror about four inches thick and 12 inches in diameter. A latticework of 2,688 one-in. slots at the focus, lets the telescope see and measure the positions of many stars at one time at high speed.

Second star mapper. Besides the main telescope capable of measuring the locations of 120,000 stars with a precision of one to two-thousands of an arc second, there also is a less-precise star mapper on board.

Assigned to position the satellite, it always is directed toward a specific area of the sky so the satellite can keep track of its own location in space. As a fringe benefit, that tracker sees and maps an additional 400,000 stars with a precision of about one-30th of an arc second, about the same precision achieved today by telescopes on the ground. It also measures the intensity of light coming from the stars in two ranges of the electromagnetic spectrum.

The satellite takes in so much information it must send its data continuously and very rapidly down to Earth. It transmits 2,000 measurements per second, eventually totaling more than 200 billion star positions. As each star will be seen and computed several times during the two to three years, its direction of movement can be traced.

Darmstadt. Data is collected by ESOC at Darmstadt and dispatched across Europe to two separate groups of scientists who will double-check each other and coordinate the mapping. The data will be processed by two different computers, which will compose the final star catalog. The finished product, showing distances to the stars and their directions of movement, is expected to be available by the end of the 1990's.

Distances between stars in binary pairs can be estimated from the data. Also, the mass of a star can be deduced from the data. Information about mass helps astronomers calculate other physical properties of stars. Previously, useful star movement data had been available for only 1,500 stars.

The telescope satellite, measuring position, proper motion and trigonometric parallax, is named for Hipparchus, a Greek astronomer who lived about 130 B.C. Hipparcos stands for High Precision Parallax Collecting Satellite. It is expected to work about two to three years in its elliptical orbit, controlled by ESA's European Space Operations Centre (ESOC) at Darmstadt, near Frankfurt, Germany.

Astrometry. Finding the distance to a star is an astrometric measurement. Hipparcos is Europe's first astrometry satellite.

Astronomers find the distance from Earth to a star by measuring the star's position from each side of Earth's orbit. The diameter of Earth's orbit around the Sun is 184 million miles. During the time it takes Earth to get from one side of its orbit to the other, a star appears to moves. It seems to move slightly—like an object held at arm's length seems to jump back and forth as you look with one eye and then the other.

This apparent motion of a star is small, but sufficient for astronomers to sketch a triangle with a 184 million mile base. Trigonometry is used to calculate the length of the other two sides of the triangle.

From its position in orbit above Earth's murky atmosphere, Hipparcos improves astrometric measurements as it spins slowly, sweeping the whole sky every ten hours.

The complete new Hipparcos Star Catalog should be available between 1995-1999. Its unprecedented accuracy will have a major impact on astronomy. Hipparcos measurements will improve man's yardstick for measuring the size of the Universe, and deepen understanding of the evolution of stars, galaxies and the Universe.

Telescopes In Orbit: Astro-1

Ultraviolet and X-ray telescopes flew to Earth orbit in the U.S. space shuttle Columbia in December 1990, capturing extraordinary images and spectra of a variety of deep space objects, despite two crippling computer failures.

Astro-1 was an array of telescopes to study the Universe in ultraviolet light. A broadband X-ray telescope known as BBXRT also was aboard. Spacelab was carried in the shuttle's cargo bay. Astro-1 photographed stars and other deep-space objects impossible to see with visible-light telescopes on the ground. Astronomers returned to Earth with a shuttle load of extraordinary photos, including ultraviolet images of a spiral galaxy, a globular cluster and a supernova remnant. The three ultraviolet instruments:

★Hopkins Ultraviolet Telescope (HUT),
★Wisconsin Ultraviolet Photo-polarimeter Experiment (WUPPE),
★Ultraviolet Imaging Telescope (UIT).
A star tracker led the instrument pointing system.

HUT. The Hopkins Ultraviolet Telescope looked for faint astronomical objects—quasars, active galactic nuclei and supernova remnants—in the little-explored ultraviolet range below 1200 Angstroms. It looked at the outer planets of the Solar System for auroras and the interactions of each planet's magnetosphere with the solar wind. HUT's mirror focused on an aperture of a spectrograph.

WUPPE. The Wisconsin photo-polarimeter measured polarization of ultraviolet light from deep-space objects—quasars, hot stars and the nuclei of galaxies. It had two mirrors and a spectro-polarimeter. WUPPE measured polarization by splitting a beam of light into perpendicular planes, passing the two beams through a spectrometer and focusing them on two arrays of detectors.

UIT. The ultraviolet imager was a telescope with two image intensifiers. Its 70 mm film transports each held 1000 frames. UIT exposed images of faint objects in broad ultraviolet bands at a wavelength of 1200 to 3200 Angstroms. It also searched out the star content and history of star formation in galaxies, the nature of spiral structure and non-thermal sources in galaxies.

BBXRT. The broad-band X-Ray telescope (BBXRT) had two imaging telescopes with cryogenically-cooled lithium-drifted silicon detectors at each focus. A two-axis pointing system (TAPS) guided BBXRT's view as it looked at active galaxies, clusters of galaxies, supernova remnants and stars. BBXRT measured the amount of energy, in electron volts, of each X-ray detected. It studied Supernova 1987a, the brightest exploding star seen by modern Earth astronomers.

Astronauts in the Columbia STS-35 flight crew were commander Vance D. Brand, pilot Guy S. Gardner, mission specialists Jeffrey A. Hoffman, John M. "Mike" Lounge, and A. Robert Parker. Payload specialists were civilian astronomers Samuel T. Durrance and Ronald A. Parise. They spent eight days orbiting 190 nautical miles above Earth in the first shuttle flight in five years dedicated entirely to science.

The four astronomers on board used the ultraviolet and X-ray telescopes to study quasars, binary stars, pulsars, black holes, galaxies and high-energy stars. Specific targets included Supernova 1987a, the nearby supergiant star Betelgeuse, radio-quiet quasar Q1821, spiral-poor galaxy cluster Abell 2256, NGC 1633, NGC 1399 in the constellation Fornax, and Q1821+64.

Despite early telescope control problems in orbit, the instruments in the cargo bay ended up examining 135 deep space targets during 394 observations. The observations included HUT, 75 targets, 101 observations; UIT, 64 targets, 89 observations; WUPPE, 70 sources, 88 observations; and BBXRT, 76 targets, 116 observations.

NASA's Space Classroom became reality as Durrance, Hoffman, Parise and Parker conducted space-to-ground TV lessons on the invisible and visible Universe, electromagnetic spectrum and how telescopes work in space. The astronauts answered questions from students at schools on the ground.

Telescopes In Orbit: SARA

SARA, a first-of-its-kind amateur radio astronomy satellite made in France, was launched to Earth orbit July 17, 1991, on an Ariane rocket.

SARA was built at ESIEESPACE, a club of amateur astronomers at France's Ecole Superieure d'Ingenieurs en Electrotechnique et Electronique (ESIEE). Previously the club had built and launched payloads for suborbital rocket and balloon flights for six years.

SARA was launched by European Space Agency from its Kourou, French Guiana, spaceport. As is the custom in amateur radio satellite work, SARA was renamed in orbit to become OSCAR 23. OSCAR stands for Orbital Satellite Carrying Amateur Radio. The shorthand version of its new name is SO-23, for SARA-OSCAR 23. SARA stands for Satellite for Amateur Radio Astronomy.

Radio astronomy. The first radio astronomer was Karl Jansky who worked at Bell Telephone Laboratories in New Jersey in 1932. He was the first to tune in radio waves coming from somewhere in the sky, beyond Earth.

If Jansky was the father of radio astronomy, amateur radio operator Grote Reber of Wheaton, Illinois, was the father of the radiotelescope. Reber built the first radiotelescope in 1937 in his backyard. He used the 31-ft. dish antenna to uncover the first discrete radio sources beyond Earth and mapped the spread of natural radio signals across the Milky Way. Reber worked alone in the field until World War II.

SARA. SARA is not a typical amateur radio satellite used for communications among hams on the ground. Rather, it is a radio astronomy satellite listening for high-frequency (HF) radio signals from Jupiter and relaying data to the ground. SARA identifies itself by the amateur callsign FXØSAT.

SARA's simple gear was conservatively designed around off-the-shelf consumer equipment, rather than space-qualified or military-specification components. It is expected to work many years in orbit.

Jupiter. Jupiter is the fifth planet out from the Sun and may itself have been a prototype Sun which wasn't big enough to become a star. Jupiter is the largest of nine planets circling the Sun. In fact, Jupiter alone is more than two-thirds of the total mass of all planets in our Solar System. It's 318 times the mass of Earth. Although a thousand times smaller than the Sun, Jupiter is a big planet—as big as 1,317 Earths.

Still, Jupiter wasn't big enough. If the planet had been several times more massive, it might have become a star as the pressure and temperature at its core triggered nuclear fusion. Jupiter is a big gas bag, with a relatively low density.

Jupiter is more than five times farther from the Sun than the Earth, at a distance of 483.6 million miles. It orbits the Sun every 11.9 years.

Astronomers have known that Jupiter somehow sends out naturally-generated radio noise in the HF part of the electromagnetic spectrum. Earth's atmosphere, and sometimes our planet itself, block the signals, making them hard to study from the ground. However, a receiver in a satellite orbiting Earth above the atmosphere can hear the radio waves from Jupiter.

Satellites have measured Jovian emissions, but not for long periods. Voyager 1 tuned in briefly as it flew by Jupiter on March 5, 1979. Unfortunately, Voyager's receiver itself was noisy and it could hear only the strongest signals coming from Jupiter.

Io. Jupiter has 16 moons: Ganymede, Callisto, Io, Europa, Amalthea, Himalia, Elara, Thebe, Metis, Adrastea, Pasiphae, Carme, Sinope, Lysithea, Ananke and Leda. The radio signals sweeping Earth are generated by the interaction between Jupiter and its moon Io. The signal arrives at Earth as a beam of energy, appearing four minutes earlier each day, taking two hours to swing across Earth.

Before SARA, no satellite had listened to Jupiter, during peak solar activity, in the HF frequency range between 2 MHz and 15 MHz.

SARA's orbit. SARA, dubbed microsat because of its small size, was a secondary payload on Ariane V-44 which took the Earth Resources Satellite (ERS-1) to orbit. Other microsats on the same trip to orbit were Orbcom-X, TUBsat and UoSAT-F.

Since SARA received a free ride to space, it had to go where engineers wanted ERS-1 to go. In its orbit, Earth sometimes gets between SARA and Jupiter. That shuts off reception of signals from Jupiter for up to 40 minutes per orbit. If SARA could have been flown to an orbit never blocked by Earth, interruptions would not occur.

Antennas. Electromagnetic energy from Jupiter actually is very strong, so SARA needs only three pairs of 15-ft. antennas perpendicular to each other to snare the radio waves. The antennas are 100-mm-wide steel tapes. They had been rolled up for the launch, then unrolled in orbit. Their perpendicular arrangement allows radiowaves to be captured and signal strength to be measured, no matter which direction the satellite is turned nor the polarity of the signal. One of the pairs doubles as a telemetry antenna, transmitting data down to Earth in the two-meter amateur radio band near 146 MHz.

Receiver. SARA's receiver listens to eight 100-KHz channels spread across the spectrum between 2 MHz and 15 MHz. The receiver switches among the eight channels and three antenna pairs many times during a 150-second period, averaging the signal strength of the eight channels. Averaging smooths out any storm peaks on Jupiter.

Transmitter. The receiver is sensitive enough to detect galactic background noise which has a constant level of strength. Galactic noise gives the amateur radio astronomers on the ground a point of reference when Jupiter and the Sun are silent.

The computerized satellite circles Earth every 100 minutes in a polar orbit 478 miles above Earth. It continuously sends data by teletype to the ground with a one-watt amplitude-modulated (AM) transmitter on 145.955 Mhz. It can be shut off by ground control. The satellite draws three watts of electricity from its solar-charged battery. Solar cells are on each side of the satellite, covering 60 percent of the 18.5-in. cube.

A thermal coating was applied to the 40 percent of the outside of SARA not covered with solar cells to keep the satellite interior at 68 degrees Fahrenheit.

Hearing SARA. When SARA is overhead, telemetry coming down from the satellite should be receivable at most locations. A listener would need a receiver which can be tuned to 145.955 MHz, such as an amateur radio two-meter radio or a police-fire scanner or monitor radio. A high outdoor antenna would work best. Warbling tones would be heard, representing 300 Baud ASCII wide-shift AFSK modulation.

Telescopes In Orbit: Solar-A

The Sun may look like a hot old yellow ball sitting out in space doing nothing, but things aren't what they seem. Our temperamental star goes through periods of intense activity followed by quiet periods. These ebb and flow over 11-year periods in what has come to be known as the Sunspot Cycle.

During periods of peak activity, the Sun fires off magnificent high-energy solar flares which effect the transmission of radio and television signals on Earth. To look at these flares from a better vantage point, Japanese and American scientists placed an eye in the sky—an 850-lb. satellite called Solar-A.

Japan's Solar-A satellite was launched on an M-3S-II rocket to a low Earth orbit on August 30, 1991, from Kagoshima Space Center. In space, following Japanese custom, it was renamed Yohkoh. That means sun ray, sun beam or sun light in Japanese.

Yohkoh will stare straight at the Sun for three years at 344 miles above Earth, looking for solar flares on the face of our star. The satellite will see X-rays and gamma rays coming from the Sun.

NASA scientists supplied the optics and detector for a soft X-ray telescope aboard Yohkoh. Other instruments were a Japanese hard X-Ray telescope, Great Britain's Bragg crystal spectrometer and a wide-band spectrometer from Japan. The Japanese built the satellite's flight electronics, computer and power system. Yohkoh is a project of Japan's Institute of Space and Astronautical Science (ISAS).

First photo. Soft X-rays are weaker X-rays containing less energy. NASA's soft X-ray telescope, one of four instruments aboard Yohkoh, snapped its first photograph of the Sun on September 30, 1991.

The photograph showed intricate filaments of the corona extending far above the Sun's surface. The Sun's magnetic field ensnares hot coronal gases in loops called magnetic bottles. The brightness of a loop reveals its temperature and gas density.

The high-resolution images of solar flares in hard X-rays were measured at 15-100 KeV, while soft X-rays were measured at 4-40 angstroms.

In addition to these experiments, two other instruments will observe the soft X-ray emission-line spectrum and obtain the time series of hard radiations from flares.

Bubble memory. The box-shaped, 3-ft. by 6-ft. Solar-A was Japan's fourteenth science satellite. It was the only dedicated solar-flare satellite during very-high activity on the Sun in 1991-92. To accomplish its objectives, the Japanese needed to design a satellite with three-axis stabilization. Solar-A was the first ISAS satellite with three-axis stabilization precise enough to point a high-resolution telescope.

The satellite circled the globe each 98 minutes. During each revolution around Earth, Yohkoh stored up to 80 megabytes of data in a magnetic-bubble recorder. Contents of the recorder are "dumped" by radio to waiting technicians on the ground every 98 minutes. The data was received by NASA's Deep Space Network and Kagoshima Center.

Solar-A was one of a series of satellites launched by Japan on M-3S-II rockets between August and September 1991, a period of very high solar activity.

Telescopes In Orbit: EUVE

NASA's four Extreme Ultraviolet Explorer (EUVE) astronomical telescopes were to be launched to orbit in December 1991 aboard a Delta 2 rocket from Cape Canaveral.

The UV telescopes were mounted in an Explorer module, a piece of standardized satellite hardware, at NASA's Goddard Space Flight Center, Greenbelt, Maryland.

EUVE was designed by astronomers wanting to make observations in an unexplored part of the electromagnetic energy spectrum—extreme ultraviolet between X-rays and ultraviolet light. Extreme Ultraviolet Explorer, during its first two years orbiting Earth, will survey the entire sky at wavelengths of 100 to 1,000 angstroms.

After that, astronomers will use EUVE to look at the brightest individual sources of extreme-ultraviolet light in deep space. A definitive sky map of the sky in the extreme-ultraviolet spectrum and a catalog of extreme-ultraviolet sources will be produced.

The science payload was built by Space Astrophysics Group, Space Sciences Lab, University of California, Berkeley. Science operations will be centered at Berkeley's Center for Extreme Ultraviolet Astrophysics. Flight control will be at Goddard.

Telescopes In Orbit: Uhuru

The U.S. satellite Uhuru detected a black hole in the constellation Cygnus on January 30, 1971, when the small astronomy satellite received X-rays and natural radio waves from the direction of a faint blue star which had a tiny, unseen, dark companion, more massive than a neutron star.

Named Explorer 42 (SAS-1) when it was launched to Earth orbit from Italy's San Marco Platform in the Indian Ocean off Kenya on December 12, 1970, the X-ray hunting satellite was renamed Uhuru, a Swahili word for "freedom," in honor of Kenya's independence day December 12. Uhuru descended into the atmosphere April 5, 1979.

Telescopes In Orbit: IUE

The 10th anniversary in 1988 of its launch to space was just another day on the job for the International Ultraviolet Explorer (IUE) space telescope.

IUE, which captures ultraviolet light from space, is one of Earth's most productive telescopes. It was launched from Cape Canaveral, Florida, January 26, 1978. The Earth satellite, originally expected to last only three years, still works well.

Astronomers noted the 10th birthday as they focused IUE on the exploding star, Supernova 1987a, 163,000 lightyears distant in the Large Magellanic Cloud galaxy.

During its first decade in space, astronomers used IUE to find halos of hot gas around our Milky Way galaxy; snapped the first pictures of Comet Halley from space; peered at volcanos on Jupiter's moon Io; and, since February 24, 1986, caught the intense blast of ultraviolet radiation from Supernova 1987a.

The telescope is controlled 16 hours a day from Goddard Space Flight Center, Greenbelt, Maryland, near Washington, D.C. For the other eight hours each day, it is controlled from Villafranca Ground Station near Madrid, Spain. IUE is a joint effort of NASA, the European Space Agency and the British Science and Engineering Research Council.

Over the years, 800 astronomers at Goddard and 750 at Villafranca have used IUE. There are many larger telescopes on the ground, but being in space above Earth's hazy atmosphere improves images.

Telescopes In Orbit: Einstein

The Einstein Observatory received X-rays from quasars billions of lightyears away. In fact, it was able to see X-rays coming from sources in deep space 1,000 times farther than previously seen. It also received X-rays from the planet Jupiter.

Also known as High Energy Astronomical Observatory 2 (HEAO-2), the Einstein Observatory was launched by NASA on November 13, 1978, into a 300-mi.-high circular orbit. Along with its X-ray telescope, it carried a television camera. The Einstein Observatory descended into the atmosphere March 25, 1982.

HEAO-1. A year before HEAO-2 was launched, HEAO-1 (High Energy Astronomical Observatory 1) was launched on an Atlas-Centaur rocket August 12, 1977. It carried telescopes to map X-ray and gamma-ray sources in deep space.

The X-ray telescope found 1,500 sources and a universal background gas enveloping galaxies. The gamma-ray telescope caught the strange galaxy Centaurus-A emitting gamma rays at one million electron volts energy. HEAO-1 descended March 15, 1979.

HEAO-3. A year after HEAO-2, NASA launched HEAO-3—the High Energy Astronomical Observatory 3—September 20, 1979. It had two cosmic-ray telescopes. HEAO-3 fell from orbit December 7, 1981.

Telescopes In Orbit: Solar Max

Solar Max was launched from Cape Canaveral, Florida, on a Delta rocket on Valentine's Day, February 14, 1980, on what then was to have been a two-year spaceflight to study the Sun's atmosphere, solar flares, and solar energy. Instead, it worked in space for ten years.

The Solar Maximum Mission (Solar Max) satellite was launched to monitor the Sun during peaks of solar activity. The satellite had gamma ray, ultraviolet and X-ray spectrographs, a coronagraph and a radiometer. The X-ray imaging spectrometer was an international project involving The Netherlands, Great Britain and the U.S.

Solar Max satellite was used to snap pictures of more than 12,500 solar flares. It exposed 250,000 photos of the Sun's corona and recorded more than 100,000 images in ultraviolet light. Astronomers using Solar Max discovered ten sun-grazing comets.

Solar Max was first to detect gamma rays from Supernova 1987a, an exploding star in a nearby galaxy. From 1986 through 1989, scientists also used Solar Max to look down to altitudes of 34 miles to 47 miles with its ultraviolet spectrometer and polarimeter to study the concentration of ozone in the Earth's mesosphere.

Solar flares are energy bursts from areas near sunspots in the Sun's corona, or outer atmosphere, blasting electrically charged particles and radiation into space. Lasting minutes to hours, solar flares release the energy of 10 billion one-megaton nuclear bombs, emissions which can disrupt communications on Earth. The Solar Maximum Mission satellite was operated by Goddard Space Flight Center, Greenbelt, Maryland.

Repair in orbit. Solar Max sent down data for nine months until a fuse blew and its attitude controls failed, leaving it unable to point correctly. Fortunately, Solar Max had been the first satellite designed to be retrieved by shuttle.

NASA launched a space repair mission April 6, 1984, sending shuttle Challenger with five astronauts to a 288-mi.-high orbit to grab the broken satellite and fix it.

The crew was commander Robert L. Crippen, pilot Francis R. "Dick" Scobee, and mission specialists James D. "Ox" Van Hoften, Terry J. Hart and George D. "Pinky" Nelson. During their third day in space, Challenger's orbit was raised to 300 miles. Scobee flew the shuttle within 200 feet of Solar Max.

Nelson and Van Hoften, wearing spacesuits, left the cabin and went out into the payload bay. Nelson donned a manned maneuvering unit (MMU) and flew out to the satellite. He tried to grasp Solar Max with a capture tool known as a Trunnion Pin Acquisition Device (TPAD). Three attempts to clamp TPAD onto the satellite failed.

Van Hoften then tried to grab the satellite using the shuttle's 60-ft. robot arm—the remote manipulating system (RMS)—but Solar Max started tumbling. The effort was called off. While the astronauts slept that night, the Goddard operating center was able to regain control of the satellite by radioing commands which ordered its magnetic torque bars to stop the tumbling. Solar Max went into a slow, regular spin.

Got it. The next day, Nelson and Van Hoften went back into the cargo hold to try again to capture Solar Max. They succeeded on the first try, using the RMS to place Solar Max on a cradle in the payload bay where they replaced the satellite's attitude control mechanism and the main electronics system of its coronagraph instrument.

The refurbished Solar Max was pushed back overboard into its own orbit the next day, ending one of the most unusual rescue and repair missions in the history of the U.S. space program. Solar Max resumed monitoring the Sun.

Just a month after that first-ever overhaul in orbit, in May 1984, Solar Max recorded one of the largest flares ever seen erupting on the Sun.

Solar Max recorded science data about the Sun for nearly 10 years, covering almost one complete solar cycle. Highlights of the Solar Max science harvest:

★One of the important findings by Solar Max was discovery of the variability of astronomy's Solar Constant.

★Solar flares are gigantic eruptions on the surface of the Sun, waxing and waning on an eleven-year cycle. Solar Max recorded more than 12,500 flares, letting scientists measure sunspot magnetic fields above the visible surface of the Sun for the first time. Solar Max confirmed the existence of neutrons in solar flares.

★The gamma ray spectrometer aboard Solar Max was first to detect gamma rays from Supernova 1987a, the distant star that was seen in 1987 exploding in a neighboring galaxy, the Magellanic Cloud.

★Solar Max discovered 10 comets skimming past or crashing into the Sun, the last September 28, 1989.

★The satellite's ultraviolet spectrometer uncovered increases in the level of high-altitude ozone above Earth north of the equator and decreases south of the equator.

Natural gamma rays are photons found in cosmic rays, probably generated by deep-space objects such as supernovas, quasars and neutron stars. Photons, produced in nuclear fission, are far more powerful than X-rays or light rays. Astronomers are not sure of the processes. Solar flares generate gamma rays. Earth's atmosphere absorbs gamma radiation before it reaches the ground.

Nuclear blindness. The USSR has built nuclear reactors into three dozen satellites since the 1960's. The reactors generate electricity for satellite electronics, often radar sweeping the oceans for U.S. warships. To save weight, satellites are designed to shield only for their own equipment. Most reactor radiation is gamma rays.

The fact that radiation from nuclear reactors in USSR spy satellites sometimes blinded Solar Max was kept secret for years by the U.S. government. The astronomy satellite was blinded eight times a day by gamma rays from the reactors. During those orbits, data could not be gathered.

Positrons. Another problem for Solar Max were electrons and positrons, which are positively-charged electrons. The charged particles often were trapped for several minutes by Earth's magnetic field. Some positrons, colliding with ordinary matter in the satellite were annihilated. That yielded even more gamma rays.

Through 1987-88, the satellite's digital data storage would become saturated by positron-generated gamma rays eight times a day. That would blind Solar Max for the rest of each orbit until the data could be transmitted to the ground.

Flux from a nearby satellite nuclear reactor would be 50 times greater than flux from celestial sources. Knowing that, scientists were able to detect a satellite nuclear reactor in orbit up to 1,500 miles altitude.

Falling. NASA knew Solar Max was falling in 1989. The satellite was down to an altitude of 206 miles October 31, to 177 miles by November 17 and 166 miles November 22, on the way to its fiery re-entry December 2.

To prepare for the drop, Goddard controllers started to put Solar Max to sleep November 17, shutting off electricity to three science instruments needing a stable orbit to work. On November 22, technicians radioed a command to jettison the high-gain antenna. The three instruments continued to record data to be sent down as Solar Max passed over ground stations. Finally, solar panels, which provided power to recharge its batteries, were jettisoned November 24, ending science operations.

"I'm kind of sad to see it come back," astronaut George D. "Pinky" Nelson told reporters as the satellite was falling from orbit.

Down. The boxy 5,000-lb. Solar Max ended ten years of duty in Earth orbit December 2, 1989, when it dived into the planet's atmosphere and burned over the Indian Ocean near Sri Lanka. Solar Max entered the atmosphere at 3.1 degrees north latitude, 88.6 degrees east longitude, far to the west of the predicted entry point over the Galapagos Islands in the Pacific Ocean 1,300 miles west of Ecuador. The drop was about 45 minutes ahead of U.S. predictions. There were no reports of debris falling on land.

Future. Scientific study of the Sun continues. The Ulysses Sun-orbiting spacecraft was launched in October 1990. A U.S. telescope to see soft X-rays was aboard Japan's Solar-A satellite launched to Earth orbit in 1991.

The Solar, Anomalous and Magnetospheric Particle Explorer (SAMPEX) was to be launched in 1992, carrying a telescope to study energy particles from the Sun. Argentina's Satellite de Aplicaciones Cientificas-B (SAC-B), to be fired to Earth orbit in 1994, will carry a hard X-ray spectrometer to study solar flares.

Telescopes In Orbit: DE-1

NASA's satellite Dynamics Explorer-1 (DE-1), launched August 3, 1981, was the first to acquire global images of Earth's aurora.

In a highly-elliptical orbit, ranging from only 300 miles above Earth out to 14,500 miles, DE-1 was used by astronomers to study the coupling of energy, electric currents and mass between the Earth's upper atmosphere, ionosphere and magnetosphere.

Dynamics Explorer-1 was designed to last three years, but lasted nearly ten. NASA ended its use of Dynamics Explorer-1 in March 1991. It had refused to accept commands since November 17, 1990.

Dynamics Explorer-2 (DE-2), also launched August 3, 1981, had descended February 19, 1983. Project scientists said the quality and quantity of data returned by the satellite companions exceeded expectations.

Telescopes In Orbit: IRAS

Results from the Infra-Red Astronomical Satellite (IRAS), one of the most sophisticated astronomy satellites orbited in the 1980's, has challenged standard theories of how galaxies formed in space.

With a superfluid helium-cooled telescope, IRAS was a joint U.S.-British-Dutch satellite launched on January 26, 1983, to an orbit 550 miles above Earth. From there, the telescope in the satellite could see infrared light coming from more than a quarter-million deep-space objects, including 15,000 galaxies. IRAS created the most comprehensive deep-sky map to that time.

A group of British and Canadian astronomers surveyed IRAS results to chart 2,000 galaxies, some 450 million light years away. It was the most comprehensive map to that time of deep space close to the Solar System.

The result: no model of galaxy formation validates the existence of large clusters of galaxies and the vast voids between the clusters.

On a map covering so much of the Universe, there simply were too many galaxy clusters, and the clusters were just too large, to fit theories to date.

One popular theory had proposed something called "cold dark matter" (CDM). Supposedly, CDM explains why much less matter is actually seen in the Universe than the amount required to keep the Universe from collapsing.

The CDM theory assumes the tug of gravity can be detected from some sort of unknown, invisible matter. Proponents say such cold, dark matter would not be like anything found on Earth or in the Sun. Unfortunately for the proponents, their CDM model doesn't explain the very large galaxy clusters.

Telescopes In Orbit: Small Explorers

Small Explorers are inexpensive 400-lb. satellites NASA plans to launch to low Earth orbit on Pegasus and Scout rockets in the 1990's.

Each Small Explorer satellite is expected to cost about $30 million. The project is operated by the special-payloads division of NASA's Goddard Space Flight Center, Greenbelt, Maryland.

Past Explorers. NASA has been launching Explorer satellites for years, with more than 75 U.S. and international-cooperation payloads sent to orbit in the program. A notable example of an Explorer-series satellite is IUE—the International Ultraviolet Explorer—which went to space in January 26, 1978, and sent down sufficient information from space for astronomers to write 1,400 articles for science journals.

The satellite's telescope, controlled from ground stations in Maryland and Madrid, Spain, has a telescope that can be used in real time, as if it were on the ground.

Tens of thousands of images have been made through the telescope, including ultraviolet light photos of planets, comets, interstellar dust and gas, supernovas, stars, galaxies and quasars.

IUE was a joint project of NASA, the European Space Agency and Great Britain's Scientific and Engineering Research Council (SERC).

Among the many other Explorers were the International Comet Explorer (ICE) and two International Sun-Earth Explorer (ISEE) satellites.

Future Explorers. NASA has six Small Explorer missions scheduled for 1991-1994. Those flights will include:

★A Small Explorer which was to be launched about June 1992 on a Scout rocket will be the Solar, Anomalous and Magnetospheric Particle Explorer (SAMPEX), carrying a telescope to study energy particles from the Sun, cosmic rays from across the Milky Way galaxy and electrons from the magnetosphere. The project was designed at the University of Maryland and will involve 10 other U.S. and West German scientists.

★Another Small Explorer will be the Submillimeter Wave Astronomy Satellite (SWAS), to be launched about June 1995 on a Pegasus rocket with a telescope to observe molecular clouds. Scientists at Harvard-Smithsonian Center for Astrophysics, Cambridge, Massachusetts, and 11 others from across the U.S. and Cologne, West Germany, hope to discover how molecular clouds collapse to form stars and planets.

★The Small Explorer known as Fast Auroral Snapshot Explorer (FAST) is to be launched about September 1994 on a Pegasus rocket. Designed at the University of California, Berkeley, it also will include scientists from Lockheed Palo Alto Research Laboratory, California, and the University of California, Los Angeles. Scientists hope to discover how aurora in Earth's upper atmosphere works.

★Two Small Explorers known as Total Ozone Mapping Spectrometer (TOMS) will be launched on Pegasus rockets in September 1993 and June 1997 to study ozone in Earth's stratosphere. It will provide daily maps of ozone around the globe, helping scientists predict ozone depletion.

Ten scientists at NASA-Goddard designed the project. The agency called TOMS a high-priority Earth-observation mission, critical for monitoring long-term depletion of ozone from the stratosphere.

★The Small Explorer known as High Energy Transient Experiment (HETE) will be launched on Pegasus in July 1994 to study gamma ray sources and their locations in space and X-ray sources and locations.

★The Small Explorer known as Satellite de Aplicaciones Cientificas-B (SAC-B), to be be fired to space on Pegasus in July 1994, will be an Argentine satellite carrying a hard X-ray spectrometer to study solar flares and cosmic transient X-ray emissions.

★NASA's schedule shows a dozen other Small Explorers being launched on Pegasus and other small rockets from 1996 to 2002.

Telescopes In Orbit: Radioastron

Radio portions of the electromagnetic spectrum will be studied after the USSR launches a radioastronomy satellite known as Radioastron to Earth orbit sometime between 1993 and 1995. The satellite will have a 33-ft. antenna to gather radio waves from deep space for five years.

The U.S. will help with tracking the satellite and lend recorders to capture data transmitted to the ground. American astronomers also will help plan the launch and work with the USSR on the project while Radioastron is in space.

Astronomers around the globe will analyze the data for new insights into galaxies, black holes, neutron stars and quasars. Seventeen other nations also are participating.

U.S. Vice President J. Danforth Quayle, chairman of the National Space Council, told the press March 8, 1990, mankind is entering a golden era in space exploration.

He said U.S. help with Radioastron is a small, but significant step in evolving cooperation in space. Quayle recalled that Soviet physicist Andrei Sakharov, before he died, had asked for U.S. involvement in the project.

Telescopes In Orbit: Italian X-ray

An X-ray astronomy satellite built for the Italian space agency—Agenzia Spaciale Italiana or ASI—and the Netherlands Agency for Space Programs is to be launched in 1994 from Cape Canaveral.

An Atlas 1 rocket will carry the 2,868-lb. satellite from launch complex 36B at Cape Canaveral Air Force Station, Florida. Manufactured by Aeritalia, the satellite would complete 2,000-3,000 X-ray observations during its lifetime.

Telescopes In Orbit: SOHO

NASA's Solar and Heliospheric Observatory (SOHO) will be launched in July 1995 on an Atlas-2AS rocket to investigate how the solar corona is formed and heated. It also will study the Sun's interior structure.

The Corona is seen from Earth as a faint haze surrounding the Sun during a total eclipse. SOHO is part of an International Solar and Terrestrial Physics Satellite program of NASA, European Space Agency (ESA) and Japan's Institute of Space and Astronautical Science (ISAS) to study the Sun and solar phenomena. Various space launches in the program are to take place from 1992 through 1995.

SOHO will be placed in a highly elliptical Earth orbit. Its on-board rocket motor will drive it one million miles from Earth. It's designed to work two years.

The SOHO project is managed by NASA's Lewis Research Center, Cleveland, Ohio.

Telescopes In Orbit: HEXT

High Energy X-Ray Telescope (HEXT) is one of NASA's sounding rocket programs. HEXT is operated by NASA's Marshall Space Flight Center, Huntsville, Alabama.

Stars and galaxies shine at many different wavelengths, depending on their chemical makeup and nuclear reactions. The higher the energy, the shorter the wavelength of light and electromagnetic energy given off. X-rays fall between ultraviolet light and gamma rays on the energy spectrum.

X-rays must be seen in space because they don't penetrate Earth's atmosphere. It does not go to orbit, but at a very high altitude in Earth's atmosphere HEXT will look for high-energy X-rays coming from space. X-rays arriving at Earth give astronomers better understanding of how stars, galaxies and the Universe itself formed.

Future U.S. Telescopes

After studying future needs for two years, a 15-member committee organized by the National Academy of Sciences' National Research Council recommended in March 1991 the U.S. government finance a few large telescope projects to study the Universe in the 1990's. The ten projects they recommended were:

★The $1.3 billion Space Infrared Telescope Facility (SIRTF) satellite, which would complete NASA's Great Observatory program. It would be almost 1,000 times more sensitive than Earth-based telescopes operating in the infrared.

★To search for Jupiter-sized planets around stars up to 500 lightyears away, the $250 million Astrometric Interferometry Mission. Improved interferometry would boost by one-thousand times the current precision in locating deep space objects.

★A $230 million, 8.2-ft. telescope mounted on a Boeing 747 jet aircraft, known as Stratospheric Observatory for Far-Infrared Astronomy.

★To provide high-resolution photos of deep-space star-birth areas and star-burst galaxies, the $115 million Millimeter Array of telescopes.

★An $80 million infrared telescope with a 26.2-ft. mirror at Mauna Kea, Hawaii.

★A $70 million Far Ultraviolet Spectroscopy Explorer satellite.

★A $55 million 26.2-foot optical telescope in the Southern Hemisphere.

★An extra $30 million for 13-foot optical telescopes.

★The $15 million Large Earth-based Solar Telescope, using adaptive optics.

★Increasing Explorer satellites launched on unmanned rockets to five.

It was not clear whether the U.S. government would finance the projects during a time of federal budget problems. NASA budgets had dwindled recently, while competition for federal dollars grew between advocates of manned vs. unmanned space research.

How Astronomers Measure Distance

Astronomers compute the distance from Earth to a nearby star by measuring its position twice, once on each side of Earth's orbit around the Sun. During the course of our year-long journey around our own star, a distant star will seem to move slightly in the same way a ball held at arm's length will seem to jump back and forth as you look at it first with one eye and then the other.

Earth orbits the Sun once each year. The diameter of Earth's circuit around the Sun is 184 million miles. The apparent motion of a star is very small because it is so far away, but sufficient for astronomers to sketch a triangle with a base 184 million miles long. Trigonometry then is used to calculate the length of the triangle's sides. Finding the distance to a star is an astrometric mesurement.

Astrometric measurements are very difficult. But, from above Earth's murky atmosphere, a telescope in a satellite such as Europe's Hipparcos can improve astrometric measurements. Hipparcos gives astronomers especially interesting results because it spins slowly, sweeping great circles across the sky, completing a rotation every two hours. The telescope scans the complete celestial sphere.

Electromagnetic Environment

The space environment—both beyond and on Earth—is being fouled by human acquisitiveness.

Electromagnetic babble from millions of man-made radio sources is drowning out whispers from the heavens and crippling research by radio astronomers who probe distant stars.

The problem is not from radio transmitters intended to provide two-way communications. Those usually are well engineered and remain on assigned frequencies. The problem comes from the machines of modern life—computers, auto engines, light dimmers, garage door openers, radar speed guns and so on—clattering away, littering the airwaves with electromagnetic rubbish.

Airwave pollution. The electromagnetic spectrum is a natural resource. As with any natural resource, it can be ruined by pollution. Radio spectrum pollution is not like water or air pollution. You can't see it, but the garbage is there. Astronomers, especially, are anxious that casual use of uncontrolled broadband radio signals be stopped.

Radio astronomers listen to the electromagnetic spectrum to overhear secrets of starbirth and galaxy accumulation millions of lightyears in the past. An individual molecule of matter in space, excited by heat from some distant sun, radiates a unique radio signal. Tuning in millions of such tiny traces, radio astronomers unlock chemical processes at points in the universe not even visible to optical telescopes. Such science has defined the shape of the Milky Way, Earth's home galaxy. It has discovered key periods in the lives of stars and confirmed the forging of complex chemicals within galaxies.

Radar splatter. But now radar signals for defense, air traffic control, even highway speed-trap guns all splatter signals across portions of the spectrum important to astronomy. Television stations do it, too.

Even space users interfere. A Russian ocean navigation system is about to black out study of the hydroxyl radical, key to the chemistry of water and organic chemicals. Listening to deep space hydroxyl radical radio signals could reveal important clues to life beyond Earth. Radio astronomers for years have listened for telltale signals of the hydroxyl radical at frequencies of 1,600 to 1,700 megahertz, but now a new system of Soviet navigation satellites transmits constantly at the same frequencies.

Controlling spectrum pollution would not be hard to accomplish if only government and industry would get with the program. In most cases, ugly sources of radio interference could be shut off with the use of inexpensive filters and metallic shields that might add only a few cents to manufacturing costs.

Use of the electromagnetic spectrum is vital in everyday life. There are important services such as police, fire, ambulances and air traffic control which need two-way radio communications.

Airwave traffic controller. The Federal Communications Commission could take measures to control unnecessary, frivolous, non-public-service use of the diminishing natural resource—the electromagnetic spectrum. The FCC could force manufacturers to shield home and office electronic gear which can generate radio signals. The FCC could refrain from turning over the valuable natural resource to anybody with a new way to make a million bucks while adding to spectrum pollution.

Light Pollution

Electromagnetic gridlock is not the only kind of pollution affecting astronomers.

Around the world, the work of science is being hampered by excessive man-made light. There aren't many really-dark places anymore. Too much light from a nearby city can render a mountaintop observatory impotent.

And science is being hampered by sunlight, reflecting from man-made junk in space, that streaks photographs. Astronomers shoot thousands of long-exposure star photos each year. Space junk swinging through the field of view blocks discoveries.

Even worse, Paris artists want to build a kite as big as a football field and send it to space as an art object. Other Frenchmen wanted to orbit a ring of 100 lights as a salute on the 100th birthday of the Eiffel Tower. A Florida mortuary wants to send up satellites, each holding the cremated remains of 10,000 people. Each of the space burial capsules would be gold coated so loved ones on the ground could see a glint of light as their relatives' ashes pass overhead.

Telescope Inventors

From time to time, it's reported that Galileo Galilei invented the reflecting telescope. Not so! Galileo died in 1642 and the reflecting telescope was conceived by James Gregory in 1663. The first one was built by Isaac Newton five years later.

In a reflector telescope, light is bounced from a mirror to an eyepiece. On the other hand, light passes through two lenses to the eyepiece in a refracting telescope. The refractor telescope was invented first.

It often is said Galileo thought up the refracting telescope, but he didn't really invent that either.

Spectacles had been invented in the 13th century, but it wasn't until the 17th century that the refracting telescope was invented. The Dutch optician Hans Lippershey came upon the principle by luck in 1608 when he looked through one lens placed in front and saw how distant objects were magnified.

Galileo Galilei. Galileo heard about the Lippershey invention in 1609 and built his own refracting telescope to make important discoveries in 1610.

Galileo found the Milky Way has separate stars, and the Moon has mountains, valleys and craters. He spotted four satellites orbiting Jupiter. This at a time when scientists thought everything there was to be discovered had been found by ancient Greeks. Galileo showed Aristotle had been wrong in claiming celestial bodies were unblemished globes. He disproved the Moon would be left behind if Earth circled the Sun. He showed Venus has phases, like the Moon, breaking down another old teaching.

Johannes Kepler improved the reflecting telescope design in 1611. All large modern telescopes are reflectors. Smaller refractors are used frequently as guide telescopes.

Schmidt Telescopes

The Schmidt telescope was invented in 1930 by Estonian optician Bernhard Voldemar Schmidt at Bergedorf Observatory, Hamburg, Germany.

Schmidt was dissatisfied with conventional parabolic telescope mirrors because a defect called coma made their fields of view too small. He found spherical mirrors unsatisfactory because they had the focusing problem spherical aberration.

To improve the situation, Schmidt designed a telescope with a concave mirror and a transparent corrector plate at its center. He combined the best of a refractor telescope and a reflector telescope. Astronomers called it the biggest advance in optics in 200 years.

Famous Schmidts. The best-known Schmidt is on Mount Palomar, California. It has a mirror diameter of 72 inches and a corrector plate of 48 inches. Operating at f/2.5, the Palomar Schmidt produced the National Geographic Atlas of the Northern Sky.

Telescopes like Palomar are at Cerro Tololo; at the European Southern Observatory (ESO), Cerro La Silla, Chile; and the UK-Schmidt telescope at Siding Spring, Australia.

At Tautenburg, Germany, the world's largest Schmidt, has a 79-in. mirror and a 52-in. corrector. Schmidt was born in 1879 and 1935.

Tech specs. Schmidt designed his telescope to take advantage of spherical aberration by incorporating a concave spherical mirror with a thin, transparent, locally-corrected corrector plate at the center of the curve.

The plate corrects for spherical aberration. The mirror is larger than the corrector plate, which must be bent to a curved focal surface. Different areas of the mirror reflect different areas of the field onto the concentric curved focal surface which is half the main mirror's radius.

Using Schmidt telescopes, astronomers can photograph star fields several degrees in diameter. Record speeds for shooting faint objects are very high, since the focal ratio is determined by corrector-plate diameter and main-mirror focal length.

Telescopes On Earth vs. Satellites

New visible-light telescopes at observatories on the ground may allow astronomers to see faraway stars even more clearly than the Hubble Space Telescope can, at less expense, according to Horace W. Babcock, former director of Carnegie Institution observatories, in 1990.

The new design, called adaptive optics, uses a mirror which is warped, or deformed, repeatedly to compensate for the distortion of the atmosphere. Such rapid changes in a telescope mirror would be ordered by a controlling computer.

Adaptive. Adaptive infrared telescopes on the ground might not match the resolution of Hubble, but visible-light telescopes on the ground would be likely to see better than Hubble. And there still would be a need for satellite observatories in orbit because Earth's atmosphere scatters ultaviolet light. Telescopes above the atmosphere would be better at collecting ultraviolet light.

The European Southern Observatory in Chile already has experimental adaptive optics to see infrared light. Research would transfer the techniques to visible light.

Installing adaptive optics in a new telescope might cost a fraction of the $1.5 billion Hubble Space Telescope price, Babcock suggested. A large eight-meter adaptive telescope might be built for $40 million.

A Hubble optics engineer at NASA's Goddard Space Flight Center confirmed adaptive optics on ground telescopes could exceed the visible-light clarity of Hubble in the future, but orbiting telescopes have the advantage of being unaffected by clouds and weather. Ground observatories can't be used on cloudy nights. Some are affected by high winds. Also orbiting telescopes can see ultraviolet light.

Important Observatories

Observing Optical Light And Infrared Light

Country:	Australia
Name:	Anglo-Australian Observatory
Location:	Siding Spring, New South Wales
Main Telescope:	3.9m reflector, 1.2m Schmidt

Country:	Canary Islands
Name:	Roque de los Muchachos Observatory
Location:	La Palma
Main Telescope:	4.2m and 2.5m reflectors

Country:	Chile
Name:	Cerro Tololo Interamerican Observatory
Location:	La Serena
Main Telescope:	4m reflector

Country:	Chile
Name:	European Southern Observatory
Location:	Cerro La Silla
Main Telescope:	3.6m, 3.5m and 2.2m reflectors, 1m Schmidt

Country:	Chile
Name:	Las Campanas Observatory
Location:	La Serena
Main Telescope:	2.5m reflector

Country:	Spain
Name:	German-Spanish Astrophysical Observatory
Location:	Calar Alto
Main Telescope:	3.5m and 2.2m reflectors

Country:	USA
Name:	Kitt Peak National Observatory
Location:	Kitt Peak, Arizona
Main Telescope:	4m, 2.3m and 2.1m relfectors, 1.5m solar

Country:	USA
Name:	Lick Observatory
Location:	Mount Hamilton, California
Main Telescope:	3m reflector

Country:	USA
Name:	Mauna Kea Observatory
Location:	Mauna Kea, Hawaii
Main Telescope:	3.8, 3.6 and 3m infrared reflectors, 10m reflector

Country:	USA
Name:	McDonald Observatory
Location:	Mount Licke, Texas
Main Telescope:	2.7m and 2.1m reflectors

Country:	USA
Name:	Mount Wilson Observatory
Location:	Mount Wilson, California
Main Telescope:	2.5m reflector

Country:	USA
Name:	Palomar Observatory
Location:	Palomar Mountain, California
Main Telescope:	5m reflector, 1.2m Schmidt

Country:	USA

Name:	Whipple Observatory
Location:	Mount Hopkins, Arizona
Main Telescope:	4.5m multiple-mirror

Country:	USSR
Name:	Crimean Astrophysical Observatory
Location:	Simeiz, Ukraine
Main Telescope:	2.6m reflector

Country:	USSR
Name:	Special Astrophysical Observatory
Location:	Zelenchukskaya, Caucasus
Main Telescope:	6m reflector

Radiotelescope Observatories

Country:	Australia
Name:	Australian National Radio Astronomy Observatory
Location:	Parkes, New South Wales
Main Telescope:	64m dish

Country:	Australia
Name:	CSIRO Solar Observatory
Location:	Culgoora, New South Wales
Main Telescope:	Australia Telescope, 3km heliograph

Country:	Australia
Name:	Molonglo Radio Observatory
Location:	Hoskinstown, New South Wales
Main Telescope:	1.6km array

Country:	Great Britain
Name:	Mullard Radio Astronomy Observatory
Location:	Cambridge, England
Main Telescope:	5km array, 1.6km array

Country:	Great Britain
Name:	Nuffield Radio Astronomy Laboratories
Location:	Jodrell Bank, Cheshire, England
Main Telescope:	76m and 38m dishes, Merlin

Country:	Japan
Name:	Nobeyama Radio Observatory
Location:	Nobeyama Highland
Main Telescope:	45m dish for millimeter wavelengths

Country:	The Netherlands
Name:	Westerbork Radio Observatory
Location:	Westerbork
Main Telescope:	2.7km array

Country:	USA
Name:	Aercibo Radio Observatory
Location:	Arecibo, Puerto Rico
Main Telescope:	305m dish

Country:	USA
Name:	National Radio Astronomy Observatory
Location:	Green Bank, West Virginia
Main Telescope:	91m dish, 43m dish

Country:	USA
Name:	National Radio Astronomy Observatory
Location:	Socorro, New Mexico
Main Telescope:	Very Large Array (VLA)

Country: USSR
Name: Special Astrophysical Observatory
Location: Zelenchukskaya, Caucasus
Main Telescope: 600m array

Country: West Germany
Name: Max Planck Institute for Radio Astronomy
Location: Bonn
Main Telescope: 100m dish

Telescopes On Earth: Mauna Kea

Mauna Kea is a dormant volcano on the big island of Hawaii with ideal viewing conditions for astronomers at its 13,800-ft. summit. The site is operated by the University of Hawaii. Many important telescopes are located atop Mauna Kea, including:

★Great Britain's 3.8-meter infrared telescope;

★the Canada-France-Hawaii 3.6-meter optical and infrared telescope;

★NASA's 3-meter infrared telescope facility;

★the University of Hawaii's 2.2-meter optical telescope;

★the British-Dutch 15-meter millimeter-wavelength radiotelescope;

★University of California Hoffman Obsrv. 10-meter reflector with mirror segments.

Telescopes On Earth: Keck

Twin huge 10-meter telescopes with segmented mirrors are under construction atop Hawaii's Mauna Kea volcano. Named Keck I and Keck II, after philanthropist W. M. Keck, the telescopes will be operated by California Institute of Technology and University of California.

The 33-ft. telescopes will be 300 feet apart on the mountain. Use of Keck I was to start in 1992. Keck II was expected to operate in 1996.

Keck I and Keck II will work in tandem, creating an optical and infrared interferometer 280 feet in diameter. They would be extraordinarily sensitive with a resolving power of 0.01 arc second. They also will be used for separate observations.

World's largest. Keck I already was billed as the world's largest and most-powerful telescope. Keck II will offer astronomers power equal to Keck I.

Keck I alone is expected to look back 12 billion years. The two Kecks may give astronomers sight back to 14 billion years ago, near a time when the Universe may have formed.

They look back by receiving light which astronomers calculate has been travelling toward Earth for billions of year. Light travels about 5.9 trillion miles or 9.5 trillion kilometers in one year.

Long way back. Astronomers measure great distances in lightyears. One lightyear is the distance one ray of light would travel in one year in a vacuum. Space is not 100 percent vacuum, but it's close. For comparison, the star nearest our Sun is Proxima Centauri, about 4.5 lightyears away. That's about 26 trillion miles.

The Keck telescopes will look for light shining from galaxies 14 billion lightyears away. That's about 3 billion times farther away than Proxima Centauri.

Great Attractor. Some scientists suggest the Universe was created 15 billion years ago in a Big Bang explosion, but that theory has been questioned recently by the discovery of an immense black hole—the Great Attractor—which is known only by the way its gravity tugs galaxies.

Astronomers say matter clumping into something as immense as the Great Attractor casts doubt on the Big Bang, which should have caused matter to be evenly distributed across the Universe. Keck astronomers will investigate whether dark matter actually is filling the seemingly empty spaces making up 90 per cent of the known Universe. The W.M. Keck Foundation, established by the late founder of Superior Oil Company, is paying most of the telescope costs.

Telescopes On Earth: Japan

The world's largest monolithic mirror blank for a telescope—eight inches thick, 28 feet in diameter, weighing 60,000 pounds—is being constructed of ultra-low expansion glass in a three-year grinding project at Canton, New York.

The mirror blank is for the Japan National Large Telescope to be built on Hawaii's Mauna Kea volcano for use starting in 1999.

The telescope is a project of the National Astronomical Observatory of Tokyo. Mitsubishi Electric Corp., a Tokyo conglomerate which is prime contractor for first work on the telescope, selected Corning Inc. to grind and polish the mirror at Canton.

Corning, having manufactured telescope mirror blanks for years, is using a "hex-seal" process for the Japan National Large Telescope. Many hexagonal pieces are joined together to form one large monolithic blank.

In 1937, Corning made the blank for the largest telescope in the United States, the five-meter reflector for the Hale telescope on Mount Palomar in Southern California.

Telescopes On Earth: SERC

A huge 50-ft. radiotelescope antenna of Great Britain's Science and Engineering Research Council (SERC) near the 14,000-ft. summit of Mount Mauna Kea, Hawaii, is designed to receive radio wavelengths shorter than one millimeter.

As temperatures change, miniature platinum thermometers monitor movement of 276 aluminum honeycomb panels forming the reflecting surface of the dish. The antenna panels were built in southern England at Rutherford Appleton Laboratory, Chilton, Didcot, Oxfordshire. The University of Hawaii and the Organization for the Advancement of Pure Science (ZWO) in The Netherlands helped in the radiotelescope project.

Telescopes On Earth: Australia

An extraordinary new radiotelescope peering out from the cotton fields of New South Wales has found yet another faraway supernova in the the Southern Hemisphere skies.

The Australia Telescope, a group of five large dish-shaped receiving antennas, was turned on in April 1990 and immediately produced results—clear images of a dead shell of a star which exploded about 1,000 years ago. Identified by astronomers as SNR 0540-693, it was the second supernova found with a spinning, radio-generating, pulsar star.

Five in one. The Australia Telescope, near Culgoora, the Southern Hemisphere's first large radiotelescope, was commissioned in 1988. Each 72.6-ft. dish weighs 266 tons. Each rolls across the windswept plain on 1.86 miles of railroad track.

Together, the dishes compose one powerful radiotelescope capable of intercepting very weak natural signals arriving at Earth from across the vast reaches of the Universe.

A sixth dish antenna, 72.6 feet wide, was hooked into the system in 1991. A seventh dish, 71.5 miles south at Coonabarabran, was to be connected in 1992.

Later, when combined with a 211-ft. dish at the older Parkes Radiotelescope 241 miles south, the giant array will equal a whopping radio antenna 200 miles wide. That will give astronomers radio images of unprecedented detail.

Even bigger. The antenna will be made even larger from time to time when it will be linked to dishes at Hobart on the southern island of Tasmania, Alice Springs in central Australia and Perth in western Australia.

Linked to Parkes, Hobart, Alice Springs or Perth, Australia Telescope could be zoomed, like a camera lens, to the electronic equivalent of a 2,400-mi.-diameter dish.

The array also will be linked electronically to a Russian radiotelescope in orbit aboard a satellite known as Radioastron to be launched between 1993-1995. That will produce the electronic equivalent of a radiotelescope with a width equal to five times the diameter of the Earth.

The competition. The Australia Telescope's approximate equal in the Northern Hemisphere is the U.S. Very Large Array (VLA) composed of 27 dish antennas on tracks in the New Mexico desert. VLA was completed in 1980.

New technology from the 1980's make Australia's six dishes about equal in sensitivity to the VLA. The former director of VLA, Ronald Ekers, now is in charge of the Australia Telescope.

The European Southern Observatory in Chile has the biggest optical telescopes in the Southern Hemisphere at La Silla, but it has no major radiotelescope array.

Most powerful. The $48.4 million radiotelescope is the most powerful astronomer's tool in the Southern Hemisphere. Australia Telescope is used to study the birth of new stars, the death of old stars and faraway galaxies like the Magellanic Clouds—actually in Earth's neighborhood at the relatively-close distance of 170,000 lightyears and a prominent feature in the southern sky.

The power of the Australia Telescope has been compared to reading a telephone directory from 660 feet. When linked with the future Radioastron telescope in space, it will be 10,000 times stronger.

Much of our own Milky Way galaxy, and its nearest neighbor galaxies, the Magellanic Clouds, can't be seen by observatories in the Northern Hemisphere. The new instrument opened a window in the southern sky, looking directly overhead at the center of the Milky Way to study black holes, pulsars and other deep space features.

Siding Spring. Siding Spring Mountain in the Australian state of New South Wales is home to the Anglo-Australian Observatory. At 3,800 feet altitude, the observatory houses the 154-in. Anglo-Australian Telescope (AAT) and Great Britain's 47-in. Schmidt Telescope.

The computer-controlled AAT, built and shared equally by the governments of Australia and Great Britain, looks at infrared as well as visible light. Through it, astronomers can see space objects as faint as magnitude 23 to 25. It could detect the weak flame of a candle at 1,000 miles distance but, of course, it's used to look at stars ten million times fainter than what can be seen with the naked eye.

The Schmidt Telescope, in use since 1973, is a remote field station of the Royal Observatory at Edinborough.

Parkes. Four hours south of Siding Spring Mountain by road, at the town of Parkes, New South Wales, is a huge wire-mesh radiotelescope dish 210 feet in diameter, used to tune in radio signals from deep space. The Parkes radiotelescope hears radio waves from distant pulsars and quasars. Astronomers on Siding Spring Mountain then use the AAT to search out and photograph visible light reaching Earth from the object.

Supernova 1987a. NASA linked its 110-ft. radiotelescope antenna at Tidbinbilla, Australia, with Australia's 210-ft. Parkes radiotelescope antenna 200 miles away for a detailed study of the exciting Supernova 1987a.

The radiotelescope antenna at Tidbinbilla is part of NASA's Deep Space Network. It was connected by microwave radio link to a CSIRO radiotelescope 200 miles away at Parkes. CSIRO is Australia's Commonwealth Scientific and Industrial Research Organization. NASA's Deep Space Network also has antennas near Madrid, Spain and at Goldstone, California. The Tidbinbilla and Parkes antennas worked together in a technique known as interferometry. Together, they created electronically an antenna bigger than either one alone. Connected in real time, they formed a receiving antenna the size of the distance between the two antennas.

Radioastronomy detects emissions from space objects at microwave radio frequencies, rather than visible light. Using interferometry, a radiotelescope can provide hundreds of times finer resolution than optical telescopes.

Voyager. The microwave link between Tidbinbilla and Parkes had been established before, for the Voyager 2 fly-by of the planet Uranus in January 1986. The radiotelescope's super-sensitive reception was critical to receiving Voyager photos and science data. A similar arrangement will link Tidbinbilla and Parkes antennas again as Voyager 2 passes Neptune in August 1989.

An even wider network, using a technique called Very Long Baseline Interferometry (VLBI), has been tested using four antennas: Tidbinbilla, Parkes, a Landsat ground station at Alice Springs in central Australia, and an 85-ft. dish antenna at the University of Tasmania, Hobart. Tasmania is an island southeast of Australia.

Radio sources. Supernovas turn into strong sources of radio waves as they explode. Natural radio signals emitted by Supernova 1987a probably were created when the expanding bubble of gases from the exploding star collided with materials nearby. Radio signals eventually penetrated the thinning cloud of material blasting out from the supernova explosion and traveled on to Earth.

The supernova, in a nearby galaxy, the Large Magellanic Cloud, 163,000 lightyears from Earth, was first detected February 23, 1987, by the reception of the sub-atomic particles known as neutrinos. It was detected optically by astronomers February 24, 1987. The Tidbinbilla-Parkes radiotelescope interferometer began looking at the supernova on February 26, 1987.

Supernova 1987a was the first such star explosion in the neighborhood of our own Milky Way galaxy since 1604. At 69 degrees south of the equator, it can be seen only from the Southern Hemisphere.

Weipa. Far north of New South Wales, on the Queensland coast, is the town of Weipa where Australia plans to build a commercial spaceport in the 1990's.

Telescopes On Earth: China

China installed in 1990 what it called the world's largest reflecting horizontal meridian circle telescope at the Chinese Academy of Sciences' Shaanxi Observatory.

The meridian circle telescope is used to determine the times at which stars cross the the meridian. It was designed by a Danish astronomer and built at the Nanjing Astronomical Instrument Factory for the Shaanxi Observatory.

The 420-mm-diameter telescope to observe distant stars is expected to provide accurate locations for more than 10,000 stars a year.

Multi-channel solar. Nanjing Astronomical Instrument Factory also built a multi-channel solar telescope for a Chinese Academy of Sciences observatory in Huariou County near Beijing. It has six lenses and 14 channels. The multi-channel solar telescope, which the Chinese said was the first of its kind, produced fine images in first tests. Leading Chinese astronomers called the telescope a solar physics breakthrough.

Mobile field lab. A mobile open-air astronomy observatory was opened by the Chinese Academy of Sciences in 1990. The field lab has a 7-ft. optical telescope, a 4-ft. infrared telescope, a solar magnetic field telescope and others.

Celestial globe. China's Purple Mountain Observatory constructed a new multifunctional globe. It not only shows the movement of celestial bodies, but also forecasts their exact positions and answers questions about astronomy. Astronomers there regard it as the best celestial globe ever produced in China. The globe is used for teaching, aviation, navigation and astronomy. The celestial globe looks like a terrestrial globe, but depicts more than 1,000 stars on its 32-cm.-diameter surface.

Telescopes On Earth: ESO

European Southern Observatory (ESO) is an international project at La Silla, Chile, 600 miles north of Santiago. Five European countries formed ESO on October 5, 1962. Participating today are Belgium, Denmark, France, Germany, the Netherlands, and Sweden. ESO is headquartered at Garching, Germany, near Munich.

Several telescopes were constructed after 1964, including a 142-in. reflecting telescope, a 59-in. spectrographic telescope, a 39-in. Schmidt, 39-in. and 20-in. photometric telescopes, and a double astrograph composed of a 10-in. refracting telescope and a 16-in. refracting telescope strapped together.

The European Southern Observatory has the biggest optical telescopes in the Southern Hemisphere at La Silla, but it has no major radio telescope array as in Australia. ESO also manages four other telescopes for national institutes.

Telescopes On Earth: USSR 6-Meter

The world's largest reflecting telescope is the USSR's Six Meter at the USSR Academy of Sciences' Special Astrophysical Observatory near Zelenchukskaya in the Caucasus Mountains. The observatory, near 7,000 feet altitude, also has a radiotelescope known as Ratan 600 as well as the 236-in. or six-meter telescope.

The Six Meter telescope was the first large optical-light telescope to have an altazimuth mount. It went into service in 1977. The telescope's first primary mirror, made of Pyrex glass, has been replaced with a glass-ceramic mirror.

A telescope primary mirror is the light-reflecting surface which faces the object being observed. Modern mirrors are made of fused quartz, such as Pyrex, or glass-ceramic compounds such as Zerodur or Cer-Vit. The back of the glass usually is coated with aluminum to create the reflecting surface. The surface of the mirror must be carefully ground to exact dimensions and highly polished. Current technology and cost limit the size of one mirror to about eight meters or 315 inches diameter. Telescopes equivalent of up to a 15-meter mirror are being built using multiple mirrors.

The Ratan 600 radiotelescope dish at Zelenchukskaya, in use since 1976, has 895 vertical metal panels, each about six feet wide and 22 feet tall, in a circle 1,890 feet across.

Telescopes On Earth: Greenwich

Laser light pulses upward day and night from the Royal Greenwich Observatory, Herstmonceaux Castle, Hailsham, East Sussex, England. Built by the Observatory and Hull University, the laser measures Earth's rotation by computing the range to a satellite as it passes overhead while circling the globe three times a day.

The Laser Geodynamic Satellite (Lageos) was launched in 1976 to an altitude near 4,000 miles. The French satellite Starlette, in a lower orbit, provides information on Earth's gravity. The observatory's laser pulse timer is accurate to 50 picoseconds. A picosecond is a billionth of a second. Only one person is needed to operate the computerized laser.

Telescopes On Earth: VLBI

Astronomers got unusually clear pictures of three quasars in 1986 when they combined two deep-space antennas in Australia and Japan with one on an orbiting satellite, creating an extraordinarily big radiotelescope.

Data from antennas on the ground was combined with data from an antenna on NASA's orbiting TDRS communications satellite in a technique known as very long baseline interferometry (VLBI).

Radiotelescope. The main element in a radiotelescope is a radio antenna, pointed at the sky to receive naturally-generated signals from space. Ground radiotelescopes in the VLBI experiment were NASA's Deep Space Network 210-ft. antenna in Australia, a 64-ft. antenna at the Institute of Space and Astronautical Sciences, Usuda, Japan, and an 80-ft. antenna at the Radio Research Laboratory, Kashima, Japan. The electronic antenna created in the experiment was equivalent to a radiotelescope bigger than Earth itself—as large as 1.4 times the diameter of Earth.

It was the first time VLBI used an orbiting satellite as one of its radiotelescope antennas. Previously, scientists had linked widely-separated antennas on the ground with VLBI techniques to create high-resolution images of deep-space objects.

Quasars. Quasars are quasi-stellar objects. They probably are the most distant objects seen by astronomers. The quasars studied in July and August 1986 in the VLBI test were 1730-130, 1741-038, and 1510-089.

Researchers said the successful experiment will lead to Quasat, an orbiting radioastronomy satellite for VLBI work to be launched by the U.S. and the European Space Agency.

Astronomers and astrophysicists involved in the VLBI project were led by NASA's Jet Propulsion Laboratory, Pasadena, California. TDRS was managed by NASA's Goddard Space Flight Center, Greenbelt, Maryland.

Scientists also were from the Massachusetts Institute of Technology, Cambridge, Massachusetts; the Haystack Observatory, Westford, Massachusetts; Bendix Field Engineering Corp., Columbia, Maryland; the TRW Space and Technology Group, White Sands, New Mexico; Australia's Commonwealth Scientific and Industrial Research Organization; Australian National University's Mount Stromlo Observatory; Japan's Institute of Space and Astronautical Science; Japan's Nobeyama Radio Observatory; and Japan's Radio Research Laboratory.

Telescopes On Earth: Hale 200-Inch

The California Institute of Technology—Caltech—paid for and operated the 200-in. Hale Telescope atop Palomar Mountain alone for more than 40 years. In 1989, Caltech took on Cornell University as a partner to help maintain the gigantic reflector.

At $500,000 a year, Cornell is paying 25 percent of the telescope's costs, building several new scientific instruments and receiving 25 percent of the observing time.

Telescopes On Earth: The Hooker

The 100-inch Hooker Telescope at the Mount Wilson Observatory in California, one of America's oldest astronomy assets, is being put back into service by a new owner.

The Carnegie Institution, which built the observatory in 1904 and operated it since, mothballed the big telescope in 1985, shifting to more powerful instruments at its new observatory in Las Campanas, Chile.

The Hooker Telescope was completed in 1917. For three decades, it was the world's largest, the prototype for modern reflecting telescopes. Its 4.5-ton 100-in. mirror was cast by a French bottle-maker.

Carnegie astronomers Edwin Hubble and Milton Humason, using the Hooker Telescope in the 1920's, established that so-called spiral nebulae actually were other galaxies of billions of stars, outside our own Milky Way galaxy. The size of the Universe then recorded by astronomers expanded a thousand-fold.

The Carnegie Institution retired the Hooker, but continued to operate three other telescopes at Mount Wilson. Now, the entire observatory has been converted to a new Mount Wilson Institution, headed by Carnegie veterans, primarily George Preston, former observatory director and now a chief scientist there. It will be financed independently, mostly from government grants.

Light pollution, encroaching from nearby Los Angeles, had destroyed the Hooker Telescope's ability to see faint, distant galaxies. Today, Preston and Arthur Vaughn expect to use the Hooker telescope, at 5,700 feet altitude, for detecting planets around stars in the Milky Way and for studying the Sun. It also will be used to study starspots, blemishes on other stars like sunspots on the Sun.

The Hooker Telescope is being re-fitted with computer controls and new instruments. Scientists are being added to five teams already there.

Telescopes On Earth: Lick

Astronomy seems an odd choice for businessman James Lick. He started out making pianos in South America. Then he turned to California real estate where he owned major chunks of downtown San Francisco and even Catalina Island.

He made a fortune and set up a trust fund, earmarking $700,000 for construction of a telescope and observatory, with concise orders, "Build a powerful telescope, superior and more powerful than any telescope yet made."

It was done, but not before Lick died in 1876—12 years before the observatory was operating. The observatory was built on California's 4,430-foot Mount Hamilton, 60 miles south of San Francisco. An immense 36-inch lens was mounted on a 57-foot tube. At that time, it was the world's biggest telescope.

It also was the first real demonstration of the value of being on a mountaintop. Before, most observatories were in cities. After Lick, mountaintop observatories became standard, including pristine sites in Hawaii, Chile and Southern California.

Lick Observatory has been the scene where:

★Edward Barnard found the fifth moon of Jupiter;

★countless distant galaxies have been discovered;

★expansion of the universe was mapped;

★scientists have studied how stars form and age.

James Lick's remains were placed in a vault at the base of the telescope on Mount Hamilton in January 1887.

Now, owned and operated by the University of California, the pioneering observatory's seven telescope domes, offices and workshops are in use. The big Lick telescope, second in size only to the 40-inch refracting telescope at Yerkes Observatory in Wisconsin, still functions.

Lick Observatory also pioneered the use of reflecting telescopes, with a mirror instead of a lens to focus light. Lick telescopes today include the 120-inch Shane reflector, a 40-inch reflector, two 24-inch reflectors and a twin 20-inch refractor.

Lick Observatory recently observed its 100th anniversary. The telescopes still are used all the time, most in demand for dark moonless nights.

Telescopes On Earth: The McMath

Dr. Robert R. McMath was a Detroit industrialist with a deep passion for the Sun and solar astronomy. A frequent visitor to the Arizona home of Kitt Peak Observatory, McMath founded the McMath-Hulbert Solar Observatory in Pontiac, Michigan.

Dr. A. Keith Pierce at the University of Michigan designed a big solar telescope in 1955. He went to Tucson, Arizona, in 1958 to begin building the telescope and an associated laboratory. In the early 1960's, industrialist Robert McMath pressured the National Science Foundation to set up a national astronomical center. He was a prime mover in pushing the National Science Foundation to find more than $2 million needed to build Pierce's major new solar telescope on Kitt Peak. The money was a big chunk of the NSF budget at the time.

The McMath. On October 30, 1962, a group of men and women gathered at Kitt Peak National Observatory, 60 miles southwest of Tucson, to dedicate a dramatic new research facility. It was Pierce's solar telescope, named the Robert R. McMath Telescope. The largest solar telescope in the world, the great size of The McMath recalled temples and pyramids of earlier times.

First light through the telescope was a photo of Jupiter, snapped October 30, 1962.

The big telescope has an unusual shape—a triangle with the light shaft inclined at 32 degrees. From Tucson's latitude, that keeps it aligned with the celestial north pole.

The entire structure is encased in copper. Coolants are piped through the telescope's skin to keep temperatures uniform. That's unusual as heat is removed from most solar telescopes by evacuating air from the light shaft. The McMath doesn't use vacuum because it's just too big. And the vacuum would make it impossible to do infrared studies.

As the telescope tracks the Sun, the solar beam is reflected by an 81-inch primary mirror to a distance of hundreds of feet into the Earth where it is reflected into an optical train that feeds it to instruments at the base of the observatory. The McMath is an all-mirror telescope which lets astronomers reconstruct images that couldn't be done if mirrors were obstructed by a secondary mirror.

Triumphs. In the early days, The McMath was used mostly to watch the Sun's magnetic field and do other Sun research. Later, it evolved into a 24-hour-a-day telescope, used by day to observe the Sun and by night to search for other Sun-like stars and examine the atmospheres of planets in our own Solar System.

Nowadays, it's operated by the National Optical Astronomy Observatories (NOAO) National Solar Observatory, with observing facilities both on Kitt Peak and on Sacramento Peak, New Mexico. The McMath has provided solar, stellar, and planetary astronomers around the world with a state-of-the-art laboratory.

The McMath Telescope is a science lab as well as an observing instrument. The facility's most important work has been in spectroscopic analysis of the Sun's atmosphere. The McMath has the most powerful spectrographs in the world for analyzing sunlight.

Sun vibes. Other important work has been discovery of previously-unseen chemical elements and isotopes in the Sun. The McMath pioneered techniques for observing small oscillations on the Sun's surface to learn more about the deep interior structure of the Sun. The lab has made extensive studies of the solar magnetic field, sunspot cycles and flare activity.

The McMath opened the way to solar-stellar research, looking for Sun-like activity on the tenuous surfaces of other stars. It even helped Apollo astronauts learn lunar geology. Today, the big telescope has a fresh coat of paint, improved instrumentation, and new computers. Scientists in NOAO's Advanced Development Program are designing flexible mirrors and other so-called adaptive optics to reduce atmospheric distortion of telescope observations. They also are working to sharpen resolution of The McMath to a small fraction of a second arc. Larger telescopes for the future are being designed.

Telescopes On Earth: Smithsonian

Is there intelligent life Out There sending signals we can detect? Every ten days or so, alarms sound as out-of-the-ordinary radio waves from deep space are received by sky sweeping receivers in Massachusetts. So far they have been false alarms.

An 84-ft. dish-shaped antenna points at the sky. Every two minutes, Earth's rotation causes the antenna to point toward a different part of the sky. Naturally-generated radio signals arriving at Earth from deep space are received around the clock and analyzed by a bank of 128 computers.

The computers divide the two-minute chunks of radio waves from space into 8.4 million separate frequencies. Each frequency is analyzed to see if it seems to contain artificial signals. Is there something in the signal not attributable to background radiation caused by natural phenomena in the sky?

The antenna, receivers and computers are part of the Harvard-Smithsonian Observatory. Once a day an operator changes the north-south elevation of the antenna just enough for it to look at different two-minute chunks of sky.

At one time or another during the year, nearly 80 percent of the entire sky passes in view of the Massachusetts site. Over seven months, the antenna sees it all.

Once in a while, every ten days or so, the computers stumble across something which might be intelligent. Human observers always examine such data but, so far, nothing intelligent has passed their tests. And the sweep goes on.

Telescopes On Earth: Townes

The Nobel-prize scientist who invented the laser 25 years ago now has developed a new infrared sky microscope which will show 100 times more detailed pictures of space than the most powerful radio telescopes on Earth.

Charles Townes, the 71-year-old physicist at Berkeley, California, who won the 1964 Nobel prize in physics for research on the laser and maser, cranked up his new telescope in 1987 and announced it will see through shrouding clouds of dust and gas.

Townes' portable system uses two infrared telescopes mounted on trucks. Computers electronically record deep-space pictures from the telescopes and display them on video screens. The extremely short wavelength of infrared light allows it to slip through clouds of dust and gas which hide much of the Universe from Earth's view.

Dust and gas clouds are particularly frustrating to astronomers when they block the view of the birth of a star. Starbirth usually is a violent cataclysm surrounded by vast banks of gas and dust. With the new dual telescopes, astronomers will be able to look through the dust at stars early in their lives.

Interferometry. Astronomers often link the antennas of two radiotelescopes, using a method known as interferometry to see deep-space objects in more detail.

A radio signal is received by two antennas at the same time. The two antennas are separated by a small distance.

The signal coming in through one antenna is mixed with the same signal received by the other antenna. Very slight differences in signals received at the separate antennas are compared by computer. The signals interfere and blend with each other. Strange as it seems, that actually enhances the final picture. The computer creates a video picture that is the electronic equivalent of visible light. That electronic image is photographed for all to see.

Townes' instruments are infrared telescopes, rather than radiotelescopes. His telescopes see infrared lightwaves that are a fraction of an inch in length, not radiowaves. By comparison, radiowaves measure from a few inches to several miles. The Townes infrared telescopes, using the same interferometry technique, reveal 100 times more detail than radiotelescopes.

Townes' telescopes are 80-in. flat mirrors moved by computer to track stars. Images of stars reflect onto a 65-inch parabolic mirror, concentrating in the focusing area. Each telescope is mounted on a truck with sliding doors.

Everything must be controlled exactly to make precise measurements. Laser beams align and calibrate the system which can be taken around the world.

Telescopes On Earth: NAAO

A Delaware, Ohio, radiotelescope won a ten-year reprieve from a fate as a golf course. Actually a meadow filled with antennas pointed toward the stars, the radiotelescope was built by Ohio State and Ohio Wesleyan universities, but Ohio Wesleyan sold the land to owners of a golf course.

Now four schools have formed the North American Astrophysical Observatory to operate the telescope. They have a ten-year lease on the land.

Telescopes On Earth: Green Bank

It was a beautiful November evening in 1988 with no weather problems in the West Virginia mountains when a 50-inch steel plate weakened and ripped, collapsing one of the world's biggest radiotelescopes as the National Radio Astronomy Observatory was using it to map the Universe.

The 26-year-old, 300-ft.-wide Green Bank antenna, a metal latticework bowl as wide as a football field and resembling a TV satellite dish, was intercepting natural radio signals from deep-space objects 10 billion lightyears away. Such signals help scientists understand the origins of the Universe.

The 600-ton radiotelescope dish, which collapsed November 15, 1988, was America's most precise astronomical instrument. A latticework support structure under the dish had obscured the stress-weakened plate each year as scientists inspected the structure. Nobody was hurt in the collapse, but the telescope was beyond repair. A control room was damaged.

Surveying the Universe, the radiotelescope covered the entire northern sky from its radio quiet zone in the West Virginia mountains, kept free by law of manmade interference.

Used by U.S. and foreign astronomers, the telescope was one of few that large which are moveable. Similar telescopes are in California, Australia and England. The Green Bank radiotelescope, which could be adjusted on a north-south axis, had taken 18 months to build when it was completed in 1962 at a cost of $850,000.

The Green Bank Observatory, opened in 1959 before the radiotelescope was completed 100 miles east of Charleston in the Monongahela National Forest, has eight smaller telescopes. The smallest is 40 feet in diameter. Observatory headquarters is in Charlottesville, Virginia.

The big radiotelescope, funded by the National Science Foundation (NSF) and operated by a group of nine universities, including Harvard, Yale and Johns Hopkins, will be rebuilt by 1993. It will cost $70 million to replace, but the new radiotelescope will be larger, sturdier and more flexible. While the replacement is under construction, a 140-ft.-wide telescope at the observatory is picking up deep-space mapping work. NSF also funds a 1,000-ft. non-moveable telescope in Arecibo, Puerto Rico.

Famous Astronomers: A to Z

Abbot. American astronomer Charles Greely Abbot, born in 1872, studied the Sun. He went to work at Smithsonian Astrophysical Observatory in 1895 and became director in 1907. His book, The Sun, was published in 1911. He died in 1961.

Abu'l-Wafa. Muhammad al-Buzjani Abu'l-Wafa, who lived from 994 to about 998, helped develop trigonometry. His Complete Book, about astronomy, and astronomical tables were widely used.

Adams. John Couch Adams, born June 5, 1819, studied at Cambridge University, England, and became a fellow, tutor, and professor of astronomy and geometry there, then was appointed director of Cambridge Observatory in 1861.

At age 24, Adams was first to predict the position of a planetary mass beyond Uranus. After the planet Neptune was discovered September 23, 1846, Adams was drawn into a dispute with French astronomer Urbain Leverrier over who predicted the planet first. The English astronomer had predicted Neptune first, but the Frenchman's prediction was published first. The dispute eclipsed Adams' notable studies of the Leonid meteor shower and the motion of the Moon. He died January 21, 1892.

Airy. British astronomer George Biddell Airy, born in England on July 27, 1801, was astronomer royal for 45 years. He modernized the Royal Greenwich Observatory. Airy was graduated from, and a professor of astronomy at, Cambridge University.

Named after the astronomer, the Airy Disk is a small image produced by diffraction of light passing through a telescope. Airy calculated the size of the disk which depended on the wavelength of the light and on the diameter of the objective lens or mirror. The larger the diameter, the smaller the disk. Airy died Jan. 2, 1892.

Aitken. American astronomer Robert Grant Aitken, born in 1864, director of the Lick Observatory in California from 1930-1935, made important observations of double stars. His New General Catalogue of Double Stars, published in 1932, listed more than 17,000. He died in 1951.

Albategnius. Muslim astronomer al-Battani, sometimes called Albategnius, lived from 858-929, observing the skies above al-Raqqa on the Euphrates River. He is credited with bringing Greek planetary theory into medieval Europe, composing tables of planetary motion and showing the Sun varies in distance from Earth. An 80-mi.-diameter walled plain on the Moon is named after Albategnius.

Al-Ma'mun. The 9th century Arab astronomer Al-Ma'mun built an observatory in 829 and outfitted it with astronomy instruments better than those of the Greeks.

Al-Sufi. The 10th century Arab astronomer Abu'l-Husayn Al-Sufi, born in 903, revised Ptolemy's star catalog from 800 years earlier, added Arab names and tables of the latest magnitudes and positions, included observational drawings of each constellation, and published it as Book on the Constellations of the Fixed Stars. Al-Sufi died in 986.

Anaxagoras. Greek philosopher Anaxagoras, who lived around 500 BC to 428 BC, said the sky is rarified, dry, hot matter, while the Sun, Moon, planets and stars had been torn from Earth, a disk of dense, cold, wet, dark material. Motion ignited the Sun, Moon, planets and stars, making them shine.

Aristarchus. Aristarchus of Samos was a Greek mathematician and astronomer living from 310-230 BC, between the times of Euclid and Archimedes. He was a student of Strato of Lampsacus, the third head of Aristotle's Lyceum.

Aristarchus advocated a Sun-centered Universe and tried to calculate the sizes and distances of the Moon and Sun. No one knows how he came up with his heliocentric cosmology which the Greeks did not accept. The theory is mentioned in Archimedes' The Sand-Reckoner.

Aristarchus' only known work is On the Sizes and Distances of the Sun and Moon, spelling out his geometry from observations in which he calculated the Sun was 20 times as far from Earth as the Moon and 20 times the size of the Moon. His computations were too small because his instruments were inaccurate.

Aristotle. Aristotle, born in 384 BC in Stagira in the northern part of Greece, may have been the most influential philosopher in the history of Western thought. His father, Nicomachus, was physician to the Macedonian court. In 367, Aristotle joined Plato's Academy at Athens, as a student, then a teacher. Men at the academy tried to organize knowledge on a theoretical basis and expand it in all directions.

In 347 BC, after Plato's death, Aristotle joined the court of Hermias of Atarneus. In 343 BC he went to the court of Philip II of Macedonia to instruct the young Alexander the Great. In 335 BC, Aristotle returned to Athens to found his own school, the Lyceum, or Peripatus. He emphasized the study of nature. Alexander the Great died in 323, feelings against Macedonia arose in Athens, and Aristotle retired to Chalcis. He died in 322 BC. The study of astronomy and other natural sciences was dominated by Aristotle for more than a millennium. Today's science of physics is a reaction to Aristotelian thinking.

Baade. Wilhelm Heinrich Walter Baade, born in Germany March 24, 1893, was graduated from the University of Gottingen, and worked eleven years at the University of Hamburg's observatory at Bergedorf. He discovered the asteroids Hidalgo and Icarus, then moved to California, in 1931 to study extragalactic astronomy with Edwin P. Hubble at the Mount Wilson Observatory, Pasadena, California.

Man-made lights in Los Angeles were blacked out during World War II, making it easier for Baade to observe light coming from the Andromeda Galaxy and other objects in deep space outside our own Milky Way galaxy. He listed two kinds of stars, today referred to as population I and population II stars. Unable to distinguish a type of variable star through the 200-in. telescope, Baade calculated that Andromeda Galaxy was more than twice as far away as had been supposed, making the Universe twice as large as had been thought. Baade died June 25, 1960.

Barnard. Edward Emerson Barnard, born at Nashville, Tennessee, December 16, 1857, studied at Vanderbilt University, worked at Lick Observatory in California, then at the University of Chicago's Yerkes Observatory at Williams Bay, Wisconsin. He was America's foremost observational astronomer of his time.

In 1916, he discovered Barnard's Star, a faint cool red dwarf. It is seen as a magnitude 9.54 star in Earth's sky in the constellation Ophiuchus. Planets orbiting Barnard's Star have been suggested. At 5.9 lightyears from Earth, it is the fourth nearest star to the Sun after three in the Alpha Centauri system. Barnard's Star has the largest proper motion.

Barnard also discovered many comets and nebulas, and the fifth moon of Jupiter. He was known for employing photography in astronomy, especially for stunning long-exposure photographs of the Milky Way. He made his last observation—of the Moon passing in front of Venus—from his bed a month before he died February 6, 1923.

Bessel. Friedrich Wilhelm Bessel, born July 22, 1784, was the German astronomer and mathematician who found the earliest method for accurately measuring stellar distances. Even though he had little formal schooling, the University of Gottingen granted Bessel the title of doctor and he was named director of Frederick William III of Prussia's Konigsberg Observatory at age 26. Bessel's work finding positions and proper motions of stars lead to the discovery in 1838 of the parallax of 61 Cygni. He also worked out the Bessel function, an indispensable mathematical analysis tool in engineering and physics. Bessel died March 17, 1846.

Boltzmann. Austrian physicist Ludwig Boltzmann, born February 20, 1844, received a doctorate from the University of Vienna in 1867, and was a professor at the universities of Vienna, Graz, Munich, and Leipzig. He used laws of probability to describe how properties of atoms determine visible properties of matter, developing the statistical mechanics branch of physics. He published a series of papers in the 1870's applying probability to atomic motion and the second law of thermodynamics. He found the Maxwell-Boltzmann distribution, showing average energy used for each direction of motion of an atom is the same, making Boltzmann the first to develop the kinetic theory of gases. Unfortunately, his work was opposed by many European physicists. He became depressed, fell into poor health, and committed suicide September 5, 1906, shortly before French scientist Jean Perrin verified his work.

Bond. William Cranch Bond, born in 1789, director of Harvard College Observatory, discovered Hyperion, the eighth satellite of the planet Saturn, in 1848 (with William Lassell), and the Crepe Ring—Saturn's C ring—in 1850. He died in 1859.

Bradley. James Bradley, born March 1693, was graduated from Balliol College, Oxford University, ordained a priest in 1719, then named Savilian professor at Oxford in 1721. He was able to make extremely accurate observations of the positions of stars. In 1729, Bradley found small changes in positions of stars due to the combined effect of the orbital motion of Earth and the velocity of light. He called the effect aberration. Later he detected the nodding of Earth's axis known as nutation.

In 1742, Bradley followed Edmond Halley as England's astronomer royal and director of the Royal Greenwich Observatory where his very accurate astrometry work was the cornerstone of 19th-century astronomy. Friedrich Bessel used it for data in his star catalogue. Bradley died July 13, 1762.

Brahe. Danish astronomer Tycho Brahe, born at Knudstrup, Scania, Denmark, December 14, 1546, was raised by Jorgen Brahe, an uncle, who sent him to study law in Copenhagen in 1559. Unfortunately for his uncle's plans, Brahe was excited by a solar eclipse in 1560 and became interested in astronomy. He studied on his own using Ptolemy's Almagest. Seeing the conjunction of Jupiter and Saturn in 1563, he uncovered the inaccuracy of existing planetary position records. His uncle died in 1565 and Brahe traveled Europe studying science at various universities. He returned to Denmark in 1571 to install a chemical lab in a relative's castle at Heridsvad Abbey. From the castle in 1572, he observed what he called a "new star" or nova in the sky in constellation Cassiopeia. In his first work, published in 1573, titled De Nova Stella, Brahe concluded the nova was a star beyond the Moon's orbit. Proving the nova a star, Brahe showed the Aristotelian world view of immutable heavens was flawed.

King Frederick II, wanting to keep Brahe in Denmark, granted the astronomer a liberal pension in 1576 and the island of Ven, or Hven, between Sweden and Denmark. Brahe constructed the observatory Uraniborg there and built instruments and made extensive observations for 21 years. He made the most accurate observations of the heavens before the telescope was invented.

Brahe's observation of a comet in 1577, and five more comets later, convinced him they were far outside the lunar orbit. The accuracy of his calculations of planetary positions led to the Copernican world view in the 17th century.

The astronomer left in 1597 after King Frederick's son, Christian IV, withdrew Brahe's pension and rights on Ven. In 1599, Holy Roman Emperor Rudolph II made Brahe imperial mathematician of the court in Prague. Some of Brahe's instruments were shipped to Prague where he was joined in 1600 by a young astronomer, Johannes Kepler. Brahe assumed correctly an annual variation in the directions of stars should be visible if the Earth revolves around the Sun but, despite precise measurements, he couldn't find the variation. In his Tychonic system, Earth was the fixed center with the Sun and Moon revolving around Earth, and other planets revolving around the Sun. When Brahe died October 24, 1601, Kepler inherited his observations, using them to write different laws of planetary motion.

Cassini. Italian astronomer Giovanni Domenico (Jean Dominique) Cassini, born June 8, 1625, headed four generations of astronomers who directed the Paris Observatory. Known for his observational astronomy, the first Jean Dominique Cassini, died September 14, 1712. He was followed by his son Jacques Cassini (1677-1756), then Cesar Francois Cassini de Thury (1714-84), and finally Cesar's son who also was named Giovanni Domenico Cassini. The last Cassini relinquished the observatory post in 1793 after the family had held it 120 years.

The first Jean Dominique Cassini was educated at Jesuit College, Genoa, then appointed in 1650 to the chair of astronomy at Bologna, Italy. For 19 years he used instruments made by Roman lens makers Eustachio Divini and Giuseppe Campani to watch seasonal changes on Mars, find bands and spots on Jupiter, and publish a table of motions for the Jovian satellites. Cassini was invited to Paris in 1669 for temporary assignment to the Academie des Sciences, but quickly took over as director of the new Paris Observatory where he discovered four satellites of Saturn, detected the division of Saturn's rings, and observed the Moon's surface and many comets. Cassini, with astronomers Jean Picard and Jean Richer, determined the parallax of Mars, unveiling the true size of the Solar System.

Jacques Cassini observed planets and satellites and followed his father in disagreeing with Newton's view of an Earth with flattened poles. Unlike his father, Jacques agreed with Copernicus. Cesar Francois Cassini was known for cartography.

The U.S. and European Space Agencies plan to send an interplanetary spacecraft to study Saturn and its moons in the 1990's. The Saturn orbiter, named Cassini, will fly by asteroids and Jupiter on the way to Saturn. It will drop a smaller probe named Huygens into Saturn's atmosphere.

Chandrasekhar. Born in India October 19, 1910, the theoretical astrophysicist Subrahmanyan Chandrasekhar was graduated from Trinity College, Cambridge University, England, moved to the United States in 1936, worked for a short while at Harvard University, then became a professor at the University of Chicago.

In 1939 Chandrasekhar's Introduction to the Study of Stellar Structure was the first broad explanation of nuclear reactions taking place in stars. Later, he discovered that the very dense kind of star known as a white dwarf cannot have a mass greater than 1.44 times the mass of the Sun. White dwarfs are so dense their electrons are squeezed into a small space. A star whose mass is greater than 1.44 times the mass of the Sun cannot become a white dwarf directly. It must lose mass to drop below the so-called Chandrasekhar Limit or end up a neutron star or black hole.

From 1952-1971, Chandrasekhar edited the Astrophysical Journal. He won the Nobel Prize for physics in 1983 for research on the atmospheres and interiors of stars and the properties of clusters of star and star systems.

Clark. Alvan Graham Clark, born July 10, 1832, was from a Massachusetts instrument making family which built the 19th century's finest telescopes. In 1897, he built the 40-in. refracting telescope for the University of Chicago's Yerkes Observatory at Williams Bay, Wisconsin. Today, that telescope remains the largest refractor. Clark was the first person to observe Sirius is a system of binary stars. He made the discovery while testing an 18.5-in. mirror. Clark died June 9, 1897.

Copernicus. Nicolaus Copernicus was born to a prosperous Polish merchant February 19, 1473, at Thorn, or Torun. After his father's death, Copernicus was raised by a maternal uncle who sent him to the University of Krakow, renowned for astronomy, mathematics and philosophy. Later he studied liberal arts at Bologna from 1496-1501, medicine at Padua, and law at the University of Ferrara. Graduated with a doctorate in canon law from Ferrara in 1503, he returned to Poland to settle within 100 miles of his birthplace at the cathedral in Frauenberg, or Frombork. His uncle had arranged Copernicus' election as a canon of the church. Copernicus studied astronomy, practiced medicine, wrote on monetary reform, and performed ecclesiastical duties.

He quietly circulated a manuscript in 1514, outlining arguments which he later supported in 1543 in On the Revolutions of the Heavenly Spheres. Opposing Aristotle, Ptolemy and a system centered around Earth, Copernicus saw Earth rotating while revolving with other planets about a stationary central Sun—explaining daily rotation of the heavens, annual movement of the Sun through the ecliptic, and periodic retrograde motion of the planets. Copernicus still believed in Aristotle's solid celestial spheres, perfect circular motion of heavenly bodies, and physics of motion, and Ptolemy's planetary motion as combinations of circles called epicycles. The heliocentric theory required a much larger Universe, although Copernicus didn't see the Universe as infinite.

Copernicus' radical theory started a scientific revolution, leading in the next century to Kepler's determination of the ellipticity of planetary orbits, Galileo's concept of motion, and Newton's theory of universal gravitation. The church didn't like Copernicus' heliocentric theory, so it wasn't published until 1540 when an enthusiastic supporter named Rheticus printed A First Account. After that, Copernicus agreed to have it printed. The astronomer is said to have received a copy on his deathbed. He died May 24, 1543.

Curtis. American spectroscopist Heber Doust Curtis, born in 1872, used novas to measure distances to galaxies. In the so-called Great Debate of 1920 with Harlow

Shapley, Curtis asserted correctly that spiral nebulas are galaxies beyond our own Milky Way galaxy. Curtis died in 1942.

D'Arrest. German astronomer Heinrich Louis D'Arrest, born in 1822, helped Johann Gottfried Galle find the planet Neptune in 1846. D'Arrest also found an asteroid and three comets, including the faint D'Arrest's Comet which he spotted from Leipzig in 1851. D'Arrest's Comet returns every 6.4 years and has been seen during 14 visits. D'Arrest, whose nebulas measurements were published in 1858 and 1867, died in 1868.

Dawes. English double-star astronomer William Rutter "Eagle Eye" Dawes, born in 1799, was nicknamed for his excellent vision. He is remembered for the Dawes Limit, a way to calculate resolving power of a telescope. Dawes died in 1868.

Eddington. English astronomer Arthur Stanley Eddington, born December 28, 1882, was graduated from Trinity College, Cambridge University. He was chief assistant at the Royal Greenwich Observatory from 1906-1913. Later he directed the Cambridge University Observatory and was Plumian professor of astronomy of Cambridge from 1913 until his death November 22, 1944.

After looking into the distribution and motion of stars in our Milky Way galaxy in his early years, Eddington turned to the structure of the stars themselves. He decided a star is held together by the balance of the forces of gravity, gas pressure, and radiation pressure.

Eddington understood Einstein's theory of general relativity and wrote a book about it in 1920, after organizing an expedition in 1919 to see an eclipse which confirmed Einstein's prediction of the bending of light in a gravitational field.

He wrote in 1926 about the relationship between star size and luminosity, the density of white dwarf stars, and the sources of stellar energy. Eddington's readable literary style brought astronomy to life for many.

Einstein. Physicist Albert Einstein was born at Ulm, Germany, March 14, 1879, to nonobservant Jews in the business of manufacturing electrical equipment. They moved from Ulm to Munich when Einstein was an infant. The business failed in 1894, the family moved to Milan, Italy, and Einstein relinquished German citizenship. He was graduated from the Swiss Federal Institute of Technology, the Zurich Polytechnic, in 1900 as a secondary school teacher of mathematics and physics.

Two years later he got a job at the Swiss patent office in Bern. While he worked there from 1902-1909, he also wrote theoretical physics papers in his spare time. Einstein submitted a paper to University of Zurich and received a Ph.D. degree in 1905. In 1908, he sent a paper to University of Bern which appointed him lecturer. In 1909, Einstein moved up to associate professor of physics at University of Zurich. Then, he held professorships at German University of Prague and at Zurich Polytechnic. In 1914 he became professor at the Kaiser-Wilhelm Gesellschaft in Berlin. He also was on faculty at University of Berlin. Einstein left Berlin in 1933 for the Institute for Advanced Study in Princeton, New Jersey.

Einstein delivered three fundamental papers in 1905. The first was about light, which is visible electromagnetic radiation. Einstein discussed the so-called light-quanta of electromagnetic energy emitted by radiating objects in quantities proportional to the frequency of the radiation. While classical theory assumed waves of electromagnetic energy flowing through an ether, Einstein saw light as bundles of radiation. Einstein depicted a photoelectric effect in which some metals illuminated by light at a given frequency would release electrons. His theory led to quantum mechanics.

Einstein's special theory of relativity was set out in the second 1905 paper. The mass of an electron increased as it approached the velocity of light. Classical theory described ether, but scientists couldn't find the ether. Einstein said the equations for calculating electron motion could describe the non-accelerated motion of any particle. The laws of physics had to apply in any frame of reference. Assuming the speed of light was the same in all frames of reference, Einstein said there was no ether. He depicted so-called time dilatation in which time is a function of velocity, like mass and length. Mass and energy were equivalent.

The third 1905 paper, about statistical mechanics, calculated the average trajectory of a microscopic particle buffeted by random collisions with molecules in a gas or fluid. Einstein's calculations accounted for brownian motion, a seemingly erratic movement of pollen in fluids. Einstein offered evidence for the existence of atom-sized molecules.

In 1907, Einstein found gravitational acceleration the same as acceleration caused by mechanical forces, thus gravitational mass was the same as inertial mass. He said if mass were equal to energy, gravitational mass would interact with the mass of electromagnetic radiation, including light.

In 1911, Einstein predicted how a ray of light from a distant star, passing near the Sun, would seem attracted to, or bent slightly, in the direction of the Sun's mass. Light radiated from the Sun would interact with the Sun's mass, changing its color toward the infrared end of the optical spectrum.

Einstein said any new theory of gravitation would have to account for a small persistent anomaly astronomers had seen in the perihelion motion of the planet Mercury. About 1912, Einstein began work on the the general theory of relativity with gravitational research in terms of tensor calculus which eased calculations in four-dimensional space-time. He published the general theory in 1915. Gravitational field equations were the same in equal frames of reference. The field equations covered the perihelion motion of Mercury. British solar eclipse expeditions in 1919 verified Einstein's predictions.

Einstein returned to Germany in 1914, but was a pacifist in the First World War. When the allies tried to keep German scientists out of international meetings after the war, Einstein—a Jewish person with a Swiss passport—was an acceptable German representative. The theory of relativity was so controversial that Einstein's 1921 Nobel prize in physics was awarded for photoelectric work rather than relativity. Einstein was attacked in the 1920s by anti-Semitic physicists, who later tried to establish a so-called Aryan physics in Germany. Conservatives called him a defeatest traitor for his pacifist views and Zionist politics. The rise of fascism in Germany led Einstein to the United States in 1933 where he agreed the Third Reich had to be stopped. In 1939, Einstein sent a letter to President Franklin D. Roosevelt asking the United States to develop an atomic bomb before Germany. The letter contributed to Roosevelt's decision to go ahead with the Manhattan Project.

Einstein continued looking for a unified field theory, in which gravity and electromagnetism could be explained by one set of equations, until his death at Princeton April 18, 1955.

Empedocles. Greek philosopher Empedocles, who may have lived from 490 BC to 430 BC, explained the material substances of the Universe as earth, fire, air and water. He said the Universe is a crystal sphere and the atmosphere is a substance, not a void.

Eratosthenes. The Greek geographer Eratosthenes of Cyrene, living from about 276 BC to 195 BC, was chief librarian at the Alexandria, Egypt, museum where he calculated Earth's circumference with good accuracy. His mathematical system, known today as Eratosthenes' Sieve, found prime numbers and determined the inclination of the ecliptic to the celestial equator.

Flamsteed. John Flamsteed, born August 19, 1646, and frail and sickly through life, received an M.A. degree from Cambridge University. He was appointed the first astronomer royal in 1675. As the first director of England's Royal Greenwich Observatory, Flamsteed is recalled today for his early catalog of stars.

Flamsteed's task was astrometry—determining accurate lunar and stellar positions as an aid to navigation. Flamsteed's astrometry was like observations made earlier by Tycho Brahe and Johannes Hevelius, but Flamsteed had a telescopic sight on his seven-ft. sextant. Thus, he achieved fifteen times better accuracy. Isaac Newton and Edmond

Halley stole some of Flamsteed's work and published it in 1712 without his permission, but the complete Historia Coelestis Britannica, recording positions of 3,000 stars, wasn't published until 1725, after Flamsteed died December 31, 1719. His Atlas Coelestis was published in 1729.

Fraunhofer. Joseph von Fraunhofer, born March 6, 1787, was a German optical craftsman who, in 1814, first mapped dark lines crossing the solar spectrum. Dark lines in star spectra were observed in 1802 by British physicist William H. Wollaston, but the lines were named for Fraunhofer after he studied them in detail starting in 1814.

The 25,000 Fraunhofer lines mapped in the Sun's spectrum come from atoms in the star's lower atmosphere. Atoms absorb light at the same wavelengths they emit. An element can be found inside a star by finding its Fraunhofer Lines. The make up of the Sun and many stars has come from the lines. Fraunhofer died June 7, 1826.

Galileo. Galileo Galilei was born near Pisa, Italy, February 15, 1564. He went to University of Pisa in 1581 as a medical student, but became interested in mathematics and left in 1585 without a degree. He taught privately at Florence, Italy, then was appointed in 1589 professor of mathematics at Pisa where he demonstrated from the Leaning Tower that Aristotelian physics was wrong in assuming speed of fall was proportional to weight.

In 1592, Galileo was appointed professor of mathematics at University of Padua. He remained there until 1610. He invented a mechanical calculator known today as a sector, worked out a mechanical explanation of tides following Copernicus' theory of the motions of Earth, and showed how machines transform power, not create it. Galileo studied motion along inclined planes and the motion of pendulums in 1602. He formulated, by 1604, the basic law of falling bodies, verified by measurements.

A very bright supernova appeared visible to the naked eye in Earth's sky in October 1604. It was seen in the constellation Ophiuchus by astronomers in Europe, China and Korea, and referred to as Kepler's Star after the astronomer who determined its position. The 1604 supernova involved Galileo in a dispute with Aristotelian philosophers who said the heavens couldn't change. Using parallax equations, Galileo said the new star was very distant, in the area Aristotelian philosophers presumed to be invariable.

After the commotion settled, Galileo returned to studying motion. He established a restricted inertial principle and determined projectiles flew across parabolic paths. He was writing about motion in 1609 when he heard of the the newly invented Dutch telescope. He set to work immediately, constructing his own telescopes. By the end of the year, Galileo had a 20-power telescope through which he could see mountains on the Moon, what he thought were planets orbiting Jupiter, and stars of the Milky Way.

In 1610, Galileo published his startling news in The Starry Messenger, churning up controversy until others made their own telescopes. The Grand Duke of Tuscany appointed Galileo court mathematician at Florence, Italy. By the end of 1610, he had recorded the phases of Venus and firmly believed in Copernicus' heliocentric world system. He was opposed by those who said the Bible required a stationary Earth. Galileo wrote a letter in 1615 to the Grand Duchess Christina asking for freedom of inquiry. He suggested mathematical proofs and sensory evidence should not be subordinated to doubtful scriptural interpretations.

The Holy Office at Rome issued an edict against Copernicanism in 1616, but Galileo's friend Maffeo Barberini became pope Urban VIII in 1623. He permitted Galileo to write a book discussing the Ptolemaic and Copernican systems. After Galileo's book Dialogue was published in 1632, he was summoned to Rome for trial by Inquisition on grounds he had been ordered in 1616 never to defend or teach Copernicanism. In 1633, Galileo was sentenced to life in prison for "vehement suspicion of heresy." Dialogue was banned and printers forbidden to publish anything by Galileo.

However, his Dialogue was translated into Latin outside Italy and read by scholars across Europe. Galileo's sentence was reduced to house arrest under custody of his friend, the Archbishop of Siena. That penalty then was reduced to house arrest at Galileo's own villa in Arcetri, near Florence, where Galileo completed his Padua studies on motion and on the strength of materials. That work was published at Leiden in 1638 as Discourses and Mathematical Demonstrations Concerning Two New Sciences.

Galileo's students included Benedetto Castelli, who founded the science of hydraulics and taught Bonaventura Cavalieri, who formulated principles leading to the development of the calculus, and Evangelista Torricelli, who explained atmospheric pressure and invented the barometer. Galileo died at Arcetri January 8, 1642.

The U.S. and European Space Agencies in 1989 launched an interplanetary spacecraft named Galileo to study Jupiter in 1995. Along the way it will have visited Venus, Earth twice and the Asteroid Belt.

Galle. German comet and meteor watcher Johann Gottfried Galle, born in 1812, discovered the C-ring around the planet Saturn in 1838, then found the planet Neptune in 1846 with the help of Heinrich D'Arrest. In 1873, Galle used observations of the asteroid Flora to calculate solar parallax. He died in 1910.

Hale. George Ellery Hale, born at Chicago, Illinois, June 29, 1868, studied at Massachusetts Institute of Technology where he invented the spectroheliograph for photographing solar prominences. After he was graduated from MIT, Hale was appointed professor of astrophysics in 1892 at the University of Chicago. Despite never earning a Ph.D., Hale founded the world-renowned Yerkes, Mount Wilson, and Palomar Mountain observatories. He is known as the father of solar observational astronomy.

Hale persuaded businessman Charles T. Yerkes to pay for a 40-in. refractor telescope for the University of Chicago's new observatory at Williams Bay, Wisconsin. Completed in 1897, it still is the world's largest refracting telescope. Hale used the Yerkes Observatory for astrophysical research on Sun and stars.

Carnegie Institution of Washington granted money to Hale to found Mount Wilson Observatory at Pasadena, California, in 1904. While Hale was director of the observatory until 1923, he discovered magnetic fields in sunspots—the first knowledge magnetic fields existed in outer space. Hale calculated a 22 to 23 year cycle of polarity reversals in the magnetic fields. The first Mount Wilson telescope was a 60-in. reflector completed in 1908. That was followed by a 100-in. reflector, named the Hooker Telescope, in use since 1919.

Experience with the 100-in. Hooker led Hale to want an even larger telescope. He obtained money needed to build the 200-in. telescope on Palomar Mountain, California. Still the largest optical telescope in the United States, the Hale Telescope saw first light in 1947. (Largest in the world is a 236-in. telescope in the USSR.)

Hale died February 21, 1938. Today, the Hale Observatories group, sponsored by California Institute of Technology and Carnegie Institution of Washington, include Mount Wilson, Palomar Mountain and Big Bear Solar Observatory, all in California, and Las Campanas Observatory in Chile.

Halley. Edmond Halley, born November 8, 1656, was a graduate of Oxford University who became a member of the Royal Astronomical Society at age 22. Halley studied archaeology and was deputy comptroller of the mint at Chester. The English astronomer is remembered for discerning that comets are periodical and discovering the proper motion of stars.

Between 1676 and 1678 on the island of Saint Helena, he catalogued positions of 350 Southern Hemisphere stars and watched a transit of Mercury. He suggested transits of Mercury and Venus could be used to calculate the distance to the Sun.

Halley worked out a theory of comet orbits, concluding the Comet of 1682—known today as Halley's Comet—is periodic. Finding it to be traveling an elliptical orbit, he calculated it was the same one seen from Earth in 1531 and 1607. He predicted correctly the comet would return in 76 years, reappearing in 1758. A modern search of historical records shows the Halley's Comet has returned every 76 years for 2,000 years.

Halley was appointed Savilian professor of geometry at Oxford in 1704. While comparing star positions in 1710 with positions listed in an ancient catalog by Ptolemy, he surmised the stars must have a slight motion of their own. Checking three stars, he found the motion, now known as proper motion.

In 1720 Halley succeeded John Flamsteed as astronomer royal. He brought the first transit instrument to the Royal Greenwich Observatory and invented a way for ships captains to determine longitude at sea by observing the Moon.

Halley supported Isaac Newton financially, calmed Johannes Hevelius in the dispute over accurate methods for measuring positions of star, and angered John Flamsteed by plotting with Newton to publish Flamsteed's observations before they were complete. Halley died January 14, 1742.

During its latest visit to the inner Solar System, astronomers used the 200-in. Hale reflector telescope on Palomar Mountain, California, to spot the returning Halley's Comet in October 1982 when it was one billion miles from Earth, beyond the orbit of Saturn.

The comet reached perihelion, closest to the Sun, February 9, 1986. It was nearest Earth—about 38 million miles distant—April 11, 1986. During the 1986 passage, Comet Halley was fainter than when last seen in 1910. In 1986, the comet was seen most readily from Earth's Southern Hemisphere.

For a closer look in 1986, Halley's Comet was visited by a number of spacecraft from Earth. The USSR's Vega 1 and Vega 2 probes, which earlier had fired probes into the atmosphere on the way by Venus, came within 5,500 miles and 5,125 miles of the core of Halley's Comet on March 6 and 9. Vega solar panels were damaged by dust particles, but both probes transmitted many color images.

On March 8, the Japanese probe Suisei passed within 94,000 miles of the comet, snapping ultraviolet images and measuring how the comet's gas coma interacted with the solar wind. Three days later the Japanese spacecraft Sakigake whistled by Halley's Comet at a greater distance of 4.3 million miles. It measured the comet's magnetic field and solar wind interactions.

The European Space Agency's probe Giotto came closest to the comet, passing within 335 miles of the core on March 14. Dust blocked its cameras shortly before closest encounter, but Giotto was able to show the core is a black, peanut-shaped ice-rock about 9.3 by 2.5 miles in size. The blackness may have been extremely fine dust embedded in Halley's ice. Jets of gas and dust shot out from the core at irregular intervals as Halley's Comet passed around the Sun.

Herschel. Caroline Lucretia Herschel, was born at Hanover, Germany in 1750, moved to England in 1772, and assisted her older brother, William Herschel, with his music and astronomy work. At Datchet and Slough, she conducted her own sky sweeps, looking for comets and nebulas. Her finds included the Sculptor Galaxy NGC-253 in 1783 and eight comets. Caroline revised Flamsteed's catalog in 1798. After William died, she returned to Hanover in 1822. The Royal Astronomical Society gave her its gold medal in 1828. She died in 1848.

Herschel. William Herschel, born in Germany November 15, 1738, escaped to England in 1757 as Hanover was occupied by the French. He worked as a professional musician for 25 years while constructing the largest reflecting telescopes of his time. The English astronomer is best known for discovering the planet Uranus while surveying the skies on the night of March 13, 1781. It was the first planet discovery since ancient times.

In 1786, William and Caroline, his sister and lifelong assistant, settled at Slough, where he lived out his life on a royal pension. He supervised construction of immense telescopes, including one 40-ft.-long tube housing a 48-in.-diameter mirror. Herschel tabulated star catalogues, found 2,400 nebulas, and even resolved some nebulas into individual stars. He assumed erroneously the brightness of a star revealed its distance, but counting stars in the field of view of his telescope as it was pointed in different directions gave him a first crude map of the Milky Way—the beginning of statistical astronomy.

Herschel studied infrared light, discovered two moons orbiting Saturn and two orbiting Uranus, and calculated the velocity and direction of the Solar System as it moves through space. He came up with a theory the Sun is like Earth and inhabited, merely enveloped in a luminous atmosphere.

Known today as the father of stellar astronomy, he was the father of astronomer John Herschel. William Herschel died August 25, 1822.

Herschel. John Frederick William Herschel, born March 7, 1792, was the only child of English astronomer William Herschel. He was graduated from Cambridge University. An outstanding astronomer, physicist, and chemist in his own right, John Herschel assisted with his father's observations.

From 1834 to 1838, John Herschel catalogued nebulas, star clusters and binary stars he saw in the skies above Earth's Southern Hemisphere from an observation point at the Cape of Good Hope. He published A Treatise on Astronomy in 1833 and Outlines of Astronomy in 1849.

John Herschel developed several photographic processes. He discovered that the chemical known today as hypo—sodium thiosulfate—is a photo fixer. He coined the photographic terms positive and negative. John Herschel died May 11, 1871.

Hertzsprung. Ejnar Hertzsprung, born October 8, 1873, was a Danish astrophysicist and professor at Gottingen who became associate director of the University of Leiden observatory in 1919 and director in 1935.

Hertzsprung published papers in 1905 and 1907 reporting discovery of giant stars and dwarf stars. He revealed a graph plotting absolute stellar magnitude against temperature. The plot is known today as the Hertzsprung-Russell Diagram. The HR diagram is named for Hertzsprung and American astronomer Henry Norris Russell. Working independently from 1911-13, they studied the relationship between the absolute magnitude of a star and its spectral type or surface temperature, which is approximately equivalent to its color. Ordinary stars, like the Sun, appear as "main sequence" stars on the HR diagram.

Hertzsprung also studied variable stars, double stars and photographic photometry. Hertzsprung died October 21, 1967.

Hipparchus. Hipparchus of Nicaea, a Greek astronomer and instrument maker living from about 190 BC to 120 BC, made astronomical observations from what now is Iznik, Turkey, and the island of Rhodes.

Our knowledge of Hipparchus comes mostly from the works of the astronomer Ptolemy of Alexandria and the Greek geographer Strabo. He is regarded as the greatest ancient Greek astronomer and the founder of systematic astronomy, even though only one minor work by Hipparchus has been found.

Hipparchus saw Earth as the center of the Universe with all celestial bodies orbiting around Earth in perfect circles. Since discrepancies in this theory could be seen, he decided the Moon and Sun travel in circular orbits eccentric to the Earth and the planets move in combined circles upon circles he called epicycles.

Hipparchus cataloged 850 stars. He found the small annual change in each star's position known today to result from the precession of Earth. He devised better methods of calculating diameters and distances of the Moon and Sun. Hipparchus also was one of the first to calculate a sine table known as a table of of chords. He used mathematics to compute longitudes and latitudes.

Hubble. Edwin Powell Hubble, born at Marshfield, Missouri, November 20, 1889, studied physics and astronomy and was a heavyweight boxer at the University of Chicago. He received a law degree from Oxford University, practiced law briefly, then went on to earn a Ph.D. in astronomy at University of Chicago in 1917.

Hubble went to Mount Wilson Observatory near Pasadena, California, after World War I. In 1923, he confirmed the Andromeda Nebula was far outside our Milky Way galaxy, establishing the so-called Island Universe theory that galaxies exist outside our own. In 1925, he composed a galaxy classification system in use today. Hubble provided convincing evidence of the expansion of the Universe.

The basic cosmological quantity Hubble's Constant stems from Hubble's Law proposed in 1929. All galaxies in the Universe are receding from us. Seeing the farthest galaxies retreating the fastest, Hubble discovered a simple relationship between velocity (v) of a galaxy and its distance (r). In the equation, H is Hubble's Constant: $v = Hr$

The exact value of Hubble's Constant is in question. One current number is 55, but recalculations of distances to galaxies, based on new computations of mass from their rotation period, and the distance of a quasar sending out fluctuating X-rays, indicate Hubble's Constant may be about 100.

Hubble died September 28, 1953. The U.S. sent the Hubble Space Telescope to Earth orbit in April 1990 on a 15-year mission to search the cosmos for new discoveries, other galaxies, and the very origins of the Universe itself. On May 20, 1990, the first HST images—crisp photos of a double star in the Carina system—were transmitted to Earth.

Huggins. Pioneer astrophysicist William Huggins was born in England February 7, 1824. He was educated at City of London School. Gustav Kirchoff had discovered in 1859 the chemical composition of the Sun could be determined from its spectral lines. Huggins then recorded the first spectra of other stars, determining they are of a chemical composition similar to the Sun.

From the late 18th century, nebulas had been considered groups of stars until 1864 when Huggins discovered spectroscopically they are luminous gas. He also used spectroscopy to determine the radial velocities of stars. Huggins died May 12, 1910.

Huygens. Dutch natural philosopher Christiaan Huygens, born at The Hague April 14, 1629, was tutored first at home by private teachers, then studied law and mathematics at the University of Leiden and the College of Orange at Breda. He went on to discover Saturn's rings, invent the pendulum clock, and compose a wave theory of light.

His scientific career was most productive from 1650-1666 while he was relatively alone at The Hague. Later, he spent 1666-1681 in Paris with some of the greatest scientists of the age. Huygens wrote about mathematical problems in 1651 and 1654, then took up a lifelong interest in lens grinding.

Looking through one of his lenses in 1655, Huygens spotted the first satellite of Saturn to be found. The next year he discovered the flattened ring which orbits Saturn. Huygens explained changes in the shape of the ring in his Systema Saturnium in 1659.

In 1656, Huygens patented the first pendulum clock, making time measurement more accurate. Published in 1673 in his Horologium Oscillatorium, Huygens explained pendulum motion and centrifugal force.

Huygens set forth a wave theory of light in his Traite de la lumiere published in 1678. He described an expanding sphere of light with each point on the wave front a new source of radiation of the same frequency and phase. The position and direction of light can be determined from the previous position and direction of the wave. Light from any point on the wave is weak, but where there is overlap, there is light. The theory explained reflection and refraction.

At the end of his life, Huygens wrote about extraterrestrial life, published posthumously in 1698 as the Cosmotheoros. Huygens died at The hague July 8, 1695.

The U.S. and European Space Agency plan to send an interplanetary probe to study Saturn in the 1990's. The Saturn orbiter Cassini will fly by asteroids and Jupiter on the way to Saturn. It will drop a smaller probe named Huygens into Saturn's atmosphere.

Innes. South African double-star astronomer Robert Thorburn Ayton Innes, born in 1861, was director of the Union/Republic Observatory in Johannesburg from 1903 to 1927. In 1915 he discovered Proxima Centauri, the star closest to our Sun. Proxima

Centauri, one of the least luminous stars ever seen, is a dwarf star and a close companion of the double-star Alpha Centauri. Proxima Centauri is a flare star and its occasional brightness has been spotted by X-ray telescope in Earth orbit. Innes died in 1933.

Isidorus. The bishop of Seville, Isidorus, living from about 570 to 637, was one of the earliest persons to spell out the difference between astrology and astronomy. Astrology is the use of positions of celestial bodies, particularly at the moment of birth, to describe an individual's personality and forecast events. Astronomy is the observation and theoretical study of celestial bodies, space, and the Universe

Janssen. French astronomer Jules Cesar "Pierre" Janssen, born in 1824, founded the Meudon Observatory in an ancient castle near Paris in 1875 and the Mont Blanc Observatory in 1895. Janssen discovered the telluric line in 1862 and helium's spectral line in 1868. He computed in 1868 a way to study prominences on the Sun without waiting for an eclipse. He photographed solar granulation in 1874. Janssen died in 1907. A low-walled, 160-mi.-diameter plain on the Moon is named for Janssen.

Jeans. English astrophysicist James Hopwood Jeans, born in 1877, popularized astronomy through books and radio broadcasts. He died in 1946.

Kepler. German astronomer Johannes Kepler, born December 27, 1571, attended seminaries at Adelberg and Maulbronn, then studied mathematics, philosophy and theology at the University of Tubingen where astronomer Michael Maestlin noticed his aptitude for science. Maestlin led Kepler toward the theories of Copernicus, although Maestlin continued officially to support Ptolemy's earlier view of the Universe. Kepler was appointed to a chair in mathematics and astronomy at Graz.

Kepler, the first strong supporter of Copernicus' heliocentric theory, conceived three laws of planetary motion. At 24, Kepler defended Copernicus' planetary system in his book Mysterium cosmographicum, or Cosmographic Mystery, published in 1596. Kepler saw the Universe ruled by the geometry of inscribed and circumscribed circles of the five regular polygons.

Tycho Brahe, mathematician at Emperor Rudolph II's court at Prague, invited Kepler in 1600 to be his assistant. With Catholics persecuting Protestants in Graz, Kepler accepted. When Brahe died in 1601, Kepler inherited Brahe's planet position data, particularly for Mars. Kepler started an intensive examination of planet orbits.

Kepler deserted the old doctrine that planets move in perfect circles. He verified the orbit of Mars is an ellipse with the Sun at one of its two focal points.

Kepler's laws of planetary motion describe the shape of the orbit, the velocities, and the distances of the planets from the Sun: the planets move in ellipses with the Sun at one focus; the line joining the Sun and a planet sweeps out equal areas in equal intervals of time; and the square of the time of revolution of a planet divided by the cube of its mean distance from the Sun gives a number that is the same for all the planets.

Kepler published his first law of planetary motion, in Astronomia nova, or New Astronomy, in 1609. In the same book, the second law, known as the law of areas, controlled planetary velocity. The third law, known as the harmonic law, governing the relationship between orbital periods and distances of planets from the Sun, was published in 1619 in Harmonices mundi, or Harmonies of the World. Kepler found beauty in the structure of the Universe and harmony in geometric figures, numbers, and music.

Kepler's notion that the Sun regulates the velocity of planets led to Isaac Newton's theory of universal gravitation. Newton generalized Kepler's laws so they apply to any motion associated with a problem in which two bodies are point masses, or small rigid spheres, and the motion of one about the other is subject to Newtonian mechanics and to Newton's law of gravitation. The possible forms of motion now are called Keplerian motion. First law orbits can be ellipses, parabolas, or hyperbolas. The third law applies only to elliptic orbits.

The laws as amended by Newton apply to a satellite revolving around a planet or to two stars forming a binary system. The amended third law is used to determine the mass of binary stars.

A bright supernova, visible to the naked eye in October 1604, was seen by astronomers in Europe, China and Korea. It sometimes is referred to as Kepler's Star because the astronomer precisely determined its position.

Kepler wrote about optics in 1604, Galileo's telescopes in 1610, and telescope lenses in 1611. Kepler's Epitome astronomiae Copernicanae, or Introduction to Copernican Astronomy, 1618-1621, was widely read in Europe. Kepler's last work, Rudolphine Tables published in 1627, compiled tables of planetary motion. In his fictional Somnium, or Dream, published posthumously in 1634, Kepler wrote of travel to the Moon and lunar people. Kepler died November 15, 1630.

Kuiper. Dutch astronomer Gerard Peter Kuiper, born December 7, 1905, was graduated from the University of Leiden in 1927 and 1933. He moved to the United States in 1933 and was appointed to the faculty of the University of Chicago in 1936.

From 1947 to 1949 and from 1957 to 1960, Kuiper headed the University of Chicago's Yerkes Observatory at Williams Bay, Wisconsin, and the University of Texas' McDonald Observatory at Mount Locke, Texas. In 1960, Kuiper founded the University of Arizona's Lunar and Planetary Laboratory.

Kuiper is known for his study of the Moon's surface, but he also discovered Miranda, a fifth satellite of the planet Uranus, in 1948 and Nereid, a second satellite of the planet Neptune, in 1949. He found a methane atmosphere on Titan and a carbon dioxide atmosphere on Mars. Kuiper died December 23, 1973.

Craters on the Moon, Mars and Mercury were named after Kuiper. NASA named its Kuiper Airborne Observatory (KAO) after the astronomer. In operation since 1975, KAO is a Lockheed C-141 transport carrying a 34-in. telescope for high-altitude infrared astronomy observations. The telescope is in the fuselage where it is subjected to low temperatures and pressures. Its Cassegrain focus is inside the aircraft cabin. Use of KAO led to discoveries of Uranus' rings and water vapor in Jupiter's atmosphere and in Halley's Comet.

Lacaille. French astronomer Nicolas-Louis de Lacaille, born in 1713, has been referred to as the father of southern astronomy after he observed the Southern Hemisphere skies above the Cape of Good Hope from 1750-1755. He named 14 new constellations. Lacaille died in 1762.

Lagrange. French physicist Joseph Louis, comte de Lagrange, born in Turin January 25, 1736, was professor of geometry at the Royal Artillery School in Turin from 1755-66. In 1757, he helped found the Royal Academy of Science there. Lagrange became director of the mathematical section of the Berlin Academy of Science in 1766, but left Berlin in 1787 to become a member of the Paris Academy of Science.

He spent the rest of his career at the Paris Academy of Science, known as the National Institute after 1795. In the 1790s he worked on the metric system, advocating a decimal base. He helped found, and taught at, the Ecole Polytechnique. Napoleon named him to the Legion of Honor and Count of the Empire in 1808.

Lagrange invented the calculus of variations and applied it to celestial mechanics, the so-called three-body problem. He contributed to numerical and algebraic solution of equations and to number theory. In his book Mecanique analytique, or Analytical Mechanics, published in 1788, he transformed mechanics into a branch of mathematical analysis. The work was noted for use of differential equations. In a 1797 book about the foundations of calculus, Lagrange stressed function and the Taylor series. Lagrange died April 10, 1813.

Lagrangian Points, named after the mathematician, are crossing points in the orbital paths of two large celestial bodies where smaller objects can remain in equilibrium. For instance, a satellite placed at a Lagrangian Point where Earth and Moon cross paths would hang there effortlessly.

Laplace. The prosperous French astronomer Pierre Simon de Laplace, born March

28, 1749, became a professor of mathematics at Paris' Ecole Militaire at the age of 19, an associate in 1773 and a pensioner in 1785 of the Paris Academy of Sciences. He helped establish the metric system, taught calculus at Ecole Normale, and became a member of the French Institute in 1795. Napoleon made him a member, then chancellor, of the Senate. Laplace received the Legion of Honor in 1805, became Count of the Empire in 1806, and marquis in 1817.

In his Exposition du systeme du monde, or Exposition of the System of the World, published in 1796, Laplace proposed a nebula theory for the origin of the Solar System in which the Solar System originated from the contracting and cooling of a large, flattened, and slowly rotating cloud of incandescent gas. From the acceleration of the Moon, the planets Jupiter and Saturn, and the inner three satellites of Jupiter, Laplace confirmed mathematically the stability of the Solar System. His five-volume Traite de mecanique celeste, or Treatise on Celestial Mechanics, published between 1799-1825, capped a century of mathematical explanations of how gravity affects the motion of Solar System bodies. Laplace found the Solar System fluctuates about a constant plane.

The astronomer contributed to the science of magnetism and electricity, but disagreed with Thomas Young's theory of light. Laplace died March 5, 1827.

Lassell. British brewer William Lassell, born June 18, 1799, was an amateur astronomer who ground many telescope lenses and built two large reflecting telescopes. With them, he discovered Triton, the largest satellite of the planet Neptune, in 1846; Hyperion, the eighth satellite of the planet Saturn, in 1848 (with William Bond); and 600 nebulas. He also verified Ariel and Umbriel, two satellites of Uranus, in 1851. Lassell died October 5, 1880.

Leavitt. American astronomer Henrietta Swan Leavitt, born July 4, 1868, was graduated in 1892 from the forerunner of Radcliffe College. She went to work in 1902 at Harvard College Observatory where she studied variable stars. Leavitt conducted Edward C. Pickering's program of photographic photometry to determine the brightness of certain stars. She found 2,400 variable stars.

Leavitt's study of the class of stars known as Cepheid variables, especially in the Small Magellanic Cloud galaxy, resulted in understanding of the Cepheid period-luminosity relationship: the longer the period of a variable star, the brighter the visual magnitude. The period-luminosity relationship opened a new way to calculate how far distant stars and galaxies are from Earth. Harlow Shapley used it to calculate the distance to globular clusters. Leavitt died December 12, 1921.

Lemaitre. Catholic priest Georges Edouard Lemaitre, born in 1894, an astronomy professor at Louvain from 1927, is remembered for studies of the expansion of the Universe. After considering Albert Einstein's General Theory of Relativity, Lemaitre's thought, that the Universe exploded from a single, radioactively-decaying primeval atom, led to the Big Bang theory. Lemaitre died in 1966.

Lemonnier. French astronomer Pierre Charles Lemonnier, born in 1715, surveyed Earth's sky in a one-degree arc along the meridian near the Arctic Circle. The meridian is an imaginary semicircle around Earth, at right angles to the equator, connecting the North Pole and South Pole. He also studied small variations, or perturbations, in the orbits of planets. Perturbation is a disturbance in the regular movement of one celestial body around another, usually caused by gravity from a nearby third body. Lemonnier died in 1799.

Lomonosov. Mikhail Lomonosov, born in 1711, son of a fisherman, had little early schooling, but was the first great Russian astronomer. In 1735, he entered the University of St. Petersburg, then went to Germany for further training. In 1745, he was appointed chemistry professor at University of St. Petersburg. Lomonosov agreed with Copernicus and observed a transit of Venus in 1761. He concluded correctly that Venus has a dense atmosphere. Lomonosov died in 1765.

Lowell. American businessman Percival Lowell, born at Boston, Massachusetts,

March 13, 1855, was travelling in the Far East when he heard of Giovanni Schiaparelli's 1877 depiction of Martian canali, or channels. He became an astronomer at the age of 39. Obsessed with the possibility of such artificial canals on the Red Planet, he founded in 1894 the Lowell Observatory at Flagstaff, Arizona, to study the surface of Mars. For more than a decade, Lowell mapped what seemed to be markings crisscrossing Mars. Most astronomers saw no canals on Mars, but Lowell stuck to his theory they had been built by intelligent beings. Finally, the unmanned interplanetary spacecraft from Earth, which probed Mars in the 1960's, 1970's and 1980's, disproved Lowell's claim.

The unmanned U.S. spacecraft Mariner 4 sent back 22 photos as it flew by Mars in July 1965. Mariner 6 passed Mars in July 1969, sending 100 photos. Mariner 7 passed Mars in August 1969, sending 100 photos. Mariner 9 arrived in orbit around the Red Planet in November 1971, the first man-made satellite to orbit a planet other than Earth. It sent back 7.329 photos and mapped the entire surface.

The USSR's unmanned Mars 2 entered Mars orbit in November 1971. Mars 3 entered Mars orbit in December 1971. Mars 5 entered Mars orbit in February 1974. Mars 6 flew past Mars in March 1974. Mars 7 flew past Mars in March 1974.

The U.S. unmanned Viking 1 arrived at Mars June 19, 1976, sending 26,000 photos to Earth and landing a science station on the rocky Martian surface. Viking 2 arrived August 7, 1976, sending 26,000 photos and landing a science station. The USSR's unmanned Phobos 2 arrived in orbit above Mars in January 1989 to map the planet.

Despite the Mars disagreement, Lowell and his observatory made significant contributions to knowledge of planets and stars. Lowel encouraged V.M. Slipher who discovered the redshifts used to measure distances to faraway galaxies. Lowell predicted where a planet would be found lurking in the far reaches of the Solar System, beyond the planet Neptune. Lowell died November 12, 1916, before his prediction was verified in 1930 when the planet Pluto was found by Clyde Tombaugh.

Ludendorff. German astrophysicist Frederich Wilhelm Hans Ludendorff, born at Koszalin, Pomerania, in 1873, studied variable star periods. He is recalled for his history of astronomy. He died in 1941.

Marius. German astronomer Simon Marius, born in 1570, said he discovered the natural satellites of Jupiter in 1609, but astronomers of his time gave the credit instead to Galileo who published news of the Jovian moons in 1610. Marius was the first to observe Andromeda Galaxy with a telescope—in 1612. Marius died in 1624.

Messier. French astronomer Charles Messier, born June 26, 1730, went to Paris at age 21 as a draftsman and recorder of astronomy observations. At the Marine Observatory, he searched for the returning Comet Halley in 1759. He wasn't the first to recover Halley, but Messier did find a different comet in 1764. Almost all comets discovered in the next 15 years were found by Messier. King Louis XV called him the "ferret of the comets." Altogether, Messier discovered 21 comets.

He is best known for his Messier Catalogue of deep-sky objects which he needed to distinguish deep-space nebulous objects from comets. Comets and nebulas have the same fuzzy look to the naked eye or through a small telescope.

Messier published his catalogue first in 1774, listing 45 deep-space objects. The 1780 edition had 68 objects. In the 1781 version, there were 103 magnificent objects seen in Earth's sky, including galaxies, star clusters and nebulas. The best-known Messier objects are the Crab Nebula, also referred to as M1 for Messier 1, and the Andromeda Galaxy known as M31.

Messier also was interested in solar and lunar eclipses, sunspots and occultations, in which one celestial object is seen passing in front of another. He died April 12, 1817.

Newton. Isaac Newton was born prematurely January 4, 1643, three months after his father's death, to a family of modest yeoman farmers in the manor house of Woolsthorpe, a hamlet near Grantham, Lincolnshire, England. Three-year-old Isaac was left with his maternal grandmother when his mother remarried and moved to a nearby village. He hated his stepfather until he died in 1656. His mother decided Isaac should be trained as a farmer to manage her estate so she removed him from grammar school in Grantham. Fortunately, she agreed when Isaac preferred attending a university.

He entered Trinity College, Cambridge University, in 1661, as a subsizar, or needy undergraduate doing menial jobs for his keep. When he was graduated in 1665, the plague closed the university and Newton returned to Woolsthorpe. Although he hadn't revealed an extraordinary ability at Trinity, his genius appeared during the next 18 months as dramatic advances in mathematics, optics, physics, and astronomy. Newton returned to Cambridge in 1667 as a Trinity fellow.

Newton worked in differential and integral calculus years before its discovery by a German mathematician. Newton's "method of fluxions" found the integration of a function (finding the area under its curve) to be the inverse of differentiating it (finding the slope of the curve at any point). Newton produced simple analytical methods to find areas, tangents and lengths of curves. Trinity College fellow and Lucasian professor of mathematics at Cambridge, Isaac Barrow, was so impressed, he resigned his chair in 1669 so the 27-year-old Newton could take his place.

Newton thought white light was a mixture of light of all colors. To prove it, Newton passed a beam of sunlight through a glass prism, projecting a long spectrum of red, yellow, green, blue and violet colors on the wall. From that, he selected light of just one color to send through a second prism. When no further spectrum breakdown occurred, and all rays of one color were bent or refracted at the same angle, Newton concluded white light is a mixture of many rays, different rays bend at different angles, and each ray is a different color. Earlier theory had said all rays of white light striking the prism at the same angle would be refracted equally. Newton showed the spectrum too long to be explained by the bending of light passing through dense media.

Telescopes of that time were refractors, using only lenses to form images. Very long, cumbersome tubes were needed for sharp images. The lenses created color fringes around bright objects—distortions known as chromatic dispersion. Newton mistakenly assumed refractors could not be made to overcome the problem.

To correct it, he designed and made the first reflecting telescope, with a mirror, by 1671. His reflector showed no color fringes and was much shorter than a refracting telescope. When he sent one of his reflectors in 1671 to the leading scientific institution of its day—the Royal Society of London—members elected him a fellow in 1672. Newton's reflector was the prototype for today's largest optical telescopes.

Newton was sensitive to criticism. When he submitted three papers to the Philosophical Transactions publication of the Royal Society in the early 1670's—on color and light in 1672, on colors in thin films of oil in 1675, and on tiny light particles or corpuscles in 1675—criticism by Robert Hooke and Christian Huygens upset him so much he retreated to Cambridge and didn't publish again for a decade. He also held back a complete report on his optical research until after Hooke died in 1703. Newton's book Opticks, published in 1704, covered light, color, colors of thin sheets, Newton's rings, and diffraction of light. Today, the rings seen in photographs as colored rings caused by sun's glare are called Newton's rings.

Newton's greatest achievements were in physics and celestial mechanics, culminating in his theory of universal gravitation. It is a myth that he discovered gravity in 1666 when he saw an apple fall from a tree in his garden. However, by 1666 Newton had conceived three laws of motion and a theory of centrifugal force.

Newton's first law of motion: unless acted upon by an outside force, an object at rest or in motion will remain at rest or in motion with the same speed and direction.

Newton's second law of motion: the change in motion of an object acted upon by an outside force is directly proportional to the force and inversely proportional to the object's mass.

Newton's third law of motion: every action force has a reaction force equal in magnitude and opposite in direction.

Newton's law of universal gravitation: force of attraction between any two particles of matter is directly proportional to the product of their masses and inversely proportional to the square of the distance between their centers of mass.

Universal gravitational constant: the constant of proportionality in Newton's law of universal gravitation.

Centrifugal force: a force away from the center of a body moves uniformly in a circular path.

However, Newton thought Earth's gravity and the motions of the planets might be caused by whirlpools, or vortices, of small corpuscles. He knew of centrifugal force, but did not understand the mechanics of circular motion. He thought it resulted from a balance between two forces—centrifugal away from the center and centripetal toward the center—rather than one centripetal force constantly deflecting the body away from its inertial path in a straight line.

Newton's great achievement in 1666 was visualizing Earth's gravity extending to the Moon, counterbalancing its centrifugal force. From his law of centrifugal force and Johannes Kepler's third law of planetary motion, Newton decided the centrifugal force, and therefore centripetal force, of the Moon or planet must decrease as the inverse square of its distance from the center of its motion.

Robert Hooke drew Newton into a discussion of orbital motion and celestial mechanics in 1679, suggesting circular motion arises from centripetal deflection of inertially moving bodies. Hooke speculated centripetal force drawing planets to the Sun should vary as the inverse square of their distances from it. Hooke couldn't prove it mathematically, but bragged he could. Newton was able to apply math and showed a body obeying Kepler's second law is acted upon by a centripetal force. In 1684, Edmond Halley tired of Hooke's boasting. He asked Newton whether he could prove Hooke's conjecture and was surprised to hear Newton had solved the problem five years earlier, but had mislaid the paper. After much urging by Halley, Newton reproduced the proofs and expanded them into a paper on laws of motion and orbital mechanics. Eventually, Halley persuaded Newton to write the full story of his new physics and astronomy. It took 18 months, but Newton finally published in 1687 the Philosophiae naturalis principia mathematica, or The Mathematical Principles of Natural Philosophy. Today it is known simply as Principia and is regarded as the greatest science book ever written.

After Principia, Newton's creative period ended. He suffered a nervous breakdown in 1693, retired from scientific research for a government job at London, became warden of the Royal Mint in 1696, and master of the Royal Mint in 1699. He conducted the major English 1690's re-coinage and hunted down counterfeiters.

Newton was elected president of the Royal Society in 1703 and re-elected every year until his death. He was the first scientist to be so honored when he was knighted in 1708 by Queen Anne. Said to be the premier scientific genius of all time, Newton made fundamental contributions to every major area of science and mathematics. He died in London March 31, 1727.

Neison. English astronomer Edmund Neison/Nevill, born in 1851, directed South Africa's old Natal Observatory from 1881 to 1910. He published a map of the Moon and an article about it in 1897. Neison died in 1938.

Oberth. Transylvania physicist Hermann Julius Oberth, born June 25, 1894, was captivated by Jules Verne's book From the Earth to the Moon when he was 11 years old. At college, he first studied for a medical career, then switched to physics. Oberth worked on combat rockets before World War I, proposing a liquid-fuel rocket in 1917.

His 92-page book Die Rakete zu den Planetenraumen, or The Rocket into Planetary Space published in 1923, probed space travel. A 423-page edition, Wege zur Raumschiffahrt, or The Road to Space Travel published in 1929, was accepted as a doctoral dissertation. Oberth worked for the German army during World War II. After the war, he worked for a time for the U.S. Army under his former student, Wernher von Braun, but returned to Germany in 1958. He is considered the founder of modern astronautics.

Olbers. The famous German doctor Heinrich Olbers, born in 1758, also was one of the so-called celestial policemen, a team of astronomers which started searching in 1800 for a planet between the orbits of Mars and Jupiter. They found none, but did discover the asteroids Pallas, Juno and Vesta from 1802-1807.

Olbers also discovered several comets and invented a way to calculate orbits of comets. Olbers' Comet is a bright body which tours the Solar System every 69.47 years. It has been seen three times, in 1815, 1887 and 1856.

Olbers' Paradox is named after the astronomer. He discussed it in 1826, although Edmond Halley had brought up the question earlier: "Why is the night sky dark?" If space is infinite and filled with stars, wherever you stare you would be looking at the surface of a star. The entire sky should be as bright as the surface of the Sun. There are various answers: if the Universe is not old enough, light from remote objects may not have reached Earth yet. And expansion of the Universe means the glow from faraway galaxies is exhausted by red-shift and can't be detected beyond some distance.

Olbers was responsible for the exceptional career of Friedrich Wilhelm Bessel, the German astronomer who found the earliest method for accurately measuring distances to stars. The orbit of Olber's Comet was calculated first by Bessel. Olbers died in 1840.

Oort. Dutch astronomer Jan Hendrik Oort, born April 28, 1900, was graduated from the University of Groningen. He worked lifelong at the University of Leiden and Leiden Observatory where, from 1945 to 1970, he was full professor and observatory director.

Oort has been best known in recent years for his theory, published about 1950, that the Sun is surrounded by a distant cloud of comets and a comet is hurled occasionally toward the center of the Solar System by gravitational perturbations from planets and nearby stars. If it exists, the so-called Oort Cloud of comets may be some 3 billion to 9 billion miles from the Sun. Earth is only about 93 million miles from the Sun.

A pioneer radio astronomer in the Netherlands, Oort advanced our understanding of the structure and rotation of our Milky Way galaxy. In the 1950's, he determined the spiral structure of the galaxy, verified a theory that the galaxy rotates as a whole, determined direction and distance from Earth to the center of the galaxy, and calculated the mass of the galaxy.

Piazzi. Italian monk Giuseppe Piazzi, born in 1746, was professor of mathematics at Palermo and director of the Palermo Observatory. He also was director of the Government Observatory at Naples. Piazzi discovered the first asteroid, Ceres, in 1801. He also compiled star catalogs. Piazzi died in 1826.

Pickering. Edward Charles Pickering, born at Boston, July 19, 1846, was graduated from Harvard in 1865, was astronomy professor there from 1876, and was director of Harvard College Observatory from 1877-1919. He led a study of magnitudes of 45,000 stars using a meridian photometer which he had invented. He compiled the Henry Draper Catalogue of Stellar Spectra. Pickering's photographic map of stars to magnitude 11 was published in 1903. He made influential contributions to stellar spectroscopy. Pickering died February 3, 1919.

Planck. German physicist Max Karl Ernst Ludwig Planck, born April 23, 1858, received a Ph.D. from University of Munich in 1879, taught at University of Kiel from 1885-1889 and at University of Berlin from 1889-1926. At University of Berlin, he also was director of the Institute of Theoretical Physics.

Planck is known for conceiving of the quantum as a fundamental increment of energy. It is basic to quantum mechanics and a cornerstone of modern physics.

He began studying blackbody radiation in 1897 and discovered at long wavelengths it did not follow classical distribution laws. In 1900, he concluded an oscillator could emit energy only in discrete quanta, contrary to classical theory. For his quantum theory, Planck received the 1918 Nobel Prize for physics.

Albert Einstein used quantum theory in 1905 to explain the photoelectric effect. Niels Bohr used it in 1913 to depict an atom with quantized electronic states. Planck died October 3, 1947.

Ptolemy. Not much is known about the life of Ptolemy, the influential Greek geographer and astronomer who lived from around A.D. 100 to 170, except he probably lived most of his life at Alexandria, Egypt, and observed the sky there from A.D. 127 to 141. He was able to write a theory of how the Universe is—a geocentric world system—which was followed for fourteen centuries.

Two important works by Ptolemy still exist—the Almagest and the Geography. The Almagest may be his earliest work. It details a mathematical theory of the motions of the Sun, Moon, and planets. Accepting a solar theory conceived 200 years earlier by Hipparchus, Ptolemy improved the lunar theory and added details of the motions of planets. Following Aristotle's Earth-centered system, Ptolemy's geometric models to predict the positions of these bodies were combinations of circles known as epicycles. The Almagest was not abandoned by scientists until a century after Copernicus published a heliocentric theory 1543.

The Geography was a map of the known world. Including the earlier works of Marinus of Tyre, Strabo and Hipparchus, Ptolemy recorded longitudes and latitudes of important places with maps and a description of mapmaking techniques. It was inaccurate, but the Geography remained popular through the Renaissance.

In two lesser works—the Optics and the Tetrabiblos—Ptolemy covered reflection, refraction and astrology.

Quenisset. French astrophotographer Ferdinand Jules Quenisset, born in 1872, worked from 1906 to 1951 at the Observatoire Camille Flammarion in Juvisy-sur-Orge, Essone. In 1934, he tried to capture the entire sky on one photographic plate, starting the fad known today as all-sky photography. Quenisset died in 1951.

Quetelet. Belgian astronomer Jacques Adolf Quetelet, born in 1796, directed the Brussels Observatory from 1833. Quetelet studied meteor showers and, in 1839, cataloged 315 different meteor displays, calling particular attention to the Perseid meteor shower which occurs about August 9 each year. Quetelet died in 1874.

Russell. Astrophysicist Henry Norris Russell, born at Oyster Bay, New York, October 25, 1877, received a doctorate in 1899 from Princeton University in 1899, then taught there most of his life. His pioneering stellar evolution work published in 1914—the Hertzsprung-Russell Diagram—related the brightness and spectral class of stars.

The Hertzsprung-Russell Diagram, also known as HR Diagram, was named for Russell and Danish astrophysicist Ejnar Hertzsprung who published a graph plotting stellar magnitude against temperature. Working independently from 1911-13, both astronomers studied how the absolute magnitude of a star relates to its spectral type or surface temperature, which is about the same as its color. The Sun and other ordinary stars are "main sequence" stars on the HR diagram. Russell died February 18, 1957.

Schmidt. Estonian optician Bernhard Voldemar Schmidt, born in 1879, was working at the Bergedorf Observatory, Hamburg, when he invented the Schmidt Telescope in 1930. The most important advance in optical astronomy in 200 years, Schmidt's telescope combined the best features of refractor and reflector telescopes.

Parabolic telescope mirrors had a small field of view and an image coma. Spherical mirrors suffered from the focusing problem spherical aberration. Schmidt incorporated a concave spherical mirror with a thin transparent plate at the center of curvature to correct for spherical aberration. Speeds for recording faint deep-space objects were high and wide-diameter star fields could be photographed. Schmidt died in 1935.

Schmidt telescopes are popular with amateur astronomers. The best-known large Schmidt telescope, on Mount Palomar, California, has a 72-in. mirror, a 48-in. corrector plate and opens to f/2.5. It photographed the popular National Geographic Atlas of the Northern Sky. The U.K. Schmidt telescope at Siding Spring, Australia, the European

Southern Observatory (ESO) Schmidt telescope at Cerro La Silla, Chile, and another at Cerro Tololo, Chile, are similar to the one at Palomar. The world's largest Schmidt telescope, with a 79-in. mirror and 52-in. corrector, is at Tautenburg, Germany.

Schmidt. Dutch astronomer Maarten Schmidt, born December 28, 1929, earned a Ph.D. from the University of Leiden in 1956 and became a professor at the California Institute of Technology. He played a pivotal role in the discovery of quasi-stellar objects, or quasars. In 1963, Schmidt recognized that odd spectra from certain intense radio sources deep in space meant those objects had enormous red shifts and must be the most distant objects known in the Universe.

Schwarzschild. German astronomer Karl Schwarzschild, born October 9, 1873, studied at the University of Strasbourg and was graduated from the University of Munich in 1896. After graduation, he worked on celestial mechanics and a new use of photography for stellar photometry.

After working at a Vienna observatory, he became a professor at the University of Gottingen in 1901, where he studied the Sun. In 1906, Schwarzschild published a report on stellar equilibrium, suggesting energy transfers upward through several layers of a star by radiation. He became director of the Potsdam Astrophysical Observatory in 1909, then enlisted in the German army in 1914 for service in World War I. Schwarzschild died two years later on May 11, 1916, from a skin disease contracted while in service.

His Schwarzschild radius, or gravitational radius, is a calculation to determine the existence of a black hole. It calculates the position of the surface of a non-rotating black hole. The space at the surface is known as the event horizon. No light can escape the event horizon. When an object undergoes gravitational collapse, contracting inside its Schwarzschild radius, it disappears from view, leaving behind a black hole.

Shapley. Astronomer Harlow Shapley, born at Nashville, Missouri, November 2, 1885, was graduated from University of Missouri in 1910 and Princeton University in 1913. After working at Mount Wilson Observatory, California, he became director of Harvard College Observatory from 1921 to 1952.

Shapley determined distances to globular clusters, then conceived a new model of our galaxy in 1918. Using the period-luminosity relation of Cepheid variables in globular clusters to calculate their distances and distribution, he concluded the Sun is far from the center of our Milky Way galaxy.

In the so-called Great Debate of 1920 with Heber Curtis, Shapley contended the Milky Way galaxy is twice as large as previously thought. He incorrectly argued that spiral nebulas are small objects near our galaxy, rather than faraway galaxies beyond our own Milky Way galaxy. Shapley died October 20, 1972.

Slipher. American astronomer Vesto Melvin Slipher, born at Mulberry, Indiana, November 11, 1875, director of Lowell Observatory, Flagstaff, Arizona, from 1916 to 1954, is known for spectroscopic studies of the Solar System and objects outside of the Milky Way galaxy. He was first to see the red-shift in faraway objects we know today as galaxies. Slipher calculated the rotation of the planets Venus, Mars, Jupiter, Saturn, and Uranus and started the final search for the planet Pluto. Slipher found the atmospheres of Saturn and Jupiter to be mostly ammonia and methane. He discovered many spiral nebulas are moving away from Earth. Slipher died November 8, 1969.

Tsiolkovsky. Russian theorist Konstantin Eduardovich Tsiolkovsky, born September 17, 1857, was the son of a Polish forester who had moved to Russia. Childhood scarlet fever left Tsiolkovsky nearly deaf, but he taught himself mathematics and became a high school teacher in the small town of Kaluga, 90 miles south of Moscow. Like Transylvanian physicist Hermann Oberth, Tsiolkovsky was captivated by Jules Verne's books at an early age. Verne's novels inspired him to design airships, planes, and rockets. Tsiolkovsky was encouraged by chemist Dmitry Mendeleyev and foreign astronautics pioneers such as Oberth.

Tsiolkovsky described the motion of a rocket in weightlessness and in a vacuum in an article, Research into Interplanetary Space by Means of Rocket Power, in 1903. From 1896 to 1913, other space science articles by Tsiolkovsky appeared infrequently in Russian journals, but were not noticed abroad.

He wrote science-fiction space adventures describing artificial satellites, spacesuits, space colonies, and asteroid mining. The science-fiction titles included On the Moon published in 1895, Dreams of the Earth and Sky in 1895, and Beyond the Earth in 1920.

Tsiolkovsky retired in 1920 because of ill health and received a government pension. But, when Oberth's works were published in Europe in 1923, Tsiolkovsky's earlier astronautics articles were expanded and republished. They were very popular in Russia and finally brought him international recognition.

Tsiolkovsky wrote about multistage rockets in 1929. He never built rockets, but encouraged young engineers who did, including Sergei Korolev who became chief designer of USSR spacecraft in the 1950's. Tsiolkovsky died September 19, 1935. A museum was built at Kaluga and a crater on the back of the Moon was named for him.

Ulugh. Muhammad Taragay, known as Ulugh Beigh or Beg, was a grandson of Tamerlane and an Islamic astronomer. He was born a prince in 1394 and grew to rule the Tamerlane empire from its chief city, Samarkand, in what more recently has been Russian Turkistan and the USSR republic of Uzbekistan. At Samarkand, Ulugh Beigh built a famous observatory noted for its astronomical tables. He died in 1449.

Van Biesbroeck. Belgian astronomer George Van Biesbroeck, born in 1880, found many double stars and three comets. In one of his binary systems, the weak companion for many years was the faintest star known. He worked at the University of Chicago's Yerkes Observatory. Van Biesbroeck died in 1974.

von Zach. Hungarian astronomer Franz Xavier von Zach, born in 1754, was director of the Seeberg Observatory. He was one of the celestial policemen, a team of astronomers which started searching in 1800 for a planet between the orbits of Mars and Jupiter. They found none, but did discover the asteroids. Franz von Zach died in 1832.

Wargentin. Swedish astronomer Pehr Wargentin, born in 1717, compiled tables of statistics on the satellites of the planet Jupiter. He died in 1783.

Wilson. English astronomer Alexander Wilson, born in 1714, first was apprenticed to a surgeon-apothecary at St. Andrews and London, then changed his plans and set up a type foundry at St. Andrews. Learning astronomy, he was appointed in 1760 to professor of practical astronomy at Glasgow.

Wilson was a sunspot watcher, finding them depressions in the surface of the Sun. He published a book in 1774, describing the Wilson Effect in which a circular sunspot seems elliptical when seen near the limb, or edge, of the Sun. In 1769, he found the penumbra farthest from the limb looks compressed, while the penumbra nearest the limb seems stretched wider—the distorted view is caused by the funnel shape of the sunspot depression. Wilson died in 1786.

Wollaston. English doctor William Hyde Wollaston, born in 1766, invented the camera lucida which he used in 1802 to observe dark lines in a spectrum from the Sun. He mistakenly took the dark lines only as boundaries between colors. A camera lucida is an instrument with a prism and mirrors used to project the image of an object on a plane surface so it can be traced. He died in 1828.

Woolley. British astronomer Richard van der Riet Woolley, born in 1906, was the tenth astronomer royal from 1956 to 1971. He worked at Mount Wilson Observatory in California, at Cambridge and at Canberra where he developed the Mount Stromlo Observatory. After 1871, he was the first director of the South African National Observatories until 1976. He studied and wrote about eclipses. Woolley died in 1986.

Wright. English astronomy teacher Thomas Wright of Durham, born in 1711, theorized in his 1750 book, An Original Theory of the Universe, that the Milky Way galaxy is nothing more than an optical illusion because the Sun is at the core of a surrounding thin hollow shell of stars or flat hollow disk of stars. Wright died in 1786.

Young. Dartmouth College astronomer Charles Augustus Young, born in 1834, was among the first to make spectroscopic observations of the Sun's corona. Watching an eclipse in 1870, he discovered the flash spectrum of the Sun's chromosphere. Young cataloged bright spectral lines in the Sun, using the data to measure the speed of rotation of the Sun. Some of his astronomy books were best sellers. Young died in 1908.

Zollner. German astronomer Johann Karl Friedrich Zollner, born in 1834, invented the polarizing photometer, which is a visual instrument used to measure the apparent magnitude of a star. The polarizing photometer has both a fixed and rotating polarizer which are used to vary the apparent brightness of an artificial star until the articifial star looks the same as a real star in the field of view. The amount of rotation reveals the apparent magnitude of the real star. Zollner theorized about the make-up of the Sun and calculated the albedos of the Moon and planets. He died in 1882.

Zhengtong. A Ming Dynasty ruler of China from 1436-1449, Zhengtong had the Ancient Beijing Observatory built at the southeast corner of the old city wall. A 46-ft.-high platform held eight Qing Dynasty bronze astronomical instruments. Two were built in 1439 and six in 1673.

Zwicky. Swiss astronomer Fritz Zwicky, born in 1898, working mostly in the United States, discovered supernovas in galaxies beyond our own Milky Way. Zwicky died in 1974.

Deep Space Glossary

Absolute magnitude. see: Magnitude.

Absorption lines. Dark lines marking the absorption and scattering of certain wavelengths of light.

Accretion disk. A large accumulation of gas, especially around a neutron star or black hole. See also: Black Hole, Event Horizon, Neutron Star.

Andromeda Galaxy. A galaxy in our neighborhood of the Universe. The spiral galaxy nearest our Milky Way. Bigger than the Milky Way with 300 billion individual stars in a disk 180,000 lightyears in diameter, 2.2 million lightyears from Earth.

Angstrom. A unit of measurement. One centimeter equals 100 million angstroms.

Anti-matter. A theoretical substance formed of particles with electrical charges opposite those found in known matter.

Apparent magnitude. see: Magnitude.

Astronomer. A person observing, studying and theorizing about celestial bodies, outer space, and the Universe. Astronomy is the observation and theoretical study of celestial bodies, outer space, and the Universe.

Astrophysics. A branch of astronomy studying the physical properties of celestial objects, outer space, and the Universe.

Astronomical unit. AU is a unit of measurement. The average distance between Sun and Earth is about 92.96 million miles. Astronomers refer to that distance as one astronomical unit (AU). The outer edge of the Solar System is estimated to be 75 to 100 times the distance from the Earth to the Sun. That is 75 to 100 AU.

AU. See: Astronomical Unit.

Barred spiral. A spiral galaxy with a bar-like formation seeming to pass through its center. Arms stem from the ends of the bar.

Big Bang. A hypothetical explosion which started the Universe evolving.

Binary stars. Double stars are two stars which appear so close together in the sky that only powerful binoculars or telescope can separate them. Often, the stars merely are in line with one star farther away than its apparent companion. On the other hand, the companions sometimes are near each other, orbiting each other around a common center of gravity. Stars which are not close but seem so in the line of sight from Earth are called optical doubles. Stars which actually are related are binary stars, or binary pairs in a binary system. A spectroscopic binary, with stars too close to each other to be resolved visually, must be studied with spectroscopy. See also: Eclipsing Binaries, Spectroscope, Variable Star.

Black hole. Possibly the end result of a total collapse of a star or several stars into a very small and extremely massive body. The matter in a black hole is crushed into something even more densely packed than a neutron star. It is so dense, making its gravity so intense, even light can't escape, making a black hole invisible. However, its presence can be guessed from its interaction with more normal space around it. For instance, a black hole might suck matter away from a nearby normal star. Vast quantities of X-rays might be given off by the matter as it crossed over onto the black hole. The event horizon is the boundary of a black hole.

Blue giant. A very hot, large, bright star.

Bolometric magnitude. See: Magnitude.

Bowshock. As a star moves through space, a large bubble of energy surrounding the star is deformed to a teardrop shape, compressed from the front with a long tail streaming behind in its wake. Ahead of the bubble is a bowshock, a disturbance something like a wave breaking on the bow of a ship moving through water.

Brown dwarf. A lump of matter in space larger than a planet, but not massive enough to become a star by producing thermonuclear reactions at its core.

Cassegrain reflector. A compact, portable reflecting telescope, first designed in the 17th century. It has a convex secondary mirror focusing light through a hole in the primary mirror.

Celestial mechanics. A branch of astronomy involving the motions and gravitational interactions of celestial bodies.

Celestial sphere, equator, poles. The celestial sphere is the imaginary inverted bowl of stars surrounding Earth. The celestial equator is a projection of Earth's equator onto the celestial sphere, equidistant from the celestial poles, dividing the celestial sphere into two hemispheres. The celestial poles are points on the celestial sphere above Earth's North Pole and South Pole. The celestial sphere seems to rotate around the celestial poles. The meridian is an imaginary line crossing the celestial sphere, passing through both Poles and the Zenith. The prime meridian is the meridian passing through the vernal equinox. Zenith is the point in the sky directly above the observer. Nadir is the point on the celestial sphere opposite Zenith. Precession is the shift of celestial poles and equinox due to the gravitational tug of the Sun and the Moon on Earth's equatorial bulge. Right ascension is the eastward angle between an object and the vernal equinox, expressed in hours-minutes-seconds.

Cepheid variables. Variable stars used by astronomers to indicate distance.

Chandrasekhar limit. The maximum mass of stable cold star beyond which it will collapse to become black hole.

Circumpolar star. A star which, at a particular latitude, does not set.

Clusters of stars. When many stars form in a large cloud of interstellar gas and dust, they are a star cluster. A galactic cluster, or open cluster, is a loose, irregularly-shaped collection of stars. An open cluster has dozens or hundreds of stars, while a globular cluster is a large but compact spherical collection of hundreds or thousands of stars.

Coelostat. A clock-driven telescope mirror to track objects across the sky.

Collimator. A telescope lens to align light rays in parallel.

Colors. See: Stars.

Constellations. The 88 constellations are arbitrary groupings of stars in Earth's sky, used by astronomers through the ages to identify particular sky locations. Seeming to form figures or patterns, they are named for mythical persons or animals. Two stars seen in one constellation may not be near each other since one could be relatively close to Earth—say, within a few lightyears distance—while the other could be thousands of lightyears beyond. For example, the star Alpha Centauri is less than five lightyears from Earth while Beta Centauri is more than 450 lightyears away. They seem close in the sky, but are not. The constellations are not equal in size of sky area they cover. The constellation Hydra covers the largest area of the sky. Crux Australis, the smallest constellation. Many different cultures have seen various mythical creatures in the constellations and attributed a variety of stories to them.

Core. The central mass and region of highest density in a star.

Corona. The faint haze surrounding the Sun during total eclipse.

Coronagraph. A disk in a telescope to create an artificial solar eclipse, permitting photographs of the Sun's corona.

Cosmic noise. Radio-frequency radiation arriving at Earth from various objects in the Milky Way, also known as galactic noise.

Cosmic rays. Energetic particles, or ions, travelling through space after being emitted by the Sun or other stars.

Cosmology. The study of origin, design and structure of the Universe. Cosmogony is a branch of cosmology studying the origin and evolution of matter.

Coude. A fixed focus telescope to observe a moving object, used mostly with a spectrograph.

Declination. The angle between a celestial object and the celestial equator.

Diffraction. Bending of light rays as they pass through slits in a grating or are deflected from an etched surface. Used to produce spectra and measure wavelengths of spectral lines.

Dispersion. Separating light into its basic wavelengths.

Doppler Effect. Change in frequency of radio or sound waves caused by motion of the observer or the source.

Double stars. See: Binary stars.

Dwarf star. A small star with low mass and low luminosity. See also: Star.

Eclipsing binaries. One member of a binary star system periodically may block our view of another in the same system. Such systems are known as eclipsing binaries. See also: Binary Stars.

Electromagnetic spectrum. Electromagnetic energy wavelengths: gamma rays, X-rays, ultraviolet rays, visible light, infrared rays, microwaves, and radio waves.

Ellipticals. Oval-shaped galaxies.

Emission lines. The colors in a spectrum produced by hot gas, revealing the chemical composition of the gas. See also: Spectral Bands.

Equatorial mount. A telescope mechanism to follow a star across the sky.

Ether. An imaginary substance once said to fill outer space.

Event horizon. The boundary of a black hole. See also: Black Hole.

Fraunhofer lines. The most prominent absorption lines in the spectrum of light from the Sun or other star.

Galactic cluster. See: Clusters of Stars.

Galaxy. A galaxy is a vast cloud of millions or billions of stars, mixed with gas and dust, attracted to each other and held close by their gravity while floating through the Universe. There may be more galaxies strewn across the Universe than grains of sand on a beach. Galaxies measure many lightyears across. Galaxies have different shapes and sizes. Most common are flat-disk pinwheel spirals, egg-shaped ellipticals, and oddly-shaped irregulars. Best-known are our own Milky Way galaxy and the nearby Large Magellanic Cloud galaxy and Andromeda Galaxy. The Sun is one star among 100 billion in the Milky Way, a flat disk 100,000 lightyears in diameter with a bulge at the center. A cluster of stars 15 billion lightyears away, known as 4C41.17, is one of the most distant galaxies. The Local Group of galaxies is a cluster of galaxies in our area of the Universe. The Milky Way is a member of the Local Group. The Milky Way was named for its faint band of light seen crossing Earth's sky—the light from billions of stars.

Galilean telescope. The first refracting telescope, with a positive objective lens and a negative eye lens, developed by Galileo Galilei in the 17th century.

Gamma ray. The shortest wavelength of electromagnetic radiation. An electromagnetic wave of extremely high frequency emitted by the nucleus of a radioactive atom. A high-energy photon emitted naturally by decay of radioactive elements in a star or celestial object. See also: Electromagnetic Spectrum, Infrared, Ultraviolet, X-Ray.

Gamma-ray astronomy. So many advances in space research have spilled out of observatories and laboratories during the last half of the 20th Century that astronomy has undergone a revolution. Many astronomers think of this as a golden age. Earth's atmosphere is a murky opaque barrier to most infrared light, ultraviolet light, gamma rays and X-rays. So, satellites orbiting above the ocean of air—as well as astronomy payloads riding in high-altitude balloons and one-shot rockets—have become popular platforms from which astronomers observe those exotic wavelengths. The U.S., USSR and others have launched many observatories to orbit since 1962 to get a better look at infrared light, ultraviolet light, gamma rays and X-rays coming from deep-space objects. Now we know a lot more about deep-space objects such as hot young stars, quasars, pulsars and black holes. The orbiting Hubble Space Telescope, the orbiting Gamma Ray Observatory, space shuttles and unmanned rockets ferrying telescopes to orbit, men in space stations, balloon flights near space, are bringing more advances.

Globular cluster. See: Clusters of Stars.

Gravitation. The force of attraction between two bodies. It brings interstellar gas and dust together to form stars, planets and galaxies.

Gravitational collapse. The force of gravity condensing matter in stars, planets and galaxies.

Halo. A bright cloud of gas around a celestial object.

Herbig-Haro object. A star not seen from Earth, but detected as it illuminates a nearby cloud of interstellar gas and dust.

Hertzsprung-Russell diagram. Stars plotted by spectral type and temperature vs. absolute magnitude and luminosities.

H I region. An area in deep space of neutral interstellar hydrogen.

H II region. An area in deep space of ionized interstellar hydrogen.

Hubble classification. Galaxies ordered by their shapes.

Hubble constant. Ratio of the velocity at which a galaxy is receding from Earth to its distance from Earth.

Hubble Space Telescope. A telescope in Earth orbit, above the planet's atmosphere, to see farther across the Universe than most telescopes on Earth's surface.

Image-intensifier tube. A device to increase the brightness of a faint image from a telescope.

Infrared. Infrared light is electromagnetic radiation with wavelength longer than red light in the visible spectrum, but shorter than microwaves. It is radiation at an extremely high frequency, emitted naturally by a star or celestial object. Human beings sense infrared radiation as heat. See also: Electromagnetic Spectrum, Gamma Ray, Ultraviolet, X-Ray.

Infrared astronomy. Scientists have known infrared light comes from deep-space objects since the 18th Century. In the early 1960's, they got down to serious infrared astronomy research. Sensitive new detectors gave astronomers eyes to explore the infrared-light portion of the electromagnetic spectrum to better understand how stars are born, how young stars evolve, and the vast dense clouds of dust surrounding stellar maternity wards. See also: Gamma-Ray Astronomy.

Infrared telescope. Equipment on Earth, aboard satellites and in interplanetary probes to sense light at infrared wavelengths.

Interstellar. Between the stars. For example, interstellar space separates the Sun and its Solar System from the nearest neighbor star, Proxima Centauri. If a probe from Earth were to leave the Solar System it would be an interstellar spacecraft.

Interstellar helium. Helium atoms moving across space, outside of the environment around any star.

Interstellar hydrogen. Hydrogen atoms moving across space, outside of the environment around any star.

Interstellar medium. The matter moving in space outside of the environment around any star. Mostly interstellar hydrogen and helium.

Interstellar space. The space outside of the environment around any star.

Interstellar travel. Theoretical manned flights to other star systems.

Interstellar wind. A stream of interstellar hydrogen and helium flowing in the space outside of the environment around any star.

Large Magellanic Cloud. A galaxy near our own Milky Way galaxy. The Large Magellanic Cloud, an irregular galaxy, caught the public's fancy in 1987 when a star there exploded as Supernova 1987a, closest supernova to Earth in 400 years. The Large Magellanic Cloud is only 163,000 lightyears from Earth.

Large Magellanic Cloud is 39,000 lightyears in diameter, less than half as wide as the Milky Way. It has a smaller companion galaxy, the Small Magellanic Cloud.

Astronomers think Large Magellanic Cloud may once have been closer to Earth as it drifted past the Milky Way Galaxy a billion years ago. A trail of gas and dust, the Magellanic Stream, extends across our sky. The Stream may be matter ripped away from the Large Cloud by Milky Way gravity. It's not clear whether the Magellanic Clouds are satellite galaxies of the Milky Way, or passers-by on a one-time encounter on their way through the Universe.

Lightyear. The distance light travels at 186,000 miles per second through space in one year, about 5.9 trillion miles. Astronomy standard length unit. See also: Parsec.

Local group. The Local Group is a cluster of galaxies in our area of the Universe. Our own Milky Way galaxy is a member of the Local Group.

Luminosity. The light emitted by a deep space object, compared with light from the Sun. The energy emitted per second by a star.

Magellanic Clouds. Two galaxies near our own Milky Way are the Large Magellanic Cloud, or LMC, and Small Magellanic Cloud. Irregular in shape and visible from Earth's Southern Hemisphere, they were named for Portuguese explorer Ferdinand Magellan. In 1987, a star in LMC exploded into the supernova closest to Earth in 400 years. LMC is 163,000 lightyears from Earth and 39,000 lightyears in diameter, less than half as wide as the Milky Way. The Small Magellanic Cloud is under 20,000 lightyears in diameter and is 196,000 lightyears from Earth. Some astronomers wonder if the Small Cloud isn't two galaxies lined up. If so, they would call the third galaxy the Mini Magellanic Cloud. It might be 230,000 lightyears from Earth. LMC once may have been closer to Earth as it drifted past the Milky Way a billion years ago. A trail of gas and dust, called the Magellanic Stream, extending across our sky, probably is matter ripped away from LMC by Milky Way gravity. It's not clear whether the Clouds are satellite galaxies of the Milky Way or merely passing by.

Magnetic field. A star generates a powerful magnetic field which dominates its environment.

Magnitude. The magnitude we assign to a deep-space object depends on the amount of light it emits and its distance from Earth. An object farther from Earth sending out a lot of light may appear brighter than a closer object not giving off much light.

The lower the magnitude number, the brighter the object appears. A magnitude 1 object seems brighter than a magnitude 2 object. As the numbers approach and pass zero, they become negative numbers on the magnitude scale. For instance, the brightest star we see is Sirius at magnitude –1.4. The very-brightest objects in our sky are the Sun at magnitude –27 and the Moon at magnitude –13. On a clear night, we see objects as faint as magnitude 5 or 6 with the naked eye. With binoculars we see objects as weak as magnitude 8 or 9. A six-inch telescope will show objects to magnitude 13. The faintest objects photographed so far through the largest telescopes are around magnitude 26. Absolute magnitude is the apparent magnitude a star would have if it were viewed from a distance of 10 parsecs. Apparent magnitude is the apparent visual brightness of a celestial object such as a star. Bolometric magnitude is the total energy at all wavelengths from a star.

Main sequence. A band of stars, on the Hertzsprung-Russell diagram, creating energy by nuclear reactions at their cores. Our Sun and most stars are main sequence stars. See also: Hertzsprung-Russell diagram.

Meridian. See: Celestial Sphere.

Milky Way. See: Galaxy.

Molecular cloud. A low-temperature loose collection of dust and gas containing molecules of carbon monoxide and formaldehyde.

Moon. A natural satellite. A body in orbit around a planet. Earth has one, Mars has two. Artificial moons are man-made satellites.

Muses. From Greek mythology, the nine daughters of Zeus and Mnemosyne who inspired artists: Calliope, heroic poetry; Clio, history; Erato, love poetry; Euterpe, music; Melpomene, tragedy; Polyhymnia, eloquence; Terpsichore, dancing; Thalia, comic, lyric poetry; and Urania, astronomy. See also: Quadrivium, Urania, Uranology.

Nadir. See: Celestial Sphere.

Nebula. A large cloud of glowing interstellar dust and gas where new stars are forming in clusters.

Neutron star. A neutron star is the small remnant of a massive star which has been destroyed in a supernova explosion.

Newtonian reflector. A kind of telescope with a spherical or parabolic primary mirror and a flat reflector angled at 45 degrees from the optical axis to focus light at the side of the telescope tube.

North Star. Polaris, the seemingly stationary, bright, supergiant binary star seen in Earth's sky near the north celestial pole. Also known as polestar and cynosure.

Nova. A star which flares in brightness, then fades to its previous brightness. It increases in brilliance by several magnitudes in a few hours due to explosive ejection of surface material. Nova stars sometimes repeat the brightening and fading cycle. Not terminal as in a supernova. See also: Pulsating Stars, RR Lyrae Stars, T-Tauri Stars.

Objective. The main light-gathering and image-forming lens in a telescope. Nearest to the object observed, it focuses light to form an image of the observed object.

Observable Universe. The part of the Universe which can be seen with a telescope or other devices.

Observatory. A structure housing a telescope and other astronomy equipment.

Open cluster. See: Clusters of Stars.

Optical astronomy. Traditional astronomers are awash in new techniques of faint-object detection including an exciting range of electronic-imaging gear, computers, video screens and more-sensitive photographic film emulsions. Using today's standard telescope, an astronomer can see fainter, more distant objects than ever before. And he no longer is limited to looking only for visible starlight. Now his instruments let him study deep space in new parts of the energy spectrum, including infrared and ultraviolet light. See also: Telescope.

Orbit. The path travelled through space by an object, often a circle or ellipse. Satellites orbit planets. Planets, comets, asteroids and satellites orbit the Sun. The Sun, and its Solar System entourage, orbits the center of the Milky Way galaxy. In a satellite's orbit around Earth, perigee is closest to the planet and apogee is farthest away. Perihelion is the point of a planet's orbit closest to the Sun. The farthest point from the Sun is aphelion.

Parallax. The apparent shift of position of a nearby object against a distant background when viewed from two places. A seeming change in position of celestial bodies with change in viewing angle. The apparent shift in position or direction of a celestial body when viewed from Earth at opposing points of its orbit around the Sun. A slight difference in a camera's field of view between lens and viewfinder. The slight difference between the image in a camera's viewfinder and the actual image as seen through the camera's lens. The apparent displacement of a moving object as it is viewed from two different points. Spectroscopic parallax is the distance to a star as based on its absolute magnitude calculated from the relative intensities of selected spectrum lines.

Parsec. The astronomer's unit of measurement equal to 3.26 lightyears, the distance required to produce a one second of arc parallax angle. See also: Lightyear.

Perturbation. Disturbance of a celestial body's orbit by another celestial body.

Photometer. A device to measure light.

Photon. A unit of light.

Planet. A body in orbit around a star such as our Sun. Earth is one of nine planets circling the Sun: Mercury, Venus, Earth, Mars, Jupiter, Saturn, Uranus, Neptune and Pluto. Astronomers say there may be a tenth planet in our Solar System plus yet-to-be-seen planets orbiting other stars.

Planetarium. A theater with a domed ceiling on which images of the Universe are projected.

Planetary nebula. A dying star, usually a red giant, at the end of its life, surrounded by a wide ring of its expelled outer layers of gas. Such an object once was thought to be forming planets. See also: Nebula, Red Giant, Star.

Population I and II stars. Stars typified by age, color and location in our Milky Way galaxy. Population I stars are blue and in the galactic plane. Population II stars are red and spread throughout the galaxy.

Precession. See: Celestial Sphere.

Prime meridian. See: Celestial Sphere.

Proper motion. The angular velocity of a star perpendicular to the line of sight.

Proton-proton reaction. The nuclear reaction converting hydrogen to helium in main sequence stars.

Protostar. A flat circular cloud of gas and dust in deep space which, it has been conjectured, could develop into a star.

Pulsar. A rapidly-spinning collapsed star of high density, known as a neutron star, it sends out regular bursts of radiation energy. A neutron star is composed mostly of neutrons whose light or natural radio emissions reach Earth as pulses, frequently at regular intervals. The result of a supernova explosion, a pulsar is so crushed by the weight of its own gravity, a thimbleful would weigh a million tons. Pulsars, like flashing beacons, emit radio bursts at regular intervals.

Pulsating star. A star varying in brightness, probably as its volume changes. See also: Nova, RR Lyrae Stars, T-Tauri Stars.

Quadrivium. In a medieval university, the name for upper-division studies, including arithmetic, astronomy, geometry and music.

Quantum. A unit of energy.

Quasar. A quasi-stellar object. A powerful source of light and natural radio energy, a quasar may be a highly luminous galaxy extremely remote from Earth at the far distant edge of the known Universe, probably undergoing violent explosion. Maybe the most distant objects in the Universe, quasars are billions of lightyears away from Earth, racing farther away at 150,000 miles per second. They spew out huge quantities of light and other radiation. Even though they seem smaller than galaxies, some emit 100 times as much energy as common galaxies. They may be violently exploding galaxies or the cores of very active galaxies.

Radar astronomy. Radar signals transmitted from Earth and from orbiting spacecraft used to measure and map Venus, Mercury and other Solar System bodies.

Radial velocity. Motion of a star or galaxy along a line of sight.

Radiation. The spreading of energy through space by electromagnetic waves such as gamma rays, X-rays, ultraviolet rays, visible light, infrared rays, microwaves, radio waves and radiant heat. See also: Electromagnetic Spectrum.

Radio astronomy. The study of sources of radio energy from deep space using reflecting radiotelescopes having one or more radio antennas to receive and measure radio waves. Radiotelescopes routinely are used to study natural electromagnetic signals from stars, planets, the dust and gas between stars in our own galaxy, and galaxies and other sources beyond our Milky Way galaxy. Radio astronomers have mapped the galaxy and discovered quasars, pulsars, and a large number of complex organic molecules in interstellar space.

Radio galaxy. A galaxy emitting strong natural radio waves. The waves may be received by a radiotelescope on Earth. See also: Radio Astronomy, Radiotelescope.

Radiometer. Device to measure microwave energy from a celestial body.

Radio star. A strong natural radio wave source of very small size.

Red dwarf. A small star with low luminosity and surface temperature.

Red giant. A large cool star late in its life, with low surface temperature and absolute magnitude near zero. Radius 15 to 30 times that of the Sun; luminosity 100 times the Sun. See also: Magnitude, Planetary Nebula.

Redshift. Astronomers find some faraway galaxies and other deep-space objects, too faint to be seen by naked eye, by the naturally-occurring radio signals they omit, often millions of times more powerful than emissions from our Sun. A galaxy's distance from Earth can be calculated from the kind of optical light it sends out. Galaxy 4C41.17, for example, transmits light characteristic of hydrogen and carbon. The wavelength of the light is stretched by movement away from Earth. Such a stretch causes the color of the light to shift toward red. By measuring this so-called red shift, astronomers calculate the distance to the galaxy. Galaxy 4C41.17 has a red shift of 3.8 which converts to 15 billion lightyears away. At 15 billion, the galaxy may be 90 percent of the distance to the edge of the Universe.

Reflector. A kind of telescope having a mirror as objective.

Refractor. A kind of telescope with a system of lenses as objective.

Resolving power. Telescope's capability to separate close objects into distinct bodies.

Right ascension. See: Celestial Sphere.

RR Lyrae stars. Stars which change in brightness, like the Cepheid variable stars used by astronomers to determine distance. See: Nova, Pulsating Stars, T-Tauri.

Satellite. A body in orbit around a planet. Natural satellites often are called moons. Earth has one Moon. Mars has two moons. Artificial satellites are man-made machines in orbit.

Schmidt telescope. A kind of reflector telescope with a spherical mirror as objective and correcting lens in front of it. Schmidt telescopes often are used for wide-angle photography.

Schwarzschild radius. The size a mass must reach to be dense enough to trap light by gravity and become a black hole. See also: Black Hole, Neutron Star.

Seyfert galaxy. A galaxy with a strong energy source at its core.

Singularity. A point in space-time at which the space-time curvature becomes infinite, for instance in a black hole. A naked singularity is a space-time singularity not surrounded by a black hole. See also: Black Hole, Event Horizon, Neutron Star.

Small Magellanic Cloud. The Large Magellanic Cloud has a smaller companion galaxy known as the Small Magellanic Cloud. The Small Cloud, under 20,000 lightyears in diameter, is 196,000 lightyears from Earth. Astronomers wonder if the Small Cloud isn't two small galaxies lined up in their telescopes. If so, they would call the third galaxy the Mini Magellanic Cloud. It would be some 230,000 lightyears from Earth.

Solar system. A star with nearby objects surrounding it in space forms a solar system. Our Solar System is composed of a star we call the Sun plus nine major planets—Mercury, Venus, Earth, Mars, Jupiter, Saturn, Uranus, Neptune and Pluto—as well as moons, asteroids, minor planets, planetoids, comets and other objects. There may be a tenth planet in our system. Distant stars also may have orbiting planets.

Spectral bands. Lines in the colors of a spectrum of light from a star, caused by absorption or emission by molecules, revealing the star's chemical composition. See also: Emission Lines.

Spectrograph. A device which collimates, or makes parallel, rays of starlight passing them through a prism so they can be photographed.

Spectroscope. An instrument to split starlight into its spectrum of different colors or wavelengths. Spectroscopy is the study of astronomical objects' spectra.

Spectroscopic binary. A binary system whose stars are too close to each other to be resolved visually and must be studied with spectroscopy. See also: Binary Stars.

Spectroscopic parallax. See: Parallax.

Spiral galaxy. A galaxy with arms extending from a central bulge, like a gargantuan luminous pinwheel floating through the Universe. See also: Galaxy.

Standard candles. Stars which astronomers use to measure standard distances.

Star. Stars are glowing bodies which shine by releasing energy from nuclear reactions at their cores. A star is a fireball of intensely hot, bubbling gas, usually orbiting the center of a galaxy. Stars have surface temperatures ranging from 3,000 to 100,000 degrees Fahrenheit. That's hot enough to melt almost anything on Earth. The color of a star tells how hot it is. A cool star, at about 3,600 degrees Fahrenheit, is red. A hot star, at about 90,000 degrees Fahrenheit, is blue. Our star, the Sun, is orbiting the center of our Milky Way galaxy. The Sun isn't a very large star, being only 865,400 miles in diameter. It is at the heart of a Solar System with nine orbiting planets, comets and asteroids. The Sun is one star among 100 billion in a blazing pinwheel of stars 100,000 lightyears in diameter. Sunspots are large storms in the atmosphere of a star. The storms generate massive explosions from the surface of a star, firing high-speed atomic particles outward.

Stargazing. A slang name for astronomy as well as astrology.

Steady state. A theory which holds that the density of the Universe remains the same as new stars form when old ones extinguish. See also: Big Bang.

Supernova. An extraordinarily powerful explosion of a massive star. The star brightens for a short time, then usually reduces to a small remnant—a neutron star. See also: Nova.

Synchrotron radiation. Naturally-occuring radio signals generated by electrons spiraling in a magnetic field.

Syzygy. The alignment of three objects in space.

Telescope. An instrument for collecting light or other energy radiated from celestial bodies, and resolving images of these objects. Best known types are reflector, refractor, Schmidt and radiotelescope. See also: Cassegrain, Coelostat, Coronagraph, Coude, Equatorial Mount, Galilean Telescope, Gamma-Ray Astronomy, Hubble Space Telescope, Image-intensifier tube, Newtonian Reflector, Objective Lens, Observatory, Optical Astronomy, Radio Astronomy, Radiotelescope, Reflector, Refractor, Schmidt Telescope, Spectroscope, Very Large Array, X-Ray Astronomy.

Trinary star. Three stars bound to each other by gravity. See also: Binary Stars.

T-Tauri stars. Stars which vary in brightness and seem to lose mass in the process. See also: Nova, Pulsating Stars, RR Lyrae stars.

Ultraviolet. Ultraviolet light is electromagnetic radiation with wavelength shorter than violet light in the visible spectrum, but longer than X-rays. It is radiation at an extremely high frequency, emitted naturally by a star or celestial object. See also:

Electromagnetic Spectrum, Infrared, Gamma Ray, X-Ray.

Ultraviolet astronomy. Earth's atmosphere is a barrier to most ultraviolet light. Orbiting satellites—as well as astronomy payloads riding high-altitude balloons and one-shot rockets—have been used as platforms from which astronomers observe ultraviolet light. The U.S., USSR and others have launched observatories to get a better look at ultraviolet light coming from deep-space objects. See also: Gamma-Ray Astronomy.

Ultraviolet telescope. Equipment on Earth, aboard satellites and in interplanetary probes to sense light at ultraviolet wavelengths.

Universe. Some astronomers think the early Universe came to life and started expanding outward with a massive explosion they call the Big Bang. The Big Bang may have happened about 20 billion years ago. Light arriving at Earth today from galaxy 4C41.17 started on its journey here 15 billion years ago. That would be only a few billion years after the Big Bang. The edge of the Universe may be 20 billion, or so, lightyears away. The light we see coming from galaxy 4C41.17 left there 15 billion years ago, so viewing the galaxy is like a window looking back in time. By comparison, the Sun and Earth probably formed recently, only 4.6 billion years ago.

Urania. From Greek mythology, the muse of astronomy. See also: Muses.

Uranography. Mapping of the heavens including the positions of stars.

Uranology. An archaic name for astronomy.

Variable star. The brightness of star, as seen from Earth, may vary with internal changes in the star. Or, sometimes the brightness of a star is diminished as another star passes between it and Earth. One member of a binary star system periodically may block our view of another in the same system. Such systems are known as eclipsing binaries. See also: Binary Stars.

Very Large Array. VLA is a large group of radiotelescope antennas planted in a Y-shape in New Mexico.

Visible-light telescope. Television camera on Earth, aboard satellites and in interplanetary probes to sense light at wavelengths visible to the human eye.

Visual binary. Two or more stars in a binary system in which separate stars can be discerned through a powerful telescope. See also: Binary Stars.

White dwarf. A small, faint, cold star, fading after gravitational collapse.

White hole. A time-reversed black hole.

X-rays. Next to the shortest wavelength of electromagnetic radiation. An electromagnetic wave of extremely high frequency, emitted naturally by a star or celestial object. X-rays are generated in high energy processes where temperatures are greater than a million degrees. Astronomers are interested in X-ray images because they highlight regions across the Universe where such high energy phenomena exist. See also: Electromagnetic Spectrum, Infrared, Gamma Ray, Ultraviolet.

X-ray astronomy. Earth's atmosphere is a barrier to most X-rays. Orbiting satellites—as well as astronomy payloads riding high-altitude balloons and one-shot rockets—have become platforms from which astronomers observe X-rays. The U.S., USSR and others have launched observatories to get a better look at X-rays coming from deep-space objects. See also: Gamma-Ray Astronomy.

X-ray burster. A source in the Milky Way of intense flashes of X-rays.

Zenith. See: Celestial Sphere.

Today In Space History
January

1   The asteroid Ceres was discovered by Giuseppe Piazzi in 1801.

2   USSR Luna 1 became the first spacecraft to leave Earth's gravity, in 1959. It missed the Moon by nearly 4,000 miles.

5   In 1969, the USSR launched Venera 5 to Venus where it landed May 16, 1969, sending back information about that planet's atmosphere.

6   In 1968, U.S. Surveyor 7 was launched to the Moon, landing near Crater Tycho January 10, 1968, for soil analysis. It sent back 3,343 photos.

7   Galileo discovered Jupiter's moons Io, Europa and Callisto in 1610.

8   In 1973, the USSR sent Luna 21 to a soft landing on the Moon January 16, 1973. The Lunokhod 2 self-propelled roving Moon car scooped up soil samples which were returned to Earth January 27.

10   In 1946, the U.S. Army Signal Corps bounced a radar beam off the Moon.

10   In 1969, the USSR sent Venera 6 to Venus where it landed May 17, 1969, sending back information about that planet's atmosphere.

10   USSR Soyuz 27 was launched in 1978 on an A-2 rocket with cosmonauts Vladimir Dzhanibekov, Oleg Makarov. The cosmonauts spent six days in space. The Soyuz 27 capsule spent 65 days. They were the first short-time visitor-crew to Salyut 6 space station. Switched individually contoured seats from Soyuz 27 to Soyuz 26 for flight home, leaving Soyuz 27 behind.

11   Uranus' mons Titania and Oberon discovered in 1787 by William Herschel.

11   USSR Soyuz 17 was launched in 1975 on an A-2 rocket for a 30-day stay at space station Salyut 4. Cosmonauts Georgi Grechko, Alexei Gubarev. It was the first cosmonaut ferry to Salyut 4. Returned in a snow storm.

11   In 1978, cosmonauts Vladimir Dzhanibekov and Oleg Makarov in Soyuz 27 successfully docked at the USSR's Salyut 6 space station, joining the crew of Soyuz 26, Georgi Grechko and Yuri V. Romanenko, to form the first three spacecraft complex in orbit. Salyut 6 was the USSR's first second-generation space station. Grechko and Romanenko had been launched in Soyuz 26 atop an A-2 rocket December 10, 1977, to break the U.S. Skylab-4 record of 84 days in space. They made the first USSR spacewalk in nine years and accomplished the first refueling in orbit from a Progress cargo freighter. Dzhanibekov and Makarov in Soyuz 27 were launched on an A-2 January 10, 1978. They were the first short-time visitors to the Salyut 6 station. They stayed six days, then switched their individually-contoured seats from Soyuz 27 to Soyuz 26 for the flight home, leaving Soyuz 27 behind. That pattern of station crew rotation continues today: long-term crew occupies station, receives short-time visitors. When short-time crew leaves, it takes old Soyuz home, leaving newer Soyuz. Grechko and Romanenko flew home in Soyuz 27 after record 96 days in orbit.

12   U.S. shuttle Columbia launched in 1986 on flight STS-61C with astronauts Robert L. "Hoot" Gibson, Charles F. Bolden Jr., Steven A. Hawley, U.S. Rep. Bill Nelson, George D. "Pinky" Nelson, Frances R. Chang-Diaz, commercial passenger Robert J. Cenker. Carried a communications sat, 12 Get Away Special cannisters (GAScans) with experiments and Material Science Lab. U.S. Congressman Bill Nelson was first member of U.S. House of Representatives in space, just 16 days before the Challenger disaster.

13   Galileo, in 1610, discovered Ganymede, a moon of Jupiter.

14   Master rocket designer Sergei P. Korolev, father of the USSR space program, died in 1966.

14   USSR Soyuz 4 was launched in 1969 on an A-2 rocket from Baikonur

Cosmodrome for a rendezvous in space with Soyuz 5. Vladimir A. Shatalov was launched in Soyuz 4. Boris V. Volynov, Alexei S. Yeliseyev, Yevgeny V. Khrunov were launched in Soyuz 5 on an A-2 rocket January 15. Soyuz 4 was the first USSR manned launch in winter. The first crew transfer between spacecraft was completed in orbit. Yeliseyev & Khrunov left Soyuz 5 in spacesuits, pulled themselves along handrails and into the Soyuz 4 airlock. The spacewalk took about an hour. The two capsules remained docked about 4.5 hours. Soyuz 4 completed 45 orbits. Soyuz 5 completed 46. Shatalov, Yeliseyev & Khrunov flew home in Soyuz 4. Volynov flew home in Soyuz 5, a day after Soyuz 4.

21	In 1960, the Mercury capsule Little Joe 4 was launched on a Little Joe rocket from Cape Canaveral carrying the monkey Miss Sam on a suborbital flight.
24	In 1985, U.S. shuttle Discovery flight STS-51C from Cape Canaveral, with crew Thomas K. Mattingly 2nd, Loren J. Shriver, Ellison S. Onizuka, James F. Buchli and Gary E. Payton, was the first secret mission. Discovery carried a military spy satellite.
24	Voyager 2 flew past Uranus in 1986.
27	A launchpad fire in 1967 killed the three-man crew of Apollo 1. Astronauts killed: Virgil I. "Gus" Grissom, Edward H. White 2nd, Roger B. Chaffee. The fire spread as they sat in an Apollo command-module capsule on the launch-pad during a pre-flight ground test at Cape Canaveral, Florida.
28	Astronomer Johannes Hevelius was born in 1611.
28	Space shuttle Challenger exploded 72 seconds into lift-off on flight STS-51L in 1986 from Cape Canaveral, Florida, killing seven astronauts: Francis R. "Dick" Scobee, Michael J. Smith, Judith A. Resnik, Ellison S. Onizuka, Ronald E. McNair, Gregory B. Jarvis, Sharon Christa McAuliffe. The flight lasted just over a minute and did not make it to orbit. Challenger exploded when a solid-fuel booster-rocket leak ignited fuel. McAuliffe was a school teacher.
31	America's first Earth satellite, Explorer 1, was launched to orbit in 1958 on a Jupiter-C rocket from Cape Canaveral. The 31-lb. satellite used a Geiger counter in its 18-lb. payload to confirm the existence of the Van Allen radiation belts surrounding Earth.
31	In 1961, Mercury capsule MR-2 was launched on a Redstone rocket from Cape Canaveral carrying the chimpanzee Ham on a 16-minute suborbital flight.
31	In 1966, the USSR launched Luna 9 to the Moon from Baikonur Cosmodrome. It arrived four days later, dropping a 220-lb. instrument capsule February 3, 1966, to the first ever soft landing on the Moon. Luna 9 sent back the first radio and television from the Moon to Earth, including 30 pictures.
31	In 1971, a Saturn V rocket lifted Alan B. Shepard Jr., Stuart A. Roosa and Edgar D. Mitchell in Apollo-14 to the Moon. The flight lasted 216 hours 01 minute 57 seconds and included 34 orbits of the Moon by the Command Service Module. Shepard and Mitchell made man's third Moon landing, collecting 96 lbs. of rock, soil samples in 33 hours 31 minutes on surface. First 6-iron golf shot on Moon. Roosa stayed in CSM.

February

1	In 1961, first Minuteman intercontinental ballistic missile (ICBM) launched.
2	In 1931, with 102 letters attached to a rocket by Friedrich Schmiedl and fired from his place to another town in Austria, the first official rocket mail was launched. The first rocket mail flight in the U.S. was five years later on February 23, 1936, from New York City to Greenwood Lake, New Jersey.
3	The USSR's Luna 9 made the first soft landing on the Moon in 1966. Its 220 lb. instrument capsule sent the first radio and television from the Moon to

Earth, including 30 pictures. Luna 9 had been launched January 31 from Baikonur Cosmodrome.

3   In 1984, U.S. shuttle Challenger flight STS-41B from Cape Canaveral, with crew Vance D. Brand, Robert L. "Hoot" Gibson, Ronald E. McNair, Robert L. Stewart and Bruce McCandless II, carried the satellites Palapa and Westar, which failed after release and had to be recovered by flight 51A in November 1984. STS-41B saw the first use of a Manned Maneuvering Unit backpack for untethered spacewalk. McCandless and Stewart used MMUs to fly free as the first human satellites. The shuttle landed at Kennedy Space Center, Florida.

4   Astronomer Clyde Tombaugh was born in 1906. He discovered Pluto 24 years later to the month, February 18, 1930.

4   In 1961, the USSR tried to launch a Venera probe to Venus on an A-1 rocket from Baikonur Cosmodrome. It failed, only reaching Earth orbit. From then the USSR referred to the spacecraft as Tyazheily Sputnik 4. The spacecraft which blasted off successfully eight days later on February 12 became Venera 1.

4   In 1967, U.S. Lunar Orbiter 3 left for the Moon, going into orbit around that natural satellite February 8. It worked until October 9, 1967, sending back photos.

5   USA Apollo 14 landed on the Moon in 1971.

7   U.S. astronaut Bruce McCandless, using a Manned Maneuvering Unit (MMU) from shuttle Challenger, made the first untethered spacewalk in 1984.

8   Leonid Kizim, Vladimir Solovyov and Oleg Atkov, in 1984 in the USSR second-generation space station, Salyut 7, set an endurance record of 237 days in space which stood through September 1987.

11   Japan became the fourth nation to launch a satellite to orbit with the flight of Ohsumi from Kagoshima Space Center in 1970. Japan orbited its second satellite, Tansei, from Kagoshima in February 16, 1971.

12   In 1961, the USSR launched Venera 1 (Venus 1) on an A-1 rocket from Baikonur Cosmodrome. It was the first interplanetary probe to escape Earth orbit. It flew within 60,000 miles of Venus on May 19, 1961.

15   Galileo was born in 1564.

16   Miranda, a moon of Uranus, was discovered in 1948 by Gerard Kuiper.

16   In 1965, the U.S. used Saturn 1 rockets to fire three large satellites to Earth orbit. The first was Pegasus 1 on February 16, followed by Pegasus 2 on May 25 and Pegasus 3 on July 30. A dummy Apollo capsule flew to space with each Pegasus. Pegasus satellites sprouted 2,300 sq. ft. panels which extended 96 ft. to measure the number of penetrations by meteoroids. Watchers with binoculars on the ground could see the satellites at night. Scientists discovered that interplanetary dust particles were considerably fewer in number than had been expected. That news allowed engineers to cut 1,000 lbs. of protective weight from each Apollo capsule.

17   In 1965, the probe Ranger 8 left the U.S. for the Moon, sending back 7,137 photos of Earth's natural satellite before crashing at Mare Tranquilitatis.

18   Discovery of the planet Pluto in 1930 by Clyde Tombaugh. Born February 4, 1906, Tombaugh was a farmer who liked to build telescopes to look at the planets. He sent his drawings of Mars and Jupiter to Arizona's Lowell Observatory in 1928. The observatory hired him to search for a ninth planet. Tombaugh used a 13-in. telescope to photograph areas of the night sky. Comparing similar photos he could see objects which had moved. It took a year to examine photo plates containing the images of two million stars, but Tombaugh found Pluto. Tombaugh's book about the discovery was called Out Of The Darkness.

19   Nicolas Copernicus was born in 1473.

20   John H. Glenn Jr., first from the USA to orbit Earth, was blasted to space in
1962 in the Mercury 6 capsule named Friendship 7 atop an Atlas D rocket. Glen
made three orbits of Earth during 4 hours 55 minutes 23 seconds in space. The
Friendship 7 capsule now is at the National Air & Space Museum, Smithsonian
Institution, Washington, D.C. Before Glenn, Alan B. Shepard Jr. and Virgil I.
"Gus" Grissom had been the first two Americans in space, but on a suborbital
flights. Shepard rode in the Mercury 3 capsule named Freedom 7 on May 5,
1961, and Grissom in Mercury 4 capsule Liberty Bell 7 on July 21, 1961.
Mercury 3 and 4 were fired to space on Redstone rockets.

20   The USSR in 1986 launched third-generation space station Mir to Earth orbit.

21   NASA blasted off six Nike-Apache sounding rockets from the Wallops Island,
Virginia, launch site in one night in 1968. To study winds in the upper
atmosphere, one rocket released a red-orange cloud of sodium seen for hundreds
of miles up and down the East Coast. Vapors from the other rockets appeared as
fainter blue and green clouds. Cameras on the ground tracked direction, speed.

23   In 1936, the first rocket mail flight in the U.S., from New York City to
Greenwood Lake, New Jersey.

23   Supernova 1987a explodes in the Large Magellanic Cloud galaxy in 1987.
Neutrinos blasted out by the supernova are recorded on Earth. On February 24,
astronomers working in the Southern Hemisphere discover Supernova 1987a,
closest supernova since 1604 and since invention of the telescope.

24   In 1969, the U.S. sent Mariner 6 to Mars. The probe came within 2,000 miles
of the surface July 31, 1969, sending back television pictures and data.

24   Reports of the first pulsars were published in 1968. Cambridge University
astronomer Antony Hewish and his associates found four sources of the
pulsating radio waves and thought they might be a communication or
navigation network operated by some distant, advanced civilization.
Unfortunately for romantics, the speed and regularity of the energy waves from
deep space led scientists to figure out they were naturally-occurring signals
from neutron stars which had been theorized since 1932. Pulsars probably are
rapidly-spinning neutron stars which have magnetic fields strong enough to
generate powerful electromagnetic-energy signals.

25   In 1969, the interplanetary probe Mariner 6 left the U.S. for Mars. It came
within 2,000 miles of the Red Planet July 31, 1969, sending back photos and
information.

26   The U.S. sent the powerful new Saturn 1B rocket on its first unmanned
suborbital test flight in 1966. It carried a dummy Apollo capsule 5,500 miles
downrange. In a second unmanned test flight July 5, 1966, the rocket second
stage went into orbit, but carried no other satellite. The third test in 1966 was a
suborbital flight August 25. The first manned test was to follow on February
21, 1967, but a fire killed astronauts Virgil I. "Gus" Grissom, Edward H. White
2nd and Roger B. Chaffee in their Apollo capsule during a simulated countdown
on the launch pad January 27, 1967. Their Apollo 204 capsule later was
renamed Apollo 1. A delay of more than a year and a half followed. There were
three unmanned tests in 1967 and 1968, using Saturn 1B and Saturn V rockets,
before a Saturn 1B hefted the manned Apollo 7 to space October 11, 1968.

31   In 1971, the U.S. moonflight Apollo 14 blasted off on a Saturn V from launch
pad 39A at Cape Canaveral, carrying astronauts Alan B. Shepard Jr., Stuart A.
Roosa and Edgar D. Mitchell. The 216-hour round-trip flight included a landing
by the lunar excursion module Antares and 34 orbits of the Moon by the
command service module Kitty Hawk. Shepard and Mitchell made man's third
Moon landing in Antares, at the Fra Mauro formation cone crater, collecting 96
lbs. of rock and soil samples in 33 hours 31 minutes on surface. They also

made the first 6-iron golf shot on Moon. Roosa stayed aloft in Kitty Hawk.

March

1   USSR Venera 3 in 1966 was the first probe to land on Venus.

2   In 1968, the USSR attempted to send Zond 4 on what probably was a flight around the Moon. It reached only parking orbit near Earth and fell back into the atmosphere March 3, 1968.

3   The U.S. National Advisory Committee for Aeronautics (NACA), NASA's predecessor, was founded in 1915 to study the science of flight.

3   In 1959, the U.S. launched Pioneer 4 to the Moon. The probe passed within 37,280 miles of the Moon.

3   In 1969, U.S. astronauts James A. McDivitt, David R. Scott and Russell L. Schweickart rode to space in Apollo 9 atop a Saturn V rocket. They flew the lunar excursion module (LEM) for the first time.

3   The 10th launch in America's series of Pioneer interplanetary spacecraft failed in 1969. That probe was to have been Pioneer 10, but became known as Pioneer E. The flight following E—11th in the series of Pioneer launches—was launched successfully March 3, 1972. It became known as Pioneer 10. The craft flew a 620 million mile passage through the Asteroid Belt to Jupiter, passing within 82,000 miles of the giant planet December 3, 1973. Today it is on its way out of the Solar System toward interstellar space.

5   The U.S. interplanetary probe Voyager 1 flew near Jupiter in 1979. Voyager 1 had been launched on a Titan-Centaur rocket from Cape Canaveral September 5, 1977, to explore Jupiter and Saturn. The craft swept past Jupiter March 5, 1979, and Saturn November 12, 1980, then pulled above the swirling disk of planets surrounding our Sun to float out of the Solar System.

6   USSR Vega 1 in 1986 sailed near Comet Halley.

7   Astronomer John Herschel was born in 1792.

8   In 1986, Japan's Suisei flies past Comet Halley, photographing its atmosphere with an ultraviolet camera, while watching the comet's nucleus rotate.

9   In 1965, the U.S. conducted the first launch of eight satellites on one rocket, a Thor-Agena from Vandenberg Air Force Base, California. Most of the satellites were military spacecraft, but one was the first active real-time amateur radio communications satellite, OSCAR 3—the first hamsat with a two-way signal repeating transponder for a radio relay in the sky. It received 146 MHz signals, transmitted with 1-watt of power at 144 MHz, and had two radio beacons. One sent a continuous signal for tracking and propagation studies. The other sent telemetry data about temperatures and battery voltages. A few solar cells were included to back-up the battery powering the beacons, thus the first use of solar power on an amateur spacecraft. But, the satellite was not fully solar powered. The solar cells allowed the beacons to work months longer than the transponder. More than 1,000 hams from 22 nations talked through OSCAR 3 during its 18 days of operation. Contacts included New Jersey to Spain, Massachusetts to Germany and New York to Alaska.

9   USSR Vega 2 in 1986 flies near Comet Halley.

10 In 1970, France launched its Dial satellite and the German Wika satellite from Kourou, French Guiana, South America.

10 Rings around the planet Uranus—the third most distant planet from the Sun were first seen by astronomers on Earth in 1977. The astronomers, aboard NASA's Kuiper Airborne Observatory (KAO) flying over the Indian Ocean, watched as Uranus passed in front of a bright star, occulting the star. They saw five rings around the planet. Then, in January 1986 while it was flying by for a close-up look at the rings, Voyager 2 saw 11 rings and uncovered strange

incomplete-circle formations now called ring arcs. The many rings and arcs varied in depth and density.

10   In 1986, Japan's Sakigake flew past Comet Halley, studying solar wind and magnetic fields, detected plasma waves.

11   In 1960, the U.S. sent the interplanetary probe Pioneer 5 to orbit the Sun between Earth and Venus, setting a record by sending radio signals 20 million miles. Pioneer 5 studied the Sun's flares, solar wind, magnetism three months.

13   The planet Uranus is discovered in 1781 by William Herschel.

13   Astronomer Percival Lowell was born in 1855.

13   The satellite China 2 was launched in 1971.

13   In 1986, European Space Agency's interplanetary probe Giotto flew by Comet Halley, closest of several probes from Earth to the comet. Giotto examined the comet's atmosphere and magnetic fields, sending back the best pictures of the comet's nucleus.

13   In 1989, U.S. shuttle Discovery flight STS-29 from Cape Canaveral, with crew Michael L. Coats, John E. Blaha, Robert C. Springer, James F. Buchli and James P. Bagian, carried NASA's TDRS-D communications satellite plus 4 rats, 32 fertilized chicken eggs, plants, crystals. Miles of IMAX film exposed Earth's environmental problems.

14   Albert Einstein was born in 1879.

14   In 1931, Johannes Winkler launched Europe's first liquid-fuel rocket at Dessau, Germany.

14   In 1956, The U.S. launched the first Jupiter rocket from Cape Canaveral.

15   Goddard Space Flight Center, Greenbelt, Maryland, was named in 1960 for Robert H. Goddard, the father of American rocketry.

16   Robert Goddard in 1926 launched America's first liquid-fueled rocket at 2:30 p.m. on a farm at Auburn, Massachusetts, outside Worcester, Massachusetts. He spoke of firing rockets to the Moon.

16   In 1961, NASA dedicated its new Goddard Space Flight center at Greenbelt, Maryland.

16   In 1966, Neil A. Armstrong and David R. Scott rode to space in the capsule Gemini 8 atop a Titan 2 rocket. They completed the first docking of one space vehicle with another when they hooked up with an Agena rocket previously launched to orbit. A broken control ended the flight, but they completed 6.5 trips around the globe in 10 hours 41 minutes 26 seconds.

18   First spacewalk, by cosmonaut Alexei A. Leonov in 1965, lasted 10 minutes. From the USSR, he was with Pavel I. Belyayev in Voskhod 2.

21   In 1965, U.S. Ranger 9 flew to the Moon, took 5,814 photos and crashed at Crater Alphonsus.

22   In 1972, the USSR's 844-lb. probe Venera 8 was launched to Venus. Arriving at Venus, the orbiting craft measured the planet with hydrogen and oxygen sensors, magnetometer, charged particle trap and cosmic ray counter, sending back data on atmosphere, pressure, temperature, magnetic field, hydrogen corona. It dropped a probe which parachuted down for an hour to the surface; measured the atmosphere with gas analyzers, thermometers, barometer, radio altimeter and atmospheric densitometer; landed on the day side of the planet July 22, 1972; analyzed soil; and transmitted information for 50 minutes to Earth. It was the second ever soft-landing on Venus, the first soft landing on the planet's day side. Venus surface temperature was 887 degrees Fahrenheit, pressure was 90 times Earth's, and soil was like granite with density of 1.5 grams/cc.

22   In 1982, Jack R. Lousma and Charles Gordon Fullerton took U.S. shuttle Columbia flight STS-3 from Cape Canaveral to space for an eight-day test

flight. Bad weather in California forced them to land at White Sands, New Mexico.

23   The first photograph of the Moon was made in 1840.

23   Virgil I. "Gus" Grissom and John W. Young in 1965 flew to space in the Gemini 3 capsule atop a Titan 2 rocket. It was the first Gemini to orbit and the first manned spacecraft to change path in orbit. Grissom ate the first corned beef sandwich in orbit as they completed three trips around the world in 4 hours 53 minutes.

23   Earth had a close call with an asteroid (1989fc) which came within 450,000 miles of our planet in 1989. Just 72 days later, a 2-mi.-wide asteroid (1989ja) passed within 8 million miles of Earth June 3. Later, another large asteroid (1989pb) came within 2.5 million miles of Earth August 24.

24   Astronomer Walter Baade was born in 1893.

24   In 1962, U.S. astronaut M. Scott Carpenter flew to Earth orbit atop an Atlas D rocket in the Mercury MA-7 capsule he named Aurora 7. It was the second U.S. manned orbital flight. Carpenter completed three revolutions around the globe in 4 hours 56 minutes 5 seconds and overshot his landing site by 250 miles. Today, Aurora 7 is at the Hong Kong Space Museum, Hong Kong.

25   Saturn's moon Titan was discovered in 1655 by Christiaan Huygens.

27   Yuri A. Gagarin, the first man in space, died in a USSR airplane crash in 1968.

27   Launched March 27, 1969, the U.S. interplanetary probe Mariner 7 flew by Mars August 5, 1969. It passed at a distance of 2,128 miles. Mariner 7 photographed the southern hemisphere of Mars and the Martian South Pole ice cap. Today Mariner 7 is in orbit around the Sun. Mariner 6 and Mariner 7 were 910-lb. twins. Launched February 24, 1969, Mariner 6 flew by Mars July 31, 1969, passing at a distance of 2,120 miles. Mariners 6 and 7 together transmitted to Earth a total of 201 television pictures of the Red Planet. They measured surface temperature, atmospheric pressure, composition and diameter of Mars and radioed the data to Earth.

27   In 1972, the USSR launched Venera 8 to Venus where it landed July 22, 1972, sending signals with atmosphere and surface data for 50 minutes. Surface pressure and heat crushed and burned the instrument capsule.

28   The asteroid Pallas, second minor planet found, discovered in 1802 by Heinrich Olbers. At 336 miles, third largest asteroid in diameter.

28   In 1935, Robert H. Goddard launched the first rocket with gyroscopic controls at Roswell, New Mexico. It flew to an altitude of 4,800 feet and, at a speed of 550 mph, to a downrange distance of 13,000 feet.

29   The asteroid Vesta, fourth minor planet found, discovered in 1807 by Heinrich Olbers. Only asteroid bright enough to be seen by naked eye. At 349 miles, second largest asteroid in diameter.

29   In 1974, the first flyby of planet Mercury by the American probe Mariner 10.

31   In 1966, the USSR's probe Luna 10 was launched to the Moon, going into orbit around that natural satellite of Earth April 3, 1966, the first spacecraft from Earth to orbit the Moon. It sent back data for two months. There were 24 craft in the Luna, or Lunik, series of unmanned probes returning photos and information on Earth's natural satellite. The Luna program included many firsts: Luna 1 in 1959 was the first-ever Moon probe from Earth. Luna 2 in 1959 was the first Earth probe to impact on the Moon. Luna 3 in 1959 made the first photographs of the Moon's far side. Luna 9 in 1966 made the first soft landing on the Moon. Luna 10 in 1966 was the first Moon orbiter. Luna 16 in 1970 returned a soil sample to Earth. The Lunokhod-1 rover vehicle flew to the Moon in Luna 17 in 1970. Between 1959 and 1976, 30 Luna and Zond spacecraft from the USSR explored the Moon.

April

1   In 1950, Wernher von Braun moved his U.S. rocket research operation from Fort Bliss, Texas, to Redstone Arsenal, Alabama.

1   Tiros 1, the first weather observation satellite, was launched from Cape Canaveral by the U.S. in 1960.

1   In 1974, the U.S. launched the Westar 1 satellite, first for communications inside the U.S.

2   The first photograph of the Sun was made in 1845.

2   In 1963, the USSR's probe Luna 4 was launched on a Moon flyby. It flew within 5,300 miles of the Moon.

2   The USSR's probe Zond 1 was launched in 1964 on a Venus flyby. It passed within 60,000 mi. of Venus July 19, 1964.

3   The first nuclear reactor to orbit Earth was Snapshot 1, also known as Snap 10A, sent to space from Vandenberg Air Force Base, California, in 1965 by NASA and the U.S. Atomic Energy Commission.

3   USSR Luna 10 in 1966 was the first spacecraft from Earth to orbit the Moon.

3   In 1973, the USSR launched its second first-generation space station, Salyut 2.

4   American Rocket Society was founded in 1930 at New York City under its first name, American Interplanetary Society.

4   The first weather satellite, Tiros 1, was launched in 1960 by the U.S.

4   In 1968, the U.S. Apollo 6 capsule rode on the last test flight of Saturn V and command module.

4   In 1983, U.S. shuttle Challenger flight STS-6 from Cape Canaveral, with crew Paul J. Weitz, Karol J. "Bo" Bobko, Donald H. Peterson and F. S. "Story" Musgrave, ferried a TDRS communications satellite to orbit. Peterson and Musgrave took the first shuttle spacewalks.

5   In 1973, the U.S. launched the interplanetary probe Pioneer 11 on a Jupiter and Saturn flyby. The spacecraft passed within 26,600 miles of Jupiter on December 2, 1974, and within 13,000 miles of Saturn on September 1, 1979. Pioneer 11 crossed the orbit of Neptune 2.8 billion miles from Earth February 3, 1990, leaving behind forever the known parts of the Solar System. It joined three other U.S. spacecraft in those far-distant unexplored spaces beyond our Solar System. Already out there were Pioneer 11's sister spacecraft Pioneer 10, the first to leave, along with Voyager 1 and Voyager 2, the second and third to fly beyond the planets.

5   In 1975, the USSR tried to launch Soyuz 18A carrying cosmonauts Vasili Lazarev and Oleg Makarov. They didn't make it to orbit. The flight was aborted to suborbital when the A-2 rocket stages didn't separate properly. Landing in Siberian cold, the crew climbed out and built a fire to stay warm.

6   In 1965, the U.S. launched Early Bird, also was known as Intelsat 1, the first commercial communications satellite, from Cape Canaveral to stationary orbit over the Atlantic Ocean.

6   In 1984, U.S. shuttle Challenger flight STS-41C from Cape Canaveral, with crew Robert L. Crippen, Francis R. "Dick" Scobee, Terry J. Hart, George D. "Pinky" Nelson and James D. Van Hoften, carried the Long Duration Exposure Facility (LDEF) satellite with 57 experiments to space. The astronauts rescued the Solar Max astronomy satellite. Nelson and Van Hoften made the first satellite repair in the shuttle's cargo bay, then released Solar Max back to orbit. Six cosmonauts were at the USSR's Salyut 7 space station while the five astronauts were in Challenger; to that date the total of 11 was the most people ever in space at one time.

7   In 1968, the USSR launched its Moon orbiter, Luna 14. The probe measured gravity from lunar orbit.

8   The U.S. fired Gemini 1 in 1964 on the first unmanned test of that two-seat space capsule.

11   In 1970, U.S. astronauts James A. Lovell Jr., Fred W. Haise Jr. and John L. Swigart Jr. were launched to Earth orbit in Apollo 13 on a Saturn V rocket. They attempted to fly on to Moon, but a service module oxygen tank ruptured. They returned safely to Earth using lunar module oxygen and power. On their way home they travelled around the Moon, but made no landing there.

11   In 1987, the astrophysics and biotechnology space laboratory, Kvant, launched March 31 from the USSR, was permanently attached to the Mir space station.

12   Man's world changed forever in 1961 when USSR cosmonaut Yuri Gagarin in the capsule Vostok 1 became the first human to orbit Earth. He lifted off from Tyuratam at 9:07 a.m. Moscow time. Vostok 1 completed one orbit in one hour 48 minutes at an altitude of 112 to 203 miles. Gagarin was killed seven years later in a March 1968 plane crash while training for the flight of Soyuz 3. On the far side of the Moon, a crater was named in his honor.

12   Robert L. Crippen and John W. Young flew America's first space shuttle, Columbia, to space from Cape Canaveral in 1981. On its maiden voyage, flight STS-1, Columbia was the first winged ship to fly to Earth orbit and return to an airport landing. It landed in California April 14.

12   In 1985, the U.S. sent the first politician to orbit—U.S. Sen. Edwin "Jake" Garn in shuttle Discovery flight STS-51D from Cape Canaveral. With him were Karol J. "Bo" Bobko, Donald E. Williams, Jeffrey A. Hoffman, S. David Griggs, Margaret Rhea Seddon and Charles D. Walker. Hoffman and Griggs made an unplanned spacewalk to try to repair their Syncom communications satellite. They failed and left it broken in orbit, to be repaired in August 1985. Discovery landed at Kennedy Space Center, Florida.

13   Transit 1B, the first navigation satellite, was launched from Cape Canaveral by the U.S. in 1960.

14   Astronomer Christiaan Huygens was born in 1629.

16   In 1946, the U.S. launched the first of its captured German V-2 rockets, at White Sands Proving Grounds, New Mexico. The rocket reached an altitude of five miles.

16   Charles M. Duke Jr., Thomas K. Mattingly and John W. Young were launched to Earth orbit in Apollo 16 atop a Saturn V rocket in 1972. They flew on to the Moon where Young and Duke made the fifth-ever Moon landing, collecting 213 lbs. of samples 71 hours.

17   In 1967, Surveyor 3 left the U.S. for the Moon where it soft landed at Oceanus Procellarum after a 65-hour flight. It scooped up lunar soil and sent test results back by radio.

18   In 1951, the U.S. launched the first Aerobee rocket from Holloman Air Force Base, New Mexico. It carried a biomedical experiment.

18   In 1965, USSR cosmonaut Aleksei Leonov took man's first-ever spacewalk, tethered to the Voskhod 2 capsule.

19   In 1971, the USSR launched man's first space station, Salyut 1.

19   Salyut 7, the second of the USSR's second-generation space stations, was launched in 1982.

20   U.S. Apollo 16 made Man's fifth landing on the Moon in 1972.

23   Molniya 1A, the USSR's first active real-time communications satellite, was launched to elliptical orbit in 1965. Molniya 1B, launched October 14, 1965.

24   In 1967, USSR cosmonaut Vladimir M. Komarov was killed in the crash landing of his capsule Soyuz 1 after 26-hour flight.

24   The Peoples Republic of China launched its first space satellite—known as

Mao 1 or China 1 or PRC-1—to orbit in 1970 on a Long March 1 rocket. That 390-lb. electronic ball floated around the Earth blaring the patriotic song, The East Is Red, making China the fifth nation with a space rocket. Before that first successful launch, the Chinese may have sustained a launch failure in 1969. They may have suffered three failures in 1974 and another in 1979. Since 1970, China has had 24 known successful firings and one known failure. Since 1984, the country has launched four of its own communications satellites. China started selling commercial space launches to foreign satellite owners in 1986 during the time U.S. shuttles and European rockets were grounded.

25   Astronauts in U.S. shuttle Discovery dropped the Hubble Space telescope overboard into orbit 380 miles above Earth. The STS-31 crew: Loren Shriver, Charles Bolden. Steven Hawley, Bruce McCandless and Kathryn Sullivan. With a 94.5-in. mirror, Hubble is the most sensitive optical telescope ever located above Earth's blurring atmosphere. Hubble sent first picture to Earth May 20.

26   In 1962, the first USSR weather satellite, Cosmos 4, was launched.

29   In 1985, U.S. shuttle Challenger flight STS-51B left Cape Canaveral with Spacelab 3 in the cargo bay for materials processing and life sciences experiments. The crew: Robert F. Overmyer, Frederick D. Gregory, Don L. Lind, Norman E. Thagard, William E. Thorton, Taylor G. Wang and Lodewijk van den Berg, of the European Space Agency, the first Dutch astronaut. The amateur satellite NUsat-1 was launched from the shuttle.

May

1   In May 1946, Wernher von Braun launched a V-2 rocket for the U.S. Army at White Sands, New Mexico.

1   Nereid, a moon of Neptune, was discovered in 1949 by Gerard Kuiper.

1   Gary Power's U-2 high-altitude spy plane shot down over the USSR in 1960.

4   In 1967, U.S. Lunar Orbiter 4 left for the Moon, going into orbit around that satellite May 7, 1967. The satellite changed orbit on command from Earth.

4   In 1989, U.S. shuttle Atlantis flight STS-30 left Cape Canaveral with crew David M. Walker, Ronald J. Grabe, Mary L. Cleave, Norman E. Thagard, M.D. and Mark C. Lee They launched Magellan, a radar-mapping probe which arrived at Venus in August 1990. It was the first interplanetary spacecraft launched from a shuttle. The crew used small video camera for Earth photos. Cleave and Lee grew crystals.

5   Alan B. Shepard Jr., flying 15 minutes 22 seconds along a sub-orbital path in 1961 in the Mercury 3 capsule launched on a Redstone rocket, was the first American in space.

7   NASA's Wallops Flight Research Facility was opened at Wallops Island, Virginia, in 1945, administered then by NASA's Langley Research Center, Virginia. The first rocket was launched there June 27, 1945.

9   In 1963, U.S. project West Ford launched metal needles to orbit and scattered them about as dipoles for passive communications antennas.

9   In 1965, the USSR's Luna 5 probe crashed on the Moon.

11   In 1973, launch of USSR's Cosmos 557 first-generation space station failed.

12   In 1950, Chuck Yeager flew the Bell X-1 on its maiden flight from Dryden Flight Research Facility, California.

14   America launched its first-generation space station, Skylab, in 1973 on a leftover Saturn V Moon rocket. It was visited three times that year and not used again. Skylab fell from orbit July 11, 1979. It was the last Saturn V rocket ever to be fired.

15   French astronomer Nicolas-Louis de Lacaille was born in 1713. He became known as the father of southern astronomy after observing skies over the Cape

of Good Hope from 1750-55 where he named 14 new constellations. Lacaille pioneered the lunar-distance method of finding longitude. He died in 1762.

15   In 1960, the USSR launched Sputnik 4, also known as Korabl Sputnik 1, to Earth orbit. It was the first unmanned space flight test of a Vostok capsule, carrying a dummy dressed in a spacesuit. The test failed when the capsule did not re-enter the atmosphere. Korabl Sputniks 1 to 7 were announced as part of the Sputnik series to prevent the West from learning of preparations for manned flight.

15   L. Gordon Cooper, 1963, Mercury-Atlas 9, stayed more than 24 hours in orbit.

15   In 1987, the USSR successfully test-fired its Energia rocket, the first flight of the world's most powerful space booster.

16   USSR Venera 5 landed on Venus in 1969, sending back information about the atmosphere. It had been launched January 5, 1969.

17   English astronomer Sir Joseph Norman Lockyer was born in 1836. He discovered helium in the Sun and founded an observatory at Sidmouth. Lockyer started Nature magazine which he edited 50 years. He died in 1920.

17   USSR Venera 6 landed on Venus in 1969, sending back information about the atmosphere. It had been launched January 10, 1969.

18   Thomas P. Stafford, Eugene A. Cernan and John W. Young, in Apollo 10 in 1969, took a lunar excursion module (LEM) for its first orbit of the Moon.

19   In 1971, The USSR launched Mars 2 which reached the Red Planet November 27, dropping a capsule to land on the surface.

20   In 1978, the interplanetary probe Pioneer 12, also known as Pioneer-Venus 1, left the U.S. for Venus, going into orbit around that planet December 4, 1978, checking the atmosphere and photographing the surface.

20   Hubble Space Telescope sent its first picture to Earth in 1990. The so-called "first light" recorded by the $1.5 billion telescope had traversed 1,260 lightyears of deep space from a an open cluster of stars known to astronomers as Theta Carina. Hubble was 381 miles above Jayapura, New Guinea, when the shutter of its electronic wide-field planetary camera snapped the first black-and white picture.

23   In 1980, the amateur radio communication satellite Phase 3-A and West Germany's Firewheel experimental satellite were lost in the Atlantic Ocean after Ariane rocket LO-2 ran of course and splashed down three minutes after lift off from the European Space Agency's launch pad at Kourou, French Guiana, South America. Two similar satellites were constructed later with Phase 3-B making it to Earth orbit on June 16, 1983, and Phase 3-C on June 15, 1988. Phase 3-B became known as OSCAR 10. Phase 3-C became OSCAR 13.

24   M. Scott Carpenter, in 1962 in Mercury-Atlas 7, overshot his intended landing site by 250 miles.

24   In 1975, the USSR launched Soyuz 18B carrying cosmonauts Pyotr Klimuk and Vitali Sevastyanov to set a USSR record of 63 days in space. Klimuk and Sevastyanov were at Salyut 4 station during the U.S./USSR Apollo-Soyuz flight in July 1975.

25   President John F. Kennedy in 1961 announced the American objective of sending a man to the Moon and back by the end of the 1960's.

25   Charles Conrad Jr., Joseph P. Kerwin and Paul J. Weitz, in 1973, made the first visit to open America's Skylab space station and repair launch damage. The original launch of the station was known as Skylab 1. The flight by Conrad, Kerwin and Weitz was Skylab 2.

28   In 1964, the U.S. first flew its powerful new Saturn 1 space rocket, on May 28 and September 18, carrying dummy Apollo capsules. Those early Saturns, with upper stages using liquid-hydrogen fuel, led to Saturn 1B and Saturn V rockets

and America's ability to launch what was called "the heaviest payload in history," the manned Apollo flights to the Moon. In 1965, Saturn 1 rockets were used to fire three large satellites to Earth orbit: Pegasus 1 on February 16, Pegasus 2 on May 25 and Pegasus 3 on July 30. A dummy Apollo capsule flew to space with each Pegasus. Pegasus satellites sprouted 2,300 sq. ft. panels which extended 96 ft. to measure the number of penetrations by meteoroids. Watchers with binoculars on the ground could see the satellites at night. Scientists discovered that interplanetary dust particles were considerably fewer in number than had been expected. That news allowed engineers to cut 1,000 lbs. of protective weight from each Apollo capsule.

28   In 1971, The USSR launched the unmanned orbiter-with-lander Mars 3 which reached the Red Planet December 2, dropping a capsule to the surface. TV signals were unexpectedly brief. Mars 3 was the first machine from Earth to soft-land on Mars.

29   In 1974, USSR launched Luna 22. It reached orbit around Moon June 2, 1974.

30   In 1966, U.S. Surveyor 1 flew to the Moon, landing June 2, 1966. It sent 10,400 photos, even after surviving a cold 14-day lunar night.

30   In 1971, the interplanetary probe Mariner 9 left the U.S. for Mars, becoming the first U.S. craft to orbit the Red Planet November 13, 1971, sending back 7,300 photos of the Martian surface and first close-ups of a Martian moon. Signals stopped October 27, 1972.

31   Robert H. Goddard launched a rocket in 1935 from Roswell, New Mexico, to an altitude of 7,500 feet.

June

1   In June 1942, Germany made the first flight test of the A-4 rocket at Peenemunde, by Wernher von Braun and Walter Dornberger. The A-4 became the V-2 missile in WWII.

1   The USSR launched Soyuz 9 in 1970 with cosmonauts Andrian G. Nikolayev and Vitali Sevastyanov. Their 18-day stay in space broke a five-year-old U.S. Gemini 7 record of 14 days.

2   In 1983, the USSR launched the Venera 15 radar-mapping orbiter to Venus. The probe orbited Venus October 10, 1983.

3   Edward H. White 2nd, in 1965 from Gemini 4, made the first American spacewalk (extravehicular activity or EVA), for 20 minutes. James A. McDivitt was with him.

3   Use of the 200-inch Hale Telescope started in 1948.

3   In 1966, the U.S. Gemini 9 capsule went to orbit carrying astronauts Thomas P. Stafford and Eugene A. Cernan.

6   Georgi T. Dobrovolsky, Vladislav N. Volkov, Viktor I. Patsayev, in 1971, flew to space in Soyuz 11, docked with Salyut space station. They stayed 23 days but died when pressure was lost during the flight back down to Earth.

6   In 1983, the USSR launched the Venera 16 radar-mapping orbiter to Venus.

6   In 1987, Pope John Paul II used various orbiting space satellites to pray and broadcast to 23 nations in 35 languages.

8   Astronomer Giovanni Cassini was born in 1625.

8   In 1959, North American Aviation pilot A. Scott Crossfield flew NASA's X-15 rocket-powered craft for the first time across the California desert at Edwards Air Force Base. The X-15 was carried by a B-52 bomber to an altitude of 38,000 feet and dropped. Crossfield made an unpowered five-minute gliding descent. The world's first hypersonic research airplane—with a skin to absorb and dissipate heat generated by re-entry into the atmosphere and instruments to measure its speed and altitude above the atmosphere—eventually reached six

times the speed of sound. X-15s flew 199 flights, until October 24, 1968. It was dubbed the first spaceplane as it flew above 67 miles altitude. Its records were 354,200 feet altitude and 4,520 mph. Twelve pilots flew the X-15 including Neil A. Armstrong who went on to Apollo program to become first man to step on Moon and Joe H. Engle who went on to fly U.S. space shuttle.

8   In 1965, the USSR launched the Luna 6 probe on a Moon flyby. It passed within 100,000 miles of the Moon.

8   In 1975, USSR launched Venera 9 probe to Venus, an orbiter with a lander. TV pictures transmitted for 115 minutes to Earth from surface October 22, 1975.

10   In 1985, USSR Vega 1 interplanetary probe, on its way for a close look at Comet Halley, drops off a package of instruments at Venus to study atmosphere and surface. It flew on, through the comet's dust particles March 6, 1986, sending back TV pictures and data.

12   In 1967, the USSR interplanetary probe Venera 4 left for Venus. It arrived October 17, 1967, its atmosphere probe sent back temperature and chemical composition information while parachuting to the planet's night side surface.

13   The American spacecraft Pioneer 10 left the Solar System in 1983.

14   In 1975, USSR Venera 10 flew to Venus, soft landed, sent TV photos.

14   In 1967, the interplanetary probe Mariner 5 left the U.S. for Venus. It flew by the planet October 19, 1967, and now is orbit around the Sun.

14   In 1985, USSR Vega 2 interplanetary probe, on its way for a close look at Comet Halley, drops off a package of instruments at Venus to study atmosphere and surface. It flew on, through the comet's dust particles March 9, 1986, sending back TV pictures and data.

16   Valentina V. Tereshkova, from the USSR, in Vostok 6 in 1963 was the first woman to travel to space. She completed 48 orbits as she and Valery F. Bykovsky in Vostok 5 composed the second tandem group flight.

17   Prince Sultan Salman Abdel Aziz Al-Saud of Saudi Arabia became the first Arab in space and the first royalty in orbit, aboard U.S. shuttle Discovery flight STS-51G from Cape Canaveral in 1985. With him in space were Daniel C. Brandenstein, John O. Creighton, Steven R. Nagel, John W. Fabian, Shannon W. Lucid and Patrick Baudry of the French space agency CNES. Discovery ferried three communications satellites to orbit—Arabsat 1-B, Morelos A, and Telstar 3-D—plus Spartan X-ray and laser satellite.

18   American astronaut Sally K. Ride in shuttle Challenger STS-7 became the first American woman in space in 1983. U.S. shuttle Challenger flight STS-7 left Cape Canaveral with crew Robert L. Crippen, Frederick H. "Rick" Hauck, John W. Fabian, Norman E. Thagard and Sally K. Ride, the first and youngest U.S. woman in space, at age 32. Challenger carried satellites Anik and Palapa. Astronauts released and recovered a pallet satellite with the shuttle's robot arm.

22   The Royal Greenwich Observatory was founded in 1675.

22   In 1960, The U.S. launched the satellites Transit 2A and Solrad 1, the first multiple payloads placed in multiple orbits.

22   In 1976, the Salyut 5 space station was launched. It was the USSR's fifth first generation station. Put to military use, it stayed in orbit until August 8, 1977.

22   Pluto's moon Charon was discovered in 1978 by James Christy.

24   In 1974, the Salyut 3 space station was launched. It was the USSR's third first generation station. Put to military use, it stayed in orbit until August 24, 1975.

25   In 1967, the Our World television program used five orbiting space satellites to bring together The Beatles in London; Marc Chagall and Joan Miro in Paris; and Van Cliburn and Leonard Bernstein in New York.

26   Charles Messier, to catalogue deep-space objects, was born in 1730.

27   The father of radioastronomy reveals his work at Bell Telephone Laboratories

for the first time in 1933. The phone company was interested in tracing static in Earth's atmosphere which sometimes interfered with calls passing through telephone lines. Since 1931, Karl Jansky had been rotating a radio antenna in an attempt to pinpoint the source of a steady hiss of static. He finally decided it was an electrical disturbance of extraterrestrial origin. The hiss Jansky recorded was the sound of radio waves from the Milky Wave galaxy.

27   The first rockets were launched in 1945 out over the Atlantic Ocean from the new Pilotless Aircraft Research Station on Wallops Island, a barrier island on Virginia's eastern shore named for the 17th century surveyor John Wallop. Five small 3.25-in. rockets were blasted off for launching experience and to test radar and optical tracking. A week later, on July 4, a larger Tiamat research rocket was launched. Congress just two months earlier on April 25 had approved construction of the research station by the Langley Aeronautical Laboratory of the National Advisory Committee for Aeronautics (NACA), NASA's predecessor. Later, after NASA was established in 1958, the research facility became independent and known as Wallops Station. In 1974 it became Wallops Flight Center, then was appended to NASA's Goddard Space Flight Center, Greenbelt, Maryland, in 1981. The station, known today as Wallops Flight Facility, expanded over the years by taking over the old Chincoteague Naval Air Station. Today it encompasses 6,000 acres. Famous American rockets such as Little Joe and Scout, many small satellites, space probes, have been launched in 15,000 blast offs from Wallops Island over the years.

27   In 1982, Thomas K. Mattingly 2nd and Henry W. Hartsfield Jr flew U.S. shuttle Columbia flight STS-4 from Cape Canaveral on last test flight. Shuttle's two boosters were lost. Trip lasted seven days, carried the first classified payload.

29   Astronomer George E. Hale was born in 1868.

30   A major explosion was caused by a tiny comet, asteroid or large meteor crashing near Tunguska in central Siberia, USSR, in 1908.

30   Cosmonauts Georgi T. Dobrovolsky, Vladislav N. Volkov and Viktor I.Patsayev died in 1971 as they returned home from space in their Soyuz 11 capsule. A USSR government commission found that, during re-entry, a faulty valve had allowed pressure to escape from the spacecraft, killing them 30 minutes before landing. Not wearing space suits for the flight home after 23 days at the Salyut 1 space station, they were found dead in the Soyuz 11 capsule when it was opened after landing.

July

2   In 1985, the European Space Agency launched the deep-space probe Giotto toward Comet Halley. It flew by the comet March 13, 1986.

4   The Crab Nebula supernova was seen for the first time in 1054.

4   In 1945, Tiamat was the first large research rocket to fly from the new Pilotless Aircraft Research Station on Wallops Island, a barrier island on Virginia's eastern shore named for the 17th century surveyor John Wallop. A week earlier, on June 27, five small 3.25-in. rockets had been blasted off for launch experience and to test radar and optical tracking for the National Advisory Committee for Aeronautics (NACA), NASA's predecessor. Today the 6,000-acre station is the Wallops Flight Facility operated by NASA's Goddard Space Flight Center, Greenbelt, Maryland.

7   In 1988, the USSR launched the ill-fated Phobos 1 probe toward Mars. The probe failed enroute August 29, 1988, and now is orbiting the Sun.

9   The American spacecraft Voyager 2 flew past Jupiter in 1979.

10   In 1962, the U.S. launched Telstar 1, the first active real-time communications satellite.

11   America's first-generation space station, Skylab, was allowed to fall into the atmosphere and burn up in 1979. It was visited only three times, all shortly after the station was launched to space in 1973.

12   In 1988, the USSR launched the Phobos 2 probe to Mars. It orbited Mars January 29, 1989, sent back photos and data about Mars and its moon Phobos, then failed March 27, 1989.

13   What today is NASA's Langley Research Center was founded at Hampton, Virginia, in 1917. Named for Samuel P. Langley, the then-small laboratory was set up to investigate the new field of aeronautics.

13   In 1969, the USSR's Luna 15 probe was launched. It arrived in Moon orbit just 48 hours before the American capsule, Apollo 11. Luna 15 was supposed to land and retrieve soil samples, but crashed on the Moon July 21.

14   The U.S. spacecraft Mariner 4 made the first Mars flyby in 1965, taking the first close-up photos of the Red Planet.

14   In 1967, U.S. Surveyor 4 flew to the Moon but contact was lost 2.5 minutes before landing.

15   In 1975, USSR cosmonauts Alexei Leonov and Valeri Kubasov were in space in Soyuz 19 six days. American astronauts Vance Brand, Thomas P. Stafford and Donald K. "Deke" Slayton were in space in Apollo 18 nine days. Their two ships formed a joint U.S.-USSR flight known as Apollo-Soyuz. Crews linked their craft in space July 18, shared meals, conducted experiments together and held a joint news conference from space. It was the last Apollo flight.

16   In 1965, the USSR launched the first Proton rocket from Baikonur Cosmodrome. The heavy-lifter, still in use, launches 50,000 lbs. to orbit.

16   Neil A. Armstrong, Edwin E. "Buzz" Aldrin Jr. and Michael Collins, in 1969, left Earth in Apollo 11 atop a Saturn V rocket bound for the Moon. The first lunar touchdown by a manned ship, the lunar excursion module Eagle, was made July 20. Armstrong became the first man to walk on the Moon; Aldrin the second. They stayed 21 hours, collecting 48 lbs. of rock and soil. Collins stayed in the lunar sky, orbiting in the command module Columbia.

16   · Pakistan's first satellite was launched in 1990. Badr-A, an amateur radio communications satellite, was launched by China on the maiden flight of the Long March 2E rocket from Xichang Launch center. Badr-A, in space 146 days, falling into the atmosphere December 9, 1990, was one of eight hamsats sent to space in 1990, the largest proliferation since Russian hams sent six in one flight in 1981.

17   The first photograph of a star was an exposure of Vega in 1850.

17   The American-Russian Apollo-Soyuz Test Project spacecraft link-up was accomplished in 1975. They remained docked with each other two days.

18   In 1965, USSR Zond 3 flew to the Moon, sending back close-up photos of three million square miles of surface. It now is orbit around the Sun.

18   In 1966, the U.S. manned Gemini 10 capsule was launched on a Titan rocket to Earth orbit carrying astronauts John W. Young and Michael Collins.

18   U.S. Apollo 18 spacecraft and USSR Soyuz 19 craft met and connected in orbit in 1975.

18   In 1980, India launched Rohini 1, its first Earth-orbiting space satellite, using its rocket known as Satellite Launch Vehicle (SLV), making India the 7th nation with a space rocket.

19   In 1967, U.S. Explorer 35 left for the Moon, arriving in orbit around that natural satellite July 22, 1967.

20   U.S. Apollo 11 carrying Neil A. Armstrong and Edwin E. "Buzz" Aldrin Jr. landed on the Moon in 1969. Armstrong was the first man to walk on the Moon. Aldrin was second. They spent 21.5 hours on the surface, including 2.5

hours outside their lunar excursion module (LEM) Eagle.

20   U.S. spacecraft Viking 1 landed on Mars in 1976.

21   In 1961, the U.S. Mercury 4 capsule atop a Redstone rocket carried astronaut Virgil I. "Gus" Grissom on America's second suborbital flight. Grissom had to be rescued from the ocean when his capsule sank after splash down.

21   In 1973, USSR launched Mars 4, which flew by the Red Planet February 1974, briefly sending back photos.

22   Astronomer Friedrich Bessel was born in 1784.

22   In 1969, the U.S. Apollo 11 lunar excursion module Eagle lifted off the Moon, returning astronauts Armstrong and Aldrin to the command module Columbia in Moon orbit, the first men launched from Moon to Earth.

23   In 1980, the USSR launched the manned capsule Soyuz 37 with cosmonauts Viktor V. Gorbatko and Pham Tuan, from Vietnam, the first non-Warsaw Pact and first third-world Intercosmos cosmonaut. The flight to the Salyut 6 station took place during Summer Olympic Games in Moscow.

24   In 1950, Bumper 8 was the first rocket to be launched from Cape Canaveral. After a series of successful test firings at White Sands, New Mexico, of captured German V-2 rockets, a V-2 was carted across the continent to the Cape on a trailer known as Meillerwagen. The V-2 was converted to become the first two stage booster, a 60-ft. rocket made up of a U.S. Army Wac Corporal second stage atop the V-2 first stage. They named it Bumper 8. The rocket flew 180 miles down range.

25   In 1973, USSR's unmanned Mars 5 probe was launched toward the Red Planet.

26   In 1963, U.S. satellite Syncom 2 was launched, the first communications sat in stationary orbit. Relayed TV pictures from 1964 Olympics in Japan.

26   David R. Scott, Alfred M. Worden and James B. Irwin flew Apollo 15, in 1971, to man's fourth landing on the Moon. Enroute they took the first ever deep space spacewalk. Scott and Irwin made the fourth-ever Moon landing July 30, collecting 170 lbs. of rocks in 6 hours. For the first time, they drove a lunar rover across the Moon's surface. The 462-lb. rover had four wheels and could range 50 miles at 8.7 mph.

28   First photograph of a total eclipse of the Sun was made in 1851.

28   In 1964, the U.S. blasted its probe Ranger 7 to the Moon. Ranger, the first NASA space program to investigate another Solar System body was a series of Moon probes in the early 1960's. At first, Rangers were to bounce onto the lunar surface with a seismometer, but the first six failed. They were redesigned for photography. The first five, launched between August 1961 and October 1962, failed. Ranger 6, launched January 30, 1964, reached its target, but its cameras broke down. Ranger was redesigned completely. Pictures were to be snapped as a Ranger approached its bulls-eye on the Moon. Ranger 7 was the first success. It impacted in an unnamed mare area later named Mare Cognitum, meaning "the sea which has become known." The level mare without craters was chosen because it might make a good landing spot for later Apollo manned flights. Ranger 7 used six sequential TV cameras to send back 4,316 photos of Earth's natural satellite during the last 15 minutes of flight, from as close as 1,000 feet above the lunar surface, before impacting near Crater Guericke 68.5 hours after launch. Those first close-ups of the Moon surface showed boulders and three-foot craters. The widest angle photo covered a square mile and showed 30-ft. craters. The most detailed pictures revealed 3-ft. craters in an area 100 ft. by 160 ft.

28   Alan L. Bean, Jack R. Lousma and Owen W. Garriott in Skylab 3 made the second visit to America's first-generation space station, Skylab, in 1973. A spider, Arabella, was able to spin her web in Skylab in space. Spacewalks

totaled 13 hours. Garriott was the first amateur radio operator to go to space.

29   In 1955, the U.S. announced its intention to launch an artificial Earth satellite for the 1958 International Geophysical Year.

29   U.S. space agency NASA was founded in 1958.

29   The U.S. administration named the Apollo Program.

29   During the lift off of flight STS-51F in 1985, Challenger was the first U.S. shuttle to have an engine failure enroute to space. But Coke and Pepsi going to space for the first time were the big news of the flight. The astronauts siphoned the carbonated beverages from special cans. There had been a public relations battle between Coca Cola and Pepsi Cola over whose can was best in space. Pepsi called its can "one giant sip for mankind," but Coke's worked better. The astronauts didn't like warm cola, however, and complained about a lack of refrigeration for the cans aboard the shuttle. The crew: Charles Gordon Fullerton, Roy D. Bridges Jr., Loren W. Acton, F. S. "Story" Musgrave, John David F. Bartoe, Anthony W. "Tony" England and Karl G. Henize, at age 58, the oldest person to fly in space up to that time. An astronomy research module known as Spacelab 2 was carried in the shuttle's cargo bay. Tony England made amateur radio contacts with hams on Earth.

30   In 1955, USSR announced its intention to launch an artificial Earth satellite.

30   U.S. Apollo 15 makes Man's fourth landing on the Moon in 1971.

31   First vehicle driven on the Moon by men.

August

1   Astronomer Maria Mitchell was born in 1818.

2   In 1967, U.S. Lunar Orbiter 5 left for the Moon, arriving in orbit there August 5. After photographing, it was sent down to crash on the Moon.

5   Neil A. Armstrong, the first man on the Moon, was born in 1930 at Wapakoneta, Ohio.

6 ·   Gherman S. Titov of the USSR made the first space flight of more than 24 hours, in Vostok 2 in 1961.

8   In 1978, the interplanetary probe Pioneer 13 (Pioneer-Venus 2) left the U.S. for Venus, going into orbit around that planet December 9, 1978, checking atmosphere, photographing surface, four probes dropped.

8   NASA's ancient mariner, the space shuttle Columbia, returned to space in 1989, its eighth trip since 1981. Columbia had been NASA's first shuttle to go to space when it first lifted into the skies over Cape Canaveral on April 12, 1981. It made two trips in 1981, three in 1982 and one in 1983. It last had flown January 12, 1986, just 16 days before the Challenger disaster. Since then, it had been cannibalized for spare parts to keep shuttles Discovery and Atlantis running and then extensively refurbished for return to duty. Following Challenger, the three remaining shuttles—Atlantis, Discovery and Columbia were rebuilt, each with more than 250 modifications improving safety and performance. The newer shuttles, Discovery and Atlantis, were fixed first. Discovery returned America to space September 29, 1988. Atlantis flew December 2, 1988. With Columbia's successful trip from August 8 to 13, 1989—flight STS-28—the space agency had three shuttles available for duty and had put the Challenger disaster behind it and returned to normal operations.

9   In 1976, the USSR sent Luna 24 to a soft landing on the Moon August 18, 1976. It brought back soil samples August 22, 1976.

10   In 1966, U.S. Lunar Orbiter 1 left for the Moon, going into orbit around Earth's natural satellite August 14, sending back 21 photos.

11   A moon of Mars, Deimos, was discovered in 1877 by Asaph Hall. He discovered Phobos, a moon of Mars, six days later.

11   In 1962, Andrian G. Nikolayev in Vostok 3 and Pavel R. Popovich in Vostok 4 made the first tandem group flight. On first orbit, Vostok 4 came within three miles of Vostok 3.

12   The giant balloon Echo 1 was launched in 1960 from Vandenberg Air Force Base, California, to Earth orbit where it was used to reflect radio signals down, including a message by U.S. President Dwight D. Eisenhower.

12   NASA hopes someday to retrieve the deep-space probe International Sun-Earth Explorer 3 (ISEE), known today as International Cometary Explorer (ICE). ISEE was launched in 1978 from Cape Canaveral to study solar wind. The name was changed to ICE in 1983 when it was sent off for a close look at Comet Giacobini-Zinner in 1985. ICE was the first spacecraft from Earth to fly through the tail of a comet. Today it is orbiting the Sun. NASA would like to bring it back to an Earth orbit in the year 2013.

17   A moon of Mars, Phobos, was discovered in 1877 by Asaph Hall. He had discovered Deimos, a moon of Mars, six days earlier.

17   In 1966, U.S. Pioneer 7 left for orbit around the Sun, ranging 92 to 102 million miles from the Sun, circling every 400 days.

17   In 1970, the USSR sent Venera 7 to Venus which it reached December 15, 1970, apparently sending data back from the surface for 58 minutes.

19   Astronomer John Flamsteed was born in 1646.

19   In 1960, the USSR launched Sputnik 5, also known as Korabl Sputnik 2, on an unmanned Vostok capsule space flight test from Baikonur Cosmodrome. The five-ton spacecraft carried dogs Belka (Squirrel in Russian) and Strelka (Little Arrow). They were ejected after re-entry, recovered after parachuting to Earth.

19   In 1960, the U.S. satellite Discoverer 14 ejected a 300-lb. data package. The capsule, equipped with a radio beacon, entered the atmosphere and descended to an altitude of 8,000 feet where it was caught by a U.S. Air Force C-119 transport plane. After that, capsules were recovered by ship and aircraft for years during the Discoverer program. Later, data was transmitted directly from satellite to Earth by radio. Meanwhile, the USSR and China have continued with capsules drops from some satellites.

20   In 1975, the U.S. sent Viking 1 to Mars, carrying cameras and life detectors. It landed on the Red Planet July 20, 1976. Designed to last 90 days, Viking 1 worked on the surface 6.5 years.

21   L. Gordon Cooper Jr. and Charles Conrad Jr. were launched in the Gemini 5 capsule atop a Titan rocket in 1965, first use of fuel cells to generate electricity in space.

24   In 1966, USSR Luna 11 left for Moon, going into orbit there August 27, 1966.

24   America's Voyager 2 interplanetary probe flew past Saturn in 1981.

24   Launched August 20, 1977, Voyager 2 sailed past the planet Neptune in the farthest reaches of the Solar System on schedule in 1989, crowning a 12-year voyage of discovery which had carried the sturdy craft past Jupiter, Saturn and Uranus. Voyager 2 flashed by Neptune just 3,000 miles above its north pole, using the planet's gravity to bend its trajectory. Neptune became the fourth planet known to have them when Voyager 2 came upon rings and partial rings or ring arcs. The probe discovered several previously-unknown small moons orbiting the giant planet. Voyager 2 found Neptune swept by the fastest winds in the Solar System, blowing frozen clouds of natural gas at 1,500 mph across the northern edge of the planet's vast Great Dark Spot hurricane. As Voyager 2 passed, Neptune was farther from the Sun—2.8 billion miles—than Pluto. Five hours later, the robot explorer sailed past Triton, a moon as big as the entire planet Pluto. Triton is Neptune's largest moon and the coldest known object in the Solar System. Astronomers had thought Triton would have a methane

atmosphere, possibly with pools of liquid nitrogen on its surface. Voyager 2 found the moon to be more reflective than expected, meaning the surface was covered with ice rather than liquid pools. Spellbinding close-up pictures disclosed a corrugated-ice surface of frozen nitrogen and methane with strange ice volcanoes. Voyager 2 photographed 5 geysers towering 5 miles above active ice volcanos on the frigid moon. The volcanoes were geysers spewing nitrogen, in the form of gas and ice, five miles high and ninety miles downwind. Today Voyager 2 is rushing out of the Solar System, looking for feeble shock waves at the boundary of the Sun's influence and interstellar space—the imperceptible edge of the Solar System. With its plutonium powered electricity generator, Voyager 2 should be able to beam data back to Earth for another 30 years.

24   A large asteroid (1989pb) came within 2.5 million miles of Earth in 1989. Earlier, Earth already had seen a close call with an asteroid (1989fc) which came within 450,000 miles of our planet March 23, 1989. A different asteroid (1989ja) had passed within 8 million miles of Earth June 3, 1989. Yet another asteroid (1990mf) flew within 3 million miles of Earth July 10, 1990.

26   Sunlight on Earth was dimmed by dust for several years after the volcano on a Dutch East Indies island, Krakatoa, erupted in 1883. The volcano tossed five cubic miles of earth 50 miles in the air—to the very edge of space. Three quarters of the island was blown away and more than 36,000 people were killed. Tidal waves reached 120 feet high, blast noise was heard 2,000 miles away.

27   In 1962, the interplanetary probe Mariner 2 left the United States for Venus, passing 21,648 miles from the beautiful planet December 14, 1962. Reported 800°F. surface temperature. Contact was lost January 3, 1963, some 54 million miles from Earth.

27   In 1985, U.S. shuttle Discovery flight STS-51I left Cape Canaveral with astronauts Joe H. Engle, Richard O. Covey, James D. "Ox" Van Hoften, John M. "Mike" Lounge and William F. Fisher. Van Hoften and Fisher took a spacewalk to repair the Syncom satellite left broken in orbit after flight STS-51D in April 1985. Discovery carried three communications satellites to orbit.

28   Saturn's moon Enceladus was discovered in 1789 by William Herschel.

30   Guion S. Bluford Jr. became the first black man to fly in space, aboard U.S. shuttle Challenger flight STS-8 from Cape Canaveral in 1983. With him were Richard H. Truly, Daniel C. Brandenstein, Dale A. Gardner and William E. Thorton. STS-8, making first night launch and landing, orbited India's Insat.

30   In 1984, U.S. shuttle Discovery flight STS-41D left Cape Canaveral with astronauts Henry W. Hartsfield Jr., Michael L. Coats, Richard M. "Mike" Mullane, Steven A. Hawley, Judith A. Resnick and Charles D. Walker, the first commercial payload specialist, who did drug research. Discovery ferried the satellites Syncom, Telstar and SBS to orbit.

September

1   In 1947, a captured WWII V-2 rocket was launched by the U.S. Navy from the carrier Midway.

1   The U.S. probe Pioneer 11 made the first flyby of the planet Saturn in 1979.

2   The asteroid Juno, the third minor planet to be discovered, was found by Karl Harding in 1894.

2   In 1971, the USSR's Luna 18 probe crashed on Moon.

3   The U.S. Viking 2 unmanned probe landed on Mars in 1976.

4   Harvest Moon, usually in September, is the full Moon nearest autumnal equinox. Hunter's Moon, usually in October, is the full Moon following the Harvest Moon.

5   In 1870, unmanned French balloon mail flies from Met across German siege.

5   In 1962, pieces of a Russian satellite landed near Manitowoc, Wisconsin.

5   In 1977, the interplanetary probe Voyager 1 left the U.S. for Jupiter and Saturn, encountering Jupiter March 5, 1979, and Saturn November 13, 1980. It now is leaving the Solar System.

5   In 1989, the planet Pluto, smallest, faintest and usually most distant from the Sun, was at perihelion—its closest point to the Sun. Pluto's average distance from the Sun is 3.75 billion miles. At perihelion of its 248-year orbit, Pluto and its moon Charon were 2.75 billion miles from the Sun. It temporarily is closer to the Sun than Neptune, until 1999.

6   In 1989, cosmonauts Alexander S. Viktorenko and Alexander A. Serebrov left the USSR in Soyuz TM-9 to reopen Mir after the space station was mothballed in orbit for 132 days from April 27.

8   Marshall Space Flight Center was opened in 1960 at Huntsville, Alabama. Mrs. George C. Marshall and U.S. President Dwight D. Eisenhower were there.

8   In 1967, U.S. Surveyor 5 left for the Moon, landing near the lunar equator September 10. Its mechanical claw dug up soil for radiological tests.

9   The Jupiter moon Amalthea was discovered by E.E. Barnard in 1892.

9   The U.S. decided to try to launch an artificial Earth satellite when the Dept. of Defense approved Project Vanguard in 1955.

9   In 1975, the U.S. sent Viking 2 to Mars, carrying life detectors and cameras. It landed on the Red Planet September 3, 1976. Designed to work 90 days, Viking 2 lasted 3.5 years.

9   In 1978, USSR Venera 11 was launched to Venus, soft landing there December 25, 1978, sending back data by radio for 95 minutes.

11   An interplanetary probe called International Cometary Explorer or ICE flew past Comet Giacobini-Zinner in 1985. The probe originally was launched in 1978 and known as ISEE-3. It was renamed and diverted by lunar gravity for the first comet fly-by in history. ICE flew on to come within 17 million miles of Comet Halley in March 1986.

12   USSR Luna 2 in 1959 was fired toward the Moon. It became the first spacecraft from Earth to hit the Moon, landing near Mare Serenitatis September 13.

12   In 1966, the U.S. Gemini 11 manned capsule was launched on a Titan rocket to Earth orbit carrying astronauts Charles Conrad Jr. and Richard F. Gordon Jr. They docked with orbiting Agena rocket for two tethered revolutions of Earth, photographed stars, galaxies, set Gemini capsule altitude record of 739.2 mi.

12   In 1970, the USSR sent Luna 16 to the Moon where it soft-landed September 20, 1970, scooping up soil samples weighing 101 grams and returning to Earth September 24, 1970. The first automated lunar sample retrieval.

13   USSR Luna 2 became the first craft to impact the Moon, in 1969.

14   In 1968, the USSR's Zond 5 probe flew from Earth to circle the Moon and return, carrying turtles, worms, flies, plants and seeds.

14   The 13th Jupiter moon Leda was discovered in 1974 by C. Kowal. At 9 miles diameter, it may be a captured asteroid.

14   In 1978, USSR Venera 12 was launched to Venus, soft landing there December 21, 1978, sending back data by radio for 110 minutes.

15   In 1989, the USSR launched Biocosmos 9, an orbiting zoo housing two rhesus monkeys, rats, fish, fish eggs, newts, snails, beetles, flies, ants, worms, seeds, plants and one-cell organisms. Also known as Cosmos 2044, the research satellite returned to Earth September 29.

19   The Saturn moon Hyperion was discovered by William Bond in 1848.

19   As the story goes, on this night in 1961 a bright flying disk stayed in sight as Barney and Betty Hill drove 30 miles through the New Hampshire's White

Mountains. Betty and Barney said later they couldn't recall two hours of the time. Under hypnosis, they described going inside the UFO where they received physical examinations. Their psychiatrist said their stories came from Barney's sensitivity to racial prejudice and Betty's nightmares about being abducted. In 1975, NBC-TV aired a movie about the UFO incident starring James Earl Jones and Estelle Parsons.

19   The Pregnant Guppy made her maiden flight in 1962. The plane was a Boeing Stratocruiser blimped up to carry large space booster rockets inside.

19   In 1988, Israel launched its first satellite, Horizon, to orbit on a Shavit rocket. Israel became the eighth nation with a space rocket, joining China, France, Great Britain, India, Japan, USSR and the U.S.

20   Wernher von Braun and other German scientists arrive in the U.S. in 1945.

20   In 1956, the U.S. first launched the Jupiter-C rocket.

20   In 1957, the U.S. first launched the Thor rocket.

20   In 1977, the interplanetary probe Voyager 2 left the U.S. for Jupiter, Saturn, Uranus and Neptune encountering Jupiter July 9, 1979, and Saturn August 26, 1981 and Uranus January 1986. It will fly within 3,000 miles of Neptune August 24, 1989, on its way out of the Solar System.

23   Neptune, the eighth major planet from the Sun, was discovered in 1846 by J.G. Galle at Berlin Observatory.

26   In 1983, USSR cosmonauts Gennadi Strekalov and Vladimir Titov narrowly escaped death in a launch pad explosion. Strekalov and Titov had boarded the capsule Soyuz T-10A atop an A-2 rocket to fly to the space station Salyut 7 when fire erupted at the base of the rocket 90 seconds before launch. The Soyuz escape rocket was triggered. The capsule separated from the booster, lifting off seconds before the A-2 exploded. Titov and Strekalov landed, uninjured, several miles away.

27   In 1945, the U.S. fired a Wac Corporal rocket from White Sands, New Mexico, to an altitude of 235,000 feet.

28   In 1971, the USSR sent Luna 19 to the Moon, which it orbited, taking measurements, making photos. It then soft-landed September 21 in the Sea of Fertility. It brought rock samples to Earth February 25, 1972.

29   In 1977, the USSR launched to orbit its Salyut 6 station, the first of two second-generation space stations.

29   In 1988, U.S. astronauts returned to space 975 days after the 1986 Challenger disaster with the launch of shuttle Discovery flight STS-26 from Cape Canaveral. The astronauts were Frederick H. Hauck, Richard O. Covey, David C. Hilmers, George "Pinky" Nelson and John M. "Mike" Lounge. Discovery carried NASA's TDRS-C communications satellite to orbit.

30   Yuri Romanenko in 1987 in the USSR's third-generation space station, Mir, surpassed the previous 237-day man-in-space endurance record.

October

1   31st anniversary of NASA, formed October 1, 1958.

3   Walter M. Schirra, in 1962, rode Mercury 8 to orbit atop an Atlas rocket and splashed down within 4.5 miles of target, best up to that date.

3   In 1971, the USSR launched the Moon-orbiter probe Luna 19.

3   In 1985, U.S. shuttle Atlantis secret military flight STS-51J left Cape Canaveral with astronauts Karol J. "Bo" Bobko, Ronald J. Grabe, Robert L. Stewart, David C. Hilmers and William A. Pailes. Atlantis carried two military communications satellites to orbit.

4   The USSR launched Sputnik 1 in 1957, the first artificial man-made satellite in Earth orbit, the start of the Space Age, beginning Man's explorations in space.

4   In 1959, the USSR launched Luna 3, a solar-powered probe to circle the Moon, snapping the first photos of the dark side October 7, 1959.

4   The first active communications satellite was the U.S. Courier 1B in 1960 with the first store-and-forward message board in space.

4   Hunter's Moon, usually in October, the full Moon following the Harvest Moon. Harvest Moon, usually in September, full Moon nearest autumnal equinox.

5   The Cepheid variable star in galaxy M-31 identified in 1923 by Edwin Hubble.

5   Kathryn D. Sullivan and Sally K. Ride, aboard U.S. shuttle Challenger flight STS-41G from Cape Canaveral in 1984, were the first two women to be in space at the same time. Sullivan made the first U.S. female spacewalk. With Ride and Sullivan in space were Robert L. Crippen, Jon A. McBride, David C. Leetsma, Marc Garneau, and Paul D. Scully-Power. Challenger carried the Earth Radiation Budget satellite to orbit. Garneau was the first Canadian to go to space. Scully-Power was the first oceanographer in orbit.

7   USSR Luna 3 photographed the far side of the Moon in 1959.

8   Danish astronomer Ejnar Hertzsprung was born in 1873. Hertzsprung, independently but at the same time as American astronomer Henry Norris Russell, plotted stars on a diagram relating brightness to color or temperature. Most stars today are called "main sequence" because they lie on a main line of the chart extending from the dimmest and coolest stars to the brightest and hottest stars. Modern astronomers call this color-magnitude diagram an H-R Diagram or Hertzsprung-Russell Diagram.

10   The Neptune moon Triton was discovered in 1846 by William Lassell.

11   The U.S. launched the Moon probe Pioneer 1 in 1958. It failed.

11   Apollo 7 launched to Earth orbit on a Saturn rocket in 1968, the first manned flight of that U.S. moonship command service module (CSM), by Walter M. Schirra Jr., Donn F. Eisele and R. Walter Cunningham.

11   In 1969, the USSR launched the Soyuz 6 manned space capsule carrying cosmonauts Georgi S. Shonin and Valery N. Kubasov. Anatoly V. Filipchenko, Vladislav N. Volkov and Viktor V. Gorbatko were launched in Soyuz 7. Aleksei Yeliseyev and Vladimir Shatalov rode Soyuz 8 to space. For the first time, three spacecraft and seven crew members were in Earth orbit at the same time. Shonin and Kubasov did the first welding of metal in the zero-gravity of space.

12   Cosmonauts Vladimir M. Komarov, Konstantin P. Feoktistov and Boris B. Yegorov, in 1964 in the Voskhod 1 capsule, made the first three-man orbital flight and the first space flight without space suits.

15   British fired V-2 rocket at Cuxhaven, Germany, Operation Clitterhouse, 1945.

17   An electrical fire damaged America's Magellan deep-space probe as it stood in crates in a warehouse at Cape Canaveral. The spacecraft was repaired in time for shuttle Atlantis to ferry it to Earth orbit May 4, 1989. From there, it was fired toward Venus where it went into orbit in August 1990, sending back radar maps of that cloud-shrouded body.

20   In 1970, the USSR's Zond 8 probe circled the Moon, sending back the first TV pictures of the Earth.

22   Man's first record of an eclipse of the Sun was made in China in 2136 BC.

22   In 1966, USSR's Luna 12 left for Moon, going into orbit October 25, 1966.

24   Uranus' moons Umbriel and Ariel were discovered in 1851 by William Lassell.

25   The Saturn moon Iapetus was discovered in 1671 by Giovanni Cassini.

25   American astronomer Henry Norris Russell was born in 1877. Russell, independently but at the same time as Danish astronomer Ejnar Hertzsprung, plotted stars on a diagram relating brightness to color or temperature. Most stars today are called "main sequence" because they lie on a main line of the chart extending from the dimmest and coolest stars to the brightest and hottest

stars. Modern astronomers call this color-magnitude diagram an H-R Diagram or Hertzsprung-Russell Diagram.

26   USSR cosmonaut Georgi T. Beregovoi, in 1968, flew his Soyuz 3 capsule to the first ever manned rendezvous, with an unmanned Soyuz 2.

26   In 1976, USSR launched first Ekran stationary-orbit communications satellite.

28   Great Britain launched its first satellite, Black Knight 1, to Earth orbit in 1971, making Great Britain the sixth nation to demonstrate a space rocket.

28   In 1974, the USSR fired its Luna 23 unmanned Moon-landing probe.

30   In 1981, the USSR's Venera 13 was launched to Venus, dropping landers onto that planet as it flew by March 1, 1982. The landers sent back data for 127 minutes, including the first X-ray fluorescence analysis of the soil of Venus.

30   In 1985, U.S. shuttle Challenger flight STS-61A left Cape Canaveral with a record crew of eight astronauts: Henry W. Hartsfield Jr., Steven R. Nagel, James F. Buchli, Guion S. Bluford Jr., Bonnie J. Dunbar, Wubbo J. Ockels from Holland, and Reinhard Furrer and Ernst Messerschmid from West Germany. Challenger carried a pressurized Spacelab-D1 in the cargo bay for West German materials processing experiments. Houston shared mission control with the West German Space Operations Center at Oberpfaffenhofen near Munich.

November

1   In 1950, German rocket expert Werner von Braun set up a U.S. Army rocket team at Redstone Arsenal, Huntsville, Alabama.

1   In 1962, the spacecraft Mars 1 was launched, first of many USSR's attempts to probe the Red Planet.

1   The 2,060th asteroid, Chiron, was discovered in 1977 by Charles Kowal.

2   Astronomer Harlow Shapley was born in 1885.

3   In 1957, Sputnik 2, the world's second space satellite ,was launched from the USSR carrying Laika, a live dog on a life-support system, part of the 1,121-lb. payload. Laika and her capsule remained attached to the spent upper stage of the converted intercontinental ballistic missile (ICBM) which blasted her to space, creating a six-ton satellite in orbit. The dog captured the hearts of people around the world as her life slipped away a few days into her journey. Later, Sputnik 2 burned as it fell into Earth's atmosphere April 14, 1958. That second satellite in orbit had confirmed the first hadn't been a fluke. Russian satellites were bigger and America hadn't been able to send even a small one to space. Laika indicated the USSR was going for man in space. Today there's still no memorial to Laika, one of the first higher lifeforms from Earth to die in space.

3   In 1973, The interplanetary probe Mariner 10 left the U.S. for Venus and Mercury. It passed Venus February 5, 1974, and arrived at Mercury March 29, 1974, the first time the gravity of one planet, Venus, was used to fling a planet toward another planet, Mercury. It was man's first visit to Mercury.

4   In 1981, USSR Venera 14 was launched to Venus, dropping landers there March 5, 1982, for X-ray fluorescence soil analysis.

7   In 1966, U.S. Lunar Orbiter 2 left for the Moon, going into orbit there November 10, 1966, sending back hundreds of photos.

7   In 1967, U.S. Surveyor 6 left for the Moon, landing in Sinus Medii November 10, 1967. It jumped eight feet to take pictures of its original landing site, among 11,524 photos it sent back.

8   Astronomer Edmond Halley, for whom the comet was named, was born in 1656.

8   The U.S. military ballistic missile program started in 1955. Intermediate range ballistic missiles (IRBM) Jupiter and Thor were approved.

8   In 1958, the American spacecraft Pioneer 2 was launched to the Moon. It failed.

8   In 1968, the U.S. Pioneer 9 probe was sent to orbit the Sun with six solar

radiation experiments aboard.

8   In 1984, U.S. shuttle Discovery flight STS-51A left Cape Canaveral with astronauts Frederick H. "Rick" Hauck, David M. Walker, Joseph P. Allen, Dale A. Gardner and Anna L. Fisher, the first mother to go to space. Allen and Gardner made the first satellite rescue, recovering Westar and Palapa which had failed after flight STS-41B in February 1984. The satellites were returned to Earth in Discovery which landed at Kennedy Space Center, Florida.

9   In 1967, the U.S. capsule Apollo 4 was launched on an unmanned test.

10   In 1968, Zond 6 left USSR on a flight out and around Moon and back to Earth.

10   In 1970, the USSR sent Luna 17 to a soft landing on the Moon, on the Sea of Rains November 17, 1970. In its first use, the Lunokhod 1 self-propelled roving Moon car analyzed soil, sent TV pictures of the lunar surface to Earth.

11   James A. Lovell Jr. and Edwin E. Aldrin Jr., 1966, Gemini 12 on a Titan rocket, made final Gemini flight in Earth orbit including record 5.5-hour spacewalk.

11   In 1982, U.S. shuttle Columbia flight STS-5 left Cape Canaveral with astronauts Vance D. Brand, Robert F. Overmyer, Joseph P. Allen and William B. Lenoir on the first operational flight. Columbia ferried the commercial communications satellites Anik and SBS to orbit. The first shuttle spacewalk was cancelled when spacesuit problems turned up.

11   In 1988, cosmonauts Vladimir Titov and Musa Manarov surpassed the previous 326-day man-in-space record at the USSR's orbiting Mir space station. The 326-day record had been held by cosmonaut Yuri Romanenko. Titov and Manarov went on to set their own records of 366 uninterrupted days in space, from December 21, 1987, through December 21, 1988.

12   In 1965, the USSR launched the probe Venera 2 on a Venus flyby.

12   In 1980, the U.S. interplanetary probe Voyager 1 flew past Saturn, the Solar System's second largest planet. TV pictures radioed back to Earth from within 77,000 miles of the ringed planet showed 900 mph winds at the Saturn's equator. Before the fly-by, astronomers thought Saturn had three large rings. Voyager 1 showed they actually were a legion of small ice particles forming thousands of tiny ringlets. Before Voyager 1, astronomers knew of ten moons orbiting Saturn. The fly-by raised the count to 17. Voyager 1 flew within 3,000 miles of the largest moon, Titan, finding a thick nitrogen atmosphere producing deep smog hiding the surface below.

12   In 1981, U.S. shuttle Columbia flight STS-2 left Cape Canaveral with astronauts Joe H. Engle and Richard H. Truly. The shuttle's robot arm was tested. The flight was cut to two days when a fuel cell failed.

13   The U.S. spacecraft Mariner 9 orbited Mars in 1971.

14   Charles Conrad Jr., Richard F. Gordon and Alan L. Bean flew Apollo 12 to the Moon in 1969. Conrad and Bean made the second-ever Moon landing November 19, collecting 74 lbs. rock samples in 31 hours.

15   Astronomer William Herschel was born in 1738 in Germany. Building his own telescopes, he discovered the planet Uranus in 1781; two moons of Uranus, Oberon and Titania, in 1787; and two moons of Saturn, Mimas and Enceladus, in 1789. William and his son John compiled the General Catalog of 5,000 deep space objects.

15   In 1988, the world's most powerful space rocket, Energia, lofted the USSR's first space shuttle, Buran, successfully to two orbits and a perfect unmanned landing. The flight lasted 3 hours 20 minutes.

15   In 1988, the 26-year-old 300-ft.-diameter Green Bank radiotelescope antenna collapsed unexpectedly 100 miles east of Charleston, in West Virginia, in the Monongahela National Forest.

16   In 1965, the USSR's spacecraft Venera 3 left for Venus, arriving in that

planet's atmosphere March 1, 1966. It was the first craft to impact Venus and first human artifact to reach surface of another planet. No data was sent back.

16   Gerald P. Carr, Edward G. Gibson and William Pogue, in 1973, rode the Skylab 4 capsule from Cape Canaveral on the final visit to America's first-generation space station, Skylab. They set the U.S. space endurance record of 84 days which still stands. The crew had a record spacewalk of seven hours, part of 22 hours of spacewalks for the flight.

18   NASA's Cosmic Background Explorer (COBE) was launched to Earth orbit in 1989, to study the history of stars forming the Universe.

19   U.S. Apollo 12 makes Man's second landing on the Moon in 1969.

20   American astronomer Edwin Powell Hubble, born in 1889 in Marshfield, Missouri, is best known for describing what galaxies are. From the time the telescope was invented in the 17th Century, astronomers had seen glowing spiral-shaped specks of light they called spiral nebulae. In 1924, Hubble discovered they were islands of stars at great distances away from our own Milky Way galaxy. This led him to define the basic structure of the Universe. Hubble had earned a law degree from Oxford University in 1910, but his heart was in astronomy. In 1919 he went on staff at Mount Wilson Observatory, California, using that institution's 100-in. Hooker Telescope, largest in the world at that time, to study spiral nebulae. Some nebulae were clouds of gas within our Milky Way, but Hubble guessed others were more-distant clouds of stars. Hubble made photos of the Andromeda spiral nebula in 1924 and, on those plates, saw Cepheid variable stars—easily-recognizable stars which fluctuate in brightness. The stars he saw were so faint, Hubble calculated the distance to the Andromeda spiral at 800,000 lightyears—eight times farther than the most distant star in the Milky Way. Hubble had discovered the Andromeda spiral nebula actually was the Andromeda Galaxy. He went on to measure the entire Universe, dying September 28, 1953.

26   The first photograph of a meteor was made in 1885.

26   France launched its first satellite, Asterix 1, to orbit on a Diamant rocket in 1965 from Hammaguir, Algeria, in the Sahara Desert. France became the third nation with space rocket. The satellite stayed up until March 25, 1967.

26   Rodolfo Neri became the first Mexican in space, aboard U.S. shuttle Atlantis flight STS-61B from Cape Canaveral in 1985. With him were Brewster H. Shaw Jr., Bryan D. O'Connor, Mary L. Cleave, Sherwood C. Spring, Jerry L. Ross and Charles D. Walker. Ross and Spring took a spacewalk to test construction methods. Atlantis carried three communications satellites to orbit—Morelos B, Satcom KU-2, and Aussat 2. Ross and Spring took a spacewalk to test construction methods.

26   The add-on module Kvant-2, as large as the original Mir itself, was launched to the USSR's orbiting Mir space station in 1989. It docked December 6.

28   The U.S. fired an intercontinental ballistic missile at full range for the first time, in 1958.

28   In 1963, the U.S. space center in Florida known as Cape Canaveral was renamed Cape Kennedy in honor of president John F. Kennedy who had been assassinated November 22. Later, local residents voted to change the name back to the original Cape Canaveral. NASA continued to call its spaceport John F. Kennedy Space Center.

28   In 1964, the interplanetary probe Mariner 4 left the U.S. for Mars. After a mid course correction in flight path, it passed behind the Red Planet July 14, 1965, for man's first fly by of Mars. Mariner 4 took 22 photos from 6,000 miles above the Martian surface.

28   The 1983 flight of U.S. shuttle Columbia, with Spacelab 1 in the cargo bay,

has been known both as STS-9 and STS-41A. It left Cape Canaveral with astronauts John W. Young, Brewster H. Shaw Jr., Owen K. Garriott, A. Robert Parker, Ulf Merbold and Byron K. Lichtenberg. Young had the widest experience with six flights on four different kinds of spacecraft. Merbold from West Germany was the first non-American to go to space in a U.S. craft. He and Lichtenberg were the first payload specialists. Garriott had amateur radio contacts with hams on Earth.

29   In 1961, the U.S. orbited Mercury 5 capsule carrying the chimp Enos to orbit.

30   In 1964, the USSR launched its Zond 2 probe on a Mars flyby.

December

2   In 1971, the USSR's Mars 3 was the first craft to soft-land on the Red Planet.

2   The U.S. interplanetary probe Pioneer 11 flew by Jupiter in 1974.

2   In 1988, U.S. shuttle Atlantis flight STS-27 left Cape Canaveral with astronauts Robert L. "Hoot" Gibson, Guy S. Gardner, Richard M. "Mike" Mullane, Jerry L. Ross and William M. Shepherd. The secret military flight carried the Lacrosse radar spy satellite to orbit.

2   The famous Solar Max fell from orbit, making its fiery re-entry into Earth's atmosphere in 1989. The 5,000-lb. Solar Maximum Mission satellite had been launched on a Delta rocket from Cape Canaveral on February 14, 1980, to study the Sun. When three of its seven instruments failed, the satellite was retrieved at a 288-mi.-high altitude and repaired in space by astronauts aboard shuttle Challenger April 6, 1984. Astronauts George D. "Pinky" Nelson and James D. "Ox" Van Hoften made the repairs in Challenger's 60-foot payload bay. Just a month after that first-ever overhaul in orbit, Solar Max recorded one of the largest flares ever seen erupting on the Sun in May 1984. Solar Max gathered data about the Sun covering almost one complete solar cycle. It recorded more than 12,500 solar flares, letting scientists measure sunspot magnetic fields above the visible surface of the Sun for the first time. Solar Max confirmed the existence of neutrons in solar flares. The gamma ray spectrometer aboard Solar Max was first to detect gamma rays from Supernova 1987a, the distant star that was seen in 1987 exploding in a neighboring galaxy, the Magellanic Cloud. Solar Max discovered 10 comets skimming past or crashing into the Sun. The satellite's ultraviolet spectrometer uncovered increases in the level of high altitude ozone above Earth north of equator and decreases south of the equator.

3   In 1958, California Institute of Technology's Jet Propulsion Laboratory became part of NASA, two months after the U.S. space agency was formed and less than a year after JPL's hastily-built *Explorer I* became the West's first Earth satellite on January 31, 1958. JPL had been an obscure secret lab, begun as a WWII Army rocket research project.

3   In 1965, the USSR's Luna 8 probe was launched from Baikonur Cosmodrome. It crashed on the Moon.

3   The U.S. spacecraft Pioneer 10 made the first flyby of Jupiter in 1973.

4   In 1959, the Mercury capsule Little Joe 3 was launched on a Little Joe rocket from Cape Canaveral carrying the monkey Sam on a suborbital flight.

4   In 1965, Frank Borman and James A. Lovell Jr. were launched in Gemini 7 atop a Titan rocket, the longest Gemini flight, more than 13 days in Earth orbit. Walter M. Schirra Jr. and Thomas P. Stafford, launched December 15 in Gemini 6, made the first space rendezvous with Borman and Lovell.

5   Iraq joined the exclusive circle of nations with rockets powerful enough for space flight, launching a 48-ton three-stage booster to orbit in 1989. Iraq did not launch a separate satellite to orbit, but the third stage of the rocket swept around the Earth for six revolutions before falling back out of orbit. A U.S. spy

satellite, intended to warn of nuclear attack, spotted the rocket's fiery exhaust as it blasted off the Al-Anbar Space Research Center launch pad 50 miles west of Baghdad. The North American Aerospace Defense Command tracked the third stage as it whirled around the globe.

6   In 1957, the U.S. Navy's Vanguard rocket failed in America's first attempt to launch a satellite.

6   In 1958, the U.S. launched the probe Pioneer 3 to the Moon. It failed, but discovered the outer Van Allen radiation layer above Earth's atmosphere.

6   The add-on module Kvant-2, as large as the original Mir itself, was attached to USSR's orbiting Mir space station in 1989. It had been launched November 26.

7   In 1972, Eugene A. Cernan, Ronald E. Evans and Harrison H. Schmitt were launched to the Moon in Apollo 17 atop a Saturn V rocket. Cernan and Schmitt made the sixth and final Moon landing December 11, collecting 243 lbs. of rocks in 75 hours.

10   In 1945, 55 German rocket experts start work at Fort Bliss, Texas, and White Sands, New Mexico. Wernher von Braun and six other German specialists join them to start building rockets for the U.S.

11   U.S. Apollo 17, carrying the first scientist to the Moon, in 1972 makes Man's sixth landing there.

12   In 1961, OSCAR 1 was the first amateur radio satellite launched to Earth orbit.

12   In 1967, U.S. Pioneer 8 was launched from Earth to an orbit around the Sun. It worked until April 28, 1968.

12   Geminid meteor shower in mid-December, up to 60 meteors an hour. The meteors are debris littered along their paths by comets orbiting the Sun. The small chunks of comets orbit the Sun until they encounter a large body, such as Earth, then vaporize when they strike the planet's atmosphere. From the ground, they show off as bright streaks of light as they plunge down into the atmosphere. Geminid meteors are thought to come from a Solar System body either ice comet or asteroid rock—discovered in 1983, named 3200 Phaethon.

13   In 1962, the Relay 1 communications satellite was launched by the U.S. It had 12 telephone circuits and one TV channel.

14   Astronomer Tyco Brahe was born in 1546.

14   The U.S. probe Mariner 2 made the first successful flyby of another planet at Venus in 1962.

15   Lunar satellite Atlas Able 5 exploded during Cape Canaveral launch in 1960.

15   In 1964, launch of the San Marco A satellite to Earth orbit made Italy the third nation to send a satellite to space. The U.S.-made Scout rocket was fired by an Italian launch team from an Italian platform in the Indian Ocean off Kenya, Africa. It was the first launch of a NASA rocket by foreign technicians.

15   In 1965, Walter M. Schirra Jr. and Thomas P. Stafford were launched to Earth orbit in Gemini 6 atop a Titan rocket. Frank Borman and James A. Lovell Jr., launched December 4 in Gemini 7, made the first space rendezvous with Schirra and Stafford .

15   The USSR's Venera 7 became the first spacecraft from Earth to soft-land on Venus in 1970.

15   In 1975, the USSR launched the first Raduga communications satellite to stationary orbit.

15   In 1984, the USSR fired its Vega 1 probe to fly by Venus and Comet Halley. Vega is an acronym for Venus-Gallei meaning Venus-Halley. Vega 1 past Venus and dropped a lander June 11, 1985, and by Halley's Comet March 6, 1986.

16   In 1965, U.S. Pioneer 6 left for a close-up view of the Sun, orbiting between Earth and Venus, from 75 to 90 million miles from the Sun in a 311-day orbit.

18   In 1958, America's Score communications satellite was launched, carrying a

Christmas message taped by U.S. President Dwight D. Eisenhower to be broadcast from space.

19   In 1960, the U.S. capsule Mercury 1 was blasted toward space on an unmanned test. It flew 235 miles and reached an altitude of 130.7 miles.

19   Astronauts Eugene A. Cernan, Ronald E. Evans and Harrison H. Schmitt in Apollo 17 returned to Earth after man's longest and last stay on the Moon, 74 hours 59 minutes by astronauts Cernan and Schmitt.

19   In 1978, the USSR launched the first Gorizont communications satellite to stationary orbit.

21   In 1965, the amateur radio satellite OSCAR 4 was launched, the first high altitude hamsat.

21   In 1966, USSR Luna 13 left for the Moon, soft landing there 80 hours later. It drove a spike into the lunar surface, used a mechanical digging arm and sent back photos.

21   In 1968, Apollo 8 was launched to the Moon on a Saturn V rocket, the first manned craft to leave Earth's gravity and the first manned craft to orbit the Moon. The launch boosted three astronauts in their capsule to 25,000 miles an hour, a speed no human had reached before. Frank Borman, James A. Lovell Jr. and William A. Anders made that first manned flight to the vicinity of the Moon in the Apollo command module, coming within 70 miles of the surface. During their 147-hour trip, they did not land, but circled the Moon ten times, sent back television pictures of the lunar surface, and returned to Earth. On their way home, the three astronauts made a worldwide broadcast on Christmas Eve reading passages from The Holy Bible, book of Genesis, while telecasting an extraordinary view of Earth from 250,000 miles out in space. The six-day space flight was the longest on record to that time. They photographed possible landing sites for Apollo 11 which landed six months later on July 20, 1969.

21   In 1984, USSR launched Vega 2 probe to fly by Venus and Comet Halley. Flew by Venus, dropped lander June 15, 1985, by Halley's Comet March 9, 1986.

21   In 1988, cosmonauts Vladimir Titov and Musa Manarov set a one-trip world record of 366 days in space as they returned to Earth from the USSR's orbiting Mir space station.

23   Astronomer Giovanni Cassini discovered the Saturn moon Rhea in 1672.

24   In 1968, three astronauts aboard Apollo 8 made a worldwide Christmas Eve broadcast reading passages from The Holy Bible, book of Genesis, while telecasting an extraordinary view of Earth from 250,000 miles out in space. "Up to that point it was the largest audience that had ever listened to a human voice," Apollo 8 command module pilot James A. Lovell Jr. recalled in 1988. With Lovell aboard Apollo 8 were Frank Borman and William A. Anders.

24   In 1979, the European Space Agency launched its first Ariane 1 space rocket from Kourou, French Guiana, carrying a CAT satellite. French Guiana, on the northeastern coast of South America, is a department of France.

25   Isaac Newton was born in 1642.

25   A Christmas message prerecorded by U.S. President Dwight D. Eisenhower was broadcast to Earth in 1958 by the orbiting Score communications satellite.

26   In 1974, the USSR launched to Earth orbit the Salyut 4 space station, fourth in the series of first-generation space stations.

27   Astronomer Johannes Kepler was born in 1571.

28   Astronomer Arthur Eddington was born in 1882.

30   Robert H. Goddard launched his first rocket from New Mexico in 1930. It reached an altitude of 2,000 feet and a speed of 500 mph.

Index
****

A

A, rocket, USSR, 250, 349, 358
A-1, rocket, USSR, 128, 349
A-2, rocket, Germany, 246, 349
A-2, rocket, USSR, 349
A-4, rocket, Germany, 246
A-9, rocket, Germany, 247
AAP, 99, 156, 192
Abbot, Charles Greely, 659
aberration, 661
abort, 240, 349
absolute magnitude, 681, 685
    red giant, 688
    spectroscopic parallax, 687
    stars, 669
    stars, Russell, 678
absorption lines, 681
    Sun, 683
Abu'l-Wafa, Muhammad al-Buzjani, 659
Academie des Sciences, 662
Academy, Plato, 660
accretion, 521
    disk, 681
achondrite, 522
acronyms, NASA, 223
Acton, Loren W., 78
Adams, John Couch, 659
Adamson, James C., 78
Advanced Launch System, 301
Advanced X-Ray Astrophysics Facility, 625
Adventure, 197
advertising, Mir, 172
Aerobee, rocket, 3, 4, 248, 283
Aerospatiale, 127
Afanasyev, Viktor, 57
Afghanistan, 77, 140
age, Universe, Olbers, 677
Agena, rocket, 8, 88
air, water, earth, fire, 665
airborne launches, 263, 358
AirBus, 127
airships, Tsiolkovsky, 679
Airy Disk, 660
Airy, George Biddell, 660
Aitken, Robert Grant, 660
Akers, Thomas D., 78
Akiyama, Toyohiro, 57, 77
Aksyonov, Vladimir V., 57
Al-Anbar, Iraq, spaceport, 276, 358
al-Battani, 660
Al-Ma'mun, 660
al-Raqqa, 660
Al-Saud, Prince Sultan Salman Abdel Aziz, 78, 124
Al-Sufi, Abu'l-Husayn Al-Sufi, 660
Alabama Space & Rocket Center, 126
Alabama, Huntsville, 4
Alaska, Poker Flats, 263
Albategnius, 660
albedo, 681
Albert, monkey, 3
Alcantara, Brazil, launch site, 280
Aldrin, Edwin E. "Buzz", Jr., 78, 135, 151
Alexander the Great, 660
Alexandrov, Alexander, 57, 77
Algeria, 358
all-sky photography, 678
Allen, Andrew M., 78
Allen, Joseph P., 78
Almagest, Ptolemy, 662

Alouette 1, satellite, Canada, 10
Alpha Centauri, 661, 682
    Innes, 670
altitude, 397
amateur radio astronomy, SARA, 635
amateur radio, 9, 74, 105, 107, 112, 113, 116, 117, 118, 119, 121, 214, 367, 368, 370-373
    shuttle, SAREX, 214
America, Apollo 17 CSM, 92
America, rockets, 244
American Radio Relay League, 372
Americas, launch sites, 260
Ames Center, 126
ammonia, Shapley, 679
AMSAT, 372
Anaxagoras, 660
Anders, William A., 78, 151
Andromeda Galaxy, 537, 661, 674, 683, 681
    Marius, 674
Andromeda Nebula, 669
angstrom, 681
Angular momentum, 396
angular velocity, 687
animals, 146
anorthosite, 522
Antares, Apollo 14 LEM, 92
antenna, dish, 398
anti-matter, 681
Apennine Mountains, 97
aphelion, 522, 686
apogee, 351, 396, 522
Apollo, 90-98, 148, 152, 192, 255
    abort, 96
    Apollo-204, 90
    Apollo Applications Program, 99
    Apollo-Soyuz, 152
    astronauts, 78-82
    capsule, 90
    Moon flights, 471, 472, 473
    Moon landing, 95
    Neil Armstrong, 135
Apollo crater, 151
Apollo group, 522
apparent magnitude, 681, 685
    star, Zollner, 681
appetite, 143
Applied Physics Lab, Johns Hopkins University, 3
Apt, Jerome "Jay", 78
Aquarius, Apollo 13 CSM, 92
Arabs, rockets, ancient, 243
Archimedes, 660
Aremir, 74
Ariane, rockets, 251, 283, 312, 358
    Ariane-5, site, 269
    flight schedule, 313
    past flights, 314
    L-5, 313
    LO-2, 313
    V-15, 313
    V-18, 313
    V-19, 315
    V-20, 318
    V-21, 319
    V-22, 319
    V-23, 321
    V-24, 322
    V-25, 323
    V-26, 324
    V-27, 325
    V-28, 326
    V-29, 327

Rohini, 358
Romanenko, Yuri V, 58 139, 153, 193
Romania, 76, 140
Roosa, Stuart A., 81
Roosevelt, Franklin D., 664
Rosat, 630
Ross, Jerry L., 81
Roswell, New Mexico, 2
Royal Academy of Science, 672
Royal Astronomical Society, 667
Royal Greenwich Observatory, 654, 660-661, 664-665, 667
Royal Society of London, 674
Royal Tern, 197
Rozhdestvensky, Valeri I., 58
RR Lyrae stars, 688
RTG, 527-528
Rukavishnikov, Nikolai, 58
Runco, Mario, Jr., 81
Russell, Henry Norris, 669, 678
Russian rocket pioneer, 2
Ryle, Martin, quasars, 10
Ryumin, Valeri V., 58

S

safety net, runaway, space shuttles, 206
Sakigake, probe, Japan, 667
Sally Ride, space shuttles, 133
Salute, 74-76
Salyut, space stations, 1-7, 74-76, 153, 156-159, 193
Sam, monkey, 6
San Marco Range, Kenya, Italian spaceport, 274
Sanger, German, rocket, 340
SAR, 398
SARA, 372, 635
Sarafanov, Gennadi, 58
SAREX, shuttle amateur radio, 214
Satellite Launch Vehicle, rocket, India, 283, 346, 358
Satellite Scoreboard, 356
satellites, 3, 351, 398, 528, 688
    Advanced X-Ray Astrophysics Facility, 625
    Alouette, Canada, 10
    amateur radio, 9, 370-373
    Ariel, Great Britain, 9
    Astro, 634
    atom bomb test, 11
    AXAF, 625
    balloon, 7
    Brazilsat, 314
    Clarke Belt, 360
    communications, 366
        early, 369
    Cosmic Background Explorer, 628
    Courier 1B, 7
    dish antenna, 398
    DSCS, 287
    Dynamics Explorer, 641
    Earth Observing System, 380
    Echo, 7, 368
    Einstein Observatory, 639
    Explorers, 2-5, 250, 351
    Extreme Ultraviolet Explorer, 638
    Gamma Ray Observatory, 622
    glossary, 396
    GOES, 362
    GRO, 113
    ham radio, 9, 370-373
    hamsats, 9, 370-373
    Helios, 314
    Hipparcos, 632
    Hubble Space Telescope, 111-112, 120, 615, 684
    Infrared Astronomical Satellite, 642
    infrared telescope, 684
    International Ultraviolet Explorer, 638
    Italian X-ray satellite, 643

Lacrosse, 108
Laika, 4
LDEF, 110
NASA, TDRS, communications, 369
Nimbus, weather, 362
NOAA, weather, 362
Orbiting Solar Observatory, 9
orbits, 358, 686
OSCAR, 9, 370-373
OSO, 9
Palapa, 296
Radioastron, 643
reconnaissance, 109, 111, 388
Relay, 10
remote-sensing, 376
Roentgen Satellite, 630
Score, 6
search and rescue, 390
Small Explorers, NASA, 642
Solar And Heliospheric Observatory, 644
Solar Max, 639
Solar-A, 637
Solidaridad, Mexico, 314
Space Infrared Telescope Facility, 627
space junk, 392, 394
space power station, 386
Spot, 376
Sputnik 1, 4-5, 249
STEPsat, 300
Syncom, 11
telescopes, 613
Television Infra-Red Orbital Satellite, 7
Telstar, 9
Tiros-1, 7, 361
Turksat, 314
UARS, 115
Uhuru, 638
uses for satellites, 360
Vanguard, 5, 249
weather, 361
    China, 364
    Europe, 364
    first, 7
    India, 364
    Japan, 364
West Ford, needles, 11
wildlife tracking, 385
Saturn, planet, 668, 670, 672
    Cassini probe, 500
    Herschel, 668
    Hyperion moon, Lassell, 673
    Laplace, 672
    Pioneer probes, 512-513
    ring C, 661, 667
    rings, Huygens, 670
    Shapley, 679
Saturn, rocket, 6, 90, 153, 193, 250, 255, 307, 352, 423, 528, 661-662
Saudia Arabia, 124, 140
Savinykh, Viktor P., 58
Savitskaya, Svetlana, 58, 133
Schiaparelli, Giovanni, 673
Schirra, Walter Marty "Wally" Jr., 6, 81, 83
Schlegel, Hans-Wilhelm, 81, 125
Schmidt telescope, 647, 679, 688-689
    Cerro La Silla, Chile, 679
    Cerro Tololo, Chile, 679
    European, 679
    Mount Palomar, 679
    Siding Spring, 679
    Tautenburg, Germany, 679
Schmidt, Bernhard Voldemar, 679
Schmidt, Maarten, quasars, 10, 679
Schmitt, Harrison H., 81
school teacher, Challenger, 124